Cherry Kerr

Mar
Unive, Durham

Władysław Narkiewicz

Elementary and Analytic Theory of Algebraic Numbers

Second Edition, Substantially Revised and Extended

Springer-Verlag
Berlin Heidelberg New York
London Paris Tokyo Hong Kong

PWN—Polish Scientific Publishers
Warszawa

Professor Władysław Narkiewicz
Wrocław University
Plac Grunwaldzki 2/4
PL-50-384 Wrocław

First edition published in the series Monografie Matematyczne
by Państwowe Wydawnictwo Naukowe, Warszawa 1974

Mathematics Subject Classification (1980): 12-02

ISBN 3-540-51250-0 Springer-Verlag Berlin Heidelberg New York
ISBN 0-387-51250-0 Springer-Verlag New York Berlin Heidelberg

Distribution rights for the non-socialist countries:
Springer-Verlag Berlin Heidelberg New York
London Paris Tokyo Hong Kong

Library of Congress Cataloging-in-Publication Data
Narkiewicz, Władysław. Elementary and analytic theory of algebraic
numbers/Władysław Narkiewicz. — 2nd ed., substantially rev. and extended.
Includes bibliographical references. ISBN 0-387-51250-0 (U.S.) 1. Algebraic number
theory. I. Title. QA247.N3332 1989 512'.74—dc20

Copyright © by PWN—Polish Scientific Publishers—Warszawa 1990
All rights reserved
No part of this publication may be reproduced, stored in retrieval system, or transmitted
in any form or by any means, electronic, mechanical, photocopying, recording or otherwise,
without the prior written permission of the copyright owner
Printed in Poland
Typesetting and printing: Drukarnia im. Rewolucji Październikowej, Warszawa
Bookbinding: K. Triltsch, Würzburg
2141/3140-543210

To my wife

Preface

The aim of this book is to present an exposition of the theory of algebraic numbers, excluding class-field theory and its consequences. There are many ways of developing this subject; the latest trend is to neglect the classical Dedekind ideal-theory in favour of local methods. However, for numerical computations necessary for the application of algebraic numbers to other areas of number theory the old approach seems more suitable, although its exposition is obviously longer. The local approach is more powerful for analytical purposes as demonstrated in Tate's thesis. Thus the author has tried to reconcile the two approaches, presenting a self-contained exposition of the classical standpoint in the first four chapters and then turning to local methods.

In the first chapter we present the necessary tools from the theory of Dedekind rings and valuation theory, including the structure of finitely generated modules over Dedekind domains. In Chapters 2, 3 and 4 the classical theory of algebraic number fields and ideals in their rings of integers is developed. The proof of the main result, the different-theorem, is based on the approach of Weil, who related it to derivations. Chapter 5 contains the fundamental notions of the theory of p-adic fields and Chapter 6 brings their application to the study of algebraic number fields. We include here Shafarevich's proof of the Kronecker–Weber theorem, and also the principal properties of adeles and ideles.

In Chapter 7 we apply analytical methods and derive the functional equations for zeta-functions, including Dedekind's zeta-functions and Dirichlet's L-functions. These functions are then applied to the study of the asymptotic distribution of ideals and prime ideals. In Chapter 8 we consider Abelian extensions of the rationals. We give a proof of the Siegel–Brauer theorem in this case, obtain the class-number formula and give an effective bound for negative quadratic discriminants with class-number one. The last chapter presents results connected with factorizations of integers into irreducibles.

Each chapter ends with a section containing comments and a short review of the relevant literature. In response to suggestions made by several colleagues, we have also provided a selection of exercises.

At the end of the book we present a choice of open problems containing some classical questions and also several problems of more recent vintage. In the first edition this list contained 35 problems, of which 9 have by now been solved. In this edition we have added a further 14.

We expect the reader to have an elementary knowledge of topological and algebraic notions, including elements of Galois theory.

For the reader's convenience we have included four appendices dealing, successively with locally compact Abelian groups, Dirichlet series, the geometry of numbers and Baker's method and giving the results which are utilized in the main text. Theorems, propositions, lemmas and formulas are numbered consecutively in each chapter. The sign □ indicates the end of a proof. We always write "iff" in place of "if and only if" and use ERH (Extended Riemann Hypothesis) to denote the statement that every conceivable zeta-function which should not have zeros in the critical strip indeed does not have them.

Several changes have been made with regard to the first edition. We have added certain theorems, including the structure of finitely generated torsion modules over Dedekind domains in Chapter 1, and an improvement of Kronecker's theorem, due to E. Dobrowolski, and Perron's theorem on approximations in Chapter 2, Remak's results on extensions with the defect of units in Chapter 3, the Pellet–Stickelberger theorem and the normal basis theorem in Chapter 4, and Moser's theorem on Minkowski's units and the density theorems of Kronecker and Frobenius in Chapter 7. Chapter 8 includes certain material on Abelian fields which was absent in the first edition, including an asymptotical formula for the number of Abelian fields with a given Galois group and a bounded conductor, and the Bundschuh–Hock proof of an upper bound for discriminants of imaginary quadratic fields with class-number one. In Chapter 9 we have omitted the results relating to arithmetic functions and instead give an elementary description of the class-group, due to J. Kaczorowski.

The notes at the end of each chapter have been rewritten to take account of the literature up to 1983 and the bibliography has been extended accordingly. Some of the literature appearing in 1984 has also been covered as have certain items appearing in 1985 and later.

Several smaller changes in the text have been made, either to clarify the argument or to remove certain flaws. I am grateful to the many friends and colleagues who commented on the first edition. Particular thanks go to Professor J. Browkin, Dr T. Nakahara and Dr T. Uehara, who sent me lists of misprints and other inaccuracies.

I could not end without thanking my family, who showed much patience during the period I was working on the typescript; I am also very grateful to my 50 year-old typewriter "Olympia".

Finally, I would like to thank PWN and Springer-Verlag for their understanding and assistance in the realization of this book.

My heartiest thanks go to Mrs J. Smólska and Dr M. Kuczma who corrected the language.

Wrocław 1989   Władysław Narkiewicz

Table of Contents

Preface VII

Chapter 1. Dedekind Rings and Valuations 1

   § 1. Dedekind Rings 1
   § 2. Valuations and Exponents 15
   § 3. Finitely Generated Modules over Dedekind Domains . . . 23
   § 4. Notes to Chapter 1 35
   Exercises to Chapter 1 41

Chapter 2. Algebraic Numbers and Integers 42

   § 1. Distribution of Integers in the Complex Plane 42
   § 2. Discriminants and Integral Bases 52
   § 3. Applications of Minkowski's Convex Body Theorem . . . 66
   § 4. Notes to Chapter 2 71
   Exercises to Chapter 2 84

Chapter 3. Units and Ideal Classes 86

   § 1. Valuations of Algebraic Number Fields 86
   § 2. Ideal Classes 94
   § 3. Units 99
   § 4. Euclidean Algorithm 120
   § 5. Notes to Chapter 3 125
   Exercises to Chapter 3 142

Chapter 4. Extensions 143

   § 1. The Homomorphisms of Injection and Norm 143
   § 2. Different and Discriminant 155
   § 3. Factorization of Prime Ideals in Extensions. More about
       the Class-Group 175
   § 4. Notes to Chapter 4 193
   Exercises to Chapter 4 206

Chapter 5. p-adic Fields 207

§ 1. Principal Properties, Integers, Units 207
§ 2. Extensions of p-adic Fields 229
§ 3. Harmonic Analysis in p-adic Fields 244
§ 4. Notes to Chapter 5 259
Exercises to Chapter 5 264

Chapter 6. Applications of the Theory of p-adic Fields to Algebraic Number Fields 265

§ 1. Arithmetic Applications 265
§ 2. Adeles and Ideles 295
§ 3. Notes to Chapter 6 317
Exercises to Chapter 6 322

Chapter 7. Analytical Methods 324

§ 1. The Classical Zeta-Functions 324
§ 2. Asymptotic Distribution of Ideals and Prime Ideals . . . 355
§ 3. Chebotarev's Theorem 379
§ 4. Notes to Chapter 7 404
Exercises to Chapter 7 418

Chapter 8. Abelian Fields 421

§ 1. Main Properties 421
§ 2. The Class-Number Formula and the Siegel–Brauer Theorem 436
§ 3. Class-Number of Quadratic Fields 448
§ 4. Notes to Chapter 8 470
Exercises to Chapter 8 488

Chapter 9. Factorization Problems 490

§ 1. Elementary Approach to Factorizations 490
§ 2. Quantitative Results 501
§ 3. Notes to Chapter 9 511
Exercises to Chapter 9 519

Appendix I. Locally Compact Abelian Groups 520
Appendix II. Dirichlet Series 534
Appendix III. Geometry of Numbers 536
Appendix IV. Baker's Method 538

Unsolved Problems 539

Bibliography 544

List of Important Symbols 713

Subject Index 715

Author Index 723

Addendum 743

Chapter 1. Dedekind Rings and Valuations

§ 1. Dedekind Rings

1. This chapter is introductory and contains the fundamental properties of Dedekind rings including their behaviour under finite extensions and the structure of finitely generated modules. Moreover, we include the elementary facts about valuations needed in the sequel.

Consider a commutative ring R without zero divisors and with a unit element and let K be its field of quotients. Any non-zero R-module I contained in K and such that for a certain non-zero $a \in R$ we have $aI \subset R$ will be called a *fractional ideal* of K. Note that this notion depends on R. Every fractional ideal contained in R is an ideal in the usual sense and the converse holds for non-zero ideals. If I_1, I_2 are fractional ideals, then their *product* is defined as the set of all sums $a_1 b_1 + \ldots + a_m b_m$ with $a_i \in I_1$ and $b_i \in I_2$. This set is also a fractional ideal. Indeed, $I_1 I_2$ is a non-zero R-module contained in K and if $x, y \neq 0$ lie in R and $xI_1 \subset R$, $yI_2 \subset R$, then for $a = \sum_m a_m b_m$ in $I_1 I_2$ we get

$$(xy)a = \sum_m (xa_m)(yb_m) \in R.$$

It is clear that the set of all fractional ideals of K forms a commutative semigroup with a unit equal to R.

Any commutative ring with a unit element, but possibly with zero-divisors, is called a *Noetherian ring* if every ascending chain of distinct ideals is necessarily finite.

Proposition 1.1. *A commutative ring with a unit element is Noetherian iff every ideal of that ring is finitely generated.*

Proof. Let R be Noetherian and let I be any of its ideals. Assume that I is not finitely generated. Select $x_1 \neq 0$ in I arbitrarily, and if the elements $x_1, x_2, \ldots, x_n$ are already selected, then choose for x_{n+1} any element of I not belonging to the ideal $x_1 R + x_2 R + \ldots + x_n R$, which choice is always possible by our assumption. But now we have an ascending chain $x_1 R, x_1 R + x_2 R, \ldots$ of distinct ideals, which is a contradiction.

Let R be a ring in which every ideal is finitely generated and consider any ascending sequence $I_1 \subset I_2 \subset \ldots$ of its ideals. The union $\bigcup_{m=1}^{\infty} I_m$ is an ideal in R and so must have a finite set of generators. But this set must already lie in some I_m, which shows that our sequence has at most m distinct terms. □

Corollary. *If in R every ideal is principal, then R is Noetherian provided it is commutative and has a unit element.*

For application we shall need also the notion of a Noetherian module. If R is a commutative ring with a unit element, and M is an R-module, then we call it a *Noetherian R-module* if every ascending chain of its submodules has only a finite number of distinct terms. In the same manner as in Proposition 1.1 we prove that an R-module M is Noetherian iff its every sub-R-module is finitely generated.

Proposition 1.2. (i) *The direct sum of a finite number of Noetherian modules is again a Noetherian module.*

(ii) *A homomorphic image of a Noetherian module is again Noetherian.*

Proof. The proof will be based on the following

Lemma 1.1. *If M is a Noetherian R-module and the sequence*

$$0 \to M_1 \to M \to M_2 \to 0 \qquad (1.1)$$

is exact (i.e., M_1 is a sub-R-module of M and $M_2 \simeq M/M_1$), then the modules M_1 and M_2 are also Noetherian. Conversely, if there exists a sequence (1.1) with Noetherian M_1, M_2 which is exact, then M is Noetherian.

Proof. Let M be Noetherian and let (1.1) be exact. As every sub-R-module of M_1 is a sub-R-module of M, the noetherianity of M_1 follows. If $\{I_m\}$ is an infinite ascending chain of distinct sub-R-modules of M_2, then the reciprocal images of I_m in M form an infinite ascending chain of distinct sub-R-modules of M, contradicting our assumption.

Assume M_1, M_2 to be Noetherian and (1.1) to be exact. Let $\{I_m\}$ be an ascending chain of sub-R-modules of M and let J_m be the image of I_m in M_2. The sequence $\{J_m\}$ is ascending, and thus for sufficiently large n we have

$J_n = J_{n+1} = J_{n+2} = \ldots$, and similarly we obtain for $L_m = I_m \cap M_1$ the equalities $L_{n_1} = L_{n_1+1} = \ldots$ for sufficiently large n_1. Let $N = \max(n, n_1)$ and let $r \geqslant N$. If $a \in I_r$, then for a certain $b \in I_N$ we have $a - b \in M_1$, whence $a - b \in I_r \cap M_1 = I_N \cap M_1 \subset I_N$; hence $a \in I_N$. Thus $I_r \subset I_N$ and the sequence $\{I_m\}$ has only a finite number of distinct terms, which implies that M is Noetherian. □

To prove our proposition, note that its part (ii) is already contained in the lemma and case (i) follows by induction owing to the observation that for any R-modules A and B the sequence

$$0 \to A \to A \oplus B \to B \to 0$$

is exact. □

From now on we shall assume that R is a commutative ring with a unit element and without zero-divisors, i.e., a domain. By K we shall denote the field of quotients of R. If I is a fractional ideal of R, then by I' we shall denote the set

$$\{x \in K : xI \subset R\}.$$

Obviously I' is a non-zero R-module. We shall show that it is a fractional ideal. To do this let us take an arbitrary non-zero element $y \in I$ and let r be a non-zero element of R satisfying $ry \in R$. Then ry lies in $R \cap I$ and for every $a \in I'$ we have $ary \in R$.

It can easily be seen that $II' \subset R$. If, for a fractional ideal I, equality $II' = R$ holds, then we say that I is *invertible* and write I^{-1} instead of I'. We now prove

Proposition 1.3. *Every principal fractional ideal (i.e., an ideal of the form aR with $a \in K$) is invertible, and the set of all invertible fractional ideals of R is a group under multiplication.*

Proof. If I is principal, say $I = aR$, then evidently $I' = a^{-1}R$, and $II' = R$. Now note that the equality $I_1 I_2 = R$ implies $I_2 = I_1^{-1}$. In fact, we have $I_2 \subset I_1'$; thus $R = I_1 I_2 \subset I_1 I_1' \subset R$; hence $I_1 I_1' = R$ and, moreover, $I_1' = I_1' R = I_1' I_1 I_2 = RI_2 = I_2$. The invertibility of the product of invertible fractional ideals and equality $I = (I^{-1})^{-1}$ being trivial, the proposition follows. □

If, in a domain R, every fractional ideal is invertible, then we say that R is a *Dedekind domain*. The first part of Proposition 1.3 implies that every principal ideal domain is a Dedekind domain and, in particular, the ring of rational integers is Dedekind. Two important properties of Dedekind domains are given by the following

Theorem 1.1. *If R is a Dedekind domain, then R is Noetherian and every non-zero prime ideal in R is maximal.*

Proof. Let I be any non-zero ideal in R. As $II^{-1} = R$ there exist elements $a_i \in I$, $b_i \in I^{-1}$ ($i = 1, 2, \ldots, m$) such that $a_1 b_1 + a_2 b_2 + \ldots + a_m b_m = 1$. Now, if $x \in I$, then $x = (xb_1)a_1 + \ldots + (xb_m)a_m$ and $xb_j \in R$, whence I is finitely generated (the system $\{a_i\}$ being a set of generators) and R is Noetherian.

Let $\mathfrak{p}$ be a non-zero prime ideal in R and let $\mathfrak{P}$ be a maximal ideal containing $\mathfrak{p}$. We have $\mathfrak{p}\mathfrak{P}^{-1} \subset \mathfrak{P}\mathfrak{P}^{-1} = R$ and so $\mathfrak{p}\mathfrak{P}^{-1}$ is an ideal of R. As $(\mathfrak{p}\mathfrak{P}^{-1})\mathfrak{P} = \mathfrak{p}$, we must have either $\mathfrak{p}\mathfrak{P}^{-1} \subset \mathfrak{p}$ or $\mathfrak{P} \subset \mathfrak{p}$ because $\mathfrak{p}$ is a prime ideal. The first possibility gives $\mathfrak{P}^{-1} \subset \mathfrak{p}^{-1}\mathfrak{p}\mathfrak{P}^{-1} \subset \mathfrak{p}^{-1}\mathfrak{p} = R$, which implies $\mathfrak{P}^{-1} = R$ and $\mathfrak{P} = R$, which is a contradiction; thus the second possibility holds and this implies $\mathfrak{p} = \mathfrak{P}$. □

2. To obtain further results on Dedekind domains we have to introduce the notion of integrality. Let R be a domain, and let L be any field containing R. There is no need for L to coincide with the field of fractions of R. An element $x \in L$ is said to be *integral over R*, or shortly *R-integral*, if there exist elements $a_0, a_1, \ldots, a_{n-1}$ of R such that the equality

$$x^n + a_{n-1} x^{n-1} + \ldots + a_0 = 0 \tag{1.2}$$

holds.

(Note that if R is a field, then the integral elements over R are exactly those which are algebraic over R, i.e. which satisfy some algebraic equation with coefficients in R.)

R-integrality may be defined in another, equivalent way, which is sometimes more suitable for applications. This is the content of

Proposition 1.4. *The following properties of an element x contained in an overfield L of a domain R are equivalent:*
  (i) *x is R-integral.*
  (ii) *The ring $R[x]$ generated by R and x is a finitely generated R-module.*
  (iii) *There exists a finitely generated and non-zero R-module M contained in L such that $xM \subset M$.*

Proof. (i) → (ii). Observe that the elements $1, x, \ldots, x^{n-1}$ with n given by (1.2) generate $R[x]$.
  (ii) → (iii). The R-module $R[x]$ may serve for M.
  (iii) → (i). Let $z_1, \ldots, z_r$ be generators of M. From $xM \subset M$ we obtain, for suitable $b_{ij} \in R$, the equalities

$$xz_i = b_{i1} z_1 + \ldots + b_{ir} z_r \quad (i = 1, 2, \ldots, r).$$

Since M is non-zero, the z_i do not all vanish, and so we get
$$\det[b_{ij} - x\delta_i^j] = 0,$$
where
$$\delta_i^j = \begin{cases} 1 & \text{if } i = j, \\ 0 & \text{if } i \neq j. \end{cases}$$
Expanding this determinant, we get an equation of the form (1.2). □

Corollary. *The set of all elements of a field L which are R-integral, where R is a given subring of L, forms a ring.*

Proof. Let $a, b \in L$ be R-integral. Choose finitely generated and non-zero R-modules M, N in L, so that $aM \subset M$, $bN \subset N$. The R-module MN is finitely generated and non-zero and, moreover, $(a \pm b)MN \subset MN$, $(ab)MN \subset MN$, whence $a+b$, $a-b$, ab are R-integral. □

The ring whose existence is asserted in this corollary is called the *integral closure* of R in L. If a domain is equal to its integral closure in its quotient field, then we call it *integrally closed*. Now we shall prove a theorem on the transitivity of integral closure.

Theorem 1.2. *Let R be a domain contained in a field K, and let L be any extension of K. If S is the integral closure of R in K, then the integral closures of R and S in L coincide.*

Proof. Let $x \in L$ be S-integral, i.e., for some $a_0, \ldots, a_{n-1}$ lying in S let $x^n + a_{n-1}x^{n-1} + \ldots + a_0 = 0$. The ring $R_1 = R[a_0, a_1, \ldots, a_{n-1}]$ generated by R and the a_i's is a finitely generated R-module, as can be seen from the consideration of the chain $R \subset R[a_0] \subset \ldots \subset R_1$ (in which every ring is a finitely generated module over its predecessor). Moreover, x is R_1-integral, whence, by Proposition 1.4, $R_1[x]$ is a finitely generated R_1-module. It follows that $R_1[x]$ is a finitely generated R-module. As it is obviously non-zero and $xR_1[x] \subset R_1[x]$, Proposition 1.4 implies the R-integrality of x. □

Corollary. *Every integral closure of a domain in a field is integrally closed.*

For the proof it suffices to apply the theorem just proved for $K = L$. □

We shall now use the notion of integrality to give a characterization of Dedekind domains.

Theorem 1.3. *A domain R is a Dedekind domain iff it satisfies the following three conditions:*
  (i) *R is Noetherian.*

(ii) *Every non-zero prime ideal of R is maximal.*
(iii) *R is integrally closed.*

Proof. *Necessity.* The necessity of (i) and (ii) is contained in Theorem 1.1. To prove (iii) take any R-integral element x of the field K of fractions of R. Then $R[x]$ is, by Proposition 1.4, a finitely generated R-module. Let $a_1, \ldots, a_m$ be its generators, and choose $b \neq 0$ so that $ba_i \in R$ for $i = 1, 2, \ldots, m$. Then $bR[x] \subset R$, showing that $R[x]$ is a fractional ideal. Moreover, $R[x]^2 = R[x]$ (as $R[x]$ is a ring), and we obtain

$$R[x] = RR[x] = R[x]R[x]R[x]^{-1} = R,$$

which implies $x \in R$. This shows that R is integrally closed.

Sufficiency. For the proof of this part of our theorem we shall need three lemmas.

Lemma 1.2. *If R is a Noetherian domain and I is a non-zero ideal in R distinct from R, then one can find prime ideals $\mathfrak{p}_1, \ldots, \mathfrak{p}_r$ of R such that $\mathfrak{p}_1 \mathfrak{p}_2 \cdots \mathfrak{p}_r \subset I \subset \mathfrak{p}_1 \cap \mathfrak{p}_2 \cap \cdots \cap \mathfrak{p}_r$.*

Proof. Let I be a maximal element of the set of all those ideals of R which do not have this property. This ideal cannot be prime, since in this case we could take $r = 1$, $\mathfrak{p}_1 = I$. Hence there exist elements $a, b \notin I$ such that $ab \in I$. Put $A = I + aR$, $B = I + bR$. Then $AB \subset I \subset A \cap B$. The ideals A, B cannot be equal to R, since, e.g., $A = R$ would imply the equality $B = I$. Hence the ideals A, B both have the property formulated in the lemma, and this implies that I has it as well. This contradiction proves the lemma. □

Lemma 1.3. *If a domain R satisfies* (i), (ii) *and* (iii), *then every non-zero prime ideal in R is invertible.*

Proof. Let $\mathfrak{p}$ be any non-zero prime ideal in R. Choose $a \neq 0$ in $\mathfrak{p}$ so that the principal ideal aR contains a product $\mathfrak{p}_1 \cdots \mathfrak{p}_k$ of the least possible number of prime ideals. Such a choice is possible by Lemma 1.2. Now one of the ideals $\mathfrak{p}_i$, say $\mathfrak{p}_1$, is contained in $\mathfrak{p}$, and so, by (ii), $\mathfrak{p} = \mathfrak{p}_1$. The product $\mathfrak{p}_2 \cdots \mathfrak{p}_k$ is not contained in aR, and so we may find $b \in \mathfrak{p}_2 \cdots \mathfrak{p}_k \setminus aR$. For such b we have $b\mathfrak{p} \subset \mathfrak{p}\mathfrak{p}_2 \cdots \mathfrak{p}_k \subset aR$ and $ba^{-1}\mathfrak{p} \subset R$, i.e., $ba^{-1} \in \mathfrak{p}' \setminus R$, showing that $R \subsetneq \mathfrak{p}'$.

The product $\mathfrak{p}\mathfrak{p}'$ is an ideal and, moreover, $\mathfrak{p} = R\mathfrak{p} \subset \mathfrak{p}\mathfrak{p}' \subset R$. We have to show that $\mathfrak{p}\mathfrak{p}' = R$. Otherwise we would have $\mathfrak{p}\mathfrak{p}' = \mathfrak{p}$ and consequently $\mathfrak{p}(\mathfrak{p}')^n = \mathfrak{p}$ for all n, which would imply for every non-zero $x \in \mathfrak{p}$ and $y \in \mathfrak{p}' \setminus R$ the inclusion $xy^n \in \mathfrak{p} \subset R$ for every n. This in turn implies $xR[y] \subset R$, and so $xR[y]$ has to be an ideal in R. By (i) it has a finite number of generators, say $a_1, \ldots, a_m$, and so the R-module $R[y]$ has $a_1 x^{-1}, \ldots, a_m x^{-1}$

as generators. Proposition 1.4 shows that y is R-integral and condition (iii) implies $y \in R$, contrary to the choice of y. □

Lemma 1.4. *If a domain R satisfies* (i), (ii) *and* (iii), *then every ideal in R except R itself equals a product of prime ideals.*

Proof. Assume that the lemma is false and choose a non-zero ideal $I \neq R$ in R which is not a product of prime ideals. Lemma 1.2 shows that I contains a product $\mathfrak{p}_1 \ldots \mathfrak{p}_k$ of prime ideals and we can assume that I is selected so that the number k is minimal. If $k = 1$, then (ii) shows that I is itself a prime ideal, whence $k \geqslant 2$. Let $\mathfrak{p}$ be a prime ideal containing I. Then, say $\mathfrak{p} = \mathfrak{p}_1$, and $\mathfrak{p}_2 \ldots \mathfrak{p}_k \subset \mathfrak{p}^{-1}I \subset \mathfrak{p}^{-1}\mathfrak{p} = R$, whence $\mathfrak{p}^{-1}I$ is an ideal in R. By the choice of I we have $\mathfrak{p}^{-1}I = \mathfrak{q}_1 \ldots \mathfrak{q}_s$ with some prime ideals $\mathfrak{q}_i$, and finally $I = \mathfrak{p}\mathfrak{q}_1 \ldots \mathfrak{q}_s$, which is a contradiction. □

Now we may prove our theorem. Let I be a non-zero fractional ideal of R and let $a \neq 0$ in R be such that $aI \subset R$. Since aI is an ideal of R, we may apply Lemma 1.4 to get $aI = \mathfrak{p}_1 \ldots \mathfrak{p}_s$ with suitable prime ideals $\mathfrak{p}_i$, which implies $I = a^{-1}\mathfrak{p}_1 \ldots \mathfrak{p}_s = (a^{-1}R)\mathfrak{p}_1 \ldots \mathfrak{p}_s$. We can see that I is a product of invertible ideals, and so it must be invertible itself. □

The theorem just proved and Lemma 1.4 imply the following

Corollary. *In a Dedekind domain every proper non-zero ideal can be represented as a product of prime ideals.*

Theorem 1.4. *In a Dedekind domain every proper non-zero ideal can be represented uniquely as a product of prime ideals (products differing in order not being regarded as distinct).*

Proof. Only the uniqueness remains to be proved. Let $I \neq 0$ be a proper ideal in a Dedekind domain, having at least two different representations as a product of prime ideals, say,

$$I = \mathfrak{p}_1 \ldots \mathfrak{p}_r = \mathfrak{q}_1 \ldots \mathfrak{q}_s.$$

We may assume that I is chosen so that the number r of prime ideal factors in one of the decompositions is the least possible. Now $\mathfrak{q}_1 \ldots \mathfrak{q}_s \subset \mathfrak{p}_1$, whence one of the $\mathfrak{q}_i$'s, say $\mathfrak{q}_1$, is contained in $\mathfrak{p}_1$. By Theorem 1.3 $\mathfrak{p}_1 = \mathfrak{q}_1$ and by Lemma 1.3 we obtain

$$\mathfrak{p}_2 \mathfrak{p}_3 \ldots \mathfrak{p}_r = \mathfrak{p}_1^{-1}\mathfrak{p}_1 \ldots \mathfrak{p}_r = \mathfrak{p}_1^{-1}\mathfrak{q}_1 \ldots \mathfrak{q}_s = \mathfrak{q}_2 \ldots \mathfrak{q}_s.$$

Thus the ideal $\mathfrak{p}_2 \ldots \mathfrak{p}_r$ has two factorizations into prime ideals which are different, as the factorizations of I were different. But the number of factors in one of them is less than r, contrary to our choice of I, which is a contradiction. □

Corollary 1. *The group of all fractional ideals of a Dedekind domain R is a free abelian group generated by the prime ideals of R.*

Proof. Let A be a fractional ideal of R and choose $a \neq 0$ in R so that $aA \subset R$. The set aA is an ideal in R, hence $A = (aR)^{-1}(aA)$ is a product of powers of prime ideals with integral, not necessarily positive, exponents. Hence the prime ideals generate the group of all fractional ideals, and Theorem 1.4 shows that they generate it freely. □

Corollary 2. *A non-zero ideal can be divisible only by finitely many distinct ideals.* □

We see thus that every fractional ideal of a Dedekind domain may be uniquely written in the form

$$I = \prod_{\mathfrak{p}} \mathfrak{p}^{a(\mathfrak{p})},$$

where the product is extended over all prime ideals of R and the exponents $a(\mathfrak{p})$ are rational integers, only a finite number of them being non-zero.

Now we regain for the ideals of R many results of elementary number theory, connected with the notions of the least common multiple or the greatest common divisor.

We shall say that an ideal A is *divisible* by B if, with a suitable ideal C, we have $A = BC$. The greatest common divisor of two ideals A and B is an ideal which divides both A and B and is divisible by every other ideal with this property. The existence and uniqueness of such an ideal follows from the product-representation. It should only be noted that if $A = \prod \mathfrak{p}^{a(\mathfrak{p})}$, $B = \prod \mathfrak{p}^{b(\mathfrak{p})}$ are ideals, then $a(\mathfrak{p})$, $b(\mathfrak{p})$ are non-negative, and A divides B iff for all $\mathfrak{p}$ we have $a(\mathfrak{p}) \leq b(\mathfrak{p})$. This clearly implies that the greatest common divisor of A and B equals $\prod \mathfrak{p}^{c(\mathfrak{p})}$ with $c(\mathfrak{p}) = \min(a(\mathfrak{p}), b(\mathfrak{p}))$.

Similarly, the least common multiple of two ideals is defined as an ideal divisible by both and dividing every other ideal with this property. It is easy to see that the least common multiple of $A = \prod \mathfrak{p}^{a(\mathfrak{p})}$ and $B = \prod \mathfrak{p}^{b(\mathfrak{p})}$ is the ideal $\prod \mathfrak{p}^{d(\mathfrak{p})}$ with $d(\mathfrak{p}) = \max(a(\mathfrak{p}), b(\mathfrak{p}))$.

In accordance with the elementary theory of integers we shall denote the greatest common divisor of ideals A and B by (A, B) and their least common multiple by $[A, B]$.

The reader may prove for himself other properties of (A, B) and $[A, B]$ which are analogous to those known from elementary number theory, e.g., the equality $(A, B)[A, B] = AB$.

We end this subsection with the following

Proposition 1.5. *If R is a Dedekind domain, then:*

(i) If A, B are fractional ideals, then the inclusion $A \subset B$ holds iff we have $A = BC$ with some ideal $C \subset R$.

(ii) If A is a fractional ideal, then there exists a principal fractional ideal aR such that $(aR)A^{-1}$ is contained in R.

(iii) If A and B are relatively prime ideals in R, then their product AB and intersection $A \cap B$ coincide.

(iv) If A and B are ideals in R, then $(A, B) = A + B$.

Proof. The implication $A = BC \to A \subset B$ in (i) is trivial. If $A \subset B$, then $AB^{-1} \subset BB^{-1} = R$, and so $C = AB^{-1}$ is an ideal in R satisfying $BC = A$.

To prove (ii) take any non-zero element a in A. Then $aR \subset A$ and we may apply (i) to get the desired conclusion.

If A and B are relatively prime, then by (i) both A and B divide $A \cap B$, and so AB divides $A \cap B$, whence $A \cap B \subset AB$. The inclusion $AB \subset A \cap B$ being trivial, (iii) follows.

Finally, to obtain (iv), note that in view of (i) $A+B$ divides (A, B) and, on the other hand, $A+B$ is the minimal ideal containing A and B, whence it has to be divisible by (A, B). □

3. This subsection is devoted to linear congruences modulo an ideal in a Dedekind domain. Once again imitating the theory of rational integers we shall take $a \equiv b \pmod{I}$ to mean $a - b \in I$, where I is an ideal in R. The problem of solubility of a linear congruence is solved by

Proposition 1.6. *If R is a Dedekind domain, I an ideal in R and $a, b \in R$, then the congruence $ax \equiv b \pmod{I}$ has a solution x in R iff the element b lies in the ideal $I + aR$.*

Proof. If the congruence $ax \equiv b \pmod{I}$ has a solution x, then for a suitable $y \in I$ we have $b = ax + y \in aR + I$, proving the necessity of the condition stated. To prove its sufficiency, observe that $b \in I + aR$ implies $b = ax_1 + y$ for suitable $x_1 \in R$ and $y \in I$, i.e. the congruence $ax \equiv b \pmod{I}$ has a solution. □

Corollary 1. *If $\mathfrak{p}$ is a prime ideal in a Dedekind domain R and $a \in R \setminus \mathfrak{p}$, then for every natural n the congruence $ax \equiv b \pmod{\mathfrak{p}^n}$ has a solution for any $b \in R$.*

Proof. It suffices to observe that $\mathfrak{p}^n + aR = R$. □

Corollary 2. *If* $\mathfrak{p}_1, \ldots, \mathfrak{p}_m$ *are distinct prime ideals in a Dedekind domain* R, *then for any given* $a_1, \ldots, a_m$ *in* R *and every natural* n *one can find a common solution of the congruences* $x \equiv a_i \pmod{\mathfrak{p}_i^n}$ *for* $i = 1, 2, \ldots, m$.

Proof. Choose b_i in $(\mathfrak{p}_1 \cdots \mathfrak{p}_{i-1} \mathfrak{p}_{i+1} \cdots \mathfrak{p}_m)^n \setminus \mathfrak{p}_i$ for $i = 1, 2, \ldots, m$ and let x_i be the solution of $b_i x_i \equiv a_i \pmod{\mathfrak{p}_i^n}$. The element $x = b_1 x_1 + \cdots + b_m x_m$ has the desired property. □

Corollary 3 (Chinese remainder-theorem). *If* $I_1, \ldots, I_m$ *are pairwise relatively prime ideals of a Dedekind domain* R *and* $a_1, \ldots, a_m$ *in* R *are given, then there exists a common solution of the congruences* $x \equiv a_i \pmod{I_i}$ *for* $i = 1, 2, \ldots, m$.

Proof. Observe that any congruence of the form $x \equiv a \pmod{I}$ is equivalent to a system of congruences $x \equiv a_i \pmod{\mathfrak{p}_i^{a_i}}$, where $I = \prod \mathfrak{p}_i^{a_i}$, and apply Corollary 2. □

Corollary 4. *Let* I *and* J *be relatively prime ideals in a Dedekind domain* R. *Then one can find an element* x *of* I *which satisfies* $(xR, J) = R$, $(xI^{-1}, I) = R$ *and is such that for every ideal* I_1 *of* R *relatively prime to* I *there exists a* $y \in I$ *with* $(yR, I_1) = R$, $(yI^{-1}, I) = R$ *and* $xR + yR = I$.

Proof. Factorize I into prime ideals

$$I = \prod_{i=1}^{m} \mathfrak{p}_i^{a_i},$$

and let $x_i \in \mathfrak{p}_i^{a_i} \setminus \mathfrak{p}_i^{a_i+1}$ for $i = 1, 2, \ldots, m$. By Corollary 3 the system of congruences

$$x \equiv x_i \pmod{\mathfrak{p}_i^{a_i+1}} \quad (i = 1, 2, \ldots, m),$$
$$x \equiv 1 \pmod{J}$$

has a solution x. We may write $xR = II_2$ for a certain I_2 satisfying $(I_2, IJ) = 1$. Applying once more Corollary 3, we obtain the existence of y in R, which satisfies the system

$$y \equiv x_i \pmod{\mathfrak{p}_i^{a_i+1}} \quad (i = 1, 2, \ldots, m),$$
$$y \equiv 1 \pmod{I_1 I_2}.$$

Now $yR = II_3$ for a suitable I_3 satisfying $(I_3, I_2) = (I_3, I_1) = R$, which is equivalent to our final assertion. □

Corollary 5. *Every ideal in a Dedekind domain* R *is generated as an R-module by at most two elements.*

Proof. This follows immediately from Corollary 4. □

We end this rather long list of corollaries with

Corollary 6. *If I and J are given ideals in a Dedekind domain R, then one can find an ideal A such that $(A, IJ) = R$ and the product AI is principal.*

Proof. Let $I = \prod \mathfrak{p}^{a(\mathfrak{p})}$, $J = \prod \mathfrak{p}^{b(\mathfrak{p})}$ and let P be the set of prime ideals dividing IJ. For every $\mathfrak{p}$ in P let $x_\mathfrak{p} \in \mathfrak{p}^{a(\mathfrak{p})} \setminus \mathfrak{p}^{a(\mathfrak{p})+1}$. Corollary 3 implies the existence of an a in R satisfying $a \equiv x_\mathfrak{p} \pmod{\mathfrak{p}^{a(\mathfrak{p})+1}}$ for $\mathfrak{p} \in P$, and for such an a we have

$$aR = \prod \mathfrak{p}^{a(\mathfrak{p})} A = IA$$

with some A relatively prime to the product IJ. □

The next theorem reduces the problem of the structure of the factor ring R/I to the case where I is a prime ideal power:

Theorem 1.5. *If I is an ideal in a Dedekind domain R, and*

$$I = \prod \mathfrak{p}^{a(\mathfrak{p})}$$

is its factorization into prime ideal powers, then the factor ring R/I is isomorphic to the direct sum

$$\oplus R/\mathfrak{p}^{a(\mathfrak{p})}.$$

Proof. Consider the homomorphism $f: R \to \oplus R/\mathfrak{p}^{a(\mathfrak{p})}$ given by

$$f(x) = [x \bmod \mathfrak{p}^{a(\mathfrak{p})}]_\mathfrak{p}.$$

Corollary 3 shows us that f is surjective, and if x lies in $\operatorname{Ker} f$, then $x \in \mathfrak{p}^{a(\mathfrak{p})}$ for all $\mathfrak{p}$, and so $x \in I$, whence $\operatorname{Ker} f$ is contained in I. On the other hand, the inclusion $I \subset \operatorname{Ker} f$ is trivial, which shows $\operatorname{Ker} f = I$ and proves the theorem. □

We finish this section with the following useful

Proposition 1.7. *If $\mathfrak{p}$ is a non-zero prime ideal in a Dedekind domain R and n is an arbitrary natural number, then the factor rings $R/\mathfrak{p}$ and $\mathfrak{p}^n/\mathfrak{p}^{n+1}$ have isomorphic additive groups.*

Proof. Choose a in $\mathfrak{p}^n \setminus \mathfrak{p}^{n+1}$ and consider the mapping $g: x \mapsto ax$ of the additive group R^+ of R into the additive group of $\mathfrak{p}^n$. Since $g(\mathfrak{p}) \subset \mathfrak{p}^{n+1}$, g induces a homomorphism $\bar{g}$ of the additive group of $R/\mathfrak{p}$ into the additive group of $\mathfrak{p}^n/\mathfrak{p}^{n+1}$. Observe that if $\bar{x}$ lies in $\operatorname{Ker} \bar{g}$ and x is any representative of $\bar{x}$ in R, then $ax \in \mathfrak{p}^{n+1}$. This implies $x \in \mathfrak{p}$, i.e., $\bar{x} = 0$ and we see that $\bar{g}$ is an injection. To prove that it is also surjective, take any $\bar{y} \in \mathfrak{p}^n/\mathfrak{p}^{n+1}$ and let y be a representative of $\bar{y}$.

Since we have $(aR, \mathfrak{p}^{n+1}) = \mathfrak{p}^n$, Propositions 1.5 (i) and 1.6 imply the existence of x in R with $ax \equiv y \pmod{\mathfrak{p}^{n+1}}$, and for the element $\bar{x}$ of $R/\mathfrak{p}$ containing x we clearly have $\bar{g}(\bar{x}) = \bar{y}$. Hence $\bar{g}$ is an isomorphism. □

4. In this subsection R will be a Dedekind domain satisfying the following additional condition:

(FN)   *For every non-zero ideal $I \subset R$ the factor ring R/I is finite.*

The number of elements in R/I we shall call the *absolute norm of I*, or the *norm*, for short, and denote it by $N(I)$. Observe that for every I the ideal $N(I)R$ is divisible by I, since the canonical image of $N(I)e$ (where e is the unit element of R) in R/I is zero and, moreover, the norm of a prime ideal is a prime power, since in this case R/I is a finite field. The main properties of the norm are given in the following theorem:

Theorem 1.6. (i) *For any non-zero ideals I, J we have $N(IJ) = N(I)N(J)$.*

(ii) *If T is a given positive number, then the number of ideals I of R satisfying $N(I) \leq T$ is finite.*

Proof. (i) By Proposition 1.7 for any prime ideal $\mathfrak{p}$ the factor ring $\mathfrak{p}^n/\mathfrak{p}^{n+1}$ has $N(\mathfrak{p})$ elements, and since $|R/\mathfrak{p}^{n+1}|/|R/\mathfrak{p}^n| = |\mathfrak{p}^n/\mathfrak{p}^{n+1}|$, we have $N(\mathfrak{p}^n) = N(\mathfrak{p})^n$ and it suffices to apply Theorem 1.5.

(ii) Consider any set of more than $T+1$ distinct elements $a_1, a_2, \ldots, a_m$ of R. For every ideal I satisfying $N(I) \leq T$ we can find distinct i, j such that $a_i \equiv a_j \pmod{I}$. The set of differences $a_i - a_j$ being finite, our assertion now follows from Corollary 2 to Theorem 1.4. □

This theorem shows that the norm is a homomorphism of the semigroup of all non-zero ideals of R into the multiplicative semigroup of natural numbers. Corollary 1 to Theorem 1.4 enables us to extend this homomorphism to the group of all fractional ideals of R, the value group being the multiplicative group of positive rationals. This extended homomorphism we shall again denote by $N(I)$ and call it the *norm*.

Now we shall prove two results generalizing the theorems of Fermat and Euler in the elementary theory of numbers.

Theorem 1.7. *If $\mathfrak{p}$ is a prime ideal in R, then for all $x \in R$ we have $x^{N(\mathfrak{p})} \equiv x \pmod{\mathfrak{p}}$. Moreover, $N(\mathfrak{p})$ is the least positive exponent n such that for all $x \in R$ the congruence $x^n \equiv x \pmod{\mathfrak{p}}$ holds true.*

Proof. If $x \in \mathfrak{p}$, then the theorem is trivial. For $x \notin \mathfrak{p}$ we have $x^{N(\mathfrak{p})-1} \equiv 1 \pmod{\mathfrak{p}}$, since $R/\mathfrak{p}$ is a field with $N(\mathfrak{p})$ elements and its multiplicative group is of order $N(\mathfrak{p})-1$. This proves the first part. To prove the second, it is enough to observe that the multiplicative group of $R/\mathfrak{p}$ is cyclic, and so, for any representative $x \in R$ of its generator, the powers $1, x, x^2, \ldots, x^{N(\mathfrak{p})-2}$ are distinct (mod $\mathfrak{p}$). □

Theorem 1.8. *If we denote by $\varphi(I)$ the number of invertible elements in R/I, then*

$$\varphi(I) = N(I) \prod_{\mathfrak{p}} (1 - N(\mathfrak{p})^{-1}),$$

where the product is extended over all prime ideals dividing I and, moreover, if $x \in R$ and $(xR, I) = R$, then

$$x^{\varphi(I)} \equiv 1 \pmod{I}.$$

Proof. Theorem 1.5 implies that the group of invertible elements in R/I equals the product of such groups in $R/\mathfrak{p}^{a(\mathfrak{p})}$ when $I = \prod \mathfrak{p}^{a(\mathfrak{p})}$ is a factorization into prime ideals. Hence it suffices to prove our theorem for $I = \mathfrak{p}^n$. For $n = 1$ the first part is evident and the second is contained in Theorem 1.7. For the general case, observe that

$$\varphi(\mathfrak{p}^n) = N(\mathfrak{p}^n) - N(\mathfrak{p}^{n-1}) = N(\mathfrak{p})^n (1 - N(\mathfrak{p})^{-1}).$$

The second part is a trivial consequence of the definition of $\varphi(I)$. □

5. We shall now investigate the behaviour of Dedekind domains under the operation of integral closure. The main classical result concerning this problem is contained in the following

Theorem 1.9. *Let R be a Dedekind domain and let K be its field of quotients. Let L/K be a finite, separable extension of K and let S be the integral closure of R in L. Then S is again a Dedekind domain. Moreover, if R satisfies the condition* (FN), *then S does so as well.*

Proof. We shall check conditions (i), (ii) and (iii) of Theorem 1.3. Since L/K is separable and finite, it can be generated over K by a single element, which may be taken from S. In fact, if a generates L over K and satisfies the equation

$$A_n X^n + A_{n-1} X^{n-1} + \ldots + A_0 = 0 \quad (A_n \neq 0)$$

with coefficients $A_i \in R$, then the element $A_n a$ satisfies the equation

$$X^n + A_{n-1} X^{n-1} + A_{n-2} A_n X^{n-2} + \ldots + A_0 A_n^{n-1} = 0,$$

and so it is integral over R, i.e. belongs to S.

Consider a fixed algebraic closure of K and let $L_1, \ldots, L_n$ ($n = [L:K]$)

be the embeddings of L into that closure. For any $x \in L$, let $x^{(i)}$ denote its image in L_i. We need a lemma which will be also used in the next chapter:

Lemma 1.5. *Let ϑ be an element of S generating the extension L/K. If $D = \det[(\vartheta^{(i)})^k]_{i,k=1,\ldots,n}$ then D^2 is a non-zero element of R and, moreover, $S \subset D^{-2}R[\vartheta]$.*

Proof. Let a be an arbitrary element of S. With suitable $a_0, a_1, \ldots, a_{n-1} \in K$ we can write

$$a = \sum_{k=0}^{n-1} a_k \vartheta^k.$$

It follows that for $i = 1, 2, \ldots, n$ we have

$$a^{(i)} = \sum_{k=0}^{n-1} a_k (\vartheta^{(i)})^k,$$

and thus $a_k = A_k/D$, where A_k is a determinant whose elements are integral over R. (Note that $D \neq 0$, since it is a Vandermonde determinant.) Observe that D^2 is invariant under automorphisms from the Galois group of the least normal extension of K containing L, whence $D^2 \in K$. But D^2 is integral over R and thus $D^2 \in R$.

Now we may write $a_k = A_k D/D^2$ ($k = 0, 1, \ldots, n-1$) and $a_k \in R$, $D^2 \in R$ show that $A_k D \in R$. Thus we can see that for every $a \in S$, $D^2 a \in R[\vartheta]$, and so, with $c = D^{-2}$ we obtain the inclusion $S \subset cR[\vartheta]$.

The mapping $f: R^n \to cR[\vartheta]$ given by

$$f: (x_1, \ldots, x_n) \mapsto c(x_1 + x_2\vartheta + \ldots + x_n\vartheta^{n-1})$$

is a homomorphism of R-modules which is surjective, and by Proposition 1.2 we see that $cR[\vartheta]$ is a Noetherian R-module and, since $S \subset cR[\vartheta]$, S must also be Noetherian. But every ideal of S is trivially an R-module, and so S is a Noetherian ring. This proves that condition (i) of Theorem 1.3 is satisfied by S. Since the quotient field of S is contained in L (in fact, it is equal to L), the Corollary to Theorem 1.2 implies that condition (iii) is also satisfied by S, and it remains to show that every prime ideal in S is maximal. For this purpose we first prove

Lemma 1.6. *Under the assumptions of Theorem 1.9 if $\mathfrak{P}$ is a prime ideal in S, then the ideal $\mathfrak{P} \cap R$ of R is also prime. Moreover, if for the prime ideals $\mathfrak{P}_1$ and $\mathfrak{P}_2$ of S we have $\mathfrak{P}_1 \subset \mathfrak{P}_2$ and $\mathfrak{P}_1 \cap R = \mathfrak{P}_2 \cap R$, then $\mathfrak{P}_1 = \mathfrak{P}_2$.*

Proof. Observe that the injection $R \to S$ carries $\mathfrak{P} \cap R$ into $\mathfrak{P}$, and so it induces a homomorphism of the factor rings $R/(\mathfrak{P} \cap R) \to S/\mathfrak{P}$, which is clearly again an injection. Since $\mathfrak{P}$ is prime, $S/\mathfrak{P}$ has no zero-divisors, and so $R/(\mathfrak{P}\cap R)$ has the same property, showing that $\mathfrak{P}\cap R$ is prime.

Now assume $\mathfrak{P}_1\cap R = \mathfrak{P}_2\cap R$, but $\mathfrak{P}_1 \subsetneq \mathfrak{P}_2$. Take any element x in $\mathfrak{P}_2\setminus\mathfrak{P}_1$ and let

$$x^n + a_{n-1}x^{n-1} + \ldots + a_0 = 0$$

be its equation over R. If all coefficients a_i lie in $\mathfrak{P}_1\cap R$, then $x^n \in \mathfrak{P}_1$, whence $x \in \mathfrak{P}_1$, which is a contradiction. Thus there is a minimal index j such that $a_j \notin \mathfrak{P}_1\cap R$. Then we have

$$x^j(x^{n-j} + a_{n-1}x^{n-j-1} + \ldots + a_j) \in \mathfrak{P}_1;$$

thus $x^{n-j} + \ldots + a_{j+1}x + a_j \in \mathfrak{P}_1 \subset \mathfrak{P}_2$. But $x^{n-j} + \ldots + a_{j+1}x$ lies in $\mathfrak{P}_2$, implying $a_j \in \mathfrak{P}_2$ and $a_j \in \mathfrak{P}_2\cap R = \mathfrak{P}_1\cap R$, contrary to the choice of j. This contradiction proves the lemma. □

Now it is easy to verify condition (ii). If $\mathfrak{P}_1$ is a non-maximal, non-zero prime ideal in S, then it is contained in a maximal (and a fortiori prime) ideal $\mathfrak{P}_2$. The lemma just proved shows that the prime ideals $\mathfrak{P}_1\cap R$ and $\mathfrak{P}_2\cap R$ of the ring R are distinct, but clearly the first of them is contained in the second, which is possible only if $\mathfrak{P}_1\cap R = 0$. However, in this case $\mathfrak{P}_1\cap R = 0\cap R$ and Lemma 1.6 gives $\mathfrak{P}_1 = 0$, a contradiction.

Now assume that R satisfies the condition (FN). In the proof of Lemma 1.6 we have seen that for an arbitrary prime ideal $\mathfrak{P}$ of S the field $S/\mathfrak{P}$ is an extension of $R/\mathfrak{p}$, with $\mathfrak{p} = \mathfrak{P}\cap R$, and $R/\mathfrak{p}$ is finite by assumption. Since every element of S satisfies an equation with coefficients from R of a degree not exceeding n, it follows that every element of $S/\mathfrak{P}$ satisfies a similar equation, and so the extension of the field of residue classes is finite. This establishes the finiteness of $S/\mathfrak{P}$ and, by Theorem 1.5 and Proposition 1.7, the property (FN) for S follows. □

§ 2. Valuations and Exponents

1. In this section we list the fundamental definitions and properties of valuations which will be needed in the sequel.

Let K be any field. A homomorphism v of its multiplicative group K^* into the group of positive reals is called a *valuation* if it satisfies the condition

$$v(x+y) \leq v(x) + v(y)$$

for all non-zero $x, y \in K$. By putting $v(0) = 0$ one extends any valuation to the whole field K.

Every valuation v induces in K a metric $d(x, y) = v(x-y)$ under which the additive and multiplicative groups of K become topological groups. In fact, continuity of addition follows immediately from the definition, and the continuity of the inverse in K^+ results from $v(-x) = v(x)$ which is implied by the formulas $v(-x) = v(-1)v(x)$ and $v(1) = v(-1)^2 = 1$. To check the continuity of multiplication choose a, b in K^* and a positive ε. If $M = \max(v(a), v(b))$ and δ does not exceed $\min(1, \varepsilon(1+2M)^{-1})$, then the inequalities $v(x-a) < \delta$ and $v(y-b) < \delta$ easily imply $v(xy-ab) < \varepsilon$. Finally, the continuity of the inverse in K^* follows from $v(x^{-1}) = v(x)^{-1}$. Hence K becomes a topological field. Note that it is not necessarily complete or locally compact, as is shown by the example of the field Q of rational numbers under the valuation $v(x) = |x|$.

The valuation defined by $v(x) = 1$ for all x in K^* is called the *trivial valuation*. It induces the discrete topology.

Two valuations are said to be *equivalent* if they define the same topology. The connection between such valuations is given in the following proposition:

Proposition 1.8. *If v and w are equivalent valuations of a field K, then with a suitable positive a one has for all x in K the equality $w(x) = v(x)^a$.*

Proof. If $v(x) = 1$ for all $x \neq 0$, then v induces the discrete topology, and so $w(x) = 1$ must hold for all $x \neq 0$ since the existence of an element x with $0 < w(x) < 1$ would imply $\lim x^n = 0$. Similarly, the existence of an element x with $w(x) > 1$ leads to a contradiction. Assume thus that v is non-trivial, choose $x_0 \neq 0$ with $v(x_0) \neq 1$ and put $a = (\log w(x_0))/(\log v(x_0))$. Note that the sets of those $x \in K$ for which $v(x)$ or $w(x)$ exceeds unity are equal, since they are formed by those elements x for which x^{-n} tends to zero in the induced topology. For any x in K let

$$b_1 = (\log w(x))/(\log w(x_0)) \quad \text{and} \quad b_2 = (\log v(x))/(\log v(x_0)),$$

and let $r = m/n$ be any rational number larger than b_1. Then $w(x_0^m) > w(x^n)$, whence $w(x_0^m x^{-n}) > 1$; thus $v(x_0^m x^{-n}) > 1$, i.e., $v(x_0^m) > v(x^n)$, which implies $r \geq b_2$ and thus $b_1 \geq b_2$. Interchanging the roles of v and w, we are led to $b_1 = b_2$, showing that the ratio $(\log w(x))/(\log v(x))$ does not depend on x, and our assertion follows. □

If a valuation v satisfies the condition

$$v(x+y) \leq \max(v(x), v(y)) \tag{1.3}$$

for all $x, y \in K$, then it is called a *non-Archimedean valuation*. All the remaining valuations are called *Archimedean*.

Proposition 1.9. *If v is a non-Archimedean valuation and $v(a) \neq v(b)$, then $v(a+b) = \max(v(a), v(b))$.*

Proof. Assume $v(a) < v(b)$. Then $v(a+b) \leq v(b)$, but on the other hand we have
$$v(b) = v((a+b)-a) \leq \max(v(a+b), v(a)),$$
and since the inequality $v(b) \leq v(a)$ is ruled out, we must have
$$v(b) \leq v(a+b). \qquad \square$$

The traditional formulation of the Archimedean axiom runs as follows.

(A) *If a and b are non-zero elements of K, then with a suitable positive integer n we have $v(na) > v(b)$.*

The non-Archimedean valuations are exactly those which disobey axiom (A). This is contained in the next proposition.

Proposition 1.10. *If v is a valuation of a field K, then the following properties are equivalent:*
   (i) *v is non-Archimedean.*
   (ii) *For every positive integer n one has $v(n) \leq 1$.*
   (iii) *For every positive integer n one has $v(n) \leq B$, where B is a constant independent of n.*
   (iv) *v does not satisfy (A).*

(Here we identify a positive integer n with the sum of n copies of the unit element of K.)

Proof. (i) $\to$ (ii). Inequality (1.3) implies $v(n) \leq v(1) = 1$.
   (ii) $\to$ (iii). Obvious.
   (iii) $\to$ (iv). For every positive integer n we have $v(na) = v(n) v(a) \leq Bv(a)$, whence if we choose b so that $v(b)$ exceeds $Bv(a)$, then (A) will fail. Such a choice is always possible for non-trivial v, since $v(x) > 1$ implies $v(x^m) \to \infty$. If, however, v is trivial then (A) clearly does not hold.
   (iv) $\to$ (ii). Choose $a, b \neq 0$ so that for all positive integers n we have the inequality $v(na) \leq v(b)$. Then $v(n) \leq v(ba^{-1})$. If, for some n_0, $v(n_0) > 1$ then with a suitable k we would have $v(n_0^k) > v(ba^{-1})$, a contradiction.
   (ii) $\to$ (i). Since
$$v((a+b)^n) \leq \sum_{k=0}^{n} v\left(\binom{n}{k} a^k b^{n-k}\right) \leq \sum_{k=0}^{n} v(a)^k v(b)^{n-k} \leq (n+1) \max(v(a)^n, v(b)^n),$$
we have
$$v(a+b) \leq (n+1)^{1/n} \max(v(a), v(b)),$$
and (1.3) follows. $\qquad \square$

Corollary. *If the field K has a non-zero characteristic then all its valuations are non-Archimedean.*

Proof. The set $\{1, 1+1, \ldots\}$ being finite, condition (iii) is trivially satisfied $\square$.

2. A valuation v is called *discrete* if the set of values of $\log v$ is discrete. Such valuations are closely connected with the *exponents* of K, i.e., homomorphisms $n\colon K^* \to \mathbb{Z}$ which are surjective and satisfy the condition

$$n(a+b) \geqslant \min(n(a), n(b)) \tag{1.4}$$

for all $a, b \neq 0$ in K.

If $0 < c < 1$ and n is an exponent of K, then $v(x) = c^{n(x)}$ is a discrete valuation of K, non-Archimedean by (1.4). Conversely, one can easily see that every discrete non-Archimedean valuation can be obtained in such a way from a suitable exponent.

There is a standard way of constructing exponents in the field K of quotients of a Dedekind domain R. Take any non-zero prime ideal P of R and let $x \neq 0$, $x \in K$. We may write $xR = P^{n(x)}I$, with $n(x) \in \mathbb{Z}$ and a fractional ideal I whose decomposition into prime ideals does not contain P. In this way we define a function n which is an exponent of K. One may define a valuation using this exponent in many ways, depending on the choice of c in $v(x) = c^{n(x)}$. However, in the case where R satisfies the (FN) condition it is convenient to make a particular choice of c, namely $c = N(P)^{-1}$. In this case we speak of a *normalized valuation* corresponding to P.

Note that if the prime ideals P_1 and P_2 are different, then the corresponding valuations are not equivalent. In fact, if $x_n \in P_1^n \setminus P_2$, then x_n tends to zero in the topology induced by P_1 but not in the topology induced by P_2.

The topology induced in K by a prime ideal P of R is called the *P-adic topology*. In the ring of rational integers the prime ideals are generated by rational primes p and the resulting topology is called the *p-adic topology*.

For an arbitrary exponent n of K put

$$R_n = \{x \in K \colon n(x) \geqslant 0\} \quad \text{and} \quad P_n = \{x \in K \colon n(x) > 0\}.$$

We now prove

Theorem 1.10. *The set R_n is a principal ideal (and hence Dedekind) domain and P_n is the unique non-zero prime ideal of R_n. It is generated by every element $a \in K$ satisfying $n(a) = 1$.*

Proof. From the definition of the exponent it follows that R_n is a ring and P_n is an ideal in R_n. If a is a non-zero element of $R_n \setminus P_n$, then $n(a) = 0$, thus $n(a^{-1}) = 0$ and $a^{-1} \in R_n$, i.e., a is invertible. This shows that P_n consists of all non-invertible elements of R_n and so is the only maximal ideal of R_n. To show that P_n is principal consider any element a satisfying $n(a) = 1$. Then obviously $aR_n \subset P_n$ and for every $b \in P_n$ with, say, $n(b) = m$ we have $ba^{-m} \in R_n$.

Thus $b = a^m(ba^{-m}) \in a^m R_n \subset aR_n$, whence $P_n \subset aR_n$ and finally $P_n = aR_n$. We see, moreover, that $P_n^k = a^k R_n = \{a \in K: n(a) = k\}$ holds for $k = 1, 2, \ldots$

The maximality of P_n implies that every proper ideal I of R_n is contained in P_n. If $I \neq 0$, then we can choose N so that $I \subset P_n^N$, $I \not\subset P_n^{N+1}$. Such a choice is possible, since the intersection of all powers of P_n is the zero ideal. Let $b \in I$ with $n(b) = N$. Then $b = a^N c$ with $c \in R_n$ and $n(c) = 0$. Thus c is invertible in R_n and this implies that the ideals $a^N R_n$ and bR_n coincide. But $P_n^N = a^N R_n = bR_n \subset I$ and thus $I = P_n^N$. It follows that every non-zero ideal of R_n is principal and a power of P_n. □

The ring R_n is called the *exponent ring of n*, and if v is any valuation induced by n, then R_n is also called the *valuation ring of v*. Similarly, P_n is called the *ideal of the exponent n* (or the *valuation ideal of v*).

Theorem 1.10 applies in particular to the case where n is the exponent induced by a prime ideal of a Dedekind domain R. We shall now consider this case more closely.

Proposition 1.11. *Let R be a Dedekind domain with the field of quotients K; let P be an arbitrary non-zero prime ideal of R and let n be the exponent induced by P in K. Then we have*
(i) $R_n = \{a/b \in K: a, b \in R, b \notin P\}$ *and*
   $P_n = \{a/b \in K: a, b \in R, a \in P, b \notin P\}$.
(ii) $P_n^m \cap R = P^m$ $(m = 1, 2, \ldots)$.
(iii) $P_n^m = P^m R_n$ $(m = 1, 2, \ldots)$.
(iv) *The rings R/P^m and R_n/P_n^m are isomorphic for $m = 1, 2, \ldots$*
(v) *The intersection of all rings R_n taken over all prime ideals of R equals R.*

Proof. (i) If $a, b \in R$, $b \notin P$, then $n(a/b) \geq 0$, hence $a/b \in R_n$. If $x \in R_n$, then we can write $xR = I/J$, where I, J are ideals of R, with $P \nmid J$. By Corollary 6 to Proposition 1.6 we can find an ideal $A \subset R$ which is not divisible by P and for which the product AJ is principal. Then $xR = (AI)/(AJ)$, the ideals AI, AJ are both principal, and if $AI = aR$, $AJ = bR$, then $b \notin P$ and $x = ac/b$ with a suitable $c \in R$. This establishes the first equality, and to obtain the second it suffices to observe that if $\pi \in P \setminus P^2$, then $P_n = \pi R_n$ follows from Theorem 1.10.

(ii) This assertion results from the observation that $P^m = \{a \in R: n(a) \geq m\}$.

(iii) Theorem 1.10 implies $P_n = \pi R_n$ for $\pi \in P \setminus P^2$ and thus $P_n \subset PR_n$. On the other hand, every x in PR_n is of the form $x = a_1 b_1 + \ldots + a_k b_k$ with $a_i \in R$, $n(a_i) \geq 1$ and $n(b_i) \geq 0$; hence $n(x) \geq \min_i n(a_i b_i) \geq 1$. This shows that $x \in P_n$ and implies that $PR_n \subset P_n$. Thus $PR_n = P_n$ follows and the equality $P_n^m = P^m R_n$ results immediately.

(iv) In view of the embedding $P^m \subset P^m R_n = P_n^m$ the embedding $R \subset R_n$ induces a homomorphism $f\colon R/P^m \to R_n/P_n^m$. We shall now show that f is an isomorphism. Let $\bar{a} \in R/P^m$, $a \in \bar{a}$ and assume that $\bar{a} \in \operatorname{Ker} f$. Since $f(\bar{a})$ is the coset mod P_n^m determined by a in R_n, we have $a \in P^m R_n \cap R = P^m$ in view of (ii) and thus $\bar{a} = 0$. So f is an embedding.

Now let $a \in R_n$. To prove that f is surjective it suffices to find an x in R satisfying $a - x \in P_n^m$. Write $a = a_1/a_2$ with $a_1, a_2 \in R$ and let $r = n(a_2)$. Then $n(a_1) \geq r$ and so a lies in the ideal $P^{m+r} + a_2 R = P^r$. Consequently, by Proposition 1.6, the congruence $a_2 x \equiv a_1 \pmod{P^{m+r}}$ has a solution in R and for this solution we have

$$n(a-x) = n(a_1 - a_2 x) - n(a_2) \geq m,$$

i.e., $a - x \in P_n^m$ as required.

(v) The inclusion $R \subset \bigcap R_n$ is obvious, and if $a \in K$ is such that for all prime ideals of R the corresponding exponents are non-negative at a, then aR is an ideal of R, whence $a \in R$. □

3. In this subsection we prove the *weak approximation theorem* for valuations, due to E. Artin and G. Whaples [45], which can be regarded as an abstract version of the Chinese remainder-theorem (Corollary 3 to Proposition 1.6), to which it reduces in the case where all valuations occurring in its statement are induced by prime ideals of a Dedekind domain R with quotient field K, and the elements a_i belong to R.

Theorem 1.11. *Let $v_1, \ldots, v_t$ be non-equivalent and non-trivial valuations of a field K and let $a_1, \ldots, a_t \in K$. To every positive ε there corresponds an element a of K such that for $i = 1, 2, \ldots, t$ one has $v(a - a_i) < \varepsilon$.*

Proof. We follow the argument of Artin and Whaples and use two lemmas.

Lemma 1.7. *Under the assumptions of the theorem one can find an element $x \in K$ for which $v_1(x) > 1$ and $v_j(x) < 1$ for $j = 2, 3, \ldots, t$.*

Proof. First consider the case of $t = 2$. Since v_1 and v_2 are not equivalent, one can find a sequence $\{x_n\}$ such that $v_2(x_n)$ tends to zero whereas $v_1(x_n) \geq \delta$ holds for a certain fixed positive δ. Select $c \in K$ so that $v_1(c) > 1/\delta$. If we take n so large that $v_2(x_n) < v_2(c)^{-1}$ holds, then for $a = cx_n$ we get

$$v_2(a) = v_2(c)v_2(x_n) < 1 \quad \text{and} \quad v_1(a) = v_1(c)v_1(x_n) > 1.$$

In the general case we proceed by induction and assume that the lemma is true for systems of $t-1$ valuations. Choose a, b in K so that the inequalities

$$v_1(a) > 1, \quad v_j(a) < 1 \quad (j = 2, 3, \ldots, t-1),$$
$$v_1(b) > 1, \quad v_t(b) < 1$$

are satisfied. If $v_t(a) \leqslant 1$, then for sufficiently large n the element $x = a^n b$ fulfils our assertion, and if $v_t(a) > 1$ then define a sequence $x_n \in K$ by

$$x_n = a^n b(a^n+1)^{-1}.$$

For $j = 2, 3, \ldots, t-1$ we obtain

$$v_j(x_n) = v_j(a)^n v_j(b)(v_j(a^n)+1))^{-1} \leqslant v_j(a)^n v_j(b)(1-v_j(a)^n)^{-1}$$

and for $j = t$ we get

$$v_t(x_n) \leqslant v_t(a)^n v_t(b)(v_t(a)^n-1)^{-1}.$$

In both cases, for sufficiently large n, the right-hand side of the resulting inequalities is less than 1 since $v_j(a)^n$ tends to zero for $j = 2, 3, \ldots, t-1$ and $v_t(a)^n(v_t(a)^n-1)^{-1}$ tends to unity. Moreover, we have

$$v_1(x_n) \geqslant v_1(a)^n v_1(b)(1+v_1(a)^n)^{-1}$$

and this is larger than 1 for a sufficiently large n. $\square$

Lemma 1.8. *Under the assumptions of the theorem one can find an element y of K such that $v_1(y-1) < \varepsilon$ and $v_j(y) < \varepsilon$ for $j = 2, 3, \ldots, t$.*

Proof. Lemma 1.7 enables us to find $a \in K$ with $v_1(a) > 1$ and $v_j(a) < 1$ for $j = 2, 3, \ldots, t$. It now suffices to put $y = a^n(1+a^n)^{-1}$ with n sufficiently large. $\square$

The theorem now follows easily. Put $M = \max_{i,j} v_i(a_j)$ and choose $x_i \in K$ in such a way that the inequalities

$$v_i(1-x_i) < \varepsilon(tM)^{-1}, \quad v_j(x_i) < \varepsilon(tM)^{-1} \quad (j \neq i)$$

are satisfied for $i = 1, 2, \ldots, t$. The element $a_1 x_1 + \ldots + a_t x_t$ then satisfies the assertion. $\square$

Corollary. *Let $v_1, \ldots, v_t$ be non-equivalent and non-trivial non-Archimedean valuations of a field K. If $a_1, \ldots, a_t \in K$ are given, then one can find an $x \in K$ such that for $j = 1, 2, \ldots, t$ one has $v_j(x) = v_j(a_j)$.*

Proof. Let, in Theorem 1.11, $\varepsilon = \min_j v_j(a_j)$ and find an $x \in K$ such that $v_j(x-a_j) < \varepsilon$ holds for $j = 1, 2, \ldots, t$. If for a certain j we had $v_j(x) \neq v_j(a_j)$ then, by Proposition 1.9, we would have

$$v_j(x-a_j) = \max(v_j(x), v_j(a_j)) \geqslant v_j(a_j) \geqslant \varepsilon,$$

contradicting the choice of x. $\square$

4. We conclude this section with a description of all valuations of the field Q of rational numbers. In Chapter 3 we shall describe all valuations of a finite extension of Q.

Theorem 1.12. *If v is a non-trivial valuation of Q, then either v is equivalent to the ordinary absolute value $|x|$ or it is equivalent to one of the p-adic valuations induced by rational primes.*

Proof. Assume first that v is Archimedean. Let $m, n > 1$ be integers and write
$$m = a_0 + a_1 n + \ldots + a_r n^r$$
with integers $0 \leq a_i < n$ and non-zero a_r. This implies
$$v(m) \leq v(a_0) + \ldots + v(a_r)v(n)^r.$$
Since for all positive rational integers N we have
$$v(N) \leq v(1) + \ldots + v(1) = N,$$
it follows that
$$v(m) \leq n(1 + v(n) + \ldots + v(n)^r) \leq n(r+1)\max(1, v(n)^r).$$
However, $n^r \leq m$ and so $r \leq (\log m)/(\log n)$, implying
$$v(m) \leq n(1 + (\log m)/(\log n))\max(1, v(n)^{(\log m)/(\log n)}).$$
Now put $m = M^k$ with a fixed M, extract the kth roots on both sides and let k tend to infinity. In this way we obtain the inequality
$$v(M) \leq \max(1, v(n)^{(\log M)/(\log n)}), \tag{1.5}$$
valid for every pair M, n of integers larger than 1. Since v is Archimedean, we may choose M with $v(M) > 1$ and obtain
$$v(n)^{(\log M)/(\log n)} > 1,$$
which proves $v(n) > 1$ for all integers $n > 1$. But this together with (1.5) implies
$$v(M) \leq v(n)^{(\log M)/(\log n)}$$
for all integers $M, n > 1$, i.e.,
$$(\log v(M))/(\log M) \leq (\log v(n))/(\log n).$$
Interchanging M and n, we finally arrive at
$$(\log v(n))/(\log n) = c$$
with a constant c, i.e., $v(n) = n^c$ for all integers $n > 1$. This implies $v(x) = |x|^c$ for all rational x, and so $v(x)$ is equivalent to $|x|$.

Now consider the case of non-Archimedean v. By Proposition 1.10 we have $v(n) \leq 1$ for all integers n. Let A be the set of all those integers n for which $v(n) < 1$. If it is $\{0\}$, then v is trivial, which is excluded. Hence A is non-zero and, since $1 \notin A$, we find from (1.3) that A is a non-zero proper ideal in Z; thus $A = mZ$ with a suitable $m > 1$. Since m is the smallest positive element of A, it must be a prime, because a factorization $m = rs$ with $r, s > 1$ would imply $1 > v(m) = v(r) v(s) = 1$, which is impossible. Let $v(m) = a$ and let n be the exponent induced by the prime ideal mZ. Then $v(x) = a^{n(x)}$, and so v is a p-adic valuation induced by $p = m$. □

Corollary. *If v is a discrete valuation of a field K, then it is non-Archimedean.*

Proof. Assume that v is Archimedean. By the Corollary to Proposition 1.10 K is of characteristic zero and thus contains Q. The restriction of v to Q must be Archimedean, and so by the theorem it must be equivalent to $|x|$, whence non-discrete. □

§ 3. Finitely Generated Modules over Dedekind Domains

1. We shall now be concerned with the structure of finitely generated modules over a Dedekind domain R with the field of quotients K. This structure is described by the following

Theorem 1.13. *Let M be a finitely generated R-module and let A be the submodule of M consisting of all torsion elements of M, i.e., of elements $x \in M$ which, for some non-zero $r \in R$, satisfy $rx = 0$. Then M can be written as a direct sum*

$$M = R^k \oplus I \oplus A,$$

where k is a natural number and I is some ideal of R.

For the proof of this theorem we shall need various results concerning projective modules over commutative rings, not necessarily Dedekind.

If R is any commutative ring with a unit element, then an R-module M is called *projective* if every diagram of the form

$$\begin{array}{c} M \\ \downarrow \\ A \to B \to 0 \end{array}$$

with exact row and arbitrary R-modules A, B can be embedded in a commutative diagram

$$\begin{array}{c} M \\ \swarrow \downarrow \\ A \to B \to 0 \end{array}$$

Proposition 1.12. *The direct sum $\oplus P_a$ of R-modules is projective if and only if its every summand is projective.*

Proof. Denote by i_a the canonical injection of P_a into $\oplus P_a$ and by p_a the projection of $\oplus P_a$ onto P_a. Assume now that $P = \oplus P_a$ is projective, the sequence $A \to B \to 0$ is exact and $f: P_a \to B$ is a homomorphism. Then $f \circ p_a$ is a homomorphism of P into B, whence, for a suitable $g: P \to A$, the diagram

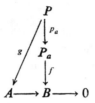

is commutative. Now it suffices to observe that the mapping $h = g \circ i_a$ makes the diagram

commutative, and so P_i is projective.

To prove the second part of the proposition assume that all modules P_a are projective, the sequence $A \to B \to 0$ is exact and a homomorphism $f: P \to B$ is given. Then $f \circ i_a$ maps P_a into B, and so, for a suitable $g_a: P_a \to A$, the diagram

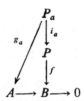

commutes. The projectivity of P follows from the observation that the mapping $h = \oplus g_a$ makes the diagram

commutative. □

Corollary. *Every free R-module is projective.*

Proof. As every free R-module is a direct sum of R-modules R, it suffices

§ 3. Finitely Generated Modules over Dedekind Domains 25

in view of the proposition just proved to establish the projectivity of R. Let $f: R \to B$ be a homomorphism, and let the sequence $A \xrightarrow{g} B \to 0$ be exact. If $f(1) = b$ and a is any element in A with $g(a) = b$, then the mapping $h: R \to A$ given by $h(x) = xa$ has the required property. □

The properties of an R-module equivalent to its projectivity are established in the following

Proposition 1.13. *The following properties of an R-module M are equivalent*:
 (i) *If the sequence $0 \to A \to B \to M \to 0$ is exact, then $A \oplus M \simeq B$.*
 (ii) *The R-module M is a direct summand of a suitable free R-module.*
 (iii) *The R-module M is projective.*

Proof. (i) → (ii). The module M is a homomorphical image of a free module F, and so, for a suitable N, the sequence $0 \to N \to F \to M \to 0$ is exact. By (i) $F \simeq M \oplus N$.

(ii) → (iii). If $M \oplus N \simeq F$ and F is free, then by Proposition 1.12 and the Corollary to it we obtain the projectivity of M.

(iii) → (i). Assume that the sequence $0 \to A \xrightarrow{i} B \xrightarrow{p} M \to 0$ is exact. Condition (iii) implies the existence of $f: M \to B$, making the composition $M \to B \to M$ an identity. Obviously f is an injection. If $x \in B$, then $f \circ p(x) = y$ lies in $\operatorname{Im} f \simeq M$. Moreover, $p(x - y) = 0$; thus $x - y \in \operatorname{Im} i$ and we may write $x = z + y$ with $z \in \operatorname{Im} i$. Finally, we see that $\operatorname{Im} f \cap \operatorname{Im} i = 0$, since for $x \in \operatorname{Im} f \cap \operatorname{Im} i$ one has $x = f(u)$ and $p(x) = 0$, giving $u = p(f(u)) = 0$; thus $x = f(0) = 0$. Hence $B \simeq \operatorname{Im} i \oplus \operatorname{Im} f$, which in turn implies $B \simeq A \oplus M$. □

Another characterization of projective modules is provided by

Proposition 1.14. *An R-module M is projective iff there exists a system $(a_t)_{t \in T}$ of elements of M and a family $(f_t)_{t \in T}$ of homomorphisms of M into R such that every element $a \in M$ may be written in the form*

$$a = \sum_{t \in T} f_t(a) a_t, \qquad (1.6)$$

where only a finite number of summands are non-zero.

Proof. Assume first that M is projective and let F be any free module whose image by a homomorphism, say f, is M. Proposition 1.13 implies that M is a direct summand of F, and so, with a suitable homomorphism $i: M \to F$, we have $f \circ i =$ the identity. Now if $a \in M$, then

$$i(a) = \sum_t f_t(a) x_t$$

for some $f_t(a) \in R$, where (x_t) is the set of free generators of F. If we put

$a_t = f(x_t)$, then we obtain

$$a = \sum_t f_t(a) a_t$$

with only a finite number of non-zero summands. Since, obviously, the mappings of M into R defined by f_t are homomorphisms, the first part of our proposition follows.

To prove the second part, assume that each $a \in M$ has the form (1.6). Let F be the free R-module with free generators (x_t) and define a homomorphism $f: F \to M$ by $f(x_t) = a_t$. If now $g: M \to F$ is given by

$$g(a) = \sum_t f_t(a) x_t$$

for $a = \sum_t f_t(a) a_t$, then the composition $M \xrightarrow{g} F \xrightarrow{f} M$ equals the identity, showing that M is a direct summand of F, which allows us to conclude, by using Proposition 1.13, that M is projective. □

Our next proposition connects the notion of projectivity with the concepts developed in Section 1.

Proposition 1.15. *If R is an integral domain and I is a non-zero ideal in R, then I is projective (as an R-module) if and only if I is invertible.*

Proof. Let K be the field of fractions of R and let I be an invertible ideal in R, i.e., $II^{-1} = R$. Then, with suitable $a_1, \ldots, a_n \in I$ and $x_1, \ldots, x_n \in I^{-1}$, we have

$$\sum_i x_i a_i = 1.$$

If we now define, for $t = 1, 2, \ldots, n$, homomorphisms f_t of I into R by $f_t(x) = xx_t$, then

$$\sum_t f_t(x) a_t = \sum_t xx_t a_t = x$$

and so, by Proposition 1.14, I is projective.

Conversely, assume I to be projective. This implies the existence of a set of elements $(a_t)_{t \in T}$ and homomorphisms $(f_t)_{t \in T}$ of I into R such that every element of I can be written in the form

$$x = \sum_t f_t(x) a_t$$

with only a finite number of non-zero summands. Observe that for x, y in I we have $y f_t(x) = f_t(yx) = f_t(xy) = x f_t(y)$ and so the ratio $x_t = f_t(x) x^{-1} \in K$ is independent of the choice of non-zero x in I. Moreover, $x_t I \subset R$;

thus $x_t \in I'$ and for any fixed $x \in I$ only finitely many elements $f_t(x) = xx_t$ are non-zero, whence only a finite number of x_t's do not vanish, say $x_1, \ldots, x_n$. Thus, for any $x \in I$, we obtain an equality of the form

$$x = \sum_{t=1}^{n} f_t(x) a_t = \sum_{t=1}^{n} xx_t a_t = x \sum_{t=1}^{n} x_t a_t,$$

which implies

$$1 = \sum_{t=1}^{n} x_t a_t,$$

and so $R \subset II' \subset R$, i.e., $R = II'$ and I is invertible. $\square$

Corollary. *In any Dedekind domain all ideals are projective.*

In fact, every non-zero ideal of R is invertible by definition. $\square$

Having proved the principal results on projective modules, we may begin the proof of Theorem 1.13. We need two lemmas.

Lemma 1.9. *Let R be a domain in which every ideal is projective. If M is a finitely generated R-module contained in a free R-module F, then M can be represented as a direct sum of a finite number of ideals of R.*

Proof. Observe first that M is contained in a finitely generated free R-module. Indeed, if $a_1, \ldots, a_m$ generate M, then the set of free generators of F occurring in the canonical form of those elements is finite and consists of, say, the elements $x_1, \ldots, x_n$. The R-module generated by $x_1, \ldots, x_n$ is obviously free and contains M.

We apply induction in n. For $n = 0$ there is nothing to prove. Assume thus the truth of our lemma for all R-modules contained in a free R-module with $n-1$ free generators. Let M be a finitely generated R-module contained in the free R-module F_n with n free generators $x_1, \ldots, x_n$ and let F_{n-1} be the free R-module generated by the first $n-1$ of them. Every element x in M can be written as $r_1 x_1 + \ldots + r_n x_n$ with $r_i \in R$, and the mapping $f: x \to r_n$ is a homomorphism of M into R. Since the sequence

$$0 \to \operatorname{Ker} f \to M \to \operatorname{Im} f \to 0$$

is exact and $\operatorname{Im} f$ is an ideal in R, projective by assumption, we may apply Proposition 1.13 to obtain $M \simeq \operatorname{Im} f \oplus \operatorname{Ker} f$. This implies that $\operatorname{Ker} f$ is finitely generated as a homomorphic image of M, and since $\operatorname{Ker} f \subset F_{n-1}$, we may apply the inductional assumption to find that $\operatorname{Ker} f$ is a direct sum of ideals of R. Since $\operatorname{Im} f$ is also an ideal, the lemma follows. $\square$

Lemma 1.10. *For every domain R, any finitely generated and torsion-free R-module M is a submodule of a free R-module.*

Proof. Write $M = Rx_1 + \ldots + Rx_n$ and let K be the field of fractions of R. Then $Kx_1 + \ldots + Kx_n = M \otimes K$ is a finite-dimensional linear space over K. If $y_1, \ldots, y_m$ is its basis, then we may write

$$x_i = \sum_{j=1}^{m} r_{ij} y_j \quad (i = 1, 2, \ldots, n; r_{ij} \in K).$$

Now take a non-zero element q in R so that $qr_{ij} \in R$ for all i and j. With such a choice we get

$$M = Rx_1 + \ldots + Rx_n \subset Ry_1/q + \ldots + Ry_m/q$$

and on the right-hand side of this inclusion we obviously have a free R-module. □

After this rather lengthy preparation we may prove our theorem. Let M be any finitely generated module over a Dedekind domain R and let A be its submodule consisting of all torsion elements of M. The factor-module $M_1 = M/A$ is clearly torsion-free and finitely generated; hence the Corollary to Proposition 1.15, Lemmas 1.9 and 1.10 imply that M_1 is a direct sum of ideals of R. The same corollary together with Proposition 1.12 shows that M_1 is projective, and so the exactness of the sequence

$$0 \to A \to M \to M_1 \to 0$$

gives, in view of Proposition 1.13, the decomposition

$$M \simeq A \oplus M_1 \simeq A \oplus I_1 \oplus \ldots \oplus I_m,$$

where $I_1, \ldots, I_m$ are ideals in R.

Now we prove that with a suitable ideal $I \subset R$ we have

$$I_1 \oplus \ldots \oplus I_m \simeq R^{m-1} \oplus I.$$

For this purpose it suffices to show that for any pair J_1, J_2 of ideals of R one may find an ideal J such that $J_1 \oplus J_2 \simeq R \oplus J$. First we show that there is an ideal J_1' of R which is isomorphic to J_1 as an R-module and satisfies $(J_1', J_2) = R$. Choose $A \subset R$ so that $J_1 A = aR$ is principal and $(A, J_2) = R$, which is possible according to Corollary 6 to Proposition 1.6. Write $A = P_1^{a_1} \ldots P_s^{a_s}$ and choose $b \in R$ so that for $i = 1, 2, \ldots, s$ one has

$$b \in P_i^{a_i} \setminus P_i^{a_i+1}$$

and $b \equiv 1 \pmod{Q}$ for every prime ideal Q dividing J_2. Then bR is divisible by A, and so we may write $bR = AJ_1'$ with some ideal J_1' relatively prime

to J_2, and we finally obtain
$$aJ_1' = J_1 AJ_1' = bJ_1,$$
which shows that $J_1 \simeq bJ_1 = aJ_1' \simeq J_1'$, as required.

Now consider the exact sequence
$$0 \to J_1' \cap J_2 \to J_1' \oplus J_2 \to J_1' + J_2 \to 0.$$
Since the ideals J_1' and J_2 are relatively prime, by Proposition 1.5, (iii) and (iv) their sum is equal to R and their intersection equals their product, whence the sequence takes the form
$$0 \to J_1' J_2 \to J_1' \oplus J_2 \to R \to 0$$
and the projectivity of R implies finally
$$J_1 \oplus J_2 \simeq J_1' \oplus J_2 \simeq R \oplus J_1' J_2$$
as asserted. As we have seen above, this proves the theorem. □

2. Now we shall consider the question of uniqueness of the direct summands occurring in Theorem 1.13. Since the torsion submodule A is clearly unique, we may assume that our module is torsion-free. In this case we prove

Theorem 1.14. *If R is a Dedekind domain and M_1, M_2 are finitely generated torsion-free R-modules written in the form*
$$M_1 = I_1 \oplus \ldots \oplus I_m, \quad M_2 = J_1 \oplus \ldots \oplus J_n,$$
where I_i, J_i are non-zero fractional ideals of R, then M_1 and M_2 are isomorphic iff $m = n$ and with a suitable a from K, the field of quotients of R, the equality
$$I_1 \ldots I_m = aJ_1 \ldots J_n$$
holds.

Proof. The sufficiency of our condition was already established in the last part of the proof of the foregoing theorem. To prove its necessity assume that M_1 and M_2 are isomorphic. The embedding of R in K induces an embedding of M_1 in K^m and of M_2 in K^n, and obviously M_1 spans K^m and M_2 spans K^n. The isomorphism of M_1 onto M_2 extends to a K-isomorphism of the spanned spaces and so $m = n$.

To prove the remaining part of the theorem we may assume that all ideals I_i, J_i contain the ring R. In fact, if I is one of those ideals, then with a suitable non-zero a in K we have, say, $R \subset aI = I'$. The mapping $x \to ax$ shows that $I \simeq I'$, whence
$$M_1 \simeq I_1' \oplus \ldots \oplus I_m', \quad M_2 \simeq J_1' \oplus \ldots \oplus J_m'.$$

If we prove the theorem in this case, then we shall have $I'_1 \ldots I'_m = cJ'_1 \ldots J'_m$ for some $c \in K$, and this obviously implies the equality $I_1 \ldots I_m = dJ_1 \ldots J_m$ with some d in K.

Now let f be an isomorphism of M_1 onto M_2 and let f_r be its restriction to I_r. If 1_r is the unit element of R, regarded as a subset of I_r, then denote its image $f_r(1_r)$ by $[a_{r1}, \ldots, a_{rm}]$ with a_{ri} lying in J_i for $i = 1, 2, \ldots, m$. We shall show that

$$J_s = a_{1s}I_1 + \ldots + a_{ms}I_m \quad (s = 1, 2, \ldots, m).$$

Accordingly, note first that if a, x, ax all lie in I_r, then $f_r(xa) = xf_r(a)$. Indeed, if $x = A/B$ with A, B in R, then $Bf_r(ax) = Bf_r(aA/B) = f_r(aA) = Af_r(a)$, hence $f_r(xa) = (A/B)f_r(a) = xf_r(a)$. If we denote by p_s the projection of M_2 onto J_s, then in view of

$$f([x_1, \ldots, x_m]) = \sum_{i=1}^{m} f_i(x_i) = \sum_{i=1}^{m} f_i(1_i)x_i,$$

we obtain the following chain of equalities:

$$\sum_{i=1}^{m} a_{is}I_i = \left\{\sum_{i=1}^{m} a_{is}x_i : x_i \in I_i\right\}$$
$$= \{p_s(f_1(1_1)x_1 + \ldots + f_m(1_m)x_m) : x_i \in I_i\}$$
$$= \{p_s(f([x_1, \ldots, x_m])) : x_i \in I_i\} = J_s.$$

Now if $C = \det[a_{ij}] = \sum_P \operatorname{sign} P \cdot A_P$ is the expansion of the determinant of $[a_{ij}]$, then, multiplying all the equalities just obtained, we get

$$J_1 \ldots J_m = \prod_{s=1}^{m} \sum_{i=1}^{m} a_{is}I_i = \sum_P A_P I_1 \ldots I_m + \ldots,$$

which shows that

$$\sum_P A_P I_1 \ldots I_m \subset J_1 \ldots J_m.$$

From this we shall now deduce the inclusion $CI_1 \ldots I_m \subset J_1 \ldots J_m$. Let P be any permutation of m letters and let $x_i \in I_i$ ($i = 1, 2, \ldots, m$). If $y_i = x_i$ for $i = 2, 3, \ldots, m$ and $y_1 = x_1 \operatorname{sign} P$, then $A_P y_1 \ldots y_m = A_P(\operatorname{sign} P)x_1 \ldots x_m \in A_P I_1 \ldots I_m \subset J_1 \ldots J_m$ and so the sum

$$\sum_P A_P(\operatorname{sign} P)x_1 \ldots x_m,$$

which equals $Cx_1 \ldots x_m$, lies in $J_1 \ldots J_m$.

If we now change the roles of M_1 and M_2, we get $C_1 J_1 \ldots J_m \subset I_1 \ldots I_m$, where C_1 is the determinant of the matrix $[b_{ij}]$ defined by

$$g_r(e_r) = [b_{r1}, \ldots, b_{rm}] \quad (b_{ri} \in I_i),$$

e_r is the unit element of R treated as a subset of J_r, and g_r is the restriction of g, the inverse mapping to f, to J_r. It is easy to see that the matrices $[a_{ij}]$ and $[b_{ij}]$ are the inverses of each other, and so $CC_1 = 1$, which at once implies the equality $I_1 \ldots I_m = C_1 J_1 \ldots J_m$. This proves our theorem. □

3. To conclude the study of finitely generated modules over Dedekind domains we shall now consider torsion modules and start with the case of a principal ideal domain.

Proposition 1.16. *If R is a principal ideal domain and M is a finitely generated non-zero torsion R-module, then there exist ideals $I_1, \ldots, I_n$ of R such that*

$$M \simeq \bigoplus_{j=1}^{n} R/I_j.$$

Proof. For any prime ideal $P \neq 0$ of R denote by $M(P)$ the submodule of M consisting of all elements of M which are annihilated by a power of P, i.e.,

$$M(P) = \{m \in M : P^r m = 0 \text{ for a certain } r \geq 1\}.$$

Since R is a principal ideal domain we have equivalently

$$M(P) = \{m \in M : p^r m = 0 \text{ for a certain } r \geq 1\},$$

where p is a generator of P. First we show that $M = \bigoplus_P M(P)$, where P runs over all prime ideals of R. Let m be a non-zero element of M and let

$$\text{Ann}(m) = \{r \in R : rm = 0\}$$

be its *annihilator*. It is a non-zero ideal in R and hence we can find irreducible elements $p_1, \ldots, p_s$ generating distinct prime ideals, and also exponents $a_i \geq 1$ ($i = 1, 2, \ldots, s$) so that $\text{Ann}(m) = p_1^{a_1} \ldots p_s^{a_s} R$. Since R is a PID, the elements

$$r_j = (p_1^{a_1} \ldots p_s^{a_s}) p_j^{-a_j} \quad (j = 1, 2, \ldots, s)$$

do not have a non-unit common divisor and thus we may find $t_1, \ldots, t_s$ in R satisfying $t_1 r_1 + \ldots + t_s r_s = 1$. This implies

$$m = t_1(r_1 m) + \ldots + t_s(r_s m) \in \sum_{j=1}^{s} M(p_j R)$$

because $r_j m$ is annihilated by $p_j^{a_j}$. This shows that $M = \sum_P M(P)$, but since only the zero element can be annihilated by two relatively prime elements, the sum $\sum_P M(P)$ is direct, and $M = \bigoplus_P M(P)$ follows. Since Proposition 1.2 implies that every finitely generated module over a Noetherian ring is itself a Noetherian module, there can be only finitely many non-zero terms $M(P)$ in the sum in question.

It follows that it suffices to consider modules of the form $M(P)$ with a suitable prime ideal P. Note that for such modules M their annihilator $\operatorname{Ann}(M) = \bigcap_{m \in M} \operatorname{Ann}(m)$ must be a power of P, because $\operatorname{Ann}(m)$ is a power of P for $m \in M$. Therefore, let $\operatorname{Ann}(M) = p^t R$, where p is a generator of P and $t \geqslant 1$. Moreover, denote by $m_1, \ldots, m_n$ a fixed set of generators of M. We use induction in the number n of generators. If $n = 1$, then M is an epimorphic image of R, the free R-module with one generator, and hence $M \simeq R/I$ with a suitable ideal I. Assume thus that our assertion holds for all modules with at most $n-1$ generators. Obviously at least one of the generators m_i satisfies $\operatorname{Ann}(m_i) = p^t R$, and we may assume that this holds for $i = n$. The factor-module $M/m_n R$ has less than n generators, whence we may write

$$M/m_n R = \bigoplus_{i=1}^{s} v(x_i) R$$

with suitable $x_1, \ldots, x_s \in M$. Here v denotes the natural map $M \to M/m_n R$. Put $\operatorname{Ann}(v(x_i)) = p^{r_i} R$ ($i = 1, 2, \ldots, s$). Then $r_i \leqslant t$ and with suitable $k_i \geqslant 0$ and $a_i \in R \setminus pR$ we have $p^{r_i} x_i = p^{k_i} a_i m_n$ ($i = 1, 2, \ldots, s$).

Because of

$$0 = p^t x_i = p^{t - t_i + k_i} a_i m_n$$

we infer that $k_i \geqslant r_i$. Putting $y_i = x_i - p^{k_i - r_i} a_i m_n$, we obtain $p^{r_i} y_i = 0$ and $v(y_i) = v(x_i)$. This gives

$$\operatorname{Ann}(v(x_i)) = p^{r_i} R \subset \operatorname{Ann}(y_i) \subset \operatorname{Ann}(v(y_i)) = \operatorname{Ann}(v(x_i)),$$

and thus $\operatorname{Ann}(v(y_i)) = \operatorname{Ann}(y_i)$. It follows that the map v restricted to $y_i R$ is an isomorphism for $i = 1, 2, \ldots, s$, and because of

$$v(y_1 R + \ldots + y_s R) = M/m_n R = \bigoplus_{i=1}^{s} v(y_i) R$$

v maps $y_1 R + \ldots + y_s R$ isomorphically onto $\bigoplus_{i=1}^{s} v(y_i) R$, and so the sum $\sum_{i=1}^{s} y_i R$ is direct. This leads to

$$M = m_n R \oplus \bigoplus_{i=1}^{s} y_i R,$$

and, applying the inductional assumption, we arrive at our assertion. □

Using the proposition just proved, we can now describe all finitely generated and torsion modules over a Dedekind domain. It turns out that their structure is not more complicated than in the case of a PID.

Theorem 1.15. *If R is a Dedekind domain and M a non-zero finitely generated and torsion R-module, then there exist ideals $I_1, \ldots, I_n$ of R such that*

$$M \simeq \bigoplus_{j=1}^{n} R/I_j.$$

§ 3. Finitely Generated Modules over Dedekind Domains

Proof. The set $I = \{r \in R: rm = 0 \text{ for all } m \in M\}$ is a non-zero ideal of R and M can be regarded as an R/I-module via $r \pmod{I} \cdot m = rm$ ($r \in R, m \in M$). Write

$$I = \prod_{j=1}^{t} P_j^{a_j}$$

with distinct prime ideals $P_1, \ldots, P_t$ and $a_j \geq 1$. Theorem 1.5 implies

$$R/I \simeq \bigoplus_{j=1}^{t} R/P_j^{a_j}$$

and to utilize this decomposition we need the following lemma.

Lemma 1.11. *If a commutative ring S with a unit is a direct sum of its subrings*

$$S = \bigoplus_{j=1}^{s} S_j,$$

then every S-module M can be written in the form

$$M = \bigoplus_{j=1}^{s} M_j,$$

where $M_1, \ldots, M_s$ are S-modules and for $i \neq j$ and $s_i \in S_i$ we have $s_i M_j = 0$.

Proof. If e denotes the unit of S and e_j is the unit of S_j ($j = 1, 2, \ldots, s$), then evidently

$$e = e_1 + e_2 + \ldots + e_s.$$

Put $M_j = e_j M$ for $j = 1, 2, \ldots, s$. Then obviously for $i \neq j$ and $s \in S_j$ we have $sM_i = 0$. Since for a in M

$$a = e_1 a + e_2 a + \ldots + e_s a \tag{1.7}$$

and $e_j a \in M_j$, the sum of the modules M_j equals M and it remains to show that this sum is direct, i.e., that the decomposition (1.7) of a into sums of elements of the M_i's is unique. This is easy to prove, since if

$$a = m_1 + m_2 + \ldots + m_s$$

(where $m_i \in M_i$; $i = 1, 2, \ldots, s$), then $m_i = e_i x_i$ for suitable $x_i \in M_i$ and thus

$$e_j a = e_j m_1 + \ldots + e_j m_s = e_j e_1 x_1 + \ldots + e_j e_s x_s$$
$$= e_j^2 x_j = e_j x_j = m_j$$

and our decomposition coincides with (1.7). □

We apply the lemma in the case of $S = R/I$, $S_j = R/P_j^{a_j}$ and obtain the decomposition

$$M = \bigoplus_{j=1}^{t} M_j$$

where each M_j can be regarded as an $R/P_j^{a_j}$-module, and for $i \neq j$ one has $(R/P_i^{a_i})M_j = 0$.

To conclude the proof it is sufficient to show that every finitely generated R/P^a-module N (where P is a prime ideal of R and $a \geq 1$) is of the form

$$N \simeq \bigoplus_{j=1}^{t} R/P^{b_j}$$

with a certain $t \geq 0$ and $1 \leq b_j \leq a$. If we denote by R_P the valuation ring induced by the P-adic valuation, then by Proposition 1.11 (iv) the rings R/P^a and $R_P/(PR_P)^a$ are isomorphic. Thus N becomes an R_P-module with the property $(PR_P)^a N = 0$. Since by Theorem 1.10 R_P is a PID, Proposition 1.16 is applicable and we see that N is a direct sum of cyclic R_P-modules, i.e.,

$$N \simeq \bigoplus_{j=1}^{t} R_P/I_j$$

with suitable ideals I_j of R_P. Theorem 1.10 implies that each I_j is a power of PR_P, and owing to $(PR_P)^a N = 0$ we must have $I_j = (PR_P)^{b_j}$ with $1 \leq b_j \leq a$. Since $R \subset R_P$, we can regard R_P/I_j as an R-module, and since the ring-isomorphism of R/P^{b_j} and $R_P/(PR_P)^{b_j}$ is also an R-module isomorphism, we finally obtain

$$N \simeq \bigoplus_{j=1}^{t} R/P^{b_j}$$

as asserted. □

4. We conclude this chapter with the introduction of the notion of the *class-group of a Dedekind domain*, which will play an important role in the sequel. Its definition is based on the following simple result.

Proposition 1.17. *If R is a Dedekind domain and $I_1 \simeq I_2$, $J_1 \simeq J_2$ are two pairs of its fractional ideals, which are isomorphic as R-modules, then the products $I_1 J_1$ and $I_2 J_2$ are also isomorphic.*

Proof. Since obviously $I_1 \oplus J_1 \simeq I_2 \oplus J_2$, Theorem 1.14 implies the existence of a non-zero $a \in K$ such that $I_1 J_1 = a I_2 J_2$, and this shows that the map $x \mapsto ax$ of $I_2 J_2$ in $I_1 J_1$ is an isomorphism. □

This proposition implies the compatibility of the multiplication of ideals with the partition of all fractional ideals into classes of isomorphic ideals and so permits us to define a multiplication in the set of these classes in the following way: if $c(I)$, $c(J)$ are classes containing I or J, then their product is defined by $c(I)c(J) = c(IJ)$. This induces a semigroup structure in the set of classes, but one can easily see that it is in fact a group structure because the existence of inverses is implied by the invertibility of fractional ideals.

The resulting group is called the *group of ideal classes of R*, or simply the *class-group of R*, and is usually denoted by $H(R)$. Its importance for the theory of Dedekind rings is explained by the following theorem.

Theorem 1.16. *If R is a Dedekind domain, then the following statements are equivalent*:
  (i) $H(R) = 1$.
  (ii) *R is a PID (principal ideal domain)*.
  (iii) *R is a UFD (unique factorization domain)*.

Proof. If $H(R) = 1$, then every non-zero ideal of R is isomorphic to R as an R-module, and so has the form aR with a certain $a \neq 0$. This establishes the implication (i) $\to$ (ii). The implication (ii) $\to$ (iii) being clear, assume that R is a UFD. We show first that every irreducible element of R (i.e., a non-zero and non-invertible element which does not have proper divisors) generates a prime ideal. If a were an irreducible element with $aR = P_1 \ldots P_s$ $(s \geq 2$, P_i—prime ideals), then by Corollary 5 to Proposition 1.6 we would have for $i = 1, 2, \ldots, s$

$$P_i = a_i R + b_i R = (a_i R, b_i R)$$

with suitable $a_i, b_i \in R$. For every $i = 1, 2, \ldots, s$ we have either $a \nmid a_i$ or $a \nmid b_i$, and we may assume that $a \nmid a_i$ holds for $i = 1, 2, \ldots, s$. However, $a_1 \ldots a_s \in P_1 \ldots P_s = aR$, and thus a divides the product $a_1 \ldots a_s$ without dividing any of the factors, which is impossible for an irreducible element in UFD.

This shows that all irreducibles generate prime ideals. If $H(R)$ were non-trivial, then there would be at least one non-principal prime ideal, say P, since otherwise all ideals would be principal. Write $P = (aR, bR)$ with suitable $a, b \in R$ and factorize a into irreducibles, say $a = a_1 \ldots a_r$. All ideals $a_i R$ are prime, and it follows that for a certain i we have $a_i R = P$, thus P is principal, a contradiction. This establishes the implication (iii) $\to$ (i). □

§ 4. Notes to Chapter 1

1. The theory of Dedekind domains was created as a generalization of results concerning rings of integers in finite extensions of the rational field, obtained mainly by R. Dedekind [71]. It was already observed by Dedekind and Weber [82] that many of these results also apply to rings of integral elements in function fields. However, the general theory had to wait for the introduction of abstract methods and concepts into algebra. In fact, the definition of an abstract ring as it is used today, appears for the first time in Fraenkel [16], and the definition of an abstract field is not much older (Steinitz [10]).

The role of the ascending chain conditions for ideals (the Noether condition) for the theory of commutative rings was emphasized by E. Noether [21]. She obtained the fundamental results of the theory of Noetherian rings, generalizing to more general situations many earlier results of D. Hilbert [90], E. Lasker [05] and F. S. Macaulay [13] for polynomial rings.

The standard proof of Proposition 1.1 which we presented uses the axiom of choice; in fact, as shown by W. Hodges [74], this cannot be avoided.

The theory of rings, now called *Dedekind rings*, originated with Noether [27a] (preceded by Noether [19]) where the condition which is both necessary and sufficient for a domain to have a unique factorization of ideals into prime ideals was given in the following form: the domain should be Noetherian and integrally closed and for every non-zero ideal the corresponding factor ring should be *Artinian* (i.e., it should satisfy the descending chain condition for ideals). The last condition is also called the *restricted minimum condition*.

The equivalence of these conditions to the definition given by us was established by N. Nakano [43], and a simple proof may be found in I. S. Cohen [50].

The proof of Theorem 1.1 shows that in a domain an invertible ideal is finitely generated, a fact first noted by W. Krull [30a].

In a domain the existence of factorizations of all non-zero proper ideals into prime ideals implies its uniqueness. This is implicit in S. Mori [40] but apparently the first explicit mention was made by K. Matusita [44]. Here again for a simple proof the reader is referred to I. S. Cohen [50].

2. Theorem 1.3 gives one of the many known characterizations of Dedekind domains. Apparently it was first formulated in this form in the second volume of van der Waerden [30], although its essence is already contained in Noether [27a]. Let us quote some other characterizations of Dedekind domains.

(i) A domain is Dedekind iff the non-zero fractional ideals form a group under multiplication. (Krull [35], p. 13, "Gruppensatz".)

(ii) A domain R is Dedekind iff every extension of a torsion R-module of bounded rank by a torsion-free R-module splits. (Kaplansky [52]—necessity, Chase [60]—sufficiency.)

This characterization is reminiscent of a similar characterization of fields conjectured by Kaplansky and proved by Rotman [60]: a domain R is a field iff every extension of a torsion R-module by a torsion-free R-module splits.

We should also mention the following result of I. Kaplansky [60] concerning Prüfer domains, i.e., domains in which every finitely generated ideal is invertible: a domain R is Prüfer iff every torsion R-module is a direct summand of every finitely generated R-module containing it. (For various properties of Prüfer domains see Camion, Levy, Mann [73] and for their characterizations cf. Gilmer [67b], Hattori [57], Jensen [63], Megibben [70], Storrer [69].)

(iii) A domain R is Dedekind iff for every proper non-zero ideal I of R the factor ring R/I is UFD (Jensen [63]).

(iv) A domain satisfying the (FN) condition is Dedekind iff Theorem 1.6 holds in it (Butts, Wade [66]).

For other characterizations of Dedekind domains see Albu [70a], Arnold [69], Butts [64], [65], Butts, Gilmer [66], Butts, Wade [66], I. S. Cohen [50], Davis [64], Gilmer [64], [67a], Gilmer, Ohm [64], Goldman [64], Hays [73], Huckaba, Papick [81], Ishikawa [69], Johnson, Lediaev [71], Koyama, Nishi, Yanagihara [74], Matlis [59], [68], Matusita [44], Moriya [40], Mott [66], Quadri, Irfan [79], and Walton [71].

Butts and Smith [66] gave a necessary and sufficient condition for a domain to have its integral closure in its field of fractions Dedekind, and R. W. Gilmer and J. L. Mott [68] characterized domains whose every proper overring is Dedekind (or Noetherian, or integrally closed). Here by an overring we understand any ring containing the given ring and contained in its field of fractions.

3. Various authors have considered rings with some properties in common with Dedekind domains. Commutative rings with a unit element in which every ideal is a unique product of prime ideals were described in Kubo [40] and Kobayashi, Moriya [41a], [41b]. Z.P.I.-rings, i.e., rings in which every ideal is a product of prime ideals, not necessarily unique, were described in S. Mori [40] (cf. Asano [51], Gilmer [63a]). A Noetherian ring has this property iff its lattice of ideals is distributive (Asano [51]). This should be compared with Jensen [63], where it is shown that a domain has a distributive lattice of ideals iff it is Prüfer (cf. Jensen [64b]).

L. Fuchs [54] characterized those ideals I of a commutative ring which can be represented as products of powers of prime ideals in a unique way, say $I = P_1^{a_1} \ldots P_r^{a_r}$, and for which every proper ideal containing I can be written in the form $P_1^{b_1} \ldots P_r^{b_r}$ where $0 \leqslant b_i \leqslant a_i$ for $i = 1, 2, \ldots, r$. If the ring in question is integrally closed, Noetherian and has a unit element, then this condition is equivalent to

$$I \cap (A+B) = I \cap A + I \cap B$$

for all ideals A, B of R (Jensen [66]). The condition holds iff I and every ideal contained in it are products of maximal ideals and, as shown in Camion, Levy, Mann [73], it is satisfied for all finitely generated ideals I iff R is Prüfer.

K. B. Levitz [72] proved that if in a commutative ring R every finitely generated ideal is a product of prime ideals, then all ideals have this property, and showed that it suffices if all ideals with at most two generators have this property. In particular, if R is a domain in which all ideals with at most two generators are products of prime ideals, then R is Dedekind.

It is a classical result of E. Noether [21] that in a Noetherian ring every ideal is an intersection of primary ideals. (An ideal I is called *primary* if $ab \in I$, $a \notin I$ implies $b^n \in I$ for a certain $n \geq 1$.) For polynomial rings over Z or over complex field this was proved in Lasker [05] (cf. Seidenberg [84]). D. D. Anderson [80a] characterized Noetherian rings in which every ideal is a product of primary ideals, as those in which every non-maximal prime ideal is a multiplication ideal. (An ideal I is a *multiplication ideal* if for every ideal $J \subset I$ one has $I \mid J$.) Rings in which all ideals are multiplication ideals are called *multiplication rings* and were already considered in Krull [25], Akizuki [32] and S. Mori [32], [33]. For characterizations of such rings cf. Anderson [76], [80b], Griffin [74], Mott [64]. Note that a multiplicative domain must be Dedekind.

A domain which obeys the cancellation law for ideal multiplication is called an *almost Dedekind domain*. This is not the original definition but an equivalent one. This notion was introduced by H. S. Butts and R. C. Phillips [65], who also gave several characterizations of almost Dedekind domains. In particular, they showed that R is almost Dedekind iff every primary ideal is a power of a maximal ideal and also iff the following three conditions hold: R is Prüfer, every prime ideal is maximal and there is no proper idempotent ideal. This implies of course that a Noetherian almost Dedekind domain is Dedekind. For a characterization of Dedekind domains among almost Dedekind domains see also Pirtle [70].

4. Commutative rings, not necessarily domains, which satisfy conditions (i)–(iii) of Theorem 1.3 have been considered by J. R. Gilbert, Jr. and H. S. Butts [68], who showed that if such a ring has no zero-divisors, then it can be embedded as an ideal in a suitable Dedekind domain.

For the history of Proposition 1.5 see Dedekind [95].

Noetherian rings in which every ideal is generated by at most k elements (with k fixed) were considered by I. S. Cohen [50] and R. Gilmer [72], [73]. E. Matlis [70] characterized domains in which every ideal is generated by at most two elements. The first example of a domain with an invertible ideal having n but not fewer generators was given by S. U. Chase. The first published example is due to R. W. Gilmer [69]. For an extensive study of invertible ideals see W. Weber [31].

Theorem 1.9 was proved in full generality in Noether [27a] and in the habilitation thesis of F. K. Schmidt in 1927 (cf. Grell [36]). F. K. Schmidt [36] gave an example marking the important distinction between the separable and the inseparable case. A newer proof of Theorem 1.9 was given by D. G. Northcott [55]. For infinite extensions Theorem 1.9 ceases to be true, as shown by N. Nakano [53].

Note that the integral closure of a Noetherian domain may be non-Noetherian (Nagata [54]). If R is Noetherian and S is a ring containing R which is a finitely generated unitary R-module, then by Proposition 1.2 S is Noetherian. P. M. Eakin, Jr. [68] proved the converse of this statement and M. Nagata [68] supplied a simple proof.

Rings with the (FN) property were studied in Chew, Lawn [70] and Levitz, Mott [72].

Of the many books concerning ideals in commutative rings we mention only the classical work of W. Krull [35] and R. W. Gilmer's more recent book [68].

5. Valuations were introduced by J. Kürschak [13] and the main results of valuation theory were established by A. Ostrowski [13], [17], [18], [35] and W. Krull [31b]. An account of this may be found in Schilling [50] (cf. also Endler [63a]).

Theorem 1.12 is due to Ostrowski [18], who also described valuations of algebraic number fields (see Theorem 3.2). The reproduced proof is that of E. Artin [32b].

6. Theorems 1.13, 1.14 and 1.15 are essentially due to E. Steinitz [12], who considered rings of integers in algebraic number fields. For the general case see Asano [50], Chevalley [36a], Franz [34], Kaplansky [52]. Cf. also Archinard [84].

Theorem 1.13 implies that if every finitely generated module over a Dedekind domain is a direct sum of cyclic modules, then this domain is a PID, hence UFD. Rings with this property are called *FGC-rings*. A. I. Uzkov [63] showed that a Noetherian FGC-domain must be UFD (cf. Bass [62]). A characterization of commutative FGC-rings appears in R. Wiegand, S. Wiegand [77] and Brandal [79]. B. Midgarden and S. Wiegand [81] considered rings whose finitely generated modules are direct sums of modules with n generators, with n fixed.

A ring R is called a *Köthe ring* if every R-module is a sum of cyclic modules. It was proved by G. Köthe [35] that every Artinian principal ideal ring with a unit is a Köthe ring. In the same paper Köthe showed that commutative Artinian Köthe rings are principal ideal rings. The assumption of Artinianity is not necessary here (Cohen, Kaplansky [51]), but commutativity is (Nakayama [40]). A characterization of rings whose non-trivial epimorphic images are Köthe rings was given by C. Faith [66].

Theorem 1.13 has its analogue for projective modules over the group ring of a finite group over a Dedekind domain. (Swan [60], Giorgiutti [60].)

Theorem 1.14 implies that if A, B are ideals of a Dedekind domain R and M is a finitely generated and torsion-free R-module such that $A \oplus M$ and $B \oplus M$ are isomorphic, then $A \simeq B$. It has been proved by W. V. Vasconcelos [67] that the same holds for all Noetherian, integrally closed domains. In the case of Dedekind domains, M need not be torsion-free. (Hsü [62], Kaplansky [52], cf. also Fuchs, Loonstra [71].) This shows that the second test-problem of Kaplansky [54] has a positive solution for Dedekind domains. However, the third test-problem has a negative solution in this case; in fact, if $H(R)$ contains an element of order two and I is a non-principal ideal such that I^2 is principal, then $I \oplus I$ and $R \oplus R$ are isomorphic but I and R are not.

For a discussion of free modules over Dedekind domains see O'Meara [56].

7. The notion of class-group essentially goes back to E. E. Kummer in the case of rings of integers of cyclotomic extensions of Q and even to C. F. Gauss in the case of quadratic extensions. Gauss dealt with classes of quadratic forms, but his theory is equivalent to the theory of ideal classes in quadratic fields, as we shall see in Chapter 8.

Certain analogues of the class-group may be considered in other, not necessarily Dedekind, domains. Cf. Claborn, Fossum [68], Fossum [73], Kennedy [80], Strooker [66].

An old conjecture that every Abelian group can serve as a class-group for a suitable Dedekind domain was settled affirmatively by L. Claborn [66]. A proof may be found in Fossum [73] and another proof was given by C. R. Leedham-Green [72]. For an extension see Claborn [68].

P. Eakin and W. Heinzer [73] showed that if the group in question is finitely generated, then one can find a suitable Dedekind domain R satisfying $Z[X] \subset R \subset Q[X]$. In the case of the trivial group this leads to an example of a non-Euclidean PID.

It has also been conjectured (see Zariski, Samuel [58]) that every Dedekind domain is an integral closure in a finite extension of the quotient field of a PID.

However, this conjecture was disproved by L. Claborn [65]. He in turn suggested that every Dedekind domain can be represented as the ring of quotients of such an integral closure with respect to a multiplicatively closed subset. However, this conjecture also fails (Leedham-Green [72]) and the question whether one can construct all Dedekind domains in a simple fashion starting with PID's remains unanswered.

EXERCISES TO CHAPTER 1

1. Let R be a commutative ring with a unit and let I, J be ideals in R. Prove that if R/I and R/J are isomorphic as R-modules, then $I = J$, and show that this implication may fail if we replace R-module-isomorphism by ring-isomorphism.

2. (Krull [35].) Show that a domain R is Dedekind iff the fractional ideals of R form a group under multiplication.

3. Prove that if R is a Noetherian ring and M is a finitely generated R-module, then M is a Noetherian module.

4. Let R be a domain and I a fractional ideal. Prove that I is invertible iff there exists a fractional ideal J such that IJ is principal.

5. Prove that if R is a domain in which all proper non-zero ideals are uniquely represented as products of prime ideals, then R is Dedekind.

6. Show that in the preceding exercise one can omit the uniqueness assumption.

7. Prove that if R is a Dedekind domain and I is a non-zero ideal in R, then the factor ring R/I is Artinian, i.e., satisfies the descending chain condition for ideals.

8. (Camion, Levy, Mann [73].) Let R be a Prüfer domain. Prove the following three assertions:

(i) If I is a finitely generated ideal of R and for certain ideals A, B of R we have $AI = BI$, then $A = B$.

(ii) If I is a finitely generated ideal of R and J an ideal contained in I, then $I|J$, i.e., there exists an ideal A such that $AI = J$.

(iii) If A, B, C are finitely generated ideals of R, then $A \cap (B+C) = A \cap B + A \cap C$.

9. Let R be a Dedekind domain and I a non-zero ideal of R, $N \geq 1$, and assume that $R^{N-1} \oplus I$ has N generators. Prove that I is a principal ideal.

10. (In the case of $d = 1$ Reiner [56], in the general case Moore [75].) Let R be a Dedekind domain. Prove that if $a_i \in R$ for $i = 1, 2, ..., n$ and $d \in Ra_1 + Ra_2 + ... + Ra_n$, then there exists an invertible $n \times n$ matrix with entries in R whose first row equals $a_1, a_2, ..., a_n$ and whose determinant equals d.

11. Let R be a Dedekind domain, $G(R)$ the group of all its fractional ideals and $\text{Pr}(R)$ the group of all principal fractional ideals. Prove that the quotient group $G(R)/\text{Pr}(R)$ is isomorphic to $H(R)$.

Chapter 2. Algebraic Numbers and Integers

§ 1. Distribution of Integers in the Complex Plane

1. In this chapter we introduce the fundamental notions of the theory and develop some of their properties. Let us start with definitions. Any complex number which is integral over the field Q of rational numbers will be called an *algebraic number*, and if it is also integral over the ring Z of rational integers, then it will be called an *algebraic integer*. The Corollary to Proposition 1.4 implies that the set of all algebraic numbers is a ring and the same holds for the set of all algebraic integers. In fact, the first of these rings is a field, since if $a \neq 0$ is algebraic then a is a root of $x^m + a_{m-1}x^{m-1} + \ldots + a_0 = 0$ with $a_0, \ldots, a_m \in Q$ and non-zero a_0, and then the number $1/a$ satisfies $x^m + a_1 a_0 x^{m-1} + \ldots + a_0^{-1} = 0$.

The minimal degree of a non-zero polynomial over Q whose root is a is called the *degree* of a over Q, or the *absolute degree* of a. We shall denote it by $\deg a$ or $\deg_Q a$. If K is any subfield of the field $\mathfrak{Z}$ of complex numbers, then $\deg_K a$, the degree of a over K, is the minimal degree of a non-zero polynomial over K having a for one of its roots. This polynomial is called the *minimal polynomial of a over K*, provided it is monic, i.e., its highest coefficient equals 1.

Any finite extension K of Q will be called an *algebraic number field*. By the corollary quoted above, the algebraic integers contained in K form a ring, which we shall denote by R_K. Its elements will be called *integers* of K, and our chief aim is to study their properties and also the properties of the ring R_K as a whole. Note that every algebraic number field K can be written in the form $K = Q(a)$ for a suitable a, which can be taken from R_K. This implies that every number from K is of the form $P(a)$, where P is a polynomial over Q of degree smaller than $\deg a = [K:Q]$, the degree of K. The naive hope that with a suitable choice of a one may be able to write every integer of K in the form $P(a)$ with $P \in Z[t]$ is unjustified, as will be shown on examples later on.

It will be useful to note at this point that R_K, being the integral closure of Z in K, is by Theorem 1.2 also the integral closure of the ring R_L for every subfield L of K.

Note also that if $[K:Q] = n$, then there exist exactly n different embeddings of K into 3. Indeed, if $K = Q(a)$ and $a_1 = a, a_2, ..., a_n$ are all roots of the minimal polynomial for a over Q, then the mappings F_i defined for $i = 1, 2, ..., n$ by

$$F_i(A_0 + A_1 a + ... + A_{n-1} a^{n-1}) = A_0 + A_1 a_i + ... + A_{n-1} a_i^{n-1}$$

(for $A_0, ..., A_{n-1} \in Q$) are all isomorphisms of K into 3 and every such isomorphism must be of this form, since the image of a under it has to coincide with one of the a_i's.

The elements $a_1, ..., a_n$ are called the *conjugates of a over Q*, and $F_i(K)$—the *fields conjugated to K*. Obviously, if K is considered as a subfield of 3 embedded in a fixed way, then one of the F_i's is the identity on K. The same terminology is applicable to a slightly more general situation, where K/k is a finite extension of degree n of a field $k \subset 3$. In this case one considers only those embeddings of K in 3 which are equal to identity on k.

An embedding F_i is called *real* if $F_i(K)$ is contained in the real field R and *complex* (or *imaginary*) otherwise. Note that if F_i is a complex embedding, then its complex conjugate is again a complex embedding distinct from F_i, and hence the number of complex embeddings is always even. One half of this number is denoted usually by $r_2 = r_2(K)$ and the number of real embeddings by $r_1 = r_1(K)$. The equality $r_1 + 2r_2 = [K:Q]$ is obvious.

Fields K with $r_2(K) = 0$ are called *totally real* and fields with $r_1(K) = 0$ *totally complex* (or *totally imaginary*). The pair $[r_1, r_2]$ is called the *signature of K*. Note that if K/Q is a Galois extension, then K is either totally real or totally complex, since all images $F_i(K)$ have to coincide.

Usually one fixes the order of the embeddings $F_1, ..., F_n$ in such a way that $F_1, ..., F_{r_1}$ are real, the remaining embeddings are complex and, moreover, for $i = r_1 + r_2 + 1, ..., n$, one has $F_i = \overline{F_{i-r_2}}$. In the sequel we shall always tacitly assume that the embeddings are ordered in this way.

An element $a \in K$ is called *totally real* if all its conjugates are real, and *totally complex* if none of them are real. This does not depend on the choice of field K containing a.

The product of r_1 copies of the multiplicative groups $\{-1, 1\}$ is called the *signature group of K* and is denoted by $\text{Sgn}(K)$. There is a homomorphism $\text{Sgn}: K^* \to \text{Sgn}(K)$ defined by $\text{Sgn}(a) = [\varepsilon_1, ..., \varepsilon_{r_1}]$ where ε_i denotes the sign of $F_i(a)$. The elements of the kernel of the signature map are called *totally positive numbers*. One writes $a \gg 0$ to indicate that a is totally positive.

Note that this notion depends on K, since in a totally complex field all non-zero numbers are totally positive and, in particular, such are all negative rationals, which are certainly not totally positive in Q. Clearly all squares of K and their sums are totally positive in K, and it follows from a result of C. L. Siegel [21b] that, conversely, every totally positive element of K is a sum of squares of elements of K.

The signature map is surjective. This is implied by the following, slightly more general, result.

Proposition 2.1. *If I is a non-zero ideal in R_K and X one of its residue classes, then there are infinitely many elements of X with an arbitrarily prescribed signature.*

Proof. If $r_1 = 0$, then $\mathrm{Sgn}(K^*) = 1$ and there is nothing to prove. So assume that $r_1 \geq 1$ and let $\varepsilon = [\varepsilon_1, \ldots, \varepsilon_{r_1}] \in \mathrm{Sgn}(K^*)$ be given. We prove first that there exists an integer b of K with $\mathrm{Sgn}(b) = \varepsilon$. For this purpose consider the additive map of R_K into the real n-space (with $n = [K:Q]$) defined by

$$\Phi: x \mapsto [F_1(x), \ldots, F_{r_1}(x), \mathrm{Re}\, F_{r_1+1}(x), \mathrm{Im}\, F_{r_1+1}(x), \ldots$$
$$\ldots, \mathrm{Re}\, F_{r_1+r_2}(x), \mathrm{Im}\, F_{r_1+r_2}(x)].$$

Obviously Φ is an embedding, since $\Phi(x) = 0$ implies that all conjugates of x vanish, whence $x = 0$. We prove that the image $\Phi(R_K)$ is an n-dimensional lattice, i.e., a free Abelian subgroup of R^n with n free generators. First observe that $\Phi(R_K)$ is not dense in R^n. Indeed, the set

$$A = \{[x_1, \ldots, x_n]: |x_i| < 1/2 \text{ for } i = 1, 2, \ldots, n\}$$

cannot contain non-zero points of $\Phi(R_K)$ since if $\Phi(x) \in A$ then $|F_i(x)| < 1/2$ holds for $i = 1, 2, \ldots, n$, implying

$$|F_1(x) \ldots F_n(x)| < 2^{-n} < 1.$$

However, the product of all conjugates of an integer $x \neq 0$ is a non-zero rational integer, hence its absolute value equals at least 1.

Thus $\Phi(R_K)$ is a discrete subgroup of R^n and hence must be isomorphic to Z^d with a suitable $0 \leq d \leq n$. Thus it suffices to exhibit n linearly independent points in $\Phi(R_K)$, and here we may take the images of $1, u, u^2, \ldots, u^{n-1}$, where u is any integer of K satisfying $\deg u = n$. In fact, if with real $A_0, \ldots, A_{n-1}$ we had

$$\sum_{j=0}^{n-1} A_j \Phi(u^j) = 0,$$

then for $k = 1, 2, \ldots, n$ we would have

$$\sum_{j=0}^{n} A_j F_k(u^j) = 0,$$

and since the determinant of $[F_k(u^j)]$ does not vanish, we obtain $A_0 = A_1 = \ldots = A_{n-1} = 0$.

It follows that $\Phi(R_K)$ has infinitely many points in each of the 2^{r_1} parts into which the hyperplanes $x_i = 0$ ($i = 1, 2, \ldots, r_1$) divide R^n. This proves the existence of $b \in R_K$ with $\text{Sgn}(b) = \varepsilon$. If now $a \in I$ is totally positive (e.g. $a = N(I)$), then the numbers $abk + x$ (with $x \in X$) are all in X and, for sufficiently large k, satisfy $\text{Sgn}(abk + x) = \text{Sgn}(ab) = \text{Sgn}(b) = \varepsilon$. □

2. For any extension L/K of fields one defines two important mappings: the norm $N_{L/K}(x)$, which equals the product of all conjugates of x over K, and the trace $T_{L/K}(x)$, equalling their sum. The following proposition exhibits their principal properties.

Proposition 2.2. (i) *The norm $N_{L/K}$ is a homomorphism of L^*, the multiplicative group of L, into K^* and the trace $T_{L/K}$ is a homomorphism of the corresponding additive groups.*

(ii) *If $R \subset K$ is a ring and S its integral closure in L, then $N_{L/K}(S) \subset R$ and $T_{L/K}(S) \subset R$. In particular, the norm and trace of an algebraic integer are again integers.*

(iii) *If L/K and M/L are finite, then*

$$N_{M/K} = N_{L/K} \circ N_{M/L}, \quad T_{M/K} = T_{L/K} \circ T_{M/L}.$$

Proof. Assertions (i) and (ii) are immediate consequences of the definitions. To prove (iii) let $\bar{M}$ be an algebraic closure of M and let $s_1, \ldots, s_m$ be the embeddings of L into $\bar{M}$ which equal the identity on K. Extend them to embeddings of M in $\bar{M}$ and let $t_1, \ldots, t_n$ be those which are equal to identity on L. If s is an arbitrary embedding of M into $\bar{M}$ equal to identity on K, then with a suitable j we have $s_j^{-1} \circ s(L) = L$, and so for a certain i we have $s_j^{-1} \circ s = t_i$. This shows that every embedding of M into $\bar{M}$ over K is of the form $s_j \circ t_i$, and hence for $a \in M$ we get

$$N_{M/K}(a) = \prod_{i,j} s_j \circ t_i(a) = \prod_j s_j \left(\prod_i t_i(a) \right)$$

$$= \prod_j s_j(N_{M/L}(a)) = (N_{L/K} \circ N_{M/L})(a),$$

as asserted. The same argument applies to the trace mapping (it suffices to replace each product by a sum and write T in place of N). □

3. The earliest result concerning the distribution of algebraic integers in the plane is due to L. Kronecker and, with the use of the notation $\boxed{a}$ for the largest absolute value of conjugates (over Q) of a, runs as follows.

Theorem 2.1. (i) *If $a \neq 0$ is an algebraic integer which is not a root of unity, then $\overline{|a|} > 1$.*

(ii) *If a is a non-zero totally real integer which is not of the form $2\cos(r\pi)$ with rational r, then $\overline{|a|} > 2$.*

Proof. (i) Let $a \neq 0$ be an algebraic integer, and assume that $\overline{|a|} \leq 1$. Let $K = Q(a)$ and $n = [K:Q]$. The numbers $a, a^2, a^3, \ldots$ all lie in R_K and therefore their minimal polynomials have rational integral coefficients and degrees not exceeding n. But all conjugates of the numbers a^k ($k \geq 1$) have their absolute values bounded by 1, whence the coefficients of their minimal polynomials do not exceed $\max\{\binom{n}{j}: 1 \leq j \leq n\}$. This shows that the sequence $a, a^2, a^3, \ldots$ contains only a finite number of distinct terms, and so for certain $i \neq j$ we must have $a^i = a^j$, i.e., a is a root of unity.

(ii) Now let a be a non-zero totally real integer whose conjugates all lie in the interval $[-2, 2]$. We may assume that $a \neq \pm 2$. The number $a^2/4 - 1$ is negative, and so $b = a/2 + (a^2 - 4)^{1/2}/2$ is not real. Since b is a root of $x^2 - ax + 1$, it is an algebraic integer, but obviously $\overline{|b|} = 1$, whence by (i) b is a root of unity, i.e., $b = \exp(2\pi i r')$ for a certain rational r'. Since $a = 2\mathrm{Re}(b)$, we get $a = 2\cos(\pi r)$ with $r = 2r'$. □

It has been observed by A. Schinzel and H. Zassenhaus [65] that for integers of a fixed degree the bound in (i) can be improved. We now give a proof of a result of this type due to E. Dobrowolski [78].

Theorem 2.2. *If a is an algebraic integer of degree $n \geq 2$ which is neither zero nor a root of unity, then*

$$\overline{|a|} \geq 1 + \frac{1}{6} \cdot \frac{\log n}{n^2}.$$

Proof. Assume the contrary and choose a rational prime in the interval $(3n, 6n)$. Such a prime exists by Bertrand's postulate. Denote by $F(x) = x^n + A_{n-1}x^{n-1} + \ldots + A_0$ the minimal polynomial of a and let

$$G(x) = \prod_{i=1}^{n}(x - a_i^p) = x^n + B_{n-1}x^{n-1} + \ldots + B_0,$$

where $a_1 = a, a_2, \ldots, a_n$ are the conjugates of a. Finally, put

$$S_k = a_1^k + a_2^k + \ldots + a_n^k \quad \text{for } k \geq 1.$$

Newton's formulas imply the following equalities for $k = 1, 2, \ldots, n$:

$$\begin{aligned} S_k + A_{n-1}S_{k-1} + \ldots + A_{n-k+1}S_1 + kA_{n-k} &= 0, \\ S_{kp} + B_{n-1}S_{(k-1)p} + \ldots + B_{n-k+1}S_p + kB_{n-k} &= 0. \end{aligned} \quad (2.1)$$

Observe now that for $k = 1, 2, \ldots, n$ we have

$$|S_{kp} - S_k| \leq 2n \lceil a \rceil^{kp} \leq 2n(1+(\log n)/6n^2)^{6kn} \leq 2n\exp(k(\log n)/n),$$

and since the function $f(t) = t^{-1}\exp(t(\log n)/n)$ has no maxima in the open interval $(1, n)$ and $f(1) \leq 2^{1/2} < 3/2$, $f(n) = e/n \leq e/2 < 3/2$, we obtain

$$|S_{kp} - S_k| \leq 3kn < kp.$$

In particular, one has

$$|S_p - S_1| < p. \tag{2.2}$$

Further, we have for $k \geq 1$ the equality

$$S_k^p = (a_1^k + \ldots + a_n^k)^p = a_1^{kp} + \ldots + a_n^{kp} + pu(p, k)$$

with an integer $u(p, k)$, which is rational, being equal to $(S_k^p - S_{kp})/p$, and thus

$$S_k \equiv S_k^p \equiv S_{kp} \pmod{p} \tag{2.3}$$

holds for all $k \geq 1$.

In particular, $S_1 \equiv S_p \pmod{p}$, and this together with (2.2) leads to $S_1 = S_p$. Utilizing (2.1), we get $A_{n-1} = B_{n-1}$. Assume now that for $j = 1, 2, \ldots, k-1$ one has $S_{jp} = S_j$ and $A_{n-j} = B_{n-j}$. Using (2.1) and (2.3), we get $S_{kp} = S_k$, and again using (2.1), we obtain $A_{n-k} = B_{n-k}$. Finally, we see that $F(x) = G(x)$ and thus the sets $\{a_1, \ldots, a_n\}$ and $\{a_1^p, \ldots, a_n^p\}$ coincide. This, however, quickly leads to a contradiction. Indeed, if K is the splitting field of F and $s \in \text{Gal}(K/Q)$ carries a onto a^p then, denoting the order of s by N, we get

$$a = s^N(a) = a^{p^N},$$

and thus a is either zero or a root of unity, contrary to our assumption. □

4. The second part of Theorem 2.1 was extended by I. Schur [18a], who proved that an interval on the real axis of length smaller than 4 can contain only a finite number of full sets of conjugates of algebraic integers. This result found its counterpart in a theorem proved by R. M. Robinson [62], which implies that Theorem 2.1 (ii) is the best possible.

Theorem 2.3. *If I is any interval on the real axis whose length is greater than 4, then one can find infinitely many full sets of conjugates of algebraic integers lying in I. Moreover, for the particular intervals $[-2-\varepsilon, 2+\varepsilon]$ with a positive ε one can find such sets not containing any number of the form $2\cos(\pi r)$ with a rational r.*

Proof. Let $I = [c-2\lambda, c+2\lambda]$ for $\lambda > 1$ and define a sequence $P_n(x)$ of polynomials by means of

$$P_n(x) = (2\lambda)^n T_n\left(\frac{x-c}{2\lambda}\right) \quad (n = 1, 2, \ldots)$$

where $T_n(x)$ is the nth Chebyshev polynomial, i.e. $T_n(x) = 2^{1-n}\cos(n\arccos x)$. To be more explicit, we have

$$T_n(x) = x^n + \sum_{k=1}^{r}(-1)^k nk^{-1}\binom{n-k-1}{k-1}4^{-k}x^{n-2k} \quad (r = [n/2]).$$

We shall need the following properties of $P_n(x)$, which are easily deducible from the corresponding properties of the Chebyshev polynomials:

(i) $$P_n(x) = x^n + \sum_{k=1}^{n} a_k^{(n)} x^{n-k},$$

where for a fixed k we have $a_k^{(n)} = F_k(n)$, which is a polynomial in n divisible by the polynomial $f(n) = n$;

(ii) $$\max_{x\in I}|P_n(x)| = 2\lambda^n,$$

and this maximal value is attained at $1+n$ points, say $u_1, \ldots, u_{n+1}$. Moreover, the products $P_n(u_i)P_n(u_{i+1})$ are negative for $i = 1, 2, \ldots, n$.

Let us now choose an integer M so as to satisfy $\lambda^M(\lambda-1) \geq 1$ and observe that for a suitable m every coefficient $a_i^{(n)}$ ($i = 1, 2, \ldots, M$) with n divisible by m is integral and even. In fact, for some rational integers $\alpha_i^{(j)}, \beta_j$ we have

$$a_k^{(n)} = F_k(n) = (\alpha_0^{(k)}n^k + \ldots + \alpha_{k-1}^{(k)}n)\beta_k^{-1},$$

and so we may take m as the least common multiple of the numbers $2\beta_1, \ldots, 2\beta_M$.

Let n be any rational integer divisible by m. Evidently we are able to find in the interval $[0, 1)$ numbers $b_{M+1}, \ldots, b_n$ such that the polynomial

$$Q_n(x) = P_n(x) + \sum_{k=1+M}^{n} b_k P_{n-k}(x) = x^n + \sum_{k=0}^{n-1} c_k x^k$$

has all the coefficients c_k integral and even, c_0 not being divisible by 4. (For the coefficients $c_{n-M}, \ldots, c_{n-1}$ this is implied already by the choice of m.)

Finally, note that for $x \in I$ we have

$$|P_n(x) - Q_n(x)| \leq \sum_{k=1+M}^{n} \max_{x\in I}|P_{n-k}(x)| = 2\sum_{k=1+M}^{n} \lambda^{n-k}$$

$$= 2\frac{\lambda^{n-M}-1}{\lambda-1} < \frac{2\lambda^n}{\lambda^M(\lambda-1)} \leq 2\lambda^n;$$

hence at all points $u_1, \ldots, u_{n+1}$ the polynomials $P_n(x)$ and $Q_n(x)$ attain values of coinciding signs. This shows that $Q_n(x)$ has exactly n zeros in I and, since it is irreducible by Eisenstein's theorem, those zeros form a full set of conjugates. This proves the first part of the theorem, since we may choose n in infinitely many ways.

To prove the second part observe that the first part shows the existence of infinitely many full sets of conjugates in the interval $J_\varepsilon = [-2-\varepsilon, 2-\varepsilon/2]$ with a positive ε. Among those sets there is an infinite number of sets not containing any number $2\cos(\pi r)$ with rational r. Indeed, every set containing such a number must contain a number $2\cos(\pi q^{-1})$ with a suitable natural q, but this is possible only for a finite number of q's, since $2\cos(\pi q^{-1})$ tends to 2 as q approaches infinity. This proves the remaining part of the theorem since the interval J_ε is contained in $[-2-\varepsilon, 2+\varepsilon]$. □

5. Now let us consider more closely the distribution of full sets of conjugates in the plane, restricting ourselves as before to algebraic integers. Since the product of all conjugates is a rational integer, it is impossible to approximate n arbitrary complex numbers by n conjugates of an algebraic integer of degree n, except when those numbers fulfil some additional conditions. However, it turns out that any $n-1$ complex numbers which lie symmetrically with respect to the real line can be approximated in this way by $n-1$ conjugates of an algebraic integer of degree n. This is the content of the following theorem, proved by Th. Motzkin [47].

Theorem 2.4. *Let $A = \{z_1, \ldots, z_{n-1}\}$ be a set of complex numbers, symmetrical with respect to the real line. To every positive ε one can find an irreducible (over Q) polynomial of degree n with integral rational coefficients whose roots $w_1, \ldots, w_n$ are algebraic integers and, after a suitable reordering, satisfy $|z_i - w_i| < \varepsilon$ for $i = 1, 2, \ldots, n-1$. Moreover, if the number z_j is real for $j = j_1, \ldots, j_t$, then the polynomial may be so chosen that the roots w_j are also real for the same values of the indices.*

The proof is based on the fact that the roots of a polynomial depend continuously on its coefficients. In a slightly more precise form this is

Lemma 2.1. *Let $P(x) = x^n + a_{n-1}x^{n-1} + \ldots + a_0$ be a polynomial with complex coefficients and let $z_1, \ldots, z_n$ be its roots. We assume that they are all distinct. Assume, moreover, that*

$$P_k(x) = x^n + a_{n-1}^{(k)} x^{n-1} + \ldots + a_0^{(k)}$$

is a sequence of polynomials over the complex field such that for $j = 0, 1, \ldots$

..., $n-1$ we have $\lim_k a_j^{(k)} = a_j$ and let $w_1^{(k)}, ..., w_n^{(k)}$ be roots of $P_k(x)$. Then it is possible to reorder those roots in such a way that for every positive ε and sufficiently large k we have the inequalities

$$|z_i - w_i^{(k)}| < \varepsilon.$$

Moreover, if the coefficients of $P(x)$ and $P_k(x)$ are real for k sufficiently large, then $P(x)$ and $P_k(x)$ have the same number of real zeros provided that k is large enough.

Proof. Observe first that all the numbers $w_i^{(k)}$ have a common bound, and then renumber them in the following way: let $w_1^{(k)}$ be that root of $P_k(x)$ which lies as near as possible to z_1. If there are more such roots, then take one arbitrarily. Of the remaining roots of $P_k(x)$ let $w_2^{(k)}$ be that one which lies as near as possible to z_2 and continue the process. Next we may take a sequence $\{k_r\}$ such that all the sequences $\{w_i^{(k_r)}\}$ ($i = 1, 2, ..., n$) are convergent and denote by $c_1, ..., c_n$ their respective limits. Then obviously

$$\lim_r P_{k_r}(z) = \lim \prod_{i=1}^n (z - w_i^{(k_r)}) = \prod_{i=1}^n (z - c_i)$$

and, since obviously $\lim_r P_{k_r}(z) = P(z)$, the sets $\{z_1, ..., z_n\}$ and $\{c_1, ..., c_n\}$ coincide.

Moreover, $|w_1^{(k_r)} - z_1| \leq |w_i^{(k_r)} - z_1|$ for $i = 2, 3, ..., n$; thus $|c_1 - z_1| \leq |c_i - z_1|$ for $i = 2, 3, ..., n$ and $c_1 = z_1$. In the same manner one proves $c_i = z_i$ for $i = 1, 2, ..., n$. Finally, note that the sequence $\{k_r\}$ could be chosen arbitrarily provided the required limits exist, and so $\lim_n w_i^{(n)} = z_i$ for $i = 1, 2, ..., n$, proving the first part of the lemma.

To prove the second, note that if the coefficients of $P(x)$ and $P_k(x)$ are real, z_i is a real root of $P(x)$ and for an infinite sequence $\{k_r\}$ the numbers $w_i^{(k_r)}$ are not real, then $\overline{w}_i^{(k_r)}$ are also roots of the P_k's, and so, by the foregoing argument, they tend to a root of $P(x)$, say z_j, distinct from z_i, which is impossible, since $z_j = \lim \overline{w}_i^{(k_r)} = \overline{z}_i = z_i$, z_i being real. □

Proof of Theorem 2.4. Let $z_1, ..., z_{n-1}$ be given complex numbers, the first t of them being real and the remaining ones being pairwise conjugate and ordered in such a way that for $i = t+1, t+2, ..., r = (n-t-1)/2$ we have the equalities

$$\overline{z}_i = z_{r+t+i}.$$

Those numbers are roots of the polynomial

$$P(z) = \prod_{i=1+t}^r (z - z_i)(z - \overline{z}_i) \cdot \prod_{i=1}^t (z - z_i) = z^{n-1} + a_1 z^{n-2} + ... + a_{n-1}$$

with real coefficients. Without restricting generality, we may assume that the numbers $1, a_1, \ldots, a_{n-1}$ are linearly independent over the rationals and that no subset of the set $\{z_1, \ldots, z_{n-1}\}$ is a full set of zeros of a monic polynomial with rational integral coefficients. This may be obtained by replacing the numbers z_i by z_i' which are sufficiently close to them and algebraically independent over Q.

Now we have to use Kronecker's theorem on diophantine approximations (see Appendix III, Theorem III), which implies the existence of rational integers $b_i = b_i(k)$ $(i = 1, 2, \ldots, n;\ k = 1, 2, \ldots)$ satisfying

$$|b_{j+1} - a_j b_1 - a_{j+1} + a_j a_1| < k^{-1} \tag{2.4}$$

for $j = 1, 2, \ldots, n-1$. (We put here $a_n = 0$.) If we define $z_n = a_1 - b_1$, $P_k(x) = x^n + b_1 x^{n-1} + \ldots + b_n$ and the numbers c_i by $(x - z_n) P(x) = x^n + c_1 x^{n-1} + \ldots + c_n$, then inequality (2.4) shows $|c_j - b_j| < k^{-1}$ for $j = 1, 2, \ldots, n$. Finally, according to Lemma 2.1, we can renumber the roots $Z_i = Z_i(k)$ of P_k in such a way as to obtain

$$\lim_k Z_i(k) = z_i \quad \text{for } i = 1, 2, \ldots, n.$$

The theorem will be proved if we show that among the polynomials $P_k(x)$ there occur an infinite number of irreducibles. Indeed, this is very simple. If, for sufficiently large k, all polynomials $P_k(x)$ were reducible, then by choosing a suitable sequence $\{k_r\}$ we would obtain a partition of the set $\{1, 2, \ldots, n\}$ into two disjoint subsets A, B such that the polynomials

$$H_r(x) = \prod_{i \in A} (x - Z_i(k_r)), \quad L_r(x) = \prod_{i \in B} (x - Z_i(k_r))$$

would have rational integral coefficients. But one of the sets A, B does not contain the number n, say the set A, and thus the polynomial

$$\prod_{i \in A} (x - z_i) = \lim_r H_r(x)$$

would have rational integral coefficients, contrary to our assumption. □

As an immediate consequence, we get the following important

Corollary. *There exist finite extensions of rationals with the signature $[r_1, r_2]$ given arbitrarily.*

Proof. If $r_1 = 0$, then let us just take the field $Q((-2)^{1/n})$ with $n = 2r_2$. If $r_1 \neq 0$, then it suffices to apply the theorem just proved to a set $z_1, \ldots, z_{n-1}$ of complex numbers, symmetrical with respect to the real axis and containing exactly $r_1 - 1$ real numbers. The nth root of the polynomial obtained will obviously be real. □

Note. This corollary can also be proved directly from Lemma 2.1 without the use of Kronecker's theorem on linear forms. Namely, take an arbitrary polynomial with r_1 real and $2r_2$ complex roots, with real coefficients and without multiple zeros. Then choose a sequence of monic polynomials with coefficients of the form $(4k+2)/(2m+1)$ (k, m—integral and rational) tending to the given polynomial. Lemma 2.1 shows that if the monic polynomials are sufficiently close to it then they have r_1 real and $2r_2$ complex roots, and Eisenstein's theorem shows that they are all irreducible, whence the fields generated by their roots have the signature $[r_1, r_2]$, as required.

§ 2. Discriminants and Integral Bases

1. To obtain some kind of classification of algebraic number fields one introduces various functions, defined on the set of all such fields. The simplest example is the degree over Q, which, however, does not suffice since there are infinitely many fields of the same degree. In this section we shall introduce another function, namely the absolute discriminant $d(K)$, which assumes non-zero integral values and has the remarkable property, noted first by C. Hermite, of assuming the same value only for a finite number of fields. Moreover, it is an invariant of isomorphisms, i.e., isomorphic fields have the same discriminants. We start with some general definitions, valid not only for number fields but also for arbitrary finite extensions of any field of zero characteristic. This slightly more general approach will be utilized in Chapter 5, where we shall speak about completions of algebraic number fields.

Assume thus that K is a field of zero characteristic and L/K is a finite extension of degree n. If $\bar{K}$ is a fixed algebraic closure of K, then there are n distinct embeddings $F_1, \ldots, F_n$ of L into $\bar{K}$ leaving K invariant. For any $v_1, \ldots, v_n \in L$ we define the *discriminant* $d_{L/K}(v_1, \ldots, v_n)$ by

$$d_{L/K}(v_1, \ldots, v_n) = (\det[F_j(v_i)]_{i,j})^2.$$

(Note that the discriminant does not depend on the ordering of the embeddings or of the v_i's, because of the squaring.)

The following proposition provides an alternative definition of the discriminant.

Proposition 2.3. *If $T_{L/K}(x) = F_1(x) + \ldots + F_n(x)$ is the trace defined for $x \in L$, then for every sequence $v_1, \ldots, v_n \in L$ we have*

$$d_{L/K}(v_1, \ldots, v_n) = \det[T_{L/K}(v_i v_j)].$$

Proof. It suffices to observe that with the notation A^T for the transposed matrix we have

$$\det[T_{L/K}(v_i v_j)] = \det\left[\sum_{k=1}^{n} F_k(v_i)F_k(v_j)\right]$$
$$= \det([F_k(v_i)][F_k(v_j)]^T) = d_{L/K}(v_1, \ldots, v_n). \qquad \square$$

If a is an arbitrary element of L, then its discriminant with respect to K, which we shall denote by $d_{L/K}(a)$, is defined by

$$d_{L/K}(a) = d_{L/K}(1, a, a^2, \ldots, a^{n-1}).$$

One can easily see that

$$d_{L/K}(a) = \prod_{i<j} (F_i(a) - F_j(a))^2, \qquad (2.5)$$

and this shows that if a generates the extension L/K then $d_{L/K}(a)$ equals the discriminant of the minimal polynomial for a over K, and thus is non-zero. If, however, a generates a proper subfield of L, then $d_{L/K}(a) = 0$.

The main properties of the discriminant are contained in the following proposition.

Proposition 2.4. *Let L/K be finite and let $v_1, \ldots, v_n$ be elements of L.*

(i) *The discriminant $d_{L/K}(v_1, \ldots, v_n)$ lies in K, and if all v's lie in the integral closure in L of a domain R contained in K, then $d_{L/K}(v_1, \ldots, v_n) \in R$. In particular, if K is an algebraic number field, then the discriminant of a set of integers is again an integer.*

(ii) *If $u_i = \sum_{j=1}^{n} a_{ij} v_j$ $(i = 1, 2, \ldots, n; a_{ij} \in K)$, then*

$$d_{L/K}(u_1, \ldots, u_n) = (\det[a_{ij}])^2 d_{L/K}(v_1, \ldots, v_n).$$

(iii) *$d_{L/K}(v_1, \ldots, v_n) = 0$ iff the system $v_1, \ldots, v_n$ is linearly dependent over K.*

(iv) *If a generates L over K and $P \in K[x]$ is its minimal polynomial, then*

$$d_{L/K}(a) = (-1)^m \det[a_{ij}] = (-1)^m N_{L/K}(P'(a)),$$

where $m = n(n-1)/2$, the elements a_{ij} are defined by

$$a^j P'(a) = \sum_{i=0}^{n-1} a_{ij} a^i \quad (j = 0, 1, \ldots, n-1),$$

P' is the formal derivative of P, and $N_{L/K}(x) = \prod_{j=1}^{n} F_j(x)$ is the norm mapping from L to K.

Proof. Part (i) results from Proposition 2.3 and (ii) is a consequence of the equality

$$\det[F_j(u_i)] = \det[a_{ij}]\det[F_j(v_i)].$$

To prove (iii) note first that if the elements $v_1, \ldots, v_n$ are linearly dependent over K, then for suitable $x_1, \ldots, x_n$ in K (not all of them zero) we have

$$\sum_{i=1}^{n} x_i F_j(v_i) = 0 \quad \text{for } j = 1, 2, \ldots, n,$$

and thus $d_{L/K}(v_1, \ldots, v_n)$ must vanish.

On the other hand, if $d_{L/K}(v_1, \ldots, v_n)$ vanishes, then the system

$$\sum_{i=1}^{n} x_i T_{L/K}(v_i v_j) = 0 \quad (j = 1, 2, \ldots, n)$$

has a non-zero solution $X_1, \ldots, X_n$ in K. If the system $v_1, \ldots, v_n$ were linearly independent over K, then $X = X_1 v_1 + \ldots + X_n v_n$ would be non-zero and $T_{L/K}(Xv_i) = 0$ would hold for $i = 1, 2, \ldots, n$. Hence for all $y \in L$ we would have $T_{L/K}(Xy) = 0$, but this leads to a contradiction, since for $y = 1/X$ we obtain $n = T_{L/K}(1) = 0$.

Finally, to obtain (iv) denote by $a_1, a_2, \ldots, a_n$ the conjugates of $a_1 = a$ and put $b_i = P'(a_i)$, all these elements lying in a fixed algebraic closure of K. Then

$$d_{L/K}(a) = \prod_{i<j} (a_i - a_j)^2 = (-1)^m \prod_{i=1}^{n} \prod_{j \neq i} (a_i - a_j)$$

$$= (-1)^m \prod_{i=1}^{n} P'(a_i) = (-1)^m b_1 b_2 \ldots b_n = (-1)^m N_{L/K}(P'(a)).$$

Moreover,

$$b_1 b_2 \ldots b_n \det[a_i^j] = \det[a_i^j b_i] = \det\left[\sum_{k=1}^{n} a_{kj} a_i^k\right] = \det[a_{kj}] \det[a_j^k],$$

and since (ii) implies $\det[a_j^k] \neq 0$, we obtain

$$b_1 b_2 \ldots b_n = \det[a_{ij}]$$

and thus

$$d_{L/K}(a) = (-1)^m \det[a_{ij}],$$

as asserted. □

2. Now we leave the abstract situation and return to the algebraic number fields to which the concepts developed in the preceding subsection apply. We shall use them for the definition of the field discriminant. Here we first need the notion of an integral basis. Let K be an algebraic number field of degree n. A system $w_1, \ldots, w_n$ of integers of K which is linearly independent over Q and generates R_K as a Z-module is called an *integral basis of K*. Note that the existence of a system $w_1, \ldots, w_n$ of elements of R_L which are linearly independent over a subfield K of L and which generate R_L as an R_K-module is equivalent to the fact that R_L is a free R_K-module. We shall see below that this always happens for $K = Q$ but may fail in the general case. The question which extensions L/K have such relative integral bases will be answered in Chapter 7 (Corollary 1 to Theorem 7.14).

Now we prove

Theorem 2.5. *Every algebraic number field has an integral basis over Q, i.e., R_K is a free Z-module for all finite extensions K/Q.*

Proof. Let $a \in R_K$ be a generator of the extension K/Q. Lemma 1.5 implies that $d_{K/Q}(a) R_K \subset Z[a]$, whence

$$R_K \subset d_{K/Q}^{-1}(a) Z[a].$$

Since $Z[a]$ is a free Z-module with finitely many generators $1, a, \ldots, a^{n-1}$ ($n = [K:Q]$), it is Noetherian by Proposition 1.2, and hence $d_{K/Q}^{-1}(a) Z[a]$ is also Noetherian. It follows that R_K is a finitely generated torsion-free Z-module, and hence Theorem 1.13 implies that, as a Z-module, R_K is isomorphic to Z^d for a certain $0 \leq d \leq n$ (because Z is PID). However, R_K spans an n-dimensional Q-space, and so we must have $d = n$. □

We shall also give an alternative proof of Theorem 2.5 which permits the construction of an integral basis of the special form needed in the sequel.

Proposition 2.5. *If $[K:Q] = n$ and the numbers $a_1, \ldots, a_n$ of R_K are linearly independent over Q, then there exists an integral basis $w_1, \ldots, w_n$ of K such that for $j = 1, 2, \ldots, n$ we have*

$$a_j = c_{j1} w_1 + \ldots + c_{jj} w_j$$

for suitable $c_{ij} \in Z$.

Proof. For $i = 1, 2, \ldots, n$ let d_{ii} be the least positive integer such that with suitably chosen rational integers $d_{i1}, d_{i2}, \ldots, d_{i,i-1}$ the number

$$w_i = d_{K/Q}(a_1, \ldots, a_n)^{-1} \sum_{j=1}^{i} d_{ij} a_j$$

lies in R_K.

The numbers $w_1, \ldots, w_n$ are linearly independent over Q, since

$$d_{K/Q}(w_1, \ldots, w_n) = (d_{K/Q}(a_1, \ldots, a_n))^{-n} \det[d_{ij}]^2 d_{K/Q}(a_1, \ldots, a_n)$$

and

$$\det[d_{ij}] = d_{11} d_{22} \ldots d_{nn} \neq 0$$

by Proposition 2.4. We show that $w_1, \ldots, w_n$ is an integral basis of K. Note first that if $c \in R_K$ can be written in the form

$$c = d_{K/Q}(a_1, \ldots, a_n)^{-1}(c_1 a_1 + \ldots + c_j a_j)$$

for $c_i \in Z$ and a certain j, then $d_{jj} | c_j$. Indeed, if $c_j = sd_{jj} + r$ with $r, s \in Z$ and $0 < r < d_{jj}$, then $c - sw_j \in R_K$ and

$$c - sw_j = d_{K/Q}(a_1, \ldots, a_n)^{-1}((c_1 - d_{j1})a_1 + \ldots + ra_j)$$

contrary to the choice of d_{jj}.

Let M_0 be the Z module generated by $w_1, \ldots, w_n$. We shall prove by recurrence that every element of R_K which has the form

$$d_{K/Q}(a_1, \ldots, a_n)^{-1}(x_1 a_1 + \ldots + x_j a_j),$$

where $x_i \in Z$, lies in M_0. For $j = n$ this will imply $R_K = M_0$.

For $j = 1$ the assertion is obvious. Assume that it holds for $j = i - 1$ and consider

$$y = d_{K/Q}(a_1, \ldots, a_n)^{-1}(x_1 a_1 + \ldots + x_i a_i) \in R_K$$

where $x_k \in Z$. Then with a suitable $A \in Z$ we have $x_i = Ad_{ii}$; thus $y - Aw_i \in R_K$ and, since $Aw_i \in M_0$, it suffices to apply the inductional assumption. $\square$

Examples of integral bases in various classes of fields will be given in the last part of this section.

3. In this subsection we define the discriminant of any Z-module with a finite index in R_K, a special case of which will be the field discriminant of K. The following simple proposition shows that all free bases of a free Z-module have the same discriminant.

Proposition 2.6. *If M is a free Z-module with n free generators $a_1, \ldots, a_n$ and $b_1, \ldots, b_n$ is another set of its generators, also free, then for some matrix $[a_{ij}]$ with elements from Z and determinant ± 1 we have*

$$b_i = \sum_{j=1}^{n} a_{ij} a_j. \qquad (2.6)$$

Conversely, if $a_1, \ldots, a_n$ generate freely a Z-module M, and the elements $b_1, \ldots, b_n$ of M are related to the a_i's through (2.6) with $\det[a_{ij}] = \pm 1$, then those elements again form a free basis of M.

Proof. If the a_i's and the b_i's are free generators, then the matrix $[a_{ij}]$ has an inverse with elements in Z, and so its determinant must equal 1 or -1.

The second part of the proposition results from the observation that under the given conditions the matrix $[a_{ij}]$ has an inverse with elements from Z. □

Corollary. *If $[K:Q] = n$ and M is a free Z-module with n free generators contained in R_K, then the discriminant of a basis of M does not depend on the choice of that basis.*

Proof. This follows from the proposition just proved and Proposition 2.4 (ii). □

The above proposition suggests the following definition: If M is a free Z-module with n free generators contained in R_K (with $n = [K:Q]$), then the discriminant $d_{K/Q}(M)$ is the discriminant of any free basis of M. In particular, if $M = R_K$, then this discriminant is called the *discriminant of the field K* and denoted by $d(K)$. It is of course the discriminant of any integral basis of K. Proposition 2.4 (ii) shows that $|d(K)|$ equals the greatest common divisor of $d_{K/Q}(v_1, ..., v_n)$ taken over all $v_1, ..., v_n \in R_K$. This fact is made more precise by

Proposition 2.7. *If $a_1, ..., a_n \in R_K$ and are linearly independent over Q, then*

$$d_{K/Q}(a_1, ..., a_n) = m^2 d(K),$$

where m is the index in R_K of the Z-module M generated by the a_i's.

Proof. Let $w_1, ..., w_n$ be an integral basis for K, and choose the numbers $b_1, ..., b_n$ from M in such a way that

$$b_i = \sum_{k=1}^{i} c_{ik} w_k \quad (c_{ik} \in Z)$$

and c_{ii} is positive and as small as possible. As in the proof of Proposition 2.5 we see that the b_i's form a set of free generators for M and that $t_1 w_1 + ... + t_i w_i$ (with $t_i \in Z$) can lie in M only if t_i is divisible by c_{ii}. This shows that the numbers

$$\alpha_1 w_1 + ... + \alpha_j w_j \quad (0 \leq \alpha_i < c_{jj}; j = 1, ..., n)$$

are all incongruent (mod M) and obviously there are $c_{11} ... c_{nn}$ of them. We shall show that all residue classes (mod M) are represented by those numbers. Indeed, let

$$\xi = \sum_{k=1}^{n} \lambda_k w_k \quad (\lambda_k \in Z)$$

be an arbitrary element from R_K, denote by μ_n the least non-negative residue of $\lambda_n \pmod{c_{nn}}$ and put $A_n = (\lambda_n - \mu_n)/c_{nn}$. Then

$$\xi = A_n b_n + \mu_n w_n + \sum_{k=1}^{n-1} (\lambda_k - A_n c_{nk}) w_k.$$

If by μ_{n-1} we denote the least non-negative residue of $\lambda_{n-1} - A_n c_{n,n-1}$ $\pmod{c_{n-1,n-1}}$ and put

$$A_{n-1} = (\lambda_{n-1} - A_n c_{n,n-1} - \mu_{n-1}) c_{n-1,n-1}^{-1},$$

then

$$\xi = A_n b_n + A_{n-1} b_{n-1} + \mu_n w_n + \mu_{n-1} w_{n-1}$$
$$+ \sum_{k=1}^{n-2} (\lambda_k - A_n c_{n,k} - A_{n-1} c_{n-1,k}) w_k.$$

Continuing this procedure, we finally obtain an equality of the form

$$\xi = \sum_{k=1}^{n} \alpha_k b_k + \sum_{k=1}^{n} \mu_k w_k \quad (0 \leq \mu_j < c_{jj}; \mu_j, \alpha_j \in Z),$$

and so

$$\xi \equiv \sum_{k=1}^{n} \mu_k w_k \pmod{M},$$

as required. Now it suffices to observe that

$$d_{K/Q}(a_1, \ldots, a_n) = d_{K/Q}(b_1, \ldots, b_n) = (c_{11} \ldots c_{nn})^2 d(K)$$

by Proposition 2.4 (ii). □

Corollary. *If a is a non-zero element of K and $I = aR_K$ is the fractional ideal generated by a, then*

$$N(I) = |N_{K/Q}(a)|.$$

Proof. Since both sides of the asserted equality are multiplicative in a, we may assume that $a \in R_K$. Let $w_1, \ldots, w_n$ be an integral basis of K. Then aR_K is the Z-module generated by $aw_1, \ldots, aw_n$, and hence the proposition implies

$$N(I)^2 = d_{K/Q}(aw_1, \ldots, aw_n) d(K)^{-1},$$

but obviously

$$d_{K/Q}(aw_1, \ldots, aw_n) = N_{K/Q}^2(a) d(K),$$

and our assertion follows. □

4. Not every non-zero rational integer is a discriminant of a suitable field. This is implied by the following result first proved by L. Stickelberger:

Theorem 2.6. *The discriminant $d(K)$ of a field is either congruent to unity (mod 4) or divisible by 4.*

Proof. Let $w_1, \ldots, w_n$ be an integral basis of K and let $w_i^{(j)}$ ($i, j = 1, 2, \ldots, n$) be all the conjugates of the w_i's. Then we have

$$d(K)^{1/2} = \det[w_i^{(j)}]$$

$$= \sum_{\substack{(1,\ldots,n \\ \mu_1,\ldots,\mu_n) \text{ even}}} w_1^{(\mu_1)} \ldots w_n^{(\mu_n)} - \sum_{\substack{(1,\ldots,n \\ \mu_1,\ldots,\mu_n) \text{ odd}}} w_1^{(\mu_1)} \ldots w_n^{(\mu_n)} = A - B.$$

If K_0 is the minimal normal extension of Q which contains K, then A and B belong to R_{K_0} and, moreover, the numbers $A+B$ and AB are rational integers, because they are invariant under the automorphisms of K_0. But we may write

$$d(K) = (A-B)^2 = (A+B)^2 - 4AB \equiv (A+B)^2 \pmod{4},$$

and so $d(K) \equiv 0, 1 \pmod{4}$. □

Our next proposition, due to A. Brill [77] determines the sign of the discriminant.

Proposition 2.8. *If $[r_1, r_2]$ is the signature of K, then $\operatorname{sign} d(K) = (-1)^{r_2}$, in other words, the discriminant of K is positive iff K has an even number of complex embeddings in the field of complex numbers that are non-conjugate (in the elementary sense).*

Proof. Once again let $w_1, \ldots, w_n$ be an integral basis of K and let $w_i^{(j)}$ be the conjugates of w_i's. Write $\det[w_i^{(j)}] = d_1 + id_2$ with real d_1, d_2. Since the change of i into $-i$ in this determinant is equivalent to the interchange of r_2 pairs of rows, we have $d_1 - id_2 = (-1)^{r_2} \det[w_i^{(j)}]$, and so $(-1)^{r_2}(d_1 + id_2) = d_1 - id_2$. If r_2 is even, this implies $d_2 = 0$ and $d(K) = d_1^2$ is positive, and if r_2 is odd, this implies $d_1 = 0$, and $d(K) = (id_2)^2 = -d_2^2$ is negative. □

To finish this subsection we prove a result of L. Kronecker, which will be strengthened later (see Chapter 4, Corollary 2 to Proposition 4.9). However, we prove it here since the proof is easy and the result will be of some use in the sequel.

Proposition 2.9. *If $Q \subset K \subset L$, then $d(K)$ divides $d(L)$.*

Proof. Let $[K:Q] = m$, $[L:K] = n$. Let $a_1, \ldots, a_m$ be an integral basis of K, and choose $a_{m+1}, \ldots, a_{mn} \in R_L$ so that the resulting set $a_1, \ldots, a_{mn}$ is linearly independent over Q. Proposition 2.5 implies the existence of an integral basis of L, say $w_1, \ldots, w_{mn}$, such that

$$a_j = c_{j1} w_1 + \ldots + c_{jj} w_j \quad (j = 1, 2, \ldots, mn)$$

with $c_{ij} \in Z$. The elements $w_1, \ldots, w_m$ of this basis lie in K and the Z-module generated by them contains an integral basis of K; hence $w_1, \ldots, w_m$ must be an integral basis of K. As usual, denote by $w_i^{(k)}$ the conjugates of the w_i's and let $F_1, \ldots, F_m$ be the embeddings of L into the complex field such that $F_k(w_1) = w_1^{(k)}$ for $k = 1, 2, \ldots, m$. Denote the remaining embeddings of L by $F_{m+1}, \ldots, F_{mn}$ and assume that $F_k(w_j) = w_j^{(k)}$ holds for all j's and $k = m+1, \ldots, mn$.

Moreover, we can assume that for $x \in K$ and $i \equiv j \pmod{m}$ we have $F_i(x) = F_j(x)$. Consequently, we get

$$d(L) = (\det[F_i(w_j)])^2$$

$$= \begin{vmatrix} w_1^{(1)} & \ldots & w_1^{(m)} & w_1^{(1)} & \ldots & w_1^{(m)} & \ldots & w_1^{(1)} & \ldots & w_1^{(m)} \\ \cdot & & & & & & & & & \cdot \\ w_m^{(1)} & \ldots & w_m^{(m)} & w_m^{(1)} & \ldots & w_m^{(m)} & \ldots & w_m^{(1)} & \ldots & w_m^{(m)} \\ w_{m+1}^{(1)} & \ldots & w_{m+1}^{(m)} & w_{m+1}^{(m+1)} & \ldots & & & & & w_{m+1}^{(mn)} \\ \cdot & & & & & & & & & \cdot \\ w_{mn}^{(1)} & \ldots & w_{mn}^{(m)} & w_{mn}^{(m+1)} & \ldots & & & & & w_{mn}^{(mn)} \end{vmatrix}^2$$

$$= \begin{vmatrix} w_1^{(1)} & \ldots & w_1^{(m)} & 0 & \ldots & & & 0 & \ldots & \\ \cdot & & & & & & & & & \cdot \\ w_m^{(1)} & \ldots & w_m^{(m)} & 0 & \ldots & & & 0 & \ldots & \\ w_{m+1}^{(1)} & \ldots & w_{m+1}^{(m)} & w_{m+1}^{(m+1)} - w_{m+1}^{(1)} & \ldots & & w_{m+1}^{(mn)} - w_{m+1}^{(m)} \\ \cdot & & & & & & & & & \cdot \\ w_{mn}^{(1)} & \ldots & w_{mn}^{(m)} & w_{mn}^{(m+1)} - w_{mn}^{(1)} & \ldots & & w_{mn}^{(mn)} - w_{mn}^{(m)} \end{vmatrix}^2$$

$$= d(K)a,$$

where a is an algebraic integer. However, $a = d(L)/d(K)$ is also a rational number, and thus it is a rational integer. $\square$

5. This subsection is devoted to examples of integral bases and discriminants in various fields K, including quadratic fields, pure cubic fields $K = Q(m^{1/3})$ and cyclotomic fields $Q(\zeta_{p^n})$.

We start with a lemma which is sometimes a very useful tool for determining the discriminant. A definition is needed: if a is an algebraic integer generating the field K, then the index of Z-module generated by $1, a, a^2, \ldots, a^{n-1}$ (n being the degree of K) in R_K will be called the *index of the number a in R_K* or, which is slightly incorrect, the *index of a in K*. (Note that this module is in fact the subring of R_K generated by a.)

Lemma 2.2. *Let a be an algebraic integer and let $K = Q(a)$ be the field generated by it. If the minimal polynomial over Q of this number is Eisensteinian with* respect to the prime number p, i.e. if it has the form $x^n + a_{n-1}x^{n-1} + \ldots + a_0$, a_i being integers divisible by p, and $p^2 \nmid a_0$, then the index of a in K is not divisible by p.

Proof. Our assumption implies that a^n/p is an algebraic integer and, moreover, $N_{K/Q}(a)$ is not divisible by p^2. Now assume that p divides the index of a. In this case one can find an integer ξ in R_K of the form

$$\xi = (b_0 + b_1 a + \ldots + b_{n-1} a^{n-1})p^{-1} \quad (b_i \in Z),$$

not all of the b_i's being divisible by p. Let j be the minimal index for which $p \nmid b_j$. Then the number

$$\eta = (b_j a^j + \ldots + b_{n-1} a^{n-1})p^{-1}$$

$$= \xi - \left(\frac{b_0}{p} + \frac{b_1}{p} a + \ldots + \frac{b_{j-1}}{p} a^{j-1}\right)$$

is an algebraic integer and so is also

$$\zeta = b_j a^{n-1} p^{-1} = \eta a^{n-j-1} - a^n(b_{j+1} + b_{j+2}a + \ldots + b_{n-1}a^{n-j-2})p^{-1}.$$

This shows that

$$p^n N_{K/Q}(\zeta) = N_{K/Q}(p\zeta) = N_{K/Q}(b_j a^{n-1}) = b_j^n N_{K/Q}(a)^{n-1}$$

and so p has to divide b_j, which is a contradiction. □

With the use of this lemma it is sometimes possible to find the exact power of a prime p dividing the discriminant of K. This happens when we have an integer a generating K and satisfying the conditions of the lemma, because then, by Proposition 2.7, the discriminant of a and the field discriminant are divisible by the same powers of p.

Another fact will also be used quite often, namely the equality $D_{K/Q}(a) = D_{K/Q}(a+n)$ with any rational integral n, which follows directly from the expression of the discriminant of any number as a product of differences.

Let us now return to our examples. To begin with we consider quadratic fields and prove

Theorem 2.7. *Let D be a rational integer not divisible by a square $\neq 1$ and let $K = Q(D^{1/2})$. (Obviously every quadratic extension of the rationals is of this shape.) If $D \equiv 1 \pmod 4$, then $d(K) = D$ and the numbers $1, (1+D^{1/2})/2$ form an integral basis of K. If $D \not\equiv 1 \pmod 4$, then $d(K) = 4D$, and an integral basis is formed by the numbers $1, D^{1/2}$.*

(It is irrelevant which of the two possible square roots of D we take here; however, it is convenient to assume that for positive D the number $D^{1/2}$ is positive, whereas for negative D the square root $D^{1/2}$ lies on the upper imaginary half-axis.)

Proof. The polynomial $x^2 - D$ is Eisensteinian for every rational prime dividing D and since the discriminant of $D^{1/2}$ equals $4D$, for even D we obtain, by Lemma 2.2 and Proposition 2.7, the equality $d(K) = 4D$. If $D \equiv 3 \pmod 4$, then the polynomial $(x+1)^2 - D = x^2 + 2x + 1 - D$ is Eisensteinian for $p = 2$. Since its root equals $D^{1/2} - 1$ and its discriminant is $4D$ in this case we have again $d(K) = 4D$.

Since in both cases considered above we have $D_{K/Q}(1, D^{1/2}) = 4D$, 1 and $D^{1/2}$ form an integral basis.

Finally, if $D \equiv 1 \pmod 4$, then the number $a = (1 + D^{1/2})/2$ is integral, since it satisfies the equation $x^2 - x + (1-D)/4 = 0$ and, moreover, its discriminant equals D, which is a square-free number, thus, by Proposition 2.7 $d(K) = D$ and 1, a is an integral basis. □

Now we consider a pure cubic field, i.e., $K = Q(m^{1/3})$ where m is not divisible by a cube of a prime. We can assume that m is positive and written in the form $m = ab^2$, where ab is square-free (this implies $(a, b) = 1$). Moreover, in the case of $3|m$ we assume that $3|a$, $3 \nmid b$ because the fields generated by $(ab^2)^{1/3}$ and $(a^2b)^{1/3}$ coincide. We now prove

Theorem 2.8. *If K is a pure cubic field of the form $K = Q(m^{1/3})$ with $m = ab^2$ as given above, then one distinguishes between three cases:*

 (i) *$m \not\equiv 1, 8 \pmod 9$. Here $d(K) = -3^3(ab)^2$ and the numbers 1, $m^{1/3}$, $m^{2/3}b^{-1}$ form an integral basis;*

 (ii) *$m \equiv 1 \pmod 9$. Here $d(K) = -3(ab)^2$ and the numbers $m^{1/3}$, $m^{2/3}b^{-1}$, $(1 + m^{1/3} + m^{2/3})/3$ form an integral basis, and*

 (iii) *$m \equiv 8 \pmod 9$. Here $d(K) = -3(ab)^2$ and an integral basis is formed by the numbers $m^{1/3}$, $m^{2/3}b^{-1}$, $(1 - m^{1/3} + m^{2/3})/3$.*

Proof. Put $A = m^{1/3}$. Using Proposition 2.4 (iv), we obtain $d_{K/Q}(A) = -3^3 m^2$. The minimal polynomial for A, $x^3 - m$, is Eisensteinian for every prime divisor of a, and thus in the case of $3|a$ we get $27a^2 | d(K)$ and in the case $3 \nmid a$ we obtain $3a^2 | d(K)$. The number $B = A^2/b$ is a root of $x^3 - a^2b$ which is Eisensteinian for all primes dividing b, and we see that b^2 divides $d(K)$. Putting all this together, we obtain $d(K) = -3^N (ab)^2$ where N is equal to 3 if $3|m$ and is equal to either 1 or 3 otherwise.

If $m \not\equiv 1, 8 \pmod 9$, then $m^3 \not\equiv m \pmod 9$ (in the case of $3|m$ this results from $3 \nmid b$). Hence $(x+m)^3 - m$ is Eisensteinian with respect to $p = 3$ and the discriminant of its root $A - m$ equals $-27m^2$. Hence by Lemma 2.2 we find $d(K) = -27(ab)^2$.

If $m \equiv 1 \pmod 9$ then the number

$$C = (1 + A + A^2)/3$$

is integral. Indeed, its trace equals 1 in view of

$$T_{K/Q}(A) = T_{K/Q}(A^2) = 0;$$

its norm equals

$$N_{K/Q}(1+A+A^2)/27 = N_{K/Q}(A^3-1)/27 N_{K/Q}(A-1) = (m-1)^2/27 \in Z$$

because the minimal polynomial for $A-1$ is $x^3+3x^2+3x+1-m$, and finally the remaining coefficient of the minimal polynomial for C equals $(1-m)/3 \in Z$.

The integrality of C implies that the index of A is divisible by 3, and Proposition 2.7 gives $d(K) = -3(ab)^2$.

The same argument also applies to the case $m \equiv 8 \pmod 9$, in which one ought to consider the number $(1-A+A^2)/3$ instead of C.

The assertion concerning integral bases can be verified directly by calculating the discriminants of the relevant sets and comparing them with the field discriminant. □

As the last example we shall consider cyclotomic fields $K_n = Q(\zeta_n)$, ζ_n being a primitive nth root of unity, in the case where n is a prime power.

Theorem 2.9. *Let ζ be any primitive p^n-th root of unity and let $K = Q(\zeta)$. The extension K/Q is then normal and its degree equals $\varphi(p^n) = p^{n-1}(p-1)$. The numbers $1, \zeta, \zeta^2, \ldots, \zeta^r$ (with $r = \varphi(p^n)-1$) form an integral basis of K and*

$$d(K) = (-1)^{p(p-1)/2} p^N$$

where $N = n\varphi(p^n) - p^{n-1}$. Finally, the Galois group of K/Q is isomorphic to the multiplicative group $G(p^n)$ of residue classes $(\bmod\, p^n)$ not divisible by p.

Proof. Since every p^nth root of unity is a power of ζ, the normality of K follows. To show that $[K:Q] = \varphi(p^n)$ it suffices to establish the irreducibility of the polynomial

$$W(x) = (x^{p^n}-1)/(x^{p^{n-1}}-1)$$

over the rationals, since $W(\zeta) = 0$. Consider $F(x) = W(x+1)$. Since easy induction shows that one can write

$$(1+x)^{jp^{n-1}} = (1+x^{p^{n-1}})^j + pW_j(x) \quad (j \geqslant 1)$$

with a suitable $W_j \in Z[x]$ of degree smaller than jp^{n-1}, it follows that

$$F(x) = pV(x) + \sum_{j=0}^{p-1}(1+x^{p^{n-1}})^j = pV(x) + \sum_{j=1}^{p}\binom{p}{j} x^{p^{n-1}(j-1)}$$

with a certain $V \in Z[x]$ of degree smaller than $(p-1)p^{n-1}$. Since $F(0) = W(1) = p$, the above equality shows that F is Eisensteinian with respect to p and hence is irreducible over Q. Clearly the same holds for W, and this proves our assertion about the degree.

To find the discriminant we use Proposition 2.4 (iv), which gives
$$d_{K/Q}(\zeta) = (-1)^{p(p-1)/2} N_{K/Q}(W'(\zeta)).$$
Since
$$W'(x)(x^{p^{n-1}}-1) + p^{n-1} W(x) x^{p^{n-1}-1} = p^n x_d^{p^n-1},$$
and $\zeta^{p^{n-1}} = \vartheta$ is a primitive pth root of unity, we get $W'(\zeta)(\vartheta-1) = p^n \zeta^{-1}$, whence
$$N_{K/Q}(W'(\zeta)) = N_{K/Q}(p^n(\vartheta-1)\zeta)^{-1} = p^{n\varphi(p^n)} N_{K/Q}^{-1}((\vartheta-1)\zeta).$$

But $N_{K/Q}(\zeta) = 1$ and since $\vartheta - 1$ is a root of the irreducible polynomial $(x+1)^{p-1} + \ldots + (x+1) + 1$, we obtain, for $K_0 = Q(\vartheta)$, the equality
$$N_{K/Q}(\vartheta - 1) = N_{K_0/Q}(N_{K/K_0}(\vartheta - 1)) = N_{K_0/Q}((\vartheta-1)^{p^{n-1}}) = p^{p^{n-1}}.$$
(Here we have utilized Proposition 2.2 (iii), which finally leads to
$$d_{K/Q}(\zeta) = (-1)^{p(p-1)/2} p^N.)$$

Finally, observe that $d_{K/Q}(\zeta) = d_{K/Q}(\zeta - 1)$ and the number $\zeta - 1$ satisfies an Eisensteinian equation with respect to p. This shows that $d_{K/Q}(\zeta) = d(K)$ and that the powers of ζ form an integral basis, as asserted.

It remains to determine $\mathrm{Gal}(K/Q)$. The irreducibility of W implies that all primitive roots of unity are conjugated, and so for $1 \leq a < p^n$, $p \nmid a$ the map $\zeta \mapsto \zeta^a$ extends to an automorphism $g_a \in \mathrm{Gal}(K/Q)$. One can easily see that the associated mapping $G(p^n) \to \mathrm{Gal}(K/Q)$ is an isomorphism, and this proves our last assertion. □

The discriminants, integral bases and Galois groups for other cyclotomic extensions will be given in Chapter 4 (see Theorem 4.10).

6. The results of the preceding subsection show that certain fields have integral bases of the form $1, a, a^2, \ldots, a^{n-1}$ with a suitable a. Such a basis is called a *power basis*. Not all fields, however, possess such bases. A necessary and sufficient condition for a field to have a power integral basis is of course the existence of an element $a \in R_K$ for which $d_{K/Q}(a) = d(K)$. It turns out that there exist fields for which the greatest common divisor of the indices of elements of R_K is different from unity. Obviously such a field cannot have a power basis. The first example of such a field was given by R. Dedekind [78], and we now reproduce it. Let $K = Q(a)$, where a is any root of the polynomial $x^3 - x^2 - 2x - 8$ which is irreducible over Q. Consider
$$b = (a^2 + a)/2.$$

§ 2. Discriminants and Integral Bases 65

It is a root of $x^3 - 3x^2 - 10x - 8$, and hence lies in R_K, and an easy computation based on Proposition 2.4 gives $d_{K/Q}(1, a, b) = -503$. Since 503 is a prime, this shows that $1, a, b$ form an integral basis of K and $d(K) = -503$. We shall now prove that for all integers $x \in R_K$ the discriminant $d_{K/Q}(x)$ is even; hence every integer of K is of even index, and there cannot exist a power basis.

Write $x = A + Ba + Cb$ with $A, B, C \in Z$. Since $b^2 = 6 + 2a + 3b$, $a^2 = 2b - a$ and $ab = 2b + 4$, we have $x^2 = (A^2 + 6C^2 + 8BC) + a(2C^2 - B^2 + 2AB) + b(2B^2 + 3C^2 + 2AC + 4BC)$, and so $d_{K/Q}(x)$ equals

$$-503 \begin{vmatrix} 1 & 0 & 0 \\ A & B & C \\ A^2+6C^2+8BC & 2C^2-B^2+2AB & 2B^2+3C^2+2AC+4BC \end{vmatrix}^2;$$

this is congruent to $-503(BC)^2(3C+B)^2 \pmod{2}$, which is obviously an even number.

Positive rational integers $\neq 1$ which divide all indices of integers from K are traditionally called *common non-essential discriminantial divisors*. We shall consider them more closely in Chapter 4 (see Theorem 4.13 and Proposition 4.17).

Even if K has no common non-essential discriminantial divisors, a power basis need not exist. To give an example we prove first a result due to K. Hensel [94b].

Proposition 2.10. *To every field of degree n over Q there corresponds a form $F(x_1, \ldots, x_{n-1})$ of degree $n(n-1)/2$ in $n-1$ variables with coefficients from Z such that the set*

$$\{|F(a_1, \ldots, a_{n-1})|: a_1, \ldots, a_{n-1} \in Z, F(a_1, \ldots, a_{n-1}) \neq 0\}$$

coincides with the set of indices of integers of K.

Proof. Proposition 2.5 implies the existence of an integral basis of K of the form $1 = w_1, w_2, \ldots, w_n$. If now

$$x = A_1 w_1 + \ldots + A_n w_n \in R_K \quad (A_i \in Z),$$

then the index of x equals the index of $x - A_1 w_1 = x - A_1$. To calculate the index explicitly, observe that for $j = 1, 2, \ldots, n-1$ we have

$$(x - A_1 w_1)^j = \sum_{k=1}^{n} F_k^{(j)}(A_2, \ldots, A_n) w_k,$$

where $F_k^{(j)}$ are forms of degree j in $n-1$ variables with coefficients in Z. Propositions 2.4 and 2.7 imply that the absolute value of the determinant

$$\det[F_k^{(j)}(A_2, \ldots, A_n)]$$

equals zero if x does not generate K and equals the index of x otherwise.

Putting now $F(X_1, \ldots, X_{n-1}) = \det[F_k^{(j)}(X_1, \ldots, X_{n-1})]$, we obtain our assertion. □

Clearly K has a power basis iff the form constructed in the last proposition represents 1 or -1. To give the promised example of a field without common non-essential discriminantial divisors and without an integral power basis, consider

$$K = Q(m^{1/3})$$

with $m = ab^2$, ab—squarefree, $3 \nmid m$ and $m \not\equiv 1, 8 \pmod 9$. By Theorem 2.8 the numbers $1, m^{1/3}, m^{2/3}b^{-1}$ form an integral basis of K, and consequently, one can see after a short computation that the form occurring in Proposition 2.10 equals $bx_1^3 - ax_2^3$ in this case. Since $(a, b) = 1$ and this form represents both a and b, there cannot be any common divisors of the indices of integers of K. However, putting $a = 7$, $b = 5$, we obtain a field without a power basis, since the form $5x_1^3 - 7x_2^3$ does not assume the values ± 1. In fact, the congruence $5X^3 \equiv \pm 1 \pmod 7$ is insoluble.

§ 3. Applications of Minkowski's Convex Body Theorem

1. In this section we shall present some results concerning evaluations of discriminants and approximations by complex integers which are obtained by means of the convex body theorem of Minkowski and its consequences, given in Appendix III. We start with a simple proof to the fact that $d(K)$ cannot be equal to 1 for an algebraic number field $\neq Q$.

Theorem 2.10. *If K is a finite extension of the rationals and $K \neq Q$, then $|d(K)| > 1$.*

Proof. Let $w_1, \ldots, w_n$ be an integral basis of K, and let $w_i^{(j)}$ $(i, j = 1, 2, \ldots, n)$ be the corresponding conjugate integers. The system

$$L_j(x_1, \ldots, x_n) = \sum_{k=1}^{n} w_k^{(j)} x_k \quad (j = 1, 2, \ldots, n)$$

of linear forms obviously satisfies the assumptions of Minkowski's theorem for linear forms, where we take for the lattice M the lattice of all points with integral rational coordinates which has $d(M) = 1$. Since $|\det[w_i^{(j)}]| = |d(K)|^{1/2}$, we obtain the existence of rational integers $X_1, \ldots, X_n$, not all equal to zero, with $|N_{K/Q}(X_1 w_1 + \ldots + X_n w_n)| = \prod_{j=1}^{n} |L_j(X_1, \ldots, X_n)| < |d(K)|^{1/2}$.

But on the left-hand side of this equation we have a norm of an integer which is surely a rational non-zero integer and so equals at least 1, i.e. $|d(K)|$ exceeds 1. □

This result may be sharpened, especially for fields of a large degree. We shall not prove here the best result known (see §4 for more information), the proof of which is quite involved, but content ourselves with a classical evaluation, proved first by H. Minkowski:

Theorem 2.11. *If the field K has the signature $[r_1, r_2]$, then*

$$|d(K)| > \left(\frac{\pi}{4}\right)^{2r_2} \left(\frac{n^n}{n!}\right)^2.$$

Proof. We shall utilize the following lemma, of which our theorem is an easy consequence:

Lemma 2.3. *Let K be a field with signature $[r_1, r_2]$ and let M be a Z-module of a finite index m in R_K. Then there is a non-zero element a in M such that*

$$|N_{K/Q}(a)| \leq \left(\frac{4}{\pi}\right)^{r_2} (n!) n^{-n} |d(K)|^{1/2}.$$

Proof. Let $k = r_1 + r_2$ and let $F_1, \ldots, F_k$ be the embeddings of K into the complex field, the first r_1 of them being real and from every pair of conjugate complex embeddings only one being taken. Consider the homomorphism F of the additive group of R_K into the n-dimensional real space defined by

$$F(x) = [F_1(x), \ldots, F_{r_1}(x), \operatorname{re} F_{r_1+1}(x), \operatorname{im} F_{r_1+1}(x), \ldots, \operatorname{re} F_k(x), \operatorname{im} F_k(x)].$$

Clearly F is injective and so $F(M)$, the image of M under F, is a lattice whose generators are the images of the generators of M. It can immediately be checked that the discriminant of this lattice equals $|d(K)|^{1/2} 2^{-r_2} m$.

Now let t be any positive real number and let X_t be the set in the n-space defined by

$$X_t = \Big\{[x_1, \ldots, x_{r_1}, y_{r_1+1}, z_{r_1+1}, \ldots, y_k, z_k] : \sum_{i=1}^{r_1} |x_i| + 2 \sum_{j=1+r_1}^{k} (y_j^2 + z_j^2)^{1/2} < t\Big\}.$$

This set is obviously convex, bounded and symmetrical about the origin. We shall prove that its volume $V(X_t)$ satisfies

$$V(X_t) = 2^{r_1} (\pi/2)^{r_2} t^n / n!. \tag{2.7}$$

In the case of $r_1 = 1$, $r_2 = 0$ or $r_1 = 0$, $r_2 = 1$ this is evident. Assume now that (2.7) holds for $r_1 = A$, $r_2 = B$. Then for $r_1 = A+1$, $r_2 = B$ we have

$$V(X_t) = \frac{2^{A-B}\pi^B}{(A+2B)!} \int_{-t}^{t} (t-x_{A+1})^{A+2B} dx_{A+1}$$

$$= \frac{2^{A+1-B}\pi^B}{(A+2B+1)!} t^{A+2B+1}.$$

Similarly, for $r_1 = A$, $r_2 = B+1$, we get

$$V(X_t) = \frac{2^{A-B}\pi^B}{(A+2B)!} \iint_{y^2+z^2 \leq t^2/4} (t-2\sqrt{y^2+z^2})^{A+2B} dy\, dz$$

$$= \frac{2^{A+1-B}\pi^{1+B}}{(A+2B+2)!} t^{A+2B+2}.$$

Now (2.7) follows by induction.

For any given positive ε, determine $t = t(\varepsilon)$ from the equality

$$t^n = \left(\frac{4}{\pi}\right)^{r_2} (n!) |d(K)|^{1/2} m + \varepsilon.$$

Then we have

$$V(X_t) > (2^{-r_2} m |d(K)|^{1/2}) 2^n,$$

and so the convex body theorem of Minkowski (Appendix III, Theorem I) implies the existence of a number $a = a(\varepsilon) \neq 0$ in M such that the point $F(a)$ lies in X_t, i.e. if $F(a) = [x_1, \ldots, z_k]$, then

$$\sum_{i=1}^{r_1} |x_i| + 2 \sum_{j=1+r_1}^{r_2} |y_j^2 + z_j^2|^{1/2} < t.$$

The inequality between the arithmetic and geometric means implies

$$|N_{K/Q}(a)|^{1/n} = \left(\prod_{i=1}^{r_1} |x_i| \prod_{j=1+r_1}^{r_2} |y_j^2+z_j^2|^{1/2}\right)^{1/n} \leq t/n,$$

whence

$$|N_{K/Q}(a)| \leq \frac{t^n}{n^n} \leq \left(\frac{4}{\pi}\right)^{r_2} \frac{n!}{n^n} |d(K)|^{1/2} m + \frac{\varepsilon}{n^n}. \tag{2.8}$$

Observe that if ε is in the interval $(0, 1)$, we have only a finite number of possibilities for $a(\varepsilon)$, and so there must exist an a_0 in M satisfying the last inequality for all positive ε's whence we obtain the assertion of the lemma. □

Proof of the theorem. Put $M = R_K$, $m = 1$ in the lemma and observe that for non-zero a in R_K we have $|N_{K/Q}(a)| \geq 1$. □

§ 3. Applications of Minkowski's Convex Body Theorem

The theorem implies the following

Corollary 1. *For any field K of degree n over the rationals we have*

$$|d(K)| \geqslant \left(\frac{11}{12}\right)^2 \left(\frac{\pi e^2}{4}\right)^n (2\pi n)^{-1}.$$

In fact, by the Stirling's formula $n^n/n! = \exp(n - \vartheta/12n)(2\pi n)^{-1/2}$ for some ϑ in $(0, 1)$, and since $\exp(1/12) \leqslant 12/11$ and $n \geqslant 1$, the evaluation stated in the corollary follows. □

Denote by $M(r_1, r_2)$ the minimal absolute value of the discriminant of a field with the signature $[r_1, r_2]$. We thus have

Corollary 2. *If $r_1 + r_2$ tends to infinity, then $M(r_1, r_2)$ does the same.* □

2. Now we are able to prove the result of C. Hermite mentioned at the beginning of § 2.

Theorem 2.12. *Only a finite number of fields can have the same discriminant.*

Proof. By Corollary 2 to Theorem 2.11 we may restrict ourselves to fields of a fixed degree, say n. Let D be a fixed natural number and let K be any field of degree n with $|d(K)| \leqslant D$. We shall show that $K = Q(a)$ with a from a fixed finite set depending only on D. Consider the mapping F used in the proof of Lemma 2.3, let $[r_1, r_2]$ be the signature of K and $r_1 + r_2 = k$. In the case of $r_1 \neq 0$ define

$$X = \{[x_1, \ldots, x_{r_1}, y_{r_1+1}, z_{r_1+1}, \ldots, z_k] : |x_i| < C_i, y_j^2 + z_j^2 < 1,$$
$$i = 1, 2, \ldots, r_1; j = r_1+1, \ldots, k\}$$

with $C_1 = (D+1)^{1/2}$ and $C_i = 1$ for $i \neq 1$.

In the case of $r_1 = 0$ we define another set. Let

$$Y = \{[y_1, z_1, \ldots, y_k, z_k] : |y_1| < 1, |z_1| < (D+1)^{1/2},$$
$$y_j^2 + z_j^2 < 1, j = 2, 3, \ldots, k\}.$$

It can easily be checked that the volumes of those sets are equal to

$$V(X) = 2^{r_1}(1+D)^{1/2} \pi^{r_2} \quad \text{and} \quad V(Y) = 2(1+D)^{1/2} \pi^{r_2-1},$$

and so the quotients $V(X)/2^{r_1}|d(K)|^{1/2}$ and $V(Y)/|d(K)|^{1/2}$ exceed 1, whence Minkowski's theorem implies the existence of non-zero points from $F(R_K)$ in X and Y, respectively. Let a be one of them. Since its conjugates are absolutely bounded by a constant depending only on D, the coefficients of the minimal polynomial of a over Q are bounded, and so we have only finitely many possibilities of choosing a.

Our theorem will be proved if we succeed in showing $K = Q(a)$. This can easily be done. In fact, if $r_1 \neq 0$, then x_1 is the only conjugate of a lying outside the unit circle (if it lay inside it or on its boundary, we would have $|N_{K/Q}(a)| < 1$), and if $r_1 = 0$, then $y_1 + iz_1$ and $y_1 - iz_1$ are the only conjugates of a which may have this property. Moreover, $z_1 \neq 0$, since otherwise every conjugate of a would lie in the unit circle. Hence in both cases there exists a conjugate of a distinct from other conjugates, and so a generates the field K. □

3. Our final application of Minkowski's convex body theorem deals with approximations of complex numbers by integers from a fixed imaginary quadratic field. We prove a result of O. Perron [33].

Theorem 2.13. *Let $K = Q(D^{1/2})$ with a square-free, negative rational integer D. For every complex number z one can then find infinitely many distinct ratios a/b with $a, b \in R_K$ such that*

$$|z - a/b| \leq c(K)^{-1} N(b)^{-1}$$

where

$$N(b) = N_{K/Q}(b) = |b|^2 \quad \text{and} \quad c(K) = \pi(2|d(K)|)^{-1/2}.$$

Proof. Fix a positive ε and observe that for all complex a, b, z the following inequality holds:

$$2|b||zb - a| \leq \varepsilon|b|^2 + \varepsilon^{-1}|zb - a|^2. \tag{2.9}$$

If a, b are arbitrary complex numbers, then for suitable real x_i, y_i we can write

$$a = x_1 + x_2 w, \quad b = y_1 + y_2 w,$$

where we put

$$w = iD^{1/2} \quad \text{if } D \not\equiv 3 \pmod 4$$

and

$$w = (1 + iD^{1/2})/2 \quad \text{if } D \equiv 3 \pmod 4.$$

Theorem 2.6 implies that $\{1, w\}$ is an integral basis for K, and hence the coefficients in this representation are integers iff the corresponding complex number is an integer of K.

Now consider the set

$$V_r = \{(x_1, x_2, y_1, y_2) \in R^4 : \varepsilon|b|^2 + \varepsilon^{-1}|zb - a|^2 \leq r^2\}$$

with a positive r. This set is compact, convex and symmetrical about the origin and an easy calculation shows that its volume equals $2r^4\pi^2|d(K)|^{-1}$. (It suffices to recall that the unit ball in R^4 has volume $\pi^2/2$ and to apply a suitable linear transformation to transform V_r, which is a four-dimensional ellipsoid in the unit ball.) If $r^2 = 2c(K)^{-1}$ then the volume of V_r equals 16, and hence by Minkowski's theorem V_r contains at least one non-zero point with integral coordinates. For such a point we have $a, b \in R_K$ and

$$\varepsilon|b|^2 + \varepsilon^{-1}|zb-a|^2 \leqslant 2c(K)^{-1},$$

and so, using (2.9), we arrive at

$$|b||zb-a| \leqslant c(K)^{-1}, \quad \text{i.e.,} \quad |z-a/b| \leqslant c(K)^{-1}N(b)^{-1}.$$

Observe finally that for ε tending to zero the value of $\varepsilon^{-1}|zb-a|$ remains bounded by $2c(K)^{-1}$; thus $|zb-a|$ must tend to zero and, since $z \neq a/b$ for $a, b \in R_K$, we get infinitely many possibilities for the ratios a/b. □

Observe that essentially the same proof leads to an analogous result concerning approximations of complex numbers by ratios $(x_1+x_2 w)/(y_1+y_2 w)$, where $x_1, x_2, y_1, y_2 \in Z$ and w is a fixed non-real number.

§ 4. Notes to Chapter 2

1. The first systematic investigation of integers lying in an algebraic number field $\neq Q$ was carried out by C. F. Gauss [32], who considered integers of $Q(i)$ and used them for the study of biquadratic reciprocity. He also suggested that numbers of the form $a+b\varrho$ ($a, b \in Z$, $\varrho^3 = 1$, $\varrho \neq 1$), i.e., integers of $Q(\varrho)$, should be used in investigations of cubic reciprocity. He never returned to this subject himself and the suggested study was carried out by G. Eisenstein [44a]. C. G. J. Jacobi [39] expressed the opinion that Gauss had been led to complex integers through his research concerning the division of the lemniscate.

The integers of $Q(\varrho)$ had been used earlier by Euler (cf. Bergmann [66a]) and the integers of $Q(5^{1/2})$ and $Q(i7^{1/2})$ by Dirichlet [28], [32b] in the proofs of Fermat's last theorem for exponents 3, 5 or 14. (Euler's proof, however, was incomplete.) They did not develop the properties of the integers considered except those few which they applied directly. A study of the connection between early research concerning algebraic numbers and Fermat's last theorem can be found in Edwards [77].

Arbitrary quadratic fields appear first in Dirichlet [32a], who noted on p. 379 that expressions of the form $t+u\sqrt{a}$ with a square-free a should be subject to theorems similar to those which concern the complex integers $a+bi$. He did not suspect at that time that the arithmetic in arbitrary quadratic fields may substantially differ from that in $Q(i)$.

The cyclotomic fields $Q(\zeta_p)$ with prime p appear in Kummer [47a], [47b] and integers of arbitrary algebraic number fields in Dirichlet [40]. However, Dirichlet's definition of integers differs from those used today since in the field $K = Q(a)$ he only regarded as integers the elements of the ring $Z[a]$ generated by a. The modern definition of algebraic integers and also the definitions of the discriminant and the integral basis are due to R. Dedekind [71].

An early survey of the beginnings of the theory of algebraic numbers may be found in Smith [94]. A good insight into the early stages of this theory may be gained through Dickson et al. [23]. A complete bibliography up to 1896 is included in Hilbert's classic work [97], which laid the foundations for modern developments.

Of the many books devoted to algebraic numbers we mention here only a selection: Artin [59], [67], Borevich, Shafarevich [64], Cassels, Fröhlich [67], Châtelet [62], Eichler [63], Gundlach [72], Hasse [49], [50a], Hecke [23], Holzer [58], Ireland, Rosen [82], Janusz [73], Landau [18e], [27a], Lang [64], [70], Long [77], Mann [55], Marcus [77], McCarthy [66], Ore [34], Ribenboim [66], [72a], [72b], Samuel [67], Stewart, Tall [79], Takagi [48], Weber [96b], Weil [67a], Weiss [63], Weyl [40].

The modern theory of cyclotomic fields is exposed in Lang [78] and Washington [82]. For a survey see Lang [82]. Expositions of class-field theory, which we shall not touch on in this book, can be found in Artin, Tate [61], Cassels, Fröhlich [67], Chevalley [54], Goldstein [71c], Hasse [26c], [67], Holzer [66], Iyanaga [75], Neukirch [67b], [69b], Weil [67a]. For surveys of this theory see Garbanati [81], Ribenboim [62b].

A new approach, which seems to be the simplest of all, was recently made by J. Neukirch [84].

A survey of computational methods in algebraic number theory with an excellent bibliography was given by H. G. Zimmer [72].

2. The definition of norm and trace is due to R. Dedekind [71]. A characterization of the norm was obtained in Artin [50b] and Flanders [53a]. Cf. also Ankeny, Rogers [51a]. For Proposition 2.2 (iii) see also Inoue [42].

C. L. Siegel [73] presented a method of finding algebraic numbers of a given norm in a given normal extension K of Q (cf. Bartels [80], Garbanati [80]). K. Győry and A. Pethö [75], [77] considered the number $F_a(t)$ of elements x of R_K which have a given norm a and satisfy $x = x_1 w_1 + \ldots + x_n w_n$ with $|x_i| \leq t$ where $w_1, \ldots, w_n$ is a fixed integral basis of K. They proved that if $F_a(t) \neq 0$, then for t tending to infinity one has

$$F_a(t) = c \log^r t + O(\log^{r-1} t)$$

with positive c and $r = r_1 + r_2 - 1$. Cf. Pethö [74].

K. Győry [73] proved that there are only finitely many algebraic integers of given degree, discriminant and norm. The case of cubic integers was considered earlier in Delaunay and Faddeev [40]. For generalizations see Győry [81a], [81b], [83], [84], Győry, Papp [77].

Multiplicative relations between norms of algebraic integers in various fields were studied in Rehm [75].

If $a_1, \ldots, a_n \in K$ are linearly independent over Q, then the form $N_{K/Q}(x_1 a_1 + \ldots + x_n a_n)$ is called a *norm-form*. The problem of representing integers by norm-forms has been studied for a long time. See, e.g., Borevich, Shafarevich [64].

For other questions concerning norm-forms see Meyer, Perlis [79] and Perlis, Schinzel [79].

A. Schinzel [75a] showed that if K is real and $f \in K[x]$, then for a certain a in K one has $T_{K/Q}(f(a)) > 0$ (cf. Bazylewicz [82]).

The book by P. E. Conner and R. Perlis [84] is devoted to the study of the quadratic form $T_{K/Q}(x^2)$, which arises when x is written as $x_1 w_1 + \ldots + x_n w_n$ where $w_1, \ldots, w_n$ is an integral basis of K. (Cf. Gallagher [85], Maurer [73], [78a], Taussky [68].)

I. Schur [18a] proved that if $c < e^{1/2} = 1.6487\ldots$, then only finitely many totally real and totally positive integers can have their trace smaller than cn, where n is their degree. C. L. Siegel [45a] showed that for $c = 3/2$ the only such integers are 1, $(3 \pm 5^{1/2})/2$ (cf. Dinghas [52], Hunter [56a]). H. Behnke [23] considered the number of totally positive numbers with a given trace in a totally real field.

3. Theorem 2.1 was first proved by L. Kronecker [57a]; another proof of the second part of it may be found in Lehmer [32]. Other proofs of the first part were given by G. Greiter [78] and J. Spencer [77]. For a generalization see Moussa, Geronimo, Bessis [84]. Kronecker's theorem obviously fails for non-integers. Indeed, there are plenty of algebraic numbers a with $|a| = 1$ which are not roots of unity, and it was shown in Blanksby, Loxton [78] that a totally complex field K is a quadratic extension of a totally real field (such fields are called *CM*-fields or *J*-fields) iff $K = Q(a)$ with $|a| = 1$, $a \neq \pm 1$.

A. Schinzel and H. Zassenhaus [65] proved that if a has $2s$ non-real conjugates, then either it is a root of unity or $|a| \geq 1 + 4^{-s-2}$, and they obtained a similar improvement of the second part of Theorem 2.1. They asked whether there exists a constant $c > 0$ such that for every integer $a \neq 0$ which is not

a root of unity we have
$$\overline{|a|} \geq 1 + c/n.$$

This question is related to another question, raised by D. H. Lehmer [33a], who considered
$$M(a) = \prod_{i=1}^{n} \max(1, |a_i|)$$
where $a = a_1, \ldots, a_n$ are all conjugates of a, and asked whether there exists a positive constant ε such that for all integers $a \neq 0$ which are not roots of unity we have
$$M(a) \geq 1 + \varepsilon.$$
If so, then the answer to the question of Schinzel and Zassenhaus is positive for $c = \log(1+\varepsilon)$.

C. J. Smyth [71] showed that the answer to Lehmer's question is affirmative for those a's whose minimal polynomials are non-reciprocal. (A polynomial P of degree N is said to be *reciprocal* if it satisfies $P(x) = x^N P(x^{-1})$.) The same holds for the Schinzel–Zassenhaus question, which, moreover, has a positive answer also in the case where a has many real conjugates (Blanksby [69]).

E. Dobrowolski [79] showed that the inequality
$$M(a) \leq 1 + c(\log\log n)^3 (\log n)^{-3}$$
where $c < 1$ implies that a is either 0 or a root of unity; D. G. Cantor, E. G. Straus [82] and U. Rausch [85] independently proved that the same holds for all $c < 2$, and R. Louboutin [83] improved this result allowing c to be any number < 2.45.

For previous results concerning this topic see Blanksby, Montgomery [71], Dobrowolski [78] (who proved Theorem 2.2) and Stewart [78]. Cf. also M. Amara [79], Bazylewicz [76], Blanksby [75], Boyd [81a], Cassels [66], Dobrowolski, Lawton, Schinzel [83], Lawton [75], Mahler [64b], Mignotte [77], Notari [78], Pathiaux [73], Schinzel [73], Smyth [80].

An analogue of Theorem 2.1 for polynomials of several variables was considered in Boyd [81b], Lawton [77], [83], and Smyth [81]. For another analogue see Montgomery, Schinzel [77].

Kronecker's constant $\varepsilon(K)$ of a field K is defined as the largest lower bound of $\overline{|a|} - 1$ where a ranges over all integers a which are neither zero nor roots of unity. T. Callahan, M. Newman, M. Sheingorn [77] proved that every field can be embedded in a field K where $\varepsilon(K) \geq 2^{1/n} - 1$ ($n = [K:Q]$). They quote the following problem of P. T. Bateman: does $\varepsilon(K) = 1$ hold for most fields K? It holds for all fields of prime degree and a sufficiently large discriminant. Cf. Robinson [65].

H. Brunotte [80], [82] studied $\overline{|a|}$ from another point of view.

4. J. Favard [29], [30] noted that if $a_1, \ldots, a_n$ are all conjugates of an integer, then $\max|a_i - a_j| > (3/2)^{1/2}$. This result was improved by C. W. Lloyd-Smith [84], who replaced the right-hand side by $3/2$, and dr P. Blanksby informed me that M. J. McAuley improved it to 1.659. The minimal value of $|a_i - a_j|$ was considered by S. M. Rump [79]. P. E. Blanksby [70] considered

$$d(a, b) = \min_i \max_j |a_i - b_j|$$

(where $a = a_1, \ldots, a_n$ and $b = b_1, \ldots, b_m$ are sets of conjugated integers), and showed that for $m = n$ and $a_i \neq b_j$ we have

$$d(a, b) \geq 0.96/nM^{n-1}$$

where $M = \max(\overline{|a|}, \overline{|b|})$.

The question of minimizing $\sum_{i<j}(a_i - a_j)^2$, where again the a_i's are conjugated integers was considered in Hunter [56a].

The case of cubic integers occurred in Brunotte, Halter-Koch [79], Godwin [60], and M. N. Gras [80].

5. There are several results connected with Theorem 2.3 and Schur's theorem which we quoted above (§ 1, subsect. 4). M. Fekete [23] generalized Schur's result, proving that every compact plane set with transfinite diameter less than 1 can contain only finitely many full conjugated sets of integers (FCS). (Since a real interval of length a has transfinite diameter $4a$, this in fact generalizes Schur's theorem.)

A plane analogue of Theorem 2.3 is due to M. Fekete and G. Szegö [55]: if E is a set on the complex plane, closed on conjugation, whose interior contains a subset with transfinite diameter equal 1, then E contains infinitely many FCS. (For a generalization see Cantor [80], where also another proof of Theorem 2.3 was given.) For real sets no similar result is true, since R. M. Robinson [64a] found sets with arbitrarily large transfinite diameters not containing any algebraic numbers at all. Cf. Robinson [64b], [64c].

V. Ennola [75a] showed that in Theorem 2.3 the resulting FCS may consist of numbers having an arbitrary sufficiently large degree. This was conjectured by Robinson [62].

Straight lines containing infinitely many FCS's were described by T. Motzkin [45] and circles with this property were described in Robinson [69], Ennola [73a], Ennola, Smyth [74], [76]. An analogous problem for arbitrary conics was studied in Smyth [82].

For similar questions see Bullig [39], Cantor [76], Ferguson [70], Jacobsthal [58], Robinson [67], Schoenberg [64].

6. In this subsection we deal with the PV-numbers (the Pisot–Vijayaraghavan numbers), named so after C. Pisot [36] and T. Vijayaraghavan [40], who studied their properties. Note, however, that they already occur in Thue [12] and Hardy [19].

A real number is a *PV-number* if it exceeds 1 and is an algebraic integer, and its remaining conjugates lie inside the unit circle. (For accounts of their theory see Cassels [57], Pisot [63a], Salem [63].)

C. Pisot [36], [38] showed that $a > 1$ is a PV-number iff for a suitable real $b \neq 0$ the series

$$\sum_{n=1}^{\infty} \sin^2(ba^n)$$

converges. For other characterizations see Coquet [77], Mendès-France [67], [76], Y. Meyer [68]. An algorithm to test whether a given number is PV was proposed by R. J. Duffin [78].

Every real field $K \neq Q$ contains PV-numbers, since the proof, presented here, of Theorem 2.12 shows the existence of PV-numbers a such that $K = Q(a)$. This result is essentially due to A. Thue [12].

Important connections between PV-numbers and the theory of trigonometrical series were discovered by R. Salem [43]. Cf. also Lohoué [70], Y. Meyer [68], [70a], [70b], [70c], Pyatetskii-Shapiro [52], Salem, Zygmund [55], Senge, Straus [71], [73].

R. Salem [44] proved that the set S of all PV-numbers is closed and that, for every k, the kth derived set of S is non-empty. (For other proofs of the first result see Dufresnoy, Pisot [53a], Salem [45].) J. B. Kelly [50] showed that the union of S and the set S_1 of all those non-real integers a whose conjugates except a and $\bar{a}$ all lie in the interior of the unit circle is also closed. For a generalization of both results see Smyth [70].

C. L. Siegel's paper [44a] started investigations concerning the smallest elements of S and its derived sets. Siegel identified the smallest two numbers in S as those roots of $x^3 - x - 1$ and $x^4 - x^3 - 1$ which exceed unity, and proved that every other PV-number exceeds $2^{1/2}$. J. Dufresnoy and C. Pisot [53a] found the next two numbers of S and showed also that $\min S' = (1+5^{1/2})/2$. For another proof see Thurnheer [83]. The value of $\min S^{(2)} = 2$ was found by M. Grandet-Hugot [61], [65a], and D. W. Boyd [79a] showed that

$$1/k \leq \min S^{(k)} \leq (2^{1/2} - 1/2 + \varepsilon_k)$$

where ε_k tends to zero.

For other results concerning small PV-numbers see Boyd [78], Dufresnoy, Pisot [55b], Lazami [78], Talmoudi [77].

G. Rauzy [69] showed that only the polynomials $P(t) = t^n$ have the property $P(S) \subset S$. See Ventadoux, Liardet [69] for a generalization.

For other results concerning PV-numbers see Boyd [78], [79b], A. Brauer [51], Cantor [65a], [77], Dufresnoy, Pisot [53b], [55a], Gerig [67], Grandet-Hugot [70], [72], Mignotte [84], Pathiaux [69], [77], Rauzy [64], K. Schmidt [80].

Similarly, one considers the set T of Salem numbers, consisting of all algebraic integers exceeding 1 which have all their remaining conjugates in the closed unit disc and which are not PV-numbers (Salem [45]). For this and other generalizations of PV-numbers see Amara [65], [66], Bertin [74], [76], [78], [80], [81], Boyd [77a], [77b], [78], [82], Cantor [62], Chamfy [57], Doubrére [55], F. Dress [68], Grandet-Hugot [62], [65a], Halter-Koch [71a], [72a], Hugot, Pisot [58], Pisot [63b], [64], Pisot, Salem [64], Prenat [75], Rejeb [77], Samet [53], K. Schmidt [80].

7. Lemma 2.1 is very old (cf. Kneser [42]). The Corollary to Theorem 2.4 occurs first in Delaunay, Faddeev [44].

For the determination of the degree of particular classes of algebraic numbers see Diviš [77], Evans, Isaacs [77], Fried [54], Isaacs [70], Kurshan, Odlyzko [81], Mostowski [55], Nagell [37].

M. Schacher and E. G. Straus [74] considered the multinomial degree of L/K defined as the smallest N such that for every $a \in L$ one can find $r \leq N$ numbers $0 < n_1 < n_2 < \ldots < n_r$ such that $1, a^{n_1}, \ldots, a^{n_r}$ are linearly dependent over K. They showed that in fields of characteristic zero the multinomial degree coincides with the usual degree. Cf. also Risman [76a], [76b].

I. Niven [51] determined all the cases where $c^{1/m}$ divides $a^{1/m} - b^{1/m}$ $(a, b, c \in Z, (a, b, c) = 1, m \geq 2, |c| \geq 2)$.

The divisibility of $a^n - b^n$ by powers of prime ideals was considered in Ennola [80], Gelfond [40], Grossman [74a], Postnikova, Schinzel [68], Rédei [58], Sachs [56], Schinzel [74a].

8. Various results concerning analogues of elementary and analytic number theory (such as residue systems, perfect numbers, various arithmetic functions, power residues etc.) may be found in Amerbaev, Pak [69], Andrukhaev [69], Bedocchi [78], [80], Bergum [71], [72], [78], Bertness, McCulloh [72], Butts, Mann [56], Carlitz [62], [63], Cheo [51], Chidambaraswamy [67], Cofré-Matta, Shapiro [77], Freitag [77], Friedlander [73a], [73b], [74], Hardman, Jordan [67], [69], Hausman, Shapiro [76], Hua [51], Jordan [67a], [67b], [67c], [68], [69], Jordan, Portratz [65], Jordan, Rabung [70], [76], Jordan, Schneider [71], Lakein [74a], McCulloh [69], McCulloh, Stout [69], McDaniel [74a],

Meissner [11], Revuz [74], Sarges, Schaal [82], Spira [61], Stout [73], Suthankar [73], Wunderlich [71], [73], Zaikina [57b].

9. The main notions and results of § 2, in particular the fundamental Theorem 2.5, are due to R. Dedekind [71]. This theorem was generalized by E. Stiemke [26], who proved that every additive group consisting of algebraic integers is free. The following result of C. U. Jensen [64a] (cf. Kulkarni [67] for a simple proof) is related to that: if L/K is an infinite algebraic extension of an algebraic number field K, then algebraic integers contained in L form a free R_K-module. We shall see later (see Proposition 7.16) that this may fail if L/K is finite.

In connection with Theorem 2.5 cf. also Nagell [31], [32a], [32b], [32c], [33], [65], Nyberg [33], Okutsu [82a].

For a generalization of Proposition 2.7 to arbitrary extensions L/K see Fuchs [48], [49].

Various methods of finding integral bases were proposed in Albert [37a], Berwick [27], Canals, Ortiz [70], Ore [25b], Petr [35], N. R. Wilson [27], [31], Zassenhaus [65]. Some of those methods are also applicable to finding Z-bases of ideals, or, more generally, sub-Z-modules of R_K. Cf. Eda [53], Mann, Yamamoto [67], McDuffee [31a], Nagell [65], Wilson [29]. Ideal bases satisfying certain inequalities were constructed by K. Mahler [64a], who used them to deduce anew various fundamental results of the theory of algebraic numbers. Cf. Luthar [66].

10. Several papers were concerned with discriminants and integral bases of particular classes of fields.

(a) *Cubic fields.* In this case algorithms and even ready-made formulas for integral bases were given in by G. Voronoi [96]. (Cf. Delaunay, Faddeev [40].) See also Albert [30b], Arai [81], Bergström [37a], Martinet, Payan [67], Mathews [93], Sergeev [73], Smadja [73], [77], Tornheim [55].

Ideal bases in cyclic cubic fields were constructed by C. Paris [72b], who used approximations to generators of the fields in various p-adic metrics, found in Paris [72a].

Tables of small discriminants are given in Delaunay, Faddeev [40], Faddeev [34], Godwin, Samet [59], Reid [01]. Larger tables are described in Angell [73], Llorente, Oneto [82]. Cf. Delaunay [30a].

Normal cubic fields are characterized by the fact that their discriminants are squares. Squares which are such discriminants were characterized by H. Hasse [48a]. An elementary proof was given by K. Girstmair [79]. Cf.

Llorente, Nart [83], Maurer [78b]. For an elementary approach to normal cubic fields see Châtelet [46].

H. Hasse [30b] described arithmetic in non-normal cubic fields from the point of view of class-field theory. He showed in particular that they have discriminants of the form Df^2 with square-free D and $f = 3^a p_1 \ldots p_m$ with distinct primes $p_i \neq 3$, satisfying $(D/p_i) \equiv p_i \pmod{3}$ and $a = 0, 1, 2$. However, not all such numbers are discriminants, and the question which of them are was settled by J. Martinet, J. J. Payan [67] and P. Satgé [81]. Cf. Venkov [34]. An elementary treatment of arithmetic in cubic fields can be found in H. Reichardt [33]. Cf. Dedekind [00], Markoff [82], Nagell [30]. In the last paper it is shown, among other things, that a cubic extension of Q is pure iff its discriminant is of the form $-3m^2$. Cf. also Carlitz [66b].

Cubic fields with given discriminants were considered in Berwick [24], Hasse [30b], Venkov [34]. Their number is unbounded, in contrast to the quadratic case. See Martinet, Payan [67] for the number of non-normal such fields.

The number of cubic fields K where $|d(K)| \leqslant x$ was evaluated by H. Davenport, H. Heilbronn [69]. For cyclic cubics this was done by H. Cohn [54]. (See Corollary 2 to Theorem 8.3.)

(b) *Quartic fields.* Tables of quartic fields with small discriminants were given in Delaunay, Billevich, Sominskii [35], Godwin [56], [57a], [57b], [62], Kwon [84] and Pohst [75a].

Discriminants and integral bases of quartic fields were considered in Albert [30a] (note, however, that some of his results are incorrect as pointed out by Zhang Xianke [84a]), Amberg [97], Grebenyuk [58], Litver [55], Smadja [77], K. S. Williams [70].

Asymptotic behaviour of the number of quartic fields K whose Galois closure has a given Galois group and $|d(K)| \leqslant x$ was considered by A. M. Baily [80].

(c) *Quintic fields.* Cyclic quintics were considered by J. J. Payan [62a], [62b]. For general quintics see Turganaliev [62]. Tables of fields with small discriminants were given by H. Cohn [55] (cf. Cartier, Roy [74]) and C. Paris [83].

(d) *Pure fields.* The discriminants and integral bases in pure fields $Q(a^{1/m})$ for $a \in Z$ and m square-free were found by W. E. H. Berwick [27]. (See also Tietze [44], Okutsu [82b].) The case of prime m was considered in Gautheron, Flexor [69], Landsberg [97], Samko [55], Wegner [32a] and Westlund [10].

Discriminants of pure extensions of prime degree of $Q(\zeta_p)$ were computed by U. Wegner [32a], [37] and integral bases for them were found by K. Komatsu [76b].

(e) *Other classes of fields.* Arithmetic in arbitrary Abelian extensions of Q is described in Leopoldt [59a], [62]. Discriminants and integral bases for arbitrary cyclotomic fields will be found in Chapter 4 (see Theorem 4.10). Cf. also Bauer [39], Nagell [64a], Rados [06], Rédei [59].

For bases and discriminants of other types of fields see Bambah, Luthar, Madan [61], Foster [70], Hancock [25], Hasse [37], Hilbert [94a], Komatsu [75], [76a], Liang, Toro [80], Llorente, Nart, Vila [84], Martinet, Payan [68], Oriat [72], Samko [49], [57], Satgé [79a], Satgé, Barrucand [76], Sergeev [78].

Relations between integral bases of various fields were considered in Miyata [68], Yokoi [60b].

Integral bases over fields other than Q, which have class-number 1, and their discriminants were considered in Goldstein [72a] and Sunley [72b].

11. Theorem 2.6 is due to L. Stickelberger [97] and the proof presented here to I. Schur [29b]. This theorem and Theorem 4.5 belong to the few results concerning the characterization of numbers which can serve as discriminants.

Using the theory of the different (see Chapter 4), we can obtain more information about discriminants. On this topic see Bauer [19a], Hensel [89], [97a], Ore [25b], [26c], [26d], Schur [32]. We quote one of the results of Ore [26c]: if $n = b_0 + b_1 p + \ldots + b_N p^N$ where $0 \leqslant b_j \leqslant p-1$ and A denotes the number of non-zero b_i's, then a prime p cannot divide the discriminant of a field of degree n in a power exceeding

$$N(n, p) = b_0 + 2b_1 p + 3b_2 p^2 + \ldots + (1+N)b_N p^N - A,$$

and this bound is attainable. W. R. Thompson [31] determined those integers k in $[0, N(n, p)]$ for which there is a field K of degree n with $p^k || d(K)$.

12. In this subsection we consider the question of a power integral basis. A field K has a power integral basis iff R is generated, as a ring, by one element. For this reason one calls such fields *monogenic*. The first example of a non-monogenic field was given by R. Dedekind [78] and we reproduced it in the main text. Proposition 2.10 is due to K. Hensel [94b]. (Cf. also M. Hall [37].) Prime divisors of the form constructed in this proposition were considered by K. Györy, Z. Z. Papp [77] and L. A. Trelina [77b]. Cf. Urazbaev [50].

Theorem 2.7 implies that all quadratic fields are monogenic. Cyclic cubic monogenic fields were described by G. Archinard, [74] M. N. Gras [73], [74], and J. J. Payan [73]. (Cf. Dummit, Kisilevsky [77], Huard [79].) Quartic cyclic monogenic fields were characterized in M. N. Gras [81]. Only two of them are complex, viz. $Q(\zeta_5)$ and $Q(\zeta_{16} - \overline{\zeta_{16}})$. Every cyclic monogenic field of prime degree $\neq 2$ must be contained in a suitable cyclotomic field $Q(\zeta_N)$ with square-free N, as shown by J. J. Payan [73]. Recently M. N. Gras [86] showed that a real cyclic field of prime degree $\neq 2, 3$ is monogenic iff it equals the maximal real subfield of a cyclotomic field. It was shown in Liang [76] that maximal real subfields of cyclotomic fields are monogenic. Cf. also Carlitz [33], Dade, Taussky [64], Kleiman [72], Nagell [30], [66], [67], [68c], Nakahara [82], [83].

A necessary and sufficient condition for a field to be monogenic was given by B. Kovács [81]. (Cf. also Gilbert [81], Kátai, Kovács [80], [81], Kátai, Szabó [75].) Another necessary condition will be proved in Chapter 4 (see Theorem 4.13).

If K is monogenic, and $R_K = Z[a]$, then for all $m \in Z$ one has also $R_K = Z[a+m]$. The resulting power bases we shall call *equivalent*. It was proved in Györy [73] that there can be only finitely many inequivalent power bases in a given field and gave an explicit bound for their number. Cf. Györy [79a], [81b], Györy, Papp [77] and a forthcoming paper by J. H. Evertse and K. Györy.

K. Uchida [77b] showed that if R is a Dedekind domain and a is integral over R, then $R[a]$ is Dedekind iff the minimal polynomial for a over R does not belong to the square of a maximal ideal in $R[X]$. Cf. Albu [79].

The minimal number of generators of R_K as a ring was determined by P. A. B. Pleasants [74].

13. From a general result of J. Lewin [67] it follows that R_K can have only a finite number of subrings of a given index. (Cf. Nagell [65].) A classification of subrings of R_K spanning K as a vector space was given in Beaumont, Pierce [61]. (See also Davis [65], Krull [28a], Skolem [23].) For a description of semifields contained in K see Koch [64a], Weinert [63].

T. Nagell [65], [67], [68c] conjectured that integers of a given field with a given discriminant form a finite number of classes, two integers lying in the same class iff they differ by a rational integer. He confirmed this for fields of degree ≤ 4, and the general case was settled by K. Györy [73] in an effective way. It follows also from the result proved in Birch, Merriman [72], which, however, is ineffective. Cf. also Györy [80a], [80b], [81a], [81b], [83], [84] Györy, Papp [77], Trelina [77a], [77b].

14. In this subsection we put together results concerning the evaluations of $d(K)$. Theorem 2.10 was stated without proof in Kronecker [82] and proved by H. Minkowski [91a], who in [91b] proved the stronger Theorem 2.11. Other proofs of Theorem 2.10 were given in Calloway [55], Landau [22], Mordell [22b], [31], Müntz [23], Odlyzko [76], Ramanathan [59], Schur [18a], Siegel [22a] (for totally real fields), Weber, Wellstein [13]. For special cases see Lubelski [39a] and Steinbacher [11].

Theorem 2.12 is due to C. Hermite [57].

Denote by $M(r_1, r_2)$ the minimal value of $|d(K)|$ for fields K with signature $[r_1, r_2]$. Evidently, $M(2, 0) = 5$ and $M(0, 1) = 3$. For cubic fields we have $M(1, 1) = 23$, realized by $K = Q(a)$, $a^3 - a - 1 = 0$, and $M(3, 0) = 49$, realized by $K = Q(a)$ where $a^3 + a^2 - 2a - 1 = 0$. (See Davenport [39], Cassels [59b].) For quartic fields the minimal discriminants were found by J. Mayer [29]: $M(0, 2) = 117$, $M(2, 1) = 275$, $M(4, 0) = 725$. For quintic fields J. Hunter [57], confirming a conjecture of H. Cohn [55], proved that $M(1, 2) = 1609$, $M(3, 1) = 4511$ and $M(5, 0) = 14641$. The last value was also found by D. K. Faddeev and V. K. Potapkin [59]. The last completely settled case is that of sextic fields. J. J. Liang, H. Zassenhaus [77] obtained $M(0, 3) = 9747$ and conjectured that $M(2, 2) = 28037$, $M(4, 1) = 92779$. This was shown to be true by M. Pohst [82], who earlier ([75b]) obtained $M(6, 0) = 300\,125$. The only known further cases are: $M(7, 0) = 20\,134\,393$ (Pohst [76], [77a]) and $M(1, 3) = 184\,607$ (Diaz y Diaz [83]). Cf. Diaz y Diaz [82].

If $r_1 + 2r_2 = n$, then for sufficiently large n we have

$$M(r_1, r_2)^{1/n} \geq 60^{r_1/n} 22^{r_2/n} - \varepsilon$$

(for every positive ε), as shown by A. Odlyzko [76]. This implies that for

$$D = \liminf_{n \to \infty} M(r_1, r_2)^{1/n}$$

one has $D \geq 22$. Assuming the Extended Riemann Hypothesis, Odlyzko (loc. cit.) showed that the constants 60 and 22 can be replaced by 180 and 41, respectively. For previous bounds see Blichfeldt [36], [39], Mulholland [60], Odlyzko [75], Rogers [50], Schur [18a], Siegel [45a]. For a description of Odlyzko's method and certain improvements see Martinet [82], Poitou [76], [77].

It had been conjectured for a long time that the true value of D is infinite; however, it was shown to be false by E. S. Golod, I. R. Shafarevich [64], who obtained $D \leq 4404.5$. This was improved by A. Brumer [65] to $D \leq 347$ and by J. Martinet [78] to $D \leq 92.4$.

For other results on small discriminants see Ankeny [51], Cohn [52a], Godwin [58], Scholz [38].

The minimal value of the discriminant is related to simultaneous diophantine approximations. See Davenport [52], [55], Furtwängler [27], Hofreiter [40a]. For another type of discriminant evaluations see Delaunay [54].

15. The first results on the approximation of complex numbers by numbers from an imaginary quadratic field appear in Hurwitz [87] and Minkowski [07]. Theorem 2.13 was proved by O. Perron [33]. The best possible value $C(d)$ of the constant $c(K)$ with $K = Q(id^{1/2})$ is called the *Hurwitz constant* of K.

It is known only in the following few cases: $C(3) = 13^{1/4}$ (Perron [31], Shiokawa, Kaneiwa, Tamura [75]), $C(4) = 3^{1/2}$ (Ford [18], [25], Perron [30]; cf. Speiser [32]), $C(7) = 8^{1/4}$ (Hofreiter [37]), $C(8) = 2^{1/2}$ (Perron [33]), $C(11) = 5^{1/2}/2$ (Descombes, Poitou [50]), $C(19) = 1$ (Poitou [53]). For evaluations of $C(d)$ see Hofreiter [35b], [37], Oppenheim [37], Perron [49], Vulah [77].

For other aspects of approximations by numbers from imaginary quadratic fields see Aral [39], Buchner [26], Cassels, Ledermann, Mahler [51], Eggan, Maier [63], Hlawka [48], Hofreiter [35b], [40b], [52], Koksma, Meulenbeld [41a], [41b], Leveque [52a], [52b], Oliwa [53], Osgood [66], Poitou [53], N. Richert [81], A. L. Schmidt [67], [75], [78], [83].

Generalizations of the algorithm of continued fractions in imaginary quadratic fields were considered in Arwin [26], Gramm [60], J. Hurwitz [02], Leveque [52b] and Sudan [67].

Approximations by biquadratic integers were considered by C. J. Hightower [75].

16. The ring P_K of polynomials from $K[X]$ assuming integral values at integers of K was studied by G. Pólya [19] and A. Ostrowski [19], who gave necessary and sufficient conditions for the existence of a sequence $F_0, F_1, \ldots$ of polynomials generating P_K as an R_K-module and satisfying $\deg F_m = m$ for $m = 0, 1, \ldots$ Fields with this property are now called *Pólya fields*. H. Zantema [82] described all cyclic Pólya fields and showed that if the Galois closure of K has S_n ($n \neq 4$) or A_n ($n \neq 3, 5$) for the Galois group, then K is a Pólya field iff $h(K) = 1$.

One can consider, more generally, the ring of all polynomials from $K[X]$ which map a subring R of K into itself or an ideal of R into another ideal. Cf. Barsky [73], Brizolis [75], [76], [79], Cahen [72], Cahen, Chabert [71], Chabert [71], [77], [78], [79a], [79b], Gunji, McQuillan [70], [78], Hensley [77], Jacob [76], McQuillan [72a], [72b].

D. Brizolis [74] showed that if $f, g \in P_K$ and for every a in R_K $f(a)$ divides $g(a)$, then f divides g in P_K. Rings with this property were studied in Gunji, McQuillan [75].

Divisors of polynomial values were studied by C. R. McCluer [71a], [71b].

It was shown in Narkiewicz [62] that if P is a polynomial which maps an infinite subset X of an algebraic number field onto itself, i.e., $P(X) = X$, then P must be linear. This was put in a more general setting by P. Liardet [71]. For similar questions see Kubota [72a], [72b], [73], Kubota, Liardet [76], Lewis [72], Liardet [72], [75], Narkiewicz [63], [65].

A. J. van der Poorten [68] proved the existence of transcendental entire functions f satisfying $f(K) \subset K$ for every algebraic number field K.

17. In this subsection we list the papers dealing with the general theory of particular classes of fields.

For cubic fields see Châtelet [46], Delaunay, Faddeev [40], Hasse [30b], Markoff [82], Reichardt [33]. For quartics see Bergström [37b], Dribin [37], Nagell [62a] and for fields with quaternion groups see Rosenblüth [34]. The relative-quadratic fields were studied in Hilbert [99] and cyclic fields in Weber [09]. See also Hull [35] for cyclic quintics, Nagell [63b] for sextics, Oriat [72] for cyclic fields of prime power degree and Payan [65] for cyclic fields of prime degree. For fields with certain metabelian groups see Porusch [33] and for Frobenius extensions see Steckel [82b].

EXERCISES TO CHAPTER 2

1. Prove that if an algebraic integer a is totally positive and all its conjugates are real then for $k = 0, 1, \ldots, n-1$ one has $(-1)^k a_k > 0$ where $f(x) = x^n + a_{n-1} x^{n-1} + \ldots + a_0$ is the minimal polynomial of a over Q.

2. Let $\varepsilon(K) = \inf\{|a|-1 : a \neq 0, a \in R_K, a \text{ not a root of unity}\}$. Determine $\varepsilon(K)$ for all quadratic fields K.

3. (Callahan, Newman, Sheingorn [77].) Prove that if K/Q is normal and the complex conjugation lies in the centre of $\mathrm{Gal}(K/Q)$, then $\varepsilon(K) \geq 2^{1/2} - 1$.

4. (Moussa, Geronimo, Bessis [84].) (i) Let $P \in Z[X]$ be monic, put $P_0(x) = x$ and for $n \geq 0$ let $P_{n+1}(x) = P(P_n(x))$. Let z be an algebraic integer and let $z_1 = z, z_2, \ldots, z_N$ be its conjugates. Prove that the sequence $P_n(z_j)$ is bounded for $j = 1, 2, \ldots, N$ iff there exist distinct integers $r \neq s$ such that $P_r(z) = P_s(z)$.

(ii) Deduce Theorem 2.1 from (i).

5. (Blanksby, Loxton [78].) Prove that a non-real field K is closed under complex conjugation iff K is generated by an element of absolute value 1. (Hint: look for a generator of the form $(a+r)/(\bar{a}+r)$ with $r \in Q$ and $\deg_Q a = [K:Q]$.)

6. (Smyth [73].) Prove that if a is a PV-number and two of its conjugates, say a_i and a_j, have the same absolute value, then $a_j = \bar{a}_i$.

7. (Cofré-Matta, Shapiro [77].) Let K be neither Q nor an imaginary quadratic field. Prove that for every ideal $I \neq 0$ of R_K and positive ε one can find a complete set $a_1, \ldots, a_N$ of residues (mod I) such that $|a_{i+1} - a_i| < \varepsilon$ holds for $i = 1, 2, \ldots, N-1$.

8. (a) Prove that if $K = Q(a)$ is normal of degree N, then $d(K)$ is the square of a rational integer iff the Galois group of K/Q treated as a permutation group of the set of conjugates of a is contained in the alternating group A_N.

(b) Prove that if K/Q is normal of odd degree, then $d(K)$ is a square.

(c) Show that if K/Q is cubic non-normal, then $d(K)$ is not a square.

9. Determine the discriminant and find an integral basis for the field $Q(i, m^{1/2})$ where m is a square-free integer $\neq \pm 1$.

10. Find the discriminant and an integral basis for the maximal real subfield of $Q(\zeta_p)$ where p is an odd prime.

11. Show, without using Theorem 2.5, that if $a_1, \ldots, a_n$ are elements of R_K (with $[K:Q] = n$), linearly independent over Q and the Z-module generated by them is a proper submodule of R_K, then there exist $b_1, \ldots, b_n \in R_K$ satisfying

$$0 < |d_{K/Q}(b_1, \ldots, b_n)| < |d_{K/Q}(a_1, \ldots, a_n)|.$$

Use this to give another proof of Theorem 2.5.

12. Show that if L/K is of degree n, then R_L can be generated, as an R_K-module, by at most $1+n$ elements. Prove also that R_L is a free R_K-module iff it has a set of n generators.

Chapter 3. Units and Ideal Classes

§ 1. Valuations of Algebraic Number Fields

1. Consider an algebraic number field K. We already know from Theorem 1.9 that its ring of integers R_K is a Dedekind domain with the finite norm property (FN), since it is an integral closure of the ring of rational integers, which is obviously a Dedekind domain with the (FN)-property, in the separable extension K/Q. This gives us the possibility of constructing discrete valuations of K via the exponents connected with different prime ideals of R_K. In this section we want to examine all valuations of K including Archimedean ones and we shall show below that every Archimedean valuation of K is generated in a suitable way by an embedding of K in the complex field, whereas every other valuation is discrete and induced by a prime ideal of R_K.

The reader should be warned that a similar result is not necessarily true for other fields. Indeed, consider $\mathfrak{Z}(X)$ the field of rational functions in one complex variable. We shall show that it contains no Dedekind domain whose prime ideals would induce all discrete valuations. The formula $n(P/Q) = \deg Q - \deg P$ (where P, Q are relatively prime polynomials) defines an exponent in $\mathfrak{Z}(X)$. Moreover, for any $z \in \mathfrak{Z}$ we have a discrete valuation n_z defined by $n_z(P/Q) = r$ where r denotes the order of P/Q at z, i.e., r equals the order of the zero z of P if $P(z) = 0$, or $-r$ equals the order of the zero z of Q if $Q(z) = 0$ or finally $r = 0$ if $P(z)Q(z) \neq 0$. Assume now that R is a Dedekind domain which has $\mathfrak{Z}(X)$ for its quotient field and whose prime ideals induce n and n_z (for all $z \in \mathfrak{Z}$). Proposition 1.11 (i) shows that R is contained in the intersection of the corresponding exponent rings, but this intersection equals $\mathfrak{Z}$ and so $\mathfrak{Z}(X)$ cannot be the quotient field of R, contrary to our assumption.

To begin with, we shall develop a little further the theory of valuations of arbitrary fields. Let K be any field with a valuation v. This valuation induces on K the structure of a metric topological space which may or may not be complete. It is an elementary result in topology that every metric space may be embedded in a complete metric space, and this can be done in an essentially unique way. It turns out that if we do this for K, then the resulting complete metric space may be given a field structure. This is the most important part of the following theorem:

Theorem 3.1. *Let K be a field with a valuation v. Then there exists a field L, with a valuation w and the following properties:*
   (i) *L is complete in the topology induced by w,*
   (ii) *K is contained in L and on K the valuations v and w coincide,*
   (iii) *K is a dense subset of L.*

Moreover, if M is any overfield of K satisfying (i)–(iii), *then the fields M and L are topologically isomorphic (in fact, this isomorphism preserves the values of the corresponding valuations) and the mapping induced on K is an identity.*

Proof. Let X be a complete metric space containing K as a dense subset on which the metric coincides with that induced by v. We shall define a field structure on X. Denote by $d(a, b)$ the metric on X. Thus for $a, b \in K$ we have $d(a, b) = v(a-b)$; in particular, $d(a, 0) = v(a)$. If $x, y \in X$, then $x = \lim_n x_n$, $y = \lim_n y_n$ with $x_n, y_n \in K$. Consider the sequences $x_n + y_n$ and $x_n y_n$. We have

$$\begin{aligned} d(x_n + y_n, x_m + y_m) &= v(x_n + y_n - x_m - y_m) \\ &\leqslant v(x_n - x_m) + v(y_n - y_m) \\ &= d(x_n, x_m) + d(y_n, y_m), \end{aligned} \quad (3.1)$$

and similarly

$$\begin{aligned} d(x_n y_n, x_m y_m) &= v(x_n y_n - x_m y_m) \\ &= v(x_n y_n + x_n y_m - x_n y_m - x_m y_m) \\ &\leqslant v(x_n) v(y_n - y_m) + v(y_m) v(x_n - x_m) \\ &= v(x_n) d(y_n, y_m) + v(y_m) d(x_n, x_m). \end{aligned} \quad (3.2)$$

But $v(x_n)$ tends to $d(x, 0)$ and $v(y_m)$ tends to $d(y, 0)$, and so for some positive constant B we have $v(x_n), v(y_m) \leqslant B$. If we now choose m and n so large as to satisfy $d(x_n, x_m) < \varepsilon$ and $d(y_n, y_m) < \varepsilon$, then this remark together with (3.1) and (3.2) gives

$$d(x_n + y_n, x_m + y_m) < 2\varepsilon, \quad d(x_n y_n, x_m y_m) < 2B\varepsilon;$$

hence the sequences $x_n + y_n$, $x_n y_n$ are both fundamental, and thus have limits in X, say w and z respectively. Now define $x + y = w$, $xy = z$. We have to show that the elements w, z do not depend on any particular choice of the auxiliary sequences x_n and y_n. This can easily be done. In fact, if, say, x'_n tends to x, y'_n tends to y and $x'_n + y'_n$ tends to w', then $d(w, w') = \lim d(x_n + y_n, x'_n + y'_n) =$

$\lim v(x_n+y_n-x'_n-y'_n) \le \lim \sup [v(x'_n-x_n)+v(y'_n-y_n)] = \lim \sup [d(x_n, x'_n)+ d(y_n, y'_n)] = 0$, and so $w = w'$. A similar argument shows that z is independent of the choice of sequences.

It is obvious that X becomes a ring. Now we prove that it is, in fact, a field, i.e., we show that every non-zero element has an inverse. Let $x \in X$, $x \ne 0$, and let $x = \lim x_n$, where $x_n \in K$. Since $v(x_n) \le v(x_n-x_m)+v(x_m)$ and also $v(x_m) \le v(x_n-x_m)+v(x_n)$, we obtain the inequality $|v(x_n)-v(x_m)| \le v(x_n-x_m)$, showing that the sequence $v(x_n)$ is fundamental and so has a limit, say A. If $A = 0$, then x_n tends to 0, whence $x = 0$, contrary to our assumption. Thus $A \ne 0$ and so $v(x_n) \ge C > 0$ holds at least for sufficiently large n. But now

$$v(x_n^{-1}-x_m^{-1}) = v(x_n-x_m)/v(x_n)v(x_m) \le v(x_n-x_m)C^{-2},$$

showing that the sequence x_n^{-1} is fundamental. Denoting its limit by y, we find $xy = \lim(x_n x_n^{-1}) = 1$, and so x has an inverse.

In the meantime we have found that the convergence of a sequence x_n implies the convergence of $v(x_n)$. It is easy to establish the independence of $\lim v(x_n)$ of everything, including the choice of x_n, except the value of $\lim x_n = x$. Thus we may define for $x = \lim x_n$ a function $w(x) = \lim v(x_n)$ which coincides on K with $v(x)$.

It only remains to prove that the valuation properties are satisfied by the function w thus defined. This follows immediately from the corresponding properties of v via the passing to the limit. The theorem will be proved if we show the uniqueness of the field constructed. This is immediate by the corresponding uniqueness result for the completion of metric spaces. One has only to note that this topological isomorphism transfers also the algebraic structure, which follows from the continuity of the algebraic operations and the density of K. □

From the proof of the theorem we now derive some simple but useful corollaries:

Corollary 1. *If the valuation v of the field K is discrete and L is the completion of K with valuation w, then w is also discrete. Moreover, the ring $R_w = \{x \in L: w(x) \le 1\}$ is the closure of the ring $R_v = \{x \in K: v(x) \le 1\}$, and similarly the prime ideal $P_w = \{x \in L: w(x) < 1\}$ is the closure of $P_v = \{x \in K: v(x) < 1\}$.*

Proof. If $x \in L$ and $\lim x_n = x$, where $x_n \in K$, then, as we saw in the proof of the theorem, $\lim v(x_n) = w(x)$, whence either $w(x) = 0$, in which case $x = 0$, or $v(x_n)$ is constant starting from a certain term. This shows that $v(K) = w(L)$, and the corollary follows immediately. □

Corollary 2. *If the valuation v of the field K is discrete and the quotient R_v/P_v is finite, then (with L as above) the mapping $R_v/P_v \to R_w/P_w$ induced by the embedding $K \to L$ is an isomorphism.*

Proof. This mapping is clearly a homomorphism. The previous corollary shows that its image is dense in R_w/P_w but it is finite, and so has to coincide with R_w/P_w. □

Corollary 3. *If the valuation v of a field K is Archimedean, so is the valuation w prolonging v in the completion L of K.*

Proof. Proposition 1.10 shows that a valuation of a field is Archimedean iff its restriction to the prime field is Archimedean and the fields K and L have the same prime field. □

We shall also need a proposition describing topologies on finite dimensional linear spaces over a complete field which are consistent with the topology of that field:

Proposition 3.1. *Let K be a complete field with the valuation v and let E be a finite dimensional linear space over K. If E has a norm topology in which addition and multiplication by scalars from K are continuous and, moreover, if it is consistent with the topology of K, i.e., induces the topology of K on one-dimensional subspaces $\{ax: a \in K\}$ (with non-zero fixed x in E), then this topology coincides with the product topology, i.e. the mapping*

$$f: [x_1, \ldots, x_n] \mapsto \sum_{j=1}^{n} x_j a_j \quad (x_i \in K),$$

with the a_i's forming a K-basis of E, is a topological isomorphism between K^n and E.

Proof. As f is obviously a continuous isomorphism, it only remains to prove the following statement:

If $\|x\|$ denotes the norm on E, then for a suitable positive constant C we have $v(x_i) \leq C \cdot \|x_1 a_1 + \ldots + x_n a_n\|$ for $i = 1, 2, \ldots, n$ and all $x_i \in K$.

Denote by E_r the subset of E consisting of all $x \in E$ such that in $x = x_1 a_1 + \ldots + x_n a_n$ at most r of the x_i's are non-zero. We prove our statement by induction in r successively for the sets $E_1, E_2, \ldots, E_n = E$. (Note that these sets are not linear spaces except E_n.) The case $r = 1$ being covered by our assumptions, we may assume that the statement is proved for $E_1, \ldots, E_{r-1}$, and thus for $x \in E_{r-1}$ we have $v(x_i) \leq C_1 \cdot \|x_1 a_1 + \ldots + x_n a_n\|$ for $i = 1, 2, \ldots, n$. If our statement were false in E_r, we could find a sequence $x^{(n)} \neq 0$ with

$$x^{(n)} = x_{i_1}^{(n)} a_{i_1} + \ldots + x_{i_r}^{(n)} a_{i_r}$$

such that the quotient
$$\frac{||x^{(n)}||}{v(x_{i_1}^{(n)})}$$
tends to zero. For $y^{(n)} = x^{(n)}/x_{i_1}^{(n)} = a_{i_1} + y_{i_2}^{(n)} a_{i_2} + \ldots + y_{i_r}^{(n)} a_{i_r}$, we thus have $||y^{(n)}|| \to 0$, and so $y^{(n)}$ tends to zero. But this leads to a contradiction. In fact, the differences $y^{(m)} - y^{(n)}$ lie in E_{r-1}, and so, by the inductive assumption, we have
$$v(y_{ij}^{(m)} - y_{ij}^{(n)}) \leqslant C_1 \cdot ||y^{(n)} - y^{(m)}|| \to 0$$
for $j = 2, \ldots, r$; thus the sequences $\{y_{ij}^{(n)}\}$ are fundamental for $j = 2, \ldots, r$, whence in view of the completeness of K they have limits, say $\lim y_{ij}^{(n)} = z_j$. But this shows that $y^{(n)}$ tends to $a_{i_1} + z_{i_2} a_{i_2} + \ldots + z_{i_r} a_{i_r}$, which is obviously a non-zero element. This produces the contradiction which we sought. □

2. Now we can prove a theorem describing all valuations of an algebraic number field, obtained first by A. Ostrowski [35].

Theorem 3.2. *Let K be an algebraic number field of degree n over the rationals. If v is a discrete valuation of K, then there exists a prime ideal $\mathfrak{p}$ of the ring R_K of integers of K such that $v(x) = a^{n(x)}$ for a certain positive a not exceeding 1, where n is the exponent induced by $\mathfrak{p}$. If $v(x)$ is an Archimedean valuation of K, then it is equivalent to $|F(x)|$, where F is an embedding of K into the complex field. Conversely, every prime ideal of R_K defines in such a way a valuation on K which is discrete, and every embedding of K into the complex field defines an Archimedean valuation. Valuations defined by different prime ideals are non-equivalent, and two valuations defined by different embeddings of K into the complex field are equivalent iff those embeddings are complex conjugated. Finally every valuation of K is either discrete or Archimedean.*

Proof. Since Theorem 1.12 covers the case $K = Q$, we may assume that $[K:Q] \geqslant 2$. We start with discrete valuations. Thus let v be a non-trivial discrete valuation of K. With a suitable exponent n and $0 < c < 1$ we may write $v(x) = c^{n(x)}$ for all $x \neq 0$, $x \in K$. Denoting by m the restriction of n to Q, observe that m does not vanish identically. In fact, in such case we would have
$$Q \subset R_n \subset \{x \in K : n(x) \geqslant 0\},$$
and since by Theorem 1.10 R_n is Dedekind and so integrally closed in K, it would contain the integral closure of Q in K, i.e., the field K, which is possible only in the case of trivial v, which we excluded. Thus m is not zero, and thus $m(x)/e$, where e is the minimal positive value of n attained in Q, is an exponent of Q, which, by Theorem 1.12, is induced by a rational prime p. Hence for

$x \in Z$ we have $n(x) = m(x) \geq 0$. If $y \in R_K \setminus Z$ satisfies
$$y^N + c_{N-1} y^{N-1} + \ldots + c_0 = 0$$
(with $N \geq 2$, $c_j \in Z$), then, using the Corollary to Theorem 1.12, we obtain
$$Nn(y) = n(y^N) \geq \min_{c_j \neq 0} \{n(c_j) + jn(y)\}.$$
This shows that $n(y) \geq 0$, since from $n(y) < 0$ we would get
$$Nn(y) \geq \min_{c_j \neq 0} \{jn(y)\} \geq (N-1)n(y)$$
and $n(y) \geq 0$, a contradiction. Thus $R_K \subset R_n$. Consider now
$$\mathfrak{p} = \{x \in R_K : n(x) > 0\} = P_n \cap R_K.$$
Obviously $\mathfrak{p}$ is a prime ideal of R_K, and we shall show that it induces n. Denote by c the exponent induced by $\mathfrak{p}$ and let x be any element with $c(x) = 0$. We may write $xR_K = A_1 A_2^{-1}$, A_1, A_2 being ideals of R_K neither of them divisible by the prime ideal $\mathfrak{p}$. Choose a_2 in $A_2 \setminus \mathfrak{p}$. Then xa_2 lies in A_1 and is not in $\mathfrak{p}$. Thus $n(a_2) = n(xa_2) = 0$ and $n(x) = 0$ follows. Now fix an element π in $\mathfrak{p} \setminus \mathfrak{p}^2$ and, for any x with $c(x) = c$, write $x\pi^{-c} = y$. Since $c(y) = 0$, we obtain by the previous reasoning $n(y) = 0$; hence $n(x) = c(x)n(\pi)$. But we may find an x_0 in K with $n(x_0) = 1$, and thus $1 = c(x_0)n(\pi)$, proving $n(\pi) = 1$. But now $n(x) = c(x)$ becomes obvious. The remaining assertions concerning discrete valuations are contained in the remarks preceding Theorem 1.10.

We may thus turn to the Archimedean valuations. Let $v(x)$ be one of them. Its restriction to the rational field is also Archimedean, and so, by Theorem 1.12 and Proposition 1.8, it equals $|x|^a$ with a suitable a.

Let L be a completion of K and let w be the extension of v to L. The field L contains the closure of Q, which is obviously a field isomorphic topologically to the field R of real numbers with the usual topology and which we shall identify with R. We shall show that L has a finite dimension over R. In fact, consider a Q-basis $w_1, \ldots, w_n$ of the linear space K. If x is an arbitrary element of L, then with a suitable choice of rational $a_k^{(m)}$ we have
$$x = \lim_m \sum_{k=1}^n a_k^{(m)} w_k.$$
Without restricting generality, we may assume that the elements $w_1, \ldots, w_r$ are linearly independent over R, whereas for $j = 1, 2, \ldots, n-r$ we have
$$w_{r+j} = \sum_{k=1}^r \lambda_k^{(j)} w_k \quad \text{with real } \lambda_k^{(j)}.$$
Thus we may write
$$w = \lim_m \sum_{k=1}^r c_k^{(m)} w_k \quad \text{with real } c_k^{(m)}. \tag{3.3}$$

Denote by M the R-subspace of L spanned by $w_1, \ldots, w_r$. As it is finite-dimensional and w serves as a norm satisfying the assumptions of Proposition 3.1, we may apply it to infer that M is closed. But (3.3) shows that M is dense in L, whence $M = L$ and L has finite dimension over R. But M is a field, and so must be equal either to R itself or to the complex field $\mathfrak{Z}$ with the product topology. This shows that w induces the usual topology on L in both cases, $L = R$ or $L = \mathfrak{Z}$. In the first case this gives the equality $w(x) = |x|^a$ at once, whereas in the second we need some more reasoning. In this case w is continuous in the usual topology of the complex plane and, moreover, $w(xy) = w(x)w(y)$. For $x = r\exp(it)$ with $r = |x|$ we obtain

$$w(x) = w(r)w(\exp(it)) = r^a w(\exp(it)).$$

But for $t = 2\pi p/q$, where $p, q \in Z$, $q > 0$, we obtain $w(\exp(it))^q = 1$, and such numbers t are dense on the circle $|z| = 1$; thus $w(e^{it}) = 1$ for all t, whence follows the equality $w(x) = r^a = |x|^a$. In the course of our proof we identified K with its image in the complex field. However, if we treat K in an abstract manner, and denote by $F_1(x), \ldots, F_n(x)$ its embedding in $\mathfrak{Z}$, then we obtain $w(x) = |F_i(x)|^a$ for a certain i.

The next lemma shows that if two embeddings of K in $\mathfrak{Z}$ define the same valuation, then they are conjugate. This will prove the part of our theorem concerning Archimedean valuations, since by taking a suitable power of one of them we may transform equivalent valuations into equal ones. We need a lemma.

Lemma 3.1. *Let K be any field and let f_1, f_2 be two embeddings of K into the complex field such that $|f_1(x)| = |f_2(x)|$ holds for all $x \in K$. Then either $f_1 = f_2$ or $f_1 = \overline{f_2}$.*

Proof. Let $a \in K$ and let $b_i = f_i(a)$ ($i = 1, 2$). Then we have

$$1 + b_1 + \overline{b_1} + b_1\overline{b_1} = |1 + b_1|^2 = |f_1(1+a)|^2 = |f_2(1+a)|^2 = 1 + b_2 + \overline{b_2} + b_2\overline{b_2};$$

thus either $b_1 = b_2$ or $b_1 = \overline{b_2}$. Now let

$$K_1 = \{a \in K : f_1(a) = f_2(a)\} \quad \text{and} \quad K_2 = \{a \in K : f_1(a) = \overline{f_2(a)}\}.$$

Obviously K_1, K_2 are subfields of K and $K = K_1 \cup K_2$. If $K_i \neq K$ ($i = 1, 2$) and $a_1 \in K_1 \setminus K_2$, $a_2 \in K_2 \setminus K_1$, then $a_1 + a_2$ belongs neither to K_1 nor to K_2 but lies in K, a contradiction. Thus either $K = K_1$ or $K = K_2$. □

Finally note that every valuation of K is either discrete or Archimedean. Indeed, if v is non-Archimedean and non-discrete, then the group $v(K)$ must be dense in the positive real half-line. If $\hat{K}$ is the completion of K and w the prolongation of v onto $\hat{K}$, then $w(\hat{K})$ is again dense. But $\hat{K}$ contains $\hat{Q}$, the completion of the rational field under $v|_Q$, which must be discrete by Theorem 1.12. One sees now that $\hat{K}$ is a finite-dimensional vector space over $\hat{Q}$ and Proposition 3.1 shows that the topology of $\hat{K}$ is the topology of $\hat{Q}^m$ for a certain finite m, but the norm in this space has a discrete image, which provides the required contradiction. □

3. We end this section with the proof of a very important although simple product formula for normalized valuations. To begin with we define the *normalized valuations* of K. As we have seen in the previous section, every valuation of K which is discrete may be written in the form $v(x) = a^{n(x)}$, where n, or more precisely n_p, is the exponent induced by a prime ideal p of R_K and a is an arbitrary real number from the interval $(0, 1)$. Note that by Theorem 1.9 the ring R_K has the (FN) property, i.e. the number of elements $N(I)$ of the factor ring R_K/I is finite for every non-zero ideal $I \subset R_K$. We shall say that the valuation v is *normalized* if we have made the following choice for a: $a = N(p)^{-1}$. In the case of an Archimedean valuation we consider the corresponding embedding F of K into the complex field. If $F(K)$ is contained in the field of real numbers, then v is normalized provided $v(x) = |F(x)|$. Otherwise things get more complicated. We adopt the convenient convention of calling every power of a valuation also a valuation, although it need not satisfy the triangle inequality. In cases where the triangle inequality becomes involved, we shall call those powers of a valuation which are not true valuations *generalized valuations*. This convention allows us to define the normalized valuation corresponding to a non-real embedding F of K in the field of complex numbers as $|F(x)|^2$.

Now we may state the theorem which gives the product formula.

Theorem 3.3. *Let V be the set of all normalized valuations of an algebraic number field K. Then for every non-zero element $a \in K$ we have*

$$\prod_{v \in V} v(a) = 1.$$

Proof. Let $a \in K$ be a non-zero element, and let $aR_K = \prod \mathfrak{P}^{n(\mathfrak{P})}$ be the factorization of the fractional ideal generated by a into prime ideals. Since only finitely many numbers $n(\mathfrak{P})$ are non-zero, denoting by $v_\mathfrak{P}$ the normalized valuation associated with the exponent generated by $\mathfrak{P}$ we see that only finitely many terms in our product are $\neq 1$. Now $v_\mathfrak{P}(a) = N(\mathfrak{P})^{-n(\mathfrak{P})}$; thus the product $\prod v(a)$ extended over all discrete valuations equals $N(aR_K)^{-1}$. On the other hand, this product extended over all normalized Archimedean valuations equals $|N_{K/Q}(a)|$, and the Corollary to Proposition 2.7 can now be used to complete the proof. □

§ 2. Ideal Classes

1. Theorem 1.9 shows that the ring R_K of integers in an algebraic number field K is always a Dedekind domain with the finite norm (FN) property. Thus we may consider the group of ideal classes of the ring R_K as defined in Chapter 1. We recall that this group consists of isomorphism classes of ideals of R_K considered as R_K-modules with multiplication induced by the usual ideal multiplication. The fact that this multiplication induces group structure was established in Proposition 1.7. The case $m = n = 1$ of Theorem 1.14 shows that the group of ideal classes is isomorphic to the quotient group $G(K)/P(K)$, where $G(K)$ is the group of all fractional ideals of K and $P(K)$ is its subgroup consisting of principal fractional ideals.

The group of ideal classes of the ring R_K will be denoted by $H(K)$. One of the principal aims of this section is to prove that it is always finite. Actually we shall consider also some generalizations of the group $H(K)$ and prove finiteness for this more general family of groups.

Let $D(K)$ be the product of $G(K)$, the group of all fractional ideals of K and the signature group $\text{Sgn}(K)$, as defined in Chapter 2. Denote by $\mathfrak{p}_{1,\infty}, \ldots, \mathfrak{p}_{r_1,\infty}$ the generators of $\text{Sgn}(K)$ and consider the homomorphism of the multiplicative group K^* of the field K into $D(K)$ defined by

$$f(a) = \mathfrak{p}_{1,\infty}^{\varepsilon_1} \cdots \mathfrak{p}_{r_1,\infty}^{\varepsilon_{r_1}} \prod \mathfrak{p}^{a(\mathfrak{p})},$$

where $\varepsilon_i = 0, 1$ and ε_i are defined by $\text{sign} F_i(a) = (-1)^{\varepsilon_i}$ and $\prod \mathfrak{p}^{a(\mathfrak{p})}$ is the factorization into prime ideals of the ideal generated by a.

The group $D(K)$ is usually called the *divisor group* of K, the elements $\mathfrak{p}_{1,\infty}, \ldots, \mathfrak{p}_{r_1,\infty}$ are called the *real infinite prime divisors* of K and the elements of the subgroup generated in $D(K)$ by the prime ideals of R_K are called the *finite divisors*. The image of f is the group of *principal divisors*; its projection on $G(K)$, which is clearly isomorphic to $P(K)$, is the group of *finite principal divisors*, finally, the intersection of the image of f with $G(K)$ is called the *group of positive principal divisors* and is denoted by $G_+(K)$.

The quotient group $G(K)/P(K)$ equals $H(K)$, but the quotient group $G(K)/G_+(K)$ is generally larger and is called the *group of ideal classes in the narrow sense* and denoted by $H^*(K)$. Both are very important for the study of multiplicative properties of R_K.

The classical definition of those groups (equivalent to the one given above) runs as follows: two ideals of R_K, say I and J, are equivalent iff, for some non-zero $a, b \in R_K$, we have $aI = bJ$. The ideals I and J are equivalent in the narrow sense iff, for some totally positive $a, b \in R_K$, we have $aI = bJ$. The set of all equivalence classes in the first sense, with multiplication induced by the usual ideal multiplication, gives $H(K)$, whereas the second kind of equivalence leads to $H^*(K)$.

The proof of the equivalence of the definitions presented here is left to the reader.

Now we make another generalization. Let $\mathfrak{f}$ be any non-zero ideal of R_K and consider the group $A_\mathfrak{f}$ of all elements of K^*, which may be represented in the form $x = ab^{-1}$ with $a, b \in R_K$, $(abR_K, \mathfrak{f}) = 1$ and $a \equiv b \pmod{\mathfrak{f}}$. Moreover, let $G_\mathfrak{f}(K)$ be the group of all fractional ideals of K which can be written as quotients of two ideals of R_K prime to $\mathfrak{f}$. We consider $G_\mathfrak{f}(K)$ as a subgroup of $D(K)$ through the embeddings $G_\mathfrak{f}(K) \subset G(K) \subset D(K)$. Finally, denote by $G_\mathfrak{f}^+(K)$ the intersection of $G_\mathfrak{f}(K)$ with the image of $A_\mathfrak{f}$ under the mapping f. We may consider the quotient groups $H_\mathfrak{f}(K) = G_\mathfrak{f}(K)/f(A_\mathfrak{f})$ and $H_\mathfrak{f}^*(K) = G_\mathfrak{f}(K)/G_\mathfrak{f}^+(K)$. The first is called the *group of ideal classes* (mod $\mathfrak{f}$) and the second — the *group of ideal classes* (mod $\mathfrak{f}$) *in the narrow sense*. Note that for $\mathfrak{f} = R_K$ we obtain $H(K) = H_\mathfrak{f}(K)$ and $H^*(K) = H_\mathfrak{f}^*(K)$.

The groups $H_\mathfrak{f}(K)$ and $H_\mathfrak{f}^*(K)$ are called the *group of ray classes* (mod $\mathfrak{f}$) and the *group of ray classes in the narrow sense* (mod $\mathfrak{f}$), respectively.

Of course we may also give here an equivalent definition, similar to that used for $H(K)$ and $H^*(K)$. One has only to consider two ideals I and J as equivalent (mod $\mathfrak{f}$) iff $aI = bJ$ holds for suitable non-zero $a, b \in R_K$ with $a \equiv b \equiv 1 \pmod{\mathfrak{f}}$, and the narrow equivalence is defined analogously with the additional requirement for a, b to be totally positive. Moreover, only ideals prime to $\mathfrak{f}$ have to be considered.

In future, speaking about ideal classes (mod $\mathfrak{f}$), we shall always have the last interpretation in mind.

2. Now we are going to prove the theorem about the finiteness of all the groups introduced in the previous section. To begin with, we show that it is enough to prove the theorem for the group $H(K)$. Indeed, we prove

Lemma 3.2. *Let $\mathfrak{f} \neq 0$ be an ideal of R_K. The homomorphism of $H_\mathfrak{f}^*(K)$ into $H(K)$ induced by the embedding $G_\mathfrak{f}(K) \subset G(K)$ has a finite kernel, and so if $H(K)$ is finite, then $H_\mathfrak{f}^*(K)$ is finite as well.*

Proof. The kernel of $H_\mathfrak{f}^*(K) \to H(K)$ consists of all those classes (mod $\mathfrak{f}$) in the narrow sense which contain a principal ideal, and so equals the quotient of the group of all principal fractional ideals by the group of all ideals generated by a totally positive element which is a quotient of two integers congruent to 1 (mod $\mathfrak{f}$). Every coset of the last quotient group contains a principal ideal generated by an integer. Indeed, if $ab^{-1}R_K$ is any principal fractional ideal with $a, b \in R_K$, relatively prime to $\mathfrak{f}$, then $c = b^{2\varphi(\mathfrak{f})}$ is totally positive and congruent to $1 \pmod{\mathfrak{f}}$ by Theorem 1.8; thus $ab^{-1}R_K$ and $ab^{2\varphi(\mathfrak{f})-1}R_K = ab^{-1}cR_K$ are in the same coset. This shows that it suffices to prove the existence of a finite set of integers in R_K, say $a_1, \ldots, a_m$, such that every integer $x \in R_K$, after multiplying by one of the a_i's, becomes totally positive and congruent to unity $(\bmod\, \mathfrak{f})$ provided $(xR, \mathfrak{f}) = 1$. This can be done for example by taking for every fixed residue class $(\bmod\, \mathfrak{f})$ in R_K a set of elements from this class with different signatures, every signature being represented, which is possible by Lemma 2.1. Since there are at most 2^{r_1} different signatures and $\varphi(\mathfrak{f})$ residue classes $(\bmod\, \mathfrak{f})$ relatively prime to $\mathfrak{f}$, the set so constructed is finite and clearly has the required property. □

So we may now consider only the simplest case, namely that of $H(K)$. We shall give two different proofs of the finiteness of $H(K)$: one due to H. Minkowski and based on his convex body theorem, which furnishes also a method of finding the number of elements in $H(K)$, and another one, due to A. Hurwitz, which avoids the use of the geometry of numbers. So we state

Theorem 3.4. *For every non-zero ideal* $\mathfrak{f} \subset R_K$ *the groups* $H_\mathfrak{f}(K)$ *and* $H_\mathfrak{f}^*(K)$ *are finite.*

Proof. As already observed, it suffices to prove the theorem for $H(K)$.

(i) *Minkowski's proof.* This proof is based on the following

Lemma 3.3. *If K is of degree n over the rationals and has signature $[r_1, r_2]$, then in every ideal class one can find an ideal of R_K with a norm not exceeding*

$$\frac{n!}{n^n} \left(\frac{4}{\pi}\right)^{r_2} \sqrt{|d(K)|}\,.$$

Proof. Let I be any ideal in R_K and choose J so that the ideal IJ becomes principal, say generated by $a \in R_K$. (This is possible by Corollary 6 to Proposition 1.6.) Since the index of J, as a Z-module in R_K, equals $N(J)$, we may apply Lemma 2.3 to obtain an element $c \neq 0$ in J with

$$|N_{K/Q}(c)| \leq \frac{n!}{n^n} \left(\frac{4}{\pi}\right)^{r_2} \sqrt{|d(K)|}\, N(J).$$

The principal ideal generated by c is divisible by J, let B be their quotient, $B = cJ^{-1}$. The ideals B and I are equivalent, belonging to the class inverse to that of J and, moreover,

$$N(B) = |N_{K/Q}(c)|/N(J) \leq \frac{n!}{n^n} \left(\frac{4}{\pi}\right)^{r_2} \sqrt{|d(K)|}$$

by the Corollary to Proposition 2.7. □

The theorem follows from Theorem 1.6 (ii). □

(ii) *Hurwitz's proof.* This proof is based on the following lemma, which is interesting in itself since it provides a kind of substitute for the Euclidean algorithm in K.

Lemma 3.4. *If a is any element of K, then one can find an integer x of this field with the property that, for some positive rational integer r not exceeding a bound depending solely on the field K, we have*

$$|N_{K/Q}(ra-x)| < 1.$$

Proof. For the time being let k be an arbitrary natural number. For $r = 0, 1, \ldots, k^n$ ($n = [K:Q]$) choose b_r in R_K so that in

$$ra - b_r = a_{1r}w_1 + \ldots + a_{nr}w_n$$

(where $w_1, \ldots, w_n$ is a fixed integral basis of K) the coefficients a_{ir} lie in the interval $[0, 1]$. Since we have $k^n + 1$ numbers $ra - b_r$, by the box principle we may find two of them, say $r_1 a - b_{r_1}$ and $r_2 a - b_{r_2}$, whose difference is of the form

$$A_1 w_1 + \ldots + A_n w_n \quad \text{with} \quad |A_i| < 1/k.$$

The norm of this difference does not exceed in absolute value the number C/k^n, where C is a positive number depending on the choice of the basis $w_1, \ldots, w_n$ exclusively, and thus, in fact, can be made bounded by a constant depending solely on K.

Taking k larger than $C^{1/n}$, we now obtain the assertion of the lemma for a certain r not exceeding k^n. □

Denote by $m(K)$ the bound on r occurring in the lemma. Now let I be an arbitrary non-zero ideal of R_K and let $a \in I$, $a \neq 0$ be chosen from the numbers of I with the least possible absolute value of the norm. Applying Lemma 3.4 to the quotient b/a, where b is an arbitrary number from I we find $r \leq m(K)$ and $c \in R_K$ such that $|N_{K/Q}(rb-ca)| < |N_{K/Q}(a)|$. Since $rb - ca$ lies in I, the choice of a implies $rb = ca$, and so $a|rb$, which in turn implies $a|m(K)!b$. Hence the ideal $m(K)!I$ is divisible by aR_K, and so for a certain ideal J we have $m(K)!I = aJ$, i.e., I and J are equivalent. But $m(K)!a \in m(K)!I = aJ$, and so $m(K)! \in J$, which shows that J divides the principal ideal generated by $m(K)!$; thus, in view of Corollary 2 to Theorem 1.4, we have only finitely many possibilities for J. □

We shall denote the number of elements in $H(K)$, $H^*(K)$, $H_\mathfrak{f}(K)$ and $H_\mathfrak{f}^*(K)$ by $h(K)$, $h^*(K)$, $h_\mathfrak{f}(K)$ and $h_\mathfrak{f}^*(K)$, respectively, and call those numbers the *class-number*, the *narrow class-number*, the *class-number* (mod $\mathfrak{f}$) and the *narrow class-number* (mod $\mathfrak{f}$), respectively. Hurwitz's proof does not give any useful information about the size of $h(K)$ but Minkowski's proof show that $h(K)$ cannot exceed the number of ideals in R_K whose norms do not exceed the number

$$\frac{n!}{n^n}\left(\frac{4}{\pi}\right)^{r_2}\sqrt{|d(K)|},$$

which we shall call the *Minkowski constant* for the field K.

For example, we can see that quadratic fields with discriminants $d(K) = -3, -4, -7, -8, 5, 8, 12, 13$ and also cubic fields with discriminants contained in the range $-81 < d(K) < 49$ have $h(K) = 1$, since only the trivial ideal has its norm bounded by the Minkowski constant. Theorem 1.16 implies that the rings of integers of those fields are unique factorization domains.

We also note two corollaries to Theorem 3.4.

Corollary 1. *If I is an ideal in R_K, then $I^{h(K)}$ is a principal ideal.*

Corollary 2. *If I is an ideal in R_K and, for a certain natural number m, the ideal I^m is principal, then either I is principal itself or $(m, h(K)) \neq 1$.*

The proofs are immediate.

The determination of the structure of class-groups is one of the main goals of the theory of algebraic numbers. We shall return to this problem later. In the next section the reader will find some results connecting $h_{\mathfrak{f}}^*(K)$ with $h(K)$, so that the problem of determination of $h_{\mathfrak{f}}^*(K)$ will be reduced to the simplest case.

Now we give an example of a field K in which the ring of integers is not a unique factorization domain, i.e., $h(K) \neq 1$. Consider $K = Q(\sqrt{-5})$ and the ideal $2R_K + (1+\sqrt{-5})R_K$. Its norm divides $N(2R_K) = 4$ and also $N((1+\sqrt{-5})R_K) = 6$; thus it equals 1 or 2. If the norm is 1, then the ideal coincides with R_K, and for some $x, y \in R_K$ we have $1 = 2x+(1+\sqrt{-5})y$. Write $x = a+bw$, $y = c+dw$ with $w = \sqrt{-5}$ and $a, b, c, d \in Z$, which is possible by Theorem 2.7. Now we obtain

$$1 = (2a+c-5d)+(2b+d+c)w,$$

and so $2a+c-5d = 1$, $2b+d+c = 0$. But this system has no integral solutions, since the second equation gives $c = -2b-d$ and so in view of the first equation 1 is even, which is clearly absurd.

Hence our ideal has norm 2. To prove that $h(K) \neq 1$, we show that there is no principal ideal with norm 2, i.e., there is no integer in K with $N_{K/Q}(a) = 2$.

(Obviously we do not have to consider the case of the norm being equal to -2, since K is quadratic imaginary and so the norm is always positive for non-zero elements.) Indeed, if $A+Bw \in R_K$, then $N_{K/Q}(A+Bw) = A^2 + 5B^2$ and this never equals 2.

§ 3. Units

1. This section is devoted to *algebraic units*, i.e. invertible elements of the ring of all algebraic integers. As in every domain, the units of this ring form a group under multiplication. The algebraic units lying in a fixed algebraic number field K form a subgroup of that group: we shall denote it by $U(K)$. Its elements will be called the *units of the field K*, which is perhaps not the best name, since every non-zero element of K is invertible in K. In fact $U(K)$ consists exactly of the invertible elements of R_K.

We shall determine the structure of the group $U(K)$; this was first done by P. G. Lejeune-Dirichlet in 1840 in a slightly different setting, since his definition of an algebraic integer does not coincide with the one we use.

We shall also consider divisibility in R_K. We say that *a divides b* and write $a \mid b$ if a, b and the quotient ba^{-1} are integers. Note that this definition does not depend on the field in which a, b are contained. If two integers a, b satisfy $a \mid b$ and $b \mid a$, then we call them *associated*. Obviously this happens iff the quotient ab^{-1} is a unit, and this means in fact that the ideals generated by a and b in any field containing these numbers coincide.

The following trivial result will be used quite often:

Proposition 3.2. *If $a, b \in R_K$ and $b \mid a$, then $N_{K/Q}(b)$ divides $N_{K/Q}(a)$.*

Proof. From $a = b(ab^{-1})$ with $ab^{-1} \in R_K$ we obtain by Proposition 2.2 (i)
$$N_{K/Q}(a) = N_{K/Q}(b) N_{K/Q}(ab^{-1}). \qquad \square$$

Now we prove some elementary results concerning the units.

Proposition 3.3. *If a is an algebraic integer, then the following six conditions are equivalent*:
  (i) *a is a unit.*
  (ii) *a divides 1.*
  (iii) *For every field K containing a, $|N_{K/Q}(a)| = 1$.*
  (iv) *There is a field K such that $a \in K$ and $|N_{K/Q}(a)| = 1$.*
  (v) *There is a monic polynomial $F(x)$ with rational integral coefficients with $F(0) = \pm 1$ and $F(a) = 0$.*

(vi) *There is a polynomial with integral coefficients $G(x) = a_n x^n + \ldots + a_0$ such that a_0 and a_n are algebraic units and $G(a) = 0$.*

Proof. (i) → (ii). This implication follows directly from the definition of a unit.

(ii) → (iii). If a divides 1, then by the preceding proposition $N_{K/Q}(a)$ divides $N_{K/Q}(1) = 1$, and so must be equal to 1 or -1.

(iii) → (iv). Trivial.

(iv) → (v). Take for F the minimal polynomial for a over Z which is monic. Then $N_{K/Q}(a)$ equals some power of $\pm F(0)$.

(v) → (vi). Trivial.

(vi) → (i). The polynomial $W(x) = a_0^{-1} x^n G(x^{-1})$ has integral coefficients and $W(a^{-1}) = 0$; thus a^{-1} is integral over R_K, K being the field $Q(a_0, a_1, \ldots, a_n)$. By Theorem 1.2 a^{-1} is integral over Z, and so a is a unit. □

Evidently every root of unity lying in K belongs to $U(K)$. All such roots of unity form a group under multiplication, which we shall denote by $E(K)$. It is clearly the maximal torsion subgroup of $U(K)$. We now prove

Proposition 3.4. *The group $E(K)$ is a finite cyclic group whose order is even and divides $2d(K)$.*

Proof. Denote by ζ_m the mth primitive root of unity which may be taken as the complex number $\exp(2\pi i/m)$. If ζ is any root of unity $\neq 1$ contained in K, then we may write

$$\zeta = \prod_{i=1}^{r} \zeta_{p_i^{a_i}}^{n_i},$$

where $p_1, \ldots, p_r$ are rational primes, $a_1, \ldots, a_r$ and $n_1, \ldots, n_r$ are positive rational integers with $p_i \nmid n_i$ for $i = 1, 2, \ldots, r$. Put

$$p_r^{a_r} = \max\{p_i^{a_i} : i = 1, 2, \ldots, r\}$$

and denote by N the product

$$p_1^{a_1} \ldots p_{r-1}^{a_{r-1}}.$$

Then the number

$$\zeta^N = \zeta_{p_r^{a_r}}^{n_r N}$$

lies in K, and, since $p_r \nmid N n_r$, it is a primitive $p_r^{a_r}$-th root of unity. But by Theorem 2.9 its degree equals $p_r^{a_r - 1}(p_r - 1)$, and so we must have the inequality

$$p_r^{a_r - 1}(p_r - 1) \leq n = [K:Q].$$

This shows that we have only a finite number of possibilities for $p_r^{a_r}$, and the same applies to other powers $p_i^{a_i}$, since they do not exceed $p_r^{a_r}$. Hence $E(K)$ is finite, and so it must be cyclic, since every finite subgroup of the group of all roots of unity is cyclic. Its order must be even, because it contains the subgroup $\{-1, 1\}$ of order two. To prove the last assertion, let m be the order of $E(K)$. Factorize m into rational primes

$$m = \prod p_i^{n_i}.$$

As before, we see that the fields $Q(\zeta_{p_i^{n_i}})$ are subfields of K, and so, by Proposition 2.9, their discriminants divide $d(K)$. But, by Theorem 2.9, they are equal to $\pm p_i^{A_i}$ with $A_i = n_i p_i^{n_i-1}(p_i-1) - p_i^{n_i-1}$, whence $d(K)$ is divisible by $\prod p_i^{A_i}$. But if $p_i \neq 2$, then $A_i \geq n_i$, and in the case of $p_i = 2$ we have $A_i \geq n_i - 1$, whence the number $p_i^{B_i}$ with

$$B_i = \begin{cases} n_i & \text{if } p_i \neq 2, \\ n_i - 1 & \text{if } p_i = 2 \end{cases}$$

divides $d(K)$, as asserted. □

2. Now we shall prove Dirichlet's unit theorem in the form due to C. Chevalley and H. Hasse. To state it we have to introduce some new concepts. Let S be any non-empty finite set of non-equivalent valuations of K (taken in the normalized form) containing the set S_∞ of all Archimedean valuations. An element $a \neq 0$ from K is called an *S-unit* if for every valuation v not in S we have $v(a) = 1$. Note that if we take for S the system of all Archimedean valuations of K, then an S-unit is an algebraic unit in the sense introduced in the previous subsection. In fact, the equality $v(a) = 1$ for all discrete valuations of K shows that, firstly, a is integral and, secondly, $v(a^{-1}) = 1$ for all such valuations, whence a^{-1} is also integral. Conversely, every algebraic unit is an S-unit for every set S.

One can easily see that the S-units form a group $U_S(K)$ under multiplication and one may ask about its structure. The answer is given by the following

Theorem 3.5 (P. G. Lejeune-Dirichlet, C. Chevalley, H. Hasse). *The group $U_S(K)$ is the direct product of the group of roots of unity $E(K)$ and a free Abelian group with $s-1$ free generators, s being equal to the number of elements in S.*

The special case where S equals the set of all Archimedean valuations gives the celebrated Dirichlet unit theorem:

Theorem 3.6. *The group $U(K)$ is the direct product of the group $E(K)$ and $r_1 + r_2 - 1$ copies of the cyclic infinite group.*

Proof of Theorem 3.5. The idea of the proof is very simple. We construct a homomorphism of $U_S(K)$ in the real s-space whose kernel equals $E(K)$ and then we prove that the image of $U_S(K)$ is a free Abelian group with $s-1$ free generators, i.e., it is an $(s-1)$-dimensional lattice. Since by the Corollary to Proposition 1.12 every free Z-module is projective (and an Abelian group and a Z-module mean the same), Proposition 1.13 (i) shows that $U_S(K)$ is indeed the direct product of $E(K)$ and $s-1$ copies of the infinite cyclic group.

Thus consider the mapping $\Phi: U_S(K) \to R^s$ given by

$$\Phi: x \to [\log v(x)]_{v \in S},$$

$v(x)$ being normalized. At our first step we prove that

(i) *The image of $U_S(K)$ under this mapping is a discrete subgroup of R^s.*

Since Φ is obviously a homomorphism, $\Phi(U_S(K))$ is a subgroup of R^s and we have only to show that there is no sequence of non-zero elements of $\Phi(U_S(K))$ converging to zero, because every subgroup of R^s is either discrete or is dense in a certain linear subspace of R^s.

Assume the contrary. Then, in particular, one can find infinitely many elements $x \in U_S(K)$ such that $|\log v(x)| < 1$ for Archimedean v, and $|\log v_\mathfrak{p}(x)| < \log N(\mathfrak{p})$ if $v_\mathfrak{p}$ is the discrete valuation associated with the prime ideal $\mathfrak{p}$ of R_K, $v_\mathfrak{p} \in S$.

But if we denote by $c_\mathfrak{p}(x)$ the exponent defined by $\mathfrak{p}$ in K, then clearly $v_\mathfrak{p}(x) = N(\mathfrak{p})^{-c_\mathfrak{p}(x)}$ and so $\log v_\mathfrak{p}(x) = -c_\mathfrak{p}(x) \log N(\mathfrak{p})$, which shows that, for our choice of x, $|c_\mathfrak{p}(x)| < 1$. Since this number is a rational integer, it must be zero; thus $x \in R_K$. Moreover, all conjugates of x are bounded by $\exp 1$, whence we can have at most a finite number of such x's, contrary to our assumption.

We have thus proved (i). Observe, moreover, that Kronecker's Theorem 2.1 implies that the kernel of Φ coincides with $E(K)$. Finally, note that the image of $U_S(K)$ is at most an $(s-1)$-dimensional lattice, since, in view of Theorem 3.3, it lies entirely in the $(s-1)$-dimensional hyperplane $X_1 + X_2 + \ldots + X_s = 0$.

Our next step is to prove that the dimension of $\Phi(U_S(K))$ equals $s-1$. We shall do this first in the simplest case, where S contains no discrete valuations, and then deduce the same in the general case, using the finiteness of the class-group $H(K)$ established in the previous section. Thus we are now going to prove that

(ii) *If $r_1 + r_2 > 1$, then there exist $r_1 + r_2 - 1$ units in $U(K)$ whose images in R^s (where $s = r_1 + r_2$) under Φ are linearly independent.*

Let $n = [K:Q]$, let $w_1, \ldots, w_n$ be a fixed integral basis of K, denote by I the set of all elements of the form $c_1 w_1 + \ldots + c_n w_n$ where $|c_i| \leq 1$, $c_i \in Q$ ($i = 1, 2, \ldots, n$) and put $g = \max\{\overline{|x|} : x \in I\}$.

§ 3. Units   103

Lemma 3.5. *Let B be a positive integer. For every $a \in K$ satisfying $|N_{K/Q}(a)| \leq B^n$ one can find a non-zero $b \in R_K$ such that for $v \in S_\infty$ we have*

$$v(ab) \leq \begin{cases} gB & \text{if } v \text{ corresponds to a real embedding,} \\ gB^2 & \text{otherwise.} \end{cases}$$

Proof. First assume that $a \in R_K$. In this case the numbers $c_1 w_1 + \ldots + c_n w_n$ ($0 \leq c_i \leq B$, $c_i \in Z$) cannot be all distinct (mod aR_K), since there are

$$(1+B)^n > |N_{K/Q}(a)| = N(aR_K)$$

of them.

Hence one of their non-zero differences must be divisible by a. Denote it by ab; then for all v's we have either $v(ab) \leq gB$ or $v(ab) \leq gB^2$, depending on the type of embedding of K into the complex field, corresponding to v. If a is not integral, then write $a = a_0/m$ with $a_0 \in R_K$ and natural m. Obviously $|N_{K/Q}(a_0)| \leq (mB)^n$ and so the preceding argument shows the existence of a non-zero b in R_K with $v(a_0 b) \leq gmB$ and $v(a_0 b) \leq g(mB)^2$, respectively. This implies the asserted inequality for $v(ab)$. □

Corollary. *There exists a constant $M > 1$ with the property that for all $a \in K$ satisfying $1/2 < |N_{K/Q}(a)| < 1$ there exists a unit $u \in U(K)$ such that $\overline{ua} \leq M$.*

Proof. Applying the lemma with $B = 1$, we obtain the existence of $b = b(a)$ satisfying $v(ab) \leq g$ for all $v \in S_\infty$. This implies $|N_{K/Q}(ab)| \leq g^n$, whence $|N_{K/Q}(b)| \leq 2g^n$. Theorem 1.6 (ii) shows now that the numbers $b(a)$ generate only a finite number of distinct ideals, and so we may find a finite set $b_1, \ldots, b_N$ such that for every a satisfying our conditions we have $b(a) = ub_i$ for a certain $i \leq N$ and a suitable unit u. Thus for every v in S_∞ we have

$$v(ua) = v(ab)v(ub^{-1}) = v(ab)v(b_i^{-1}) \leq M$$

for a certain constant M which does not depend on a. □

Now let $v_1, \ldots, v_{r_1+r_2}$ be all Archimedean valuations of K ordered so that $v_1, \ldots, v_{r_1}$ correspond to the real embeddings of K into $\mathfrak{Z}$. Moreover, let $0 < a_i < b_i$ ($i = 1, 2, \ldots, r_1+r_2$) be given real numbers. We claim that it is possible to find an element a in K (not necessarily integral) such that $a_i < v_i(a) < b_i$. In fact, define

$$A_i = \begin{cases} a_i & \text{for } i = 1, 2, \ldots, r_1, \\ a_i^{1/2} & \text{for } i = r_1+1, \ldots, r_1+r_2 \end{cases}$$

and

$$B_i = \begin{cases} b_i & \text{for } i = 1, 2, \ldots, r_1, \\ b_i^{1/2} & \text{for } i = r_1+1, \ldots, r_1+r_2. \end{cases}$$

Then our conditions take the form

$$A_i < |F_i(a)| < B_i \quad \text{for } i = 1, 2, \ldots, r_1 + r_2,$$

where F_i is the embedding corresponding to v_i. Now take an arbitrary integral basis of K, say $w_1, \ldots, w_n$, and consider the system of linear equations

$$x_1 F_i(w_1) + \ldots + x_n F_i(w_n) = h_i \quad (i = 1, 2, \ldots, n),$$

where for $i = r_1 + r_2 + 1, \ldots, n$ the embeddings F_i are defined by $F_i(x) = \overline{F_{i-r_2}(x)}$, and h_i are real numbers with $A_i < h_i < B_i$, for $i = 1, 2, \ldots, r_1 + r_2$ and $h_i = h_{i-r_2}$ for i larger than $r_1 + r_2$.

By Cramer's rule this system has a real solution. If we now choose rational numbers $y_1, \ldots, y_n$ with the differences $|x_i - y_i|$ small enough to satisfy the inequalities

$$A_i < |y_1 F_i(w_1) + \ldots + y_n F_i(w_n)| < B_i \quad (i = 1, 2, \ldots, n)$$

(where for $i = r_1 + r_2 + 1, \ldots, n$ we define $A_i = A_{i-r_2}$, $B_i = B_{i-r_2}$), then the number $a = y_1 w_1 + \ldots + y_n w_n$ gives what is required.

This being done, we may now find $a_1, a_2, \ldots, a_{r_1+r_2}$ in K satisfying the inequalities

$$M < v_i(a_j) < cM \quad (i \neq j),$$

$$1/(2M^{r_1+r_2-1}) < v_j(a_j) < 1/(cM)^{r_1+r_2-1},$$

where the constant $c > 1$ is chosen so that it satisfies $c^{r_1+r_2-1} < 2$.

These numbers satisfy $1/2 < |N_{K/Q}(a_i)| < 1$, and so, by the Corollary to Lemma 3.5 we obtain for certain units u_i the inequality $v_j(u_i a_i) < M$ for all j. In particular, for $i \neq j$ we get $v_j(u_i) < 1$, i.e., $\log v_j(u_i) < 0$. Now consider the matrix $[\log v_j(u_i)]$ with $1 \leq i, j \leq r_1 + r_2 - 1$. Statement (ii) will follow once we prove that it is non-singular. For this purpose we note that Theorem 3.3 gives

$$\sum_{j=1}^{r_1+r_2-1} \log v_j(u_i) = -\log v_{r_1+r_2}(u_i) > 0$$

for $i = 1, 2, \ldots, r_1 + r_2 - 1$, and so (ii) results from the following lemma:

Lemma 3.6. *If $[a_{ij}]$ is a real matrix with $a_{ij} < 0$ for $i \neq j$ and $a_{i1} + \ldots + a_{in} > 0$ $(i = 1, 2, \ldots, n)$, then $\det[a_{ij}]$ is non-zero.*

Proof. Otherwise there exist a system $x_1, \ldots, x_n$ of numbers not all equal to zero and such that

$$\sum_{j=1}^{n} a_{ij} x_j = 0 \quad (i = 1, 2, \ldots, n).$$

If $\max_i |x_i| = x_N$ (we may assume that x_N is positive, multiplying, if necessary, all numbers x_i by -1), then we obtain a clear contradiction:

$$0 = \sum_{j=1}^{n} a_{Nj} x_j = a_{NN} x_N + \sum_{j \neq N} a_{Nj} x_j \geq x_N \sum_{j=1}^{n} a_{Nj} > 0. \qquad \square$$

So we see that the dimension of $\Phi(U(K))$ is $r_1 + r_2 - 1$. Now we can prove that

(iii) *There exist $s-1$ units in $U_S(K)$ whose images in R^s are linearly independent.*

Since $U(K)$ is contained in $U_S(K)$, we may consider $\Phi(U(K))$ as a sublattice of $\Phi(U_S(K))$. Moreover, $\Phi(U(K)) \simeq U(K)/E(K)$, $\Phi(U_S(K)) \simeq U_S(K)/E(K)$; thus $\Phi(U_S(K))/\Phi(U(K)) \simeq U_S(K)/U(K)$ is a free Abelian group with at most $s - 1 - (r_1 + r_2 - 1) = s - r_1 - r_2 = m$ free generators by (ii). We have to prove that it has exactly m free generators; in order to do this it suffices to show the existence of m elements in the group for which no product of their integral powers can be equal to the unit element except the trivial case where all the exponents are zero. Let $v_1, \ldots, v_m$ be the discrete valuations in S and let $\mathfrak{P}_1, \ldots, \mathfrak{P}_m$ be the corresponding prime ideals in R_K. By Corollary 1 to Theorem 3.4 the ideals $\mathfrak{P}_i^{h(K)}$ are principal, and thus of the form $a_i R_K$, where a_i's are all S-units. We show that they are independent in $U_S(K)/U(K)$. In fact, if for some rational integers $m_1, \ldots, m_m$ we have $a = a_1^{m_1} \ldots a_m^{m_m} \in U(K)$, then the ideal $aR_K = (\mathfrak{P}_1^{m_1} \ldots \mathfrak{P}_m^{m_m})^{h(K)}$ equals R_K, and thus by the unique factorization of ideals we obtain $m_1 = \ldots = m_m = 0$ as asserted. This proves (iii).

As we already said at the beginning, this proves also the theorem, for $\Phi(U_S(K))$ is shown to be a free Z-module, and so we have an exact sequence

$$0 \to E(K) \to U_S(K) \to \Phi(U_S(K)) \to 0$$

with a free, and thus projective, last non-zero term. Proposition 1.13 now gives

$$U_S(K) = E(K) \oplus \Phi(U_S(K))$$

and the second term here is a free Abelian group with $s-1$ free generators. $\square$

Corollary 1. *Every S-unit u can be written in a unique way in the form*

$$u = \zeta \varepsilon_1^{n_1} \ldots \varepsilon_{s-1}^{n_{s-1}} \qquad (n_i \in Z, \zeta \in E(K)),$$

where $\varepsilon_1, \ldots, \varepsilon_{s-1}$ are certain fixed S-units. $\square$

Every system of S-units $\varepsilon_1, \ldots, \varepsilon_{s-1}$ having the property stated in this corollary is called a *fundamental system of S-units*. In the case $S = S_\infty$ we speak simply about *fundamental units of K.*

For any system $u_1, \ldots, u_r$ of units of K (with $r = r_1+r_2-1$) which are not roots of unity we define the *regulator* $R(u_1, \ldots, u_r)$ as the absolute value of the determinant of the matrix $[\log v(u_i)]$, where v runs over all valuations from S_∞ except one, which can be chosen arbitrarily.

Corollary 2. (i) *The value of $R(u_1, \ldots, u_r)$ does not depend on the deleted valuation.*

(ii) *The regulator $R(u_1, \ldots, u_r)$ vanishes iff the units $u_1, \ldots, u_r$ are multiplicatively dependent, i.e., for certain $n_1, \ldots, n_r \in Z$, not all vanishing, we have $\prod_{i=1}^{r} u_i^{n_i} = 1$.*

(iii) *If $u_1, \ldots, u_r$ and $u'_1, \ldots, u'_r$ are two fundamental systems of units of K, then*
$$R(u_1, \ldots, u_r) = R(u'_1, \ldots, u'_r).$$

If we denote this common value by $R(K)$ and $a_1, \ldots, a_r$ is any system of multiplicatively independent units of K, then
$$R(K) \leq R(a_1, \ldots, a_r).$$

Proof. All the assertions follow directly from the definition of the regulator, the elementary properties of the determinants and the equality $\sum_{v \in S} \log v(u) = 0$ for u in $U(K)$. □

The regulator $R(K)$ is called the *regulator of the field K*, and $r = r_1+r_2-1$ is the *rank of the units of K*.

3. Using Dirichlet's theorem, we now give a description of fields which have a proper subfield with the same rank of units.

Proposition 3.5. *If $K \subset L$, $K \neq L$, then the groups*
$$U(L)/E(L) \quad \text{and} \quad U(K)/E(K)$$
are isomorphic iff K is totally real, L is totally complex and $[L:K] = 2$.

Proof. If $U(L)/E(L)$ and $U(K)/E(K)$ are isomorphic, then their ranks coincide and Theorem 3.5 gives
$$r_1(K)+r_2(K) = r_1(L)+r_2(L).$$
Putting $N = [L:K]$, we get
$$r_1(L)+2r_2(L) = [L:Q] = N[K:Q] = Nr_1(K)+2Nr_2(K),$$
and thus
$$r_2(L) = (N-1)r_1(K)+(2N-1)r_2(K)$$

and
$$r_1(L) = r_1(K)+r_2(K)-r_2(L) = (2-N)r_1(K)+(2-2N)r_2(K).$$

Since $r_1(L) \geqslant 0$, the last equality implies $N \leqslant 2$; thus $N = 2$ and hence $r_1(L) = -2r_2(K)$, which is possible only if $r_1(L) = r_2(K) = 0$, whence L is totally complex and K is totally real. The converse implication is obvious. □

Corollary 1. *If K is a proper subfield of L, then the factor group $U(L)/U(K)$ is finite iff L is a totally complex quadratic extension of K and K is totally real.*

Proof. If $U(L)/U(K)$ is finite, then $U(K)/E(K)$ is of a finite index in $U(L)/E(L)$, and since they are both free Abelian groups, their ranks coincide, whence they are isomorphic. Now we can apply the proposition to obtain the first implication. To prove the converse, observe that $U(K)/E(K) \subset U(L)/E(L)$ and our assumptions imply that both groups are free Abelian groups of the same rank, whence their quotient must be finite. □

Corollary 2. *An extension K/Q is generated by a unit except possibly in the case where K is a totally complex quadratic extension of a totally real field.* □

Note that in the exceptional case of the last corollary one can give examples with or without unit generators. E.g., $K = Q(i)$ has a unit generator, whereas $K = Q(i5^{1/2})$ has not, since its only units are ± 1 in view of $1 = x^2+5y^2 = N(x+yi5^{1/2})$ having only $x = \pm 1$, $y = 0$ for solutions.

If L and K satisfy the conditions given in the last proposition, then K equals the maximal real subfield of L, i.e., $K = L \cap R$. This follows from the observation that K is real, L is non-real and $[L:K] = 2$ is a prime. Usually one denotes the maximal real subfield of a field L by L^+. A field L is said to have the *defect of the units*, if it has the property given in the last proposition and, moreover, if $U(L) = U(K)E(L)$, i.e., if every unit of L is a product of a unit of K and a root of unity, then L is said to have the *strong defect of units*. This happens iff L has a system of fundamental units lying in K. We now prove that this situation arises in the case where L/Q is a complex cyclic extension and $K = L^+$.

Theorem 3.7. *If L/Q is cyclic, then one can find in L a fundamental system of real units.*

Proof. If L itself is real, then there is nothing to prove. So assume that L is totally complex, i.e., that its signature equals $[0, n/2]$ for $n = [L:Q]$. Let $r = n/2-1$ and denote by $\varepsilon_1, \ldots, \varepsilon_r$ a fixed system of fundamental units of L and by s the generator of the Galois group $G(L/Q)$. Finally, let ζ be the generator of $E(L)$ and N its order.

We shall consider a matrix $A = [a_{ij}]$ whose elements are uniquely defined by

$$s(\varepsilon_i) = \zeta^{a_i} \prod_{j=1}^{r} \varepsilon_j^{a_{ij}} \quad (i = 1, 2, \ldots, r; a_i, a_{ij} \in Z),$$

and we shall prove that it satisfies the matrix equation

$$X^r + X^{r-1} + \ldots + X + E = 0,$$

E being the unit matrix.

Simple induction shows that if, for natural k, we put

$$s^k(\varepsilon_i) = \zeta^{a_i(k)} \prod_{j=1}^{r} \varepsilon_j^{a_{ij}^{(k)}} \quad (i = 1, 2, \ldots, r),$$

then $[a_{ij}^{(k)}] = A^k$, which in turn implies

$$\varepsilon_i s(\varepsilon_i) \ldots s^r(\varepsilon_i) = \zeta^{b_i} \prod_{j=1}^{r} \varepsilon_j^{b_{ij}} \quad (i = 1, 2, \ldots, r)$$

with $[b_{ij}] = A^r + A^{r-1} + \ldots + A + E$; thus we have to show that all b_{ij}'s vanish or, which means the same, that all products $T_i = \varepsilon_i s(\varepsilon_i) \ldots s^r(\varepsilon_i)$ are roots of unity. Since $s^{r+1} = s^{n/2}$ is the only element of order two in the Galois group $G(L/Q)$, it coincides with the complex conjugation, and we obtain

$$s^{n/2}(\varepsilon_i) s^{n/2+1}(\varepsilon_i) \ldots s^{n-1}(\varepsilon_i) = s^{n/2}(\varepsilon_i s(\varepsilon_i) \ldots s^{n/2-1}(\varepsilon_i)) = s^{n/2}(T_i) = \overline{T}_i,$$

which shows that

$$N_{K/Q}(\varepsilon_i) = \varepsilon_i s(\varepsilon_i) \ldots s^{n-1}(\varepsilon_i) = T_i \overline{T}_i = |T_i|^2.$$

Proposition 3.3 now gives $|T_i| = 1$ for $i = 1, 2, \ldots, r$. Moreover, we obtain $|s^k(T_i)| = 1$ for $k = 1, 2, \ldots, n-1$. In fact, for $k = 1$ we have

$$s(T_i) = s(\varepsilon_i) \ldots s^{r+1}(\varepsilon_i) = T_i \varepsilon_i^{-1} \overline{\varepsilon}_i, \quad \text{whence } |s(T_i)| = 1,$$

and if $|s^j(T_i)| = 1$ holds for $j = 1, 2, \ldots, k-1$, then the equalities

$$s^k(T_i) = s^{k-1}(s(T_i)) = s^{k-1}(T_i \varepsilon_i^{-1} \overline{\varepsilon}_i)$$
$$= s^{k-1}(T_i) s^{k-1}(\varepsilon_i^{-1}) \overline{s^{k-1}(\varepsilon_i)}$$
$$= s^{k-1}(\overline{T_i}) \overline{s^{k-1}(\varepsilon_i)} / s^{k-1}(\varepsilon_i)$$

show that $|s^k(T_i)| = 1$. Theorem 2.1 (i) enables us to conclude that all T_i's are roots of unity, and so A satisfies the equation stated above. In particular, we can see that $A^{n/2} = E$, and so, for $i = 1, 2, \ldots, r$, the equalities

$$s^{n/2}(\varepsilon_i) = \zeta^{t_i} \varepsilon_i$$

hold if

$$t_i = \sum_{j=1}^{r} a_j(\delta_{ij}\beta^r + a_{ij}\beta^{r-1} + a_{ij}^{(2)}\beta^{r-2} + \ldots + a_{ij}^{(r)}), \qquad (3.4)$$

where β is defined by $1 \leq \beta \leq N-1$ and $s(\zeta) = \zeta^\beta$, and δ_{ij} is the Kronecker symbol.

By Proposition 3.4 the number N is even, and so, in view of

$$\zeta^{-1} = \bar{\zeta} = s^{n/2}(\zeta) = \zeta^{\beta^{n/2}}, \qquad \beta^{n/2} + 1 \equiv 0 \pmod{N},$$

we may conclude that β is odd.

Applying once more the fact that $A^r + \ldots + A + E = 0$ and using the oddness of β, we can see that the expressions in parentheses in (3.4) are all even, and thus the same holds for the t_i's. Putting $t_i = 2u_i$ and $\eta_i = \zeta^{u_i}\varepsilon_i$ for $i = 1, 2, \ldots, r$, we can see that the η_i's form a fundamental system of units; moreover, by

$$\bar{\eta} = s^{n/2}(\eta_i) = \zeta^{-u_i}s^{n/2}(\varepsilon_i) = \zeta^{-u_i} \cdot \zeta^{t_i}\varepsilon_i = \zeta^{u_i}\varepsilon_i = \eta_i,$$

they are all real. $\square$

Corollary. *If p is an odd prime and $n \geq 1$, then every unit ε of the field $L = Q(\zeta_{p^n})$ can be written in the form $\varepsilon = \zeta \cdot \varepsilon_0$ where ζ is a root of unity and ε_0 a real unit, both lying in L.*

Proof. It suffices to note that owing to Theorem 2.9 the extension L/Q is cyclic and to apply the preceding theorem. $\square$

A similar but weaker result is true for a fairly large class of normal extensions of the rationals:

Proposition 3.6. *Let K/Q be a normal complex extension and assume that K^+/Q is also normal. Then every unit ε of K can be written in the form $\varepsilon = \zeta \cdot \varepsilon_0$ with a root of unity ζ whose square lies in K and a real unit ε_0 whose square lies in K^+.*

Proof. We need a lemma:

Lemma 3.7. *Under the assumptions of the theorem the complex conjugation acts trivially on $U(K)/E(K)$, i.e., for every unit ε of K we have $\bar{\varepsilon} = u\varepsilon$, with u lying in $E(K)$.*

Proof. Observe first that the normality of K^+/Q implies that the complex conjugation s lies in the centre of $G = \text{Gal}(K/Q)$. Indeed, the group $\{e, s\}$ corresponds to K^+ by the Galois theory and so it is a normal subgroup of G. Thus for g in G we must have either $gsg^{-1} = e$ or $gsg^{-1} = s$. However, the first equality gives $s = e$, a contradiction, and thus $gs = sg$. If ε is a unit of K,

then for $u = \bar{\varepsilon}\varepsilon^{-1}$ we have $|u| = 1$, whence for all g in G we get
$$1 = g(u\bar{u}) = g(u)g(\bar{u}) = g(u)\overline{g(u)};$$
thus $|g(u)| = 1$ and $|\bar{u}| = 1$ follows. By Theorem 2.1 (i) u must be a root of unity and the lemma follows. □

If ε lies in $U(K)$, then by the lemma we have $\bar{\varepsilon} = u\varepsilon$, with a suitable root of unity u lying in K. The number $|\varepsilon|$ is a real unit, being a root of $X^2 - \varepsilon\bar{\varepsilon}$. If we put $\zeta = \varepsilon/|\varepsilon|$, then $\zeta^2 = \varepsilon^2/|\varepsilon|^2 = \varepsilon\bar{\varepsilon}^{-1} = u^{-1}$, and hence ζ is a root of unity. Since $\varepsilon\bar{\varepsilon}$ lies in K^+, our assertion follows with $\varepsilon_0 = |\varepsilon|$. □

4. Some subgroups of the group $U(K)$ of units are also of importance. Let $\mathfrak{f}$ be any ideal of R_K and denote by $U(K, \mathfrak{f})$ the subgroup of $U(K)$ consisting of all units of K congruent to unity (mod $\mathfrak{f}$). Similarly, let $U^+(K, \mathfrak{f})$ be the group of all totally positive units of K, congruent to unity (mod $\mathfrak{f}$). In the particular case of $\mathfrak{f} = R_K$ we shall simply write $U^+(K)$ for this group. The structure of those groups is described by the following

Proposition 3.7. *If K has the signature $[r_1, r_2]$ and $r = r_1 + r_2 - 1$ is not zero, then the group $U^+(K, \mathfrak{f})$ can be written as the product of the cyclic group consisting of all roots of unity congruent to unity (mod $\mathfrak{f}$) which are totally positive and r copies of the cyclic infinite group. The group $U(K, \mathfrak{f})$ is the product of the cyclic group consisting of all roots of unity congruent to unity (mod $\mathfrak{f}$) and r copies of the cyclic infinite group.*

Proof. Since the $2\varphi(f)$th power of every element of $U(K)$ lies in $U^+(K, \mathfrak{f})$ and $U(K)/E(K)$ is the rth power of the cyclic infinite group, the same holds for $U^+(K, \mathfrak{f})/(E(K) \cap U^+(K, \mathfrak{f}))$ and the result follows. The same argument applies to $U(K, \mathfrak{f})$. □

This proposition shows that for a given ideal $\mathfrak{f}$ one can find units $\eta_1, \ldots, \eta_r$ in $U^+(K, \mathfrak{f})$ and in $U(K, \mathfrak{f})$ such that every unit in $U^+(K, \mathfrak{f})$ or $U(K, \mathfrak{f})$ may be uniquely written in the form
$$\zeta \cdot \eta_1^{n_1} \ldots \eta_r^{n_r}$$
with rational integral exponents, ζ being a suitable root of unity contained in $U^+(K, \mathfrak{f})$ or $U(K, \mathfrak{f})$. The units $\eta_1, \ldots, \eta_r$ are called the *fundamental totally positive units* of K (mod $\mathfrak{f}$) or the *fundamental units* (mod $\mathfrak{f}$), respectively.

The regulators of a system of fundamental units (mod $\mathfrak{f}$), and of a system of fundamental totally positive units (mod $\mathfrak{f}$) are denoted by $R_\mathfrak{f}(K)$ and $R_\mathfrak{f}^+(K)$, respectively.

Now we shall establish the connections between the class numbers $h(K)$, $h^*(K)$, $h_{\mathfrak{f}}(K)$ and $h_{\mathfrak{f}}^*(K)$. To do this we have to introduce some notation. Consider the mapping $f\colon K^* \to D(K)$ as defined in subsection 1 of § 2 and specifically the images of $U(K)$ and of $U(K, \mathfrak{f})$ under this mapping. They obviously lie entirely in the 2-group generated by the infinite prime divisors; thus they are finite and we may write, say, $|f(U(K))| = 2^s$ and $|f(U(K, \mathfrak{f}))| = 2^t$. For our immediate purpose it is important to observe that $|f(U(K))|$ equals the number of possible signatures of units and $|f(U(K, \mathfrak{f}))|$ equals the number of possible signatures of units congruent to 1 (mod $\mathfrak{f}$).

Moreover, denote by $\psi(\mathfrak{f})$ the number of residue classes (mod $\mathfrak{f}$) which can be represented by units from K.

Theorem 3.8. *Let $\mathfrak{f}$ be any non-zero ideal of R_K. Then we have:*
  (i) $H(K) \simeq H_{\mathfrak{f}}(K)/P_{\mathfrak{f}}^{(0)}(K)$, $h_{\mathfrak{f}}(K) = h(K)\varphi(\mathfrak{f})\psi^{-1}(\mathfrak{f})$,
  (ii) $H_{\mathfrak{f}}(K) \simeq H_{\mathfrak{f}}^*(K)/P_{\mathfrak{f}}^*(K)$, $h_{\mathfrak{f}}^*(K) = 2^{r_1-t}h_{\mathfrak{f}}(K)$,
  (iii) $H(K) \simeq H^*(K)/P_1^*(K)$, $h^*(K) = 2^{r_1-s}h(K)$,
where $P_{\mathfrak{f}}^*(K)$ and $P_{\mathfrak{f}}(K)$ denote the groups of classes of $H_{\mathfrak{f}}^*(K)$ and $H_{\mathfrak{f}}(K)$, respectively, consisting of principal ideals having a generator congruent to unity (mod $\mathfrak{f}$), $P_1^*(K)$ stands for $P_{R_K}^*(K)$, and $P_{\mathfrak{f}}^{(0)}(K)$ denotes the subgroup of $H_{\mathfrak{f}}(K)$ consisting of classes containing principal ideals.

Proof. If two ideals are in the same class in $H_{\mathfrak{f}}(K)$, then they are also in the same class in $H(K)$. This simple remark shows the existence of a canonical homomorphism of $H_{\mathfrak{f}}(K)$ into $H(K)$ which carries every class of $H_{\mathfrak{f}}(K)$ to the class of $H(K)$ containing each of its ideals. Corollary 6 to Proposition 1.6 shows that every ideal class in $H(K)$ contains an ideal relatively prime to $\mathfrak{f}$, and so the homomorphism just defined is in fact an epimorphism. Its kernel being obviously $P_{\mathfrak{f}}^{(0)}(K)$, we arrive at the first assertion of (i). To prove the second assertion we have to evaluate $P_{\mathfrak{f}}^{(0)}(K)$. It consists of a union of $h_{\mathfrak{f}}(K)/h(K)$ classes of $H_{\mathfrak{f}}(K)$, and this is the number which we have to find. Let A be any ideal in X_0 with $(A, \mathfrak{f}) = R_K$. If a is any generator of A, then all the other generators of A have the form εa for some ε in $U(K)$, and so with every such ideal A we may associate a system $\Lambda_A = \{\lambda_1, \ldots, \lambda_k\}$ (with $k = \psi(\mathfrak{f})$) of residue classes (mod $\mathfrak{f}$), namely of those which have a representative generating the ideal A. Obviously, two ideals A, B from X_0 with $(AB, \mathfrak{f}) = R_K$ are in the same class of $H_{\mathfrak{f}}(K)$ iff the systems Λ_A and Λ_B coincide. But two such systems either are disjoint or coincide, and so we have exactly $\varphi(\mathfrak{f})/\psi(\mathfrak{f})$ of them, and finally we obtain the equality
$$h_{\mathfrak{f}}(K)/h(K) = \varphi(\mathfrak{f})/\psi(\mathfrak{f}),$$
which proves (i).

The proof of (ii) follows the same pattern. We consider the canonical mapping $H_{\mathfrak{f}}^*(K) \to H_{\mathfrak{f}}(K)$, which is trivially surjective, with kernel $P_{\mathfrak{f}}^*(K)$.

Consider the principal class of $H_\mathfrak{f}(K)$, which consists of all principal ideals of the form aR_K with $a \equiv 1 \pmod{\mathfrak{f}}$. We have to find the number of classes of $H_\mathfrak{f}^*(K)$ formed by such ideals. If $a \equiv 1 \pmod{\mathfrak{f}}$ and a generates an ideal I, then every other generator of I congruent to 1 $\pmod{\mathfrak{f}}$ must be equal to εa for some ε in $U(K, \mathfrak{f})$. Hence we may associate with every such ideal a system $\Gamma = \langle \gamma_1, \ldots, \gamma_m \rangle$ (with $m = 2^t$) of signatures, namely of those which are signatures of generators of this ideal congruent to unity $\pmod{\mathfrak{f}}$. Obviously, two ideals from the principal class of $H_\mathfrak{f}^*(K)$ are in the same class of $H_\mathfrak{f}(K)$ iff their systems of signatures coincide.

By Proposition 2.1 every possible signature may be represented by an integer congruent to unity $\pmod{\mathfrak{f}}$, whence there are exactly 2^{r_1-t} systems Γ, proving (ii).

Finally, (iii) is a special case of (ii), with $\mathfrak{f} = R_K$. $\square$

Corollary 1. *The equality $H(K) = H^*(K)$ holds iff in K there exist units of every possible signature.* $\square$

Corollary 2. *If K is quadratic with a positive discriminant, and the norm of the fundamental unit is positive, then $h^*(K) = 2h(K)$. Otherwise $h^*(K) = h(K)$.*

Proof. For such fields we have $r_1 + r_2 - 1 = 1$, and so there is only one fundamental unit. The factor 2^{r_1-s} equals either 1 (iff there are units of every signature) or 2 (otherwise), and obviously the first case arises iff there is a unit with a negative norm. But this implies that the fundamental unit has a negative norm. $\square$

Corollary 3. *If K is totally real, then $H(K) = H^*(K)$ holds iff every totally positive unit of K is a square in K, i.e., $U^+(K) = U^2(K)$.*

Proof. Theorem 3.6 and Proposition 3.4 imply that the quotient $U(K)/U^2(K)$ is isomorphic to C_2^n, where $n = [K:Q]$, and since the number of possible signatures equals 2^n and $U^2(K) \subset U^+(K)$, it follows that the quotient

$$U^+(K)/U^2(K) \simeq (U(K)/U^2(K))/(U(K)/U^+(K))$$

is trivial iff $|U(K)/U^+(K)| = 2^n$. However, the index of $U^+(K)$ in $U(K)$ equals the number of unit signatures in K, and it remains to apply Corollary 1. $\square$

To state the next corollary, denote by $e_q(A)$ (for a prime-power q) the number of the invariants of the finite Abelian group A which are divisible by q, i.e. the number of cyclic factors of an order divisible by q in a decomposition of A into cyclic summands. (Note that this number does not depend on the decomposition.)

Corollary 4. *For all $n \geq 1$ we have*
$$e_{2^n}(H(K)) = e_{2^n}(H^*(K)) + A(n) - A(n-1),$$
where $2^{A(n)}$ equals the number of classes in $P_1^*(K)$ which are 2^n-th powers of elements of $H^*(K)$.

Proof. For brevity, write H, H^* and P for $H(K)$, $H^*(K)$ and $P_1^*(K)$. For every Abelian finite group A and $n \geq 1$ we have
$$e_{2^n}(A) = \dim_k A^{2^{n-1}}/A^{2^n},$$
where k denotes the 2-element field, since every group C_2^N can be regarded as a linear space over k and $A^{2^{n-1}}/A^{2^n}$ has this form.

The natural map $H^* \to H$ induces the exact sequence
$$0 \to X \to (H^*)^{2^{n-1}}/(H^*)^{2^n} \to H^{2^{n-1}}/H^{2^n} \to 0$$
(with $X = (P \cap H^*)^{2^{n-1}}/(P \cap H^*)^{2^n}$) of linear spaces.

Hence
$$e_{2^n}(H^*) = e_{2^n}(H) + \dim_k X = e_{2^n}(H) + \log_2 |X| = e_{2^n}(H) + A(n-1) - A(n). \quad \square$$

Corollary 5. *The 2-ranks of $H(K)$ and $H^*(K)$ coincide iff every class of $H^*(K)$ consisting of principal ideals is a square.*

Proof. This follows from the case $n = 1$ of the preceding corollary. $\quad \square$

5. In the case of a normal extension K/Q, one may ask about the possibility of finding a system of fundamental units which are all conjugate, or at least a system of conjugate independent units having $r = r_1 + r_2 - 1$ elements, where independence means the linear independence of their images in the real n-space under the mapping considered in the proof of Dirichlet's theorem. The main result of this type is due to H. Minkowski, who proved the following

Theorem 3.9. *If K is a normal extension of the rational field, then one can find a system of $r_1 + r_2 - 1$ independent units which are all conjugate. Here $[r_1, r_2]$ denotes the signature of K.*

Proof. Let $[K:Q] = n$ and denote by $F_1, \ldots, F_n$ the embeddings of K into the complex field. Moreover, let G be the Galois group of the extension K/Q. Note that every embedding F_i may be written in the form $F_i = F_1 \circ g$, where g, defined by $g(x) = F_1^{-1}(F_i(x))$, is an element of G, since the normality of K implies $F_1(K) = F_2(K) = \ldots = F_n(K)$.

If K is totally complex, i.e., $r_1 = 0$, then denote by Ω the maximal real subfield of K, which is obviously the fixed field of the subgroup $H = \{1, \tau\}$

of G, τ being the complex conjugation. In the case of a totally real field K we simply put $\Omega = K$ and $\tau = 1$.

Note that every Archimedean valuation of K is of the form $v_g(x) = v_1(g^{-1}(x))$ with $v_1(x) = |F_1(x)|$ and $g \in G$, where g is determined only up to a factor from H, since v_g and $v_{\tau g}$ are equal.

As in the proof of Dirichlet's Theorem 3.5, we may find a unit ε with $v_1(\varepsilon) > 1$ and $v_i(\varepsilon) < 1$ for $v_i \neq v_1$, i.e., with $v_1(g^{-1}(\varepsilon)) < 1$ for all $g \neq 1, \tau$.

Now define $\eta = \varepsilon$ if K is totally real and $\eta = \varepsilon\tau(\varepsilon)$ if K is totally complex, and note that $v_1(\eta) > 1$ and $v_i(\eta) < 1$ for $i \neq 1$. In fact, we have already proved this in the first case; in the second we have $v_1(\eta) = v_1(\varepsilon) v_1(\tau(\varepsilon)) = v_1^2(\varepsilon) > 1$, and for $i \neq 1$, $v_i(\eta) = v_1(g^{-1}(\eta)) = v_1(g^{-1}(\varepsilon))v_1(g^{-1}(\tau(\varepsilon))) < 1$, since for $g \neq 1, \tau$ also $g^{-1} \neq 1, \tau$. Note also that, for all $g \in G$, we have $g(\tau(\eta)) = g(\eta)$. If we now consider a system A of representatives of the cosets $G \pmod{H}$ and the set $\{g(\eta) : g \in A\}$, then this set turns out to have the following property:

$$v_g(h(\eta)) < 1 \quad \text{if } g \neq h, g \neq \tau h,$$
$$v_g(h(\eta)) > 1 \quad \text{if } g = h \text{ or } g = \tau h.$$

Indeed, we have

$$v_g(h(\eta)) = v_1(g^{-1}h(\eta)) < 1 \quad \text{if } g \neq h \text{ and } g \neq \tau h$$

and

$$v_g(g(\eta)) = v_1(\eta) > 1.$$

Now the reasoning used in the proof of Theorem 3.5 shows that the constructed set of units is independent, and of course they are all conjugate. □

The last theorem gives some information about the action of the Galois group on the group of units. To be more precise, let $Z[G]$ be the group-ring of a given finite group over Z, i.e., a free Abelian group generated by elements of G in which we additionally define multiplication by means of

$$\sum_{g \in G} a_g g \cdot \sum_{h \in G} b_h h = \sum_{t \in G} \Big(\sum_{gh=t} a_g b_h\Big) t.$$

If K/Q is normal with the Galois group G, then every subgroup D of the multiplicative group of K, which is G-invariant, acquires the structure of a $Z[G]$-module, where an element $A = \sum_{g \in G} a_g g$ of $Z[G]$ acts on $x \in K^*$ by

$$Ax = \prod_{g \in G} g(x)^{a_g}.$$

We shall be interested in the case $D = U(K)$. Since $E(K)$ is G-invariant, the factor group $U_0(K) = U(K)/E(K)$ is also a $Z[G]$-module. Observe now that if $a \in U(K)$ is such that r_1+r_2-1 of its conjugates are multiplicatively independent (as in Theorem 3.9), then the submodule of $U_0(K)$ generated by the image $\bar{a}$ of a is of finite index. Indeed, both $U_0(K)$ and its submodule generated by $\bar{a}$ are free Abelian groups of the same rank. We may thus state Theorem 3.9 in the following way: if K/Q is normal with the Galois group G, then the $Z[G]$-module $U_0(K)$ contains a cyclic submodule of finite index.

We may ask, under what circumstances $U_0(K)$ will itself be cyclic. If this happens, then any representative of the generator of $U_0(K)$ in $U(K)$ is called a *Minkowski unit of* K. Moreover, if there exists a unit $u \in U(K)$ such that certain conjugates of u form a system of fundamental units, then u is called a *strong Minkowski unit*.

Proposition 3.8. *A strong Minkowski unit is a Minkowski unit, and if K is real, then these notions coincide.*

Proof. If $\varepsilon_1, \ldots, \varepsilon_r$ is a system of fundamental units, all conjugated to $\varepsilon_1 = \varepsilon$, then the image of ε in $U_0(K)$ generates $U_0(K)$, whence ε is a Minkowski unit.

If K is real and ε is a Minkowski unit, then every unit u of K can be written in the form

$$u = \zeta \prod_{g \in G} g(\varepsilon)^{a_g}$$

with a root of unity ζ and $a_g \in Z$ $(g \in G)$.

Fix g_0 in G and observe that

$$\prod_{g \in G} g(\varepsilon) = N_{K/Q}(\varepsilon) = \pm 1.$$

This implies

$$g_0(\varepsilon) = \pm \prod_{g \neq g_0} g(\varepsilon)^{-1},$$

whence

$$u = \pm \zeta \prod_{g \neq g_0} g(\varepsilon)^{a_g - a_{g_0}},$$

i.e., the set $\{g(\varepsilon): g \neq g_0\}$ is a set of generators of $U(K)$, and since it contains $[K:Q]-1 = r_1+r_2-1$ elements, it is a set of fundamental units. □

The problem in which fields there exist Minkowski units or strong Minkowski units is unsolved, except in certain special cases. We now present one of them.

Theorem 3.10. *If p is an odd prime such that the cyclotomic field $Q(\zeta_p)$ has class-number 1, and K/Q is a normal extension of degree p, then K has a strong Minkowski unit.*

Proof. We start with an easy lemma:

Lemma 3.8. *Let K/Q be normal with Galois group G and let*
$$N = \sum_{g \in G} g \in Z[G].$$
Then the principal ideal of $Z[G]$, generated by N equals NZ and if Λ denotes the factor ring $Z[G]/NZ$, then $U_0(K)$ is a Λ-module in a canonical way.

Proof. The ideal generated by N obviously equals $NZ[G]$ and contains NZ. If now $a = Nb \in NZ[G]$ with $b = \sum_{g \in G} b_g g \in Z[G]$, then $a = (\sum_{g \in G} b_g) N \in NZ$, and $NZ[G] = NZ$ follows. The second assertion follows from the observation that for any unit $u \in U(K)$ we have
$$Nu = \prod_{g \in G} g(u) = N_{K/Q}(u) = \pm 1 \in E(K);$$
thus N acts trivially on $U_0(K)$, and hence the action of $Z[G]$ on it induces canonically the action of $Z[G]/NZ = \Lambda$. □

The theorem now follows without much effort. Indeed, in our case the group G is cyclic, and so denote by g any of its generators. Then the homomorphism
$$\varphi: Z[G] \to R = Z[\zeta_p]$$
defined by $\varphi(g) = \zeta_p$ is surjective and its kernel equals $NZ = (e+g+\ldots+g^{p-1})Z$. Indeed, NZ lies in the kernel because of $1+\zeta_p+\ldots+\zeta_p^{p-1} = 0$, and if $a = \sum_{j=0}^{p-1} a_j g^j$ lies in the kernel, then
$$0 = \varphi(a) = \sum_{j=0}^{p-1} a_j \zeta_p^j = \sum_{j=0}^{p-2} (a_j - a_{p-1}) \zeta_p^j;$$
thus by Theorem 2.9 we get $a_0 = a_1 = \ldots = a_{p-1}$, i.e., $a \in NZ$. It follows that R and $Z[G]/NZ$ are isomorphic rings and the lemma shows that $U_0(K)$ is an R-module which is torsion-free and finitely generated. Since by Theorem 2.9 R is the ring of integers of $Q(\zeta_p)$, it is Dedekind and by assumption it has class-number 1, i.e., it is a PID. Theorem 1.13 shows that with a suitable $m \geq 0$ we have $U_0(K) \simeq R^m$. By Theorem 3.6 the Z-rank of $U_0(K)$ equals $p-1$, and since the Z-rank of R^m equals $m(p-1)$, we obtain $m = 1$; thus $U_0(K)$ is isomorphic to R as an R-module and a fortiori as a $Z[G]$-module. But R is cyclic, and thus K has a Minkowski unit. Since K is real, this unit has to be a strong Minkowski unit by Proposition 3.8. □

In Chapter 7 we shall establish the existence of Minkowski units for all complex extensions of Q with a dihedral Galois group of $2p$ elements with prime p. (See Theorem 7.9.)

6. In this section we shall determine explicitly the units of quadratic fields. We shall first deal with the case of imaginary quadratics, which is nearly trivial, since then $r_1+r_2-1 = 0$ and so the group $U(K)$ coincides with the group of roots of unity $E(K)$ contained in K and the structure of $E(K)$ is in this case readily determined:

Proposition 3.9. *Let* $K = Q(\sqrt{-D})$, *where D is a square-free and positive rational integer. Then* $E(K) = \{-1, 1\}$ *except the following two cases*:

(i) $D = 1$, *where* $E(K) = \{1, -1, i, -i\}$

and

(ii) $D = 3$, *where* $E(K) = \{1, -1, \frac{1}{2}(\pm 1 \pm i\sqrt{3})\}$.

Proof. To begin with, let $-D$ be congruent to unity (mod 4). Then, by Theorem 2.7, the numbers $1, w = (1+\sqrt{-D})/2$ form an integral basis of K, and

$$N_{K/Q}(x+yw) = (x+y/2)^2 + Dy^2/4.$$

Since the norm is positive, the norm of a unit equals unity; hence if $\varepsilon = x+yw$ is a unit, we must have

$$(2x+y)^2 + Dy^2 = 4.$$

If D exceeds 4, then $y = 0$, $2x+y = \pm 2$, whence $\varepsilon = \pm 1$. However, if $D = 3$, then either $y = 0$, which again gives $\varepsilon = \pm 1$, or $y = \pm 1$, giving $(2x+1)^2 = 3$; thus $x = 0$ or $x = -1$, which shows that in this case $E(K)$ has the form given in (ii).

If $-D$ is not congruent to unity (mod 4), then by Theorem 2.7 the numbers $1, \sqrt{-D}$ form an integral basis, and so every unit $\varepsilon = x+y\sqrt{-D}$ has to satisfy

$$x^2 + Dy^2 = 1.$$

If D exceeds 1, then $y = 0$, and so $\varepsilon = \pm 1$. However, if $D = 1$, then either $y = 0$, which again gives $\varepsilon = \pm 1$, or $y = \pm 1$ and $x = 0$, giving $\varepsilon = \pm i$. □

If the quadratic field K is real, then, contrary to the previous case, the group $E(K)$ contains only ± 1, and since $r_1+r_2-1 = 1$, every unit may be written as $\pm \varepsilon_0^n$ for a suitable fundamental unit ε_0 and rational integral n. Note that if ε_0 is a fundamental unit, then every other such unit must be equal either to $-\varepsilon_0$ or to $\pm \varepsilon_0^{-1}$. This shows that we can always select a fundamental unit which exceeds unity, this selection being unique. In the sequel, when speaking about the fundamental unit in a real quadratic field, we shall have this particular choice in mind. Note that this fundamental unit can also be defined as the smallest unit of K that exceeds 1.

Theorem 2.7 shows that if $K = Q(D^{1/2})$ with a square-free and positive D,

then every integer of K may be written in the form
$$a = (x+yD^{1/2})/2,$$
where x, y are rational integers which in the case of $D \equiv 2, 3 \pmod 4$ are both even and in the case of $D \equiv 1 \pmod 4$ are of the same parity. This shows that the units of K satisfy the equation
$$X^2 - DY^2 = \pm 4 \quad (\varepsilon = (X+YD^{1/2})/2) \tag{3.5}$$
with $X \equiv Y \pmod 2$ and $X \equiv Y \equiv 0 \pmod 2$ if $D \equiv 2, 3 \pmod 4$.

When we speak about the solutions of (3.5), we shall always tacitly assume that X, Y satisfy those congruence conditions.

It is evident that finding units in real quadratic fields is the same thing as solving a Pellian or a non-Pellian equation (the latter name actually seems to be more accurate for both, as Pell had nothing to do with either of them).

The following proposition shows that the minimal solution of such an equation gives the fundamental unit of K.

Proposition 3.10. *Let $K = Q(D^{1/2})$, where D is square-free and positive. Let A, B be the minimal positive solutions of (3.5). Then $\varepsilon = (A+BD^{1/2})/2$ is the fundamental unit of K.*

Proof. The number ε is evidently a unit, since it satisfies $\varepsilon^2 - A\varepsilon \pm 1 = 0$. Denote by ε_0 the fundamental unit of K exceeding 1. Then $\varepsilon = \varepsilon_0^n$ for a certain natural n, since ε also exceeds unity. We are going to show that $n = 1$. Let us put
$$\varepsilon_0^m = (a_m + b_m D^{1/2})/2$$
for $m = 1, 2, \ldots$ and observe that the numbers a_m, b_m just defined are natural. In fact, if we denote by ε_0' the number conjugate with ε_0 (i.e., $\varepsilon_0' = N_{K/Q}(\varepsilon_0)/\varepsilon_0$), then
$$a_m = \varepsilon_0^m + (\varepsilon_0')^m, \quad b_m = D^{-1/2}(\varepsilon^m - (\varepsilon_0')^m)$$
and $|\varepsilon_0'| < 1 < \varepsilon_0$. Our proposition will be proved if we show that, for every m, $a_{m+1} > a_m$ and $b_{m+1} > b_m$.

Since
$$(a_{m+1}^2 - a_m^2)(b_{m+1}^2 - b_m^2) = D > 0,$$
it is sufficient to prove for each m one of these inequalities. If ε_0 has a positive norm, then $\varepsilon_0' = \varepsilon_0^{-1}$ and
$$D^{1/2}(b_{m+1} - b_m) = \varepsilon_0^m(\varepsilon_0 - 1) - \varepsilon_0^{-m}(\varepsilon_0^{-1} - 1) > 0,$$
and if the norm of ε_0 is negative, then $\varepsilon_0' = -\varepsilon_0^{-1}$, whence we obtain
$$D^{1/2}(b_{m+1} - b_m) = \varepsilon_0^m(\varepsilon_0 - 1) - \varepsilon_0^{-m}(\varepsilon_0^{-1} + 1) > 0$$

for even m, and
$$a_{m+1} - a_m = \varepsilon_0^m(\varepsilon_0 - 1) + \varepsilon_0^{-m}(\varepsilon_0^{-1} + 1) > 0$$
for odd m. □

Using this proposition and some facts from the elementary theory of numbers, we are now able to find the fundamental unit of any real quadratic field.

Theorem 3.11. *Let D be a square-free, positive rational integer and let $K = Q(D^{1/2})$. Denote by ε_0 the fundamental unit of K which exceeds unity, by s the period of the continued fraction for $D^{1/2}$, and by P/Q the $(s-1)$-th convergent of it.*
If $D \not\equiv 1 \pmod 4$ or $D \equiv 1 \pmod 8$, then
$$\varepsilon_0 = P + QD^{1/2}.$$
However, if $D \equiv 5 \pmod 8$, then either
$$\varepsilon_0 = P + QD^{1/2}$$
or
$$\varepsilon_0^3 = P + QD^{1/2}.$$
Finally, the norm of ε_0 is positive if the period s is even and negative otherwise.

Proof. We have to use some results from the elementary theory of Pell's equation as presented for example in W. Sierpiński [64], p. 307. Namely, we use the following theorem:

If the period s is even, then the equation $X^2 - DY^2 = -1$ has no solutions and the smallest natural solution of $X^2 - DY^2 = 1$ is given by $X = P$, $Y = Q$, whereas if s is odd, then the smallest natural solution of $X^2 - DY^2 = -1$ also equals $X = P$, $Y = Q$.

Note that if $D \not\equiv 5 \pmod 8$, then every unit of the field K has the form $X + YD^{1/2}$ for $X, Y \in Z$. In fact, if $D \not\equiv 1 \pmod 4$, then this results from the form of the integral basis as given by Theorem 2.7 and in the case of $D \equiv 1 \pmod 8$ one should note that if $(X + YD^{1/2})/2$ (with X, Y in Z) is a unit, then
$$X^2 - DY^2 = \pm 4,$$
whence $X^2 - Y^2$ is congruent to 4 (mod 8), which is possible only if both X and Y are even.

This remark together with the preceding proposition settles the case of $D \not\equiv 5 \pmod 8$.

Now turn to $D \equiv 5 \pmod 8$. The number $\varepsilon = (2P + 2QD^{1/2})/2$ is the least unit exceeding 1 where the coefficients in the numerator are both even. This is guaranteed by the above mentioned result about Pell's equation. If it is also a fundamental unit, this proves the case in question. Otherwise $\varepsilon = \varepsilon_0^N$ for

suitable natural N and the fundamental unit ε_0 must have the form
$$\varepsilon_0 = (A+BD^{1/2})/2$$
with odd A and B. Note that in
$$\varepsilon_0^2 = ((A^2+DB^2)/2+ABD^{1/2})/2$$
the coefficients in the numerator are also both odd, but in ε_0^3 those coefficients are even, and so $\varepsilon_0^3 = R+SD^{1/2}$ with R, S in Z is clearly the least unit exceeding unity with the above property, whence it equals ε. Since the norms of ε and ε_0 coincide, this proves the theorem. □

This theorem gives a method of finding a fundamental unit which is not very practicable, since it involves computing the full period of the expansion of $D^{1/2}$ into a continued fraction, which may sometimes be rather awkward. In particular, it gives no simple answer to the question whether K has units with negative norms. Many papers have been concerned with this problem. We shall deal with them in the last section of this chapter.

§ 4. Euclidean Algorithm

1. We shall now be concerned with fields K for which the rings of integers are Euclidean. It is well known that every such ring is necessarily UFD, and so such fields have trivial class-groups. We recall the definition:

A domain R is said to be a *Euclidean domain* if there exists a function $p(x)$ defined on R whose values are non-negative rational integers satisfying the following condition:

If $a, b \in R$ and b is non-zero, then one can find $c, r \in R$ such that
$$a = bc+r,$$
where either $r = 0$ or $p(r) < p(b)$. Moreover, $p(x) = 0$ iff $x = 0$.

In the earlier treatments of this subject usually the additional condition
$$p(ab) \geq p(b) \quad \text{for non-zero } a, b$$
was made; however, G. R. Veldkamp [60] demonstrated that it is redundant, since it does not reduce the class of the rings concerned.

We shall say that an algebraic number field K is *Euclidean* if the ring R_K is Euclidean with $p(x) = |N_{K/Q}(x)|$. The following proposition is helpful in establishing the euclidicity of a given field:

Proposition 3.11. *A field K is Euclidean iff for every element $a \in K$ there exists an integer b of the same field with $|N_{K/Q}(a-b)| < 1$.*

Proof. In our case the condition for euclidicity takes the form

$$|N_{K/Q}(a-bc)| < |N(b)|$$

for given $a, b \neq 0$ and suitable c, all of them integers in K, and in view of the multiplicativity of the norm we may write it as

$$|N_{K/Q}(ab^{-1}-c)| < 1.$$

It remains to observe that we may put every element of K in the form ab^{-1} with a, b in R_K. □

Corollary. *An imaginary quadratic field K is Euclidean iff $d(K) = -3, -4, -7, -8, -11$.*

Proof. First we show that all the fields listed are indeed Euclidean. If $d = -4$ or $d = -8$, then every integer of K has the form $x+yD^{1/2}$, where $x, y \in Z$ and $D = d/4$. Now if $a+bD^{1/2}$ $(a, b \in Q)$ is an arbitrary element of K and we choose $x, y \in Z$ with $|a-x| \leq 1/2$, $|b-y| \leq 1/2$, then

$$|N_{K/Q}(a+bD^{1/2}-(x+yD^{1/2}))| = (a-x)^2+(b-y)^2|D| \leq 1/4+|D|/4 < 1$$

since in our cases $D = -1$ or -2.

If $d = -3, -7$ or -11, then every integer of K is of the form $(x+yd^{1/2})/2$ with $x, y \in Z$ and of the same parity. If $a+bd^{1/2}$ $(a, b \in Q)$ lies in K, then choosing first $y \in Z$ so that $|y/2-b| \leq 1/4$ and afterwards x so as to fulfil $|x/2-a| \leq 1/2$ and $x \equiv y \pmod 2$, we obtain

$$|N_{K/Q}(a+bd^{1/2}-(x/2+(y/2)d^{1/2}))| = (a-x/2)^2+(b-y/2)^2|d|$$
$$\leq 1/4+|d|/16 < 1.$$

Now assume that K is a Euclidean imaginary quadratic field and again consider first the case $d(K) \equiv 0 \pmod 4$. Put $D = d/4$ and let $a = D^{1/2}/2$. Since K is Euclidean, we may find an integer $x+yD^{1/2}$ in K with

$$|N_{K/Q}(x+(y-1/2)D^{1/2})| < 1,$$

i.e.

$$(y-1/2)^2|D|+x^2 < 1;$$

but for every $y \in Z$ we have $(y-1/2)^2 \geq 1/4$, whence $|D|/4 < 1$, and so $D = -1$ or -2, which means that $d(K) = -4$ or -8.

If $4 \nmid d(K)$ then consider $a = 1/4+d^{1/2}/4$. For certain $x, y \in Z$ we have

$$|N_{K/Q}(a-x/2-yd^{1/2}/2)| < 1,$$

and thus

$$(1/4-x/2)^2+(1/4-y/2)^2|d| < 1.$$

However, for all $m \in Z$ we have $|1/4 - m/2| \geq 1/4$, whence $|d| + 1 < 16$, and this leaves us with $d = -3, -7$ and -11. □

2. The determination of all real quadratic fields which are Euclidean is much more difficult. H. Heilbronn [38a] proved that their number is finite, and later, through the efforts of many authors, it was proved that their discriminants are equal to 5, 8, 12, 13, 17, 21, 24, 28, 29, 33, 37, 41, 44, 57, 73 and 76. The final step was made by H. Chatland and H. Davenport [50]. We shall not present the proof of this result, which rather belongs to the geometry of numbers, and prove only the following partial result:

Theorem 3.12. *The real quadratic fields with discriminants $d = 5, 8, 12, 13, 17, 21, 24, 28$ and 29 are all Euclidean.*

Proof. We need a simple lemma:

Lemma 3.9. *If $0 \leq a < 2$ and $a \neq 5/4$, then to every real x there corresponds a real y such that $x - y \in Z$ and $|y^2 - a| < 1$.*

Proof. If $a = 0$, then a suitable number from the interval $(-1/2, 1/2]$ will do. The assertion will hold also with a suitable $y \in [1/2, 1]$ if $0 < a < 5/4$ and with a suitable $y \in [1, 3/2]$ if $5/4 < a < 2$. □

Consider now $d = 8, 12, 14$ and 28. The integers of $Q(d^{1/2})$ then have the form $x + yD^{1/2}$ with $D = d/4$ and $x, y \in Z$. By Proposition 3.11 it suffices to find for every $a, b \in Q$ a pair x, y of rational integers satisfying

$$|(a-x)^2 - D(b-y)^2| < 1.$$

Choose $y \in Z$ with $|b - y| \leq 1/2$. Then

$$D(b-y)^2 \leq D/4 \leq 7/4 \quad \text{and} \quad D(b-y)^2 \neq 5/4,$$

since $D \neq 5$. We may thus use Lemma 3.9 to obtain the required inequality. The remaining fields have integers of the form $x + yw$ with $w = (1 + d^{1/2})/2$, $x, y \in Z$ and so we have to show that for every pair of rational a, b we may find $x, y \in Z$ such that $|(a - x - y/2)^2 - d(b - y/2)^2| < 1$.

Choose y in Z with $|2b - y| \leq 1/2$. Then $d(b - y/2)^2 \leq d/16 \leq 29/16$ and, since $d(b - y/2)^2 \neq 5/4$ (for if $d = 5$, then $d(b - y/2)^2$ cannot exceed $5/16$), we may apply Lemma 3.9 as before. □

3. In the definition of the euclidicity of K we assumed that the Euclidean norm $p(x)$ coincides with $|N_{K/Q}(x)|$. One may ask whether this restriction is essential, i.e. whether there exist algebraic number fields K for which the rings of integers are Euclidean under a norm $p(x)$ but not under $|N_{K/Q}(x)|$.

If one assumes the Extended Riemann Hypothesis (ERH), then the answer is positive. In fact, it was shown by P. J. Weinberger [72a] that if ERH holds, $h(K) = 1$ and K is neither Q nor an imaginary quadratic field, then K has a Euclidean Algorithm under a norm possibly distinct from $|N_{K/Q}(x)|$. However, for imaginary quadratic fields this is not possible:

Proposition 3.12. *If K is an imaginary quadratic field whose ring of integers is Euclidean under a suitable norm, then K is one of the fields listed in the Corollary to Proposition 3.11.*

Proof. Assume that $|d(K)| > 11$, and let $p(x)$ be a Euclidean norm in R_K. Choose $t \neq 0, -1, 1$ in R_K so that

$$p(t) = \min\{p(x) : x \neq 0, 1, -1\}.$$

Then for every $b \in R_K$ we can find $q \in R_K$ with $b - qt$ equal to either 0 or ± 1, whence $N_{K/Q}(t) \leq 3$ by the Corollary to Proposition 2.7.

If D is the square-free kernel of $d(K)$ (i.e. $D = d(K)/4$ in the case where $d(K)$ is divisible by 4, and $D = d(K)$ otherwise) and $D \not\equiv 1 \pmod{4}$, then $t = x + yD^{1/2}$ where $x, y \in Z$, whence

$$x^2 + |D|y^2 \leq 3,$$

which is possible for $|d(K)| > 11$, i.e., for $|D| \geq 3$ and $t \neq 0, 1, -1$ only when $D = -3 \equiv 1 \pmod{4}$, giving a contradiction.

Similarly, if $D \equiv 1 \pmod 4$, then

$$t = (x + yD^{1/2})/2 \quad (x, y \in Z \text{ and } x \text{ and } y \text{ are of the same parity}).$$

If x and y are both even, then the foregoing argument leads to a contradiction. But if x and y are both odd, then

$$12 \geq N_{K/Q}(x + yD^{1/2}) = x^2 + y^2|D| \geq 1 + |D|,$$

and so $|d| = |D| \leq 11$, again giving a contradiction. □

In Chapter 4 we shall prove that the field $Q((-19)^{1/2})$ has class-number 1, and so its ring of integers may serve as an example of a Dedekind domain which is a UFD without being a Euclidean domain.

To end this chapter we shall prove a result which in the case of rings of integers goes back to R. Dedekind [31] and was obtained in its most general form by H. Hasse [28]; it gives a necessary and sufficient condition for a domain to be a PID, i.e., a principal ideal domain. In the case of rings of integers UFD and PID means the same, and so we obtain another characterization of fields with a trivial class-group.

We have included this result in the section dealing with the Euclidean algorithm, as there is a great formal resemblance between the conditions of the theorem which follows and the definition of a Euclidean domain.

Theorem 3.13. *Let R be an integral domain and assume that there is a function $f(x)$ defined in R with values in the set of natural numbers such that*
   (i) $f(x) = 0$ iff $x = 0$,
   (ii) $f(xy) = f(x)f(y)$,
   (iii) *if $0 < f(y) \leq f(x)$ and y does not divide x, then for suitable $a, b \in R$*

$$0 < f(ax - by) < f(y).$$

Then R is a principal ideal domain.

Conversely, if in R every finitely generated ideal is principal and $f(x)$ is a function defined on R whose values are natural numbers and which satisfies (i), (ii) *and the condition*
   (iv) $f(x) = 1$ iff x has an inverse in R,
then $f(x)$ satisfies also (iii).

Note that in R every finitely generated ideal is principal iff every two elements of R generate a principal ideal.

Proof. Assume that f satisfies (i), (ii) and (iii), and let I be any non-zero ideal of R. Choose $y \in I$ so that

$$f(y) = \min\{f(t): t \neq 0, t \in I\},$$

and let x be any non-zero element of I. Then $0 < f(y) \leq f(x)$, and if y did not divide x, then by (iii) we could find $a, b \in R$ with $0 < f(ax - by) < f(y)$, contrary to the choice of y, since $ax - by$ lies in I. Hence $x = ry$ for some $r \in R$, and so I is principal, generated by y.

Now assume that in R every two elements generate a principal ideal and let $f(x)$ be any function satisfying (i), (ii) and (iv). Let $x, y \in R$, $0 < f(y) \leq f(x)$ and assume that y does not divide x. The ideal $xR + yR$ must be principal, say equal to zR; hence $y = rz$ for a certain $r \in R$ which is not invertible in R, since otherwise y would divide z, and so it would divide x, contrary to our assumption. Hence (ii) and (iv) imply $f(z) < f(y)$. Moreover, for suitable $a, b \in R$ we have $ax - by = z$, and so by (i) we obtain

$$0 < f(ax - by) = f(z) < f(y),$$

as asserted. □

Corollary. *An algebraic number field K has trivial class-group iff the function $f(x) = |N_{K/Q}(x)|$ satisfies* (iii).

Observe that a function satisfying (i), (ii) and (iv) need not exist in a ring which is not a Dedekind domain. (For a Dedekind domain which is a PID the existence is trivial, since we may put $f(0) = 0$, $f(\varepsilon) = 1$ for invertible elements ε, and $f(p) = 2$ for elements which generate prime ideals, and extend this definition to R by multiplicativity.)

To give an example, take any domain in which every finitely generated ideal is principal, but which is not a PID. Then the existence of such a function would imply by the first part of Theorem 3.13 that this domain is actually a PID.

§ 5. Notes to Chapter 3

1. It has been assumed for a long time that the theory of ideals in rings of integers owes its existence to the unsuccesful attempts of E. E. Kummer to prove Fermat's Last Theorem in its full generality. Recent research however, has shed some doubt on this story and now it seems more likely that Kummer's work on ideal numbers, which later led Dedekind to introduce ideals, had been motivated by his research concerning reciprocity laws for power residues. See Edwards [77] and Neumann [81a] on this topic.

Ideal numbers, whose purpose was to restore unique factorization in rings of algebraic integers, were introduced by E. E. Kummer [47a], [47b], first in the case of cyclotomic integers in $Q(\zeta_p)$ with a prime p. He did the same for arbitrary cyclotomic fields in [56] and for Kummerian extensions $Q(\zeta_p, a^{1/p})$ in [59]. For subfields of cyclotomic fields his theory was developed by L. Fuchs [63], [66]. Kummer's ideal numbers lay outside the field in question, but, adjoined to it, provide a set of numbers closed under multiplication and having the unique factorization property. In modern language these numbers correspond roughly to the ideals of R_K. In the set of ideal numbers Kummer introduced the notion of equivalence and was led in this way to the classnumber of cyclotomic fields.

The modern treatment of ideal numbers was introduced by E. Hecke [18], and in the form due to him they are still used in certain parts of the analytic theory of algebraic numbers.

Ideal theory in rings of integers was formulated by R. Dedekind [71]. His method was applicable in a uniform way to all algebraic number fields and that was its advantage over Kummer's approach. His fundamental result, the theorem about unique factorization of ideals into prime ideals, was proved in Dedekind [71], and other proofs were later supplied by L. Kronecker [82] and D. Hilbert [94b].

L. Kronecker [82] founded his theory of algebraic numbers on the theory of forms. Its outline may be found in Hilbert [97] and Weber [96b]. Kronecker's theory had a much better reception than that of Dedekind, although Kronecker's paper was rather difficult to decipher whereas Dedekind wrote very clearly.

Recently a manuscript of Dedekind was published, commenting upon Kronecker's approach. (See Edwards, Neumann, Purkert [82].) For a modern exposition of Kronecker's theory see Edwards [80], Flanders [60], Weyl [40]. (The paper of Edwards contains also an exposition of the history of the creation of the ideal theory. Cf. Edwards [83].) Kronecker's theory can be translated into the language of ideals simply by associating with every form F the ideal generated by its coefficients, the content of F. An axiomatical characterization of the content of a form was given by F. Krakowski [65b]. Cf. Dedekind [92], Hurwitz [94], [95a], Krakowski [65a], Mertens [94].

Some ideas of Kronecker were developed by K. Hensel, who introduced p-adic numbers and used them for studying algebraic number fields. We shall say more about that in Chapter 5.

Another approach was adopted by E. Zolotarev [80]. It is based on the notion of p-integral numbers and p-divisibility. For its exposition see Chebotarev [25], [30], [37a], [47]. Cf. Grave [24a], [24b].

A method based on a kind of ideal numbers determined by infinite systems of linear congruences every finite subsystem of which is solvable was introduced by H. Prüfer [25], [32]. Cf. v. Neumann [26], Walton [50].

Krull's method (Krull [51]) was based on the consideration of homomorphisms of the multiplicative group of a field into free Abelian groups. A fresh exposition may be found in Borevich, Shafarevich [64]. Cf. also Koch [66]. For further developments see Frey, Geyer [72], Skula [70].

For other approaches see Delaunay, Faddeev [40], Fields [24], Furtwängler [21], Żyliński [32].

The theory of infinite algebraic extensions of Q was initiated by E. Stiemke [26] and developed in papers of W. Krull [28b], [28c] and J. Herbrand [31c], [32b]. (See also Gut [37], Moriya [34a], Scholz [43].)

2. Theorem 3.1 is due to J. Kürschak [13] and Theorem 3.2 to A. Ostrowsk (the Archimedean case in [18], the non-Archimedean case in [35]). Cf. Artin [32b], McLane [36], Ore [27]. For a constructivistic approach in the Archimedean case see Mines, Richman [81], [84].

The importance of the product-formula (Theorem 3.3) was stressed by E. Artin and G. Whaples [45], [46], who used it for an axiomatical characterization of algebraic number fields and also for developing anew the fundamental results of their theory, including Dirichlet's theorem on units and the finiteness of the class-number.

3. The notion of an ideal class is essentially due to Kummer [47b] in the case of cyclotomic fields. In the case of quadratic fields it is closely connected with classes of binary quadratic forms over Z, with a given determinant, under the action of SL(2, Z), as already studied by J. L. Lagrange [73] and C. F. Gauss [01]. We shall establish this relation in Chapter 8.

Ideal classes (mod $\mathfrak{f}$) were introduced by H. Weber [97] and ideal classes in the narrow sense, which form $H_\mathfrak{f}^*(K)$, by E. Landau [18b]. Theorem 3.8 is due to them.

An algorithm checking the equivalence of two ideals was given by K. K. Billevich [49a], [62].

Cf. Arwin [29] for the case of cubic fields.

Class-numbers of subrings of R_K were considered by Dedekind [77]. Cf. Dade, Taussky [64], Dade, Taussky, Zassenhaus [61], [62].

The class-number can also be defined in terms of matrices. See Châtelet [11], Fueter [33], Hurwitz [95b]. There are several relations between matrix theory and the theory of algebraic numbers. See Albert [37b], Bender [68a], [68b], Bennett [23], Bhandari, Nanda [79], Châtelet [50], Chowla, Cowles, Cowles [80], Faddeev [74], Gorshkov [41], Latimer, McDuffee [33], Taussky [49], [51], [57], [62], [77a], [77b], [80], Taussky, Todd [40], Wagner [69]. Further references may be found in a survey paper by O. Taussky [60].

4. Theorem 3.4 in the quadratic case goes back to Lagrange [73]. In the cubic case it was established by G. Eisenstein [44c] (who used also the language of forms). For cyclotomic fields it was proved in Kummer [47b], [56] and the general case is due to R. Dedekind [71] and L. Kronecker [82]. The proofs of A. Hurwitz [95c] and H. Minkowski [91a] were presented in the main text. Other proofs were given in Artin, Whaples [45], Mahler [64a]. A proof in which $H(K)$ is shown to be at the same time a continuous image of a compact group and a discrete group can be found in Cassels, Fröhlich [67].

In Chapter 4 we shall give an effective method of constructing all ideals of norm not exceeding a given bound, and this will lead to a procedure for the determination of $H(K)$.

It is not known whether every finite Abelian group can serve as $H(K)$ or $H^*(K)$ for suitable K. As shown by S. Chowla [34b] (see Corollary to Proposition 8.8), the answer is negative if we restrict K to be imaginary quadratic. Chowla's result is not effective, but D. Shanks [69a] showed that $H(K) = C_p^2$ is impossible for such fields if $p = 5, 7$ or 11.

A. Fröhlich [62a] and H. Hasse [69d] showed that every finite Abelian group is a homomorphic image and also a subgroup of $H(K)$ for infinitely many K. Cf. Frey, Geyer [72]. This result was strengthened by G. Cornell [71], who proved that one can always find a cyclotomic field K with this property. J. Sonn [83] showed that in Fröhlich's result one can replace the word "subgroup" by "direct summand". This had been known previously for cyclic groups (Sonn [79]) and for powers of cyclic groups (Iimura [81a]).

O. Yahagi [78] established that any given finite Abelian p-group is the Sylow p-subgroup of a suitable $H(K)$. Cf. Gerth [75b], Iimura [79b], [81a].

5. In this subsection we shall indicate the connection of the theory of algebraic numbers with K-theory. (For an introduction to K-theory see Bass [68], Lam, Siu [75], Milnor [71].)

It follows from Theorems 1.13 and 1.14 that if R is a Dedekind domain, then $K_0(R) \simeq H(R) \oplus Z$. (See, e.g., Milnor [71], § 1.) In fact, those theorems show that $H(R)$ is isomorphic to the projective Grothendieck group $\Gamma(R)$, defined as the quotient group of the free group generated by equivalence classes of finitely generated projective R-modules by the subgroup generated by the equivalence classes of free R-modules. The class-groups appear also naturally in the study of $\Gamma(R)$ for other classes of rings. Cf. Heller, Reiner [65], Rim [59], Swan [63].

The group $K_1(R_K)$ coincides with $U(K)$ (see Milnor [71], § 16); however, there is no simple description of $K_2(R_K)$ or the higher K-groups either for K or for R_K. It was shown by D. Quillen [73] that all groups $K_m(R_K)$ are finitely generated and it follows from the results of A. Borel [72], [74] that for $m \geqslant 2$ the group $K_m(R_K)$ is finite iff either K is totally real and $m \equiv 3 \pmod 4$ or K is arbitrary and m is even.

The main point of interest about $K_2(R_K)$ lies in the Birch–Tate conjecture, which asserts that for totally real K we have

$$|K_2(R_K)| = w_2(K)|\zeta_K(-1)|,$$

where $\zeta_K(s)$ denotes the Dedekind zeta-function, which will be studied in Chapter 7, and $w_2(K)$ denotes the maximal integer N such that every non-unit element of $\text{Gal}(K(\zeta_N)/K)$ is of order two. For K/Q Abelian, the Birch–Tate conjecture is true up to 2-torsion, being a consequence of the Main Conjecture of the Iwasawa theory, proved recently by B. Mazur and A. Wiles [84]. (Cf. Coates [81].) For a certain class of real quadratic fields as well as for the maximal real subfields of the cyclotomic fields K_{2^n} and K_{3^n} the full Birch–Tate conjecture is true. See Hettling [85], Hurrelbrink [82a], Hurrelbrink, Kolster [86], Kolster [84], Urbanowicz [84]. Cf. also Gebhart [77], Hurrelbrink [82b].

For other results on K_2 for R_K and related rings see Browkin [82a], [82b], Browkin, Hurrelbrink [84], Browkin, Schinzel [82], Coates, Sinnott [74a], Dennis, Stein [72], [75], Dunwoody [76], G. Gras [82c], [86], Hoffmann, Hettling, Browkin [84], Hurrelbrink [83], v. d. Kallen [81], Moore [69], Tate [76].

An analogue of the Birch–Tate conjecture for higher K-groups was stated by S. Lichtenbaum [73]. See Harris, Segal [75], Hurrelbrink [82a], Soulé [79], [82] on this subject.

On higher K-groups for R_K see also Coates, Sinnott [74a], Dwyer, Friedlander [82].

A description of $K_2(K)$ for algebraic number fields K was obtained by H. Matsumoto [69]. (See Bass [69], [71], Birch [69a], Coates [72a], Milnor [71], Tate [70].) For other results on $K_2(K)$ see Brinkhuis [84a], Garland [71], Lenstra [76], Tate [77a]. For an introduction to the theory of K_2 the reader may also consult Jaulent [83].

6. Lemma 3.3 is due to H. Minkowski [91a]. (Cf. Minkowski [96a], [07].) If we denote by $C(r_1, r_2)$ the lower bound of numbers C with the property that for every field of signature $[r_1, r_2]$ we can find in every class of $H(K)$ an ideal with norm not exceeding $C|d(K)|^{1/2}$, then for sufficiently large $r_1 + 2r_2$ we have

$$C(r_1, r_2) \leq (50.7)^{-r_1/2}(19.9)^{-r_2},$$

as shown by R. Zimmert [80], who improved upon the previous results of C. A. Rogers [50] and H. P. Mulholland [60]. An improvement of Minkowski's bound for cubic fields can be found in Delaunay, Faddeev [40]. On the other hand, E. A. Anfert'eva and N. G. Chudakov [68], [70] proved that for imaginary quadratic fields K we have

$$\max_{X \in H(K)} \min_{I \in X} N(I) \geq B|d|^{1/2}(\log|d|\log\log^T|d|)^{-1}$$

with $d = d(K)$, and B, T are positive numbers, depending only on $h(K)$.

Generalizations of Lemma 3.3 to other base fields were given in Kuroda [43a], Lakein [69], Mordell [69c], Y. Môri [33], Nymann [67]. Lemma 3.4 is due to A. Hurwitz [95c], [19].

A bound for the smallest norm of an element in a residue class (mod I) was obtained by H. Davenport [52]. Cf. also Egami [80], Rieger [58d], Tatuzawa [73a].

7. Algebraic units appear first in Gauss [32] in the case $K = Q(i)$. The general notion was worked out side by side with that of an integer. The main result of the theory of units, Theorem 3.6, was proved by P. G. Dirichlet [46] for the units of $Z[a]$, with an integer a; however, his proofs can be extended to cover also the general case. In [41c] he treated the cubic case and earlier, in [40], he showed that for $r_1(K) \geq 1$ there are infinitely many units. (Cf. Bachmann [64].) In the special case of $K = Q(\zeta_p)$ with prime p Dirichlet's theorem was proved independently by Kronecker [45a]. (Cf. Kummer [51].) According to Minkowski the idea of the proof occurred to Dirichlet as he was listening to the Easter concert in the Sistine Chapel.

Other proofs can be found in Artin, Whaples [45], Hermite [50], Iwasawa [53a], Kronecker [83], [84] (cf. Molk [83]), Minkowski [96a], v.d. Waerden [28]. An elementary proof for pure cubic fields can be found in Christofferson [57].

The more general Theorem 3.5 was first published by C. Chevalley [40]. For another proof see Mahler [64a], and for generalizations see Bass [66], Roquette [57], Samuel [66]. W. May [70] obtained an analogue for infinite Abelian extensions of algebraic number fields.

Note that Theorem 3.5 can be deduced from Theorem 3.6 with the aid of the finiteness of the class-number. (Cf. Joly [70a].)

From the proof of Theorem 3.6 one can extract effective procedures for the determination of fundamental units. Such procedures were considered in Avanesov [79], Billevich [64], Nagell [32b], [32c], Pohst, Weiler, Zassenhaus [82], Pohst, Zassenhaus [77], [82], Rudman, Steiner [78], Steiner [77]. Cf. also Benson, Weber [73].

The last part of Proposition 3.4 is due to O. Ore [24]. Cf. Norris, Vélez [80].

The Corollary to Lemma 3.5 gives in the case of totally real K certain information about the action of $U(K)$ on R^n (with $n = [K:Q]$) defined by

$$\varepsilon[x_1, \ldots, x_n] = [F_1(\varepsilon)x_1, \ldots, F_n(\varepsilon)x_n]$$

for ε in $U(K)$. This induces an action of $U^+(K)$ on the cone in R^n consisting of elements with non-negative coordinates. The fundamental domain for this action was given by T. Shintani [76b] and his construction was made explicit by K. Nakamula [77] in the case of a cyclic cubic field.

In Shintani [81] a similar construction was carried out for fields of arbitrary signature.

Theorem 3.6 shows that $U(K)$ is finitely generated. H. Zassenhaus [72] proved this for the group of units of any commutative ring with a unit whose additive group is isomorphic to Z^n for a certain n. (He even showed that in this case the unit group is finitely presentable.) C. Ayoub [68] proved that a subring R of an algebraic number field K has a finitely generated group of units iff the quotient $R^+/(R \cap R_K)^+$ of additive groups has non-zero p-components for finitely many primes p.

T. Skolem [48] deduced from Theorem 3.6 that K^* is the direct product of $E(K)$ and a free Abelian group with denumerably many free generators. This was extended by E. Schenkman [64] to the following form: if K_n is the field generated by all algebraic numbers of degree $\leq n$, then K_n is a direct product of cyclic groups. This implies that every multiplicative group generated by a set of algebraic numbers of bounded degree is a product of cyclic groups. Cf. Charin [54], Grishin [73], Iwasawa [53c], May [72], Meyer [78].

To the problem of the structure of K^* the following result of A. Brandis [65] is related: if K is an infinite field and $K \neq L$, $K \subset L$, then the group L^*/K^* cannot be finitely generated. Cf. Gay, Vélez [81], Kataoka [79], May [79].

8. Units of quadratic fields.

It is clear that every result about the Pellian equation $X^2 - DY^2 = \pm 4$ can be regarded as a result concerning the units in real quadratic fields. This approach permits us to treat J. L. Lagrange [66] as the author of Theorem 3.6 in this case. For a survey of older results on the Pellian equation the reader should refer to **Chapter XII** of Dickson [19].

For certain quadratic fields one can give the fundamental unit explicitly. The first formulas of this type were obtained by C. Richaud [66] and rediscovered by G. Degert [58]. They showed that if D is square-free and $D = n^2 + r$ with $r|4n$, $-n < r \leq n$, then the number

$$\varepsilon = \begin{cases} n + D^{1/2} & \text{if } r = \pm 1, D \neq 5, \\ (n + D^{1/2})/2 & \text{if } r = \pm 4, \\ (2n^2 + r + 2nD^{1/2})/|r| & \text{if } r \neq \pm 1, \pm 4 \end{cases}$$

is the fundamental unit of $Q(D^{1/2})$. For similar results see Bernstein, Hasse [75], Kutsuna [74], Nakahara [70], Neubrand [81], Nordhoff [74], Yokoi [68a], [70a]. Such explicit formulas are often helpful in establishing $h(K) \neq 1$. Cf. Ankeny, Chowla, Hasse [65], Nordhoff [74].

Methods of finding the fundamental unit were given in Bernstein [76a], Denenberg [75], Hunter [56b], Kuroda [43b], Lehmer [26a], Nakahara [74], Pohst, Zassenhaus [79]. D. Shanks [72a] gave a method of obtaining a good approximation of it.

Tables of solutions of Pellian equation were published by A. M. Legendre [98], C. F. Degen [17], C. E. Bickmore [93], E. E. Whitford [12], covering jointly the range [2, 1700]. The list of errors in them was made up by D. H. Lehmer [26b]. The tables of W. Patz [41], giving continued fractions for $D^{1/2}$ ($D < 10^4$), may also be useful. Cf. also Cohn [62a], v. Thielmann [26], Williams, Broere [76], Williams, Buhr [79], Yokoi [68a].

Several authors have studied the sign of $N_{K/Q}(\varepsilon)$ for the fundamental unit ε, trying to express it in terms of various invariants of K, since its dependence on the parity of the period of the continued fraction given in Theorem 3.11 is not very useful. A quick algorithm for determining this sign was found by J. C. Lagarias [80a]. In a long series of papers L. Rédei [32a], [34c], [35], [38], [53a], [53b] used class-field theory to deduce a necessary and sufficient condition for $N_{K/Q}(\varepsilon) = -1$. (Cf. Morton [79].) For various sufficient or necessary conditions see Brown [72a], [83], Despujols [45], Epstein [34], Furuta [59a], Jensen [62a], [62c], v. Lienen [78], Nagell [55], Pall [69], Perott [88], Pumplün [68], Scholz [35], Tano [89], Trotter [69].

Let $L(D)$ be the length of the continued fraction of $D^{1/2}$. It was shown by T. Vijayaraghavan [27] that for all positive ε's we have $L(D) = O(D^{1/2+\varepsilon})$ and L. K. Hua [42] improved this bound to $O(D^{1/2}\log D)$. In Pen, Skubenko [69] and Podsypanin [79] this result was extended to all quadratic irrationalities, and in the last paper it was shown that the Extended Riemann Hypothesis implies the bound $O(D^{1/2}\log\log D)$. For numerical evidence in favour of this bound see H. C. Williams [81a]. The best lower bounds are due to J. C. Lagarias [80a], who obtained $L(D) \geq \frac{1}{3}D^{1/2}\log^{-1} D$ for infinitely many non-square free D's and to Y. Yamamoto [71], who, in the square-free case, obtained $L(D) \geq c\log^3 D$ infinitely often for a certain $c > 0$. Numerical investigations made by M. D. Hendy [75] lead to the conjectural bound

$$L(D) \geq C(\varepsilon)D^{1/2-\varepsilon}$$

for infinitely many square-free D and all positive ε's.

For other results concerning $L(D)$ and its generalization for arbitrary quadratic irrationalities see J. H. E. Cohn [77], Danilov, Danilov [75], Golubeva [84], Hendy [74a], Hickerson [73], Hirst [72], Skubenko [62], Stanton, Sudler, Williams [76], Stanton, Williams [76].

J. C. Lagarias [78] noted an interesting relation between signatures and congruence classes (mod 4) of units and, more generally, of integers a prime to 2. He showed that if $d(K)$ is a sum of 2 squares, then a (mod 4) determines the signature of a. J. S. Sunley [79] proved that the same holds for all totally real fields K with odd $h^*(K)$ and J. C. Lagarias [80b] proved that a totally real field K has this property iff the 2-ranks of $H(K)$ and $H^*(K)$ coincide. Cf. Haggenmüller [82].

Quadratic, cubic and higher characters of units were investigated in Brandler [73], Buell, Leonard, Williams [81], Evans [84], Furuta, Kaplan [81], E. Lehmer [71], [72], [73], [74], Leonard, Williams [77], [80], Parry [76], Scholz [35], Weinberger [72b], K. S. Williams [80].

N. C. Ankeny, E. Artin, S. Chowla [52] conjectured that if $p \equiv 1$ (mod 4) is a prime and $(T+Up^{1/2})/2$ is the fundamental unit of $Q(p^{1/2})$ then $p \nmid U$.

It was shown by L. J. Mordell [60a] in the case of $p \equiv 5 \pmod 8$ and by N. C. Ankeny and S. Chowla [62] for the remaining primes $\equiv 1 \pmod 4$ that $p|U$ iff the $(p-1)/2$-th Bernoulli number is divisible by p. (We recall that the nth Bernoulli number B_n is defined by

$$t(e^t-1)^{-1} = \sum_{n=0}^{\infty} B_n t^n/n! ; \qquad (3.6)$$

thus $B_3 = B_5 = \ldots = B_{2k+1} = \ldots = 0$.) So far no such p has been found and the conjecture is known to be true for all $p \leq 6\,270\,713$. (Beach, Williams, Zarnke [71].) For a generalization of this conjecture to other primes and to composite integers see Beach, Williams, Zarnke [71], H. Lang [73b], Mordell [60a], Slavutsky [65a], Taussky, Todd [60].

For other results concerning units of quadratic fields see Carlitz [55], Chowla [61a], [61b], [65a], Hafner [68], Iyanaga [35], Jensen [62b], Morikawa [72], [79], Motoda [77], Nagell [54], Rédei, Reichardt [34], Sprindzhuk [74a], Stender [79a], [79b], Stolt [52], Teege [21], Thérond [77].

For papers dealing with congruences relating the fundamental unit to the class-number see Chapter 8, § 4.

9. *Units of cubic fields.*

A theory of units in cubic fields was outlined by C. Hermite in a letter to Jacobi (Hermite [50]). His ideas were used for the actual construction of units by E. Zolotarev [69] in the case of pure cubic fields. The general case was treated by G. F. Voronoi [96]. Cf. Brentjes [81], Delaunay, Faddeev [40], Uspensky [31], H. C. Williams [80], [81b] and Williams, Cormack, Seah [80].

Other methods and algorithms and also explicit formulas for various classes of cubic fields may be found in Appelgate, Onishi [82], Arwin [29], Avanesov, Billevich [81], Bergmann [66b], [66c], Bernstein [76b], Berwick [13], [32], [34], Billevich [56], Brunotte, Halter-Koch [79], [81a], Bullig [36], [38], Cohn [52b], Cusick [82], [84b], Endô [78], Godwin [60], [84], Güting [77], Hasse [48a], Ishida [73], Jeans, Hendy [78], Minkowski [96b], Morikawa [74], Nagell [26], Nakamula [81], Pierce [26], Rudman [73], Steiner, Rudman [76], Stender [69], [72], [75], [77], Vel'min [51], Watabe [83], H. C. Williams [76], Yokoi [74]. Cf. also Thomas [79], where units of subrings of R_K with K cubic were constructed.

Tables of fundamental units for cubic fields were given (or described) in Angell [73], [76], Beach, Williams, Zarnke [71], Billevich [56], Brentjes [81], Cassels [50], Cohn, Gorn [57], Delaunay [28], M. N. Gras [75] (cf. Godwin [83]), Reid [01], Selmer [55], Sved [70], Wada [70c].

10. Various methods of determining fundamental units in quartic fields were given in H. Amara [81], Berwick [32], Frei [81a], [82], Hasse [48a], Lakein [71], Levesque [81], [82], Nagell [62a], Nakamula [81], Podsypanin [49], R. Scharlau [80a], Stein [27], Stender [73], [75], [77], [78], [83], Wada [66]. Cf. also Kubota [56b], Kuroda [43a]. Tables are given in Billevich [56], Cohn [71], M. N. Gras [79], Pohst [75b]. Units of sextic fields were dealt with in Bergmann [65], Frei [81b], Hasse [48b], Iimura [80], Mäki [80] (where also a table was given), Nakamula [79], [81], Setzer [78], Stender [74], [75], [77], [78]. For octic fields see Cohn [71] (with a table), Levesque [82], Stender [83]. Fundamental units in cyclotomic fields were found by P. Dénes [53b] and I. Yamaguchi [77] and in arbitrary Abelian fields a method of H. W. Leopoldt [56] is applicable. Cf. Gras and Gras [79].

For the construction of units in some classes of fields one can use the Jacobi–Perron algorithm which goes back to Jacobi [68] and Perron [07]. On this subject see Bernstein [67], [70], [71], [72], [74a], [74b], [75a], Bernstein, Hasse [65], [69], David [49], [50], [51], [56], Dubois, Paysant-le-Roux [71], [75], Elsner, Hasse [67], Levesque [79], Raju [76].

For other classes of fields see Bernstein [75b], [75c], [77], [78a], [78b], Cohn [76], Frei, Levesque [79], [80], Greiter [80], Halter-Koch [75], [82], Halter-Koch, Stender [74], Neubrand [78], Oozeki [78], [79].

11. A totally complex field which is a quadratic extension of a totally real field is called a *CM-field* or, sometimes, a *J-field*. Proposition 3.5 characterizes such fields in terms of the unit group. It is due to R. Remak [52], who studied fields with the defect of the units. Cf. Dénes [51], Remak [54].

Several characterizations of *CM*-fields were given in Györy [75]. Cf. also Blanksby, Loxton [78], McCluer, Parry [75], Parry [75a]. *CM*-fields of a particular form were considered in Ennola [78].

Theorem 3.7 was stated in Kummer [50a] and proved in [51] for $K = Q(\zeta_p)$. (For another proof in this case see McCluer, Parry [75].) The general case is due to C. G. Latimer [34]. Cf. Lednev [39], Weiss [36]. Proposition 3.6 is due to P. Dénes [51] and the proof to McCluer, Parry [75]. Lemma 3.6 for $K = Q(\zeta_p)$ appears already in Kummer [51].

For the Corollary 4 to Theorem 3.8 see Kaplan [74]. A sufficient condition for the existence of units of all signatures was given by O. Neumann [77b]. By the Corollaries 2 and 3 to Theorem 3.8 this occurs for a totally real field iff $U^+(K) = U^2(K)$. It was proved by D. A. Garbanati [76] that the above condition holds for the maximal real subfield of the *m*th cyclotomic field (where $m \not\equiv 2 \pmod 4$) iff m is a prime power. For $m = p$ a prime, this result is due to Kummer [70] and for $m = 2^k$ to H. Weber [96b]. H. Hasse [52a] showed that $U^+(K) = U^2(K)$ holds for all real Abelian extensions of K of degree 2^k provided $|d(K)|$ is a prime power. If K is of a prime-power degree p^k with odd p, the order of 2 (mod p) is even, and $2 \nmid h(K)$, then we have also $U^+(K) = U^2(K)$ as shown by J. V. Armitage and A. Fröhlich [67]. Cf. Hughes, Mollin [83].

12. Theorem 3.9 was proved by H. Minkowski [00] and later extended by J. Herbrand [30b], [31b] to relative extensions. A simplified proof of Herbrand's result was given by E. Artin [32a] and a further generalization can be found in Lednev [39]. Theorem 3.10 is due to A. Zeinalov [65] (for $p = 3$ see Brunotte, Halter-Koch [79], Delaunay, Faddeev [40], Godwin [60], M. N. Gras [80], Halter-Koch [82], Hasse [48a]) and A. Brumer [69] proved that the condition $h(Q(\zeta_p)) = 1$ can be replaced by the assumption that all ideals of norm $h(K)$ in $Q(\zeta_p)$ are principal. For a generalization see Gillard [80a].

N. Moser [83] proved that every imaginary extension K/Q of degree $2p$ (p prime) and dihedral Galois group has a Minkowski unit. We shall prove her result in Chapter 7. (See Theorem 7.9.) For other classes of fields no such complete results are known. It was shown in N. Moser [79a] that if K is real then it has a Minkowski unit iff $U(K)/E(K)$ is isomorphic with $Z[G]/NZ[G]$, where $N = \sum_{g \in G} g$, as a $Z[G]$-module. Cf. Latimer [34], Payan [81], Weiss [36].

A description of possible actions of $G = \mathrm{Gal}(K/Q)$ on $U(K)$ was given by B. Setzer [80b] for K real with $G = C_2^2$, who also treated in [78] the case of $G = S_3$, K complex. (Cf. Hasse [30b], Ishida [73].) Further results concerning the action of G on $U(K)$ are contained in Hasse [48b] for $G = C_4$, Bouvier, Payan [75] for $G = C_{p^n}$, Bouvier, Payan [79], Duval [81], Pollaczek [29] for $G = C_p^2$, and in Jaulent [79], N. Moser [78], [79a], [79b], Nakamula [82a] for certain classes of non-Abelian G's. Most of these papers give conditions for the existence of a Minkowski unit.

For a new approach to the study of unit groups under the action of G see Chinburg [83b].

The cohomology groups $H^i(G, U(L))$ with $G = \mathrm{Gal}(L/K)$ are related to ambiguous ideal classes, i.e., classes of $H(L)$ which are invariant under G. (Iwasawa [56], cf. Matsumura [72], McQuillan [73], Rosen [66].)

The action of $\mathrm{Gal}(L/K)$ on the group $U(L/K)$ of relative units, introduced in Leopoldt [56] and defined as the set of all $u \in U(L)$ for which $N_{L/M}(u) \in E(M)$ holds for all $K \subset M \subset L$, was considered by H. Brunotte [78].

For the existence of a conjugated system of generators in $U^+(K)$ see Hasse [48a], Morikawa [68].

13. Theorem 3.8 shows that if K is real, then it has totally positive fundamental units iff $h^*(K) = 2^{r_1-1}h(K)$. See Armitage, Fröhlich [67], Taylor [75].

The results of W. E. H. Berwick [32] imply that every totally real cubic field has a fundamental system of units which are PV-numbers. E. D. Zlebov [66] showed that the same holds for all real fields. Cf. Brunotte, Halter-Koch [81b].

14. For a given field K one can consider the subgroup $U'(K)$ of $U(K)$ generated by units of proper subfields of K, or, more generally, the subgroup $U'_S(K)$ generated by those units of $U_S(K)$ which already lie in a proper subfield of K. J. H. Smith [70a] considered the case of $S_\infty \subset S$, K/Q normal and $U_S(K)$ invariant under $\text{Gal}(K/Q)$ and characterized the situation when $U'_S(K)$ is of a finite index in $U_S(K)$. In the case of $S = S_\infty$ (and K not necessarily normal) this index was computed or evaluated in Barrucand, Cohn [71], Berwick [34], Brunotte, Klingen, Steurich [77], Halter-Koch [76], [78a], Ishida [73], Kubota [56b], Kuroda [43a], Liang [72], N. Moser [79b], Nakamula [80], [82a], Parry [77a], [81], Scholz [33], Setzer [78]. (Certain authors considered instead the index of $U'(K)E(K)$ in $U(K)$.)

15. *Cyclotomic units.*

Let p be an odd prime and $K_p = Q(\zeta_p)$. It was shown by E. E. Kummer [50a], [50b], [51] (cf. Hilbert [97], Satz 142, S. Lang [78], Washington [82]) that the numbers

$$\left\{ \frac{(1-\zeta^g)(1-\zeta^{-g})}{(1-\zeta)(1-\zeta^{-1})} \right\}^{1/2} = \zeta_1 \frac{1-\zeta^g}{1-\zeta}$$

(where g is a fixed primitive root (mod p), ζ runs over all primitive pth roots of unity and $\zeta_1 \in E(K_p)$) are units and generate a subgroup $C(K_p)$ of $U(K_p)$ of index equal to $h(K_p^+)$ (K^+ denoting the maximal real subfield of a field K). Thus they form a fundamental system of units iff $h(K_p^+) = 1$. The last equality occurs, e.g., for all $p \leq 73$ (v.d. Linden [82a]) but it was already known to Kummer ([70]) that for $p = 163, 229, 257$ and 937 it fails. Since that time many other such examples have been found, and it is known that the ERH implies $h(K_p^+) = 1$ for all $p < 163$ (Ankeny, Chowla, Hasse [65], v.d. Linden [82a]), however, the computation of $h(K_p^+)$ still presents difficulties, even for small primes, and the same applies to the more general question of computing $h(K_m^+)$ with $K_m = Q(\zeta_m)$. (Cf. Cohn [60a], [61a] for the case of $m = 2^n$.)

The elements of $C(K_p)$ are called the *cyclotomic* or *circular units of* K_p. The reader should consult Washington [82] (Ch. 8) for their main properties and S. Lang [82] for an extensive survey of their theory and applications, including the generalizations which we describe below.

The analogues of cyclotomic units were defined also in other Abelian fields, and this led to an exact analogue of Kummer's index formula for cyclotomic fields K_{p^n} with prime p and to suitably adjusted formulas in the general case. There are in fact two definitions of cyclotomic units, which agree for K_{p^n}. According to the first, the group of cyclotomic units of an Abelian field K equals the intersection of $U(K)$ with the group generated by $E(K_f)$ and $1 - \zeta_f^a$ ($a = 2, 3, \ldots, f-1$), where K_f is the minimal cyclotomic field containing K. (Such a field always exists by the Kronecker-Weber theorem, which we shall prove in Chapter 6, Theorem 6.5.) This definition is commonly used for cyclotomic fields. For other Abelian fields one defines the group of cyclotomic units as the intersection of $U(K)$ with the group generated by -1 and all numbers of the form $N_{K_n/L_n}(1 - \zeta_n^a)$ with $L_n = K \cap K_n$ and $n = 2, 3, \ldots; n \nmid a$.

Let $C_0(K)$ and $C(K)$ be the groups so defined. W. Sinnott [78] computed $[U(K):C_0(K)]$ for cyclotomic K and in [80] obtained a formula for $[U(K):C(K)]$ in the general case.

In Hasse [52a] still another definition was used, leading to a finite index only for K_{p^n}, in which case the resulting group coincided with $C(K_{p^n})$. Its rank was computed by K. Feng [82a].

If p and $q = (p-1)/2$ are both primes and either $p \leqslant 4703$ or 2 is a primitive root (mod q), then every square in $C(K_p^+)$ is totally positive, as proved by D. Davis [78], who also conjectured that the bound on p may be removed. This question is related to the parity of $h(K_p^+)$. Cf. Armitage, Fröhlich [67], Garbanati [75a], G. Gras [75], Gras, Gras [75], Hasse [52a].

For the structure of the p-Sylow subgroup of $U(K)/C(K)$ in the case of real K and $p \nmid [K:Q]$ see Gillard [77], G. Gras [77a], [77b], Greenberg [75], Mazur, Wiles [84].

For other results concerning cyclotomic units see Ennola [71], [72], Ishida [82], Milgram [81], Montouchet [71], Ramachandra [66], C. G. Schmidt [78], [80a], Vandiver [25], [29a], [30a], [30b], [34a] (who extended a result of Kummer [52] concerning the character values of certain cyclotomic units).

Analogues of cyclotomic units in Abelian extensions of imaginary quadratic fields are provided by elliptic units, defined with the use of singular values of modular functions. They were introduced by G. Robert [73a], [73b], but in special cases appeared already in Fueter [10], Siegel [61] and Novikov [67]. For their properties and applications see Coates, Wiles [78], Gillard [79a], [79b], [79c], [80a], [80b], Gillard, Robert [79], Kersey [80], Kubert, Lang [79], Nakamula [82b], Robert [78], [79]. Cf. also D. S. Kubert and S. Lang [81].

16. If p is an odd prime not dividing $h(K_p)$ (such primes are called regular), then every unit of K_p congruent (mod p) to a rational integer is the pth power of some unit in K_p. This was proved by E. E. Kummer [50b]. (Cf. Hilbert [97] (Satz 156), Leahey [66], Mirimanoff [91], Vandiver [34c].) A similar result was proved for arbitrary fields K by F. K. Schmidt [30]: for any positive integer n there exists an integer $m = m(n, K)$ such that every unit of K congruent to unity (mod m) is an nth power in K. An explicit determination of $m(n, K)$ for a prime power n was made in Jehne [61]. For generalizations see Bass [65], Chevalley [51], J. H. Smith [70b].

It was shown by A. Thue that the linear form $ax+y$ with an algebraic a is a unit for infinitely many $x, y \in Z$ if a lies in a real quadratic field and is irrational. A proof may be found in Chabauty [38]. If a, b are algebraic integers generating the unit ideal in R_K (with $K = Q(a, b)$), then $ax+b$ is a unit for infinitely many algebraic integers x.

A similar result holds also for polynomials in several variables over Z. (Skolem [35]. Cf. Cantor, Roquette [84], Dade [63], Lagarias, Lenstra [81], Steinitz [12], Watson [63].)

Let W be the class of fields in which every integer is a sum of distinct units. B. Jacobson [64] proved that $Q(2^{1/2})$ and $Q(5^{1/2})$ lie in W and J. Śliwa [74] showed that W does not contain other quadratic fields and that no pure cubic field belongs to W. P. Belcher [74], [75] found infinitely many cubic and quartic fields in W.

It was proved by J. H. Evertse [84] that the equation $ax+by = 1$ (with $a, b \in K$, $ab \neq 0$) has at most $3 \cdot 7^{N+2s}$ solutions in S-units $x, y \in K$, with $N = [K:Q]$ and $s = |S|$. Cf. Györy [79b]. The particular case of $a = b = 1$, $S = S_\infty$ had been considered earlier in Chowla [61c] and Nagell [59], [60], [64b], [68b], [69a], [69b].

Units lying in a hyperplane were considered by C. Sirovich [69].

For other equations involving units see Chabauty [36], [37], Ennola [73b], [75b], [75c], Grossman [76], [77], Jacobsthal [13], Loxton [74b], Mahler [50], Mordell [63], Nagell [28] (cf. Delaunay [30b]), Newman [71], [74a], Rados [90], Siegel [21a], Watabe [82].

R. M. Robinson [65] proposed several problems concerning roots of unity. See Cassels [69b], Davenport, Schinzel [67], Jones [68], Loxton [72], [74c], Newman [71], Ojala [75], Schinzel [66b].

Units in arithmetic progressions were considered by M. Newman [74b]

Only finitely many units of given degree can have the same discriminant, as shown in Györy [73]. For cubic units this was proved by V. A. Tartakovskii (see Delaunay, Faddeev [40]) and for quartic units by T. Nagell [68c].

An interval I on the real axis is called *critical* if every larger interval contains infinitely many full systems of conjugated units but no proper subinterval of I

has this property. R. M. Robinson [62] showed that every interval $[c-2, c+2]$ with $|c| \leq 2$ is critical and in [64c] he proved that the length of a critical interval is ≥ 4. (Cf. Madan, Pal [77], Robinson [77].) For an analogous problem for closed subsets of the complex plane see Cantor [80], Robinson [67].

Hilbert's Theorem 90 (Hilbert [97]) asserts that if K/Q is cyclic and s is a generator of $\text{Gal}(K/Q)$, then every element of norm 1 in K is of the form $s(a)/a$. Such elements for other extensions were studied in Pitti [70], [71], [72], [73a], [73b]. Cf. Connell [65].

For miscellaneous questions involving units see Benzaghou [68], [69], Dénes [53a], [53b], [55], Grossman [74b], Moriya [33], Newman [82], Shapiro, Sparer [73], Washington [79b], Yokoi [60a].

17. *Regulator evaluations.*

The first general upper bound for $R(K)$ was obtained by E. Landau [18d], who proved

$$R(K) \leq O(D^{1/2}\log^{n-1}D) \quad \text{where} \quad D = |d(K)|.$$

(See Corollary 4 to Theorem 7.1.) He obtained also an upper bound for $\overline{|u|}$, $u \in U(K)$. Cf. Landau [18a], Remak [31], Siegel [69a]. Bounds for the product $h(K)R(K)$ will be considered in Chapter 8.

Lower bounds for $R(K)$ were first established by R. Remak [31], [32], [52], [54], who showed, among other things, that if K is totally real, then $R(K) \geq 10^{-3}$. A substantial improvement was made here by M. Pohst [77b], [78], who proved $R(K) \geq 0.315$ and for sufficiently large $N = [K:Q]$ obtained $R(K) \geq \exp(4N/5)$, for a totally real K. For cubic and quartic fields see Cusick [84a]. Upper bounds for the number of fields of a given degree satisfying $R(K) \leq x$ were obtained by V. G. Sprindzhuk [74b].

For real quadratic fields the regulator equals $\log \varepsilon$, where $\varepsilon > 1$ is the fundamental unit. Its best evaluation is due to A. F. Lavrik [70a], who proved that for sufficiently large $d = d(K)$ we have

$$h(K)\log \varepsilon \leq (.263+c)d^{1/2}\log d$$

for every positive c. Cf. Chowla [64], J. H. E. Cohn [77], Hua [42], Perron [14], Remak [13], Schmitz [16], Schur [18b], Stephens [72], Takaku [71], Wang Yuan [64], Yokoi [70b], and the papers quoted in subsection 8 on $L(D)$.

The lower bound $\log \varepsilon \geq B\log^3 d$, valid for infinitely many d with a suitable positive B, was established in Yamamoto [71]. Cf. Hooley [84].

18. The Euclidean Algorithm in real quadratic fields was first considered by L. E. Dickson [23] who showed that fields K with $d(K) = 5, 8, 12$ and 13 are Euclidean and asserted that there are no other such fields. This was shown to be false by O. Perron [32], who proved Theorem 3.12 and showed that $Q(44^{1/2})$ is also Euclidean. Later more Euclidean real quadratic fields were found, namely those with discriminants $d = 33, 37, 41, 57, 73$ and 76 (Berg [35], Hofreiter [35c], Oppenheim [34], Rédei [41a], Remak [34]). A uniform proof that all those fields are Euclidean was given by P. Varnavides [52]. Cf. Barnes [50], Cassels [48], Ennola [58a].

The finiteness of the set of Euclidean real quadratic fields was established by H. Heilbronn [38a] and for $Q(p^{1/2})$ with prime p by P. Erdös and Chao Ko [38]. The final step in showing that all such fields are among those listed above was made by H. Chatland and H. Davenport [50]. (Cf. Davenport [51].) The intermediate work was done in Barnes, Swinnerton-Dyer [52], Behrbohm, Rédei [36], Berg [35], A. Brauer [40], Chatland [49], Cugiani [48], [50], Hofreiter [35a], [35c], Hua [44], Hua, Min [44], Hua, Shih [45], Inkeri [47], [48], Min [47], Oppenheim [34], Rédei [41a], [41b], Schuster [38].

H. Davenport [49], [50a] showed that there are only finitely many Euclidean cubic fields with $d(K) < 0$ and they all satisfy $|d(K)| \leq 64 \cdot 10^{26}$. He also proved, in [50b], the finiteness of the set of totally complex quartic Euclidean fields. The bound given in that paper was improved by J. W. S. Cassels [52] to $d \leq 24\,846\,000$.

V. G. Cioffari [79] showed that the only pure cubic Euclidean fields are $Q(D^{1/3})$ with $D = 2, 3$ and 10. For the pure quartic case see Cioffari [79], Egami [79].

H. Heilbronn [50] showed that there are only finitely many cyclic cubic Euclidean fields and later showed ([51]) that there are only finitely many cyclic Euclidean fields of a fixed degree and prime-power discriminant. This seems to be the only general result concerning Euclidean fields of an arbitrary degree.

To find Euclidean fields one usually applies geometrical methods. H. W. Lenstra [77a] used them to formulate arithmetical criteria for euclidicity. Define $L(K)$ (the Lenstra constant of K) as the largest integer M such that there exist M distinct integers in K whose all non-zero differences are all units. Lenstra showed that if $L(K)$ is sufficiently large (e.g., if it exceeds the Minkowski constant of K) then K is Euclidean. Using this result he was able to produce several new Euclidean fields (Lenstra [77a], [78]) and the same approach was utilized in Leutbecher [85], Leutbecher, Martinet [82a], [82b], Martinet [79a] and Mestre [81] to enlarge considerably the list of known Euclidean fields. According to Leutbecher [84] 576 such fields are now known, all of degrees

≤ 10. For 42 of them the degree equals 10, e.g. for $Q(\zeta_{11})$ which was shown to be Euclidean in Lenstra [75]. Other papers searching for Euclidean fields are: Clarke [51], Davenport [39], [47], Godwin [65a], [67], J. R. Smith [69], [71], Swinnerton-Dyer [71], Taylor [76] in the cubic case, Godwin [65b], Lakein [72], v.d. Linden [84] in the quartic case and Godwin [65b] in the quintic case. Cyclotomic Euclidean fields were considered in Lenstra [75], [78], Masley [75a], Ojala [77] and Uspensky [06], [09].

The number of steps in the Euclidean algorithm in the fields listed in the Corollary to Proposition 3.11 was dealt with by N. Baldisseri [75]. Replacing in Proposition 3.11 the right-hand side of the inequality by $N(I)$ and adding the condition $b \in I$, where I is an ideal of R_K, we arrive at the notion of a Euclidean ideal, introduced by H. W. Lenstra [79]. He showed that this property depends only on the class of I in $H(K)$ and showed that such a class, if it exists, must be unique and generate $H(K)$, and so $H(K)$ must be cyclic.

Let S be a finite set of non-Archimedean valuations of $A = R_K$ and denote by A_S the intersection of all valuation rings of valuations outside $S \cup S_\infty$. O. T. O'Meara [65] and C. S. Queen [73a] proved that for suitable S the ring A_S is Euclidean. (Cf. also Markanda [75].) The same assertion holds also for function fields in one variable over a finite field of constants. (For this case cf. also Madan, Queen [73], Queen [73b], [74].) F. J. van der Linden [82b] determined all Euclidean rings of the form A_S with $|S| = 2$.

Under ERH every algebraic number field K with $h(K) = 1$ which is not complex quadratic is Euclidean under a suitable norm, as proved in Weinberger [72a]. (See Lenstra [77b], Samuel [71] for an analogue in the case of A_S.) This does not hold for imaginary quadratic fields, as shown by the example $K = Q(i19^{1/2})$. (Dubois, Steger [58], Lemmlein [54], Motzkin [49], K. S. Williams [75].)

Theorem 3.13 is due to H. Hasse [28] and the Corollary to it was rediscovered in Kutsuna [80]. For similar results see Bougaut [80], Cooke [76], Cooke, Weinberger [75], Dedekind [31], Kantz [55a], [55b], Krull [31a], Leutbecher [77], Rabinowitsch [13].

One can also consider a modification of the definition of a Euclidean domain, allowing the norm p to assume values in an arbitrary well-ordered group. This modification was made by P. Samuel [71], and M. Nagata [78] gave an example showing that it extends the class of the domains concerned.

A substitute for the Euclidean algorithm in algebraic number fields was considered in McDuffee [31b] and McDuffee, Jenkins [35].

For other results on euclidicity cf. Arpaia [68], Pollák [59], [62], Rodosskii [80], Tena Ayuso [77].

Exercises to Chapter 3

1. Prove that a non-real field K is a *CM*-field (i.e., it is totally complex and is a quadratic extension of a totally real field) iff K is closed under complex conjugation s, and s commutes with all embeddings of K in $\mathfrak{Z}$.

2. Call two ideals I, J of R_K strongly equivalent, provided there exist non-zero a, b in R_K such that $aI = bJ$ and $N_{K/Q}(a/b)$ is positive. Prove that the equivalence classes form a group under multiplication and find a relation between its cardinality and $h(K)$.

3. (Skolem [48].) Prove that the multiplicative group of an algebraic number field K is the direct product of $E(K)$ and a free abelian group of denumerably many free generators.

4. Determine the fundamental units of $Q(17^{1/2})$ and $Q(19^{1/2})$.

5. (Yokoi [68a].) Prove that a real quadratic field K has its fundamental unit of negative norm iff $K = Q(a^{1/2})$ with $a = t^2+4$ for a certain $t \in Z$.

6. (Kronecker [57b].) Prove that every unit of a cyclotomic field can be written as a product of a root of unity and a real unit, and this unit may lie outside K.

7. Let K/Q be complex and normal with Galois group G. Prove that if $g \neq e$ is an element of G which acts trivially on $U(K)/E(K)$, then g equals the complex conjugation.

8. Under the assumptions of the preceding exercise show that if the complex conjugation acts trivially on $U(K)/E(K)$ then it lies in the centre of G.

9. Still under the same assumptions show that the complex conjugation acts trivially on $U(K)/E(K)$ iff the maximal real subfield of K is normal over Q.

10. (Garbanati [76].) Let $Q \subset K \subset L$ and assume that L/Q is Abelian and real. Prove that if $U^+(K) = U^2(K)$ then $U^+(L) = U^2(L)$.

11. (Garbanati [76].) Let $Q \subset K \subset L$, let L/Q be Abelian and assume that $\mathrm{Gal}(K/Q)$ is isomorphic with the 2-Sylow subgroup of $\mathrm{Gal}(L/Q)$. Prove that

$$N_{L/Q}(U(L)) = N_{K/Q}(U(K)).$$

12. A unit u is called *exceptional* if $1-u$ is also a unit. Prove that in a real quadratic field either there are no exceptional units or there are exactly six of them.

13. Prove that the Lenstra constant of a field of degree n (as defined in subsection 18 of § 5) cannot exceed 2^n.

Chapter 4. Extensions

§ 1. The Homomorphisms of Injection and Norm

1. This chapter is devoted to the connections between arithmetic in an algebraic number field K and a finite extension L of K. Such an extension L/K we shall call traditionally a *relative extension* and in the case of $K = Q$ we shall speak about *absolute extension*. The same applies to other notions which will arise in the sequel, and so we shall speak about, say, a *relative discriminant* of an extension, whereas by the *absolute discriminant* we shall mean the discriminant $d(K)$ defined in Chapter 2.

The fundamentals of the theory can be proved in a more general setting, and at least at the start we shall work with an arbitrary Dedekind domain R with the field of quotients K. Let L be a finite separable extension of K and S let be the integral closure of R in L. By Theorem 1.9 the ring S is also a Dedekind domain and, moreover, if R has the (FN) property, then S shares this property. We shall denote by $G(K)$ and $G(L)$ the groups of fractional ideals of R and S, respectively, and by $I(K)$, $I(L)$ the semigroups of integral ideals, i.e., of ideals in the usual sense of R and S, respectively. Our aim is to describe the relations between $G(K)$ and $G(L)$ and, in particular, we shall deal with two maps defined in a canonical way—the injection map acting from $G(K)$ in $G(L)$ and the norm map acting in the opposite direction.

We start with the injection map $i_{L/K}: G(K) \to G(L)$ defined by

$$i_{L/K}(A) = AS$$

for $A \in G(K)$.

One can immediately see that $i_{L/K}(A)$ is the smallest S-module containing A. It is also clear that this map preserves inclusion and maps principal ideals into principal ideals. Moreover, the equality $ABS = (AS)(BS)$ shows that it is a homomorphism. Its main properties are given in

Proposition 4.1. *The injection map $i_{L/K}$ is a monomorphism which maps $I(K)$ into $I(L)$.*

Proof. Note first that the intersection of S and K equals R. In fact, the inclusion $R \subset S \cap K$ is trivial and, since every element of that intersection is integral over R and lies in K and thus belongs to R, we obtain the opposite inclusion $S \cap K \subset R$. Now if $i_{L/K}(A) = S$, then $SA = S$, whence $R = S \cap K = SA \cap K \supset A$ and so A must lie in $I(K)$. But $i_{L/K}(A^{-1}) = S^{-1} = S$, and we may repeat this argument to obtain $A^{-1} \in I(K)$. This can happen only if A is the unit element, i.e., $A = R$ and the kernel of $i_{L/K}$ is trivial.

If $A \in I(K)$, then $A \subset R$; thus $i_{L/K}(A) \subset i_{L/K}(R) = S$, whence $i_{L/K}(A) \in I(L)$. □

This proposition allows us to treat every ideal of R as an ideal of S if we identify the ideals A and $i_{L/K}(A)$. Actually, this idea can be extended so as to make it possible to treat ideals belonging to integral closures of R in various finite extensions of K "in abstracto" without referring to a particular extension. Observe first that if M/L and L/K are both finite and separable extensions, then $i_{M/K} = i_{M/L} \circ i_{L/K}$. Now consider the system of groups $\{G(L)\}$, where L runs over all finite separable extensions of a fixed field K which is the field of quotients of a Dedekind domain R, and the group $G(L)$ is the group of fractional ideals attached to the integral closure of R in L. This system can be partially ordered by $G(L) \prec G(M)$ holding iff $L \subset M$, all fields concerned being subfields of a fixed separable algebraic closure of K. Since the mappings $i_{M/L}$ are compatible with this ordering, we may consider the direct limit G of the system $(G(L), i_{M/L})$. The elements of G are fractional ideals of all fields L which are finite and separable extensions of K, two such ideals, say $A \in G(L)$ and $B \in G(M)$, being regarded as equal iff there is a common extension N of the fields L and M such that $i_{N/L}(A) = i_{N/M}(B)$.

In the same way one can consider also ideals of integral closures of R (i.e., integral ideals) and principal ideals in various fields without referring to the field in which they actually lie.

The most important case, at least for us, is the one arising when we put $K = Q$ and $R = Z$ and consider all finite extensions of the rationals contained in the complex field.

Let us return to the injection map. If $\mathfrak{p}$ is a prime ideal in R, then there is no need for the ideal $i_{L/K}(\mathfrak{p})$ to be prime in S; however, we can always write

$$i_{L/K}(\mathfrak{p}) = \mathfrak{P}_1^{e_1} \ldots \mathfrak{P}_s^{e_s},$$

where the $\mathfrak{P}_i$'s are prime ideals of S and the exponents e_i are positive rational integers. We shall say that the ideals $\mathfrak{P}_1, \ldots, \mathfrak{P}_s$ lie *above* $\mathfrak{p}$ (and $\mathfrak{p}$ lies *below* any of them) and the exponent e_i will be called the *ramification index* of $\mathfrak{P}_i$ over K and denoted by $e_{L/K}(\mathfrak{P}_i)$. The ideal $\mathfrak{P}_i$ is said to be *ramified* in L/K provided $e_{L/K}(\mathfrak{P}_i) > 1$ and *unramified* otherwise. If $e_{L/K}(\mathfrak{P}_i)$ is divisible by the characteristic of the field $S/\mathfrak{P}_i$, then $\mathfrak{P}_i$ is called *wildly ramified in L/K* and ramified prime ideals which are not wildly ramified are called *tamely ramified*. Finally, if $\mathfrak{P}_i$ is either unramified or tamely ramified, then it is said to be *at most tamely ramified*.

One says also that a prime ideal $\mathfrak{p}$ of R is *ramified in L/K* if at least one of the prime ideals of S lying above $\mathfrak{p}$ is ramified. Similarly, $\mathfrak{p}$ is called, respectively *unramified* and *tamely ramified* if all prime ideals of S above $\mathfrak{p}$ are such.

An extension L/K is called, respectively *unramified* and *tamely ramified* (or simply *tame*) if all prime ideals of S are unramified resp. at most tamely ramified. Note that an unramified extension is tamely ramified by definition.

If L is an algebraic number field and $K \subset L$, then the extension L/K is called *unramified at infinity* if no real embedding of K into $\mathfrak{Z}$ can be prolonged to a complex embedding. In particular, every extension of a totally complex field is unramified at infinity.

Proposition 4.2. *If $\mathfrak{P}$ is a prime ideal of S, then there is exactly one prime ideal $\mathfrak{p}$ of R lying below $\mathfrak{P}$. Moreover, $\mathfrak{p} = \mathfrak{P} \cap R$.*

Proof. Let I be the intersection of $\mathfrak{P}$ and R. It is clearly an ideal in R distinct from R since it does not contain 1, and non-zero since for every $a \in \mathfrak{P}$ it contains its norm $N_{L/K}(a)$. The embedding $R \to S$ carries I into $\mathfrak{P}$ and thus induces a homomorphism of R/I into $S/\mathfrak{P}$, which is actually an embedding as $I = \mathfrak{P} \cap R$. Thus R/I is a subring of a field, whence it cannot possess zero-divisors, and so I is a prime ideal. Moreover, $i_{L/K}(I) = S(\mathfrak{P} \cap R) \subset S\mathfrak{P} \cap SR = \mathfrak{P}$, whence $\mathfrak{P}$ lies above I. If J is another prime ideal of R lying below $\mathfrak{P}$, then it must be contained in $\mathfrak{P} \cap R = I$ and hence equal to I. □

We see thus that there is an embedding of the field $R/\mathfrak{p}$ into $S/\mathfrak{P}$ provided that $\mathfrak{P}$ lies above $\mathfrak{p}$. The extension $(S/\mathfrak{P})/(R/\mathfrak{p})$ is finite and its degree cannot exceed $[L:K]$ because if a is any element of S and

$$a^n + c_{n-1}a^{n-1} + \ldots + c_0 = 0 \quad (c_i \in R, \; n = [L:K]),$$

then the class $\bar{a}$ of a (mod $\mathfrak{P}$) satisfies

$$\bar{a}^n + \bar{c}_{n-1}\bar{a}^{n-1} + \ldots + \bar{c}_0 = 0,$$

where $\bar{c}_i$ are in $R/\mathfrak{p}$.

We shall denote the degree $[S/\mathfrak{P} : R/\mathfrak{p}]$ by $f_{L/K}(\mathfrak{P})$ and call it the *degree of $\mathfrak{P}$ over the field K*.

The definitions of $e_{L/K}$ and $f_{L/K}$ immediately imply the following proposition:

Proposition 4.3. *If $K \subset L \subset M$, T is the integral closure of S in M and $\mathfrak{p} \subset \mathfrak{P}$ are prime ideals of S and T respectively, then*

$$e_{M/K}(\mathfrak{P}) = e_{M/L}(\mathfrak{P})e_{L/K}(\mathfrak{p})$$

and

$$f_{M/K}(\mathfrak{P}) = f_{M/L}(\mathfrak{P}) f_{L/K}(\mathfrak{p}). \qquad \Box$$

Corollary. *If L/K and M/L are both unramified, or tamely ramified, so is M/K.*
$\Box$

2. Our subsequent results will concern the relations between the ramification indices and ideal degrees. Although they are true in the most general situation (under the only assumption of finiteness and separability of L/K), we shall prove them only in those cases which are really needed for our purposes. This will enable us to shorten the proofs. Our standing assumptions now are:

(i) L/K is a finite and separable extension, R is a Dedekind domain with the quotient field K and S is its integral closure in L.

(ii) The ring R (and hence also S) has the (FN) property.

(iii) The class-numbers of R and S are finite.

In particular, the rings R_K satisfy these assumptions and this is the most important case for us.

We now prove

Theorem 4.1. *If $\mathfrak{p}$ is a prime ideal in R and*

$$i_{L/K}(\mathfrak{p}) = \mathfrak{p}S = \mathfrak{P}_1^{e_1} \ldots \mathfrak{P}_s^{e_s},$$

where $\mathfrak{P}_1, \ldots, \mathfrak{P}_s$ are prime ideals in S, then

$$e_1 f_1 + \ldots + e_s f_s = n,$$

where $f_i = f_{L/K}(\mathfrak{P}_i)$ and $n = [L:K]$.

Proof. For simplicity, let us denote the field $R/\mathfrak{p}$ by k. The embedding $R \to S$ induces a homomorphism of k into the ring $S/\mathfrak{p}S$, which is actually an embedding since it is non-zero and k is a field. This induces in $S/\mathfrak{p}S$ the structure of a linear k-space, and the same applies to $S/\mathfrak{P}_i^{e_i}$. We prove

Lemma 4.1. $\dim_k S/\mathfrak{p}S = e_1 f_1 + \ldots + e_s f_s$.

Proof. By Theorem 1.5 it suffices to prove the equality

$$\dim_k S/\mathfrak{P}_i^{e_i} = e_i f_i$$

for $i = 1, 2, \ldots, s$.

Moreover, we have $(S/\mathfrak{P}_i^m)/(S/\mathfrak{P}_i^{m-1}) \simeq \mathfrak{P}_i^{m-1}/\mathfrak{P}_i^m$ and this, jointly with Proposition 1.7, implies (note that we use here the (FN) property)

$$\dim_{S/\mathfrak{P}_i} S/\mathfrak{P}_i^{e_i} = e_i.$$

Consequently,
$$\dim_k S/\mathfrak{P}_i^{e_i} = (\dim_{S/\mathfrak{P}_i} S/\mathfrak{P}_i^{e_i})(\dim_k S/\mathfrak{P}_i) = e_i f_i,$$
as required. □

Let v be the exponent induced by $\mathfrak{p}$ in K and consider its *exponent-ring* R' and the R'-module $N = R'S \subset L$. If $x \in N$, then $x = x_1 y_1 + \ldots + x_m y_m$ with $x_j \in R'$, $y_j \in S$. According to Proposition 1.11 (i) we can write $x_j = c_j/b$ for $j = 1, 2, \ldots, m$ with $c_j \in R$ and $b \in R \setminus \mathfrak{p}$. Hence $x = a/b$ with a certain $a \in S$. If $P(t)$ is the minimal polynomial for a over K and $r = \deg_K a$, then the polynomial $P(bt)b^{-r}$ has all its coefficients in R' and is minimal for x. Thus for all $x \in N$ we get $T_{L/K}(x) \in R'$.

Now choose in N an element w generating L/K and denote by $w_1, \ldots, w_n$ its conjugates over K in a fixed algebraic closure of K. Then $\det[w_i^j] \neq 0$, and so we may find $v_1, \ldots, v_n \in L$ satisfying $T_{L/K}(w^i v_j) = 1$ if $i+1 = j$ and $T_{L/K}(w^i v_j) = 0$ otherwise, for $i = 0, 1, \ldots, n-1$ and $j = 1, 2, \ldots, n$. Since the v_i's are K-linearly independent, we can write every $x \in N$ in the form $x = x_1 v_1 + \ldots + x_n v_n$ ($x_i \in K$). Since $xw^i \in N$, we get $T_{L/K}(xw^i) \in R'$ for $i = 0, 1, \ldots, n-1$ but

$$T_{L/K}(xw^i) = \sum_{j=1}^{n} x_j T_{L/K}(v_j w^i) = x_{i-1} \quad (i = 1, 2, \ldots, n);$$

thus all x_i's lie in R' and we infer that N is a submodule of the free R'-module, freely generated by $v_1, \ldots, v_n$. By Theorem 1.10, R' is a PID, and thus N cannot have more than n generators, and since it is torsion-free, Theorem 1.13 shows that N is a free R'-module with at most n free generators. However, N contains n elements K-linearly independent, and thus it has exactly n free generators.

Thus $N = \bigoplus_{i=1}^{n} u_i R'$ for suitable $u_1, \ldots, u_n$. Now take an arbitrary $a \in \mathfrak{p} \setminus \mathfrak{p}^2$ and consider $aN = \bigoplus_{i=1}^{n} au_i R'$. The mapping $N \to (R'/aR')^n$ defined by

$$\sum_{i=1}^{n} a_i u_i \mapsto [a_1 (\operatorname{mod} aR'), \ldots, a_n (\operatorname{mod} aR')]$$

is clearly surjective and its kernel equals aN. Thus N/aN and $(R'/aR')^n$ are isomorphic as Abelian groups. Applying Proposition 1.11 (iv), we get $|R'/aR'| = |k|$, whence $|N/aN| = |k|^n$ and

$$\dim_k N/aN = n. \tag{4.1}$$

On the other hand, the embedding $S \to N$ induces an isomorphism $S/pS \to N/aN$, and so, by Lemma 4.1,

$$\dim_k N/aN = e_1 f_1 + \ldots + e_s f_s.$$

Comparing this with (4.1), we obtain the assertion of Theorem 4.1. □

Corollary 1. *If $\mathfrak{p}$ is a prime ideal in R and $\mathfrak{P}$ lies above $\mathfrak{p}$, then $e_{L/K}(\mathfrak{P})$ and $f_{L/K}(\mathfrak{P})$ do not exceed $[L:K]$.*

Corollary 2. *There are at most $[L:K]$ distinct prime ideals in S lying above a fixed prime ideal in R.*

If $\mathfrak{p}$ is an ideal in R such that there is only one prime ideal $\mathfrak{P}$ over $\mathfrak{p}$ in S and it satisfies $e_{L/K}(\mathfrak{P}) = [L:K]$, then $\mathfrak{p}$ is called *totally* (or *fully*, or *completely*) *ramified by the extension L/K*. This notion is specially important in the study of the extension of fields complete under a discrete valuation. (See Chapter 5.)

If $\mathfrak{p}S$ is a product of distinct prime ideals of degree 1, then we say that the prime ideal $\mathfrak{p}$ *splits completely in L/K*.

3. In this subsection we assume that our extension L/K is normal. If I is any fractional ideal in L and s is an element of the Galois group $G(L/K)$, then $s(I)$ is also a fractional ideal. All fractional ideals obtained in this way from a given I will be called *conjugated to I*. The notion of conjugated ideals has a meaning also in the case where L/K is not normal—let L_i be the images of L under its different embeddings into a fixed algebraic closure of K, and let $s_i: L \to L_i$ be the corresponding isomorphism, leaving K fixed. Now if I is a fractional ideal of L, then $s_i(L)$ are the conjugated ideals lying in L_i.

In the case of normal extensions the foregoing theorem can be extended as follows:

Theorem 4.2. *If L/K is normal and $\mathfrak{p}$ is a prime ideal in R, then all prime ideals of S which lie above $\mathfrak{p}$ are conjugated and have the same ramification index e and degree f. Moreover, if g denotes their number, then $efg = n$.*

Proof. Once we get the equality of all ramification indices and of all degrees, the last assertion follows immediately from Theorem 4.1. Moreover, it suffices to show that the prime ideals lying above $\mathfrak{p}$ are conjugated, since then the equality of degrees is immediate (the corresponding fields being isomorphic) and the equality of ramification indices follows by applying a suitable $g \in G(L/K)$ to

$$\mathfrak{p}S = \mathfrak{P}_1^{e_1} \ldots \mathfrak{P}_s^{e_s}. \tag{4.2}$$

In fact, if $\mathfrak{P}_i = g(\mathfrak{P}_1)$, then $pS = g(pS) = g(\mathfrak{P}_1)^{e_1} \ldots g(\mathfrak{P}_s)^{e_s}$ and so $e_i = e_1$ by the unique factorization.

So we have to prove that all $\mathfrak{P}_i$'s in (4.2) are conjugated. If g is an arbitrary element of $G(L/K)$, then g acts as a permutation group on the set $\mathfrak{P}_1, \ldots, \mathfrak{P}_s$, and we have to show that it acts transitively. We shall use here the finiteness of the class-number of S. This fact implies the existence of $a \in S$ such that $\mathfrak{P}_1^{h(S)} = aS$. Clearly a lies in $\mathfrak{P}_1$. Denote by $a_1, \ldots, a_n$ the conjugates of $a = a_1$ over K. Since $a \in \mathfrak{P}_1$, we have also $N_{L/K}(a) = a_1 \ldots a_n \in \mathfrak{P}_1$; hence the principal ideal $N_{L/K}(a)S$ is contained in pS and so is divisible by it. This shows that the product $a_1 \ldots a_n$ is divisible by $\mathfrak{P}_i$, where $\mathfrak{P}_i$ is a fixed prime ideal lying above p, and so, say, $a_i \in \mathfrak{P}_i$. Let g be that element in $G(L/K)$ which carries a onto a_i. Then $g(\mathfrak{P}_1)^{h(S)} = g(a)S = a_i S$, but $a_i \in \mathfrak{P}_i$; thus $g(\mathfrak{P}_1)^{h(S)} \subset \mathfrak{P}_i$ and $g(\mathfrak{P}_1) = \mathfrak{P}_i$ because both $\mathfrak{P}_i$ and $g(\mathfrak{P}_1)$ are prime ideals. So $G(L/K)$ indeed acts transitively. □

4. This subsection is devoted to the norm homomorphism, which will be a mapping from $G(L)$ to $G(K)$, defined for every extension L/K satisfying our standing assumptions. Let $\mathfrak{P}$ be any prime ideal of S and let p be the prime ideal in R which lies below $\mathfrak{P}$. We define

$$N_{L/K}(\mathfrak{P}) = p^{f_{L/K}(\mathfrak{P})}$$

and extend the definition of $N_{L/K}$ to the full group $G(L)$ by multiplicativity, i.e., we put

$$N_{L/K}(I) = \prod p^{a_\mathfrak{P} f_{L/K}(\mathfrak{P})}$$

for $I = \prod \mathfrak{P}^{a_\mathfrak{P}}$ in $G(L)$.

Note that this definition is meaningless if L/K does not satisfy assumptions (i) and (ii) of subsection 2. However, it may be applied also in those cases which violate (iii).

We now state some simple properties of the norm map just defined.

Proposition 4.4. (i) *If the field K is the field of rational numbers, and $I \subset S$, then $N_{L/K}(I)$ is the principal ideal generated by $N(I)$.*

(ii) *If $I \subset S$, then $N_{L/K}(I) \subset R$.*

(iii) *If $I \subset J$, then $N_{L/K}(I) \subset N_{L/K}(J)$.*

(iv) *If I, J are ideals in S such that I and $N_{L/K}(J)S$ are relatively prime, then the norms $N_{L/K}(I)$ and $N_{L/K}(J)$ are also relatively prime.*

(v) *For every I in $G(K)$, we have*

$$N_{L/K} \circ i_{L/K}(I) = I^n,$$

where $n = [L:K]$.

(vi) *For any chain $K \subset L \subset M$ of fields, we have*
$$N_{L/K} \circ N_{M/L} = N_{M/K}.$$

Proof. (i) It suffices to check the assertion for a prime ideal $\mathfrak{P}$. If $\mathfrak{p}$ is the prime ideal of Z lying below $\mathfrak{P}$, then $\mathfrak{p}$ is generated by a rational prime p and $N_{L/K}(\mathfrak{P}) = \mathfrak{p}^{f_{L/K}(\mathfrak{P})}$. But in this case $N(\mathfrak{P}) = p^{f_{L/K}(\mathfrak{P})}$ and our assertion follows.

(ii) This is immediate from the definition of the norm map.

(iii) This follows from (ii) and Proposition 1.5 (i).

(iv) Assume the contrary. Then there is a prime ideal $\mathfrak{p}$ in R dividing $N_{L/K}(I)$ and $N_{L/K}(J)$. This shows that $\mathfrak{p}S$ divides $N_{L/K}(J)S$; thus $\mathfrak{p}S$ and I are relatively prime, and so no prime ideal lying above $\mathfrak{p}$ occurs in the factorization of I. But this means that $N_{L/K}(I)$ cannot be divisible by $\mathfrak{p}$, giving a contradiction.

(v) It suffices to consider the case of $I = \mathfrak{p}$, a prime ideal. If $i_{L/K}(\mathfrak{p}) = \mathfrak{P}_1^{e_1} \ldots \mathfrak{P}_s^{e_s}$, then
$$N_{L/K} \circ i_{L/K}(\mathfrak{p}) = N_{L/K}(\mathfrak{P}_1)^{e_1} \ldots N_{L/K}(\mathfrak{P}_s)^{e_s} = \mathfrak{p}^{e_1 f_1 + \ldots + e_s f_s} = \mathfrak{p}^n$$
by Theorem 4.1.

(vi) Immediate by Proposition 4.3. □

The next proposition shows that one can regard the norm of an ideal as the product of all ideals conjugated to it. This makes the norm mapping of ideals similar to the norm mapping of elements.

Proposition 4.5. *Let M be the minimal normal extension of K containing L and let $A \in G(L)$. Then*
$$N_{L/K}(A)T = \prod_{g \in H} g(AT),$$

where T is the integral closure of R in M and H is a subset of the Galois group $G(M/K)$ consisting of representatives of cosets of this group relative to its subgroup acting trivially on L.

In particular, if L/K is itself a normal extension, then
$$N_{L/K}(A)S = \prod_{g \in G(L/K)} g(A).$$

Proof. We first prove our proposition in the case where L/K is normal. It is obviously sufficient to prove it for a prime ideal, say $A = \mathfrak{P}$. Let $\mathfrak{p}$ be the prime ideal of R lying below $\mathfrak{P}$. According to Theorem 4.2 we may write
$$\mathfrak{p}S = (\mathfrak{P}_1 \ldots \mathfrak{P}_g)^e,$$
$\mathfrak{P}_1, \ldots, \mathfrak{P}_g$ being prime ideals in S and $\mathfrak{P}_1$ equal to $\mathfrak{P}$. Let G_i (for $i = 1, 2, \ldots, g$) be the set of $g \in G(L/K)$ carrying $\mathfrak{P}_1$ onto $\mathfrak{P}_i$. Clearly G_1 is a group and the G_i's are its cosets, whence we obtain
$$|G_1| = n/g = ef$$

with $f = f_{L/K}(\mathfrak{P})$ and $n = [L:K]$. This equality in turn implies

$$\prod_{g \in G(L/K)} g(\mathfrak{P}) = \prod_{i=1}^{g} \mathfrak{P}^{ef} = (pS)^f = N_{L/K}(\mathfrak{P})S,$$

as asserted.

In the general case it again suffices to consider $A = \mathfrak{P}$, a prime ideal. Observe that

$$|H| = |G(M/K)/G(M/L)| = [L:K]$$

and, moreover, $g^{-1}g_1 \in G(M/L)$ implies $g(\mathfrak{P}T) = g_1(\mathfrak{P}T)$; hence

$$\left(\prod_{g \in H} g(\mathfrak{P}T)\right)^{[M:L]} = \prod_{g \in G(M/K)} g(\mathfrak{P}T) = N_{M/K}(\mathfrak{P}T)T$$

$$= N_{L/K}(N_{M/L}(\mathfrak{P}T))T = N_{L/K}(\mathfrak{P})^{[M:L]}T = (N_{L/K}(\mathfrak{P})T)^{[M:L]}$$

(using Proposition 4.4 (v), (vi), and the part of our proposition already proved) and this implies our assertion. □

Corollary. *The mapping $N_{L/K}$ maps a principal ideal onto a principal ideal; more precisely, if A is a principal fractional ideal generated by an element a in L, then $N_{L/K}(A)$ is the principal fractional ideal generated by $N_{L/K}(a)$.*

Proof. Denoting by M the least normal extension of K containing L and by T the integral closure of R in M, we obtain from the foregoing proposition

$$N_{L/K}(A)T = \prod_{g \in H} g(AT) = \prod_{g \in H} g(aT) = \prod_{g \in H} g(a)T = N_{L/K}(a)T;$$

hence by Proposition 4.1

$$N_{L/K}(A) = N_{L/K}(A)T \cap S = N_{L/K}(a)T \cap S = N_{L/K}(a)S.$$ □

The above corollary enables us to present a characterization of the ideals $N_{L/K}(A)$ which is often used as a definition of the norm mapping.

Theorem 4.3. *For any fractional ideal I of L the norm $N_{L/K}(I)$ is the smallest fractional ideal of K which contains all norms $N_{L/K}(a)$ when a ranges over all elements of I.*

Proof. The Corollary to Proposition 4.5 shows that $N_{L/K}(I)$ contains all norms $N_{L/K}(a)$ ($a \in I$). To prove that such norms generate $N_{L/K}(I)$ consider first the case of integral I, i.e., $I \subset S$. By Corollary 4 to Proposition 1.6 we may find a, b in I such that $I = aS + bS$ and the ideals bS and $N_{L/K}(aI^{-1}S)S$ are relatively prime. If we define Q_1 and Q_2 by $aS = IQ_1$ and $bS = IQ_2$, then

$N_{L/K}(aS) = N_{L/K}(I)N_{L/K}(Q_1)$ and $N_{L/K}(bS) = N_{L/K}(I)N_{L/K}(Q_2)$, but Proposition 4.4 (iv) shows that the norms of Q_1 and Q_2 are relatively prime; thus

$$N_{L/K}(aS) + N_{L/K}(bS) = N_{L/K}(I),$$

as asserted.

Now if I is an arbitrary fractional ideal of L, then we may write it in the form $I = A/aS$ with a in R and $A \subset S$. In fact, we can write $I = A_1/bS$ with $A_1 \subset S$ and $b \in S$, but the element $N_{L/K}(b)/b = c$ lies in S and $I = cA_1/cbS = A/a$ with $A = cA_1 \subset S$ and $a = N_{L/K}(b) \in R$. Having done this, observe that for $x \in I$ we have $ax = x_1 \in A$, and so the fractional ideal generated by all elements $N_{L/K}(x)$ with $x \in I$ coincides with the product of a^{-n} by the smallest fractional ideal containing all elements $N_{L/K}(x_1)$ with x_1 in A. But this equals $a^{-n}N_{L/K}(A) = N_{L/K}(a^{-1}A) = N_{L/K}(I)$. □

5. From Corollary 2 to Theorem 4.1 one can deduce an evaluation of $h(K)$, due to E. Landau [18a]. Although it is rather simple, no better evaluation of $h(K)$ holding for all K is known so far.

We start with

Lemma 4.2. *Let $F(n)$ be the number of distinct ideals with the norm $N(I)$ equal to n in a given algebraic number field K. Then the function $F(n)$ is multiplicative, i.e., the equality $F(mn) = F(m)F(n)$ holds for relatively prime m, n and, moreover, satisfies the inequality*

$$F(n) \leqslant d_N(n) = O(n^\varepsilon)$$

for every positive ε, where $d_N(n)$ denotes the number of factorizations of the number n into N factors taking into account their order, and $N = [K:Q]$.

Proof. The multiplicativity is an immediate consequence of the fact that every ideal of norm mn with $(m, n) = 1$ is a unique product of ideals with the norms m and n respectively. Now let $n = p^a$ be a prime power and let $pR_K = \mathfrak{P}_1^{e_1} \ldots \mathfrak{P}_s^{e_s}$ where $\mathfrak{P}_i$ are prime ideals in R_K.

If I is an ideal of norm $N(I) = p^a$ and $\mathfrak{P}$ a prime ideal dividing I, then by Theorem 1.6 $N(\mathfrak{P})|N(I)$, whence $N(\mathfrak{P})$ must be a power of p, thus $p \in \mathfrak{P}$ and we see that $\mathfrak{P}$ coincides with one of the ideals $\mathfrak{P}_1, \ldots, \mathfrak{P}_s$. This shows that $I = \mathfrak{P}_1^{b_1} \ldots \mathfrak{P}_s^{b_s}$ with suitable $b_i \geqslant 0$.

It follows that every such I induces a factorization of p^a into s factors: $p^a = N(I) = N(\mathfrak{P}_1^{b_1}) \ldots N(\mathfrak{P}_s^{b_s})$; and if J is another ideal of norm p^a inducing the same factorization, then $J = \mathfrak{P}_1^{c_1} \ldots \mathfrak{P}_s^{c_s}$, and obviously $N(\mathfrak{P}_i^{b_i}) = N(\mathfrak{P}_i^{c_i})$ holds for $i = 1, 2, \ldots, s$, whence $b_i = c_i$ and $I = J$. Thus $F(p^a)$ cannot exceed $d_s(p^a)$ and, since by Corollary 2 to Theorem 4.1 s does not exceed N, we obtain

$F(p^a) \leq d_N(p^a)$. By multiplicativity the lemma follows, the last evaluation being standard. □

The announced evaluation of $h(K)$ runs as follows:

Theorem 4.4. *If* $[K:Q] = N > 1$,
$$h(K) = O(D^{1/2}\log^{N-1} D)$$
where $D = |d(K)|$.

Proof. By Lemma 3.3 we have
$$h(K) \leq \sum_{n \leq cD^{1/2}} F(n)$$
where c is the Minkowski constant of K, and so, by the last lemma,
$$h(K) \leq \sum_{n \leq cD^{1/2}} d_N(n).$$
However, it is a standard result in elementary number theory, which can easily be proved by recurrence in N, that
$$\sum_{n \leq x} d_N(n) = O(x\log^{N-1} x),$$
and so the theorem follows. □

For further reference we point out a special case:

Corollary. *If K is an imaginary quadratic field and $D = |d(K)|$, then*
$$h(K) \leq D^{1/2}\log D(\pi^{-1} + .35(\log D)^{-1}).$$

Proof. It suffices to note that
$$\sum_{n \leq x} d_2(n) = \sum_{n \leq x} \sum_{d|n} 1 = \sum_{d \leq x} [xd^{-1}] \leq x + x\log x.$$ □

6. We conclude this section with the proof of the Pellet–Stickelberger theorem concerning the quadratic character of polynomial and field discriminants. We could not give this proof in the preceding chapter, where we treated discriminants, since it uses an observation proved at the beginning of this section, namely that if $\mathfrak{P}$ lies over p, then $S/\mathfrak{P}$ is a finite extension of $R/\mathfrak{p}$.

Theorem 4.5. *If $f \in Z[X]$ is monic and irreducible and D denotes its discriminant, then for any odd prime p not dividing D we have*
$$\left(\frac{D}{p}\right) = (-1)^{n-r_p},$$

where n is the degree of f and r_p denotes the number of irreducible factors of $\bar{f}$, the reduction of $f \pmod{p}$, over $GF(p)$.

Proof. Put $k_0 = GF(p)$, let $x_1, \ldots, x_n$ be the roots of f, denote by k the splitting field of $\bar{f}$ over k_0 and let $\bar{x}_1, \ldots, \bar{x}_n$ be the roots of $\bar{f}$ in k. Since $p \nmid D$, these roots are distinct. Observe now that if $\bar{D} \equiv D \pmod{p}$, then

$$\bar{D} = \prod_{i<j} (\bar{x}_i - \bar{x}_j)^2.$$

Indeed, if L is the splitting field of f over Q and $\mathfrak{P}$ is a prime ideal of R_L lying over p, then $R_L/\mathfrak{P}$ is a finite extension of k_0 containing k. Applying the map $s: R_L \to R_L/\mathfrak{P}$ and noting that its restriction to Z maps Z onto k_0 and moreover we can assume $s(x_i) = \bar{x}_i$ (for $i = 1, 2, \ldots, n$), we arrive at the formula asserted.

Write $\bar{f} = F_1 \ldots F_r$, where the F_i's are irreducible over k_0 and $r = r_p$. If we denote by $A_i = \{u_1^{(i)}, \ldots, u_{n_i}^{(i)}\} \subset k$ the set of all roots of F_i for $i = 1, 2, \ldots, r$, then we can write

$$\bar{D} = \prod_{i=1}^{r} D(F_i) \prod_{\substack{s \\ s>i}} B_{i,s}^2,$$

where $D(F_i)$ is the discriminant of F_i and

$$B_{i,s} = \prod_{\substack{\bar{x} \in A_i \\ \bar{y} \in A_s}} (\bar{x} - \bar{y}).$$

Since $B_{i,s}$ is invariant under the action of $\text{Gal}(k/k_0)$, we have $B_{i,s} \in k_0$ and with a certain non-zero $B \in k_0$ we obtain

$$\prod_{i=1}^{n} \prod_{\substack{s \\ s>i}} B_{i,s}^2 = B^2.$$

Let n_i be the degree of F_i. If for a certain i this degree is odd, then

$$D(F_i) = \prod_{t<s} (u_t^{(i)} - u_s^{(i)})^2$$

is a square in the splitting field k_i of F_i over k_0, but since k_i/k_0 is cyclic of an odd degree, $D(F_i)$ is a square already in k_0. This shows that

$$\left(\frac{D}{p}\right) = +1 \quad \text{iff} \quad \prod_{\substack{i \\ 2|n_i}} D(F_i) \text{ is a square in } k_0. \tag{4.3}$$

To decide when this happens observe that $D(F_i)$ is a square in k_0 precisely

when
$$\prod_{t<s}(u_t^{(i)}-u_s^{(i)}) \in k_0,$$
which occurs iff this product is invariant under $\operatorname{Gal}(k_i/k_0)$. Since for $g \in \operatorname{Gal}(k_i/k_0)$ the ratio
$$\frac{\prod_{t<s}(g(u_t^{(i)})-g(u_s^{(i)}))}{\prod_{t<s}(u_t^{(i)}-u_s^{(i)})}$$
equals $(-1)^{\operatorname{sgn} g}$ (if we treat $\operatorname{Gal}(k_i/k_0)$ as a permutation group on A_i), $D(F_i)$ is a square in k_0 iff all elements of $\operatorname{Gal}(k_i/k_0)$ are even permutations. However, this cannot happen for even n_i. Indeed, the group $\operatorname{Gal}(k_i/k_0)$ is cyclic and acts transitively on A_i; thus its generator must be a cycle of even length n_i, which is an odd permutation. This observation, jointly with (4.3), shows that $\left(\frac{D}{p}\right) = +1$ holds iff the number of even n_i's is even. Since
$$n = \sum_{i=1}^{r} n_i = r + \sum_{i=1}^{r}(n_i-1) \equiv r + \sum_{\substack{i \\ 2\mid n_i}} 1 \pmod{2},$$
we arrive at our assertion. □

The following corollary results immediately:

Corollary. *Let a be an algebraic integer generating the field K, f its minimal polynomial and p a rational prime not dividing the index of a. Then either $p \mid d(K)$, or $\left(\frac{d(K)}{p}\right) = (-1)^{n-r_p}$, where $n = \deg a$ and r_p equals the number of irreducible factors of $f \pmod{p}$.* □

§ 2. Different and Discriminant

1. In this section we shall define the discriminant for a relative extension. This cannot be done in the same way as in Chapter 2 for extensions of Q, since we used in that chapter the existence of integral bases, which is assured only if the ring of integers of the ground field is a PID. Even if it is, and we imitate the procedure used for absolute extensions, we find that the discriminants of the various integral bases may differ by a square of a unit, which is not necessarily equal to 1. So we have to follow another procedure. Since we want to use our construction also for completions of algebraic number fields, we shall not restrict ourselves to finite extensions of Q, but proceed as in the preceding section, which means that L/K will be a finite separable extension, R a Dedekind domain having K for its field of fractions and S the integral closure of R in K. We still assume that the condition (FN) and the finiteness of the class-number are satisfied.

Let $A \subset L$ be a non-zero R-module. The set

$$A^* = \{x \in L: T_{L/K}(xA) \subset R\}$$

will be called the *codifferent of A over K*. It is an R-module, which may even be equal to the zero module; this happens if A is sufficiently large, e.g. $A = L$. To obtain non-trivial properties of the codifferent we have to make some additional assumptions on A. The most obvious thing to do is to assume that A is an S-module and a fractional ideal in L. In this case the following result holds:

Proposition 4.6. *If A is a fractional ideal in L, then its codifferent A^* is also a fractional ideal in L and $AA^* = S^*$. Moreover, if A is an ideal in S, then $(A^*)^{-1}$ is also an ideal in S.*

Proof. If $x_1, x_2 \in A^*$ and $b_1, b_2 \in S$, then

$$T_{L/K}((b_1 x_1 + b_2 x_2)A) \subset T_{L/K}(x_1 A) + T_{L/K}(x_2 A) \subset R,$$

whence A^* is a S-module.

Since with some non-zero $a \in S$ we have $aA \subset S$ and every such element a lies in A^*, we can see that A^* is non-zero. It remains to show that with some non-zero $q \in S$ we have $qA^* \subset S$. For this purpose let $w_1, \ldots, w_n$ be any K-basis of L contained in S and let b be any non-zero element of $A \cap R$ (it can be any element of the form $N_{L/K}(c)$ with some $c \in A \cap S$, $c \neq 0$). We shall prove that the element $q = b \det[T_{L/K}(w_i w_j)]$, which is non-zero (as L/K is separable) and lies in R, is good for us. Take any $x = c_1 w_1 + \ldots + c_n w_n$ in A^* with $c_k \in K$ and observe that all elements bw_j lie in S; thus $T_{L/K}(bxw_j) \in R$. But

$$T_{L/K}(bxw_j) = b \sum_{k=1}^{n} c_k T_{L/K}(w_j w_k),$$

whence by Cramer's rule we infer that all elements $c_k b \det[T_{L/K}(w_i w_j)]$ lie in R and so $qx \in S$. This shows $qA^* \subset S$, as required. The equality $AA^* = S^*$ results from the following chain of equivalences

$$a \in A^* \Leftrightarrow T_{L/K}(aA) \subset R \Leftrightarrow T_{L/K}(aAS) \subset R \Leftrightarrow aA \subset S^* \Leftrightarrow a \in A^{-1}S^*.$$

To prove the final assertion observe that $A \subset S$ implies $S \subset A^*$, whence $(A^*)^{-1} = S(A^*)^{-1} \subset A^*(A^*)^{-1} = S$. □

If A is an arbitrary fractional ideal in L, then $(A^*)^{-1}$, which is a fractional ideal by the above proposition (and in the case of $A \subset S$ even an integral ideal) will be called the *different of A over K*. We shall denote it by $D_{L/K}(A)$.

The different of the unit ideal S will be called the *different of the extension L/K*. We shall denote it by $D_{L/K}$.

Note that this notion depends not only on the extension L/K but also on the choice of R. However, in all cases which we shall consider the choice of R will be obvious, so that no ambiguity will arise. Note also that the map $G(L) \to G(L)$ given by $A \to D_{L/K}(A)$ is in general not a homomorphism. Now we prove some properties of the differents.

Proposition 4.7. (i) *If $A \in G(L)$, then $D_{L/K}(A) = AD_{L/K}$.*

(ii) *If $K \subset L \subset M$, then $D_{M/K} = D_{M/L} D_{L/K}$.*

(iii) *If L/K is normal, then for every g in the Galois group $G(L/K)$ we have $g(D_{L/K}) = D_{L/K}$, i.e., $D_{L/K}$ is an invariant ideal.*

(iv) *If $A \in G(K)$, then the conditions*

$$T_{L/K}(B) \subset A$$

and

$$B \subset AD_{L/K}^{-1}$$

are equivalent for every $B \in G(L)$.

Proof. (i) By the previous proposition we have

$$(D_{L/K}(A))^{-1} AD_{L/K} = A^* AD_{L/K} = S^* D_{L/K} = S$$

and (i) results.

(ii) The chain of equivalences

$$a \in D_{M/L}^{-1} \Leftrightarrow T_{M/L}(a) \in S \Leftrightarrow D_{L/K}^{-1} T_{M/L}(a) \subset D_{L/K}^{-1}$$
$$\Leftrightarrow T_{L/K}(D_{L/K}^{-1} T_{M/L}(a)) \subset R \Leftrightarrow T_{M/K}(aD_{L/K}^{-1}) \subset R$$
$$\Leftrightarrow aD_{L/K}^{-1} \subset D_{M/K}^{-1} \Leftrightarrow a \in D_{L/K} D_{M/K}^{-1}$$

gives $D_{M/L}^{-1} = D_{L/K} D_{M/K}^{-1}$, which is equivalent to our assertion.

(iii) For any x in S^* we have

$$T_{L/K}(g(x)S) = T_{L/K}(xg^{-1}(S)) = T_{L/K}(xS) \subset R;$$

hence $g(S^*) \subset S^*$ for all $g \in G(L/K)$ and thus also $g^{-1}(S^*) \subset S^*$, i.e., $S^* \subset g(S^*)$, and finally $g(S^*) = S^*$. This proves (iii).

(iv) It is enough to consider the sequence of equivalences

$$T_{L/K}(B) \subset A \Leftrightarrow A^{-1} T_{L/K}(B) \subset R \Leftrightarrow T_{L/K}(A^{-1}B) \subset R$$
$$\Leftrightarrow A^{-1}B \subset D_{L/K}^{-1} \Leftrightarrow B \subset AD_{L/K}^{-1}.$$

□

The notion of the different allows us to characterize those extensions in which the mapping of S into R induced by the trace is surjective and, more generally, to identify the set of all traces of elements from S.

Corollary 1. *The largest divisor of the ideal $D_{L|K}$ which lies in the image of $i_{L/K}$ equals $T_{L/K}(S)S$, and so $T_{L/K}(S)$ is the least common multiple of integral ideals from R whose images under $i_{L|K}$ divide the different.*

Proof. Obviously $T_{L/K}(S)$ is an ideal in R and so the corollary follows by putting $B = S$ in (iv). □

Corollary 2. *The mapping of S into R induced by the trace is surjective iff the different $D_{L|K}$ does not have any divisor $\neq S$ lying in the image of $i_{L|K}$, i.e. being an ideal from R.*

Proof. Immediate by the previous corollary. □

2. In the number-theoretic case we may prove some more results about the different and the codifferent:

Proposition 4.8. *Assume that K is the rational field Q. If A is a fractional ideal in L with a Z-basis $a_1, \ldots, a_n$ and $b_1, \ldots, b_n$ are defined by*

$$T_{L/K}(a_i b_j) = \begin{cases} 1, & i = j, \\ 0, & i \neq j, \end{cases}$$

then the Z-module generated by $b_1, \ldots, b_n$ equals A^*.

Moreover, the absolute norm of $D_{L/Q}(A)$ equals $N(A)|d(L)|$; in particular

$$N(D_{L/Q}) = |d(L)|.$$

Proof. Any element $x \in L$ may be written as $x = x_1 b_1 + \ldots + x_n b_n$ ($x_i \in Q$), and if $y \in A$, then with some $y_i \in Z$ we have $y = y_1 a_1 + \ldots + y_n a_n$.

Now

$$T_{L/Q}(xy) = \sum_{i,j} y_i x_j T_{L/Q}(a_i b_j) = x_1 y_1 + \ldots + x_n y_n,$$

and the last sum will be integral for every choice of $y \in A$ iff all x_i's are integral. This shows that A^* coincides with the Z-module generated by $b_1, \ldots, b_n$.

(Note that the same argument applies in the more general situation where R is a principal ideal domain.)

To prove the second assertion it suffices, in view of Proposition 4.7 (i), to consider the case of $A = R_L$. Let $w_1, \ldots, w_n$ be an integral basis of L. By the already proved part of our proposition the ideal $D_{L/Q}^{-1}$ has a Z-basis $b_1, \ldots, b_n$ such that

$$T_{L/Q}(w_i b_j) = \begin{cases} 0, & i \neq j, \\ 1, & i = j. \end{cases}$$

Choose a natural m such that all numbers $c_i = mb_i$ are integral and consider the ideal $I = mD_{L/Q}^{-1} \subset R_L$. Using Proposition 2.7 and its corollary, we obtain

$$N(D_{L/Q}^{-1})^2 = N(I)^2 N_{L/Q}^{-2}(m) = d_{L/Q}(c_1, \ldots, c_n) d^{-1}(L) m^{-2n}$$
$$= d_{L/Q}(b_1, \ldots, b_n) d(L)^{-1}.$$

Finally observe that the product $[w_i^{(j)}][b_i^{(j)}]^T$ equals $[T_{L/Q}(w_i b_j)]$ and thus coincides with the unit matrix; hence

$$d_{L/Q}(b_1, \ldots, b_n) = d_{L/Q}^{-1}(w_1, \ldots, w_n) = d(L)^{-1},$$

which gives $N(D_{L/Q})^2 = d^2(L)$. □

Now we define the discriminant of an extension L/K. In contrast to the discriminant $d(K)$, it will not be an element of the ring R_K but an ideal in this ring; however, for the extension L/Q it will coincide with the ideal generated by $d(L)$ in Z. A definition of a discriminant which resembles more the absolute discriminant will be given in terms of ideles in Chapter 6.

Our definition is very simple: the *discriminant of the extension L/K* equals $N_{L/K}(D_{L/K})$. We shall denote it by $d(L/K)$. Proposition 4.8 immediately implies that $d(L/Q)$ is the ideal generated by $d(L)$. Moreover, Proposition 4.7 (ii) implies the following

Proposition 4.9. *If* $K \subset L \subset M$, *then*

$$d(M/K) = d(L/K)^{[M:L]} N_{L/K}(d(M/L)).$$

Corollary 1. *If K/Q is an extension with a square-free discriminant $d(K)$, then there is no field between Q and K.*

Proof. If L were such a field, then by Theorem 2.10 $d(L)$ would have a prime divisor and by the last proposition the discriminant $d(K)$ would be divisible by the $[K:L]$ th power of that prime. □

Corollary 2. *If L/Q is finite and $Q \subset K \subset L$, then the discriminant $d(L)$ is divisible by $d(K)^{[L:K]}$.* □

This improves Proposition 2.9.

3. Let us return to the general case. For any element $a \in S$ generating the extension L/K consider its minimal polynomial

$$F(t) = t^n + a_{n-1}t^{n-1} + \ldots + a_0$$

over R and define the *different* $\delta_{L/K}(a)$ as the value of the derivative $F'(t)$ at a. Our aim is the following:

Theorem 4.6. *The different $D_{L/K}$ (as an ideal of S) is generated by the set of all differents $\delta_{L/K}(a)$ with a running over S.*

For convenience we define also the different $\delta_{L/K}(a)$ for all elements of S, not necessarily generating L/K, by putting $\delta_{L/K}(a) = 0$ in the case yet undefined.

For the proof we need some results concerning subrings of S. Let A be such a subring containing R. The greatest common divisor of ideals of S contained in A will be denoted by $\mathfrak{f}_A$ and called the *conductor* of A. Another characterization of the conductor is contained in

Proposition 4.10. *If $R \subset A \subset S$ and A is a ring, then its conductor $\mathfrak{f}_A$ equals $\{x \in L: xA^* \subset S^*\}$ and is contained in A.*

(Note that A in the case of $A \neq S$ is not an S-module and so its codifferent A^* is only an R-module and not an S-module.)

Proof. Put $I = \{x \in L: xA^* \subset S^*\}$. For $x \in I$, $y \in S$ we get $yxA^* \subset yS^* \subset S^*$ and so I is an S-module. Moreover, $IA^* \subset S^*$, which gives $T_{L/K}(IA^*) \subset T_{L/K}(S^*) \subset R$ and $A^* \subset I^*$, implying $I \subset A \subset S$. Consequently, I is an ideal of S contained in A and hence $\mathfrak{f}_A$ divides I.

Now if J is an ideal of S contained in A and y lies in A^*, i.e., $T_{L/K}(yA) \subset R$, then $T_{L/K}(JyS) \subset T_{L/K}(yA) \subset R$; hence $yJ \subset S^*$, proving $J \subset I$. Consequently, I divides all ideals lying in A and so must divide $\mathfrak{f}_A$. Hence $\mathfrak{f}_A = I \subset A$. $\square$

Of particular importance are the rings $R[a]$ generated by R and an element a of S which generates the extension L/K. Obviously, we may write for such a's

$$R[a] = \bigoplus_{j=0}^{n-1} a^j R,$$

where $n = [L:K]$. For such rings it is possible to give an explicit set of generators for the codifferent, treated as an R-module:

Proposition 4.11. *If $A = R[a]$ and $f(t)$ is the minimal monic polynomial over R for a, then the codifferent A^* is generated as an R-module by the set $a^j/f'(a)$ ($j = 0, 1, \ldots, n-1$).*

Proof. Let B be the R-module generated by those elements. Applying the Lagrange interpolation formula successively to the polynomials $x, x^2, \ldots, x^{n-1}$ and $x^n - f(x)$, we obtain the identities

$$\sum_{j=1}^{n} \frac{a_j^k}{f'(a_j)} \cdot \frac{f(x)}{x - a_j} = \begin{cases} x^{1+k}, & k = 0, 1, \ldots, n-2, \\ x^n - f(x), & k = n-1, \end{cases}$$

in which a_j ($j = 1, 2, \ldots, n$) denote the conjugates of a, and $a = a_1$. Putting $x = 0$, we immediately obtain $T_{L/K}(a^j/f'(a)) \in R$, whence $a^j/f'(a) \in A^*$, i.e. $B \subset A^*$.

To obtain the converse inclusion, consider any b in A^*. If we put

$$P(x) = \sum_{i=1}^{n} b_i f(x)/(x - a_i),$$

where the b_i's are all conjugates of b, then we obtain

$$P(x) = \sum_{j=1}^{n} c_j \sum_{k=0}^{j-1} x^k T_{L/K}(ba^{j-k-1}),$$

where c_i are defined by $f(t) = c_n t^n + \ldots + c_0$ ($c_n = 1$). Since, by our choice of b, $T_{L/K}(bA) \subset R$, we can see that the coefficients of $P(x)$ lie in R. But now $bf'(a) = P(a) \in R[a]$, i.e., $b \in B$ and thus $A^* \subset B$. □

Corollary. $f'(a) S \subset R[a]$.

Proof. $S \subset A^*$, and thus $f'(a) S \subset f'(a) A^* \subset R[a]$. □

The above proposition allows us to find a simple formula for the conductor of $R[a]$ and also to obtain a relation between the different $D_{L/K}$ and the differents of elements, $\delta_{L/K}(a)$.

Proposition 4.12. *Let $A = R[a]$. Then*
  (i) $\mathfrak{f}_A = \delta_{L/K}(a) D_{L/K}^{-1}$,
  (ii) $\mathfrak{f}_A = \{x \in A : xS \subset A\}$.

Proof. (i) Since $S^* \subset A^*$, the preceding proposition implies

$$\delta_{L/K}(a) D_{L/K}^{-1} = f'(a) S^* \subset R[a] \subset S;$$

thus the ideal $\delta_{L/K}(a) D_{L/K}^{-1}$ is integral and is divisible by the conductor $\mathfrak{f}_A$.

Moreover, we have seen in the proof of the preceding proposition that $T_{L/K}(A/f'(a)) \subset R$; hence $A/f'(a) \subset S^*$ and so $\mathfrak{f}_A/f'(a) \subset S^*$, giving finally

$$\mathfrak{f}_A \subset f'(a) S^* = \delta_{L/K}(a) D_{L/K}^{-1}.$$

(ii) If $x \in \mathfrak{f}_A$, then $xS \subset \mathfrak{f}_A S = \mathfrak{f}_A \subset A$, showing that $\mathfrak{f}_A$ is contained in $\hat{A} = \{x \in A : xS \subset A\}$, and since $\hat{A}$ is an S-ideal contained in A, we get $\hat{A} \subset \mathfrak{f}_A$. □

For the proof of Theorem 4.6 we need two further lemmas.

Lemma 4.3. (i) *Let $A = R[a]$, where a is an element of S generating L/K. If $\mathfrak{P}$ is a prime ideal of S not dividing $\mathfrak{f}_A$, then the embedding of A in S gives rise to isomorphisms of the quotient rings $S/\mathfrak{P}^m$ and $A/(A \cap \mathfrak{P}^m)$ for $m = 1, 2, \ldots$, i.e., every residue class (mod $\mathfrak{P}^m$) in S can be represented by an element from A.*

(ii) *Let $\mathfrak{P}$ be a prime ideal of S and let $\mathfrak{p}$ be the prime ideal of R lying below $\mathfrak{P}$. If $\mathfrak{p}S = \mathfrak{P}^e I$, where I is not divisible by $\mathfrak{P}$, and a is an element in $\Gamma \backslash \mathfrak{P}$ which generates L/K and, with $A = R[a]$, the factor rings $S/\mathfrak{P}^m$ and $A/(A \cap \mathfrak{P}^m)$ are isomorphic for $m = 1, 2, \ldots$, the isomorphisms being induced by the canonical embedding of A in S, then $\mathfrak{P}$ does not divide $\mathfrak{f}_A$.*

Proof. The first part is very easy. Let b be any element of $\mathfrak{f}_A \backslash \mathfrak{P}$ and $c \in S$. Proposition 4.12 (ii) shows that for some $W(t) \in R[t]$ we have $c = W(a)/b$. But for suitable k we have $b^k \equiv 1 \pmod{\mathfrak{P}^m}$, and since $b = V(a)$ for some $V \in R[t]$, we finally obtain

$$c = W(a) b^{k-1} \equiv W(a) V(a)^{k-1} \pmod{\mathfrak{P}^m};$$

thus all residue classes (mod $\mathfrak{P}^m$) are represented by elements from A.

The second part is more complicated. First observe that every element of S may be written in the form $x = P(a)/D$ with $P(t) \in R[t]$ and $D \in R$, D not equal to 0 and independent of x. Indeed, $A \subset S$ implies $S \subset S^* \subset A^*$, and thus by Proposition 4.11 we obtain $S \subset R[a]/f'(a)$, where f is the minimal polynomial of a over R. However, $1/f'(a) = g(a)$ for a suitable $g \in K[X]$, whence $1/f'(a) = V(a) D^{-1}$ with $V \in R[X]$ and $D \neq 0$, $D \in R$.

Let $\mathfrak{p}^m$ be the highest power of $\mathfrak{p}$ dividing D. By assumption we can write every element x of S in the form $x = b + x_1$ with $b \in A$ and $x_1 \in \mathfrak{P}^{em}$. Then $x_1 a^m \in \mathfrak{P}^{em} I^m = \mathfrak{P}^m S \subset D^{-1} \mathfrak{p}^m A$. Since $\mathfrak{p} \nmid D\mathfrak{p}^{-m}$, we can find $c \in D\mathfrak{p}^{-m} \backslash \mathfrak{p} \subset R$ and thus $x_1 a^m \in c^{-1} A$. Write $x_1 a^m = c^{-1} r$ with $r \in A$. Then $x_1 = r/ca^m = V(a)/ca^m$ with suitable $V \in R[X]$, implying $ca^m x \in A$, whence $ca^m \in \mathfrak{f}_A$. However, $ca^m \in A \backslash \mathfrak{P}$ and so $\mathfrak{P}$ does not divide $\mathfrak{f}_A$, as asserted. □

Lemma 4.4. *Let $\mathfrak{P}$ be a prime ideal of S and let I be an ideal of S not divisible by $\mathfrak{P}$. Then there exists an element $a \in I$ such that to every $x \in S$ and natural n one can find an y_n in $R[a]$ such that $x \equiv y_n \pmod{\mathfrak{P}^n}$.*

Proof. Let $\mathfrak{p}$ be the prime ideal of R lying below $\mathfrak{P}$ and let $f = f_{L/K}(\mathfrak{P})$ be the degree of $\mathfrak{P}$. Write, for short, $k = S/\mathfrak{P}$ and $k_0 = R/\mathfrak{p}$ and denote by $\bar{a}$ the image in k of $a \in S$. The extension k/k_0 is of degree f, and since k is finite, generated by a single element $\bar{y}$. Let $F(t)$ be a polynomial over R of the form $t^f + b_{f-1} t^{f-1} + \ldots + b_0$ such that the minimal polynomial for $\bar{y}$ over k_0 equals $t^f + \bar{b}_{f-1} t^{f-1} + \ldots + \bar{b}_0$. Now choose y in S so that $y \pmod{\mathfrak{P}} = \bar{y}$ and $F(y) \notin \mathfrak{P}^2$. (This can be obtained by replacing y by $y + c$, where $c \in \mathfrak{P} \setminus \mathfrak{P}^2$ if necessary.) Finally, let $a \in S$ be any element satisfying $a \equiv 0 \pmod{I}$, $a \equiv y \pmod{\mathfrak{P}^2}$. The existence of such elements is assured by Corollary 3 to Proposition 1.6. We show now that every such a satisfies the assertion of the lemma. Let $x \in S$. Then for some polynomial $V(t) \in k_0[t]$ we have $\bar{x} = V(\bar{a})$ and if $V_1(t)$ is a polynomial over R which after reduction $(\bmod\ \mathfrak{p})$ coincides with $V(t)$, then obviously $x \equiv V_1(a) \pmod{\mathfrak{P}}$. This proves our assertion for $n = 1$. Assume its truth for some natural number n; thus, for every $x \in S$ with some $V_n(t) \in R[t]$ we have $x \equiv V_n(a) \pmod{\mathfrak{P}^n}$. The principal ideal generated by $F(a)$ in S is divisible by $\mathfrak{P}$; thus $F(a)S = \mathfrak{P}J$ for a certain J, which, moreover, is not divisible by $\mathfrak{P}$ by our choice of a. Let u be an element of $J \setminus \mathfrak{P}$, $u'u \equiv 1 \pmod{\mathfrak{P}}$ and choose a polynomial $W(t)$ over R to fulfil $u' \equiv W(a) \pmod{\mathfrak{P}}$. The element $c = (x - V_n(a)) u^n / F^n(a)$ lies in S, and thus $c \equiv T(a) \pmod{\mathfrak{P}}$ with a $T(t)$ from $R[t]$, and since $x = V_n(a) + c F^n(a)/u^n$, we finally obtain

$$x \equiv V_n(a) + T(a) F^n(a) W^n(a) \pmod{\mathfrak{P}^{n+1}}. \qquad \square$$

Proof of Theorem 4.6. By Proposition 4.12 (i) every different $\delta_{L/K}(a)$ is contained in $D_{L/K}$, and since $\mathfrak{f}_A D_{L/K} = \delta_{L/K}(a)$ holds with $A = R[a]$, it suffices to choose an a with $\mathfrak{f}_A$ not divisible by a given prime ideal $\mathfrak{P}$ of S. However, the previous two lemmas show that such a choice is possible. $\qquad \square$

4. It was observed by A. Weil [43] that the different $D_{L/K}$ is related to abstract differentiation in commutative rings. If R is any commutative ring and M is an R-module, then every homomorphism of the additive group R^+ into the additive group M^+ is called a *derivation* provided it satisfies

$$f(xy) = xf(y) + yf(x).$$

In the case where M is a commutative ring the derivation f is called *essential* provided its image $f(R)$ contains at least one element which is not a zero-divisor. The connection between the existence of essential derivations and the different

$D_{L/K}$ is given by the following

Theorem 4.7. *The different $D_{L/K}$ equals the least common multiple of ideals I of S for which there exists an essential derivation of S into S/I which equals zero on R.*

Proof. We need a lemma:

Lemma 4.5. *If $\mathfrak{P}$ is a prime ideal of S and $m \geqslant 1$, then for every derivation $D\colon S \to S/\mathfrak{P}^m$ we have $D(x) = 0$ for all $x \in \mathfrak{P}^{1+m}$.*

Proof of the lemma. For $x \in \mathfrak{P}^{m+1}$ and $\pi \in \mathfrak{P} \setminus \mathfrak{P}^2$ write $x = \pi^{m+1}A/B$ with $A, B \in S$, $B \notin \mathfrak{P}$, which is possible by Proposition 1.11 (i). This leads to

$$D(Bx) = BD(x) + xD(B) = AD(\pi^{m+1}) + \pi^{m+1}D(A),$$

but the equality $D(uv) = uD(v) + vD(u)$ implies

$$D(u^r) = ru^{r-1}D(u),$$

and so we get

$$D(Bx) = BD(x) = 0.$$

But $B \notin \mathfrak{P}$ and so $D(x) = 0$. □

Now observe that it suffices to consider only ideals I of the form $I = \mathfrak{P}^m$, $\mathfrak{P}$ being a prime ideal. Indeed, if $D\colon S \to S/I$ is an essential derivation equal to zero on R and $I = \mathfrak{P}_1^{m_1} \ldots \mathfrak{P}_t^{m_t}$, then $D_i \colon S \to S/\mathfrak{P}_i^{m_i}$ defined by $D_i(x) = D(x) \pmod{\mathfrak{P}_i^{m_i}}$ is also an essential derivation vanishing on R. Conversely, if for $i = 1, 2, \ldots, t$ we have essential derivations $D_i\colon S \to S/\mathfrak{P}_i^{m_i}$ vanishing on R, then by putting $D(x) = [D_1(x), \ldots, D_t(x)]$ we obtain a derivation of S into the direct sum of the rings $S/\mathfrak{P}_i^{m_i}$ which is isomorphic to S/I, and so D induces a derivation D' of S in S/I which vanishes on R. It has to be shown that D' is essential. Choose $x_1, \ldots, x_t \in S$ so that $D_i(x_i)$ is not a divisor of zero in $S/\mathfrak{P}_i^{m_i}$ and let x be an element of S for which $x \equiv x_i \pmod{\mathfrak{P}^{m_i+1}}$ for $i = 1, 2, \ldots, t$ holds. Such an element exists by the Chinese remainder-theorem, and Lemma 4.5 implies that $D_i(x) = D_i(x_i)$ for $i = 1, 2, \ldots, t$, whence $D(x)$ is not a zero-divisor.

Thus we have to prove that an essential derivation of S into $S/\mathfrak{P}^m$ (with a prime ideal $\mathfrak{P}$) exists iff $\mathfrak{P}^m$ divides $D_{L/K}$. Assume first that D is such a derivation and let a be chosen in S in such a way that for $A = R[a]$ we have $\mathfrak{P} \nmid \mathfrak{f}_A$, which is possible by Lemmas 4.3 and 4.4.

If $x \in S$ and $x \equiv W(a) \pmod{\mathfrak{P}^{m+1}}$, then, by Lemma 4.5 and the definition of a derivation, we obtain $D(x) = D(W(a)) = W'(a)D(a)$. If $D(a)$ were a divisor of zero, then $D(x)$ would be a zero-divisor for all $x \in S$ — a case we

§ 2. Different and Discriminant

excluded by assumption. Thus $D(a)$ is not a zero-divisor, and if $f(t)$ is the minimal polynomial for a over R, then $f(a) = 0$ implies $0 = D(0) = D(f(a)) = f'(a)D(a)$; hence $f'(a) \equiv 0 \pmod{\mathfrak{P}^m}$ and thus, by Proposition 4.12 (i), $\mathfrak{P}^m$ divides $D_{L/K}$ since $\mathfrak{P}$ does not divide the conductor $\mathfrak{f}_A$.

Now assume that $\mathfrak{P}^m | D_{L/K}^!$ and choose an $a \in S$ so that the conductor of the ring $R[a]$ is not divisible by $\mathfrak{P}$. Moreover, let b be any element of this conductor which does not lie in $\mathfrak{P}$. Then every $x \in S$ can be written as

$$x = V(a)/b,$$

where $V(t)$ is a polynomial over R. The element b can be put in the form $b = W(a)$ with $W(t) \in R[t]$. Since it does not belong to $\mathfrak{P}$, it has an inverse in $S/\mathfrak{P}^m$. Hence we may find $c \in S$ such that $bc \equiv 1 \pmod{\mathfrak{P}^m}$. Now we are able to define our derivation, the form of which is prompted by the familiar rule of differentiation of ratios:

For $x = V(a)/b$ we put

$$D(x) = (V'(a)W(a) - V(a)W'(a))c^2 \pmod{\mathfrak{P}^m}.$$

This definition makes sense because if we have two expressions for x, say $x = V_i(a)/b$ ($i = 1, 2$), then $V_1(a) - V_2(a) = 0$; hence the polynomial $V_1(t) - V_2(t)$ is a multiple of the minimal polynomial for a, say $g(t)$, and so we may write

$$V_1(t) - V_2(t) = g(t)h(t) \quad (h(t) \in R[t]);$$

consequently

$$V_1'(a) - V_2'(a) = h(a)g'(a) \equiv 0 \pmod{\mathfrak{P}^m}$$

since by Theorem 4.6 $g'(a)S$ is divisible by $D_{L/K}$ and a fortiori by $\mathfrak{P}^m$.

The linearity of D is obvious. Moreover, if $x = x_1 x_2$, $x = P(a)/b$, $x_i = V_i(a)/b$ ($i = 1, 2$), then

$$\begin{aligned}
x_1 D(x_2) + x_2 D(x_1) &= cV_1(a)[V_2'(a)W(a) - V_2(a)W'(a)]c^2 \\
&\quad + cV_2(a)[V_1'(a)W(a) - V_1(a)W'(a)]c^2 \\
&= c^3([V_1(a)V_2'(a) + V_1'(a)V_2(a)]W(a) \\
&\quad - 2V_1(a)V_2(a)W'(a)),
\end{aligned}$$

all equalities being understood $\pmod{\mathfrak{P}^m}$.

However, $V_1(a)V_2(a)$ equals $P(a)W(a)$, whence with a suitable polynomial $A(t)$ over R we obtain $V_1(t)V_2(t) = P(t)W(t) + g(t)A(t)$, which easily gives

$$V_1(a)V_2'(a) + V_1'(a)V_2(a) \equiv P'(a)W(a) + P(a)W'(a) \pmod{\mathfrak{P}^m}.$$

Finally, we arrive at

$$x_1 D(x_2) + x_2 D(x_1) = c^2(P'(a)W(a) - P(a)W'(a)) \pmod{\mathfrak{P}^m},$$

which gives $x_1 D(x_2) + x_2 D(x_1) = D(x_1 x_2)$. Hence D is indeed a derivation, obviously vanishing on R and, moreover, it is essential, since $D(a)$ is the unit element of $S/\mathfrak{P}^m$. □

As an application of the theorem obtained, we prove the main result of the theory of the different, namely the different theorem:

Theorem 4.8. *If $\mathfrak{P}$ is a prime ideal of S, $\mathfrak{p}$ is a prime ideal of R lying below $\mathfrak{P}$ and $\mathfrak{p}S = \mathfrak{P}^e A$ with $\mathfrak{P} \nmid A$, then the ideal $\mathfrak{P}^{e-1}$ divides $D_{L/K}$. Moreover, if the number e is relatively prime to the absolute norm of $\mathfrak{p}$ (or $\mathfrak{P}$, which means the same), then $\mathfrak{P}^e$ does not divide $D_{L/K}$.*

Proof. The first part is easy, and we shall obtain it without the use of Theorem 4.7. Denote by M the least normal extension of K containing L in the algebraic closure of L, and let $L = L_1, \ldots, L_n$ be the fields conjugated to L in M. Let x be any element of $\mathfrak{P}A$ and let q be the rational prime dividing the absolute norm of $\mathfrak{P}$. For sufficiently large N we certainly have

$$x^{q^N} \in \mathfrak{P}^{q^N} A \subset \mathfrak{p}S,$$

and similarly the elements $(x^{q^N})^{(i)}$ of the integral closure S_i of R in L_i which are conjugated to x^{q^N} lie in $\mathfrak{p}S_i$ for $i = 1, 2, \ldots, n$. This proves

$$T_{L/K}(x^{q^N}) \in \mathfrak{p}S_0 \cap R = \mathfrak{p},$$

where S_0 is the integral closure of R in M. But by the choice of q the difference

$$T_{L/K}(x^{q^N}) - T_{L/K}(x)^{q^N}$$

lies in $\mathfrak{p}$; thus, finally, $T_{L/K}(x) \in \mathfrak{p}$. But the resulting inclusion $T_{L/K}(\mathfrak{P}A) \subset \mathfrak{p}S$ leads us, by Proposition 4.7 (iv) to $\mathfrak{P}A \subset \mathfrak{p}D_{L/K}^{-1}$, $\mathfrak{P}AD_{L/K} \subset \mathfrak{p}S = \mathfrak{P}^e A$ and finally $D_{L/K} \subset \mathfrak{P}^{e-1}$, proving the first part of our theorem.

To prove the second part, assume that e is relatively prime to the absolute norm of $\mathfrak{p}$, i.e., it is not divisible by the characteristic of $R/\mathfrak{p}$. In view of Theorem 4.7 it suffices to prove that every derivation of S into $S/\mathfrak{P}^e$ vanishing on R has exclusively zero-divisors in its image. Let D be such a derivation. Let $\pi \in \mathfrak{p}\setminus\mathfrak{p}^2$ and $\Pi \in \mathfrak{P}\setminus\mathfrak{P}^2$. Then, for suitable $a, b \in S$, $a, b \notin \mathfrak{P}$, we have $\pi = \Pi^e a/b$, i.e., $\pi b = \Pi^e a$, which gives

$$0 = \pi D(b) + bD(\pi) = D(\pi b) = e\Pi^{e-1}D(\Pi) + \Pi^e D(a)$$
$$= e\Pi^{e-1}D(\Pi);$$

hence $D(\Pi)$ is a zero-divisor in $S/\mathfrak{P}^e$. Now if $x \in \mathfrak{P}^k$ ($k \geq 2$), then $x = \Pi^k a/b$, where $a, b \in S$, $b \notin \mathfrak{P}$, whence

$$bD(x) + xD(b) = k\Pi^{k-1}D(\Pi)a + \Pi^k D(a),$$

and all the terms of this equality except possibly the first lie in $\mathfrak{P}/\mathfrak{P}^e$, whence the first lies there as well. But this shows that $D(x)$ is a zero-divisor in $S/\mathfrak{P}^m$.

§ 2. Different and Discriminant   167

Finally, let $x \notin \mathfrak{P}$. Then by Theorem 1.8 we have $x^{N(\mathfrak{P})-1} = 1+a$ with $a \in \mathfrak{P}$. This gives
$$(N(\mathfrak{P})-1)x^{N(\mathfrak{P})-2}D(x) = D(x^{N(\mathfrak{P})-1}) = D(1+a) = D(a),$$
which is a zero-divisor. Since $(N(\mathfrak{P})-1)x^{N(\mathfrak{P})-2} \notin \mathfrak{P}$, it follows that $D(x)$ is also a zero-divisor. □

Corollary 1. *The prime ideals of S which are ramified in L/K are exactly those which divide the different L/K. It follows that there is only a finite number of such ideals.* □

Corollary 2 (*The discriminant theorem*). *The prime ideals of R which are ramified by the extension L/K are exactly those which divide the discriminant* $d(L/K)$. □

Corollary 3. *There is no unramified extension of the rational field Q except the trivial extension Q/Q.* (Apply Theorem 2.10 and the preceding corollary.) □

It should be noted that other fields may have non-ramified extensions. To give an example, consider two square-free and relatively prime natural numbers A, B such that $A, B \equiv 1 \pmod 4$. Let
$$K = Q((AB)^{1/2}) \quad \text{and} \quad L = Q(A^{1/2}, B^{1/2}).$$
Then obviously $L = K(A^{1/2}) = K(B^{1/2})$. The differents of $A^{1/2}$ and $B^{1/2}$ over K are equal to $2A^{1/2}$ and $2B^{1/2}$, respectively, and so, by Theorem 4.6, $2 \in D_{L/K}$ as $(A, B) = 1$. Now choose natural numbers C and D satisfying $C^2 - 4D = A$. The root of $X^2 + CX + D = 0$ is an algebraic integer lying in L, and its different over K equals $A^{1/2}$, whence $A = (A^{1/2})^2$ lies in $D_{L/K}$. But $(A, 2) = 1$ and so the different contains two relatively prime natural numbers, whence it is equal to R_L and $d(L/K) = R_K$, showing that L/K is unramified.

5. Here we shall prove certain results concerning the discriminant of a composite of two extensions of the same ground-field.

Proposition 4.13. *If the extension L/K is the composite of K_1/K and K_2/K, i.e., L is the minimal field containing K_1, K_2 in a fixed common algebraic closure of K_1 and K_2, then the sets of prime ideals dividing $d(L/K)$ and $d(K_1/K)d(K_2/K)$ coincide.*

Proof. Proposition 4.9 shows that every prime ideal dividing $d(K_1/K)d(K_2/K)$ divides also $d(L/K)$. Assume now that p is a prime ideal of R which divides $d(L/K)$ but does not divide $d(K_1/K)$. We have to prove that it divides $d(K_2/K)$.

168   4. Extensions

The definition of the discriminant implies that there is an ideal $\mathfrak{P}$ of S lying over p and dividing the different $D_{L/K}$. This ideal cannot divide $D_{K_1/K}S$ since this would imply, in view of

$$\mathfrak{p}^f = N_{L/K}(\mathfrak{P})|N_{L/K}(D_{K_1/K}S) = N_{K_1/K}(N_{L/K_1}(D_{K_1/K}S))$$
$$= N_{K_1/K}(D_{K_1/K}^{[L:K_1]}) = d(K_1/K)^{[L:K_1]},$$

the divisibility of $d(K_1/K)$ by p. Since $D_{L/K} = D_{L/K_1}D_{K_1/K}$, this shows that $\mathfrak{P}$ divides D_{L/K_1}.

Now let a be any element of K_2 integral over R and generating the extension K_2/K. Let $G(t)$ be the minimal polynomial for a over K_1 and $F(t)$ its minimal polynomial over K. Then $L = K_1(a)$ and $F(t) = G(t)H(t)$ with some polynomial H over K_1. Then $F'(a) = G'(a)H(a)$, and so $F'(a)$ lies in the ideal generated in S by $G'(a)$. By Theorem 4.6, $G'(a) \in D_{L/K_1} \subset \mathfrak{P}$; thus $F'(a) \in \mathfrak{P}$. The same Theorem 4.6 gives now $D_{K_2/K} \subset \mathfrak{P}$, showing that p divides $d(K_2/K)$. □

Corollary 1. (i) *If* p *is a prime ideal of R unramified in K_1/K and K_2/K, then it is also unramified in K_1K_2/K.*

(ii) *If the extensions K_1/K and K_2/K are both unramified, then so is the composite extension K_1K_2/K.*   □

Corollary 2. *If M/K is the minimal normal extension of K containing L/K, then the discriminants $d(L/K)$ and $d(M/K)$ have the same prime ideal divisors. In particular, L/K is non-ramified iff M/K is.*   □

In the case where the ground field equals Q we can sometimes obtain more precise results. As an example of those we prove

Theorem 4.9. *Let K_i/Q ($i = 1, 2$) be finite extensions of degrees n_1, n_2 with relatively prime discriminants $d(K_1)$, $d(K_2)$ and let $L/Q = K_1K_2/Q$ be their composite. Then the degree of L/Q equals n_1n_2, the discriminant $d(L)$ equals $d(K_1)^{n_2}d(K_2)^{n_1}$, and if $w_1, \ldots, w_{n_1}$ is an integral basis of K_1 and $v_1, \ldots, v_{n_2}$ is an integral basis of K_2, then the set $\{w_iv_j\}$ forms an integral basis of L.*

Proof. Let K be the minimal normal extension of the rationals containing K_1. By Corollary 2 to Proposition 4.13 we have $(d(K), d(K_2)) = 1$. Let $K = Q(a)$ and let $f(t)$ be the minimal polynomial for a over Q. We shall show that it is irreducible over K_2. Consider $g(t)$, the minimal polynomial for a over K_2. Obviously it divides $f(t)$, and to show that $f(t)$ and $g(t)$ are equal it suffices to prove that the coefficients of $g(t)$ are all rational. Those coefficients lie in K because they are rational functions of some conjugates of a over Q, whence the field k generated by them lies in the intersection $K \cap K_2$. By Proposition 2.9 we infer that $d(k)$ divides both $d(K)$ and $d(K_2)$ and hence equals ± 1. But this can happen only if $k = Q$ (Theorem 2.10) and so $f(t) = g(t)$, whence $[L:K_2] = n_1$ and the first part of the theorem follows.

To obtain the remaining parts, observe that the discriminant of the system $\{w_i v_j\}$ equals $d(K_1)^{n_2} d(K_2)^{n_1}$, and so the discriminant of L is a divisor of that number. Moreover, Corollary 2 to Proposition 4.9 shows that $d(L)$ is divisible by $d(K_1)^{n_2}$ and also by $d(K_2)^{n_1}$, and to conclude the proof one has only to note that those numbers are relatively prime, whence $d(L)$ has to be divisible by their product. □

6. We now apply Theorem 4.9 to a description of the principal properties of cyclotomic fields. Let $K_m = Q(\zeta_m)$ be the mth cyclotomic field. Observe that for odd m the fields K_m and K_{2m} coincide in view of $\zeta_{2m} = -\zeta_m$, whence it suffices to deal in the sequel only with the case $m \not\equiv 2 \pmod 4$.

Theorem 4.10. (i) *For any $m, n \geq 1$ we have $K_m K_n = K_{[m,n]}$.*
(ii) *For any $m \geq 1$ we have $[K_m : Q] = \varphi(m)$,*

$$d(K_m) = \prod_{p^a || m} d(K_{p^a})^{\varphi(mp^{-a})},$$

and $1, \zeta_m, \zeta_m^2, \ldots, \zeta_m^{\varphi(m)-1}$ form an integral basis. Moreover, K_m/Q is normal with the Galois group isomorphic to $G(m)$, the multiplicative group of residue classes $(\bmod\ m)$, prime to m, the isomorphism being given by

$$r \mapsto g_r, \qquad g_r(\zeta_m) = \zeta_m^r.$$

Finally, the minimal polynomial of ζ_m over Q equals

$$F_m(x) = \prod_{\substack{1 \leq j < m \\ (j,m)=1}} (x - \zeta_m^j) = (x^m - 1) \bigg/ \bigg(x^m - 1, \prod_{j<m} (x^j - 1)\bigg).$$

(iii) *If $m \not\equiv 2 \pmod 4$, then $K_m \subset K_n$ holds iff $m|n$; thus for distinct m, n both incongruent to $2 \pmod 4$ the fields K_m and K_n are distinct.*

(iv) *If $m \not\equiv 2 \pmod 4$ and $m|n$, then the subgroup of $G(n)$, corresponding to K_m according to the Galois theory equals*

$$\{r \in G(n) : r \equiv 1 \pmod m\}.$$

(v) *For any m, n we have $K_m \cap K_n = K_{(m,n)}$.*

Proof. (i) If $[m, n] = s$, then

$$\zeta_m = \zeta_s^{s/m}, \qquad \zeta_n = \zeta_s^{s/n};$$

thus $K_m K_n \subset K_s$. To get the converse inclusion, solve $mx + ny = (m, n) = mn/s$ in rational integers x, y and observe that

$$\zeta_m^x \zeta_n^y = \zeta_s;$$

thus $K_s \subset K_m K_n$.

(ii) If $m = \prod_{p|m} p^{a_p}$ is the canonical factorization of m, then by (i) K_m is the composite of the fields $K_{p^{a_p}}$, which by Theorem 2.9 have their discriminants pairwise relatively prime. Theorem 4.9 gives us now the form of the discriminant and the statement about the degree. Proposition 2.9 and Theorem 2.10 imply that the intersection of $K_{p^{a_p}}$ and $K_{mp^{-a_p}}$ equals Q, whence from Theorem 2.9 it follows that K_m/Q is normal and has the Galois group isomorphic to $G(m)$. Observe finally that if $(j, m) = 1$, then $Q(\zeta_m^j) = K_m$ since $jr \equiv 1 \pmod{m}$ implies $(\zeta_m^j)^r = \zeta_m$. This shows that ζ_m^j are conjugated to ζ_m for $(j, m) = 1$, which proves the assertion about F_m. The statement about the form of isomorphism between $\mathrm{Gal}(K_m/Q)$ and $G(m)$ now results immediately.

(iii) If m divides n, then $\zeta_m = \zeta_n^{n/m}$, whence $K_m \subset K_n$ results. Conversely, if we have $K_m \subset K_n$, then denoting by d the order of $E(K_n)$ we get with a suitable a the equality $\zeta_n = \zeta_d^a$, implying $n|d$. Obviously $K_n = K_d$ and so (ii) gives $\varphi(n) = \varphi(d)$. Because of $n|d$ this is possible if either $d = n$ or n is odd and $d = 2n$. In the first case ζ_m is a power of $\zeta_d = \zeta_n$ and thus $m|n$, and in the second case the same argument leads to $m|2n$. In the third case, if m is odd, then $m|n$ results, and if m is even, then $m \equiv 2 \pmod{4}$, which is excluded by our assumption.

(iv) Denote by H the subgroup of $G(n)$, corresponding to K_m. By (iii) we have $m | n$. Clearly $r \in H$ iff $g_r(\zeta_m) = \zeta_m$; however, with $q = n/m$, we have $\zeta_m = \zeta_n^q$ and $g_r(\zeta_m) = g_r(\zeta_n^q) = \zeta_n^{rq}$, showing that $r \in H$ iff $rq \equiv q \pmod{n}$, which is equivalent to $r \equiv 1 \pmod{m}$.

(v) Let $d = (m, n)$ and $D = [m, n]$. By (i) the fields K_d, K_m, K_n are contained in K_D. If H_r is the subgroup of $G(D)$ corresponding to K_r (for $r | D$), then $K_m \cap K_n$ corresponds to

$$H_m H_n = \{s \pmod{D}: s = r_1 r_2, r_1 \equiv 1 \pmod{m}, r_2 \equiv 1 \pmod{n}\}$$

by (iv). We have to show that $H_m H_n = H_d$. The inclusion $H_m H_n \subset H_d$ being immediate, take $s \in H_d$ and solve the systems of congruences

$$\begin{cases} x \equiv s \pmod{m} \\ x \equiv 1 \pmod{n} \end{cases} \quad \text{and} \quad \begin{cases} y \equiv 1 \pmod{m} \\ y \equiv s \pmod{n}, \end{cases}$$

which is possible in view of $s \equiv 1 \pmod{d}$. Now

$x \pmod{D} \in H_n, \quad y \pmod{D} \in H_m \quad$ and $\quad xy \equiv s \pmod{D}$,

proving $s \in H_m H_n$. □

Corollary. *One has*

$$|E(K_m)| = \begin{cases} m & \text{if } 2|m, \\ 2m & \text{if } 2 \nmid m. \end{cases}$$

Proof. If $|E(K_m)| = T$ then $T \geq m$ and $K_T \subset K_m$ (because $E(K_m)$ is cyclic by Proposition 3.4) and applying (iii) we infer that $T \neq m$ can happen only in the case $T = 2t$, $2 \nmid t$; thus $K_t = K_T \subset K_m$, and so $t | m$ and $T | 2m$. But in this case $T > m$, and we get $T = 2m$ as asserted. □

The polynomial F_m occurring in the theorem is called the *m-th cyclotomic polynomial*. Its irreducibility follows from (ii); however, this can also be established in a much simpler way. From the various proofs we select one, due to K. Grandjot [24], and based on Dirichlet's prime number theorem:

Let $V \in Z[X]$ have ζ_m for one of its roots. It suffices to show that for every j prime to m the number ζ_m^j is also a root of V. For every prime $p \equiv j \pmod{m}$ we have

$$0 = V(\zeta_m)^p \equiv V(\zeta_m^p) \equiv V(\zeta_m^j) \pmod{pR_{K_m}}$$

and this shows that the algebraic integer $V(\zeta_m^j)$ is divisible by infinitely many rational primes, whence it vanishes as asserted.

7. If L/K is normal and $G = \mathrm{Gal}(L/K)$, then the question arises of determining the structure of L as $K[G]$-module and of R_L as an $R_K[G]$-module, the group rings acting in a natural way by

$$\left(\sum_{g \in G} a_g g\right) x = \sum_{g \in G} a_g g(x).$$

The structure of L as a $K[G]$-module is described by the following theorem of E. Noether:

Theorem 4.11. *If K is an infinite field and L/K its finite separable extension then L is isomorphic to $K[G]$ as an $K[G]$-module.*

(For finite K this is also true, but the proof of it is different.)

Proof. (We follow W. C. Waterhouse [79].) We start with Dedekind's theorem on linear independence of automorphisms:

Lemma 4.6. *If $g_1, \ldots, g_N$ are distinct automorphisms of a field L and for certain $c_1, \ldots, c_N \in L$ we have*
$$c_1 g_1(x) + \ldots + c_N g_N(x) = 0$$
for all $x \in L$, then $c_1 = \ldots = c_N = 0$.

Proof. Assume that the assertion is false and select a sum

$$\sum_{j=1}^{N} c_j g_j(x) \qquad (4.4)$$

with non-zero $c_j \in L$, which vanishes identically and has the number N of summands minimal. Obviously we must have $N \geq 2$. Choose y in L with $g_1(y) \neq g_2(y)$. Then

$$0 = \sum_{i=1}^{N} c_i g_i(xy) = \sum_{i=1}^{N} c_i g_i(y) g_i(x),$$

and since obviously $g_1(y) \neq 0$, we get

$$\sum_{i=1}^{N} c_i g_i(y) g_1(y)^{-1} g_i(x) = 0.$$

Subtracting this equality from (4.4), we arrive at an identically vanishing linear combination of automorphisms with fewer non-zero terms, contradicting the choice of N. □

Now let $g_1 = e, g_2, \ldots, g_n$ be all the elements of G and observe that the set of all vectors of the form

$$[g_1(x), g_2(x), \ldots, g_n(x)]$$

with $x \in L$ generate L^n, because the lemma shows that they cannot all lie in a proper subspace. Having this in mind, we may find an integer N and elements $c_1, \ldots, c_N, x_1, \ldots, x_N$ in L satisfying

$$c_1 x_1 + \ldots + c_N x_N = 1$$

and

$$\sum_{i=1}^{N} c_i g_j(x_i) = 0 \quad \text{for } j = 2, 3, \ldots, n.$$

This choice implies that the polynomial P defined by

$$P(X_1, \ldots, X_n) = \det\left[\sum_{i=1}^{N} X_i h^{-1} g(x_i)\right]_{g, h \in G}$$

does not vanish identically, since $P(c_1, \ldots, c_N) = 1$. Hence there are elements $u_1, \ldots, u_N$ in K such that $P(u_1, \ldots, u_N) \neq 0$, and this, for $A = u_1 x_1 + \ldots + u_N x_N \in L$, gives

$$\det[h^{-1} g(A)]_{g, h \in G} = P(u_1, \ldots, u_N) \neq 0.$$

This implies that the set $\{g(A) : g \in G\}$ is K-linearly independent. Indeed, if we had

$$\sum_{g \in G} d_g g(A) = 0$$

with $d_g \in K$, not all zero, then for every $h \in G$ we would obtain

$$\sum_{g \in G} d_g h^{-1} g(A) = 0,$$

and so $\det[h^{-1}g(A)]_{g,h \in G} = 0$, a contradiction.

To prove the theorem observe that the map $K[G] \to L$ defined by

$$\sum_{g \in G} a_g g \mapsto \sum_{g \in G} a_g g(A)$$

is K-linear, preserves the action of G and is surjective, since its image contains n K-linearly independent elements and $\dim_K L = n$, whence it is an isomorphism of $K[G]$-modules. □

One can ask whether an analogous result is true also for the ring R_L of integers, i.e., whether R_L is isomorphic to $R_K[G]$ as an $R_K[G]$-module. The answer, however, is negative in general. First of all the isomorphism of R_L and $R_K[G]$ would imply that R_L is a free R_K-module, and this may fail if $h(K) \neq 1$. (See the Corollary to Proposition 7.16.) However, even if $K = Q$ then R_L is not necessarily a free $Z[G]$-module, as the example $L = Q(i)$ shows. We now give a necessary condition for $R_L = R_K[G]$ to hold and do this in a more general setting, using the notation set up at the beginning of this section:

Proposition 4.14. *If L/K is normal with the Galois group G and $S \simeq R[G]$, then the trace map $T_{L/K}: S \to R$ is surjective.*

Proof. If $f: R[G] \to S$ is an isomorphism of $R[G]$-modules and $a = f(e)$, e being the unit element of G, then for $x \in R$ we have with suitable a_g ($g \in G$)

$$x = f\left(\sum_{g \in G} a_g g\right) = \sum_{g \in G} a_g g(a);$$

thus for all $h \in G$

$$x = h(x) = \sum_{g \in G} a_g hg(a) = \sum_{g \in G} a_{h^{-1}g} g(a),$$

and we infer that the coefficients a_g are independent of g, whence

$$x = a_e \sum_{g \in G} g(a) = a_e T_{L/K}(a) \in T_{L/K}(S),$$

and the trace is surjective. □

Corollary 1. *If L/K is normal with the Galois group G and $S \simeq R[G]$, then $d(L/K)$ cannot be divisible by an n-th power of a prime ideal, with $n = [L:K]$.*

Proof. Applying Corollary 2 to Proposition 4.7, we find that $D_{L/K}$ cannot have non-trivial divisors which are ideals of R_K. By Proposition 4.7 (iii) we obtain

$$d(L/K)S = N_{L/K}(D_{L/K})S = D_{L/K}^n;$$

thus if $\mathfrak{p}$ were an ideal with $\mathfrak{p}^n | d(L/K)$, then $\mathfrak{p}^n S | D_{L/K}^n$ and $\mathfrak{p} S | D_{L/K}$, a contradiction. □

Corollary 2. *If K/Q is normal of degree n and has a normal integral basis, then $d(K)$ cannot be divisible by the n-th power of a prime.*

Proof. The assertion results from the preceding corollary and Proposition 4.7. □

We shall see later that if L/K is normal, then the trace map from R_L to R_K is surjective iff L/K is tame. (See Corollary 3 to Proposition 6.2.)

The condition given in Corollary 2 above is in general not sufficient for the existence of a normal integral basis. In fact, no necessary and sufficient conditions for that are known, except for special types of Galois groups. (See § 4, subsection 6 of this chapter.)

We now prove certain simple results which permit us to obtain normal integral bases in some extensions of Q.

Proposition 4.15. (i) *If L/Q is normal and has a normal integral basis, and K/Q is normal with $K \subset L$, then K/Q also has a normal integral basis. In fact, if the conjugates of a form a normal basis of L/Q, then the conjugates of $T_{L/K}(a)$ form a normal integral basis of K/Q.*

(ii) *If K_i/Q $(i = 1, \ldots, m)$ have pairwise relatively prime discriminants, are all normal and have normal integral bases, then their composite $L = K_1 \ldots K_m$ also has a normal integral basis.*

Proof. (i) Let $G = \text{Gal}(L/Q)$ and let $\{g(a): g \in G\}$ be a normal integral basis of L. Let H be the subgroup of G fixing K. An integer $x = \sum_{g \in G} A_g g(a)$ $(A_g \in Z)$ lies in R_K iff for all $h \in H$ we have $h(x) = x$, and since

$$h(x) = \sum_{\in G} A_{h^{-1}g} g(a),$$

this occurs precisely when $A_{hg} = A_g$ holds for all h in H. It follows that if $X_1 = H, X_2, \ldots, X_m$ are the cosets (mod H) in G, then every $x \in R_K$ can be uniquely written in the form

$$x = A_1 w_1 + \ldots + A_m w_m$$

with

$$w_j = \sum_{\in X_j} g(a) \quad \text{and} \quad A_j \in Z,$$

and thus $w_1, \ldots, w_m$ is an integral basis of K. Since the w_j's are all conjugated it is a normal integral basis, and the last assertion follows from $w_1 = T_{L/K}(a)$.

(ii) An easy induction, based on Theorem 4.9 and Proposition 4.13, shows that if $w_j^{(i)}$ ($j = 1, 2, \ldots, [K_i:Q]$) is a normal basis of K_i ($i = 1, 2, \ldots, m$) then the set of all products $w_{i_1}^{(1)} \ldots w_{i_m}^{(m)}$ forms an integral basis for the composite $K_1 \ldots K_m$ and it suffices to observe that this set is invariant under the action of the Galois group. □

As a corollary we can now characterize cyclotomic fields having normal integral bases.

Corollary. *The m-th cyclotomic field K_m has a normal integral basis iff m is square-free.*

Proof. From Theorem 2.9 it follows that in the case of $m = p$, a prime, the set $\zeta_p, \zeta_p^2, \ldots, \zeta_p^{p-1}$ is an integral basis which is normal, and so the existence of a normal integral basis in K_m for square-free m results from part (ii) of the proposition.

If m is not square-free and for a certain prime p we have $p^2|m$, then $K_{p^2} \subset K_m$ and in view of part (i) of the proposition it suffices to show that K_{p^2} does not have a normal integral basis. Now Theorem 2.9 gives $[K_{p^2}:Q] = p^2 - p$ and $|d(K_{p^2})| = p^a$ with $a = 2p^2 - 3p$. Since $2p^2 - 3p \geq p^2 - p$, Corollary 2 to Proposition 4.14 implies that K_{p^2} cannot have a normal integral basis. □

We shall prove in Chapter 6 (Theorem 6.5) that every Abelian extension of the rationals is a subfield of a suitable cyclotomic field, and from this result it will follow that an Abelian field K has a normal integral basis iff it is contained in a cyclotomic field K_m with m square-free.

§3. Factorization of Prime Ideals in Extensions. More about the Class-Group

1. We shall now consider the question of establishing effectively the decomposition of a prime ideal $\mathfrak{p}$ of R into prime ideals of S. In many cases this can be done with the use of Theorem 4.12 below, which goes back to Kummer in special cases and was proved by Dedekind for all rings of algebraic integers.

We need first an analogue of Lemma 4.3:

Lemma 4.7. *Let $\mathfrak{p}$ be a prime ideal of R, a an element of S generating the extension L/K, $A = R[a]$ and $\mathfrak{f}$ the conductor of A. Then the following conditions are equivalent*:

(i) $\mathfrak{p}S \cap A = \mathfrak{p}[a]$.

(ii) *The embedding of A in S induces an isomorphism $S/\mathfrak{p}S \to A/(A \cap \mathfrak{p}S)$, i.e., every coset of S (mod $\mathfrak{p}S$) contains an element of A.*

(iii) *The embedding of A in S induces isomorphisms $S/\mathfrak{p}^m S \to A/(A \cap \mathfrak{p}^m S)$ for $m = 1, 2, \ldots$*

(iv) *For $m = 1, 2, \ldots$ we have $\mathfrak{p}^m S \cap A = \mathfrak{p}^m[a]$.*

(v) *The prime ideal $\mathfrak{p}$ does not divide $N_{L/K}(\mathfrak{f})$.*

Proof. (i) → (ii). Since $|S/\mathfrak{p}S| = N(\mathfrak{p}S) = (N\mathfrak{p})^n$ with $n = [L:K]$ and $|A/(A \cap \mathfrak{p}S)| = |A/\mathfrak{p}[a]| = (N\mathfrak{p})^n$, the induced homomorphism of $A/(A \cap \mathfrak{p}S)$ in $S/\mathfrak{p}S$ is surjective, and since it is obviously injective, we obtain (ii).

(ii) → (iii). For $m = 1$ the assertion is true by assumption. Assume thus that (iii) holds for a certain m and choose $c \in \mathfrak{p} \setminus \mathfrak{p}^2$; write $cR = \mathfrak{p}I$ (where $\mathfrak{p} \nmid I$), let $d \in I \setminus \mathfrak{p}$, define d' by $dd' \equiv 1 \pmod{\mathfrak{p}^m}$, and finally let $b \in S$. By assumption there is a polynomial V over R such that $b \equiv V(a) \pmod{\mathfrak{p}^m S}$, whence $b_1 = (b - V(a))d^m c^{-m} \in S$. If we now choose $V_1 \in R[X]$ so that $b_1 \equiv V_1(a) \pmod{\mathfrak{p}S}$, then finally

$$b = V(a) + b_1 c^m d^{-m} \equiv V(a) + (cd')^m V_1(a) \pmod{\mathfrak{p}^{1+m}S},$$

whence (iii) holds for $m+1$.

(iii) → (iv). It suffices to note that $\mathfrak{p}^m[a]$ is contained in $A \cap \mathfrak{p}^m S$ and the indices of those two rings in S are both equal to $N(\mathfrak{p})^{mn}$ by (iii).

(iv) → (i). This implication is obvious.

We have thus proved the equivalence of the first four conditions. To show that also the fifth is equivalent to them we prove two implications:

(iii) → (v). Let f be the minimal polynomial for a over R, write $N_{L/K}(f'(a))R = \mathfrak{p}^m I$ ($\mathfrak{p} \nmid I$) and let $b \in I \setminus \mathfrak{p}$. By (iii) for every $c \in S$ there is an element $d \in A$ such that $b(c-d)/N_{L/K}(f'(a))$ lies in S. Thus $b(c-d)/f'(a) \in S$, whence $b(c-d) \in f'(a)S \subset A$ by the Corollary to Proposition 4.11. Hence $bd \in A$ and we obtain $bS \subset A$, which implies $b \in \mathfrak{f}$. Since $b \notin \mathfrak{p}$, we also obtain $N_{L/K}(b) = b^n \notin \mathfrak{p}$, but $N_{L/K}(b) \subset N_{L/K}(\mathfrak{f})$, and so $\mathfrak{p}$ does not divide $N_{L/K}(\mathfrak{f})$.

(v) → (ii). If $\mathfrak{p} \nmid N_{L/K}(\mathfrak{f})$ and $\mathfrak{p}S = \prod_i \mathfrak{P}_i^{e_i}$, then no $\mathfrak{P}_i$ can divide $\mathfrak{f}$. The Chinese remainder-theorem gives the existence of $b \in S$ satisfying $b \equiv 0 \pmod{\mathfrak{f}}$, $b \equiv 1 \pmod{\mathfrak{P}_i}$ for all $\mathfrak{P}_i$'s dividing $\mathfrak{p}$. Having such a $b = V(a)$ (with $V \in R[X]$), we may put every $c \in S$ in the form $c = W(a)/b = W(a)/V(a)$ for suitable $W \in R[X]$. Since $(bS, \mathfrak{p}S) = 1$, we have for a suitable positive integer r the congruence $V(a)^r \equiv 1 \pmod{\mathfrak{p}S}$, whence

$$c \equiv W(a)V(a)^{r-1} \pmod{\mathfrak{p}S}. \qquad \square$$

Now we can state and prove the main result of this section:

Theorem 4.12. *Let $a \in S$ be a generating element for the extension L/K and let $f \in R[X]$ be its minimal polynomial. Moreover, let $\mathfrak{p}$ be a prime ideal of R satisfying the equivalent conditions of the lemma. Denote by φ the canonical map $R[t] \to R[t]/\mathfrak{p}[t] = k[t]$ resulting from the application of the residue map $R \to R/\mathfrak{p} = k$ to every coefficient. If*

$$\varphi(f) = f_1^{a_1} \ldots f_m^{a_m}$$

where the f_i's are distinct monic and irreducible polynomials over k, then

$$\mathfrak{p}S = \mathfrak{P}_1^{a_1} \ldots \mathfrak{P}_m^{a_m},$$

where $\mathfrak{P}_i$ are distinct prime ideals of S, $f_{L/K}(\mathfrak{P}_i) = n_i = \deg f_i$ and finally

$$\mathfrak{P}_i = \mathfrak{p}S + F_i(a)S \quad (i = 1, 2, \ldots, m),$$

where F_i is an arbitrary polynomial from $R[X]$ satisfying $\varphi(F_i) = f_i$.

Proof. Consider the residue maps

$$\varphi_i \colon k[t] \to k[t]/f_i k[t] = k_i$$

for $i = 1, 2, \ldots, m$. The composition $\varphi_i \circ \varphi$ will be denoted by Φ_i. Observe that the kernel of this composition is the ideal $I_i = \mathfrak{p}[t] + F_i(t)R[t]$ of $R[t]$. In fact, I_i is evidently contained in this kernel and if $V(t) \in \operatorname{Ker} \Phi_i$, then $\varphi(V) \in \operatorname{Ker} \varphi_i$, whence for a suitable $g \in k[t]$ we have $\varphi(V) = g(t)f_i(t)$, i.e. $V(t) = A(t) + B(t)F_i(t)$ for some $A, B \in \mathfrak{p}[t]$, and thus $V \in I_i$.

The principal ideal generated by f in $R[t]$ is divisible by I_i, whence the homomorphism Φ_i induces a map of $R[t]/fR[t]$ into k_i. But $R[t]/fR[t]$ is isomorphic to $R[a]$ and so, using this isomorphism, we obtain a map

$$\psi_i \colon R[a] \to k_i,$$

which is easily seen to be surjective. Its kernel equals $\mathfrak{p}[a] + F_i(a)R[a]$ because it consists of elements $V(a)$ with $V \in I_i$.

After these preparations, we may define a mapping of S into k_i in the following way:

Let $b \in S$ and choose (using (ii) of Lemma 4.7) a polynomial $V \in R[t]$ such that $b \equiv V(a) \pmod{\mathfrak{p}S}$. Then define

$$\Psi_i(b) = \psi_i(V(a)).$$

This is well defined since from $V(a) \equiv 0 \pmod{\mathfrak{p}S}$ we obtain, by condition (i) of Lemma 4.7, $V(a) \in \mathfrak{p}S \cap A = \mathfrak{p}[a]$, and so $\psi_i(V(a)) = 0$.

The mapping so defined is surjective because ψ_i was such. This shows that the ideal $\mathfrak{P}_i = \operatorname{Ker}\Psi_i$ is a non-zero prime ideal in S. Note that $\mathfrak{P}_i = \mathfrak{p}S + F_i(a)S$. In fact, we have

$$\mathfrak{P}_i = \{c \in S: c \equiv V(a) \pmod{\mathfrak{p}S}, V \in R[t], V(a) \in \mathfrak{p}[a] + F_i(a)A\}$$
$$= \mathfrak{p}S + \mathfrak{p}[a] + F_i(a)A = \mathfrak{p}S + F_i(a)A = \mathfrak{p}S + F_i(a)S,$$

since condition (ii) of Lemma 4.7 implies $S = A + \mathfrak{p}S$, which in turn gives $F_i(a)S \subset F_i(a)A + \mathfrak{p}F_i(a)S \subset F_i(a)A + \mathfrak{p}S$.

If we had $\mathfrak{P}_i = \mathfrak{P}_j$ for $i \neq j$, then in view of $(f_i, f_j) = 1$ there would exist polynomials $A, B \in R[t]$ and $C \in \mathfrak{p}[t]$ such that $A(t)F_i(t) + B(t)F_j(t) = 1 + C(t)$. (Remember that k is a finite field!) But $F_i(a) \in \mathfrak{P}_i$ and $F_j(a) \in \mathfrak{P}_j$; thus putting $t = a$ in the resulting equality, we can see that $1 \in \mathfrak{P}_i$, which is nonsense.

Since the element

$$F_1(a)^{a_1} \ldots F_m(a)^{a_m} \tag{4.5}$$

belongs to $\mathfrak{p}S$, the prime ideals $\mathfrak{P}_1, \ldots, \mathfrak{P}_m$ are the only possible prime ideals dividing $\mathfrak{p}S$, whence for some $e_1, \ldots, e_m$ we must have $\mathfrak{p}S = \mathfrak{P}_1^{e_1} \ldots \mathfrak{P}_m^{e_m}$. The degree $f_{L/K}(\mathfrak{P}_i)$ equals $[k_i : k]$ and this is equal to n_i. It remains to show that $a_i = c_i$ holds for $i = 1, 2, \ldots, m$. The ideal $\mathfrak{P}_i^{e_i}$ divides the product (4.5) but, as we have seen, $\mathfrak{P}_i$ can divide only the factor $F_i(a)^{a_i}$ of that product, and thus $\mathfrak{P}_i^{e_i}$ divides $F_i^{a_i}(a)S$. It follows that the ideal $\mathfrak{p}S + F_i(a)^{a_i}S$ is divisible by $\mathfrak{P}_i^{e_i}$ but it divides $\mathfrak{P}_i^{a_i} = \mathfrak{p}^{a_i}S + F_i(a)^{a_i}S$; thus $a_i \geq c_i$ for $i = 1, 2, \ldots, m$. Finally, observe that by Theorem 4.1 $e_1 n_1 + \ldots + e_m n_m = n$ and obviously $a_1 n_1 + \ldots + a_m n_m = n$, which together with the inequalities just obtained shows $a_i = e_i$. □

Observe that condition (v) of Lemma 4.7 shows that the conditions of the theorem proved are satisfied by almost all prime ideals of R, i.e. by all with finitely many exceptions. In the case of a rational ground-field the exceptional primes are, in view of Lemma 4.7 (v), exactly those which divide the index of a. By varying a we may apply the theorem proved above to all primes except those which divide the indices of all integers in the field, the possibility of which situation was demonstrated in Chapter 2, § 2.5.

2. Now let us turn to applications of Theorem 4.12 to algebraic number fields. We start with a necessary and sufficient condition for a rational prime to divide all indices of integers from a given field K. It is useful to introduce the following definition: the greatest common divisor of indices of all integers of K will be called the *index of the field* K. The index of K will be denoted by $i(K)$.

If $a \in R_K$, then the definition of the index of a implies immediately that a rational prime p divides this index iff there exists a polynomial over Z with a degree not exceeding $n-1$ with $n = [K:Q]$ whose coefficients are not all divisible by p but which satisfies $V(a) \equiv 0 \pmod{p}$. Having this in mind, we prove the following result:

Theorem 4.13. *Let p be a rational prime (which is convenient to identify with the prime ideal in Z generated by it) and let*

$$pR_K = \mathfrak{P}_1^{e_1} \ldots \mathfrak{P}_t^{e_t}$$

be the factorization into prime ideals in K. Moreover, let $f_i = f_{K/Q}(\mathfrak{P}_i)$. Then the number p does not divide the index of the field K iff there exist distinct polynomials $V_1, \ldots, V_t$ over the finite field $Z/pZ = GF(p)$ irreducible over that field, their degrees being equal to $f_1, \ldots, f_t$ respectively.

Proof. If $p \nmid i(K)$, one can find an integer of K with index not divisible by p and then Theorem 4.12 implies the existence of polynomials with the required properties. To prove the converse implication let $W_1, \ldots, W_t$ be polynomials over Z which after reduction $(\bmod\ p)$ coincide with $V_1, \ldots, V_t$. For each $i = 1, 2, \ldots, t$ choose an $a_i \in R_K$ satisfying the following conditions:

$$W_i(a_i) \equiv 0 \pmod{\mathfrak{P}_i}, \quad W_i(a_i) \not\equiv 0 \pmod{\mathfrak{P}_i^2}, \quad a_i \equiv 0 \pmod{I_i},$$

where $I_i = pR_K/\mathfrak{P}_i^{e_i}$. From the proof of Lemma 4.4 we infer that the conductor of $Z[a_i]$ is not divisible by $\mathfrak{P}_i$. It follows immediately that if $a \equiv a_i \pmod{\mathfrak{P}_i^2}$ for $i = 1, 2, \ldots, t$ then the conductor of $Z[a]$ is prime to $\mathfrak{P}_1 \mathfrak{P}_2 \ldots \mathfrak{P}_t$. Put $\mathfrak{Q}_i = pR_K + W_i(a)R_K$. We shall show that $\mathfrak{P}_i = \mathfrak{Q}_i$. In fact, $\mathfrak{P}_i | \mathfrak{Q}_i$, $\mathfrak{P}_i^2 \nmid \mathfrak{Q}_i$ and moreover for $i \neq j$ we have $\mathfrak{P}_j \nmid \mathfrak{Q}_i$ since otherwise we would have $W_i(a) \equiv 0 \pmod{\mathfrak{P}_j}$ and thus $W_i(a_j) \equiv 0 \pmod{\mathfrak{P}_j}$, but in view of $W_j(a_j) \equiv 0 \pmod{\mathfrak{P}_j}$ this would imply that V_i and V_j have a common root in the field $R_K/\mathfrak{P}_j$, which is impossible, since they are relatively prime. Hence we have $\mathfrak{P}_i = \mathfrak{Q}_i$ for $i = 1, 2, \ldots, t$. Now consider the polynomial

$$R = W_1^{e_1} \ldots W_t^{e_t}.$$

Clearly $W(a) \equiv 0 \pmod{pR_K}$. If $M = W_1^{a_1} \ldots W_t^{a_t}$ is a divisor of R which satisfies $M(a) \equiv 0 \pmod{pR_K}$, then $\mathfrak{P}_i^{e_i}$ divides $M(a)R_K$, and since $\mathfrak{P}_i$ and $W_j(a)R_K$ are relatively prime for $i \neq j$, we get $\mathfrak{P}_i^{e_i} | W_i^{a_i}(a)$, i.e. $a_i = e_i$ and $M = R$.

Write $n = [K:Q]$ and observe that if $p | i(K)$, then for a suitable $S \in Z[X]$ with coefficients not all divisible by p and $\deg S \leq n-1$ we would have $S(a) \equiv 0 \pmod{pR_K}$. But the polynomial $(R(X), S(X))$ divides R and has the same property as S, whence by the previous remark it equals R, i.e. $R | S$, whence finally $n = \deg R \leq \deg S \leq n-1$, a contradiction. □

180   4. Extensions

To apply this theorem one has to know the number $r_p(n)$ of non-associated irreducible polynomials over $GF(p)$ of degree n. This number was found by Gauss, and we now prove his formula:

Proposition 4.16. *For every prime p and $n \geq 1$ one has*

$$r_p(n) = \frac{1}{n} \sum_{d|n} \mu(d) p^{n/d}, \qquad (4.6)$$

where $\mu(d)$ is the familiar Möbius function.

Proof. Consider the field $GF(p^n)$ which has degree n over $GF(p)$. Every irreducible polynomial over $GF(p)$ of degree d dividing n has exactly d distinct roots in $GF(p^n)$ since this is the unique extension of $GF(p)$ of degree n. Noting that non-associated irreducible polynomials cannot have common zeros, we obtain

$$p^n = |GF(p^n)| = \sum_{d|n} d r_p(d),$$

and the application of the Möbius inversion formula gives (4.6).   □

The formula obtained permits us to bound the prime divisors of $i(K)$ from above:

Proposition 4.17. *All prime divisors of $i(K)$ are smaller than $n = [K:Q]$.*

Proof. If $p | i(K)$ and $p \geq n$, then among the numbers $f_{K/Q}(\mathfrak{P})$ (where $\mathfrak{P}$ ranges over all prime ideals dividing pR_K) there can be at most n/k equal to k, whence by Theorem 4.14 there must exist k_0 such that $r_p(k_0) < n/k_0 \leq p/k_0$. However, the sum on the right-hand side of (4.6) is divisible by p and non-zero, and this implies $r_p(k_0) \geq p/k_0$, a contradiction.   □

Corollary. *If K is a cubic extension of Q and $p|i(K)$, then $p = 2$. This number is an index-divisor iff $2R_K = \mathfrak{P}_1 \mathfrak{P}_2 \mathfrak{P}_3$ with $\mathfrak{P}_i$ distinct.*

Proof. Observe that there are two irreducible linear polynomials over $GF(2)$.   □

Our next application of Theorem 4.12 deals with splitting primes:

Theorem 4.14. *If L/K is finite, then there are infinitely many prime ideals of R_K splitting completely in L/K.*

Proof. It suffices to prove the assertion in the case $K = Q$ and L/Q normal.

In fact, if M/Q is a normal extension containing L, then Proposition 4.3 shows that if pZ splits in M/Q then every prime ideal of R_K lying above pZ splits also in L/K. Thus, using Theorem 4.12, we can reduce our assertion to the following form: If $P \in Z[X]$ is non-constant, then for infinitely many primes p the congruence $P(x) \equiv 0 \pmod{p}$ has a solution. Assume this to be false for a certain P. Then its values can be divisible only by finitely many primes, say $p_1, \ldots, p_r$. Select x_0 with $|P(x_0)| \geq 2$. Then with suitable non-negative $a_1, \ldots, a_r$ we can write

$$|P(x_0)| = p_1^{a_1} \ldots p_r^{a_r}.$$

If $M = p_1^{a_1+1} \ldots p_r^{a_r+1}$ and $x \equiv x_0 \pmod{M}$, then with suitable b_i we have

$$p_1^{b_1} \ldots p_r^{b_r} = |P(x)| \equiv |P(x_0)| \pmod{M},$$

which implies $a_i = b_i$ for $i = 1, 2, \ldots, r$; thus for infinitely many x we have $|P(x)| = |P(x_0)|$, which shows that P must be constant, a contradiction. $\square$

One more application of Theorem 4.12, which is very helpful in factorizing prime ideals, is based on an extension of the notion of an Eisensteinian polynomial: If P is a polynomial over some R_K, $P(x) = a_n x^n + \ldots + a_0$, and for a certain prime ideal $\mathfrak{p}$ of R_K we have $a_n \notin \mathfrak{p}$, $a_{n-1}, \ldots, a_0 \in \mathfrak{p}$, $a_0 \notin \mathfrak{p}^2$, then P is called $\mathfrak{p}$-*Eisensteinian*, or *Eisensteinian with respect to* $\mathfrak{p}$. One proves in the same way as for the case of $R_K = Z$ that such a polynomial is irreducible over K. We now prove

Proposition 4.18. *If $a \in R_L$ generates the extension L/K and its minimal polynomial $f(x) = x^n + a_{n-1} x^{n-1} + \ldots + a_0$ over R_K is $\mathfrak{p}$-Eisensteinian for a certain prime ideal $\mathfrak{p}$ of R_K, then that prime ideal $\mathfrak{p}$ ramifies completely in L/K, i.e. $\mathfrak{p} R_L = \mathfrak{P}^n$ for a certain prime ideal $\mathfrak{P}$ of R_L.*

Proof. Let $\mathfrak{P}$ be any prime ideal of R_L lying above $\mathfrak{p}$ and let $\mathfrak{P}^e$ be its highest power dividing $\mathfrak{p} R_L$. Since e cannot exceed n by Corollary 1 to Theorem 4.1, it suffices to show that e is at least equal to n. Since $a^n \in \mathfrak{p} R_L$, we have $a \in \mathfrak{P}$. Moreover, the coefficients $a_0, a_1, \ldots, a_{n-1}$ belong to $\mathfrak{P}^e$, and if we had $e \leq n-1$, then the ideal $\mathfrak{P}^{1+e}$ would divide the ideal generated by the number $a_0 = -(a^n + \ldots + a_1 a)$ and so $\mathfrak{p}^2$ would contain a_0, giving a contradiction. $\square$

3. Finally, we turn to the principal applications of Theorem 4.12, namely to explicit factorization of prime ideals. We start with quadratic fields and prove

Theorem 4.15. *Let K be a quadratic extension of the rationals with the discriminant $d(K) = d$ and let p be a rational prime.*

If p is odd, then

$$pR_K = \begin{cases} \mathfrak{P}_1\mathfrak{P}_2, & \text{iff } \left(\dfrac{d}{p}\right) = 1, \\ \mathfrak{P}, & \text{iff } \left(\dfrac{d}{p}\right) = -1. \end{cases}$$

If $p = 2$, then

$$2R_K = \begin{cases} \mathfrak{P}_1\mathfrak{P}_2, & \text{iff } d \equiv 1 \pmod{8}, \\ \mathfrak{P}, & \text{iff } d \equiv 5 \pmod{8}. \end{cases}$$

Finally, if p divides d, then irrespective of its parity we have

$$pR_K = \mathfrak{P}^2.$$

Proof. If D is the square-free part of d, i.e., $D = d$ if $d \equiv 1 \pmod 4$ and $D = d/4$ if d is divisible by 4, then the number $D^{1/2}$ generates K and its index is, in view of Theorem 2.7, either 1 or 2, the last case arising if $d \equiv 1 \pmod 4$. As the minimal polynomial for $D^{1/2}$ equals $x^2 - D$, our assertion concerning odd primes p follows immediately from Theorem 4.12 and the same happens for $p = 2$ provided d is divisible by 4. The remaining case requires the observation that $(1+D^{1/2})/2$ has index 1 and its minimal polynomial equals $x^2 - x + (1-D)/4$, which is irreducible over $GF(2)$ iff D is congruent to 5 (mod 8). □

The next class of fields with which we shall deal consists of cyclotomic fields $Q(\zeta_m)$ with ζ_m equal to a primitive mth root of unity. In this case we prove

Theorem 4.16. *Let $K = Q(\zeta_m)$. If the rational prime p does not divide m and f is the least natural number such that $p^f \equiv 1 \pmod m$, then $pR_K = \mathfrak{P}_1 \ldots \mathfrak{P}_g$ with $g = \varphi(m)/f$, all $\mathfrak{P}_i$'s distinct and of degree f.*

However, if p divides m, $m = p^a m_1$ with $p \nmid m_1$ and f is the least natural number such that $p^f \equiv 1 \pmod{m_1}$, then $pR_K = (\mathfrak{P}_1 \ldots \mathfrak{P}_g)^e$ with $e = \varphi(p^a)$, $g = \varphi(m_1)/f$, all $\mathfrak{P}_i$'s being distinct and of degree f.

Proof. We start with a lemma which is also applicable in a more general situation:

Lemma 4.8. *Let ζ_m be a primitive m-th root of unity and let K be any algebraic number field. Assume that $\mathfrak{p}$ is a prime ideal of R_K not containing m and of the first degree over the rational field Q. If f is the least natural number such that $N(\mathfrak{p})^f \equiv 1 \pmod m$, then in the field $L = K(\zeta_m)$ we have the decomposition*

$$\mathfrak{p}R_L = \mathfrak{P}_1 \ldots \mathfrak{P}_g,$$

all prime ideals $\mathfrak{P}_i$ being distinct and of degree f over K and $g = [L:K]/f$.

§ 3. Factorization of Prime Ideals in Extensions   183

Proof of the lemma. Since the extension L/K is always normal, we have only to show that if t is the degree of any prime ideal $\mathfrak{P}$ of R_L which lies over $\mathfrak{p}$, then $t = f$. Observe first that the conductor $\mathfrak{f}_A$ of the ring $A = R_K[\zeta_m]$ divides the ideal mR_L. Indeed, if we denote by $f(x)$ the minimal polynomial of ζ_m over K, then for a certain $g(x) \in R_K[x]$ we have $x^m - 1 = f(x)g(x)$. (The coefficients of $g(x)$ lie in R_K because all roots of $g(x)$ are algebraic integers.) This shows that

$$\delta_{L/K}(\zeta_m) g(\zeta_m) = m\zeta_m^{m-1},$$

and since ζ_m is a unit, we can see that the ideal $\delta_{L/K}(\zeta_m) R_L$ divides mR_L and so does $\mathfrak{f}_A$ in view of Proposition 4.12 (i). But this show that $N_{L/K}(\mathfrak{f}_A)$ divides $m^{[L:K]}R_K$, and so our $\mathfrak{p}$ satisfies condition (v) of Lemma 4.7; hence for every x in R_L we may find a polynomial F over R_K such that

$$x \equiv F(\zeta_m) \pmod{\mathfrak{p} R_L}. \tag{4.7}$$

From the properties of multinomial coefficients it follows that

$$x^{N(\mathfrak{p})^f} \equiv F(\zeta_m)^{N(\mathfrak{p})^f} \equiv F(\zeta_m^{N(\mathfrak{p})^f}) \equiv F(\zeta_m) \pmod{\mathfrak{p} R_L}$$

and also (mod $\mathfrak{P}$), but Theorem 1.7 shows that $N(\mathfrak{P}) = N(\mathfrak{p})^t$ is the minimal exponent r for which $x^r \equiv x \pmod{\mathfrak{P}}$ holds identically in R_L; hence t cannot exceed f. If we had $N(\mathfrak{p})^t \not\equiv 1 \pmod{m}$, then $\zeta_m^{N(\mathfrak{p})^t}$ would be a primitive mth root of unity $\neq \zeta_m$, and so the difference $\zeta_m^{N(\mathfrak{p})^t} - \zeta_m$ would be an element of $\mathfrak{P}$. If we put $K_m = Q(\zeta_m)$, then this shows that the discriminant $d_{K_m/Q}(\zeta_m)$ lies in $\mathfrak{P}$, and hence in $\mathfrak{P} \cap Z$, which is generated by a rational prime not dividing m. But this would contradict Theorem 4.10. Hence $N(\mathfrak{p})^t \equiv 1 \pmod{m}$, and so f cannot exceed t and $t = f$ follows, as asserted. □

Our assertion concerning primes not dividing m follows immediately. The case of $p|m$ is more complicated. Denote by K_1 the field $Q(\zeta_{m_1})$. Since $[K:Q] = \varphi(m)$ and $[K_1:Q] = \varphi(m_1)$, it follows that

$$[K:K_1] = \varphi(m)/\varphi(m_1) = \varphi(p^a) = [Q(\zeta_{p^a}):Q],$$

and we can see that the p^ath cyclotomic polynomial $F_{p^a}(x)$ is irreducible over K_1. Since p does not divide m_1, we may apply the already proved part of the theorem to the field K_1 and obtain

$$pR_{K_1} = \mathfrak{p}_1 \ldots \mathfrak{p}_g$$

for distinct prime ideals $\mathfrak{p}_1, \ldots, \mathfrak{p}_g$ and $g = \varphi(m_1)/f$. In K we have the factorization

$$pR_K = (\mathfrak{P}_1 \ldots \mathfrak{P}_r)^e$$

with some r and e yet to be determined. But we know that $ref_{K/Q}(\mathfrak{P}_i) = [K:Q] = \varphi(m_1)$, and since p is not ramified in K_1/Q, we must have

$$e|[K:K_1] = \varphi(p^a).$$

Consider the following factorization of the rational prime p in K:

$$p = F_{p^a}(1) = \prod_k (1-\zeta_{p^a}^k) = (1-\zeta_{p^a})^{\varphi(p^a)} \prod_k (1+\zeta_{p^a}+ \ldots +\zeta_{p^a}^{k-1}),$$

where all products are taken over k not divisible by p from the interval $[1, p^a]$. The number $1-\zeta_{p^a}$ is obviously not a unit, whence the last product occurring in the above equality must be a unit, since p cannot be factorized into more than $\varphi(p^a)$ non-unit factors in the field $Q(\zeta_{p^a})$. Denoting this unit by ε, we finally arrive at the equality

$$p = \varepsilon(1-\zeta_{p^a})^{\varphi(p^a)},$$

and so

$$pR_K = (1-\zeta_{p^a})^{\varphi(p^a)}R_K = (\mathfrak{P}_1 \ldots \mathfrak{P}_r)^e,$$

and since e divides $\varphi(p^a)$, we obtain $e = \varphi(p^a)$. Now we determine r. The inequality $r \leq g$ is obvious. On the other hand, we have

$$r\varphi(p^a)f_{K/Q}(\mathfrak{P}_i) = \varphi(m) = \varphi(m_1)\varphi(p^a);$$

thus

$$rf_{K/Q}(\mathfrak{P}_i) = \varphi(m_1),$$

and also

$$fg = \varphi(m_1), \quad f_{K/Q}(\mathfrak{P}_i) = f_{K/K_1}(\mathfrak{P}_i)f.$$

Putting these equalities together, we get $rf_{K/K_1}(\mathfrak{P}_i) = g$ and thus $g \leq r$. □

For further reference we state separately the factorization of the prime p in $Q(\zeta_{p^a})$ obtained in the proof of the theorem:

Corollary. *If p is a prime and $n \geq 1$, then for $K = Q(\zeta_{p^n})$ we have $pR_K = \mathfrak{P}^e$, where $e = \varphi(p^n)$ and $\mathfrak{P}$ is the principal prime ideal generated by $1-\zeta_{p^n}$.* □

As a third example we consider the Kummerian extensions, i.e. extensions L/K, where $L = K(a)$, a being a root of an irreducible polynomial over K of the form $x^n - c$ with all nth roots of unity contained in K. Such extensions are always cyclic, and every cyclic extension of degree n of a field containing the nth roots of unity must necessarily be of this form. The proof of this fact may be found in any textbook of the Galois theory.

We prove

Theorem 4.17. *Let L/K be a Kummerian extension of degree m, generated by $a \in R_L$ with $a^m = c \in R_K$. Moreover, let $\mathfrak{p}$ be a prime ideal of R_K not containing mc, and define r as the maximal natural divisor of m for which the congruence*

$$x^r \equiv c \pmod{\mathfrak{p}}$$

has a solution in R_K. Then $\mathfrak{p}R_L = \mathfrak{P}_1 \ldots \mathfrak{P}_r$ with all prime ideals $\mathfrak{P}_i$ distinct.

Proof. We need first a simple lemma:

Lemma 4.9. *Let k be the field $R_K/\mathfrak{p}$ and let k_1 be the field $R_L/\mathfrak{P}$, where $\mathfrak{P}$ is any prime ideal of R_L lying above $\mathfrak{p}$. Denote by $\bar{a}$ the image of a in k_1, and finally let $tu = m$. Then the congruence*
$$x^t \equiv c \pmod{\mathfrak{p}}$$
has a solution in R_K iff the element $(\bar{a})^u$ lies in k.

Proof of the lemma. If $(\bar{a})^u \in k$, then obviously every element of R_K lying in the residue class $(\bar{a})^u$ satisfies our congruence. Conversely, if $x^t \equiv c \pmod{\mathfrak{p}}$ and $\bar{x}$ is the image of x in k, then $\bar{c} = (\bar{a})^m = (\bar{a})^{tu}$, whence $(\bar{x}/\bar{a}^u)^t = 1$. But the finite field k contains all tth roots of unity (from a fixed algebraic closure of k) since by our assumption K contains the splitting field of $x^m - 1$ and the same applies to k. Thus $\bar{x}/\bar{a}^u \in k$ and $\bar{a}^u \in k$. □

In the proof of the theorem we retain the notation of the lemma. Let f be the minimal positive exponent such that $\bar{a}^f = \bar{b}$ lies in k. Observe that the polynomial $x^f - \bar{b}$ is irreducible over k. In fact, k contains primitive mth roots of unity; so let us denote one of them by z, and if $m = Af + B$, where $0 \leq B < f$, then $\bar{a}^B$ lies in k; thus $B = 0$, f divides m and k contains also fth roots of unity. Since
$$x^f - \bar{b} = x^f - \bar{a}^f = \prod_{i=0}^{f-1} (x - \bar{a}w^i),$$
w being a primitive fth root of unity in k, the constant term of any presumable factor of $x^f - \bar{b}$ would have the form $\bar{a}^u y$ for an $y \in k$ and a $u < f$. But our choice of f gives $u = f$, which proves the irreducibility of the polynomial concerned. Since a root of this polynomial generates the extension k_1/k, we obtain $f_{L/K}(\mathfrak{P}) = [k_1:k] = f$, and the application of Lemma 4.9 shows the truth of our assertion. □

4. The knowledge of factorization of prime ideals of Z in an extension K/Q is very helpful in determining the class-number of K and the structure of its class-group $H(K)$. We shall illustrate this on the examples which follow.

We first consider imaginary quadratic fields and prove

Proposition 4.19. *Quadratic imaginary fields with discriminants $d = -3, -4, -7, -8, -11, -19, -43, -67$ and -163 have a trivial class group.*

Proof. For fields with $d = -3, -4, -7, -8$ we obtain this immediately from Lemma 3.3. In the remaining cases we again use this lemma. According to it in every class of ideals in a quadratic imaginary field there must occur an ideal whose abolute norm does not exceed $(2/\pi)|d|^{1/2} < 0.63662|d|^{1/2}$. Applying this bound to $d = -11, -19, -43, -67, -163$, we obtain in every class an ideal with absolute norm equal at most to 2, 2, 4, 5 and 8, respectively. Since our discriminants are congruent to 5 (mod 8), Theorem 4.15 shows that the number 2 generates a principal prime ideal, and so there is only one ideal, namely R_K, with absolute norm not exceeding 2. This proves that $h = 1$ in the first two cases. For $d = -43$ we have to consider also ideals with norms 3 and 4. Since $\left(\frac{-43}{3}\right) = -1$, $3R_K$ is a prime ideal of norm 9, and thus there are no ideals of norm 3, and there is only one ideal of norm 4, namely the principal ideal $2R_K$. This proves that $h = 1$ for $d = -43$. In the case of $d = -67$ the required result follows from $\left(\frac{-67}{3}\right) = \left(\frac{-67}{5}\right) = -1$ and in the case of $d = -163$ from $\left(\frac{-163}{p}\right) = -1$ for $p = 3, 5, 7$. □

It is much more difficult to prove that those fields are the only imaginary quadratic fields with class-number one. (Cf. Theorem 8.10.)

To obtain a less trivial example of $H(K)$ consider the field $Q((-14)^{1/2})$ with the discriminant -56. Lemma 3.3 shows that every class of ideals contains an ideal with absolute norm equal to at most four. Theorem 4.15 implies $2R_K = \mathfrak{p}_2^2$ with a prime ideal $\mathfrak{p}_2$ which is not principal, since its norm equals 2, and the quadratic form $x^2 + 14y^2 = N_{K/Q}(x+y(-14)^{1/2})$ does not represent 2; so there are no principal ideals with that norm. By the same theorem we get $3R_K = \mathfrak{p}_3\mathfrak{p}_3'$ with non-principal $\mathfrak{p}_3, \mathfrak{p}_3'$ for the same reason. Finally, the only ideal with norm 4 is $2R_K$, which is principal. Let X be the ideal class containing $\mathfrak{p}_2$ and Y the class containing $\mathfrak{p}_3$. If we had $X = Y$, then in view of $X^2 = E$, E being the principal class, the product $\mathfrak{p}_2\mathfrak{p}_3$ would be principal, but its norm is 6 and there is no such principal ideal. Thus $X \neq Y$. Since $h(K)$ cannot exceed 4, $H(K)$ contains an element of order two and finally $h(K) \geq 3$; thus we must have $h(K) = 4$, and $H(K)$ consists of E, X, Y, Y^{-1} with $Y \neq Y^{-1}$ and so is cyclic.

Similar reasoning can be applied to real quadratic fields. However, in this case additional difficulties arise connected with testing whether there are principal ideals of the given norm. We illustrate this on the example of $K = Q((10)^{1/2})$. Here the discriminant equals 40, and so we have to test ideals with norms not exceeding 3. By Theorem 4.15 we have $2R_K = \mathfrak{p}_2^2, 3R_K = \mathfrak{p}_3\mathfrak{p}_3'$. But there are no principal ideals of norm 2, since the form $x^2 - 10y^2$ does not represent 2, 2 being a quadratic nonresidue (mod 5), and the same applies to number 3. Thus $\mathfrak{p}_2, \mathfrak{p}_3$ and $\mathfrak{p}_3'$ are non-principal. However, they all lie in the same class, since the only ideals with norm 6 are $\mathfrak{p}_2\mathfrak{p}_3'$ and $\mathfrak{p}_2\mathfrak{p}_3$, and the principal ideals generated by $4+10^{1/2}$ and $4-10^{1/2}$ have norm 6 and are distinct, the ratio $(4+10^{1/2})/(4-10^{1/2})$ not being integral. This shows that $\mathfrak{p}_3$ and $\mathfrak{p}_3'$ lie in the same class as $\mathfrak{p}_2$. Finally, we get $H(K) = C_2$.

To give some examples of class-determination in pure cubic fields, we shall consider fields $K = Q(m^{1/3})$ with m square-free and not congruent to 1 or 8 (mod 9). Theorem 2.8 shows that $d(K) = -27m^2$ and the numbers $1, m^{1/3}, m^{2/3}$ form an integral basis for K; thus the index of $m^{1/3}$ equals unity. A trivial but tedious computation shows that

$$N_{K/Q}(x+ym^{1/3}+zm^{2/3}) = x^3 + my^3 + m^2 z^3 - 3xyzm$$

and Lemma 3.3 gives the existence in every class of an ideal with a norm not exceeding $3m/2$.

If $m = 2$, we have to check only ideals with norm 2, but the polynomial $X^3 - 2$ is a third power (mod 2); hence $2R_K = \mathfrak{p}^3$, and the prime ideal $\mathfrak{p}$ is principal, generated by $m^{1/3}$. Thus $h = 1$.

In the case of $m = 3$ we deal with ideals with norms equal to 2, 3 and 4. Since $x^3 - 3 \equiv (x-1)(x^2+x+1) \pmod{2}$, both factors being irreducible, we see that $2R_K = \mathfrak{p}\mathfrak{q}$ with $N(\mathfrak{p}) = 2$ and $N(\mathfrak{q}) = 4$. Those prime ideals are principal, since $N_{K/Q}(m^{1/3}-1) = 2$ and $\mathfrak{p}$ is the only ideal with norm 2. Moreover, $x^3 - 3$ is a third power (mod 3), and so $3R_K = \mathfrak{p}_1^3$, $\mathfrak{p}_1$ being generated by $3^{1/3}$. Hence again we obtain $h = 1$. By the same method one shows that for $m = 5$ and $m = 6$ one obtains fields with a trivial class-group. Finally, consider $m = 7$. Here we have to check ideals with norms at most equal to 10. Factorizing the polynomial $x^3 - 7$ in $GF(p)$ for $p = 2, 3, 5$ and 7, we can see that

$$2R_K = \mathfrak{p}_2\mathfrak{q}_2 \quad \text{with } N(\mathfrak{p}_2) = 2, N(\mathfrak{q}_2) = 4,$$
$$3R_K = \mathfrak{p}_3^3 \quad \text{with } N(\mathfrak{p}_3) = 3,$$
$$5R_K = \mathfrak{p}_5\mathfrak{q}_5 \quad \text{with } N(\mathfrak{p}_5) = 5, N(\mathfrak{q}_5) = 25,$$
$$7R_K = \mathfrak{p}_7^3 \quad \text{with } N(\mathfrak{p}_7) = 7.$$

Now observe that every cube is either divisible by 7 or congruent to 1 or $-1 \pmod{7}$, and so every norm of an integer of K is congruent to $0, -1$ or $1 \pmod{7}$. This shows that the ideals $\mathfrak{p}_2, \mathfrak{q}_2, \mathfrak{p}_3, \mathfrak{p}_5, \mathfrak{q}_5$ are not principal. The ideal $\mathfrak{p}_7$, however, is obviously principal since it is generated by $7^{1/3}$. Denote by X the class of ideals containing $\mathfrak{p}_3$. Then $X^3 = E$, E being the unit class. Putting in our expression of the norm first $x = 1, y = -1, z = 0$ and then $x = 2, y = 1, z = 0$, we can see that there are principal ideals of the norms 6 and 15 which have to be equal to $\mathfrak{p}_2\mathfrak{p}_3$ and $\mathfrak{p}_3\mathfrak{p}_5$, respectively. But this shows that $\mathfrak{p}_2 \in X^{-1} = X^2$, $\mathfrak{p}_5 \in X^2$; thus $\mathfrak{q}_2 \in X$, $\mathfrak{q}_5 \in X^2$, and so every listed prime ideal lies in the cyclic subgroup of $H(K)$ generated by X. But every other ideal yet to be tested is a product of those prime ideals, and so $H(K)$ coincides with this subgroup, i.e. $h = 3$, and $H(K) = C_3$.

Now we turn to cyclotomic fields and prove

Proposition 4.20. *The cyclotomic fields $Q(\zeta_m)$ with $m = 3, 4, 5, 6, 7, 8$ have a trivial class-group.*

Proof. The cases of $m = 3$, $m = 4$ and $m = 6$ are covered by the preceding proposition. For $m = 5$ we have to check only ideals with norm 2, but by Theorem 4.16 we have $2R_K = \mathfrak{p}$, which is a prime ideal, and so there are no ideals with norm 2 at all, and thus $h = 1$. For $m = 7$ we have to consider ideals with norms at most equal to 5. Since the numbers 3 and 5 generate prime ideals, the prime 2 remains, and for it we get $2R_K = \mathfrak{p}\mathfrak{q}$ with $N(\mathfrak{p}) = N(\mathfrak{q}) = 8$, whence again $h = 1$. Finally, in the case of $m = 8$ we have the same bound on norms of ideals to be tested. We easily get the decompositions

$$2R_K = \mathfrak{p}_2^4, \qquad N(\mathfrak{p}_2) = 2,$$
$$3R_K = \mathfrak{p}_3\mathfrak{q}_3, \qquad N(\mathfrak{p}_3) = N(\mathfrak{q}_3) = 9,$$
$$5R_K = \mathfrak{p}_5\mathfrak{q}_5, \qquad N(\mathfrak{p}_5) = N(\mathfrak{q}_5) = 25,$$

and it follows that R_K, $\mathfrak{p}_2$ and $\mathfrak{p}_2^2$ are the only ideals in which we are interested. But observe that the number $1 - i^{1/2}$ of our field satisfies the irreducible (over Q) equation $x^4 - 4x^3 + 6x^2 - 4x + 2 = 0$, and so we have a principal ideal of norm 2, which is necessarily equal to $\mathfrak{p}_2$. Thus also in this case the class-group is trivial. □

5. We conclude this section with some results concerning the behaviour of the class-group $H(K)$ under an extension. As we already saw at the beginning of this chapter, the embedding $K \to L$ induces a monomorphism $i_{L/K}$ of the group of fractional ideals $G(K)$ into $G(L)$ which maps principal ideals into principal ones. This in turn induces a homomorphism of $H(K)$ into $H(L)$, which we shall denote by $i^*_{L/K}$. Immediately from the definition follows

Proposition 4.21. (i) *The mapping $i^*_{L/K}$ is injective iff no non-principal ideal of K is mapped on a principal ideal of L by the embedding $i_{L/K}$.*

(ii) *The mapping $i^*_{L/K}$ is surjective iff to every fractional ideal I of L there correspond a non-zero element a of L and an ideal J of R_K such that $I = aJR_L$.*

(iii) *The mapping $i^*_{L/K}$ is trivial iff every fractional ideal of K becomes principal in L.* □

The following theorem, in a quite different formulation, was used by Kummer for founding his theory of ideal numbers:

Theorem 4.18. *Every algebraic number field K has an extension L/K with $[L:K]$ not exceeding $h(K)$ such that $i^*_{L/K}$ is trivial.*

Proof. Let
$$H(K) = C_{h_1} \times \ldots \times C_{h_r}$$
be a factorization of $H(K)$ into a product of cyclic groups, and let $X_1, \ldots, X_r$ be classes of $H(K)$ generating those factors. For each $i = 1, 2, \ldots, r$ choose an ideal A_i in X_i. Then the ideal $A_i^{h_i}$ is principal and we may choose one of its generators, say a_i. Denote by L the field obtained by adjunction to K of one root of each of the polynomials $X^{h_i} - a_i$. Let $b_1, \ldots, b_r$ be those roots. The degree of L over K evidently does not exceed $h(K) = h_1 \ldots h_r$. Now take any non-principal ideal A in R_K and let
$$X = X_1^{m_1} \ldots X_r^{m_r}$$
be the class to which it belongs. Then the fractional ideal $AA_1^{-m_1} \ldots A_r^{-m_r}$ is principal, generated by, say, $a \in K$. But for $i = 1, 2, \ldots, r$ we have
$$(b_i R_L)^{h_i} = a_i R_L = A_i^{h_i} R_L,$$
and thus $A_i R_L = b_i R_L$. This implies
$$A R_L = a A_1^{m_1} \ldots A_r^{m_r} R_L = a b_1^{m_1} \ldots b_r^{m_r} R_L,$$
and so A becomes principal in L. □

Now we define the norm-homomorphism for ideal classes. Since the norm $N_{L/K}$ carries principal ideals into principal ones, it induces a mapping of the corresponding class groups, which we shall again denote by $N_{L/K}$. By Proposition 4.4 (v) we see that $N_{L/K} \circ i^*_{L/K}(X) = X^n$, where n is the degree of L/K. Moreover, we immediately obtain from the definition

Proposition 4.22. (i) *The mapping $N_{L/K}$ is injective iff every non-principal ideal has a non-principal norm.*

(ii) *The mapping $N_{L/K}$ is surjective iff for every ideal A of R_K one can find a non-zero element a in K such that the ideal aA is the norm of a fractional ideal of L.*

(iii) *The mapping $N_{L/K}$ is trivial iff the norm of every ideal of R_L is principal.* □

In a similar way one can define analogous homomorphisms of the corresponding groups $H^*(K)$ and $H^*(L)$.

A simple sufficient condition for the injectivity of the mapping $i^*_{L/K}$ is a consequence of the following

Proposition 4.23. *Let n be the degree of the extension L/K and let p be any rational prime not dividing n. If we denote by $H_p(K)$ and $H_p(L)$ the p-components of the corresponding class-groups, then the map $i^*_{L/K}$ restricted to $H_p(K)$ is an embedding into $H_p(L)$.*

Proof. If the class X lies in $H_p(K) \cap \operatorname{Ker} i^*_{L/K}$, then $X^n = N_{L/K}(i^*_{L/K}(X)) = E$, E being the unit class; thus $X = E$ in view of $p \nmid n$. □

Corollary. *If the class-number $h(K)$ is prime to the degree of L/K, then the mapping $i^*_{L/K}$ is injective and, in particular, the class-number of L is divisible by $h(K)$.* □

A result dual to the proposition just proved is contained in the following

Proposition 4.24. *Under the assumptions of Proposition 4.23 the map $N_{L/K}$ restricted to $H_p(L)$ maps this group onto $H_p(K)$.*

Proof. Let X be any class of $H_p(K)$ and let p^r be its order. Since p does not divide n, we can find A, B in Z such that $Ap^r + Bn = 1$. Then $N_{L/K}(i^*_{L/K}(X^B)) = X^{nB} = X$. □

Corollary. *If p does not divide n, then the group $H_p(L)$ is the direct product of $H_p(K)$ and $\operatorname{Ker} N_{L/K} \cap H_p(L)$. If $(n, h(K)) = 1$, then $H(L) = H(K) \times \operatorname{Ker} N_{L/K}$.* □

In the case where $H(K)$ is cyclic we can obtain an improvement of Proposition 4.24 yielding the following counterpart of Theorem 4.18.

Theorem 4.19. *If $H(K)$ is cyclic and L is a field containing K in which every ideal of K becomes principal, then the degree n of L/K equals at least $h(K)$.*

Proof. We need an auxiliary result:

Lemma 4.10. *Let p be a prime, and let C_{p^a} be a cyclic factor of $H_p(K)$. If L/K is an extension of degree $n = p^b N$ with $p \nmid N$, then $i^*_{L/K}(C_{p^a})$ is cyclic, consisting of at least p^{a-b} elements.*

Proof. Let X be a generator of C_{p^a}. Obviously $i^*_{L/K}(X)$ is a generator of $i^*_{L/K}(C_{p^a})$; assume that its order equals p^c. Then plainly we have

$$X^{Np^{b+c}} = X^{np^c} = N_{L/K}(i^*_{L/K}(X^{p^c})) = 1,$$

and so $b+c \geqslant a$, $c \geqslant a-b$ as asserted. □

§ 3. Factorization of Prime Ideals in Extensions

The assumptions of the theorem imply that for every prime p the image $i^*_{L/K}(H_p(K))$ is trivial. Writing $n = p^b N$ with $p \nmid N$ and $H_p(K) = C_{p^a}$, we obtain by the lemma the inequality $a \leqslant b$; thus $h(K)$ divides n, which is even more than we asserted. □

It should be pointed out that Theorem 4.19 may fail if $H(K)$ is not cyclic. We demonstrate this on an example due to P. Furtwängler [16]: consider the field $K = Q((-21)^{1/2})$. We shall show that $H(K) = C_2 \times C_2$; thus $h(K) = 4$ and every ideal of K becomes already principal in suitable quadratic extensions of K, namely in $K((-3)^{1/2})$, $K((-7)^{1/2})$ and $K((21)^{1/2})$. The discriminant of K equals -84, and so we have to test ideals with norms not exceeding 6. By Theorem 4.15 we obtain $2R_K = \mathfrak{p}_2^2$, $3R_K = \mathfrak{p}_3^2$ and $5R_K = \mathfrak{p}_5 \mathfrak{q}_5$. Since the norms of the resulting prime ideals cannot be expressed in the form $x^2 + 21y^2$, none of them is principal. For the same reason the ideals $\mathfrak{p}_2 \mathfrak{p}_3$, $\mathfrak{p}_2 \mathfrak{p}_5$, $\mathfrak{p}_2 \mathfrak{q}_5$, $\mathfrak{p}_3 \mathfrak{p}_5$ and $\mathfrak{p}_3 \mathfrak{q}_5$ are non-principal. If X is the ideal class containing $\mathfrak{p}_2$, Y the class containing $\mathfrak{p}_3$ and $\mathfrak{p}_5 \in Z$, $\mathfrak{q}_5 \in T$, then $X^2 = Y^2 = ZT = E$, $X \neq Y, X \neq Z, Y \neq Z$; thus $h(K) \geqslant 4$. On the other hand, evidently $h(K) \leqslant 5$, and since $h(K)$ is even, X being of order 2, $h(K) = 4$ and $H(K)$ is non-cyclic. It remains to prove that in the three fields mentioned above the ideals $\mathfrak{p}_2$ and $\mathfrak{p}_3$ become principal; this is implied by the following decompositions:

$$2 = (8+3 \cdot 7^{1/2})(3-7^{1/2})^2, \quad 3 = -((-3)^{1/2})^2 \quad \text{in } K((-3)^{1/2}),$$

$$2 = (2-3^{1/2})(1+3^{1/2})^2, \quad 3 = (3^{1/2})^2 \quad \text{in } K((-7)^{1/2}),$$

$$2 = -i(1+i)^2, \quad 3 = -\left(\frac{5-21^{1/2}}{2}\right)\left(\frac{3+21^{1/2}}{2}\right)^2 \quad \text{in } K(21^{1/2}).$$

One should only observe that $8 + 3 \cdot 7^{1/2}$, $2 - 3^{1/2}$, $-i$ and $-(5-21^{1/2})/2$ are units.

Finally, we consider a normal extension L/K with the Galois group G. This group acts on the group of all fractional ideals of L and the subgroup consisting of all principal ideals is invariant, which shows that G acts also on the group $H(K)$ of ideal classes. A class A of $H(K)$ is called *ambiguous* if it is invariant under G, and obviously the set of all ambiguous classes forms a group which we shall denote by $\text{Am}(L/K)$ and call the *group of ambiguous classes*. For any prime p we shall denote by $\text{Am}_p(L/K)$ the p-component of $\text{Am}(L/K)$.

Observe that the image of $H(K)$ in $H(L)$ under the map $i^*_{L/K}$ is contained in $\text{Am}(L/K)$; however, it may happen that this image does not cover the full group $\text{Am}(L/K)$, as may be seen from the following example: Let $K = Q$ and $L = Q((-14)^{1/2})$. In subsection 5 we showed that $H(L)$ is a cyclic group of order 4, and the only prime ideal dividing $2R_L$ lies in a class X which generates H. This ideal is invariant under the action of the Galois group, and so the class X is ambiguous, but it does not lie in the image of the (trivial) class-group of K.

In some cases it is possible to say more about $\mathrm{Am}(L/K)$.

Proposition 4.25. *Let L/K be normal of degree n. Then*

(i) *If $(n, h(K)) = 1$, then the image of $H(K)$ (which is, by Corollary to Proposition 4.23 isomorphic with $H(K)$) is a direct factor of $\mathrm{Am}(L/K)$.*

(ii) *If $(n, h(K)) = (n, h(L)) = 1$, then $\mathrm{Am}(L/K) = i^*_{L/K}(H(K)) \simeq H(K)$.*

(iii) *If $n = p^a$ is a prime power, p does not divide $h(K)$ and we have $\mathrm{Am}(L/K) = i^*_{L/K}(H(K))$, then p does not divide $h(L)$ and the quotient $h(L)/h(K)$ is a rational integer congruent to unity $(\mathrm{mod}\, p)$.*

Proof. (i) From the proof of Proposition 4.24 one can readily see that our assumptions imply
$$N_{L/K} \circ i^*_{L/K}(H(K)) = H(K).$$
We infer in turn that $f = i^*_{L/K} \circ N_{L/K}: \mathrm{Am}(L/K) \to \mathrm{Am}(L/K)$ is an endomorphism onto $i^*_{L/K}(H(K))$. The kernel of f equals $\{X \in \mathrm{Am}(L/K): X^n = E\}$ because for ambiguous X we have $f(X) = i^*_{L/K}(X^n)$, and since $i^*_{L/K}(H(K))$ and $\mathrm{Ker}\, f$ have relatively prime orders, we can see that the extension
$$1 \to \mathrm{Ker}\, f \to \mathrm{Am}(L/K) \to i^*_{L/K}(H(K)) \to 1$$
splits.

(ii) It suffices to observe that $\mathrm{Ker}\, f$ must be trivial in our case and apply (i).

(iii) Observe first that if H is any finite Abelian group on which another finite group G acts as a group of endomorphisms and G is a p-group, then $|H|$ is congruent $(\mathrm{mod}\, p)$ to the number of its elements which are under G invariant. To see this denote by $o(L)$ the orbit of an element $h \in H$ under G, i.e., the set $\{g(h): g \in G\}$. If h is invariant under G, then $|o(h)| = 1$ and otherwise the number $|o(h)|$ is divisible by p, and so
$$|H| \equiv \sum_{\substack{h \\ o(h)=1}} 1 \,(\mathrm{mod}\, p).$$

Applying this remark to our situation, we see first that $h(L) \equiv |\mathrm{Am}(L/K)|$ $(\mathrm{mod}\, p)$, and since $|\mathrm{Am}(L/K)| = h(K)$, we can see that $h(L)$ cannot be divisible by p. The quotient $h(L)/h(K)$ is a rational integer by the Corollary to Proposition 4.23, and for the same reason $H(K)$ is a subgroup of $H(L)$. Now the quotient group $H(L)/H(K)$ has no non-trivial elements invariant under the Galois group G; hence, by the preceding argument, its order has to be congruent to unity $(\mathrm{mod}\, p)$. □

Note. If $(n, h(K))$ is not equal to 1, then similar results hold for the groups $H_p(K)$, $H_p(L)$ and $\text{Am}_p(L/K)$ for every prime p not dividing n. In particular, one gets the following

Corollary. *If L/K is normal of degree n and $p \nmid n$, then the groups $H_p(K)$ and $H_p(L)$ are isomorphic iff every class of $H_p(L)$ is ambiguous.* □

§ 4. Notes to Chapter 4

1. The main results of this chapter are due to R. Dedekind [71], [78]. The proof of Proposition 4.1 is due to H. Flanders [60]. For the minimal field K for which a given ideal A of R_L satisfies $A \in i_{L/K}(I(K))$ see Mann [50]. A simple deduction of Proposition 4.5 in the case of a normal extension was given by M. Bauer [27].

Theorem 4.4 is due to E. Landau [18a]. The evaluation of $h(K)$ is closely connected with that of $R(K)$ since, as proved by C. L. Siegel [35] for quadratic fields and by R. Brauer [47a] in the general case, for fields K of a given degree one has the inequalities

$$|d(K)|^{1/2-\varepsilon} \leqslant h(K)R(K) \leqslant |d(K)|^{1/2+\varepsilon}$$

for all positive ε and sufficiently large $|d(K)|$. For Abelian fields we shall establish this result in Chapter 8 (see Theorem 8.5). In that chapter more information will be found about the asymptotical behaviour of the class-number.

For another bound see Newman [65].

2. Theorem 4.5 was first proved by A. Pellet [78] and later rediscovered by L. Stickelberger [97] and G. Voronoi [04]. See also Carlitz [53c], Cvetkov [83], Dalen [55], Hensel [05b], Herbrand [32d], Lasker [16], Skolem [52], Swan [62].

3. The definition and principal properties of the different and the conductor appear in Dedekind [78], containing, in particular, Theorem 4.6, Lemma 4.3 and Proposition 4.12 (i). For other definitions of the different see Hilbert [97], Noether [30], [50]. Generalizations to other classes of rings were considered in Auslander, Buchsbaum [59], Berger [60], [62], [64], Fitting [36], Fossum [66], Grell [35], Kähler [53], Kuniyoshi [58], Kunz [60], [68].

A characterization of ideals which can serve as conductors of $R[a]$ was given by R. Dedekind [82]. Cf. Furtwängler [19], Grell [27]. For other results on conductors see Bauer [36], Hensel [87], Ore [27].

Proposition 4.7 (iv) appears first in Hecke [17b]. For the Corollary 2 to it see Halter-Koch [67], Yokoi [60b]. The trace map $S \to R$ is surjective iff L/K is tame. We shall establish this for normal extensions in Chapter 6 (see Corollary 3 to Proposition 6.2). Cf. Ullom [69a], where it is shown that if L/K is tame, then for every ideal I of R_K one has $T_{L/K}(I) = I \cap R_K$.

For Proposition 4.9 see Keller [39]. Proposition 4.11 goes back essentially to L. Euler. Possible forms of $d(L/K)$ were determined in Ore [27], Thompson [31].

4. Theorem 4.7 was stated by A. Weil [43] and a proof was published by Y. Kawada [51]. The proof given here, as well as that of Theorem 4.8, is a modification of that of Kawada. Cf. Kinohara [52], [53], [54], Moriya [53], Narkiewicz [69], Neukirch [67a]. The first proof of Theorem 4.8 was given in Dedekind [82]. For other proofs of Theorem 4.8 or its Corollary 2 see Artin [59], [67], Bauer [23], Chebotarev [37b], Hecke [23], Hensel [94a], Hilbert [97], Narkiewicz, Schinzel [69]. For a generalization see Noether [27b].

The different theorem does not shed light on the highest power $\mathfrak{P}^a$ of a prime ideal $\mathfrak{P}$ dividing it in the wildly ramified case. An upper bound for it was obtained by K. Hensel [02], who confirmed a conjecture of R. Dedekind [82]. (See Proposition 6.3.) See also Bauer [19a], [21b], Ore [25b], [25c], [26a], [26d], [27], Yokoi [66].

As regards Theorem 4.9 and other properties of the differents and discriminants of composite fields see Bauer [21c], Hensel [89], [93], [97d], [97e], [99], Kataoka [80], Liang [73], Pumplün [66], Rados [90], [30], Tôyama [55]. For a special case see Motoda [75].

The analogue of Theorem 2.10 fails for relative extensions, although it holds in certain particular cases. Examples of fields with this property were given by Kuroda [62] and Nakahara [73]. Class-field theory shows that all such fields must have $h^*(K) = 1$, since otherwise there exist Abelian unramified extensions of K. In fact, such a maximal extension has its Galois group over K isomorphic to $H^*(K)$. For non-Abelian unramified extensions see Fujisaki [57], J. H. Smith [69], Taussky [37b], Uchida [70], Y. Yamamoto [70].

5. The irreducibility of the mth cyclotomic polynomial F_m for prime m was established by C. F. Gauss [01]. Cf. Kronecker [45a], [45b], [56c], Plemelj [33]. The first proof for arbitrary m was given by L. Kronecker [54]. Other proofs may be found in Arndt [58b], Bauer [17] (for prime-powers), Dedekind [57], Grandjot [24] (whose proof we have reproduced), Landau [29], Lebesgue [59], Levi [31], Schur [29a], Skolem [49], Späth [27], Toepken [37]. M. Ruthinger [07] gave a survey of early proofs.

For characterizations of the cyclotomic polynomials see Bauer [04a] and Wegner [31]. A conjecture of P. Lévy [37] stating that for prime p one cannot write F_p as a product of polynomials with real non-negative coefficients was proved by M. Krasner, B. Ranulac [37] and D. A. Raikov [37].

Let A_m be the maximal absolute value of coefficients of F_m, i.e., the height of F_m. For infinitely many m one has

$$\log \log A_m \geq C \log m / \log \log m$$

with a positive constant C (Erdös [46]). Cf. Erdös [57], E. Lehmer [36], Vaughan [74]. Erdös stated also that

$$\log \log A_m = O(\log m / \log \log m),$$

and this was established by P. T. Bateman [49]. See also Bateman, Pomerance, Vaughan [85], Beiter [68], [78], Bloom [68], Endo [74], Erdös, Vaughan [74], Justin [69], Möller [70], [71].

The irreducibility of F_m in extensions of Q was considered in Kronecker [54], Nagell [64a], Petersson [55], [59], Pumplün [63], Weisner [28]. For the factorization of F_m (mod p) (which can also be obtained from Theorems 4.12 and 4.16) see Ballieu [54], Chowla, Vijayaraghavan [44], Guerrier [68].

For other questions concerning F_m see Apostol [70a], [75], Bochkarev [49], [54], Carlitz [66a], [67], Conway, Jones [76], Diederichsen [40], Ennola [80], Felsch, Schmidt [68], Golomb [78], Hölder [36], Hsia [73], Jacobsthal [58], Kanold [49], [52], Kazandzidis [63a], [63b], [64a], [64b], Klobe [49], Kostandi [43], D. H. Lehmer [33a], [66], E. Lehmer [30], McDaniel [74b], Mollin [83b], Rédei [54], Richter [72], [74], Rogers [65], Schoenberg [64], Torelli [24], Zeitlin [68].

For Theorem 4.10 see also Bauer [39], Nagell [19], [64a], Rados [06].

6. *Additive Galois-module structure.*

Theorem 4.11 is due to E. Noether [32] and the proof presented here to W. C. Waterhouse [79]. For other proofs see Berger, Reiner [75], Cassels, Wall [50], Chevalley [33a], Deuring [32], Stauffer [36], Winter [72]. Special cases were considered in Brinkhuis [84b], Hensel [89], Perlis [42]. Theorem 4.11 implies the existence of an element $w \in L$ with $L = K[G]w$. The norms of possible elements w were studied by C. J. Bushnell [77b], [79], [83]. F. Halter-Koch, F. Lorenz [81] characterized the case where one can find $w \in U(L)$. Generators w for certain extensions of $Q(i)$ were given in Okada [80].

M. Deuring [32] utilized Theorem 4.11 to deduce the main theorems of the Galois theory and the theorem on the primitive element.

We have already noted that for normal extensions the surjectivity of the trace map, occurring in Proposition 4.14, is equivalent to tame ramification. (This is essentially due to A. Speiser [16]. See Corollary 3 to Proposition 6.2.) It follows that only tame extensions of Q may have normal integral bases (NIB). Tame ramification, however, is not sufficient here, as shown by J. Martinet [71a], who gave an example of a tame extension K/Q with $\text{Gal}(K/Q) = H_8$, the quaternion group, without NIB. Other examples of tame extensions without NIB may be found in Brinkhuis [81a], Cougnard [83a].

For certain large classes of fields tame ramification is sufficient for the existence of NIB. This holds, e.g., for Abelian fields (Hilbert [97], Fueter [45a], Fröhlich [76b], Leopoldt [62]; see the Corollary to Proposition 8.1), for fields with the dihedral Galois group D_n (T. Miyata [80]; for n = odd prime, Martinet [68], [69], [71b]; for $n = 2^k$, Fröhlich, Keating, Wilson [74]; for $n = p^k$, Taylor [78a]) and for extensions with a generalized quaternion group of order 2^n with $n \geqslant 4$ (Fröhlich, Keating, Wilson [74]). The same is true for all normal K/Q with a square-free degree (Ph. Cassou-Noguès [77], [78], Taylor [81a], [82a]). For other classes of fields see Cougnard [72], [73], [74], [80], Fröhlich [74a], [75], [76a], [76b], Jaulent [81a], Queyrut [72].

M. Newman, O. Taussky [58] and R. C. Thompson [62] described extensions K/Q which cannot have more than one NIB apart from those obtained by permutations and sign changes. These results can also be deduced from Higman [40].

Confirming a conjecture of J. P. Serre, A. Fröhlich [72] proved that if $G = H_8$ and K/Q is tame, then NIB exists iff the so-called "root-number" $W(\chi)$ corresponding to the unique irreducible symplectic character χ of H_8 equals unity. (Cf. Fröhlich [76b], Martinet [77b].) This in turn is equivalent to the non-vanishing of Artin's L-function corresponding to χ at $s = 1/2$ (Armitage [72]). This condition can also be expressed in an elementary way (Fröhlich [72], Martinet [77b]).

The discovery of the connection between NIB and the root-numbers lead to an extensive theory, which is presented in a recent book by A. Fröhlich [83a]. For earlier expositions see Cougnard [83b], Fröhlich [74c], [77a], Martinet [73].

Denote by $\text{Cl}(R_K[G])$ the group of classes of stably-isomorphic locally free rank one $R_K[G]$-modules, i.e. finitely generated modules M whose completions under valuations induced by prime ideals $\mathfrak{p}$ of R_K are free $(R_K)_\mathfrak{p}[G]$-modules with one free generator, where $(R_K)_\mathfrak{p}$ is the closure of R_K in the corresponding completion of K. Two modules M, N are stably isomorphic if with suitable free modules F_1, F_2 one has $M \oplus F_1 \simeq N \oplus F_2$.

It was proved in Noether [32] that R_L is a locally free $R_K[G]$-module iff L/K is tame. Hence, for every normal and tame extension L/K, we can consider the class $[R_L]$ of R_L in $\text{Cl}(R_K[G])$. Clearly, $[R_L] = 1$ is necessary for the existence of a NIB, but in general it is not sufficient. Fröhlich conjectured that $[R_L]$ is related to Artin's L-functions associated with L/K, and this was proved in Taylor [81a]. (Cf. Cassou-Noguès, Taylor [83].) Fröhlich's conjecture implies in particular $[R_L]^2 = 1$, which in the case of $K = Q$ was conjectured in Fröhlich [76b]. For previous results in this direction see Ph. Cassou-Noguès [78], [79], [81], Fröhlich [74a], [76b], [80], Taylor [79], [80a].

There is another class-group of the group-ring $Z[G]$ used in this context. Theorem 4.11 shows that if K/Q is normal with group G, then $K = Q[G]w$ with a suitable $w \in K$, and it follows that there exists an invertible fractional $Z[G]$-ideal $I \subset Q[G]$ with $R_K = Iw$. Now it is clear that K has a NIB iff I is principal. Hence, if we introduce the class-group cl($Z[G]$) as the factor group of the group of all invertible fractional $Z[G]$-ideals contained in $Q[G]$ by the subgroup of all such principal ideals, then K will have a NIB iff the image (R_K) of I in cl($Z[G]$) equals 1. If, for a given G, we denote by R the set of all classes in cl($Z[G]$) which have the form (R_K) for a suitable tame extension K/Q with group G, then the question arises how large R can be. In the case of $G = C_p^n$ (p prime, $n \geq 1$) R was described in McCulloh [82]. (Cf. also Childs [81], McCulloh [77] for previous work.)

On class-groups of group rings the reader is advised to consult Reiner [76] and the literature quoted there.

The structure of R_K as a $Z[G]$-module for Abelian $G = \text{Gal}(K/Q)$ was determined by H. W. Leopoldt [62]. (Cf. Rud'ko [68] for the cyclotomic case.) M. J. Taylor [78b] showed that if L/K is Abelian and tame, then R_L is a free $Z[G]$-module. For the structure of R_L in the case of particular Abelian Galois groups see Bertrandias [79], Chatelain [70], [73], Childs [77], M. N. Gras [83], Yokoi [60b]. For Kummerian extensions see Childs [80], Fröhlich [62b], McCulloh [77], Taylor [80b].

The structure of R_L in the case of a wildly ramified extension L/K was considered in Bergé [78], Cassou-Noguès, Queyrut [82], Fröhlich [78], Queyrut [81a], [81b], [82], Wilson [80].

Let O be a maximal order in $Q[G]$, i.e. a maximal subring containing the unit element. The structure of R_K as an O-module was considered in Bergé [72], [78], [81], Chatelain [81], Cougnard [75], [76], [77], Fröhlich [76b], [77c], Jacobinski [63], Jaulent [81a], [81b], Leopoldt [59a], Martinet [72], Taylor [81b].

New necessary conditions for NIB were found by J. Brinkhuis [81a], [81b], [83], [84b], who related the Galois structure of R_K to the embedding problem for fields.

Other results concerning the Galois structure of R_K and NIB may be found in Chinburg [83b], Cougnard [82], Kleiman [73] (where, however, the term normal basis is used in two different meanings), Martinet, Payan [67], Maurer [78b], Taylor [82b], [83], Ullom [80], [81], Vostokov [77].

The Galois structure of ideals was studied in Jaulent [81b], Ullom [69a], [69b], [74a].

7. E. E. Kummer [47b], [56] actually defined ideal prime numbers in a cyclotomic field as formal factors of a prime p, imitating the factorization (mod p) of the corresponding cyclotomic polynomial. This allows us to consider him as the originator of Theorem 4.12, which first appeared in its modern form in Dedekind [78]. (Cf. Zolotarev [74], Engstrom [30b].) L. Kronecker [82] related the factorization of primes in extensions of Q to the factorization of the corresponding norm forms into factors irreducible (mod p), and his method was also used later in Bauer [19b], Hensel [94a], Ore [26d].

O. Ore [27] employed a method which is p-adic in nature and based on the behaviour of the defining polynomial (mod p^N) for sufficiently large N. (Cf. Bauer [36], Bauer, Chebotarev [28].)

The simple proof of Lemma 4.7 was communicated to the author by A. Schinzel.

Theorem 4.13 is due to R. Dedekind [78]. Cf. also Hensel [94b], [97b] and Ore [27]. Proposition 4.16 was first proved in Gauss [01]. See also Schönemann [46], Schwarz [61].

Proposition 4.17 was proved in Żyliński [13], and Y. Nakamura [74] generalized it to relative extensions. Unramified prime divisors of $i(K)$ were described by K. Hensel in his dissertation [84]. Cf. Hensel [94b]. The maximal power p^a of a given prime p which can divide $i(K)$ in the case of $[K:Q] = n$ was in certain special cases determined by H. T. Engstrom [30a]. For $n = 3$, $p = 2$ his result gives $a = 3$. Another proof of this fact was given by L. Tornheim [55]. J. Śliwa [82a] determined a for all unramified primes p. The determination of a presented in Sukallo [55] does not seem to be quite convincing.

For other results on $i(K)$ see Bauer [07], Bungers [36], Carlitz [33], [52], Kleiman [72], Nagell [66], Nart [85], Ore [27].

8. In using Theorems 4.12 and 4.13 it is useful to have tables of polynomials f irreducible over $GF(p)$. For $p = 2$ such tables were given for $\deg f \leq 20$ in Church [35], Garakov [70], Marsh [57], Mossige [72]. For $p = 3, 5, 7$ see the already quoted papers of Church (for $\deg f \leq 7, 5, 4$ resp.) and Garakov (for $\deg f \leq 8, 6, 5$ resp.). Garakov deals also with $p = 11$, $\deg f \leq 4$. For tables of irreducible polynomials of a special kind see Chang, Godwin [69], Conway [68], McEliece [69], Rodemich, Rumsey [68], Stahnke [73], E. J. Watson [62], Zierler [69], [70], Zierler, Brillhart [68]. Cf. Popovici [59].

General methods of factoring polynomials (mod p) were given in Berlekamp

[67], [70], Kempfert [69], Knuth [69] (vol. II), McEliece [69]. See also Leonard [69] for quartics.

9. The proof of Theorem 4.14 presented here is due to T. Nagell. We shall give an analytical proof in Chapter 7. Other proofs, some of which apply also to more general situations, can be found in Dress [64], Dujčev [56], Moriya [50], Nagata, Nakayama, Tuzuku [53]. A counterpart to Theorem 4.14 is also true: there exist infinitely many prime ideals which do not split completely in a given extension L/K. (See Corollary 5 to Proposition 7.9.)

Class-field theory determines the prime ideals splitting in an Abelian extension L/K as those whose classes lie in a certain subgroup of $H_\mathfrak{f}^*(K)$ with a suitable $\mathfrak{f}$. This property characterizes Abelian extensions. (See, e.g., Cassels, Fröhlich [67].) G. Bruckner [66], [68] characterized normal extensions in which the splitting primes are exactly those which are representable by binary quadratic forms from a given set. It was shown by F. Gauthier [78] that the set of primes splitting in K/Q contains an arithmetic progression iff K/Q is Abelian. Cf. Schulze [76a]. Further results about sets of splitting primes will be given in Chapter 7, where Chebotarev's theorem will be available.

H. Hasse [26a], [26b] (cf. Ore [26c]) proved that if there is a given finite set of prime ideals of R_K, then one can find infinitely many extensions L/K of a prescribed degree in which every prime from that set factorizes into prime ideals in a given way, subject only to conditions resulting from Theorem 4.1. (We shall return to this subject in Chapter 6.) For generalizations see Endler [63b], Krull [59], Ribenboim [59], [62a].

10. Theorem 4.15 for the field $Q(i)$ was already known to Gauss [32]. In this case, as for all other fields with $h(K) = 1$, the factorization law for prime ideals describes in reality the classes of associated irreducible integers, and this description can be carried out for certain cases in an elementary way. (Cf. Hardy, Wright [60], Ch. XII.) For irreducibles in $Q(\zeta_3)$ see v.d. Pol, Speziali [51], where a graphical representation of them is given. An analogue of Theorem 4.15 for relative quadratic extensions is given in Hilbert [99].

Theorem 4.16 goes back to Kummer [47b], [56]. (See also Bachman [66], Bhaskaran [71], Wegner [31].) Also Theorem 4.17 was proved by Kummer [59] in a special case. The general case is due to Hilbert [97]. Cf. Hensel [18], [21a], Rella [24a].

Factorizations in cubic fields were considered in Agou [71], Arai [81], Dedekind [00], Engstrom [30c], Hasse [30b], Latimer [29], Llorente, Nart [83], Markoff [82], [91], Martinet, Payan [67], Nowlan [26], Reichardt [33], Wahlin [22], Westlund [13]. A table for pure cubic fields was given by E. S. Selmer [55].

Factorizations in Abelian extensions are described by the class-field theory; however, there is no comparable approach in the non-Abelian case, and our knowledge is reduced to extensions of a rather special form. On this topic see Bruckner [68], Büsser [44], H. Cohn [79a], Dribin [37], Fröhlich [60a], Furuta [58], [59b], [61], [77], Gut [32], [33], [34b], Halter-Koch [71b], [84a], S. Kuroda [51], S. N. Kuroda [70a], Mann, Vélez [76], Satgé, Barrucand [76], Sato [81], Vélez [77], [78], v.d. Waall [74a]. A connection with modular forms was found by G. Shimura [66]. Cf. Chowla, Cowles [77], Hiramatsu [82], Ito [77], Moreno [80], Serre [68].

For other methods of factorizations see Gut [34a], Hensel [13], [18], [21a], [21b], McLane [35], Ore [23], [25a], [28a], Rella [24a], Speiser [22].

Relations between factorizations in two extensions and their composite were studied in Bauer [16b], [20], [21a], [40a], [40b], Herbrand [31a], Maus [67], Ore [26b], Vassiliou [32].

A numerical study of decompositions of primes in 63 fields of degrees $\leqslant 5$ was made in Cartier, Roy [73].

For other questions concerning factorizations of prime ideals see Barrucand, Laubie [82], Bhaskaran [74], Maurer [73], McCluer [71b], Parry [71a], [71b], [71c], Reichardt, Wegner [37], Rella [19]. For a survey see Hasse [51d].

11. Proposition 4.19 is very old. A proof of it without using ideal theory was given by E. Lánczi [65]. (Cf. Zaupper [83].) Elementary proofs in special cases may be found in Gauss [32], Popovici [55], Rudin [61]. The converse to that proposition was proved by H. M. Stark [67a], [67b]. (See Theorem 8.11.)

In principle, knowledge of prime ideals with small norms suffices for the determination of the class-number and the class-group; however, the relevant procedure may be very awkward. An algorithm valid for real Abelian fields was given by G. Gras and M. N. Gras [79].

For arbitrary fields a method of finding the number of invariants of $H(K)$ divisible by a given prime power was given by L. Rédei [44]. For cyclic fields of degree p this was done for invariants divisible by a power of p by E. Inaba [40]. Cf. Fröhlich [52], [54a], [54b], [57], Inaba [41], Latimer [33a].

D. A. Garbanati [78a] considered invariants of a certain homomorphic image of $H^*(K)$ for Abelian K, the central class-group of K.

Methods of determining $h(K)$ for various classes of fields were presented in Aigner [53], H. Amara [81], Billevich [49b], Brown, Parry [77], Hasse [75], Hayashi [84], Lagarias [80a], Lakein [71], [74b], Lenstra [82], Nakamula [81], [82b], Pohst [75c], Shanks [69b], [71], [72b], Smadja [73], [77], Vel'min [51]. Some of these papers deal only with certain Sylow subgroups of $H(K)$ and their cardinalities.

For surveys of earlier approaches to the problem see Shanks [76a] and Zimmer [72].

The first tables of $H(K)$ for quadratic fields were given by A. Cayley [62] and A. E. Cooper [25], who used the language of quadratic forms. For later tables, mostly giving only $h(K)$, see Bitimbaev [77], Bouvier [72], Buell [76], [77], Gupta [42], Hendy [75], Ince [34], Juricic [65], Müller [78], Schaffstein [28], Urazbaev, Alibaev [72], Wada [70b], Williams, Broere [76]. Cf. also H. Cohn [62b].

Tables for cubic fields were given in Angell [73], [76], Barrucand, Williams, Baniuk [76], Beach, Williams, Zarnke [71], Cassels [50], Cohn [57], Eisenbeis, Frey, Ommerborn [78], Godwin [61], Godwin, Samet [59], Gras, Moser, Payan [73], Hasse [48a], Reid [01], Shanks [74], H. C. Williams [77], [81b], Williams, Shanks [79].

For quartic fields see H. Cohn [58], M. N. Gras [79], Hasse [48a], Lakein [74b], Pohst [75b], [75c]. Cf. Gudiev [57], where certain cases of $h(K) = 1$ were found.

For various Abelian fields see Bauer [69], M. N. Gras [77], v.d. Linden [82a], Montouchet [71].

Certain conjectures based on numerical results which concern the distribution of imaginary quadratic fields with certain types of the class-group were presented in Cohen [83] and Cohen, Lenstra [84], where also heuristic support was given.

12. An analogue of Theorem 4.18 for a more general situation was obtained by H. Chatland and H. B. Mann [49]. The argument used in its proof was utilized by E. Hecke [18] (part II) in the construction of Hecke's ideal numbers, which were useful in the theory of Hecke's zeta-functions, which we shall consider in Chapter 7. After the invention of ideles and adeles they slowly disappeared.

Proposition 4.23 is due to P. Furtwängler [08] and has been rediscovered several times. Proposition 4.25 appears in Yokoi [68b] and Yokoyama [65], [66]. Other results on these topics may be found in Chevalley [31], Cornell, Rosen [80], Dénes [52a], Draxl [70], Fröhlich [52], Iwasawa [55a], Kitaoka [77], Ohta [78], Okamoto [76], Scharlau [73], Schipper [77], Yokoi [67], Yokoyama [65].

13. The structure of $H(K)$ as a $Z[G]$-module in the case where K/Q is normal with the Galois group G remains still largely unknown. One of the central problems here is the determination of the ideal of relations of $H(K)$ in $Z[G]$, i.e. the set of all $\sum_g a_g g \in Z[G]$ such that for any class X in $H(K)$ one has

$$\prod_g g(X)^{a_g} = 1.$$

It follows from Propositions 4.4 (i) and 4.5 that this ideal always contains the element $\sum_g g$, and in the case of imaginary Abelian extensions K/Q other non-trivial relations are provided by the Stickelberger ideal S defined by

$$S = Z[G] \cap AQ[G],$$

where $A = f^{-1}\sum_a a g_a^{-1}$, f being the conductor of K, i.e., the smallest integer with $K \subset K_f = Q(\zeta_f)$ (which exists by the Kronecker–Weber theorem; see Theorem 6.5) and g_a being the restriction to K of the automorphism of K_f given by $\zeta_f \mapsto \zeta_f^a$ for $1 \leq a < f$, $(a,f) = 1$. The fact that S annihilates $H(K)$ was established in the case of a cyclotomic field by L. Stickelberger [90], and for the field $Q(\zeta_p)$ it follows immediately from a result in Kummer [47b]. For other proofs see Childs [81], Fröhlich [77b]. See also Grant [34], McKenzie [52]. For the general case see Lang [78], Washington [82], where further references will be found. Here we would like to draw attention to papers concerning the determination of the index $[Z[G]^-:S^-]$, where for any $Z[G]$-module M we denote by M^- the set of all $m \in M$ satisfying $s(m) = -m$, where $s \in G$ acts on K as the complex conjugation. K. Iwasawa [62] showed that for $K = Q(\zeta_q)$ with a prime-power q, this index equals $h(K)/h(K^+)$, K^+ being the maximal real subfield of K. For another proof see Skula [81]. This result was generalized by W. Sinnott [78] to arbitrary cyclotomic fields, and C. G. Schmidt [79] and K. Iimura [81b] did the same for arbitrary complex Abelian fields. Cf. Kimura, Horie [82].

An analogue of Stickelberger's ideal annihilating $H_i^*(K)$ for cyclotomic K was constructed by C. G. Schmidt [82].

For other results on annihilators of $H(K)$ see G. Gras [79a], [79b], Kobayashi [82], Oriat [81].

The group $\text{Am}(L/K)$ was extensively studied in the case of a cyclic extension of prime degree in G. Gras [72b], [73a], [73b], [74a], [78], Moriya [30]. Cf. also Chevalley [33a]. For other fields see Furuya [82], Jaulent [81c], Payan [73]. An analogue of $\text{Am}(L/K)$ for non-normal extensions was considered in Walter [79a].

The maximal factor group of $H(K)$ on which G operates trivially is called the *central ideal class group of K*. See Fröhlich [54a], [83b], Furuta [71], [76], [77], Garbanati [78a], Shirai [75], [78], [79].

For other results on the action of the Galois group on $H(K)$ see Cornell, Rosen [81], Dénes [52a], Gerth [75c], S. N. Kuroda [64b], Taussky [69].

14. The class-field theory gives a canonical extension of K having the property stated in Theorem 4.18, namely the Hilbert class-field $\overline{K}$ of K, defined as the maximal Abelian extension of K which is unramified (also at infinity). This was first proved by P. Furtwängler [30] and is called the *Principal Ideal Theorem* (Hauptidealsatz). Other proofs were given in Artin [30a], Borevich [57], Iyanaga [34a], Magnus [34], Schumann [37], Taketa [32]. For generalizations and analogues see Furuya [77], Herbrand [32c], Iyanaga [31], [39], Kempfert [62], Kuniyoshi, Takahashi [53], Miyake [80a], Takahashi [64], [65], Tannaka [33a], [33b], [34], [49], [50], [56], [58], Tannaka, Terada [49], Taussky [32], Terada [50], [52], [53], [54a], [54b], [55], [71], Zink [75]. A cohomological interpretation was given by Y. Kawada [68].

If a class $X \in H(K)$ lies in the kernel of $i^*_{L/K}$, then X is said to *capitulate in L*. Thus the principal ideal theorem asserts that all classes capitulate in $\bar{K}$. The first result concerning capitulation is the Theorem 94 of Hilbert [97], which states that if L/K is cyclic of degree N and unramified, then the order of the kernel of $i^*_{L/K}$ is divisible by N. (Cf. Cornell, Rosen [80], Taussky [71], Zink [75].) Capitulation in cyclic extensions was also considered in Furuya [77], M. N. Gras [79], Heider, Schmithals [82], Kisilevsky [70], Scholz, Taussky [34], Terada [71]. For other cases see Bond [81], Chang [77], Chang, Foote [80], Heider [84], Iyanaga [34b], Jehne [77a], Miyake [80b].

Hasse [31a], Hsü [63], Reichardt [58], Terada [54a] dealt with explicit generators in $\bar{K}$ for ideals of an imaginary quadratic field K. Cf. Furtwängler [32], Pollaczek [24], Schäfer [29], Scholz, Taussky [34], Taussky [71].

15. If $h(\bar{K}) \neq 1$, then one can consider the Hilbert class-field of $\bar{K}$ and continue this process. It has been conjectured that in this way one will be led to an extension L/K with $h(L) = 1$ (the *class-field-tower problem*; Klassenkörperturmproblem (see Hasse [26c])). However, E. S. Golod and I. R. Shafarevich [64] proved that the class-field tower may be infinite. This happens, e.g., if K is a quadratic field with sufficiently many ramified primes. (Previously A. Scholz [29] noted that this tower may be arbitrarily long.) For expositions of the Golod–Shafarevich theorem see Panella [66], Roquette [67], Serre [66]. Cf. Koch [69], [75], Kostrikin [65], Vinberg [65].

It was shown by L. V. Kuzmin [69] that if K/Q is normal and at least 8 primes ramify in it, then the class-field-tower of K is infinite. Examples of quadratic fields with an infinite class-field-tower and few ramified primes were given in Martinet [78], Matsumura [77] and Schmithals [80a]. The record-holder is $K = Q(d^{1/2})$ with $d = -191.40961$ given in the paper of N. Matsumura. For other classes of fields see Brumer [65], Cornell [83a], [83b], Cornell, Rosen [80], Fröhlich [52], Furuta [72], Lamprecht [67], Nehrkorn [33], Shafarevich [63], T. Takeuchi [79], [80].

For the study of the class-field-tower see also Browkin [63a], Fröhlich

[54d], [62d], Kisilevsky [76], Koch [64b], Moriya [36], Taussky [37a], [37b], Venkov, Koch [74].

An explicit determination of the Hilbert class-field or related Abelian extensions was obtained for particular fields in Barrucand, Cohn [73], H. Cohn [79a], [79b], [81a], [81b], [83], Cohn, Cooke [76], Deuring [52], Eichler [56], Gogia, Luthar [78a], G. Gras [72c], Gut [43a], [43b], Hasse [33], [47], [64], Hasse, Liang [69], Hecke [12], [13], Herz [66], Hilbert [99], Honda [60a], Kaplan [77a], Liang, Zassenhaus [69], Madden, Vélez [80], Rédei [53a], Schertz [78a].

The *ray-class-fields* (mod $\mathfrak{f}$) of a given field K are defined as the maximal Abelian extensions of K in which only prime ideals dividing a given ideal $\mathfrak{f}$ are ramified. Their Galois group over K equals $H_\mathfrak{f}(K)$ if no ramification at infinity is allowed, and equals $H_\mathfrak{f}^*(K)$ otherwise. For imaginary quadratic fields K the ray-class-fields are described by the theory of complex multiplication, which goes back to H. Weber [97] (cf. also [96b], vol. III) and R. Fueter [07], [11], and which assumed a modern form in Hasse [27]. (See also Fueter [05], [14], [33], [45b], Hasse [31b], [31c].) Hasse's paper [27] gives a pair of generators for every such field. A major improvement here was made by K. Ramachandra [64], who gave in each case one generator, which is the value of a point in K of a holomorphic function, independent of K. This is fully analogous to the case of $K = Q$, where the Kronecker-Weber theorem (see Chapter 6) shows that one can take $f(z) = \exp(2\pi i z)$, the ray-class-field (mod mZ) (with infinite ramification allowed) being equal to the mth cyclotomic field. For an introduction to complex multiplication see Borel *et. al.* [66].

A generalization of the classical theory of complex multiplication, leading to the construction of class-fields for a large class of fields was given by G. Shimura [62] and G. Shimura and Y. Taniyama [61]. See also Shimura [68], [71a], [71b], [72].

Fields which are Hilbert class-fields for suitable subfields were considered by K. Győry and W. Leahey [77], who showed that there can be only finitely many such fields of a given degree which are (*CM*)-fields. See also Hamamura [81].

One of the main problems of algebraic number theory is the construction of an analogue of the class-field theory for non-Abelian extensions. An attempt to do this was made by M. Krasner [47]. A way towards its solution is indicated in the Langlands program, which predicts deep connections between number theory and infinite-dimensional group representations. The reader is advised to read the excellent introduction to it written by S. Gelbart [84], where also further literature on the subject is listed.

16. The use of cohomology in algebraic number theory was initiated by

G. Hochschild and T. Nakayama [52] and T. Nakayama [52]. This approach turned out to be very successful in restating the class-field theory in a modern fashion. For its exposition see Artin, Tate [61], and the lectures of J. P. Serre and J. Tate in Cassels, Fröhlich [67]. Actually the first cohomological result about algebraic number fields goes back to D. Hilbert [97], whose Theorem 90 gives the triviality of $H^1(G, L^*)$ in the case of normal L/K with Galois group G, assumed to be cyclic, and Theorem 97, asserting the non-triviality of $H^1(G, U(L))$ if $\operatorname{Gal}(L/K)$ is cyclic of prime order $\neq 2$.

Several important results concerning cohomological properties of number fields were stated by J. Tate [62]. (Cf. Haberland [78], T. Takahashi [68], Uchida [69].)

The cohomology of rings of integers was studied by H. Yokoi [60b], [62a], [62b], [64], who proved in [64] that if L/K is cyclic with Galois group G, then all groups $H^q(G, R_L)$ are of the same order. The same was shown by M. Rosen [66], who found, moreover, that if $[L:K]$ is square-free, then all these groups are isomorphic. (For $[L:K]$ prime this was proved in Yokoi [62b].) An analogous result fails for non-cyclic extensions. (See Lee, Madan [69].)

For other results concerning cohomology in algebraic number fields see Furuta, Sawada [68], Kawada [51], Kuniyoshi [58], McQuillan [76], Nakayama [51], S. Takahashi [53], Tate [52], Ullom [69b].

17. Several authors devoted their attention to maximal extensions of the rationals or their finite extensions with prescribed properties. T. Kubota [57] determined Ulm's invariants for the Galois group of the maximal Abelian extension of an algebraic number field. H. Koch's book [70] contains a study of Galois groups of maximal p-extensions of algebraic number fields, with given ramified primes.

If $\hat{K}$ denotes the algebraic closure of K, K/Q and L/Q are both finite and normal and $\operatorname{Gal}(\hat{K}/K) \simeq \operatorname{Gal}(\hat{L}/L)$, then, as shown by J. Neukirch [69a], one has $K = L$. One cannot replace here the algebraic closure by the maximal Abelian extension (Onabe [76]). J. Neukirch conjectured that $\operatorname{Gal}(\hat{Q}/Q)$ has only inner automorphisms, and this was shown to be true by M. Ikeda [75b], [77] and K. Uchida [76a], [77a]. A simpler proof was given by J. Neukirch [77]. See also Ikeda [75a], Kanno [73], Komatsu [74]. For surveys of results concerning $\operatorname{Gal}(\hat{Q}/Q)$ see Geyer [78] and Neukirch [74a].

For other results concerning these topics see Bond [79], Browkin [63b], Brumer [66], Cvetkov [80], G. Gras [82a], [83a], Hironaka–Kobayashi [76], Hoechsmann [66], Iwasawa [53b], [53c], Kawada [54], Koch [63a], [65a], [65b], [67], [68], [74], [77], Kuzmin [72], [75a], [75b], Markshaitis [63], Miki

[78a], Neukirch [69c], Neumann [75], [77a], Shafarevich [62], [63], J. Smith [65], Uchida [82].

EXERCISES TO CHAPTER 4

1. Prove that if L/K is normal, then the kernel of the trace map equals the additive group generated by $(1-g)a$, g running over $\text{Gal}(L/K)$ and $a \in L$.

2. Show that if L/K is normal of odd degree then it is unramified at infinity, and prove that one cannot remove here the normality assumption.

3. Let A_m be the maximal absolute value of the coefficients of the mth cyclotomic polynomial. Prove that
$$\log \log A_m = O(\log m).$$

4. Let K be a quadratic field and $a \in R_K$. For $A = Z[a]$ determine the conductor and A^*.

5. Compute the different of a pure cubic extension of Q.

6. Prove Theorem 4.11 for finite K.

7. (McCulloh.) Prove that if $K = Q(i5^{1/2})$ and $L = K(i)$, then L/K has a relative integral basis but does not have a relative normal integral basis.

8. (i) Let K/Q be normal with group G and assume that it has a normal integral basis $\{g(w)\}_{g \in G}$. Prove that there exists $w' \in R_K$ not of the form $\pm g(w)$ with $g \in G$, such that its conjugates again form a normal integral basis iff the group ring $Z[G]$ has invertible elements not of the form $\pm g$ ($g \in G$).

(ii) (Newman, Taussky [58].) Prove that if, moreover, G is cyclic of order 2, 3, 4 or 6, then K can have at most one normal integral basis (up to permutations and sign changes).

(iii) Show on an example that (ii) fails in the case of $G = C_5$ and $G = C_8$.

9. Determine the factorization of ramified primes in pure cubic fields.

10. Let K be a cubic field. Prove that $i(K) \neq 1$ iff 2 splits completely in K/Q.

11. Let K/Q be a normal extension of degree 4 with a non-cyclic Galois group. Prove that with suitable $a, b \in Z$ we have $K = Q(a^{1/2}, b^{1/2})$, determine the discriminant $d(K)$ and find the law of decomposition of prime ideals in K/Q.

12. Show that if $f_1, \ldots, f_m$ are non-constant polynomials over Z, then there exist a prime p and integers $N_1, \ldots, N_m$ such that p divides $f_i(N_i)$ for $i = 1, 2, \ldots, m$.

13. Prove that if K is a quadratic field with $|d(K)| = D$, then for all complex K with $D > e^{24}$ and for all real K we have
$$h(K) < \tfrac{1}{3} D^{1/2} \log D.$$

Chapter 5. $\mathfrak{p}$-adic Fields

§ 1. Principal Properties, Integers, Units

1. In this chapter we shall consider fields which are completions of algebraic number fields under a discrete valuation. According to Theorem 3.1 every valuation gives rise to a complete field which is uniquely determined by it up to a topological isomorphism. By Theorem 3.2 every discrete valuation v of an algebraic number field K is induced by a prime ideal $\mathfrak{p}$ of its ring of integers. The completion of K under v will be denoted by $K_\mathfrak{p}$ or K_v and called the $\mathfrak{p}$-*adic field*. In the case of the rational field Q we shall not distinguish between the prime p and the prime ideal in Z which it generates, and we shall write Q_p for the field which is the completion of Q under the valuation induced by pZ.

We shall not give here the full story of p-adic fields, and we shall develop their theory only to such an extent as is needed for our immediate purposes. The interested reader should refer to J. P. Serre's monograph [62].

We start with defining the integers and units of a p-adic field. Let K be any algebraic number field, $\mathfrak{p}$ a prime ideal of its ring R_K of integers, v the valuation of K associated with $\mathfrak{p}$, $K_\mathfrak{p}$ the completion of K under v, and k the quotient field $R_K/\mathfrak{p}$. The prolongation of v to $K_\mathfrak{p}$ will be denoted by the same letter v. By Corollary 1 to Theorem 3.1 we can see that v is discrete, the ring $R_\mathfrak{p} = \{x \in K_\mathfrak{p} : v(x) \leqslant 1\}$ is the closure of the ring $\{x \in K: v(x) \leqslant 1\}$, $\mathfrak{P} = \{x \in K_\mathfrak{p} : v(x) < 1\}$ is a prime ideal of $R_\mathfrak{p}$ which is the closure of the prime ideal $\{x \in K: v(x) < 1\}$ of that ring, and the proof of Theorem 3.1 shows that the sets $v(K)$ and $v(K_\mathfrak{p})$ coincide. This enables us to define a prolongation of the exponent corresponding to $\mathfrak{p}$ defined in K to the whole field $K_\mathfrak{p}$ by putting

$$n_\mathfrak{p}(x) = \log_a v(x)$$

for non-zero x in $K_\mathfrak{p}$, where the number a of the interval $(0, 1)$ is chosen in such a way as to satisfy

$$v(x) = a^{n_\mathfrak{p}(x)}$$

for x in K.

The ring $R_\mathfrak{p}$ will be called the *ring of integers of* $K_\mathfrak{p}$, and its elements the *integers of* $K_\mathfrak{p}$. We shall also write Z_p for the ring of integers of Q_p. The invertible elements of $R_\mathfrak{p}$ form a group $U(K_\mathfrak{p})$, which is the *group of units of* $K_\mathfrak{p}$.

Having established the notation, we prove

Proposition 5.1. (i) *The ring $R_\mathfrak{p}$ is a Dedekind domain with a trivial class-group, $\mathfrak{P}$ is its unique non-zero prime ideal, and the quotient field $R_\mathfrak{p}/\mathfrak{P}$ is isomorphic with $R_K/\mathfrak{p}$.*

(ii) *The ring $R_\mathfrak{p}$ is the closure of R_K in $K_\mathfrak{p}$ and, more generally, $\mathfrak{P}^m$ is the closure of $\mathfrak{p}^m$ for each m. Moreover, the intersection $\mathfrak{P}^m \cap R_K$ equals $\mathfrak{p}^m$.*

(iii) $U(K_\mathfrak{p}) = R_\mathfrak{p} \setminus \mathfrak{P}$ *and if π is any fixed element of $\mathfrak{P} \setminus \mathfrak{P}^2$ (which may be chosen from R_K), then every element x of $K_\mathfrak{p}$ can be uniquely written in the form $x = a\pi^m$ with $a \in U(K_\mathfrak{p})$ and $m = n_\mathfrak{p}(x)$.*

Proof. Part (i) follows immediately from Theorem 1.10, Proposition 1.11 and Corollary 2 to Theorem 3.1. To prove (ii) consider x in K with $n_\mathfrak{p}(x) = m$. We can write $x = a/b$ with a and b in R_K, $a \in \mathfrak{p}^m$, $b \notin \mathfrak{p}$. If we now choose for every natural n an element z_n in $\mathfrak{P}^m$ such that $a - bz_n \equiv 0 \pmod{\mathfrak{P}^n}$, which is possible by Proposition 1.6, then we can see that the resulting sequence converges to x. This shows that $\mathfrak{P}^m \cap K$ lies in the closure of $\mathfrak{p}^m$ and so does $\mathfrak{P}^m$. But $\mathfrak{P}^m$ is closed and thus it must be equal to the closure of $\mathfrak{p}^m$. The rest of (ii) is obvious. Finally, (iii) follows from (i) and the observation that if, in a domain, there is a unique non-zero prime ideal then it consists of all non-invertible elements. □

An explicit form for elements of $K_\mathfrak{p}$ is provided by the following theorem:

Theorem 5.1. *Let K be a field of zero characteristic, complete under a discrete valuation with a finite residue class field R/P, where R is the valuation ring and P its prime ideal. Let A be a system of representatives of R/P containing the zero element, and let t_n ($n = 0, \pm 1, \ldots$) be a sequence of elements of K such that $t_1 \in P \setminus P^2$, and $t_{n+1}/t_n \in P \setminus P^2$ for all n. Then every element x of K can be uniquely represented as the sum of convergent series*

$$x = a_N t_N + a_{N+1} t_{N+1} + \ldots \qquad (5.1)$$

for suitable $N \in Z$ and $a_N, a_{N+1}, \ldots$ in A, with $a_N \neq 0$ in the case of $x \neq 0$.

Proof. Assume first that $x \in R$. If $x = 0$, then there is no question of existence. Assume thus that $x \neq 0$. Choose an $a_0 \in A$ so that $x \equiv a_0 t_0 \pmod{P}$. Then the element $x_1 = (x-a_0 t_0)/t_1$ lies in R. Choose in turn an $a_1 \in A$ so that $x_1 \equiv a_1 (\mathrm{mod}\, P)$ and let $x_2 = (x_1-a_1)t_1/t_2 \in R$. If the elements x_i and a_i are already chosen for $i \leqslant r$ and satisfy $x_i \equiv a_i \pmod{P}$, then put $x_{r+1} = (x_r-a_r)t_r/t_{r+1} \in R$ and define $a_{r+1} \in A$ by $x_{r+1} \equiv a_{r+1} \pmod{P}$. Observe that for every r we have

$$x \equiv a_0 t_0 + a_1 t_1 + \ldots + a_r t_r \pmod{P^{r+1}},$$

and so the series (5.1) converges to x.

If x is an arbitrary element of K, then write $x = \varepsilon t_m$, ε being a unit of K and m being suitably chosen. Applying the proved part of our theorem to the number ε and the sequence $t'_n = t_{m+n}/t_m$, we obtain (5.1) for our x.

To prove the uniqueness of representation (5.1) it suffices to prove that $0 = a_m t_m + a_{m+1} t_{m+1} + \ldots$ implies the vanishing of all coefficients a_i. Assume that a_m is the first non-vanishing coefficient and observe that $0 = t_m^{-1}(a_m t_m + a_{m+1} t_{m+1} + \ldots) = a_m + a_{m+1} t_{m+1} t_m^{-1} + \ldots$. In this series all terms except possibly the first lie in P and so $a_m \in P$, but a_m lies in A and so it equals zero, which gives a contradiction. □

The problem of convergence criteria for series in K_p is easily solved by the following proposition:

Proposition 5.2. *If L is a field complete under a non-Archimedean valuation v then the series $\sum_{n=1}^{\infty} a_n$ ($a_n \in L$) converges iff a_n tends to zero.*

Proof. The necessity being obvious, we turn to the sufficiency. Assume thus that the sequence $\{a_n\}$ converges to zero, i.e. that $v(a_n)$ tends to zero. Then for any $M \geqslant N$ we have

$$v(a_M + a_{M+1} + \ldots + a_N) \leqslant \max_{M \leqslant i \leqslant N} v(a_i),$$

and the right-hand side tends to zero with M^{-1} independently of the behaviour of N, showing that the sequence of partial sums of our series is fundamental, and hence convergent by completeness. □

The topological properties of fields K considered in Theorem 5.1 are described in

Theorem 5.2. *The additive and multiplicative groups of K are both locally compact topological groups; the ring R and also the ideals P^m ($m = 1, 2, \ldots$) of it are compact. Those ideals, moreover, are open in R and form a basis of neighbourhoods of the zero-element; thus R is a totally disconnected topological space.*

Proof. We need a lemma.

Lemma 5.1. *If for $j = 1, 2, \ldots$ the ring-homomorphisms $f_j: R/P^{1+j} \to R/P^j$ are defined by*

$$f_j: x \mapsto x \pmod{P^j},$$

then the inverse limit space $X = \liminv R/P^j$ is topologically isomorphic with R.

Proof of the lemma. Let $x \in R$ and let x_j be the image of x in R/P^j under the residue class map. We define a mapping of R into X by putting

$$f: x \mapsto [x_1, x_2, \ldots].$$

It is obvious that f is a homomorphism and its kernel is trivial. Moreover, if $[a_1, a_2, \ldots]$ is any element of X, then for m and n larger than a given r we have $a_m \equiv a_n \pmod{P^r}$; thus, if we consider a sequence $A_1, A_2, \ldots$ of elements of R for which the class $A_n \pmod{P^n}$ equals a_n, then this sequence is fundamental, and for its limit A we have $f(A) = [a_1, a_2, \ldots]$. Thus f is surjective and so it is an algebraic isomorphism. It remains to show that both f and f^{-1} are continuous, and since those mappings are homomorphisms and the underlying space is homogeneous (being the carrier of a topological group), it suffices to consider continuity at the zero element. If a sequence a_n in R converges to zero, then a_n lies in P^N from a certain n on for any given N; hence $f(a_n)$ is of the form $[0, \ldots, 0, x_{N+1}, \ldots]$, i.e. tends to zero. Conversely, if $y_n = [a_1^{(n)}, a_2^{(n)}, \ldots]$ tends to zero, then from a certain n on we have $a_i^{(n)} = 0$ for $i = 1, 2, \ldots, N$ for any given N; hence if $f(x_n) = y_n$, then x_n also tends to zero. □

To prove the theorem, observe that R is compact as the inverse limit of finite sets, and so are ideals P^m as closed subsets of R. Moreover, if $a \in P^m$ and $a - b \in P^{1+m}$, then the element b also lies in P^m, showing that the ideals P^m are open. The set P is a compact neighbourhood of the zero element in the additive group of K_p and $1 + P$ has the same property in the multiplicative group, whence those sets are locally compact. The remaining assertions are now evident. □

Corollary. *Let $F_1(x_1, \ldots, x_r), \ldots, F_k(x_1, \ldots, x_r)$ be polynomials over R. The system of equations*

$$F_i(x_1, \ldots, x_r) = 0 \quad (i = 1, 2, \ldots, k)$$

has a solution in R iff the system of congruences

$$F_i(x_1, \ldots, x_r) \equiv 0 \pmod{P^m} \quad (i = 1, 2, \ldots, k)$$

has a solution in R for every m.

If $F_1, \ldots, F_k$ are forms, and for every m the system of congruences has a solution with some $x_i \notin P$, then the corresponding system of equations has a solution with some $x_i \neq 0$.

Proof. The necessity is trivial. To prove the sufficiency, let $x_1^{(m)}, \ldots, x_r^{(m)} \in R$ be elements forming a solution of the system of congruences $(\bmod P^m)$. As R is a compact metric space, we can find a sequence m_k such that all sequences $\{x_i^{(m_k)}\}$ are convergent, say, to $x_i^{(0)}$ $(i = 1, 2, \ldots, r)$. It is obvious that the resulting elements satisfy our system of equations. If $x_i^{(m_k)}$ are outside P, then the same applies to their limit points $x_i^{(0)}$ because P is open. □

2. One of the main tools used in the study of p-adic fields is the theorem which we are now going to prove and which is commonly known as "Hensel's lemma".

Theorem 5.3. *Let K_p be a p-adic field, R_p its ring of integers and $\mathfrak{P}$ the prime ideal of R_p. For x in R_p denote by $\bar{x}$ the image of x in the residue class field $R_p/\mathfrak{P}$, which we shall simply denote by k. Similarly, for any polynomial $W(t)$ over R_p we shall denote by $\overline{W}(t)$ the polynomial over k obtained from $W(t)$ by replacing each of its coefficients by the corresponding residue in k.*

If $\deg \overline{W} = \deg W$ and the polynomial $\overline{W}(t)$ can be represented as a product of two relatively prime polynomials $f(t)$ and $g(t)$ over k of positive degrees, then there exist two relatively prime polynomials $F(t)$ and $G(t)$ over R_p such that $\deg F = \deg f$, $\deg G = \deg g$, $\overline{F}(t) = f(t)$, $\overline{G}(t) = g(t)$ and $W(t) = F(t)G(t)$. Moreover, if $f(t)$ is monic, then we can choose also a monic $F(t)$.

Proof. Let m be the degree of $f(t)$ and n the degree of $g(t)$. We shall define inductively two sequences, $f_k(t)$ and $g_k(t)$ of polynomials over R_p enjoying the following properties:
  (a) $\deg f_k(t) = m$, $\deg g_k(t) = n$.
  (b) All coefficients of the polynomial $W(t) - f_k(t)g_k(t)$ lie in $\mathfrak{P}^{k+1}$.
  (c) All coefficients of $f_k(t) - f_{k-1}(t)$ and $g_k(t) - g_{k-1}(t)$ lie in $\mathfrak{P}^k$.
  (d) If the polynomial $f(t)$ is monic, so is $f_k(t)$.

The existence of such sequences will immediately imply the theorem. In fact, conditions (b) and (c) show that the coefficients of $f_k(t)$ and $g_k(t)$ form convergent sequences, and by (a) the polynomials $f_k(t)$ and $g_k(t)$ converge to some polynomials $F(t)$ and $G(t)$, respectively, in the topology defined in the set A of polynomials with degree not exceeding $\max(m, n)$ by the family $H(t) + P^k R_p[t] \cap A$ with $H(t) \in A$ and $k = 1, 2, \ldots$. From (b) follows the equality $W(t) = F(t)G(t)$, (c) implies $\overline{F}(t) = f(t)$, $\overline{G}(t) = g(t)$ and since $f(t)$ and $g(t)$ are relatively prime, so are $F(t)$ and $G(t)$. Moreover, if $f(t)$ is monic, then by (d) $F(t)$ is monic as well.

So let us now find the required sequences. Let $f_0(t)$ be any polynomial of degree m over $R_\mathfrak{p}$ for which $\overline{f}_0(t) = f(t)$ and which is monic if $f(t)$ is such. Similarly, let $g_0(t)$ be any polynomial of degree n over $R_\mathfrak{p}$ with $\overline{g}_0(t) = g(t)$. Assume now that we have already chosen $f_0(t), \ldots, f_N(t), g_0(t), \ldots, g_N(t)$ subject to (a)–(d) for an $N \geq 0$, and let π be any element of $\mathfrak{P} \setminus \mathfrak{P}^2$. Then the polynomial $c_N(t) = (W(t) - f_N(t)g_N(t))\pi^{-N-1}$ has all its coefficients in $R_\mathfrak{p}$ by (b), and since $f(t)$ and $g(t)$ are relatively prime, we are able to determine $a_N(t)$ and $b_N(t)$ from $R_\mathfrak{p}[t]$ which are of degree not exceeding $m-1$ and $n-1$, respectively, and which satisfy

$$\overline{c}_N(t) + \overline{a}_N(t)g(t) + \overline{b}_N(t)f(t) = 0,$$

i.e.

$$c_N(t) + a_N(t)g_N(t) + b_N(t)f_N(t) \equiv 0 \;(\mathrm{mod}\; \mathfrak{P}).$$

Now observe that the polynomials

$$f_{N+1}(t) = f_N(t) + a_N(t)\pi^{N+1},$$
$$g_{N+1}(t) = g_N(t) + b_N(t)\pi^{N+1}$$

satisfy (a)–(c) and if $f(t)$ was monic, also (d). □

Corollary 1. *If $W(t)$ is a polynomial over $R_\mathfrak{p}$ such that the polynomial $\overline{W}(t)$ has a simple zero u in $R_\mathfrak{p}/\mathfrak{P}$, then there exists an element b in $R_\mathfrak{p}$ such that $W(b) = 0$ and $\overline{b} = u$.*

Proof. We can write $\overline{W}(t) = (t-u)V(t)$ with relatively prime $t-u$ and $V(t)$; thus the theorem shows the existence of a factorization $W(t) = (t-b)V_1(t)$ with $\overline{b} = u$. □

Corollary 2. *If the residue class field $R_\mathfrak{p}/\mathfrak{P}$ has p^f elements, then in every non-zero residue class $(\mathrm{mod}\; \mathfrak{P})$ there is a $(p^f - 1)$-th root of unity.*

Proof. Consider the polynomial $t^{p^f - 1} - 1$. In the field $R_\mathfrak{p}/\mathfrak{P}$ every non-zero element is its root and hence it has $p^f - 1$ distinct roots in it and we can apply the preceding corollary. □

Corollary 3. *If p is a rational odd prime and m is a rational integer not divisible by p, then the p-adic field Q_p contains a square root of m iff $\left(\dfrac{m}{p}\right) = 1$.*

Proof. Apply Corollary 1 to the polynomial $t^2 - m$. □

Corollary 4. *If $W(t)$ is a polynomial over $R_\mathfrak{p}$ which is irreducible over $K_\mathfrak{p}$, then $\overline{W}(t) = V(t)^m$ for a certain $m \geq 1$, $V(t)$ being irreducible over $R_\mathfrak{p}/\mathfrak{P}$.*

Proof. Otherwise $\overline{W}(t)$ would have a factorization into two relatively prime and non-trivial factors and Theorem 5.2 would imply the existence of such a factorization for $W(t)$. □

3. The aim of this section is to show that every finite extension of a p-adic field is also a p-adic field and, moreover, that p-adic fields can be characterized as those fields of zero characteristic which are complete with respect to a discrete valuation and have a finite residue class field R/P, where R is the valuation ring and P the valuation ideal. It is obvious that the p-adic fields have those properties, and so we assume up to the end of this subsection that the field K considered satisfies the conditions:

(i) char $K = 0$.

(ii) K is complete with respect to a discrete valuation $v(x)$.

(iii) The residue class field $k = R/P$, where R and P are the valuation ring and the valuation ideal of K, respectively, is finite.

Note that not all of those conditions will actually be used in every proof, but this will not be of any importance to us. The results developed in this subsection will also have applications in later subsections and not only to the characterization of p-adic fields.

We start with

Proposition 5.3. *If L/K is a finite extension of a field satisfying (i)–(iii), then there exists exactly one valuation $w(x)$ of L which coincides with $v(x)$ on K. It is equal to*

$$w(x) = (v(N_{L/K}(x)))^{1/n},$$

where $n = [L:K]$. Moreover, L satisfies also (i)–(iii).

Proof. Denote by S the integral closure of R in L. By Theorems 1.9 and 1.10 S is Dedekind, whence if P_1 is a prime ideal of S lying above P, then the valuation associated with it extends v after a suitable normalization. Call this extension w. If w' is another extension of v to L, then by Proposition 3.1 they both induce the product topology in L, and so Proposition 1.8 implies $w = w'$. So we have both existence and uniqueness. By Theorem 1.9 the ring S has the (FN) property and Proposition 1.11 (iv) shows that the same holds for the valuation ring of w, giving (iii). Since (i) and (ii) are obviously satisfied in view of L being complete in the product topology, it remains to show that w has the asserted form. To do this let M/K be the minimal normal extension containing L in a fixed algebraic closure of L and denote by u the unique extension of v to M. If $s \in \text{Gal}(M/K)$, then putting $f(x) = u(s(x))$ for $x \in M$, we obtain again a prolongation of v to M, and thus $f = u$. Since

$$N_{L/K}(x) = \prod_{i=1}^{n} s_i(x),$$

where s_i ranges over a subset of $\text{Gal}(M/K)$, we obtain for $x \in L$

$$v(N_{L/K}(x)) = w(N_{L/K}(x)) = u(x)^n = w(x)^n. \qquad \square$$

Corollary 1. *If K satisfies* (i)–(iii) *and L/K is finite, S is the valuation ring of the extension of v to L, and $a \in L$, then the conditions $a \in S$ and $N_{L/K}(a) \in R$ are equivalent.* $\qquad \square$

Corollary 2. *If K satisfies* (i)–(iii) *and L/K is finite and normal, then every automorphism $g \in \text{Gal}(L/K)$ is continuous in the topology induced by w.*

Proof. If for $x \in L$ we define $w'(x) = w(g(x))$, then by the proposition $w' = w$, i.e. g is an isometry. $\qquad \square$

Our next result is simple but very useful. It is known as "Krasner's lemma".

Proposition 5.4. *Let K be a field satisfying* (i)–(iii) *and L a finite normal extension of it. Let w be the prolongation of v to L. If a lies in L and $M = K(a)$, then every element b of L satisfying $w(a-b) < w(s(b)-b)$ for every s with $s(b) \neq b$ from the Galois group $G(L/K)$ lies in M.*

Proof. If the last statement were not true, then for some s in $G(L/K)$ we would have $s(b) \neq b$, $s(a) = a$. But then

$$w(b-s(b)) = w(b-a+a-s(b)) \leq \max(w(b-a), w(a-s(b)))$$
$$= \max(w(b-a), w(s(a)-s(b)))$$
$$= w(b-a) < w(s(b)-b),$$

giving a contradiction. $\qquad \square$

With the aid of Krasner's lemma we shall now show that two monic irreducible polynomials of the same degree whose corresponding coefficients are sufficiently close to each other define the same extension. Before doing this, consider any algebraic closure $\hat{K}$ of a field K satisfying (i)–(iii) and observe that the valuation v can be extended in a unique way to a valuation of $\hat{K}$, which, however, will not be discrete; also the field $\hat{K}$ will not be complete in it, but since we shall not need this fact in the sequel, the proof is omitted. The interested reader may look for it in Ostrowski [17]. To obtain such a valuation on $\hat{K}$, take any element of this field, say a. The field $K(a)$ is a finite extension of K, and so there is a unique prolongation of v to it. Define $V(a)$ as equal to the value of this prolongation at the element a. If we consider a as an element of any other finite extension of K, we always obtain the same value of the valuation at that element, whence V is a valuation defined unambiguously in the whole field $\hat{K}$, which extends v. Having this in mind, we prove

Proposition 5.5. *Let K be any field satisfying (i)–(iii), and let $F(t) = t^n + a_{n-1}t^{n-1} + \ldots + a_0$ be a polynomial irreducible in K with coefficients in R. If $G(t) = t^n + b_{n-1}t^{n-1} + \ldots + b_0$ is another polynomial over R, whose coefficients are sufficiently close to the corresponding coefficients of $F(t)$, then $G(t)$ is also irreducible over K and, moreover, to every root a of $F(t)$ there exist a root b of $G(t)$ such that the fields $K(a)$ and $K(b)$ coincide. (We regard a and b as elements of a fixed algebraic closure of K.)*

Proof. We shall denote the extension of v to the field $\hat{K}$ by the same symbol v. First we prove that if $\max_i v(a_i - b_i)$ is sufficiently small, then to every root a of $F(t)$ there exists a root b of $G(t)$ such that $v(a-b)$ is smaller than a given positive ε.

Let $b_1, \ldots, b_n$ be the roots of $G(t)$ in $\hat{K}$, each of them occurring in this sequence as many times as its multiplicity. Assume that $v(a-b_i) \geq \varepsilon$ for $i = 1, 2, \ldots, n$. Since $G(t) = (t-b_1) \ldots (t-b_n)$, we have

$$v(G(a)) = \prod_{i=1}^{n} v(a-b_i) \geq \varepsilon^n \quad \text{and} \quad v(F(a) - G(a)) = v(G(a)) \geq \varepsilon^n.$$

On the other hand, we have

$$F(a) - G(a) = (a_{n-1} - b_{n-1})a^{n-1} + \ldots + (a_1 - b_1)a + (a_0 - b_0);$$

thus

$$v(F(a) - G(a)) \leq \max_{\substack{j \\ a_j \neq b_j}} v(a_j - b_j) v(a^j),$$

and finally

$$\varepsilon^n \leq B \max v(a_j - b_j)$$

with $B = \max_j v(a)^j$. However, the last inequality is impossible if $v(a_j - b_j)$ is sufficiently small for $j = 0, 1, 2, \ldots, n-1$.

Now let $a_1, a_2, \ldots, a_n$ be the set of all roots of $F(t)$ and put $M = \min_{i \neq j} v(a_i - a_j)$. To every a_i we can find a root b_i of $G(t)$ provided the coefficients of $G(t)$ are sufficiently near to those of $F(t)$. Applying Proposition 5.4 to the splitting field of $F(t)$ over K, we obtain $K(a_i) \subset K(b_i)$. But $K(a_i)$ is of degree n over K and $[K(b_i):K]$ cannot exceed n, whence $K(a_i) = K(b_i)$ for all $i = 1, 2, \ldots, n$.

The irreducibility of $G(t)$ is now obvious. □

Corollary. *A field K satisfying* (i)–(iii) *can have only finitely many extensions of a given degree.*

Proof. Let $n \geqslant 2$ be given and consider R^n with the product topology. For a monic polynomial

$$F(t) = t^n + a_{n-1} t^{n-1} + \ldots + a_0 \in R[t]$$

denote by $f(F)$ the element $[a_{n-1}, \ldots, a_0]$ of R^n. The proposition implies that the set V of all monic irreducible polynomials is open, and hence compact in R^n. For a given extension L/K of degree n let V_L be the set of all those polynomials from V whose roots include one which generates L/K. By the proposition each of the sets $f(V_L)$ is open, and since their union covers $f(V)$ and they are disjoint, there can be only finitely many of them. □

In Propositions 5.12 and 5.13 we shall enumerate the quadratic extensions of Q_p as well as those with Galois group C_2^2.

Now we prove a characterization of p-adic fields:

Theorem 5.4. *Let K be a field with a valuation $v(x)$. The following properties are then equivalent*:

(a) *K is a p-adic field with the p-adic valuation.*
(b) *K satisfies conditions* (i)–(iii).
(c) *K is a finite extension of a certain Q_p.*

Proof. The implication (a) → (b) is obvious. Assume that K satisfies (i), (ii) and (iii). The restriction of v to the field Q of rational numbers is a discrete valuation of Q; thus by Theorem 1.12 it is equivalent to some p-adic valuation.

Since K is complete, it contains by Theorem 3.1 an isomorphic copy of some field Q_p, and v, restricted to it, coincides with its usual valuation after a suitable renorming if required. We may simply assume that $Q_p \subset K$. Let n be the exponent of K induced by v and let $e = n(p)$, e being a non-zero integer. Thus $pR = \mathfrak{P}^e$. Note that if we apply Theorem 5.1 to K, taking for $\{t_n\}$ any sequence satisfying the required assumptions and such that $t_{je} = p^j$ holds for $j = 0, \pm 1, \ldots$, then we find that R is a finitely generated free module over the ring Z_p of integers of Q_p and so K has as a linear Q_p-space a finite dimension. This shows (c).

Finally, assume that K is a finite extension of Q_p, and let a be any generator of K/Q_p lying in R. The coefficients of its minimal monic polynomial over the ring Z_p of integers of Q_p may be approximated arbitrarily close by rational integers. If we replace the coefficients by those approximants, we obtain a polynomial over Z which, by Proposition 5.5, will be irreducible over Q_p, and one of its roots, say b, will generate $Q_p(a) = K$ over Q_p. Now consider the field $Q(b)$, which is a finite extension of Q and lies in K. Its closure in K contains Q_p and also b, and thus it coincides with K. Now observe that the valuation of K induces in $Q(b)$ a discrete valuation which, by Theorem 3.2, is induced by a prime ideal of the ring of integers of $Q(b)$, and obviously K is the completion of $Q(b)$ under it. □

4. The following theorem enables us to translate many problems concerning finite extensions of algebraic number fields into the language of p-adic fields. Its applications will be discussed in the next chapter.

Theorem 5.5. *Let K be an algebraic number field, $\mathfrak{p}$ a non-zero prime ideal of R_K, and $K_\mathfrak{p}$ the completion of K corresponding to $\mathfrak{p}$. Moreover, let L/K be a finite extension, $\mathfrak{P}$ a prime ideal of R_L lying above $\mathfrak{p}$, and $L_\mathfrak{P}$ the corresponding completion. Finally, let $R_\mathfrak{p}$ and $R_\mathfrak{P}$ be the corresponding rings of integers in $K_\mathfrak{p}$ and $L_\mathfrak{P}$. Then we have:*

(i) $[L_\mathfrak{P}:K_\mathfrak{p}] = e_{L/K}(\mathfrak{P})f_{L/K}(\mathfrak{P})$.

(ii) *The field $L_\mathfrak{P}$ is generated by L and $K_\mathfrak{p}$, i.e., $L_\mathfrak{P} = K_\mathfrak{p}L$.*

(iii) *The ring $R_\mathfrak{P}$ is the integral closure of $R_\mathfrak{p}$ in $L_\mathfrak{P}$.*

(iv) *If $\bar{\mathfrak{p}}$ and $\bar{\mathfrak{P}}$ are the prime ideals in $R_\mathfrak{p}$ and $R_\mathfrak{P}$, respectively, then $\bar{\mathfrak{P}}$ lies above $\bar{\mathfrak{p}}$ and $e_{L_\mathfrak{P}/K_\mathfrak{p}}(\bar{\mathfrak{P}}) = e_{L/K}(\mathfrak{P})$, $f_{L_\mathfrak{P}/K_\mathfrak{p}}(\bar{\mathfrak{P}}) = f_{L/K}(\mathfrak{P})$. $\bar{\mathfrak{P}}$ is the only prime ideal in $R_\mathfrak{P}$ lying above $\bar{\mathfrak{p}}$.*

Proof. The closure of K in $L_\mathfrak{P}$ is complete, and so, in view of Theorem 3.1, we may identify it with $K_\mathfrak{p}$. Let $a_1, \ldots, a_n$ be a basis of the K-space L and put

$$L_1 = a_1 K_\mathfrak{p} + a_2 K_\mathfrak{p} + \ldots + a_n K_\mathfrak{p}.$$

L_1 is a $K_\mathfrak{p}$-space contained in $L_\mathfrak{P}$ and containing L. Moreover, $\dim_{K_\mathfrak{p}} L_1$ cannot exceed $\dim_K L = n$. By Proposition 3.1, L_1 has the product topology; hence it is closed, and this implies $L_1 = L_\mathfrak{P}$. But L_1 is generated by $K_\mathfrak{p}$ and L; hence we get (ii). Proposition 5.1 shows that $\bar{\mathfrak{P}}$ is indeed the only prime ideal of $R_\mathfrak{P}$ and so it must lie above $\bar{\mathfrak{p}}$; hence, by Theorem 4.1 (which is applicable in our case), we get $[L_{\mathfrak{P}}:K_\mathfrak{p}] = e_{L_{\mathfrak{P}}/K_\mathfrak{p}}(\overline{\mathfrak{P}})f_{L_{\mathfrak{P}}/K_\mathfrak{p}}(\overline{\mathfrak{P}})$. The equality $f_{L_{\mathfrak{P}}/K_\mathfrak{p}}(\overline{\mathfrak{P}}) = f_{L/K}(\mathfrak{P})$ follows from Proposition 5.1 (i), and if $e = e_{L_{\mathfrak{P}}/K_\mathfrak{p}}(\overline{\mathfrak{P}})$, $e_1 = e_{L/K}(\mathfrak{P})$, then $\mathfrak{p}R_L = \mathfrak{P}^{e_1}\mathfrak{Q}$ with $\mathfrak{P} \nmid \mathfrak{Q}$. Taking closures, we get $\mathfrak{p}R_{\mathfrak{P}} = \overline{\mathfrak{P}}^{e_1} \cdot \mathfrak{Q}_1$, $\mathfrak{Q}_1$ being the closure of $\mathfrak{Q}$. But since $\mathfrak{Q}$ is not divisible by $\mathfrak{P}$, it contains an element which does not lie in $\mathfrak{P}$ and it is a unit in $R_{\mathfrak{P}}$, whence $\mathfrak{Q}_1 = R_{\mathfrak{P}}$, and we finally get $\overline{\mathfrak{P}}^e = \overline{\mathfrak{p}}R_{\mathfrak{P}} = \overline{\mathfrak{P}}^{e_1}$; thus $e = e_1$. This proves (iv) and (i) follows immediately. It remains to prove (iii). Let $a \in R_{\mathfrak{P}}$. By Proposition 5.1 (ii) we can find a sequence $x_1, x_2, \ldots$ of elements of R_L converging to a. Let

$$F_m(t) = t^n + c_{n-1}^{(m)} t^{n-1} + \ldots + c_0^{(m)}$$

be the minimal polynomial for x_m over R_K. From every sequence $\{c_j^{(m)}\}$ we can extract a convergent subsequence and let $c_j \in R_\mathfrak{p}$ be its limit. If now $F(t) = t^n + c_{n-1} t^{n-1} + \ldots + c_0$, then obviously $F(a) = 0$; thus a is integral over $R_\mathfrak{p}$. This shows that $R_{\mathfrak{P}}$ is contained in the integral closure of $R_\mathfrak{p}$ in L, but it is integrally closed and contains $R_\mathfrak{p}$, and thus (iii) follows. □

We point out a useful corollary:

Corollary. *If L_1, L_2 are extensions of an algebraic number field K and a prime ideal $\mathfrak{p}$ of K splits completely in L_1 and L_2, then it splits also in their composite $L_1 L_2$.*

Proof. Let $\mathfrak{P}$ be a prime ideal lying over $\mathfrak{p}$ in $L = L_1 L_2$ and put $\mathfrak{P}_1 = \mathfrak{P} \cap R_{L_1}$, $\mathfrak{P}_2 = \mathfrak{P} \cap R_{L_2}$. Then by (i) we have the equality

$$[(L_1)_{\mathfrak{P}_1} : K_\mathfrak{p}] = [(L_2)_{\mathfrak{P}_2} : K_\mathfrak{p}] = 1,$$

whence both L_1 and L_2 are contained in $K_\mathfrak{p}$. Thus $L = L_1 L_2 \subset K_\mathfrak{p}$, whence $L_{\mathfrak{P}}$, being the closure of L, also lies in $K_\mathfrak{p}$, leading to $[L_{\mathfrak{P}} : K_\mathfrak{p}] = 1$, which by (i) implies

$$e_{L/K}(\mathfrak{P}) = f_{L/K}(\mathfrak{P}) = 1,$$

as asserted. □

We see that the behaviour of the prime ideals $\mathfrak{p}$ and $\mathfrak{P}$ is the same as that of $\overline{\mathfrak{p}}$ and $\overline{\mathfrak{P}}$. Moreover, one can easily see that $\overline{\mathfrak{P}} = \mathfrak{P}R_{\mathfrak{P}}$ and $\overline{\mathfrak{p}} = \mathfrak{p}R_\mathfrak{p}$, and so we can dispense with the pedantic distinguishing of $\mathfrak{p}$ and $\overline{\mathfrak{p}}$, $\mathfrak{P}$ and $\overline{\mathfrak{P}}$. In the sequel we thus use the same letter for a prime ideal of an algebraic number field and the prime ideal in its completion.

Note, moreover, that as there is only one prime ideal in the ring of integers of a p-adic field, we can freely write $e(L_{\mathfrak{P}}/K_\mathfrak{p})$ for $e_{L_{\mathfrak{P}}/K_\mathfrak{p}}(\mathfrak{P})$ and $f(L_{\mathfrak{P}}/K_\mathfrak{p})$

for $f_{L_\mathfrak{P}/K_\mathfrak{p}}(\mathfrak{P})$ and speak about the ramification index or the residue class field degree of the extension $L_\mathfrak{P}/K_\mathfrak{p}$ rather than that of a prime ideal.

5. Note that to an extension $L_\mathfrak{P}/K_\mathfrak{p}$ we may apply the theory developed in Chapter 4, and since the class-group of $R_\mathfrak{p}$ is trivial, we can even obtain more. In particular, we can prove the following

Proposition 5.6. *The ring $R_\mathfrak{P}$ is a free $R_\mathfrak{p}$-module with $n = [L_\mathfrak{P}:K_\mathfrak{p}]$ free generators. Moreover, we can find an element a in $R_\mathfrak{P}$ such that the $R_\mathfrak{p}$-module generated freely by $1, a, a^2, \ldots, a^{n-1}$ coincides with $R_\mathfrak{P}$.*

Proof. Lemmas 4.3 and 4.4 show that there is an element a in $R_\mathfrak{P}$ such that the conductor of the ring $R_\mathfrak{p}[a]$ is not divisible by $\mathfrak{P}$. But then this conductor equals $R_\mathfrak{P}$ and thus contains the unit element and, by Proposition 4.12 (ii), $R_\mathfrak{p}[a]$ and $R_\mathfrak{P}$ coincide. □

Now we shall define for finite extensions of p-adic fields the numerical discriminant $\partial(L_\mathfrak{P}/K_\mathfrak{p})$, which will be very useful for the intended applications to algebraic number fields. According to the previous proposition, $R_\mathfrak{P}$ is a free $R_\mathfrak{p}$-module, and so we can consider its free generators, say $w_1, \ldots, w_n$, and define the discriminant $D_{L_\mathfrak{P}/K_\mathfrak{p}}(w_1, \ldots, w_n)$ of this set by

$$D_{L_\mathfrak{P}/K_\mathfrak{p}}(w_1, \ldots, w_n) = \det[T_{L_\mathfrak{P}/K_\mathfrak{p}}(w_i w_j)],$$

or, equivalently, by

$$D_{L_\mathfrak{P}/K_\mathfrak{p}}(w_1, \ldots, w_n) = (\det[w_i^{(j)}])^2,$$

where $w_i^{(j)}$ are the conjugates to w_i in an algebraic closure of $L_\mathfrak{P}$. In the same way as in Chapter 2 in the case of algebraic number fields, we see that this discriminant is non-zero and the discriminants of different sets of free generators differ by the square of a unit of the ring $R_\mathfrak{p}$. Since in Q the only square of a unit was the unit element, we could then define the field discriminant as the common value of discriminants of generators; however, now it is no longer possible to proceed in the same way. Instead, we define the discriminant $\partial(L_\mathfrak{P}/K_\mathfrak{p})$ as the class in the factor group $K_\mathfrak{p}^*/U(K_\mathfrak{p})^2$ which contains all discriminants of sets of free generators.

Principal properties of the discriminant defined in this way are contained in

Proposition 5.7. *Let $L_\mathfrak{P}/K_\mathfrak{p}$ be a finite extension of a p-adic field $K_\mathfrak{p}$. Then*
    (i) The ideal in $R_\mathfrak{p}$ generated by any representative of $\partial(L_\mathfrak{P}/K_\mathfrak{p})$ equals the discriminant $d(L_\mathfrak{P}/K_\mathfrak{p})$ defined in Chapter 4.
    (ii) If the extension $L_\mathfrak{P}/K_\mathfrak{p}$ is unramified, then $\partial(L_\mathfrak{P}/K_\mathfrak{p})$ lies in $U(K_\mathfrak{p})/U(K_\mathfrak{p})^2$.

(iii) *If $M/L_\mathfrak{P}$ is a finite extension of degree n, then*

$$\partial(M/K_\mathfrak{p}) = \partial(L_\mathfrak{P}/K_\mathfrak{p})^n N_{L_\mathfrak{P}/K_\mathfrak{p}}(\partial(M/L_\mathfrak{P})),$$

where the norm mapping is to be understood as acting from $L_\mathfrak{P}^*/U(L_\mathfrak{P})^2$ to $K_\mathfrak{p}^*/U(K_\mathfrak{p})^2$, which is allowed since evidently the norm of a square of a unit is also a square of a unit.

Proof. (i) Since every representative of $\partial(L_\mathfrak{P}/K_\mathfrak{p})$ generates the same ideal, we can consider a set of generators of the form $1, a, a^2, \ldots, a^{m-1}$ with $m = [L_\mathfrak{P} : K_\mathfrak{p}]$, which exists according to Proposition 5.6. In the same way as in the proof of Proposition 2.4 (iii) we find that the discriminant of this set generates the same ideal as the element $N_{L_\mathfrak{P}/K_\mathfrak{p}}(f'(a))$, where $f(t)$ is the minimal monic polynomial for a over $R_\mathfrak{p}$. By the Corollary to Proposition 4.5 and the observation that the different $D_{L_\mathfrak{P}/K_\mathfrak{p}}$ equals $f'(a) R_\mathfrak{P}$ assertion (i) follows. Assertion (ii) follows from (i) and the discriminant theorem, and so it remains to prove (iii). To do this let $a_1, \ldots, a_m$ be a set of free generators for $R_\mathfrak{P}$ over $R_\mathfrak{p}$, and $b_1, \ldots, b_n$ the set of free generators of the ring S of integral elements of M over $R_\mathfrak{P}$. One can easily see that the products $a_i b_j$ ($i = 1, 2, \ldots, m$; $j = 1, 2, \ldots, n$) form a set of free generators for S over $R_\mathfrak{p}$ and this will allow us to compute the discriminant $\partial(M/K_\mathfrak{p})$. Let N be a normal extension of $K_\mathfrak{p}$ containing M, let T_1 be the set of embeddings of $L_\mathfrak{P}$ in N (over $K_\mathfrak{p}$), extended to the embeddings of M in N in an arbitrary way, and let T_2 be the set of embeddings of M in N (over $L_\mathfrak{P}$). If $A = [t(a_i)]$, $B = [s(b_j)]$, where $t \in T_1$, $s \in T_2$, $i = 1, 2, \ldots, m$ and $j = 1, 2, \ldots, n$, then we have

$$\partial(L_\mathfrak{P}/K_\mathfrak{p}) = (\det A)^2 (\bmod U(K_\mathfrak{p})^2)$$

and

$$\partial(M/L_\mathfrak{P}) = (\det B)^2 (\bmod U(L_\mathfrak{P})^2).$$

Moreover, if $C = [t(a_i)(st)(b_j)]$, where i, j, s and t are as above, then

$$\partial(M/K_\mathfrak{p}) = (\det C)^2 (\bmod U(K_\mathfrak{p})^2).$$

To find the relations between the determinants of A, B and C write

$$A_1 = [t(a_i) \delta_k^j]_{(i,j),(t,k)},$$

where $i = 1, 2, \ldots, m$; $j = 1, 2, \ldots, n$; $k = 1, 2, \ldots, n$ and

$$B_1 = [(st)(b_k) \delta_u^t]_{(k,t),(s,u)},$$

where $k = 1, 2, \ldots, n$; $u \in T_2$, and observe that the product $A_1 B_1$ equals C. But

$$\det A_1 = \pm (\det A)^n,$$

$$\det B_1 = \pm N_{L_\mathfrak{P}/K_\mathfrak{p}}(\det B);$$

hence we have $(\bmod U(K_\mathfrak{p})^2)$ the following chain of equalities:

§ 1. Principal Properties, Integers, Units   221

$$\partial(M/K_\mathfrak{p}) = \det C^2 = \det A_1^2 \det B_1^2 = (\det A)^{2n} N_{L_\mathfrak{P}/K_\mathfrak{p}}^2(\det B)$$
$$= (L_\mathfrak{P}/K_\mathfrak{p})^n N_{L_\mathfrak{P}/K_\mathfrak{p}}(\partial(M/L_\mathfrak{P})),$$

as asserted. □

To give an example we consider quadratic extensions of a p-adic field and prove

Theorem 5.6. *Let K be a p-adic field, R its ring of integers and a an element of R which is not a square in K and which does not lie in the square of the prime ideal $\mathfrak{p}$ of R. Moreover, denote by π any element from $\mathfrak{p} \setminus \mathfrak{p}^2$. If $L = K(a^{1/2})$ and S is the ring of integers of L, then the following possibilities arise:*

(i) *If π does not divide $2a$, then*

$$S = R[1, a^{1/2}] \quad \text{and} \quad \partial(L/K) \equiv a \pmod{U(K)^2}.$$

(ii) *If π divides a, then*

$$S = R[1, a^{1/2}] \quad \text{and} \quad \partial(L/K) \equiv 4a \pmod{U(K)^2}.$$

(iii) *If π divides 2 but does not divide a and k is the largest natural number such that π^k divides 2 and the congruence $x^2 \equiv a \pmod{\mathfrak{p}^{2k}}$ has a solution b in R, then*

$$S = R[1, (b + a^{1/2})\pi^{-k}] \quad \text{and} \quad \partial(L/K) \equiv 4a\pi^{-2k} \pmod{U(K)^2}.$$

Proof. We start with case (i). By Proposition 5.6 we have $S = [1, c]$ for a certain c in S, which in any case can be written in the form

$$c = (x + ya^{1/2})\pi^{-k} \quad (x, y \in R, k \geqslant 0). \tag{5.2}$$

Without restricting generality we may assume that $(x, y, \pi) = 1$. Note that the norm $N_{L/K}(c) = (x^2 - ay^2)\pi^{-2k}$ lies in R, and thus if k is positive, then the congruence

$$X^2 \equiv aY^2 \pmod{\mathfrak{p}}$$

has solutions in R with $X, Y \notin \mathfrak{p}$, but this shows that

$$X^2 \equiv a \pmod{\mathfrak{p}}$$

has a solution in R. By Corollary 1 to Theorem 5.3 this implies (in view of $2a \notin \mathfrak{p}$) that a is a square in K, contrary to our assumption. Thus $k = 0$, and $c = x + ya^{1/2}$ with x and y from R. But this gives

$$S = R[1, x + ya^{1/2}] \subset R[1, a^{1/2}] \subset S;$$

thus $S = R[1, a^{1/2}]$ and obviously $\partial(L/K) \equiv 4a \pmod{U(K)^2}$. But 4 is the square of a unit, and thus $\partial(L/K) \equiv a \pmod{U(K)^2}$.

Now consider case (ii). As before, we have $S = R[1, c]$, c being of the form (5.2), where again $(x, y, \pi) = 1$. Denote by n the exponent associated with the prime ideal $\mathfrak{p}$ in K. Since $n(a) \geq 1$ and $\pi^2 \nmid a$, we have $n(a) = 1$; thus $n(x^2 - ay^2) = \min(2n(x), 1+2n(y))$ (by Proposition 1.9), whence $2k \leq \min(2n(x), 1+2n(y))$ results from $N_{L/K}(c) \in R$. Since one of the numbers $n(x), n(y)$ has to be zero, we again obtain $k = 0$, and proceeding in the same way as in case (i) we arrive at $S = R[1, a^{1/2}]$ and $\partial(L/K) = 4a (\mathrm{mod}\, U(K)^2)$. (This time 4 need not be a square of a unit.)

We are left with case (iii). We start as before with $S = R[1, c]$ and c given by (5.2), $(x, y, \pi) = 1$ and the condition $(x^2 - ya^2)\pi^{-2k} \in R$. First let k be zero. Then again $S = R[1, a^{1/2}]$ and $\partial(L/K) = 4a$. We have to show that the congruence $X^2 \equiv a\,(\mathrm{mod}\,\mathfrak{p}^2)$ is insoluble in R. Indeed, if X were its solution, then the element $(X + a^{1/2})\pi^{-1}$ would have its norm in R and so, by Corollary 1 to Proposition 5.3 it would lie in S, but it does not belong to $R[1, a^{1/2}]$, which gives a contradiction.

Now let $k \neq 0$. We show that $n(xy) = 0$. In fact, if $n(x) = 0$, then from $n(x^2) = 0$ and $n(x^2 - ay^2) \geq 2k$ it follows that $n(ay^2) = 0$ and $n(y) = 0$. A similar argument applies in the case of $n(y) = 0$. We have $\partial(L/K) = 4y^2 a\pi^{-2k} \equiv 4a\pi^{-2k}(\mathrm{mod}\,U(K)^2)$, and it remains to show that k is the largest natural number not exceeding $n(2)$ for which the congruence

$$X^2 \equiv a\,(\mathrm{mod}\,\mathfrak{p}^{2k})$$

has a solution in R. Since π^{2k} divides $4ay^2$ and $\pi \nmid ay$, π^{2k} divides 4, π^k divides 2, i.e. $n(2)$ is at least equal to k. Moreover, $X = xy^{-1}$ satisfies our congruence. Finally we prove that no number m exceeding k has both these properties. In fact, if for a $t \in R$ we have $t^2 \equiv a\,(\mathrm{mod}\,\mathfrak{p}^{2m})$, then $(t + a^{1/2})\pi^{-m}$ lies in S; thus for suitable A, B in R we have

$$(t + a^{1/2})\pi^{-m} = A + B(x + ya^{1/2})\pi^{-k};$$

hence

$$\pi^{k-m} = By \in R$$

and $k \geq m$. □

Note that this theorem actually covers all quadratic extensions of a p-adic field since every such extension has a generator $a^{1/2}$ with either $n(a) = 0$ or $n(a) = 1$, because if $a = \varepsilon\pi^m$ for an $m \geq 2$, then $a^{1/2}$ defines the same extension as $\varepsilon^{1/2}$ or $(\varepsilon\pi)^{1/2}$ depending on the parity of m.

Corollary. *Let K be a p-adic field and let $L = K(a^{1/2})$, where $n(a) = 0$ or 1. Then the extension L/K is unramified iff either K is not an extension of Q_2 and a is a unit, or K is an extension of Q_2, a is a unit and the congruence $X^2 \equiv a(\mathrm{mod}\,4)$ has a solution in the ring R of integers of K.*

§ 1. Principal Properties, Integers, Units   223

Proof. By the discriminant theorem and Proposition 5.7 (i) L/K is unramified iff $\partial(L/K)$ lies in $U(K)/U(K)^2$, and the verification that this happens only in the two cases listed above is immediate by the theorem proved above. □

6. Now let us have a closer look at the group of units. Let K be a p-adic field, R its ring of integers with the prime ideal $\mathfrak{p}$, v the valuation of K and n the exponent associated with $\mathfrak{p}$, p the characteristic of the residue class field $k = R/\mathfrak{p}$ (thus $Q_p \subset K$), e the ramification index of K over Q_p, f the degree of $\mathfrak{p}$ over Q_p, i.e. $f = [k:GF(p)]$, and finally $U = U(K)$ the group of units of K. In the last group we mark out two subgroups—the group $E(K)$ of all roots of unity contained in K and the group $U_1 = U_1(K)$ (called the *group of principal units*) equal to $1+\mathfrak{P}$. Those groups may have a non-trivial intersection, as shown by the example $K = Q_2(i)$, where $i \in E(K) \cap U_1(K)$.

However, the group $E_1(K)$ of all $(p^f - 1)$th roots of unity, which according to Corollary 2 to Theorem 5.3 is a subgroup of $E(K)$, does not contain any principal unit except 1, since its elements lie in different residue classes (mod $\mathfrak{p}$). This group is the uninteresting factor in the decomposition occurring in

Proposition 5.8. $U(K) = E_1(K) \times U_1(K)$, $K^* = E_1(K) \times U_1(K) \times C_\infty$.

Proof. Apply Corollary 2 to Theorem 5.3, the preceding remark and Proposition 5.1 (iii). □

Corollary. $U(K)/U_1(K) \simeq k^*$.

Proof. Both groups are cyclic groups of $p^f - 1$ elements. □

Now consider $U_1(K)$. We distinguish in it a descending sequence of subgroups $U_m = U_m(K) = 1+\mathfrak{p}^m$ forming a basis for the open neighbourhoods of 1 in K^*. All those groups are open and compact and the factor groups U_m/U_{m+1} are described by

Proposition 5.9. *For $m = 1, 2, \ldots$ we have an isomorphism between U_m/U_{m+1} and the additive group of $GF(p^f)$.*

Proof. Define a mapping of the additive group of $\mathfrak{p}^m$ into U_m/U_{m+1} by putting
$$f(x) = 1+x \pmod{U_{m+1}}.$$
Since $f(x+y) = 1+x+y \pmod{U_{m+1}} = (1+x)(1+y) \pmod{U_{m+1}} = f(x)f(y)$, f is a homomorphism which is evidently surjective and has $\mathfrak{p}^{m+1}$ for the kernel. Thus U_m/U_{m+1} and the quotient $\mathfrak{p}^m/\mathfrak{p}^{m+1}$ of additive groups are isomorphic and by Proposition 1.7 the latter group is isomorphic to $k^+ = GF(p^f)^+$. □

Corollary. *The quotient U_m/U_{m+1} is the f-th power of C_p, the cyclic group of p elements.*

Proof. Observe that k is an f-dimensional linear space over $GF(p)$. □

Now we shall prove a result describing the structure of U_1 as a Z_p-module, but first we have to show that it really is such a module. Let ε be any principal unit and let $a = a_0 + a_1 p + \ldots$ be an element of Z_p with $a_0, a_1, \ldots$ in Z. Consider the sequence

$$\varepsilon_n = \varepsilon^{a_0 + a_1 p + \ldots + a_n p^n}$$

and observe that for $n > m$ we have

$$v(\varepsilon_n - \varepsilon_m) = v(\varepsilon^{a_{m+1} p^{m+1} + \ldots + a_n p^n} - 1);$$

a short calculation shows that the last expression converges to zero with $1/m$. Thus ε_n converges to an element ε_0 of U_1, and we may define the action of Z_p on U_1 by

$$\varepsilon^a = \varepsilon_0.$$

Since $1/k$ lies in Z_p for $k \in Z$, $p \nmid k$, it follows that if ε lies in U_1 then so does $\varepsilon^{1/k}$.

It is now easy to check that in this way U_1 acquires the structure of a Z_p-module. To prove the structure theorem we need some more auxiliary results.

Lemma 5.2. *If m exceeds $e/(p-1)$ and a belongs to U_{m+e}, then the polynomial $x^p - a$ has a root in U_m.*

Proof. Write $a = 1 + b\pi^{m+e}$ with $b \in R$ and $\pi \in \mathfrak{p} \setminus \mathfrak{p}^2$ and let $c = 1 + d\pi^m$ be an arbitrary element of U_m, $d \in R$. Then

$$c^p - a = (1 + d\pi^m)^p - (1 + b\pi^{m+e}) = \sum_{j=1}^{p} \binom{p}{j} d^j \pi^{mj} - b\pi^{m+e}.$$

Since $n(p) = e$, we have

$$n\left(\binom{p}{j} d^j \pi^{mj}\right) \geq e + mj \geq m + e;$$

hence writing $p = \varepsilon \pi^e$ for a certain unit ε, we finally obtain

$$(c^p - a)\pi^{-m-e} = V(d) + \varepsilon d - b,$$

where $V(t)$ is a polynomial whose coefficients all lie in $\mathfrak{p}$. It follows that in order to solve $x^p - a = 0$ in U_m it suffices to find a root of $V(y) + \varepsilon y - b$ in R. But after reducing the last polynomial (mod $\mathfrak{p}$) we get a non-constant linear polynomial over k, and the application of Corollary 1 to Theorem 5.3 gives what is needed. □

Corollary. *Under the assumption $m > e/(p-1)$ the groups U_m and U_{m+e} are isomorphic. This isomorphism is established by the mapping $g\colon U_m \to U_{m+e}$ defined by $g(x) = x^p$.*

Proof. The lemma shows that g is surjective. If $x \in U_m$ and x is distinct from 1, then $x = 1 + a\pi^m$ for a certain non-zero a in R and we obtain

$$x^p - 1 = \sum_{j=1}^{p} \binom{p}{j} a^j \pi^{jm},$$

but the minimal value of $n\left(\binom{p}{j} a^j \pi^{jm}\right)$ is attained only for $j = 1$; thus $n(x^p - 1)$ is finite, and $x^p \neq 1$, and so x does not lie in the kernel of g. □

The next lemma is of a general nature.

Lemma 5.3. *Let $A = A_0 \supset A_1 \supset A_2 \supset \dots$ and $B = B_0 \supset B_1 \supset B_2 \supset \dots$ be two descending sequences of Abelian groups and assume that $A \simeq \liminv A/A_n$, $B \simeq \liminv B/B_n$, the isomorphisms being given by the canonical maps*

$$a \mapsto [a \bmod A_1, a \bmod A_2, \dots] \quad (a \in A),$$
$$b \mapsto [b \bmod B_1, b \bmod B_2, \dots] \quad (b \in B).$$

Let $f\colon A \to B$ be a homomorphism mapping every A_n into the corresponding B_n and assume that all the induced homomorphisms $f_n\colon A_n/A_{n+1} \to B_n/B_{n+1}$ are injective, or surjective, or bijective. Then the same holds for f.

Proof. Assume first that all f_n are injective and let $x \in \operatorname{Ker} f$. Our assumption shows that the intersection of all A_i's and also of all B_i's is a trivial group; thus if x is non-zero, then $x \in A_n \setminus A_{n+1}$ for a suitable n. Since $f(x) = 0$, we have $f_n(x \pmod{A_{n+1}}) = 0$, and in view of the injectivity of f_n, we get $x \in A_{n+1}$, which is a contradiction. Thus $\operatorname{Ker} f = 0$ and f is injective.

Now assume that all f_n are surjective and let $y \in B$. Our assumptions imply that both A and B are complete topological groups in the topology induced by the families A_n and B_n, respectively, of open neighbourhoods of the zero element. This means that if a_m is a sequence of elements of A such that for $m \leqslant n$ and m sufficiently large the difference $a_m - a_n$ lies in a given A_N, then it is convergent, and the same holds for B. If our element y equals zero, then $f(0) = y$. Otherwise we can find a natural number n such that

$y \in B_n \setminus B_{n+1}$. By the surjectivity of f_n we can find x_1 in A_n and y_1 in B_{n+1} such that $y = f(x_1) + y_1$. Similarly, we get x_2 in A_{n+1}, y_2 in B_{n+2} such that $y_1 = f(x_2) + y_2$ and, proceeding in this way further, we obtain a sequence $x_1, x_2, \ldots$ of elements of A such that $x_k \in A_{n+k-1}$ and also a sequence $y_1, y_2, \ldots$ of elements of B such that $y_k \in B_{n+k}$. Those two sequences are related by

$$y_k = f(x_{k+1}) + y_{k+1} \quad (k = 1, 2, \ldots)$$

and

$$y = f(x_1) + y_1.$$

Now let $z_k = x_1 + \ldots + x_k$ and observe that this sequence is fundamental in A; hence it is convergent to an element x of A. Note also that the mapping f is continuous, since for every neighbourhood B_n of zero $f(x) \in B_n$ implies $f(x + A_n) \subset B_n$. Thus we get

$$y = \lim_k (y - y_k) = \lim_k (f(x_1) + \ldots + f(x_k)) = \lim_k f(z_k) = f(x),$$

and so f is surjective. □

Now we can prove

Theorem 5.7. *If K does not contain a primitive p-th root of unity, then $U_1(K)$ is a free Z_p-module of rank $[K:Q_p]$. However, if K does contain a primitive p-th root of unity, then $U_1(K)$ is a direct sum of a cyclic group of order p^s for a certain $s \geqslant 1$ on which Z_p acts by $x^a = x^{a \pmod{p^s}}$ and a free Z_p-module of rank $[K:Q_p]$.*

(The exponent s is called the *index of irregularity* of K. K is called a *regular field* if $s = 0$ and otherwise it is called *irregular*.)

Proof. Let $m > e/(p-1)$. Since $f(U_m) \subset U_{m+e}$ for $f(x) = x^p$, U_m/U_{m+e} can be regarded as a vector space over $GF(p)$ (an element a of $GF(p)$ acting on x by $a \cdot x = x^a$), and since it has p^{ef} elements, this implies $U_m/U_{m+e} = C_p^n$ ($n = [K:Q_p]$).

Let $x_1, \ldots, x_n$ be the generators of $U_m \pmod{U_{m+e}}$ and consider the mapping $f: Z_p^n \to U_m$ defined by

$$f([a_1, \ldots, a_n]) = x_1^{a_1} \ldots x_n^{a_n}.$$

If $[a_1, \ldots, a_n] \in (p^k Z_p)^n$ for a certain k, then

$$f([a_1, \ldots, a_n]) = x_1^{b_1 p^k} \ldots x_n^{b_n p^k}$$

with suitable $b_i \in Z_p$; thus f maps $(p^k Z_p)^n$ in U_{m+ke} and the Corollary to Lemma 5.2 shows that the homomorphisms

$$f_k: (p^k Z_p)^n/(p^{k+1} Z_p)^n \to U_{m+ke}/U_{m+(k+1)e}$$

induced by f are well defined and surjective. Since each f_k is also injective and thus bijective, applying Lemma 5.3 to our situation, we find that f is an isomorphism, and a little reflection shows that it is also a Z_p-module isomorphism. Thus, for $m > e/(p-1)$, U_m is a free Z_p-module of rank n. Since U_1/U_m is finite, this shows that U_1 is a finitely generated Z_p-module, and so by Theorem 1.13 we obtain

$$U_1 \simeq Z_p^s \oplus H$$

with a suitable $s \geq 0$ and a torsion Z_p-module H. Since U_1 contains U_m, we have $s \geq n$ and, using again the finiteness of U_1/U_m, we get $s = n$.

Now we show that H is a finite p-group. Since H is finitely generated and Z_p has the (FN) property, Theorem 1.15 implies that H must be finite, and thus, being a subgroup of the multiplicative group of a field, it consists of roots of unity. If H were not a p-group, then it would contain a primitive root of unity of a prime order $q \neq p$, say u, but since $1/q \in Z_p$, every element of U_1 has a unique qth root lying in U_1, and this implies $u = 1$. Thus H consists of roots of unity of order p^s with a suitable $s \geq 0$. This number s is positive iff U_1 contains a primitive pth root of unity, but this occurs exactly when K contains such a root. □

Corollary. *If $[K:Q_p] = n$, then there exist $a_1, \ldots, a_n$ in U_1 such that for $\pi \in \mathfrak{p} \setminus \mathfrak{p}^2$ fixed but arbitrary, z_1 being the primitive $(p^f - 1)$-th root of unity and z_2 being the primitive p^s-th root of unity (with s as in the theorem), every element of K^* can be uniquely written in the form*

$$z_1^{m_1} z_2^{m_2} a_1^{l_1} \ldots a_n^{l_n} \pi^N,$$

where $0 \leq m_1 < p^f - 1$, $0 \leq m_2 < p^s$, $l_1, \ldots, l_n \in Z_p$ and $N \in Z$.

Proof. Combine the theorem just proved with Proposition 5.8. □

Because of Proposition 5.6 and Theorem 5.7 in the case of a regular field K the Z_p-modules $U_1(K)$ and R and also $\mathfrak{p}, \mathfrak{p}^2, \ldots$ are all isomorphic. In some cases it is possible to establish a particularly simple isomorphism between $U_1(K)$ and $\mathfrak{p}$. This is contained in

Proposition 5.10. (i) *The series*

$$\exp x = \sum_{n=0}^{\infty} x^n/n!$$

converges for every x in the domain $n(x) > e/(p-1)$ and defines in it a continuous function.

(ii) *The function* $\exp x$ *provides an isomorphic map from the additive group* $\{x\colon n(x) > e/(p-1)\}$ *onto the multiplicative group* $\{x\colon n(1-x) > e/(p-1)\}$.

(iii) *If e does not exceed $p-1$, then* $\exp x$ *provides an isomorphic map of* $\mathfrak{p}$ *onto* $U_1(K)$, *both treated as Z_p-modules.*

Proof. To show (i) we have to prove that for x satisfying $n(x) > e/(p-1)$ the general term $x^k/k!$ of our series tends to zero. Now $n(x^k/k!) = kn(x) - n(k!)$ and by elementary number theory $k!$ is divisible exactly by the c_kth power of p with

$$c_k = [k/p] + [k/p^2] + \ldots + [k/p^r],$$

where r is determined from $p^r \leqslant k < p^{r+1}$. Moreover,

$$n(k!) = ec_k \leqslant ek(p^{-1} + p^{-2} + \ldots + p^{-r}) = ek(1-p^{-r})/(p-1),$$

and since for every x in our region we have $n(x) = e/(p-1) + a$ for a positive a, it follows that $n(x^k/k!) > ka$; thus $x^k/k!$ indeed tends to zero. It can immediately be verified that for x, y in the set of convergence of our series we have the identity $\exp(x+y) = \exp x \exp y$, which shows that $\exp$ is a homomorphism. The values of $\exp x$ clearly lie in $U_1(K)$ and the continuity at $x = 0$ is evident. This implies (i).

Now observe that the minimal value of $n(x^k/k!)$ is for $x \neq 0$ in the range of convergence attained for $k = 1$ only. Indeed, if $kn(x) - n(k!) \leqslant n(x)$, then $n(x) \leqslant ec_k/(k-1) \leqslant ek(1-p^{-r})/(p-1)(k-1) < e/(p-1)$, which is a contradiction in the case of $k \neq 1$. This shows that $n(\exp x - 1) = n(x) < e/(p-1)$. Moreover, this implies the injectivity of $\exp x$, since from $x \neq y$, $\exp x = \exp y$ we obtain $\exp(x-y) = 1$, but $n(\exp(x-y)-1) = n(x-y)$, a finite number. Finally, we have to show that $\exp x$ is surjective, i.e. that every y of the form $y = 1+z$ with $n(z) > e/(p-1)$ can be put in the form $y = \exp x$. To obtain this consider the series

$$x - x^2/2 + x^3/3 - x^4/4 + \ldots + (-1)^{n+1} x^n/n + \ldots,$$

which is convergent for x with $n(x) > 1$, and observe that if we denote its sum by $\log(1+x)$ for obvious reasons, we obtain for an element $1+y$ with $n(y) > e/(p-1)$ the identity

$$\exp(\log(1+y)) = 1+y,$$

and so $1+y$ lies in the image of $\exp x$. This proves (ii).

To prove (iii) observe first that the additional assumptions and (ii) show the mapping $\exp\colon \mathfrak{p} \to U_1$ to be a group-isomorphism, and it has to be checked whether $\exp$ commutes with the action of Z_p. However, for elements in Z this follows from the homomorphism-property, and the general case follows by continuity. $\square$

§ 2. Extensions of p-adic Fields

1. In this section L/K will be a finite extension of a p-adic field K of degree $n = ef$, where e is the ramification index of L/K and f is the degree of the residue class field extension k_L/k_K, where k_L is the residue class-field of L and k_K that of K. By p we shall denote the characteristic of k_K; thus $Q_p \subset K$. Moreover, R and S will denote the rings of integers in K and L respectively, $\mathfrak{p}$ will be the prime ideal of R and $\mathfrak{P}$ that of S, whence $\mathfrak{p}S = \mathfrak{P}^e$; by π we shall denote a fixed element from $\mathfrak{p} \setminus \mathfrak{p}^2$ and by Π such an element from $\mathfrak{P} \setminus \mathfrak{P}^2$. Finally, n_K, n_L will be the exponents associated with $\mathfrak{p}$, $\mathfrak{P}$ and v_K, v_L the corresponding valuations, which will be normalized so as to satisfy

$$v_K(\pi) = p^{-f(K/Q_p)}, \quad v_L(\Pi) = p^{-f(L/Q_p)}.$$

As before, we shall call L/K *unramified* iff $e = 1$ and, moreover, we shall say that L/K is *totally* (or *fully*) *ramified* iff $f = 1$, i.e. iff $e = n$, and that it is *tamely ramified* iff p does not divide e. An extension which is not tamely ramified will be called, as would be expected, *wildly ramified*.

We shall now describe all unramified extensions of a given p-adic field. It will turn out that there exists exactly one unramified extension of a given degree, which is necessarily normal, and its Galois group is isomorphic to the Galois group of the extension of the corresponding residue class-field. To obtain this we first prove a lemma.

Lemma 5.4. *A finite extension L/K is unramified iff there exists an $a \in S$ such that $L = K(a)$ and the image of a in k_L is a simple root of a polynomial $\varphi(x)$ over k_K which is obtained from a monic polynomial $F(x)$ over R such that $F(a) = 0$ by reduction (mod $\mathfrak{p}$) of its coefficients.*

Proof. If L/K is unramified, then $n = f = [k_L : k_K]$. Let $\bar{a}$ be a generator for the extension k_L/k_K and denote by $\varphi(x)$ its minimal polynomial. Moreover, let $F(x)$ be a monic polynomial of degree n over R whose reduction (mod $\mathfrak{p}$) equals $\varphi(x)$. By Corollary 1 to Theorem 5.3 $F(x)$ has a root in L, say a. Moreover, $F(x)$ is irreducible over K because of $[L:K] = [k_L:k_K] = [K(a):K]$ (since k_L is the residue class field for $K(a)$), whence $L = K(a)$.

To prove the converse, observe first that we can assume the irreducibility of $F(x)$ over K. Corollary 4 to Theorem 5.3 shows that φ is a power of an irreducible polynomial; thus it must be irreducible itself since it has a simple root. Hence

$$n = \deg F = \deg \varphi = [k_L : k_K] \leqslant n;$$

thus $f(L/K) = n$ and L/K is unramified. $\square$

Corollary 1. *If L/K is unramified and M/Q_p is finite, then LM/KM is again unramified.*

Proof. Let $L = K(a)$ with a as in the lemma. Then $LM = KM(a)$ and the image of a in the field k_{LM} is a simple root of the polynomial φ, minimal for the image of a in k_L over k_K; hence by the lemma LM/KM is unramified. □

Corollary 2. *If L/K and M/K are unramified, then LM/K is also unramified.*

Proof. Apply the multiplicative property of ramification indices and the preceding corollary. □

Corollary 3. *If M is the composite of all unramified extensions of K contained in L, then M/K is unramified, and L/M is fully ramified. Hence every finite extension can be decomposed into two consecutive extensions—the first unramified and the second fully ramified.*

Proof. By the preceding corollary M/K is unramified. Denote by f_1 the degree of k_L/k_M. As every finite field, the field k_L may be written as $GF(p)(\zeta_r)$ for a certain r not divisible by p, ζ_r being a primitive rth root of unity. Thus the equation $x^r = 1$ has r solutions in k_L, and so, by Corollary 1 to Theorem 5.3, it has r solutions in L. If we had $f_1 \neq 1$, then not all of them would lie in M, and so the splitting field N of $x^r - 1$ over K would be larger than M and contained in L. Now we shall show that N/K is unramified, which will contradict the maximality of M. Thus we have to prove that if $p \nmid r$ and the field K_1 is obtained from K by adjoining to it a certain root of $x^r = 1$, then K_1/K is unramified. To get this, we consider the minimal polynomial $\varphi(x)$ of such a root a. We certainly have $x^r - 1 = \varphi(x)g(x)$ with a suitable $g(x)$; hence the different of K_1/K divides ra^{r-1}, which is a unit in K. Thus by the different-theorem K_1/K cannot be ramified. □

Now we can prove

Theorem 5.8. *If K is a p-adic field, then to every finite extension k/k_K there exists a unique non-ramified extension L/K such that k_L and k are isomorphic. This extension is normal and its Galois group is isomorphic with that of k/k_K.*

Proof. Let a be a generator of k/k_K with a minimal polynomial $\varphi(x)$ over k_K. Choose a monic polynomial $F(x)$ over K such that $F(x)(\mathrm{mod}\,\mathfrak{p}) = \varphi(x)$ and put $L = K(b)$, where b is any root of $F(x)$, taken as usual from a fixed algebraic closure of Q_p. We obtain

$$[L:K] = \deg_K b \leq \deg F = \deg \varphi = [k:k_K] \leq [k_L:k_K] \leq [L:K]$$

because the image of b in k_L is a root of $\varphi(x)$, and so k is contained in k_L. The resulting chain of inequalities shows that $[k_L:k_K] = [L:K]$; thus L/K is unramified and $[k:k_K] = [k_L:k_K]$; hence $k = k_L$. Thus L/K satisfies our first assertion. It remains to prove its uniqueness and normality. Let L_1 be another field, unramified over K and with $k_{L_1} = k$. By Hensel's lemma the polynomial $F(x)$ has a root b_1 in L_1 and we have $K(b_1) \simeq K(b) = L$, but $[K(b_1):K] = [k:k_K] = [L_1:K]$; thus $L_1 = K(b_1)$ and L_1 is indeed isomorphic to L.

Now let us turn to the normality. The extension k/k_K is normal and thus k is a splitting field of some polynomial $h(x)$ over k. Choose $H(x)$ over K so that $h(x) = H(x) \pmod p$. By Hensel's lemma $H(x)$ splits into linear factors in L, and the preceding argument shows that one of its roots generates L over K, i.e. that L is the splitting field of $H(x)$ over K and so is normal. If g is any element of the Galois group G of L/K and for any a in S we denote by $\bar{a}$ its image in k, then the formula $\bar{g}(\bar{x}) = \overline{g(x)}$ defines an automorphism $\bar{g}$ of k/k_K. We shall prove that the mapping $g \mapsto \bar{g}$ is bijective. Since the groups of L/K and k/k_K have the same number of elements, it suffices to show that this mapping is surjective. Let a, b, $\varphi(x)$ and $F(x)$ have the same meaning as at the beginning of the proof and let $\bar{b} = a$. If s is an element of the Galois group $G(k/k_K)$, then $s(a) = a_1$ is again a root of $\varphi(x)$, and Hensel's lemma implies the existence of b_1 in S such that $F(b_1) = 0$ and $\bar{b}_1 = a_1$. Such an element is unique because φ has in k as many roots as F has in L, and so all roots of F have distinct images in k. Now if g is an element of G which takes b into b_1, then $\bar{g} = s$; hence follows the required surjectivity. □

Corollary 1. *There exists exactly one non-ramified extension of a $\mathfrak{p}$-adic field with a given degree.*

Proof. There is only one extension of the finite field k_K with this property. □

Corollary 2. *If L/K is unramified, then its Galois group is cyclic.*

Proof. Every finite extension of a finite field is cyclic. □

To finish with the subject of unramified extensions we give an explicit description of them.

Theorem 5.9. *If K is a $\mathfrak{p}$-adic field, then its finite extension L/K is unramified iff $L = K(\zeta_m)$, where ζ_m is a primitive m-th root of unity, where m is not divisible by p.*

Proof. In the proof of Corollary 3 to Lemma 5.4 we have already seen that $K(\zeta_m)/K$ is unramified provided $p \nmid m$. Conversely, if L/K is unramified, then $k_L = k_K(\overline{\zeta}_m)$ with $m = p^f - 1$, where $f = f(L/Q_p)$. Applying Hensel's lemma, we see that $x^m - 1$ has m roots in L, and if L_1 is the field generated over K by one of them, which we can take to be primitive, then L_1 has the same residue class field as L, whence, since L/L_1 has to be unramified, we obtain $L = L_1$. □

Corollary. *If $L = K(\zeta_m)$, where m is not divisible by the characteristic p of the residue class field, then every subfield of L containing K has the form $K(\zeta_r)$ with r not divisible by p.*

Proof. Every subfield of a unramified field over K is also unramified. □

2. Now we turn to fully ramified extensions L/K, i.e. such extensions that $f(L/K) = 1$. We shall show that every such extension can be generated over K by a root of an Eisensteinian polynomial, i.e. a polynomial whose coefficients, all except the leading one (which is equal to unity), lie in the prime ideal $\mathfrak{P}$ of R and the free term lies in $\mathfrak{P} \setminus \mathfrak{P}^2$. One shows in the same way as in the classical case that every such polynomial is irreducible over K. We prove

Theorem 5.10. *If K is a p-adic field and L/K is its fully ramified extension, then there exists an Eisensteinian polynomial $f(x)$ over K whose root a generates L/K and, moreover, $S = R[a]$. Conversely, every extension of K which is generated by a root of an Eisensteinian polynomial is fully ramified.*

Proof. Let L/K be fully ramified. We can use Theorem 5.1 with the set A of representatives of $S/\mathfrak{P}$ contained in R to get the existence of an element a in $\mathfrak{P} \setminus \mathfrak{P}^2$ such that $S = R[a]$. (Namely, choose b in R with $n_K(b) = 1$ and define $t_k = b^i a^j$ for $k = ie + j$ with $j = 0, 1, \ldots, e-1$.) Let $x^e + c_{e-1} x^{e-1} + \ldots + c_0$ be its minimal polynomial over R. Since

$$a^e + c_{e-1} a^{e-1} + \ldots + c_1 a + c_0 = 0,$$

we see that c_0 lies in $\mathfrak{p}$ and thus $n_L(c_0) > 0$. Now assume that $c_0, c_1, \ldots, c_{j-1}$ all lie in $\mathfrak{p}$; thus $c_i = bb_i$ with $b_i \in R$ and

$$-c_j a^j = a^e + \ldots + c_{j+1} a^{j+1} + b(b_{j-1} a^{j-1} + \ldots + b_0);$$

thus, in view of $a^e | b$, we see that c_j is divisible by a, and consequently lies in $\mathfrak{p}$.

It follows by induction that all c_i's lie in $\mathfrak{p}$. Moreover, $n_L(c_j a^j) > e$ and $n_L(a^e) = e$, whence $n_L(c_0) = e$, i.e. $n_K(c_0) = 1$, showing that our polynomial is indeed Eisensteinian.

To prove the converse assume that L/K is generated by a root a of an Eisensteinian polynomial $c_0 + \ldots + c_{n-1} x^{n-1} + x^n$. Obviously $n_L(a^n)$ has to be positive and thus $n_L(a)$ is also positive. This shows that

$$n n_L(a) = n_L(a^n) = n_L(c_0) = e,$$

whence n cannot exceed e, and finally $n = e$ showing that L/K is fully ramified. □

Corollary. *Every p-adic field has fully ramified extensions of any prescribed degree.*

Proof. In fact, the field L generated over K by any root of $x^n + \pi$ is fully ramified over K. □

An analogue of Corollary 2 to Lemma 5.4 would be false for fully ramified extensions, as the following example shows:

Let $K = Q_3$, $K_1 = Q_3(3^{1/2})$, $K_2 = Q_3((-3)^{1/2})$ and $L = K_1 K_2$. The extensions K_1/K and K_2/K are, by the preceding theorem, fully ramified with $e(K_1/K) = e(K_2/K) = 2$. Since -1 is not a square (mod 3), it cannot be square in Q_3 and the fields K_1 and K_2 are certainly distinct. The extension L/K is of degree 4; however, L/K_1 is unramified by Theorem 5.9 since $L = K_1(i)$, and so $e(L/K) = e(L/K_1) e(K_1/K) = 2$ showing that $f(L/K) = 2$; thus L/K is not fully ramified.

Now we shall use Theorem 5.10 and Corollary 3 to Lemma 5.4 to obtain a refinement of the different-theorem for p-adic fields, which will be extended later (see Chapter 6) also to algebraic number fields.

Proposition 5.11. *If L/K is a finite extension of a p-adic field which is wildly ramified, i.e. $\mathfrak{p}$ divides e, then the different $D_{L|K}$ is divisible by $\mathfrak{P}^e$.*

Proof. By Corollary 3 to Lemma 5.4 we can restrict our attention to fully ramified extensions L/K. Thus let Π be chosen in S in such a way that $S = R[\Pi]$, $n_L(\Pi) = 1$ and the minimal polynomial $f(x) = x^e + a_{e-1} x^{e-1} + \ldots + a_0$ for Π over R is Eisensteinian. Then we obviously have

$$n_L(f'(\Pi)) = n_L\left(e\Pi^{e-1} + \sum_{k=1}^{e-1} k a_k \Pi^{k-1}\right)$$

$$\geq \min(n_L(e\Pi^{e-1}), n_L(a_1), n_L(2 a_2 \Pi), \ldots, n_L((e-1) a_{e-1} \Pi^{e-2})),$$

and since $n_L(e\Pi^{e-1}) = n_L(e)+e-1 \geq e$ and, for every $k = 1, \ldots, e-1$, $n_L(ka_k\Pi^{k-1}) \geq n_L(a_k)+k-1 \geq e+k-1 \geq e$, we have

$$n_L(f'(a\Pi)) \geq e,$$

and so finally $D_{L/K} = f'(\Pi)S$ is divisible by $\mathfrak{P}^e$. □

Corollary. *If L/K is an extension of degree n of a p-adic field, then the following conditions are equivalent:*
  (i) L/K *is tame.*
  (ii) $\mathfrak{p}^n \nmid d(L/K)$.
  (iii) $T_{L/K}(S) = R$.

Proof. The equivalence of (i) and (ii) results from Corollary 1 to Theorem 4.8 and the proposition just proved. To prove the equivalence of (i) and (iii) we use Corollary 2 to Proposition 4.7, according to which (iii) holds iff $D_{L/K}$ has no divisors AS with $A \subsetneq R$. Since $D_{L/K} = \mathfrak{P}^m$ with $m \leq e-1$ if L/K is tame and $m \geq e$ otherwise (by Theorem 4.8 and the above proposition) and $\mathfrak{P}^e$ is the minimal power of $\mathfrak{P}$ of the form AS with $A \subsetneq R$, the assertion follows. □

3. Now we shall look closely at the tame extensions. Since every such extension can be decomposed in two consecutive extensions—the first unramified and the second fully and tamely ramified (Corollary 3 to Lemma 5.4), and since we already know the structure of unramified extensions, we shall restrict our attention to fully and tamely ramified extensions. A description of them is given by

Theorem 5.11. *A finite extension L/K is fully and tamely ramified iff $L = K(a)$ where a is a root of $x^n - b$ with $n_K(b) = 1$ and, moreover, n is not divisible by p.*

Proof. If $L = K(a)$ and $x^n - b$ is the minimal polynomial of a over R, with $n_K(b) = 1$, and $p \nmid n$, then by Theorem 5.10 L/K is fully ramified and its ramification index equals n, whence it is not divisible by p, which means that L/K is tamely ramified.

To prove the converse we need a lemma:

Lemma 5.5. *Let L/K be a finite extension of a p-adic field K, m a natural number not divisible by $\mathfrak{p}$ and a an element of R. If b is an element of S such that $n_L(b^m) = n_L(a)$ and the quotient $b^m a^{-1}$ is congruent to a unit of K (mod $\mathfrak{P}$), then there exists an element c in R which differs from a only by a unit factor such that the equation $x^m = c$ has a solution in $K(b) \subset L$.*

Proof. If $b^m a^{-1} \equiv \varepsilon \pmod{\mathfrak{P}}$, where ε is a unit in K, then put $c = a\varepsilon$. If $x_1, \ldots, x_m$ denote all roots of $f(x) = x^m - c$ in a fixed algebraic closure of L then, denoting by v the extension of the valuation of L to the field generated by $x_1, \ldots, x_m$, we have

$$\prod_{i=1}^{m} v(b - x_i) = v(b^m - c) = v(a)v(b^m a^{-1} - ca^{-1})$$
$$= v(a)v(b^m a^{-1} - \varepsilon) < v(a) = v(b)^m;$$

thus for a certain i_0 we have $v(b - x_{i_0}) < v(b)$. Moreover,

$$v(b)^{m-1} = v(x_{i_0})^{m-1} = v(f'(x_{i_0})) = \prod_{i \neq i_0} v(x_i - x_{i_0}),$$

but for $i \neq i_0$ we have $v(x_i - x_{i_0}) \leq \max(v(x_i), v(x_{i_0})) = v(b)$; hence for $i \neq i_0$ we should have $v(x_i - x_{i_0}) = v(b)$. Finally, we obtain, for $i \neq i_0$, the inequality $v(b - x_{i_0}) < v(x_i - x_{i_0})$, and we may apply Krasner's lemma (Proposition 5.4). □

Now assume that L/K is fully and tamely ramified of degree $n = e$. With a suitable unit ε of L we have $\pi = \varepsilon \Pi^e$. Since in our case $f = 1$, we can find a unit η of K congruent to $\varepsilon \pmod{\mathfrak{P}}$. Applying the lemma proved above (for $m = e$, $a = \pi$, $b = \Pi$), we obtain the existence of c in K with $n_K(c) = 1$ for which the polynomial $x^e - c$ has in L a solution, say x_0. Since this polynomial is Eisensteinian, it is irreducible; thus the degree of $K(x_0)$ over K equals e and so $K(x_0) = L$. □

Corollary 1. *If L/K is tamely ramified and M/K is finite, then the extension LM/M is also tamely ramified.*

Proof. Let L_0 be the maximal unramified extension of K contained in L. The extension L/L_0 is then fully and tamely ramified; hence by Theorems 5.9 and 5.11 we have $L = L_0(a)$, $L_0 = K(\zeta_r)$, where a is a root of $x^m - b$ for a certain b in L_0 with $n_{L_0}(b) = 1$, $m = e(L/L_0) = e(L/K)$ and ζ_r is a root of unity with r not divisible by p. Put $M_1 = M(\zeta_r, \zeta_m)$, $N_1 = M(a)$, and denote by M_0 the maximal unramified extension of M contained in M_1. Since $LM \subset N_1$, it suffices to show that N_1/M is tamely ramified. Observe that N_1/M_0 is fully ramified and, moreover, $N_1 = M_0(a)$. Since M_0 contains all mth roots of unity, it follows that a is a root of $x^{m_1} - b_1$ for a certain b_1 from M_0 and m_1 dividing m, and this polynomial is irreducible over M_0. Indeed, we have

$$x^m - b = \prod_{j=0}^{m-1} (x - \zeta_m^j a),$$

and if $f(x) = \prod_{j \in J}(x - \zeta_m^j a)$ is an irreducible factor over M_0 of $x^m - b$ (J being a subset of $\{0, 1, \ldots, m-1\}$), then a^r lies in M_0, r being equal to the number of elements in J. Now choose an irreducible factor $f(x)$ with the minimal possible value of r and observe that such an r has to divide m. If $a^r = b_1$, then $f(x)$ equals $x^r - b_1$. Indeed, $x^r - b_1$ has to be divisible by $f(x)$ since the last polynomial is irreducible, they have a common root a, and the equality of degrees implies the equality of polynomials.

Finally, we obtain $e(N_1/M_0) = [N_1 : M_0] = m_1 | m$, and thus N_1/M_0 is tamely ramified. But $e(N_1/M) = e(N_1/M_0)$, and so N_1/M is also tamely ramified. □

Corollary 2. *If L/K and M/K are tamely ramified, so is LM/K.*

Proof. Apply the preceding corollary and the multiplicativity of ramification indices at consecutive extensions. □

Corollary 3. *If L/K is finite and M is the composite of all tamely ramified extensions of K contained in L, then M/K is tamely ramified and L/M is a wildly and fully ramified extension whose degree is a power of p. We thus have $e(L/K) = e_1 p^m$ with e_1 not divisible by p, $e_1 = e(M/K)$ and $p^m = [L:M]$.*

Proof. The ramification of M/K results from the preceding corollary. Assume now that $[L:M] = e(L/M) = e_0 p^m$ for a certain e_0 not divisible by p. If we apply Lemma 5.5 for $m = e_0$, $a = \pi_M$, $b = \pi_L^{p^e}$ (where π_M and π_L are the generators of the prime ideals in the rings of integers in M and L respectively), then we infer that the equation $x^{e_0} = \varepsilon \pi_M$ has a solution x_0 in L with a suitable unit ε of M. The extension $M(x_0)/M$ is, according to Theorem 5.11, tamely ramified; thus x_0 lies in M and $e_0 = 1$ follows. □

Corollary 4 (Abhyankar's lemma). *If L/K is tame, M/K is finite and $e(L/K)$ divides $e(M/K)$, then LM/M is unramified.*

Proof. Let L_0 be the maximal subfield of L unramified over K. By Corollary 3 to Lemma 5.4 L/L_0 is fully ramified. Since Corollary 1 to Lemma 5.4 implies that ML_0/M is unramified, we get $e(ML_0/K) = e(M/K)$, which leads to

$$e(M/K) = e(ML_0/K) = e(ML_0/L_0) e(L_0/K) = e(ML_0/L_0).$$

Since L/L_0 is fully and tamely ramified, Theorem 5.11 shows that we may write $L = L_0(\pi)$, π being a root of $x^e - A$, where $e = e(L/K)$, $A \in L_0$ and $n_{L_0}(A) = 1$. If $B \in ML_0$ satisfies $n_{ML_0}(B) = 1$, then with a certain unit u of

ML_0 we have

$$A = uB^{e(M/K)}.$$

Since $e|e(M/K)$, we get, with $e' = e(M/K)/e$,

$$\pi^e = uB^{ee'};$$

thus $(\pi B^{-e'})^e = u$, leading to $LM = ML_0(u^{1/e})$.

Using Proposition 5.8, we can write $u = zu_1$ with a root of unity z and $u_1 \in U_1(ML_0)$. Because of $p \nmid e$, the number $1/e$ is a p-adic integer; thus $u_1^{1/e} \in ML_0$ by Theorem 5.7. Hence $LM = ML_0(z^{1/e})$, but now we may invoke Theorem 5.9 to prove that LM/ML_0 is unramified. Finally, we arrive at

$$e(LM/M) = e(LM/ML_0)e(ML_0/M) = 1,$$

and thus LM/M is unramified, as asserted. □

4. In this subsection we give some examples. We first consider quadratic extensions of a given p-adic field K, enumerate them and give generators for them.

Proposition 5.12. *Let K be of degree n over Q_p. If $p \neq 2$, then K has three quadratic extensions, namely $K(\pi^{1/2})$, $K((\pi\zeta)^{1/2})$ and $K(\zeta^{1/2})$, where ζ is the primitive $(p^f - 1)$-th root of unity with $f = f(K/Q_p)$ and π is a generator of the prime ideal in K. However, if $p = 2$, then K has $2^{n+2} - 1$ distinct quadratic extensions, each of them of the form $K(a^{1/2})$, $K((a\pi)^{1/2})$ with*

$$a = \zeta^{a_0}\varepsilon_1^{a_1} \ldots \varepsilon_n^{a_n}, \tag{5.3}$$

where ζ is the primitive 2^s-th root of unity with s equal to the irregularity index of K, $\varepsilon_1, \ldots, \varepsilon_n$ are the free generators of the free summand of $U_1(K)$ as described in Theorem 5.7, $a_0, a_1, \ldots, a_n$ are either 0 or 1, and finally π is as in the first case. In both cases one extension is unramified and all the others are fully ramified (in the first case tamely and in the second wildly).

Proof. Note that for any field K (not necessarily p-adic) of characteristic $\neq 2$ the quadratic extensions are in a one-to-one correspondence with the elements of $K^*/(K^*)^2$, where we denote by $(K^*)^2$ the group of all non-zero squares of the field K. Indeed, if $a, b \in K^*$ lie in the same coset $(\mathrm{mod}(K^*)^2)$, then for a certain c we have $a = bc^2$ and thus $K(a^{1/2}) = K(b^{1/2})$. Conversely, if $K(a^{1/2}) = K(b^{1/2})$, then $a^{1/2} = A + Bb^{1/2}$ for some $A, B \in K$, and a short computation shows that a and b lie in the same coset. Finally, observe that any quadratic extension of a field K of characteristic $\neq 2$ is of the form $K(a^{1/2})$.

Since in view of Proposition 5.8 we have $K^* = E_1(K) \times U_1(K) \times C_\infty$, we must consider the factor groups $E_1(K)/E_1^2(K)$, $U_1(K)/U_1^2(K)$ and $C_\infty/2C_\infty$. The last group being trivially C_2, we are left with the first two. Assume first that $p \neq 2$. Then $U_1(K) = U_1^2(K)$ since $1/2$ is an integral element of Q_p; hence every principal unit has a unique quadratic root in K, which is necessarily also a principal unit. Moreover, $E_1(K)$ is a cyclic group of even order $p^f - 1$; thus the factor group $E_1(K)/E_1^2(K)$ is C_2 and the equality $K^*/(K^*)^2 = C_2 \times C_2$ results. The representatives of the cosets may be taken to be equal to 1, ζ, π, $\zeta\pi$, and hence we have three quadratic extensions of K, the first of which is unramified and the others fully and tamely ramified.

Now take $p = 2$. In this case $E_1(K)$ is a cyclic group of odd order, and hence every element of it is a square; so it remains to look at $U_1(K)$. By Theorem 5.7 we can write every element of this group in the form

$$a = \zeta^{a_0} \varepsilon_1^{a_1} \ldots \varepsilon_n^{a_n}$$

with $0 \leq a_0 < 2^f - 1$ and $a_1, \ldots, a_n \in Z_2$. But obviously, for each i we may also write $a_i = A_i + 2B_i$ with $A_i = 0, 1$ and $B_i \in Z_2$; hence our a takes the form

$$a = \zeta^{A_0} \varepsilon_1^{A_1} \ldots \varepsilon_n^{A_n} \varepsilon^2,$$

where ε is a principal unit. Hence every coset of $U_1(K)$ (mod $U_1^2(K)$) contains an element of the form (5.3) and one can easily see that they all lie in distinct cosets. Thus there are 2^{1+n} of them, and finally $K^*/(K^*)^2 = C_2^{n+2}$ as asserted. The remaining assertions are now immediate. □

In the case of $K = Q_2$ we can be more explicit. We already know that it has seven quadratic extensions. Observe now that in view of the insolubility of $x^2 + 1 \equiv 0 \pmod 4$ no primitive fourth root of unity lies in Q_2, whence -1 is the primitive 2^s-th root of unity contained in Q_2 with s maximal. Moreover, 2 is obviously the generator of the prime ideal, and hence in order to find the generators for the quadratic extensions of Q_2 we have only to determine a generator of the free summand of

$$U_1(Q_2) = Z_2[\varepsilon_1].$$

To find ε_1 observe that, according to the proof of Theorem 5.7, it is given by a generator of the cyclic group U_2/U_3 because in our case the Corollary to Lemma 5.2 applies from $m = 2$ on. Since this group is a cyclic group of two elements, we can take for ε_1 any element from $U_2 \setminus U_3$, e.g. $\varepsilon_1 = 5$. It follows that the quadratic extensions of Q_2 are the following: $Q_2(5^{1/2})$, $Q_2(i)$, $Q_2(2^{1/2})$, $Q_2(i5^{1/2})$, $Q_2(i2^{1/2})$, $Q_2((10)^{1/2})$ and $Q_2(i(10)^{1/2})$. Since the congruence $x^2 \equiv 5 \pmod 4$ has a solution in Z_2, the Corollary to Theorem 5.6 shows that $Q_2(5^{1/2})$ is the only unramified extension of Q_2 which is quadratic, whereas the remaining ones are fully and wildly ramified.

Now consider biquadratic extensions which are normal and have the Galois group equal to $C_2 \times C_2$. If L/K is such an extension, then there are three quadratic extensions of K in L, and so we get

Proposition 5.13. *Let K be of degree n over Q_p. If $p \neq 2$, then K has exactly one normal biquadratic extension with the Galois group $C_2 \times C_2$, namely $L = K(\pi^{1/2}, \zeta^{1/2})$, where π and ζ have the same meaning as in the preceding proposition. We have $e(L/K) = f(L/K) = 2$, and thus L/K is ramified and tamely ramified. However, if $p = 2$, then there are $\dfrac{(2^{n+2}-1)(2^{n+1}-1)}{3}$ such extensions, each of them of the form $K(a^{1/2}, b^{1/2})$, where*

$$a = \zeta^{a_0} \varepsilon_1^{a_1} \ldots \varepsilon_n^{a_n}, \quad b = \zeta^{b_0} \varepsilon_1^{b_1} \ldots \varepsilon_n^{b_n}$$

with $a_0, \ldots, a_n, b_0, \ldots, b_n = 0, 1$; $\varepsilon_1, \ldots, \varepsilon_n, \zeta$ have the same meaning as in the preceding proposition and not all a_i's are equal to the corresponding b_i's. Not all such extensions are distinct and all possible equalities between them are of the form

$$K(a^{1/2}, b^{1/2}) = K(a^{1/2}, c^{1/2}) = K(b^{1/2}, c^{1/2})$$

with

$$c = \zeta^{c_0} \varepsilon_1^{c_1} \ldots \varepsilon_n^{c_n},$$

where $c_i = 0, 1$; $c_i \equiv a_i b_i \pmod{2}$. All such extensions are wildly ramified. Moreover, if one of the fields $K(a^{1/2})$, $K(b^{1/2})$, $K(c^{1/2})$ is unramified over K, then for $L = K(a^{1/2}, b^{1/2})$ we have $e(L/K) = f(L/K) = 2$. Otherwise $e(L/K) = 4$, $f(L/K) = 1$, and thus L/K is fully ramified.

Proof. First we deal with the case of $p \neq 2$. Since L contains three subfields quadratic over K and by the preceding proposition there are only three of them, we must have $L = K(\pi^{1/2}, \zeta^{1/2})$ as asserted and there is no other possibility for L. Since L contains $K(\zeta^{1/2})$, which is unramified, $e(L/K)$ cannot exceed 2. On the other hand, the remaining two quadratic subfields of L are ramified, and thus L/K must be ramified, whence $e(L/K) = 2$ and this implies $f(L/K) = 2$. This settles the first case. Now let $p = 2$. By the preceding proposition we have $2^{n+2} - 1$ quadratic extensions of K and every pair of such extensions generates a biquadratic extension with the group $C_2 \times C_2$. Let $c_1, \ldots, c_r$ (with $r = 2^{n+2} - 1$) be the generators of all possible quadratic extensions of K. Every biquadratic extension of the form in question contains exactly three of the c_i's and every c_i will occur in $(r-1)/2$ distinct fields; thus there are $r(r-1)/6$ distinct fields. The remaining parts are immediate—it suffices to observe that none of our fields can be unramified, since the Galois group is not cyclic, and if a field does not contain an unramified subfield $\neq K$, then it has to be fully ramified. □

5. Now we shall consider normal extensions of p-adic fields. Our notation will be the same as in the previous subsections of this section; moreover, by G we shall denote the Galois group of L/K. To emphasize its dependence on the extension we shall also write $G(L/K)$ for it. We shall define a sequence of subgroups of G, the study of which leads to some important results concerning the structure of finite extensions of p-adic fields, and, as we shall see in the next chapter, also of algebraic number fields.

First we define a mapping of $G(L/K)$ onto $G(k_L/k_K)$. Let g be any element of $G(L/K)$ and $a \in S$. Then the element $\overline{g(a)}$ (where by $\bar{x}$ we denote the canonical images of elements of S and R in k_L and k_K respectively) depends only on $\bar{a}$. Indeed, if $\bar{a} = \bar{b}$, then $n_L(a-b) \geq 1$, and thus $n_L(g(a)-g(b)) \geq 1$ and $\overline{g(a)} = \overline{g(b)}$. If we now define a mapping $\bar{g}: \bar{a} \mapsto \overline{g(a)}$ of k_L into k_L, then one can immediately see that it is an automorphism of k_L which leaves k_K fixed and thus belongs to $G(k_L/k_K)$. Moreover, we have

Proposition 5.14. *The mapping $f: g \mapsto \bar{\ }$ is a homomorphism of $G(L/K)$ onto $G(k_L/k_K)$ whose kernel G_0 consists of all those $g \in G$ which, for all y in S, satisfy $g(y) \equiv y \pmod{\mathfrak{P}}$.*

Proof. The surjectivity of f was proved in the last part of the proof of Theorem 5.8, where the assumption that L/K is unramified was not used. The remaining assertions are clear. □

The group G_0 occurring in the last proposition is called the *inertia group* of the extension L/K. Now we define, for $i = 1, 2, \ldots$, the *ramification groups* G_i of L/K by

$$G_i = \{g \in G_0: g(x)x^{-1} \in U_i(L) \text{ for all non-zero } x \text{ in } S\}.$$

Let us check that the G_i's are indeed groups. Take g_1, g_2 in G_i and let x be any non-zero element of S. Then

$$(g_1 g_2)(x) x^{-1} = (g_1(g_2(x))(g_2(x))^{-1})(g_2(x) x^{-1}) \in U_i;$$

hence $g_1 g_2$ lies in G_i and, since G_i is finite, it is a group.

Note that for i large enough the ith ramification group is trivial, since any element g which lies in all G_i's has to satisfy $g(x)x^{-1} \in 1 + \mathfrak{P}^i$ for all non-zero elements of S, and this can happen only if $g = 1$. The last non-trivial ramification group we shall denote by G_t.

The main properties of the sequence G_i are given in

Theorem 5.12. *Let L/K be a normal extension of a p-adic field K and let G be its Galois group.*

(i) *The maximal unramified extension L_0/K of K contained in L corresponds by the Galois theory to the group G_0 (the inertia group) of L/K. This group G_0 is a normal subgroup of G of order $e(L/K)$ and the factor group G/G_0 is cyclic of order $f(L/K)$. Moreover, the ramification groups of L/K and L/L_0 coincide if we identify $G(L/L_0)$ and G_0.*

(ii) *The maximal tamely ramified extension L_1/K of K contained in L corresponds by the Galois theory to the first ramification group G_1 of L/K. The group G_1 is a normal subgroup of G; it is a p-group and the factor-group G_0/G_1 is cyclic, its order is not divisible by p and there is an embedding of it into the multiplicative group of the field k_L.*

(iii) *For $i = 1, 2, \ldots, t$ the ramification groups G_i are normal subgroups of G. The factor-groups G_i/G_{i+1} can be isomorphically embedded in U_i/U_{i+1}. The corresponding isomorphisms are induced by*

$$f_i \colon G_i \to U_i/U_{i+1},$$

where $f_i(g) = g(\Pi)\Pi^{-1} \pmod{\mathfrak{P}^i}$.

Proof. (i) The group corresponding to L_0 can be identified with $G(L/L_0)$. Let g be any element of it. As every element of k_L has a representative in the field L_0 (for $k_{L_0} = k_L$) and this is left invariant under g, g belongs to G_0 and so $G(L/L_0)$ is contained in G_0. To prove that those groups are equal it remains to show that they are of the same order. The isomorphism $G(L_0/K) \simeq G/G(L/L_0)$ shows that $|G(L/L_0)| = |G|/|G(L_0/K)|$ and, on the other hand, Theorem 5.8 and Proposition 5.14 give $G/G_0 \simeq G(k_L/k_K) \simeq G(L_0/K)$, whence $|G_0| = |G|/|G(L_0/K)|$ and thus $G_0 = G(L/L_0)$. Since G/G_0 is equal to the Galois group of a non-ramified extension, it is cyclic by Corollary 2 to Theorem 5.8. The statements about the orders now become evident and the last assertion follows from the fact, already used, that G_0 leaves L_0 invariant.

(ii) Consider the mapping $\Phi \colon G_0 \to k_L^*$ defined by

$$\Phi(g) = g(\Pi)\Pi^{-1} \pmod{\mathfrak{P}}.$$

This mapping does not depend on the choice of Π since for every unit ε of L we have

$$g(\varepsilon\Pi)(\varepsilon\Pi)^{-1} = g(\varepsilon)\varepsilon^{-1}g(\Pi)\Pi^{-1} = g(\Pi)\Pi^{-1} \pmod{\mathfrak{P}},$$

in view of $g(\varepsilon)\varepsilon^{-1} \equiv 1 \pmod{\mathfrak{P}}$ for g in G_0. Moreover, Φ is a homomorphism since

$$\Phi(gh) = g(h(\Pi))\Pi^{-1} = g(h(\Pi))(h(\Pi))^{-1}h(\Pi)\Pi^{-1} = \Phi(g)\Phi(h),$$

in view of $h(\Pi) = \varepsilon\Pi$ for a suitable unit ε. It is clear that G_1 is the kernel of Φ. If e_1 is the order of the group Im Φ, then p cannot divide e_1, and if we denote by M the field corresponding to G_1, then $G_1 \subset G_0$ shows that M contains L_0, which is the maximal unramified extension of K contained in L. Moreover, $[L:M] = |G_1| = e(L/K)/e_1$, and since L/M is fully ramified (for $L_0 \subset M$), we obtain $e_1 = e(L/K)/e(L/M) = e(M/K)$ hence M/K is tamely ramified, i.e. $M \subset L_1$. We now have the following situation

$$K \underset{f}{\subset} L_0 \underset{e_1}{\subset} M \underset{e_2}{\subset} L_1 \underset{p^k}{\subset} L,$$

where the subscripts indicate the respective degrees, $e(L/K) = e_0 p^k$ with e_0 not divisible by p and $e_2 = e_0/e_1$.

Since $G(L/L_1)$ is the maximal p-subgroup of $G(L/M)$ and it is unique (if there were another maximal p-subgroup of $G(L/M)$, we could take the field corresponding to it and compose it with L_1 to get a tamely ramified extension of K larger than L_1 and contained in L), it has to be normal; thus L_1/M is normal, tamely and fully ramified. By Theorem 5.11 we can write $L_1 = M(a)$ with integral $a^{e_1} = b \in M$. Now let g be in G_1. Then we can write $g(a)a^{-1} = 1 + \Pi A$ with $A \in S$, by the definition of G_1. But this gives

$$b = g(b) = g(a^{e_1}) = g(a)^{e_1} = a^{e_1}(1+\Pi A)^{e_1} = b(1+\Pi A)^{e_1};$$

thus $(1+\Pi A)^{e_1} = 1$, and since e_1 is not divisible by p, its inverse lies in Z_p; hence elements of $U_1(L)$ have unique e_1 st roots, and consequently $1+\Pi A = 1$ and finally $g(a) = a$. Thus g leaves L_1 invariant, whence $L_1 = M$.

All conjugates of L_1 in L are equal to L, and this implies the normality of G_1. We have seen already that the mapping Φ induces an embedding of G_0/G_1 in the cyclic group k_L^*, and this implies that G_0/G_1 is cyclic itself.

(iii) It suffices to check that f_i are homomorphisms independent of the choice of Π, and this can be accomplished in the same way as in (ii). The equality $\text{Ker} f_i = G_{i+1}$ is also evident. To obtain the normality of G_i in G note that G_i is the maximal subgroup of G which acts trivially on the quotient S/P^{i+1}, and it is clear that if g lies in G and h acts trivially on S/P^{i+1}, then ghg^{-1} does the same. □

Corollary 1. *If we denote by F the residue class field degree of L over Q_p, then the quotients G_i/G_{i+1} are (for $i = 1, 2, ..., t-1$) embeddable in C_p^F.*

Proof. This corollary follows from (iii) and the Corollary to Proposition 5.9. □

Corollary 2. *Let L_i be the field corresponding by the Galois theory to the group G_i for $i = 0, 1, 2, ..., t$. Then $K \subset L_0 \subset ... \subset L_t \subset L$, and if we write $e =$*

$e_0 p^k$ with e_0 not divisible by p, then $e(L_0/K) = 1$, $e(L_1/L_0) = e_0$ and for $i = 1, 2, \ldots, t-1$ we have $e(L_{i+1}/L_i) = p^{a_i}$ with a suitable positive a_i except possibly the case of $L_{i+1} = L_i$. Finally $e(L/L_t) = p^{a_t}$ for a positive a_t, $a_1 + a_2 + \ldots + a_t = k$, $f(L_0/K) = f(L/K)$, $f(L/L_0) = 1$.

Proof. Clear. □

Corollary 3. *The Galois group of every normal finite extension of a p-adic field is solvable.*

Proof. By the preceding corollary every such extension can be obtained in three consecutive steps—first two cyclic extensions and then a p-extension, i.e. an extension whose Galois group is a p-group. Since at every step we have a solvable group, the same holds for $G(L/K)$. □

Now we can determine the different of a normal extension in terms of its ramification groups. Before doing this, we prove a simple lemma which allows us to determine whether a given element of G_0 belongs to the ith ramification group.

Lemma 5.6. *If $S = R[a]$, then an element g of G_0 belongs to G_i iff $n_L(g(a)-a) \geq 1+i$.*

Proof. The necessity of this condition being evident, we turn to its sufficiency. Let x be a non-zero element of S. We can write it as $x = a_0 + a_1 a + \ldots + a_{n-1} a^{n-1}$ with $a_i \in R$; thus

$$g(x) = a_0 + a_1 g(a) + \ldots + a_{n-1} g(a)^{n-1} \equiv x \pmod{\mathfrak{P}^{i+1}}$$

and $n_L(g(x)x^{-1}-1) \geq i$, and so $g \in G_i$. □

Corollary. *If $S = R[a]$, then $g \in G_i \setminus G_{i+1}$ iff $n_L(g(a)-a) = 1+i$.*

Proof. Clear. □

Now we can prove the result concerning $D_{L/K}$.

Theorem 5.13. *If L/K is normal, then*

$$D_{L/K} = \mathfrak{P}^A, \quad \text{where } A = \sum_{i=0}^{t}(|G_i|-1).$$

Proof. In view of Theorem 5.12 (i) and the different-theorem it suffices to prove our result for fully ramified extensions only. Assume thus that L/K is fully ramified. Then by Theorem 5.10 we have $S = R[a]$ for some a with

$n_L(a) = 1$. If $f(t)$ is the minimal polynomial for a over R, then $D_{L/K} = f'(a)S$; hence A equals $n_L(f'(a))$. Now, if g runs over all elements of $G(L/K)$ except the unit element 1, then

$$f'(a) = \prod_g (a - g(a));$$

thus

$$n_L(f'(a)) = \sum_g n_L(a - g(a))$$

$$= \sum_{j=0}^{t-1} \sum_{g \in G_j \setminus G_{j+1}} n_L(a - g(a)) + \sum_{\substack{g \in G_t \\ g \neq 1}} n_L(a - g(a)) + \sum_{g \notin G_0} n_L(a - g(a)).$$

But for $g \in G_j \setminus G_{j+1}$ we have, by the corollary to the preceding lemma, $n_L(a - g(a)) = j n_L(a) = j$, and for $g \notin G_0$ we have $n_L(a - g(a)) = 0$; hence finally we get

$$n_L(f'(a)) = \sum_{i=0}^{t-1} i(|G_i| - |G_{i+1}|) + t(|G_t| - 1) = \sum_{i=0}^{t} (|G_i| - 1),$$

as asserted. □

Corollary. *If L is a normal q-extension of a p-adic field K (with a prime q), then $D_{L/K}$ is a $(q-1)$-st power.*

Proof. Since every q-extension can be obtained by consecutive extensions with group C_q and its different is the product of the differents of those extensions, it suffices to consider the case of $G(L/K) = C_q$. If $q \neq p$, then $G_0 = C_q$ in the case of $e(L/K) \neq 1$ and $G_0 = 1$ otherwise, whence in the first case $A = q-1$ and in the second $A = 0$. (We could also quote the different-theorem here.) If $q = p$, then the groups G_i equal C_q or 1. If $G_0 = 1$, then L/K is unramified and $A = 0$. Finally, if $G_0 = G_1 = \ldots = G_t = C_q$, then $A = t(q-1)$. □

§ 3. Harmonic Analysis in p-adic Fields

1. In this section we apply results from the theory of locally compact Abelian groups to the study of additive and multiplicative groups of a p-adic field K and also of its group of units. The principal facts of that theory may be found in Appendix I.

Observe first that, by Theorem 5.3, K^+ and K^* are locally compact and the same holds for $U_m(K)$. The latter groups are even compact. In this subsection we describe the duals of those groups, and later we shall consider the Mellin transform in K^*. The principal result of this section is a theorem about the functional equation for the Mellin transform, proved in the thesis of J. Tate [50].

First we determine the dual group $\hat{K}^+$ of K^+.

Theorem 5.14. *The group $\hat{K}^+$ is topologically isomorphic with K^+ and every character of K^+ has the form $X(ax)$, where a is an element of K^+ and $X(x)$ is a fixed non-trivial character of K^+. The map $f: a \mapsto X(ax)$ gives the required isomorphism.*

Proof. Fix a non-trivial character $X(x)$. Then $X(ax)$ is obviously a character of K^+, and it is clear that f is an algebraic homomorphism. Let $t \in K^+$ be chosen so that $X(t) \neq 1$. For a in the kernel of f we have $X(ax) = 1$ for all x, whence also for $x = ta^{-1}$ provided $a \neq 0$. But this gives the contradiction $X(t) = 1$, whence $a = 0$ and f is injective. Now we prove that f is a homeomorphism of K^+ into its dual group. If X_1 is a character of K^+, then the sets

$$U(\varepsilon, C) = \{\chi : |\chi(x) - X_1(x)| < \varepsilon \text{ for } x \text{ in } C\}$$

(with a positive ε and a compact C) form the fundamental system of neighbourhoods of the character X_1, and so to prove that f is continuous it suffices to observe that the set $f^{-1}(U)$ is open, being equal to

$$\{a : |X_1(x) - X(ax)| < \varepsilon \text{ for } x \text{ in } C\}.$$

On the other hand, if U is an arbitrary open subset of K^+, then $f(U)$ equals

$$\{\chi : \chi(x) = X(ax), a \in U\}.$$

Let $X_1(x) = X(ax)$ be an arbitrarily fixed character from $f(U)$. We prove that a suitable neighbourhood of it in the set of characters having the form $X(bx)$ ($b \in K^+$) with the topology induced from $\hat{K}^+$ is entirely contained in $f(U)$. Let $v(x)$ be the valuation of K and again denote by t a fixed element with $X(t) \neq 1$. Moreover, put

$$\varepsilon_0 = |1 - X(t)|, \quad M = 2v(t)/\inf_{x \notin U} v(a-x), \quad C = \{x \in K^+ : v(x) \leq M\}.$$

The neighbourhood $U(\varepsilon_0, C) \cap f(K^+)$ of X_1 will satisfy our requirement. In fact, if $X(bx)$ belongs to it, then for every x in C we have

$$|X(bx) - X(ax)| < \varepsilon_0,$$

i.e.

$$|1 - X((b-a)x)| < |1 - X(t)|,$$

showing that the element $t(b-a)^{-1}$ does not belong to C, thus $v(t(b-a)^{-1})$ exceeds M, and we finally obtain the inequality

$$v(b-a) < v(t)/M = \tfrac{1}{2}\inf_{x\notin U} v(a-x).$$

This implies $b \in U$; hence $f(U)$ is open in $f(K^+)$ and we see that K^+ and $f(K^+)$ are homeomorphic. Now the local compactness of K^+ implies the local compactness of $f(K^+)$, and thus the last set is complete and consequently closed in $\hat{K}^+$.

Now observe that if $X(ax) = 1$ holds for all a in K^+, then $xK^+ \neq K^+$ and thus $x = 0$. Hence $f(K^+)$ is dense in $\hat{K}^+$, and since it is closed, our assertion follows. $\square$

In the sequel we shall use a standard choice of the character $X(x)$. First let K be the p-adic field Q_p. Observe that to every x in Q_p we can find a unique rational number $\lambda(x)$ with $0 \leq \lambda(x) < 1$ whose denominator is a power of p and for which the difference $\lambda(x)-x$ is a p-adic integer. Indeed, every element of Q_p can, in view of Theorem 5.1, be written uniquely in the form

$$x = a_{-N}p^{-N} + \ldots + a_0 + a_1 p + a_2 p^2 + \ldots$$

with $0 \leq a_i < p$, a_i being integers, and so $\lambda(x) = a_{-N}p^{-N} + \ldots + a_{-1}p^{-1}$ if $x \notin Z_p$ and $\lambda(x) = 0$ otherwise. The function $\lambda(x)$ is continuous and additive (mod 1), i.e. $\lambda(x+y) - \lambda(x) - \lambda(y)$ is a rational integer; hence if we put

$$X(x) = \exp(2\pi i \lambda(x)),$$

we obtain a non-trivial character of Q_p^+. In the general case assume that K contains Q_p and define the standard character by

$$X(x) = \exp(2\pi i \lambda(T_{K/Q_p}(x))).$$

To see that it is non-trivial write $[K:Q_p] = p^m q$, with q not divisible by p and consider $x = p^{-m-1}$. Then

$$T_{K/Q_p}(x) = qp^{-1}$$

and obviously $\lambda(qp^{-1})$ is not a rational integer.

We can now determine the character group of the additive group R^+ of integers of K. If $D = D_{K/Q_p}$, then the definition of the different shows that $T_{K/Q_p}(x)$ lies in Z_p iff $x \in D^{-1}$, and thus the character $X(ax)$ is trivial on R iff a lies in D^{-1}, and we obtain

Proposition 5.15. *The group of characters of the additive group R^+ of integers of K equals K^+/D^{-1}.* $\square$

The reader may check that the above quotient group is topologically isomorphic to $(Q_p/Z_p)^n$ with $n = [K:Q_p]$.

Theorem 5.8 reduces the study of characters of K^* to the study of those of $U_1(K)$, and in order to find $\widehat{U_1(K)}$ we apply Theorems IV and V from Appendix I, which lead to the following result:

Theorem 5.15. *The dual group of U_1 is a discrete p-group. To every character X of U_1 there corresponds a positive rational integer m such that X trivializes on U_m; hence X may be regarded as a character of U_1/U_m.*

Proof. By Theorem V of Appendix I the group U_1 can be represented as the inverse limit of U_1/U_i with obvious maps. Hence $\hat{U}_1$ is the direct limit of $\widehat{U_1/U_i}$, which are finite p-groups. Since the mappings of this system are in fact embeddings, the theorem follows. □

Let $m = m(X)$ be the smallest natural number for which the character X of U_1 trivializes on U_m. The ideal $\mathfrak{f}_X = \mathfrak{P}^m$ is called the *conductor* of X. We shall also use the definition of the conductor of an arbitrary character of K^*. Namely, Theorem 5.15 jointly with Proposition 5.8 implies that every such character χ is determined by a character X of U_1, a natural number n not exceeding $p^f - 2$ with $f = f(K/Q_p)$ and a complex number z on the unit circle. If $x = \varepsilon \zeta^r \pi^m$ is a non-zero element of K with ε in U_1, ζ being the primitive $(p^f - 1)$th root of unity, $0 \leqslant r < p^f - 1$, π in $\mathfrak{P} \setminus \mathfrak{P}^2$ and m in Z, then

$$\chi(x) = \chi(\varepsilon) \zeta^{nr} z^m.$$

The *conductor of χ* is by definition the conductor of X. In the same way one may speak about conductors of characters of any subgroup of K^* containing U_1.

Now we turn to quasicharacters and prove

Theorem 5.16. (i) *Every quasicharacter of K^+ is a character.*
(ii) *Every quasicharacter $q(x)$ of K^* is determined by a pair $[X, z]$, where X is a character of $U(K)$ and z is a non-zero complex number. If $x = \varepsilon \pi^n$ is the canonical form of an element of K^*, then $q(x) = X(\varepsilon) z^n$.*

Proof. (i) It suffices to show that every quasicharacter of K^+ is bounded. To do this observe that the restriction of such a quasicharacter $q(x)$ to the ring of integers must be a character since this ring is compact. Hence for integral x we have $|q(x)| = 1$. But every element of K may be put in the form x/m with an integral x and a natural m, and obviously $|q(x/m)|^m = |q(x)| = 1$, and thus $|q(x/m)| = 1$.

(ii) Observe that $U(K)$ is compact, and so every quasicharacter on it must be a character. □

By the preceding theorems every quasicharacter of K^* is trivial on a certain U_m. If it is trivial on U, we call it *unramified*, and if it is not and m is the minimal natural number for which it is trivial on U_m, then m is called its *ramification degree* and the ideal $\mathfrak{P}^m$—its *conductor*. For convenience we call $\mathfrak{P}^0 = R$ the conductor of any unramified quasicharacter.

2. In this section we shall fix a Haar measure in K^+ in order to make the inversion formula for the Fourier transform look as simple as possible; then we shall determine a Haar measure in K^*.

First a triviality—we extend the absolute norm of ideals as defined in Chapter 1 to all fractional ideals by multiplicativity, and observe that the following analogue of Proposition 4.4 (i) holds for p-adic fields:

Proposition 5.16. *If K contains Q_p, then the absolute norm of a fractional ideal I of K generates in Z_p the ideal $N_{K/Q_p}(I)$.*

Proof. Repeat the argument of Proposition 4.4 (i), replacing Z by Z_p. □

Through the remainder of this section we shall denote by R the ring of integers of K, by D the different D_{K/Q_p}, by $\mathfrak{P}$ the prime ideal of R and by v the valuation of K normalized by the requirement that for elements of $\mathfrak{P}\setminus\mathfrak{P}^2$ it has the value $(N(\mathfrak{P}))^{-1}$, i.e.

$$v(x) = (N(\mathfrak{P}))^{-n(x)},$$

where n is the exponent associated with $\mathfrak{P}$.

We now choose the Haar measure $m(E)$ in K^+ so that the ring R acquires measure $N(D)^{-1/2}$. Integrals with respect to $m(E)$ will be written simply as $\int f(x)\,dx$. The reason for such a choice becomes obvious when we prove

Proposition 5.17. *If $f(x)$ is a continuous function in $L_1(K^+)$ whose Fourier transform*

$$\hat{f}(y) = \int_{K^+} f(x) X^{-1}(xy)\,dx$$

belongs to $L_1(K^+)$, then the following inversion formula holds:

$$f(x) = \int_{K^+} \hat{f}(y) X(xy)\,dy = \hat{\hat{f}}(-x).$$

Proof. We begin with a lemma which is very useful for integration in K^+.

Lemma 5.7. *If a is a non-zero element of K, then the measure $m(aR)$ of the fractional ideal generated by it equals $N(aR)^{-1}(N(D))^{-1/2}$.*

Proof. Choose a natural number m so that $ma = b \in R$. The additive group of bR has in aR the index $N(mR) = m^n$ with $n = [K:Q_p]$, and since all cosets have the same measure, we get $m(aR) = m^n m(bR)$. Moreover, the additive group bR has the index $N(bR)$ in the additive group of R; thus $m(bR) = m(R)/N(bR) = N(D)^{-1/2} N(bR)^{-1}$ and finally

$$m(aR) = m^n N(D)^{-1/2} N(bR)^{-1} = N(D)^{-1/2} N(aR)^{-1}. \qquad \Box$$

Corollary 1. *If I is a fractional ideal of K, then $m(I) = N(I)^{-1} N(D)^{-1/2}$.* $\qquad \Box$

Corollary 2. *If I is a fractional ideal of K generated by a, then $m(I) = v(a) N(D)^{-1/2}$.* $\qquad \Box$

It suffices to check our assertion for one arbitrarily chosen non-zero function $f(x)$. We shall take the characteristic function of R, which is continuous since R is clopen and lies in $L_1(K^+)$, the measure of R being finite. Its Fourier transform equals

$$\hat{f}(y) = \int_R X^{-1}(xy) dx,$$

and since the integrand is a character and we integrate over a compact subgroup, the integral vanishes except when $X^{-1}(xy)$ equals unity for all x in R, in which case $\hat{f}(y)$ equals the measure of R. But we already know that $X^{-1}(xy)$ is trivial on R iff $y \in D^{-1}$, and so ultimately we arrive at

$$\hat{f}(y) = \begin{cases} 0 & \text{if } y \notin D^{-1}, \\ N(D)^{-1/2} & \text{if } y \in D^{-1}. \end{cases}$$

The resulting function lies in $L_1(K^+)$ since D^{-1} is compact. Now we find its inverse transform

$$\int_{K^+} \hat{f}(y) X(xy) dy = N(D)^{-1/2} \int_{D^{-1}} X(xy) dy,$$

and again the integrand is a character whereas we integrate over a subgroup which is compact. An analogous argument as before leads to the conclusion that the integrand is trivial on D^{-1} iff x lies in R; hence taking into account the equality $m(D^{-1}) = N(D)^{1/2}$, resulting from Corollary 1 to Lemma 5.7, we obtain

$$N(D)^{-1/2} \int_{D^{-1}} X(xy) dy = f(x). \qquad \Box$$

Having a Haar integral in K^+ it is easy to obtain a Haar integral in the multiplicative group of K through

Proposition 5.18. *If* $I(f) = \int_{K^+} f(x)\, dx$ *is a Haar integral in* K^+, *then the functional*

$$C(f) = \int_{K\setminus\{0\}} f(x) v^{-1}(x)\, dx$$

is a Haar integral in K^*.

Proof. Indeed,

$$\int_{K\setminus\{0\}} f(xy) v^{-1}(x)\, dx = \int_{K\setminus\{0\}} f(xy) v^{-1}(xy) v(y)\, dx$$

$$= \int_{K\setminus\{0\}} f(xy) v^{-1}(xy)\, d(xy)$$

$$= \int_{K\setminus\{0\}} f(x) v(x)^{-1} dx,$$

the crucial point here lying in $d(xy) = v(y)\, dx$, i.e. $m(Ey) = v(y) m(E)$ — a fact which follows from Lemma 5.7. □

In the sequel we shall utilize a multiple of the Haar measure in K^* resulting from the Haar integral just obtained, namely

$$d^*x = \frac{N(\mathfrak{P})}{N(\mathfrak{P})-1} \frac{dx}{v(x)}.$$

One immediately verifies that this measure gives for the group $U(K)$ of units the measure $N(D)^{-1/2}$ and for the group $U^m(K)$ the measure

$$(N(\mathfrak{P})-1)^{-1} N(D)^{-1/2} N(\mathfrak{P})^{-m+1}.$$

In fact,

$$\int_{1+\mathfrak{P}^m} dx^* = N(\mathfrak{P})/(N(\mathfrak{P})-1) \int_{1+\mathfrak{P}^m} dx$$

$$= N(\mathfrak{P})/(N(\mathfrak{P})-1) \int_{\mathfrak{P}^m} dx$$

$$= (N(\mathfrak{P})-1)^{-1} N(D)^{-1/2} N(\mathfrak{P})^{-m+1}$$

by Corollary 1 to Lemma 5.7 and the invariance of dx under addition.

3. Now we shall consider the Mellin transform in K^* and prove the functional equation for it. The results of this section will be applied in Chapter 7 to the deduction of functional equations for classical analytical functions connected with algebraic number fields.

First we put the quasicharacters $q(x)$ in a convenient form, related to the specific choice of $v(x)$. Namely, by Theorem 5.16 (ii), we have, for $x = \varepsilon \pi^n$,

$q(x) = X(\varepsilon) z^n$ with an X in $\hat{U}$ and a non-zero complex z. Now $v(x) = N(\mathfrak{P})^{-n}$; thus $n = \log v(x)/\log N(\mathfrak{P})$ and $z^n = v(x)^s$ with $s = \log z/\log N(\mathfrak{P})$, where the complex number s is determined only up to a rational integral multiple of $2\pi i/\log N(\mathfrak{P})$. We thus obtain

$$q(x) = X(\varepsilon) v(x)^s,$$

and we can always assume that $0 \leq \operatorname{Im} s < 2/\log N(\mathfrak{P})$. The real part of s is uniquely determined by $q(x)$; we shall call it the *exponent* of the quasicharacter $q(x)$ and write $e(q)$ for it. Observe that the equality $e(q) = 0$ is both necessary and sufficient for q to be a character.

Two quasicharacters, $q(x) = X(\varepsilon) v(x)^s$ and $q_1(x) = X_1(\varepsilon) v(x)^{s_1}$ are *equivalent* if $X = X_1$. If the character X is fixed, then all quasicharacters in the equivalence class of X are in a one-to-one correspondence with points of the Riemann surface obtainable by identifying, on the complex plane, points whose difference is a rational integral multiple of $2\pi i/\log N(\mathfrak{P})$. This correspondence is obviously bicontinuous, and so we get an analytical structure in any fixed equivalence class of quasicharacters. In particular, we can consider analytical functions defined on an equivalence class: if $F(q)$ is a complex-valued function defined for quasicharacters in an equivalence class, then $F(q)$ may be written as a function of s alone, and we say that $F(q)$ is analytic at a point $q(x) = X(\varepsilon) v(x)^s$ provided it is analytic at s.

Let $f(x)$ be a continuous complex-valued function on K which lies in $L_1(K^+)$ and assume that for every positive t the functions $f(x) v(x)^t$ are in $L_1(K^*)$. If $q(x)$ is a quasicharacter of K^* with positive $e(q)$, then the function $f(x) q(x)$ lies in $L_1(K^*)$, and so the formula

$$Z(f, q) = Z(f, X, s) = \int_{K^*} f(x) q(x) dx^* \tag{5.4}$$

defines, for fixed f and X in $q(x) = X(\varepsilon) v(x)^s$, a function of the complex variable s in the half-plane $\operatorname{Re} s > 0$. This function will be called the *zeta-function* associated with f and the equivalence class of quasicharacters determined by $X(\varepsilon)$.

To give some examples we shall compute the zeta-functions of the following functions:
(a) The characteristic function χ_{U_m} of U_m ($m = 1, 2, \ldots$),
(b) The characteristic function χ_U of U,
(c) The function

$$f(x) = \begin{cases} X(x), & x \in \mathfrak{P}^{-m-N}, \\ 0, & x \notin \mathfrak{P}^{-m-N}, \end{cases}$$

where $X(x)$ is the additive character defined after the proof of Theorem 5.14,

$\mathfrak{P}^m$ equals the different $D = D_{K/Q_p}$, and $\mathfrak{P}^N$ is the conductor of the quasicharacter $q(x)$. In this case, we shall compute the zeta-function only for one equivalence class of quasicharacters.

Cases (a) and (b) are simple. Let $\mathfrak{P}^N$ be the conductor of $q(x) = X(\varepsilon)\, v(x)^s$ and observe that it remains the same for all quasicharacters of the same equivalence class. Now on U_m we have $x = \varepsilon$ and $q(x) = X(\varepsilon)$, and thus

$$Z(\chi_{U_m}, X, s) = \int_{U_m} X(\varepsilon)\, d\varepsilon^* = \begin{cases} \int_{U_m} d\varepsilon^* & \text{if } m \geq N, \\ 0 & \text{if } m = 1, \ldots, N-1. \end{cases}$$

But we already know that the measure of U_m equals

$$(N(\mathfrak{P}) - 1)^{-1} N(D)^{-1/2} (N(\mathfrak{P}))^{-m+1},$$

and so we finally arrive at

$$Z(\chi_{U_m}, X, s) = \begin{cases} (N(\mathfrak{P}) - 1)^{-1} N(D)^{-1/2} N(\mathfrak{P})^{1-m}, & m \geq N, \\ 0, & 1 \leq m \leq N-1; \end{cases} \quad (5.5)$$

thus, in our case, the zeta-function is constant on each equivalence class of quasicharacters. In particular, although it is originally defined only in the half-plane $\mathrm{Re}\, s > 0$, it can be extended to an entire function.

In case (b) we proceed similarly and obtain

$$Z(\chi_U, X, s) = \begin{cases} N(D)^{-1/2} & \text{if } X = 1, \\ 0, & \text{if } X \neq 1; \end{cases} \quad (5.6)$$

and again the zeta-function does not depend on s.

However, in case (c) the situation is different. If we write $x = \varepsilon_x \pi^{n(x)}$, then

$$Z(f, X, s) = \int_{(D\mathfrak{P}^N)^{-1}} X(x) X(\varepsilon_x) v(x)^s dx^*$$

$$= \sum_{j=-m-N}^{\infty} N(\mathfrak{P})^{-js} \int_{\pi^j U} X(x) X(\varepsilon_x)\, dx^*$$

$$= \sum_{j=-m-N}^{\infty} N(\mathfrak{P})^{-js} \int_U X(\pi^j x) X(x)\, dx^*.$$

For $j \geq -m$ we have $\pi^j U \subset D^{-1}$; hence in this case

$$I_j = \int_U X(\pi^j x) X(x)\, dx^* = \int_U X(x)\, dx^*$$

$$= \begin{cases} 0 & (N > 0, \text{ i.e. } X \neq 1), \\ N(D)^{-1/2} & (N = 0, \text{ i.e. } X = 1) \end{cases}$$

and thus

$$Z(f, X, s) = \sum_{j=-m-N}^{-m-1} N(\mathfrak{P})^{-js} I_j + C(X) N(D)^{s-1/2} N(\mathfrak{P})^{sN} (1 - N(\mathfrak{P})^{-s})^{-1},$$

where $C(X)$ equals 1 if the character X is trivial and equals zero in all other cases.

We can now dispose of the case of $X = 1$. In fact, we then have $N = 0$, and thus the sum occurring in the above expression for the zeta-function is void and $C(X) = 1$, i.e. we obtain

$$Z(f, 1, s) = N(D)^{s-1/2} (1 - N(\mathfrak{P})^{-s})^{-1}. \tag{5.7}$$

It remains to consider what happens if X is non-trivial, and to get some insight into this case we have to compute the integrals I_j for $-m-N \leq j \leq -m-1$. It will turn out that they all vanish with the exception of I_{-m-N}.

Observe first that the value of the additive character $X(t)$ depends only on the coset $\mod D^{-1}$ determined by t. Note, moreover, that the set $\pi^j U$ consists of full cosets of this kind because if a lies in $\pi^j U$, i.e., $n_K(a) = j$ and b differs from a by an element of D^{-1}, then in view of $j < -m$ we have $n_K(b) = j$.

Now we restrict ourselves to $j \neq -m-N$. We can find suitable $a_1, \ldots, a_k$ in $\pi^j U$ such that

$$\pi^j U = \bigcup_{i=1}^{k} (a_i + D^{-1})$$

and I_j takes the form

$$I_j = \sum_{i=1}^{k} X(a_i) \int_{A_i} X(x) \, dx^*$$

with $A_i = (a_i + D^{-1}) \pi^{-j} = a_i \pi^{-j} (1 + \mathfrak{P}^{-m-j}) = a_i \pi^{-j} U_{-m-j}$. But

$$\int_{A_i} X(x) \, dx^* = X(a_i \pi^{-j}) \int_{U_{-m-j}} X(x) \, dx^*,$$

and since X is non-trivial on U_{-m-j} (owing to $-m-j < N$), its integral vanishes. We thus see that our zeta-function equals

$$N(\mathfrak{P})^{(m+N)s} I_{-m-N},$$

and so our last step will consist in a reduction of I_{-m-N} to a handy form. Let $\varepsilon_1, \ldots, \varepsilon_r$ be the full set of representatives of the cosets $\mod U_N$ in the group U. Then, since the character X is constant on those cosets, we obtain

$$I_{-m-N} = \sum_{i=1}^{r} X(\varepsilon_i) \int_{\varepsilon_i U_N} X(\pi^{-m-N} x) \, dx^*;$$

but

$$\pi^{-m-N}x \in \pi^{-m-N}\varepsilon_i U_N$$

implies

$$\pi^{-m-N}x \in \pi^{-m-N}\varepsilon_i + D^{-1};$$

hence $X(\pi^{-m-N}x) = X(\pi^{-m-N}\varepsilon_i)$ holds for x from $\varepsilon_i U_N$, and it follows that

$$I_{-m-N} = \sum_{i=1}^{r} X(\varepsilon_i) X(\pi^{-m-N}\varepsilon_i) \int_{\varepsilon_i U_N} dx^* = \tau_0(X) \int_{U_N} dx^*$$
$$= \tau_0(X)(N(\mathfrak{P})-1)^{-1} N(D)^{-1/2} N(\mathfrak{P})^{-N+1},$$

where

$$\tau_0(X) = \sum_{i=1}^{r} X(\varepsilon_i) X(\pi^{-m-N}\varepsilon_i). \tag{5.8}$$

Hence the final result looks as follows:

$$Z(f, X, s) = \tau_0(X) N(D)^{s-1/2} N(\mathfrak{P})^{N(s-1)+1} (N(\mathfrak{P})-1)^{-1} \tag{5.9}$$

for $X \neq 1$.

Now we can present the proof of the promised functional equation for zeta-functions. The precise formulation of this result runs as follows:

Theorem 5.17. *Let f be a complex-valued function on K, continuous and belonging to $L_1(K^+)$. Assume, moreover, that its Fourier transform*

$$\hat{f}(y) = \int_{K^+} f(x) \overline{X(yx)} dx$$

(where dx indicates integration with respect to the Haar measure in K^+ chosen in subsection 2) also belongs to $L_1(K^+)$. Finally assume that for all positive t the functions $f(x) v(x)^t$ and $\hat{f}(y) v(y)^t$ lie in $L_1(K^*)$. If we denote by $\hat{q}(x)$ the quasicharacter of K^* equal to $q^{-1}(x) v(x)$, where $q \in \tilde{K}^*$, then for all quasicharacters with $0 < e(q) < 1$ the following functional equation involving the zeta-function holds:

$$Z(f, q) = \varrho(q) Z(\hat{f}, \hat{q}),$$

where

$$\varrho(q) = \begin{cases} N(D)^{s-1/2}(1-N(\mathfrak{P})^{s-1})(1-N(\mathfrak{P})^{-s})^{-1} & \text{if } N = 0, \\ \tau_0(X) N(D)^{s-1/2} N(\mathfrak{P})^{N(s-1)} & \text{if } N \neq 0 \end{cases}$$

for $q(x) = X(\varepsilon) v(x)^s$, $\tau_0(X)$ as defined in (5.8) and $\mathfrak{P}^N$ equal to the conductor of q.

Proof. Let $f(x)$ and $g(x)$ be two functions on K, both of them satisfying the conditions imposed on f in our theorem, and let $0 < e(q) < 1$. This condition ensures the existence of all integrals occurring below owing to the evident equality $e(\hat{q}) = 1 - e(q)$. Now

$$Z(f, q) Z(\hat{g}, \hat{q}) = \iint_{K^* \times K^*} f(x) \hat{g}(y) q(xy^{-1}) v(y) dx^* dy^*$$

$$= \iint_{K^* \times K^*} f(x) g(y) q(y^{-1}) v(xy) dx^* dy^*$$

$$= \int_{K^*} \left(\int_{K^*} f(x) g(xy) v(x) dx^* \right) q(y^{-1}) v(y) dy^*.$$

We prove that the inner integral here is symmetrical in f and g, and this will produce the equality

$$Z(f, q) Z(\hat{g}, \hat{q}) = Z(g, q) Z(\hat{f}, \hat{q}). \tag{5.10}$$

In fact, we have

$$\int_{K^*} f(x) \hat{g}(xy) v(x) dx^* = \int_{K^*} f(x) v(x) \int_{K^+} g(u) \overline{X(xyu)} du dx^*$$

$$= \iint_{K^+ \times K^*} f(x) g(u) \overline{X(xyu)} du dx$$

$$= \int_{K^*} g(u) v(u) \int_{K^+} f(x) X(xyu) dx du^*$$

$$= \int_{K^*} g(u) \hat{f}(xy) v(u) du^*.$$

Now let us fix an equivalence class C_X of quasicharacters determined by the character X of U, i.e. $C_X = \{q: q(x) = X(\varepsilon_x) v(x)^s\}$, and let $\mathfrak{P}^N$ be the conductor of quasicharacters from C_X. We shall now specify our function $g(x)$:

$$g(x) = \begin{cases} X(x) & \text{for } x \text{ in } (D\mathfrak{P}^N)^{-1}, \\ 0 & \text{for the remaining } x. \end{cases}$$

We then have

$$\hat{g}(y) = \int_{D^{-1}\mathfrak{P}^{-N}} X(x) X(-xy) dx = \int_{D^{-1}\mathfrak{P}^{-N}} X(x(1-y)) dx,$$

which in the case of $N \neq 0$ equals $N(D)^{1/2} N(\mathfrak{P})^N$ for y in U_N and zero for y outside U_N. Similarly in the case of $N = 0$ the last integral equals $N(D)^{1/2}$ for y in the ring R of integers of K and zero otherwise. The zeta-functions for g and $\hat{g}$ are already known to us (see (5.5), (5.7) and (5.9)) with the exception of $Z(\hat{g}, q)$ in the case of $N = 0$. But in this case $q(x) = v(x)^s$ and we obtain

5. p-adic Fields

$$Z(\hat{g}, q) = \int_{R \setminus \{0\}} \sqrt{N(D)} v(x)^s dx^*$$

$$= \sum_{j=0}^{\infty} \int_{v(x)=j} \sqrt{N(D)} v(x)^s dx^*$$

$$= \sum_{j=0}^{\infty} N(\mathfrak{P})^{-js} \sqrt{N(D)} \int_{\pi^j U} dx^* = (1 - N(\mathfrak{P})^{-s})^{-1}.$$

It follows that

$$Z(\hat{g}, \hat{q}) = N(\mathfrak{P})/(N(\mathfrak{P}) - 1) \quad \text{for } N \neq 0$$

and

$$Z(\hat{g}, \hat{q}) = (1 - N(\mathfrak{P})^{s-1}) \quad \text{for } N = 0,$$

and we can see that the quotient $Z(g, q)/Z(\hat{g}, \hat{q})$ equals $\varrho(q)$. This, in view of equality (5.10), implies the theorem. □

Corollary. *Under the assumptions of the theorem the zeta-function $Z(f, q)$ is, for every fixed equivalence class of quasicharacters, a regular function of s in the half-plane $\mathrm{Re}\, s > 0$ and can be prolonged analytically to a meromorphic function.*

Proof. We begin with the regularity. Let $\mathrm{Re}\, s = \sigma$ be positive and consider the difference

$$(Z(f, X, s+h) - Z(f, X, s))/h - \int_{K^*} f(x) X(x) v(x)^s \log v(x) dx^*$$

$$= \int_{K^*} f(x) X(x) v(x)^s ((v(x)^h - 1)/h - \log v(x)) dx^*,$$

where h is a complex number not exceeding unity in absolute value. We have to prove that this difference tends to zero with h. We can write it in the form

$$h \int_{K^*} f(x) X(x) v(x)^s \sum_{j=2}^{\infty} h^{j-2} \log^j v(x)/(j!) \, dx^*,$$

and so it does not exceed

$$|h| \int_{K^*} |f(x)| v(x)^\sigma \left(\sum_{j=2}^{\infty} |h|^{j-2} |\log^j v(x)|/j! \right) dx^*.$$

We split the last integral in two, the first, I_1, extended over the set $\{x: v(x) \geq 1\}$ and the second, I_2, over the remaining part of K^*.

Now

$$I_1 \leq |h| \int_{v(x) \geq 1} |f(x)| v(x)^\sigma \sum_{j=2}^{\infty} (\log^j v(x)/j!) \, dx^*$$

$$\leq |h| \int_{v(x) \geq 1} |f(x)| v(x)^{1+\sigma} dx^*,$$

and this obviously tends to zero with h. To evaluate I_2 choose a constant $B \geq 4/\sigma^2$ and let $|h| \leq (\sigma/2)^3$. Then for $j = 2, 3, \ldots$ we have $|h|^{j-2} \leq B(\sigma/2)^j$ and thus the integral I_2 does not exceed

$$|h| \int_{v(x)<1} |f(x)| v(x)^\sigma B \exp(-\tfrac{1}{2}\sigma \log v(x)) \, dx^* = B|h| \int_{v(x)<1} |f(x)| v(x)^{\sigma/2} dx^*,$$

which again tends to zero.

Now observe that as $\varrho(q) = \varrho(X, s)$ with a fixed X is a meromorphic function of s, the functional equation

$$Z(f, q) = \varrho(q) Z(\hat{f}, \hat{q})$$

can be used for the prolongation of $Z(f, q)$ to a meromorphic function, since for $\mathrm{Re}\, s < 1$ the right-hand side of this equation is well-defined and regular up to the poles of $\varrho(q)$ in it. In view of our theorem the two definitions of $Z(f, q)$ in the strip $0 < \mathrm{Re}\, s < 1$ (one direct and the other via the functional equation) agree with each other. □

It should be pointed out that the zeta-functions can have their poles only at those points at which the function $\varrho(q)$ has them. Moreover, the theorem shows the importance of the sum $\tau_0(X)$ for the general theory and not only as a factor of the zeta-function of a particular function. In the next chapter we shall say more about this sum, and now let us return to the factor $\varrho(q)$ occurring in the functional equation.

We prove

Proposition 5.19. *If for a quasicharacter $q(x)$ we denote by $\hat{q}(x)$, as before, the quasicharacter $q^{-1}(x) v(x)$ and the bar denotes complex conjugation, then we have*
  (i) $\varrho(\hat{q}) = q(-1) \varrho(q)^{-1}$;
  (ii) $\varrho(\bar{q}) = q(-1) \overline{\varrho(q)}$;
  (iii) *If $e(q) = 1/2$, then $|\varrho(q)| = 1$, whence $|\tau_0(\chi)| = N(\mathfrak{P}^N)^{1/2}$, where $\mathfrak{P}^N$ is the conductor of X.*

Proof. (i) For an arbitrary function $f(x)$ which satisfies the assumptions of the foregoing theorem and whose zeta-function $Z(f, q)$ does not vanish at q we obtain from the functional equation the equality

$$Z(f, q) = \varrho(q) Z(\hat{f}, \hat{q}) = \varrho(q)\varrho(\hat{q}) Z(\hat{\hat{f}}, \hat{\hat{q}}).$$

But $\hat{\hat{f}}(x) = f(-x)$ and $\hat{\hat{q}}(x) = q(x)$, whence

$$Z(\hat{\hat{f}}, \hat{\hat{q}}) = \int_{K^*} f(-x) q(x) dx^* = q(-1) \int_{K^*} f(x) q(x) dx^* = q(-1) Z(f, q),$$

giving in turn

$$Z(f, q) = \varrho(q)\varrho(\hat{q}) q(-1) Z(f, q),$$

and finally

$$\varrho(q)\varrho(\hat{q}) q(-1) = 1,$$

i.e. (in view of $q(-1) = \pm 1$)

$$\varrho(q)\varrho(\hat{q}) = q(-1).$$

(ii) This time let $f(x)$ be a function satisfying the assumptions of the preceding theorem and such that the zeta-function of its Fourier transform $\hat{f}(x)$ does not vanish at $\hat{q}$. Then we obviously have

$$\overline{Z(f, q)} = Z(\bar{f}, \bar{q}) = \varrho(\bar{q}) Z(\hat{\bar{f}}, \hat{\bar{q}}).$$

But

$$\hat{\bar{f}}(y) = \int_{K^*} \overline{f(y)} \overline{X(xy)} dy = \overline{\int_{K^*} f(y) X(xy) dy} = \overline{\hat{f}(-y)}$$

and $\overline{\hat{q}}(x) = \hat{\bar{q}}(x)$; thus

$$Z(\hat{\bar{f}}, \hat{\bar{q}}) = \int_{K^*} \overline{\hat{f}(x)} \hat{\bar{q}}(x) dx^* = \int_{K^*} \overline{\hat{f}(-x)} \hat{q}(x) dx^*$$

$$= \hat{q}(-1) \overline{\int_{K^*} \hat{f}(x) \hat{q}(x) dx^*} = q(-1) \overline{Z(\hat{f}, \hat{q})},$$

whence by the functional equation we obtain

$$\overline{Z(f, q)} = \varrho(\bar{q}) q(-1) \overline{Z(\hat{f}, \hat{q})}.$$

But we also have

$$\overline{Z(f, q)} = \overline{\varrho(q)} \overline{Z(\hat{f}, \hat{q})},$$

and so $\varrho(\bar{q}) q(-1) = \overline{\varrho(q)}$ and $\varrho(\bar{q}) = q(-1) \overline{\varrho(q)}$, proving (ii).

(iii) If $e(q) = 1/2$, then $|q(x)| = v(x)^{1/2}$ and thus

$$q(x)\bar{q}(x) = v(x) = q(x)\hat{q}(x),$$

i.e., $\bar{q}(x) = \hat{q}(x)$. But applying (i) and (ii), we obtain

$$q(-1)\varrho(q)^{-1} = \varrho(\bar{q}) = \varrho(\hat{q}) = q(-1)\overline{\varrho}(q) \quad \text{and} \quad \varrho(q)^{-1} = \overline{\varrho}(q),$$

showing the truth of (iii). □

§ 4. Notes to Chapter 5

1. The p-adic and p-adic fields were introduced by K. Hensel in a remarkable sequence of papers starting with [97c] (Hensel [02], [04], [05a], [09]) and culminating in two books (Hensel [08], [13]). He defined p-adic numbers as formal power series in p with suitably defined arithmetic operations. The valuation-theoretical approach is due to J. Kürschak [13] and an axiomatical approach was made by A. Fraenkel [12].

With regard to Theorem 5.2 let us remark that if K is a topological field with a topology induced by a non-Archimedean valuation, then it will be locally compact iff it is complete, the valuation is discrete and the residue class-field is finite. In this case the valuation ring must be compact. The theorem of H. Hasse and F. K. Schmidt [33] implies that every such field is either p-adic or a finite extension of the field of formal power series over a finite field. The original proof was incomplete, as pointed out by S. McLane [39a]. (Cf. Chao [51].) Other proofs were given in McLane [39b], Teichmüller [36], [37], Witt [36]. Part of this theorem is our Theorem 5.4.

Fields with a topology induced by a valuation were described by I. R. Shafarevich [43]. (Cf. Dürbaum, Kowalsky [53], Fleischer [53], Kaplansky [47], Zelinsky [48].) A survey of the theory of topological fields was given by W. Więsław [82].

2. Hensel's lemma (Theorem 5.3) appears first in Hensel [04] for Q_p and Hensel [05a] for K_p in a form equivalent to that given by us, which appears first in Hensel [18]. For other forms of Hensel's lemma see Hensel [08], [13], Rella [24b].

Fields with a non-Archimedean valuation in which the analogue of Hensel's lemma holds are called *relatively complete*. They were considered by A. Ostrowski [35], who showed that a field K is relatively complete iff it is closed under separable algebraic extensions in its completion. Examples of relatively complete fields which are not complete were given in Ostrowski [13]. It was shown in Rim [57] that a field is relatively complete iff its valuation has a unique extension to the algebraic closure. Cf. Inaba [52], Nagata [53], Neukirch [68], Rayner [57], [58], Schilling [43], Zassenhaus [54].

3. The corollaries to Hensel's lemma are Hensel's. For further applications see Hensel [08], Thurston [43]. Corollary 3 implies that Q_p contains subfields K, normal over Q with the Galois group equal to C_2^m ($m = 1, 2, ...$). H. Kleiman [71] proved that one can find such real fields K with $\mathrm{Gal}(K/Q) = S_n$ for $n = 2, 3, ..., p$. Cf. Frey, Geyer [72], Iwata [72]. J. W. S. Cassels [76] deduced from Hensel's lemma that every finitely generated extension of Q is embeddable in infinitely many fields Q_p. For finite extensions this can be deduced from Theorem 4.14.

4. Proposition 5.4 (Krasner's lemma) was proved in Krasner [46]. The Corollary to Proposition 5.5 appears in Hensel [09]. (Cf. Albert [40].)

M. Krasner considered in [37b], [38], [52], [62], [66] the number of extensions of a given p-adic field with prescribed properties. We quote two of his many results:

(i) The number of fully ramified extensions of degree n is divisible by n.

(ii) Let $N_K(n)$ denote the number of all extensions of degree n of a p-adic field K, containing Q_p, and write $n = hp^m$ with $p \nmid h$ and $N = n[K:Q_p]$. Then

$$N_K(n) = \sigma(h)((p^{m+1}-1)/(p-1)+S),$$

where $\sigma(h)$ is the sum of positive divisors of h,

$$S = \sum_{s=1}^{\infty} (p^{m+s+1}-p^{2s})(p^{\varepsilon(s)N}-p^{\varepsilon(s-1)N})/(p-1),$$

and $\varepsilon(s)$ equals $1/p+1/p^2+ ... +1/p^s$ if s is positive and vanishes otherwise.

On this topic see also Feit [59], Henniart [77], Krasner [79], Massy, Nguyen [75], Pavlova [68], Payan [65], Serre [78a], Shafarevich [47].

Conditions for an isomorphism of two extensions of Q_p were given in Hensel [09]. (Cf. Krasner [37a], [37c].) S. McLane [40], and F. K. Schmidt [33] proved that two fields complete under certain valuations which are algebraically isomorphic must also be topologically isomorphic.

Proposition 5.6 is due to Hensel [18]. For other integral bases see Okutsu [82a], Toro [80]. The discriminant $\partial(L/K)$ and also Proposition 5.7 and Theorem 5.6 are due to A. Fröhlich [60b].

5. For the construction of the algebraic closure $\hat{Q}_p$ of Q_p see Bauer [24] and Strassmann [26]. The valuation of Q_p has a unique extension to $\hat{Q}_p$; however, $\hat{Q}_p$ is not complete under it (Ostrowski [13], [17]). As shown in Kürschak [13] and Rychlik [24], the completion of $\hat{Q}_p$ is both algebraically closed and complete. There is an extensive literature concerning the theory of functions in this field. See Koblitz [77], Mahler [73], Washington [82], where further references are given.

For a p-adic field K let $G(K) = \text{Gal}(\hat{Q}_p/K)$. A fully satisfactory description of the structure of $G(K)$ was given for odd p by U. Janssen, K. Wingberg [82]. (For $p = 2$ and $\zeta_4 \in K$ see Diekert [84].) Cf. Janssen [82]. Let $G_p(K)$ be the Galois group of the minimal extension of K closed under p-extensions. If K is regular, then, as shown by I. R. Shafarevich [46], $G_p(K)$ is a free topological group with $1 + [K:Q_p]$ free generators. If K is irregular, then $G_p(K)$ has $m = 2 + [K:Q_p]$ topological generators and one relation (Kawada [54]). Cf. Faddeev, Skopin [59], Miki [76]. S. P. Demushkin [59], [61] showed that for odd p one can choose the generators $g_1, \ldots, g_m$ so that the defining relation takes the form

$$g_2^{p^s}[g_1, g_2][g_3, g_4] \cdots [g_{m-1}, g_m] = 1,$$

where s is the exponent of irregularity of K. For $p = 2$ see Demushkin [63], Labute [67]. The groups arising in this way were studied in Andozhskii [68], Labute [65], [66a], [66b], Sonn [74].

Previous results on $G(K)$ and $G_p(K)$ are contained in Borevich [56], Iwasawa [55b], Jakovlev [68], [78a], Koch [61], [62], [65c], [68], [78], Kuzmin [75b], Watanabe [77], Zelvenskii [72], [73], [78]. A full story of p-extensions is presented in H. Koch [70].

Extensions K_1, K_2 of Q_p with $G(K_1)$ isomorphic to $G(K_2)$ were constructed in Yamagata [76]. Cf. Jarden, Ritter [79], Ritter [78]. Automorphisms of $G(K)$ were studied in Jarden, Ritter [80], Wingberg [82].

6. Theorem 5.7 was proved in Hensel [14b]. See also Guan-Chi-Wen [63], Halter-Koch [72b], Hasse [49], Hasse, Hensel [23], Hensel [15], [16a], [16b], [17], [21c], [27], Holzapfel [67], Rella [20], Wahlin [16], [32]. We shall show in Chapter 6 that $U(K)/U_N(K)$ is isomorphic to the group of invertible elements in R/P^N and we shall describe all situations in which it is cyclic. For an interpretation of U_1/U_N in terms of algebraic groups see Kambayashi [75].

The p-adic logarithm occurring in the proof of Proposition 5.10 was introduced by K. Hensel [15]. For another definition of it, which easily leads to its main properties, see Leopoldt [61]. Cf. Disse [25], [26], Pollaczek [46]. The function exp gives a continuous homomorphism of an open subgroup of K in K^*. All such homomorphisms were described in Gout [70a], [70b]. Explicit isomorphisms of M^+ with $(1+M)^*$ in certain local rings with maximal ideal M were constructed by C. Ayoub [72], [81].

An important class of functions defined on Q_p and the completion of $\hat{Q}_p$

form the p-adic L-functions, introduced by T. Kubota and H. W. Leopoldt [64], which we shall consider in the notes to Chapter 7. Analogues of the Γ-function were introduced in Morita [75], [78]. Cf. Amice [79], Barsky [77a], Leopoldt [75]. For other p-adic generalizations of various classical functions see Diamond [77], [79a], [79b], Koblitz [78], Morita [77], Naito [82]. A survey was given by J. I. Manin [76].

7. Most of the results of Section 2 are due to K. Hensel, in particular Theorems 5.9, 5.10 and 5.11 (Hensel [08]). A discussion of tame extensions is contained in Albert [40], where also a new proof of Corollary 3 to Theorem 5.12 is given. Another proof was given by M. Bauer [22]. Ramified cyclic extensions of degree p of fields containing Q_p were treated in McKenzie, Whaples [56]. For non-normal ramified extensions of degree p this was done in Amano [58], [71]. Cf. Tran [77].

The ramification groups were defined by D. Hilbert [94a] for finite normal extensions of algebraic number fields. In the p-adic case they were apparently first used in Ore [28b]. Ramification groups of composite extensions were considered in Maus [67]. In Maus [68] one finds a partial characterization of the sequence of ramification groups. Cf. Maus [72], Miki [77].

For other results on ramification groups and their generalizations see Arf [39], Davis, Wishart [71], Deuring [31], Fontaine [71], Gordeev [77], Hasse [34], Herbrand [31a], Idt [74], Jacobinski [61], Kawada [53], Krasner [37c], [37d], [40], [44], [49], [50], Krull [30b], [52], Laubie [80], Marshall [71], Maruno [76], Maus [73], Miki [81], Y. Miyata [71], Nagata [53], Nguyen [75], Satake [52], Schilling [40], Sen [69], Sen, Tate [63], Tamagawa [51], Wyman [69], Yamamoto [68].

8. If L/K is normal, then it is tamely ramified iff $S \simeq R[G]$ as $R[G]$-modules, as shown by E. Noether [32]. (Cf. Lesev [66b].) Hence the problem of the existence of a normal integral basis is solved completely for p-adic fields, in contrast to the case of algebraic number fields. Normal integral bases for ideals were considered in Ferton [73], Ullom [70], Vostokov [74], [76a], [76b].

The structure of S as an $R[G]$-module or an $O[G]$-module with $O = \{a \in K[G]: aS \subset S\}$ was studied in Bergé [75], [78], Bertrandias [78], [79], F. Bertrandias, J. P. Bertrandias, Ferton [72], Bertrandias, Ferton [72], Borevich, Vostokov [73], Ferton [72], [74], [75], Martel [74], Y. Miyata [74], [79], [80].

9. If L/K is normal and K contains Q_p then one can consider $U_1(L)$ as an $Z_p[G]$-module, with $G = \text{Gal}(L/K)$. In fact, if $u \in U_1(L)$ and $A = \sum_g a_g g \in$

$Z_p[G]$, then the action of A upon u can be defined by

$$u^A = \prod_g g(u)^{a_g}$$

which is well-defined, since G preserves the prime ideal P and so $g(U_1) = U_1$ holds for all g in G. If $U_1(L)$ is, as an $Z_p[G]$-module, the direct sum of a finite module and a free module, then L/K is said to *have a normal basis for units*. M. Krasner [39] proved for regular L, that tame ramification of L/K is sufficient for the existence of such a basis and D. Gilbarg [42] proved the converse. In this case no finite summand arises and so $U_1(L)$ is free iff L/K is tame. In the irregular case M. Krasner [39] and Z. I. Borevich [65a] demonstrated that tame ramification is neither necessary and sufficient and the full solution of the problem was obtained by Z. I. Borevich, A. I. Skopin [65]. We quote their result only in the case of L/K tame and $p \neq 2$. In this case L/K has a normal basis for units iff $p \nmid [L:K(z)]$, where z is the primitive p^sth root of unity, with s being the index of irregularity of L.

The structure of $U_1(L)$ as a $Z_p[G]$-module was treated for various classes of extensions in Arutyunyan [77], Borevich [64], [65a], [65b], [65c], [66], [67], Borevich, Gerlovin [76], David [78], Gerlovin [69], Iwasawa [55b], [60], Jakovlev [70], Krasner [36], Lesev [64], [65], [66a], Pieper [72], [73], Rosenbaum [66], [70], [78], Wahlin [32], Wingberg [79]. (The last paper determines the structure in the regular case.)

The structure of the groups $L^*/(L^*)^{p^n} = U^{(n)}$ and $\lim\mathrm{inv}_n U^{(n)}$ as $Z_p[G]$-modules was considered in Bashmakov, Keshtov [76], Borevich, El-Musa [73], Faddeev [60], Jakovlev [75], [78b], Janssen, Wingberg [79], [80], Koch [63b], [72], Lesev [73], Nguyen [75], [76], [82], Sueyoshi [78], Wintenberger [80].

10. The results of Section 3 are due to J. Tate [50]. For a simple proof of Theorem 5.14 see Washington [74]. More about Gaussian sums defined by (5.8) will be said in the next chapter.

For the class-field theory in p-adic fields see Iwasawa [80], Serre [62] and the exposition of Serre in Cassels, Fröhlich [67]. We quote here only the principal papers on this subject: Chevalley [33a], [33b], Hasse [30c], Hochschild [50], Lubin, Tate [65], Schilling [61], Schmidt [30]. For new approaches to it see Hazewinkel [69], [75] and Neukirch [84]. The class-field theory establishes a one-to-one correspondence between Abelian extensions L/K of a given p-adic field K and subgroups H of finite index in K^*. This correspondence implies $H = N_{L/K}(L^*)$ and $\mathrm{Gal}(L/K) \simeq K^*/H$.

The groups K_2 of p-adic fields and their rings of integers were considered in Carroll [73] and Dennis, Stein [72], [75].

Exercises to Chapter 5

1. Prove that the algebraic closure of Q_p is not complete.

2. Prove that the completion of the algebraic closure of Q_p is algebraically closed.

3. Show that every automorphism of Q_p is continuous.

4. Prove that for every prime p and $n \geq 3$ one can find a polynomial irreducible over Z of degree n which has a root in Q_p but does not split there into linear factors.

5. Let L/K be a finite extension of a p-adic field, and let $R \subset K$, $S \subset L$ be the corresponding rngs of integers. Prove that the set of all $a \in S$ such that $S = R[a]$ is open.

6. (a) Let K be a p-adic field containing Q_p and let n be an integer not divisible by p. Determine the Galois group of the splitting field of $x^n - 1$ over K.

   (b) Show that if n is a power of p and $K = Q_p(\zeta_n)$, ζ_n being a primitive nth root of unity, then $\mathrm{Gal}(K/Q_p)$ is isomorphic to the multiplicative group of residue classes (mod n) prime to n.

7. Prove that if L/K is unramified, then the norm map $N_{L/K}$ maps $U_n(L)$ onto $U_n(K)$ for $n = 1, 2, \ldots$

8. Prove that if $[K:Q_p] = n$, and $D = D_{K/Q_p}$, then the groups K^+/D^{-1} and $(Q_p/Z_p)^n$ are topologically isomorphic.

9. (Hensel [94a].) Prove that if K is a p-adic field and $D_{L/K} = P^m$, then $m \leq e(L/K) + n_L(e_{L/K}) - 1$, where n_L is the exponent in L.

10. Let K/Q be finite. Show that for infinitely many primes p there exists an embedding of K in Q_p.

11. Describe all cubic extensions of the fields Q_2 and Q_3.

12. Let R be the ring of integers in a finite extension K/Q_p and let $A_N = R/p^N$, where p is the prime ideal of R. Moreover, let $\pi \in \mathfrak{p} \setminus \mathfrak{p}^2$.
    (i) Prove that for $N = 1, 2, \ldots, e = e(K/Q_p)$ one has $pA_N = 0$.
    (ii) Prove that if $N = 1, 2, \ldots, e$, then
    $$A_N^+ \simeq C_p^{fN}$$
    with $f = f(K/Q_p)$.
    (iii) Show that if $m > n$, then $\pi^n A_m^+ \simeq A_{m-n}^+$.
    (iv) Let $k \geq 1$. Prove that if $(k-1)e < N \leq ke$, then
    $$A_N^+ \simeq C_{p^{k-1}}^a \oplus C_{p^k}^b,$$
    where $a = (ke-N)f$, $b = (N-ke+e)f$.

Chapter 6. Applications of the Theory of $\mathfrak{p}$-adic Fields to Algebraic Number Fields

§ 1. Arithmetic Applications

1. In this chapter we shall give various applications of the results obtained in Chapter 5 to the study of algebraic numbers. In the first section we give direct arithmetic applications, in the second we introduce the ring of adeles and the group of ideles of an algebraic number field, both of them being equipped with a suitable topology, we study their principal properties and finally we perform some harmonic analysis, including the deduction of the functional equation for suitably defined zeta-functions.

The method of transferring results concerning p-adic fields is, broadly speaking, as follows: if L/K is a finite extension of an algebraic number field K, $\mathfrak{p}$ is a prime ideal in R_K and $\mathfrak{P}_1, \ldots, \mathfrak{P}_r$ are the prime ideals of R_L lying above $\mathfrak{p}$, then we can consider the p-adic fields $K_\mathfrak{p}$, $L_{\mathfrak{P}_1}, \ldots, L_{\mathfrak{P}_r}$. By Theorem 5.5 the fields $L_{\mathfrak{P}_i}$ are finite extensions of $K_\mathfrak{p}$ and, moreover, by the same theorem, part (iv), the corresponding ramification indices and also prime ideal degrees are not affected by this localization. Now, having a problem concerning L/K, we can try to solve the corresponding problem in $L_{\mathfrak{P}_i}/K_\mathfrak{p}$ for all possible $\mathfrak{p}$, which may be easier than in the original setting, since we have additional topological means at our disposal. Having done this, we can again try to put those local solutions together in some way and this may happen to yield a solution of the "global" problem.

We start with

Proposition 6.1. *Let L/K be an extension of degree n of an algebraic number field K, let $a \in R_L$ be a generator of this extension and let $f(t)$ be its minimal polynomial over R_K. If $\mathfrak{p}$ is a prime ideal of R_K and we have, in the $\mathfrak{p}$-adic field*

$K_\mathfrak{p}$, a factorization $f(t) = f_1(t) \ldots f_m(t)$ into polynomials irreducible in $K_\mathfrak{p}$, then

$$\mathfrak{p} R_L = \mathfrak{P}_1^{e_1} \ldots \mathfrak{P}_m^{e_m},$$

where $\mathfrak{P}_i$ are distinct prime ideals of R_L and for each i the product $e_i f_{L/K}(\mathfrak{P}_i)$ equals the degree of $f_i(t)$.

Moreover, for $i = 1, 2, \ldots, m$ we have $L_{\mathfrak{P}_i} \simeq K_\mathfrak{p}(b_i)$ with $f(b_i) = 0$, and finally there is a topological isomorphism between $\oplus L_{\mathfrak{P}_i}$ and $L \otimes_K K_\mathfrak{p}$, treated as $K_\mathfrak{p}$-spaces.

Proof. For any $i = 1, 2, \ldots, m$ consider a field M_i generated over $K_\mathfrak{p}$ by an element which is a root of $f_i(t)$. Call this element b_i. By Propositions 3.1 and 5.3 the field M_i is complete under the unique extension of the valuation of $K_\mathfrak{p}$ to M_i. Moreover, M_i must contain an isomorphic copy of L, namely $K(b_i)$. Finally M_i coincides with the closure of $K(b_i)$ because this closure contains $K_\mathfrak{p}$ as well as b_i. If n is the exponent of the field M_i induced by its valuation, then its restriction to $K(b_i) = L$ corresponds to a prime ideal $\mathfrak{P}_i$, which lies above $\mathfrak{p}$, and so M_i is isomorphic with $L_{\mathfrak{P}_i}$ under a valuation-preserving isomorphism. This together with Theorem 5.5 (i) proves all but the last assertion of our proposition. To prove the last one, note that

$$\dim_{K_\mathfrak{p}} \oplus L_{\mathfrak{P}_i} = [L:K] = \dim L \otimes_K K_\mathfrak{p}$$

by the already proved part and Theorem 4.1, and apply Proposition 3.1. □

Corollary. *If $x \in L$, then*

$$N_{L/K}(x) = \prod_{i=1}^m N_{L_{\mathfrak{P}_i}/K_\mathfrak{p}}(x),$$

and similarly

$$T_{L/K}(x) = \sum_{i=1}^m T_{L_{\mathfrak{P}_i}/K_\mathfrak{p}}(x).$$

Proof. If x generates L/K, then the asserted equalities are an immediate consequence of Proposition 6.1. However, if $M = K(x) \subsetneq L$, $\mathfrak{Q}_1, \ldots, \mathfrak{Q}_s$ are the prime ideals of R_M lying above $\mathfrak{p}$ and for each $j = 1, 2, \ldots, s$ we denote by $\mathfrak{P}_{1j}, \ldots, \mathfrak{P}_{k_j j}$ the prime ideals of R_L lying above $\mathfrak{Q}_j$, then we can write

$$N_{L/K}(x) = N_{M/K}(N_{L/M}(x)) = (N_{M/K}(x))^{[L:M]} = \prod_{i=1}^s (N_{M_{\mathfrak{Q}_i}/K_\mathfrak{p}}(x))^m;$$

but, on the other hand, we have

$$N_{L_{\mathfrak{P}_{ij}}/M_{\mathfrak{Q}_j}}(x) = x^{n_{ij}}$$

with $n_{ij} = e_{L/M}(\mathfrak{P}_{ij}) f_{L/M}(\mathfrak{P}_{ij})$ by Theorem 5.5 (i) and so, using Theorem 4.1, we arrive at the equality

$$\prod_{i=1}^{k_j} N_{L_{\mathfrak{P}_{ij}}/M_{\mathfrak{Q}_j}}(x) = x^{[L:M]},$$

which gives

$$\prod_{j=1}^{s} \prod_{i=1}^{k_j} N_{L_{\mathfrak{P}_{ij}}/K_\mathfrak{p}}(x) = \prod_{j=1}^{s} \prod_{i=1}^{k_j} N_{M_{\mathfrak{Q}_j}/K_\mathfrak{p}}(N_{L_{\mathfrak{P}_{ij}}/M_{\mathfrak{Q}_j}}(x))$$

$$= \prod_{j=1}^{s} N_{M_{\mathfrak{Q}_j}/K_\mathfrak{p}}(x)^{[L:M]} = N_{L/K}(x),$$

as required. A similar argument, in which products are replaced by sums and exponents by coefficients, applies to traces. □

2. In this subsection we shall consider the connections between the different of an extension L/K of an algebraic number field and the corresponding local differents.

Proposition 6.2. *If K is an algebraic number field and L/K is finite, then*

$$D_{L/K} = \prod_{\mathfrak{P}} (D_{L_\mathfrak{P}/K_\mathfrak{p}} \cap R_L),$$

where the product is taken over all prime ideals of R_L, and for every such $\mathfrak{P}$ we denote by $\mathfrak{p}$ the prime ideal of R_K lying below $\mathfrak{P}$.

Proof. The assertion results immediately from the following lemma:

Lemma 6.1. *If L/K is a finite extension of an algebraic number field K, $\mathfrak{p}$ is a prime ideal of R_K and $\mathfrak{P}$ lies above $\mathfrak{p}$, then the different of L/K is of the form*

$$D_{L/K} = \mathfrak{P}^m \mathfrak{Q} \quad (\mathfrak{P} \nmid \mathfrak{Q}),$$

$\mathfrak{P}^m R_{L_\mathfrak{P}}$ *being equal to the different of the corresponding $\mathfrak{p}$-adic extension $L_\mathfrak{P}/K_\mathfrak{p}$.*

Proof. After identifying the prime ideal of $L_\mathfrak{P}$ with $\mathfrak{P}$ write $D_{L_\mathfrak{P}/K_\mathfrak{p}} = \mathfrak{P}^n$. By Theorem 4.6 we can find an integer a in L such that if $f(t)$ is its minimal polynomial over R_K, then $f'(a)$ lies in $\mathfrak{P}^m \setminus \mathfrak{P}^{m+1}$. If $f(t) = f_1(t) \ldots f_r(t)$ is the factorization of $f(t)$ into monic factors irreducible in $K_\mathfrak{p}$ with integral coefficients, then for an i we have $f_i(a) = 0$, and obviously a generates the extension $L_\mathfrak{P}/K_\mathfrak{p}$. Differentiating, we obtain the equality $f'(a) = f_i'(a) g(a)$ with a certain polynomial $g(t)$ with integral coefficients, whence, using the fact that $f_i'(a)$ generates the different $D_{L_\mathfrak{P}/K_\mathfrak{p}}$, we obtain $m \geq n$.

Assume that m exceeds n. According to Theorem 4.7 there exists an essential derivation $d: R_{L_\mathfrak{p}} \to R_{L_\mathfrak{p}}/\mathfrak{P}^n$ vanishing on $R_{K_\mathfrak{p}}$. Since its restriction to R_L is also a derivation $d^*: R_L \to R_L/\mathfrak{P}^n$ vanishing on R_K, the same theorem implies that it is not essential, i.e. that the image $d^*(R_L)$ contains only zero-divisors. We show now that this leads to a contradiction. Let A be the set of all a's in $R_{L_\mathfrak{P}}$ for which $d(a)$ is a zero-divisor. Lemma 4.5 shows that d is continuous, and since $R_{L_\mathfrak{P}}/\mathfrak{P}^n$ is finite, the set A must be closed. But it contains R_L and so has to coincide with $R_{L_\mathfrak{P}}$. This contradicts the essentiality of d. □

The proposition follows by considering separately all prime ideals dividing $D_{L/K}$ and applying the lemma. □

Corollary 1. *A prime ideal $\mathfrak{P}$ of R_L is unramified and tamely ramified in L/K iff the corresponding p-adic extension is unramified and tamely ramified, respectively.*

Proof. This follows from the proposition and Corollary 1 to Theorem 4.8. □

Corollary 2. *A prime ideal $\mathfrak{P}$ of R_L is wildly ramified in L/K iff $\mathfrak{P}^e | D_{L/K}$ with $e = e_{L/K}(\mathfrak{P})$.*

Proof. Apply the proposition, Proposition 5.11 and Theorem 4.8. □

Corollary 3. *A normal extension L/K of algebraic number fields is tame iff the trace map $T_{L/K}: R_L \to R_K$ is surjective.*

Proof. If L/K is normal, then using Corollary 2 to Proposition 4.7 and Theorem 4.2 one finds that $T_{L/K}(R_L) \neq R_K$ holds iff there is a prime ideal $\mathfrak{P}$ of R_L such that $\mathfrak{P}^e | D_{L/K}$ with $e = e_{L/K}(\mathfrak{P})$, and the preceding corollary shows that this is equivalent to wild ramification. □

The same argument shows that if L/K is tame, but not necessarily normal, then the trace map from R_L to R_K is surjective. However, the converse implication fails in the non-normal case. In fact, let $K/\mathbf{Q}$ be a cubic extension in which 3 is unramified, and $2R_K = \mathfrak{P}_1^2 \mathfrak{P}_2$. Then $\mathfrak{P}_1$ is the only wildly ramified prime ideal, because, in view of Theorem 4.1, in a cubic extension only divisors of $2R_K$ and $3R_L$ may wildly ramify. Assume now that $T_{K/\mathbf{Q}}(R_K) \neq \mathbf{Z}$. Then by Corollary 2 to Proposition 4.7 there is a prime p such that if

$$pR_K = \prod_{j=1}^{g} \mathfrak{P}_j^{e_j}$$

§ 1. Arithmetic Applications   269

s $\mathfrak{P}_i^{e_i} | D_{L/K}$, and thus, by Theorem 4.8 all prime
lly ramified. In our case the only choice for p is
:ase is tame, whence we have $T_{K/Q}(R_K) = Z$.
occur in nature, consider $K = Q(a)$, where a is
$+15X^2+20X+30$, which is irreducible, being
he prime 5. Proposition 2.4 (iv) easily leads to
is 3 is unramified and 2 ramified, and we have
$_K = \mathfrak{P}^3$. But this is easy. Indeed, if $2R_K = \mathfrak{P}^3$,
$_{/Q}$, whence $2^2 = N_{K/Q}(\mathfrak{P}^2) | d(K)$. In view of
pes not divide the index of 2, and we may apply

$\cdot 30 \equiv X^3 + X^2 \equiv X^2(X+1) \pmod{2}$,

th $\mathfrak{P}_1 \neq \mathfrak{P}_2$, contrary to our assumption.

$i = 1, 2)$ and $L = K_1 K_2$, then $d(L/K)$ divides

$d(K_1/K)^{n_2} d(K_2/K)^{n_1}$.

n R_L and $\mathfrak{p}_1, \mathfrak{p}_2, \mathfrak{p}$ the prime ideals in R_{K_1}, R_{K_2},
R_K which lie below it. Assume that $\mathfrak{p}_1$ divides the different $D_{K_1/K}$ with exponent
a. Consider the corresponding p-adic fields $K_\mathfrak{p}$, $L_\mathfrak{P}$, $k_1 = (K_1)_{\mathfrak{p}_1}$ and $k_2 = (K_2)_{\mathfrak{p}_2}$. Then we have

$$D_{k_1/K_\mathfrak{p}} = \mathfrak{p}_1^a = \delta(c) R_{k_1}$$

for a suitable c in k_1. Since $L = K_1 K_2$, we have also $L_\mathfrak{P} = k_1 k_2$ and thus
$L_\mathfrak{P} = k_2(c)$. If $F(t)$ and $G(t)$ are the minimal polynomials for c over $K_\mathfrak{p}$ and
k_2, respectively, then for a certain $H(t)$ over k_2 with integral coefficients we
have $F(t) = G(t)H(t)$, which gives $F'(c) = G'(c)H(c)$, and we see that the
different $D_{L_\mathfrak{P}/k_2}$ divides the ideal generated by $F'(c)$, which equals $D_{k_1/K_\mathfrak{p}} R_{L_\mathfrak{P}}$.
Applying our proposition, we obtain

$$D_{L/K_2} | D_{K_1/K} R_L;$$

hence

$$D_{L/K} = D_{L/K_2} D_{K_2/K} | D_{K_1/K} D_{K_2/K},$$

and taking norms we arrive at

$$d(L/K) = N_{L/K}(D_{L/K}) | N_{L/K}(D_{K_1/K}) N_{L/K}(D_{K_2/K})$$
$$= d(K_1/K)^{[L:K_1]} d(K_2/K)^{[L:K_2]}.$$

It suffices now to note that $[L:K_1] \leq n_2$ and $[L:K_2] \leq n_1$. □

Errata

Page/line	For	Read
268_3	$3R_L$	$3R_K$
270_8	$\mathfrak{P}_i^{s_i+1}$	$\mathfrak{P}_i^{s_i+1}$
276_{12}	$G(1^2)$	$G(\mathfrak{P}^2)$
278_{17}	C_m	C_{m_i}
289_{11}	C_i	$C_{p^n_i}$

Finally, we prove a bound for the maximal power of a prime ideal which can divide the different, generalizing the second part of Theorem 4.8 to wildly ramified prime ideals:

Proposition 6.3. *If K is an algebraic number field, L/K its finite extension, $\mathfrak{p}$ a prime ideal of R_K such that*

$$\mathfrak{p} R_L = \mathfrak{P}_1^{e_1} \ldots \mathfrak{P}_r^{e_r}$$

with distinct $\mathfrak{P}_1, \ldots, \mathfrak{P}_r$, then the different $D_{L/K}$ cannot be divisible by $\mathfrak{P}_i^{e_i+s_i}$, where s_i is defined by $e_i \in \mathfrak{P}_i^{s_i} \setminus \mathfrak{P}_i^{s_i+1}$.

Proof. By Proposition 6.2 it suffices to prove this result for finite extensions of p-adic fields, in our case for $L_{\mathfrak{P}_i}/K_\mathfrak{p}$. Denote by M the maximal tamely ramified extension of $K_\mathfrak{p}$ contained in $L_{\mathfrak{P}_i}$. Then for the prime ideal $\mathfrak{P}$ of R_M we have $\mathfrak{P} R_{L_{\mathfrak{P}_i}} = \mathfrak{P}_i^{e_i}$ and $L_{\mathfrak{P}_i}/M$ is fully ramified. Using Theorem 5.10, we can write $L_{\mathfrak{P}_i} = M(\pi)$, where π is a generator of $\mathfrak{P}_i$ which satisfies an Eisensteinian equation over M, say $f(\pi) = 0$ with $f(t) = t^{e_i} + \ldots + a_{e_i}$, and for which $f'(\pi) R_{L_{\mathfrak{P}_i}}$ equals the different of $L_{\mathfrak{P}_i}/M$. Denoting by $n_i(x)$ the exponent corresponding to $\mathfrak{P}_i$ in $L_{\mathfrak{P}_i}$, we see that for $r = 1, 2, \ldots, e_i$ we have the congruence

$$n_i(a_r(e_i - r) \pi^{e_i - r - 1}) \equiv -r - 1 \pmod{e_i},$$

because $n_i(a_r)$ is divisible by e_i and the same applies to $n_i(e_i - r)$. Hence all the summands of the sum

$$f'(\pi) = e_i \pi^{e_i - 1} + a_1 (e_i - 1) \pi^{e_i - 2} + \ldots + a_{e_i - 1}$$

give distinct values to the exponent n_i, and since the first summand gives $n_i(e_i \pi^{e_i - 1}) = s_i + e_i - 1$, the different cannot be divisible by a higher power of $\mathfrak{P}_i$ than $\mathfrak{P}_i^{s_i + e_i - 1}$. □

3. If the extension L/K is normal, then it is possible to apply the results obtained in subsection 5, Section 2, of the preceding chapter and, in particular, to define the ramification groups of such an extension. This is made possible by

Theorem 6.1. *If L/K is a normal extension of an algebraic number field K, $G = G(L/K)$ is its Galois group, $\mathfrak{p}$ is a prime ideal of R_K and $\mathfrak{P}$ is any prime ideal of R_L which lies above $\mathfrak{p}$, then the corresponding extension $L_\mathfrak{P}/K_\mathfrak{p}$ of p-adic fields is normal and there is a canonical embedding of its Galois group $G(L_\mathfrak{P}/K_\mathfrak{p})$ into G. The index of $G(L_\mathfrak{P}/K_\mathfrak{p})$ in G equals the number of prime ideals lying above $\mathfrak{p}$ in L. (This makes sense provided we identify $G(L_\mathfrak{P}/K_\mathfrak{p})$ with its image in G.)*

Proof. By the preceding proposition we have $L_\mathfrak{P} = K_\mathfrak{p}(a)$, where a is a generator of L/K. The conjugates of a over $K_\mathfrak{p}$ form a subset of the set of conjugates to a over K, and so they all lie in $L \subset L_\mathfrak{P}$, whence $L_\mathfrak{P}/K_\mathfrak{p}$ must be normal.

The above observation shows also that the restriction of any $s \in G(L_\mathfrak{P}/K_\mathfrak{p})$ to L is an element of G, and this restriction mapping is in fact injective, because if $s_{|L}$ is the identity map, then $s(a) = a$ and s is the identity map on the whole of $L_\mathfrak{P}$. Since $G(L_\mathfrak{P}/K_\mathfrak{p})$ has $e_{L/K}(\mathfrak{P})f_{L/K}(\mathfrak{P})$ elements by Theorem 5.5(i), the application of Theorem 4.2 proves the last assertion of the theorem. □

The image of $G(L_\mathfrak{P}/K_\mathfrak{p})$ in $G(L/K)$ is called the *decomposition group* of the ideal $\mathfrak{P}$. In the sequel we shall not distinguish between this group and $G(L_\mathfrak{P}/K_\mathfrak{p})$. Hence the inertia group and the ramification groups of the corresponding local extension become subgroups of $G(L/K)$ and are usually called, respectively, the *inertia group* and the *ramification groups* of the prime ideal $\mathfrak{P}$ over K. Those groups were originally defined by D. Hilbert without the use of p-adic fields. The equivalence of his definition with the one given above is proved in the following proposition:

Proposition 6.4. *Let L/K be a normal extension of an algebraic number field K and let G be its Galois group. Moreover, let $\mathfrak{P}$ be a prime ideal of R_L. If we now define a sequence $G_{-1}(\mathfrak{P}), G_0(\mathfrak{P}), G_1(\mathfrak{P}), \ldots$ of subgroups of G by*

$$G_{-1}(\mathfrak{P}) = \{g \in G : g(\mathfrak{P}) = \mathfrak{P}\},$$

$$G_i(\mathfrak{P}) = \{g \in G : \text{for all } a \in R_L, g(a) - a \in \mathfrak{P}^{1+i}\} \quad (i = 0, 1, \ldots),$$

then $G_{-1}(\mathfrak{P})$ equals the decomposition group of the ideal P, $G_0(\mathfrak{P})$ equals its inertia group and, for $i = 1, 2, \ldots, G_i(\mathfrak{P})$ equals its i-th ramification group.

Proof. Observe first that every element of the decomposition group fixes the ideal $\mathfrak{P}$. Indeed, we can write in it the form $g = \bar{g}_{|L}$, where $\bar{g}$ is an automorphism of the extension $L_\mathfrak{P}/K_\mathfrak{p}$, p being the prime ideal of R_K lying below $\mathfrak{P}$. This gives first the equality

$$\bar{g}(\mathfrak{P}R_{L_\mathfrak{P}}) = \mathfrak{P}R_{L_\mathfrak{P}}$$

and secondly

$$g(\mathfrak{P}) = \bar{g}(\mathfrak{P}) = \bar{g}(\mathfrak{P}R_{L_\mathfrak{P}} \cap R_L) \subset \mathfrak{P}R_{L_\mathfrak{P}} \cap R_L = \mathfrak{P}.$$

But $g(\mathfrak{P})$ is again a prime ideal and $g(\mathfrak{P}) = \mathfrak{P}$ follows.

Secondly note that $G_{-1}(\mathfrak{P})$ and the decomposition group have the same order. In fact, the cosets of G with respect to G_{-1} consist of those elements which map $\mathfrak{P}$ onto a fixed prime ideal conjugated with $\mathfrak{P}$ and so the index of G_{-1} in G equals the number of such prime ideals. It remains to apply the last part of Theorem 6.1 to obtain the equality of G_{-1} and the decomposition group.

Now let us look at the remaining groups from our sequence. Directly from the definitions it follows that the inertia group of $\mathfrak{P}$ is contained in $G_0(\mathfrak{P})$ and the ith ramification group is contained in $G_i(\mathfrak{P})$. To prove that we have equalities here, note that by Corollary 2 to Proposition 5.3 every element of $\text{Gal}(L_\mathfrak{P}/K_\mathfrak{p})$ is continuous. Let $g \in G_i$ for some $i \geq 0$, and let $\bar{g}$ be the restriction of $\bar{g} \in \text{Gal}(L_\mathfrak{P}/K_\mathfrak{p})$. We show that $\bar{g}(a) - a$ lies in $\mathfrak{P}^{i+1}$ for all a in $R_{L_\mathfrak{P}}$ (although by Lemma 5.6 it would suffice to do so for a particularly chosen element a). Now a is a limit of a sequence $\{a_n\}$ of elements of R_L, and since $\mathfrak{P}^{i+1}$ is open, for n large enough we have $a_n \in a + \mathfrak{P}^{i+1}$. But by our assumption $g(a_n) - a_n \in \mathfrak{P}^{i+1}$; hence $g(a_n) \in a + \mathfrak{P}^{i+1}$ for large n. Finally $a + \mathfrak{P}^{i+1}$ contains $\lim g(a_n) = \bar{g}(a)$ since it is closed. $\square$

The field corresponding to the decomposition group of $\mathfrak{P}$ will be denoted by $K_{-1}(\mathfrak{P})$ and called the *decomposition field* of $\mathfrak{P}$ or the *splitting field* of $\mathfrak{P}$. Similarly, the fields which correspond to the inertia group and the ith ramification group will be denoted by K_0 and K_i, respectively, and called the *inertia field* and the *i-th ramification field* of $\mathfrak{P}$. We have the following

Proposition 6.5. *Denote by $\mathfrak{P}_i$ the prime ideal of R_{K_i} lying below $\mathfrak{P}$ ($i = -1, 0, 1, \ldots$) and let $\mathfrak{p}$ be the prime ideal of K lying below $\mathfrak{P}$. Then in K_{-1} we have $\mathfrak{p}R_{K_{-1}} = \mathfrak{P}_{-1}\mathfrak{Q}$ where $\mathfrak{P}_{-1}$ does not divide $\mathfrak{Q}$ and $f_{K_{-1}/K}(\mathfrak{P}_{-1}) = 1$. Moreover, K_{-1} is the maximal subfield of L having these properties.*

In the field K_0 we have $\mathfrak{p}R_{K_0} = \mathfrak{P}_0 \mathfrak{Q}_0$ with $\mathfrak{P}_0$ not dividing $\mathfrak{Q}_0$, $f_{K_0/K}(\mathfrak{P}_0) = f_{L/K}(\mathfrak{P})$ and again K_0 is the maximal subfield of L with these properties.

Finally, in K_1 we have $\mathfrak{p}R_{K_1} = \mathfrak{P}_1^{e_0} \mathfrak{Q}_1$ where $\mathfrak{P}_1$ does not divide $\mathfrak{Q}_1$, $f_{K_1/K}(\mathfrak{P}_1) = f_{L/K}(\mathfrak{P})$, e_0 defined by $e(L/K) = e_0 p^m$, $p \nmid e_0$, p being the characteristic of $R_L/\mathfrak{P}$, and also here K_1 is the maximal subfield of L with these properties.

The extension K_0/K_{-1} is cyclic of degree $f_{L/K}(\mathfrak{P})$, K_1/K_0 is cyclic of degree e_0, and finally L/K_1 is a p-extension of degree e/e_0.

Proof. Clearly $K_\mathfrak{p} \subset (K_{-1})_{\mathfrak{P}_{-1}} \subset L_\mathfrak{P}$, and since the group $\text{Gal}(L_\mathfrak{P}/K_\mathfrak{p})$ fixes K_{-1}, it must fix, by continuity, also its closure $(K_{-1})_{\mathfrak{P}_{-1}}$, which shows that $(K_{-1})_{\mathfrak{P}_{-1}} = K_\mathfrak{p}$. We obtain in turn the equalities

$$e_{K_{-1}/K}(\mathfrak{P}_{-1}) = 1, \quad f_{K_{-1}/K}(\mathfrak{P}_{-1}) = 1$$

by Theorem 5.5, and so $\mathfrak{p}R_{K_{-1}} = \mathfrak{P}_{-1}\mathfrak{Q}$ where $\mathfrak{P}_{-1}$ does not divide $\mathfrak{Q}$, as asserted. Conversely, if $K \subset K' \subset L$ and for the prime ideal $\mathfrak{P}'$ of $R_{K'}$ lying below $\mathfrak{P}$ we have $e_{K'/K}(\mathfrak{P}') = f_{K'/K}(\mathfrak{P}') = 1$, then $K'_{\mathfrak{P}'} = K_\mathfrak{p}$, K' is fixed by G_{-1}, i.e., $K' \subset K_{-1}$. The remaining assertions are immediate by Theorems 5.5, 5.12 and the definition of the groups $G_i(\mathfrak{P})$. $\square$

Corollary. *If L/K is a normal extension of an algebraic number field K, $\mathfrak{P}$ is a prime ideal of R_L and $\mathfrak{p}$ is the prime ideal of R_K which lies below $\mathfrak{P}$, then*
$$e_{L/K}(\mathfrak{P}) = |G_0(\mathfrak{P})|, \quad f_{L/K}(\mathfrak{P})e_{L/K}(\mathfrak{P}) = |G_{-1}(\mathfrak{P})|;$$
hence $\mathfrak{p}R_L$ is a product of $[L:K]/|G_{-1}(\mathfrak{P})|$ distinct prime ideals of R_L.

Proof. Evident. □

This corollary shows that the knowledge of the sequence $G_i(\mathfrak{P})$ allows one to determine the decomposition of the prime ideal lying below $\mathfrak{P}$. The next lemma shows that this sequence determines also the decomposition of this prime ideal in every subfield of L containing K.

Lemma 6.2. *If L/K is a normal extension of an algebraic number field K, M a subfield of L containing K and H the corresponding subgroup of $G(L/K)$, then the subgroup $G_i(\mathfrak{P})$ of $G(L/M) = H$ which corresponds to the prime ideal $\mathfrak{P}$ equals the intersection of the subgroup $G_i(\mathfrak{P})$ of $G(L/K)$ with H for $i = -1, 0, 1, ...$*

Proof. Immediate from the definition of the groups G_i. □

We conclude this subsection with the determination of the maximal power of a prime ideal which divides the different of a normal extension:

Proposition 6.6. *If L/K is a finite and normal extension of an algebraic number field K and $\mathfrak{P}$ is a prime ideal of R_L, then its maximal power dividing the different $D_{L/K}$ equals $\mathfrak{P}^A$ with*
$$A = \sum_{i=0}^{t} (|G_i| - 1),$$
where G_0 is the decomposition group and $G_1, G_2, ..., G_t$ are the non-trivial ramification groups of L/K.

Proof. Apply Proposition 6.2 and Theorem 5.13. □

Corollary 1. *If L/K is normal of odd degree N and we denote by T_N the greatest common divisors of all numbers $(p-1)/2$ for $p|N$, p prime, then $d(L/K)$ is a $2T_N$-th power.*

Proof. Let $\mathfrak{p}$ be a prime ideal of R_K ramified in L/K and let $\mathfrak{p}R_L = (\mathfrak{P}_1 ... \mathfrak{P}_g)^e$. Let p be the rational prime lying below $\mathfrak{p}$. The proposition shows that $d(L/K)$

is divisible exactly by p^{efA}, where $f = f_{L/K}(\mathfrak{P}_1)$ and A is given in the proposition. Since e is an odd divisor of N, its every prime divisor is congruent to unity $(\mathrm{mod}\,2T_N)$, and hence $e \equiv 1(\mathrm{mod}\,2T_N)$, leading to $|G_0| \equiv 1(\mathrm{mod}\,2T_N)$ by the Corollary to Proposition 6.4. If $|G_1| = 1$, then $A = |G_0| - 1$ and thus $2T_N | A$. However, if $|G_1| \neq 1$, then $p|e$ and hence $p|N$. By Proposition 6.4 and Theorem 5.12 the ramification groups are all p-groups; thus for $i = 1, 2, \ldots$ we have $|G_i| \equiv 1(\mathrm{mod}(p-1))$, but since p divides N, $p-1$ is divisible by $2T_N$ and we conclude that also in this case $2T_N$ divides A. This shows that $d(L/K)$ is a $2T_N$th power of an ideal in R_K. □

Corollary 2. *If L/K is a normal extension of a prime degree q, then $d(L/K)$ is the $(q-1)$-st power of an ideal of R_K.*

Proof. This is a special case of the preceding corollary. □

4. In this subsection we shall consider some properties of characters $(\mathrm{mod}\,I)$ which will later be utilized in the theory of Gaussian sums.

Let us start with definitions. Consider an algebraic number field K and an ideal I of its ring of integers. We shall always assume that I is non-zero; however, the case of $I = R_K$ will not be excluded. Denote by $G(I)$ the multiplicative group of residue classes $(\mathrm{mod}\,I)$ which are relatively prime to I and consider any character χ of that group. It is customary to treat this character as a function defined for all integers of K by means of

$$\chi(a) = \begin{cases} \chi(a\,\mathrm{mod}\,I), & \text{if } a \bmod I \text{ lies in } G(I), \\ 0 & \text{otherwise.} \end{cases}$$

Observe that the function so defined preserves the multiplicative property of χ; however, it is not a group character in the usual sense.

A character of $G(I)$ is called *primitive* if there is no ideal $J \neq I$ dividing I and having the property that if $(xR_K, I) = 1$ and $x \equiv 1(\mathrm{mod}\,J)$ then $\chi(x) = 1$. It is obvious that if I is a prime ideal then every non-trivial character of $G(I)$ is primitive and the trivial character is primitive only in $G(1)$. Observe, moreover, that every character of $G(I)$ can be regarded as a primitive character of $G(J)$, J being a suitable divisor of I, owing to the isomorphism between the subgroup of the group of characters of $G(I)$ which are trivial on the residue class $1(\mathrm{mod}\,I_1)$ and the group of those characters of $G(J)$ where $J = I/I_1$, for any I_1 dividing I. In fact, take I_1 to be equal to the least common multiple of all ideals I_0 dividing I for which the character considered is trivial at the residue class $1(\mathrm{mod}\,I_0)$. The resulting ideal J is called the *conductor* of our character.

In the case of $K = Q$ the unique positive generator of J is called the *conductor of χ*. Clearly no confusion can arise here.

For every character χ we have $\chi^2(-1) = 1$, whence $\chi(-1) = \pm 1$. Characters χ satisfying $\chi(-1) = 1$ are called *even* characters, and the remaining characters are called *odd*.

Corollary 3 to Proposition 1.6 shows that if $I = I_1 \ldots I_r$ is a factorization of I into factors relatively prime in pairs, then we have also the factorization $G(I) = G(I_1) \times \ldots \times G(I_r)$ of corresponding groups. Thus every character of $G(I)$ is a product of characters of $G(I_j)$, i.e., if we define for a given character χ of $G(I)$ the characters χ_i of $G(I_i)$ by $\chi(x \bmod I_i) = \chi(u)$, where $u \equiv x \pmod{I_i}$ and $u \equiv 1 \pmod{I/I_i}$, then

$$\chi(x \bmod I) = \prod_{i=1}^{r} \chi(x \bmod I_i).$$

We prove

Proposition 6.7. *If χ is a character of $G(I)$ and $I = \prod_{i=1}^{r} I_i$ is a factorization into pairwise co-prime factors, then χ is primitive iff every factor χ_i in the factorization $\chi = \prod_{i=1}^{r} \chi_i$ is primitive.*

Proof. Assume that for $i = 1, 2, \ldots, r$ the character χ_i has the ideal J_i, dividing I_i for the conductor. Then for x congruent to unity $(\bmod J)$ with $J = \prod_{i=1}^{r} J_i$ we have $\chi(x) = \prod_{i=1}^{r} \chi_i(x) = 1$. It follows that if a certain factor χ_i is not primitive, which means that $I_i \neq J_i$, then $J \neq I$ and χ is not primitive either. Conversely, if χ is not primitive, say $\chi(x) = 1$ for $x \equiv 1 \pmod{J}$ with a certain $J \neq I$ dividing I, then $J = \prod_{i=1}^{r} J_i$ with J_i dividing I_i and for a certain i we must obtain the inequality $I_i \neq J_i$. But then $\chi_i(x) = 1$ for $x \equiv 1 \pmod{J_i}$, showing that χ_i is not primitive. □

To determine the characters of $G(I)$ the knowledge of the structure of this group is very useful. In particular, it is important to know in which cases the group $G(I)$ is cyclic. If it is cyclic, then we say that there exist *primitive roots* $(\bmod I)$, every such root being a generator of $G(I)$. In the case of the rational field it is a classical result that the group of residue classes $(\bmod M)$ prime to M under multiplication is cyclic iff either M is a power of an odd prime or M is the double of such a power, or, finally, if $M = 1, 2, 4$. In the general case things become more complicated, as we shall now show:

Theorem 6.2. *The group $G(I)$ is cyclic iff one of the following cases holds*:
 (i) $I = \mathfrak{P}$, *a non-zero prime ideal.*
 (ii) $I = \mathfrak{P}^2$ *provided that $f_{K/Q}(\mathfrak{P}) = 1$.*

(iii) $I = \mathfrak{P}^N$ for arbitrary N, provided that $\mathfrak{P}$ does not contain 2, and, moreover, $f_{K/Q}(\mathfrak{P}) = e_{K/Q}(\mathfrak{P}) = 1$, i.e. the corresponding local extension $K_\mathfrak{P}/Q_p$ is trivial.

(iv) $I = \mathfrak{P}^3$ provided 2 lies in $\mathfrak{P}$, $f_{K/Q}(\mathfrak{P}) = 1$ and $e_{K/Q}(\mathfrak{P}) \neq 1$.

(v) $I = \mathfrak{P}^N Q_1 \ldots Q_k$, where $2 \notin \mathfrak{P}$, $2 \in Q_i$ ($i = 1, 2, \ldots, k$) and with the notation $f = f_{K/Q}(\mathfrak{P})$, $f_i = f_{K/Q}(\mathfrak{P}_i)$ ($i = 1, 2, \ldots, k$), p being the unique rational prime contained in $\mathfrak{P}$, we have either $N = 1$ and the integers $p^f - 1$, $2^{f_1} - 1, \ldots, 2^{f_k} - 1$ are pairwise relatively prime, or $N = 2$, $f = 1$, and the numbers $p(p-1)$, $2^{f_1} - 1, \ldots, 2^{f_k} - 1$ are pairwise relatively prime, or finally, $N \geq 3$, $f = e_{K/Q}(\mathfrak{P}) = 1$ and $p(p-1)$, $2^{f_1-1}, \ldots, 2^{f_k-1}$ are pairwise relatively prime.

Proof. We start with a simple lemma which translates the whole problem into the language of p-adic fields:

Lemma 6.3. *For any prime ideal $\mathfrak{P}$ the group $G(\mathfrak{P}^N)$ is isomorphic to the factor group $U(K_\mathfrak{P})/U_N(K_\mathfrak{P})$.*

Proof. The embedding of K into $K_\mathfrak{P}$ maps the integers prime to $\mathfrak{P}$ into the group of units U, and since every coset of $U \pmod{U_N}$ contains elements of R_K, the induced map $G(\mathfrak{P}^N) \to U(K_\mathfrak{P})/U_N(K_\mathfrak{P})$ is bijective. □

We deal first with prime ideal powers $I = \mathfrak{P}^N$. The case of $N = 1$ is trivial, since $G(I) = G(\mathfrak{P}) = (R_K/\mathfrak{P})^* = GF(p^f)^*$. Moreover, by the Corollary to Proposition 5.9 the group U_1/U_2 equals C_p^f with $f = f(K/Q_p)$; hence it can be cyclic only if $f = 1$. But if U/U_N is cyclic, then U_1/U_2 is cyclic as well, and so $G(\mathfrak{P}^N)$ can be cyclic for $N \geq 2$ only if $f(K/Q) = 1$. If this condition is satisfied, then $U_1/U_2 = C_p$, and since $U/U_1 = C_{p-1}$, in view of $(p, p-1) = 1$ we obtain $G(I^2) = U/U_2 = C_{p(p-1)}$. This settles cases (i) and (ii).

Since we already know that in the case of $f_{K/Q}(\mathfrak{P}) \neq 1$ the group $G(\mathfrak{P}^N)$ can be cyclic only for $N = 1$, we may assume in the sequel that $f_{K/Q}(\mathfrak{P}) = 1$. Since

$$(1+\pi a)^p = 1 + p\pi a + \ldots + p\pi^{p-1}a^{p-1} + \pi^p a^p$$

(where π is a generator of the prime ideal $\mathfrak{P}$ in $K_\mathfrak{P}$), we can see that the mapping $x \mapsto x^p$ maps U_1 into U_r with $r = \min(p, 1+e)$, whence U_1/U_r is a cyclic group only in the case of $r = 2$, for in all other cases all elements are of order p; however, the group itself is of a larger order. The case of $r = 2$ arises if either $p = 2$ or $e = 1$.

In the remaining cases the group U_1/U_r equals C_p^{r-1}, whence U_1/U_3 cannot be cyclic, and so $G(\mathfrak{P}^N)$ for $N \geq 3$ cannot be cyclic either.

Now in the case of $e = 1$ we have $K_\mathfrak{P} = Q_p$, as $f_{K/Q}(\mathfrak{P}) = 1$, and thus $G(\mathfrak{P}^N) = G(p^N)$, which is cyclic for all N and odd p and non-cyclic for $p = 2$ and $N = 3, 4, \ldots$ This settles (iii).

In the case of $p = 2$, $e \neq 1$ we see that U_1/U_3 is cyclic with the generator $1+\pi$, since $(1+\pi)^2 = 1+\pi^2+2\pi \in U_2 \setminus U_3$ and U_1/U_4 is non-cyclic because every element of it has the order at most equal to 4 owing to

$$(1+a\pi)^4 = 1+4\pi a+6\pi^2 a^2+4\pi^3 a^3+\pi^4 a^4 \in U_4,$$

whereas $|U_1/U_4| = 8$. Since in this case $U = U_1$, (iv) follows.

To establish (v) one has to observe that a product of cyclic groups is cyclic iff their orders are relatively prime in pairs and to apply (i)–(iv). We leave the easy but tedious computations to the reader. □

Lemma 6.3 shows that every character of $G(\mathfrak{P}^N)$ is in fact a character of $U(K_\mathfrak{P})/U_N(K_\mathfrak{P})$ and one can easily see that the two definitions of the conductor which we have at this moment are in perfect agreement.

We end this subsection with the determination of all ideals I for which the group $G(I)$ has a real primitive character, i.e. a primitive character assuming the values 1 and -1 exclusively. We prove

Proposition 6.8. *The group $G(I)$ has real primitive characters iff I is of the form*

$$I = \mathfrak{Q}_1^{a_1} \ldots \mathfrak{Q}_t^{a_t} \mathfrak{P}_1 \ldots \mathfrak{P}_r,$$

where $\mathfrak{Q}_1, \ldots, \mathfrak{Q}_t$ are the prime ideals containing 2, $\mathfrak{P}_i$ are distinct prime ideals not containing 2 and $a_i = 0, 2, 4, \ldots, 2e_i-2, 2e_i, 2e_i+1$ with $e_i = e_{K/Q}(Q_i)$ ($i = 1, 2, \ldots, t$).

Proof. Owing to Proposition 6.7 we may restrict our attention to prime ideal-powers, $I = \mathfrak{P}^N$. First let us assume that $2 \notin \mathfrak{P}$ and let χ be a real character on $G(\mathfrak{P}^N)$. It is convenient to consider χ as a character of U/U_N which is allowed by Lemma 6.3. Since χ is real, for every x we have $\chi(x^2) = \chi^2(x) = 1$; thus χ is trivial on all squares. But we have also

$$U/U_N \simeq U/U_1 \times U_1/U_N,$$

because the orders of the groups involved are relatively prime, and since in U_1 every element is a square (1/2 being integral in the completion of K), we see that our character is trivial on U_1/U_N and hence its conductor equals $\mathfrak{P}$ or $\mathfrak{P}^0$. This shows that in the case of $2 \notin \mathfrak{P}$ only $G(\mathfrak{P})$ can have a primitive real character, and indeed it does have one, being equal to C_{p^f-1} with $f = f_{K/Q}(\mathfrak{P})$.

Now let us turn to the case of $2 \in \mathfrak{P}$. If χ is a real character of $G(\mathfrak{P}^N)$, then again it is trivial on every square. By the Corollary to Lemma 5.2 the map $x \mapsto x^2$ maps U_{1+e} onto U_{1+2e}; hence our character has to be trivial on U_{1+2e} and finally we find that there are no real primitive characters of $G(\mathfrak{P}^N)$ for $N \geqslant 2(1+e)$, where, as usual, $e = e_{K/Q}(\mathfrak{P})$. Hence let $1 \leqslant N \leqslant 2e+1$. Observe first that since U/U_1 is of odd order, the group $G(\mathfrak{P}^N) \simeq U/U_N$ has a primitive real character iff there is a real character of $V_1 = U_1/U_N$ which is non-trivial on $V_2 = U_{N-1}/U_N$. Since the kernel of a non-trivial real character is of index 2 and every subgroup of index 2 induces such a character, this will happen iff there is a subgroup of index 2 in V_1, not containing V_2, i.e. V_2 is not contained in the intersection A of all subgroups of index 2 in V_1. Note that A consists of all squares of elements of V_1. Indeed, V_1 is a product of cyclic 2-groups:

$$V_1 = \prod_{i=1}^{r} C_{m_i}$$

with $m_i = 2^{a_i}$, $a_i \geqslant 1$, $i = 1, 2, \ldots, r$. If y_i is the generator of C_{m_i} and $x = [x_1, \ldots, x_r] \in A$ ($x_i \in C_{m_i}$), then x lies in every subgroup $V_1^{(j)}$ defined by

$$V_1^{(j)} = \left(\prod_{i \neq j} C_m\right) \times C_{m_j/2},$$

$C_{m_j/2}$ being the group generated by y_j^2.

This implies that for every j we must have $x_j = y_j^{2n_j}$ with a suitable n_j, and thus x is a square. Conversely, every square evidently lies in A.

Let now N be odd and not exceeding $2e-1$. If π is a generator of $\mathfrak{P}$ in $K_\mathfrak{P}$ and $1+c\pi^{N-1} \in U_{N-1}$, then, determining $a \pmod{\mathfrak{P}}$ from $a^2 \equiv c \pmod{\mathfrak{P}}$ (which is possible, since $|G(\mathfrak{P})| = N(\mathfrak{P})-1$ is odd in view of $2 \in \mathfrak{P}$), we get

$$1+c\pi^{N-1} \equiv 1+a^2\pi^{N-1} \pmod{\mathfrak{P}^N},$$

and in view of

$$2a\pi^{(N-1)/2} \equiv 0 \pmod{\mathfrak{P}^{e+(N-1)/2}}$$

and $e+(N-1)/2 \geqslant N$, we obtain

$$1+c\pi^{N-1} \equiv (1+a\pi^{(N-1)/2})^2 \pmod{\mathfrak{P}^N},$$

i.e. $V_2 \subset A$, showing that in this case there is no primitive real character $\pmod{\mathfrak{P}^N}$.

If $1 \leqslant N \leqslant 2e$ and N is even, then the image of $1+\pi^{N-1}$ in V_2 does not lie in A. In fact, otherwise, for a certain unit u and $M \geqslant 0$, we would have the congruence

$$1+\pi^{N-1} \equiv (1+u\pi^M)^2 \pmod{\mathfrak{P}^N};$$

thus, if u_1 is defined by $2 = u_1 \pi^e$,
$$1+\pi^{N-1} \equiv 1+uu_1\pi^{M+e}+u^2\pi^{2M} \pmod{\mathfrak{P}^N}.$$
If $M \geqslant e$, then this gives $1+\pi^{N-1} \equiv 1 \pmod{\mathfrak{P}^{2e}}$ and hence $2e \leqslant N-1$, a contradiction. However, if $M < e$, then
$$1+\pi^{N-1} \equiv 1+\pi^{2M}(u^2+uu_1\pi^{e-M}) \pmod{\mathfrak{P}^N}$$
leads to $N-1 = 2M$, which is not possible, because $N-1$ is odd. This establishes the existence of primitive real characters $(\bmod\, \mathfrak{P}^N)$ for $N = 0, 2, 4, \ldots$ $\ldots, 2e$.

There remains the case of $N = 1+2e$. Assume that every element $1+a\pi^{2e}$, a being a unit, is a square $(\bmod\, U_{1+2e})$ of a principal unit $1+b\pi$. Writing $b = \varepsilon\pi^n$ with a suitable unit ε, we obtain
$$\begin{aligned}1+a\pi^{2e} &\equiv 1+2\pi b+\pi^2 b^2 \\ &\equiv 1+\varepsilon\varepsilon_1\pi^{1+e+n}+\varepsilon^2\pi^{2n+2} \pmod{\mathfrak{P}^{1+2e}},\end{aligned} \quad (6.1)$$
with $2 = \varepsilon_1\pi^e$. Now we distinguish three cases:

(i) $e-1 < n$. In this case $\varepsilon\varepsilon_1\pi^{1+e+n}$ and $\varepsilon^2\pi^{2n+2}$ both lie in $\mathfrak{P}^{1+2e}$ and hence $1+a\pi^{2e} \equiv 1 \pmod{\mathfrak{P}^{1+2e}}$, which gives a contradiction.

(ii) $e-1 > n$. In this case $\varepsilon\varepsilon_1\pi^{1+e+n}$ lies in $\mathfrak{P}^{2n+3}$ and hence (6.1) leads to $2n+2 = 2e$, $n+1 = e$, which again gives a contradiction.

(iii) $e-1 = n$. Here we get $1+a\pi^{2e} \equiv 1+(\varepsilon\varepsilon_1+\varepsilon^2)\pi^{2e} \pmod{\mathfrak{P}^{1+2e}}$, and hence
$$a \equiv \varepsilon\varepsilon_1+\varepsilon^2 \pmod{\mathfrak{P}},$$
and we see that the polynomial $X^2+X\bar{\varepsilon}_1$ (where $\bar{\varepsilon}_1$ is the image of ε_1 in $k = R_{K\mathfrak{P}}/\mathfrak{P}$) maps k^* onto itself, i.e. for every non-zero y in k the polynomial $X^2+X\bar{\varepsilon}_1+y$ is reducible in k. Writing $X^2+X\bar{\varepsilon}_1+y = (X-X_1)(X-X_2)$, we obtain $X_1+X_2 = \bar{\varepsilon}_1$, $X_1 X_2 = y$. Our assumption implies that the map
$$\varphi: X_1 \to X_1(\bar{\varepsilon}_1-X_1)$$
satisfies $\varphi(k^*) = k^*$ and thus it is injective, but $\varphi(X_1) = \varphi(\bar{\varepsilon}_1-X_1)$ shows that this is impossible. So there must be non-squares in V_2 and our proof is complete. □

5. Now we introduce Gaussian sums in any algebraic number field. They will be seen to be intimately connected with the sums $\tau_0(\chi)$ considered in the previous chapter. In fact, in the most important cases they will be equal, and in general our sums will be products of various $\tau_0(\chi)$.

Let I be a non-zero ideal in R_K and let a be a fixed number lying in the fractional ideal $I^{-1}D_{K/Q}^{-1}$. If χ is any character of $G(I)$, then the *Gaussian sum*

$\tau_a(\chi)$ corresponding to the pair $\langle \chi, a \rangle$ is defined by

$$\tau_a(\chi) = \sum_{x(\bmod I)} \chi(x)\exp(2\pi i T_{K/Q}(ax)). \tag{6.2}$$

Observe that this sum does not depend on the choice of the residue system $x\,(\bmod I)$ because for $x \equiv y\,(\bmod I)$, $x, y \in R_K$ we have $\chi(x) = \chi(y)$, and since $ax = ay + a(x-y)$, $T_{K/Q}(ax) = T_{K/Q}(ay) + T_{K/Q}(a(x-y))$, $a(x-y) \in D_{L/K}^{-1}$, by Proposition 4.7 (iv) we get $T_{K/Q}(a(x-y)) \in Z$, and so $\exp(2\pi i T_{K/Q}(ax)) = \exp(2\pi i T_{K/Q}(ay))$.

In the special case of $K = Q$ we can write $I = nZ$, $a = k/n$ with suitable natural n and $k \in Z$, and obtain the usual Gaussian rational sum

$$\tau_a(\chi) = \sum_{x(\bmod n)} \chi(x)\exp(2\pi i k x/n).$$

The case of $k = 1$ is specially important, and we shall denote the corresponding Gaussian sum simply by $\tau(\chi)$.

We now prove

Proposition 6.9. (i) *If b is an integer of K relatively prime to I, then $\tau_{ab}(\chi) = \overline{\chi(b)}\,\tau_a(\chi)$. In particular, if $K = Q$, $I = nZ$ with natural n and $a = k/n$ with $(k, n) = 1$, then $\tau_a(\chi) = \overline{\chi(k)}\,\tau(\chi)$.*

(ii) *Let $I = I_1 \ldots I_r$ be a factorization of I into factors relatively prime in pairs and let $\chi = \chi_1 \ldots \chi_r$ be the corresponding factorization of a character χ of $G(I)$. If a lies in $D_{K/Q}^{-1}I^{-1}$ and for $i = 1, 2, \ldots, r$ the elements a_i belong to $D_{K/Q}^{-1}I_i^{-1}$ and satisfy $a - a_i \in J_i^{-1}D_{K/Q}^{-1}$ with $J_i = II_i^{-1}$, then*

$$\tau_a(\chi) = \prod_{i=1}^{r} \tau_{a_i}(\chi_i).$$

Proof. (i) Under our assumptions we have

$$\tau_{ab}(\chi) = \sum_{x \in G(I)} \chi(x)\exp(2\pi i T_{K/Q}(xab));$$

but if x runs over $G(I)$, then xb does the same and we obtain

$$\tau_{ab}(\chi) = \overline{\chi(b)} \sum_{x \in G(I)} \chi(bx)\exp(2\pi i T_{K/Q}(xab)) = \overline{\chi(b)}\,\tau_a(\chi).$$

(ii) For every $x \in R_K$ which is relatively prime to I choose $x_i \in R_K$ with $x_i \equiv x\,(\bmod I_i)$ and $x_i \equiv 1\,(\bmod I_j)$ ($j \neq i$). Moreover, let $y_1, \ldots, y_r$ be integers of K such that $y_i \equiv 1(\bmod I_i)$ and $y_i \in I_j$ for $j \neq i$. Then we have

$x_1 y_1 + \ldots + x_r y_r \equiv x \pmod{I}$ and thus

$$\tau_a(\chi) = \sum_{\substack{x_i \in G(I_i) \\ i=1,\ldots,r}} \chi_1(x_1) \ldots \chi_r(x_r) \exp(2\pi i T_{K/Q}(ax))$$

$$= \prod_{i=1}^{r} \sum_{x_i \in G(I_i)} \chi_i(x_i) \exp(2\pi i T_{K/Q}(ax_i y_i))$$

$$= \prod_{i=1}^{r} \sum_{x_i \in G(I_i)} \chi_i(x_i) \exp(2\pi i T_{K/Q}(a_i x_i y_i))$$

since $ax_i y_i - a_i x_i y_i \in D_{K/Q}^{-1}$, and this equals

$$\prod_{i=1}^{r} \tau_{a_i}(\chi_i)$$

because of (i) and $\chi_i(y_i) = 1$. □

Corollary. *Let $n = n_1 \ldots n_r$ be a factorization of a natural number n into mutually prime factors, let χ be a character of $G(nZ)$ (which group we shall for convenience denote by $G(n)$) and let $\chi = \chi_1 \ldots \chi_r$ be its factorization into characters of $G(n_i)$. Then we have the equality*

$$\tau(\chi) = \prod_{i=1}^{r} \chi_i(n/n_i) \prod_{i=1}^{r} \tau(\chi_i).$$

Proof. For $i = 1, 2, \ldots, r$ the congruence

$$xn/n_i \equiv 1 \pmod{n_i}$$

is solvable and we shall denote by b_i one of its solutions. Then the numbers $a_i = b_i/n_i$ satisfy the assumptions of (ii), whence

$$\tau(\chi) = \prod_{i=1}^{r} \tau_{a_i}(\chi_i). \tag{6.3}$$

Applying part (i) of the proposition, we obtain

$$\tau_{a_i}(\chi_i) = \overline{\chi_i(b_i)} \tau(\chi_i);$$

however, $b_i(n/n_i)$ is congruent to unity $\pmod{n_i}$ and so $\overline{\chi_i(b_i)} = \chi_i(n/n_i)$, which together with (6.3) and the above equation proves our assertion. □

The proposition proved above shows that Gaussian sums connected with characters of $G(I)$ can be decomposed into products of such sums resulting from characters of $G(\mathfrak{P}^m)$. We are mainly interested in Gaussian sums $\tau_a(\chi)$

for primitive characters χ and we are now going to establish the relations between them and the sums $\tau_0(\chi)$.

Proposition 6.10. *Let $\mathfrak{P}$ be a non-zero prime ideal of R_K and χ a primitive character of $G(\mathfrak{P}^N)$. Let $\mathfrak{P} = \mathfrak{P}_1, \mathfrak{P}_2, \ldots, \mathfrak{P}_g$ be all the prime ideals of R_K which lie above $pZ = \mathfrak{P} \cap Z$; denote by $\mathfrak{P}^m$ the maximal power of $\mathfrak{P}$ dividing the different $D_{K/Q}$, and choose an element Π of K such that $\Pi R_K = \mathfrak{P}_1 I$, $I \subset R_K$, $\mathfrak{P}_1 \nmid I$. If we denote the character of $U(K_\mathfrak{P})/U_N(K_\mathfrak{P})$ induced by χ by the same letter and put $a = \Pi^{-m-N}$, then the Gaussian sum $\tau_a(\chi)$ coincides with the sum $\tau_0(\chi)$ as defined by (5.8).*

Proof. By the Corollary to Proposition 6.1 we have for u in K the equality

$$T_{K/Q}(u) = \sum_{i=1}^{g} T_{K_i/Q_p}(u), \qquad (6.4)$$

where K_i is the completion of K with respect to $\mathfrak{P}_i$. Moreover, for every integer x of K for which $(xR_K, \mathfrak{P}_2 \ldots \mathfrak{P}_g) = 1$ the quotient x/Π^{m+N} is integral in K_i for $i = 2, 3, \ldots, g$; thus

$$\exp(2\pi i \lambda(T_{K_i/Q_p}(x\Pi^{-m-N}))) = 1 \qquad (6.5)$$

for $i = 2, 3, \ldots, g$. (Here $\lambda(x)$ denotes the function introduced in Chapter 5, § 3.1.) Observe that $T_{K/Q}(x\Pi^{-m-N})$ is a rational number whose denominator is a power of p, whence $\lambda(T_{K/Q}(x\Pi^{-m-N}))$ coincides with the fractional part of $T_{K/Q}(x\Pi^{-m-N})$. In view of (6.4) and (6.5) this leads us to

$$\exp(2\pi i T_{K/Q}(x\Pi^{-m-N})) = \exp(2\pi i \lambda(T_{K_1/Q_p}(x\Pi^{-m-N}))).$$

Finally, using the Chinese remainder-theorem, choose a full system $x_1, \ldots, x_k$ of representatives of $G(\mathfrak{P}^N)$ satisfying

$$(x_i R_K, \mathfrak{P}_2 \ldots \mathfrak{P}_g) = 1;$$

thus for our $a = \Pi^{-m-N}$ we obtain by the preceding remarks the equality

$$\tau_a(\chi) = \sum_{i=1}^{k} \chi(x_i) \exp(2\pi i \lambda(T_{K_1/Q_p}(x_i \Pi^{-m-N}))).$$

But Lemma 6.3 shows that the x_i's form at the same time a full system of representatives of $U(K_1)/U_N(K_1)$, and our assertion becomes evident. □

Corollary 1. *Let χ be a primitive character of $G(\mathfrak{P}^N)$, let $\mathfrak{P}^m$ be the maximal power of $\mathfrak{P}$ which divides $D_{K/Q}$ and let a be any element of $\mathfrak{P}^{-m-N} \setminus \mathfrak{P}^{-m-N+1}$. Then the Gaussian sum $\tau_a(\chi)$ differs from $\tau_0(\chi)$ only by a factor which is a root of unity.*

Proof. Choose Π as in the proposition just proved and observe that, by Proposition 6.9 (i), if b is an element of K such that both b and b^{-1} are integral in $K_{\mathfrak{P}}$, then for every c in $\mathfrak{P}^{-m-N}$ the sums $\tau_c(\chi)$ and $\tau_{bc}(\chi)$ differ by a factor which is a root of unity. But our choice of a shows that a/Π^{-m-N} can be taken as b and Π^{-m-N} as c. $\square$

Corollary 2. *Let $I = \prod_{i=1}^{r} \mathfrak{P}_i^{N_i}$ and for any $i = 1, 2, \ldots, r$ let $\mathfrak{P}_i^{m_i}$ be the maximal power of $\mathfrak{P}_i$ which divides $D_{K/Q}$. Moreover, let $n_i(t)$ be the exponent corresponding to the prime ideal $\mathfrak{P}_i$. If a is an element of $I^{-1} D_{K/Q}^{-1}$, which, moreover, satisfies $n_i(a) = -m_i - N_i$ for $i = 1, 2, \ldots, r$ and χ is a primitive character of $G(I)$, then*

$$|\tau_a(\chi)| = N(I)^{1/2}.$$

Proof. Choose for our a elements $a_1, \ldots, a_r$ which satisfy the assumptions of Proposition 6.9 (ii) and for which $n_i(a_i) = -m_i - N_i$. Then the sum $\tau_a(\chi)$ equals the product of $\tau_{a_i}(\chi_i)$ ($i = 1, 2, \ldots, r$) and the preceding corollary together with Theorem 5.17 and Proposition 5.19 (iii) implies our assertion. $\square$

Similarly, this time using Propositions 5.19 (ii) and 6.9 (ii), we get the next corollary:

Corollary 3. *If χ is a primitive character of $G(I)$, then*

$$\tau_a(\bar{\chi}) = \chi(-1) \overline{\tau_a(\chi)}. \qquad \square$$

6. Now we shall have a closer look at Gaussian sums for real characters in the field of rational numbers. In this case Proposition 6.9 (i) shows that it suffices to deal with $\tau(\chi)$, and the Corollaries 2 and 3 to Proposition 6.10 imply that if χ is real, then $\tau(\chi)^2 = \chi(-1) M$, where M is the conductor of χ, i.e. $\tau(\chi)$ differs by a fourth root of unity from $M^{1/2}$. The determination of the suitable root forms the content of the following theorem:

Theorem 6.3. *If χ is a real primitive character of $G(M)$, then*

$$\tau(\chi) = M^{1/2} \quad \text{if } \chi \text{ is even,}$$

and

$$\tau(\chi) = iM^{1/2} \quad \text{if } \chi \text{ is odd.}$$

Proof. The crucial point of the proof lies in the case of $M = p$, p being an odd prime, and we start with the consideration of this case. The group $G(p)$, being cyclic of order $p-1$, has exactly one subgroup H of index 2, which consists of all squares. If χ is a primitive real character of $G(p)$, then $\mathrm{Ker}\,\chi$ must equal H, and so we see that $\chi(x)$ is equal to $\left(\dfrac{x}{p}\right)$, the Legendre symbol. In particular, $\chi(-1) = (-1)^q$ with $q = (p-1)/2$.

The simple reasoning which now follows is due to W. C. Waterhouse [70].

Consider the space of all complex-valued functions defined on $G(p)$. It is a $(p-1)$-dimensional linear space and has two obvious bases: one consisting of characteristic functions of points, i.e., $f_k(x) = 1$ if $x = k$ and 0 otherwise for $k = 1, 2, \ldots, p-1$, and another, consisting of all characters $\chi_1, \ldots, \chi_{p-1}$ of $G(p)$. Define now a linear transformation of this space by means of the formula

$$B(g)(n) = \sum_{k=1}^{p-1} g(k) \zeta_p^{kn}.$$

Using Proposition 6.9 (i), one can immediately see that

$$B(\chi_j) = \tau(\chi_j) \cdot \bar{\chi}_j.$$

Since the equality $\chi = \bar{\chi}$ holds only for the trivial and quadratic characters, we see that the matrix of B in the basis $\chi_1, \ldots, \chi_{p-1}$ consists of blocks

$$\begin{pmatrix} 0 & \tau(\psi) \\ \tau(\bar{\psi}) & 0 \end{pmatrix}$$

corresponding to pairs $\psi, \bar{\psi}$ of conjugate characters which are non-trivial and non-quadratic, and of two diagonal entries, equal to -1 (which corresponds to the trivial character) and $\tau(\chi)$, χ being the quadratic character. The determinant of B thus equals

$$(-1) \prod_{\psi} (-\tau(\psi)\tau(\bar{\psi})) \cdot \tau(\chi)$$

the product being taken over all pairs $(\psi, \bar{\psi})$ of conjugate characters not equal to the trivial or the quadratic one, and since Corollaries 2 and 3 to Proposition 6.10 imply the equality

$$-\tau(\psi)\tau(\bar{\psi}) = -p\psi(-1),$$

we obtain

$$\det B = (-1)(-p)^{(p-3)/2} \tau(\chi) \prod_{\psi} \psi(-1)$$

$$= (-1)(-p)^{(p-3)/2} \tau(\chi) (-1)^{[(p-1)/4]}$$

there being exactly $[(p-1)/4]$ characters ψ with $\psi(-1) = -1$ when we count the conjugate characters only once.

Now we compute det B, utilizing the matrix of B in the basis $f_1, \ldots, f_{p-1}$, and obtain

$$\det B = \det(\zeta_p^{kn}),$$

which is a Vandermonde determinant. Since

$$B^2 = \begin{pmatrix} p & 0 & \ldots & 0 \\ 0 & 0 & \ldots & p \\ \ldots & \ldots & \ldots & \ldots \\ 0 & p & \ldots & 0 \end{pmatrix}$$

we immediately see that $\det B^2 = (-1)^{p(p-1)/2} p^p$; thus

$$\det B = \pm i^{p(p-1)/2} p^{p/2}.$$

To compute the sign, write $x = \cos(\pi/p) + i\sin(\pi/p)$. We thus obtain

$$\det B = \prod_{0 \leqslant l < k \leqslant p-1} (\zeta_p^k - \zeta_p^l) = \prod_{l < k} x^{l+k}(x^{k-l} - x^{-(k-l)})$$

$$= \prod_{l<k} x^{l+k} \prod_{l<k} \left(2i\sin\frac{(k-l)\pi}{p}\right).$$

Because of

$$\sum_{l<k} (k+l) = 2p((p-1)/2)^2$$

the first factor equals 1 and the second equals $i^{p(p-1)/2}$ times a positive quantity. Finally, we arrive at

$$\det B = i^{p(p-1)/2} p^{p/2},$$

and comparing this with the previous expression for B, we obtain the assertion of the theorem in the case of $M = p$, p being an odd prime.

Now we prove the theorem by direct verification in the cases of $M = 4$ and $M = 8$ and then reduce the general case to the cases already settled. (Note that $M = 2$ cannot arise since then $G(M)$ is trivial and so there is no primitive character at all.) For $M = 4$ our character is given by $\chi(1) = 1$, $\chi(3) = -1$, and the corresponding Gaussian sum equals $i - i^3 = 2i$. In the case of $M = 8$ we have two characters to consider: one given by $\chi_1(1) = \chi_1(3) = 1$, $\chi_1(5) = \chi_1(7) = -1$ and the other by $\chi_2(1) = \chi_2(7) = 1$, $\chi_2(3) = \chi_2(5) = -1$. The corresponding Gaussian sums are equal to $2^{3/2}i$ and $2^{3/2}$, respectively, and so the theorem is true for $M = 4$ and $M = 8$.

Proposition 6.8 shows that real primitive characters of $G(M)$ exist if and only if $M = 2^a p_1 \ldots p_r$, where $a = 0, 2$ or 3 and $p_1, \ldots, p_r$ are distinct odd primes; so suppose that M is of this form and that χ is a primitive character of $G(M)$ assuming the values 1 and -1 only. Write $\chi = \chi_0 \cdot \chi_1 \cdots \chi_r$, where χ_0 is 1 in the case of $a = 0$ and is a primitive real character of $G(2^a)$

otherwise, and $\chi_i(t) = \left(\dfrac{t}{p_i}\right)$ are the only real primitive characters of $G(p_i)$ for $i = 1, 2, \ldots, r$. Using the Corollary to Proposition 6.9, we can write

$$\tau(\chi) = \prod_{i=0}^{r} \chi_i(M/p_i)\tau(\chi_i),$$

where for simplicity we write $p_0 = 2^a$. By the already proved part of our theorem we have

$$\tau(\chi_i) = \begin{cases} p_i^{1/2} & \text{if } \chi_i \text{ is even}, \\ ip_i^{1/2} & \text{if } \chi_i \text{ is odd}, \end{cases}$$

and since, for $i \geqslant 1$, χ_i is odd iff $p_i \equiv 3 \pmod 4$, we obtain

$$\tau(\chi) = i^{s+\varepsilon} M^{1/2} \prod_{i=0}^{r} \chi_i(M/p_i),$$

where s equals the number of prime divisors of M congruent to $3 \pmod 4$ and $\varepsilon = 1$ if χ is odd and equals 0 otherwise. Using the quadratic reciprocity law, we obtain

$$\prod_{i=0}^{r} \chi_i(M/p_i) = \chi_0(p_1 \cdots p_r) \prod_{i=1}^{r} \left(\frac{p_0 p_1 \cdots p_{i-1} p_{i+1} \cdots p_r}{p_i}\right)$$

$$= \chi_0(p_1 \cdots p_r) \prod_{i=1}^{r} \prod_{\substack{0 \leqslant j \leqslant r \\ j \neq i}} \left(\frac{p_i}{p_j}\right)$$

$$= \chi_0(p_1 \cdots p_r) \prod_{j=1}^{r} \left(\frac{p_0}{p_j}\right) \prod_{1 \leqslant i < k} \left(\frac{p_i}{p_k}\right)\left(\frac{p_k}{p_i}\right)$$

$$= \chi_0(p_1 \cdots p_r) \prod_{j=1}^{r} \left(\frac{p_0}{p_j}\right) \prod_{1 \leqslant l < k} (-1)^{(p_i-1)(p_k-1)/4}$$

$$= \chi_0(p_1 \cdots p_r) \prod_{j=1}^{r} \left(\frac{p_0}{p_j}\right)(-1)^{s(s-1)/2}$$

and the equality

$$\tau(\chi) = i^{s+\varepsilon} M^{1/2} (-1)^{s(s-1)/2} \chi_0(p_1 \cdots p_r) \prod_{j=1}^{r} \left(\frac{p_0}{p_j}\right)$$

follows. There are now three cases to be considered separately, namely $p_0 = 1$, $p_0 = 4$ and $p_0 = 8$. In the first case $\varepsilon = 0$ and we see that $\tau(\chi) = iM^{1/2}$ if s is odd and $\tau(\chi) = M^{1/2}$ if s is even. But $\chi(-1) = (-1)^s$, and so our form of $\tau(\chi)$ agrees with the asserted one.

§ 1. Arithmetic Applications   287

In the second case we have $\varepsilon = 1$, $\left(\dfrac{p_0}{p_i}\right) = 1$ for all i and $\chi_0(p_1 \ldots p_r) = (-1)^s$ because $\chi_0(m) = (-1)^{(m-1)/2}$, and by putting these equalities together we easily confirm the truth of the theorem also in this case.

Finally we are left with $p_0 = 8$. For $i = 1, 3, 5, 7$ denote by t_i the number of primes congruent to i (mod 8) which divide M. With this notation we have

$$\prod_{j=1}^{r}\left(\dfrac{p_0}{p_j}\right) = \prod_{j=1}^{r}(-1)^{(p_j^2-1)/8} = (-1)^{t_3+t_5},$$

$$\chi_0(p_1 \ldots p_r) = (-1)^{t_5+t_7} \quad \text{if } \varepsilon = 1, \text{ i.e. if } \chi_0(1) = \chi_0(3) = 1,$$

and

$$\chi_0(p_1 \ldots p_r) = (-1)^{t_3+t_5} \quad \text{if } \varepsilon = 0, \text{ i.e. if } \chi_0(1) = \chi_0(7) = 1.$$

(We already know that there are exactly those two possibilities for a primitive real character of $G(8)$.) Moreover, $s = t_3 + t_7$ and

$$\chi(-1) = \chi_0(-1)\prod_{i=1}^{r}\chi_i(-1) = \chi_0(-1)\prod_{i=1}^{r}\left(\dfrac{-1}{p_i}\right) = \chi_0(-1)(-1)^s.$$

Finally $\chi_0(-1) = 1$ iff $\varepsilon = 0$. Having established this, we can now easily obtain the assertion of the theorem by checking the resulting cases:

(a) $s \equiv 0, 1 \pmod 4$, $\varepsilon = 1$. Here

$$\tau(\chi) = i^s i M^{1/2}(-1)^{t_5+t_7+t_3+t_5} = i^{s+1}(-1)^s M^{1/2},$$

and this equals $iM^{1/2}$ for $s \equiv 0 \pmod 4$ and $M^{1/2}$ for $s \equiv 1 \pmod 4$.

(b) $s \equiv 0, 1 \pmod 4$, $\varepsilon = 0$. Here

$$\tau(\chi) = i^s M^{1/2}$$

again in accordance with our assertion.

(c) $s \equiv 2, 3 \pmod 4$, $\varepsilon = 1$. Here

$$\tau(\chi) = i^s i M^{1/2}(-1)^{t_5+t_7+t_3+t_5+1} = i^{s+1} M^{1/2}(-1)^{1+s},$$

also in accordance with the theorem, which can be seen after a short computation, and finally

(d) $s \equiv 2, 3 \pmod 4$, $\varepsilon = 0$, in which case

$$\tau(\chi) = i^s M^{1/2}(-1)^{t_3+t_5+t_3+t_5+1} = -i^s M^{1/2},$$

which is equal to $M^{1/2}$ for $s \equiv 2 \pmod 4$ and $iM^{1/2}$ for $s \equiv 3 \pmod 4$. □

We conclude this subsection with a counterpart to the last theorem:

Theorem 6.4. *Let p be an odd prime and let χ be a primitive character of $G(p)$.*

If the quotient $\tau(\chi)/p^{1/2}$ is a root of unity, then $\chi(t) = \left(\dfrac{t}{p}\right)$.

Proof. Let χ be a primitive character of $G(p)$ (which in this case means only that χ is non-trivial) and denote by χ_1 the real character $\left(\dfrac{t}{p}\right)$. Assume that

$$\tau(\chi) = \varepsilon p^{1/2} \tag{6.6}$$

where ε is a root of unity.

Obviously the number $\tau(\chi)$ lies in the field K generated over Q by the values of χ and the pth primitive root of unity ζ_p. To get more information about it observe that the values of χ form a subgroup of the group of all $(p-1)$th roots of unity, and so we may write $\chi(k) = \omega^m$ with $m = m(k)$ for $k = 1, 2, \ldots, p-1$, where ω is a $(p-1)$th root of unity.

By Theorem 6.3 we have $\tau(\chi_1) = \eta p^{1/2}$ with $\eta = 1$ or i depending on the residue of $p \pmod 4$; hence

$$\tau(\chi) = \varepsilon \eta^{-1} \tau(\chi_1)$$

and since both Gaussian sums occurring here lie in K, we can infer that $\varepsilon \eta^{-1} \in K = Q(\omega, \zeta_p)$. Now note that $Q(\omega) \subset Q(\zeta_{p-1})$, and so, by Theorems 2.9 and 4.10, the fields $Q(\omega)$ and $Q(\zeta_p)$ have relatively prime discriminants, which in view of Theorem 4.9 leads to $K = Q(\omega \zeta_p)$. (Just check the degrees over Q.) Finally, we find that $\varepsilon \eta^{-1}$ lies in K and, applying Theorem 4.10, we obtain the equality

$$\varepsilon \eta^{-1} = \pm \omega^k \zeta_p^k$$

with suitable k. Now let g_a be that element of the Galois group of K/Q which maps ζ_p into ζ_p^a ($1 \leq a \leq p-1$) and fixes ω. We have $g_a(\chi(x)) = \chi(x)$, and now, using Proposition 6.9 (i), we obtain

$$\pm \omega^k \zeta_p^{ak} = g(\pm \omega^k \zeta_p^k) = g(\varepsilon \eta^{-1}) g(\tau(\chi))/g(\tau(\chi_1))$$
$$= \overline{\chi(a)} \overline{\chi_1(a)} \tau(\chi)/\tau(\chi_1) = \pm \overline{\chi(a)} \overline{\chi_1(a)} \omega^k \zeta_p^k,$$

which in turn implies $\zeta_p^{ak} = \overline{\chi(a)} \overline{\chi_1(a)} \zeta_p^k$, i.e.,

$$\chi(a) = \zeta_p^{(a-1)k} \left(\dfrac{a}{p}\right) \in Q(\zeta_p)$$

for all a in $[1, p-1]$, showing that ω lies in $Q(\zeta_p)$. But since the discriminants of the fields generated by ω and ζ_p are relatively prime, this can happen only if ω lies in Q, which means that χ must be a real character. But $\left(\dfrac{t}{p}\right)$ is the only real primitive character of $G(p)$. □

7. We conclude this rather long section with the proof of the Kronecker–Weber theorem:

Theorem 6.5. *If K is a normal extension of the rationals with an Abelian Galois group, then K is contained in a suitable cyclotomic field $K_m = Q(\zeta_m)$.*

(This result belongs properly to the class-field theory; however, we have decided to include it here because its proof, due to Shafarevich is a very illuminating example of the application of p-adic methods to the theory of algebraic number fields.)

Proof. Observe first that it suffices to prove the theorem for cyclic extensions of the rationals with a prime-power degree. Indeed, if K/Q is Abelian and $G = \prod_{i=1}^{r} C_{r_i}$ is the factorization of its Galois group into cyclic groups of prime-power orders, then the fields K_j ($j = 1, 2, \ldots, r$) corresponding to the subgroups generated by the product of all factors except $C_{p_j^{n_j}}$ have the latter groups for their Galois groups, and their composite equals K since it has to correspond to the unit group.

Now fix a prime-power q^m and let $S = \{p_1, \ldots, p_r\}$ be a finite set of primes. Denote by $K = K_{q^m, S}$ the composite of all cyclic extensions of Q whose degrees divide q^m and whose discriminants can be divisible only by primes from S. Observe that K/Q is finite. In fact, if L/Q is cyclic of degree q^r ($r \leq m$) and $d(L/Q) = \prod_{i=1}^{r} p_i^{m_i}$ ($m_i \geq 0$), then, using Proposition 6.3, we immediately infer that the m_i's cannot exceed a certain bound depending on q^m only, and so by Theorem 2.12 there can be only finitely many such L's. Hence their composite K/Q is finite. Note also that the Galois group of K/Q is a subgroup of a suitable power of C_{q^m}. Obviously it suffices to prove that every field $K_{q^m, S}$ is contained in a cyclotomic field.

For every prime $p \neq q$ denote by $Z = Z_{q^m, p}$ the maximal subfield of $Q(\zeta_p)$ such that the order of each element of its Galois group divides q^m. Moreover, for $p = q \neq 2$ let Z be the unique subfield of degree q^m of $Q(\zeta_{q^{1+m}})$ and for $p = q = 2$ put $Z = Q(\zeta_{2^{m+2}})$. We obviously have $Z_{q^m, p} \subset K_{q^m, S}$ if $p \in S$, and our aim will now be to prove the following fact:

(*) *If $S = \{p_1, \ldots, p_r\}$, then $K_{q^m, S}$ coincides with the composite $Z'_{q^m, S}$ of the fields Z_{q^m, p_i} ($i = 1, 2, \ldots, r$).*

This will prove the theorem since the fields Z_{q^m, p_i} are by definition contained in suitable cyclotomic fields and so is their composite.

Since in the extension $Z_{q^m, p}/Q$ only the prime p ramifies, a repeated application of Theorem 4.9 leads us to the equality

$$[Z'_{q^m,S}:Q] = \prod_{i=1}^{r}[Z_{q^m,p_i}:Q],$$

but in any case Z is a subfield of $Q(\zeta_{p_i^{m+2}})$, whence p_iZ becomes in it a power of a prime ideal of the first degree by Theorem 4.16 and Proposition 4.3.

Thus $[Z_{q^m,p_i}:Q]$ equals the ramification index e_i of p_i in this extension; hence the degree of $Z'_{q^m,S}$ over Q equals $e_1 \ldots e_r$ and, in view of $Z'_{q^m,S} \subset K_{q^m,S}$, in order to prove assertion (*) it suffices to obtain the inequality $[K_{q^m,S}:Q] \leq e_1 \ldots e_r$.

Assume for a moment that the following lemma is true:

Lemma 6.4. *If $L_{q^m,p}$ is the composite of all cyclic extensions of Q_p whose degrees divide q^m, then $L_{q^m,p}$ coincides with the composite of $Z_{q^m,p}$ and the unique unramified extension $W_{q^m,p}/Q_p$ of degree q^m.*

We shall deduce from the above lemma assertion (*) and then we shall give a proof of it. It is clear that for every p in S the field $K_{q^m,S}$ lies in $L_{q^m,p}$, and so by the lemma we obtain $K_{q^m,S} \subset Z_{q^m,p}W_{q^m,p}$.

But W_{q^m,p_i} is unramified over Q_{p_i} and the closure of Z_{q^m,p_i} in L_{q^m,p_i} has the ramification index over Q_{p_i} equal to e_i; hence the ramification index of p_i in $K_{q^m,S}/Q$ cannot exceed e_i by Theorem 5.5 (iv) and the multiplicativity of ramification indices.

At this point we need another auxiliary result, which is sometimes called the *monodromy theorem for algebraic number fields*:

Lemma 6.5. *If L/K is a normal finite extension of an algebraic number field with the Galois group G, then the subgroup H of G generated by all the inertia subgroups of G determined by the prime ideals of R_L corresponds to the maximal subfield of L which is unramified over K. In particular, if $K = Q$, then the inertia subgroups of G generate G.*

Proof. For every prime ideal $\mathfrak{P}$ of R_L we have $G_0(\mathfrak{P}) \cap H = G_0(\mathfrak{P})$. If M is the field corresponding to H, then by Lemma 6.1 and the Corollary to Proposition 6.5 we have

$$e_{M/K}(\mathfrak{P} \cap R_M) = |G_0(\mathfrak{P})|/|G_0(\mathfrak{P}) \cap H| = 1;$$

hence no prime ideal ramifies in M/K, and so M is contained in the maximal subfield of L unramified over K. Conversely, if M is unramified over K and contained in L, then for every prime ideal $\mathfrak{P}$ of R_L w have $G_0(\mathfrak{P}) = G_0(\mathfrak{P}) \cap H_0$, where H_0 is the subgroup of G corresponding to L; thus $G_0(\mathfrak{P}) \subset H_0$ and $H \subset H_0$, showing that M lies in the field corresponding to H.

If $K = Q$, then by Corollary 3 to Theorem 4.8 $H = G$. □

This lemma implies that the Galois group of $K_{q^m,S}$ is generated by its inertia subgroups and, as it is Abelian, its order cannot exceed the product of the orders of those subgroups which are at most equal to $e_1, \ldots, e_r$ respectively. Hence the degree of $K_{q^m,S}$ over Q does not exceed $e_1 \ldots e_r$, and we have already noticed that this implies (*).

Thus everything has been reduced to the proof of Lemma 6.4, which we will now give.

Proof of Lemma 6.4. The inclusion $WZ \subset L$ is obvious, and hence it suffices to show the equality of the degrees (over Q_p) of the two fields involved. We have to consider separately the cases of $p \neq q$, $p = q \neq 2$ and $p = q = 2$.

The first case: $p \neq q$. The Galois group L/Q_p is a subgroup of some power of the cyclic group of order q^m, and hence it can contain no cyclic subgroup except those with orders dividing q^m. This shows that the maximal unramified subfield of L equals W since it has to be cyclic by Corollary 2 to Theorem 5.8, and so its degree divides q^m and W is unramified by definition. It follows that L/W is fully ramified of degree $e(L/Q_p)$, and as it is also tamely ramified (since p does not divide its degree, which is a power of q), Theorem 5.12 shows that $G_1(L/W) = 1$ and L/W must be cyclic. However, its Galois group is a subgroup of G and so we get

$$e(L/Q_p) = [L:W] \mid q^m. \tag{6.7}$$

Now denote by π any generator of the prime ideal of L which generates the ring R_L as an R_W-module whose existence follows from Theorem 5.10. Moreover, let s be the generator of the group $G_0(L/Q_p)$ (which is equal to the Galois group of L/W) and let t be that automorphism of G which induces the map $x \mapsto x^p$ in the residue class field. The existence of t follows from Proposition 5.14. We can write $s(\pi) \equiv a\pi \pmod{\pi^2}$ for a certain unit a of L. Note that $s^k(\pi) \equiv a^k\pi \pmod{\pi^2}$ holds for every k and Lemma 5.5 shows that from $s^k(\pi) \equiv \pi \pmod{\pi^2}$ follows the divisibility of k by $e(L/W) = e(L/Q_p)$; hence a congruence of the form $a^k \equiv 1 \pmod{\pi}$ can hold only if $e(L/Q_p)$ divides k. But we have

$$a^p t(\pi) \equiv t(a\pi) \equiv t(s(\pi)) \equiv s(t(\pi)) \pmod{\pi^2}$$

since our extension is Abelian. If $t(\pi) \equiv b\pi \pmod{\pi^2}$, then $s(t(\pi)) \equiv s(b)a\pi \equiv ba\pi \equiv at(\pi) \pmod{\pi^2}$, i.e.

$$a^{p-1}t(\pi) \equiv t(\pi) \pmod{\pi^2}$$

and $a^{p-1} \equiv 1 \pmod{\pi}$, giving

$$e(L/Q_p) \mid p-1. \tag{6.8}$$

From (6.7) and (6.8) we get $e(L/Q^p) \mid (p-1, q^m)$. Since the polynomial $(x^p - 1)/(x - 1)$ is irreducible over Q_p, the extension ZQ_p/Q_p is of degree

$(p-1, q^m)$, and since $ZQ_p \cap W = Q_p$ (the first field being fully ramified and the second unramified), we see that ZW/Q_p is of degree

$$f(K/Q_p)(p-1, q^m) \geqslant f(K/Q_p)e(K/Q_p) = [K:Q_p],$$

and thus $K = ZW$ as asserted.

The second case: $p = q \neq 2$. Here the Galois group of the composite ZW equals $C_{p^m} \times C_{p^m}$; hence we only have to show that the Galois group G of K/Q_p has at most two independent generators. To do this it suffices to prove that the Galois group of $K_{p,p}/Q_p$ has at most two independent generators. Indeed, assume this fact and let r be the number of independent generators of G. Then there are subfields $K_1, \ldots, K_r$ of K which are independent in the following sense: for each $i = 1, 2, \ldots, r$ the intersection of K_i with the field generated by the remaining K_i's equals Q_p. But selecting in each such K_i a subfield of degree p over Q_p, we obtain r independent subfields of $K_{p,p}$, and thus $r \leqslant 2$. Hence we have to prove $[K_{p,p}:Q_p] \leqslant p^2$. We first need a simple lemma:

Lemma 6.6. *Let $M = Q_p(\zeta_p)$ and let s be a generating element of the Galois group of M/Q_p. Define k by $s: \zeta_p \to \zeta_p^k$ and let a be any non-zero element of M. Then the extension $M(a^{1/p})/Q_p$ is Abelian iff there exists an element c in M such that $s(a) = a^k c^p$.*

Proof. Assume first that $M(a^{1/p})/Q_p$ is Abelian and lift s to an automorphism of this extension. If $b = s(a)$, then $b^{1/p}$ lies in $M(a^{1/p})$, and so, g being the generator of the cyclic extension $M(a^{1/p})/M$, we can write $g(a^{1/p}) = \zeta_p a^{1/p}$ and $g(b^{1/p}) = \zeta_p^r b^{1/p}$. This shows that the element $a^{r/p} b^{-1/p}$ is invariant under g and hence belongs to M, i.e., we have $b^{1/p} = c a^{r/p}$ with a suitable c in M. This in turn implies $b = a^r c^p$, and if we now define two automorphisms g_1 and g_2 of $M(a^{1/p})/Q_p$ by assuming g_1 to be equal to the identity on M and $g_1(a^{1/p}) = \zeta_p a^{1/p}$, whereas $g_2|_M = s$ and $g_2(a^{1/p}) = c a^{r/p}$, then owing to the abelianity of our extension we get $g_1 g_2 = g_2 g_1$, and finally we obtain

$$c \zeta_p^r a^{r/p} = g_1(c a^{r/p}) = g_1 g_2(a^{1/p})$$
$$= g_2 g_1(a^{1/p}) = g_2(\zeta_p a^{1/p}) = \zeta_p^k c a^{r/p},$$

which leads us to $k = r$, and proves the "only if" part of the lemma. Conversely, if a satisfies $s(a) = a^k c^p$ with $c \in M$, then we immediately see that $M(a^{1/p})/Q_p$ is normal and, repeating our previous argument in the reversed order, we obtain its abelianity. □

(Note that the lemma and its proof hold for all fields whose characteristic is not divisible by p; however, we shall have no opportunity to use this.)

Now if L/Q_p is a cyclic extension of degree p, then for $M = Q_p(\zeta_p)$ the extension ML/M is also cyclic, and we can write $ML = M(a^{1/p})$ for a certain a in M. The previous lemma shows that every such extension is determined by an element of H/H^p, where H is the group of non-zero elements a of M for which $s(a) = a^k c^p$ holds, the notation being the same as in the lemma.

Now H/H^p can be regarded as a vector-space over $GF(p)$, and if the fields $L_1, \ldots, L_j$ are independent in the sense defined above, then the corresponding elements of H/H^p are also linearly independent. In fact, otherwise we could find independent fields $L_1, \ldots, L_j$ such that with $L_i M = M(a_i^{1/p})$ the images x_i of a_i in H/H^p would satisfy $x_1 = x_2^{n_2} \ldots x_j^{n_j}$. This would imply $a_1 = a_2^{n_2} \ldots a_j^{n_j} b^p$ with a suitable b, so $ML_1 \subset ML_2 \ldots L_j$. But the independence of the L_i's gives $[L_1 \ldots L_j : Q_p] = p^j$ and $[L_2 \ldots L_j : Q_p] = p^{j-1}$, which leads us to $[ML_1 \ldots L_j : Q_p] = (p-1)p^j$ and $[ML_2 \ldots L_j : Q_p] = (p-1)p^{j-1}$; thus ML_1 cannot be contained in $ML_2 \ldots L_j$.

Observe that every class of H/H^p has a representative in $U_1(M)$. Indeed, if n is the exponent of the field M, $a \in H$ and $a = a_1 \pi^t$, where $a_1 \in U(M)$ and π is a generator of the prime ideal $\mathfrak{P}$ of M, which may be chosen equal to $1 - \zeta_p$, then we have $kt + pn(c) = n(s(a)) = n(a) = t$; thus $(k-1)t$ is divisible by p. Since p does not divide $k-1$, it has to divide t, whence $a^{1/p}$ and $a_1^{1/p}$ determine the same extension of M and so lie in the same coset of H (mod H^p). This shows that every class of H/H^p has a representative in $U(M)$, but since $U(M)$ is the direct product of $U(M)/U_1(M)$ and $U_1(M)$ and the first factor is of order not divisible by p, every element in it is a pth power, and so we find that in every class of H/H^p there are indeed elements of $U_1(K)$ as asserted.

Now we show that $U_1(M)^p$ coincides with $U_{1+p}(M)$. Since in our case $e(M/Q_p) = p-1$, the Corollary to Lemma 5.1 shows that every element of $U_{1+p}(M)$ is a pth power, and so it remains to show that every pth power of a principal unit lies in $U_{1+p}(M)$. To obtain this, write $x = 1 + a(1 - \zeta_p) + b(1 - \zeta_p)^2$, a being a rational integer from the interval $[0, p-1]$ and $b \in M$ being an integer, and note that

$$x^p = (1 + a(1-\zeta_p) + b(1-\zeta_p)^2)^p$$
$$= \sum_{i_1 + i_2 + i_3 = p} p!/(i_1! i_2! i_3!) a^{i_2} b^{i_3} (1-\zeta_p)^{i_2 + 2i_3}.$$

The only terms of this expansion which do not lie in $\mathfrak{P}^{1+p}$ are 1, $pa(1-\zeta_p)$ and $a^p(1-\zeta_p)^p$; hence

$$x^p \equiv 1 + pa(1-\zeta_p) + a^p(1-\zeta_p)^p \pmod{\mathfrak{P}^{1+p}}.$$

But $p = \varepsilon(1-\zeta_p)^{p-1}$, where ε is a unit, equal to

$$\prod_{k=1}^{p-1} (1 + \zeta_p + \ldots + \zeta_p^{k-1})$$

(cf. the proof of Theorem 4.16), and since $\zeta_p \equiv 1(\mathrm{mod}(1-\zeta_p))$, we infer that ε is congruent to $(p-1)!$ and thus to $-1(\mathrm{mod}(1-\zeta_p))$ by Wilson's theorem. This shows finally that $pa(1-\zeta_p)+a^p(1-\zeta_p)^p$ lies in $\mathfrak{P}^{1+p}$ and so $x^p \in U_{1+p}$. The result obtained shows that $|H/H^p|$ does not exceed the number of those residue classes $a \pmod{\mathfrak{P}^{1+p}}$ for which $s(a) \equiv a^k \pmod{\mathfrak{P}^{1+p}}$. Denote by H_1 the group of those residues. Obviously ζ_p belongs to H_1 and so does $1+(1-\zeta_p)^p = A$. Indeed, we have $A^k \equiv 1+k(1-\zeta_p)^p \pmod{\mathfrak{P}^{1+p}}$ and, on the other hand, $s(A) = 1+(1-\zeta_p^k)^p = 1+(1-\zeta_p)^p(1+\zeta_p+\ldots+\zeta_p^{k-1}) \equiv 1+k(1-\zeta_p)^p \pmod{\mathfrak{P}^{1+p}}$; thus $A \in H_1$. It remains to prove that ζ_p and A generate the whole group H_1 since then $|H_1|$ cannot exceed p^2.

First let $a \in H_1 \cap U_2$ ($a \neq 1$) so that

$$a \equiv 1+a_1(1-\zeta_p)^i \pmod{\mathfrak{P}^{1+i}}$$

with a rational integer a_1 not divisible by p and $p \geq i \geq 2$. Then

$$s(a) \equiv 1+a_1 s((1-\zeta_p)^i) \equiv 1+a_1 k^i(1-\zeta_p)^i \pmod{\mathfrak{P}^{1+p}},$$

but, on the other hand,

$$s(a) \equiv a^k \equiv 1+a_1 k(1-\zeta_p)^i \pmod{\mathfrak{P}^{1+p}},$$

showing that $k^{i-1} \equiv 1 \pmod{\mathfrak{P}}$; hence $i = p$ because k is a primitive root $(\mathrm{mod}\, p)$. Thus $H_1 \cap U_2 \subset H_1 \cap U_p$. Moreover, for any $a \in H_1 \setminus U_2$, say $a \equiv 1+b(1-\zeta_p) \pmod{\mathfrak{P}^2}$, one easily obtains $a\zeta_p^b \in U_2 \cap H_1$, and so every $a \in H_1$ can be written in the form

$$a \equiv \zeta_p^b(1+c(1-\zeta_p)^p) \pmod{\mathfrak{P}^{1+p}},$$

and since $A^c = (1+(1-\zeta_p)^p)^c \equiv 1+c(1-\zeta_p)^p \pmod{\mathfrak{P}^{1+p}}$, we obtain $a \equiv \zeta_p^b A^c \pmod{\mathfrak{P}^{1+p}}$, and thus indeed H_1 is generated by ζ_p and A. As already observed, this settles the case under consideration.

The third case: $p = q = 2$. In this case $Q_2 Z/Q_2$ has the group $C_2 \times C_{2^m}$ for its Galois group, and so the composite WZ has its Galois group over Q_2 equal to $C_2 \times C_{2^m} \times C_{2^m}$. To obtain the equality $WZ = L$ we have to show that the Galois group of L/Q_2 has three generators at least one of which is of order two. As in the previous case, the first assertion reduces to $m = 1$, and so we have to prove that the Galois group of $L_{2,2}$ has three generators. But we know that Q_2 has 7 quadratic extensions (cf. Proposition 5.12) generated by the square roots of $-1, 2, -2, 5, -5, 10$ and -10, and one can immediately see that the fields corresponding to $-1, 2$ and 5 form a maximal independent system. This proves that the group of $L_{2,2}$ has three generators.

Finally, if the group of $L_{2^m, 2}$ had all generators of orders at least equal to 4, then the group of $L_{2^2, 2}$ would be equal to $C_4 \times C_4 \times C_4$, and so every quadratic extension of Q_2 would be contained in a cyclic quartic extension. But this fails already for the field $Q_2(i)$. Indeed, assume that it is contained in a field

$K = Q_2(i, (a+bi)^{1/2})$ whose Galois group over Q_2 equals C_4. Since K is normal, it is generated over $Q_2(i)$ also by $(a-bi)^{1/2}$, and so with suitable A, B in Q_2 we have

$$a - bi = (a+bi)(A+Bi)^2.$$

Observe that by putting $s((a+bi)^{1/2}) = (a+bi)^{1/2}(A+Bi)$, $s(i) = -i$ we define an automorphism of K/Q_2 whose order equals four and so it generates the group of K/Q_2. Moreover, s^2 is the only non-trivial element of the Galois group of $K/Q_2(i)$ and thus takes $(a+bi)^{1/2}$ into $-(a+bi)^{1/2}$; hence we get

$$-(a+bi)^{1/2} = s^2((a+bi)^{1/2}) = s((a+bi)^{1/2}(A-Bi))$$
$$= (A^2 + B^2)(a+bi)^{1/2}$$

and $-1 = A^2 + B^2$. But the latter equality is impossible in Q_2, and so we get a contradiction. This proves Lemma 6.4 in all cases. □

§ 2. Adeles and Ideles

1. In this section K will be an algebraic number field and V the set of all its non-equivalent valuations, Archimedean and non-Archimedean. For any v in V denote by K_v the completion of K under v and put $R_v = K_v$ in the case of Archimedean v, and R_v equal to the ring of integers of K_v in the case of non-Archimedean v. The restricted direct product of the additive groups K_v^+ with respect to the subgroups R_v^+ is called the *adele group* of K, and its elements are called *adeles* of K. Incidentally, this group also has a ring structure induced by multiplication in the factors. The resulting ring is called the *adele ring* of K and will be denoted by A_K. Since the continuity of multiplication is obvious, we conclude by Lemma 1 of Appendix I that A_K is a locally compact topological ring. For any finite subset $S \subset V$ denote by A_S the group $\prod_{v \in S} K_v^+ \times \prod_{v \notin S} R_v^+$. From the definition of the restricted product it follows that the topology induced in A_S by the topology in the adele ring coincides with the product topology and that every group A_S is open in A_K. Obviously the union of all groups A_S exhausts the full adele group.

The invertible elements of A_K are called *ideles* and their set—the *idele group* of K. However, it is not convenient to give that group the topology induced by the topology of A_K and we shall endow it with another restricted product topology. To do this observe that in the multiplicative group K_v^* we can select for almost all v (which means: all except finitely many v's) the subgroup $U(K_v)$ of units, which is compact and open, and it can immediately be seen that the elements of the restricted product of the K_v^*'s with respect to $U(K_v)$ are exactly the ideles of K. The group of ideles with the topology given by this restricted product will be denoted by I_K, and only this topology on the idele group will be used in the sequel. If we put for a finite set $S \subset V$

$$I_S = \prod_{v \in S} K_v^* \times \prod_{v \notin S} U(K_v),$$

then again the I_S get the product topology and are open in I_K. One particular selection of S is of importance: S equal to the set of all Archimedean valuations of K. In this case we denote I_S by U_K and call its elements *unit ideles*.

In Chapter 3 we considered a particular choice of the system V of non-equivalent valuations of K, and with this system we shall now work constantly. We recall that for $v \in V$ we have $v(x) = N(\mathfrak{p})^{-n_\mathfrak{p}(x)}$ if v is p-adic and corresponds to the prime ideal $\mathfrak{p}$ of R_K with exponent $n_\mathfrak{p}(x)$ and $v(x) = |F_i(x)|^{\varepsilon_i}$ for Archimedean v, where F_i is the embedding of K into the complex field corresponding to v, and ε_i equals unity if $F_i(K)$ is real and equals 2 otherwise.

It is convenient to arrange the embeddings F_i (considering only one from each pair of conjugated complex embeddings) so that the first r_1 of them are real and the remaining ones are complex. Later on we shall use this arrangement without further comments.

Observe now that for every x in K we have $x \in R_v$ for almost all v and also, for every non-zero $x \in K$, we have $x \in U(K_v)$ for almost all v. This allows us to define a canonical embedding i of K in A_K and of K^* in I_K by means of $i(x) = (x_v)$, where $x_v = x$ for non-Archimedean v and $x_v = F_i(x)$ if v is Archimedean and corresponds to the embedding F_i. Put $i(K) = A_0$, $i(K^*) = I_0$ and call these images the *ring of principal adeles* and the *group of principal ideles*, respectively.

For any idele $x = (x_v)$ define its volume $V(x)$ by

$$V(x) = \prod_{v \in V} v(x_v).$$

This infinite product is convergent since for every x it contains only finitely many terms distinct from 1, and so this definition makes sense. Theorem 3.3 implies that the volume of any principal idele equals unity; hence I_0 lies in the kernel J_K of the map $V: I_K \to R_+^*$ (the multiplicative group of positive real numbers). This simple fact implies the following

Proposition 6.11. *The ring A_0 is a discrete subring of A_K.*

Proof. It suffices to find a neighbourhood of zero in A_K which does not contain non-zero principal adeles. Let us take as that neighbourhood the set $U = \{(a_v): v(a_v) \leqslant 1 \text{ for all } v, v(a_v) < 1 \text{ for Archimedean } v\}$. If a non-zero principal adele $i(a)$ lies in U, then it is an idele and its volume is smaller than 1, which is impossible. □

A similar result holds also for ideles:

Proposition 6.12. *The group I_0 is a discrete subgroup of I_K.*

Proof. Let U be the neighbourhood of unity in I_K consisting of all those ideles (a_v) which for Archimedean v satisfy $v(a_v - 1) < 1$ and lie in the open subgroup U_K. Now observe that a principal idele lies in U_K iff it is a unit of K. Thus, if we had a principal idele in U, say $i(a)$, then a would be a unit of K and so $a - 1$ would be an integer. However, its norm equals

$$|N_{K/Q}(a-1)| = \prod_{v \text{ Archimedean}} v(a_v - 1) < 1$$

in absolute value, which is possible only for $a = 1$. □

The factor group A_K^+/A_0^+ is called the *adele class group*, and similarly the group I_K/I_0 is called the *idele class group*. We shall denote the latter group by $C(K)$. About these groups we now prove

Proposition 6.13. *The adele class group is compact whereas the idele class group is not.*

Proof. Let (a_v) be an arbitrary adele. We shall prove the existence of a compact set B in A_K^+ independent of (a_v) and such that in the class mod A_0^+ determined by (a_v) there is an adele from B. First choose a rational integer m so that for all v we have $ma_v \in R_v$. Denote by P the set of all prime ideals of R_K which divide mZ, and for any $\mathfrak{p} \in P$ let $n_\mathfrak{p}$ denote the corresponding exponent. Put $N = \max_{\mathfrak{p} \in P} n_\mathfrak{p}(m)$ and let $x \in R_K$ be a solution of the system

$$ma_{v_\mathfrak{p}} \equiv x \pmod{\mathfrak{p}^N} \quad (\mathfrak{p} \in P)$$

of congruences. For $b_v = a_v - x/m$ we have, for any non-Archimedean v, the inequality

$$v(b_v) = v(ma_v - x)/v(m) \leq 1,$$

and so the idele (b_v) which lies in the same class as (a_v) has all its coordinates integral. Now let $v_1, \ldots, v_r$ form the set of all Archimedean valuations of K (with $r = r_1 + r_2$) and observe that since the points $[F_1(t), \ldots, F_r(t)]$ form an r-dimensional lattice in the Euclidean r-space as t runs over R_K, we can select a suitable t_0 in R_K so that the point $[b_{v_1} - F_1(t_0), \ldots, b_{v_r} - F_r(t_0)]$ lies in a compact set independent of (a_v). But this gives an evaluation $v(b_v - t_0) < C$ for a certain positive C; so the idele $(b_v - t_0)$ lies in a compact set independent of (a_v), and of course this idele is in the same class as (a_v). This proves the first assertion; to prove the second it suffices to observe that $V(x)$, the volume of x, is an unbounded and continuous function on the idele class-group, and so this group cannot be compact. □

2. There is a canonical homomorphism of the idele group I_K onto the group of all fractional ideals of K, which we denote by $G(K)$. To define it consider an arbitrary idele $(a_v) = a$, and, for any non-Archimedean valuation v, let the corresponding exponent be denoted by n_v. If we now put

$$f(a) = \prod_v \mathfrak{p}_v^{n_v(a_v)},$$

where $\mathfrak{p}_v$ denotes the prime ideal of K_v, then the resulting mapping $f: I_K \to G(K)$ is clearly a surjective homomorphism. Its kernel equals U_K, and thus I_K/U_K is isomorphic to $G(K)$. One can also easily see that f is continuous.

Similarly, the group $H(K)$ of ideal classes in K is isomorphic to the quotient $I_K/U_K I_0$. To prove this, consider the map $g: G(K) \to H(K)$, which attributes to every ideal the class to which it belongs. Then $g \circ f$ maps I_K onto $H(K)$ and the kernel of this map equals $\{x: f(x) \text{ is principal}\} = \{x: f(x) \in f(I_0)\} = \{x: x \in I_0 \cdot \text{Ker} f\} = I_0 U_K$.

We now generalize these results and prove:

Proposition 6.14. *Let $\mathfrak{f}$ be any integral ideal in R_K and denote by $I_\mathfrak{f}$ and $I_\mathfrak{f}^*$ the subgroups of I_K consisting, respectively, of all principal ideles (a) such that a is congruent to unity (mod $\mathfrak{f}$) and of all principal ideles (a) such that a is congruent to unity (mod $\mathfrak{f}$) and totally positive. Then*

$$H_\mathfrak{f}(K) \simeq I_K/I_\mathfrak{f} U_K \quad \text{and} \quad H_\mathfrak{f}^*(K) \simeq I_K/I_\mathfrak{f}^* U_K.$$

Proof. Let g be the canonical map of $G(K)$ onto $H_\mathfrak{f}(K)$ and onto $H_\mathfrak{f}^*(K)$ and consider $g \circ f$. It maps I_K onto the group of ideal-classes under consideration and its kernel equals $I_\mathfrak{f} U_K$ and $I_\mathfrak{f}^* U_K$, respectively. □

Corollary. *Every open subgroup of the idele class-group $C(K)$ is of finite index; in other words, every discrete continuous image of $C(K)$ is finite.*

Proof. Let G be an open subgroup of $C(K)$ and let H be that subgroup of I_K which is mapped on G by the canonical map. Obviously H is open and so it contains a neighbourhood of unity of the form

$$\{(x_v): v(x_v - 1) < \varepsilon \text{ for Archimedean } v, x_v \in U(K_v) \text{ for non-Archimedean } v\}$$

and also the subgroup generated by it, which turns out to be U_K since it has a non-void interior. So it suffices to show that the image of U_K has a finite index in $C(K)$. But this follows from Proposition 6.14 and the finiteness of the class-number. □

If we did not know that the groups $H_f(K)$ and $H_f^*(K)$ are finite and gave them the topology induced by the resulting isomorphisms, then it would follow that they are discrete since U_K is open. There is a way of proving their finiteness by showing that they are also compact, since every topological space which is compact and discrete is necessarily finite. Cf. Cassels, Fröhlich [67]. Now we prove a result of importance in the class-field theory:

Theorem 6.6. *The quotient group J_K/I_0 is compact.*

Proof. We shall use both the finiteness of the ideal class-group $H(K)$ and Dirichlet's theorem on the structure of $U(K)$, i.e. the group of units of K. Let $A_1, \ldots, A_N$ be ideals from $G(K)$ representing the various classes from $H(K)$. Then every ideal A of $G(K)$ can be written as $A = aA_i$ with a non-zero a in K and a suitable $1 \leq i \leq N$. Now let $a = (a_v)$ be any idele from J_K, i.e. an idele having unit volume $V(a)$, and let $A = f(a)$ be the corresponding ideal. Moreover, fix the ideles $b_1, \ldots, b_N$ from J so that $f(b_i) = A_i$ for $i = 1, 2, \ldots, N$. Such ideles exist since the equation $f(b) = A_i$ restricts only the non-Archimedean coordinates of b and we can adjust the Archimedean coordinates so that $V(b) = 1$ holds. Since $A = cA_i$ for some i and $c \in K^*$, we obtain $f(ac^{-1}b_i^{-1}) = 1$; thus the idele $ac^{-1}b_i^{-1}$ lies in $J_K \cap U_K$.

Now let $v_1, \ldots, v_{r+1}$ be all the Archimedean valuations of K and define a homomorphism of $U_K \cap J_K$ in the Euclidean r-space R^r by means of

$$L((u_v)) = [\log v_1(u_1), \ldots, \log v_r(u_r)].$$

This gives actually an epimorphism since, given $[t_1, \ldots, t_r] \in R^r$, we can easily find $u_1, \ldots, u_r$ so that $v_i(u_i) = \exp t_i$ holds for $i = 1, 2, \ldots, r$, and then assuming $v_{r+1}(u_{r+1}) = \exp(-(t_1 + \ldots + t_r))$ and putting $u_v = 1$ for non-Archimedean v, we get an idele $u = (u_v)$ in $J_K \cap U_K$ with $L(u) = [t_1, \ldots, t_r]$.

From Theorem 3.6 (cf. its proof) we infer that the images under L of the fundamental units $\varepsilon_1, \ldots, \varepsilon_r$ of K are linearly independent and thus span the full space. Denote by P the set of all ideles in $J_K \cap U_K$ whose images in R^r can be written in the form $x_1 L(\varepsilon_1) + \ldots + x_r L(\varepsilon_r)$ where $0 \leq x_i \leq 1$ ($i = 1, 2, \ldots, r$). This set is obviously compact and, moreover, for every idele u in $J_K \cap U_K$ we can find a suitable unit ε in K so that the product εu gets into P. Applying this observation to the idele $ac^{-1}b_i^{-1}$, we see that a can be written as $a = \varepsilon c b_i a_1$ with a_1 in P, and thus every idele from J_K has in its class (mod I_0) an idele from $\bigcup_{i=1}^N b_i P$, which is a fixed compact set. But this proves the compactness of J_K/I_0. □

Corollary. *If $D(K)$ is the connected component of the unit element of $C(K)$, then the group $C(K)/D(K)$ is compact and totally disconnected.*

Proof. The second assertion is trivial and to prove the first it suffices to show that every coset of $C(K)$ (mod $D(K)$) contains an element of J_K/I_0. Take any idele (x_v) and let t be its volume. Since the group of ideles (a_v) with $a_v = 1$ for non-Archimedean v is connected, the class determined by the idele (y_v) with $V((y_v)) = t$, $y_v = 1$ for non-Archimedean v lies in $D(K)$; hence the classes determined by (x_v) and $(x_v y_v^{-1})$ are in the same coset (mod $D(K)$), but it is obvious that $(x_v y_v^{-1})$ determines a class in J_K/I_0. □

3. The general theory of restricted direct products (see Appendix I) gives a method of constructing a standard Haar measure in the group of adeles and ideles once the Haar measures in K_v^+ and K_v^* have been fixed. We recall that in the case of non-Archimedean v we fix that Haar measure dx_v on K_v^+ which gives to the ring of integers R_v the measure $N(D_v)^{-1/2}$, where D_v denotes the different of K_v/Q_p, p being the characteristic of the residue class field of K_v. On the multiplicative group K_v^* we choose the measure

$$dx_v^* = N(P)(N(P)-1)^{-1} v^{-1}(x) dx_v,$$

$P = P_v$ being the prime ideal in R_v. This measure gives to the group of units the measure $N(D_v)^{-1/2}$. In the case of Archimedean v we proceed as follows: if v is real, i.e. if K_v equals the field of real numbers, then we take for dx_v the usual Lebesgue measure and for dx_v^* the usual Haar measure equal to $dx_v/v(x) = dx_v/|x|$. However, if v is complex, then dx_v will be the double Lebesgue plane measure and $dx_v^* = dx_v/v(x) = dx_v/|x|^2$. Note that these Haar measures dx_v are self-dual provided that in the Archimedean case we made the slightly artificial identification between K_v^+ and its dual shown in the appendix.

The resulting Haar measures in A_K and I_K will be denoted by m_A and m_I, respectively. Note that the first of them is self-dual. Observe also that for every idele a we have $dm_A(ax) = V(a) dm_A(x)$, which results from the corresponding result for the measures dx_v, which in turn is easily established through Corollary 2 to Lemma 5.7.

Now we are going to prove the strong approximation theorem. Let us first establish an auxiliary result:

Proposition 6.15. *There exists a positive constant C such that if (a_v) is an idele with volume exceeding C, then one can find a principal idele (b_v) such that for all v we have $v(b_v) \leqslant v(a_v)$.*

Proof. We need a lemma from the theory of topological groups:

Lemma 6.7. *Let G be a locally compact Abelian group and H its countable discrete subgroup. Moreover, assume that the quotient group G/H is compact. If A is an open subset of G with a sufficiently large measure, then there are elements x_1, x_2 in A lying in the same coset with respect to H.*

Proof. Let U be any neighbourhood of the unit element of G which satisfies $U \cap H = \{e\}$ and has a finite measure. By the compactness of G/H we can find suitable elements $x_1, \ldots, x_r$ of G such that with the notation φ for the canonical map $G \to G/H$ we have

$$G/H = \varphi\left(\bigcup_{i=1}^{r} x_i U\right).$$

Put $V = \bigcup_{i=1}^{r} x_i U$, let B be the measure of V and let A be any open subset of G all elements of which lie in distinct cosets. Then the set $\varphi^{-1}(\varphi(A)) \cap V = A_1$ is open because φ is open and continuous, and its measure does not exceed B. Every element of A_1 can be uniquely put in the form ah, where $a \in A$, $h \in H$. Write $X_h = \{a \in A: ah \in A_1\}$. Those sets are all open as intersections of $A_1 h$ with A, and we have also $A = \bigcup_h X_h$, $A_1 = \bigcup_h hX_h$, both sums being disjoint. This gives the equality of measures of A and A_1, and so the measure of A does not exceed a fixed constant. □

(Note that if H is not discrete, then every non-void open set contains elements from the same coset regardless of the remaining assumptions.)

Now consider the set $X = \{(x_v): v(x_v) < \frac{1}{2}v(a_v)$ for Archimedean v and $v(x_v) \leq v(a_v)$ for non-Archimedean $v\}$, which is open in the adele group of K. Let us compute its measure.

Write S for the set consisting of all Archimedean valuations of K, all non-Archimedean valuations v satisfying $v(a_v) > 1$ and all non-Archimedean valuations v such that K_v is ramified over the completion of Q. Then $X \subset A_S$ and the definition of our measure gives

$$m_A(X) = \prod_{v \in S} \int_{X_v} dx_v,$$

where X_v is the projection of X on K_v. But for real v we see that X_v has the measure $v(a_v)$ and for complex v the measure $\pi v(a_v)$, whereas for non-Archimedean v it equals $v(a_v)N(D_v)^{-1/2}$; thus finally

$$m_A(X) = \pi^{r_2} V((a_v)) |d(K)|^{-1/2}$$

by Propositions 6.9 and 4.7.

In view of Propositions 6.11 and 6.13 we may now apply Lemma 6.7 and find that if the volume of the adele (a_v) is large enough there must be two distinct adeles in X whose difference is a principal adele, say (b_v). But obviously this adele satisfies $v(b_v) \leq v(a_v)$ for all v's, and since it is non-zero, it is an idele. □

This proposition allows us to prove the strong approximation theorem:

Theorem 6.7. *Let v_0 be any fixed valuation from V. Then for every finite set $V_1 \subset V$ not containing v_0, every system α_v ($v \in V_1$) of elements of K_v and every positive ε there exists a principal adele (b_v) such that*

$$v(b_v - \alpha_v) < \varepsilon \quad \text{for } v \in V_1$$

and

$$v(b_v) \leq 1 \quad \text{for } v \notin V_1, v \neq v_0.$$

(A more sophisticated formulation of this theorem reads as follows: the restricted direct product of the additive groups K_v^+ ($v \neq v_0$) with respect to R_v^+ is dense in the adele group.)

Proof. Let B be a compact set in A_K^+, whose existence was demonstrated in Proposition 6.13, such that every class of adeles has in B its representative. As every compact subset of the adele group, it must be contained in a set of the shape $\{(x_v): v(x_v) \leq c_v\}$ with suitable c_v almost all of which equal unity. Now take a non-zero principal adele (d_v) so that

$$v(d_v) \leq \varepsilon/c_v \quad \text{for } v \in V_1,$$
$$v(d_v) \leq 1/c_v \quad \text{for } v \notin V_1, v \neq v_0,$$

which is possible by Proposition 6.15. (Choose a_v with $0 < v(a_v) \leq \varepsilon/c_v$ for $v \in V_1$, $0 < v(a_v) \leq 1/c_v$ for $v \notin V_1$, $v \neq v_0$ and finally with $v_0(a_{v_0})$ sufficiently large so that the volume $V((a_v))$ will exceed C.) Now define α_v also for $v \notin V_1$ simply by putting $\alpha_v = 0$. We can write

$$(\alpha_v d_v^{-1}) = (\beta_v) + (\gamma_v),$$

where (β_v) is in B and (γ_v) is principal, and thus

$$v(\alpha_v - d_v \gamma_v) = v(d_v \beta_v)$$

and it easily follows that the adele $(d_v \gamma_v)$ satisfies our assertion. □

4. Now we present the proof of a theorem of Hecke concerning the discriminant $d(L/K)$. To prove it we have to introduce a new kind of discriminant of a finite extension of an algebraic number field based on the discriminants $\partial(L_\mathfrak{P}/K_\mathfrak{p})$ defined in Chapter 5, § 1.5, for the extension of p-adic fields. First we prove a simple

Lemma 6.8. *The factor-group I_K/U_K^2 is algebraically (not topologically) isomorphic with that subgroup of the product $\prod_v K_v^*/U(K_v)^2$ which consists of all elements (x_v) satisfying $x_v \in U(K_v)/U(K_v)^2$ for almost all v's.*

Proof. The embedding of the idele group I_K into the product of all groups K_v^* obviously induces such an isomorphism. □

Now we shall define the discriminant $\partial(L/K)$, which will be an element of the factor group I_K/U_K^2. Let $v \in V$ and let $w_1, \ldots, w_m$ form the full system of non-equivalent valuations of L which are normalized and whose restriction to K is equivalent to v. If V' is the set of all valuations of L obtained in this way, then Theorem 3.2 shows that V' coincides with the set of all normalized non-equivalent valuations of L. We define

$$\partial(L/K) = (\partial_v(L/K)) \in \prod_v K_v^*/U(K_v)^2,$$

where the components $\partial_v(L/K)$ are defined as follows:

First look at the Archimedean v. If K_v is the complex field, then we simply put $\partial_v(L/K) = 1$ and if K_v is the real field and r is the number of complex fields among the K_{w_i} ($i = 1, 2, \ldots, m$), then $\partial_v(L/K)$ equals $(-1)^r$. For non-Archimedean v we put

$$\partial_v(L/K) = \prod_{i=1}^m \partial(L_{w_i}/K_v).$$

Proposition 5.7 (ii) shows that for almost all v the component $\partial_v(L/K)$ lies in $U(K_v)/U(K_v)^2$, and so by Lemma 6.8 we can regard our discriminant as an element of I_K/U_K^2.

The principal properties of the discriminant $\partial(L/K)$ are given in the following

Proposition 6.16. (i) *If g is the homomorphism of I_K/U_K^2 into $I_K/U_K = G(K)$ induced by the embedding of U_K^2 into U_K, then $g(\partial(L/K))$ equals $d(L/K)$, i.e. the usual discriminant of the extension L/K.*

(ii) *The extension L/K is unramified iff $\partial(L/K)$ lies in U_K/U_K^2.*

(iii) *If $K \subset L \subset M$, then*

$$\partial(M/K) = \partial(L/K)^{[M:L]} N_{L/K}(\partial(M/L)), \tag{6.9}$$

where $N_{L/K}: I_L/U_L^2 \to I_K/U_K^2$ is the mapping induced by the norm maps

$$\prod_w N_{L_w/K_v}$$

(where w ranges over all extensions of v to L) at all components v.

Proof. (i) results from Proposition 5.7 (i) and the observation that the value of g does not depend on the Archimedean components at all and (ii) follows immediately from (i) and the discriminant theorem (Corollary 2 to Theorem 4.8). Finally, for the non-Archimedean components, (iii) results from Proposition 5.7 (iii), and to obtain the same for Archimedean components we have to examine them more carefully. If K_v is complex, then there is nothing to prove since the v-components of both sides of (6.9) reduce to 1; so assume that K_v is real, which fact we simply express by saying that v is real. Let $w_1, \ldots, w_r$ be the real valuations of L extending v and let $w_{r+1}, \ldots, w_m$ be the complex ones. By Theorem 3.2 they are all determined by the embeddings of L into the complex field which map K onto a fixed subfield of the reals and are all equal on it. This shows that $r + 2(m-r) = [L:K]$. Similarly every real valuation w_i has, say, a_i real and b_i complex extensions to M with $a_i + 2b_i = [M:L]$, and every complex w_i has exactly $[M:L]$ complex extensions to M and of course no real ones. This shows that our valuation v has $b = b_1 + \ldots + b_r + (m-r)[M:L]$ complex extensions to M, and so the v-component of the left-hand side of (6.9) equals $(-1)^b$. Now look at the right-hand side. The v-component of $\partial(L/K)^{[M:L]}$ equals $(-1)^{(m-r)[M:L]}$, and for $i = 1, 2, \ldots, r$ the w_i-component of $\partial(M/L)$ equals $(-1)^{b_i}$, whereas for the remaining i's it equals 1. This shows that the v-component of the norm is $(-1)^{b_1 + \ldots + b_r}$ because the extension L_{w_i}/K_v for real w_i is trivial, and for complex w_i we have to find the norm of 1, which is 1. Finally, we see that the v-components of both sides of (6.9) are equal for all v's. □

Now we can prove

Theorem 6.8. *If L/K is a finite extension of the algebraic number field K and $d(L/K)$ is its discriminant (in the sense of Chapter 4), then its class in $H(K)$ is a square.*

Proof. We prove first that $\partial(L/K)$ always lies in $I_K^2 I_0 / U_K$. The easiest method of doing this requires some simple facts about algebras which we shall now indicate. Let M be a field and let

$$E = \bigoplus_{i=1}^{r} N_i,$$

where each N_i is a finite separable extension of M contained in a fixed algebraic closure of M. For $i = 1, 2, \ldots, r$ let the set A_i form an M-basis for N_i and put

$$A = \bigcup_{i=1}^{r} A_i.$$

Then A is an M-basis for E and we define its discriminant $d(A)$ as the product of the discriminants $d_{N_i/M}(A_i)$ (as defined in Chapter 2, § 2.1). Note that if we start with other bases $B_1, \ldots, B_r$ of $N_1, \ldots, N_r$ and obtain a basis B of E, then the discriminants $d(B)$ and $d(A)$ will differ by a factor which is a square in M in view of Proposition 2.4 (ii).

We now apply this setup to our situation. Proposition 6.1 shows that for every non-Archimedean valuation $v \in V$ we have an isomorphism

$$L \underset{K}{\otimes} K_v \simeq \bigoplus_{i=1}^{m} L_{w_i},$$

where $w_1, \ldots, w_m$ are the distinct extensions of v to L. Now choose in each field L_{w_i} a K_v-basis consisting of elements from L, which can be done since L spans L_{w_i} over K_v, and put them all together to yield a K_v-basis $a_1, \ldots, a_n$ of $L \otimes_K K_v$ over K_v. Its elements lie in L and so their discriminant D belongs to K. Now let (D_v) be any idele representing $\partial(L/K)$. For non-Archimedean v we may write

$$D_v = \prod_{i=1}^{m} D'_{w_i},$$

where D'_{w_i} is a representative of $\partial(L_{w_i}/K_v)$. If we choose in each field L_{w_i} a K_v-basis with discriminant D'_{w_i} and put them together, we obtain a basis of $L \otimes_K K_v$ which has D_v for its discriminant. But this shows that $D_v D^{-1}$ lies in $(K_v^*)^2$ at least for all non-Archimedean v. Thus it remains to show that for real Archimedean v the quotient D_v/D is positive (for complex v $K_v^* = (K_v^*)^2$), but here $D_v = (-1)^s$, where s is the number of complex extensions of v to L, and the argument used in the proof of Proposition 2.8 shows that the sign of D is also equal to $(-1)^s$.

Thus we have obtained the inclusion $\partial(L/K) \in I_K^2 I_0/U_K^2$; hence, by the canonical map of I_K onto $I_K/I_0 U_K = H(K)$, every representative of $\partial(L/K)$ becomes a square, and Proposition 6.16 (i) now shows that $d(L/K)$ lies in a class which is a square. □

The discriminant $\partial(L/K)$ introduced in this subsection gives more information about the extension L/K than the usual discriminant $d(L/K)$, which is an ideal; in fact, $\partial(L/K)$ is the true generalization of the numerical discriminant $d(K)$. In particular, in terms of $\partial(L/K)$ it is possible to describe completely the structure of R_L as an R_K-module and to give a necessary and sufficient condition for the existence of a relative integral basis for L/K. However, in the proofs of those results some auxiliary propositions, which need analytical tools, are used, and so we have to postpone their exposition until Chapter 7. (See Theorem 7.14.)

5. We conclude this section with the proof of the functional equation for the zeta-functions of I_K, which we shall define below after introducing some results on characters and quasicharacters of the adele and idele groups.

We already know that A_K^+ is self-dual because all groups K_v^+ are such, but in the argument which will follow an explicit form of this self-duality will be needed. We state it as

Proposition 6.17. *Let* $X(t) = \prod_v X_v(t_v)$ $(t = (t_v))$ *be a character of* A_K^+ *and write*

$$X_v(t_v) = \begin{cases} \exp(2\pi i \lambda_v(T_{K_v/Q_p}(a_v t_v))), & \text{non-Archimedean } v, \\ \exp(-2\pi i a_v t_v), & \text{Archimedean real } v, \\ \exp(-4\pi i \operatorname{Re}(a_v t_v)), & \text{Archimedean complex } v \end{cases}$$

for suitable $a_v \in K_v$. *Here* λ_v *denotes the function occurring in the canonical form of a character of* K_v. *If* $a = (a_v)$, *then there is a biunique correspondence between* A_K^+ *and its dual given by*

$$a \leftrightarrow X_a(t) = X(at).$$

Proof. Clear. □

Corollary. *For any two adeles* a *and* b *we have* $X_{ab}(t) = X_a(bt)$. □

Let us now recall that any continuous homomorphism of I_K into the multiplicative group of the complex field is called a *quasicharacter* of I_K. (See Appendix I.) We shall mostly be interested in those quasicharacters of I_K which are trivial on I_0. They are described completely by the following

Proposition 6.18. (i) *If* $q(x)$ *is a quasicharacter of* I_K *trivial on* J_K, *then* $q(x) = V(x)^s$ *with a suitable complex* s, *and conversely, for every complex* s, $V(x)^s$ *is a quasicharacter of* I_K *trivial on* J_K. *Such a quasicharacter is a character iff* $\operatorname{Re} s = 0$, *i.e. iff* $q(x) = V(x)^{it}$ *for a real* t.

(ii) *If* $q(x)$ *is a quasicharacter of* I_K *trivial on* I_0, *then* $q(x) = X(x)V(x)^s$, *where* s *is a complex number and* $X(x)$ *is a character of* I_K *which is trivial on* I_0. *Conversely, every pair* $(X(x), s)$ *determines a quasicharacter* $X(x)V(x)^s$ *of* I_K *trivial on* I_0.

Two pairs (X, s), (X_1, s_1) *determine the same quasicharacter iff* $\operatorname{Re} s = \operatorname{Re} s_1$ *and we have* $X_1(x) = X(x)V(x)^{it}$ *for* $t = \operatorname{Im}(s - s_1)$. *In particular one can always select a real* s.

Proof. If $q(x)$ is a quasicharacter of I_K trivial on J_K, then q depends only on $V(x)$ and so $q(x) = f(V(x))$, where $f(t)$ is a quasicharacter of the multiplicative group of positive reals. But every such $f(t)$ equals t^s for suitable complex s, and so $q(x) = V(x)^s$. The remaining parts of (i) are trivial if one remembers that for Re$s \ne 0$ the function $V(x)^s$ is unbounded. To prove (ii) we use Theorem 6.6, according to which J_K/I_0 is compact, and so for every quasicharacter $q(x)$ of I_K which is trivial on I_0 its restriction to J_K is in fact a character. Denote it by $X(x)$, and since J_K is a closed subgroup of I_K, we can extend $X(x)$ to a character of I_K. If $X_1(x)$ and $X_2(x)$ are two such extensions, then $X_1 X_2^{-1}$ is trivial on J_K, and so by (i) we must have $X_1(x) X_2^{-1}(x) = V(x)^{it}$ for a real t, i.e.

$$X_2(x) = X_1(x) V(x)^{it}. \tag{6.10}$$

Moreover, $q(x) X_1^{-1}(x)$ is trivial on J_K and again by (i) we infer that $q(x) = X_1(x) V(x)^s$. Equality (6.10) shows that the character $X_1(x)$ is determined up to the factor $V(x)^{it}$, and so the real part of s is unaffected by the change of $X_1(t)$. The remaining assertions are trivial. □

The number Res, which, by Proposition 6.18, is uniquely determined by a quasicharacter of I_K trivial on I_0, is called the *exponent* of that quasicharacter and denoted by $\exp q$.

Denote by Q_0 the group of all quasicharacters of I_K which are trivial on I_0 and let An J_K denote the annihilator of J_K in Q_0, i.e. the subgroup of Q_0 formed by all quasicharacters which are trivial on J_K. It is possible to make a convenient choice of the set of representatives of $Q_0/\text{An}\, J_K$ owing to the following

Lemma 6.9. *In every coset of Q_0 mod An J_K there is a unique character $X(x)$ of I_K such that for $x = (x_v)$ we have*

$$X(x) = \prod_v X_v(x_v),$$

where X_v is a character of K_v^*, trivial on $U(K_v)$ for almost all v, which in the case of Archimedean v has the form

$$X_v(x_v) = (x_v/|x_v|)^{m_v} v(x_v)^{it_v}$$

(with real t_v, integral m_v in the case of complex v and $m_v = 0, 1$ in the case of real v), where $\sum t_v = 0$, the sum being taken over all Archimedean v's

Proof. The form of $X(x)$ results from Theorem VIII of Appendix I since every character of I_K must be of such form. Moreover, the previous proposition shows that in every coset mod An J_K there are characters and every two characters lying in the same coset differ by a factor $V(x)^{it}$ for a real t. Let $X(x)$ be any character lying in a given coset and denote the sum $\sum t_v$ corresponding to it by T. Then for the character

$$X(x)V(x)^{it}$$

this sum is equal to $T+t(r_1+r_2)$ (where r_1 denotes as usual the number of real Archimedean valuations and r_2 the number of complex Archimedean valuations), and so with a suitable choice of t we can make the sum vanish. Evidently such a choice of t is unique. □

Characters satisfying the conditions of this lemma will be called *normalized characters*.

Let $\{q_a\}$ be the set of characters obtained according to the lemma and denote by Q_a the coset $\mod \mathrm{An}\, J_K$ determined by q_a. Every quasicharacter from this coset can be uniquely written in the form

$$q(x) = q_a(x)V(x)^s \qquad (6.11)$$

with a suitable complex s. In this way the coset Q_a can be described by a complex parameter s and the group Q_0 can be treated as a family of $|Q_0/\mathrm{An}\, J_K|$ copies of the complex plane.

There is another way of describing the normalized characters q_a, which we shall now present.

Let J'_K be that subgroup of I_K which consists of all ideles (a_v) whose non-Archimedean components equal unity and whose Archimedean components are of the form $a_v = t^{\varepsilon_v/n}$, where t is positive and independent of v, $n = [K:Q]$, $\varepsilon_v = 1$ for real v and $\varepsilon_v = 1/2$ for complex v. It is easy to see that J'_K is topologically isomorphic with the multiplicative group of positive reals with the usual topology, this isomorphism being given by

$$t \mapsto [t^{1/n}, \ldots, t^{1/2n}, \ldots, 1, 1, \ldots].$$

Moreover, we have the topological isomorphism $I_K = J_K \times J'_K$, and so it is possible to choose in the group J_K a Haar measure $dj(x)$ so that our Haar measure dm_I on I_K equals the product of that dj and the standard Haar measure dt/t on the positive reals transferred onto J'_K, dt being the usual Lebesgue measure on the real line.

Proposition 6.19. *A character X of I_K is normalized iff X is trivial on J'_K.*

Proof. If X is normalized, then for Archimedean v we have, with $X = \prod_v X_v$,

$$X_v(x_v) = (x_v/|x_v|)^{m_v} v(x_v)^{it_v},$$

where the sum of the t_v's vanishes. If $(x_v) = [t^{1/n}, \ldots]$ lies in J'_K, then

$$X((x_v)) = t^{iA}$$

with $A = \sum_v t_v/n = 0$, and thus X trivializes on J'_K.

Conversely, if X trivializes on J_K', then for every positive t we must have

$$n^{-1}\left(\sum_v t_v\right) \log t = 2N_t \pi i$$

with a certain integral N_t, which shows that $N_t/\log t$ is constant. But this can happen only if $N_t = 0$, and thus the sum of the t_v's vanishes. □

Now let f be any complex-valued function defined on I_K about which we only assume that for all quasicharacters $q \in Q_0$ with $\exp q > 1$ the Mellin transform

$$\tilde{f}(q) = \int_{I_K} f(x) q(x) dm_I(x)$$

is well-defined. If we now fix the coset Q_a and consider only the quasicharacters from this coset, then owing to (6.11) we can consider $\tilde{f}(q)$ as a function of one complex variable s defined in the open half-plane $\operatorname{Re} s > 1$. To stress this fact we shall also write

$$\tilde{f}(q) = Z(f, s, q_a)$$

or

$$\tilde{f}(q) = Z(f, s)$$

and call $\tilde{f}(q)$ the *zeta-function associated with* $f(x)$. The zeta-functions so defined are meromorphic and satisfy a functional equation, provided the function $f(x)$ satisfies some regularity assumptions. This result forms the main part of the following

Theorem 6.9. *Let $f(x)$ be a complex-valued function defined on the adele-group A_K^+ which satisfies the following conditions:*

(i) *$f(x)$ and its Fourier transform $\hat{f}(\hat{x})$ are both continuous and integrable.*
(ii) *For every adele x and idele a the two series*

$$\sum_{t \in A_0} f(a(x+t)) \quad \text{and} \quad \sum_{t \in A_0} \hat{f}(a(x+t))$$

are uniformly convergent for (a, x) lying in an arbitrary but fixed compact subset of the product $I_K \times A_K$.

(iii) *If g is the restriction of f to I_K and g_0 is the restriction of $\hat{f}$ to I_K, then for every $t > 1$ the functions $g(x)V(x)^t$ and $g_0(x)V(x)^t$ are integrable in I_K.*

Then for each fixed coset Q_a of Q_0 mod $\operatorname{An} J_K$ the integral

$$Z(g, s) = \int_{I_K} g(x) q(x) dm_I(x)$$

(with q as in (6.11)) represents a regular function of s in the open half-plane $\operatorname{Re} s > 1$. This function can be continued analytically to a meromorphic function in the open plane. If $Q_a = \operatorname{An} J_K$ is the zero-coset, i.e. $q_a(x) = 1$, then $Z(g, s)$ can have at most two single poles at $s = 0$ and $s = 1$ and its residues there are equal to $-h\varkappa \hat{f}(0)$ and $h\varkappa \hat{f}(0)$, respectively, where h is the class-number of K and the constant $\varkappa$ equals

$$\varkappa = 2^{r_1}(2\pi)^{r_2} R(K) |d(K)|^{-1/2} w^{-1}(K), \tag{6.12}$$

where r_1 and r_2 have their usual meaning, $R(K)$ is the regulator of K, $d(K)$ is its discriminant and $w(K)$ is the number of roots of unity lying in K.

However, if Q_a is not the zero-coset, then $Z(g, s)$ is entire.

Finally, if $q_{a^{-1}}$ denotes the representative of the coset inverse to Q_a (which obviously equals q_a^{-1}), then for all s the following functional equation holds:

$$Z(g, s, q_a) = Z(g_0, 1-s, q_{a^{-1}}). \tag{6.13}$$

Moreover, in every strip $c_1 \leq \operatorname{Re} s \leq c_2$ the function

$$s(s-1) Z(g, s, q_a)$$

is bounded.

Proof. Assumption (iii) implies the existence of the integral defining $Z(g, s)$ for every s in the half-plane $\operatorname{Re} s > 1$ as $|g(x)q(x)| = |g(x)| V(x)^{\exp q}$. Denote by A_1 and A_2 the sets of ideles with $V(x) \geq 1$ and $V(x) \leq 1$, respectively, and for $i = 1, 2$ let B_i be equal to the integral

$$\int_{A_i} g(x) q(x) dm_I(x).$$

Evidently $Z(g, s) = B_1 + B_2$ as $J_K = A_1 \cap A_2$ is of measure zero. Observe that for $\operatorname{Re} s > 1$ integrals B_1 and B_2 are both regular functions of s. Indeed, under this assumption we have for $i = 1, 2$ the equality

$$k^{-1}\left(\int_{A_i} g(x) q_a(x) V(x)^{s+k} dm_I(x) - \int_{A_i} g(x) q_a(x) V(x)^s dm_I(x)\right)$$
$$= \int_{A_i} g(x) q_a(x) V(x)^s (V(x)^k - 1) k^{-1} dm_I(x),$$

and since the integrand does not exceed

$$|g(x)| V(x)^{1 + \operatorname{Re} s} |\log V(x)|$$

in absolute value for k sufficiently small, we again find, using (iii), that B_i are differentiable for $\operatorname{Re} s > 1$. Note also that the same argument proves the differentiability of B_1 in the remaining part of the plane; thus B_1 is entire and it remains to show that B_2 can be analytically continued to a meromorphic function.

Applying Proposition 6.19, we can write

$$B_2 = \int_0^1 t^{-1} dt \left(\int_{J_K} g(tj) q(tj) \, dj \right)$$

$$= \int_0^1 t^{-1} dt \int_{J_K} g(tj) q_a(j) t^s \, dj = \int_0^1 t^{s-1} \int_{J_K} g(tj) q_a(j) \, dj \, dt.$$

To evaluate the inner integral in the last term of the above equality let us consider the set $E \subset J_K$ defined in the following manner:

In Theorem 6.6 we proved the existence of a compact set B in J_K which contained representatives of every class from $C(K)$; it had the form

$$B = \bigcup_{i=1}^h b_i P,$$

where $b_1, \ldots, b_h$ were ideles whose images in $H(K) = I_K/U_K I_0$ represented all ideal classes and P was the set of all ideles from U_K whose images in R^r through the map

$$L: (u_v) \mapsto [\log v_1(u_{v_1}), \ldots, \log v_r(u_{v_r})]$$

($v_1, \ldots, v_r$ being all Archimedean valuations of K except one) had the shape

$$\sum_{j=1}^r x_j L(\varepsilon_j), \quad 0 \leq x_j \leq 1, \tag{6.14}$$

where the ε_j's were the fundamental units of K.

Note that if P' is the set of all ideles from U_K which, under the map L, have the form (6.14), $0 \leq x_j < 1$, then again the set

$$B' = \bigcup_{i=1}^h b_i P'$$

represents all classes of $C(K)$ although it is no longer compact. However, also in this case there are members of the same class of $C(K)$ which belong to B'. Let us find out under what circumstances this happens. Let a_1, a_2 be two elements of $b_i P'$ which represent the same class of $C(K)$. (Note that elements from distinct summands $b_i P'$ and $b_j P'$ cannot lie in the same class of $C(K)$ since they induce ideals lying in different classes of $H(K)$.) We can write $a_1 = b_i p_1$, $a_2 = b_i p_2$ with p_1, p_2 in P' and principal $p_1 p_2^{-1}$. But

$$L(p_1 p_2^{-1}) = \sum_{j=1}^r x_j L(\varepsilon_j), \quad \text{where } -1 < x_j < 1,$$

and since $p_1 p_2^{-1}$ is principal and induces the unit ideal, it must be a unit, whence the x_j's are rational integers and we see that they must vanish, i.e. $L(p_1 p_2^{-1}) = 0$. Hence $v(p_1 p_2^{-1}) = 1$ holds for every valuation with the exception of one, and the product formula shows that the same must be true also for the excepted valuation; hence, by Theorems 3.2 and 2.1 (i), we see that $p_1 p_2^{-1}$ must be a root of unity, and so finally $a_1 = \zeta_w^m a_2$ holds for a suitable m. Conversely, if two ideles from P' differ by a factor which is a root of unity, then they lie in the same class of $C(K)$.

In the definition of the mapping L one of the Archimedean valuations was not used. It is convenient to have it complex if there exist complex Archimedean valuations of K at all, and we shall make this choice whenever possible. Let v_0 be that valuation. Let us now define

$$P_1 = \{(x_v) \in P' : 0 \leqslant \arg x_{v_0} < 2\pi/w\},$$

which definition in the case of real v reduces to

$$P_1 = \{(x_v) \in P' : x_{v_0} > 0\}.$$

The foregoing argument shows that the set

$$E = \bigcup_{i=1}^{h} b_i P_1$$

contains exactly one element from every class of $J_K/I_0 \subset C(K)$. We now prove

Lemma 6.10. *The measure $j(E)$ of the set E equals $h\varkappa$ ($\varkappa$ defined by (6.12)).*

Proof. From the definition of E we readily obtain

$$j(E) = \sum_{i=1}^{h} j(b_i P_1) = hj(P_1),$$

and in addition $j(P_1) = j(P')/w$; hence we have to find the measure of P'. An idele (x_v) lies in P' iff it satisfies the following conditions:

$$v(x_v) = 1 \quad \text{for all non-Archimedean } v,$$
$$V((x_v)) = 1$$

and

$$0 \leqslant \sum_{\substack{v \neq v_0 \\ \text{Archimedean}}} A_{v,j} \log v(x_v) < 1 \quad (j = 1, 2, \ldots, r),$$

where $[A_{v,j}]$ is the matrix inverse to $[\log|F_v(\varepsilon_j)|]$. Now let Z be the product of P' and the interval $[1, e]$, considering the latter as a subset of J'_K; thus $Z \subset I_K$ and any pair (a, t) with $a \in P'$, $t \in [1, e]$ can be considered as the product at', t' being the idele $[t^{1/n}, \ldots]$. By the definition of the measure j and the fact that $[1, e]$ is of unit measure in J'_K we see that $j'(P) = m_I(Z)$. To find $m_I(Z)$ observe that if $S \subset V$ consists of all Archimedean valuations and those non-Archimedean valuations for which the extension K/Q is ramified, then our set Z lies in A_S and so

$$j(P') = \int_Z dx_S.$$

Now an idele (u_v) lies in Z iff it satisfies the following conditions: $v(u_v) = 1$ for all non-Archimedean v, $1 \leq V((u_v)) < e$, and if $V((u_v)) = t$ and we put $t_v = t^{e_v/n}$ for Archimedean v and $t_v = 1$ for non-Archimedean v, then for $j = 1, 2, \ldots, r$ we have

$$0 \leq \sum_{v \neq v_0 \text{ Archimedean}} A_{v,j} \log v(u_v/t_v) < 1. \qquad (6.15)$$

Moreover, by the definition of the measure dx_S, the measure of Z equals the product of two integrals, one taken over the product of $U(K_v)$ for non-Archimedean v and lying in S and the other taken over the projection Z_{Arch} of Z into the product of K_v^* for Archimedean v, in both cases the integrand being equal to unity. But the first integral equals the product of $N(D_{K_v/Q_p})^{-1/2}$ taken over all v's at which K/Q is ramified; hence it equals $|d(K)|^{-1/2}$, and so we are left with the second integral. It is equal to the multiple integral in $R^{r_1} \times \mathfrak{Z}^{r_2}$ of the function

$$\prod_{v \text{ Archimedean}} v(u_v)^{-1}$$

taken over the set of all sequences $(u_v)_{v \text{ Archimedean}}$ which for a certain $1 \leq t < e$ satisfy (6.15) and

$$\prod_{v \text{ Archimedean}} v(u_v) = t,$$

and one sees immediately that it differs by $R(K)$ from the integral S of the same function taken over the set of sequences $(u_v)_{v \text{ Archimedean}}$ which for a certain $1 \leq t < e$ satisfy

$$1 \leq v(u_v/t_v) < e \quad \text{for } v \neq v_0$$

and

$$\prod_{v \text{ Archimedean}} v(u_v) = t;$$

to deal with S we make the following change of variables:

$$u'_v = v_v \quad \text{for } v \neq v_0,$$

$$t = u_{v_0} / \left(\prod_{v \neq v_0 \text{ Archimedean}} u_v \right)$$

after which S takes the form

$$\int_1^e \left(\prod_{v \neq v_0 \text{ Archimedean}} \int_{B_{v,t}} v(u'_v)^{-1} du'_v \right) t^{-1} dt,$$

where

$$B_{v,t} = \{u'_v : t^{1/n} \leq v(u'_v) < et^{1/n}\}$$

and so equals $2^{r_1}(2\pi)^{r_2}$. Putting all this together, we obtain the assertion of our lemma. □

Now we return to the integral B_2, which, as we have seen, equals

$$\int_0^1 t^{s-1} \int_{J_K} g(tj) q_a(j) \, dj \, dt, \qquad (6.16)$$

and using the fact that every element of J_K can be uniquely written as bj with $b \in I_0$ and $j \in E$, we can put the inner integral in the form

$$\int_E \sum_{b \in I_0} g(btj) q_a(j) \, dj = \int_E q_a(j) \sum_{b \in I_0} g(btj) \, dj.$$

But

$$\sum_{b \in I_0} g(btj) = \sum_{b \in A_0} f(btj) - f(0),$$

and so we can apply

Lemma 6.11. *If the function f satisfies the assumptions of the theorem, then we have*

$$\sum_{b \in A_0} f(bx) = V^{-1}(x) \sum_{b \in A_0} \hat{f}(b/x)$$

for every idele x.

Proof. We want to apply Poisson's formula for the adele group and its subgroup consisting of principal adeles, and to do that we have to check the assumptions. We need a fundamental set D with unit measure. From the proof of Proposition 6.13 we infer that if $w_1, \ldots, w_n$ is an integral basis for K/Q and $w'_1, \ldots, w'_n$ denote the projection of the principal adeles generated by the w_i's into the product $\prod_{v \text{ Archimedean}} K_v^+$, then for D we can take the set of all adeles (a_v) such that for non-Archimedean v's one has $v(a_v) \leq 1$, whereas the element $(a_v)_{v \text{ Archimedean}}$ of $\prod_{v \text{ Archimedean}} K_v^+$ belongs to the set

$$\left\{ \sum_{i=1}^n x_i w'_i : 0 \leq x_i < 1 \right\}.$$

Using the definition of the discriminant and Corollary 1 to Lemma 5.7, one can easily see that the measure of D in fact equals 1. Now define the Haar measure in the adele class-group via the bi-unique mapping between D and A/A_0; so all the assumptions of the Poisson formula are satisfied. We now apply it to the function $f_1(t) = f(xt)$.

First we need its Fourier transform:

$$\hat{f}_1(y) = \int_{A_K^+} f_1(t) X_y(-t) dm_A(t)$$

$$= \int_{A_K^+} f(xt) X_y(-t) dm_A(t)$$

$$= \int_{A_K^+} f(t) X_y(-tx^{-1}) V^{-1}(x) dm_A(t)$$

$$= \int_{A_K^+} f(t) X_{y/x}(-t) V^{-1}(x) dm_A(t)$$

$$= V^{-1}(x) \hat{f}(y/x)$$

by the Corollary to Proposition 6.16, and this leads us immediately to the asserted equality in view of Theorem VII of Appendix I. □

Thus the inner integral in (6.16) becomes

$$\int_E q_a(j) V(tj)^{-1} \left(\sum_{b \in A_0} \hat{f}(b/tj) - f(0) \right) dj$$

$$= t^{-1} \int_E q_a(j) \sum_{b \in I_0} g_0(b/tj) dj + t^{-1} \hat{f}(0) \int_E q(j) dj - f(0) \int_E q_a(j) dj,$$

and so B_2 turns out to be the sum of three integrals:

$$B_{21} = \int_0^1 t^{s-2} \int_E q_a(j) \sum_{b \in I_0} g_0(b/tj) dt\, dj,$$

$$B_{22} = \hat{f}(0) \int_0^1 t^{s-2} dt \int_E q_a(j) dj$$

and

$$B_{23} = -f(0) \int_0^1 t^{s-1} dt \int_E q_a(j) dj,$$

which we shall now evaluate separately. The first of them is similar in nature to B_1. Indeed, it equals

$$\int_0^1 t^{s-2} \int_{J_K} g_0(1/tj) q_a(j) dj\, dt = \int_{A_2} g_0(1/x) V(x)^{s-1} q_a(x/V(x)) dm_1(x)$$

$$= \int_{A_1} g_0(x) V(x)^{1-s} q_{a^{-1}}(x/V(x)) dm_1(x)$$

and thus has the same form as B_1 and for the same reasons defines an entire function. To evaluate B_{22} and B_{23} observe that q_a is a character of J_K which trivializes on I_0 and E is a full set of representatives of J_K/I_0; thus if $q_a|_{J_K} = 1$ then

$$\int_E q_a(j)\,dj = j(E) = h\varkappa$$

by Lemma 6.10, and in all other cases this integral vanishes. It should be noted that with our choice of the q_a's the equality $q_a|_{J_K} = 1$ implies $q_a = 1$ on I_K since we have

$$q_a(x) = V(x)^{it} = \prod_v v(x_v)^{it} \quad \text{for a suitable } t,$$

but in view of our normalization (Lemma 6.9) this gives $t = 0$. We thus arrive at the equality

$$Z(g, s, q_a) = \int_{V(x) \geq 1} g(x) V(x)^s q_a(x)\,dm_1(x)$$
$$+ \int_{V(x) \geq 1} g_0(x) V(x)^{1-s} q_{a^{-1}}(x)\,dm_1(x)$$
$$+ \varepsilon(\hat{f}(0)/(s-1) - f(0)/s) h\varkappa, \tag{6.17}$$

where the first two terms are integral functions of s and $\varepsilon = 1$ if $q_a = 1$ and $\varepsilon = 0$ otherwise.

This proves all the assertions of the theorem save the functional equation (6.13) because the boundedness of $Z(g, s, q_a)$ in every fixed vertical strip is immediate. But to obtain the functional equation it suffices to observe that the equality

$$Z(g_0, 1-s, q_{a^{-1}}) = \int_{V(x) \geq 1} g_0(x) V(x)^{1-s} q_{a^{-1}}(x)\,dm_1(x)$$
$$+ \int_{V(x) \geq 1} g(-x) V(x)^s q_a(-x)\,dm_1(x)$$
$$+ \varepsilon(-f(0)/s + \hat{f}(0)/(s-1)) h\varkappa$$

follows from (6.17), and since the second integral remains unchanged after we substitute x for $-x$, one can immediately see that this zeta-function coincides with $Z(g, s, q_a)$. □

In the next chapter we shall use this functional equation for various functions $f(x)$ to obtain the classical results of B. Riemann and E. Hecke.

§ 3. Notes to Chapter 6

1. The first to apply the p-adic approach to algebraic numbers was K. Hensel, who as early as 1894 demonstrated the power of this approach while solving a problem of Dedekind's concerning the discriminant (Hensel [94a], [94b], [97a], [97b]). A systematic application of this method was outlined in Hensel [97c], [05a], [07], [13], [18]. Quadratic fields were treated in Hensel [14a]. Cf. Bauer [22], Wahlin [15a]. Applications to the factorization of prime ideals in various extensions can be found in Bauer [36], Bauer, Chebotarev [28], Hensel [21a], [21b], Ore [27], Rella [24a], Wahlin [15b], [22].

2. The applicability of p-adic numbers to quadratic forms was established by H. Hasse [23a], [23b], [24a], [24b] (cf. Hasse [62a] for historical remarks), who proved that two such forms over an algebraic number field K are equivalent iff they are equivalent over every completion of K and, moreover, a given quadratic form represents an element a of K iff it represents it in every completion of K. (Cf. Cassels [59a], O'Meara [63], Siegel [41], Springer [57].)

It is said that for a given sentence the *Hasse principle holds* provided the sentence is true in a field K iff it is true in every completion of K. E. Witt [35a] noted that the Hasse principle fails for the solvability of $x^2 + y^2 = a$ in fields of algebraic functions. It fails also for zeros of cubic forms, even over Q, as shown first by E. S. Selmer [51], who produced the example $3x^3 + 4y^3 + 5z^3$, which has zeros in every completion of Q but not in Q itself. See also Bremner [78], Cassels, Guy [66], Mordell [65], Selmer [53], Swinnerton-Dyer [62].

For quintic forms the Hasse principle also fails (Fujiwara [72]), and also for certain other classes of forms (Birch, Swinnerton-Dyer [75], Fujiwara, Sudo [76], Manin [71]). V. A. Iskovskikh [71] showed that it fails also for common zeros of two quadratic forms; however, as shown in Waterhouse [76], it holds for the equivalence of such pairs.

Several cases of the applicability of Hasse's principle were given in Cantor, Roquette [84], Colliot-Thélène, Coray, Sansuc [80], Colliot-Thélène, Sansuc [82], Dress [65], Hasse [32], Hijikata [63], Landherr [36], Noether [33], O'Meara [59], Opolka [81], Waterhouse [77], [78].

An important example of the validity of this principle is the Hasse Norm Theorem (HNT) (Hasse [31d]; in special cases Furtwängler [04], [09], Hilbert [97] (Satz 167), [99]), which states that if L/K is cyclic, then $N_{L/K}(L)$ equals the intersection of N_{L_w/K_v} where v ranges over all valuations of K and w is a prolongation of v to L. Another proof is given in Gold [77]. For the history of HNT see Jehne [82]. For non-cyclic extensions HNT may fail, as noted already in Hasse [31d]; however, there are classes of extensions for which it is true. It is valid, e.g., for all extensions of prime degree, not necessarily normal (Bartels [81b]; for the cubic case see Tasaka [70]), for the extensions with a generalized quaternion Galois group of order $\equiv 0 \pmod 8$ (Arnaudon [76]), and also for all extensions whose normal closure has a dihedral Galois group (Bartels [81a]). For other classes of fields see Garbanati [75b], [75c], [77], [78a], Gerth [77a], [77b], [78], Gurak [78a], [78b], [80], Platonov, Drakokhrust [85], Scholz [36].

A. Scholz [36] considered the factor group of the group of elements in K^* which are norms in all local extensions associated with L/K by $N_{L/K}(L^*)$. This factor group is called the *number-knot* of L/K and was studied by W. Jehne [79], who considered also several analogues of it. (Cf. Heider [80], Steckel [82a].) It is canonically isomorphic to the Galois group of a certain extension of K considered by K. Masuda [59]. Cf. Razar [77].

F. Lorenz [80] proved that if L/K is normal then there exists a normal extension M/K such that $L \subset M$ and all local norms of M/K become global norms in L/K. Cf. Lorenz [82], Opolka [80a], [80b], [82].

Cohomological obstructions to the validity of HNT were given in Amano [79], Chevalley [54], Gurak [78a], Platonov, Drakokhrust [85], Razar [77], Tate (see Cassels, Fröhlich [67]).

For other results on HNT see Amano [77], Gurak [78a], Ono [63]. An application to diophantine equations was given by J. H. Smith [75]. For analogous questions concerning infinite extensions see Heider [81] and for units see Gold [77], Nakagoshi [75].

Elements which are norms in all Abelian extensions of a given field were studied in Ax [62].

3. The ramification groups were introduced by D. Hilbert [94a] directly, without p-adic fields. In addition to papers quoted in Chapter 5, let us mention Herbrand [30a], Krasner [35], Speiser [19], Tomás [73], v.d. Waerden [34], Wegner [35] on that topic.

Corollary 1 to Proposition 6.6 is due to L. R. McCulloh [66] and Corollary 2 to E. Netto [83], [84].

Theorem 6.2 was first proved by A. Wiman [99] and rediscovered later by J. Westlund [12]. A new proof was given by V. S. Albis-González [73]. In the first edition the treatment of case (v) was not complete.

The structure of the group $G(\mathfrak{P}^n)$ was determined in Hensel [16b], [17], Nakagoshi [79], Takenouchi [13]. Cf. Wolff [05], Hensel [37]. For particular cases see Cross [83], Dirichlet [41a], Halter-Koch [72b], Ranum [10]. For a related group see Loh [77]. The first study of $G(\mathfrak{P})$ occurs in Hensel [87], [88]. The structure of the multiplicative semigroup of the ring R_K/I was described in Rieger [64a].

Generators of the group $G(I)$ in the case where it is cyclic are called *primitive roots* $(\bmod I)$, and thus Theorem 6.2 describes ideals having primitive roots. An analogue of Artin's conjecture on primitive roots was considered by S. Egami [81] and J. Hinz [84], [86]. The primitive root $(\bmod \mathfrak{P})$ of smallest norm was evaluated by J. Hinz [83a], [83b], [83c]. The bounds obtained by him are of the same order of magnitude as those in the classical case $K = Q$. On primitive roots in linear recurrences see Nechaev, Stepanova [65].

Finite commutative rings with a cyclic group of units were described by R. W. Gilmer, Jr. [63b]. For other proofs see C. W. Ayoub [69], Pearson, Schneider [70], Raghavendran [70].

Proposition 6.8 had in the first edition a wrong formulation. This was noticed by J. Lewittes [83], who computed the number of real primitive characters $(\bmod \mathfrak{P}^k)$.

4. Gaussian sums in the case of $K = Q$ go back to Gauss and Lagrange. In the general case they were introduced by E. Hecke [19] and Hasse [51a]. See also Hasse [52b], [54a]. A general theory of the Gaussian sum in rings was developed by E. Lamprecht [53], [57].

Theorem 6.3 is due to C. F. Gauss [11], and the proof we quote — to W. C. Waterhouse [70]. There are more than 25 proofs of that theorem, all listed in the survey by B. C. Berndt and R. J. Evans [81], and so we quote only the following few: Carlitz [56], [68b], Cauchy [40], Dirichlet [35], Estermann [45], Kronecker [56a], [89], [90], Landau [28], Mertens [96], Mordell [18b], [62a], Schur [21].

Theorem 6.4 was obtained independently by S. Chowla [62] and L. J. Mordell [62b]. We reproduced the proof of the latter. See Evans [77], Yokoyama [64] for generalizations. For characterizations of Gaussian sums with real characters among cyclotomic integers see Cavior [64], Rédei [47].

Explicit formulas for Gaussian sums with cubic and quartic characters in terms of elliptic functions were given by C. R. Matthews [79]. Cf. Loxton [74a], [78], McGettrick [72a], K. Yamamoto [65].

There was a conjecture, originating in Kummer [42], [46], which stated that if p runs over all primes congruent to unity $(\bmod 3)$, then the arguments of the Gaussian sums associated with a cubic character $(\bmod p)$ are non-uniformly distributed on the unit circle. Actually this distribution is uniform as shown by D. R. Heath-Brown, S. J. Patterson [79]. (See Venkov, Proskurin [82] for an exposition of the proof.) The same holds also for Gaussian sums of higher orders (Patterson [81]). For previous work on Kummer's conjecture see Beyer [54], Cassels [70], Fröberg [74], Krätzel [67], Kubota [68], E. Lehmer [56], Loxton [74a], [78], McGettrick [72b], Moreno [74], v. Neumann, Goldstine [53], Patterson [78], Reshetukha [70], [75].

Gaussian sums related to characters (mod p^n) with $n \geq 2$ are easier to handle and can be evaluated explicitly, as shown by R. W. K. Odoni [73a]. See also Mauclaire [83].

Various relations between Gaussian sums, including important reciprocity theorems, were given in Barner [67], Davenport, Hasse [34], Hecke [19], Kubota [60], [61], [63], Kunert [35], Mordell [22a], Schmid [36], K. Yamamoto [66].

Gaussian sums were used by H. Hasse [40] and H. Bergström [44] for the study of class-numbers. The role of Gaussian sums in the arithmetic of Abelian extensions of Q was emphasized by H. W. Leopoldt [62], who employed them for a systematic development of that theory.

Closely related to the Gaussian sums are the Ramanujan sums, introduced in algebraic number fields by G. J. Rieger [60a].

For other results on Gaussian sums and their generalizations see Aposto [70b], Auslander, Tolimieri, Winograd [82], Cassels [69a], Hölder [35], Joris [77], Kloosterman [30], Kubota [64], Landsberg [93], Shiratani [64b]. Cf. also the literature quoted in Berndt, Evans [81].

5. Theorem 6.5 was stated by L. Kronecker [53], [77] and the first proof (not quite complete) was published by H. Weber [86]. The proof we give is that of I. R. Shafarevich [51]. For other proofs see Chebotarev [24a], Delaunay [23], Greenberg [74], Hilbert [96], [97], Lubelski [39a], Mertens [06], Speiser [19], Steinbacher [11], Weber [07], Yamamoto, Onuki [75], Zassenhaus [68]. O. Neumann [81b] gave two proofs along the lines of Kronecker and Weber, discussing also other proofs. For a special case see Neiss [31].

An analogue for Abelian extensions of imaginary quadratic fields had been conjectured by Kronecker and proved, in a modified form, by H. Weber [96b], [97] and R. Fueter [05]. (Cf. Fueter [11], [14], [24], Hasse [27], Hecke [17b], Ramachandra [64], Takagi [20].) In other cases no analogue is known, although in the parallel problem of Abelian extension of function fields the situation is much better. D. R. Hayes [74] solved the problem for the field of rational functions in one variable over a finite field, and V. G. Drinfel'd [74] obtained a solution for all its finite extensions. His proof was simplified in Hayes [79]. Cf. also the survey of D. Goss [80].

Every Abelian extension of Q_p is contained in a cyclotomic extension. Abelian extensions of arbitrary p-adic fields were described in Lubin, Tate [65]. Cf. Gold [81], Hazewinkel [69], [75], Lubin [81], Rosen [81].

For Lemma 6.5 see Chebotarev [29].

6. The question of the existence of a cyclic extension L/K of an algebraic number field K with prescribed local extensions $L_\mathfrak{P}/K_\mathfrak{p}$ for $\mathfrak{p}$ in a given finite set was first considered by W. Grunwald [33]. He asserted that such an extension always exists; however, it was pointed out by S. Wang [48] that there is a case to which Grunwald's argument does not apply. An improved version was obtained independently by H. Hasse [50b], and S. Wang [50a], and the final result is called the Grunwald–Hasse–Wang theorem. For other proofs see Artin, Tate [61], Neukirch [74b], Whaples [42] (where again there are objections with regard to a special case). An analogous result holds for arbitrary Abelian extensions (Wang [50b]). Cf. Miki [78b], Neukirch [73], [74c], Saltman [82], [84].

7. Ideles were introduced by C. Chevalley [36b], who used them for his reformulation of the class-field theory (Chevalley [40]). Adeles occur first in Artin, Whaples [45], where they are called valuation vectors.

The ring of adeles was characterized by K. Iwasawa [53a] in the following way: if R is a semi-simple commutative locally compact ring with a unit element, neither compact nor discrete, and there is a discrete field $K \subset R$ with the same unit element such that R/K is compact, then R is the ring of adeles over an algebraic number field or over a field of algebraic functions in one variable over a finite field of constants. (In the latter case one defines the ideles in a completely analogous way.)

For the topological properties of adeles and ideles see Artin, Tate [61], Iwasawa [53a]. The structure of $D(K)$ was determined by A. Weil [51]. For other proofs see Artin [56], M. Kobayashi [83]. One proves in the class-field theory that $C(K)/D(K)$ is topologically isomorphic to the Galois group of K^{ab}/K, where K^{ab} is the maximal Abelian extension of K.

A characterization of characters of $C(K)$ among those of I_K was given in Gurevich [71]. More about characters of $C(K)$ will be said in the next chapter.

For the cohomological properties of adeles and ideles cf. any modern account of the class-field theory, e.g., J. Tate's lecture in Cassels, Fröhlich [67]. See also Albu [70b], Katayama [81]. Certain subgroups of I_K were studied in Furuta [66], Masuda [57], Miyake [80b].

The idele class-groups of imaginary quadratic fields were described in Onabe [78]. The same paper contains examples of distinct fields K with isomorphic idele class-groups but distinct $H(K)$. (E.g., $K = Q(id^{1/2})$ with $d = 8$, 20, 23, 47 and 71.) Examples of non-isomorphic fields whose adele rings are isomorphic were given by K. Komatsu [78], who showed in [84] that there are arbitrarily large finite sets of such fields. A criterion for the isomorphism of adele rings in terms of norm-forms was given by W. Meyer, R. Perlis [79]. K. Komatsu [74] showed that if $\text{Gal}(\hat{Q}/K) \simeq \text{Gal}(\hat{Q}/L)$ then $I_K \simeq I_L$, and in [76c] he produced examples to show that the converse implication fails for adele rings in place of I_K.

An exposition of adeles was given by A. Robert [74].

8. The discriminant $\partial(L/K)$ was introduced by A. Fröhlich [60b] (cf. Fröhlich [61]), who applied this notion in [60c], [60d]. Proposition 6.16 is due to him. D. Maurer [78a], [79] characterized normal extensions L/K with $\partial(L/K) \in I_K^2/U_K^2$ as those for which the 2-Sylow subgroup of $\text{Gal}(L/K)$ is non-cyclic. Cf. Gallagher [85].

E. Hecke [23] proved that the class of the different $D_{L/K}$ in $H(L)$ is a square. This immediately implies Theorem 6.8. Other proofs were given by Armitage [67], Fröhlich [60b], Knebusch, Scharlau [71], Weil [67a]. For a special case see Herbrand [32d]. Our proof is that of Fröhlich. Theorem 6.8 holds for an arbitrary Dedekind domain (Armitage [67], [72]); however, it was shown in Fröhlich, Serre, Tate [62] that the class of the different is not necessarily a square for suitable Dedekind domains.

The results of § 2.5, in particular Theorem 6.9, are due to J. Tate [50]. More general zeta-functions were considered by T. Ono [70].

9. A p-adic analogue of the regulator matrix was introduced by H. W. Leopoldt [62], who conjectured that its rank is independent of p and coincides with the rank of the usual regulator matrix. This was shown to be true for Abelian fields by A. Brumer [67] (cf. Ax [65]) and other special cases were settled in Bertrandias, Payan [72], G. Gras [72a], Kuzmin [81], Miyake [82], Wingberg [83]. In the general case M. Waldschmidt [81] proved that the rank of the p-adic regulator is $\geq (r_1+r_2-1)/2$.

For other results concerning Leopoldt's conjecture see Gillard [79d], Iwasawa [73a], Kramer, Candiotti [78], Schneider [79].

Diophantine approximations and the geometry of numbers in the ring of adeles and also analogues of PV-numbers in them were studied in Bertrandias [65], Cantor [65b], Decomps-Guilloux [65a], [65b], [70], Grandet-Hugot [65b], [66], [75a], [75b], McFeat [71], Pisot [66], Senge [67].

EXERCISES TO CHAPTER 6

1. Characterize those extensions L/K which are wildly ramified, but the trace map $T_{L/K}$: $R_L \to R_K$ is surjective in terms of factorization of prime ideals.

2. Determine the inertia group and the ramification groups for prime ideals in the following extensions of the rationals:

(a) K/Q quadratic,
(b) $K = Q(\zeta_p)$ with prime p,
(c) $K = Q(i^{1/2})$.

3. Determine the structure of the group $G(\mathfrak{P}^2)$ where $\mathfrak{P}$ is a prime ideal in R_K and $f_{K/Q}(\mathfrak{P}) \neq 1$.

4. Determine all those rational positive integers n for which the group $G(nZ)$ has primitive characters of a given order k.

5. Prove that if K/Q_p is Abelian, then K is contained in a cyclotomic extension of Q_p.

6. Let L/K be a finite extension of an algebraic number field and define the norm mapping by the formula:

$$N_{L/K}((a_v)_v) = \Big(\prod_{v \text{ over } w} N_{L_v/K_w}(a_v) \Big)_w.$$

Prove that $N_{L/K}: I_L \to I_K$ is a homomorphism which, on principal ideals, agrees with the usual norm map from L^* to K^*. Show that $N_{L/K}$ is continuous.

7. Prove an analogue of the preceding exercise for an appropriately defined trace map from A_L to A_K.

8. (Weil [39].) Estimate the number of principal adeles $(x_v)_v$ satisfying $v(x_v) \leq T_v$, where T_v is a given family of non-negative real numbers of which only finitely many are non-zero.

9. (Weil [39].) Let $a \in K$ and let $(a_v)_v$ be the corresponding principal adele. For infinite v define

$$T_v(a) = \begin{cases} a_v & \text{if } v \text{ is real,} \\ a_v + \bar{a}_v & \text{if } v \text{ is complex} \end{cases}$$

and for finite v put $T_v(a) = T_{K_v/Q_p}(a_v)$, where p is the unique rational prime with $v(p) \neq 1$. Moreover define $t_v(a)$ for finite v as any rational number r for which $T_v(a) - r \in R_v$, and put $t_v(a) = -T_v(a)$ for v infinite.

Prove that for all $a \in K$ the number $\sum_v t_v(a)$ is a rational integer.

Chapter 7. Analytical Methods

§ 1. The Classical Zeta-Functions

1. In this section we introduce the Dirichlet series defining the Dedekind zeta-function and also some other zeta-functions (including Dirichlet's *L*-functions) and we derive the functional equation for those functions. Our reasoning is based on the results of the preceding chapter. Subsequent sections are devoted to asymptotic distribution of ideals and prime ideals. Our main tool will be the Tauberian theorem of H. Delange, an account of which is to be found in Appendix II.

We start with Dedekind's zeta-function of an algebraic number field K, defined by

$$\zeta_K(s) = \sum_I N(I)^{-s} \tag{7.1}$$

in the open half-plane $\operatorname{Re} s > 1$, summation spreading over all ideals of R_K. We first prove

Proposition 7.1. *In the half-plane* $\operatorname{Re} s > 1$ *the series* (7.1) *is absolutely convergent. The convergence is uniform in every compact subset of this half-plane and so* $\zeta_K(s)$ *is regular there. Moreover, in the same half-plane the infinite product*

$$\prod (1 - N(\mathfrak{p})^{-s})^{-1}$$

taken over all prime ideals $\mathfrak{p}$ *of* R_K *converges to* $\zeta_K(s)$.

Proof. Let n be the degree of K over Q. Since there can be at most n prime ideals in R_K with a given norm and every such norm must be a prime power, the series $\sum N(\mathfrak{p})^{-s}$ is absolutely convergent for $\operatorname{Re} s > 1$. Its convergence is uniform in every compact set since it is majorized by the series

$$\sum_q n q^{-\operatorname{Re} s},$$

where q runs over all prime-powers in $\mathbb{Z}$. If T is any positive real number, then obviously

$$\sum_{N(I)\leq T}|N(I)^{-s}| \leq \prod_{N(\mathfrak{p})\leq T}(1+N(\mathfrak{p})^{-\operatorname{Re}s}+N(\mathfrak{p})^{-2\operatorname{Re}s}+\ldots).$$

But the series in brackets on the right-hand side does not exceed $1+3N(\mathfrak{p})^{-\operatorname{Re}s}$ and so

$$\sum_{N(I)\leq T}|N(I)^{-s}| \leq \prod_{N(\mathfrak{p})\leq T}(1+3N(\mathfrak{p})^{-\operatorname{Re}s}) \leq B,$$

B being a constant, because the series $3\sum_{\mathfrak{p}}N(\mathfrak{p})^{-\operatorname{Re}s}$ converges in our range uniformly in every compact set. This proves everything except the infinite product representation of $\zeta_K(s)$. To get this we need an analogue of Euler's product expansion.

Lemma 7.1. *Let $f(I)$ be a complex-valued function defined on the set of all non-zero ideals of R_K and assume that it is multiplicative, i.e. $f(IJ) = f(I)f(J)$, for relatively prime ideals I and J. Moreover, assume that the series*

$$G(f,s) = \sum_I f(I)N(I)^{-s}$$

converges absolutely in the half-plane $\operatorname{Re} s > C$. Then we have the equality

$$G(f,s) = \prod_{\mathfrak{p}} \sum_{m=0}^{\infty} f(\mathfrak{p}^m)N(\mathfrak{p})^{-ms},$$

holding in the half-plane $\operatorname{Re} s > C$, with $\mathfrak{p}$ ranging over all non-zero prime ideals of R_K.

Proof. Fix a real positive number T and denote by A_T the set of all ideals of R_K such that all prime ideals dividing them have norms not exceeding T. Then obviously

$$\prod_{N(\mathfrak{p})\leq T}\sum_{m=0}^{\infty} f(\mathfrak{p}^m)N(\mathfrak{p})^{-ms} = \sum_{I\in A_T} f(I)N(I)^{-s},$$

and so the difference

$$\left|\prod_{N(\mathfrak{p})\leq T}\sum_{m=0}^{\infty} f(\mathfrak{p}^m)N(\mathfrak{p})^{-ms} - \sum_{N(I)\leq T} f(I)N(I)^{-s}\right|$$

does not exceed

$$\sum_{N(I)>T}|f(I)N(I)^{-s}|,$$

and this tends to zero in $\operatorname{Re} s > C$, according to our assumptions. □

The application of this lemma to $f(I) = 1$ proves the proposition. □

Corollary 1. *The function $\zeta_K(s)$ does not vanish in the half-plane $\operatorname{Re} s > 1$.*

Proof. No term of the infinite product for $\zeta_K(s)$ can vanish in this half-plane. □

Corollary 2. *The function $\zeta_K(s)^{-1}$ is regular for $\operatorname{Re} s > 1$ and is there equal to the series $\sum_I \mu(I)N(I)^{-s}$, where $\mu(I) = (-1)^k$ if I is the product of k distinct prime ideals, $\mu(I) = 0$ if I is divisible by the square of a prime ideal and $\mu(R_K) = 1$. Moreover, in the same range we have*

$$|\zeta_K(s)^{-1}| \leq \zeta_K(\operatorname{Re} s).$$

Proof. Regularity follows from the preceding corollary. Since the function $\mu(I)$ satisfies the equality $\sum_{I|J} \mu(I) = \varepsilon(J)$ (with $\varepsilon(R_K) = 1$, $\varepsilon(J) = 0$ for $J \neq R_K$), which can be proved in exactly the same way as the corresponding result for the classical Möbius-function, we obtain

$$\sum_I \mu(I) N(I)^{-s} \cdot \zeta_K(s) = 1,$$

as asserted. The final inequality results from $|\mu(I)| \leq 1$. □

Corollary 3. *If we write*

$$\zeta_K(s) = \sum_{n=1}^{\infty} F(n) n^{-s},$$

then $F(n)$ is the number of ideals of R_K with norm n and for every positive ε we have $F(n) = o(n^\varepsilon)$. Moreover, for $\operatorname{Re} s > 1$ we have

$$|\zeta_K(s)| \leq \zeta(\operatorname{Re} s)^N,$$

where $N = [K:\mathbb{Q}]$ and $\zeta(s) = \zeta_\mathbb{Q}(s)$ is the Riemann zeta-function.

Proof. The first statement is obvious and the second is contained in Lemma 4.2. To obtain the last assertion observe that if $d_N(n)$ denotes the number of representations of n as a product of N factors, then Lemma 4.2 implies $F(n) \leq d_N(n)$, hence

$$|\zeta_K(s)| \leq \sum_{n=1}^{\infty} F(n) n^{-\operatorname{Re} s} \leq \sum_{n=1}^{\infty} d_N(n) n^{-\operatorname{Re} s} = \zeta(\operatorname{Re} s)^N.$$

□

2. Now we prove the principal result concerning the zeta-function:

Theorem 7.1. *The function $\zeta_K(s)$ can be continued analytically to a meromorphic function with a simple pole at $s = 1$, where it has the residue $h\varkappa$ with $\varkappa$ defined by (6.12). Moreover, the function*

$$\Phi(s) = A^s \Gamma(s/2)^{r_1} \Gamma(s)^{r_2} \zeta_K(s)$$

(where $A = 2^{-r_2}\pi^{-n/2}|d(K)|^{1/2}$, $n = r_1+2r_2 = [K:Q]$), satisfies the functional equation $\Phi(s) = \Phi(1-s)$ and in every vertical strip $c_1 \leqslant \operatorname{Re} s \leqslant c_2$ we have
$$|s(s-1)\Phi(s)| \leqslant B(c_1, c_2) \quad \text{for a positive } B(c_1, c_2).$$

Proof. We are going to apply Theorem 6.9 to the function $f(x) = \prod_v f_v(x_v)$ ($x = (x_v) \in I_K$), where for non-Archimedean v the function f_v is the characteristic function of the ring R_v of integers of K_v, whereas for Archimedean v
$$f_v(x_v) = \begin{cases} \exp(-\pi x_v^2), & v \text{ real,} \\ \exp(-2\pi v(x_v)), & v \text{ complex.} \end{cases}$$

(Remember that for a complex number z and complex valuation v we have $v(z) = |z|^2$.)

Let us now compute the Fourier transforms and zeta-functions corresponding to the quasicharacter $v(x_v)^s$ of the functions $f_v(x_v)$. In the real Archimedean case we have
$$\hat{f}_v(y_v) = \int_{-\infty}^{\infty} \exp(-\pi x_v^2 + 2\pi i x_v y_v) dx_v = \exp(-\pi y_v^2) = f_v(y_v)$$

and
$$\int_{K_v^*} f_v(x_v) v(x_v)^s dx_v^* = \int_{K_v^*} \hat{f}_v(y_v) v(y_v)^s dy_v^* = \pi^{-s/2}\Gamma(s/2), \tag{7.2}$$

which is easily checked.

Similarly, in the complex Archimedean case we get
$$\hat{f}_v(y_v) = f_v(u+iv) = 2\int_{-\infty}^{\infty}\int_{-\infty}^{\infty} \exp(-2\pi(x^2+y^2)+4\pi i(xu-yv)) dx dy$$
$$= \exp(-2\pi(u^2+v^2)) = f_v(y_v),$$

where $x_v = x+iy$ and $y_v = u+iv$ with real x, y, u, v and
$$\int_{K_v^*} f_v(x_v) v(x_v)^s dx_v^* = 2\int_{-\infty}^{\infty}\int_{-\infty}^{\infty} \exp(-2\pi(x^2+y^2))(x^2+y^2)^{s-1} dx dy$$
$$= 2\int_0^{\infty} dr \int_0^{2\pi} \exp(-2\pi r^2) r^{2s-1} d\varphi = \Gamma(s)/(2\pi)^{s-1}$$
$$= \int_{K_v^*} \hat{f}_v(y_v) v(y_v)^s dy_v. \tag{7.3}$$

For non-Archimedean v we obtain
$$\hat{f}_v(y_v) = \int_{K_v} f_v(x_v) X_{y_v}(-x_v) dx_v = \begin{cases} N(D_v)^{-1/2}, & y_v \in D_v^{-1}, \\ 0, & \text{otherwise,} \end{cases}$$

where $X_{y_v}(t) = X(ty_v)$, $X(t)$ is the standard character of K_v^+ and D_v is the different of the extension K_v/Q_p. Also

$$\int_{K_v^*} f_v(x_v) v(x_v)^s dx_v^* = \frac{N(p_v)}{N(p_v)-1} \int_{K_v} f_v(x_v) v(x_v)^{s-1} dx_v$$

$$= \frac{N(p_v)}{N(p_v)-1} \int_{R_v} v(x_v)^{s-1} dx_v$$

holds for $\operatorname{Re} s > 1$, p_v being the prime ideal of R_v. If we denote by π_v any of its generators, we have

$$\int_{R_v} v(x_v)^{s-1} dx_v = \sum_{j=0}^{\infty} \int_{\pi^j U(K_v)} v(x_v)^{s-1} dx_v$$

$$= \sum_{j=0}^{\infty} N(p_v)^{-js} \int_{U(K_v)} v(x_v)^{s-1} dx_v$$

$$= (1 - N(p_v)^{-s})^{-1} \int_{U(K_v)} dx_v$$

$$= (N(p_v)-1) N(p_v)^{-1} N(D_v)^{-1/2} (1 - N(p_v)^{-s})^{-1},$$

because the equality

$$N(D_v)^{-1/2} = \int_{R_v} dx_v = \sum_{j=0}^{\infty} \int_{\pi^j U(K_v)} dx_v$$

$$= \sum_{j=0}^{\infty} N(p_v)^{-j} \int_{U(K_v)} dx_v$$

implies

$$\int_{U(K_v)} dx_v = (N(p_v)-1) N(p_v)^{-1} N(D_v)^{-1/2}.$$

We have thus obtained

$$\int_{K_v^*} f_v(x_v) v(x_v)^s dx_v^* = N(D_v)^{-1/2} (1 - N(p_v)^{-s})^{-1}. \tag{7.4}$$

Similarly,

$$\int_{K_v^*} \hat{f}_v(y_v) v(y_v)^s dy_v^* = N(D_v)^{s-1} (1 - N(p_v)^{-s})^{-1}; \tag{7.5}$$

the details of computation are left to the reader.

Now we check that the function $f(x)$ satisfies the assumptions of Theorem 6.9. Since all the functions $f_v(x_v)$ are continuous and integrable in K_v^+ and since almost all of them are equal to 1 on R_v, Lemma 2 of Appendix I shows that f

is continuous and, moreover, the equality

$$\int_{K_v} |f_v(x_v)| \, dx_v = N(D_v)^{-1/2}$$

for non-Archimedean v shows that we can apply the Corollary to Lemma 2 of Appendix I to obtain the integrability of $f(x)$. Moreover, Lemma 3 and the Corollary to Lemma 4 of the same appendix together with the equality

$$\int_{K_v} |\hat{f}_v(y_v)| \, dy_v = 1$$

show that the Fourier transform $\hat{f}(y)$ is continuous and integrable and that the following formula holds:

$$\hat{f}(y) = \prod_v \hat{f}_v(y_v) \quad (y = (y_v)).$$

Hence, condition (i) of Theorem 6.9 is surely satisfied and we may turn to condition (ii). Readers acquainted with Hecke's classical proof of our theorem, which was based on the properties of theta-functions, will recognize them hidden in the background of the verification of (ii), which now follows.

Let $a = (a_v) \in I_K$, $x = (x_v) \in A_K$ and let $t \in K = A_0$. We can factorize the value of $f(a(x+t))$ into two factors:

$$\prod_{v \text{ Archimedean}} f_v(a_v(x_v+t)) \quad \text{and} \quad \prod_{v \text{ non-Archimedean}} f_v(a_v(x_v+t)).$$

For the moment we shall call the first factor P_1 and the second P_2. It is immediate that

$$P_2 = \begin{cases} 1 & \text{if for all } v \text{ non-Archimedean } a_v(x_v+t) \in R_v, \\ 0 & \text{if this is not the case.} \end{cases}$$

On the other hand, if we denote by F_v, as before, the embeddings of K in K_v and put $\eta_v = 1$ for real v and $\eta_v = 2$ for complex v, we get

$$P_1 = \prod_{v \text{ Archimedean}} \exp(-\eta_v \pi |a_v(x_v+F_v(t))|^2),$$

and thus

$$\sum_{t \in K} f(a(x+t)) = \sum_{t \in \Omega(a,x)} \exp\left(-\pi \sum_{v \text{ Archimedean}} \eta_v |a_v(x_v+F_v(t))|^2\right),$$

where $\Omega(a, x)$ denotes the set of all elements t of K such that $a_v(x_v+t) \in R_v$ for all non-Archimedean v's. Now let C_1, C_2 be compact subsets of I_K, A_K respectively and let

$$\Omega = \bigcup_{a \in C_1} \bigcup_{x \in C_2} \Omega(a, x).$$

The compactness of C_1 and C_2 implies the existence of positive constants c_1, c_2, c_3 and a finite set of valuations $S \subset V$ such that for any (a, x) in the product $C_1 \times C_2$ we have $v(a_v) = 1$, $v(x_v) \leq 1$ outside S and $c_1 \leq v(a_v) \leq c_2$, $v(x_v) \leq c_3$ for $v \in S$. Writing n_v for the exponent of K_v in the case of non-Archimedean v, we get $b_1 \leq n_v(a_v) \leq b_2$, $n_v(x_v) \geq b_3$ for non-Archimedean $v \in S$, with suitable b_1, b_2, b_3. Now let

$$J = \prod_{\text{non-Archimedean } v \in S} p_v^{\min(b_3, -b_2)}$$

and observe that the set Ω is contained in the fractional ideal J. In fact, if t is any element of Ω, then for some $(a_v) \in C_1$, $(x_v) \in C_2$ we have the inequality

$$n_v(a_v(x_v + t)) \geq 0$$

for non-Archimedean v; hence, for such v we get

$$n_v(t) = n_v(x_v + t - x_v) \geq \min(n_v(x_v), n_v(x_v + t))$$
$$\geq \min(n_v(x_v), -n_v(a_v)) \geq \min(b_3, -b_2)$$

if $v \in S$ and similarly $n_v(t) \geq 0$ if $v \notin S$.

It remains to prove that the series

$$\sum_{t \in J} \exp\left(-\pi \sum_{v \text{ Archimedean}} \eta_v |a_v(x_v + F_v(t))|^2\right)$$

is uniformly convergent for $(a, x) \in C_1 \times C_2$. For this purpose let $w_1, w_2, \ldots, w_n$ be any Z-basis of J so that every element t of J can be uniquely written as

$$t = y_1 w_1 + \ldots + y_n w_n \quad (y_i \in Z).$$

Our series becomes

$$\sum_{y_1, \ldots, y_n \in Z} \exp\left(-\pi \sum_{v \text{ Archimedean}} \eta_v \left|a_v\left(x_v + \sum_{j=1}^{n} y_j F_v(w_j)\right)\right|^2\right).$$

But our assumptions imply that $|a_v|$ is bounded from below by a positive constant and $|x_v|$ is bounded from above; hence, the inner sum exceeds a constant multiple of $\sum_{j=1}^{n} y_j^2$ except for a finite number of terms depending only on C_2; and finally we see that our series is majorized by a constant multiple of

$$\left(\sum_{y=-\infty}^{\infty} \exp(-By^2)\right)^n,$$

which is obviously convergent.

In the same way we deal with the second series occurring in condition (ii) of Theorem 6.9.

Condition (iii) is in our case a consequence of the estimates

$$\sup_S \prod_{v \in S} \int_{K_v^*} |f_v(x_v)| v(x_v)^t dx_v^*$$

$$\leqslant (\pi^{-t/2}\Gamma(t/2))^{r_1}((2\pi)^{1-t}\Gamma(t))^{r_2}$$

$$\times \prod_{v \text{ non-Archimedean}} (1-N(\mathfrak{p}_v)^{-t})^{-1} |d(K)|^{-1/2}$$

and

$$\sup_S \prod_{v \in S} \int_{K_v^*} |\hat{f}_v(y_v)| v(y_v)^t dy_v^*$$

$$\leqslant (\pi^{-t/2}\Gamma(t/2))^{r_1}((2\pi)^{1-t}\Gamma(t))^{r_2}$$

$$\times \prod_{v \text{ non-Archimedean}} (1-N(\mathfrak{p}_v)^{-t})^{-1} |d(K)|^{t-1},$$

which follow from (7.2), (7.3), (7.4) and (7.5). The convergence of the product

$$\prod_{v \text{ non-Archimedean}} (1-N(\mathfrak{p}_v)^{-t})^{-1}$$

is ensured by Proposition 7.1.

So, all conditions of Theorem 6.9 are satisfied by our function $f(x)$. If we define a function $\Psi(s)$ for $\text{Re}\, s > 1$ by

$$\Psi(s) = \prod_{v \text{ Archimedean}} \int_{K_v^*} f_v(x_v) v(x_v)^s dx_v^* = \prod_{v \text{ Archimedean}} \int_{K_v^*} \hat{f}_v(y_v) v(y_v)^s dy_v^*,$$

then, again using equalities (7.2)–(7.5), we obtain

$$Z(f, s) = \Psi(s) \zeta_K(s) |d(K)|^{-1/2}$$

and

$$Z(\hat{f}, s) = \Psi(s) \zeta_K(s) |d(K)|^{s-1}.$$

By Theorem 6.9 these two zeta-functions can be analytically continued to meromorphic functions, whose only poles can lie at $s = 0$ or $s = 1$. Furthermore, Ψ satisfies the functional equation

$$\Psi(s) \zeta_K(s) |d(K)|^{-1/2} = \Psi(1-s) \zeta_K(1-s) |d(K)|^{-s}. \tag{7.6}$$

We check without difficulty that this is just the equation which occurs in the formulation of the theorem. At $s = 1$ the function $Z(f, s)$ has a simple pole, and since $\Psi(1) = 1$, we have $\hat{f}(0) = |d(K)|^{-1/2}$, which shows that $\zeta_K(s)$ has at $s = 1$ a simple pole with residue $h\kappa$. At $s = 0$ the function $\Psi(s)\zeta_K(s)$ has a simple pole but $\Psi(s)$ has a pole of order r_1+r_2 and so $\zeta_K(s)$ is regular at $s = 0$ (having a zero of order r_1+r_2-1 there, provided $r_1+r_2 \neq 1$). The final assertion follows immediately from the corresponding assertion of Theorem 6.9. □

Corollary 1. *The following equality holds:*

$$\int_{V(x)\geq 1} (\hat{f}(x)V(x)^s + f(-x)V(x)^{1-s})dm_I(x) + h\varkappa\left(\frac{1}{s-1} - \frac{1}{s|d(K)|^{1/2}}\right)$$
$$= \zeta_K(s)|d(K)|^{s-1}\Gamma(s/2)^{r_1}\Gamma(s)^{r_2}\pi^{-sr_1/2}(2\pi)^{r_2(1-s)}.$$

Proof. Apply the equality (6.17) to the function $\hat{f}(y)$ and $q_a = 1$, remembering that $\hat{\hat{f}}(x) = f(-x)$. □

Corollary 2. *If $s > 1$, then the residue $h\varkappa$ of $\zeta_K(s)$ fulfils the estimate*

$$h\varkappa \leq s(s-1)D^{1/2}(sD^{1/2}+1-s)^{-1}\zeta_K(s)D^{s-1}\Gamma(s/2)^{r_1}\Gamma(s)^{r_2}\pi^{-sr_1/2}(2\pi)^{r_2(1-s)},$$

with $D = |d(K)|$.

Proof. Apply the preceding corollary and note that the integrand is non-negative. □

Corollary 3. *For $0 < a \leq 1$ we have*

$$h\varkappa \leq 2^{1+n}(D\pi^{-n/2})^a a^{1-n} \leq 2^{1+n}D^a a^{1-n}$$

with $n = [K:Q]$ and $D = |d(K)|$.

Proof. Put in the preceding corollary $s = 1+a$ and observe that for $0 < a \leq 1$ we have

$$\Gamma\left(\frac{1+a}{2}\right)^{r_1}\Gamma(1+a)^{r_2}\pi^{-(1+a)r_1/2}(2\pi)^{-ar_2} \leq \pi^{-an/2}2^{-ar_2} < (\pi^{n/2})^{-a},$$

because

$$\Gamma\left(\frac{1+x}{2}\right) \leq \Gamma(1/2) = \pi^{1/2}$$

and

$$\Gamma(1+x) \leq \Gamma(2) = 1 \quad \text{for } x \in (0, 1].$$

Now, Corollary 3 to Lemma 7.1 shows that $\zeta_K(x) \leq \zeta(x)^n$ holds for $x > 1$, and in view of the inequality

$$\zeta(x) = \sum_{n=1}^{\infty} n^{-x} \leq 1 + \int_1^{\infty} t^{-x}dt = \frac{x}{x-1} \leq \frac{2}{x-1},$$

holding for $1 < x \leq 2$ we get $\zeta_K(x) \leq 2^n(x-1)^{-n}$.

The assertion now results by observing that

$$s(s-1)D^{1/2}(sD^{1/2}+1-s)^{-1} \leq 2(s-1). \qquad \square$$

§ 1. The Classical Zeta-Functions

Corollary 4. *For all fields of fixed degree n we have*
$$h(K)R(K) = O(D^{1/2}\log^{n-1}D)$$
with $D = |d(K)|$.

Proof. Put $a = \log^{-1}D$ in the last corollary. This leads to $h\varkappa \leqslant C\log^{n-1}D$ with a certain $C = C(n)$. Using the equality
$$h(K)R(K) = h(K)\varkappa D^{1/2}w(K)2^{-r_1-r_2}\pi^{-r_2}$$
(where $w(K)$ is the number of roots of unity contained in K) we arrive at our assertion, because there are only finitely many roots of unity of bounded degree. □

Corollary 5. *If K and L are two algebraic number fields having the same Dedekind zeta-function, then they have equal degrees and discriminants, and moreover*
$$\frac{h(K)R(K)}{w(K)} = \frac{h(L)R(L)}{w(L)}.$$

Proof. If $\zeta_K(s) = \zeta_L(s)$, then Corollary 3 to Lemma 7.1 shows that for every n the numbers of ideals of norm n in R_K and R_L coincide. For prime n these numbers are bounded by the corresponding degrees and by Theorem 4.14 the bounds are attained. This implies $[K:Q] = [L:Q]$. If now Φ_K and Φ_L are the functions appearing in Theorem 7.1, corresponding to K and L, and A_K, A_L are the respective constants, then
$$\Phi_K(s) = \Phi_K(1-s) = A_K^s \Gamma(s/2)^{r_1(K)}\Gamma(s)^{r_2(K)}\zeta_K(s)$$
and similarly
$$\Phi_L(s) = \Phi_L(1-s) = A_L^s \Gamma(s/2)^{r_1(L)}\Gamma(s)^{r_2(L)}\zeta_L(s).$$

Division yields
$$(A_K/A_L)^{2s-1}\Gamma(s/2)^{r_1(K)-r_1(L)}\Gamma(s)^{r_2(K)-r_2(L)}$$
$$= \Gamma((1-s)/2)^{r_1(K)-r_1(L)}\Gamma(1-s)^{r_2(K)-r_2(L)}.$$

The right-hand side is regular in a neighbourhood of $s = 0$, and since $\Gamma(s)$ has a pole there, we must have $r_1(K)+r_2(K) = r_1(L)+r_2(L)$; and since the degrees of K and L coincide,
$$r_1(K) = r_1(L) = r_1, \quad r_2(K) = r_2(L) = r_2.$$

It follows that
$$(A_K/A_L)^{2s-1} = 1$$
for all s, and the equality $A_K = A_L$ results, which in turn implies $|d(K)| = |d(L)|$.

By Proposition 2.8 the signs of the discriminants coincide; hence $d(K) = d(L)$. The last assertion follows by comparing the residues of ζ_K and ζ_L at $s = 1$. □

Corollary 6. *If K is neither Q nor a quadratic imaginary field, then ζ_K has a zero at $s = 0$ of order r_1+r_2-1. In the two cases just mentioned ζ_K is regular and non-zero at $s = 0$.*

Proof. Apply the functional equation and use the fact that $\Gamma(s)$ has a simple pole at $s = 0$. □

3. We now introduce a class of multiplicative complex-valued functions defined on the group $G(K)$ of all fractional ideals of K, which we shall call *Hecke characters*. Let X be any quasicharacter of the idele-class-group $C(K)$, i.e. a quasicharacter of I_K trivial on I_0. According to Theorem IX of Appendix I we can write

$$X((x_v)) = \prod_v X_v(x_v),$$

where X_v is for each v a quasicharacter of K_v^* equal to unity on $U(K_v)$ for almost all v's.

Let S be a finite subset of V containing S_∞, the set consisting of all Archimedean valuations and also of all those v for which X_v is non-trivial on $U(K_v)$. For $v \notin S$ the value $X_v(x_v)$ depends only on $v(x_v)$ and thus we can consider X_v as a quasicharacter of the group of all fractional ideals of K_v, which is a cyclic infinite group. We now define a function χ on $G(K)$ by putting for all prime ideals $\mathfrak{P}$

$$\chi(\mathfrak{P}) = \begin{cases} X_v(\mathfrak{p}_v) & \text{if } \mathfrak{P} = \mathfrak{p}_v, v \notin S, \\ 0 & \text{otherwise} \end{cases}$$

($\mathfrak{p}_v$ denoting the prime ideal of R_v) and extending it to $G(K)$ by multiplicativity:

$$\chi(I_1 I_2) = \chi(I_1)\chi(I_2).$$

The restriction of χ to the subgroup of $G(K)$ generated by the prime ideals $\mathfrak{p}_v$ ($v \notin S$) is a quasicharacter. Although χ itself is not a quasicharacter (except when $S = S_\infty$), since it can assume value zero, we call it a *Hecke character*. If X is a character of $C(K)$, then χ is called a *proper Hecke character*. The set S is called the *exceptional set of χ*.

If X is a character which satisfies the normalization condition stated in Lemma 6.9, then χ is called a *normalized Hecke character*. Note that, in view of (6.11), if χ is an arbitrary Hecke character, we can represent it uniquely as a product

$$\chi(I) = \chi_1(I) N(I)^s \qquad (7.7)$$

with a suitable normalized (and thus proper) Hecke character χ_1 and a complex number s. Moreover, the character χ is proper iff $\operatorname{Re} s = 0$.

The formula (7.7) shows that in most questions one can restrict attention to proper Hecke characters.

Our first aim is to give an intrinsic characterization of Hecke characters. To this purpose we consider a slightly more general situation. Let G be a complete metric Abelian group and let S be a finite subset of V containing all Archimedean valuations of K. Denote by $G(K; S)$ the group of all fractional ideals of K whose factorizations into prime ideals contain no factors $\mathfrak{p}_v$ with $v \in S$. Moreover, for any fractional ideal I of K denote by I_S the S-free part of I, i.e. the unique fractional ideal of $G(K; S)$ for which we have $I = I_S I'$, all prime factors of I' being in S. A homomorphism $f: G(K; S) \to G$ will be called *admissible* if to every neighbourhood U of unity in G there is a positive ε such that for every x in K^* satisfying $v(x-1) < \varepsilon$ for $v \in S$ the element $f((xR_K)_S)$ lies in U.

It will turn out that proper Hecke characters are exactly those homomorphisms of $G(K)$ into T which are induced by admissible homomorphisms of groups $G(K; S)$. This is implied by the following result, which we shall later use also in the case of finite G:

Theorem 7.2. *Let G be a complete metric Abelian group possessing a neighbourhood of the unit element which does not contain any non-trivial subgroup of G. Then a homomorphism $f: G(K; S) \to G$ is admissible iff there is a continuous homomorphism F of I_K into G, trivial on I_0, such that for every idele (x_v) with $x_v = 1$ for $v \in S$ we have*

$$F((x_v)) = f \circ g((x_v)),$$

where g is the canonical map $I_K \to I_K/U_K = G(K)$.

If such an F exists, then it is necessarily unique.

Proof. We deal with the sufficiency part first. Let F satisfy all conditions stated and let us write it as

$$F(x_v) = \prod_v F_v(x_v),$$

by Theorem IX of Appendix I. Here F_v are continuous homomorphisms of K_v^* into G, almost all of them being trivial on $U(K_v)$. Since G does not have arbitrarily small subgroups, there corresponds to each non-Archimedean $v \in S$ an ideal $\mathfrak{f}_v$ of R_v such that $x_v \equiv 1 \pmod{\mathfrak{f}_v}$ implies $F_v(x_v) = 1$; and so, if $v(x-1)$

is sufficiently small, for each non-Archimedean $v \in S$, then

$$\prod_{\text{non-Archimedean } v \in S} F_v(x_v) = 1.$$

But then we have for $x \in K^*$

$$1 = F(x) = \prod_{v \text{ Archimedean}} F_v(x) \cdot f((xR_K)_S),$$

i.e.

$$f((xR_K)_S) = \prod_{v \text{ Archimedean}} F_v(x^{-1}).$$

If $v(x-1)$ is small for Archimedean v, then $F_v(x)$ and hence also $F_v(x^{-1})$ are close to 1. Choosing for a given neighbourhood U of unity in G a suitable ε, we can ensure $f((xR_K)_S) \in U$ provided $v(x-1) < \varepsilon$ for $v \in S$.

This shows sufficiency, and now we prove necessity. Let f be an admissible homomorphism of $G(K; S)$ into G. We have to define $F((x_v))$ so that for any idele with $x_v = 1$ for $v \in S$ it coincides with $f \circ g((x_v))$, is continuous and trivial on I_0. Thus let $x = (x_v)$ be any given idele. The strong approximation theorem (Theorem 6.7) yields the existence of a sequence a_n of principal ideles such that

$$\lim_n v(x^{-1} - a_n) = 0 \qquad (7.8)$$

holds for each $v \in S$. Let I_n denote the ideal $g(a_n x)$ and let $A_n = (I_n)_S$. For arbitrary natural m, n we have

$$f(A_n)/f(A_m) = f((a_n a_m^{-1} R_K)_S)$$

and this tends to 1 for m, n tending to infinity (since (7.8) implies $v(a_n a_m^{-1} - 1) \to 0$ for $v \in S$); thus the sequence $f(A_n)$ is convergent to a limit which is independent of the choice of a_n provided (7.8) is satisfied. (Remember that G is complete!) Now put $F(x) = \lim_n f(A_n)$. The resulting homomorphism of I_K into G is obviously trivial on I_0 (put $a_n = x^{-1}$ for principal x) and it remains to check its continuity. Let U_1 be a given neighbourhood of the unit element in G. Take a positive ε such that for any principal idele (x_v) the inequalities $v(x_v - 1) < \varepsilon$ for $v \in S$ imply $f(g(x_v)) \in U_1$. Consider the following neighbourhood of unity in I_K:

$$U = \prod_{v \in S} O_v \times \prod_{v \notin R} U(K_v),$$

where $O_v = \{x_v : v(x_v - 1) < \varepsilon\}$. Any (x_v) in U satisfies $x_v \in U(K_v)$ for v outside S and thus

$$F((x_v)) = \lim_n f((a_n R_K)_S). \qquad (7.9)$$

§ 1. The Classical Zeta-Functions   337

Moreover, for sufficiently large n we must have $v(a_n - 1) < \varepsilon$ for $v \in S$, which leads to $f((a_n R_K)_S) \in U_1$ and (7.9) shows that $F((x_v))$ lies in the closure of U_1. Now let U_0 be any neighbourhood of unity in G; choose another neighbourhood U_1 of unity so that its closure lies in U_0 and apply the preceding argument. This gives continuity. Finally, we prove the uniqueness of F. Let S' be the complement of S and for any idele x choose $a_n \in I_0$ satisfying (7.8). We then have

$$F(x) = F(a_n x) = F((a_n x)_1) F((a_n x)_2),$$

where $(a_n x)_1$ and $(a_n x)_2$ denote the ideles whose components at $v \in S$ and $v \in S'$, respectively, agree with those of $a_n x$, whereas the remaining components are equal to 1; it follows that

$$F(x) = \lim_n F((a_n x)_1) \lim_n F((a_n x)_2) = \lim_n f(g(a_n x))$$

because of $(a_n x)_1 \to 1$, and so $F(x)$ is determined by the homomorphism f. □

Corollary 1. *If S is a finite subset of V containing all Archimedean valuations and f is any admissible homomorphism of $G(K; S)$ into T, then the function f_1 defined on $G(K)$ by*

$$f_1(I) = \begin{cases} f(I) & \text{if } I \in G(K; S), \\ 0 & \text{if } I \notin G(K; S) \end{cases}$$

is a proper Hecke character. Conversely, every proper Hecke character can be obtained in this way.

Proof. If f is admissible, then Theorem 7.2 shows that there exists a character of I_K trivial on I_0, i.e. a character X of $C(K)$ such that $X((x_v)) = f \circ g((x_v))$ for any ideals (x_v) with $x_v = 1$ for v in S. This shows that the components X_v of X are trivial on $U(K_v)$ if $v \notin S$, and it is clear that in this case f_1 coincides with the Hecke character induced by X and S. The converse is trivial. □

A large class of Hecke characters is induced by characters of the groups $H_f^*(K)$. Namely, we have

Corollary 2. *Let χ be a character of $H_f^*(K)$ and let φ be the canonical mapping of $G(K; S)$ onto $H_f^*(K)$ (S being the set of $v \in V$ which are either Archimedean or for which the prime ideal $\mathfrak{p}_v$ of R_v appears in the factorization of $\mathfrak{f}$). Then the function*

$$f(I) = \begin{cases} \chi(\varphi(I)) & \text{if } I \in G(K; S), \\ 0 & \text{otherwise} \end{cases}$$

is a proper Hecke character, and the character X of $C(K)$ inducing $\mathfrak{f}$ is of finite order.

Conversely, if X is a character of finite order of C(K), then every Hecke character induced by X arises in this way from a character of $H_{\mathfrak{f}}^*(K)$ with a suitable $\mathfrak{f}$.

Proof. Write

$$\mathfrak{f} = \prod_v \mathfrak{p}_v^{a_v}$$

and consider the set A of all ideles (x_v) satisfying $v(x_v - 1) < 1/2$ for real Archimedean v, and $x_v \equiv 1 \pmod{\mathfrak{p}_v^{a_v}}$ for $\mathfrak{p}_v$ dividing $\mathfrak{f}$. Every principal idele from this set is totally positive and congruent to unity (mod $\mathfrak{f}$); hence the principal ideal generated by x lies in the principal class of $H_{\mathfrak{f}}^*(K)$ and so $\chi(\varphi((xR_K)_S)) = 1$, i.e. the composition $\chi \circ \varphi$ is admissible.

If X is the character of the idele-class-group inducing $f(I)$, then for any idele $x = (x_v)$ lying in A which additionally satisfies the equality $v(x_v) = 1$ for $v \notin S$ we have

$$X(x) = f(g(a_n x)_S)$$

for large n, with a_n as in (7.8), because the image of f must be discrete. However,

$$f(g(a_n x)_S) = f(g(a_n)_S)$$

since, outside S, $(a_n x)_v$ and $(a_n)_v$ generate the same ideal. Finally, we see that $X(x) = 1$ holds because the a_n's ultimately fall into A. This proves the triviality of X on the subgroup $\bar{A}$ of $C(K)$ generated by the image of A. However, this subgroup has a non-void interior and hence is open and by Corollary to Proposition 6.14 it is of finite index. But then X must be of finite order.

Now assume that X is a character of $C(K)$ of finite order. If $X = \prod_v X_v$, then each character X_v is also of finite order; in particular, we must have $X_v(x_v) = 1$ for complex Archimedean v and either $X_v(x_v) = 1$ or $X_v(x_v) = \operatorname{sgn} x_v$ for real Archimedean v. If $f(I)$ is a Hecke character induced by X and S is its exceptional set, it suffices to show that with a suitable ideal $\mathfrak{f}$ the conditions $x \equiv 1 \pmod{\mathfrak{f}}$, x being totally positive, imply $f((xR_K)_S) = 1$. But if, for $v \in S$ (v being non-Archimedean), $\mathfrak{f}_v$ is the conductor of X_v and $\mathfrak{f} = \prod_v \mathfrak{f}_v$, then this $\mathfrak{f}$ satisfies our needs. □

Corollary 3. *If χ is a character of the group $G(N)$ of residue classes (mod N) prime to N and we extend its definition by putting*

$$\chi(nZ) = \begin{cases} \chi(|n| \bmod N) & \text{if } (n, N) = 1, \\ 0 & \text{if } (n, N) \neq 1, \end{cases}$$

then the resulting function is a proper Hecke character.

Proof. The group $G(N)$ is canonically isomorphic to $H^*_{NZ}(Q)$ and we may apply the preceding corollary. □

Later we shall have an oportunity to use also the following

Corollary 4. *The canonical map φ of $G(K; S)$ onto $H^*_\mathfrak{f}(K)$, with S as in Corollary 2, is induced by a continuous homomorphism $F: C(K) \to H^*_\mathfrak{f}(K)$ which trivializes on the image of J'_K.*

Proof. The admissibility of φ is evident and this provides the needed F. Since J'_K is topologically isomorphic to the multiplicative group of positive reals, it cannot have proper open subgroups, and so every continuous homomorphism into a finite group must necessarily be trivial. □

We now establish a relation between the system of all groups $H^*_\mathfrak{f}(K)$ and the factor group $C(K)/D(K)$, where $D(K)$ is, as before, the unit component of $C(K)$. This relation is of importance in the class-field theory, which provides a proof of the fact that the group $C(K)/D(K)$ is canonically isomorphic to the Galois group of the maximal Abelian extension K_{ab} over K with the Krull topology (in which subgroups corresponding to finite extensions of K form a fundamental system of neighbourhoods of the unit element).

Proposition 7.2. (i) *A character X of $C(K)$ induces a character of $H^*_\mathfrak{f}(K)$ for some $\mathfrak{f}$ iff X is trivial on $D(K)$.*

(ii) *Let $H^\S(K)$ be the inverse limit of the system of all groups $H^*_\mathfrak{f}(K)$ in which the defining maps are defined for all pairs $\mathfrak{f}_1, \mathfrak{f}_2$, with $\mathfrak{f}_1$ dividing $\mathfrak{f}_2$, by $t_{\mathfrak{f}_2,\mathfrak{f}_1}: H^*_{\mathfrak{f}_2}(K) \to H^*_{\mathfrak{f}_1}(K)$ ($t_{\mathfrak{f}_2,\mathfrak{f}_1}$ maps the class containing an ideal I onto the class containing this ideal in the second group); then the groups $H^\S(K)$ and $C(K)/D(K)$ are topologically isomorphic.*

Proof. (i) By Corollary 2 to Theorem 7.2 the character X of $C(K)$ which induces a character of $H^*_\mathfrak{f}(K)$ is of finite order and conversely; so it remains to show that the characters of finite order of $C(K)$ coincide with those which are trivial on $D(K)$. If X is of finite order, then the image of $D(K)$ under X is a proper connected subgroup of T, i.e., $\{1\}$. By Corollary to Theorem 6.6 $C(K)/D(K)$ is compact and totally disconnected; hence every its character is of finite order.

(ii) Observe that the mapping

$$g: \widehat{(C(K)/D(K))} \to \text{lim dir } \widehat{H^*_\mathfrak{f}(K)},$$

which maps every character X of $C(K)$ trivial on $D(K)$ onto an element of the right-hand side determined by the set of all characters of $H^*_\mathfrak{f}(K)$ induced by X is a well-defined homomorphism. The uniqueness assertion in Theorem 7.2 jointly with Corollary 2 to it shows that g is an injection and (i) proves its surjectivity. The mapping is evidently continuous since both groups are discrete and it remains to apply the duality theorem. □

Now we define the conductor of a Hecke character. Let χ be such a character defined via a quasicharacter $X = \prod_v X_v$ of $C(K)$ with an exceptional set S. For any non-Archimedean v let $\mathfrak{f}_v$ be the conductor of X_v and observe that for the ideal $\mathfrak{f} = \prod_v \mathfrak{f}_v$ the following proposition is true:

Proposition 7.3. (i) *If $x \in K^*$, $x \equiv 1 \pmod{\mathfrak{f}}$, then for the ideal $I = xR_K$ we have*

$$\chi(I) = \prod_{v \text{ Archimedean}} X_v(x^{-1}).$$

Moreover, if X is of finite order, then also

(ii) *If $x \in K^*$, $x \equiv 1 \pmod{\mathfrak{f}}$ and is totally positive, then for the ideal $I = xR_K$ we have $\chi(I) = 1$.*

(iii) *If $\mathfrak{f}_1$ is an ideal of R_K for which (ii) holds, then $\mathfrak{f}_1$ is divisible by $\mathfrak{f}$.*

Proof. (i) For our x we have

$$1 = X(x) = \prod_{v \in S} X_v(x) \prod_{v \notin S} X_v(x) = \prod_{v \in S} X_v(x) \chi(I),$$

and so

$$\chi(I) = \prod_{\in S} X_v(x^{-1}).$$

For non-Archimedean v in S we have $x^{-1} \equiv 1 \pmod{\mathfrak{f}_v}$, which gives $X_v(x) = 1$, and so in the last formula we may omit the factors corresponding to non-Archimedean v's.

(ii) Since X_v is of finite order, it equals 1 or $\operatorname{sgn} x_v$ for real v's and 1 for complex v's. Hence, for totally positive x the product

$$\prod_{v \text{ Archimedean}} X_v(x^{-1})$$

is equal to 1.

(iii) Suppose there is a prime ideal $\mathfrak{p}_{v_0}$ such that $\mathfrak{p}_{v_0}^{n_{v_0}}$ divides $\mathfrak{f}$ but not $\mathfrak{f}_1$. Then $\mathfrak{p}_{v_0}^{n_{v_0}}$ divides $\mathfrak{f}_{v_0}$ and there is an element a_{v_0} in $K_{v_0}^*$ such that $X_{v_0}(a_{v_0}) \neq 1$ although $a_{v_0} \equiv 1 (\operatorname{mod} \mathfrak{p}_{v_0}^{n_{v_0}-1})$. Choose a totally positive x in K^* such that $x \equiv a_{v_0} \pmod{\mathfrak{p}_{v_0}}$, $x \equiv 1 \pmod{\mathfrak{p}_v}$ for every $\mathfrak{p}_v \neq \mathfrak{p}_{v_0}$ ($v \in S$) and every $\mathfrak{p}_v \neq \mathfrak{p}_{v_0}$ dividing $\mathfrak{f}_1$. Then evidently $x \equiv 1 \pmod{\mathfrak{f}_1}$; hence $\chi(I) = 1$. On the other hand

$$1 = X(x) = \prod_{v \in S} X_v(x)\chi(I) = X_{v_0}(a_{v_0}) \neq 1,$$

a contradiction. □

This proposition justifies calling $\mathfrak{f}$ the conductor of the character χ. It is easily verified that in the case covered by Corollary 3 to Theorem 7.2 this definition of conductor perfectly agrees with the usual one.

Our definition of Hecke characters does not formally coincide with the classical one, used by Hecke, and we are now going to show that the two notions are in fact the same. This is the content of

Proposition 7.4. *Let $\mathfrak{f}$ be any non-zero ideal of R_K, let $\varepsilon_1, \ldots, \varepsilon_r$ be a system of fundamental units of $U^+(K, \mathfrak{f})$ and let ξ be a generator of the group of all roots of unity congruent to unity $(\bmod \mathfrak{f})$ and contained in K. For Archimedean v choose numbers $n_v \in \mathbb{Z}$ so as to satisfy*

$$\prod_{v \text{ Archimedean}} F_v(\xi)^{n_v} = 1; \tag{7.10}$$

and for each Archimedean v let a real number a_v be given so that

$$\sum_v a_v \log|F_v(\varepsilon_j)| + \sum_{v \text{ complex}} n_v \arg F_v(\varepsilon_j)$$

is for $j = 1, 2, \ldots, r_1 + r_2 - 1 = r$ an integral rational multiple of 2π.

Now choose ideals $J_1, \ldots, J_s$ in R_K whose classes in the group $H_{\mathfrak{f}}^*(K)$ generate independent cyclic factors of it with orders $h_1, \ldots, h_s$ and let $\psi(I)$ be an arbitrary character of that group extended to a function defined in the set $G(K)$ of all ideals. For every $x \in K^*$ which is totally positive and congruent to unity $(\bmod \mathfrak{f})$ put

$$f(x) = \prod_{v \text{ Archimedean}} |x_v|^{ia_v} \prod_{\substack{v \text{ Archimedean} \\ \text{complex}}} (x_v/|x_v|)^{n_v}. \tag{7.11}$$

For any ideal of the form $I = xJ_1^{b_1} \ldots J_s^{b_s}$ (with x totally positive and congruent to unity $(\bmod \mathfrak{f})$, $0 \leq b_i < h_i$) define

$$\chi(I) = f(x)w_1^{b_1} \ldots w_s^{b_s}\psi(I), \tag{7.12}$$

where for $i = 1, 2, \ldots, s$ the numbers w_i are h_i-th roots of $f(x_i)$ arbitrarily chosen, x_i being a fixed generator of the principal ideal $J_i^{h_i}$, totally positive and congruent to unity $(\bmod \mathfrak{f})$.

The function $\chi(I)$ defined by (7.12) for ideals I relatively prime to $\mathfrak{f}$ and by $\chi(I) = 0$ for other ideals is a proper Hecke character; and conversely, every proper Hecke character is obtainable in this way.

Proof. First we check that $\chi(I)$ is well defined, i.e. depends only on I and not on the choice of x. It suffices to show that $f(\varepsilon) = 1$ for every unit ε which is totally positive and congruent to unity (mod $\mathfrak{f}$), and of course it is enough to do this for $\varepsilon = \xi, \varepsilon_1, \ldots, \varepsilon_r$.

Now, (7.10) implies $f(\xi) = 1$ and from (7.11) we obtain

$$\log f(\varepsilon_i) \equiv i \sum_{v \text{ Archimedean}} a_v \log|F_v(\varepsilon_i)| + \sum_{\substack{v \text{ Archimedean} \\ \text{complex}}} n_v L_v(\varepsilon_i) \pmod{2\pi i},$$

where we write $L_v(\varepsilon_i) = \log(F_v(\varepsilon_i)/|F_v(\varepsilon_i)|)$ for simplicity. But our choice of the a_v's implies $\log f(\varepsilon_i) \equiv 0 \pmod{2\pi i}$ and so $f(\varepsilon_i)$ is indeed equal to unity. Now observe that $\chi(I)$ is an admissible homomorphism of $G(K; S)$, where S consists of all Archimedean valuations and of those non-Archimedean ones for which the corresponding prime ideal divides $\mathfrak{f}$; this is accomplished by the observation that if for $v \in S$ we have $v(x-1)$ sufficiently small, then x is totally positive and congruent to unity (mod $\mathfrak{f}$) and x_v, for Archimedean v, is close to unity. Thus $\chi(xR_K) = f(xR_K)$ is close to unity, owing to (7.11). Now Corollary 1 to Theorem 7.2 shows that $\chi(I)$ is a Hecke character.

To prove the converse take any Hecke character $\chi(I)$ and let $\mathfrak{f}$ be its conductor. Take any element x of K^* which is totally positive and congruent to unity (mod $\mathfrak{f}$) and let A be the principal ideal generated by it. If now $X = \prod_v X_v$ is the character of the idele-class-group inducing χ, then

$$X((x_v)) = \prod_{v \in S} X_v(x_v) \chi(g((x_v))_S)$$

(S denoting as usual the finite set of valuations used to define χ); and since $X((x_v)) = 1$ for the principal idele determined by x, we obtain

$$\chi(A) = \prod_{v \in S} X_v(x_v^{-1}).$$

However, the congruence $x \equiv 1 \pmod{\mathfrak{f}}$ and the total positivity of x show that for non-Archimedean v in S we have $X_v(x_v) = 1$ and for Archimedean v we have

$$X_v(x_v) = |x_v|^{ia_v}(x_v/|x_v|)^{n_v}$$

for some real a_v and rational integral n_v equal to zero in the case of real v. Moreover, for $x = \xi, \varepsilon_1, \ldots, \varepsilon_r$ we must have $\chi(A) = 1$ and this leads to (7.10) and the form of a_v given in the statement of the theorem. Thus, if $f(x)$ is the value of $\chi(A)$ in the case of x totally positive and congruent to unity (mod $\mathfrak{f}$), then f has the form (7.11). Now let $J_1, \ldots, J_s, x_1, \ldots, x_s$ be as in the theorem. If $I = xJ_1^{b_1} \ldots J_s^{b_s}$, x being congruent to unity (mod $\mathfrak{f}$) and totally positive, then $\chi(I) = f(x)\chi(J_1)^{b_1} \ldots \chi(J_s)^{b_s}$ and obviously $\chi(J_i) = w_i$ satisfies

$w_i^{h_i} = f(x_i)$. It remains to note that $\psi(I) = \chi(I)f(x)^{-1}w_1^{-b_1}\ldots w_s^{-b_s}$ is a multiplicative function equal to unity on the set of all ideals which have a generator congruent to unity (mod $\mathfrak{f}$) and totally positive. But such a function must necessarily be a character of $H_\mathfrak{f}^*(K)$ and our proof is complete. □

Note that the last part of the proof shows that the character is normalized iff $\sum_v a_v = 0$.

4. Now we introduce zeta-functions associated with Hecke characters. As a special case we obtain the classical Dirichlet L-functions. For an arbitrary Hecke character χ we define the *Hecke zeta-function* of χ by the formula

$$\zeta(s, \chi) = \sum_I \chi(I) N(I)^{-s},$$

the sum being taken over all non-zero ideals of R_K.

Proposition 7.5. *If $\chi(I) = \chi_1(I) N(I)^w$ with a normalized Hecke character χ_1 and a complex w, then the series defining $\zeta(s, \chi)$ converges absolutely in the half-plane $\mathrm{Re}\, s > 1 + \mathrm{Re}\, w$ and defines there a regular function. Moreover, in that half-plane we have*

$$\zeta(s, \chi) = \prod_\mathfrak{p} (1 - \chi(\mathfrak{p}) N(\mathfrak{p})^{-s})^{-1}$$

and

$$\zeta(s, \chi) = \zeta(s - w, \chi_1).$$

Proof. It suffices to observe that the series for $\zeta(s, \chi)$ is majorized by the series for $\zeta_K(\mathrm{Re}(s-w))$ and to apply Proposition 7.1 and Lemma 7.1. □

This proposition shows that it suffices to study the behaviour of Hecke zeta-functions for normalized (hence proper) Hecke characters. This is accomplished in the following theorem:

Theorem 7.3. *Let $X(x)$ be a normalized character of $C(K)$ and let $\chi(I)$ be the induced Hecke character with the "exceptional" set $S \subset V$. If $X(x) = 1$, then the zeta-function of χ equals*

$$\zeta_K(s) \prod_{\text{non-Archimedean } v \in S} (1 - N(\mathfrak{p}_v)^{-s}),$$

and so it is a meromorphic function with a simple pole at $s = 1$, where it has the residue

$$h\varkappa \prod_{\text{non-Archimedean } v \in S} (1 - N(\mathfrak{p}_v)).$$

If $X(x) \neq 1$, then $\zeta(s, \chi)$ can be analytically continued to an entire function satisfying the equation

$$\zeta(s, \chi) \prod_{v \notin S} N(D_v)^{-1/2} \prod_{v \in S} \varrho_v(X_v v^s) = \zeta(1-s, \bar\chi) \prod_{v \notin S} (N(D_v)^{-s} \chi(D_v)),$$

where $\varrho_v(q)$ is the function defined on quasicharacters of K_v^* in Theorem 5.17 and in Appendix I.

Proof. The first part results immediately from Theorem 7.1 and in the second part we shall imitate the proof of that theorem. Consider the function $f((x_v)) = \prod_v f_v(x_v)$ whose components are defined in the following way: For $v \notin S$ we take f_v to be the characteristic function of R_v. For non-Archimedean $v \in S$ we first look at the conductor of X_v. If it equals $\mathfrak{p}_v^{N_v}$ ($N_v \geq 0$), then we put $f_v(x_v)$ equal to the standard character $X(x_v)$ of K_v^+ for x_v in $(D_v \mathfrak{p}_v^{N_v})^{-1}$ and zero otherwise. Notice that this function was already used in the proof of Theorem 5.17. Finally, let v be Archimedean and consider the real case first. If $X_v = v(x_v)^{it_v}$, then we put $f_v(x_v) = \exp(-\pi x_v^2)$, and in the remaining case we put $f_v(x_v) = x_v \exp(-\pi x_v^2)$. However, if v is complex and $X_v(x_v) = (x_v/|x_v|)^{N_v} v(x_v)^{it_v}$, then we put

$$f_v(x_v) = \begin{cases} \bar{x}_v^{N_v} \exp(-2\pi v(x_v)) & \text{if } N_v \geq 0, \\ x_v^{-N_v} \exp(-2\pi v(x_v)) & \text{if } N_v < 0. \end{cases}$$

These are the same functions as those used in Appendix I for the proof of the functional equation for zeta-functions in the real and complex field.

To check that our function f satisfies the assumptions of Theorem 6.9 we have to know the Fourier transforms $\hat{f}_v(y_v)$. Luckily, they were already computed in the proofs of Theorems 5.17, 7.1 and in the Appendix I, where we obtained the following results:

$$\hat{f}_v(y_v) = \begin{cases} N(D_v)^{-1/2}, & y_v \in D_v^{-1}, \\ 0, & y_v \notin D_v^{-1}, \end{cases} \quad \text{if } v \notin S,$$

$$\hat{f}_v(y_v) = \begin{cases} N(D_v)^{1/2} N(\mathfrak{p}_v)^{N_v}, & y_v \equiv 1 \pmod{\mathfrak{p}_v^{N_v}}, \\ 0, & \text{otherwise}, \end{cases} \quad \text{if } v \in S, N_v \neq 0,$$

$$\hat{f}_v(y_v) = \begin{cases} N(D_v)^{1/2}, & y_v \in R_v, \\ 0, & \text{otherwise}, \end{cases} \quad \text{if } v \in S, N_v = 0,$$

$$\hat{f}_v(y_v) = \begin{cases} f_v(y_v), & v \text{ real}, X_v = v^{it_v}, \\ i f_v(y_v), & v \text{ real}, X_v \neq v^{it_v}, \\ i^{|N_v|} f_v(y_v), & v \text{ complex}. \end{cases}$$

Proceeding as in the proof of Theorem 7.1 we find that our function f indeed satisfies the assumptions of Theorem 6.9 and so we may apply it. Write now $Z_v(f_v, s)$ for the integral

$$\int_{K_v^*} f_v(x_v) X_v(x_v) v(x_v)^s dx_v^*$$

and look at it in the case of $v \notin S$ first. We obtain

$$Z_v(f_v, s) = \int_{R_v} X_v(x_v) v(x_v)^s dx_v^*$$

$$= \sum_{j=0}^{\infty} \int_{\pi_v^j U(K_v)} X_v(x_v) v(x_v)^s dx_v^*$$

$$= \sum_{j=0}^{\infty} X_v(\pi_v)^j N(\mathfrak{p}_v)^{-js} \int_{\pi_v^j U(K_v)} dx_v^*$$

$$= \sum_{j=0}^{\infty} N(\mathfrak{p}_v)^{-js} X_v(\pi_v)^j \int_{U(K_v)} dx_v^*$$

$$= N(D_v)^{-1/2} \sum_{j=0}^{\infty} X_v(\pi_v)^j N(\mathfrak{p}_v)^{-js},$$

and so

$$Z_v(f_v, s) = N(D_v)^{-1/2} (1 - \chi(\mathfrak{p}_v) N(\mathfrak{p}_v)^{-s})^{-1} \qquad (7.13)$$

holds for $v \notin S$.

We do the same for the Fourier transform $\hat{f}_v(y_v)$, which—as we already saw in the proof of Theorem 7.1—equals the characteristic function of D_v^{-1} multiplied by $N(D_v)^{-1/2}$. Proceeding as before, we get for

$$Z_v(\hat{f}_v, 1-s) = \int_{K_v^*} \hat{f}_v(y_v) X_v(y_v) v(y_v)^{1-s} dy_v^*$$

the following equalities:

$$Z_v(\hat{f}_v, 1-s) = N(D_v)^{-1/2} \sum_{j=-m}^{\infty} \int_{\pi_v^j U(K_v)} X_v(y_v) v(y_v)^{1-s} dy_v^*$$

$$= N(D_v)^{-1/2} \sum_{j=-m}^{\infty} N(\mathfrak{p}_v)^{-j(1-s)} X_v(\pi_v)^j \int_{U(K_v)} dy_v^*$$

$$= N(D_v)^{-1} N(\mathfrak{p}_v)^{m(1-s)} X_v(\pi_v)^{-m} (1 - X_v(\pi_v) N(\mathfrak{p}_v)^{s-1})^{-1}$$

(where we put $D_v = \mathfrak{p}_v^m$), and so

$$Z_v(\hat{f}_v, 1-s) = N(D_v)^{-s} \chi(D_v) (1 - \overline{\chi(\pi_v)} N(\mathfrak{p}_v)^{s-1})^{-1}. \qquad (7.14)$$

For the zeta-function $Z(f, s)$ of f we obtain, in view of (7.13),

$$Z(f, s) = \prod_v Z_v(f_v, s) = \zeta(s, \chi) \prod_{v \notin S} N(D_v)^{-1/2} \prod_{v \in S} Z_v(f_v, s),$$

and similarly, for the zeta-function $Z(\hat{f}, 1-s)$, equation (7.14) gives

$$Z(\hat{f}, 1-s) = \prod_v Z_v(\hat{f}_v, 1-s)$$

$$= \zeta(1-s, \bar{\chi}) \prod_{v \notin S} (N(D_v)^{-s} \chi(D_v)) \prod_{v \in S} Z_v(\hat{f}_v, 1-s).$$

Now we may use Theorem 5.17 for non-Archimedean $v \in S$ and Theorem VII of Appendix I for Archimedean v, to obtain

$$Z_v(f_v, s) = \varrho_v(X_v v^s) Z_v(\hat{f}_v, 1-s) \quad (v \in S),$$

and this together with the last two equalities proves the theorem. □

For later use we now write explicitly the zeta-function

$$Z(f, s) = \prod_v Z_v(f_v, s)$$

occurring in the proof of the last theorem. We state this as

Corollary. *For the function f used in the proof above we have*

$$Z(f, s) = \zeta(s, \chi) |d(K)|^{-1/2} \prod{}_1 \prod{}_2 \prod{}_3 \prod{}_4,$$

where

$$\prod{}_1 = \prod_{\substack{v \in S \\ f_v \neq 1}} \tau_0(X_v) N(\mathfrak{f}_v D_v)^s N(\mathfrak{f}_v)^{-1} (1 - N(\mathfrak{p}_v)^{-1})^{-1},$$

$$\prod{}_2 = \prod_{\substack{v \in S \\ f_v = 1}} (1 - N(\mathfrak{p}_v)^{-s})^{-1} N(D_v)^s,$$

$$\prod{}_3 = \prod_{v \text{ real}} \Gamma\left(\frac{s + it_v + N_v}{2}\right) \pi^{-(s+it_v+N_v)/2},$$

and

$$\prod{}_4 = \prod_{v \text{ complex}} \Gamma\left(s + it_v + \frac{|N_v|}{2}\right) (2\pi)^{s + it_v - 1 - |N_v|/2}.$$

Here $\mathfrak{f}_v$ is the conductor of X_v and the numbers t_v and N_v are determined for Archimedean v's from

$$X_v(x_v) = (x_v/|x_v|)^{N_v} v(x_v)^{it_v}$$

($N_v \in \mathbb{Z}$ for complex v and $N_v = 0, 1$ for real v).

Proof. Combine the expressions for $Z_v(f_v, s)$ given in (5.7), (5.9) and (7.13) for non-Archimedean v, and for Archimedean v use the results from Appendix I. □

Our first application concerns the L-functions of Dirichlet. However, before turning attention to them we introduce primitive Hecke characters; this notion will enable us to put the results in a convenient form. It is clear that the same quasicharacter of $C(K)$ defines many distinct Hecke characters with the help of distinct finite subsets of V. If S is chosen so as to be the smallest possible (which means that it consists of all Archimedean valuations and also of those non-Archimedean v's for which X_v is non-trivial on the unit group), then we call the resulting character a *primitive Hecke character*. To every Hecke character there corresponds a unique primitive character induced by the same quasicharacter X and the smallest possible set S. It is evident that for almost all prime ideals these two Hecke characters coincide and for the remaining ideals the non-primitive character equals zero whereas the primitive one attains a non-zero value.

We now establish connection between this notion of primitivity and the usual one in the case of Hecke characters defined with the use of a character $(\bmod m)$ in the field of rational numbers, according to Corollary 3 to Theorem 7.2. First note that distinct characters, say $\chi_1(x)$ and $\chi_2(x)$, the first defined $(\bmod m)$ and the other $(\bmod n)$, may be used to define the same Hecke character $\chi(xZ)$ in the case where m and n have the same prime divisors and there are a number r, again with the same set of prime divisors, $r|(m, n)$, and a character $\chi_3(x) (\bmod r)$ such that

$$\chi_3(f_1(x)) = \chi_1(x), \quad \chi_3(f_2(y)) = \chi_2(y)$$

holds for $x \in G(m)$, $y \in G(n)$, f_1, f_2 being the canonical homomorphisms of $G(m)$, $G(n)$ onto $G(r)$. Clearly enough, this is the only case in which the equality of Hecke characters may arise.

After these remarks we may prove

Proposition 7.6. *If X is a primitive Hecke character induced by a character χ (mod m), then one can find a primitive character ψ (mod n), with a certain n dividing m, which also induces X. Conversely, every primitive character (mod m) induces a primitive Hecke character.*

Proof. If χ (mod m) induces X, then the conductor of χ equals that of X, as was observed after Proposition 7.4. If X_0 is the primitive character associated with X and ψ is the primitive character associated with χ with the conductor n ($n|m$), then obviously ψ induces X_0. The converse is also immediate. □

If χ_1 is the primitive Hecke character induced by the same quasicharacter X of $C(K)$ as a given Hecke character χ and S_1, S are the corresponding exceptional sets, then we have

$$\zeta(s, \chi) = \zeta(s, \chi_1) \prod_{v \in S \setminus S_1} (1 - \chi(\mathfrak{p}_v) N(\mathfrak{p}_v)^{-s}), \tag{7.15}$$

and so the properties of $\zeta(s, \chi)$ may be obtained from the corresponding properties of $\zeta(s, \chi_1)$. This allows us to restrict our attention to primitive characters.

Now we turn to examples and prove

Proposition 7.7. *Let $f(m)$ be a primitive character of the group $G(N)$ in the rational number field Q and let $\chi(I)$ be the Hecke character which corresponds to it. Assume $N \neq 1$ and write*

$$N = \prod_{i=1}^{r} p_i^{a_i}$$

with all exponents a_i positive. Then the function

$$L(s, \chi) = \sum_{n=1}^{\infty} \chi(n) n^{-s} \quad (\mathrm{Re}\, s > 1)$$

is an integral function which satisfies the functional equation

$$L(s, \chi) N^{s-1} \left(\prod_{i=1}^{r} \tau(f_i) \right) \Gamma(s/2) \pi^{-s/2} = L(1-s, \bar{\chi}) \Gamma((1-s)/2) \pi^{(s-1)/2}$$

if f is even and

$$iL(s, \chi) N^{s-1} \prod_{i=1}^{r} \tau(f_i) \Gamma((1+s)/2) \pi^{-(1+s)/2} = L(1-s, \bar{\chi}) \Gamma((2-s)/2) \pi^{(s-2)/2}$$

if f is odd.

Here $f = f_1 \ldots f_r$ is the factorization of f into primitive characters of $G(p_i^{a_i})$ and $\tau(f_i)$ are the corresponding Gaussian sums.

Proof. First we determine the Archimedean component $X_0(x_0)$ of the character X inducing χ. Corollary 2 to Theorem 7.2 shows that X_0 is of finite order; hence it equals either 1 or $\mathrm{sgn}\, x_0$. To determine its actual value we use Proposition 7.3 (i) and take an $x \in K^*$ congruent to unity (mod N) and negative. We obtain

$$X_0(x^{-1}) = f(|x| \bmod N)$$

and $x \equiv 1 \pmod N$ gives $|x| \equiv -1 \pmod N$; thus

$$X_0(x^{-1}) = f(-1)$$

and so $X_0(x) = X_0(x^{-1}) = \operatorname{sgn} x$ if f is odd and equals 1 if f is even. As regards the non-Archimedean components, their conductors are only of interest for us, and they are equal to $p_i^{a_i}$ for $p_i | N$ and to 1 otherwise. Using Theorem 7.3 (observe that X is necessarily normalized) and also the explicit form of ϱ_v provided by Theorem 5.17 and Theorem VII of Appendix I, we obtain the functional equation asserted. □

Corollary. *If χ is an even character $\pmod N$ with $N \neq 1$, which is primitive, then $L(s, \chi)$ has a simple zero at $s = 0$. If χ is odd, then $L(0, \chi) \neq 0$.*

Proof. Use the functional equation and the fact that the gamma-function has a simple pole at $s = 0$. □

In the same manner we can obtain the functional equation for the zeta-function of any primitive Hecke character which is in addition assumed to be normalized. Denote by $\mathfrak{f}$ its conductor and let $X = \prod_v X_v$ be the character of $C(K)$ inducing our character. Write $\mathfrak{f}_v$ for the conductor of X_v for non-Archimedean v, and for Archimedean v write

$$X_v(x_v) = (x_v/|x_v|)^{N_v} v(x_v)^{it_v}, \quad N = \sum_v N_v, \quad T = \sum_{v \text{ complex}} t_v.$$

Then we have

Proposition 7.8. *Under the above assumptions and notation we have the following equation for $\zeta(s, \chi)$:*

$$\zeta(s, \chi) \pi^{-n(s-1/2)} 2^{-2r_2(s-1/2)-2iT} |d(K)|^{s-1/2} N(\mathfrak{f})^{s-1}$$

$$\times \prod_{\mathfrak{p}_v | \mathfrak{f}} \tau_0(X_v) i^{-N} \prod_{v \text{ real}} \Gamma((N_v+s+it_v)/2) \Gamma((N_v+1-s-it_v)/2)$$

$$\times \prod_{v \text{ complex}} \Gamma(|N_v|/2+s+it_v)/\Gamma(|N_v|/2+1-s-it_v)$$

$$= \prod_{v \notin S} \chi(D_v) \zeta(1-s, \bar{\chi}).$$

Proof. Apply Theorem 7.3 using Theorem 5.17 and Theorem VII of Appendix I, which give the explicit form of ϱ_v, and remember that the normalized characters have $\sum_v t_v = 0$. □

Corollary. *If χ is a character induced by a character of some group $H_{\mathfrak{f}}^*(K)$ of the conductor $\mathfrak{f}$, then for its zeta-function we have*

$$\zeta(s,\chi)\pi^{-n(s-1/2)}2^{-2r_2(s-1/2)}|d(K)|^{s-1/2}N(\mathfrak{f})^{s-1}i^{-N}$$

$$\times \prod_{\mathfrak{p}_v | \mathfrak{f}} \tau_0(X_v)(\Gamma(s/2)/\Gamma((1-s)/2))^A$$

$$\times (\Gamma((1+s)/2)/\Gamma((2-s)/2))^{r_1-A}(\Gamma(s)/\Gamma(1-s))^{r_2}$$

$$= \prod_{\mathfrak{p}_v \nmid \mathfrak{f}} \chi(D_v)\zeta(1-s,\bar\chi),$$

A being the number of real v's with $X_v = 1$.

Proof. Since by Corollary 2 to Theorem 7.2 the characters X_v are of finite order, we can apply the last proposition with $t_v = 0$ for all v's and $N_v = 0$ for complex v's. □

Sometimes the following, slightly less precise, form of the functional equation is sufficient for applications (still under the assumptions of Proposition 7.7):

$$\zeta(s,\chi) = W(\chi)\zeta(1-s,\bar\chi)\pi^{n(s-1/2)}2^{2r_2(s-1/2)}$$

$$\times (N(\mathfrak{f})|d(K)|)^{1/2-s} \prod_{v \text{ real}} \Gamma((N_v+1-s-it_v)/2)/\Gamma((N_v+s+it_v)/2)$$

$$\times \prod_{v \text{ complex}} \Gamma(|N_v|/2+1-s-it_v)/\Gamma(|N_v|/2+s+it_v), \quad (7.16)$$

where $|W(\chi)| = 1$.

To prove this equality apply the preceding proposition and remember that by Proposition 5.19 (iii) the product

$$\prod_{\mathfrak{p}_v | \mathfrak{f}} \tau_0(X_v)$$

is of absolute value $N(\mathfrak{f})^{1/2}$.

Now we work out some evaluations of our zeta-functions, which will find applications in subsequent sections. We prove

Theorem 7.4. *Let χ be a normalized and primitive Hecke character with conductor $\mathfrak{f}$. Assume that it is induced by a character of $H_{\mathfrak{f}}^*(K)$. Write D for $|d(K)|N(\mathfrak{f})$ and let $a < 0$, $b = 1-a > 1$ be real numbers. Then for $s = \sigma+it$ ($a \leqslant \sigma \leqslant b$) we have the evaluation*

$$|\zeta(s,\chi)| \ll (b-1)^n D^{1/2-a}|t|^{(1/2-a)n}$$

for $|t|$ tending to infinity, $|t| \geqslant \varepsilon > 0$. The constant involved depends on ε and a, but not on χ, its conductor or even the field K.

§ 1. The Classical Zeta-Functions 351

Proof. A lemma is needed first.

Lemma 7.2. *Under the assumptions of the theorem we have in every fixed region $c_1 \leqslant \sigma \leqslant c_2$, $|t| \geqslant c_3 > 0$ the equality*

$$\zeta(s,\chi) = W_1(\chi)(\zeta(1-s,\bar{\chi})N(\mathfrak{f})^{1/2-s}\zeta_K(s)/\zeta_K(1-s))(1+O(\exp(-c_4 t)))$$

for a positive c_4 and $|W_1(\chi)| = 1$. The constant implied by the symbol O depends only on c_1 and c_2.

Proof. By equality (7.16) we have

$$\zeta(s,\chi)/\zeta(1-s,\bar{\chi}) = W(\chi)\pi^{n(s-1/2)}2^{2r_2(s-1/2)}D^{1/2-s}$$
$$\times (\Gamma((1-s)/2)/\Gamma(s/2))^A$$
$$\times (\Gamma((2-s)/2)\Gamma((1+s)/2))^{r_1-A}$$
$$\times (\Gamma(1-s)/\Gamma(s))^{r_2}$$
$$= W(\chi)N(\mathfrak{f})^{1/2-s}(\Gamma((2-s)/2)\Gamma(s/2)$$
$$\times \Gamma^{-1}((1-s)/2)\Gamma^{-1}((1+s)/2))^{r_1-A}$$
$$\times \zeta_K(s)/\zeta_K(1-s),$$

with A as in the Corollary to Proposition 7.8.
But $\Gamma(z)\Gamma(1-z) = \pi/\sin(\pi z)$, and so we get

$$\Gamma((2-s)/2)\Gamma(s/2)\Gamma^{-1}((1-s)/2)\Gamma^{-1}((1+s)/2)$$
$$= \cos(\pi s/2)/\sin(\pi s/2)$$
$$= i(\exp(\pi si/2)+\exp(-\pi si/2))(\exp(\pi si/2)-\exp(-\pi si/2))^{-1}$$
$$= -i(1+\exp(\pi si))(1-\exp(\pi si))^{-1}$$
$$= -i+O(\exp(-\pi t)),$$

which proves the lemma. □

To deduce the theorem let us consider the function

$$f(s) = \zeta(s,\chi)D^{a-1/2}s^{n(a-1/2)}(b-1)^{-n}$$

in the region $\{s: a \leqslant \sigma \leqslant b, |t| \geqslant \varepsilon\}$. Our aim is to show that $f(s)$ is bounded in it and to do this we use the Phragmén–Lindelöf theorem, applying it to both components of our region. It clearly suffices to consider only $\Omega = \{s: a \leqslant \sigma \leqslant b, t \geqslant \varepsilon\}$, since the general case will follow if we replace our zeta-function by the zeta-function attached to $\bar{\chi}$. So we have to prove first

$$|f(s)| \ll \exp(c_4 t) \qquad (7.17)$$

for a certain c_4 and all $s \in \Omega$ and then

$$|f(s)| \ll 1 \qquad (7.18)$$

for $s = a+it$ and $s = b+it$, the bound being independent of χ, f and K.

To obtain (7.17) note that the Corollary to Theorem 7.3 and the last sentence of Theorem 6.9 imply the following evaluation, holding for all s in Ω:

$$|\zeta(s,\chi)| \ll |d(K)|^{1/2} N(\mathfrak{f})^{\sigma-1/2} \prod_{\mathfrak{p}_v | \mathfrak{f}} N(D_v)^\sigma (1-N(\mathfrak{p}_v)^{-1})$$
$$\times |\Gamma((1-s)/2)^{-A}\Gamma((2-s)/2)^{A-r_1}\Gamma(s)^{-r_2}|\pi^m 2^{m_1},$$

where $m = (1-\sigma)A/2 + (2-\sigma)(r_1-A)/2 + (1-\sigma)r_2$ and $m_1 = (1-\sigma)r_2$; one has only to remember that all functions

$$|\Gamma(s)|^{-1}, \quad |\Gamma((1-s)/2)^{-1}|, \quad |\Gamma((2-s)/2)^{-1}|$$

are $\ll \exp(c_4 t)$. Indeed, more precise evaluations of the Γ-function are provided by

$$|\Gamma(s)| \ll |t|^{\sigma-1/2} \exp(-\pi|t|/2),$$
$$|\Gamma(s)^{-1}| \ll |t|^{1/2-\sigma} \exp(\pi|t|/2). \tag{7.19}$$

They will be useful in the proof of (7.18). To begin with, let $s = b+it$. Then $|\zeta(s,\chi)| \ll |\zeta_K(b)|$ because $b > 1$. Now use Corollary 2 to Proposition 7.1 to get

$$|\zeta(s,\chi)| \leq \zeta(b)^n \ll (b-1)^n,$$

since by Theorem 7.1 $\zeta(s)$ has a simple pole at $s = 1$ and is bounded in every half-plane $\operatorname{Re} s \geq 1+\varepsilon > 1$. This settles (7.18) for $\operatorname{Re} s = b$. To obtain (7.18) for $\operatorname{Re} s = a$ write

$\zeta(a+it, \chi)$
$= W_1(\chi)\zeta(b-it, \bar{\chi}) N(\mathfrak{f})^{1/2-s} \zeta_K(a+it)/\zeta_K(b-it)(1+O(\exp(-c_4 t)))$

(by Lemma 7.2) and apply Theorem 7.1 to get (with the notation used in it)

$\zeta_K(a+it)/\zeta_K(b-it)$
$= A^{1-2(a+it)}(\Gamma((b-it)/2)/\Gamma((a+it)/2))^{r_1}(\Gamma(b-it)/\Gamma(a+it))^{r_2}.$

This, in view of (7.19) leads to

$$|\zeta(s,\chi)| \ll (b-1)^n N(\mathfrak{f})^{1/2-a} A^{1-2a} |t|^{(1/2-a)n}$$

for our s (remember that $\zeta(b-it, \bar{\chi})$ is $\ll (b-1)^n$), and since A does not exceed $|d(K)|^{1/2}$, the bound (7.18) follows. By the Phragmén–Lindelöf theorem the function $f(s)$ is bounded throughout Ω, independently of χ, $\mathfrak{f}$ and K. □

5. We conclude this section with the proof of a result concerning the zeros of $\zeta(s,\chi)$, which will be very useful in the next section, concerning the distribution of prime ideals.

§ 1. The Classical Zeta-Functions

Theorem 7.5. *If χ is an arbitrary Hecke character, then its zeta-function does not vanish in the closed half-plane $\operatorname{Re} s \geq 1 + \operatorname{Re} w$, where w is given in Proposition 7.5. In particular, if χ is proper, then $\zeta(s, \chi) \neq 0$ for $\operatorname{Re} s \geq 1$.*

Proof. By Proposition 7.5 we may assume that χ is normalized. For $\operatorname{Re} s > 1$ our assertion follows from the fact that the Eulerian product of $\zeta(s, \chi)$ does not vanish. So let us assume that $1 + it$ is a zero of $\zeta(s, \chi)$ and exclude for the time being the case of $t = 0$, $\chi^2(I) = 1$ for all I prime to the conductor of χ. Denote by χ_0 the character with the same exceptional set S as χ, generated by the trivial character of $C(K)$. In the open half-plane $\operatorname{Re} s > 1$ we may write

$$\zeta(s, \chi_0) = \exp\left(\sum_{v \notin S} \sum_{j=1}^{\infty} j^{-1} N(\mathfrak{p}_v)^{-js}\right),$$

$$\zeta(s, \chi) = \exp\left(\sum_{v \notin S} \sum_{j=1}^{\infty} j^{-1} \chi(\mathfrak{p}_v) N(\mathfrak{p}_v)^{-js}\right)$$

and

$$\zeta(s, \chi^2) = \exp\left(\sum_{v \notin S} \sum_{j=1}^{\infty} j^{-1} \chi^2(\mathfrak{p}_v) N(\mathfrak{p}_v)^{-js}\right).$$

Now, the series

$$\sum_{v \notin S} \sum_{j=2}^{\infty} f(\mathfrak{p}_v) N(\mathfrak{p}_v)^{-js}$$

with $f(\mathfrak{p}_v)$ equal to 1, $\chi(\mathfrak{p}_v)$ or $\chi^2(\mathfrak{p}_v)$ is absolutely and uniformly convergent in the closed half-plane $\operatorname{Re} s \geq 1/2 + \varepsilon$ for any fixed positive ε and so it defines a function regular there. Indeed, it is majorized by

$$\sum_{\mathfrak{p}} \sum_{j=2}^{\infty} N(\mathfrak{p})^{-j \operatorname{Re} s} = \sum_{\mathfrak{p}} (N(\mathfrak{p})^{2 \operatorname{Re} s} - N(\mathfrak{p}^{\operatorname{Re} s}))^{-1}$$

$$\leq n \sum_{p} (p^{2 \operatorname{Re} s} - p^{\operatorname{Re} s})^{-1},$$

and this series is evidently convergent. We can thus write

$$\zeta(s, \chi_0) = H_0(s) g_0(s),$$
$$\zeta(s, \chi) = H_1(s) g_1(s),$$
$$\zeta(s, \chi^2) = H_2(s) g_2(s),$$

where $g_0(s), g_1(s), g_2(s)$ are functions regular and non-vanishing in the open half-plane $\operatorname{Re} s > 1/2$ and the functions $H_i(s)$ are defined by

$$H_i(s) = \exp\left(\sum_{v \notin S} f_i(\mathfrak{p}_v) N(\mathfrak{p}_v)^{-s}\right),$$

where $f_0 = \chi_0, f_1 = \chi$ and $f_2 = \chi^2$.

Put $F(s) = F(\sigma+it) = H_0^3(\sigma)H_1^4(s)H_2(\sigma+2it)$, and look at $\lim F(s)$, where s tends to $1+it$, the assumed zero of $\zeta(s, \chi)$, keeping $\operatorname{Im} s = t$ fixed. We assert that

$$\lim_{\sigma \to 1} F(\sigma+it) = 0. \tag{7.20}$$

Indeed, $H_1(s)$ vanishes for $1+it$, and so $H_1(s)(\sigma-1)^{-1}$ tends, as σ approaches 1, to a finite limit, say a. The function $H_0(s)$ has at $s = 1$ a simple pole, in view of the first part of Theorem 7.3, and so $H_0(\sigma)(\sigma-1)$ tends also to a finite limit, say b, when σ tends to 1. Finally, $H_2(s)$ is regular at $1+2it$ because otherwise we would have $\chi^2 = \chi_0$ and $t = 0$, the case which we excluded for the time being. But now

$$F(\sigma+it) = H_0^3(\sigma)(\sigma-1)^3 H_1^4(s)(\sigma-1)^{-4} H_2(\sigma+2it)(\sigma-1),$$

and as σ tends to 1 this expression tends to

$$a^3 b^4 H_2(1+2it) \lim_{\sigma \to 1}(\sigma-1) = 0.$$

This settles (7.20). On the other hand, we have

$$F(s) = \exp\left(\sum_{v \notin S} (3+4\chi(\mathfrak{p}_v) + \chi^2(\mathfrak{p}_v)) N(\mathfrak{p}_v)^{-s}\right);$$

hence

$$|F(s)| = \exp\left(\sum_{v \notin S} \operatorname{Re}\left((3+4\chi(\mathfrak{p}_v) + \chi^2(\mathfrak{p}_v)) N(\mathfrak{p}_v)^{-s}\right)\right)$$

$$= \exp\left(\sum_{v \notin S} N(\mathfrak{p}_v)^{-\sigma} (3+4\operatorname{Re}(\chi(\mathfrak{p}_v)N(\mathfrak{p}_v)^{-it}) + \operatorname{Re}(\chi^2(\mathfrak{p}_v)N(\mathfrak{p}_v)^{-2it}))\right),$$

and if we temporarily denote $\operatorname{Re}(\chi(\mathfrak{p}_v)N(\mathfrak{p}_v)^{-it})$ by $\cos\vartheta$, then we obtain

$$|F(s)| = \exp\left(\sum_{v \notin S} N(\mathfrak{p}_v)^{-\sigma}(3+4\cos\vartheta+\cos 2\vartheta)\right)$$

$$= \exp\left(2\sum_{v \notin S} N(\mathfrak{p}_v)^{-\sigma}(1+\cos\vartheta)^2\right) \geq 1,$$

which contradicts (7.20).

It remains to prove our theorem in the case of $\chi^2 = \chi_0$ and $t = 0$. Of course, only the case $\chi \neq \chi_0$ is interesting since $\zeta(s, \chi_0)$ has a pole at $s = 1$. Thus assume that $\chi \neq \chi_0$ and that $\zeta(1, \chi) = 0$. In this case we can write

$$\zeta(s, \chi)\zeta(s, \chi_0) = \exp G(\chi, s) \quad \text{for } \operatorname{Re} s > 1,$$

where

$$G(\chi, s) = \sum_{v \notin S} \sum_{j=1}^{\infty} j^{-1}(1+\chi(\mathfrak{p}_v)^j) N(\mathfrak{p}_v)^{-sj}.$$

The assumption $\chi^2 = \chi_0$ shows that χ assumes only the values 0, 1 and -1, and thus $1+\chi(\mathfrak{p}_v)^j$ is always non-negative. Moreover, if s is real and exceeds $1/2$, then for any fixed T we have

$$\sum_{\substack{v \notin S \\ N(\mathfrak{p}_v) \leq T}} \sum_{j=1}^{\infty} j^{-1}(1+\chi(\mathfrak{p}_v)^j)N(\mathfrak{p}_v)^{-sj} \geq \tfrac{1}{2} \sum_{\substack{v \notin S \\ N(\mathfrak{p}_v) \leq T}} N(\mathfrak{p}_v)^{-2s},$$

and with a suitable choice of T and $s > 1/2$ we can make the right-hand side arbitrarily large (by Proposition 7.1 and Theorem 7.1), which shows that the series defining $G(\chi, s)$ cannot be convergent at $s = 1/2$. Now observe that this series is a Dirichlet series with non-negative coefficients, and so in view of a theorem of Landau (see Appendix II, Theorem II) the function $G(\chi, s)$ must have a singularity on the real axis. But then $\exp G(\chi, s)$ would have a singularity as well (because $G(\chi, s)$ is non-negative to the right of this singularity). The only possible point, where $\exp G(\chi, s)$ can have a singularity is $s = 1$; but there $\zeta(s, \chi_0)$ has a simple pole and $\zeta(s, \chi)$ has a zero; thus $\exp G(\chi, s)$ is regular at that point, which gives a contradiction. □

Corollary. *The zeta-function $\zeta_K(s)$ does not vanish on the line $\operatorname{Re} s = 1$, and neither do the Dirichlet L-functions $L(s, \chi)$.* □

§ 2. Asymptotic Distribution of Ideals and Prime Ideals

1. We are now going to apply the analytical results obtained in §1 to the problem of distribution of ideals and prime ideals. We shall also use various results concerning Dirichlet series including a Tauberian theorem of Ikehara–Delange, an account of which may be found in Appendix II, and also the method of complex integration.

To avoid endless repetitions we adopt the following conventions in the sequel: by $g(s)$ (with or without indices) we denote functions which are regular in the closed half-plane $\operatorname{Re} s \geq 1$, not always the same even in the same chain of formulas. (So we write e.g. $x^s + s^{-1} = x^s + g(s) = g(s)$.) If S is a subset of V, then we write $\mathfrak{p} \in S$ to mean $\mathfrak{p} = \mathfrak{p}_v$, $v \in S$. If A is a set of prime ideals and an equality of the form

$$\sum_{\mathfrak{p} \in A} N(\mathfrak{p})^{-s} = a \log \frac{1}{s-1} + g(s)$$

holds for $\operatorname{Re} s > 1$ for a certain real a, we say that A is a *regular set of prime ideals* and we call a its *Dirichlet density*. Sometimes it is convenient to speak about the Dirichlet density of non-regular sets as well; to say that a is the density of such a set A means that for real $s > 1$ the quotient

$$\left(\sum_{\mathfrak{p} \in A} N(\mathfrak{p})^{-s}\right) : \left(a \log \frac{1}{s-1}\right)$$

tends to unity as s approaches 1. The existence of important regular sets will be demonstrated later on.

Finally, the symbols $\sum_I$, $\sum_\mathfrak{p}$ denote sums taken over all ideals and over all prime ideals of R_K, respectively. The letter $\mathfrak{p}$ always denotes a prime ideal.

Almost everything that follows is based on

Proposition 7.9. *If $\chi(I)$ is a normalized Hecke character and S is its exceptional set, then for $\operatorname{Re} s > 1$ we have*

(i) $\zeta(s, \chi) = \sum_I \chi(I) N(I)^{-s} = \alpha(\chi)/(s-1) + g(s)$,

(ii) $\sum_\mathfrak{p} \chi(\mathfrak{p}) N(\mathfrak{p})^{-s} = \beta(\chi) \log \dfrac{1}{s-1} + g(s)$, *where*

$$\alpha(\chi) = \beta(\chi) = 0 \quad \text{if } \chi \text{ is not the trivial character}$$

and

$$\alpha(\chi) = h\varkappa \prod_{\mathfrak{p} \in S}(1 - N(\mathfrak{p})^{-1}), \quad \beta(\chi) = 1 \quad \text{otherwise.}$$

Proof. Part (i) is an immediate consequence of Theorem 7.3, since in view of Theorem 7.1 the residue of $\zeta_K(s)$ at $s = 1$ equals $h\varkappa$. To prove (ii) write for $\operatorname{Re} s > 1$

$$\zeta(s, \chi) = \exp\left(\sum_{\mathfrak{p} \notin S} \chi(\mathfrak{p}) N(\mathfrak{p})^{-s}\right) g(s),$$

where

$$g(s) = \exp\left(\sum_{\mathfrak{p} \in S} \sum_{j=2}^{\infty} j^{-1} \chi(\mathfrak{p})^j N(\mathfrak{p})^{-js}\right)$$

is regular and not vanishing for $\operatorname{Re} s > 1/2$. This leads to

$$\sum_\mathfrak{p} \chi(\mathfrak{p}) N(\mathfrak{p})^{-s} = \log \zeta(s, \chi) - \log g(s) = \log \zeta(s, \chi) + g(s),$$

and the right-hand side of this equality is by Theorem 7.5 regular for $\operatorname{Re} s \geq 1$, with the possible exception of $s = 1$ in the case of trivial χ. As regards the trivial character, we have by Theorems 7.3 and 7.4

$$\zeta(s, \chi) = g(s)/(s-1),$$

with $g(s)$ non-vanishing on the line $\operatorname{Re} s = 1$. Taking logarithms we arrive at (ii). □

§ 2. Asymptotic Distribution of Ideals and Prime Ideals

Corollary 1. *The set of all prime ideals of K is regular and its Dirichlet density equals* 1. □

Corollary 2. *The set of all prime ideals of the first degree over Q is regular and its Dirichlet density equals* 1.

Proof. Denote this set by A. Clearly

$$\sum_{\mathfrak{p} \in A} N(\mathfrak{p})^{-s} = \sum_{\mathfrak{p}} N(\mathfrak{p})^{-s} - \sum_{\mathfrak{p} \notin A} N(\mathfrak{p})^{-s},$$

and the series $\sum_{\mathfrak{p} \notin A} N(\mathfrak{p})^{-s}$ is majorized by

$$[K:Q] \sum_{\substack{p \\ k \geq 2}} p^{-k \operatorname{Re} s},$$

which converges uniformly for $\operatorname{Re} s \geq 3/4$, and so the proposition just proved implies

$$\sum_{\mathfrak{p} \in A} N(\mathfrak{p})^{-s} = \log(1/(s-1)) + g(s). \qquad \square$$

Corollary 3. *If L/K is a finite extension and A is the set of all prime ideals in L which are of the first degree over K, then A is regular and its Dirichlet density is* 1.

Proof. Let B be the set of all prime ideals of L which are of the first degree over Q. Clearly $B \subset A$ and

$$\sum_{\mathfrak{p} \in A \setminus B} N(\mathfrak{p})^{-s} = g(s),$$

this being a subseries of $\sum_{\mathfrak{p} \notin B} N(\mathfrak{p})^{-s}$, which—as we have just seen—converges uniformly for $\operatorname{Re} s \geq 3/4$. Hence, using the preceding corollary, we obtain our assertion. □

Corollary 4. *Let L/K be a finite normal extension of K and let A be the set of all prime ideals of K which split in L/K, i.e. which become products of $[L:K]$ distinct prime ideals of the first degree in L. Then A is regular and its Dirichlet density is $1/[L:K]$.*

Proof. Since L/K is normal, a prime ideal of K splits in L/K iff it has at least one prime divisor in L which is of first degree over K and is not ramified. Hence

$$\sum_{\mathfrak{p} \in A} N(\mathfrak{p})^{-s} = [L:K]^{-1} \sum_{\substack{\mathfrak{P} \\ f_{L/K}(\mathfrak{P})=1 \\ e_{L/K}(\mathfrak{P})=1}} N(\mathfrak{P})^{-s},$$

but as the number of ramified prime ideals is finite, the preceding corollary gives

$$\sum_{\mathfrak{p}\in A} N(\mathfrak{p})^{-s} = [L:K]^{-1}\log(1/(s-1))+g(s). \qquad \square$$

Corollary 5. *If L/K is finite and $L \neq K$, then there exist infinitely many prime ideals of K which do not split in L/K. If moreover L/K is normal, then they form a regular set of Dirichlet density $1-[L:K]^{-1}$.*

Proof. If L/K is normal, then the assertions follow from Corollaries 1 and 4. If L/K is not normal and $L = L_1, \ldots, L_r$ are all conjugates to L over K, then their composite field M is normal over K. If $\mathfrak{p}$ splits in L/K then by Corollary to Theorem 5.5 it splits also in M/K, hence we may apply the part already proved. $\qquad \square$

It should be noted that one cannot replace the words "do not split" by "remain prime", even for $K = Q$. In fact, every normal non-cyclic field can serve as a counter-example, as we will see later. Now we only show that the cyclotomic field $Q(\zeta_8)$ has the property that no prime ideal of the rational field remains prime in it. Indeed, by Theorem 4.16 we see that an odd prime generates in $Q(\zeta_8)$ a prime ideal iff it is a primitive root (mod 8), but there is no such prime. As regards $p = 2$, it does not remain prime even in $Q(i) \subset Q(\zeta_8)$.

Finally we prove

Corollary 6. *Let $\mathfrak{f}$ be any ideal of R_K and let X be a class from $H_\mathfrak{f}^*(K)$. Then the set of all prime ideals which lie in X is regular and has $1/h_\mathfrak{f}^*(K)$ for its density*

Proof. Let $\mathscr{X}$ be the set of all characters of $H_\mathfrak{f}^*(K)$. By the orthogonality property of characters of a finite Abelian group we obtain (for $\text{Re}\, s > 1$)

$$\sum_{\mathfrak{p}\in X} N(\mathfrak{p})^{-s} = (h_\mathfrak{f}^*(K))^{-1}\sum_{\chi\in\mathscr{X}}\overline{\chi(X)}\sum_{\mathfrak{p}}\chi(\mathfrak{p})N(\mathfrak{p})^{-s},$$

which equals $(h_\mathfrak{f}^*(K))^{-1}\log(1/(s-1))+g(s)$ by Proposition 7.9 if we recall that every character of $H_\mathfrak{f}^*(K)$ can be considered as a Hecke character. $\qquad \square$

Corollary 7. *In every class of $H_\mathfrak{f}^*(K)$ there are infinitely many prime ideals, even of the first degree.*

Proof. Apply Corollary 6 and note that by Corollary 2 every set of prime ideals of degree $\neq 1$ is regular and its density is 0. $\qquad \square$

Note that in the special case of $K = Q$ the last corollary coincides with

§ 2. Asymptotic Distribution of Ideals and Prime Ideals 359

the Dirichlet prime number theorem in its qualitative form. Observe also that Corollary 4 provides another proof of Theorem 4.14.

Let us now turn to quantitative results concerning ideals and prime ideals. The application of Theorem I of Appendix II leads to

Proposition 7.10. (i) *If χ is a normalized Hecke character, then*

$$\sum_{N(I) \leq x} \chi(I) = \alpha(\chi) x + o(x)$$

and

$$\sum_{N(\mathfrak{p}) \leq x} \chi(\mathfrak{p}) = \beta(\chi) x / \log x + o(x/\log x),$$

where $\alpha(\chi)$, $\beta(\chi)$ are defined in the preceding proposition.

(ii) *If A is a regular set of prime ideals and c is its density, then for the number $N(x)$ of prime ideals in A with norms not exceeding x the following asymptotic formula holds:*

$$N(x) = cx/\log x + o(x/\log x).$$

Proof. Part (i) in the case of the trivial character and part (ii) in the case of $c \neq 0$ result immediately from Theorem I of Appendix II.

If the character χ is non-trivial, put

$$a_m = \sum_{N(I)=m} 1 \quad \text{and} \quad b_m = 2a_m + \sum_{N(I)=m} (\chi(I) + \overline{\chi(I)}) \geq 0.$$

The sum of the series

$$\sum_{m=1}^{\infty} b_m m^{-s} = 2\zeta_K(s) + \zeta(s, \chi) + \zeta(s, \overline{\chi}),$$

which converges for $\operatorname{Re} s > 1$, equals by Proposition 7.9, to

$$2h\varkappa(s-1)^{-1} + g(s),$$

with a function g regular in $\operatorname{Re} s \geq 1$, and so Theorem I of Appendix II implies

$$\sum_{m \leq x} b_m = (2h\varkappa + o(1))x.$$

However, the same theorem implies also

$$\sum_{m \leq x} 2a_m = (2h\varkappa + o(1))x,$$

and thus

$$\left| \sum_{N(I) \leq x} (\chi(I) + \overline{\chi(I)}) \right| = o(x),$$

leading to

$$\left|\text{Re} \sum_{N(I)\leqslant x} \chi(I)\right| = o(x).$$

Considering now the sequence

$$c_m = 2a_m - i \sum_{N(I)\leqslant x} (\chi(I) - \overline{\chi(I)})$$

we obtain in the same way

$$\left|\text{Im} \sum_{N(I)\leqslant x} \chi(I)\right| = o(x)$$

and the first assertion of (i) follows. The second assertion follows similarly; just we have to replace the function $\zeta_K(s)$ by

$$\sum_{\mathfrak{p}} N(\mathfrak{p})^{-s}.$$

Finally, to show (ii) in the case of $c = 0$ observe that by Corollary 4 to Proposition 7.9 we can find for every positive ε a regular set of prime ideals with a positive density smaller than ε. If A_ε is such a set, then $A \cup A_\varepsilon$ is regular and has the same density as A_ε. Applying statement (ii) (already proved for $c \neq 0$) to the set $A \cup A_\varepsilon$ we get

$$N(x) \leqslant (\varepsilon + o(1))x/\log x,$$

i.e.,

$$\limsup N(x)\log x/x \leqslant \varepsilon,$$

and since ε was arbitrarily small, we get our assertion. □

Corollary 1 (Prime Ideal Theorem). *If $\pi_K(x)$ denotes the number of prime ideals in K with norms not exceeding x, then*

$$\pi_K(x) = (1+o(1))x/\log x. \qquad \square$$

Corollary 2. *If $\pi'_K(x)$ denotes the number of prime ideals in K of the first degree over Q with norms not exceeding x, then*

$$\pi'_K(x) = (1+o(1))x/\log x. \qquad \square$$

Corollary 3. *If L/K is a finite normal extension of degree n and $N(x)$ is the number of prime ideals of K which split in L/K and have norms not exceeding x, then*

$$N(x) = (1/n + o(1))x/\log x. \qquad \square$$

Corollary 4 (The Prime Ideal Theorem for ideal classes). *If X is any class of $H_1^*(K)$ and $\pi_X(x)$ is the number of prime ideals in X with norms not exceeding*

x, then
$$\pi_X(x) = (h_{\mathfrak{f}}^*(K)^{-1} + o(1))x/\log x.$$

(For $K = Q$ this gives the quantitative form of Dirichlet's prime number theorem in its weak form, i.e. without any evaluation of the remainder term.)

Applying the same Tauberian theorem to Proposition 7.9 (i), we get

Theorem 7.6. *If X is any class of $H_{\mathfrak{f}}^*(K)$ and $M_X(x)$ is the number of ideals in X with norms not exceeding x, then*
$$M_X(x) = (\varphi(\mathfrak{f})h(K)\varkappa/N(\mathfrak{f})h_{\mathfrak{f}}^*(K) + o(1))x.$$

Proof. We proceed in the same manner as in the proof of Corollary 6 to the last proposition and again we write X for the set of all characters of $H_{\mathfrak{f}}^*(K)$. If $F_X(m)$ is the number of ideals in X with norm m, then for $\operatorname{Re} s > 1$ we get by Proposition 7.9 the equality

$$\sum_{m=1}^{\infty} F_X(m)m^{-s} = \sum_{I \in X} N(I)^{-s} = (h_{\mathfrak{f}}^*(K))^{-1} \sum_{\chi \in X} \overline{\chi(X)} \zeta(s, \chi)$$

$$= \left(h(K)\varkappa(h_{\mathfrak{f}}^*(K))^{-1} \prod_{\mathfrak{p}|\mathfrak{f}} (1 - N(\mathfrak{p})^{-1}) \right) / (s-1) + g(s)$$

$$= (\varphi(\mathfrak{f})h(K)\varkappa/h_{\mathfrak{f}}^*(K)N(\mathfrak{f}))/(s-1) + g(s),$$

and we may apply the Tauberian theorem of Appendix II. □

We point out the following corollary, which in fact was already obtained in the course of proving Proposition 7.9:

Corollary (Ideal Theorem). *If $M(x)$ is the number of ideals in R_K whose norms do not exceed x, then $M(x) = (h(K)\varkappa + o(1))x$.*

Proof. Sum $M_X(x)$ over all classes of $H_1^*(K)$. □

In the same manner one can obtain various evaluations of the number of prime ideals or ideals with norms $N(I) \leq x$ having prescribed properties. Here we prove only the following

Proposition 7.11. *Let K be a normal extension of the rationals of degree n and denote by $F(x)$ the number of those natural numbers not exceeding x which are norms of ideals from R_K. Then we have for some positive constant $C = C(K)$*
$$F(x) = (C + o(1))x(\log x)^{(1-n)/n}.$$

Proof. Denote by P_j ($j = 1, 2, \ldots, n$) the set of those rational primes whose jth power is the norm of a suitable prime ideal in K. If m is the norm of an ideal of R_K, then a prime from P_j can occur in the factorization of m only with an exponent divisible by j. The normality of K/Q implies that, conversely, every m satisfying this demand is in fact the norm of a suitable ideal. Denote by A the set of all such m's. Then for $\operatorname{Re} s > 1$ we have

$$\sum_{m \in A} m^{-s} = \prod_{j=1}^{n} \prod_{\in P_j} (1 + p^{-js} + p^{-2js} + \ldots) = G(s) \exp\left(\sum_{p \in P_1} p^{-s}\right),$$

where

$$G(s) = \prod_{j=2}^{n} \prod_{p \in P_j} (1 + p^{-js} + p^{-2js} + \ldots) G_1(s)$$

with

$$G_1(s) = \exp\left(\sum_{j=2}^{n} \sum_{p \in P_1} j^{-1} p^{-js}\right).$$

Obviously, $G(1) \neq 0$ and $G(s) = g(s)$. Moreover, P_1 differs from the set of all primes which split in K/Q only by a finite set of primes, and so by Corollary 4 to Proposition 7.9 we get

$$\sum_{p \in P_1} p^{-s} = n^{-1} \log(1/(s-1)) + g(s),$$

which gives

$$\sum_{m \in A} m^{-s} = g(s)/(s-1)^{1/n} \quad \text{with } g(1) \neq 0.$$

An application of the Tauberian theorem of Appendix II leads to the required result. □

This proposition has an immediate application to the theory of quadratic forms:

Corollary. *Let $K = Q(D^{1/2})$ with a square-free D and assume that the class-number $h(K)$ equals unity. Denote by $f(x)$ the number of natural numbers not exceeding x which can be represented in the form*

$$\pm(a^2 - Db^2) \quad (\text{if } D \equiv 2, 3 \pmod{4})$$

or

$$\pm(a^2 + ab - (D-1)b^2/4) \quad (\text{if } D \equiv 1 \pmod{4})$$

with rational integral a, b. Then there is a positive constant $C = C(K)$ such that

$$f(x) = (C + o(1)) x/(\log x)^{1/2}.$$

Proof. Since $h(K) = 1$, every natural number which is the norm of an ideal of R_K is also the absolute value of the norm of a suitable integer in K. Now note that in the case of $D \equiv 2, 3 \pmod{4}$ the norms of integers are of the form $a^2 - Db^2$ and in the case of $D \equiv 1$ (mod 4) they are of the form $a^2 + ab - (D-1)b^2/4$; it suffices to apply the proposition proved above. □

2. The asymptotics results obtained in the preceding section do not tell us anything about the size of the remainder term. In this subsection we show that in the case of the Prime Ideal Theorem and its analogue for ideal classes (Corollaries 1 and 4 to Proposition 7.10) this size is intimately connected with the zero-free regions for zeta-functions $\zeta(s, \chi)$. In particular, we prove the following

Theorem 7.7. *Let $\mathfrak{f}$ be a fixed ideal from R_K, X a class of $H_\mathfrak{f}^*(K)$ and $\pi_X(x)$ the number of prime ideals in this class with norms not exceeding x. Let $Z_\mathfrak{f}$ be the set of all functions $\zeta(s, \chi)$ associated with the characters of $H_\mathfrak{f}^*(K)$ and assume that in the region*

$$\Omega = \{\sigma + it : |t| \geq t_0, \sigma \geq 1 - A(\log|t|)^{-a}(\log\log|t|)^{-b}\}$$
$$\cup \{\sigma + it : |t| < t_0, \sigma \geq 1 - A(\log t_0)^{-a}(\log\log t_0)^{-b}, |\sigma + it - 1| \geq \varepsilon > 0\}$$

(where $a, b \geq 0$, $A > 0$, $t_0 > e$, $\varepsilon > 0$ are suitable constants) we have
  (i) $\zeta(s, \chi) \neq 0$ *for* $\zeta(s, \chi) \in Z_\mathfrak{f}$,
  (ii) $\zeta'(s, \chi)/\zeta(s, \chi) = O((\log|\operatorname{Im} s|)^N)$ *for* $\zeta(s, \chi) \in Z_\mathfrak{f}$ *with a suitable positive N.*

Then the following asymptotic equality holds:

$$\pi_X(x) = (h_\mathfrak{f}^*(K))^{-1} \operatorname{li} x + O(x \exp(-B(\log x)^u (\log\log x)^v)),$$

where B is a positive constant, $u = 1/(1+a)$, $v = b/(1+a)$, and

$$\operatorname{li} x = \int_0^{1-} + \int_{1+}^{x} dt/\log t.$$

Proof. Let $S(x)$, $T(x)$ be the sums

$$\sum_{\substack{N(\mathfrak{p}) \leq x \\ \mathfrak{p} \in X}} \log N(\mathfrak{p}), \quad \sum_{\substack{N(\mathfrak{p}) \leq x \\ \mathfrak{p} \in X}} \log N(\mathfrak{p}) \log(x/N(\mathfrak{p})),$$

respectively. We shall reduce our problem to the determination of the behaviour of $T(x)$ as x tends to infinity, and this problem will be approached by the method of complex integration. We first prove

Lemma 7.3. *Assume that for x tending to infinity we have an equality of the form*

$$S(x) = \alpha x + O(R(x)) \quad (\alpha > 0),$$

where $R(x)$ is a positive function and $R(x)/(\log x)^2$ increases monotonically from some point x_0 on. Then we have

$$\pi_X(x) = \alpha \operatorname{li} x + O(R(x)/\log x).$$

7. Analytical Methods

Proof. We begin with a modified form of partial summation:

Consider $S(x)/\log x - \pi_X(x)$, which may also be written in a slightly artificial form as

$$S(x)/\log x - \sum_{\substack{N(\mathfrak{p}) \leq x \\ \mathfrak{p} \in X}} \log N(\mathfrak{p})/\log N(\mathfrak{p})$$

$$= \sum_{\substack{N(\mathfrak{p}) \leq x \\ \mathfrak{p} \in X}} \log N(\mathfrak{p})(1/\log x - 1/\log N(\mathfrak{p}))$$

$$= \sum_{\substack{N(\mathfrak{p}) \leq x \\ \mathfrak{p} \in X}} \int_{N(\mathfrak{p})}^{x} \log N(\mathfrak{p}) \frac{d}{dt}(1/\log t)\, dt = \int_{2}^{x} S(t) \frac{d}{dt}(1/\log t)\, dt.$$

The validity of the last equality can be verified as follows: order the prime ideals $\mathfrak{p} \in X$ with $N(\mathfrak{p}) \leq x$ so that $i \leq j$ implies $N(\mathfrak{p}_i) \leq N(\mathfrak{p}_j)$ and observe that

$$\sum_{\substack{N(\mathfrak{p}) \leq x \\ \mathfrak{p} \in X}} \int_{N(\mathfrak{p})}^{x} \log N(\mathfrak{p}) \frac{d}{dt}(1/\log t)\, dt$$

$$= \sum_{\substack{N(\mathfrak{p}_j) \leq x \\ \mathfrak{p}_j \in X}} \log N(\mathfrak{p}_j) \sum_{i > j} \int_{N(\mathfrak{p}_{i-1})}^{N(\mathfrak{p}_i)} \frac{d}{dt}(1/\log t)\, dt$$

$$= \sum_{\substack{N(\mathfrak{p}_i) \leq x \\ \mathfrak{p}_i \in X}} \int_{N(\mathfrak{p}_{i-1})}^{N(\mathfrak{p}_i)} \sum_{j \leq i-1} \log N(\mathfrak{p}_j) \frac{d}{dt}(1/\log t)\, dt$$

$$= \int_{2}^{x} S(t) \frac{d}{dt}(1/\log t)\, dt.$$

This shows that

$$\pi_X(x) = S(x)/\log x - \int_{2}^{x} S(t) \frac{d}{dt}(1/\log t)\, dt,$$

and applying our assumption about $S(x)$ we get

$$\pi_X(x) = \alpha x/\log x + O(R(x)/\log x) - \int_{2}^{x} \alpha t \frac{d}{dt}(1/\log t)\, dt$$

$$+ O\left(\int_{2}^{x} |R(t)| \left|\frac{d}{dt}(1/\log t)\right| dt\right).$$

Integrating $t\frac{d}{dt}(1/\log t)$ by parts and evaluating the last term by $O(1)+O(R(x)/\log x)$ (since the integrand equals $|R(t)|/t\log^2 t$ and so equals $(|R(x)|/\log^2 x)t$ at most), we obtain the assertion of the lemma. □

Our second step consists in the reduction of the problem of the asymptotic behaviour of $S(x)$ to that of $T(x)$:

Lemma 7.4. *Assume that for $T(x)$ the asymptotic formula*

$$T(x) = \alpha x + O(x\exp(-c(\log x)^{c_1}(\log\log x)^{c_2}))$$

holds for suitable positive c and non-negative c_1, c_2, at least one of those constants being positive. Then the same evaluation holds for $S(x)$ as well, perhaps with a smaller c.

Proof. Let $\delta = \delta(x) = \exp(-(\log x)^{c_1/2}(\log\log x)^{c_2/2})$ and consider the difference $T(x(1+\delta)) - T(x)$. It obviously equals

$$S(x)\log(1+\delta) + \sum_{\substack{x<N(\mathfrak{p})\leqslant x(1+\delta) \\ \mathfrak{p}\in X}} \log N(\mathfrak{p})\log(x(1+\delta)/N(\mathfrak{p})),$$

and since the second summand is non-negative, we obtain

$$S(x) \leqslant (T(x(1+\delta)) - T(x))/\log(1+\delta). \qquad (7.21)$$

On the other hand, the same difference can be written as

$$S(x(1+\delta))\log(1+\delta) + \sum_{\substack{x<N(\mathfrak{p})\leqslant x(1+\delta) \\ \mathfrak{p}\in X}} \log N(\mathfrak{p})\log(x/N(\mathfrak{p})),$$

and since the second summand is negative, we get, replacing x by $x/(1+\delta)$,

$$S(x) \geqslant (T(x) - T(x/(1+\delta)))/\log(1+\delta). \qquad (7.22)$$

Applying to (7.21) and (7.22) our assumption about $T(x)$, we get

$$\frac{\alpha\delta}{(1+\delta)\log(1+\delta)}x + O\left(\frac{R_1(x)}{\log(1+\delta)}\right) \leqslant S(x)$$

$$\leqslant \frac{\alpha\delta}{\log(1+\delta)}x + O\left(\frac{R_1(\delta x)}{\log(1+\delta)}\right),$$

where by $R_1(x)$ we denote the remainder term of the evaluation of $T(x)$. Since $\delta(x)$ tends to zero as x goes to infinity, we immediately obtain the evaluation

$$S(x) = \alpha x + O(x\delta^2) + O(R_1(x)/\delta) + O(\delta x),$$

and our choice of δ implies that the three remainder terms here are majorized by $O(x\exp(-c'(\log x)^{c_1}(\log\log x)^{c_2}))$ with a suitable constant c'. □

366 7. Analytical Methods

So we are left with the task of evaluating $T(x)$ with a good remainder term and the simplest way of doing this consists in complex integration. Define a function $K(s) = K_X(s)$ for $\operatorname{Re} s > 1$ by putting

$$K(s) = \sum_{\mathfrak{p} \in X} \log N(\mathfrak{p})/N(\mathfrak{p})^s.$$

Its relevance to our problem is explained by the first part of

Lemma 7.5. (i) *If I_2 denotes the line $\operatorname{Re} s = 2$, then*

$$T(x) = (2\pi i)^{-1} \int_{I_2} x^s K(s) s^{-2} ds.$$

(ii) *The function $K(s)$ is regular for $\operatorname{Re} s > 1$ and can be analytically continued to the region Ω, where it is regular except for a simple pole at $s = 1$. Moreover, in that region we have*

$$K(s) = O((\log|\operatorname{Im} s|)^N) \quad \text{for } |\operatorname{Im} s| \geq \varepsilon,$$

where ε is an arbitrarily fixed positive number.

Proof. To prove (i) observe that

$$\int_{I_2} x^s K(s) s^{-2} ds = \sum_{\mathfrak{p} \in X} \log N(\mathfrak{p}) \int_{I_2} (x/N(\mathfrak{p}))^s s^{-2} ds,$$

and the series

$$\sum_{\mathfrak{p} \in X} \log N(\mathfrak{p})/s^2 N(\mathfrak{p})^s$$

is uniformly convergent to a function integrable on I_2; it remains to note that for a positive real a we have

$$\int_{I_2} a^s s^{-2} ds = \begin{cases} 2\pi i \log a, & a > 1, \\ 0, & 0 < a < 1. \end{cases}$$

To obtain the second assertion write for $\operatorname{Re} s > 1$

$$K(s) = (h_1^*(K))^{-1} \sum_{\chi} \overline{\chi(X)} \sum_{\mathfrak{p}} \log N(\mathfrak{p}) \chi(\mathfrak{p}) N(\mathfrak{p})^{-s}, \qquad (7.23)$$

where χ runs over all characters of $H_1^*(K)$. In the same half-plane we have also for every character χ of $H_1^*(K)$

$$\zeta(s, \chi) = \exp\left(\sum_{\mathfrak{p}} \chi(\mathfrak{p}) N(\mathfrak{p})^{-s} + \sum_{\mathfrak{p}} \sum_{j=2}^{\infty} j^{-1} \chi^j(\mathfrak{p}) N(\mathfrak{p})^{-js}\right),$$

which easily implies the equality

$\zeta'(s, \chi)/\zeta(s, \chi)$

$$= -\sum_{\mathfrak{p}} \chi(\mathfrak{p}) \log N(\mathfrak{p}) N(\mathfrak{p})^{-s} - \sum_{\mathfrak{p}} \sum_{j=2}^{\infty} \chi^j(\mathfrak{p})(\log N(\mathfrak{p})) N(\mathfrak{p})^{-js},$$

where the second term is regular and bounded for $\operatorname{Re} s \geq 1/2+\varepsilon$, and hence also in Ω. This fact in conjunction with (7.23) leads to the equality

$$K(s) = -(h_f^*(K))^{-1} \sum_{\chi} \overline{\chi(x)} \zeta'(s, \chi)/\zeta(s, \chi) + G(s), \qquad (7.24)$$

where $G(s)$ is regular and bounded in Ω. This gives the required continuation of $K(s)$ to Ω. Its evaluation as asserted is a consequence of our assumption (ii). □

Now let $T \geq t_0$ be a positive real number (whose exact value we shall fix later) and consider the boundary Γ of the set

$$\{z = \sigma+it\colon z \in \Omega, \sigma \leq 2, |t| \leq T\},$$

which can be written as

$$\Gamma = \bigcup_{j=0}^{5} \Gamma_j,$$

where

$\Gamma_0 = \{2+ti\colon |t| \leq T\},$
$\Gamma_1 = \{\sigma+Ti\colon 1-\varphi(T) \leq \sigma \leq 2\},$
$\Gamma_2 = \{\sigma+ti\colon T \geq |t| \geq t_0, \sigma = 1-\varphi(t)\},$
$\Gamma_3 = \{1-\varphi(t_0)+it\colon |t| \leq t_0\}$

and Γ_4, Γ_5 are the symmetric images of Γ_2, Γ_1 with respect to the real axis. The function $\varphi(t)$ equals $A(\log|t|)^{-a}(\log\log|t|)^{-b}$.

We now prove that $\int_{\Gamma_2} x^s K(s) s^{-2} ds$ is for large T well approximated by the integral over Γ_0, and to deal with the last integral we utilize Cauchy's integral theorem. This gives the following evaluation:

Lemma 7.6.

(i) $\int_{\Gamma_2} x^s K(s) s^{-2} ds = \int_{\Gamma_0} x^s K(s) s^{-2} ds + O(x^2/T),$

(ii) $(2\pi i)^{-1} \int_{\Gamma_0} x^s K(s) s^{-2} ds = (h_f^*(K))^{-1} x - (2\pi i)^{-1} \int_{\Gamma_2 \cup \Gamma_4} x^s K(s) s^{-2} ds$
$\qquad + O(x^{1-\varphi(t_0)}) + O(x^2 (\log T)^N/(\log x) T^2).$

Proof. (i) For $s \in I_2$ the function $K(s)$ does not exceed $K(2)$ in absolute value, and thus the difference between the integrals occurring in (i) is

$$O\left(\int_T^\infty (x^2/(4+t^2))\,dt\right) = O(x^2 T^{-1}).$$

(ii) Equality (7.24) shows that the only singularity of $x^s K(s) s^{-2}$ in Ω occurs at $s = 1$, where it has a simple pole with the residue $h_1^*(K)^{-1}$, since the functions $\zeta'(s, \chi)/\zeta(s, \chi)$ are for $\chi \neq \chi_0$ (the trivial character) regular in Ω and $\zeta'(s, \chi_0)/\zeta(s, \chi_0)$ has a simple pole at $s = 1$ with the residue -1. Hence

$$(2\pi i)^{-1} \int_{\Gamma_0} x^s K(s) s^{-2} ds = (2\pi i)^{-1} \int_\Gamma x^s K(s) s^{-2} ds - (2\pi i) \sum_{j=1}^{5} \int_{\Gamma_j} x^s K(s) s^{-2} ds$$

$$= (h_1^*(K)) x^{-1} - (2\pi i) \sum_{j=1}^{5} \int_{\Gamma_j} x^s K(s) s^{-2} ds.$$

Now, on Γ_1 we have $|K(s)| = O((\log T)^N)$ and $|s|^2 \gg T^2$; thus the integral over Γ_1 is

$$O\left((\log T)^N T^{-2} \int_{1-\varphi(T)}^{2} x^u\, du\right) = O(x^2 (\log T)^N (\log x)^{-1} T^{-2}),$$

and the same applies to the integral over Γ_5. Moreover, on Γ_3 we have $|K(s) x^s s^{-2}| = O(x^{1-\varphi(t_0)})$, and so the corresponding integral is $O(x^{1-\varphi(t_0)})$. □

Finally, we have to evaluate the integrals over Γ_2 and Γ_4. There is nearly no difference between these two cases and we consider only the integral over Γ_2 in detail. Obviously

$$\int_{\Gamma_2} x^s K(s) s^{-2} ds \ll \int_{t_0}^{T} x^{1-\varphi(t)} t^{-2} (\log t)^N \left|\frac{d}{dt}(1-\varphi(t)+it)\right| dt;$$

but

$$\frac{d}{dt}(1-\varphi(t)+it) \ll t^{-1}(\log t)^{-a-1}(\log\log t)^{-b}$$

and so our integral is

$$\ll x \int_{t_0}^{T} \exp(-\varphi(t) \log x) t^{-2} dt.$$

To deal with the last integral we partition the interval $[t_0, T]$ into two subintervals $[t_0, U]$, $[U, T]$ with

$$U = \exp(A(\log x)^{1/(1+a)} (\log\log x)^{-b/(1+a)}).$$

Since $\varphi(t)$ decreases and is positive, we get

$$x \int_{t_0}^{U} \exp(-\varphi(t)\log x) t^{-2} dt \ll x \exp(-\varphi(U)\log x)$$

$$\ll x \exp(-A_1 (\log x)^{1/(1+a)} (\log\log x)^{-b/(1+a)})$$

for a suitable positive A_1 and

$$x \int_{U}^{T} \exp(-\varphi(t)\log x) t^{-2} dt \ll x \int_{U}^{\infty} t^{-2} dt \ll xU^{-1}$$

$$\ll x \exp(-A(\log x)^{1/(1+a)} (\log\log x)^{-b/(1+a)}).$$

Using Lemmas 7.5 (i) and 7.6 we obtain

$$T(x) = (h_{\mathfrak{f}}^*(K))^{-1} x + O(x^2 T^{-1}) + O(x^{1-\varphi(t_0)})$$
$$+ O(x^2 (\log T)^N (\log x)^{-1} T^{-2})$$
$$+ O(x \exp(-A_2 (\log x)^{1/(1+a)} (\log\log x)^{-b/(1+a)})),$$

where the choice of an appropriate T is yet to be made. Putting $T = x^2$, one sees that the last remainder term dominates all the other terms; thus

$$T(x) = (h_{\mathfrak{f}}^*(K))^{-1} x + O(x \exp(-A_2 (\log x)^{1/(1+a)} (\log\log x)^{-b/(1+a)}))$$

and now Lemmas 7.2 and 7.4 give the asserted formula for $\pi_X(x)$. □

To make the contents of Theorem 7.7 non-void we now prove that its assumptions are satisfied for a certain region Ω with $a = 1$, $b = 0$; this leads us to the following

Corollary. (i) *If X is any class of $H_{\mathfrak{f}}^*(K)$, then*

$$\pi_X(x) = (h_{\mathfrak{f}}^*(K))^{-1} \mathrm{li}\, x + O(\exp(-B(\log x)^{1/2}))$$

holds for a certain B depending on the ideal $\mathfrak{f}$.

(ii) *For the number $\pi_K(x)$ of prime ideals in the field K whose norms do not exceed x we have the formula*

$$\pi_K(x) = \mathrm{li}\, x + O(x \exp(-B(\log x)^{1/2})).$$

Proof. We first verify the assumption concerning the zeros of our zeta functions. Since $\overline{\zeta(\bar{s}, \bar{\chi})} = \zeta(s, \chi)$ (which is obvious for $\operatorname{Re} s > 1$ and follows for other s by analytic continuation), we may restrict our attention to those zeros which lie in the upper half-plane. Moreover, we will not bother about zeros with small imaginary parts, since by a change of the constant A in the definition of Ω we can keep them outside our region. Thus let $z_0 = \xi_0 + i\eta_0$ be a root of

$\zeta(s, \chi)$, where $\xi_0 \geq 7/8$, $\eta_0 \geq 3$ and write $\xi_0 = 1 - c/\log\eta_0$. We have to show that c must exceed a positive constant, which is independent of the chosen zero z_0 but which may depend on our function $\zeta(s, \chi)$, since we are not interested in uniform result. (In fact, we shall need such a result for all zeta-functions associated with the characters of $H_f^*(K)$, but as their number is finite, we can always take the minimal value of the constants c obtained and still assert that it is positive.) We apply Theorem III of Appendix II taking $f(s) = \zeta(s, \chi)$, $s_0 = \sigma + i\eta_0$ with $\sigma = 1 + a/\log\eta_0$, where $0 < a < 1/2$ (the precise value of a will be fixed later) and finally $r = 1/2$. Since $\eta_0 \geq 3$, the only possible singular point of $\zeta(s, \chi)$, viz. $s = 1$, lies outside $|s - s_0| \leq 1/2$, and, moreover, $\zeta(s, \chi)$ does not vanish for $\mathrm{Re}(s - s_0) \geq 0$ since $\mathrm{Re}\, s_0$ exceeds 1. So we need only to have a bound for the ratio $|\zeta(s, \chi)/\zeta(s_0, \chi)|$ in $|s - s_0| \leq 1/2$. Now, this is provided by the following reasoning: Theorem 7.4 gives for every positive ε the evaluation

$$|\zeta(s, \chi)| = O(|\mathrm{Im}\, s|^{n/2 + \varepsilon})$$

(with $n = [K:Q]$) and we have also for $\mathrm{Re}\, s > 1$ the identity

$$\zeta(s, \chi)^{-1} = \sum_I \chi(I)\mu(I)N(I)^{-s},$$

which is easily proved by multiplying out the series occurring in it, and this implies the evaluation

$$|\zeta(s_0, \chi)^{-1}| \ll \zeta_K(\mathrm{Re}\, s_0) = O(\log\eta_0/a).$$

The resulting evaluations yield

$$|\zeta(s, \chi)/\zeta(s_0, \chi)| \ll \eta_0^{n/2 + \varepsilon}\log\eta_0/a$$

and applying the said theorem we obtain

$\mathrm{Re}(\zeta'(s_0, \chi)/\zeta(s_0, \chi))$
$\geq -8(\log B + \log(1/a) + (n/2 + \varepsilon)\log\eta_0 + \log\log\eta_0) + (\sigma - \xi_0)^{-1}$
$\geq -B_1(\log(1/a) + \log\eta_0) + (\sigma - \xi_0)^{-1}$

for some positive B, B_1.

We apply the same procedure to the point $\sigma + 2i\eta_0$ again with $r = 1/2$ but this time with the function $f(s) = \zeta(s, \chi^2)$, and similarly we obtain the evaluation

$$\mathrm{Re}(\zeta'(\sigma + 2i\eta_0, \chi^2)/\zeta(\sigma + 2i\eta_0, \chi^2)) \geq -B_2\left(\log\frac{1}{a} + \log\eta_0\right)$$

for a suitable positive B_2.

§ 2. Asymptotic Distribution of Ideals and Prime Ideals

Now observe that for $\sigma > 1$
$$\operatorname{Re}\left\{3\frac{\zeta'(\sigma, \chi_0)}{\zeta(\sigma, \chi_0)} + 4\frac{\zeta'(\sigma+i\eta_0, \chi)}{\zeta(\sigma+i\eta_0, \chi)} + \frac{\zeta'(\sigma+2i\eta_0, \chi^2)}{\zeta(\sigma+2i\eta_0, \chi^2)}\right\} \leq 0.$$

In fact, $\zeta'(s, \chi)/\zeta(s, \chi)$ equals
$$-\sum_I \chi(I)a(I)N(I)^{-s} \quad \text{for Re}\, s > 1,$$

where $a(I)$ is a non-negative function independent of χ and equal to $\log N(\mathfrak{p})$ if $I = \mathfrak{p}^k$ is a power of a prime ideal $\mathfrak{p}$, and 0 otherwise, which follows at once if we take the logarithmic derivative of the Eulerian product for our zeta-function. Having done this, we substitute the resulting series into the left-hand side of the required inequality, thus obtaining the expression

$$-\sum_I \frac{a(I)}{N(I)^\sigma} \operatorname{Re}\{3\chi_0(I) + 4\chi(I)N(I)^{-i\eta_0} + \chi^2(I)N(I)^{-2i\eta_0}\}$$
$$= -\sum_I \frac{a(I)}{N(I)^\sigma}(3 + 4\cos\alpha_I + \cos 2\alpha_I)$$
$$= -2\sum_I \frac{a(I)}{N(I)^\sigma}(1+\cos\alpha_I)^2 \leq 0,$$

where $\alpha_I = \operatorname{Arg}\{\chi(I)N(I)^{-i\eta_0}\}$. The resulting inequalities give
$$-\operatorname{Re}(3\zeta'(\sigma, \chi_0)/\zeta(\sigma, \chi_0)) \geq 4/(\sigma-\xi_0) - B_3(\log(1/a) + \log\eta_0).$$

On the other hand, using the expansion of $\zeta'(s, \chi_0)/\zeta(s, \chi_0)$ at $s = 1$, we can write
$$\operatorname{Re}(3\zeta'(\sigma, \chi_0)/\zeta(\sigma, \chi_0)) = -3(1/(\sigma-1) + c_0 + c_1(\sigma-1) + \ldots),$$

and this shows that for every positive ε we have the inequality
$$\operatorname{Re}(3\zeta'(\sigma, \chi_0)/\zeta(\sigma, \chi_0)) > -(3+\varepsilon)/(\sigma-1),$$

given that $\sigma - 1$ is sufficiently small. Now we have two inequalities for the same expression and we shall prove their inconsistency for sufficiently small c with the constant a chosen in a suitable way. Indeed, the comparison of our inequalities yields
$$(3+\varepsilon)/(\sigma-1) \geq 4/(\sigma-\xi_0) - B_3(\log(1/a) + \log\eta_0),$$

but $\sigma - \xi_0 = (a+c)\log\eta_0$ and $\sigma - 1 = a/\log\eta_0$. Thus
$$(3+\varepsilon)\log\eta_0/a + B_3(\log(1/a) + \log\eta_0) \geq 4\log\eta_0/(a+c),$$

and we finally arrive at
$$B_3\log(1/a) > \log\eta_0(4/(a+c) + (3+\varepsilon)/a + B_3)$$
$$\geq (4/(a+c) + (3+\varepsilon)/a + B_3)\log 3.$$

Now, for a fixed and sufficiently small we have
$$B_3 \log(1/a) \leq 1/a;$$
but if in the preceding inequality we let c tend to zero, then we get
$$1/a > (4\log 3 + (3+\varepsilon)\log 3)/a + B_3 \log 3,$$
which is an obvious contradiction. We have thus obtained a zero-free region for $a = 1$, $b = 0$ and a constant $A > 0$. It remains to prove the estimate $O(\log_a^N t)$ (where $t = |\operatorname{Im} s|$) for the function $\zeta'(s, \chi)/\zeta(s, \chi)$ in a region Ω' contained in Ω and defined in the same way but perhaps with another constant A. We can freely restrict ourselves to large positive values of $\operatorname{Im} s$ and so we do. By what has already been proved our zeta-function does not vanish in Ω and so $f(s) = \log \zeta(s, \chi)$ is regular there. Now, take for Ω' the region defined similarly to Ω but with $4A$ instead of A. Let C be a large positive number and let $z = \sigma + it$ be a point of Ω' with $t \geq C$, $\sigma \leq 1$. (It obviously suffices to get our evaluation for such z's.) Take also $s_0 = 2 + it$ and consider the circle $|s - s_0| \leq 1 + 1/(2A\log t)$. It lies entirely in Ω and we may thus apply Theorem IV of Appendix II to our function $f(s)$ and s_0 with $R = 1 + 1/(2A\log t)$ and $r = 1 + 1/(3A\log t)$. Since $R \leq 2$, $|f(s_0)| = O(1)$ and
$$\max_{|s-s_0| \leq R} \operatorname{Re} f(s) \leq \log \max \{|\zeta(s, \chi)| : 0 \leq \operatorname{Re} s \leq 4, t-2 \leq \operatorname{Im} s \leq t+2\}$$
$$= O(\log t)$$
by Theorem 7.4, the said theorem of the appendix implies the formula
$$|f(s)| = O(\log^2 t),$$
valid for all s in $|s - s_0| < 1 + 1/(3A\log t)$. Finally, note that for any point s with $|s - s_0| = R$ we have $|z - s| \geq 1/(12A\log t)$; hence
$$|f'(z)| = \left| \frac{1}{2\pi i} \int_{|s-s_0|=R} \frac{f(z)}{(z-s)^2} ds \right| \leq R\log^2 t \cdot 144 A^2 \log^2 t = O(\log^4 t)$$
and this is all we want. $\square$

Observe that the corollary just proved implies the following evaluations:
$$\pi_K(x) = \operatorname{li} x + O(x/\log^N x) \quad \text{for all } N,$$
and
$$\pi_K(x) = x/\log x + O(x/\log^2 x).$$

3. Corollary 4 to Proposition 7.10 and Theorem 7.6 can be regarded as results concerning uniform distribution and may be obtained by the methods of that theory. We now prove a fact which yields those two results, and also some other theorems describing the distribution of ideals and prime ideals.

§ 2. Asymptotic Distribution of Ideals and Prime Ideals

Proposition 7.12. *Let G be a compact Abelian group and let f be a continuous homomorphism of the idele-group I_K onto G which satisfies the following two conditions:*

(i) $f(x) = 1$ *for every principal idele x,*
(ii) f *trivializes on J'_K.*

For every non-Archimedean v choose in K_v an element π_v generating the prime ideal $\mathfrak{p}_v$ and define a mapping g of the set of all prime ideals of K into G by

$$g(\mathfrak{q}) = f((x_v^{(\mathfrak{q})})),$$

where the idele $(x_v^{(\mathfrak{q})})$ has component π_v if $\mathfrak{q} = \mathfrak{p}_v$ and all other components equal to 1. Extend this mapping by multiplicativity to the semi-group of all ideals of R_K and denote this extension also by g. Then the sequence $\{g(I)\}$ ordered in accordance with the sequence of norms of ideals is uniformly distributed in G and the same holds for the sequence $\{g(\mathfrak{q})\}$, where $\mathfrak{q}$ ranges over all prime ideals of K arranged according to their norms. (The ordering of ideals with equal norms is irrelevant.)

Proof. We consider first the sequence $\{g(I)\}$. It is necessary to establish the estimate

$$\sum_{N(I) \leq x} \chi(g(I)) = o\left(\sum_{N(I) \leq x} 1 \right) = o(x)$$

for every non-trivial character χ of G, the last equality resulting from the Ideal Theorem. The composition $\chi \circ f$ is a character of the idele-class-group and condition (ii) shows that it is normalized. Let $\psi(I)$ be the primitive Hecke character induced by $\chi \circ f$, S its exceptional set, A the set of all ideals of R_K having no prime ideal divisor from S, and B the set of all ideals of R_K whose prime ideal divisors lie all in S. We can write

$$\sum_{N(I) \leq x} \chi(g(I)) = \sum_{\substack{I \in A \\ N(I) \leq x}} \chi(g(I)) + \sum_{\substack{I \notin A \\ N(I) \leq x}} \chi(g(I)). \tag{7.25}$$

Since for every ideal I in A we have $\chi(g(I)) = \psi(I)$, the first summand in the last formula is $o(x)$ by Proposition 7.10 and we have to show the same for the second summand. For this purpose note that every ideal I can be uniquely written as $I = I_1 I_2$, where $I_1 \in A$, $I_2 \in B$. Hence, for $\operatorname{Re} s > 1$ we get the identity

$$\sum_{I \notin A} \chi(g(I)) N(I)^{-s} = \zeta(s, \psi) \sum_{I_2 \in B} \chi(g(I_2)) N(I_2)^{-s};$$

the second factor of the expression on the right equals

$$\prod_{\mathfrak{p} \in S} (1 - \chi(g(\mathfrak{p})) N(\mathfrak{p})^{-s})^{-1}$$

and so is regular for $\operatorname{Re} s \geq 1$.

Since χ is non-trivial, Theorem 7.3 implies that the function $\sum_{I \notin A} \chi(g(I))N(I)^{-s}$ can be continued to a function regular for $\operatorname{Re} s \geq 1$, and so the Tauberian theorem gives the evaluation $o(x)$ for the second summand in (7.25).

We now turn to the part of our proposition concerning prime ideals. Again we have to prove that the evaluation

$$\sum_{N(\mathfrak{p}) \leq x} \chi(g(\mathfrak{p})) = o\left(\sum_{N(\mathfrak{p}) \leq x}' 1\right) = o(x/\log x)$$

holds for every non-trivial character χ of G. Let ψ be the Hecke character used in the first part of the proof. Then the sum

$$\sum_{N(\mathfrak{p}) \leq x} \chi(g(\mathfrak{p}))$$

differs from $\sum_{N(\mathfrak{p}) \leq x} \psi(\mathfrak{p})$ only by a finite number of terms, and Proposition 7.10 shows that it is $o(x/\log x)$. □

We apply this proposition to the solution of a problem concerning the distribution of principal prime ideals of a given imaginary quadratic number field on the complex plane.

Theorem 7.8. *Let K be an imaginary quadratic number field, $h = h(K)$ its class-number and w the number of roots of unity contained in K. Finally, let X be a fixed class of $H(K)$. If $N(x) = N(a, b, x, X)$ denotes the number of integers t of K with norms not exceeding x, which generate the h-th power of a prime ideal of X and satisfy*

$$a \leq \operatorname{Arg} t^h < b$$

(Arg$u$ denoting the argument of u on the complex plane under a fixed embedding of K therein), then

$$N(x) = ((b-a)w/2\pi h + o(1)) x^{1/h}/\log x.$$

Moreover, if $M(x) = M(a, b, x)$ is the number of integers t of K whose norms do not exceed x and which satisfy $a \leq \operatorname{Arg} t < b$ and generate a prime ideal, then

$$M(x) = ((b-a)w/2\pi h + o(1)) x/\log x.$$

Proof. For every idele x denote its Archimedean component by x_0 and consider the embedding of K in the complex plane which was used above to define Arg$u$. Further, for every idele x denote by $g = g(x)$ any generator of the principal ideal A^h, where A is the ideal generated by x through the canonical map $I_K \to I_K/U_K$. If we now put

$$f(x) = \exp(i \operatorname{Arg}(g/x_0^h)^w), \quad f_1(x) = \text{class of } A \text{ in } H(K)$$

and $F(x) = [f(x), f_1(x)]$, then we obtain a homomorphism of I_K into the product $T \times H(K)$, which is a compact group. Observe that F trivializes on J'_K and also on I_0 (the last property being ensured by the fact that $F(x)$ does not depend on the particular choice of the generator g of A, since all such generators differ by a root of unity and so their wth powers are equal) and, moreover, is surjective. The last assertion can be checked as follows: let $z \in T$ and $X \in H(K)$ be arbitrarily given, let y be any idele generating an ideal B which lies in X and let g be any generator of the principal ideal B^h. Now put $x_v = y_v$ for non-Archimedean v, and for Archimedean v make x_v equal to any solution of $u^{wh} = \bar{z} g^w / |g^w|$. The idele $x = (x_v)$ is mapped by F onto the pair $[z, X]$. Proposition 7.1 shows that if $G(I) = [\exp(i \operatorname{Arg} g^w(I))$, class of $I]$ (where $g(I)$ is any generator of I^h), then the sequence $G(\mathfrak{p})$ is uniformly distributed in $T \times H(K)$ and this gives the first part of our theorem in the following form:

The number of prime ideals $\mathfrak{p}$ with norms not exceeding x which lie in the class X and satisfy $a \leqslant \operatorname{Arg} g_\mathfrak{p}^w < b$ (where $g_\mathfrak{p}$ is the generator of $\mathfrak{p}^h$) is asymptotically equal to $((b-a)/2\pi h) x/\log x$.

But we have to count the $g_\mathfrak{p}$'s occurring in the class X and not the ideals $\mathfrak{p}$, and hence in this asymptotic formula we have to change x into $x^{1/h}$, since $N(g_\mathfrak{p}) = N(\mathfrak{p})^h$, and multiply it by w, since to every $\mathfrak{p}$ there are exactly w distinct $g_\mathfrak{p}$'s at our disposal.

The second part follows immediately by considering only principal prime ideals. □

It is seen without much difficulty that the second part of the theorem just proved can be regarded as a statement concerning the distribution of rational primes of the form $N_{K/Q}(z)$, where z runs over integers from a given imaginary quadratic field which lie in a given sector in the plane. Since $N_{K/Q}$ is in fact a definite quadratic form in x, y when z is written as $z = x + yw$ (1, w forming an integral basis), we obtain a result concerning the distribution of primes represented by that form.

4. Now we can apply Corollary 7 to Proposition 7.9 to a result concerning Minkowski's units. Let us recall that if K/Q is normal, with Galois group G, then a unit of K is called a *weak Minkowski unit*, if its image in $U(K)/E(K)$ generates $U(K)/E(K)$ as a $Z[G]$-module. In Theorem 3.10 we considered the case where G was cyclic of prime order and now we turn to the dihedral group D_{2n}, which can be defined as the group of motions of the regular n-gon, or equivalently, as the group generated by two elements σ, τ with relations

$$\sigma^n = \tau^2 = 1, \qquad \tau\sigma = \sigma^{-1}\tau.$$

Note that for τ one can choose any element of order 2 in D_{2n}.

Theorem 7.9 (N. Moser [83]). *Every complex normal extension K/Q with Galois group G isomorphic to D_{2p} (with p being an odd prime) has a weak Minkowski unit.*

Proof. Denote by R the ring of integers of the pth cyclotomic field $Q(\zeta_p)$ and let A be the R-module consisting of all linear polynomials in one variable with coefficients in R. Defining multiplication in A by

$$(a+bX)(c+dX) = (ac+b\bar{d}) + (ad+b\bar{c})X$$

(with $\bar{z}$ being the complex conjugate of z) we make A into a non-commutative ring. The following lemma implies that A is an epimorphic image of $Z[G]$ and hence can be regarded as a $Z[G]$-module:

Lemma 7.7. *The map $\sigma \mapsto \zeta_p$, $\tau \mapsto X$ can be extended to a surjective ring homomorphism $\varphi: Z[G] \to A$, whose kernel equals $N_H Z[G]$, where H is the subgroup of G generated by σ and*

$$N_H = \sum_{h \in H} h = 1 + \sigma + \sigma^2 + \ldots + \sigma^{p-1}.$$

Proof. Every element of D_{2p} can be uniquely written in the form σ^k or $\sigma^k \tau$ (with $0 \leq k \leq p-1$), and hence the formula

$$\varphi\left(\sum_{j=0}^{p-1}(a_j \sigma^j + b_j \sigma^j \tau)\right) = \sum_{j=0}^{p-1} \zeta_p^j (a_j + b_j X), \quad (a_j, b_j \in Z),$$

defines a linear mapping of $Z[G]$ onto A. It can be easily checked that it is a ring-homomorphism.

If

$$u = \sum_{j=0}^{p-1}(a_j \sigma^j + b_j \sigma^j \tau)$$

lies in the kernel of φ, then

$$\sum_{j=0}^{p-1} a_j \zeta_p^j = \sum_{j=0}^{p-1} b_j \zeta_p^j = 0.$$

Hence

$$\sum_{j=0}^{p-2}(a_j - a_{p-1})\zeta_p^j = 0.$$

Since $1, \zeta_p, \ldots, \zeta_p^{p-2}$ is an integral basis of R, we get $a_0 = a_1 = \ldots = a_{p-1}$ and similarly $b_0 = b_1 = \ldots = b_{p-1}$, i.e., $u = (a_0 + b_0 \tau) N_H \in Z[G]N_H = N_H Z[G]$, because N_H lies in the centre of $Z[G]$. In fact, it suffices to check that N_H

commutes with every $g \in G$; but this follows from the observation that H is of index two in G and thus is a normal subgroup.

This shows that $\operatorname{Ker} \varphi \subset N_H Z[G]$ and the converse inclusion results from $\varphi(N_H) = 0$. □

Before proceeding further, observe that the complex conjugation is an element of order 2 in G and so we may choose the generator τ so that it corresponds to the complex conjugation. We shall use this in the proof of the next lemma, which will allow us to induce the structure of an A-module on $U(K)/E(K)$.

Lemma 7.8. *If k is the subfield of K corresponding by the Galois theory to H, then k is quadratic imaginary and thus $U(k)/E(k) = 1$.*

Proof. Since H is of index 2, k must be quadratic. The field k_0 corresponding to the subgroup H_0 generated by τ is the maximal real subfield of K. Because of $H \cap H_0 = 1$, we have $K = k k_0$, and if k were real, K would be real as well, contrary to our assumption. Hence k is imaginary quadratic and the last assertion follows immediately. □

Corollary 1. *The ideal $N_H Z[G]$ of $Z[G]$ annihilates $U(K)/E(K)$, i.e., for every $a \in N_H Z[G]$ and $u \in U(K)/E(K)$ we have $au = 1$.*

Proof. It suffices to observe that N_H acts on a unit u of K as follows:

$$u \mapsto \prod_{j=0}^{p-1} \sigma^j(u) = N_{K/k}(u) \in U(k) = E(k) \subset E(K).$$

□

Corollary 2. *The action of $Z[G]$ induces on $U(K)/E(K)$ the structure of an A-module.*

Proof. This follows from Lemma 7.7 and the preceding corollary. □

Since A contains R as a subring, the last corollary shows that $U(K)/E(K)$ is a torsion-free and finitely generated R-module. We may thus invoke Theorem 1.13 to obtain the existence of an ideal J of R such that $U(K)/E(K)$ is isomorphic as an R-module to $R^s \oplus J$, with a suitable $s \geq 0$. Comparing the Z-ranks we get $p - 1 = (s+1)(p-1)$, showing that $s = 0$ and so $U(K)/E(K) \simeq J$. This isomorphism induces on J an A-module structure and so we may regard J as a $Z[G]$-module by Lemma 7.7. To prove the theorem it now suffices to show that J is a cyclic $Z[G]$-module, i.e., has one generator.

Theorem 1.14 shows that if I is an ideal belonging to the same class in $H(Q(\zeta_p))$ as J, then $U(K)/E(K) \simeq I$ as an R-module. Thus in the following argument we may freely replace J by any such ideal I.

The action of A on J is determined by the action of X, since the elements of R act on J by multiplication. We now show that there is an element c in $Q(\zeta_p)$ of absolute value 1, such that for all $z \in J$ we have $X(z) = c\bar{z}$, where $\bar{z}$ is the complex conjugate of z and $X(z)$ denotes the result of the action of X on z. If $r \neq 0$, $r \in R$, $z \neq 0$, $z \in J$, then

$$X(rz) = (Xr)z = (\bar{r}X)z = \bar{r}X(z),$$

and hence, if $r, z \in J$, $rz \neq 0$, then

$$\bar{r}X(z) = X(rz) = X(zr) = \bar{z}X(r),$$

implying $X(r)/\bar{r} = X(z)/\bar{z} = c$, with a certain c independent of r and z. This leads to $X(z) = c\bar{z}$ for all z in J, since for $z = 0$ this equality is evident. If now z_0 is the image of a real unit in J, then $X(z_0) = z_0$ (because on $U(K)/E(K)$ the element X acts as complex conjugation); thus $c\bar{z}_0 = z_0$ and we get $|c| = 1$.

Since $c\bar{J} = X(J)$ and $X(J) \subset J$, we obtain

$$\bar{J} = c\bar{c}\bar{J} \subset \bar{c}J \subset \bar{J}, \quad \text{i.e. } \bar{c}J = \bar{J},$$

and thus $X(J) = J = c\bar{J}$. We will use this to prove that in the class of J there is an ideal I satisfying $\bar{I} = I$. Lemma 4.6 implies that the function $x + \bar{c}x$ does not vanish identically on R; thus we can find $x_0 \in R$ such that $a = x_0 + \bar{c}x_0$ is non-zero. Since $\bar{a} = \bar{x}_0 + cx_0$, we obtain $\bar{c}\bar{a} = a$; hence $\bar{c} = a/\bar{a}$. Consider the ideal $I = qaJ$, where q is a positive rational integer such that $qa \in R$. Then obviously I and J lie in the same class of ideals and moreover $\bar{I} = q\bar{a}\bar{J} = I$, as needed.

We may thus replace J by I or, which is simpler, assume that $\bar{J} = J$. This gives $cJ = J$ and consequently c is a unit of R. Now, choose a prime ideal $\mathfrak{P}$ of R in such a way that $\mathfrak{P}J = xR$ is a principal ideal, $\mathfrak{P}$ is unramified and of the first degree. Such a choice is possible in view of Corollary 7 to Proposition 7.9. Note that, in particular, we have $\mathfrak{P} \neq \bar{\mathfrak{P}}$ and hence $(\mathfrak{P}, \bar{\mathfrak{P}}) = 1$.

If now

$$u = \sum_{i=0}^{p-1} (a_i \sigma^i + b_i \sigma^i \tau) \quad (a_i, b_i \in Z)$$

is an arbitrary element of $Z[G]$, then

$$ux = \sum_{i=0}^{p-1} a_i \zeta_p^i x + \sum_{i=0}^{p-1} b_i \zeta_p^i \bar{x}.$$

Hence $Z[G]x = Rx + R\bar{x} = \mathfrak{P}J + \overline{\mathfrak{P}}J = J(\mathfrak{P} + \overline{\mathfrak{P}}) = JR = J$, showing that J is generated by x as a $Z[G]$-module. □

§ 3. Chebotarev's Theorem

1. In Chapter 6 we have seen that to every prime ideal $\mathfrak{P}$ in a normal extension L/K of an algebraic number field K there corresponds a canonically determined subgroup of the Galois group $G(L/K)$, namely, the decomposition group $G_{-1}(\mathfrak{P})$ of $\mathfrak{P}$, which in turn is isomorphic in a natural way to the Galois group of the corresponding p-adic extension $L_\mathfrak{P}/K_\mathfrak{p}$ (where p is the prime ideal in K lying below $\mathfrak{P}$). Let us look more closely at the case of unramified $\mathfrak{P}$. Theorem 5.8 shows that in this case the group $G(L_\mathfrak{P}/K_\mathfrak{p})$ is isomorphic to the Galois group of the corresponding extension of residue class-fields, which is necessarily cyclic and, moreover, possesses a distinctive generator s acting as follows: $s(a) = a^{p^F}$, where $F = f(K_\mathfrak{p}/Q_p)$. We can trace down this generator to $G_{-1}(\mathfrak{P})$ through our isomorphisms, and so we obtain an automorphism $s_\mathfrak{P}$ of L/K which satisfies

$$s_\mathfrak{P}(x) \equiv x^{N(\mathfrak{p})} \pmod{\mathfrak{P}}$$

for all x in R_L. Note also that this property determines $s_\mathfrak{P}$ uniquely. Traditionally $s_\mathfrak{P}$ is denoted by

$$\left[\frac{L/K}{\mathfrak{P}} \right]$$

and called the *Frobenius automorphism* associated with $\mathfrak{P}$ or the *Frobenius symbol*.

If we take another prime ideal, say $\mathfrak{P}_1$, which lies over p, then the automorphisms $s_\mathfrak{P}$ and $s_{\mathfrak{P}_1}$ may differ; however, in this case we have

Proposition 7.13. *If t is an element of the Galois group $G(L/K)$, then*

$$\left[\frac{L/K}{t(\mathfrak{P})} \right] = t \circ \left[\frac{L/K}{\mathfrak{P}} \right] \circ t^{-1}.$$

Proof. Denote $t(\mathfrak{P})$ by $\mathfrak{P}_1$ and let x be an arbitrary element of R_L. Then $s_\mathfrak{P}(x) - x^{N(\mathfrak{p})}$ lies in $\mathfrak{P}$ and so,

$$t \circ s_\mathfrak{P} \circ t^{-1}(x) - t \circ t^{-1}(x^{N(\mathfrak{p})}) \in t(\mathfrak{P}) = \mathfrak{P}_1,$$

i.e.

$$t \circ s_\mathfrak{P} \circ t^{-1}(x) - x^{N(\mathfrak{p})} \in \mathfrak{P}_1,$$

proving the proposition. □

Corollary 1. *If L/K is Abelian, then the Frobenius automorphism of $\mathfrak{P}$ depends only on the prime ideal $\mathfrak{p}$ which lies below $\mathfrak{P}$ in K.* □

Corollary 2. *If L/K is normal and for any element g in $G(L/K)$ we denote by $\mathrm{Cl}(g)$ the class of all elements conjugate to g, then by*

$$F_{L/K}: \mathfrak{p} \mapsto \mathrm{Cl}(s_{\mathfrak{P}}),$$

where $\mathfrak{P}$ is an arbitrary prime ideal of L lying above $\mathfrak{p}$, we define a mapping of the set of all prime ideals of K unramified in L/K into the set of all conjugacy classes in $G(L/K)$. □

The following theorem gives the main properties of the mapping $F_{L/K}$.

Theorem 7.10. (i) *If L/K is a normal extension and $\mathfrak{p}$ is a prime ideal of K unramified in it, then*

$$\mathfrak{p} R_L = \mathfrak{P}_1 \ldots \mathfrak{P}_g, \quad f_{L/K}(\mathfrak{P}_i) = f$$

holds with f equal to the order in $G(L/K)$ of any element of $F_{L/K}(\mathfrak{p})$ and $g = n/f$.

(ii) *If $K \subset L \subset M$, M/K and L/K are normal, $\mathfrak{p}$ is a prime ideal in K unramified in M/K and $R = R_{M/L}: G(M/K) \to G(L/K)$ is the restriction mapping, then $R(F_{M/K}(\mathfrak{p})) = F_{L/K}(\mathfrak{p})$.*

(iii) *If L/K and M/K are normal and $\mathfrak{p}$ is a prime ideal on K unramified in LM/K, then the inclusion*

$$F_{LM/K}(\mathfrak{p}) \subset F_{L/K}(\mathfrak{p}) F_{M/K}(\mathfrak{p})$$

holds, provided we identify the group $G(LM/K)$ with its image in the product $G(L/K) \times G(M/K)$ under the map $g \mapsto [g|_L, g|_M]$.

(iv) *If L/K is normal, M/K is arbitrary finite, $\mathfrak{p}$ is a prime ideal of K unramified in L/K and $\mathfrak{q}$ is a prime ideal of M lying above $\mathfrak{p}$, then $\mathfrak{q}$ is unramified in LM/M and we have the inclusion*

$$F_{LM/M}(\mathfrak{q}) \subset F_{L/K}(\mathfrak{p})^f.$$

(v) *Let L/K be finite and let M/K be the minimal normal extension of K containing L. Let $\mathfrak{p}$ be a prime ideal of K unramified in M/K, $\mathfrak{P}$ a prime ideal of M lying above $\mathfrak{p}$, $\mathfrak{Q}$ the prime ideal of L lying below $\mathfrak{P}$, $s = \left[\dfrac{M/K}{\mathfrak{P}}\right]$ and H the subgroup of $G(M/K)$ generated by s. Then*

$$f_{L/K}(\mathfrak{Q}) = |H|/|H \cap U|$$

where U is the subgroup of $G(M/K)$ corresponding to L, according to Galois theory.

Moreover, the set of prime ideals of M dividing $\mathfrak{Q}$ coincides with $\{us^k(\mathfrak{P}): u \in U, k \geq 0\}$.

Proof. (i) results from the definition of $\left[\dfrac{L/K}{\mathfrak{P}}\right]$ as a generator of the decomposition group of $\mathfrak{P}$ and the Corollary to Proposition 6.5, the equality $e_{L/K}(\mathfrak{P}) = 1$ being taken into account.

(ii) If $s \in F_{M/K}(\mathfrak{p})$ and $s_1 = R(s)$, then for every x in R_L we have $s_1(x) \equiv x^{N(\mathfrak{p})} (\bmod \mathfrak{P})$, where $\mathfrak{P}$ is a suitable prime ideal in M lying above p. But this implies $s_1(x) \equiv x^{N(\mathfrak{p})} (\bmod \mathfrak{P}_1)$, where $\mathfrak{P}_1$ is the prime ideal in L lying below $\mathfrak{P}$ and $x \in R_L$, and so s_1 belongs to $F_{L/K}(\mathfrak{p})$. Hence R maps $F_{M/K}(\mathfrak{p})$ into $F_{L/K}(\mathfrak{p})$, and since R is surjective and preserves conjugacy classes, we have $R(F_{M/K}(\mathfrak{p})) = F_{L/K}(\mathfrak{p})$.

(iii) If, for any g in $G(LM/K)$ we define g_L and g_M by $g_L = g|_L$, $g_M = g|_M$, then the mapping $g \mapsto [g_L, g_M]$ is an injection of $G(LM/K)$ into the product $G(L/K) \times G(M/K)$, and the composition of this injection with projection onto $G(L/K)$ and $G(M/K)$ coincides with $R_{LM/L}$ and $R_{LM/M}$, respectively. Moreover, the image of g equals $[g_L, 1] \cdot [1, g_M]$. Applying this to g in $F_{LM/K}(\mathfrak{p})$, we get

$$g = [g_L, g_M] = [R_{LM/L}(g), R_{LM/M}(g)] \subset F_{L/K}(\mathfrak{p}) F_{M/K}(\mathfrak{p})$$

by (ii).

(iv) Write $L = K(a)$, $LM = M(a)$. If g is an element of $G(LM/M)$, then $g_L = g|_L$ lies in $G(L/L \cap M)$ and the mapping $g \mapsto g_L$ is an isomorphism. We can thus identify $G(LM/M)$ with $G(L/L \cap M)$. Applying Corollary 1 to Lemma 5.4 to the corresponding p-adic extension and using Theorem 5.5 (iv), we see that q is unramified under LM/M. This proves the first part of our assertion. To prove the second, consider any g in $F_{LM/M}(\mathfrak{q})$ and let $\mathfrak{P}_1$ be the prime ideal in LM which lies over q and satisfies

$$\left[\dfrac{LM/M}{\mathfrak{P}_1}\right] = g.$$

If $\mathfrak{P}$ is the prime ideal in L lying below $\mathfrak{P}_1$, then for any x in R_L we have

$$x^{N(\mathfrak{q})} \equiv g(x) \;(\bmod \mathfrak{P}_1) \text{ and also } (\bmod \mathfrak{P}).$$

If $g_1 \in F_{L/K}(\mathfrak{p})$ is chosen so that for $x \in R_L$ the congruence

$$x^{N(\mathfrak{p})} \equiv g_1(x) \;(\bmod \mathfrak{P})$$

is satisfied, then in view of $N(\mathfrak{q}) = N(\mathfrak{p})^f$

$$x^{N(\mathfrak{q})} \equiv g_1^f(x) \;(\bmod \mathfrak{P})$$

holds, and this leads to

$$g(x) \equiv g_1^f(x) \;(\bmod \mathfrak{P}),$$

which can hold for each x in R_L only if g and g_1^f coincide. But this shows that $F_{LM/M}(\mathfrak{q}) \subset F_{L/K}(\mathfrak{p})^f$, as required.

(v) First observe that by Corollary 2 to Proposition 4.12 $\mathfrak{p}$ is unramified under M/K. The elements s^k ($k \in Z$) belong to the decomposition group of $\mathfrak{P}$ over K; thus $s^k(\mathfrak{P}) = \mathfrak{P}$, and if u belongs to U and $\mathfrak{P}_1$ is the prime ideal $u(\mathfrak{P})$, then the inclusion $\mathfrak{Q} R_M \subset \mathfrak{P}$ implies $u(\mathfrak{Q} R_M) \subset \mathfrak{P}_1$, which in view of $u(\mathfrak{Q}) = \mathfrak{Q}$ ($\mathfrak{Q} \subset R_L$!) gives $\mathfrak{Q} R_M \subset \mathfrak{P}_1$; thus $\mathfrak{P}_1$ lies over $\mathfrak{Q}$. This shows that all ideals $(us^k)(\mathfrak{P})$ are prime ideals lying over $\mathfrak{Q}$. We now prove that no other prime ideal can lie over $\mathfrak{Q}$. In fact, if $\mathfrak{P}'$ lies over $\mathfrak{Q}$, then $\mathfrak{P}' = u(\mathfrak{P})$ with a suitable $u \in G(M/L) = U$.

Finally, Lemma 6.2 shows that the decomposition group of $\mathfrak{P}$ over K equals $H \cap U$, and so we obtain

$$f_{L/K}(\mathfrak{Q}) = f_{M/K}(\mathfrak{P})/f_{M/L}(\mathfrak{P}) = |H|/|H \cap U|.$$
□

Corollary 1. *If L/K is normal, then a prime ideal $\mathfrak{p}$ of K splits in L/K iff $F_{L/K}(\mathfrak{p}) = 1$.*

Proof. Immediate from (i). □

Corollary 2. *If L/K is normal and $\mathfrak{p}$ is a prime ideal in K, then $\mathfrak{p}$ remains a prime ideal in L iff the Galois group $G(L/K)$ is cyclic and generated by $F_{L/K}(\mathfrak{p})$.*

Proof. The ideal $\mathfrak{p}$, prime in K, remains so in L iff it is unramified and $f_{L/K}(\mathfrak{P}) = [L:K]$ holds for any prime ideal $\mathfrak{P}$ over $\mathfrak{p}$; this, by (i), is equivalent to our assertion. □

(This corollary gives an explanation of the example presented after Corollary 5 to Proposition 7.9.)

Corollary 3. *If L/K is finite and M/K is the minimal normal extension of K containing L, then the type of factorization (i.e. the number and degrees of the factors) of any prime ideal $\mathfrak{p}$ of K unramified in L/K depends solely on $F_{M/K}(\mathfrak{p})$ and $G(M/L)$.*

Proof. Apply (v) and (i). □

2. The principal result concerning the Frobenius automorphisms is the density theorem of Chebotarev, which we first present in its weaker form:

Theorem 7.11. *If L/K is normal of degree n and A is an arbitrary class of conjugate elements in the Galois group $G(L/K)$, then the set*

$$P_A = \{\mathfrak{p}: F_{L/K}(\mathfrak{p}) = A\}$$

is infinite and has Dirichlet density $|A|/n$.

The stronger form requires for its proof the Artin reciprocity theorem, which lies outside the scope of this book. Chebotarev's theorem in its stronger form reads as follows:

Theorem 7.11*. *If L/K is normal of degree n and A is an arbitrary class of conjugate elements in the Galois group $G(L/K)$, then the set $P_A = \{\mathfrak{p}: F_{L/K}(\mathfrak{p}) = A\}$ is regular and its density equals $|A|/n$. If $N_A(x)$ is the number of prime ideals in P_A whose norms do not exceed x, then $N_A(x) = (|A|/n + o(1)) x/\log x$.*

In the case of $K = Q$, L/Q Abelian, or more generally, when L is contained in a cyclotomic extension of K, one can obtain an elementary proof, and this will be the first step toward the proof of Theorem 7.11. In the next step we shall reduce the problem to the case of cyclic extensions and finally, for a given cyclic extension we shall combine various cyclic extensions contained in cyclotomic extensions to obtain Theorem 7.11. In this last step we shall approximate the sum

$$\sum_{\mathfrak{p} \in P_A} N(\mathfrak{p})^{-s}$$

and our approach will not be strong enough to yield the regularity of P_A.

Proof of Theorem 7.11. First we establish Theorem 7.11* for cyclotomic extensions:

Lemma 7.9. *If $L = K(\zeta_m)$ where ζ_m is a primitive m-th root of unity, then Theorem 7.11* holds for the extension L/K.*

Proof. We apply Theorem 7.10 (iv) to the normal extension $Q(\zeta_m)/Q$ and the extension K/Q, not necessarily normal. Let $\mathfrak{q}$ be a prime ideal of K and p the prime ideal in Q below $\mathfrak{q}$, and assume that, firstly, p does not ramify in $Q(\zeta_m)$, and secondly, $f_{K/Q}(\mathfrak{q}) = 1$. Since the extensions $Q(\zeta_m)/Q$ and L/K are both Abelian, the quoted theorem implies $F_{L/K}(\mathfrak{q}) = F_{Q(\zeta_m)/Q}(p)$ if we consider the Galois group $G(L/K)$ as a subgroup of $G(Q(\zeta_m)/Q)$. Now note that every automorphism of $Q(\zeta_m)$ is determined by its action on ζ_m and hence by a residue class $r \pmod{m}$ prime to m. Denote by g_r the automorphism satisfying $g_r(\zeta_m) = \zeta_m^r$ ($1 \leq r < m$, $(r, m) = 1$) and let p be the positive generator of p. The definition of the Frobenius automorphism gives

$$F_{Q(\zeta_m)/Q}(p) = g_r \quad \text{iff} \quad p \equiv r \pmod{m}.$$

This shows, in turn, that if P_r denotes the set of all prime ideals $\mathfrak{q}$ with $f_{K/Q}(\mathfrak{q}) = 1$ and $F_{L/K}(\mathfrak{q}) = g_r$, then

$$P_r = \{\mathfrak{q}: f_{K/Q}(\mathfrak{q}) = 1, N(\mathfrak{q}) \equiv r \pmod{m}\},$$

and thus
$$\sum_{F_{L/K}(\mathfrak{q})=g_r} N(\mathfrak{q})^{-s} = \sum_{\mathfrak{q}\in P_r} N(\mathfrak{q})^{-s} + g(s)$$
$$= \sum_{\substack{\mathfrak{q} \\ N(\mathfrak{q})\equiv r(\mathrm{mod}\,m)}} N(\mathfrak{q})^{-s} + g(s),$$

since the prime ideals $\mathfrak{q}$ with $f_{K/Q}(\mathfrak{q}) \neq 1$ contribute only to $g(s)$ in the equalities above.

Now observe that if a is an integer of K congruent to unity $(\mathrm{mod}\,m)$ and totally positive, then $N_{K/Q}(a)$ is congruent to $1(\mathrm{mod}\,m)$, as well. Indeed, if m divides $a-1$, then, for every automorphism g of K, m divides $g(a)-1$ because of $g(m) = m$, and this implies $g(a) \equiv 1(\mathrm{mod}\,m)$ and $N_{K/Q}(a) \equiv 1(\mathrm{mod}\,m)$. In view of the total positivity of a we have $N(aR_K) = N_{K/Q}(a)$. Hence the set $\{I: N(I) \equiv 1(\mathrm{mod}\,m)\}$ is a sum of, say, N classes from $H_\mathfrak{m}^*(K)$, where $\mathfrak{m}$ is the ideal in R_K generated by m, and this shows that each set $\{I: N(I) \equiv r(\mathrm{mod}\,m)\}$ either is void or contains N classes of $H_\mathfrak{m}^*(K)$; thus by Corollary 6 to Proposition 7.9 we get

$$\sum_{\substack{\mathfrak{q} \\ N(\mathfrak{q})\equiv r(\mathrm{mod}\,m)}} N(\mathfrak{q})^{-s} = \varepsilon_r N(h_\mathfrak{m}^*(K))^{-1}\log\frac{1}{s-1} + g(s), \qquad (7.26)$$

where ε_r equals 0 or 1. Put $N_1 = \varepsilon_1 + \ldots + \varepsilon_{m-1}$. The last equality implies
$$N_1 N(h_\mathfrak{m}^*(K))^{-1} = 1;$$
but N_1 cannot exceed the number of elements in $G(L/K)$ and thus
$$nN \geq h_\mathfrak{m}^*(K) \qquad (7.27)$$
must hold. On the other hand, every prime ideal $\mathfrak{q}$ of the first degree in K whose norm is congruent to unity $(\mathrm{mod}\,m)$ splits in L/K. In fact, in this case $F_{Q(\zeta_m)/Q}(\mathfrak{p}) = 1$ and $F_{L/K}(\mathfrak{q}) = 1$. Now (7.26) shows that the density of the set of prime ideals of K which split in L/K equals at least $N(h_\mathfrak{m}^*(K))^{-1}$, and by Corollary 3 to Proposition 7.10 it equals $1/n$; thus
$$nN \leq h_\mathfrak{m}^*(K)$$
and (7.27) implies that we have an equality here. So (7.26) becomes
$$\sum_{\substack{\mathfrak{q} \\ N(\mathfrak{q})\equiv r(\mathrm{mod}\,m)}} N(\mathfrak{q})^{-s} = \varepsilon_r n^{-1}\log\frac{1}{s-1} + g(s),$$
and we see that exactly n of the sets P_r are non-void. But this proves that every element of $G(L/K)$ is the Frobenius automorphism for prime ideals forming a regular set of density $1/n$. □

Using this lemma we easily obtain Theorem 7.11* for subextensions of cyclotomic extensions:

Lemma 7.10. *If $K \subset M \subset L$ and L/K satisfies the assumptions of the previous lemma, then Theorem 7.11* is also true for the extension M/K.*

Proof. Denote by φ the restriction map of $G(L/K)$ onto $G(M/K)$ and let p be any prime ideal of K which does not ramify under the extension L/K. By Theorem 7.10 (ii) we have

$$F_{M/K}(\mathfrak{p}) = \varphi(F_{L/K}(\mathfrak{p})),$$

and so for any g in $G(M/K)$ we obtain

$$\sum_{\substack{\mathfrak{p}\\ F_{M/K}(\mathfrak{p})=g}} N(\mathfrak{p})^{-s} = \sum_{\substack{\mathfrak{p}\\ F_{L/K}(\mathfrak{p})\in\varphi^{-1}(g)}} N(\mathfrak{p})^{-s} + g(s).$$

Lemma 7.9 implies that the last sum equals

$$|\varphi^{-1}(g)|m^{-1}\log\frac{1}{s-1} + g(s),$$

where $m = [L:K]$, but obviously $|\varphi^{-1}(g)| = [L:M]$, and so

$$\sum_{\substack{\mathfrak{p}\\ F_{M/K}(\mathfrak{p})=g}} N(\mathfrak{p})^{-s} = [M:K]^{-1}\log\frac{1}{s-1} + g(s). \qquad \square$$

Corollary. *Theorem 7.11* holds for all Abelian extensions of the rationals.*

Proof. Apply Theorem 6.5 and the lemma. $\square$

The next lemma reduces the proof of Theorem 7.11 to the case of cyclic extensions.

Lemma 7.11. *If Theorem 7.11 or 7.11* is true for all cyclic extensions of every algebraic number field, then it is true for all normal extensions.*

Proof. Let L/K be an arbitrary normal extension. Select an arbitrary element g in $G(L/K)$. We consider the set P_g of those prime ideals in K which do not ramify in L/K and for which $F_{L/K}(\mathfrak{p})$ is the conjugacy class containing g. For every prime ideal $\mathfrak{p}$ in P_g there is a prime ideal $\mathfrak{P}$ lying over $\mathfrak{p}$ in L such that

$$\left[\frac{L/K}{\mathfrak{P}}\right] = g. \tag{7.28}$$

Let $Z(g)$ be the centralizer of g (i.e. the set of all elements of $G(L/K)$ which commute with g), $C(g)$ the cyclic group generated by g and $m(g)$ the index of $C(g)$ in $Z(g)$. Observe that $m(g)$ coincides with the number of those prime ideals $\mathfrak{P}$ over a given $\mathfrak{p} \in P_g$ which satisfy (7.28), this number being independent of p. In fact, if $t \in G(L/K)$ and $\mathfrak{P}$ satisfies (7.28), then the equality

$$\left[\frac{L/K}{t(\mathfrak{P})}\right] = g$$

holds by Proposition 7.13 iff $tgt^{-1} = g$, i.e. iff t lies in $Z(g)$ and there are exactly $|C(g)|$ distinct prime ideals of the form $t(\mathfrak{P})$, where $t \in G(L/K)$, since obviously $C(g)$ is the decomposition group of P.

Now, let M be the subfield of L corresponding to $C(g)$ and denote by $\mathfrak{q}$ the prime ideal of M lying below $\mathfrak{P}$, which is a fixed prime ideal over $\mathfrak{p}$ satisfying (7.28). Proposition 6.5 shows that $\mathfrak{q}R_L = \mathfrak{P}$ and $N_{M/Q}(\mathfrak{q}) = N_{K/Q}(\mathfrak{P})$ (because of $f_{M/K}(\mathfrak{q}) = 1$) and so the conditions

$$g(x) \equiv x^{N(\mathfrak{q})} \pmod{\mathfrak{P}}$$

and

$$g(x) \equiv x^{N(\mathfrak{p})} \pmod{\mathfrak{P}}$$

are equivalent; hence $\left[\frac{L/M}{\mathfrak{P}}\right] = g$, and also $F_{L/M}(\mathfrak{q}) = g$. We thus see that if $F_{L/K}(\mathfrak{p})$ is the conjugacy class containing g, then $\mathfrak{p}$ has $m(g)$ prime ideal divisors $\mathfrak{q}$ in M satisfying $N_{M/Q}(\mathfrak{q}) = N_{K/Q}(\mathfrak{p})$ and $F_{L/M}(\mathfrak{q}) = g$. Note that also conversely, if $F_{L/M}(\mathfrak{q}) = g$ and $N_{M/Q}(\mathfrak{q}) = N_{K/Q}(\mathfrak{p})$, then $g \in F_{L/K}(\mathfrak{p})$ and the set of the remaining prime ideals $\mathfrak{Q}$ from M with $F_{L/M}(\mathfrak{Q}) = g$ is regular and has density zero, since every such ideal is of degree exceeding 1 over the rationals. We can thus write

$$m(g) \sum_{g \in F_{L/K}(\mathfrak{p})} N(\mathfrak{p})^{-s} = \sum_{F_{L/M}(\mathfrak{q})=g} N(\mathfrak{q})^{-s} + g(s). \tag{7.29}$$

Now we use the assumed validity of Theorem 7.11 for all cyclic extensions in the case of the extension L/M. This leads us to

$$\sum_{F_{L/M}(\mathfrak{q})=g} N(\mathfrak{q})^{-s} = [L:M]^{-1}\log\frac{1}{s-1} + o\left(\log\frac{1}{s-1}\right)$$

$$= |C(g)|^{-1}\log\frac{1}{s-1} + o\left(\log\frac{1}{s-1}\right),$$

as s tends to 1 over real numbers larger than 1; so, by (7.29),

$$\sum_{g \in F_{L/K}(\mathfrak{p})} N(\mathfrak{p})^{-s} = |Z(g)|^{-1}\log\frac{1}{s-1} + o\left(\log\frac{1}{s-1}\right),$$

and it suffices to remember that $|Z(g)| = |G(L/K)|/|\{tgt^{-1}\}|$.

If we assume that Theorem 7.11* holds for all cyclic extensions, then in the above arguments we can replace $o(\log 1/(s-1))$ by $g(s)$, which yields Theorem 7.11* for our extension L/K. □

So we are left with cyclic extensions only. We need a technical lemma:

Lemma 7.12. *If L/K is a given finite extension and G is a given Abelian group, then there exists a normal extension M/K, which has the Galois group isomorphic to G and which can be embedded in a cyclotomic extension N/K and satisfies $L \cap M = K$.*

Proof. Let $G = C_{p_1^{a_1}} \times \ldots \times C_{p_r^{a_r}}$ be the factorization of G into primary cyclic factors and consider cyclotomic extensions of K having the form $K(\zeta_m)$ with $m = q_1 \ldots q_r$, where the q_i's are distinct primes satisfying the conditions

$$q_i \equiv 1 + p_i^{a_i} \pmod{p_i^{a_i+1}} \quad (i = 1, 2, \ldots, r)$$

and not dividing the absolute discriminant $d(K)$ of K. The existence of such primes follows from the theorem of Dirichlet (Corollary 7 to Proposition 7.9). Theorem 2.9 and Corollary to Theorem 4.9 show that all prime divisors of $d(Q(\zeta_m))$ divide m, and so by Theorem 4.9 we obtain $[K(\zeta_m):K] = \varphi(m)$. Since for relatively prime m_1 and m_2 we have $K(\zeta_{m_1}, \zeta_{m_2}) = K(\zeta_{m_1 m_2})$, in this case

$$[K(\zeta_{m_1}, \zeta_{m_2}):K] = \varphi(m_1 m_2) = \varphi(m_1)\varphi(m_2)$$
$$= [K(\zeta_{m_1}):K][K(\zeta_{m_2}):K];$$

hence $K(\zeta_{m_1}) \cap K(\zeta_{m_2}) = K$. This shows that if v is the number of subfields of L which contain K, then from every system of $v+1$ fields $K(\zeta_{m_i})$ one can extract one, say $K(\zeta_m) = N$, for which $N \cap L = K$. If $m = q_1 \ldots q_r$, $q_i = 1 + b_i p_i^{a_i}$ ($i = 1, 2, \ldots, r$), then the Galois group $G(N/K)$ equals

$$\prod_i C_{b_i p_i^{a_i}} = \prod_i (C_{b_i} \times C_{p_i^{a_i}})$$

since $(b_i, p_i^{a_i}) = 1$. Now, take for M the field which corresponds to the subgroup $\prod_i C_{b_i}$ of $G(N/K)$. Obviously $G(M/K)$ is isomorphic to G and $L \cap M = K$. □

Now, as all necessary preparations have been made, we can proceed with the proof of Theorem 7.11. Let L/K be a cyclic extension with group G and let M/K be an extension contained in a suitable cyclotomic extension such that $L \cap M = K$. Let G_1 be the Galois group of M/K and observe that by Lemma 7.12 we can take for G_1 any finite Abelian group. Choose an element g in G and let t be an element of G_1 whose order f_1 is divisible by the order f of g, if such an element at all exists. Let U be the cyclic subgroup of the product $G \times G_1$ generated by the pair (g, t) and observe that, since this product equals $G(LM/K)$, we may consider the field $N \subset LM$ corresponding to U. The extension LM/N is cyclic with group U and MN/N is also cyclic with group

$U/([g] \cap U) \simeq [t]$ (where $[x]$ denotes the cyclic group generated by x) and contained in a cyclotomic extension of N. Lemma 7.10 implies that if $\bar{t}$ is the coset determined by t in $U/([g] \cap U)$, then

$$\sum_{F_{MN/N}(q)=\bar{t}} N(q)^{-s} = (1/f_1)\log\frac{1}{s-1} + g(s). \tag{7.30}$$

Now let q be any prime ideal in N of the first degree over K and let p be the prime ideal of K lying below it. By Theorem 7.10 (iv) we have $F_{MN/N}(q) = F_{M/K}(\mathfrak{p})$, and thus (7.30) implies

$$\sum_{\substack{F_{M/K}(\mathfrak{p})=t \\ \mathfrak{p} \text{ splits in } N}} N(\mathfrak{p})^{-s} = [N:K]^{-1} \sum_{F_{MN/N}(q)=\bar{t}} N(q)^{-s}$$

$$= ([N:K]f_1)^{-1}\log\frac{1}{s-1} + g(s). \tag{7.31}$$

Because of $G(N/K) = G/U$, Corollary 1 to Theorem 7.10 implies that a prime ideal p from K splits in N/K iff $F_{L/K}(\mathfrak{p}) \in U$. Hence, the set A_t of all those prime ideals from K which split in N/K and for which $F_{M/K}(\mathfrak{p}) = t$ is identical with

$$\{\mathfrak{p}: F_{M/K}(\mathfrak{p}) = t, F_{LM/K}(\mathfrak{p}) = (gt)^r \text{ for some } r\}.$$

Now Theorem 7.10 (iii) shows that the condition $F_{LM/K}(\mathfrak{p}) = (gt)^r$ is equivalent to the conjunction of $F_{L/K}(\mathfrak{p}) = g^r$, $F_{M/K}(\mathfrak{p}) = t^r$; thus, for $\mathfrak{p} \in A_t$, we must have $t^r = t$, i.e. $r \equiv 1 \pmod{f_1}$ and also $r \equiv 1 \pmod{f}$ because f divides f_1. Finally, we arrive at

$$A_t = \{\mathfrak{p}: F_{M/K}(\mathfrak{p}) = t, F_{L/K}(\mathfrak{p}) = g\},$$

and (7.31) leads to

$$\sum_{\substack{F_{L/K}(\mathfrak{p})=g \\ F_{M/K}(\mathfrak{p})=t}} N(\mathfrak{p})^{-s} = ([N:K]f_1)^{-1}\log\frac{1}{s-1} + g(s). \tag{7.32}$$

Denote by $N(f)$ the number of those elements t in G_1 whose orders are divisible by f and add equalities (7.32) over all such elements. This gives

$$\sum_{F_{L/K}(\mathfrak{p})=g} N(\mathfrak{p})^{-s} \geq (N(f)/[LM:K])\log\frac{1}{s-1} + O(1),$$

as s tends to 1 over real numbers exceeding 1, because of

$$f_1[N:K] = [LM:N][N:K] = [LM:K].$$

Now we use the fact that G_1 can be an arbitrary finite Abelian group. We choose it in such a way that the ratio $N(f)/[M:K]$ falls between $1-\varepsilon$ and 1, where ε is an arbitrary positive number. We have to show that such a choice is in fact possible. Factorize $f = p_1^{a_1} \ldots p_r^{a_r}$ and note that if G_1 is a cyclic group of the order $p_1^{b_1} \ldots p_r^{b_r}$, where $b_i \geq a_i$ ($i = 1, 2, \ldots, r$), then G_1 contains

$$\prod_{i=1}^{r} (p_i^{b_i} - p_i^{a_i-1})$$

elements whose orders are divisible by f, and so for such a group the ratio $N(f)/|G_1|$ equals

$$\sum_{i=1}^{r} (1 - p_i^{a_i - b_i - 1});$$

and thus can be made arbitrarily close to 1 by a suitable choice of the b_i's. Hence

$$\sum_{F_{L/K}(\mathfrak{p}) = g} N(\mathfrak{p})^{-s} \geq ([L:K]^{-1} + o(1)) \log \frac{1}{s-1}, \tag{7.33}$$

and adding the resulting equalities for $g \in G$, we get

$$\log \frac{1}{s-1} + O(1) = \sum_{g \in G} \sum_{F_{L/K}(\mathfrak{p}) = g} N(\mathfrak{p})^{-s} \geq (1 + o(1)) \log \frac{1}{s-1},$$

which is possible only if in (7.33) the equality sign occurs. □

Thus the proof of Theorem 7.11 is complete. Now we deduce the stronger Theorem 7.11* from the reciprocity theorem of Artin. First we have to state the latter. There are many equivalent formulations of this theorem, which is fundamental for the class-field theory; we choose the version which is most suitable for our purpose.

Artin's Reciprocity Theorem. *If L/K is Abelian and G is its Galois group, then there exists an ideal $\mathfrak{f}$ in R_K such that the prime ideals ramified under L/K coincide with the prime ideals dividing $\mathfrak{f}$, and if $\mathfrak{p}_1$ and $\mathfrak{p}_2$ are two unramified prime ideals in the same class in $H_\mathfrak{f}^*(K)$, then $F_{L/K}(\mathfrak{p}_1) = F_{L/K}(\mathfrak{p}_2)$. The map of $H_\mathfrak{f}^*(K)$ into G induced in this way by $F_{L/K}$ turns out to be a surjective homomorphism.*

Theorem 7.11* is an immediate corollary. Indeed, in view of Lemma 7.11 it suffices to deal with the case where L/K is cyclic and hence Abelian, and the reciprocity theorem shows that the set of all prime ideals $\mathfrak{p}$ in K for which $F_{L/K}(\mathfrak{p}) = g$, g being a given element of the group $G(L/K)$, is non-void and consists of all prime ideals lying in a suitable set of classes of $H_\mathfrak{f}^*(K)$, the number of these classes being independent of g. Now it is enough to apply Corollary 6 to Proposition 7.9. □

3. Now we turn to applications of Chebotarev's Density Theorem. We start with two old results of Frobenius and Kronecker, which formed the first steps towards Theorem 7.11.

Proposition 7.14. (i) *If L/K is normal of degree n and for a given g in $G(L/K)$ we denote by $A(g)$ the union of conjugacy classes of $g, g^2, g^3, \ldots$, then the set of all prime ideals $\mathfrak{p}$ of K which satisfy $F_{L/K}(\mathfrak{p}) \in A(g)$ is infinite and its Dirichlet density equals $|A(g)|/n$.*

(ii) *Let L/K be finite of degree n. For $m = 0, 1, 2, \ldots, n$ denote by P_m the set of all prime ideals of K having exactly m prime divisors of the first degree in L. Then each of the sets P_m has a Dirichlet density, say d_m, and moreover*

$$d_0 + d_1 + \ldots + d_n = d_1 + 2d_2 + 3d_3 + \ldots + nd_n = 1.$$

Proof. Assertion (i) follows immediately from Theorem 7.11, since every $A(g)$ is a union of disjoint conjugacy classes.

To prove (ii) observe that the existence of the densities d_m is a consequence of Corollary 3 to Theorem 7.10 and Theorem 7.11. Since the sets P_m are disjoint and their union contains every prime ideal of K, we get $d_0 + d_1 + \ldots + d_n = 1$. To obtain the second equality we use Corollary 3 to Proposition 7.9, getting for $\operatorname{Re} s > 1$

$$(1 + o(1)) \log \frac{1}{s-1} = \sum_{\substack{\mathfrak{P} \\ f_{L/K}(\mathfrak{P})=1}} N(\mathfrak{P})^{-s}$$

$$= \sum_{m=0}^{n} \sum_{\substack{\mathfrak{P} \\ f_{L/K}(\mathfrak{P})=1 \\ \mathfrak{P} \cap R_K \in P_m}} N(\mathfrak{P})^{-s}$$

$$= \sum_{m=0}^{n} m \sum_{\mathfrak{p} \in P_m} N(\mathfrak{p})^{-s} + g(s)$$

$$= \left(\sum_{m=1}^{n} md_m + o(1) \right) \log \frac{1}{s-1},$$

whence

$$\sum_{m=1}^{n} md_m = 1. \qquad \square$$

Let L/K be finite of degree n and let M/K be the smallest normal extension of K containing L with Galois group G. Let $F_1, \ldots, F_n$ be all embeddings of L in $\mathfrak{Z}$ equal to identity on K and let F_1 be the identity on L. Put $L_i = F_i(L)$ and observe that G acts as a permutation group on the sequence $\{L_1, \ldots, L_n\}$

by $g(L_i) = L_j$ iff $g \circ F_i = F_j$. In the next proposition we keep this interpretation of G in mind.

Proposition 7.15. *The set of all prime ideals $\mathfrak{p}$ of K unramified in L/K and satisfying*

$$\mathfrak{p}R_L = \mathfrak{P}_1 \ldots \mathfrak{P}_r$$

with fixed $f_{L/K}(\mathfrak{P}_i) = f_i$ ($i = 1, 2, \ldots, r$) has a density, which equals the relative frequence in G of the set of all permutations which are products of r disjoint cycles of orders $f_1, f_2, \ldots, f_r$, respectively.

Proof. We need an auxiliary result:

Lemma 7.13. *If $\mathfrak{p}$ is a prime ideal of K unramified in L/K, $g \in F_{M/K}(\mathfrak{p})$ and $g = g_1 \ldots g_r$ is the factorization of g into disjoint cycles, then*

$$\mathfrak{p}R_L = \mathfrak{P}_1 \ldots \mathfrak{P}_r$$

with

$$f_{L/K}(\mathfrak{P}_i) = |g_i| \quad (i = 1, 2, \ldots, r).$$

Proof. Let U be the subgroup of G corresponding to L. Fix i, $1 \leqslant i \leqslant r$, let L_a be an element of the cycle g_i, choose t_i in G mapping L_a onto L_1 and consider $h_i = t_i g t_i^{-1}$. By Theorem 7.10 (v) there exists a prime ideal $\mathfrak{P}_i$ lying over $\mathfrak{p}$ with $f_{L/K}(\mathfrak{P}_i) = |H_i|/|H_i \cap U|$, where H_i denotes the subgroup generated by h_i. Observe that in the factorization of h_i into disjoint cycles L_1 lies in a cycle of length $f_i = |g_i|$. Since the remaining cycles leave L_1 invariant, they must belong to U and it follows that f_i is the smallest positive integer with $h^{f_i} \in U$ and so $f_{L/K}(\mathfrak{P}_i) = f_i$ results. Since for distinct i's we get distinct prime ideals, we are ready. □

Proposition 7.15 follows immediately from the lemma and Theorem 7.11. □

4. Now we give an application of Theorem 7.11 to the question to what extent L is determined by the set $P(L/K)$ of all unramified prime ideals of K which have at least one prime ideal divisor in L of the first degree over K. It is clear that if L_1 and L_2 are conjugate over K, then the sets $P(L_1/K)$ and $P(L_2/K)$ coincide; thus $P(L/K)$ can determine L only up to an isomorphism over K. However, even this may fail, as we shall see below. Let us call an extension L/K a *Bauerian extension* if for every extension M/K the inclusion $P(M/K) \subset P(L/K)$ implies that M contains a subfield L_1 isomorphic to L over K. It is convenient at this point to consider two sets of prime ideals which differ only in a finite number of elements as identical, since this allows us to forget about all ramified prime ideals. So in this section an inclusion $A \subset B$ for sets of prime ideals will mean that the difference $A \setminus B$ is finite, and an equality $A = B$ will mean that the symmetric difference $A \div B$ is finite. The solution to our problem is given in the following theorem:

Theorem 7.12. *Let L/K and M/K be two arbitrary extensions and let N/K be the minimal normal extension of K which contains L. Denote by H and U the subgroups of $G(N/K)$ which correspond to L and $M \cap N$, respectively, and let $H_1 = H, H_2, \ldots, H_r$ be the subgroups of $G(N/K)$ conjugate to H. Then*

$$P(L/K) \supset P(M/K)$$

holds iff $U \subset \bigcup_{i=1}^{r} H_i$.

Proof. We start with a lemma:

Lemma 7.14. *Let L/K be a normal extension and M/K an arbitrary finite extension. Put $N = L \cap M$ and denote by U the subgroup of $G(L/K)$ corresponding to N. Moreover, let C be an arbitrary conjugacy class in $G(L/K)$. Then, for the existence of infinitely many prime ideals $\mathfrak{p}$ in K which have at least one prime ideal divisor of the first degree in M/K and for which $F_{L/K}(\mathfrak{p}) = C$, the condition $C \cap U \neq \emptyset$ is necessary and sufficient. If N/K is normal, then this condition may also be written in the form $C \subset U$.*

Proof. Denote by R/K the minimal normal extension of K containing M and let A be the set of all those unramified prime ideals from K which have at least one prime ideal divisor of the first degree in M. If U' is the subgroup of $G(R/K)$ which corresponds to M and $\mathfrak{P}$ is a prime ideal of R unramified in R/K, then the conditions

$$\left[\frac{R/K}{\mathfrak{P}}\right] \in U'$$

and $f_{M/K}(\mathfrak{P}') = 1$ (where $\mathfrak{P}'$ is the prime ideal of M below $\mathfrak{P}$) are equivalent by Theorem 7.10 (v). Thus a prime ideal $\mathfrak{p}$ of K lies in A iff the intersection $F_{R/K}(\mathfrak{p}) \cap U'$ is non-void. Consider S, the composite of L and R. Its Galois group over K can be considered as the subgroup of the product $G(L/K) \times G(R/K)$ consisting of all pairs (g_1, g_2) for which the restrictions of g_1 and g_2 to $L \cap R$ agree. Observe that if C_1 and C_2 are conjugacy classes in $G(L/K)$ and $G(R/K)$, respectively, then there exists a prime ideal $\mathfrak{p}$ (and in fact an infinity of them) with $F_{L/K}(\mathfrak{p}) = C_1$, $F_{R/K}(\mathfrak{p}) = C_2$ iff to every g_2 in C_2 one can select a g_1 in C_1 so that $g_1 g_2 \in G(S/K)$. Indeed, if $g_1 g_2$ lies in $G(S/K)$, then by Theorem 7.11 we can find a suitable $\mathfrak{p}$ (and even infinitely many of them), where $g_1 g_2 \in F_{S/K}(\mathfrak{p})$, and then Theorem 7.10 (iii) gives $g_1 \in F_{L/K}(\mathfrak{p})$, $g_2 \in F_{R/K}(\mathfrak{p})$. Conversely, if such a prime ideal $\mathfrak{p}$ exists, then by Theorem 7.10 (ii) the restrictions of C_1 and C_2 to $L \cap R$ coincide, both being equal to $F_{L \cap R/K}(\mathfrak{p})$, and so our assertion becomes obvious.

Thus we see that $\mathfrak{p} \in A$ iff the intersection $CU' \cap G(S/K)$ is not void. Since the intersection of $G(S/K)$ with $G(L/K)U'$ equals $G(S/M)$, we see that the set of all those $g_1 \in G(L/K)$ for which there exists a $g_2 \in U'$, where $g_1 g_2 \in G(S/K)$, equals the restriction of $G(S/M)$ to L, which in turn coincides with $G(L/N) = U$, and so finally we find that the set $A \cap \{\mathfrak{p}: F_{L/K}(\mathfrak{p}) = C\}$ is non-void iff there is a g_1 in C lying in U, i.e. $C \cap U \neq \emptyset$. If N/K is normal, then U is a normal subgroup of $G(L/K)$, and then the conditions $C \cap U \neq \emptyset$ and $C \subset U$ are equivalent. □

To prove the theorem, first assume that $P(M/K) \subset P(L/K)$. Let C be a conjugacy class in $G(N/K)$ which has elements in common with U. By the lemma the set $\{\mathfrak{p}: F_{N/K}(\mathfrak{p}) = C, \mathfrak{p} \in P(M/K)\}$ is infinite and our assumption implies that the same holds for $\{\mathfrak{p}: F_{N/K}(\mathfrak{p}) = C, \mathfrak{p} \in P(L/K)\}$, whence by the same proposition $C \cap H \neq \emptyset$. Now, if g lies in U, then one of its conjugates lies in H, hence $g \in \bigcup_{i=1}^r H_i$ proving $U \subset \bigcup_{i=1}^r H_i$.

To prove the second part assume that $U \subset \bigcup_{i=1}^r H_i$. Then all subgroups conjugate to U are also contained in $\bigcup_{i=1}^r H_i$. But for every $\mathfrak{p} \in P(M/K)$ the union of all subgroups conjugate to U contains $F_{N/K}(\mathfrak{p})$; thus $F_{N/K}(\mathfrak{p}) \cap H$ is non-void, proving that $\mathfrak{p} \in P(L/K)$. □

Corollary 1. *If L/K is a normal extension, then for every extension M/K the conditions $L \subset M$ and $P(L/K) \supset P(M/K)$ are equivalent, i.e. L/K is Bauerian.*

Proof. If $L \subset M$, then the multiplicativity of prime ideal degrees leads to $P(M/K) \subset P(L/K)$. However, if $P(M/K) \subset P(L/K)$, then we apply the theorem just proved with $N = L$, $H = \{1\}$; thus $U = \{1\}$, $L \cap M = L$ and so $L \subset M$. □

Corollary 2. *If L/K is normal, L_1/K arbitrary finite and for every prime ideal $\mathfrak{p}$ of R_K the sets $\{L_\mathfrak{P}\}_{\mathfrak{P} \text{ over } \mathfrak{p}}$ and $\{(L_1)_{\mathfrak{P}_1}\}_{\mathfrak{P}_1 \text{ over } \mathfrak{p}}$ contain fields pairwise isomorphic, then $L = L_1$.*

Proof. The assumption implies that $P(L/K) = P(L_1/K)$, and so by the preceding corollary we must have $L \subset L_1$; but these fields have the same degree and $L = L_1$ follows. □

Corollary 3. *Let L/K be an arbitrary extension and let M/K be the minimal normal extension of K containing L. Denote by H the subgroup of $G(M/K)$ corresponding to L and let $H_1 = H, H_2, \ldots, H_r$ be the subgroups conjugate to H. Then the extension L/K is Bauerian iff for every subgroup G of $G(M/K)$ it follows from $G \subset \bigcup_{i=1}^{r} H_i$ that G is contained in a certain H_i.*

Proof. Immediate from Theorem 7.12 and the definition of a Bauerian extension. □

This corollary allows us to present an example of a non-Bauerian extension. Let K be the rational field Q and take for L any field of fifth degree over Q for which the minimal normal extension of the rationals containing it has the symmetrical group on 5 letters for its Galois group. (E.g. one can take $L = Q(a)$, where a is any root of $2x^5 - 32x + 1$.) The subgroup H corresponding to L consists of all permutations fixing a suitable letter, say 1. Then H_i consists of all permutations fixing i ($i = 1, 2, \ldots, 5$) and so if we take $G = \{e, (123), (132), (12)(45)\}$, then G is contained in the union of the groups H_i without being contained in one of them. Thus L/Q cannot be Bauerian.

5. Our next application concerns power residues. Let p be an arbitrary rational prime and K any field of finite degree over Q containing all pth roots of unity. We shall call a set $(a_1, \ldots, a_n)$ of elements of K *p-independent* if the product

$$a_1^{x_1} \ldots a_n^{x_n}$$

with rational integral $x_1, \ldots, x_n$ can be a pth power of an element of K only in the case where all exponents x_i are divisible by p. Equivalently, a set $(a_1, \ldots, a_n)$ is p-independent if its image in $K^*/(K^*)^p$ treated as a linear space over $GF(p)$ forms a linearly independent set.

Let $\mathfrak{P}$ be a prime ideal in K which does not divide pR_K and is of first degree over Q. Then its norm in $Q(\zeta_p)$ is also of first degree over Q, and thus $N_{K/Q}(\mathfrak{P}) = q$ is a rational prime congruent to unity (mod p) by Theorem 4.16.

Now, for any non-zero a in R_K and any such prime ideal $\mathfrak{P}$ we can define the pth power residue symbol $\left(\dfrac{a}{\mathfrak{P}}\right)_p$ as that pth root of unity which is congruent to $a^{(q-1)/p}$ (mod $\mathfrak{P}$). This definition makes sense, since the identity

$$a^{q-1} - 1 = \prod_{j=0}^{p-1} (a^{(q-1)/p} - \zeta_p^j) \equiv 0 \,(\mathrm{mod}\, \mathfrak{P})$$

shows that $a^{(q-1)/p}$ is congruent to a certain pth root of unity, and indeed only to one of them, since they all are distinct (mod $\mathfrak{P}$), $\mathfrak{P}$ being unramified.

We are going to prove the following result:

Theorem 7.13. *Let K be an algebraic number field containing all p-th roots of unity, where p is a fixed prime. Assume that $a_1, \ldots, a_n \in R_K$ are given, forming a p-independent set, and $z_1, \ldots, z_n$ are given p-th roots of unity.*

Then one can find infinitely many prime ideals $\mathfrak{P}$ of the first degree over Q for which

$$\left(\frac{a_i}{\mathfrak{P}}\right)_p = z_i$$

holds for $i = 1, 2, \ldots, n$.

Proof. For $i = 1, 2, \ldots, n$ write $z_i = \zeta_p^{\varepsilon_i}$, where $0 \leq \varepsilon_i \leq p-1$ and let K_i be the field $K(a_i^{1/p})$. The extension K_i/K is normal with the cyclic group C_p as its Galois group. Let t_i be its generator acting on $\vartheta_i = a_i^{1/p}$ by $t_i(\vartheta_i) = \zeta_p \vartheta_i$. If now $\mathfrak{P}$ is a prime ideal in K of the first degree over Q then from the definition of the Frobenius automorphism we infer that $F_{K_i/K}(\mathfrak{P}) = t_i^j$ is equivalent to $\left(\frac{a_i}{\mathfrak{P}}\right)_p = \zeta_p^j$. In fact, if $q = N(\mathfrak{P})$, then $F_{K_i/K}(\mathfrak{P}) = t_i^j$ holds iff $\zeta_p^j \vartheta_i \equiv \vartheta_i^q$ (mod $\mathfrak{P}$), and this in turn is equivalent to $\zeta_p^j \equiv \vartheta_i^{q-1} \equiv a_i^{(q-1)/p}$ (mod $\mathfrak{P}$), i.e. to $\left(\frac{a_i}{\mathfrak{P}}\right)_p = \zeta_p^j$. Note that it suffices to prove the equality $[K(\vartheta_1, \ldots, \vartheta_n):K] = p^n$, since then the set of all prime ideals in K of the first degree over Q satisfying the assertion of our theorem coincides with the set of all those prime ideals of the first degree over Q for which

$$F_{K(\vartheta_1, \ldots, \vartheta_n)/K}(\mathfrak{P}) = [t_1^{\varepsilon_1}, \ldots, t_n^{\varepsilon_n}] \in \prod_{j=1}^{n} G(K(\vartheta_j)/K),$$

and this set, according to Theorem 7.11, is infinite. We will obtain this result as a special case of the following more general lemma:

Lemma 7.15. (i) *Let K be any field of characteristic zero and q a rational prime. If a_1, a_2 are two elements of K such that the fields $K(a_1^{1/q})$ and $K(a_2^{1/q})$ coincide (here $a^{1/q}$ denotes any q-th root of a), then one can find an element b in K and a rational integer r with $1 \leq r \leq q-1$ such that $a_1 = b^q a_2^r$.*

(ii) *If $a_1, \ldots, a_s$ are q-independent elements of a field K of zero characteristic, q being a rational prime, then the degree of the extension $K(a_1^{1/q}, \ldots, a_s^{1/q})/K$ equals q^s.*

Proof. (i) If $q = 2$, the assertion is clear; so assume $q > 2$. If one of the a_i's is a qth power in K, then the proof becomes trivial; so let us assume that this is not the case.

Let $L = K(a_1^{1/q}) = K(a_2^{1/q})$ and write $M_0 = K(\zeta_q)$, $M = M_0(a_1^{1/q}) = L(\zeta_q)$, where ζ_q is any fixed primitive qth root of unity. The extension M/M_0 is generated by a root of $X^q - a_1$, so either $a_1^{1/q} \in M_0$ or M/M_0 is cyclic of degree q. In the first case we would have

$$q = \text{degree of } a_1^{1/q} \text{ over } K \leq \text{degree of } \zeta_q \text{ over } K \leq q-1$$

giving a contradiction; thus only the second case has to be considered.
Write

$$a_2^{1/q} = \sum_{j=0}^{q-1} \lambda_j (a_1^{1/q})^j$$

with certain $\lambda_j \in M_0$ and let g be that automorphism of M/M_0 which maps $a_1^{1/q}$ onto $\zeta_q a_1^{1/q}$. Then clearly, for a certain r ($1 \leq r \leq q-1$), we must have $g(a_2^{1/q}) = \zeta_q^r a_2^{1/q}$. But then

$$\zeta_q^r a_2^{1/q} = \sum_{j=0}^{q-1} \lambda_j \zeta_q^r a_1^{j/q} = \sum_{j=0}^{q-1} \lambda_j \zeta_q^j a_1^{j/q},$$

and so necessarily $\lambda_j = 0$ for $j \neq r$; hence $a_2^{1/q} = \lambda_r a_1^{r/q}$ and $a_2 = \lambda_r^q a_1^r$, where $\lambda_r \in M_0$. Since λ_r^q lies in K, either $\zeta_q^s \lambda_r \in K$ for a certain s or the degree of λ_r over K equals q, but the latter is impossible in view of $[M_0:K] \leq q-1$, and so we arrive at our first assertion.

(ii) Write $K_0 = K$, $K_j = K(a_1^{1/q}, \ldots, a_j^{1/q})$ for $j = 1, 2, \ldots, s$ and observe that it suffices to show that $[K_{j+1}:K_j] = q$ for $j = 0, 1, \ldots, s-1$. If this were false for a certain j, the element $c = a_{j+1}^{1/q}$ would lie in K_j. Let t be the minimal index for which $c \in K_t$. As $t \neq 0$, we have $K_t = K_{t-1}(c)$ and part (i) of this lemma gives

$$a_{j+1} = \lambda^q a_t^r$$

for a certain λ in K_{t-1} and $0 \leq r \leq q-1$. Write t_1 for the minimal index for which $\lambda \in K_{t_1}$. Then $K_{t_1} = K_{t_1-1}(\lambda) = K_{t_1-1}(a_{t_1}^{1/q})$ and again we get $\lambda^q = \lambda_1^q a_{t_1}^{r_1}$ for a certain λ_1 in K_{t_1-1} and $0 \leq r_1 \leq q-1$; thus $a_{j+1} = \lambda_1^q a_{t_1}^{r_1} a_t^r$. Proceeding in this way we finally arrive at

$$a_{j+1} = \lambda_z^q a_t^r a_{t_1}^{r_1} \ldots a_{t_z}^{r_z}$$

with $0 \leq r, r_1, \ldots, r_z \leq q-1$ and λ_z in K, which contradicts the q-independence of the a_i's. □

We point out one of the consequences of this lemma:

Corollary. *If the rational integers $n_1, \ldots, n_s$ are q-independent and the field K contains $n_i^{1/q}$ ($i = 1, 2, \ldots, s$), then the degree of K/Q is divisible by q^s. Moreover, if K/Q is normal and $s \neq 0$, then its degree must be divisible by $(q-1)q^s$.*

Proof. The proof follows immediately from part (ii) of the lemma just proved if one recalls that a normal K containing a pure field of degree q must contain also all qth roots of unity. □

The theorem now follows immediately. □

Corollary. *Let K be a finite extension of the rationals containing the p-th roots of unity, p being a rational prime. Then the equality $I_0^p = I_0 \cap I_K^p$ holds, i.e. every element of K^* which is a p-th power in each completion K_v is necessarily a p-th power in K^*.*

Proof. Let $(x_v) \in I_0 \cap I_K^p$. Then $x_v = x \in K^*$ and $x = y_v^p$ holds for every v with a certain y_v in K_v. Multiplying x, if necessary, by a suitable pth power, we may assume that it already lies in R_K. Now, let $\mathfrak{P}$ be any prime ideal of K of the first degree over Q and not containing x. Then the congruence $x \equiv y_{\mathfrak{P}}^p \pmod{\mathfrak{P}}$ holds for a certain $y_{\mathfrak{P}} \in R_K$. This implies

$$\left(\frac{x}{\mathfrak{P}}\right)_p \equiv x^{(N(\mathfrak{P})-1)/p} \equiv y_{\mathfrak{P}}^{N(\mathfrak{P})-1} \equiv 1 \pmod{\mathfrak{P}},$$

and thus

$$\left(\frac{x}{\mathfrak{P}}\right)_p = 1$$

holds for almost all prime ideals $\mathfrak{P}$ of the first degree over Q. By the theorem just proved this can happen only if the set $\{x\}$ is not p-independent, but this means that x is a pth power in K^*. □

6. Our final application concerns the structure of R_L as an R_K-module for every finite extension L/K.

In the case of $K = Q$ the description is simple, namely, R_L is a free Z-module with $[L:Q]$ free generators and the same happens if $h(K) = 1$, by Theorem 1.13. In the general case this theorem shows that R_L is isomorphic, as an R_K-module, to $R_K^{n-1} \oplus I$, where $n = [L:K]$ and I is an ideal of R_K. Thus by Theorem 1.14 the isomorphism type of R_L is determined by n and the class in $H(K)$ to which I belongs. This class is called the *Steinitz class of L/K* and is denoted by $C_K(L)$. We now describe it in terms of the discriminant $\partial(L/K)$; this will give us the possibility to resolve the problem when R_L is a free R_K-module. To state the theorem we first define a homomorphism of $I_K^2 I_0/U_K^2$ onto the class-group $H(K)$. Denote by t the isomorphism of $I_K/I_0 U_K$ onto $I_K^2/I_0^2 U_K^2$ induced by the map $a \mapsto a^2$ in I_K; let t_1 be the surjective homomorphism of $I_K^2 I_0/U_K^2$ onto $I_K^2 I_0/U_K^2 I_0$ induced by $(x_v) \mapsto (x_v) \bmod I_0$; finally, define a homomorphism $u_0: I_K^2 I_0/U_K^2 I_0 \to I_K^2/U_K^2 I_0^2$ in the following way: for $(x_v^2 y) \in I_K^2 I_0$ put

$$v_0((x_v^2 y)) = (x_v^2) \pmod{I_0^2}.$$

This is well defined because if $(x_v^2 y) = (X_v^2 Y)$ with (x_v), (X_v) in I_K, (y) and (Y) in I_0, then $((x_v X_v^{-1})^2) = (Y y^{-1}) \in I_K^2 \cap I_0 = I_0^2$ by Corollary to Theorem 7.13; thus the class $(x_v^2) \pmod{I_0^2}$ is determined by $(x_v^2 y)$ uniquely. Observe that v_0 gives an isomorphism of $I_K^2 I_0/I_0$ and I_K^2/I_0^2 because of $\operatorname{Ker} v_0 = I_0$ and we denote the induced map of $I_K^2 I_0/U_K^2 I_0$ onto $I_K^2/U_K^2 I_0^2$ by u_0.

Denote by u the homomorphism of $I_K^2 I_0/U_K^2$ onto $H(K)$ obtained by composition of the mappings below:

$$I_K^2 I_0/U_K^2 \xrightarrow{t_1} I_K^2 I_0/U_K^2 I_0 \xrightarrow{u_0} I_K^2/U_K^2 I_0^2 \xrightarrow{t^{-1}} I_K/U_K I_0 \xrightarrow{t_2} H(K),$$

where t_2 is the natural isomorphism. Now we can state

Theorem 7.14. *For any finite extension L/K we have $C_K(L) = u(\partial(L/K))$.*

Proof. Let M be an arbitrary torsion-free and finitely generated R_K-module and denote by V the linear space over K spanned by M. Write n for the dimension of V and take an arbitrary torsion-free R_K-module N spanning the same space V. If v is any non-Archimedean valuation of K and R_v the ring of integers of K_v, then we can consider the R_v-modules M_v and N_v generated by M and N, respectively, in the n-dimensional K_v-space $V_v = V \otimes_K K_v$. Since R_v is a PID, the resulting modules are free, and so there is an automorphism of V_v mapping N_v onto M_v. Let us denote by d_v the determinant of that automorphism and observe that it is determined by M and N up to a unit factor form R_v. It follows that the ideal in R_v generated by d_v is determined uniquely by M and N. Let us write $[N:M]_v$ for this ideal.

Lemma 7.16. *If M, N, N_1 are torsion-free and finitely generated R_K-modules spanning the same linear space over K and v is a non-Archimedean valuation of K, then*
    (i) $[M:M]_v = R_v$.
    (ii) $[M:N]_v [N:N_1]_v = [M:N_1]_v$.
    (iii) *The inclusion $M_v \subset N_v$ implies $[N:M]_v \subset R_v$.*
    (iv) *For almost all v's $[M:N]_v = R_v$.*

Proof. Assertions (i) and (ii) are immediate. To prove (iii) it suffices to note that the elements of a free R_v-basis of M_v are linear combinations over R_v of the elements of a similar basis of N_v, which implies $d_v \in R_v$. Finally, to prove (iv), consider non-zero elements a and b of R_K such that $aM \subset N$ and $bN \subset M$. Now (ii) and (iii) imply

$$[N:M]_v[M:aM]_v \subset R_v, \quad [M:N]_v[N:bN]_v \subset R_v.$$

By definition, $[M:aM]_v = a^n R_v$, $[N:bN]_v = b^n R_v$, which is equal to R_v for almost all v's. Hence, for those v's we have $[M:N]_v \subset R_v$ and $[M:N]_v^{-1} = [N:M]_v \subset R_v$ showing that $[M:N]_v = R_v$, as asserted. □

Part (iv) of the above lemma implies that the product $\prod_v [M:N]_v$ is a fractional ideal of K. Let us denote it by $[M:N]$. Its principal properties are given in the following lemma:

Lemma 7.17. *Under the same assumptions as in the last lemma we have:*
  (i) $[M:M] = R_K$.
  (ii) $[M:N][N:N_1] = [M:N_1]$.
  (iii) *If* $M \subset N$, *then* $[N:M] \subset R_K$.
  (iv) *If M and N are fractional ideals in K, then* $[M:N] = M^{-1}N$.

Proof. Assertions (i)–(iii) follow immediately from the previous lemma. To prove (iv) observe that if M and N are fractional ideals in K, then M_v and N_v are fractional ideals in K_v, and so the automorphism mapping N_v onto M_v has the form $x \mapsto a_v x$ for a certain $a_v \in K_v^*$; hence $[M:N]_v = a_v R_v$ and so $[M:N]_v = N_v M_v^{-1}$, and we may apply the definition of $[M:N]$. □

Now let F_1 and F_2 be two free R_K-modules spanning the same K-space V as M. By assertion (ii) of the last lemma we have $[F_1:M] = [F_1:F_2][F_2:M]$. Since F_1, F_2 are free, we can find an isomorphism of F_1 onto F_2 which can be extended to an automorphism f of the space V and also, for each non-Archimedean v, to an automorphism f_v of the corresponding K_v-space V_v. Now note that if d is the determinant of f, then its images in K_v are the determinants of f_v; thus $[F_1:F_2]_v = dR_v$ and $[F_1:F_2] = dR_K$, dR_K being a principal ideal. We thus see that the classes of $[F_1:M]$ and $[F_2:M]$ in $H(K)$ are the same and depend only on M. Denote this common class by $\mathrm{Cl}(M)$. We now prove

Lemma 7.18. (i) *If I is a fractional ideal in K, then $\mathrm{Cl}(I)$ coincides with that class which contains I.*
  (ii) $\mathrm{Cl}(M_1 \oplus M_2) = \mathrm{Cl}(M_1)\mathrm{Cl}(M_2)$.
  (iii) *If $M = R_K^{n-1} \oplus I$ (where I is an ideal in R_K), then $\mathrm{Cl}(M)$ coincides with the class which contains I.*

Proof. (i) In this case R_K is a free R_K-module spanning the same space as I; thus $\mathrm{Cl}(I)$ is the class containing $[R_K:I]$. But Lemma 7.17 (iv) shows that $[R_K:I] = I$.

(ii) Let F_1 and F_2 be free R_K-modules spanning the same spaces V_1, V_2 as M_1 and M_2, respectively. Then the free R_K-module $F_1 \oplus F_2$ and $M_1 \oplus M_2$ span $V_1 \oplus V_2$ and the corresponding fact is true also for $(F_1 \oplus F_2)_v$ and $(M_1 \oplus M_2)_v$. Moreover, if $(d_i)_v$ is the determinant of the automorphism of $(V_i)_v$ induced by the isomorphism $(F_i)_v \to (M_i)_v$ ($i = 1, 2$), then the determinant of the automorphism of $(V_1 \oplus V_2)_v$ induced by $(F_1 \oplus F_2)_v \to (M_1 \oplus M_2)_v$ equals $(d_1)_v (d_2)_v$, which gives

$$[F_1 \oplus F_2, M_1 \oplus M_2]_v = [F_1, M_1]_v [F_2, M_2]_v,$$

and this implies (ii).

Finally, (iii) easily follows from (i) and (ii). □

We can now conclude the proof of our theorem. Let $a_1, \ldots, a_n$ be an arbitrary K-basis of L and let $A \in K^*$ be its discriminant. Let M be the free R_K-module generated by this basis and choose in I_K an idele $[x_v]$ which represents the discriminant $\partial(L/K)$ (which is an element of I_K/U_K^2 and even, as the proof of Theorem 6.8 shows, of $I_K^2 I_0/U_K^2$). Now, for every valuation v of K which is non-Archimedean consider a basis $\{b_j\}$ of the R_v-module

$$\bigoplus_{w \in E_v} R_{L_w},$$

where E_v is the set of non-equivalent extensions of v to L and R_{L_w} the ring of integers of L_w. Denote its discriminant by x_v. We can express the b_j's in terms of the a_i's; let A_v be the determinant of the matrix involved. Then

$$x_v = A_v^2 A. \tag{7.34}$$

Now we apply the homomorphism u to $\partial(L/K)$, looking at the behaviour of $[x_v]$ under the mappings defining u. Under t_1 $[x_v]$ goes into $[x_v]$ (mod $U_K^2 I_0$); however, by (7.34) this equals $[A_v^2]$ (mod $U_K^2 I_0$), since A is principal. Applying u_0, we obtain $[A_v^2]$ (mod $U_K^2 I_0^2$) and it remains to show that this element is carried into $C_K(L)$ by t^{-1}. To this end observe that $[M:R_L]$ equals the ideal induced by the idele $[A_v]$, $C_K(L)$ is the class in $H(K)$ in which $[M:R_L]$ lies, and the image of this class under t is the coset of $U_K^2 I_0^2$ in I_K^2 which contains $[A_v^2]$. □

Note that the proof of this theorem gives also the following explicit form of $u(\partial(L/K))$:

If $\partial(L/K) = aI^2 U_K^2$, where $a \in I_0$, $I \in I_K$, then $u(\partial(L/K))$ is the class in $H(K)$ containing the ideal induced by the idele I.

Corollary 1. *The ring R_L is a free R_K-module iff $u(\partial(L/K)) = 1$, i.e. if $\partial(L/K)$ has a representative in I_0.* □

If this happens, we say that L/K has a relative integral basis.

Corollary 2. *If C is the class in $H(K)$ containing $d(L/K)$, then $C = C_K(L)^2$.* (This gives an explicit formulation of Theorem 6.8.)

Proof. Note that C contains the ideal induced by any representative of $\partial(L/K)$, $C_K(L)$ contains the ideal induced by the idele $[A_v]$ and apply (7.34). □

Corollary 3. *If the class-number $h(K)$ is odd, then the extension L/K has a relative integral basis iff the discriminant $d(L/K)$ is principal.*

Proof. If L/K has a relative integral basis, then $C_K(L) = 1$ and so by the preceding corollary $d(L/K)$ is principal. This does not depend on the parity of $h(K)$. Conversely, if $d(L/K)$ is principal, then the oddness of $h(K)$ and the last corollary imply $C_K(L) = 1$. □

We know from Chapter 1 that if $h(K) \neq 1$, then there exist non-free, finitely generated and torsion-free R_K-modules. We now show that the rings R_L for suitable L/K have this property. More precisely, we prove:

Proposition 7.16. *If M is an arbitrary torsion-free R_K-module with two generators, which is not generated by a single element, then there exists a quadratic extension L/K such that R_L, as an R_K-module, is isomorphic to M.*

Proof. We need a lemma, which describes explicitly the class $C_K(L)$ for quadratic L/K:

Lemma 7.19. *Let $L = K(a^{1/2})$ with $a \in R_K$ and write $aR_K = \prod_{\mathfrak{p}} \mathfrak{p}^{a_{\mathfrak{p}}}$. For every $\mathfrak{p}$ with even (or zero) $a_{\mathfrak{p}}$ we define a number $s_{\mathfrak{p}}$ by $s_{\mathfrak{p}} = \max\{s: \mathfrak{p}^s | 2R_K,\ x^2 \equiv a\pi_{\mathfrak{p}}^{-a_{\mathfrak{p}}} \pmod{\mathfrak{p}^{2s}}$ is solvable$\}$, where $\pi_{\mathfrak{p}}$ is any element in K which generates the prime ideal in the completion $K_{\mathfrak{p}}$. For remaining ideals $\mathfrak{p}$ we put $s_{\mathfrak{p}} = 0$. Then the class $C_K(L)$ contains the ideal*

$$2 \prod_{\mathfrak{p}} \mathfrak{p}^{-s_{\mathfrak{p}} - m_{\mathfrak{p}}},$$

where $m_{\mathfrak{p}}$ equals the integral part of $a_{\mathfrak{p}}/2$.

Proof. Let us look at the components $\partial_v(L/K)$ of $\partial(L/K)$. If v is Archimedean, then

$$\partial_v(L/K) = \begin{cases} \operatorname{sgn} F_v(a) & \text{if } v \text{ is real,} \\ 1 & \text{if } v \text{ is complex} \end{cases}$$

(F_v denoting the embedding of K in K_v). If v is non-Archimedean and the prime ideal $\mathfrak{p}_v$ corresponding to v splits in L/K, then $\partial_v(L/K) = 1 \pmod{U^2(K_v)}$. In all other cases we have $\partial_v(L/K) = \partial(K_v(a^{1/2})/K_v)$ and we can utilize Theorem 5.6. Let v be fixed and let $\mathfrak{p}$ be the corresponding prime ideal. If $\mathfrak{p}$ does not divide the ideal generated by $2a$, then Theorem 5.6 (i) gives

$$\partial_v(L/K) = a \pmod{U^2(K_v)} = (2\pi^{-m_\mathfrak{p}-s_\mathfrak{p}})^2 a \pmod{U^2(K_v)},$$

because in this case $m_\mathfrak{p} = s_\mathfrak{p} = 0$ and 2 is a unit in K_v. If $\mathfrak{p}$ does not divide the ideal generated by 2 and the exponent $a_\mathfrak{p}$ is even and positive, then we write $a = \varepsilon(\pi^{m_\mathfrak{p}})^2$ with ε in $U(K_v)$ to see that $K_v(a^{1/2}) = K_v(\varepsilon^{1/2})$. Theorem 5.6 (i) gives

$$\partial_v(L/K) = \varepsilon \pmod{U^2(K_v)} = (2\pi^{-m_\mathfrak{p}-s_\mathfrak{p}})^2 a \pmod{U^2(K_v)},$$

since again we have $s_\mathfrak{p} = 0$ and $2 \in U(K_v)$.

If $a_\mathfrak{p}$ is odd, then the distinction between the cases $\mathfrak{p}|2R_K$ and $\mathfrak{p}\nmid 2R_K$ is irrelevant; writing $a = \varepsilon\pi(\pi^{m_\mathfrak{p}})^2$ with ε in $U(K_v)$, we get $K_v(a^{1/2}) = K_v((\pi\varepsilon)^{1/2})$, and applying Theorem 5.6 (ii), we arrive at

$$\partial_v(L/K) = 4\varepsilon\pi \pmod{U^2(K_v)} = 4a(\pi^{-m_\mathfrak{p}-s_\mathfrak{p}})^2 \pmod{U^2(K_v)}.$$

It remains to consider the case of $\mathfrak{p}|2R_K$ and even $a_\mathfrak{p}$. Here $a = \varepsilon\pi^{2m_\mathfrak{p}}$ (ε in $U(K_v)$); thus $K_v(a^{1/2}) = K_v(\varepsilon^{1/2})$ and by Theorem 5.6 (iii)

$$\partial_v(L/K) = 4\varepsilon\pi^{-2s_\mathfrak{p}} \pmod{U^2(K_v)} = a(2\pi^{-s_\mathfrak{p}-m_\mathfrak{p}})^2 \pmod{U^2(K_v)}.$$

The results obtained show that $\partial(L/K) = a\xi^2 \pmod{U_K^2}$, where ξ is the idele $[x_v]$ with $x_v = 2\pi^{-s_\mathfrak{p}-m_\mathfrak{p}}$ for non-Archimedean v and $x_v = 1$ for Archimedean v. Indeed, it remains only to examine the v-component for those v for which $\mathfrak{p}$ splits in L/K. If $\mathfrak{p}\nmid 2R_K$, then $s_\mathfrak{p} = 0$ and a is a square in K_v, say $a = b^2$; hence $ax_v^2 = 4b^2\pi^{-m_\mathfrak{p}} \pmod{U^2(K_v)}$. Obviously $2b\pi^{-m_\mathfrak{p}}$ lies in $U(K_v)$, and so $ax_v^2 = 1 \pmod{U^2(K_v)}$ and hence $ax_v^2 \pmod{U^2(K_v)}$ coincides with $\partial_v(L/K)$. If $\mathfrak{p}|2R_K$, then $a\pi^{-a_\mathfrak{p}}$ is a square, say b^2, in K_v. Thus $\mathfrak{p}^{s_\mathfrak{p}}|2R_K$ and $\mathfrak{p}^{s_\mathfrak{p}+1}\nmid 2R_K$. Now $ax_v^2 = 4a(\pi^{-s_\mathfrak{p}-m_\mathfrak{p}})^2 \pmod{U^2(K_v)} = (2b\pi^{-s_\mathfrak{p}})^2 \pmod{U^2(K_v)} = 1 \pmod{U^2(K_v)}$, as required.

Finally, observe that by Theorem 7.14 $C_K(L) = u(\partial(L/K))$, which equals the class in $H(K)$ containing the ideal induced by the idele ξ, and this ideal coincides with

$$2\prod_{\mathfrak{p}} \mathfrak{p}^{-s_\mathfrak{p}-m_\mathfrak{p}}.$$

$\square$

To prove the proposition it remains to show that to every class C in $H(K)$ there exists a quadratic extension L/K, where $C_K(L) = C$. To this end write

$$A = \prod_{2 \in \mathfrak{p}} \mathfrak{p}$$

and let $\mathfrak{q}_1$ be any prime ideal not dividing A which lies in the class C^{-1}. Select a prime ideal $\mathfrak{q}_2$ which does not divide $A\mathfrak{q}_1$ and for which the product $A\mathfrak{q}_1^2\mathfrak{q}_2$ is a principal ideal, say, equal to aR_K. Now put $L = K(a^{1/2})$. According to the lemma, the class $C_K(L)$ contains the ideal

$$\prod_{\mathfrak{p}|aR_K} \mathfrak{p}^{-s_\mathfrak{p}-m_\mathfrak{p}},$$

and for $\mathfrak{p}$ dividing aR_K we have $s_\mathfrak{p} = m_\mathfrak{p} = 0$ with the single exception of $\mathfrak{p} = \mathfrak{q}_1$, in which case $s_\mathfrak{p} = 0$, $m_\mathfrak{p} = 1$. Hence $\mathfrak{q}_1^{-1}$ lies in $C_K(L)$, i.e. $C = C_K(L)$. □

Corollary. *The equality $h(K) = 1$ holds iff for any quadratic extension L/K there exists a relative integral basis.* □

The last proposition shows that the Steinitz classes $C_K(L)$ ranging over quadratic extensions of K cover the class-group $H(K)$. An analogous statement for normal extensions of a fixed odd degree fails to hold in general, as can be seen from the following result of L. R. McCulloh [66]:

Proposition 7.17. *Let N be an odd integer and denote by T_N the greatest common divisor of $\{(p-1)/2: p|N, p \text{ prime}\}$. Assume that T_N has an odd prime divisor q, and choose a field K with $q|h(K)$. Then there is a class in $H(K)$ which is not the Steinitz class of any normal extension L/K of degree N.*

Proof. By Corollary 1 to Proposition 6.6 we have $d(L/K) = I^{2T_N}$ with a suitable ideal I of K. If X denotes the class of I in $H(K)$, then by Corollary 2 to Lemma 7.18 we get

$$C_K(L)^2 = X^{2T_N}.$$

Since $q|(T_N, h(K))$ and $q \neq 2$, we see that $C_K(L)$ must be a qth power in $H(K)$ and thus lies in a proper subgroup of $H(K)$. □

Note that, in view of a result of T. Nagell [22] (which we shall prove in the next chapter; see Theorem 8.9), one can take for K a suitable imaginary quadratic field.

§ 4. Notes to Chapter 7

1. The first application of Dirichlet series to number theory was given by Dirichlet in his research concerning the occurrence of primes in arithmetic progressions. He used L-functions corresponding to Dirichlet characters. The first systematic study of $\zeta(s)$ was made by B. Riemann. Dedekind's zeta-function was introduced by R. Dedekind [71], although in certain special cases it appeared already in Eisenstein [44b]. Dedekind established the existence of the limit $\lim_{s\to 1}(s-1)\zeta_K(s)$ and its connection with $h(K)$. E. Landau [03a] was the first to continue $\zeta_K(s)$ beyond the line $\operatorname{Re} s = 1$ to the open half-plane $\operatorname{Re} s > 1 - [K:Q]^{-1}$. E. Hecke [17a] proved that Dedekind's zeta-function can be extended to a meromorphic function in the plane and found the functional equation for this function (our Theorem 7.1). For some special fields, including pure cubic and Abelian extensions of Q, this had been known before, in part implicitly. (Cf. Dedekind [00], Hurwitz [82].) Other proofs of Theorem 7.1 were given in Müntz [24], Siegel [22b] and Tate [50]. The proof given by us follows that of Tate.

2. A simple product formula for the residue of $\zeta_K(s)$ was given by A. Wintner [46b]. A new proof of the formula for this residue can be found in P. Cassou-Noguès, J. Fresnel [79]. Corollary 4 to Theorem 7.1 was improved by A. F. Lavrik [70a], A. F. Lavrik, Zh. Edgorov [75] and C. L. Siegel [69a]. Lavrik showed that for large n one can have the constant in Corollary 4 equal to $(2/3)^n$. For pure cubic fields H. Cohn [56b] obtained the bound $h(K)R(K) = O(\log D \log\log D)$. This result is related to the diophantine equation $X^2 = Y^3 - D$. (Cf. Cassels [50], Edgar [66], Nagell [23], Selmer [56].)

For explicit bounds for $h\varkappa$ see also Hoffstein [79a], [79b].

A discussion of various ways to generalize results about Riemann zeta-function to Dedekind zeta-function is given in Lang [71].

3. Zeta-functions with arbitrary Hecke characters (called by him "Grössencharakters") were introduced by E. Hecke [18], who also gave the functional equation for them. He utilized the language of ideal numbers. The idelic approach is due to A. Weil [36], [51]. Cf. Hasse [54b], Tatuzawa [60].

In the special case of Dirichlet's L-functions, proofs of the functional equation have been given in Apostol [70c], Ayoub [67a], Berndt [73a], Hamburger [21], Speiser [39]. For the theory of Dirichlet L-functions see e.g. Chudakov [47a], Prachar [57]. From among numerous results of this theory we mention here only one, due to T. Apostol [72], who proved that Dirichlet

L-functions with non-primitive characters do not have a functional equation of the type occurring in Proposition 7.7.

4. Our proof of Theorem 7.3 is modelled upon Tate [50]. For an interpretation of this result in terms of distributions see Weil [66].

The analogues of Dedekind zeta-functions for maximal orders of finite-dimensional algebras have been constructed by F. Hey [29] and E. Witt [35b]. For this and related generalizations see Andrianov [68a], [68b], Bushnell, Reiner [80], [81a], [81b], [81c], Fujisaki [62], Galkin [73], Godement, Jacquet [72], Jenner [69], Leptin [55], Radford [71], Ramanathan [59], Solomon [77], Tamagawa [63].

Functions satisfying functional equations similar to that of zeta-functions have been considered in Berndt [69a], [69b], [71b], Chandrasekharan, Joris [73], Chandrasekharan, Narasimhan [60], [61a], [61b], [61c], [62a], [62b], [64], [68b], Gurevich [71], Hamburger [21], Hecke [36], [44], Marke [37], Siegel [22d], Weil [67b].

J. Popken [66] proved that $\zeta_K(s)$ cannot satisfy an algebraic difference-differential equation and studied also the algebraic independence of sets of Dedekind zeta-functions of quadratic fields over the ring of polynomials with complex coefficients.

S. M. Voronin [75] showed that if the fields K_i ($i = 1, 2, \ldots, m$) are either all quadratic or all cyclotomic of the form $Q(\zeta_r)$ with r square-free, then for every choice of continuous functions $F_1, \ldots, F_m$, not all vanishing, the sum

$$\sum_{k=0}^{m} s^k F_k(\zeta_{K_1}(s), \ldots, \zeta_{K_m}(s))$$

cannot vanish identically.

A study of sequences of Dedekind zeta-functions can be found in Gut [39], [40].

5. Corollary 5 to Theorem 7.1 is due to R. Perlis [77a]. Fields satisfying the assumptions of that corollary are called arithmetically equivalent; the first examples were constructed by F. Gassmann [26], and were of degree 180. It was shown by Perlis (*loc. cit.*) that there do not exist arithmetically equivalent fields of degree smaller than seven and this bound cannot be improved.

It is not known whether arithmetically equivalent fields have the same class-number, but it was shown in Perlis [78] that the *p*-parts of the corresponding class-groups must be isomorphic for all primes p not dividing the degree.

For further results concerning arithmetically equivalent fields see Komatsu [76c], [78], [84], Perlis [77b], Perlis, Schinzel [79], Roggenkamp, Scott [82]. For a generalization to relative extensions see Klingen [78].

6. If $\chi(I) = \chi_1(I)N(I)^s$ is a Hecke character, χ_1 is proper, s is a rational number and in the formula (7.11) applied to χ_1 we have $a_v = 0$ for all v, then χ is called a *character of type* (A). If, moreover, for totally positive x congruent to unity (mod $\mathfrak{f}$) (with $\mathfrak{f}$ being the conductor of χ) we have

$$\chi(x) = \pm \prod_{v \in S_\infty} x_v^{r_v} \bar{x}_v^{-s_v}$$

with $r_v, s_v \in Z$, then χ is called a *character of type* (A_0).

These notions were introduced by A. Weil [56], who showed that characters of type (A) have algebraic values and that the values of a character of type (A_0) lie in a finite extension of the rationals. The converse theorem was established by M. Waldschmidt [81]. There are deep relations between zeta-functions of such characters and certain zeta-functions of Abelian varieties, established by Y. Taniyama [57].

A. Weil [52a] showed that Jacobi sums (introduced by C. G. J. Jacobi [46] in certain special cases) are in fact Hecke characters. They are defined as follows: let $m > 1$, $K = Q(\zeta_m)$, $\mathfrak{p}$ an unramified prime ideal in K and $q = N(\mathfrak{p})$. Let χ be the mth power character (mod $\mathfrak{p}$), i.e.

$$\chi(x)^m = 1 \quad \text{and} \quad \chi(x) \equiv x^{(q-1)/m} \pmod{\mathfrak{p}}.$$

Put for any $a = [a_1, \ldots, a_r]$ (with $a_i \in Z/mZ$)

$$J_a(\mathfrak{p}) = (-1)^{1+r} \sum \chi(x_1)^{a_1} \ldots \chi(x_r)^{a_r},$$

summation extended over all $x_1, \ldots, x_r$ (mod $\mathfrak{p}$) satisfying $x_1 + \ldots + x_r \equiv -1$ (mod $\mathfrak{p}$), and extend J_a by multiplicativity to all ideals I prime to mR_K. The result of Weil states that for all $a \neq [0, \ldots, 0]$, $J_a(I)$ is a Hecke character. Cf. Brattström [82], Iwasawa [75], Jensen [60], Kubert [85], Kubert, Lichtenbaum [83], Schmidt [80b], Weil [74].

Hecke's characters were used by K. Iwasawa [77] to obtain class-number relations in a *CM*-field. They have also been utilized to the study of elliptic curves. (See e.g. Birch, Swinnerton-Dyer [65], Gross [80].)

For other results on Hecke characters and their zeta-functions see Barner [81], Masuda [61], Schmidt [79], Weil [52b], Yagi [72].

For the proof of Theorem 7.2 see Tate [50], Tatuzawa [73b].

7. If χ_1, χ_2 are Hecke characters of K_1 and K_2 respectively, then the function

$$Z(s, \chi_1, \chi_2) = \sum_{N(I_1) = N(I_2)} \chi_1(I_1) \chi_2(I_2) N(I_1)^{-s}$$

is called the *scalar product* of the zeta-functions of χ_1 and χ_2. Similarly one defines scalar product in the case of more characters. A. I. Vinogradov [65] continued this function to the half-plane $\operatorname{Re} s > 1/2$ and N. Kurokawa [78a],

[78b] obtained a necessary and sufficient condition for the scalar product to be meromorphic in the case when all characters are of finite order. For the analogue in the general case see Moroz [82]. Cf. also Draxl [71], Fomenko [72], Gaigalas [75], [77], [79], Moroz [63], [64], [65a], [65b], [68], [80], [84].

8. A new class of functions of importance in algebraic number theory was introduced by E. Artin [24], [30b] and they are now called *Artin's L-functions*. Every such function is associated with a normal extension L/K and a character X of a finite-dimensional representation of $\text{Gal}(L/K)$ and is denoted accordingly by $L(s, X, L/K)$, or simply $L(s, X)$. If L/K is Abelian and X is a character of an irreducible representation, then $L(s, X, L/K)$ coincides with a suitable zeta-function of a Hecke character of finite order.

Accounts of the theory of Artin's L-functions may be found in Cassels, Fröhlich [67] (Heilbronn's lecture), Hasse [26c] and Martinet [77a], and so we limit ourselves just to a few remarks. It was shown by R. Brauer [47b] that every Artin's L-function is a product of powers (with exponents from Z) of zeta-functions corresponding to Hecke's characters of finite order in various fields, and thus, by Theorem 7.3, is meromorphic. Artin conjectured that Artin's functions are entire, with some explicitly given exceptions. Classfield theory implies the validity of this conjecture for characters of one-dimensional representations. In the two-dimensional case the occurring representations can be partitioned into four families—dihedral, tetrahedral, octahedral and icosahedral representations. Artin's conjecture for dihedral characters was established by Artin himself (Artin [24]). For tetrahedral representations it was proved by R. P. Langlands [80] (for expositions of his proof see Gelbart [77], Gérardin, Labesse [79]), who also showed its truth for certain octahedral representations and J. Tunnell [81] proved it for all octahedral representations. For certain classes of icosahedral representations for which the conjecture holds see Buhler [78].

In the general case Artin's conjecture would follow from Langlands conjectures dealing with automorphic representations of GL_n. See Gelbart [84] on this topic.

It has been shown by H. Yoshida [77] that there are no obstructions to Artin's conjectures which are related to the group structure of $\text{Gal}(L/K)$. For other results on Artin's conjecture see Aramata [39], Sato [77], Uchida [75], Vinogradov [71], [73], van der Waall [73].

Artin's conjecture is related to the question, considered in a special case already in Dedekind [00] and stated by E. Artin [23], whether the quotient $\zeta_L(s)/\zeta_K(s)$ is entire in case $K \subset L$. If L/K is normal then the answer is positive, as shown by H. Aramata [31], [33] and R. Brauer [47a]. Earlier Z. Suetuna

[30] showed that if ζ_L does not vanish in a half-plane $\operatorname{Re} s > a$, then the same applies to ζ_K. For further results see Aramata [39], Artin [23], Brauer [73], Ishida [57], Sato [82], v.d. Waall [74b], [75], [82].

The set of all Artin's L-functions associated with an extension L/K does not characterize this extension; this was shown by T. Funakura [79], who also proved that there are only finitely many normal extensions of Q with the same set of Artin's L-functions.

The functional equation for Artin's L-functions, established by E. Artin [24], contains a term $W(X)$ which is independent of s and whose absolute value is 1. It is called the *Artin's root-number*. H. Hasse [54a] posed the problem of factorization of $W(X)$ into local factors, corresponding to various completions of the field concerned. A solution was given by B. Dwork [55], [56], up to a sign determination, and a complete solution was announced in Langlands [70]. Another proof was found by P. Deligne [73]. (See Tate [77b] for a variant of it.) Cf. also Deligne [76], Lakkis [66a], [66b], [67].

If X is real-valued, then $W(X) = \pm 1$ and it was shown by A. Fröhlich and J. Queyrut [73] that if X is a character of a real representation, then $W(X) = 1$, but otherwise $W(X) = -1$ may also occur. In fact, $W(X) = -1$ is equivalent to $L(1/2, X) = 0$; examples of this situation have been found by J. V. Armitage [72] and J. P. Serre. (Cf. Kuroda [74].) For connection with the problem of normal integral bases see the notes to Chapter 4. Cf. also Fröhlich [72], [74a], [74b], [75], [83c], Gechter [76], Queyrut [72], Ullom [80].

For analogues of Gaussian sums associated with the local factors of the root-number see Fröhlich [83c], Fröhlich, Taylor [80], Martinet [77a]. Analogues of Jacobi sums have been considered in Bushnell [77a], Fröhlich [76b], [77d].

H. M. Stark [75a] proposed certain conjectures concerning the behaviour of Artin's L-functions at $s = 0$ and $s = 1$. They are true for Abelian extensions of Q and of imaginary quadratic fields and he established them also for all L-functions with rational characters. From results of C. L. Siegel [70] their validity follows for all Abelian extensions of a totally real field. Cf. also Chinburg [83a], Shintani [77c], [78], Stark [76a], [77a], [77b], [80], Tate [81a], [81b].

For other results on Artin's L-functions see Barrucand [71], Goldfeld [73], Lagarias, Odlyzko [79], Odlyzko [77], Serre [71a], Suetuna [35], [36], [37], Weinstein [79], [80].

A still more general class of zeta-functions was introduced by A. Weil [51], the Artin-Hecke functions, attached to characters of representations of a group $G_{L/K}$ defined for every normal extension L/K as a certain extension of $C(L)$ by $\operatorname{Gal}(L/K)$. T. Tamagawa [53] found a functional equation for them. See Jehne [54], Lakkis [66b], [67], [69], Tamagawa [51], Weil [71], [72].

9. Zeta-functions satisfy also certain approximate functional equations. For the Riemann zeta-function such an equation was given by G. H. Hardy, J. E. Littlewood [21], [23], [29], but actually it goes back to Riemann. (See Siegel [32].) For Dirichlet's L-function it was obtained by Z. Suetuna [25c]. Products of $\zeta(s)$ and Dirichlet's L-functions (which case, as we shall see in the next chapter, covers all Dedekind zeta-functions of Abelian fields) were treated by A. O. Gelfond [60]. Approximate functional equations for a large class of zeta-functions including all Dedekind zeta-functions have been obtained by K. Chandrasekharan, R. Narasimhan [63] and A. F. Lavrik [68b]. Cf. Kaufman [78b] for another proof of Lavrik's result. Previous work for particular classes of zeta-functions has been done in Apostol, Sklar [57], Bulota [62], [64a], Chudakov [47b], Davies [65], Fischer [52], Fogels [69], Lavrik [65a], [65b], [66a], [66b], [67a], [67b], [68a], Linnik [46], Matuljauskas [69], [71], Potter [36], Suetuna [32], Tatuzawa [52], Titchmarsh [38], Wiebelitz [52], Wilton [30].

The approximate functional equation leads to several results concerning the behaviour of zeta-functions. K. Chandrasekharan, R. Narasimhan [63] deduced from it the evaluations

$$\zeta_K(1+it) = O(\log|t|)$$

(for $|t|$ tending to infinity), and

$$\sum_{m \leqslant x} F(m)^2 = (C+o(1))x\log^{n-1}x$$

(where $F(m)$ denotes the number of ideals of norm m and C is positive).

The first result had been previously obtained in another way by E. Landau [03a] and later A. V. Sokolovskii [68] improved the bound to $O(\log^{2/3}|t|)$. Cf. Bohr, Landau [10], Walfisz [27].

The second result was then extended to higher powers of $F(m)$ in Chandrasekharan, Good [83].

For evaluations of the Dedekind zeta-function and certain other zeta-functions in the critical strip $0 < \operatorname{Re} s < 1$ see Bartz [80], Hinz [79], Kalnin [68], Kaufman [78a], [78b], [79], Motohashi [70], Staś [61a], [76], [79], Suetuna [24], [25b], Weinstein [77], Wieczorkiewicz [79]. Such evaluations have also been obtained in several papers concerning the zeros of zeta-functions quoted in the next subsection.

10. Theorem 7.5 for Dirichlet's L-functions is due to Dirichlet [38]. Other proofs in this case can be found in Bateman [59], Chowla, Mordell [61],

Ingham [30], Narasimhan [68], Stoll [50], Teege [11]. In the general form it is due to E. Hecke [18]. Another proof, based on class-field theory, may be found in Weil [67a] (Ch. XIII, Th.11).

The Extended Riemann Hypothesis (ERH) asserts that no zeta-function associated with a proper Hecke character can vanish in the strip $1/2 < \text{Re}\, s < 1$.

The largest zero-free region for Dedekind zeta-function was obtained by T. Mitsui [68] and A. V. Sokolovskii [67], [68]. They proved that $\zeta_K(s) \neq 0$ in the region

$$\text{Re}\, s \geq 1 - C(\log t)^{-2/3}(\log \log t)^{-1/3},$$

$$|t| = |\text{Im}\, s| \geq t_0$$

with suitable constants $C > 0$, t_0, depending on K. It was shown by K. Bartz [78] that one can take $t_0 = 4$ and $C = c_1 n^{-11}|d(K)|^{-3}$ with a positive constant c_1, independent of K and $n = [K:Q]$. For previously known zero-free regions for $\zeta_K(s)$ see Landau [19b], [24a] (cf. Fryska [79]), Sokolovskii [66a], Walfisz [27].

The result of Mitsui and Sokolovskii was extended to zeta-functions of arbitrary Hecke characters of finite order by J. G. Hinz [76a], [80], who extended previous work of E. Fogels [62a] and T. Mitsui [56]. In this case the constants C and t_0 depend additionally on the conductor of the character.

11. Denote by $N_K(\sigma, T)$ the number of zeros $a+bi$ of $\zeta_K(s)$ with $|b| \leq T$ and $a \geq \sigma$, counted according to their multiplicity. From ERH the equality $N_K(\sigma, T) = 0$ follows for every $\sigma > 1/2$ and so it is of interest to obtain upper unconditional bounds for $N_K(\sigma, T)$. The best result here is due to D. R. Heath-Brown [77], who obtained

$$N_K(\sigma, T) = O(T^{(n+\varepsilon)(1-\sigma)}\log^C T)$$

for every positive ε, with $C = C(\varepsilon, K)$ and $n = [K:Q] \geq 3$, uniformly for $1/2 \leq \sigma \leq 1$, and got also a similar result in the case of $n = 2$. The Density Hypothesis, which is weaker than ERH, claims that

$$N_K(\sigma, T) = O(T^{(2+\varepsilon)(1-\sigma)}\log^C T)$$

holds uniformly for $1/2 \leq \sigma \leq 1$ with $C = C(\varepsilon, K)$. M. Jutila [77] proved the Density Hypothesis for $K = Q$ and $\sigma \geq 11/14$ and also for all Abelian fields and $\sigma \geq 21/26$. Cf. Heath-Brown [77], Sokolovskii [66b], Wieczorkiewicz [80].

Analogous problems for various classes of Hecke's zeta-functions have been studied in Bulota [63], [64b], Fogels [65], [71], [72], Hilano [74a], [74b], Hinz [76b], Huxley [68], Johnson [79], Kondakova [71], Koval'chik [74],

[75a], [75b], Kubilius [52], Maknys [75a], [76]. For computations see Davies [61].

12. It follows from a result of H. S. A. Potter and E. C. Titchmarsh [35] that for every class $X \in H(K)$ the function

$$Z(s, X) = \sum_{I \in X} N(I)^{-s}$$

(which in view of Theorem 7.3 can be prolonged to a meromorphic function) has in the case of imaginary quadratic K infinitely many zeros on the line $\mathrm{Re}\,s = 1/2$. (Cf. Hecke [37].) For Riemann's zeta-function the same assertion was proved by G. H. Hardy [14], and for Dirichlet's L-function by E. Landau [15]. For all quadratic fields K this was established in Chandrasekharan, Narasimhan [68a] (cf. Berndt [71a]) and for certain other classes of fields with small degrees by B. C. Berndt [70]. Cf. Czarnowski [82].

It is an old unsolved question, whether the Dedekind zeta-functions of quadratic fields may have real zeros. This problem is related to the magnitude of the class-number (see Chapter 8). Cf. Çallial [80], P. Chowla [74], Chowla, Erdös [51], Chowla, Goldfeld [76], Chowla, Hartung [74b], Chowla, de Leon [74], Chowla, de Leon, Hartung [73], Low [68], Rosser [49].

For fields of higher degrees the equality $\zeta_K(1/2) = 0$ may well happen, as the example of $K = Q(a, b, c)$ with $a^2 = 5$, $b^2 = 41$, $c^2 = (5+a)(41+ab)$ shows (Armitage [72]). The fact that zeta-functions of a large class of Hecke's characters do not vanish at $s = 1/2$ has been established in Montgomery, Rohrlich [82], Rohrlich [80a], [80b], [80c], [82], [84a], [84b].

It was shown by H. Heilbronn [72], [73] that if L/K is normal then every real simple zero of $\zeta_L(s)$ is already a zero of the Dedekind zeta-function of the composite of all quadratic extensions of K contained in L.

E. Landau [19b] showed that there exists a $C = C(n)$ such that for any real T all Hecke's zeta-functions of any field of degree n have a zero in the rectangle $1/2 \leqslant \mathrm{Re}\,s < 1$, $T \leqslant \mathrm{Re}\,s \leqslant T+C$, and later C. L. Siegel [72a] showed that C may be taken uniformly for all fields. Cf. Hoffstein [79a]. For further results on the zeros of Hecke functions see Fogels [63], Fujii [77], Haselgrove [51], Kubilius [51], Marinina [56].

13. Let X be a given class in $H(K)$ and put

$$a_0(X) = \lim_{s \to 1} ((s-1) Z(s, X) - \varkappa),$$

with $Z(s, X)$ defined as in subsection 12. An explicit formula for the value of $a_0(X)$ was in the case of imaginary quadratic K given by L. Kronecker [85] ("Kronecker's limit formula"). See Siegel [61] for a proof and various applications. Cf. also Koshlyakov [37], Ramachandra [64], [69], Shintani [80], Zagier [75a]. For real quadratic fields an analogous result has been proved in Hecke [17d], Herglotz [23], Novikov [80], Shintani [76a], [77a], Zagier [75a], for certain biquadratic fields in Katayama [66] and for CM-fields in Konno [65]. Finally, an analogue of Kronecker's formula for all algebraic number fields was obtained by L. J. Goldstein [74].

14. Much attention has been attracted by the values of various zeta-functions at particular points, mostly at integers and rationals; it seems that many algebraic properties of algebraic number fields are encoded in those values. As an example of this philosophy one can take the interrelation between the vanishing of $\zeta_K(s)$ at $s = 1/2$ and the existence of a normal integral basis in the case when K/Q is tame with the quaternion Galois group; we discussed this in the notes to Chapter 4.

The values of $\zeta(s)$ for even positive integers s have been computed already by Euler and the analogous question for $\zeta_K(s)$ with $K = Q(i)$ has been dealt with by A. Hurwitz [99]. Later E. Hecke [21a] (part II, Satz 3) considered real quadratic K and asserted that for even values of $s > 0$ and for every class X in $H^*(K)$ the equality

$$Z(s, X) = r(s, X)\pi^{2s}d(K)^{1/2}$$

holds with a certain non-zero rational $r(s, X)$. This clearly implies a similar assertion about $\zeta_K(s)$ and from the functional equation it follows that for $k = 1, 2, \ldots$ the value $\zeta_K(1-2k)$ is rational. Hecke's assertion was established by C. L. Siegel [37]; analogues for $Z(s, X)(X \in H_f^*(K))$ have been obtained for totally real fields by H. Klingen [62] and C. L. Siegel [69b], [70], Cf. Barner [68], [69], Borel [77], H. Cohen [74], [76], Gundlach [65], [73], Hida [78], Katayama [76], H. Lang [68], [72], [73a], [73b], [75], Meyer [67], Shintani [76b], [77b], [81], Siegel [68a], [75], Uehara [76], [78], [82], Zagier [76], [77].

Values of Dirichlet's L-functions at negative integers are expressible in terms of generalized Bernoulli numbers, introduced for real characters by A. Berger [91] and in the general case by H. W. Leopoldt [58a]. (See Washington [82], ch.IV; cf. also Grosswald [73].) For properties and applications of these numbers see Carlitz [59a], [59b], Fresnel [67], Hasse [62b], Kimura [79b], Kubota, Leopoldt [64], Leopoldt [59b], [60], [62], Shanks, Wrench [63], Shiratani [72], Washington [82].

J. P. Serre [71b] posed two conjectures concerning $\zeta_K(1-2k)$ ($k = 1, 2, \ldots$) for totally real fields K. The first asserts that if p is a rational prime and $\mathfrak{P}$ is a prime ideal lying over p in K, then the denominator of

$$(N(\mathfrak{P})^{2k} - 1)\zeta_K(1-2k)2^{-n}$$

(with $n = [K:Q]$) is a power of p, and the second states that the ratio $\zeta_K(1-2k)/\zeta(1-2k)$ is integral and divisible by an explicitly given power of two. For K Abelian these conjectures were proved by J. Fresnel [71] and J. Coates, S. Lichtenbaum [73]. Cf. also K. S. Brown [74], P. Cassou-Noguès [78], [79], Coates, Sinnott [77], Deligne, Ribet [80], Greenberg [73a], Ribet [75].

The first conjecture follows from more general conjectures of S. Lichtenbaum [72] and J. Coates, S. Lichtenbaum [73], which, in particular, express the exponent of a prime p occurring in the factorization of $\zeta_K(1-2k)$ (and, more generally, of $L(1-2k, \chi, L/K)$) in terms of p-adic cohomologies. These conjectures were deduced by P. Bayer, J. Neukirch [79] from the "Main Conjecture" (see Coates [77]), which is now a theorem, due to the work of B. Mazur, A. Wiles [84] (see Coates [81] for an exposition), and so is Serre's conjecture. For previous and/or related results see Coates [72a], [72b], Lichtenbaum [73], [75], Schneider [79], Soulé [79].

The value of Dedekind zeta-function at $s = -1$ is related to the theory of quaternion algebras. See Guého [72a], [72b], [74a], [74b], Vignéras [74], [75a], [75b]. A formula for this value in terms of subgroups of $SL_2(R_K)$ was obtained by K. S. Brown [74] and F. Hirzebruch [73]. For K quadratic, the numbers $\zeta_K(1/2)$ and $\zeta_K(1/3)$ were considered in Toyoizumi [81], [82]. For a numerical study of values of $\zeta_K(s)$ see Cartier, Roy [73]. The values at integers of Artin's L-functions were studied by M. J. Taylor [81c].

For other results concerning zeta-functions see Bartz [76], Flett [51], Hinz [77], Landau [19a], Rieger [59a], Terras [76], [77a], [77b], Walfisz [32].

15. Proposition 7.9 is due to E. Landau [18f], as well as most corollaries to it. Corollary 5 got an elementary proof in Babaev [71]. Corollary 6 with $g(s)$ replaced by $O(1)$ occurs in Furtwängler [07] and in the general form (for classes of $H_f(K)$) in Hecke [17c], as well as Corollary 7. This last corollary in the case of $K = Q$ gives the Dirichlet prime number theorem, and in the case of $K = Q(i)$ it gives another result of Dirichlet [41b]. For $K = Q(\zeta_3)$ see Fanta [01].

An analogue of Linnik's theorem on primes in progressions, giving a more precise version of Corollary 7 to Proposition 7.9 was obtained by E. Fogels [61a], [61b], [62a], [62b], [65], [66a] and G. J. Rieger [61a] in the following form: in every class of $H_f^*(K)$ there exists a prime ideal with norm not exceeding $cN(\mathfrak{f})^b$, with certain constants b, c depending on K. See also Lagarias, Odlyzko [77], Weiss [83]. Cf. Kalnin [65a] for the quadratic case.

For evaluations of character sums occurring in Proposition 7.10 (i) see Barban, Levin [68], Fogels [65], Friedlander [73a], [74], Goldstein [70a], Hinz [83a], [83b], Jordan [67c], Lee [79].

16. Corollary 1 to Proposition 7.10 (usually called the Prime Ideal Theorem) was conjectured by E. Landau [03a] and proved in Landau [03b]. Previously, upper and lower bounds for $\pi_K(x)$ had been known, similar to those obtained by Chebyshev in the case of $K = Q$. See Landau [03a], Phragmén [92], Poincaré [92], Torelli [01].

Other proofs of the Prime Ideal Theorem were given in Ahern [64] and Rieger [59b], both covering also Corollary 4. The first proof not utilizing analytic tools was given by H. N. Shapiro [49], who used an analogue of Selberg's lemma. For other proofs of this type see Bredikhin [58], Eda, Nakagoshi [67], Forman, Shapiro [54]. For Selberg's lemma in this contex see Ahern [65], Ayoub [55], G. L. Cohen [75b], Rieger [58a], [58b], [61d], [62d], [63b], Yamamoto [58].

Landau's proof gave for the remainder term the bound $O(x\exp(-\log^{1/13} x))$. The remainder given by us in Corollary to Theorem 7.7 is due to E. Landau [18f]. Further improvements have been given in Landau [24a] and Sokolovskii [66a] and the best evaluation at this moment is that obtained by T. Mitsui [68] and A. V. Sokolovskii [67], [68], namely

$$O(x\exp(-C(\log x)^{3/5}(\log\log x)^{-1/5}))$$

with a certain positive C. The dependence of the bound on the field has been investigated in Goldstein [70a] and Wiertelak [78]. Cf. Friedlander [80], Révesz [83], Sokolovskii [71], Suetuna [25a], Walfisz [27] for other results concerning $\pi_K(x)$.

Theorem 7.7 shows that an enlargement of the zero-free region of zeta-functions leads to an improved bound for the remainder term for the corresponding counting function. This dependence can be reversed. In the case of $K = Q$ this has been shown by P. Turán [50], [53; §§ 9, 13] and for arbitrary K this result is due to W. Staś [59], [60], [61b] and W. Staś, K. Wiertelak [73], [75a], [75b], [76a].

Corollary 4 to Proposition 7.10 was proved for classes in $H(K)$ by E. Landau [07]. His proof was based on a result of P. Furtwängler [07] giving the existence of the Hilbert class-field of K, since at that time the analytic continuation of the zeta-functions was not known. For classes of $H_f(K)$ this result was obtained by E. Hecke [17c] and for classes of $H_f^*(K)$ again by E. Landau [18f]. A uniform evaluation for $N(\mathfrak{f}) = O(\exp(c\log\log x/\log x))$ was obtained by J. Hinz [76a], [80]. Cf. Kalnin [65a], Staś, Wiertelak [76b]. This theorem generalizes Dirichlet's theorem about primes in progressions and is often called "Hecke's theorem about progressions". For early results on primes in progressions in $Q(i)$ and $Q(\zeta_3)$ see Fanta [01], Weber [05].

17. The Ideal Theorem (Corollary to Theorem 7.6) is due to R. Dedekind

[71], and Theorem 7.6 to E. Landau [18f] in its most general form. The last paper, which we have already quoted several times, contains a detailed treatment of zeta-functions associated with characters of $H_{\mathfrak{f}}^*(K)$. For the Ideal Theorem see also Weber [96a] and Wintner [45].

E. Landau [18e], [18f] showed that the error term in Theorem 7.6 (and hence also in the Ideal Theorem) is $O(x^{1-c})$ with $c = 2/(1+n)$, $n = [K:Q]$. (Previously (Landau [12a]) he had an extra factor x^ε.) In this result the dependence on K of the constants involved was made explicit in the case of $\mathfrak{f} = 1$ by E. Fogels [66a]. Cf. Dzhiemuratov [68a]. T. Tatuzawa [73a] improved the error term to $O(x^{1-1/n})$, making the constants explicit. Better bounds are known for quadratic fields (Richert [57]) and cyclotomic fields (Karatsuba [72]).

It follows from Landau [24b] that the error term in the Ideal Theorem cannot be $O(x^{1-(n+1)/2n-\varepsilon})$ with positive ε. The factor $x_{\mathfrak{d}}^{-\varepsilon}$ was later removed in K. Chandrasekharan, R. Narasimhan [62a] and H. Joris [70], [72], who replaced it by a function tending to infinity. Cf. also R. G. Ayoub [58], [68], Berndt [69c], Chandrasekharan, Good [83], Chandrasekharan, Narasimhan [64], v.d. Corput [23], Dzhiemuratov [68b], Linnik, Vinogradov [66], Rieger [58c], Szegö, Walfisz [27], Tatuzawa [77], Walfisz [26], Warlimont [67].

Proposition 7.11 is due to E. Wirsing (see Ostmann [68], vol. II, p. 67) and its analogues for non-normal extensions have been obtained by R. W. K. Odoni [75a]. Cf. Heupel [68], Landau [08], Luthar [67], Odoni [73b], [75b], [77a], [77b], [78], Schmid, Shanks [66], Shanks [64], Wintner [46a]. For other questions related to norms of integers and ideals see Evteev [70], [73], Gurak [77a].

18. For Proposition 7.12 see Lang [64], Mautner [53]. Theorem 7.8 is a particular case of a result of E. Hecke [18]. Cf. Knapowski [69] for another proof. Evaluations of the remainder term were provided by I. P. Kubilius [52] and T. Mitsui [56] made them uniform. Cf. also Ankeny [52b], Babaev [60], [63], Babaev, Korchagina [72], Bulota [64b], Fogels [66b], Kalnin [65b], Kaufman [77], Korchagina [79], Környei [62], Koval'chik [74], [75b], Kubilius [50], [51], [55], Maknys [75a], [75b], [76], [80], Rademacher [35], [36a], [36b], Schulz-Arenstorff [57], Urbelis [64], [65a], [65b].

The problem of counting Gaussian primes in sufficiently regular subsets of the plane was considered by I. V. Chulanovskii [56] and his results were extended by G. L. Cohen [75a], [75b]. Cf. also Hensley [76].

19. The function $F_{L/K}(I)$ was introduced by E. Artin [24] and is usually called the *Artin's symbol*. The first step towards Chebotarev's Density Theorem was done by G. Frobenius [96], who proved Proposition 7.14 (i). Other proofs have been given in Chebotarev [26], Hasse [26c] (vol. II, § 24), Hurwitz [26], Schreier [27]. Part (ii) of this proposition is essentially due to L. Kronecker [80], who tacitly assumed the existence of the densities d_m.

The first proof of Theorem 7.11 was given by N. G. Chebotarev [23b], [26]. (Cf. also Chebotarev [27], [37c].) It was simplified in Deuring [35a], McCluer [68], Scholz [31] and Schreier [27]. A purely algebraic proof in a special case was given in Wójcik [75]. Proofs leading to explicit evaluation of the remainder term in Theorem 7.11* and the smallest norm of a prime ideal in a conjugacy class have been obtained in Lagarias, Montgomery, Odlyzko [79], Lagarias, Odlyzko [77], Schulze [72] (III).

For a generalization of Chebotarev's theorem see Odoni [77a] and for a conjecture related to it cf. Lenstra [77b]. The analogue of Chebotarev's theorem for infinite extensions was shown by M. Moriya [34a] to be false in general; however, J. P. Serre [81] obtained such a generalization in special cases. The paper of Serre contains also several applications of Theorem 7.11* to the theory of elliptic curves and modular forms.

For other questions around Chebotarev's theorem see Jarden [74]. For Artin's reciprocity law, used in the proof of Theorem 7.11* see Artin [27], Cassels, Fröhlich [67], Hasse [26c].

20. Proposition 7.15 is due to E. Artin [23] and Lemma 7.14 to H. Hasse [30a]. Lemma 7.13 in a more precise form, which takes into account also ramified primes, occurs in v.d. Waerden [34]. Cf. Bauer [37], Wegner [35]. The notion of a Bauerian field and Theorem 7.12 are due to A. Schinzel [66a]. Corollary 1 to that theorem was obtained earlier by M. Bauer [16a] and in a special case in Bauer [04b]. For another proof see Deuring [35b] and for applications see Ankeny, Rogers [51b], Flanders [53b] and Mann [54], [55].

The first example of a non-Bauerian extension was given by F. Gassmann [26]. For other examples see Gerst [70], Lewis, Schinzel, Zassenhaus [66] and Schinzel [66a]. In the last paper it is also shown that all fields of degree $\leqslant 4$ over Q are Bauerian. For similar problems see Bilhan [81], Nakatsuchi [68], [70], [72], [73], [75].

It follows from Theorem 4.12 that if $f \in Z[x]$ is the minimal polynomial of a generator of L/K, then $P(L/K)$ differs from $P(f)$, the set of all primes p for which the congruence $f(x) \equiv 0 \pmod{p}$ is solvable, only by finitely many elements. A survey of properties of the set $P(f)$ was given by I. Gerst, J. Brillhart [71], so we quote here only a few papers on this subject: Fjellstedt [55], Grölz [69], Hornfeck [70], Nagell [68a], Schinzel [68], Schulze [72], [73], [76a], [76b].

Two extensions L_1/K and L_2/K are called *Kronecker-equivalent*, if the sets $P(L_1/K)$ and $P(L_2/K)$ differ only by finitely many elements. The resulting equivalence classes are called *Kronecker classes*. This notion is due to W. Jehne [77b], who showed that all minimal elements of a Kronecker class have the same Galois hull over K (i.e. the smallest normal extension of K, containing such a field) and thus are finite in number. He gave also examples of infinite Kronecker classes. Towers $K_1 \subset K_2 \subset \ldots$ of fields within the same Kronecker class were studied by N. Klingen [78]. (Cf. Komatsu [82].) For other problems on Kronecker classes see Jehne [77c], Klingen [79], [80], [83], Schulze [81].

In Corollary 2 to Theorem 7.12 the assumption of normality is essential (Gassmann [26]). Its analogue for infinite extensions is not valid (Whaples [47]). Corollary 3 was proved in Schinzel [66a].

Theorem 7.13 is due to D. Hilbert [97], and the proof given by us is that of Chebotarev [23a]. Cf. Chebotarev [37c], Rabung [70], Wójcik [69]. Generalizations of it are contained in Elliott [70a], Mills [63], Shafarevich [54].

A generalization of Lemma 7.15 (i) to composite q was for $K = \mathbb{Q}$ obtained by I. Gerst [70] and for arbitrary fields K of characteristic not dividing q by A. Schinzel [75b]. Note that the generalization given in Nagell [39] is inexact, as the example $K = \mathbb{Q}$, $n = 8$, $a = -1$, $b = 16$ shows. (Gerst [70], Schinzel [75b].) For generalizations of part (ii) of Lemma 7.15 see Besicovitch [40], Halter-Koch [80], Kneser [75], Mordell [53], Richards [74], Schinzel [75b], Siegel [72b], Ursell [74]. Cf. also Mostowski [55], Roth [71].

21. A conjecture concerning the density of the set of all primes which do not split completely in fields from a certain infinite family closed under composition was stated by L. J. Goldstein [68], [70b]. Although it turned out to fail in general (Weinberger [72c]), it holds nevertheless in certain interesting cases (Goldstein [71a], [73a], Ram Murty [84]).

22. Theorem 7.14 is due to A. Fröhlich [60b], [60d] and our exposition is based on his work. Corollary 3 to it fails to hold in the case of even $h(K)$, as shown by S. Pierce [74].

The first necessary and sufficient condition for the existence of a relative integral basis (RIB) was given by E. Artin [50a]: if $L = K(a)$ with $a \in R_L$, then the fractional ideal $d_{L/K}^{-1}(a) d(L/K)$ is the square of the fractional ideal I and RIB exists iff I is principal. (Cf. Hecke [12] for the quadratic case, and Fujisaki [74].) For simple examples of extensions without RIB see Edgar [79], McKenzie, Scheunemann [71].

Corollary to Proposition 7.16 is due to V. Hanly, H. B. Mann [58]. Previously, H. B. Mann [58] proved it in a weaker form.

Conditions for the existence of RIB in particular classes of extensions have been considered in Bird, Parry [76], Edgar, Peterson [80], Feng Keqin,

Zhang Xianke [83], Martinet, Payan [67], [68], McCulloh [63], [71], Wada [70a], Washington [76a], Zhang Xianke [84d], [84e].

Proposition 7.17 shows that Steinitz classes of normal extensions of a given degree need not cover $H(K)$. Denote by $R(K, G)$ the set of Steinitz classes of normal extensions of K with $\text{Gal}(L/K) \simeq G$. This set was for cyclic p-groups determined by R. L. Long [71], [75]. (Cf. Long [72].)

23. We conclude this section with a short survey of p-adic analogues of L-functions. T. Kubota and H. W. Leopoldt [64] defined a p-adic L-function $L_p(s, \chi)$ associated with a non-principal Dirichlet character χ as the unique continuous function on Q_p which for all positive integers n, divisible by $p-1$ if $p \neq 2$ and even if $p = 2$, satisfies

$$L_p(1-n, \chi) = (1 - \chi(p)p^{n-1})L(1-n, \chi)$$

and proved its existence. Other proofs have been given in Amice, Fresnel [72], Barsky [78], Fresnel [66], [67], Iwasawa [69], Katz [77], Osipov [79], [80], Washington [76b]. Accounts of the theory of p-adic L-functions are given in Iwasawa [72a], Koblitz [77] and Washington [82], where also further references may be found.

The values of $L_p(s, \chi)$ at $s = 1$ have been examined in Cartier, Roy [73], Koblitz [79], Leopoldt [62], Shiratani [74], Washington [77], [79a], and the values at other positive integers in Diamond [79b], Hatada [79], Koblitz [79], Shiratani [77]. A formula for $L_p'(0, \chi)$ was given in Ferrero, Greenberg [78].

Using his theory of p-adic modular forms, J. P. Serre [73] constructed p-adic zeta-functions of arbitrary totally real fields. (Cf. Serre [78b].) P-adic L-functions over real quadratic fields have been constructed by J. Coates, W. Sinnott [74b], over imaginary quadratic fields by N. M. Katz [76] and M. M. Vishik, J. I. Manin [74], and over arbitrary totally real fields by D. Barsky [77b], P. Cassou-Noguès [79] and P. Deligne, K. A. Ribet [80]. Cf. also Katz [78], [81], Lichtenbaum [80], Queen [77], Ribet [79b].

For other results on p-adic L-functions see Coates, Wiles [78], Diamond [79a], G. Gras [79c], [82b], Greenberg [75], Gross [81], Martin [83], Metsänkylä [78a], Morita [78], [79], Ribet [78], [79a], Slavutsky [69a], Uehara [75], Vishik [77], Washington [78a], [81a], [81b].

EXERCISES TO CHAPTER 7

1. Compute $\zeta_K(0)$ for imaginary quadratic K.

2. For any two complex-valued functions f, g defined on the set of all non-zero ideals of R_K define their *Dirichlet convolution* by

$$(f * g)(I) = \sum_{J|I} f(J)g(IJ^{-1}).$$

Prove that the set of all such functions forms a commutative ring with unit under usual addition and Dirichlet convolution as multiplication. Prove also that f has an inverse in this ring iff $f(R_K) \neq 0$.

3. A function $f(I)$ is called *multiplicative* if $f(IJ) = f(I)f(J)$ or $(I, J) = 1$. Prove that the Dirichlet convolution of multiplicative functions is again multiplicative and the inverse of a multiplicative function is also multiplicative.

4. Let $h = f * g$ and assume that the two Dirichlet series

$$F(s) = \sum_I f(I) N(I)^{-s}, \quad G(s) = \sum_I g(I) N(I)^{-s}$$

are absolutely convergent in a half-plane. Prove that the series

$$H(s) = \sum_I h(I) N(I)^{-s}$$

onverges absolutely in that half-plane and $H(s) = F(s)G(s)$.

5. Let $d_m(I)$ be the number of representations of an ideal I as a product of m factors (representations differing by the order of factors are viewed as distinct). Prove that for $\text{Re } s > 1$ the series

$$\sum_I d_m(I) N(I)^{-s}$$

converges absolutely and its sum equals $\zeta_K(s)^m$.

6. Let X be a class in $H_{\mathfrak{f}}^*(K)$. Prove that for x tending to infinity

$$\sum_{\substack{N(I) \leq x \\ I \in X}} d_m(I) = (c + o(1)) x \log^{m-1} x$$

where $c = h(K)^m \varkappa^m \varphi(\mathfrak{f})^m / h_{\mathfrak{f}}^*(K) (m-1)! N(\mathfrak{f})^m$.

7. Let X be a class in $H_{\mathfrak{f}}^*(K)$. Prove that

$$\sum_{\substack{\mathfrak{P} \in X \\ N(\mathfrak{P}) \leq x}} N(\mathfrak{P})^{-1} = h_{\mathfrak{f}}^*(K)^{-1} \log\log x + B + O(1/\log x)$$

with a certain $B = B(X)$.

8. Prove that the function $Z(s, X)$ defined in subsection 12 of § 4 can be prolonged to a meromorphic function with a simple pole at $s = 1$ and compute its residue there.

9. Determine explicitly all Hecke characters of Q and $Q(i)$.

10. Show that if X is a quasicharacter of I_K trivial on I_0, and Ω is the group consisting of all ideles (x_v) with $x_v = 1$ for non-Archimedean v, then the equality

$$X((x_v)) = \prod_{v \text{ Arch.}} (x_v/|x_v|)^{n_v} v(x_v)^{s+ia_v}$$

holds on Ω, with suitable rational integers n_v, real a_v and a complex number s.

11. (Weil [56].) (a) Prove that every Hecke character of type (A) (as defined in subsection 6 of § 4) has algebraic values.

(b) Prove that all values of a Hecke character of type (A_0) lie in a finite extension of the rationals.

(c) Let χ be a character of $H_{\mathfrak{f}}^*(K)$ for a certain ideal $\mathfrak{f}$ and let $r \in Q$. Show that $X(I) = \chi(I)N(I)^r$ is a Hecke character of type (A).

(d) Let K be a CM-field. Prove the existence of Hecke characters of type (A) which are not of the form given in (c).

(e) Prove that all Hecke characters of type (A) have the form given in (c) iff K does not contain any CM-field.

12. Prove that all prime ideals lying in $P(L/K)$ split completely iff L/K is normal.

13. (Hecke [17c].) Prove that for every ideal I and $a \in R_K$ satisfying $(aR_K, I) = 1$ there exist infinitely many non-associated elements c generating prime ideals of the first degree and lying in the residue class $a \pmod{I}$.

14. (Bilhan [81].) Let $P_0(L/K)$ be the set of all prime ideals of K which split completely in L. Prove that if $P(L/K) \subset P_0(M/K)$, then $M \subset L$.

15. Prove that the set of all prime ideals of K which have at least one prime ideal factor of the first degree in L/K has Dirichlet density $\geq 1/[L:K]$, equality holding iff L/K is normal.

16. Prove that if $h(K) = 3$, then every cyclic extension L/K of degree 7 has a relative integral basis.

Chapter 8. Abelian Fields

§ 1. Main Properties

1. This chapter is devoted to the arithmetic of Abelian extensions of the rationals, i.e. normal extensions L/K with an Abelian Galois group. According to the Kronecker–Weber theorem (Theorem 6.5) every such extension is contained in a suitable cyclotomic extension $K_n = Q(\zeta_n)$. The least integer f with the property $K \subset K_f$ is called the *conductor of K* and denoted by $f(K)$. The main properties of the conductor are listed in the following proposition:

Proposition 8.1. *If K/Q and L/Q are Abelian, then:*
  (i) $K \subset K_m$ *holds iff* $f(K)|m$,
  (ii) $f(K \cap L)|(f(K), f(L))$,
  (iii) $f(KL) = [f(K), f(L)]$,
  (iv) *a prime p ramifies in K/Q iff* $p|f(K)$,
  (v) *a prime p is tamely ramified in K/Q iff* $p|f(K)$ *and* $p^2 \nmid f(K)$.

Proof. (i) If $K \subset K_m$, then writing $f = f(K)$ and using Theorem 4.10 (v) we get $K \subset K_m \cap K_f = K_{(m,f)}$. Hence

$$f \leq (m, f) \leq f$$

and $f(K) | m$ follows. The converse implication is obvious.

  (ii) Since $K \cap L \subset K_{f(K)} \cap K_{f(L)} = K_{(f(K),f(L))}$ by Theorem 4.10 (v), the assertion results from (i).

  (iii) Write $f = f(KL)$. In view of $K \subset KL$, $L \subset KL$ and $KL \subset K_f$, we infer from (i) that f is divisible both by $f(K)$ and $f(L)$, so that $[f(K), f(L)] | f$. On the other hand, we have $KL \subset K_{f(K)} K_{f(L)} = K_{[f(K),f(L)]}$ and so (i) gives $f | [f(K), f(L)]$.

  (iv) If $p \nmid f = f(K)$, then by Theorem 4.16 p does not ramify in K_f/Q and since $K \subset K_f$, it cannot ramify in K/Q. If $p|f$ and p does not ramify in K/Q, then write $f = p^a m$ with $a \geq 1$, $p \nmid m$, and observe that in view of Corollary 2 to Lemma 5.4, Theorem 4.16 and Corollary 1 to Proposition 6.2 p is unramified in KK_m/Q. Writing N for the degree of KK_m/K_m, we get, again using Theorem 4.16,

$$\varphi(p^a) = e_{K_f/Q}(P) = e_{K_f/KK_m}(P) \leq [K_f : KK_m]$$
$$= [K_f : K_m]/N = \varphi(f)/\varphi(m)N = \varphi(p^a)/N$$

where P is any prime ideal above p in K_f. Thus $N = 1$ and $K \subset K_m$, contrary to the choice of f.

(v) Write again $f = f(K) = p^a m$ with $a \geq 0$, $p \nmid m$, and let P be a prime ideal lying over p in K_f. If $p | f$, $p^2 \nmid f$, then Theorem 4.16 shows that $p \nmid e_{K_f/Q}(P)$. Thus p is tamely ramified in K_f/Q and hence also in K/Q. Conversely, if p is tamely ramified in K/Q, then by (iv) we have $a \geq 1$ and by Corollary 2 to Lemma 5.5, Theorem 4.16 and Corollary 1 to Proposition 6.2 p is tamely ramified in KK_{pm}/Q. Theorem 4.16 gives

$$e_{K_f/Q}(P) = \varphi(p^a) = p^{a-1}(p-1)$$

and we get $p^{a-1} | e_{K_f/KK_{pm}}(P)$. Writing $M = [KK_{pm} : K_{pm}]$ we obtain

$$p^{a-1} | e_{K_f/KK_{pm}}(P) | [K_f : KK_{pm}] = [K_f : K_{pm}]/M = \varphi(f)/M\varphi(pm) = p^{a-1}/M.$$

Thus $M = 1$, and we arrive at the inclusion $K \subset K_{pm}$, which implies $f | pm$ and $p^2 \nmid f$. □

Corollary. *If K/Q is Abelian, then it has a normal integral basis iff it is tamely ramified.*

Proof. Necessity follows from Proposition 4.14 and Corollary 3 to Proposition 6.2. If K/Q is Abelian and tame, then (v) shows that $f(K)$ is square-free, and the existence of a normal integral basis results from Proposition 4.15 (i) and the corollary to it. □

It should be pointed out that one cannot expect equality in (ii), as the example of $K = K_3$, $L = Q(3^{1/2})$ shows. Indeed, $f(K) = 3$, and since $d(L) = 12$, we get from (iv) that $f(L) = 2^a 3^b$ with certain positive a, b (actually $f(L) = 12$), and so $(f(K), f(L)) = 3$. However, $K \cap L = Q$ and $f(K \cap L) = 1$.

2. Now we are going to determine the factorization of an arbitrary prime in an Abelian extension of Q. According to Theorem 4.10 (ii) the Galois group of K_m/Q can be identified with the group $G(m)$ of residue classes (mod m) prime to m, a residue class $a \pmod{m} \in G(m)$ acting on K_m by $\zeta_m \mapsto \zeta_m^a$. If K/Q is Abelian and $K \subset K_m$, then according to Galois theory K corresponds to a subgroup H of K_m. The following theorem shows that it is enough to know m and H to obtain the factorization laws for every rational prime in K.

Theorem 8.1. *Let $K/\mathbb{Q}$ be Abelian and let K_m be an arbitrary cyclotomic field containing K. Then for every rational prime p we have*
$$pR_K = (\mathfrak{P}_1 \ldots \mathfrak{P}_g)^e, \quad f_{K/\mathbb{Q}}(\mathfrak{P}_i) = f \quad (i = 1, 2, \ldots, g),$$
where the numbers e, f, g are determined as follows:

If $p \nmid m$, then $e = 1$, f equals the order of $p \pmod{m}$ in the factor-group $G(m)/H$ and $g = [K:\mathbb{Q}]/ef$.

If $p | m$, then write $m = p^a m_1$ with $p \nmid m_1$, denote by N the number of residue classes $r \pmod{m} \in H$ which satisfy $r \equiv 1 \pmod{m_1}$ and let N_1 be the number of residue classes $r \pmod{m} \in H$ for which $r \pmod{m_1}$ lies in the cyclic group generated by $p \pmod{m_1}$ in $G(m_1)$. Then
$$e = \varphi(p^a)/N, \quad f = FN/N_1, \quad g = [K:\mathbb{Q}]/ef$$
where F is the order of $p \pmod{m_1}$ in $G(m_1)$.

Proof. We first determine the decomposition group, the inertia group and the first ramification group of an arbitrary prime ideal of K_m lying over p, and then apply Lemma 6.2 to descend to K. The first step is contained in the following lemma, and the second is immediate.

Lemma 8.1. *Let $m = p^a m_1$ with $p \nmid m_1$ and let $\mathfrak{P}$ be an arbitrary prime ideal of K_m lying above $p\mathbb{Z}$. Then the decomposition group $G_{-1}(\mathfrak{P})$ consists of those residue classes $x \pmod{m}$ for which $x \pmod{m_1}$ lies in the cyclic group generated by $p \pmod{m_1}$, the inertia group $G_0(\mathfrak{P})$ consists of $x \pmod{m}$ satisfying $x \equiv 1 \pmod{m_1}$ and the first ramification group $G_1(\mathfrak{P})$ is the maximal p-subgroup of $G_0(\mathfrak{P})$.*

Proof. First assume that $\alpha = 0$, i.e. that m is not divisible by p, and denote by f the order of $p \pmod{m}$ in the group $G(m)$ of residue classes $\pmod{m}$ relatively prime to m. By Theorem 4.16 p is unramified and $f_{K_m/K}(\mathfrak{P}) = f$; hence by Proposition 6.5 the group $G_{-1}(\mathfrak{P})$ has f elements and is cyclic. We now show that it contains the automorphism g_p mapping ζ_m onto ζ_m^p, and this will confirm our assertion concerning $G_{-1}(\mathfrak{P})$ in this case, since evidently g_p is of order f. Observe that if $x = P(\zeta_m) \equiv 0 \pmod{\mathfrak{P}}$ (with $P(t) \in \mathbb{Z}[t]$), then
$$0 \equiv P^p(\zeta_m) \equiv P(\zeta_m^p) \pmod{\mathfrak{P}},$$
and so $g_p(x) = P(\zeta_m^p) \in \mathfrak{P}$, i.e., $g_p \in G_{-1}$.

Secondly, consider the case of $m_1 = 1$, i.e., $m = p^\alpha$ with non-zero α. In this case Theorem 4.16 yields $e_{K_m/\mathbb{Q}}(\mathfrak{P}) = \varphi(p^\alpha)$, and $f_{K_m/\mathbb{Q}}(\mathfrak{P}) = 1$. Since G_{-1} has $e_{K_m/\mathbb{Q}}(\mathfrak{P})$ elements and this number equals $[K_m : \mathbb{Q}]$, it coincides with the full Galois group and so our assertion about G_{-1} is true also in this case.

In the general case observe that K_m is the composite of $K_{m_1} = L$ and $K_{p^\alpha} = M$ and the Galois group $G(K_m/Q) = G(m)$ is the product of $G(L/Q) = G(m_1)$ and $G(M/Q) = G(p)$. These two factors are embedded in $G(K_m/Q)$ as follows: $G(L/Q)$ is the group which fixes every element of M and $G(M/Q)$ fixes every element of L. Let $\mathfrak{p}_1$, $\mathfrak{p}_2$ be the prime ideals of R_L, R_M which lie below $\mathfrak{P}$ and let g be an arbitrary element of $G_{-1}(\mathfrak{P})$. We can write $g = [g_1, g_2]$, where $g_1 \in G(L/Q)$ is the restriction of g to L lifted back to an automorphism of K_m, trivial on M, and g_2 is an element of $G(M/Q)$ obtained in the same way, M being replaced by L. Note that for $i = 1, 2$ $g_i(\mathfrak{p}_i)$ is conjugated with $\mathfrak{p}_i$, and since $g_i(\mathfrak{p}_i)R_{K_m} \subset g(\mathfrak{P}) = \mathfrak{P}$, we have $g_i(\mathfrak{p}_i) = \mathfrak{p}_i$, i.e. g_i lies in $G_{-1}(\mathfrak{p}_i)$. This shows that $G_{-1}(\mathfrak{P})$ is contained in the product of $G_{-1}(\mathfrak{p}_1)$ and $G_{-1}(\mathfrak{p}_2)$. Conversely, if for $i = 1, 2$, the elements g_i lie in $G_{-1}(\mathfrak{p}_i)$, then for $g = [g_1, g_2]$ we have $g(\mathfrak{P}) = \mathfrak{P}$. For if we had $g(\mathfrak{P}) = \mathfrak{Q} \neq \mathfrak{P}$, then $\mathfrak{p}_1 \subset \mathfrak{P}$ would imply $g_1(\mathfrak{p}_1) = g(\mathfrak{p}_1) \subset \mathfrak{Q}$, and thus $\mathfrak{P}\mathfrak{Q}$ would divide $\mathfrak{p}_1 R_{K_m}$; but this is impossible because in the extension K_m/L the prime ideal factors of pZ do not factorize into a product of two or more distinct prime ideals. (This can be inferred from Theorem 4.16.) Thus we have arrived at the equality $G_{-1}(\mathfrak{P}) = G_{-1}(\mathfrak{p}_1) \times G_{-1}(\mathfrak{p}_2) = G_{-1}(\mathfrak{p}_1) \times G(M/Q)$, and it suffices to translate this result into the language of residue classes to obtain our assertion concerning G_{-1} in the general case. To describe the group $G_0(\mathfrak{P})$, we recall that according to Proposition 6.5 it corresponds to the maximal subfield of K_m in which p is unramfied. But this subfield equals $K_{m_1} = L$. In fact, Theorem 4.16 implies that p does not ramify in L/Q, and we have also the equalities

$$[K_m : L] = \varphi(p^\alpha) = e_{K_m/L}(\mathfrak{P}),$$

showing that $f_{K_m/L}(\mathfrak{P}) = 1$, and so the corresponding p-adic extension is fully ramified. Hence p ramifies at every extension of Q containing L and contained in K_m. Consequently $G_0(\mathfrak{P})$ is the Galois group of K_m/L, which can be identified with $\{r: g_r(\zeta_{m_1}) = \zeta_{m_1}\}$. Now, in view of

$$\zeta_{m_1} = \zeta_m^p,$$

we have $g_r(\zeta_{m_1}) = \zeta_m^{rp}$, and this equals ζ_{m_1} if and only if $rp \equiv p \pmod{m}$, i.e. $r \equiv 1 \pmod{m_1}$, as asserted. Finally, the statement concerning $G_1(\mathfrak{P})$ follows from Proposition 6.5. □

The theorem now follows immediately. □

Corollary 1. *If K/Q is Abelian and $f = f(K)$, then $P(K/Q)$ coincides with the set of all primes p with $p \pmod{f} \in H$.*

Proof. Apply the theorem with $m = f$, remembering that p lies in $P(K/Q)$ iff $e = f = 1$ and noting that by Proposition 8.1 (iv) no prime from $P(K/Q)$ can divide f. □

Corollary 2. *If K/Q is Abelian and $f = f(K)$, then the map*
$$F: p \mapsto F_{K/Q}(p)$$
defined for all primes p not dividing f induces an isomorphism of $G(f)/H$ onto $\mathrm{Gal}(K/Q)$.

The existence of an isomorphism between these groups is immediate by Galois theory. The importance of this corollary lies in the fact that it provides an explicit isomorphism.

Proof. We first show that F indeed induces a well-defined map from $G(f)/H$ onto $\mathrm{Gal}(K/Q)$. The surjectivity of F results from Theorem 7.11 and so it suffices to show that if $p_1 \pmod{f}$ and $p_2 \pmod{f}$ lie in the same coset mod H then $F_{K/Q}(p_1) = F_{K/Q}(p_2)$. By Theorem 7.10 (ii), $F_{K/Q}(p)$ equals the restriction of $F_{K_f/Q}(p)$ to K, and since $F_{K_f/Q}(p) = p \pmod{f}$, $F_{K/Q}(p)$ equals the coset $\pmod H$ in which $p \pmod f$ lies. Thus indeed the map induced by F is well-defined.

To show that it is a homomorphism, it suffices to show that if $p \equiv p_1 p_2 \pmod{f}$, then $F_{K/Q}(p) = F_{K/Q}(p_1) F_{K/Q}(p_2)$; but this is obvious for $K = K_f$ and the general case follows by restriction to K. Now the assertion results immediately; injectivity is a consequence of the equality $|G(f)/H| = |\mathrm{Gal}(K/Q)|$, which follows from Galois theory. □

Corollary 3. *If m is a given positive integer and H is a subgroup of $G(m)$, then there exists an Abelian extension K/Q such that $P(K/Q)$ differs from the set of all primes p with $p \pmod{m} \in H$ by only finitely many primes.*

Proof. Let K be the subfield of K_m corresponding to H. An application of Corollary 1 shows that the two sets occurring in the assertion differ only by those primes which divide m but not $f(K)$. □

The three corollaries proved above constitute a significant part of the class-field theory for Abelian extensions of Q. Corollary 1 shows that every Abelian field over Q is a class-field in the sense of H. Weber [96b, 2nd ed.], [97], and Corollary 2 is a form of Artin's Reciprocity Law for Abelian extensions of Q. Corollary 3 is called the Existence Theorem. The three statements have analogues concerning Abelian extensions of an arbitrary fixed algebraic number field; they, however, belong properly to class-field theory, which lies outside the scope of this book.

3. Let K/Q be Abelian and let $K \subset K_m$. As before, denote by H the subgroup of $G(m)$ which corresponds to K by Galois theory. Let $X(K)$ be the group

of all characters of $G(m)$ which are equal to unity on H. Extend each of these characters first to a Dirichlet character $(\mod m)$ and then to a primitive character. For any χ in $X(K)$ denote by $f(\chi)$ its conductor and by χ' the corresponding primitive Dirichlet character.

Note that if $X(K)$ and $X'(K)$ are the groups of characters associated with K which correspond to the embeddings of K in K_f (with $f = f(K)$) and in K_m, then every character of $X'(K)$ is lifted from a character of $X(K)$. This induces an isomorphism between $X(K)$ and $X'(K)$, which preserves conductors and induced primitive characters; thus we may identify $X(K)$ with $X'(K)$. Under this convention the equality $X(K) = X(L)$ implies $K = L$.

Proposition 8.2. *If m is fixed, then the map $K \mapsto X(K)$ defined for all subfields K of K_m induces a one-to-one correspondence between subfields of K_m and subgroups of the character group of $G(m)$. This map has the following properties:*

(i) $K \subset L$ *holds iff* $X(K) \subset X(L)$,
(ii) $X(K \cap L) = X(K) \cap X(L)$,
(iii) $X(KL)$ *equals the group generated by* $X(K) \cup X(L)$,
(iv) $X(K) = G(m)$ *holds iff* $K = K_m$ *and* $X(K) = 1$ *holds iff* $K = Q$,
(v) $|X(K)| = [K:Q]$.

Proof. This is simply a restatement of the fundamental theorem of Galois theory for K_m in terms of characters. □

Proposition 8.3. *If K is a real Abelian field, then all characters of $X(K)$ are even, and if K is a complex Abelian field of degree n, then $X(K)$ contains $n/2$ even and $n/2$ odd characters. The even characters form a subgroup of $X(K)$ equal to $X(K^+)$, with K^+ being the maximal real subfield of K.*

Proof. Since the element $-1 \pmod{m}$ of $G(m)$ acts as complex conjugation, the field K is real iff $-1 \pmod{m}$ acts on K trivially, i.e. -1 lies in H, and this occurs iff all characters of $X(K)$ are even. This argument also shows that if K is complex, then $X(K)$ contains at least one odd character. Since the even characters in this case form a subgroup X' of index two in $X(K)$, we see that $X(K)$ contains the same number of odd and even characters. X' is the maximal subgroup of $X(K)$ consisting of even characters. Therefore the first part of the proposition jointly with Proposition 8.2 show that $X' = X^+(K)$. □

4. We are now going to use the group $X(K)$ to express the Dedekind zeta-function $\zeta_K(s)$ of an Abelian field K as a product of Dirichlet's L-functions.

This formula will be used later in deriving an explicit formula for the class number of Abelian fields and also for the proof of the Siegel–Brauer theorem.

Theorem 8.2. *If K/Q is Abelian, then*
$$\zeta_K(s) = \prod_{\chi \in X(K)} L(s, \chi').$$

(Recall that χ' denotes the primitive Dirichlet character induced by χ.)

Proof. Since both sides of the asserted equality are meromorphic, it suffices to establish it for s in the half-plane $\mathrm{Re}\, s > 1$. Now, for $\mathrm{Re}\, s > 1$ we have
$$\zeta_K(s) = \prod_p \prod_{\mathfrak{p}\,\text{over}\,p} (1 - N(\mathfrak{p})^{-s})^{-1}$$
and
$$L(s, \chi') = \prod_p \prod_\chi (1 - \chi'(p)p^{-s})^{-1}$$

(where p runs over all rational primes and $\chi'(p) = 0$ for primes dividing the conductor of χ), the two products being absolutely convergent. To prove the theorem, it is enough to establish the equality
$$\prod_{\mathfrak{p}\,\text{over}\,p} (1 - N(\mathfrak{p})^{-s}) = \prod_\chi (1 - \chi'(p)p^{-s}) \tag{8.1}$$

for every prime p. This is what we now do. First observe that if
$$pR_K = (\mathfrak{P}_1 \ldots \mathfrak{P}_g)^e$$
and $f = f_{K/Q}(\mathfrak{P}_1) = \ldots = f_{K/Q}(\mathfrak{P}_g)$, then the left-hand side of (8.1) equals $(1 - p^{-fs})^{-g}$. We shall show that the right-hand side of (8.1) has the same value. We start with the case of $p \nmid m$. In this case $e = 1$. If $\chi \in X(K)$, then χ' equals unity on H, and since Theorem 8.1 shows that f is equal to the order of p (mod m) in $G(m)/H$, we get $\chi'(p)^f = \chi'(p^f) = 1$ and hence all possible values of $\chi'(p)$ are fth roots of unity. Moreover, if A is the group of characters of $G(m)/H$ which are equal to 1 on the coset (mod H) determined by p (mod m), then the set of those characters χ' which attain at p a fixed fth root of unity forms a coset (mod A) in the group of all characters of G/H. It follows that this set has $|G/H|f^{-1} = g$ elements and we can write
$$1 - \chi'(p)p^{-s} = \prod_{j=0}^{f} (1 - \zeta_f^j p^{-s})^g.$$

Now note that for all x we have the identity
$$\prod_{j=0}^{f} (1 - \zeta_f^j x) = 1 - x^f.$$

Putting $x = p^{-s}$ we obtain

$$\prod_{j=0}^{f} (1 - \zeta_f^j p^{-s}) = 1 - p^{-fs}$$

and

$$\prod_\chi (1 - \chi'(p) p^{-s}) = (1 - p^{-fs})^g,$$

implying (8.1).

If p divides m, then write $m = p^a m_1$ with $p \nmid m_1$. It follows from Proposition 8.1 that $L = K \cap K_{m_1}$ is the maximal subfield of K in which p remains unramified and Proposition 6.5 implies that if $\mathfrak{p}$ is an arbitrary prime ideal of L lying over p, then $f_{L/Q}(\mathfrak{p}) = f$, $g_{L/Q}(\mathfrak{p}) = g$.

Since $p | f(\chi)$ implies $\chi'(p) = 0$, we have

$$\prod_{\chi \in X(K)} (1 - \chi'(p) p^{-s}) = \prod_{\substack{\chi \in X(K) \\ f(\chi) | m_1}} (1 - \chi'(p) p^{-s}).$$

But $f(\chi) | m_1$ holds iff $\chi \in X(K_{m_1})$ and so

$$\{\chi \in X(K) : f(\chi) | m_1\} = X(K) \cap X(K_{m_1}) = X(L)$$

by Proposition 8.2 (ii). It follows that

$$\prod_{\chi \in X(K)} (1 - \chi'(p) p^{-s}) = \prod_{\chi \in X(L)} (1 - \chi'(p) p^{-s});$$

but $p \nmid m_1$ and so applying the case already considered we obtain

$$\prod_{\chi \in X(K)} (1 - \chi'(p) p^{-s}) = (1 - p^{-f's})^{g'}$$

with $f' = f_{L/Q}(\mathfrak{p})$, $g' = g_{L/Q}(\mathfrak{p})$. We have already shown that $f' = f$, $g' = g$ and thus (8.1) results also in this case. $\square$

Our next result expresses the discriminant and the conductor of an Abelian field in terms of the associated character group.

Proposition 8.4. *Let K/Q be Abelian and let $X(K)$ be the associated group of characters. Then*

$$d(K) = (-1)^u \prod_{\chi \in X(K)} f(\chi),$$

where u denotes the number of odd characters in $X(K)$ and

$$f(K) = \mathrm{LCM} \{f(\chi) : \chi \in X(K)\}.$$

Proof. The preceding theorem implies

$$1 = \frac{\zeta_K(1-s)}{\zeta_K(s)} \prod_{\chi \in X(K)} \frac{L(s,\chi')}{L(1-s,\bar{\chi}')}.$$

Using Theorem 7.1 and Proposition 7.7 we arrive (in the case of K real) at

$$1 = |d(K)|^{1/2-\sigma} \prod_\chi f(\chi)^{\sigma-1/2} \tag{8.2}$$

with $\sigma = \mathrm{Re}\,s$. Putting $s = 0$ and using Proposition 2.8 and Proposition 8.3 we get the first assertion.

If K is complex, then the same approach leads to

$$1 = 2^{-r_2(1-2\sigma)} \pi^{-r_2+\sigma r_2} |d(K)|^{1/2-\sigma} \prod_\chi f(\chi)^{\sigma-1/2}$$

$$\times \left\{ \frac{\Gamma(1-s)\,\Gamma(s/2)\,\Gamma((1+s)/2)}{\Gamma(s)\,\Gamma((1-s)/2)\,\Gamma(1-s/2)} \right\}^{r_2}.$$

Using the classical formula

$$\Gamma(z)\Gamma(z+1/2) = \pi^{1/2} 2^{1-2z} \Gamma(2z)$$

we again get (8.2) and the first assertion results as in the first case.

To obtain the second assertion, we embed K in K_f with $f = f(K)$ and note that for χ in $X(K)$ we have $f(\chi)|f$. Thus $M = \mathrm{LCM}\,\{f(\chi): \chi \in X(K)\}$ divides f. Since all characters of $X(K)$ can be regarded as characters (mod M), we get $K \subset K_M$ and so $f|M$. Thus finally $f = M$. □

Corollary. *We have*

$$\left| \prod_{\substack{\chi \in X(K) \\ \chi \neq 1}} \tau(\chi') \right| = |d(K)|^{1/2}.$$

Proof. The equality $|\tau(\chi')| = f(\chi)^{1/2}$ results from the corollaries to Proposition 6.10 and so the assertion follows from the first part of the proposition. □

5. The results just presented permit us to determine arithmetic properties of an Abelian field in a rather quick way, provided we know its conductor f and its position in the lattice of subfields of K_f. To illustrate this, let us consider subfields of K_{13} as an example. Since 13 is a prime, the group $G(13)$ is cyclic of order 12 and has the following subgroups: $H_1 = G(13)$, $H_2 = \{1, 3, 4, 9, 10, 12\}$, $H_3 = \{1, 5, 8, 12\}$, $H_4 = \{1, 3, 9\}$, $H_6 = \{1, 12\}$ and $H_{12} = \{1\}$. (Here we denote the residue m (mod 13) simply by m.) Accordingly, we have five fields of conductor 13, the field corresponding to H_1 being equal to Q. Let L_r be the field corresponding to H_r ($r = 1, 2, 3, 4, 6, 12$). Proposition 8.4

gives $d(L_r) = (-1)^{u_r} 13^{a_r}$, where u_r is the number of odd characters equal to unity on H_r and

$$a_r = |X(L_r)| - 1 = |G(13)/H_r| - 1 = 12/|H_r| - 1 = [L_r : Q] - 1 = r - 1.$$

Since 2 is a primitive root (mod 13), every character of $G(13)$ is determined by its value at 2 and we easily obtain a complete list of those characters (we invite the reader to prepare such a list by himself), from which it follows immediately that $u_2 = u_3 = u_6 = 0$, $u_4 = 2$, $u_{12} = 6$. Thus L_2, L_3, L_6 are real whereas L_4 and L_{12} are complex. In all cases the discriminant is positive and we get $d(L_r) = 13^{r-1}$ for all r.

Factorization of rational primes in L_r follows immediately from Theorem 8.1. The only ramified prime is $p = 13$ and we get $13R_{L_r} = \mathfrak{P}_r^r$ for all r, and for other primes we have

$$pL_r = \mathfrak{P}_1 \ldots \mathfrak{P}_g$$

with $g = r/t$, t being the least positive integer such that $p^t \pmod{13} \in H_r$.

To conclude the example, let us determine explicitly the fields involved. Clearly, $L_{12} = K_{13}$ and $L_2 = Q((13)^{1/2})$, since L_2 is a quadratic field of discriminant 13. Since the maximal real subfield of K_{13} is of degree 6, it must be equal to L_6; thus $L_6 = Q(\cos(2\pi/13))$. This leaves us with L_3 and L_4. Observe that every element of K_{13} can be uniquely written in the form

$$x = \sum_{j=1}^{12} A_j \zeta_{13}^j \quad (A_j \in Q)$$

and $x \in L_r$ holds iff $A_j = A_{js \pmod{13}}$ for every j and every $s \in H_r$. Consequently $w_1 = \zeta_{13} + \zeta_{13}^5 + \zeta_{13}^8 + \zeta_{13}^{12}$ lies in L_3, and since w_1 is not in Q, it generates L_3/Q. It follows easily from Theorem 2.9 that x is integral iff $A_j \in Z$ for $j = 1, 2, \ldots, 12$ and hence w_1, w_2, w_3 form an integral basis of L_3 with

$$w_2 = \zeta_{13}^2 + \zeta_{13}^{10} + \zeta_{13}^3 + \zeta_{13}^{11} \quad \text{and} \quad w_3 = \zeta_{13}^4 + \zeta_{13}^7 + \zeta_{13}^6 + \zeta_{13}^9.$$

The same argument shows that the numbers

$$\omega_1 = \zeta_{13} + \zeta_{13}^3 + \zeta_{13}^9, \quad \omega_2 = \zeta_{13}^2 + \zeta_{13}^6 + \zeta_{13}^5,$$
$$\omega_3 = \zeta_{13}^4 + \zeta_{13}^{12} + \zeta_{13}^{10}, \quad \omega_4 = \zeta_{13}^7 + \zeta_{13}^8 + \zeta_{13}^{11}$$

form an integral basis of L_4. Since L_4 has no proper non-real subfield and ω_1 is non-real, it must generate L_4/Q. We leave to the reader the dull task of obtaining minimal polynomials for the generators of L_3 and L_4.

6. Now we are in position to obtain information on asymptotic properties of the number of Abelian extensions K/Q with a given Galois group and $f(K) \leq x$, when x tends to infinity.

We shall do this in a slightly more general setting. Let k be a fixed field, and let for each integer $N \geqslant 1$ an extension L_N/k be given, which is assumed to be finite and Abelian. Let $A(N)$ denote the Galois group of L_N/k.

The family $\{L_N\}$ will be called *multiplicative*, if it satisfies the following conditions:

(i) $L_M L_N = L_{[M,N]}$, for all M, N,
(ii) if $(M, N) = 1$, then $A(MN) = A(M) \oplus A(N)$,
(iii) $L_M \cap L_N = L_{(M,N)}$ for all M, N.

Note that if $k = Q$ and L_N is the Nth cyclotomic field then by Theorem 4.10 the family L_N is multiplicative.

We also consider the family F of all extensions K/k which satisfy $K \subset K_N$ for a suitable N depending on K. For a given finite Abelian group A let $N_A(x)$ be defined as the number of all fields K belonging to F which satisfy $\mathrm{Gal}(K/k) \simeq A$ and which are contained in a suitable field L_N with $N \leqslant x$. If $k = Q$ and $L_N = K_N$, then $N_A(x)$ counts fields with Galois group A and conductor not exceeding x.

We now prove that, under certain restrictions on $A(q)$ for prime powers q, an asymptotic formula for $N_A(x)$ can be obtained.

Theorem 8.3. *Let $\{L_N\}$ be a multiplicative family of finite Abelian extensions of a field k and let A be a fixed Abelian finite group of order $\neq 1$. For any finite Abelian group H let $n_H(N)$ be the number of homomorphisms of $A(N)$ into H. Further, assume that for every subgroup B of A there exist non-negative constants $a(B)$ and $c(B)$ such that for all prime powers q the inequality*

$$n_B(q) \leqslant c(B), \tag{8.3}$$

is satisfied and for all complex s satisfying $\mathrm{Re}\, s > 1$ we have the equality

$$\sum_p n_B(p) p^{-s} = a(B) \log \frac{1}{s-1} + g(s) \tag{8.4}$$

with p running over all rational primes and $g(s)$ being regular for $\mathrm{Re}\, s \geqslant 1$. The function g may depend on B.

If, moreover,

$$a(B) < a(A) \tag{8.5}$$

for every proper subgroup B of A, then the asymptotic equality

$$N_A(x) = (C(A) + o(1)) x (\log x)^{a(A)-2}$$

holds for x tending to infinity, with a certain positive constant $C(A)$.

Proof. First observe that for every fixed B the function $n_B(N)$ is multiplicative in N. Hence, in view of (8.3), we can expand the function

$$F_B(s) = \sum_{N=1}^{\infty} n_B(N) N^{-s}$$

into the Euler product

$$F_B(s) = \prod_p \left(1 + \sum_{j=1}^{\infty} n_B(p^j) p^{-js}\right)$$

in the half-plane $\operatorname{Re} s > 1$. Since $F_B(s)$ can vanish only at zeros of those factors of the Euler product which correspond to primes not exceeding $1+c(B)$ (this being ensured by (8.3)), we can write for $\operatorname{Re} s > 1$

$$F_B(s) = g_1(s) \exp\left(\sum_p n_B(p) p^{-s}\right)$$

with $g_1(s)$ being regular for $\operatorname{Re} s \geq 1$ and $g_1(1) \neq 0$.

Now, (8.4) leads to the equality

$$F_B(s) = g_2(s)(s-1)^{-a(B)} \tag{8.6}$$

valid for $\operatorname{Re} s > 1$, with $g_2(s)$ being regular in $\operatorname{Re} s \geq 1$ and $g_2(1) \neq 0$.

Denote by $m_B(N)$ the number of surjective homomorphisms of $G(N)$ onto B. Then obviously

$$\sum_{G \subset B} m_G(N) = n_B(N)$$

(with G ranging over all subgroups of B), and this relation permits us to calculate $m_B(N)$, in view of the following lemma:

Lemma 8.2. *One can assign to every pair of finite Abelian groups $B \subset A$ a complex number $t(A, B)$ so that, whenever f, g are complex-valued functions defined in the set of all subgroups of an arbitrary finite Abelian group G which satisfy*

$$f(A) = \sum_{B \subset A} g(B) \quad (A \subset G),$$

then $t(A, A) = 1$ and

$$g(A) = \sum_{B \subset A} t(A, B) f(B) \quad (A \subset G).$$

Proof. Let $\{1\} = H_1, H_2, \ldots, H_T = A$ be all subgroups of a group $A \subset G$, ordered in such a way that $H_i \subset H_j$ implies $i \leq j$ (but not necessarily conversely) and put

$$\varepsilon(i, j) = \begin{cases} 1 & \text{if } H_i \subset H_j, \\ 0 & \text{otherwise.} \end{cases}$$

Then
$$f(H_j) = \sum_{i=1}^{j} \varepsilon(i,j) g(H_i) \quad (j = 1, 2, \ldots, T).$$

The matrix $[\varepsilon(i,j)]$ is invertible and its inverse $[e(i,j)]$ satisfies $e(i,i) = 1$ and $e(i,j) = 0$ for $j > i$; therefore

$$g(H_j) = \sum_{i=1}^{j} e(i,j) f(H_i) \quad (j = 1, 2, \ldots, T).$$

Putting $j = T$ and $t(H_T, H_j) = e(j, T)$ we get our assertion. □

This lemma implies
$$m_B(N) = \sum_{G \subset B} t(B, G) n_G(N)$$

and this leads to the equality

$$\sum_{N=1}^{\infty} m_A(N) N^{-s} = F_A(s) + \sum_{\substack{B \subset A \\ B \neq A}} t(A, B) F_B(s)$$

holding for $\operatorname{Re} s > 1$. Using (8.5) and (8.6) we get, with a certain r, the equality

$$\sum_{N=1}^{\infty} m_A(N) N^{-s} = g_2(s)(s-1)^{-a(A)} + \sum_{j=1}^{r} h_j(s)(s-1)^{-t_j},$$

where $\operatorname{Re} s > 1$, $t_1, \ldots, t_r$ are real numbers smaller than $a(A)$ and $h_1, \ldots, h_r$ are functions regular in the half-plane $\operatorname{Re} s \geq 1$.

The kernel of any homomorphism counted by $m_A(N)$ corresponds by Galois theory to a field K satisfying

$$k \subset K \subset L_N \quad \text{and} \quad \operatorname{Gal}(K/k) \simeq A. \tag{8.7}$$

If $V(A)$ denotes the number of automorphisms of A, then to every field satisfying (8.7) there correspond $V(A)$ different homomorphisms counted by $m_A(N)$. This shows that if $c_A(N)$ denotes the number of such fields K, then

$$c_A(N) = m_A(N)/V(A).$$

Finally, let $b_A(N)$ be the number of fields K satisfying (8.7), which are not contained in L_M for any $M \underset{\neq}{\subset} N$. Our assumptions imply that $K \subset L_M$, $K \subset L_N$ leads to $K \subset L_{(M,N)}$ and hence every field counted by $c_A(N)$ is counted by $b_A(d)$ for exactly one divisor d of N.

This gives

$$\sum_{d \mid N} b_A(d) = c_A(N)$$

and so we arrive at the equality

$$\sum_{N=1}^{\infty} b_A(N)N^{-s} = \zeta^{-1}(s) \sum_{N=1}^{\infty} c_A(N)N^{-s}$$

$$= V(A)^{-1} g_2(s)(s-1)^{1-a(A)} + \sum_{j=1}^{r} V(A)^{-1} h_j(s)(s-1)^{-t_j},$$

valid for Re $s > 1$.

Note that our assumptions imply $a(A) > 1$. Indeed, for $B = \{e\}$ we have $n_B(p) = 1$ for all primes p, and (8.5) gives $a(A) > a(B) = 1$. We may thus invoke the Tauberian theorem of Delange (Theorem I of Appendix II) to obtain the asserted formula for

$$N_A(x) = \sum_{N \leq x} b_A(N). \qquad \square$$

Corollary 1. *If A is a finite Abelian group, then the number of extensions K/Q with Galois group A and $f(K) \leq x$ equals*

$$(C(A) + o(1)) x (\log x)^{a(A)-2}$$

(for x tending to infinity), where $C(A)$ is a positive number and

$$a(A) = \sum_{d \mid M} r_A(d) \varphi(d)^{-1}$$

with $M = |A|$ and $r_A(d)$ being the number of elements of A of order d.

Proof. We apply the theorem with $k = Q$ and L_N being the Nth cyclotomic field. Condition (8.3) follows from the fact that in this case the group $A(q)$ has for prime-powers q at most two generators. Moreover, it is easily seen that for prime p the equality

$$n_B(p) = \sum_{d \mid p-1} r_B(d)$$

holds for every finite Abelian group B. Hence, formula (8.4) with

$$a(B) = \sum_{d \mid |B|} r_B(d) \varphi(d)^{-1}$$

results from

$$\sum_{p} n_B(p) p^{-s} = \sum_{d \mid |B|} r_B(d) \sum_{p \equiv 1 \pmod{d}} p^{-s} = a(B) \log \frac{1}{s-1} + g(s)$$

with a function $g(s)$ regular in Re $s \geq 1$; the last equality follows from Corollary 6 to Proposition 7.9 applied to $K = Q$. It remains to verify (8.5). This is immediate, because if B is a proper subgroup of A, then for at least one value of d dividing $|A|$ we must have $r_B(d) < r_A(d)$. Thus all assumptions of the theorem are satisfied, and the assertion follows. □

In certain cases one can deduce from the last corollary asymptotic results on the number of fields with a given Galois group and bounded discriminant. We can see this on the example of cyclic fields of prime degree:

Corollary 2. *The number of cyclic extensions of Q of a prime degree p and discriminants not exceeding x equals $(C+o(1))x^a$ with $a = 1/(p-1)$ and C positive and depending on p.*

Proof. If K/Q is cyclic of degree p, then $X(K)$, the associated group of characters, contains $p-1$ non-principal characters, all of the same conductor, because each of them generates $X(K)$ and if χ_1 is a power of χ_2 then $f(\chi_1)$ divides $f(\chi_2)$. Denoting this conductor by f, we have by Proposition 8.4 $d(K) = f^{p-1}$, $f(K) = f$, and so the conditions $d(K) \leqslant x$ and $f(K) \leqslant x^a$ are equivalent. The statement now results from Corollary 1. □

The next corollary shows that it is also possible to obtain asymptotics for extensions counted in Corollary 1 with additional restrictions on ramification.

Corollary 3. *Let A be a finite Abelian group of order D and let P_0 be a set of rational primes such that, for every divisor d of D, the equality*

$$\sum_{\substack{p \in P_0 \\ p \equiv 1 \pmod{d}}} p^{-s} = \alpha(d) \log \frac{1}{s-1} + g(s; d)$$

holds for $\operatorname{Re} s > 1$, with positive $\alpha(d)$ and $g(s; d)$ regular for $\operatorname{Re} s \geqslant 1$. Then the number of extensions K/Q with Galois group A which satisfy $f(K) \leqslant x$ and are ramified only at primes from P_0 equals to

$$(C(A, P_0) + o(1)) x (\log x)^{u(A)-2}$$

where $C(A, P_0)$ is positive and

$$u(A) = 1 + \sum_{\substack{d \mid D \\ d \neq 1}} r_A(d) \alpha(d),$$

$r_A(d)$ *having the same meaning as in Corollary 1.*

Proof. The assertion follows from the theorem by taking for L_N the nth cyclotomic field, with n being the maximal divisor of N composed of primes from outside P_0. We leave the details to the reader. □

§ 2. The Class-Number Formula and the Siegel–Brauer Theorem

1. In this section we give a formula for the class-number of an arbitrary Abelian extension K/Q in terms of the values of L-functions associated with characters from $X(K)$ and other invariants of K; and then we prove the Siegel–Brauer theorem concerning the asymptotic behaviour of the product $h(K)R(K)$ for such fields. The proof of this theorem in the general case requires class-field theory and therefore lies outside the scope of this book.

Theorem 8.4. *If K/Q is Abelian and $w(K)$ denotes the number of roots of unity contained in K, then*

$$h(K) = \frac{w(K)|d(K)|^{1/2}}{2^{r_1+r_2}\pi^{r_2}R(K)} \prod_{\substack{\chi \in X(K) \\ \chi \neq 1}} L(1, \chi').$$

Proof. By Theorem 8.2 we have

$$\zeta_K(s) = \prod_{\chi \in X(K)} L(1, \chi').$$

Comparing the residues at $s = 1$, using Theorem 7.1 and noting that $L(s, \chi'_0) = \zeta(s)$ for the principal character χ_0, we obtain the assertion. □

Corollary 1. *If $m \not\equiv 2 \pmod 4$, then for the m-th cyclotomic field K_m we have*

$$h(K_m)R(K_m) = \varepsilon_m |d(K_m)|^{1/2}(2\pi)^{-\varphi(m)/2} \prod_{\chi \neq 1} L(1, \chi')$$

with

$$\varepsilon_m = \begin{cases} 1 & \text{if } 2|m, \\ 2 & \text{if } 2 \nmid m \end{cases}$$

and χ running over all non-principal characters $\pmod m$.

Proof. Follows directly from Theorems 8.4 and 4.10. □

Corollary 2. *If K is a quadratic field of discriminant d, w is the number of roots of unity contained in K, $h(d)$ is the class-number of K and, in case of $d > 0$, $\varepsilon > 1$ denotes the fundamental unit of K, then*

$$h(d) = \begin{cases} w|d|^{1/2}L_d(1)/2 & \text{if } d < 0, \\ d^{1/2}L_d(1)/2\log\varepsilon & \text{if } d > 0, \end{cases}$$

where $L_d(s) = L(s, \chi_d)$ is the L-function associated with the Kronecker's extension $\chi_d(x) = \left(\dfrac{d}{x}\right)$ of the Legendre symbol.

§ 2. The Class-Number Formula and the Siegel-Brauer Theorem   437

Proof. In view of the last theorem, it suffices to establish that the primitive character induced by the unique non-trivial character $\chi \in X(K)$ equals χ_d. Since $|X(K)| = [K:Q] = 2$, χ is real and Proposition 8.4 gives sgn $d = \chi(-1)$, $f(\chi) = |d|$. Thus χ' is a real primitive character (mod $|d|$), satisfying $\chi'(-1) =$ sgn d. Write

$$d = (\text{sgn } d)p_1^a p_2 \cdots p_t$$

with $p_1 = 2, p_2 < p_3 < \ldots < p_t$ odd primes, and $a = 0, 2$ or 3. (This representation of d results from Proposition 6.8.) By Proposition 6.7 we may write $\chi' = \chi_1 \cdots \chi_t$ where χ_i is a real primitive character (mod p_i) for $i = 2, 3, \ldots, t$, resp. (mod 2^a) for $i = 1$. Thus, for $i = 2, 3, \ldots, t$ we have $\chi_i(x) = \left(\dfrac{x}{p_i}\right)$. If $a = 0, 1$ then $\chi' = \chi_d$ results immediately. If $a = 2$, then $\chi_1(x)$ equals $(-1)^{(x-1)/2}$, the unique primitive character (mod 4). For $a = 3$ we have a choice, since there are two distinct real primitive characters (mod 8), namely $(-1)^{(x^2-1)/8}$ and $(-1)^{(x^2-1)/8 + (x-1)/2}$; but the condition $\chi'(-1) = \text{sgn } d$ determines χ_1 uniquely also in this case. It is immediate that in all cases χ' equals $\left(\dfrac{d}{x}\right)$. (The reader not acquainted with the Kronecker character $\left(\dfrac{d}{x}\right)$ may take the above construction for its definition.)   □

Corollary 3. *Let p be an odd prime and let K_p^+ be the maximal real subfield of the p-th cyclotomic field K_p. Put*

$$h_p^+ = h(K_p^+) \quad \text{and} \quad h_p^- = h(K_p)/h^+(K_p).$$

Then

$$h_p^+ = p^{(p-3)/4} R(K_p)^{-1} \prod_{\chi \text{ even}} L(1, \chi)$$

and

$$h_p^- = 2^{-(p-3)/2} p^{(p+3)/4} \pi^{-(p-1)/2} \prod_{\chi \text{ odd}} L(1, \chi),$$

products taken over non-principal characters (mod p), even in the first formula, and odd in the second.

Proof. Since K_p/Q is cyclic, Theorem 3.7 shows that there is a fundamental system of units of K_p lying in K_p^+, where it obviously also forms a system of fundamental units. This immediately implies the equality

$$R(K_p) = 2^{(p-3)/2} R(K_p^+).$$

In view of the equalities $w(K_p^+) = 2$, $|d(K_p^+)| = p^{(p-3)/2}$ (by Proposition 8.4) the formula for h_p^+ results. Applying Corollary 1 to $m = p$ and dividing the expression thus obtained by h_p^+ we get the second formula.   □

2. The number h_p^- occurring in the last corollary is always an integer. This was first noted by E. E. Kummer [50a]; we present a proof due to L. Kronecker [63]:

Proposition 8.5. *If p is an odd prime, then the canonical map $i^*_{K_p/K_p^+}$ is injective, i.e. $H(K_p^+) \subset H(K_p)$ and $h(K_p) = h_p^+ h_p^-$ is a factorization into integers.*

Proof. Write $L = K_p$, $L^+ = K_p^+$ and let I be an ideal in R_{L^+} such that IR_L is principal. In view of Proposition 4.21 (i) it suffices to show that I itself is principal. Write $IR_L = aR_L$ and observe that $\bar{I} = I$ implies $\bar{a} = ua$ with a suitable unit u of L. By Corollary to Theorem 3.7 we can write $u = zu_1$ with u_1 lying in $U(L^+)$ and a root of unity $z \in E(L)$. From

$$|a| = |\bar{a}| = |zu_1 a| = |u_1 a|$$

we get $|u_1| = 1$; hence $u_1 = \pm 1$. Thus $\bar{a} = z_0 a$ with $z_0 \in E(L)$. By Corollary to Theorem 4.10 with a suitable integer m we can write $z_0 = \pm \zeta_p^m$; hence $\bar{a} = \pm \zeta_p^m a$. Let s be a solution of $2s \equiv m \pmod{p}$ and put $b = \zeta_p^s a$. Then

$$\bar{b} = \zeta_p^{-s} \bar{a} = \pm \zeta_p^{m-s} a = \pm \zeta_p^s a = \pm b$$

and $IR_L = bR_L$. If $b = \bar{b}$, then $b \in L^+$ and so I is principal. Thus it remains to show that the equality $\bar{b} = -b$ is impossible.

Let $\pi = 1 - \zeta_p$. By Corollary to Theorem 4.16 π generates a prime ideal in L and so we may write

$$b = \pi^r c$$

with $(cR_L, \pi R_L) = 1$ and a suitable integer $r \geqslant 0$. Since, by the same Corollary

$$pR_L = (\pi R_L)^{p-1}, \quad f_{L/Q}(\pi R_L) = 1,$$

we get $pR_{L^+} = \mathfrak{P}^{(p-1)/2}$ with a suitable prime ideal $\mathfrak{P}$. Thus $\mathfrak{P} R_L = (\pi R_L)^2$ and it follows that the exponent r must be even, say, $r = 2t$. This leads to

$$-\pi^{2t} c = -b = \bar{b} = \bar{\pi}^{2t} \bar{c} = \pi^{2t} \zeta_p^{-2t} \bar{c}$$

and $\bar{c} = -\zeta_p^{2t} c$, and finally $\bar{c} \equiv -c \pmod{\pi}$ because of $\zeta_p \equiv 1 \pmod{\pi}$. This congruence, however, is impossible. For if we write

$$c = \sum_{j=0}^{p-2} A_j \zeta_p^j$$

with $A_j \in Z$, then $c \equiv A_0 + A_1 + \ldots + A_{p-2} \equiv \bar{c} \pmod{\pi}$ and thus $2c \equiv 0 \pmod{\pi}$, which is excluded by our choice of c. □

3. The class-number formula given in Theorem 8.4 contains values of Dirichlet's L-functions at $s = 1$. We now express these values by finite sums of elementary functions.

§ 2. The Class-Number Formula and the Siegel–Brauer Theorem

Proposition 8.6. *If χ is a primitive character $(\mod m)$ with $m > 2$, then*

$$L(1, \chi) = -\frac{\tau(\chi)}{m} \sum_{x=1}^{m-1} \overline{\chi(x)} \left\{ \log\left(2\sin\frac{\pi x}{m}\right) - i\frac{\pi x}{m} \right\}.$$

If moreover χ if odd, then

$$L(1, \chi) = i\pi \frac{\tau(\chi)}{m^2} \sum_{x=1}^{m-1} x\overline{\chi(x)}.$$

Proof. For $|x| < 1$ define

$$f(x) = \sum_{n=1}^{\infty} \chi(n) n^{-1} x^n.$$

Since the series defining $L(s, \chi)$ converges at $s = 1$ (the sum $\sum_{n \leq x} \chi(n)$ being bounded by m and the sequence $(1/n)$ decreasing to zero), we have

$$\lim_{x \to 1} f(x) = L(1, \chi).$$

Now,

$$f(x) = \sum_{r=1}^{m-1} \chi(r) \sum_{n \equiv r \,(\mod m)} n^{-1} x^n = \sum_{r=1}^{m-1} \chi(r) \sum_{n=0}^{\infty} (r+nm)^{-1} x^{r+nm}.$$

Hence

$$f'(x) = -\left(\sum_{r=1}^{m-1} \chi(r) x^{r-1}\right)(x^m - 1)^{-1}$$

and we get, in view of $f(0) = 0$, the equality

$$f(x) = -\int_0^x \sum_{r=1}^{m-1} \chi(r)(t^r - 1)(t^m - 1)^{-1} dt.$$

The integrand can be written in the form

$$\sum_{j=0}^{m-1} a_j (t - \zeta_m^j)^{-1}$$

and the constants $a_0, a_1, \ldots, a_{m-1}$ occurring here can be determined by putting $t = \zeta_m^j$. This gives

$$\sum_{r=1}^{m-1} \chi(r) \zeta_m^{jr-1} = a_j m \zeta_m^{j(m-1)} = a_j m \zeta_m^{-j}$$

and thus $a_j = \tau_j(\chi)/m$.

Finally we get

$$L(1,\chi) = -\frac{1}{m}\int_0^1 \sum_{j=0}^{m-1} \tau_j(\chi)(t-\zeta_m^j)^{-1}dt. \tag{8.8}$$

A direct computation shows that

$$\int_0^1 (t-\zeta_m^j)^{-1}dt = \log\left(2\sin\frac{\pi j}{m}\right) + i\left(\frac{\pi}{2} - \frac{\pi j}{m}\right)$$

and thus the first assertion follows from Proposition 6.6 (i) and the following lemma:

Lemma 8.3. *If χ is a primitive Dirichlet character (mod m) and $(j,m) \neq 1$, then $\tau_j(\chi) = 0$.*

Proof. Write $(j,m) = m_1$, $k = m/m_1$ and choose an integer a, prime to m and satisfying $a \equiv 1 \pmod{k}$, for which $\chi(a) \neq 1$. If there were no such integer, $m = f(\chi)$ would divide k, and thus $(j,m) = 1$, contrary to our assumption. Now,

$$\tau_j(x) = \sum_{r=1}^{m-1} \chi(ar)\zeta_m^{arj} = \chi(a)\sum_{r=1}^{m-1} \chi(r)\zeta_m^{arj} = \chi(a)\tau_j(\chi),$$

because ζ_m^j is a primitive kth root of unity, $ar \equiv r \pmod{k}$ and thus $\zeta_m^{arj} = \zeta_m^{rj}$. Finally, we see that $\tau_j(\chi) = 0$, as asserted. □

To obtain the second assertion note that if χ is odd, then $\bar\chi(m-x) = \bar\chi(-x) = -\bar\chi(x)$ and thus

$$\sum_{x=1}^{m-1} \bar\chi(x)\log\left(2\sin\frac{\pi x}{m}\right)$$

$$= \frac{1}{2}\sum_{x=1}^{m-1}\left\{\bar\chi(x)\log\left(2\sin\frac{\pi x}{m}\right) + \bar\chi(m-x)\log\left(2\sin\frac{\pi(m-x)}{m}\right)\right\} = 0. \quad □$$

Corollary 1. *Dirichlet's class-number formulas.*

If K is a quadratic number field of discriminant d, w is the number of roots of unity in K, $h(d)$ is the class-number of K and $\varepsilon > 1$ denotes, in the case of positive d, the fundamental unit of K, then

§ 2. The Class-Number Formula and the Siegel-Brauer Theorem

$$h(d) = \begin{cases} \dfrac{-w}{2|d|} \sum_{x=1}^{|d|} \left(\dfrac{d}{x}\right) x & \text{if } d < 0, \\ -\dfrac{1}{\log \varepsilon} \sum_{0 < x < d/2} \left(\dfrac{d}{x}\right) \log \left(\sin \dfrac{\pi x}{d}\right) & \text{if } d > 0. \end{cases}$$

Proof. For $d < 0$ we use Theorem 6.3 to get

$$L(1, \chi_d) = \operatorname{Re} L(1, \chi_d) = -|d|^{-3/2} \sum_{x=1}^{|d|} x \chi_d(x)$$

and it suffices to apply Corollary 2 to Theorem 8.4.

For positive d we obtain similarly

$$L(1, \chi_d) = \operatorname{Re} L(1, \chi_d) = -d^{-1/2} \sum_{x=1}^{d-1} \chi_d(x) \log \left(2 \sin \dfrac{\pi x}{d}\right)$$

$$= -d^{-1/2} \sum_{x=1}^{d-1} \chi_d(x) \log \left(\sin \dfrac{\pi x}{d}\right).$$

Since by Proposition 8.3 χ_d is even, we get $\chi_d(d-x) = \chi_d(x)$ and applying Corollary 2 to Theorem 8.4 we get the claim. □

Our second application of the proposition proved above consists in finding an upper bound for h_p^-, as defined in Corollary 3 to Theorem 8.4.

Corollary 2. *If p is an odd prime, then*

$$h_p^- < 2p(p/24)^{(p-1)/4}.$$

Proof. The proposition and Corollary 3 to Theorem 8.4 give

$$h_p^- = p^{(p+3)/4} 2^{-(p-3)/2} p^{-p+1} \prod_{\chi \text{ odd}} |\tau(\chi)| \prod_{\chi \text{ odd}} \left|\sum_{x=1}^{p-1} x \bar{\chi}(x)\right|,$$

and since Corollary to Proposition 8.4 implies

$$\prod_{\chi \text{ odd}} |\tau(\chi)| = |d(K_p)/d(K_p^+)|^{1/2},$$

we get

$$h_p^- = (2p)^{-(p-3)/2} \prod_{\chi \text{ odd}} \left|\sum_{x=1}^{p-1} x \bar{\chi}(x)\right|.$$

It remains to evaluate the product appearing in the last equality.

Since
$$\sum_{\chi \text{ odd}} \chi(m)\bar{\chi}(n) = \begin{cases} (p-1)/2 & \text{if } p \nmid mn, m \equiv n \pmod{p}, \\ -(p-1)/2 & \text{if } p \nmid mn, m \equiv -n \pmod{p}, \\ 0 & \text{in all other cases,} \end{cases}$$

we have
$$\sum_{\chi \text{ odd}} \left|\sum_{x=1}^{p-1} \bar{\chi}(x)x\right|^2 = \sum_{\chi \text{ odd}} \sum_{x,y=1}^{p-1} \bar{\chi}(x)\chi(y)xy$$

$$= \sum_{j=1}^{p-1}\left(\sum_{x\equiv y\equiv j \pmod{p}} xy(p-1) - \sum_{x\equiv -y\equiv j \pmod{p}} xy(p-1)/2\right)$$

$$= \frac{p-1}{2}\left(\sum_{j=1}^{p-1} j^2 - \sum_{j=1}^{p-j} j(p-j)\right) = p(p-1)^2(p-2)/12.$$

Applying the inequality between the arithmetic and geometric means we arrive finally at

$$\sum_{\chi \text{ odd}} \left|\sum_{x=1}^{p-1} \bar{\chi}(x)x\right|^{4/(p-1)} \leq p(p-1)(p-2)/6 < p^3/6,$$

and this leads to

$$h_p^- < (2p)^{-(p-3)/2}(p^3/6)^{(p-1)/4} = 2p(p/24)^{(p-1)/4}. \qquad \square$$

4. Now we shall be concerned with the asymptotic behaviour of $h(K)R(K)$ for Abelian K/Q. An upper estimate of this product is provided by Corollary 4 to Theorem 7.1, which implies

$$\limsup \log(h(K)R(K))/\log|d(K)| \leq 1/2$$

when K runs over all fields of a fixed degree. The Siegel–Brauer theorem, which we now prove for K/Q Abelian, states that in this relation lim sup can be replaced by lim and the inequality by equality:

Theorem 8.5. *If K runs over all Abelian extensions of a fixed degree of the rationals, then*

$$\lim \log(h(K)R(K))/\log|d(K)| = 1/2.$$

Proof. We start with a lower bound for the residue of the Dedekind zeta-function:

Lemma 8.4. *There is a constant $B = B(n) > e^{-13n} > 0$ such that for every field K with $[K:Q] = n$ and every s_0 with $0 < s_0 < 1$ the inequality $\zeta_K(s_0) \leq 0$ implies $h(K)\varkappa(K) \geq Bs_0(1-s_0)|d(K)|^{s_0-1}$.*

§ 2. The Class-Number Formula and the Siegel-Brauer Theorem

Proof. By Corollary 1 to Theorem 7.1 the inequality $\zeta_K(s_0) \leq 0$ implies

$$\int_{V(x) \geq 1} (\hat{f}(x) V(x)^{s_0} + f(x) V(x)^{1-s_0}) \, dm_I(x)$$

$$\leq h(K) \varkappa(K) \left(\frac{1}{s_0 |d(K)|^{1/2}} + \frac{1}{1-s_0} \right) \leq h(K) \varkappa(K) / s_0 (1-s_0), \qquad (8.9)$$

where $f(x)$ is the function used in the proof of Theorem 7.1. Let us recall the definition of f:

$$f((x_v)) = \prod_v f_v(x_v),$$

where for Archimedean v

$$f_v(x_v) = \begin{cases} \exp(-\pi x_v^2) & \text{if } v \text{ is real}, \\ \exp(-2\pi |x_v|^2) & \text{if } v \text{ is complex}, \end{cases}$$

and for non-Archimedean v, $f_v(x_v)$ is the characteristic function of the ring R_v of integers of K_v. It was shown in the proof of Theorem 7.1 that for Archimedean v we have $\hat{f}_v(x_v) = f_v(x_v)$ and for non-Archimedean v the function $\hat{f}_v(x_v)$ equals $N(D_v)^{-1/2}$ times the characteristic function of D_v^{-1}, D_v being the different of K_v/Q_p.

Since $f(x)$ and $\hat{f}(x)$ are both non-negative, (8.9) leads to

$$h(K) \varkappa(K) \geq s_0(1-s_0) \int_{V(x) \geq 1} \hat{f}(x) V(x)^{s_0} dm_I(x)$$

$$\geq s_0(1-s_0) \int_{\prod_v P_v} \hat{f}(x) V(x)^{s_0} dm_I(x)$$

$$= s_0(1-s_0) \prod_v \int_{P_v} \hat{f}_v(x_v) v(x_v)^{s_0} dx_v^*$$

$$= s_0(1-s_0) \prod_v I_v, \qquad (8.10)$$

where $P_v = \{x_v : v(x_v) \geq 1\}$ for non-Archimedean v and $P_v = \{x_v : 1 \leq v(x_v) \leq 2\}$ for Archimedean v. We now estimate the integrals I_v from below. In the non-Archimedean case we have

$$I_v = (N(D_v))^{-1/2} \int_{D_v^{-1} \cap P_v} v(x_v)^{s_0} dx_v^*$$

$$= (N(D_v))^{-1/2} \sum_{m=-M}^{0} \int_{\pi_v^m U_v} v(x_v)^{s_0} dx_v^*$$

$$\geq (N(D_v))^{-1/2} \int_{\pi_v^{-M} U_v} v(x_v)^{s_0} dx_v^*,$$

where π_v is a fixed generator of the prime ideal p_v in K_v, U_v is the group of units of K_v and M is defined by $D_v = \pi_v^M R_v$. This leads to

$$I_v \geq (N(D_v))^{-1/2} N(p_v)^{Ms_0} \int_{\pi_v^{-M} U_v} dx_v^* = (N(D_v))^{s_0-1}.$$

In the real case we get

$$I_v = \int_1^2 \exp(-\pi x^2) x^{s_0} \frac{dx}{x} \geq \exp(-4\pi) \int_1^2 \frac{dx}{x}$$

$$= \log 2 \exp(-4\pi)$$

because $x^{s_0} \geq 1$, and in the complex case we get

$$I_v = 2 \iint_{1 \leq y_1^2 + y_2^2 \leq 2} \exp(-2\pi(y_1^2+y_2^2))(y_1^2+y_2^2)^{s_0-1} dy_1 dy_2$$

$$\geq 2\exp(-4\pi) \int_0^{2\pi} d\varphi \int_1^{2^{1/2}} r^{2s_0} \frac{dr}{r} = 2\pi \log 2 \exp(-4\pi).$$

This shows that

$$\prod_v I_v \geq B \prod_{v \text{ non-Archimedean}} N(D_v)^{s_0-1} = B|d(K)|^{s_0-1},$$

where

$$B = B(n) = \inf_{r_1 + 2r_2 = n} ((2\pi)^{r_2} (\log 2 \exp(-4\pi))^{r_1+r_2}) > e^{-13n}$$

and this jointly with (8.10) proves the lemma. □

Our next lemma gives an evaluation of the value of Dirichlet L-functions at $s = 1$.

Lemma 8.5. *For every $m \geq 3$ and every non-principal character $\chi \pmod{m}$, not necessarily primitive, we have*

$$|L(1, \chi)| \leq \log m + 2 < 3 \log m.$$

Proof. For every N and every non-trivial character $\chi \pmod{m}$

$$\left| \sum_{n=1}^N \chi(n) n^{-1} \right| = \left| \sum_{n=1}^m \frac{\chi(n)}{n} + \frac{1}{N} \sum_{n=1+m}^N \chi(n) + \right.$$

$$\left. + \sum_{M=1+m}^{N-1} \left(\frac{1}{M} - \frac{1}{M+1} \right) \sum_{n=1+m}^M \chi(n) \right|$$

$$\leq \sum_{n=1}^{m} \frac{1}{n} + \frac{m}{N} + \sum_{M=1+m}^{N-1} \left(\frac{1}{M} - \frac{1}{M+1}\right) m$$

$$= \sum_{n=1}^{m} \frac{1}{n} + \frac{m}{m+1} \leq \log m + 2 < 3 \log m. \qquad \square$$

Utilizing the two preceding lemmas we now give a lower bound for the residue of $\zeta_K(s)$ in terms of $|d(K)|$. This will be achieved by a suitable choice of the parameter s_0 in Lemma 8.4:

Lemma 8.6. *To each positive ε there corresponds a positive constant $B_1 = B_1(n, \varepsilon)$ such that for every Abelian extension K/Q of degree n*

$$h(K)\varkappa(K) \geq B_1 |d(K)|^{-\varepsilon}.$$

Proof. Let a positive ε be given, which may be assumed to be less than $1/2$. First suppose that for all Abelian extension K/Q of degree n we have

$$\zeta_K(s) \neq 0$$

for $1 - \varepsilon/2n < s < 1$. Since Corollary 1 to Theorem 7.1 implies $\zeta_K(s) < 0$ for some interval $(1-a, 1)$, the inequality

$$\zeta_K(1 - \varepsilon/2n) \leq 0$$

results for all fields in consideration. Lemma 8.4 (with $s_0 = 1 - \varepsilon/2n$) now gives

$$h(K)\varkappa(K) \geq |d(K)|^{-\varepsilon/2n} \geq |d(K)|^{-\varepsilon},$$

and the proof is complete.

Now assume that there is an Abelian extension K_0/Q of degree n such that for a certain s_0 in $(1 - \varepsilon/2n, 1)$ we have

$$\zeta_{K_0}(s_0) = 0.$$

Let us fix such an $s_0 = s_0(\varepsilon)$ and let K/Q be an arbitrary Abelian extension of degree n. Moreover, let L be the minimal field containing K and K_0. It is obviously Abelian over Q and its degree does not exceed n^2. Denote by $K_m = Q(\zeta_m)$ a cyclotomic field containing L and let H_0, H_1, H_2 be the subgroups of its Galois group which correspond to K_0, K and L, respectively. Finally, for $i = 0, 1, 2$, let A_i be the set of characters $(\bmod\, m)$ which are trivial on H_i and, as before, denote by χ' the primitive character induced by χ. According to Theorem 8.2 we can write

$$\zeta_L(s) = \prod_{\chi \in A_2} L(s, \chi'),$$

$$\zeta_{K_0}(s) = \prod_{\chi \in A_0} L(s, \chi'),$$

and this together with $A_2 \supset A_0$ implies $\zeta_L(s_0) = 0$. We can now apply Lemma 8.4, which yields
$$h(L)\varkappa(L) \geqslant B's_0(1-s_0)|d(L)|^{s_0-1},$$
where $B' = \min_{m \leqslant n^2} B(m)$.

Corollary 4 to Proposition 6.2 gives
$$|d(L)| \leqslant |d(K_0)d(K)|^n,$$
and we obtain
$$h(L)\varkappa(L) \geqslant B_2|d(K)|^{n(s_0-1)} \geqslant B_3|d(K)|^{-\varepsilon/2}$$
(with B_2, B_3 depending on n, ε and s_0). In view of the equality
$$h(K)\varkappa(K) = h(L)\varkappa(L) \prod_{\chi \in A_2 \setminus A_1} L(1, \chi')^{-1}$$
we may write
$$h(K)\varkappa(K) \geqslant B_3|d(K)|^{-\varepsilon/2} \prod_{\chi \in A_2 \setminus A_1} L(1, \chi')^{-1}.$$
Using Lemma 8.5, which gives
$$\prod_{\chi \in A_2 \setminus A_1} L(1, \chi') \leqslant B_4 \prod_{\chi \in A_2 \setminus A_1} \log m_\chi \leqslant B_5 \prod_{\chi \in A_2 \setminus A_1} m_\chi^{\varepsilon/2}$$
(with $m_\chi = f(\chi)$), we finally obtain
$$h(K)\varkappa(K) \geqslant B_6|d(K)|^{-\varepsilon/2} \prod_{\chi \in A_2 \setminus A_1} m_\chi^{-\varepsilon/2} \geqslant B_7|d(K)|^{-\varepsilon},$$
since in view of Proposition 8.4 the product
$$\prod_{\chi \in A_2 \setminus A_1} m_\chi$$
equals $|d(L)/d(K)|$. □

The last lemma shows that
$$\log h(K)R(K) \geqslant \log(|d(K)|^{1/2-\varepsilon}) + O(1)$$
and now it suffices to apply Corollary 4 to Theorem 7.1 to obtain the assertion. □

Corollary 1. *For imaginary quadratic fields K we have*
$$\log h(K) = (1/2 + o(1))\log|d(K)|$$
as $|d(K)|$ tends to infinity.

Proof. It suffices to observe that in this case $R(K) = 1$. □

Corollary 2. *There is only a finite number of imaginary quadratic fields with unique factorization.*

Proof. Apply Theorem 1.16 and the preceding corollary. □

This corollary is not effective and does not lead to an explicit bound for the largest $|d(K)|$ with K imaginary quadratic and $h(K) = 1$. An effective proof of this corollary will be given in the last section of this chapter (see Theorem 8.11).

Corollary 3. *For any given n, the ratio $h(K)/h(K^+)$ tends to infinity, when K runs over all totally complex Abelian fields of degree n, arranged according to the absolute value of $|d(K)|$. More precisely,*

$$\liminf \log(h(K)/h(K^+))/\log|d(K)| \geq 1/4.$$

Proof. Proposition 3.5 implies that any fundamental system of units of K^+ generates a subgroup of finite index in $U(K)$. This shows that the regulator R' of such system, taken in K, is not smaller than $R(K)$, according to Corollary 2 (iii) to Theorem 3.5. Since obviously $R' = 2^r R(K^+)$ (with $r = n/2 - 1$), this leads to

$$\log R(K) \leq r \log 2 + \log R(K^+), \qquad (8.11)$$

and thus

$$\log h(K)/h(K^+) \geq \log(h(K)R(K)) - \log(h(K^+)R(K^+)).$$

Let ε be a fixed positive number. It follows from the theorem that if $|d(K^+)|$ is sufficiently large, say $|d(K^+)| \geq T$, then

$$\log(R(K)h(K)) \geq (1/2 - \varepsilon)\log|d(K)| \geq (1 - 2\varepsilon)\log|d(K^+)|$$

and

$$\log(h(K^+)R(K^+)) \leq (1/2 - \varepsilon)\log|d(K^+)|.$$

(We used here the inequality $|d(K)| \geq |d(K^+)|^2$ resulting from Corollary 2 to Proposition 4.9.)

Subtracting, we get

$$\log(h(K)/h(K^+)) \geq (1/2 - 3\varepsilon)\log|d(K^+)| \geq (1/4 - 3\varepsilon/2)\log|d(K)|.$$

It remains to consider those fields K for which $|d(K^+)| \leq T$. By Theorem 2.12 this gives only finitely many possibilities for K^+; thus $R(K^+)$ lies between two positive constants and (8.11) gives $R(K) = O(1)$. By Theorem 8.5 we have

$$\log h(K) = (1/2 + o(1))\log|d(K)| - \log R(K).$$

Now, in the case of $R(K) \geq 1$ we have $\log R(K) = O(1)$ and in the case of $R(K) < 1$ the number $\log R(K)$ is negative, and thus

$$\log h(K) \geq (1/2 + o(1)) \log |d(K)|.$$

Since in our case $1 \leq h(K^+) = O(1)$, we are done. □

Corollary 4. *There are only finitely many totally complex Abelian extensions of Q with given degree and class-number.* □

§ 3. Class-Number of Quadratic Fields

1. The problem of determining the class-number of quadratic fields goes back to C. F. Gauss [01], who actually considered rather the number of classes of binary quadratic forms $ax^2 + 2bxy + cy^2$ with integral rational coefficients a, b, c satisfying $(a, b, c) = 1$, under the action of $SL(Z, 2)$. Those two numbers are intimately connected, as we shall see below. However, in contrast to Gauss, we shall consider primitive binary quadratic forms $ax^2 + bxy + cy^2$ with rational integral coefficients ("primitive" means $(a, b, c) = 1$). For any such form f we define its discriminant by $d(f) = b^2 - 4ac$.

A matrix

$$M = \begin{bmatrix} A & B \\ C & D \end{bmatrix} \in SL(Z, 2)$$

(i.e. a matrix having rational integral entries and unit determinant) acts on f by means of the formula

$$Mf = g(x, y) = a(Ax + By)^2 + b(Ax + By)(Cx + Dy) + c(Cx + Dy)^2$$
$$= f(Ax + By, Cx + Dy).$$

Since M^{-1} also belongs to $SL(Z, 2)$, we obtain a partition of the set of all forms of the same discriminant into classes, each class consisting of equivalent forms, two forms being considered equivalent if there is a matrix M in $SL(Z, 2)$ mapping one of them onto the other.

The theorem which we now prove settles the connection (mentioned above) between the number of equivalence classes (relative to partition just defined) and the class-number of a quadratic field.

Theorem 8.6. *Let I be an arbitrary ideal of the quadratic field K of discriminant d and let a_1, a_2 be a Z-basis of I chosen in such a way that the expression*

$$a_1 a_2' - a_2 a_1'$$

§ 3. Class-Number of Quadratic Fields

either is positive or lies on the upper half of the imaginary axis. With I we associate a binary quadratic form, putting

$$f_I(x, y) = N(I)^{-1}(a_1 x + a_2 y)(a'_1 x + a'_2 y).$$

(By c' we denote the conjugate of c in K.)

We now assert:

(i) *The form $f_I(x, y)$ has integral rational coefficients, its discriminant equals d, it is primitive and in case of negative d it is positive definite.*

(ii) *The mapping $I \mapsto f_I$ of the set of all ideals of R_K into the set of all primitive binary quadratic forms with rational integral coefficients (which, in the case of $d < 0$, are positive definite) is surjective.*

(iii) *If two ideals I, J lie in the same class of $H^*(K)$, then the forms f_I, f_J are equivalent, and conversely.*

Proof. First note that $a_1 a'_2 - a'_1 a_2$ equals $N(I) d^{1/2}$ (where by $d^{1/2}$ we mean either the positive surd or that which lies on the upper part of the imaginary axis) and so, changing, if necessary, the sign of a_1 we can find a basis of I in the required form. Observe, moreover, that the coefficients of f_I are equal to $N_{K/Q}(a_1)/N(I)$, $(a_1 a'_2 + a_2 a'_1)/N(I)$ and $N_{K/Q}(a_2)/N(I)$, respectively. Since a_1 and a_2 both belong to I, we get $N(I) | N_{K/Q}(a_i)$ ($i = 1, 2$), and we see that the number $a_1 a'_2 + a_2 a'_1$ is a rational integer lying in $II' = N(I) R_K$, and hence is divisible by $N(I)$. This shows that our form has integral coefficients. Note that the discriminant of f_I equals d and so this form is primitive. Indeed, suppose that, for a prime number p, the form $p^{-1} f_I$ has integral coefficients; then its discriminant D is equal to d/p^2 and thus $p = 2$, and since every discriminant of a form is congruent to 0 or 1 (mod 4), we have $d = 4D$ with $D \equiv 0, 1$ (mod 4), which is obviously impossible. To obtain the last part of (i) it suffices to observe that in the case of a negative d the coefficient of x^2 in f_I is positive.

To prove (ii) consider a primitive binary quadratic form $F(x, y) = Ax^2 + Bxy + Cy^2$ with discriminant d, which is assumed to be positive definite in the case of negative d. If C is positive, consider the ideal $I = (B + d^{1/2}) Z + 2CZ$, and if C is negative, consider $I = (B - d^{1/2}) Z + 2CZ$, the surd $d^{1/2}$ being chosen in the same way as before. (The case of $C = 0$ cannot arise, for then d would be a square.) Easy checking shows that $f_I = F$, thus proving (ii).

Assume that the ideals $I = Za_1 + Za_2$ and $J = Zb_1 + Zb_2$ lie in the same class of $H^*(K)$ (in particular, the a's and b's may constitute distinct bases of the same ideal, subject to the assumption concerning signs stated above), and let c_i be totally positive integers of K (for $i = 1, 2$) for which $c_1 I = c_2 J$ holds. Then $c_1 a_1$, $c_1 a_2$ and $c_2 b_1$, $c_2 b_2$ are two bases for $c_1 I_1$ and thus we can find a matrix $M = [m_{ij}] \in GL(Z, 2)$ with

$$c_1 a_i = m_{i1} c_2 b_1 + m_{i2} c_2 b_2 \quad (i = 1, 2). \tag{8.12}$$

Now, to the ideal $c_1 I$ there corresponds, via the basis $c_1 a_1, c_1 a_2$, the form
$$F(x, y) = (c_1 a_1 x + c_1 a_2 y)(c_1' a_1' x + c_1' a_2' y) N(I)^{-1} N_{K/Q}(c_1^{-1})$$
and via the basis $c_2 b_1, c_2 b_2$ the form
$$G(x, y) = (c_2 b_1 x + c_2 b_2 y)(c_2' b_1' x + c_2' b_2' y) N(J)^{-1} N_{K/Q}(c_2^{-1}).$$
Since evidently $F = f_I$, $G = f_J$ and $G = MF$, we have $f_J = M f_I$, and so it suffices to show that det $M = 1$. Now,
$$N(c_2 J) d^{1/2} = c_2 b_1 c_2' b_2' - c_2' b_1' c_2 b_2 = N(c_1 I) d^{1/2} = c_1 a_1 c_1' a_2' - c_1' a_1' c_1 a_2$$
$$= (m_{11} c_2 b_1 + m_{12} c_2 b_2)(m_{21} c_2' b_1' + m_{22} c_2' b_2')$$
$$- (m_{11} c_2' b_1' + m_{12} c_2' b_2')(m_{21} c_2 b_1 + m_{22} c_2 b_2)$$
$$= \det M(c_2 b_1 c_2' b_2' - c_2' b_1' c_2 b_2),$$

and this shows that indeed det $M = 1$. Finally, we prove that equivalent forms correspond to ideals from the same class in $H^*(K)$. Let $F(x, y) = (a_1 x + a_2 y)(a_1' x + a_2' y) N(I)^{-1}$ and $G(x, y) = (b_1 x + b_2 y)(b_1' x + b_2' y) N(J)^{-1}$ be two forms associated with ideals $I = a_1 Z + a_2 Z$ and $J = b_1 Z + b_2 Z$, respectively, and assume that F and G are equivalent, i.e. for a certain $M = [m_{ij}]$ in $SL(Z, 2)$ we have
$$G(x, y) = F(m_{11} x + m_{12} y, m_{21} x + m_{22} y).$$
Comparing the coefficients, we see that either $b_1 b_2^{-1}$ or $(b_1 b_2^{-1})'$ equals $(a_1 m_{11} + a_2 m_{21})(a_1 m_{12} + a_2 m_{22})^{-1}$, since these numbers differ only in sign from the solutions of $G(1, X) = 0$.

Observe that the second case is impossible. Indeed, in this case we would have, for a suitable t in K, the equalities
$$a_1 m_{11} + a_2 m_{21} = t b_1', \quad a_1 m_{12} + a_2 m_{22} = t b_2',$$
and so
$$(b_1 b_2' - b_1' b_2) N_{K/Q}(t) = -\det M(a_1 a_2' - a_1' a_2) = -(a_1 a_2' - a_1' a_2),$$
showing that $N_{K/Q}(t)$ is negative. But on the other hand,
$$G(x, y) = (t b_1' x + t b_2' y)(t' b_1 x + t' b_2 y) N(I)^{-1}$$
$$= N(t) N(I)^{-1} G(x, y) N(J),$$
whence $N_{K/Q}(t) = N(I) N(J)^{-1}$ must be positive, a contradiction.

Hence the first case must hold and so for a suitable t in K we have
$$a_1 m_{11} + a_2 m_{21} = t b_1, \quad a_1 m_{12} + a_2 m_{22} = t b_2,$$
and the previous argument gives $N_{K/Q}(t) > 0$. Since $\det[m_{ij}] = 1$, we see that $t b_1 Z + t b_2 Z = I$; thus $tJ = I$ and so I and J lie in the same class in

$H^*(K)$, because t is either totally positive or totally negative, and the last case can be avoided by the change of the signs of a_1, a_2, which does not affect the argument. □

2. Using Theorem 8.6 and also results concerning the reduction of binary quadratic forms one can find the value of $h^*(K)$ and also obtain upper bounds for it. (See e.g. Narkiewicz [86], Ch.V.) However, using Corollary 2 to Theorem 8.4 one can obtain much better bounds in the case of real K, and this is what we now show. To simplify notation, from this point on we write $h(d)$ for $h(K)$ whenever K is a quadratic field of discriminant d.

Proposition 8.7. *For positive d we have $h(d) < d^{1/2}$.*

The proof will be based on a lemma of Loo-Keng Hua, which in this case improves Lemma 8.5:

Lemma 8.7. *If d is a positive discriminant, then for*

$$L_d(s) = L(s, \chi_d) \quad \text{with} \quad \chi_d(n) = \left(\frac{n}{d}\right)$$

we have

$$L_d(1) < 1 + \tfrac{1}{2}\log d.$$

Proof. Consider the function $S(n)$ defined on positive integers by

$$S(n) = \sum_{a=1}^{n}\sum_{m=1}^{a} \chi_d(m)$$

and put for convenience $S(-1) = S(0) = 0$. In view of the equality

$$S(n) - 2S(n-1) + S(n-2) = \chi_d(n) \quad (n = 1, 2, \ldots)$$

we may write

$$L_d(1) = \sum_{n=1}^{\infty} \chi_d(n) n^{-1} = 2 \sum_{n=1}^{\infty} S(n)/n(n+1)(n+2)$$

and split the last sum into two:

$$S_1 = \sum_{n=1}^{A-1} S(n)/n(n+1)(n+2),$$

$$S_2 = \sum_{n=A}^{\infty} S(n)/n(n+1)(n+2),$$

where $A = [d^{1/2}]+1$. Since $|S(n)|$ does not exceed $n(n+1)/2$, S_1 is in absolute value not larger than

$$\sum_{n=1}^{A-1} (n+2)^{-1} \leq \log(1+A) - \log 2 \leq \log d^{1/2}.$$

To evaluate the sum S_2 we have to prove the inequality

$$|S(j)| \leq j d^{1/2}/2 \quad \text{for } j \geq d^{1/2} \tag{8.13}$$

first. Clearly,

$$d^{1/2} S(j) = d^{1/2} \sum_{a=0}^{j} \sum_{n=1}^{a} \chi_d(n) = \tfrac{1}{2} d^{1/2} \sum_{a=0}^{j} \sum_{n=-a}^{a} \chi_d(n)$$

as $\chi_d(-n) = \chi_d(n)$. Theorem 6.3 combined with Proposition 6.9 (i) gives the equality

$$\chi_d(n) d^{1/2} = \sum_{x=1}^{d} \chi_d(x) \exp(2\pi i x n/d)$$

(which holds also for $(d, n) > 1$, both sides being 0); hence

$$d^{1/2} S(j) = \tfrac{1}{2} \sum_{a=0}^{j} \sum_{n=-a}^{a} \sum_{x=1}^{d} \chi_d(x) \exp(2\pi i x n/d)$$

$$= \tfrac{1}{2} \sum_{x=1}^{d} \chi_d(x) \sum_{a=0}^{j} \sum_{n=-a}^{a} \exp(2\pi i x n/d)$$

implying

$$|d^{1/2} S(j)| \leq \tfrac{1}{2} \sum_{x=1}^{d} \left| \sum_{a=0}^{j} \sum_{n=-a}^{a} \exp(2\pi i x n/d) \right|.$$

In view of the identity

$$\sum_{a=0}^{j} \sum_{n=-a}^{a} \exp(i\alpha n) = (\sin(\alpha(j+1)/2)/\sin(\alpha/2))^2$$

holding, for real $\alpha \neq 2\pi m$ ($m \in \mathbb{Z}$), we have

$$\sum_{x=1}^{d} \left| \sum_{a=0}^{j} \sum_{n=-a}^{a} \exp(2\pi i x n/d) \right|$$

$$= \sum_{x=1}^{d} (\sin(\pi x(j+1)/d)/\sin(\pi x/d))^2$$

$$= \sum_{x=1}^{d} (\sin(\pi x(j'+1)/d)/\sin(\pi x/d))^2$$

$$= \sum_{x=1}^{d} \sum_{a=0}^{j'} \sum_{n=-a}^{a} \exp(2\pi ixn/d)$$

$$= \sum_{a=0}^{j'} \sum_{n=-a}^{a} \sum_{x=1}^{d} \exp(2\pi ixn/d) = (j'+1)d - (j'+1)^2,$$

where j' denotes the least non-negative residue of $j \pmod{d}$, and this leads to

$$|S(j)| \le \tfrac{1}{2}(1+j')(d^{1/2}-(j'+1)d^{-1/2}). \tag{8.14}$$

Now, if $d^{1/2} \le j < d$, then $j' = j$ and $\tfrac{1}{2}(1+j)d^{1/2}-\tfrac{1}{2}(j+1)^2 d^{-1/2}$ does not exceed $jd^{1/2}/2$, which implies (8.13) in this case; and in the case of $j > d$ we obtain from (8.14) by trivial estimation the inequality

$$|S(j)| \le \tfrac{1}{2}(1+d)d^{1/2} \le jd^{1/2}/2,$$

which again gives (8.13). Finally,

$$|S(d)| = |S(d-1)| \le (d-1)d^{1/2}/2.$$

Inequality (8.13) being established, we can finish the proof. In fact, the inequality implies

$$|S_2| \le d^{1/2} \sum_{j=A}^{\infty} ((j+1)(j+2))^{-1} = d^{1/2}/(A+1) < 1,$$

and this shows

$$|S| \le |S_1| + |S_2| \le 1 + \log d^{1/2},$$

as required. □

Proof of Proposition 8.7. We use Corollary 2 to Theorem 8.4. The required evaluation of $L_d(1)$ is provided by the lemma just proved and we need only to establish a convenient estimation of $\log \varepsilon$ from below. For this purpose write the fundamental unit $\varepsilon > 1$ in the form

$$\varepsilon = \tfrac{1}{2}(T + Ud^{1/2})$$

with integral T, U and $T^2 - dU^2 = \pm 4$. It is clear that

$$T^2 = dU^2 \pm 4 \ge (d-4)U^2,$$

whence

$$T \ge U(d-4)^{1/2}$$

and

$$\varepsilon > \tfrac{1}{2}U(d^{1/2} + (d-4)^{1/2}) \ge (d-3)^{1/2},$$

i.e. $\log \varepsilon > \frac{1}{2}\log(d-3)$. This evaluation will suit our purpose. Indeed,

$$h(d) = d^{1/2}L_d(1)/(2\log \varepsilon) \leq d^{1/2}(1+\tfrac{1}{2}\log d)/(\log(d-3)) < d^{1/2}$$

for $d \geq 17$. But for $d < 17$ we have $h(d) = 1$, since otherwise by Lemma 3.3 there would exist a non-principal ideal of norm not exceeding $d^{1/2}/2$. However, if d is a discriminant < 17, then $d \leq 13$, so the only ideal in question is the unit ideal, obviously principal. □

3. A lower bound for $h(d)$ is provided for negative d by Theorem 8.5, which implies $h(d) > d^{1/2-\varepsilon}$ for every positive ε and sufficiently large $|d|$. No comparable result is known for positive d. In fact, it is not known whether the equality $h(d) = 1$ can hold infinitely often. We shall now give a lower bound valid for infinitely many positive discriminants, which is a particular case of the Ankeny–Brauer–Chowla theorem:

Theorem 8.7. *For every positive ε one can find infinitely many real quadratic fields $Q(d^{1/2})$ with $h(d)$ exceeding $d^{1/2-\varepsilon}$.*

Proof. By Theorem 8.5 we have in our case

$$h(d) > d^{1/2-\varepsilon}/\log E(d)$$

holding for every positive ε and sufficiently large d; $E(d)$ stands for the fundamental unit of $Q(d^{1/2})$, which is to be taken larger than 1. We have thus to estimate $\log E(d)$ from above and any evaluation of the form

$$\log E(d) = O(\log d)$$

holding for infinitely many fields will do.

We need an elementary lemma first:

Lemma 8.8. *For infinitely many n the number n^2+1 is square-free.*

Proof of the lemma. First observe that n^2+1 is never divisible by 4. For any odd prime p consider $A_p(T)$, the number of integers $n \leq T$ for which n^2+1 is divisible by p^2, and note that $A(T)$, the number of $n \leq T$ for which n^2+1 is square-free is at least equal to

$$T - \sum_{3 \leq p < T} A_p(T).$$

Obviously,

$$A_p(T) = \sum_{\substack{n \leq T \\ n^2 \equiv -1 (\bmod\, p^2)}} 1 \leq (1+[T/p^2]) \sum_{\substack{x (\bmod\, p^2) \\ x^2 \equiv -1 (\bmod\, p^2)}} 1 \leq 2+2T/p^2,$$

and so
$$A(T) \geqslant T - 2\sum_{p \leqslant T} 1 - 2T \sum_{3 \leqslant p \leqslant T} p^{-2} = (1 - 2\sum_{p \geqslant 3} p^{-2})T + o(T).$$
Further,
$$\sum_{p \geqslant 3} p^{-2} \leqslant \sum_{n=1}^{\infty} n^{-2} - 1 - 1/4 = \pi^2/6 - 5/4 \leqslant 5/12,$$
and so finally
$$A(T) \geqslant T/6 + o(T). \qquad \square$$

Now consider the field $K = Q((m^2+1)^{1/2})$ with m chosen in such a way that m^2+1 is square-free. Its discriminant d equals either m^2+1 or $4(m^2+1)$. The number $\eta = m + (m^2+1)^{1/2}$ is a unit in K, exceeding 1, and so $\eta = E^N$ with some natural N. It follows that
$$\log E = N^{-1} \log \eta \leqslant \log \eta = O(\log m) = O(\log d),$$
and this suffices to our purpose. $\qquad \square$

4. Not much is known about the structure of $H(K)$ or $H^*(K)$, in general. We now prove an old result, going back to C. F. Gauss [01], which determines he number of even invariants of $H^*(K)$ for quadratic K.

Let K be an arbitrary quadratic field. The factor-group $\mathfrak{G}(K) = H^*(K)/H^*(K)^2$ is called the *genus group of* K and cosets mod $H^*(K)^2$ in $H^*(K)$ are called the *genera*. Denote by $g(K)$ the cardinality of $\mathfrak{G}(K)$. Obviously all non-unit elements of $\mathfrak{G}(K)$ are of order 2; thus $g(K)$ is a power of 2. The structure of $G(K)$ is determined by the following theorem:

Theorem 8.8. *If the discriminant d of K has t distinct prime divisors, then the group of genera $\mathfrak{G}(K)$ is the product of $t-1$ copies of C_2.*

Proof. We may assume that $d \neq -3, -4$, since in those cases the theorem is trivially true. Write $d = \pm p_1 \ldots p_t$ with prime p_i and observe that by the law of decomposition in K (Theorem 4.15) $p_i R_K = \mathfrak{P}_i^2$ with suitable prime ideals $\mathfrak{P}_i$ for $i = 1, 2, \ldots, t$. Denote by X_i the class in $H^*(K)$ which contains $\mathfrak{P}_i$ and let $[X_i]$ be its image in $\mathfrak{G}(K)$ for $i = 1, 2, \ldots, t$. The ideal $\mathfrak{P}_1 \ldots \mathfrak{P}_t$ is principal, for it is generated by $d^{1/2}$. Therefore
$$[X_t] = [X_1]^{-1} \ldots [X_{t-1}]^{-1} = [X_1] \ldots [X_{t-1}]. \qquad (8.15)$$
The theorem will be proved if we show that every element of $\mathfrak{G}(K)$ can be written in a unique way in the form
$$[X_1]^{\varepsilon_1} \ldots [X_{t-1}]^{\varepsilon_{t-1}}, \quad \text{where } \varepsilon_i = 0, 1. \qquad (8.16)$$

Let I be an ideal in R_K such that I^2 lies in the principal class in $H^*(K)$ and denote by J its conjugate. Then I and J lie in the same class; hence $I = aJ$ with some totally positive a in K. Since $N(I) = N(J)$, we infer that $N_{K/Q}(a) = 1$. If a' denotes the element conjugate to a, we obtain the equality

$$a = (1+a)/(1+a'),$$

and so the ideal $(1+a)^{-1}I$ is equal to its conjugate. But every such ideal must be a product of prime ideals dividing the discriminant and a positive rational number. Thus

$$I = (1+a)r\mathfrak{P}_1^{c_1} \ldots \mathfrak{P}_t^{c_t}$$

with rational positive r and integral c_i. This shows that the image of I in $\mathfrak{G}(K)$ is of the form (8.16) because $r(1+a)$ is totally positive. But the reciprocal image of any element of $\mathfrak{G}(K)$ in the ideal group contains an ideal whose square is principal, and so every element of $\mathfrak{G}(K)$ has the form (8.16). It remains to show that the elements $[X_i]$ ($i = 1, 2, \ldots, t-1$) are independent, i.e. none of them is generated by the others. Assume the contrary. Renumbering, if necessary, the ideals $\mathfrak{P}_i$, we may assume that for a certain $s \leq t-1$ we have

$$[X_1] \ldots [X_s] = 1.$$

Then the product $\mathfrak{P}_1 \ldots \mathfrak{P}_s$ must be principal and must have a totally positive generator, say a. We have $N_{K/Q}(a) = p_1 \ldots p_s$, and, moreover, $a^2 R_K = p_1 \ldots p_s R_K$; thus

$$a^2 = \varepsilon p_1 \ldots p_s = \varepsilon a a',$$

with a suitable unit ε; and finally

$$a = \varepsilon a'.$$

Now we can get rid of the case of negative d. In fact, as $d \neq -3, -4$, the last equality implies $a = \pm a'$ and we shall see that both signs are impossible. If $a = a'$, then a lies in Q, its norm is a square and so $p_1 \ldots p_s$ must be a square, which is not the case. If $a = -a'$, then $T_{K/Q}(a) = 0$, and

$$0 = a^2 + N_{K/Q}(a) = a^2 + p_1 \ldots p_s;$$

thus $Q((-p_1 \ldots p_s)^{1/2}) = Q(a) = Q(d)^{1/2}$ because a is irrational and lies in $Q(d^{1/2})$; but this implies that $p_1 \ldots p_s$ is divisible by d, which also is absurd.

Hence, we are left with the case of positive discriminant. First assume that the fundamental unit ε_0 is of negative norm. In this case $\varepsilon = \varepsilon_0^{2n}$ with a suitable $n \in Z$ and

$$((a/\varepsilon_0 d^{1/2})^n)' = (a\varepsilon^{-1})/(\varepsilon_0^{-1}d^{1/2})^n = (a\varepsilon_0^{-2n})/(\varepsilon_0^{-1}d^{1/2})^n = a/(\varepsilon_0 d^{1/2})^n,$$

showing that the number $a/(\varepsilon_0 d^{1/2})^n$ is invariant under the Galois group of K/Q and so must be rational. We thus have

$$a = r(\varepsilon_0 d^{1/2})^n$$

with some rational r. Thus

$$p_1 \ldots p_s = N_{K/Q}(a) = r^2(-1)^n(-d)^n = r^2 d^n,$$

and so $p_1 \ldots p_n$ differs from d by a square rational factor or is a square itself, which is a contradiction.

Finally, let d be positive and $N_{K/Q}(\varepsilon_0) = 1$. In this case we have $\varepsilon = \varepsilon_0^n$ for a suitable integral n and the number $a/(1+\varepsilon_0)^n$ must be rational. Indeed,

$$(a/(1+\varepsilon_0)^n)' = a'/(1+\varepsilon_0^{-1})^n = a\varepsilon_0^{-n}/(1+\varepsilon_0^{-1})^n = a/(1+\varepsilon_0)^n.$$

We can thus write $a = r(1+\varepsilon_0)^n$ with a rational r and obtain $p_1 \ldots p_s = r^2 N_{K/Q}(1+\varepsilon_0)^n$. This implies that n has to be odd, since otherwise $p_1 \ldots p_s$ would be a square. Now consider the ideal $I = \mathfrak{P}_{s+1} \ldots \mathfrak{P}_t$. It is principal, because it is equal either to $d^{1/2} R_K/\mathfrak{P}_1 \ldots \mathfrak{P}_s$ or to $d^{1/2} R_K/2\mathfrak{P}_1 \ldots \mathfrak{P}_s$, according to the residue class of $d \pmod 4$. If b is the generator of I, then repeating our argument with b instead of a we get

$$p_{s+1} \ldots p_t = r_1^2 N_{K/Q}(1+\varepsilon_0)^m$$

with a suitable rational r_1 and an odd m. But this implies

$$p_1 \ldots p_t = (rr_1)^2 N_{K/Q}(1+\varepsilon_0)^{m+n}$$

and on the right-hand side we obtain a square, whereas the left-hand side is square-free and $\neq 1$, which is a contradiction. □

Corollary 1. *For a quadratic field K the group $H^*(K)$ has $t-1$ even invariants, t being the number of distinct prime divisors of the discriminant of K.*

Proof. If $H^*(K) = \prod_i C_{n_i}$ is its factorization in cyclic groups, then by the canonical map of $H^*(K)$ onto $\mathfrak{G}(K)$ every factor with an odd n_i trivializes, whereas every factor with an even n_i becomes a non-trivial cyclic factor of $\mathfrak{G}(K)$, and so the number of even n_i's equals the number of non-trivial cyclic factors of $\mathfrak{G}(K)$ (remember that for distinct i, j with even n_i, n_j the images of C_{n_i} and C_{n_j} cannot coincide). □

Corollary 2. *The narrow class-number $h^*(K)$ of a quadratic number field K is odd if and only if the discriminant of K is a prime-power, i.e. if $d = -4$, -8, $-p$ (p being a prime congruent to $3 \pmod 4$), 8, p (p being a prime congruent to unity $\pmod 4$).*

Proof. A trivial deduction from the preceding corollary. □

As an application of the theorem just proved we now present examples of quadratic fields in which the class group $H(K)$ is a direct factor of $H^*(K)$ and also of fields in which this is not the case.

First consider $K = Q((34)^{1/2})$. Here $d(K) = 2^3 \cdot 17$, $t = 2$; hence $H^*(K)$ has one even invariant. Moreover, the fundamental unit has a positive norm because the unit $35 + 6(34)^{1/2}$ has a positive norm but is not a square in K. Consequently $h^*(K) = 2h(K)$. By Lemma 3.4 in every ideal class of $H(K)$ there must be an ideal with norm not exceeding 5; a routine examination of the ideals involved leads us to $h(K) = 2$, whence $H(K) = C_2$. Since $h^*(K) = 2h(K) = 4$, we see that $H^*(K) = C_4$ and so $H(K)$ is not a direct factor of $H^*(K)$.

Now look at the field $K = Q((15)^{1/2})$. Here $d(K) = 2^2 \cdot 3 \cdot 5$, $t = 3$. Again the norm of the fundamental unit is positive, since the equation $x^2 - 15y^2 = -1$ has no solutions even (mod 3). Consequently $h^*(K) = 2h(K)$ and the usual inspection of ideals with small norms gives $h(K) = 2$; thus $H(K) = C_2$ and $H^*(K) = C_2 \times C_2$, and in this case $H(K)$ is a direct factor of $H^*(K)$.

We conclude this subsection with an asymptotic result concerning the number $g(d)$ of genera for a negative d.

Proposition 8.8. *If d tends to $-\infty$, then the quotient $g(d)/h(d)$ tends to zero.*

Proof. Assume that for a sequence d_k of discriminants we have $h(d_k) \leq Bg(d_k)$ with some fixed B. Corollary 1 to Theorem 8.5 implies $g(d_k) \to \infty$, and so by the last theorem the number t_k of distinct prime divisors of d_k tends to infinity with k. Observe that

$$\log |d_k| \geq \sum_{p | d_k} \log p \geq \sum_{p \leq p_{t_k}} \log p,$$

and by an elementary result of Chebyshev the last sum exceeds $C_1 p_{t_k} \geq C_2 t_k \log t_k$ for some positive C_1, C_2. Using again Corollary 1 to Theorem 8.5, we get for large k the inequality

$$h(d_k) \geq |d_k|^{1/4} \geq \exp(C_3 t_k \log t_k)$$

with a suitable positive C_3, and so finally we arrive at

$$\exp(C_3 t_k \log t_k) \leq Bg(d_k) = B \exp((t_k - 1) \log 2),$$

which is a clear contradiction for t_k large enough. □

Corollary. *There is no imaginary quadratic field whose class group is the product of N copies of C_2, provided N is sufficiently large.*

Proof. If we have $H(K) = H^*(K) = C_2^N$ for $K = Q(d^{1/2})$, then $g(K) = h(K)$ and by the proposition just proved there is only a finite number of such fields. □

5. It follows from Theorem 8.8 that for any power of 2 there exist infinitely many quadratic fields K, real and imaginary, with $h(K)$ divisible by it. We now present a result of T. Nagell [22], which asserts that the same holds for any positive integer, if we restrict attention to imaginary quadratic fields.

Theorem 8.9. *If n is an arbitrary positive rational integer, then there exist infinitely many imaginary quadratic fields with class-number $h(K)$ divisible by n.*

The proof will be based on the following auxiliary result:

Lemma 8.9. *Let D be a square-free positive rational integer and $q = p^a$ a prime-power, p being an odd prime. Assume that the equation*
$$x^2 + Dy^2 = z^q$$
has a solution x, y, z with $(x, y) = 1$, z odd and x divisible by p but not by p^2. Then for the field $K = Q((-D)^{1/2})$ we have $q|h(K)$.

Proof of the lemma. Our equation may be written in the form
$$JJ' = N(J) R_K = z^q R_K,$$
where J is the principal ideal generated by $x + y(-D)^{1/2}$. As our assumptions imply $(J, J') = 1$, it follows that J must be a qth power of an integral ideal, say I. Assume, contrary to our assertion, that $p^b \| h(K)$ and b is strictly less than a. Then $(p^a, h(K)) = p^b$, and so with suitable $A, B \in Z$ we obtain
$$Ap^a + Bh(K) = p^b,$$
showing that the ideal $I^{p^b} = I^{Ap^a} I^{Bh(K)}$ is principal. It follows that the number $x + y(-D)^{1/2}$ is a pth power of an integer in K. We now have to distinguish between two cases:

(a) $D \equiv 1$ or $2 \pmod{4}$

and

(b) $D \equiv 3 \pmod{4}$.

In the case (a) we may write
$$x + y(-D)^{1/2} = (u + v(-D)^{1/2})^p$$
with suitable $u, v \in Z$, which are relatively prime in view of $(x, y) = 1$. Comparing the rational parts on both sides of this equality we get
$$x = u^p - \binom{p}{2} u^{p-2} v^2 D + \ldots + \binom{p}{p-1} uv^{p-1}(-D)^{(p-1)/2};$$
thus
$$0 \equiv x \equiv u \pmod{p},$$
and this shows $u^p \equiv 0 \pmod{p^2}$ and $x \equiv 0 \pmod{p^2}$, giving a contradiction.

In the case (b) we proceed similarly. We may write

$$x+y(-D)^{1/2} = \left(u + \frac{v}{2} + \frac{v}{2}(-D)^{1/2}\right)^p$$

with suitable relatively prime rational integers u, v. Putting $t = 2u+v$ and proceeding as in the previous case we easily get

$$2^p x = t^p - \binom{p}{2}t^{p-2}v^2 D + \ldots + \binom{p}{p-1}tv^{p-1}(-D)^{(p-1)/2},$$

and from $p|x$ we infer that $p|t$. Hence $p^2|2^p x$ and $p^2|x$, which is a contradiction. □

Corollary. *Let D be a square-free positive rational integer and n an odd integer. Assume that the equation $x^2 + Dy^2 = z^n$ has a solution x, y, z with $(x, y) = 1$, z odd and (x, n^2) equal to n^*, the product of all distinct primes dividing n. Then the class-number of the field $Q((-D)^{1/2})$ is divisible by n.*

Proof. Immediate from the lemma. □

Proof of the theorem. Let n be a given odd number and let x be an integer chosen so that (n, x) equals n^*. Then the polynomial $z^n - x^2$ is irreducible over Q because x^2 is not a n_1th power, for any non-trivial divisor n_1 of n. As we saw in the proof of Theorem 4.14, this implies that the congruence $z^n \equiv x^2 \pmod{p}$ has solutions in z for infinitely many primes p. We can even obtain solutions with $z^n \not\equiv x^2 \pmod{p^2}$ replacing z, if necessary, by $z + p$. By the Chinese remainder-theorem we can now obtain, for any m, infinitely many sets $A = \{p_1, \ldots, p_m\}$ of m primes such that $p_i \| z_A^n - x^2 > 0$ $(i = 1, 2, \ldots, m)$, and p_i does not divide $z_A x$, $z_A \in Z$. Write $z_A^n - x^2 = Dy^2$ with a square-free D. Clearly the product $p_1 \ldots p_m$ divides D; thus by Theorem 8.8 the class-number of $K = Q((-D)^{1/2})$ is divisible by 2^{m-1}. To apply the corollary to the last lemma observe that we can always choose z_A to be odd (replacing z_A, if necessary, by $z_A + (p_1 \ldots p_m)^2$), and, moreover, that the condition $(x, y) = 1$ can also be fulfilled. Indeed, if q is a prime dividing (x, y) and $q \notin A$, then replacing z_A by $z_A + u(p_1 \ldots p_m)^2$ with suitable $u \in Z$, we can obtain $(z_A, q) = 1$. However, if $q \in A$, then $(z_A, q) = 1$ by construction. But q divides x, y and so has to divide z_A, which is contradiction. Thus we may indeed apply the last corollary to get the divisibility of $h(K)$ by n. We have obtained an infinity of imaginary quadratic fields with class-number divisible by $2^{m-1}n$, n being an arbitrarily given odd number and m being arbitrary. This proves the theorem. □

6. Theorem 8.8 implies that if a quadratic field has sufficiently many ramified primes, then its class-number is larger than any prescribed positive number.

An analogue of this result for arbitrary normal extensions of a fixed degree will be deduced from the following theorem, which gives a sufficient condition for the existence of subgroups of the form C_q^N (with prime q) in $H(K)$:

Theorem 8.10. *There exists a function $c(n)$ with the following property: for any prime q and any field K of degree n, the class-group $H(K)$ contains C_q^N as a subgroup, with*

$$N \geqslant t_q - c(n),$$

where $t_q = t_q(K)$ denotes the number of rational primes p such that the ramification indices of all prime ideals lying in R_K over p are divisible by q.

Moreover, if $q \neq 2$, $r(K)$ is the rank of the group of units of K, $\overline{K}$ is the field complex conjugated to K and for any field L we denote by $A_q(L)$ the degree of the composite of all pure fields $Q(a^{1/q})$ (with $a \in Q$) contained in L, then $N \geqslant t_q - \max\{c_1(K) : [K : Q] = n\}$, where

$$c_1(K) = \begin{cases} r(K) + \log A_q(K)/\log q & \text{if } \zeta_q \notin K, \\ r(K) + \log A_q((K\overline{K})^+)/\log q & \text{if } \zeta_q \in K. \end{cases}$$

Proof. The maximal subgroup of the form C_q^T of $H(K)$ obviously equals H_q/H_q^q, where H_q denotes the q-Sylow subgroup of $H(K)$. Let N be the maximal value of T. Denote by I_q the group of all non-zero elements of K which generate qth powers of fractional ideals and let $P = (K^*)^q$, $U = U(K)$ and $E = E(K)$. Then $H_q/H_q^q \simeq I_q/UP$ and the sequence

$$1 \to UP/EP \to I_q/EP \to I_q/UP \to 1$$

is exact. Since all its terms are vector spaces over $k = GF(q)$, we get

$$N = \dim_k I_q/UP = \dim_k I_q/EP - \dim_k UP/EP.$$

Theorem 3.6 shows that

$$\dim_k UP/EP = \dim_k(U/EU^q) = r(K)$$

and so it remains to obtain a good lower bound for $\dim_k I_q/EP$. Let $t = t_q$ and let $p_1, \ldots, p_t$ be rational primes whose all prime ideal divisors have their ramification indices divisible by q. These primes lie all in I_q and we consider the subspace X of I_q/EP generated by their images. If s is the dimension of X, then there are $M = t - s$ independent linear relations between our generators of X. This shows that with suitable $0 \leqslant x_{ij} \leqslant q-1$, $A_j \in K^*$ and $z_j \in E$ ($i = 1, 2, \ldots, t; j = 1, 2, \ldots, M$) we have

$$b_j = \prod_{i=1}^{t} p_i^{x_{ij}} = z_j A_j^q \quad (j = 1, 2, \ldots, M), \tag{8.17}$$

and $\operatorname{rank}_k [x_{ij}] = M$.

Observe that for any c and $M \geqslant M(c,n)$ at least $c+1$ numbers z_j are equal, say $z_1 = z_2 = \ldots = z_{c+1}$, and then $b_j/b_{1+c} = B_j^q$ holds for $j = 1, 2, \ldots, c$ with suitable $B_j \in K^*$. Thus K contains the field generated by $(b_j b_{1+c}^{-1})^{1/q}$ for $j = 1, 2, \ldots, c$. Further, observe that the numbers $b_j b_{1+c}^{-1}$ are q-independent in Q. Indeed, otherwise we would have (with suitable $0 \leqslant y_1, \ldots, y_j < q$, not all vanishing, and an $R \in Z$) the equality

$$\prod_{j=1}^{c} (b_j b_{1+c}^{-1})^{y_j} = R^q.$$

Thus

$$\prod_{j=1}^{t} p_i^{a_i} = R^q$$

with

$$a_i = \sum_{j=1}^{c} y_j(x_{ij} + x_{i,c+1}(q-1)).$$

However,

$$0 \equiv a_i \equiv \sum_{j=1}^{c} y_j(x_{ij} - x_{i,c+1}) \pmod{q};$$

and since $\mathrm{rank}_k[x_{ij} - x_{i,c+1}] = c$, all y_j's must vanish.

The Corollary to Lemma 7.15 now implies $q^c | n$; thus $c \leqslant \log n / \log q$ and hence M must be bounded. This gives the first assertion in view of the equality $N = t - M - r(K) = t + O(1)$, since for fields of fixed degree the rank of units is bounded.

If q is odd, the proof can be shortened and one obtains a much better upper bound for M proceeding as follows: taking the complex conjugate of both sides of (8.17) and multiplying pairwise we get

$$b_j^2 = B_j^q$$

with $B_j = A_j \bar{A}_j \in (K\bar{K})^+$ for $j = 1, 2, \ldots, M$. Since the b_j's are q-independent in Q and $q \neq 2$, so are the b_j^2's. Consequently the field $(K\bar{K})^+$ contains the composite of all fields $Q((b_j^2)^{1/q})$ which is of degree q^M and thus $q^M \leqslant A_q((K\bar{K})^+)$. If K does not contain ζ_q, all b_j's are qth powers in K and the same argument leads to $q^M \leqslant A_q(K)$. This establishes the last assertion. □

Corollary 1. *If q is an odd prime and K/Q is a pure extension of degree q, then $H(K)$ has at least $t_q - q$ invariants divisible by q.*

Proof. Here $A_q(K) = q$ and $\zeta_q \notin K$. Thus $c_1(K) = r(K) + 1 = q$. □

Corollary 2. *If K/Q is Abelian then $H(K)$ has at least $t_q - r(K)$ invariants divisible by q, provided q is an odd prime.*

Proof. In this case $A_q(K) = 1$. □

(Note that this corollary fails to hold for $q = 2$ in the case of imaginary buadratic fields, as shown in Theorem 8.8.)

Corollary 3. *If K/Q is cyclic of an odd prime degree q, then $H(K)$ has at least $\omega(d(K)) - q + 1$ invariants divisible by q, with $\omega(m)$ denoting the number of distinct prime divisors of m.*

Proof. In this case every ramified prime is counted by t_q. Thus $t_q = \omega(d(K))$ and $r(K) = q - 1$, so the assertion follows from the preceding corollary. □

Corollary 4. *If $K_1, K_2, \ldots$ is a sequence of normal extensions of Q of a given degree n, and $\omega(d(K_j))$ tends to infinity, then $\lim_j h(K_j) = \infty$.*

Proof. Since every ramified prime in K/Q is counted by $t_q(K_i)$ for a certain prime divisor q of n, the number

$$\max \{t_q(K_i) : q | n\}$$

tends to infinity, and we may apply the theorem. □

7. Finally we turn to imaginary quadratic fields with class-number one. We start with a curious result of Frobenius and Rabinowitsch, which displays a connection between these fields and the problem of representing primes by quadratic polynomials.

Proposition 8.9. *Let K be an imaginary quadratic field with discriminant $d \neq -3, -4, -8$. Then $h(K) = 1$ holds if and only if d is congruent to unity (mod 4) and for $x = 1, 2, \ldots, (1-d)/4$ the polynomial*

$$x^2 - x + (1-d)/4$$

assumes prime values only.

Proof. We start with a general lemma, which is often helpful in establishing $h(K) \neq 1$.

Lemma 8.10. *Let K be an algebraic number field with $h(K) = 1$. Assume that p is a rational prime such that for a certain a in R_K we have $p^s || N_{K/Q}(a)$. Then* there must be an irreducible element b in R_K with $|N_{K/Q}(b)| = p^t$ and $t \leq \max(s, n)$, where n is the degree of K/Q.

Proof of the lemma. Let b be any irreducible divisor of a with $N_{K/Q}(b)$ divisible by p. Since $h(K) = 1$, b generates a prime ideal. Hence $|N_{K/Q}(b)| = p^t$ for a certain t, and it is immediately seen that b must divide p; so we get

$$p^t | N_{K/Q}(b) | N_{K/Q}(p) = p^n,$$

and thus $t \leq n$. The inequality $t \leq s$ being obvious, our proof is complete. □

Now assume that K is imaginary quadratic, $h(K) = 1$ and $d \neq -3, -4, -8$. We first dispose of the possibility of $d \not\equiv 1 \pmod{4}$. In this case we would have $d = 4D$ with D square-free and congruent to 2 or 3 (mod 4); so either $|D| = N_{K/Q}(D^{1/2})$ or $1 + |D| = N_{K/Q}(1 + D^{1/2})$ would be divisible by 2 but not by 4, whence in view of the lemma there would be an irreducible element b with $N_{K/Q}(b) = 2$; but this is clearly impossible, since the equation $X^2 + DY^2 = 2$ cannot have integral solutions.

So d is congruent to unity (mod 4). Put $w = (1 + d^{1/2})/2$ and note that $1, w$ is an integral basis for our field. Obviously we have

$$N_{K/Q}(x + w) = x^2 - x + (1 - d)/4,$$

and for $x = 1, 2, \ldots, (1-d)/4$ we have $N_{K/Q}(x+w) < (1-d)^2/16$. Note that for x in this range the numbers $x + w$ are all irreducible, independently of the assumption that $h(K) = 1$. Indeed, from $x + w = ab$ we infer that $N_{K/Q}(a) \times N_{K/Q}(b) < (1-d)^2/16$, and so the norm of one of the factors, say b, does not exceed $(1-d)/4$. But if $r + sw$ is an arbitrary irrational element of R (i.e. $s \neq 0$), then

$$N_{K/Q}(r + sw) = r^2 - rs + (1-d)s^2/4 = (r - s/2)^2 + |d|s^2/4,$$

and this is larger than $(1-d)/4$. Hence b must be rational and it suffices to remark that $x + w$ cannot have rational integral divisors except ± 1, and so $x + w$ must be irreducible.

Now we use the assumption that $h(K) = 1$ to infer that the elements $x + w$ generate prime ideals, and so their norms are either primes or squares of primes. The last possibility could arise only if $x + w$ happened to be rational, which is impossible, and so $N_{K/Q}(x + w) = x^2 - x + (1-d)/4$ are rational primes for $x = 1, 2, \ldots, (1-d)/4$, as asserted.

The first part of our proposition is already proved. Now assume that $d \equiv 1 \pmod{4}$ (thus $|d| > 7$) and that for $x = 1, 2, \ldots, (1-d)/4$ the values of $x^2 - x + (1-d)/4 = f(x)$ are prime. Assume, moreover, that $h(K) \neq 1$ and choose a non-principal ideal I with the least possible norm. Obviously, it must be a prime ideal and its norm has to be a rational prime, say p, not exceeding $2|d|^{1/2}/\pi$ in view of Lemma 3.3.

As the index of $w = (1+d^{1/2})/2$ equals unity and $f(x)$ is its minimal polynomial, Theorem 4.12 shows that $pR_K = IJ$, $I = pR_K+(w-w_1)R_K$, $J = pR_K+(w-w_2)R_K$, where w_1, w_2 are roots of $f(x)$ (mod p), which may be taken from the interval $[0, p-1]$. Clearly $I = pR_K+(w+p-w_1)R_K$. Look at the number $w+p-w_1$. Its norm equals $(p-w_1)^2-(p-w_1)+(1-d)/4$ and since

$$0 < p-w_1 \leqslant p-1 \leqslant 2|d|^{1/2}/\pi < (1-d)/4$$

for $|d| \geqslant 7$, in view of our assumption this norm is a rational prime. But $w-w_1$ lies in I and hence its norm must be equal to p. Hence $w-w_1$ generates I, which is impossible, because I was assumed to be non-principal. □

8. We conclude this section with an upper bound for $|d(K)|$, where K is an imaginary quadratic field with class-number one. We follow here P. Bundschuh and A. Hock [69], who utilized Baker's method.

Theorem 8.11. *If K is an imaginary quadratic field with $h(K) = 1$, then $|d(K)| \leqslant \exp 6352 < 10^{2759}$.*

Proof. According to Corollary 2 to Theorem 8.8, if $|d(K)| \geqslant 10$ and $h(K) = 1$, then $d(K) = -p$, where p is a prime congruent to 3 (mod 4). We now show that p must be congruent to 3 (mod 8).

Lemma 8.11. *If $d = d(K) = -p < -10$ and $h(K) = 1$, then $p \equiv 3 \pmod 8$.*

Proof. Assume, to the contrary, that $p \equiv 7 \pmod 8$. Since $(p+1)/4$ is an even integer exceeding 2, we can write $(p+1)/4 = 2a$ with $a > 1$. Consider the quadratic form $f(x, y) = 2x^2+xy+ay^2$, which is of discriminant d. Since $h(K) = 1$, this form must be, according to Theorem 8.6, equivalent to every positive definite and primitive form of discriminant d, and so, in particular, it must be equivalent to x^2+2ay^2. Since the latter represents 1, the same must be true for f. However, this is not possible, because $f(x, 0) = 2x^2 \neq 1$ and for $y \neq 0$ we get

$$f(x, y) = 2(x+y/4)^2+py^2/8 \geqslant p/8 > 1. \qquad \square$$

The next lemma provides an analytic identity, on which the proof of the theorem rests. We use the following notation: $Q(x, y)$ is the quadratic form $x^2+xy+\frac{1}{4}(p-1)y^2$ of discriminant d, X denotes the primitive character induced by the unique non-principal character in $X(K)$ (as we have seen in Corollary 2 to Theorem 8.4, $X(x) = \left(\dfrac{d}{x}\right)$, and for prime $q \neq 2$ we denote by X_q the primitive real character (mod q). In what follows we assume that $|d|$ is sufficiently large, say, $|d| > 200$.

Lemma 8.12. *If $q \equiv 1 \pmod 4$ is a prime, then the following identity holds for $\operatorname{Re} s > 1$:*

$$\begin{aligned} L(s, X_q)&L(s, XX_q) \\ &= \zeta(2s)(1-q^{-2s}) \\ &\quad + (p/4)^{1/2-s}\pi^{1/2}q^{-1}\Gamma(s-1/2)\Gamma^{-1}(s)(q^{2-2s}-1)\zeta(2s-1) \\ &\quad + R_q(s), \end{aligned} \qquad (8.18)$$

where

$$R_q(s) = q^{-1}(p/4)^{1/2-s}\sum_{y=1}^{\infty} y^{1-2s} \sum_{m=0}^{q-1} X_q(Q(m,y)) \sum_{k\neq 0} T_k(m, y, s),$$

$$T_k(m, y, s) = \exp(2\pi i k(m+y/2)q^{-1}) J_2(ky, s)$$

and

$$J_2(N, s) = \int_{-\infty}^{\infty} \exp(-\pi i N p^{1/2} u/q)(1+u^2)^{-s} du.$$

Proof. We start with the identity

$$L(s, X_q)L(s, XX_q) = \sum_{m,n=1}^{\infty} X_q(mn)X(m)(mn)^{-s}$$

$$= \sum_{r=1}^{\infty} X_q(r) r^{-s} \sum_{d|r} X(d)$$

valid for $\operatorname{Re} s > 1$. To evaluate the inner sum we use Theorem 8.2, which implies $\zeta_K(s) = \zeta(s)L(s, X)$. Comparing the coefficients of the Dirichlet series on both sides of this equality we see that

$$F(r) = \sum_{d|r} X(d)$$

equals the number of integral ideals of K with norm r. By assumption, $h(K) = 1$; thus every ideal in K is principal and since K contains only two roots of unity, $F(r)$ equals half the number of integers of K with norm $r \neq 0$. By Theorem 2.7 the numbers 1 and $\vartheta = (1+d^{1/2})/2$ form an integral basis of K, and since

$$N_{K/Q}(x+y\vartheta) = Q(x, y),$$

we see finally that $F(r)$ equals half the number of representations of a positive integer r by the form $Q(x, y)$. (This is the only point of the proof, where the assumption $h(K) = 1$ is used.)

It follows that

$$L(s, X_q)L(s, XX_q)$$
$$= \tfrac{1}{2}\sum_{x\neq 0} X_q(Q(x,0))Q(x,0)^{-s} + \tfrac{1}{2}\sum_{y\neq 0} X_q(Q(x,y))Q(x,y)^{-s}.$$

The first summand equals

$$\tfrac{1}{2}\sum_{x\neq 0} X_q(x^2)x^{-2s} = \sum_{\substack{x\geq 1 \\ q\nmid x}} x^{-2s} = \zeta(2s)(1-q^{-2s});$$

hence it agrees with the first summand in (8.18). The second summand is more complicated and requires more powerful tools. Let us call this summand S and write it in the form

$$S = \sum_{y=1}^{\infty} \sum_{x=-\infty}^{\infty} X_q(Q(x,y))Q(x,y)^{-s}$$
$$= \sum_{y=1}^{\infty} \sum_{m=0}^{q-1} X_q(Q(m,y)) \sum_{x=-\infty}^{\infty} Q(qx+m,y)^{-s}$$

(applying the equality $Q(-x,-y) = Q(x,y)$). Now we use Poisson's formula (Appendix I, Theorem VIII in the case where G = additive group of the reals, $H = \mathbb{Z}$), obtaining

$$S = \sum_{y=1}^{\infty} \sum_{m=0}^{q-1} X_q(Q(m,y)) \sum_{k=-\infty}^{\infty} \int_{-\infty}^{\infty} Q(qt+my)^{-s}\exp(-2\pi ikt)\,dt$$
$$= q^{-1}(p/4)^{1/2-s} \sum_{y=1}^{\infty} y^{1-2s} \sum_{m=0}^{q-1} X_q(Q(m,y))G(m,y)$$

where

$$G(m,y) = \sum_{k=-\infty}^{\infty} \exp\left(\frac{2\pi ik}{q}\left(m+\frac{y}{2}\right)\right) J_2(ky,s),$$

after substituting $m+qt+y/2 = p^{1/2}yu/2$ in the inner integral and noting that $Q(x,y) = (x+y/2)^2 + py^2/4$.

In the equality thus obtained we put $z = ky$ and isolate the term corresponding to $z = 0$, which equals

$$S_0 = q^{-1}(p/4)^{1/2-s} \int_{-\infty}^{\infty} (1+u^2)^{-s}du \sum_{y=1}^{\infty} y^{1-2s} \sum_{m=0}^{q-1} X_q(Q(m,y))$$
$$= q^{-1}(p/4)^{1/2-s}\pi^{1/2}\Gamma(s-1/2)\Gamma^{-1}(s) \sum_{y=1}^{\infty} y^{1-2s} \sum_{m=0}^{q-1} X_q(Q(m,y)).$$

Since $q \equiv 1 \pmod 4$, the character X_q is even and Proposition 6.9 (i) and Theorem 6.3 imply
$$X_q(a) = \tau_a(X_q)/\tau(X_q) = \tau(X_q)q^{-1/2}.$$
Thus
$$\sum_{m=0}^{q-1} X_q(Q(m,y)) = q^{-1/2} \sum_{j=1}^{q-1} X_q(j) \sum_{m=0}^{q-1} \exp(2\pi i j Q(m,y)/q).$$
Since
$$Q(m,y) \equiv (m+My)^2 + pM^2 y^2 \pmod q$$
where $M = (q+1)/2$, and since Proposition 6.9 (i) and Theorem 6.3 imply
$$\sum_{n=0}^{q-1} \exp(2\pi i j n^2/q) = \left(\frac{j}{q}\right) q^{1/2} = X_q(j) q^{1/2},$$
we arrive at
$$\sum_{m=0}^{q-1} X_q(Q(m,y)) = q^{-1/2} \sum_{j=1}^{q-1} X_q(j) \exp(2\pi i j p M^2 y^2/q) \sum_{n=0}^{q-1} \exp(2\pi i j n^2/q)$$
$$= \sum_{j=1}^{q-1} \exp(2\pi i j p M^2 y^2/q) = \begin{cases} q-1 & \text{if } q \mid y, \\ 0 & \text{if } q \nmid y. \end{cases}$$
This gives
$$\sum_{y=1}^{\infty} y^{1-2s} \sum_{m=0}^{q-1} X_q(Q(m,y)) = \sum_{\substack{y > \\ q \mid y}} (q-1) y^{1-2s} - \sum_{\substack{y > \\ q \nmid y}} y^{1-2s}$$
$$= (q^{2-2s} - 1) \zeta(2s-1)$$
and finally we see that S_0 coincides with the second term on the right-hand side of (8.18). It is immediately seen that the remaining part of S equals $R_q(s)$, and so equality (8.18) follows. □

The formula for the product of two L-functions, which we have obtained, has the advantage over other possible formulas of that type that the remainder term $R_q(s)$ is very small at $s = 1$, and this will enable us to apply Baker's method. An evaluation of $R_q(1)$ is contained in the next lemma:

Lemma 8.13. *The series defining $R_q(s)$ converges for $s = 1$ to a function continuous at this point, and if $q = 5$ or 13 and $p \geqslant 200$, then*
$$|R_q(1)| \leqslant 14 p^{-1/2} \exp(-\pi p^{1/2} q^{-1}).$$

Proof. The first assertion follows from the estimate
$$|J_2(ky, s)| \leqslant C/k^2 y^2,$$

§ 3. Class-Number of Quadratic Fields

which holds for s in a neighbourhood of 1 and non-zero k with a suitable constant C, and which can be obtained from the definition of J_2 by iterated partial integration. To obtain the second assertion we have to compute $J_2(ky, 1)$. Write

$$A = -\pi k p^{1/2} y q^{-1} \quad \text{and} \quad f(z) = \exp(iAz)(1+z^2)^{-1}.$$

When $k > 0$, we integrate $f(z)$ over the upper half of the circle $|z| = R$, and if $k < 0$, we integrate over the lower half of that circle. Since the integral over the half-circle tends to zero when R goes to infinity, the residue theorem leads to

$$J_2(ky, 1) = \pi z^{|ky|}$$

with $z = \exp(-\pi p^{1/2} q^{-1}) < 1$.
This yields

$$|R_q(1)| \leq 2q^{-1} p^{-1/2} \sum_{y \geq 1} \sum_{m=0}^{q-1} \sum_{k \neq 0} |J_2(ky, 1)| \leq 2\pi p^{-1/2} \sum_{y \geq 1} \sum_{k \neq 0} z^{|ky|}$$

$$\leq 4\pi p^{-1/2} \sum_{y \geq 1} \sum_{k \geq 1} z^{ky} = 4\pi p^{-1/2} \sum_{y \geq 1} z^y (1-z^y)^{-1}$$

$$\leq 4\pi p^{-1/2} (1-z)^{-1} \sum_{y \geq 1} z^y = 4\pi p^{-1/2} z (1-z)^{-2}.$$

Using the inequalities $p \geq 200$ and $q \leq 13$ we get $(1-z)^{-2} \leq 1.07$, and in view of $4\pi(1.07) < 14$ our assertion follows. □

Letting in (8.18) s tend to 1, which is allowed by the lemma, we get

$$L(1, X_q) L(1, XX_q) = \frac{\pi^2}{6}(1-q^{-2}) + \frac{2\log q}{p^{1/2} q} + R_q(1).$$

Putting here first $q = 5$ and multiplying by $1 - 13^{-2}$ and then putting $q = 13$ and multiplying by $1 - 5^{-2}$, we obtain by subtraction

$$\tfrac{168}{169} L(1, X_5) L(1, XX_5) - \tfrac{24}{25} L(1, X_{13}) L(1, XX_{13}) + 2\pi p^{-1/2} (\tfrac{168}{845} \log 5 - \text{og } 13)$$

$$= \tfrac{168}{169} R_5(1) - \tfrac{24}{25} R_{13}(1).$$

The values of $L(1, X_q)$ are computed with use of Proposition 8.6, which gives

$$L(1, X_5) = (\log a) 5^{-1/2}, \quad L(1, X_{13}) = (\log b) 13^{-1/2}$$

with $a = (1 + 5^{1/2})/2$ and $b = (3 + 13^{1/2})/2$.
Corollary 2 to Theorem 8.4 shows that

$$L(1, XX_5) = h(K_1)/(5p)^{1/2}, \quad L(1, XX_{13}) = h(K_2)/(13p)^{1/2}$$

with $K_1 = Q((-5p)^{1/2})$, $K_2 = Q((-13p)^{1/2})$ and thus taking $c = 5^{1680}13^{-624}$ we obtain

$$840h(K_1)\log a - 312h(K_2)\log b + \log c = p^{1/2}\pi^{-1}(4200R_5(1) - 4056R_{13}(1)).$$

Corollary to Theorem 4.4 implies that for $p > 10^{11}$ we have

$$h(K_1) < \frac{5^{1/2}}{3}p^{1/2}\log(5p) < \tfrac{4}{3}p^{1/2}\log p$$

and similarly

$$h(K_2) < \tfrac{4}{3}p^{1/2}\log p.$$

Hence, applying Lemma 8.12 we get

$$|x_1 \log a + x_2 \log b + \log c| \leq 36791\exp(-\pi p^{1/2}/5) \leq \exp(11 - \pi p^{1/2}/5)$$

with $x_1, x_2 \in Z$, $|x_i| \leq 672p^{1/2}\log p$. Now, $Q(a, b, c)$ is of degree 4. Denoting by $H(u)$ the height of an element u, as defined in Appendix IV, we have $H(a) = 1$, $H(b) = 3$, $H(c) = \max(5^{1680}, 13^{624}) \leq \exp 2704$. Moreover, $x_1 \log a + x_2 \log b + \log c$ does not vanish, since otherwise we would have $a^{x_1}b^{x_2}c = 1$, implying $b^{x_2} \in Q(5^{1/2})$, which is impossible. We can thus apply Baker's theorem in the form given in Appendix IV, with $A_1 = A_2 = 4$, $A_3 = \exp 2704$, $C = 192^{600} < \exp 3155$, $M = 700p^{1/2}\log p < p$, $D = \log^2 4 \cdot 2704 < 5197$, $D' = \log^2 4 < 2$ to get

$$|x_1 \log a + x_2 \log b + \log c| \geq \exp(-e^{3166}\log p).$$

Hence

$$e^{3166}\log p \geq p^{1/2}/5 - 11 \geq 0.625 p^{1/2},$$

which implies $p \leq \exp 6352 < 10^{2759}$, as asserted. □

§ 4. Notes to Chapter 8

1. The results presented at the beginning of this chapter are specializations of the class-field theory to the case of the rational ground-field, in which case they are elementary consequences of the theorems of Kronecker–Weber and Chebotarev. Cf. Carlitz [33].

Corollary to Proposition 8.1 is essentially due to D. Hilbert [97].

An analogue of Theorem 8.2 is valid for arbitrary extensions L/K of an algebraic number field K, the quotient $\zeta_L(s)/\zeta_K(s)$ being in this case a product of suitable L-functions of Artin. If L/K is Abelian, then this quotient is the product of zeta-functions associated with characters of a suitable group $H_{\mathfrak{f}}^{*}(K)$. See Cassels, Fröhlich [67], Hasse [26c], Lang [70]. For a kind of converse result see Ankeny [52a], Iwasaki [52].

2. The formulas given in Proposition 8.4 are known as "the conductor-discriminant formula". An analogue of this proposition holds for all Abelian L/K. See Hasse [30d], [34], Tatuzawa [73c], Vassiliou [33]. For the non-Abelian case see Artin [31].

It seems that Theorem 8.3 and its Corollary 1 have never been stated explicitly, although all ingredients of the proof are contained in Kubota [56a], on which our approach is modelled. Lemma 8.2 is due to S. Delsarte [48] and the proof given by us is that of R. Wiegandt [59].

It was shown by G. Sarbasov [67] that in Corollary 2 to Theorem 8.3 the remainder term is $O(x^a)$, where a is any number larger than $1/(p-1) - 3/(p-1)(p+3)$, for $p \geq 5$. For $p = 3$ one can take here any $a > 0.25$ (Urazbaev [54a]). Asymptotics for the number of fields with Galois group C_p^N over Q and bounded discriminant was determined by Zhang Xianke [84c]. It can be also obtained from Corollary 1 to Theorem 8.3. For particular cases see Baily [80], [81], Zhang Xianke [82], [84b].

Several papers by B. M. Urazbaev and his collaborators dealt with analogous problems for various classes of Abelian fields. See Asenov [76], Bobrovskii [71a], [71b], [72], Bulenov [69], Hushvaktov [72], [75], [76], [77a], [77b], [77c], Karibaev [72a], [72b], Kenzhebaev [70], [71], [72], [79], Kurmanalin [66], Maikotov [69], Omarov [68], Ramazanov, Bobrovskii [72], Sarbasov [66], Sarbasov, Urazbaev [66], Urazbaev [54b], [55], [62], [64], [67], [69], [72], [77], [81].

A complete solution to the problem of asymptotics for the number of Abelian fields with given Galois group and bounded discriminant was recently given by S. Mäki [85].

K. Haberland [74] obtained asymptotics for the number of Abelian fields with given Galois group, with all prime factors of the discriminant bounded by x.

Similar questions for non-Abelian extensions are much harder to handle; answers have been settled only in few special cases. See Baily [80], Davenport, Heilbronn [69], Steckel [83].

3. Corollary 1 to Theorem 8.4 goes back to E. E. Kummer [50a], [61], [63]. Other special cases of Theorem 8.4 were given in Fuchs [66] and Fueter [17] and the general case occurs first in Beeger [19], [20] and Gut [29]. A thorough study of the class-number formula given in that theorem is accomplished in the important book of H. Hasse [52a].

There exist class-number formulas also for other classes of fields. For pure cubic fields such a formula was given by R. Dedekind [00]. For Abelian extensions of imaginary quadratic fields see Fueter [10], Hasse [51b], Meyer [57], Novikov [62], [67], Ramachandra [64], [69], Robert [73a], Schertz [77] and also Chapter XIII of Kubert, Lang [81]. E. Hecke [21b] obtained a formula for $h(L)/h(K)$ in the case when K is totally real and $L = K(a^{1/2})$ with a totally positive; he conjectured that there might exist an elementary formula for this quotient also in the case of totally negative a. This conjecture was confirmed by L. J. Goldstein [73b] (except certain particular cases) and T. Shintani [76b]. Cf. Goldstein, de la Torre [75], Reidemeister [22].

In the case of real Abelian K one can express $h(K)$ by indices of certain subgroups of $U(K)$. See Leopoldt [53b], Schertz [79]. These formulas are particularly simple for $K = K_p^+$; namely, $h(K_p^+) = [U(K_p):C(K_p)]$, where $C(K_p)$ is the group of cyclotomic units of K_p. (Kummer [50a], [50b], [51], cf. Washington [82].) No algebraic proof of this formula is known. See Iwasawa [76], Segal [68].

M. Deuring [69] utilized a method of Siegel, developed by K. G. Ramanathan [59], to express $h(K)$ for arbitrary K by Bessel functions. Modular forms were used in Kiselev [55a] to get a formula for $h(K)$ in the case of cubic fields K of negative discriminant.

It was shown by E. Landau [04] that the Dirichlet series of $\zeta_K(s)/\zeta(s)$ converges at $s = 1$ and this leads to a formula for the product $h(K)R(K)$.

4. The class-number formulas for quadratic fields (Corollary 2 to Theorem 8.4 and Corollary 1 to Proposition 8.6) as well as Proposition 8.6 are due to Dirichlet [38], [39] (whose result covers only the case of $4|d(K)$), and Kronecker [85]. Both of them used the language of quadratic forms. The old problem of finding an elementary proof was solved in the case of $d < 0$ by H. L. S. Orde [78], and in special cases earlier in Davis [76] and Venkov [28], [31]. An exposition of Orde's proof is given in Narkiewicz [86] (Ch. V). L. J. Mordell [18a] found another analytical proof of Dirichlet's formulas and E. Fogels [50], [52] gave a variant of the classical proofs, replacing infinite series by finite approximations. For other results on this subject the reader should consult the third volume of Dickson [19].

For further papers concerning formulas for the class-number of quadratic fields see Adibaev [72], Barkan [75], Bergström [44], Berndt [73b], Berndt, Evans [77], Bitimbaev [68], Bölling [79], Eichler [55], Goldstein, Razar [76],

Hasse [40], Hecke [25], [30], [39], Lerch [05], McQuillan [62], Mordell [60b], [64].

For short proofs of the fact that the sum appearing in Dirichlet's formulas does not vanish for negative d see Metsänkylä [77], Ullom [74b].

Analogues of Proposition 8.6 for zeta-functions of Hecke characters of finite order in a quadratic field were obtained by C. L. Siegel [61]. (Cf. Rideout [73].) For other fields see King [68], Shintani [77b].

5. Proposition 8.5 is valid for all cyclotomic fields K_n and leads to the factorization $h(K_n) = h_n^+ h_n^-$ with $h_n^+ = h(K_n^+)$. Kummer utilized $h^*(K)$ to obtain an analogous result for prime powers. For composite n the property $h^*(K_n^+)|h^*(K_n)$ may fail to hold, as can be seen e.g. on the example of $n = 100$. The number h_n^- is sometimes called the first factor of $h(K_n)$ and h_n^+ the second factor. Tables of h_n^- were prepared for all n with $\varphi(n) \leqslant 256$ by G. Schrutka v. Rechtenstamm [64] (cf. Washington [82]), for prime $p < 200$ by M. Newman [70] and for prime $p \leqslant 521$ by D. H. Lehmer, J. M. Masley [78]. Cf. D. H. Lehmer [77], Metsänkylä [69b]. Approximate values for $p \leqslant 1097$ were given by S. Pajunen [76]. The structure of $H(K_m)$ for all m with $h(K_m) < 10^4$ and of $H(K_p)/H(K_p^+)$ for primes $p \leqslant 227$ with certain exceptions was determined by K. Tateyama [82a]. Cf. Gerth [80].

Corollary 2 to Proposition 8.6 is due to T. Lepistö [69] and T. Metsänkylä [72] and the proof given by us to Metsänkylä [74]. Cf. Carlitz [61], Masley [78b]. For an improvement see Feng Keqin [82b].

Bounds for $h_{p^n}^-$ were given in Metsänkylä [67a] and Feng Keqin [82b].

K. Iwasawa [62] interpreted h_n^- for prime-power n in terms of indices of subgroups in the group ring $Z[C_p]$. For another proof see Skula [81].

An expression for h_p^- by an elementary determinant was given in Turnbull [41] and Carlitz, Olson [55]. Cf. Kühnova [79], Masley [78b], Metsänkylä [67a], [84], Tateyama [82b], Wang [84]. Another elementary determinant giving h_p^- was given by V. M. Galkin [72].

6. A congruence relating $h_p^- \pmod{p^n}$ to Bernoulli numbers was given by H. S. Vandiver [18] and a simpler proof was presented by I. S. Slavutskii [69b]. (Cf. Hasse [66], Inkeri [55].) We recall that the sequence B_n of Bernoulli numbers is defined by (3.6).

Tables of Bernoulli numbers for $n \leqslant 2 \cdot 62$ can be found in Washington [82]. For $2 \cdot 63 \leqslant n \leqslant 2 \cdot 92$ such tables were prepared by Serebrennikov (see Davis [35]) and for $2 \cdot 91 \leqslant n \leqslant 2 \cdot 110$ by D. H. Lehmer [36].

For generalizations of Vandiver's congruence see Shiratani [71], Slavutskii [72a], [72b], Uehara [75].

L. Carlitz [68a] determined explicitly an integer $g(p)$ for odd primes p such that

$$g(p)h_p^+ \equiv h_p^- \pmod{p}.$$

This implies, in particular, that $p|h_p^+$ yields $p|h_p^-$, a result of Kummer [50a]. A new proof was given by T. Metsänkylä [70a], [73], who also generalized Carlitz's congruence to subfields of K_p and regained certain congruences for class-numbers of quadratic fields obtained earlier in Ankeny, Artin, Chowla [52] and Kiselev [48]. For these and other congruences for the class-number of quadratic fields see Carlitz [53a], [53b], [55], P. Chowla [68], P. Chowla, S. Chowla [68], S. Chowla [60a], Cohn, Cooke [76], Gut, Stünzi [66], Hayashi [77], Hurwitz [95d], Kaplan [77a], [81], Kaplan, Williams [82a], [82b], Kimura [79b], Kiselev [55b], [59], Kiselev, Slavutskii [59], [62], [64], Lang, Schertz [76], Lerch [05], Pizer [76], Pumplün [65], [68], Rédei [28], Schertz [73], Slavutskii [60], [61], [66], K. S. Williams [79], [81a], [81b], [81c], [82], Williams, Currie [82].

Similar congruences for class-number of other types of fields have been considered in Carlitz [54a], Kudo [75b], Slavutskii [72a].

7. It was conjectured by Kummer that h_p^- equals asymptotically $L(p) = 2p(p^{1/2}/2\pi)^{(p-1)/2}$ and it was shown by N. C. Ankeny and S. Chowla [49], [51] that

$$\log h_p^- = \log L(p) + o(\log p).$$

Cf. Hyyrö [67], Lepistö [63], [66], [68a], [74], Masley, Montgomery [76], Metsänkylä [67a], [70a], Pajunen [76], Siegel [64], Tatuzawa [53].

If q runs over all prime powers, then, as shown in Goldstein [73c],

$$\log h_q^- = (0.25 + o(1))(1 - p^{-1})q \log q$$

(with q being a power of p).

The result of Ankeny and Chowla implies, in particular, that $h_p^- = 1$ can hold only for a finite number of primes p. K. Uchida [71] proved that this implies $p < 2400$; J. M. Masley and H. L. Montgomery [76] showed that this holds iff $p \leq 19$; they also determined all integers $m \not\equiv 2 \pmod{4}$ with $h_m^- = 1$. (There are 29 of them, the largest being $m = 84$.) Moreover, they proved that $h_p^- > 10^{20}$ for $p \geq 137$ and T. Lepistö [74] obtained for $p > 229$ the lower bound $h_p^- > 31 \cdot 10^{45}$. Cf. Hoffstein [79b].

The ratio $k(p^n) = h_{p^n}^-/h_{p^{n-1}}^-$ is always an integer, as shown by J. Westlund [03]. T. Lepistö [66], [67] proved that for large fixed p, $k(p^n)$ increases with n.

For other results on $k(p^n)$ see Metsänkylä [69a], [72], Morishima [33], [34], Pollaczek [24], Shiratani [67].

8. Primes p with $p \nmid h_p^-$ are called *regular* and the remaining primes are called *irregular*. Kummer [50a] proved that p is irregular iff p divides the numerator of a non-zero Bernoulli number with index $\leq p-3$. Cf. Kronecker [56b].

For similar results concerning other classes of fields see Adachi [73], K. S. Brown [74], Coates, Wiles [77], Greenberg [73a], Kudo [75a], Novikov [69], Ribet [76], Robert [74], [78], Yager [82].

A similar criterion for $p | h_p^-$, $p^2 \nmid h_p^-$ was given by Kummer [57]. Cf. Vandiver [20].

A table of subfields K of K_p ($p < 125\,000$) with $h(K)/h(K^+)$ divisible by p was prepared by K. Selucký and L. Skula [81].

K. A. Ribet [76] strengthened Kummer's criterion, relating the divisibility of Bernoulli numbers by p to the action of the Galois group on the Sylow p-subgroup A of $H(K_p)$. There is a canonical decomposition of A into direct summands A_χ, each of them associated to a character $\chi \pmod{p}$. Let X be the unique character (mod p) which satisfies for all $a \not\equiv 0 \pmod{p}$ the congruence $X(a) \equiv a \pmod{pR_K}$, with $K = K_{p-1}$. Ribet's theorem states that for even k with $2 \leq k \leq p-3$ the conditions $p | n(B_k)$ (with $n(x)$ being the numerator of $x \in Q$) and $M_{X^{1-k}} \neq 0$ are equivalent. The easier part of this equivalence was proved by J. Herbrand [32a]. Another proof of Ribet's theorem was found by V. Snaith [82]. It follows from the results of B. Mazur and A. Wiles [84] that one can express the maximal power of p dividing $|M_{X^k}|$ by generalized Bernoulli numbers. Cf. Coates [81].

For previous work on this topic see Wiles [80], S. Yamamoto [72].

The number $d(p)$ of Bernoulli numbers B_{2k} ($2 \leq 2k \leq p-3$) with $p | n(B_{2k})$ is called the *irregularity index* of p. It follows from Ribet's theorem that $H(K_p)$ contains $C_p^{d(p)}$.

The estimate $d(p) \leq p/4$ holds true and the value of $d(p)$ can be found by solving a certain system of congruences (Skula [80]). If $d(p) < [p^{1/2}] - 1$, then Fermat's Last Theorem is true for the exponent p in the first case (Skula [77], Uehara [78]). For $d(p) = 0$ this was shown by Kummer [50c]. Cf. Eichler [65], Fueter [22], Hecke [10], Hilbert [97], Krasner [34], Kummer [53], [57], Vandiver [19], [26], [34b].

The largest known index of irregularity equals 5 and is attained by the primes $p = 78\,233$ and $p = 94\,693$ (Wagstaff [78]).

It is not known whether there are infinitely many regular primes. On the other hand, the number of irregular primes, even of special form, is infinite. This was first established by K. L. Jensen [15]. Cf. Carlitz [54b], Gandhi,

Gadia [78], Metsänkylä [71a], [76], Montgomery [65], Siegel [64], Slavutsky [63], Yokoi [75].

One can also consider a generalization of regular primes using in place of Bernoulli numbers their analogues, associated with Dirichlet characters. The sets of primes thus obtained have also connections with the divisibility of class-numbers of cyclotomic fields. See Carlitz [54b], Ernvall [75], [79], Ernvall, Metsänkylä [78], Gut [51a], Kleboth [55], Slavutskii [72a].

Divisibility of h_p^- by primes $\neq p$ was considered in Metsänkylä [67b], [68a], [71b].

9. A prime p is called *pseudo-regular*, if $H(K_p)$ contains C_p, but not C_p^2. L. Skula [75] proved that for such primes we have $p \nmid h_p^+$; earlier ([72]) he showed that for pseudo-regular primes the first case of Fermat's Last Theorem is true. Yet earlier H. S. Vandiver [34b] had asserted that for this the condition $p \nmid h_p^+$ is sufficient, but his proof seems to be incomplete (see Ribenboim [79], p. 188). Cf. Dénes [52b], Vandiver [29c].

No prime satisfying $p|h_p^+$ is known, and a conjecture of Vandiver says that this never happens. Consequences of this conjecture are discussed in Washington [82] (Ch.X). Cf. Vandiver [39a], [39b], [41]. S. Wagstaff [78] showed that it is true for all primes $p < 125\,000$. Vandiver's conjecture is equivalent to the statement that for all real Abelian K with $f(K) = p$ one has $p \nmid h(K)$. This was shown to be true for cyclic fields of degrees 2, 3, 4 and 6 in C. Moser [81] and Moser, Payan [81]. The conjecture that $h_p^+ < p$ always holds was destroyed first by G. Cornell, L. Washington [85] under the assumption of the Extended Riemann Hypothesis and then by E. Seah, L. Washington, K. S. Williams [83] with the example of $p = 11\,290\,018\,777$.

It was conjectured that $h_{2^n}^+ = 1$ for all n. See Cohn [60a] for numerical results on this question.

For other results concerning h_n^+ see Ankeny, Chowla, Hasse [65], Dénes [55], Gerth [83a], Hasse [55], S. D. Lang [77], Metsänkylä [69a], Mirimanoff [91], Morishima [66], Vandiver [29b], Yamaguchi [71], [75].

10. The Iwasawa theory of Z_p-extensions, developed by K. Iwasawa in [58], [59a], [59b], [59c], [59d], [73a], brought new life into the theory of cyclotomic fields. For its exposition the reader is referred to Washington [82] (Ch.XIII) and here we point out certain of its highlights. Let K be an algebraic number field. L/K is called a Z_p-*extension* (with prime p), if L/K is an infinite algebraic extension with $\text{Gal}(L/K)$ isomorphic to the additive group of p-adic integers (as a topologic group with Krull topology). L/K is called a *cyclotomic* Z_p-

extension, if L is the fixed field of the torsion subgroup of $\text{Gal}(M/K)$, where M is obtained by adjoining to K all p^nth roots of unity ($n = 1, 2, \ldots$). If $K = K_p$ ($p \neq 2$), then its cyclotomic Z_p-extension coincides with the union of K_{p^n} ($n \geq 1$).

Iwasawa proved that if L/K is a Z_p-extension, L_n is the unique subfield of L of degree p^n over K, and p^{e_n} is the exact power of p dividing $h(L_n)$, then

$$e_n = \lambda n + \mu p^n + \nu \qquad (8.19)$$

holds for sufficiently large n, with certain constants λ, μ, ν depending on L/K.

These constants are still subject to intensive investigations and their behaviour is not yet completely understood. It was shown by B. Ferrero and L. C. Washington [79] that if K/Q is Abelian and L/K is a cyclotomic Z_p-extension, then μ vanishes. Another proof can be found in Sinnott [84]. For earlier results see Candiotti [74], Ferrero [78], [80], Iwasawa [72b], Iwasawa, Sims [66], Gold [72], [76b], Johnson [73], [75]. V. A. Babaitsev [80], [81] showed that for fixed K the coefficient μ is bounded by a constant, not depending on p or L. Cf. Gerth [79b], Greenberg [73b]. On the other hand, Iwasawa [73b] proved that one can find cyclotomic Z_p-extensions with μ arbitrary large.

For other results on Iwasawa coefficients see Bloom [79], Candiotti [74], Carroll, Kisilevsky [81], Cuoco [80], Ferrero [77], [78], [80], Gerth [79a], Gillard [76], Gold [74a], [76b], Greenberg [75], [76], [78a], Iwasawa [81], Jehne [59], Kida [80], [82], Metsänkylä [74], [75a], [75b], [78b], [83], Shiratani [64c], Washington [76c].

L. C. Washington [75], [78b] considered divisibility of $h(L_n)$ in a cyclotomic Z_p-extension by primes $q \neq p$ and showed that for sufficiently large n the maximal power of q dividing $h(L_n)$ remains constant. This need not be the case for non-cyclotomic Z_p-extensions.

Analogues of (8.19) have also been obtained for composites of Z_p-extensions, either for fixed p (Cuoco [82], Cuoco, Monsky [81], Monsky [83]) or for variable p (Friedman [82a], [82b]).

For other questions concerning Iwasawa's theory see Babaitsev [76], Bloom, Gerth [81], Carroll [75], Carroll, Kisilevsky [76], Coates [73], Federer, Gross [81], Gerth [77c], Gillard [79c], Gold [74b], G. Gras [83b], Greenberg [78b], [78c], Iwasawa [65], [66b], [83], Jaulent [81d], Kramer, Candiotti [78], Kudo [78], Kuroda [68], [71], Kuzmin [72], Miki [78a].

11. Theorem 8.5 was proved by C. L. Siegel [35] for quadratic fields and by R. Brauer [47a] for all extensions of a fixed degree (not necessarily Abelian or normal). In the second part of his paper Brauer obtained the same assertion for any sequence $\{K_n\}$ of fields, satisfying $[K_n:Q] = o(\log|d(K_n)|)$. A simple proof of the Siegel–Brauer theorem was given in Pintz [74b]. For other proofs of Siegel's theorem see Chowla [50], Chudakov [42], Estermann [48], Goldfeld [74], Heilbronn [38b], Knapowski [68], Linnik [43], [50], Pintz [74a], [76c], [77c], Ramachandra [80], Rodosskii [56], Tatuzawa [51].

As observed by A. Walfisz [36], Siegel's theorem is equivalent to the non-vanishing of $L(s, X)$ for a real character X of conductor D in the interval $(1 - c(\varepsilon)D^{-\varepsilon}, 1)$. Unfortunately, all known proofs of the Siegel–Brauer theorem are ineffective, and so is the constant $c(\varepsilon)$ for $\varepsilon < 1/2$. In the case of $\varepsilon = 1/2$ one can get effective results, as shown in Goldfeld, Schinzel [75], Haneke [73], Pintz [76b], [77c]. (Cf. also Davenport [66], Hoffstein [80], Page [35], Pintz [76c], [77b], Ramachandra [75], Tatuzawa [51].) H. M. Stark [74] proved that the Siegel–Brauer theorem can be effectivized for a large class of fields, including all fields of a bounded degree which do not have a quadratic subfield. Cf. Stark [75c].

The error term in Theorem 8.5 (and more generally, in the Siegel–Brauer theorem) can be improved by taking into account the possible real zeros of $\zeta_K(s)$. This was proved by A. I. Vinogradov [62], [63a], who showed that if c_K denotes the largest real zero of $\zeta_K(s)$, then for all fields K of a fixed degree we have

$$\log h(K) R(K) = \tfrac{1}{2}\log|d(K)| + \log(1 - c_K) + O(\log\log|d(K)|).$$

Cf. Levin, Tulyaganova [66], Vinogradov [63b]. For more recent results on c_K see Goldfeld [75], Goldfeld, Schinzel [75].

In the Abelian case Theorem 8.5 was made more precise by T. Lepistö [70], who earlier ([68b], [68c]) treated cyclotomic fields.

An analogue of Theorem 8.5 for function fields was obtained in Gogia, Luthar [78b].

12. Corollary 1 to Theorem 8.5 shows that for imaginary quadratic K, the class-number $h(K)$ tends to infinity with $|d(K)|$. This was conjectured essentially by C. F. Gauss [01] and the first step towards it was made by E. Hecke (see Landau [18c]), who deduced it from the Extended Riemann Hypothesis. (Essentially the same result is contained implicitly in Gronwall [13].) Cf. Grosswald [66], Landau [19b], [27b], Mahler [34], Pintz [76a], [77c]. Next M. Deuring [33] deduced Corollary 2 to Theorem 8.5 from the falsehood of Riemann Hypothesis and under the same assumption L. J. Mordell [34] deduced $h(K) \to \infty$. S. Chowla [34a] and E. Landau [18b] showed that if there are infinitely many quadratic imaginary fields with a given class-number, then they must be very rare, and finally H. Heilbronn [34] established the limit relation $h(K) \to \infty$. Cf. Chowla [34d].

13. All imaginary quadratic fields with even discriminant and class-number 1 were determined by E. Landau [03a] and another proof was given by M. Lerch [03]. The first important step towards the determination of such fields in the general case was done by H. Heilbronn and E. Linfoot [34], who proved that apart of those listed in Proposition 4.19 there can be at most one such field. (See Ayoub [67b] for another proof.) This follows also from a more recent result of T. Tatuzawa [51], which states that if $|d(K)|$ exceeds $2100k^2\log^2(3k)$, then $h(K) > k$ with at most one exception. (Cf. Landau [36], Pintz [77c], Ramachandra [75].) Evaluations of the possible tenth discriminant (Dickson [10], D. H. Lehmer [33b], Stark [66]) have strengthened the belief that there is no such field, especially in view of the fact that this is implied by ERH. Cf. also Bateman, Grosswald [62], Chowla, Erdös [51], Gelfond, Linnik [48], Grimm [32], Grosswald [66], Low [68], Rosser [49], Selberg, Chowla [49], [67], Turán [59].

The expected proof of the non-existence of the tenth discriminant was found by A. Baker [66] and H. M. Stark [67a], [67b]. Baker's paper indicated only the method, which was later succesfully applied by P. Bundschuh and A. Hock [69]. We adopted their approach in the proof of Theorem 8.11.

One should note that an earlier proof by K. Heegner [52] was known; however, it was regarded as erroneous until M. Deuring [68] and H. M. Stark [69a] provided the needed clarifications. Cf. Birch [69b], Meyer [70], Schertz [76].

For other proofs see Chowla [70a], Cohn [81b], Chudakov [69], Feldman, Chudakov [72], Siegel [68b].

In Stark's proof zeta-functions and L-functions attached to quadratic forms appear. Their relevance to this problem was demonstrated already in Davenport, Heilbronn [36]. On this topic see Callahan, Smith [76], Chowla [67a], [67b], Deuring [37], Epstein [03], Kenku [75], Kitaoka [71a], Mordell [29], Rosen [80], Schertz [73], Selberg, Chowla [49], [67], Stark [68], [71b], Udrescu [69].

For expositions of the class-number one problem see Baker [71a], Stark [69b], [69c], [69d], [71a], [71d], [73].

14. An effective way of finding all quadratic imaginary fields with $h = 2$ was established by A. Baker [71b] and H. M. Stark [71c]. In the second part of Stark's paper it is shown that for such fields K one has $|d(K)| < 10^{1030}$; fields in that range have been dealt with in Montgomery, Weinberger [74] (using the approach of Weinberger [75]) and Stark [75b]. There are 18 such fields, all satisfying $|d(K)| \leqslant 427$. The case of even discriminants was settled already by Weinberger [69]. (See also Baker [69], Ellison et al. [71], Kenku

[70]].) For an approach utilizing Heegner's method see Abrashkin [74], Antoniadis [83], Meyer [75]. For previous work on the $h = 2$ problem see Goldstein [72d], Iseki [51a], [51b], [52].

The list of all 16 known imaginary quadratic fields with $h = 3$ and of 54 such fields with $h = 4$ is given in Buell [77] (cf. Masley [79]) and it is believed that it is complete. The more general question of finding an effective method of determining all imaginary quadratic fields with a given class-number was reduced by D. M. Goldfeld [76], [77] to a problem in the theory of elliptic curves, which in turn was solved by B. Gross and D. Zagier [83], [86], who proved the evaluation

$$h(K) \geqslant C(\varepsilon)\log^{1-\varepsilon}|d(K)|$$

for every positive ε with an effective constant $C(\varepsilon)$.

Various conjectures leading to an effective lower bound for $h(K)$ have been considered in Chowla [34e], [66], Chowla, Friedlander [76], Chowla, Hartung [74b], Chowla, de Leon [74], Chowla, de Leon, Hartung [73], Friedlander [78].

15. Corollary 4 to Theorem 8.5 shows that there can be only finitely many complex Abelian fields of given degree and class-number. A much stronger result was proved by K. Uchida [71], who removed the restriction on the degree. It follows, in particular, that there can be only finitely many cyclotomic fields with a given class-number. All fields K_m with $m \not\equiv 2 \pmod 4$ and $h(K_m) = 1$ have been found by J. M. Masley, H. L. Montgomery [76]. (A proof is given in Washington [82], Ch.XI.) Cf. Hoffstein [79b], Lepistö [74], Masley [75b], [76], [77], [78a], [79], Metsänkylä [70b].

A list of 73 totally complex fields with Galois group of the form C_2^N and class-number one was given by K. Uchida [72], who showed that there can be only one more. Later it was shown that the exceptional field does not exist (Brown, Parry [74]). In all cases one has $N \leqslant 3$. For the case $N = 2$ see also Buell, Williams, Williams [77] and Goldstein [71b]. The quoted paper of Uchida contains also bounds for the conductor of imaginary Abelian fields with $h = 1$ which either are of degree 6 or are cyclic quartic. Using this bound C. B. Setzer [80a] determined all such cyclic quartic fields. Thus all Abelian complex quartic fields with $h = 1$ are known.

An effective bound for the discriminants of CM-fields with given class-number and given maximal real subfield was given by J. S. Sunley [72a], [73] (with at most one exception) and L. J. Goldstein [72b], [72c]. The same problem with $h(K)$ replaced by $h(K)/h(K^+)$ was solved in Odlyzko [75] and Uchida [73]. Cf. Friedlander [76], Mallik [79], Uchida [71].

It seems that for fields of large degree one cannot have $h = 1$. This is true at least for certain classes of fields, as we have seen above. Cf. also Hoffstein [79b]. J. Martinet [79b] found two fields of degree 116 with $h = 1$ and showed that under ERH there are such fields of degree 480.

16. Theorem 8.6 is classical (see Dedekind [71]). The proof given by us follows that of E. Hecke [23]. For other proofs see Jones [49], König [13], Mitchell [26]. (Cf. also Butts, Pall [68], Lubelski [36b].) Essentially the same argument leads to a correspondence between classes of binary quadratic forms of arbitrary discriminant and classes of ideals of an order in a suitable quadratic field. (A subring of an algebraic number field K is called an *order*, if it contains 1 and spans K as a linear space over Q. Every order in K must be contained in R_K and this explains the name "maximal order" used in older literature for the ring of integers of K. Many results concerning R_K can be carried through to orders, e.g. the analogue of Dirichlet's unit theorem or the finiteness of the class-number. Orders are very useful in the study of representations of rational integers by forms and a good exposition of their properties and applications can be found in Borevich, Shafarevich [64].)

The literature on the class-number of binary quadratic forms is quite formidable and has been reviewed, up to 1922, in the third volume of Dickson [19] by G. H. Cresse. From newer papers on this subject, which are not directly connected with quadratic fields, we quote only Butts, Estes [68], Kaplansky [68], Kneser [82], Shyr [79] and Towber [80] with generalizations of the classical theory, as well as Leonard, Williams [73], [74] with interesting applications. A generalization of Theorem 8.6 to forms over certain *CM*-fields was given in Cohn [83].

17. Proposition 8.7 appears in Slavutskii [65a], [65b], and Lemma 8.7, on which its proof rests, is due to Loo-Keng Hua [42]. (Cf. Kanemitsu [77].) From prime d the bound in Lemma 8.7 can be improved. In fact, for sufficiently large primes p one has

$$|L_p(1)| \leqslant 0.19674 \log p,$$

as shown by P. J. Stephens [72]. Another proof was given by J. Pintz [77b]. For previous results see Ankeny, Chowla [60], Burgess [66], Chowla [65b], Gut [63], Newman [65].

It is conjectured that there is a positive constant C such that for all d we have $L_d(1) > C/\log d$. This inequality follows from the Extended Riemann Hypothesis, as shown by Hecke (see Landau [18c], Mahler [34]). The best unconditional result is

$$L_d(1) > C(\varepsilon)d^{-\varepsilon}$$

resulting from Siegel's theorem. If $0 < \varepsilon < 0.0723$, then for all d with at most one exception (which satisfies $d \geq \exp(\varepsilon^{-1})$) we have

$$L_d(1) > \min(0.125/\log d, 2.865\varepsilon d^{-\varepsilon}),$$

as proved by J. Hoffstein [80]. Cf. Gelfond [53], Metsänkylä [70b], Pintz [71], [77c], Tatuzawa [51].

J. E. Littlewood [28] showed that under the Extended Riemann Hypothesis each of the inequalities

$$L_d(1) \leq C \log \log d$$

and

$$L_d(1) \geq C_1/\log \log d$$

holds for infinitely many d with suitable positive C, C_1. Later S. Chowla [34c] proved the first result unconditionally and the second was established by Yu. V. Linnik [42] and A. Walfisz [42]. Cf. Barban [64], Bateman, Chowla, Erdös [50], Chowla [43], [47], [49], Elliott [69], Joshi [70], Shanks [73].

For other questions concerning the distribution of $L_d(1)$ and $L_d(s)$ see Barban [62], Chowla, Erdös [51], Elliott [70b], [73], Fainleib [69], [72], Fluch [64], Joly, Moser [79], Lavrik [70b], Pintz [76b], Stankus [75].

The mean value of the class-number of imaginary quadratic fields was considered already by C. F. Gauss [01], [63] in the language of quadratic forms. Let $H(d)$ denote the class-number of primitive binary totally positive quadratic forms of determinant d. Gauss conjectured that

$$\sum_{d \leq N} H(-d) = (4\pi/21\zeta(3))N^{3/2} + o(N^{3/2}).$$

This was established by F. Mertens [74] and a similar assertion for positive determinants was obtained by C. L. Siegel [44b]. A further work of I. M. Vinogradov [18], [49], [55], [63] and Chen-Jing-Run [62], [63] led to the error term $-2N/\pi^2 + O(N^{2/3+\varepsilon})$. Cf. Ayoub [64], Fomenko [61].

Mean value of $H^k(-d)$ and $h^k(-d)$ for $k \geq 2$ has been considered in Barban [62], [67], Barban, Gordover [66], Fainleib, Saparniyazov [75], Jutila [73], Lavrik [59], [71a], [71b], Pan [63], Saparniyazov [65], Stankus [76], Warlimont [71], Wolke [69], [71], [72].

Mean value of $h(d)$ for positive d has been considered by C. Hooley [84].

18. Theorem 8.7 is a particular case of the result of N. C. Ankeny, R. Brauer, S. Chowla [56], who proved that for every positive ε one can find infinitely many fields K of given degree and signature, with $h(K) \geq |d(K)|^{1/2-\varepsilon}$. This was strengthened in Sprindzhuk [74b], where it is shown that this inequality holds for almost all fields of a given degree. For real quadratic K, H. L. Montgomery and P. J. Weinberger [77] obtained the estimate $h(K) \geq d^{1/2}\log\log d/\log d$ holding for infinitely many $d = d(K)$. If ERH is true, then this inequality cannot be improved. (Cf. also Mallik [81b], Nagell [38].) In another direction, Y. Yamamoto [71] proved $h(K) \leq cd^{1/2}\log^{-2}d$ for infinitely many $d = d(K)$ with a certain positive c. V. G. Sprindzhuk [74c], [76], [77] constructed sequences of quadratic fields of special forms with class-number tending to infinity. (Cf. Chowla, Cowles, Cowles [80].) For a similar result for cubic fields see Watabe [83].

19. Theorem 8.8 is due to Gauss [01]. See also Arndt [58a], Bebbe [66], Gogia, Luthar [79], Goldstein [72a], Kronecker [64], Mertens [05], Reiner [45], Shyr [79]. A class-field interpretation of the group of genera was given by H. Hasse [51c], who showed that the maximal unramified extension of K which is Abelian over Q coincides with that extension of K which according to the class-field theory corresponds to the group of genera of K. This was generalized to cyclic K/Q in Iyanaga, Tamagawa [51] and to arbitrary Abelian K/Q in Leopoldt [53a]. (Cf. Gold [75], Halter-Koch [71c], Hasse [69a].) In this generalization the role of the principal genus is played by $\{X^{1-s}: X \in H^*(K), s \in \text{Gal}(K/Q)\}$. For non-Abelian extensions of the rationals a theory of genera was given by A. Fröhlich [59], [83b]. The genus field of K is defined as the maximal extension of K of the form KL with L/Q Abelian, which is unramified at all finite primes. One can also demand that the genus field be unramified at infinity, which leads to a parallel theory. (See Halter-Koch [78b], Horie [83], Stark [76b].) For the cubic case see Barrucand, Cohn [70], [71], Cohn [77]. Constructions of the genus fields for certain classes of fields have been given in Cornell [82], Fröhlich [59], Ishida [74], [75a], [76], [77], [80] and Nakagoshi [84b]. For other results on the genus theory see Furuta [67], [70], Gold [76a], Gurak [77b], Halter-Koch [72c], [79], Hamamura [81], Kubokawa [77], Takase [82], Takeuchi [82].

20. Using class-field theory one can obtain an analogue of Corollary 1 to Theorem 8.8 for the group $H(K)$ of quadratic K: $H(K)$ has $t-1$ even invariants, except the case when K is real and at least one prime $p \equiv 3 \pmod{4}$ is ramified; in this case $H(K)$ has $t-2$ even invariants, t being as in Theorem 8.8. (See e.g. Herz [66].) An elementary proof of Corollary 2 was supplied by A. Takaku [75]. It was shown by H. W. Leopoldt [53a] that if K/Q is cyclic of a prime degree p, then $p \nmid h^*(K)$ iff $|d(K)|$ is a prime power. Cf. Fröhlich [54e], Kuroda [64b], Moriya [30].

A simple formula for the number e_4 of invariants of $H^*(K)$ divisible by 4, for quadratic K, was given by L. Rédei and H. Reichardt [34]: let $F(d)$ be the number of factorizations $d = d_1 d_2$, with $d = d(K)$ and d_i being discriminants of quadratic fields, for which

$$\left(\frac{d_1}{q}\right) = \left(\frac{d_2}{p}\right) = 1$$

holds for all primes $p|d_1$ and $q|d_2$. Then $F(d) = 2^{e_4}$. Cf. Kisilevsky [82], Lagarias [80c], Rédei [34a], [34b]. For a similar description of the number of invariants divisible by higher powers of 2 see Reichardt [34]. For other results concerning the 2-part of $H^*(K)$ with quadratic K see Bauer [71], [72], Endô [73a], Halter-Koch [84b], Hasse [69b], [69c], [70a], [70b], Kaplan [72], [73a], [73b], [74], [76], [77b], Lagarias [80a], Moine [72], Oriat [77], [78], Rédei [32a], [32b], [34c], [36], [38], Reichardt [70], Scholz [35], Waterhouse [73].

Far less is known about the number $e_p(H)$ of invariants of $H(K)$ divisible by an odd prime p, even in the simplest case of quadratic fields. In fact, it is even not known, whether or not the set of possible values of $e_p(H)$ is unbounded when H ranges over class-groups of quadratic fields and p over all odd primes. In the case of $p = 3$ it was shown by M. Craig [77] that $e_3(H) \geq 4$ holds for infinitely many imaginary quadratic fields and the same assertion for real quadratic fields was established by F. Diaz y Diaz [78b]. Cf. Craig [73], Diaz y Diaz [74], [78a], Diaz y Diaz, Shanks, Williams [79], Neild, Shanks [74], Shanks [72c], [76b], Shanks, Serafin [73], Shanks, Weinberger [72], Y. Yamamoto [70].

It follows from Nakano [84a] that $e_p(H) \geq 2$ holds infinitely often for every prime p and infinitely many imaginary quadratic fields. Formerly this had been known only for $p = 5$ and $p = 7$ (Mestre [83]).

Examples are known of imaginary quadratic fields with $e_5(H) \geq 3$ and one of them, with $d(K) = -258\,559\,351\,511\,807$, has $e_5(H) = 4$ (Schoof [83]). (Cf. Solderitsch [85].)

An interesting conjecture concerning the structure of H^* in relation to arithmetics in certain extensions was proposed by H. Cohn and J. C. Lagarias [83]. Cf. Morton [82a], [82b], [83].

21. A formula for $e_3(H)$ for cubic fields was given by F. Gerth [76c] and, in the case of pure cubic fields, earlier in Gerth [73], [75a] and Kobayashi [73]. Cf. Kobayashi [71], [77]. This was generalized by G. Gras [74b] to fields of degree p, whose Galois closure has a dihedral Galois group. Cf. Kobayashi [74].

Invariants of $H(K)$ for arbitrary fields were studied in Rédei [44] and the number $e_{p^k}(H)$ for cyclic fields of prime degree p in Inaba [40], [41]. Cf. Fröhlich [54b].

For other classes of fields see Armitage, Fröhlich [67], Cornell [83a], Gerth [76a], [76d], G. Gras [74b], Gras, Moser, Payan [73], Halter-Koch [78c], Kuroda [70b], Moriya [30], Oriat [76], Shanks [74], Taylor [75].

22. Proposition 8.8 is due to S. Chowla [34b]. It shows, in particular, that the equality $g(d) = h(d)$ can hold only for finitely many discriminants $d < 0$. Such discriminants are intimately connected with idoneal numbers considered by L. Euler. (See Grosswald [63], Grube [74], Steinig [66].) Conditions equivalent to $g(d) = h(d)$ were given in Kitaoka [71b] and Papkov [44]. For other results on this topic see Briggs, Chowla [54], N. A. Hall [37], [39], Hendy [74a], Möller [76a], Swift [48].

A. Baker and A. Schinzel [71] proved that every genus in an imaginary quadratic field contains an ideal of norm not exceeding $B(\varepsilon)|d(K)|^{3/8+\varepsilon}$ for every positive ε, and the exponent was reduced to $1/4+\varepsilon$ by D. R. Heath-Brown [79].

A discriminant d is called *regular*, if the principal genus in $Q(d^{1/2})$ is cyclic. Apparently it is not known, whether or not there are infinitely many regular discriminants. See Gauss [01], Lippmann [63].

Denote by $m(d)$ the exponent of $H^*(K)$ for $K = Q(d^{1/2})$, i.e. the order of its biggest cyclic subgroup. The Corollary to Proposition 8.8 implies that for negative d the equality $m(d) = 2$ can hold only in finitely many cases, and it was shown by D. W. Boyd, H. Kisilevsky [72] and P. J. Weinberger [73a] that the same is true for discriminants d with $m(d) = 3$. A. G. Earnest, D. R. Estes [81] proved the same for $m(d) = 4$ and A. G. Earnest, O. H. Körner [82] generalized this by replacing 4 by any power of 2. The same assertion holds also for totally imaginary quadratic extensions of a totally real field with $h = 1$.

It was shown by P. J. Weinberger [73a] under ERH that

$$m(d) \geqslant C \log|d|/\log\log|d|$$

for negative d, with a suitable positive C.

23. Theorem 8.9 is due to T. Nagell [22], whose proof we have reproduced here. Other proofs are given in Ankeny, Chowla [55], Humbert [40] and Kuroda [64a]. The last mentioned proof shows that in this theorem one can also assert that $d(K)$ is divisible by a given integer. Cf. Cowles [80].

The analogue of Theorem 8.9 for real quadratic fields was established by Y. Yamamoto [70]. (See also Ankeny, Chowla [55], Nakahara [78], Weinberger [73b].) A. Fröhlich [57] obtained the same assertion for cyclotomic fields. (Cf. Metsänkylä [68b].) For cyclic cubics this was done by K. Uchida [74] and for pure cubics by S. Nakano [83a].

It was recently shown by S. Nakano [84a], [85a] that there exist infinitely many fields of given degree and signature with $h(K)$ divisible by any given integer. Previously, T. Azuhata, H. Ichimura [84] had proved this for not totally real fields.

Divisibility of the class-number of quadratic fields by powers of 2 is closely connected with representations of certain divisors of the discriminant by quadratic forms. The oldest result of this type is contained in a letter of Gauss to Dirichlet (Gauss [28]), where it is shown that if $p \equiv 1 \pmod 8$ is a prime and $p = x^2 + y^2$, then $8 \mid h(-4p)$ iff $x+y$ is congruent to 1 or 7 (mod 8). On this topic see Barrucand, Cohn [69], Bauer [72], Brown [72b], [73], [74a], [74b], [75], [81], Brown, Parry [73], Hasse [69b], [69c], [70a], [70b], Kaplan [72], [73a], Koch, Zink [72], Leonard, Williams [82], Pall [69], Reichardt [70], K. S. Williams [76].

For other results concerning the divisibility of $h(K)$ for quadratic K see Chowla, Hartung [74a], Endô [73b], Glaisher [03], Hartung [74a], [74b], Hayashi [77], Honda [68], E. Lehmer [72], Oriat [78], Parry [75b], Queen [76], Slavutskii [75]. Similar questions for fields of higher degrees were considered in Barrucand, Cohn [70], Callahan [76], Cohn [56a], Cornell, Rosen [84], Endô [76], Feng Keqin [82c], Frey, Geyer [72], Fröhlich [54a], [54e], [59], [62c], Furuta [72], Furuya [71], Gerth [76b], Gold, Madan [78], G. Gras [75], Gras, Gras [75], Gut [51b], [54], [73], Holzer [50], Honda [71], Iimura [71], [79a], Ishida [69], [70], [71], [73], [74], Kobayashi [79], [80], Madan [70], Mollin [83a], Montouchet [71], Moriya [30], [34b], Morton [83], Nakano [83b], Neumann [73], Ohta [72], [81], Parry [75b], [75c], [75d], [78], [80], Parry, Walter [76], Satgé [79a], [79b], Schertz [78b], [81], Uchida [76b], Uehara [82], Wada [70a], Walter [80], Watabe [78], Yokoyama [67].

Asymptotics for cyclic fields of prime degree p, with class-number divisible exactly by p^s, was found by F. Gerth [82]. The cases $s = 0, 1$ were treated earlier in Gerth, Graham [85]. Cf. Gerth [83b], [83c], [83d]. A similar problem for cubic fields was considered in Gerth [86].

24. Theorem 8.10 is due to A. Brumer [65]. It was strengthened by P. Roquette and H. Zassenhaus [69], who also gave an elementary proof. Further improvement was done by I. G. Connell, D. Sussman [70]. The value of $c(n)$ given in our proof is, in the case of normal K/Q and $q \neq 2$, equal to that obtained by Roquette and Zassenhaus. Corollary 1 was earlier obtained by A. Fröhlich [62c] and Corollary 4 was proved by A. Brumer and M. Rosen [63]. Cf. Cornell, Rosen [80], Halter-Koch [81], Kobayashi [71], Schmithals [80b].

Infinitely many fields of a given degree and signature with $C_N^{1+r_2} \subset H(K)$ were constructed by S. Nakano [84a], [85]. Previously T. Azuhata, H. Ichimura [84] had r_2 in place of $1+r_2$. Cf. also Ichimura [82], Iimura [79b], Ishida [75b], Madan [72], Nakano [84b], [86].

Another method for getting lower bounds for N in $C_q^N \subset H(K)$ was found by K. Iwasawa [66a].

25. Relations between 3-ranks of the class-groups of $Q(m^{1/2})$ and $Q(im^{1/2})$ were obtained by A. Scholz [32]. Other proofs of his result were given in Martinet, Payan [67] and Oriat [76]. (Cf. Shanks [72c].) Oriat's proof is based on the "reflection theorem" proved by H. W. Leopoldt [58b], which generalizes previous work of E. Hecke [10], F. Pollaczek [24] and T. Takagi [27]. (Cf. Oriat, Satgé [79], Shiratani [64a].) For other applications of the reflection theorem see G. Gras [72c], [77b], [79b], Kudo [72], Oriat [78].

Relations between 4-ranks of $H(K)$ for quadratic fields were considered in Damey, Payan [70], Halter-Koch [84b], Taussky [77b]. For 8-ranks see Bouvier [71] and for 3-ranks in the case of cubic fields see Callahan [74].

26. The first result connecting class-numbers of various fields is due to Dirichlet [42], who showed that if $k_1 = Q(m^{1/2})$, $k_2 = Q(im^{1/2})$ and $K = k_1k_2$, then

$$h(K) = ah(k_1)h(k_2),$$

with $a = 1$ or $1/2$. Cf. Amberg [97], Bachmann [64], Halter-Koch [72c], Herglotz [22], Hilbert [94c], Kubota [53], [56b], Kuroda [43a], Lang, Schertz [74], Lubelski [36], Reichardt [72].

Analogues for composites of extensions of prime degree have been proved in Kuroda [50], Litver [49], [59], Nehrkorn [33], Pollaczek [29]. For similar results in other classes of fields see Halter-Koch, Moser [78], Inaba [35], N. Moser [79a], Parry [77a], Schertz [74a], [74b], Schertz, Stender [79], Scholz [30], [33], Värmon [30].

Relations between the class-groups of a field and its subfield have been studied in Castela [78], G. Gras [74b], Halter-Koch [77], N. Moser [75] and Oriat, Satgé [79]. (Cf. J. Smith [69].) R. Brauer [51] obtained very general relations between the class-numbers of subfields of a given field. See also Jaulent [81e], [82], Jehne [77a], Kuroda [50], Nakagoshi [81], [84a], Rehm, Happle [74], Shyr [75], Walter [77], [79b], [79c].

27. E. E. Kummer [50a] asserted that if p is an odd prime and $K \subset L \subset K_p$, then $h^*(K)$ divides $h^*(L)$. His proof was incorrect and the first proof of this result was given by P. Furtwängler [08] (cf. Chebotarev [24b]). It was shown by J. Herbrand [32a] that one can replace here h^* by h.

Another important result of P. Furtwängler [11] states that if $p \nmid h^*(K_p)$ then $p \nmid h^*(K_{p^n})$ for all $n \geqslant 1$. Cf. Miki [78a]. For other results concerning divisibility of $h(L)$ by $h(K)$ in case of $K \subset L$ see Adachi [73], Fröhlich [57], Honda [60a], [60b], Inaba [37], Iwasawa [55a], H. Lang [73c], [77], Latimer [33b], Yokoi [67], [68b], Yokoyama [67].

28. Proposition 8.9 is due to G. Frobenius [12] and G. Rabinowitsch [13]. Our proof follows that of I. G. Connell [62], who also proved Lemma 8.10. For other proofs see Ayoub, Chowla [81] and Szekeres [74]. Generalizations, dealing mostly with $h = 2$ and $h = g$, have been given in Haneke [69], Hendy [74b], Kutsuna [80], Möller [76b] and Papkov [44].

For various conditions related to $h = 1$ in quadratic fields see Behrbohm, Rédei [36], Chowla [61b], [70b], Ennola [58b], Inkeri [48], Lu [79], Mallik [81a], Mitchell [26], Nagell [22], Popovici [55], [57], Rédei [60].

The old question of Gauss [01], whether $h = 1$ holds for infinitely many positive quadratic discriminants, is still open. The existing tables (see e.g. Kloss [65]) seem to confirm it. Cf. Takhtayan, Vinogradov [80], [82] for a heuristical approach to this problem. In Ankeny, Chowla, Hasse [65], Çallial [80], Hasse [65], H. Takeuchi [81] many fields $Q(p^{1/2})$ with $h \neq 1$ were produced; however, it is not known whether this approach leads to infinitely many such fields. Cf. Yamaguchi, Oozeki [72].

Interesting relations between the class-number of $Q(ip^{1/2})$ and the elements in the period of the continued fraction expansion of $p^{1/2}$ were for prime p obtained by F. Hirzebruch and D. Zagier. (See Hirzebruch [76], Zagier [75a], [75b], [81] (§ 14).) For related conjectures and results see Chowla, Chowla [72], [73], H. Lang [76], Schinzel [74b].

For diverse results on the class-number of quadratic fields see Ankeny, Chowla [50], Lemmlein [68], Linnik [54], Papkov [38].

EXERCISES TO CHAPTER 8

1. Let K/Q be Abelian, let $X(K)$ be the associated group of characters and let p be a prime.

(i) Prove that the ramification index of any prime ideal lying in K over p equals the number of those characters whose conductor is a power of p and which appear in the canonical factorization of at least one character of $X(K)$ into characters with prime power conductors.

(ii) Determine the decomposition and inertia groups of a prime ideal in K in terms of $X(K)$.

2. For a given $k \geq 3$ determine all prime powers p^N for which there exist primitive characters $(\bmod\, p^N)$ of order k and find the number of such characters.

3. Characterize those integers which are discriminants of normal cubic extensions of Q.

4. Find the form of discriminants of Abelian quartic extensions of Q.

5. Determine all subfields of K_N, find their discriminants, conductors and generators, with N being your favourite number.

6. Let K be a CM-field. Prove that the index of $E(K)U(K^+)$ in $U(K)$ equals either 1 or 2.

7. Prove that if $K = K_n$, the nth cyclotomic field, then the index in the previous exercise equals unity iff n is a prime power, assuming that $n \not\equiv 2 \pmod 4$.

8. Prove Proposition 8.5 for all cyclotomic fields.

9. (Dirichlet [39], Mordell [61], Honda [75].)
  (i) Prove that if $p \equiv 3 \pmod 4$ is a prime $\neq 3$ and $K = Q(ip^{1/2})$, then
  $$h(K) = a^{-1} \sum_{x=1}^{b} \left(\frac{x}{p}\right),$$
  where $a = 2 - \left(\dfrac{2}{p}\right)$ and $b = (p-1)/2$.
  (ii) Prove that if $u_p = \pm 1$ satisfies
  $$\left(\frac{p-1}{2}\right)! \equiv u_p \pmod p,$$
  then $u_p = (-1)^N$, where N is the number of $r \in [1, (p-1)/2]$ with $\left(\dfrac{r}{p}\right) = -1$.
  (iii) Deduce Jacobi's conjecture (Jacobi [32]):
  $$h(K) \equiv -u_p \pmod 4.$$

10. Prove that if $f(x, y)$ is a quadratic form of discriminant d lying in the class of forms which, according to Theorem 8.6, corresponds to a class X of ideals in $Q(d^{1/2})$, then the set of integers represented by f coincides with the set of norms of integral ideals belonging to X^{-1}.

11. Under assumptions of the preceding exercise show that if d is negative and $d \neq -3, -4$, then the number of representations of an integer m by $f(x, y)$ equals double the number of integral ideals in X^{-1} having norm m.

12. Find a bound for $c(n)$ in Theorem 8.9 in the case of $q = 2$.

13. Let f be a quadratic polynomial over Z of discriminant D and assume that it represents primes for T consecutive integers. Prove that either the roots of f generate a quadratic field with $h = 1$ or $T \leq c|D|^{1/2}$ with a certain absolute constant c.

Chapter 9. Factorization Problems

§ 1. Elementary Approach to Factorizations

1. We already know that the condition $h(K) = 1$ is both necessary and sufficient for the uniqueness of factorization in R_K. This shows that fields with trivial class-group can be characterized arithmetically in terms of factorization properties. The discovery made by L. Carlitz [60] that one can similarly characterize in a simple way those fields which have class-number 2 gave rise to the thought that it might be possible to obtain an analogous characterization of fields with a given class-number or class-group. This is indeed the case and we give a proof of this fact in this section. We start with the result of Carlitz. To be able to state it, we have to introduce the notion of the length of a factorization. If $a \in R_K$ is neither zero nor a unit and $a = a_1 \ldots a_k$ is a factorization into numbers irreducible in R_K, then k is called the *length* of this factorization.

Theorem 9.1. *If K is an algebraic number field with $h(K) \neq 1$, then for every non-zero and non-unit element a of R_K all factorizations of a have the same length iff $h(K) = 2$.*

Proof. First assume that $h(K) = 2$ and let a be a given integer of K, $a \neq 0$ and $a \notin U(K)$. Let

$$aR_K = \mathfrak{P}_1^{a_1} \ldots \mathfrak{P}_t^{a_t} \tag{9.1}$$

with distinct prime ideals $\mathfrak{P}_i$ and assume that $\mathfrak{P}_1, \ldots, \mathfrak{P}_s$ are principal ($0 \leq s \leq t$), whereas the remaining $\mathfrak{P}_i$'s are not. Then every factorization of a into irreducibles must be of the form

$$a = c_1^{a_1} \ldots c_s^{a_s} d_1 \ldots d_u,$$

where $c_i R_K = \mathfrak{P}_i$ ($i = 1, 2, \ldots, s$) and every number d_i generates an ideal of the form $\mathfrak{P}_i \mathfrak{P}_j$ with $i, j > s$. Hence, the length of such a factorization equals

$$a_1 + \ldots + a_s + u = a_1 + \ldots + a_s + (a_{s+1} + \ldots + a_t)/2$$

and this number obviously depends only on a, as asserted.

To get the converse implication, assume that $h(K) \neq 1, 2$. We have to show that there is an integer in K with factorizations of distinct lengths. First consider the case when $H(K)$ has elements of order $g \geq 3$. Let X be such an element. Corollary 7 to Proposition 7.9 shows that there is a prime ideal $\mathfrak{P} \in X$ and a prime ideal $\mathfrak{P}_1 \in X^{-1}$. Then the ideals $\mathfrak{P}^g$, $\mathfrak{P}_1^g$, $\mathfrak{P}\mathfrak{P}_1$ are all principal, their corresponding generators a, b, c are irreducible, and since $ab = uc^g$, with a suitable unit u, the number ab has factorizations of lengths g and 2. If $H(K)$ is of the form C_2^N and $N \geq 2$, then there are distinct classes X, Y in $H(K)$ of order 2. Choose prime ideals $\mathfrak{P}_1 \in X$, $\mathfrak{P}_2 \in Y$, $\mathfrak{P}_3 \in XY$ and observe that $\mathfrak{P}_1 \mathfrak{P}_2 \mathfrak{P}_3$ and $\mathfrak{P}_i^2$ ($i = 1, 2, 3$) are principal and generated by irreducible elements a, b_1, b_2, b_3, respectively. Since $a^2 = u b_1 b_2 b_3$, with a suitable unit u, the number a^2 has factorizations of different lengths and we are ready. □

2. Now we prove a purely arithmetic characterization of the class-group, due to J. Kaczorowski [84]; to state it, we have to introduce certain definitions. If a is an irreducible element of R_K such that all of its powers have unique factorization, then a is called an *absolutely irreducible element*. For every such element we define its *order*, $\mathrm{ord}\, a$, as the maximal rational integer m with the property that with a suitable $b \in R_K$ we have $a \mid b^m$, $a \nmid b^{m-1}$.

Theorem 9.2. *If $a_1, a_2, \ldots, a_r$ are non-associated absolutely irreducible elements of R_K such that $a_1 a_2 \ldots a_r$ has unique factorization and the sum*

$$\mathrm{ord}\, a_1 + \mathrm{ord}\, a_2 + \ldots + \mathrm{ord}\, a_r$$

is maximal, then

$$H(K) \simeq \bigoplus_{i=1}^{r} C_{n_i}$$

with $n_i = \mathrm{ord}\, a_i$ ($i = 1, 2, \ldots, r$).

Proof. We start with a lemma, which translates the definitions of an absolutely irreducible element and its order into ideal-theoretic language:

Lemma 9.1. *An element $a \in R_K$ is absolutely irreducible iff $a R_K = \mathfrak{P}^m$ holds for a certain prime ideal $\mathfrak{P}$ lying in a class of $H(K)$ of order m. If this condition is satisfied, then $\mathrm{ord}\, a = m$. In particular, a generates a prime ideal iff $\mathrm{ord}\, a = 1$.*

Proof. If $aR_K = \mathfrak{P}^m$ holds with a certain prime ideal $\mathfrak{P}$, whose class is of order m, then a is irreducible and obviously all powers of a have unique factorization, since they cannot be divisible by an irreducible element not associated with a. To show that $\text{ord}\, a = m$, choose b in $\mathfrak{P}\setminus\mathfrak{P}^2$ and note that $a\,|\,b^m$, $a\nmid b^{m-1}$, so that $\text{ord}\, a \geqslant m$. If now $n > m$ and $c \in R_K$ is such that $a\,|\,c^n$, $a \nmid c^{n-1}$, we define s by $\mathfrak{P}^s \,\|\, cR_K$ and observe that $\mathfrak{P}^m\,|\,(cR_K)^n$, $\mathfrak{P}^m \nmid (cR_K)^{n-1}$, which leads to

$$(n-1)s < m \leqslant ns.$$

This implies $s \geqslant 1$ and we get

$$n-1 \leqslant (n-1)s \leqslant m-1 < n-1,$$

a contradiction. Thus $\text{ord}\, a = m$.

Now let a be absolutely irreducible and let (9.1) be the canonical factorization of aR_K, with distinct $\mathfrak{P}_i$'s and $t \geqslant 2$. Let g be the order of the class containing $\mathfrak{P}_1$. Then $\mathfrak{P}_1^g$ is principal and generated by b, say. If $c_1 \ldots c_t = a^g/b$ is a factorization into irreducibles, then $a^g = bc_1 \ldots c_t = a \ldots a$ has at least two factorizations, contrary to our assumption. This shows that necessarily $t = 1$, and so $aR_K = \mathfrak{P}_1^m$, with a certain $m \geqslant 1$. Since a is irreducible, it follows that $m = g$. □

The proof of the theorem will result immediately from the next lemma:

Lemma 9.2. *There exists a constant $B(K)$ with the following property: if $a_1, \ldots, a_r$ are non-associated absolutely irreducible elements of R_K such that their product has a unique factorization and their orders $\text{ord}\, a_i$ all exceed 1, then*

$$\sum_{i=1}^{r} \text{ord}\, a_i \leqslant B(K),$$

and the class-group $H(K)$ contains a subgroup isomorphic to

$$\bigoplus_{i=1}^{r} C_{n_i} \qquad (9.2)$$

where $n_i = \text{ord}\, a_i$.

Conversely, if $H(K)$ contains a subgroup of the form (9.2), X_i is a generator of C_{n_i}, $\mathfrak{P}_i$ is a prime ideal in X_i and

$$\mathfrak{P}_i^{n_i} = a_i R_K \qquad (i = 1, 2, \ldots, r),$$

then $a_1, \ldots, a_r$ are non-associated absolutely irreducible integers and their product has a unique factorization.

Proof. To obtain the first assertion, we use the preceding lemma, which implies that $a_i = \mathfrak{P}_i^{n_i}$ ($i = 1, 2, \ldots, r$) with suitable prime ideals $\mathfrak{P}_i$. Denote by X_i the class of $\mathfrak{P}_i$ and observe that, factorization $a_1 \ldots a_r$ being unique,

$$X_1^{c_1} \ldots X_r^{c_r} \quad (0 \leqslant c_i < n_i)$$

can be equal to the principal class only if all exponents c_i vanish, and hence the classes $X_1, \ldots, X_r$ generate a subgroup of $H(K)$ of the form (9.2). In particular, we get

$$\sum_{i=1}^r \operatorname{ord} a_i = \sum_{i=1}^r n_i \leqslant \prod_{i=1}^r n_i \leqslant h(K).$$

To get the second assertion it suffices, in view of Lemma 9.1, to show that $a_1 \ldots a_r$ has a unique factorization. But this follows from the observation that if c is an irreducible factor of $a_1 \ldots a_r$ non-associated with $a_1, a_2, \ldots, a_r$, then

$$cR_K = \mathfrak{P}_1^{b_1} \ldots \mathfrak{P}_r^{b_r}$$

holds with $0 \leqslant b_i \leqslant n_i$ and for at least one index i we have $0 < b_i < n_i$. If now $A_i = 0$ when $b_i = n_i$ and $A_i = b_i$ otherwise, then

$$X_1^{A_1} \ldots X_r^{A_r} = 1,$$

which is impossible by our assumptions. □

Corollary. *The class-group of K is cyclic of N elements iff there exists in K an absolutely irreducible element of order N and for any absolutely irreducible elements $a_1, \ldots, a_r$ whose product has a unique factorization one has*

$$\operatorname{ord} a_1 + \ldots + \operatorname{ord} a_r \leqslant N. \qquad \square$$

3. Several factorization properties in R_K can be expressed by using elementary combinatorics in finite Abelian groups. We shall now do this for the irreducibility.

Let A be a finite Abelian group written additively. A non-empty finite system $b = (g_1, g_2, \ldots, g_n)$ of elements of A is called a *block*, if $g_1 + g_2 + \ldots + g_n = 0$; the number n is then called the *length* of b. In the set $B(A)$ of all such blocks multiplication is defined as juxtaposition, i.e.

$$(g_1, \ldots, g_m)(h_1, \ldots, h_n) = (g_1, \ldots, g_m, h_1, \ldots, h_n).$$

B is thus given the structure of a commutative semi-group. (Two systems differring in the order are considered as identical.)

A block b is called *irreducible*, if it cannot be written as a product of two blocks. The relevance of this notion to that of irreducible integers in R_K is made clear in the following easy proposition:

Proposition 9.1. *An element a of R_K such that*
$$aR_K = \mathfrak{P}_1 \ldots \mathfrak{P}_s$$
(with $\mathfrak{P}_i$ being prime ideals, not necessarily distinct, $s \geq 1$) is irreducible iff the block $(X_1, X_2, \ldots, X_s)$ in $B(H(K))$ is irreducible, X_i denoting the ideal class of $\mathfrak{P}_i$.

Proof. It suffices to observe that any factorization of a, say $a = a_1 a_2$, with
$$a_1 R_K = \mathfrak{P}_{i_1} \ldots \mathfrak{P}_{i_r}, \quad a_2 R_K = \mathfrak{P}_{j_1} \ldots \mathfrak{P}_{j_s}$$
induces a factorization of the corresponding block,
$$(X_1, \ldots, X_s) = (X_{i_1}, \ldots, X_{i_r})(X_{j_1}, \ldots, X_{j_s}),$$
and conversely. □

The Davenport's constant $D(A)$ of a finite Abelian group A is defined as the smallest integer m with the property that from any sequence of m elements of A one can extract a subsequence with zero sum. This constant is finite and does not exceed the cardinality N of A; indeed, if $a_1, \ldots, a_N \in A$, then either all sums $a_1, a_1+a_2, a_1+a_2+a_3, \ldots, a_1+a_2+ \ldots +a_N$ are distinct, and since there are N of them, one must be zero, or two of them are equal, and by subtraction we obtain a non-empty zero-sum subsequence.

Proposition 9.2. *The maximal length of an irreducible block in $B(A)$ equals $D(A)$.*

Proof. Clearly, no irreducible block can have its length greater than $D(A)$; so assume that the maximal length of such a block is $n < D(A)$. Let $a_1, a_2, \ldots, a_n$ be a sequence in A without a subsequence with vanishing sum and put $h = -a_1 - a_2 - \ldots - a_n$. Then
$$b = (a_1, \ldots, a_n, h)$$
is a block of length $> n$; hence it cannot be irreducible. Thus $b = b_1 b_2$ with certain blocks b_1, b_2. But then one of the b_i's must be of the form $(a_{i_1}, \ldots, a_{i_r})$, and thus $a_{i_1} + \ldots + a_{i_r} = 0$ which is impossible. □

Corollary (H. Davenport). *The maximal number of prime ideal factors of an irreducible element of R_K equals $D(H(K))$.*

Proof. Apply Propositions 9.1 and 9.2. □

4. An explicit formula for $D(A)$ is in the general case unknown. We now prove a result of J. E. Olson giving such a formula for p-groups.

§ 1. Elementary Approach to Factorizations

Theorem 9.3. *If p is a prime and*

$$A = \bigoplus_{i=1}^{t} C_{p^{n_i}},$$

then

$$D(A) = 1 + \sum_{i=1}^{t}(p^{n_i}-1).$$

Proof. The theorem will result from the following lemma concerning group-rings of finite Abelian p-groups:

Lemma 9.3. *If A is a finite Abelian p-group written multiplicatively, which is a product of cyclic groups of orders $p^{n_1}, \ldots, p^{n_t}$, and if $g_1, \ldots, g_k$ are elements of A with*

$$k \geq 1 + \sum_{i=1}^{t}(p^{n_i}-1),$$

then the element

$$(1-g_1)(1-g_2)\ldots(1-g_k) = \sum_{g \in A} c_g g$$

of $Z[A]$ has all its coefficients c_g divisible by p.

Proof. If $g_i = 1$ for some i, then the assertion is evident; so assume that $g_i \neq 1$ for $i = 1, 2, \ldots, k$. Denote by x_i a fixed generator of the group $C_{p^{n_i}}$. Then every $g \in A$ can be written uniquely in the form

$$g = \prod_{j=1}^{t} x_j^{a_j}, \quad \text{with } 0 \leq a_j < p^{n_j}.$$

Define

$$F(g) = a_1 + \ldots + a_t.$$

We prove by induction in $\max_j F(g_j)$ that taking suitable $h_j \in A$, M, $f_{ij} \geq 0$ with $\sum_{j=1}^{M} f_{ij} = k$ we have

$$\prod_{j=1}^{k}(1-g_j) = \sum_{j=1}^{M} h_j(1-x_1)^{f_{1j}}\ldots(1-x_t)^{f_{tj}}. \tag{9.3}$$

If $\max_j F(g_j) = 1$, then for $j = 1, 2, \ldots, k$ we have $g_j = x_{i_j}$ with suitable i_j and the assertion holds with $M = 1$, $h_1 = 1$, $f_{ij} = 1$. In the general case we can write $g_j = x_{i_j} t_j$ for $j = 1, 2, \ldots, k$, with suitable i_j and $t_j \in A$ such that $F(t_j) = F(g_j) - 1$. In view of the equality

$$1 - g_j = (1-x_{i_j}) + x_{i_j}(1-t_j)$$

we get

$$\prod_{j=1}^{k}(1-g_j) = \prod_{j=1}^{k}((1-x_{i_j})+x_{i_j}(1-t_j))$$

and the last product is a sum of terms of the form

$$t\prod_{i=1}^{k}(1-u_i)$$

(with $t, u_i \in A$), where $\max_i F(u_i) < \max_j F(g_j)$.

We can thus apply the inductive assumption, except when some of the u_i's are equal to 1; but in this case we can just omit the corresponding terms.

The lemma now follows easily. In fact, by the assumption imposed on the lower bound for k, at least one f_{ij} for every j in (9.3) satisfies $f_{ij} \geq p^{n_i}$; but since

$$(1-x_i)^{p^{n_i}} = 1 + \sum_{k=1}^{p^{n_i}-1}\binom{p^{n_i}}{k}(-x_i)^k + (-x_i)^{p^{n_i}} = \sum_{j=1}^{p^{n_i}-1}s_j x_i^j$$

with $p \mid s_j$, we infer that $(1-x_i)^{f_{ij}}$ also has all its coefficients divisible by p. Thus the same applies to $(1-g_1) \ldots (1-g_k)$, in view of (9.3). □

Now observe that

$$(1-g_1) \ldots (1-g_k) = \sum_{g \in A} c_g g,$$

where

$$c_g = \sum_{\substack{g_1 \ldots g_r = g \\ 2 \mid r}} 1 - \sum_{\substack{g_1 \ldots g_r = g \\ 2 \nmid r}} 1 + \varepsilon_g$$

with

$$\varepsilon_g = \begin{cases} 1 & \text{if } g = 1, \\ 0 & \text{if } g \neq 1. \end{cases}$$

The lemma implies that $p \mid c_1$ and if no subsequence of $g_1, \ldots, g_k$ had the unit product, we would have $c_1 = 1$, a contradiction. Since the g_i's were arbitrary, we get

$$D(A) \leq 1 + \sum_{i=1}^{t}(p^{n_i}-1).$$

To obtain the converse inequality, it suffices to consider the sequence $x_1, \ldots, x_t$ where x_i is, as in the proof of the lemma, a fixed generator of $C_{p^{n_i}} \subset A$. □

5. It is also possible to obtain a combinatorial interpretation of the uniqueness of factorization. Consider a block $b = (g_1, \ldots, g_k) \in B(A)$ and fix the ordering of its elements. If $\alpha: b = b_1 \ldots b_t$ is a factorization of b, then we can associate with it a surjective mapping

$$\Phi_\alpha: \{1, 2, \ldots, k\} \to \{1, 2, \ldots, t\}$$

defining: $\Phi_\alpha(i) = j$ if g_i appears in the block j. (From a formal point of view one should consider here rather the sequence of pairs (g_i, i) in place of $\{g_i\}$ but we will not adhere to this pedantical formulation.) Two factorizations α, β of b will be called *equivalent* if they have the same number t of factors and, moreover, there is a permutation P of $1, 2, \ldots, t$ such that the sets

$$\{i: \Phi_\alpha(i) = j\} \quad \text{and} \quad \{i: \Phi_\beta(i) = P(j)\}$$

coincide. A block b is said to *have a unique factorization* if all its factorizations into irreducible blocks are equivalent. Note that this property is independent of the ordering of the elements of b.

Proposition 9.3. *If $(X_1, \ldots, X_k)$ is a block in $B(H(K))$ which has a unique factorization, then for every choice of prime ideals $\mathfrak{P}_i \in X_i$ ($i = 1, 2, \ldots, k$) any generator of the principal ideal $\mathfrak{P}_1 \mathfrak{P}_2 \ldots \mathfrak{P}_k$ has unique factorization in R_K. Moreover, if $a \in R_K$ has unique factorization and $aR_K = \mathfrak{P}_1 \ldots \mathfrak{P}_m$ with distinct prime ideals $\mathfrak{P}_i$, then the block in $B(H(K))$ formed by the classes of $\mathfrak{P}_i$ ($i = 1, 2, \ldots, k$) has a unique factorization.*

Proof. If $aR_K = \mathfrak{P}_1 \ldots \mathfrak{P}_k$ with $\mathfrak{P}_i \in X_i$, then every factorization of the block $b = (X_1, \ldots, X_k)$ induces a factorization of a. In fact, if α is a factorization of b, then $a = a_1 \ldots a_r$ with

$$a_j R_K = \prod_{m=1}^{c_j} \mathfrak{P}_{m(j,k)},$$

where for every fixed j we have

$$\{m(j, k): 1 \leq k \leq c_j\} = \{i: \Phi_\alpha(i) = j\}.$$

It is immediately seen that every factorization of a into irreducibles is induced in this way by a factorization of b, and different factorizations of a are induced by non-equivalent factorizations of b. If all prime ideals $\mathfrak{P}_i$ are distinct then, conversely, non-equivalent factorizations of b induce distinct factorizations of a. □

Note that the last assertion may fail to hold if aR_K is not a product of distinct prime ideals. For instance, if $H(K) = C_2$ and $b = (a_1, a_2, a_3, a_4)$ with $a_1 = a_2 = a_3 = a_4$ being the non-unit element of C_2, then $b =$

$(a_1 a_2)(a_3 a_4) = (a_1 a_3)(a_2 a_4)$ has non-equivalent factorizations; however, if $\mathfrak{P}$ is a non-principal prime ideal in R_K and $aR_K = \mathfrak{P}^2$, then $b = a^2$ (with $bR_K = \mathfrak{P}^4$) has unique factorization.

Denote by $a_1(A)$ the maximal length of a block in $B(A)$ which has unique factorization and does not contain the unit element e. Note that $a_1(A)$ is finite, because if $b = b_1 \ldots b_s$ is a unique factorization of b into irreducible blocks, then no element can appear in two b_i's; thus $s \leq |A|$ and since the length of each b_i does not exceed $D(A)$, we get $a_1(A) \leq D(A)|A|$. If $a \in R_K$, $a \neq 0$, we denote by $s(a)$ the number of distinct non-principal prime ideals dividing aR_K.

Corollary. *For all elements $a \in R_K$ with unique factorization we have $s(a) \leq a_1(H(K))$, and there is an element a with unique factorization, for which equality holds.*

Proof. The second part of the assertion results from the proposition, as well as the first part in the particular case when the ideal aR_K is square-free, i.e. is a product of distinct prime ideals. It remains to show that the inequality $s(a) \leq a_1(H(K))$ holds also for these elements a with unique factorization which generate a non-square-free ideal.

Let a be such an element and let (9.1) be the factorization into prime ideals of the ideal generated by a. We may assume that all ideals $\mathfrak{P}_i$ are non-principal, i.e. $t = s(a)$. Assume that the ideals $\mathfrak{P}_j^{q_j}$ are principal for $j = 1, 2, \ldots, r$ and non-principal for $j \geq 1+r$ and denote by X_j the class of $H(K)$ containing $\mathfrak{P}_j$ (if $j \leq r$) resp. $\mathfrak{P}_j^{q_j}$ (if $j \geq r+1$). Thus all classes X_i are non-principal. Choose prime ideals $q_i \in X_i$ ($i \leq t$) and $q_i' \in X^{-1}$ ($i \leq r$) so that they are all distinct, and observe that the ideal $q_1 \ldots q_t q_1' \ldots q_r'$ is principal. Let a' be one of its generators. To prove our assertion, it suffices to show that a' has unique factorization, since the ideal generated by a' is square-free; the assertion being proved for that particular case, we get

$$s(a) = t \leq t+r = s(a') \leq a_1(H(K)).$$

If π_i is a generator of $\mathfrak{P}_i^{m_i}$ ($i = 1, 2, \ldots, r$), with m_i being the order of X_i, and $A_i = a_i/m_i$, then the number

$$\prod_{i=1}^{r} \pi_i^{A_i}$$

has unique factorization, being a divisor of a, and this shows that $X_1, \ldots, X_r$ generate independent cyclic subgroups of $H(K)$. Indeed, supposing that

$$\prod_{i=1}^{r} X_i^{\alpha_i} = 1,$$

with suitable $0 \leqslant \alpha_i < m_i$, we see that all α_i must vanish, since otherwise $\prod_{i=1}^{r} \mathfrak{P}_i$ would be principal, and none of its generators would be of the form $\prod_{i=1}^{r} \pi_i^{B_i}$ with $0 \leqslant B_i \leqslant A_i$.

Further, if G denotes the subgroup of $H(K)$ generated by $X_1, \ldots, X_r$, then no product $X_{i_1} \ldots X_{i_k} \neq 1$ with $r+1 \leqslant i_1 < i_2 < \ldots < i_k \leqslant t$ can lie in G. Indeed, if this were the case, then

$$\prod_{i=1}^{r} X_i^{b_i} = X_{i_1} \ldots X_{i_k}$$

with suitable $0 \leqslant b_j < m_j$ ($j = 1, 2, \ldots, r$) and thus the ideal $\mathfrak{P}_1^{a_1-b_1} \ldots \mathfrak{P}_r^{a_r-b_r} \mathfrak{P}_{i_1}^{a_{i_1}} \ldots \mathfrak{P}_{i_k}^{a_{i_k}}$ would be principal; hence its generator, a divisor of a, would have unique factorization. This however forces $m_j | a_j - b_j$ for $j = 1, 2, \ldots, r$, and thus $m_j | b_j$ for $j = 1, 2, \ldots, r$, implying $b_1 = \ldots = b_r = 0$.

Now the uniqueness of factorization of a' follows immediately. □

6. We conclude this section with introducing another combinatorial constant associated with a finite Abelian group, which is related to the question of unique factorization of rational integers in quadratic fields.

Let A be a finite Abelian group and let $M(A)$ be the maximal cardinality of a subset $\{a_1, \ldots, a_n\}$ of A with the property that all sums

$$\sum_{j=1}^{n} \varepsilon_j a_j \quad (\varepsilon_j = 0, 1; j = 1, 2, \ldots, n)$$

are distinct.

To establish the link between $M(A)$ and factorizations, consider a quadratic number field K and observe that if $X \in H(K)$, then the orbit of X under the action of the Galois group C_2 of K/Q equals (X, X^{-1}). Indeed, if $\mathfrak{P}$ is a prime ideal of the first degree contained in X and $\mathfrak{P}'$ is its conjugate, then by Theorems 4.2 and 4.15 $\mathfrak{P}' \in X^{-1}$. From each orbit $\neq (E, E)$ (with E being the principal class) choose one class, let $X = (X_1, \ldots, X_t)$ be the set of all classes obtained in this way and let $O_i = (X_i, X_i^{-1})$. If p is a rational prime such that $pR_K = \mathfrak{P}_1 \mathfrak{P}_2$ with $\mathfrak{P}_1 \in X_i$, $\mathfrak{P}_2 \in X_i^{-1}$, then we say that p belongs to the orbit O_i and write $p \sim O_i$.

Proposition 9.4. *Suppose that $O_{i_1}, \ldots, O_{i_s}$ are distinct orbits $\neq (E, E)$ and for $j = 1, 2, \ldots, s$ we have $p_j \sim O_{i_j}$; then the number $m = p_1 p_2 \ldots p_s$ has a unique factorization in K iff all products*

$$X_{i_1}^{\varepsilon_1} \ldots X_{i_s}^{\varepsilon_s} \tag{9.4}$$

with $\varepsilon_j = 0$ or 1 are distinct.

Proof. By Proposition 9.3 m has unique factorization iff the block $b = (X_{i_1}, X_{i_1}^{-1}, \ldots, X_{i_s}, X_{i_s}^{-1})$ has unique factorization, and since the blocks (X_i, X_i^{-1}) are irreducible, this happens iff b has no irreducible factors of length exceeding 2. Now, observe that if

$$(X_{k_1}, \ldots, X_{k_m}, X_{r_1}^{-1}, \ldots, X_{r_n}^{-1}) \tag{9.5}$$

with $k_1, \ldots, k_m, r_1, \ldots, r_n \subset \{i_1, \ldots, i_s\}$ is such a factor, then $k_a \neq r_b$ for all a, b and

$$X_{k_1} \ldots X_{k_m} = X_{r_1} \ldots X_{r_n}, \tag{9.6}$$

showing that not all products (9.4) are distinct. Conversely, if two products of the form (9.4) are equal, then after suitable cancellation we arrive at an equality of the form (9.6), showing that (9.5) is a block which does not have an irreducible factor of length 2, i.e. of the form (X, X^{-1}). □

Corollary. *If m is a square-free rational integer having unique factorization in an quadratic number field K, then m can have at most $M(H(K))$ prime factors which do not generate prime ideals and do not split into principal factors in K. Moreover, there are integers m, for which this bound is attained.*

Proof. First observe that if $m = p_1 \ldots p_s$ has a unique factorization in K and the primes p_i are distinct, then they must belong to different orbits. (If e.g. p_1 and p_2 belong to the same orbit (X, X^{-1}), then by Proposition 9.3 $p_1 p_2$ cannot have unique factorization.) We may thus apply the last proposition to obtain $s \leq M(H(K))$. To show that this bound is attained, observe that if $M = M(H(K))$ and $Y_1, \ldots, Y_M$ are classes in $H(K)$ such that all products

$$\prod_{j=1}^{M} Y_j^{\varepsilon_j}$$

with $\varepsilon_j = 0$ or 1 are distinct, then for $i \neq j$ we have $Y_i Y_j \neq E$; thus all orbits (Y_i, Y_i^{-1}) are distinct. If now $p_i \sim (Y_i, Y_i^{-1})$ for $i = 1, 2, \ldots, M$, then by the proposition the product $p_1 \ldots p_M$ has a unique factorization. □

Our last result gives certain information on the size of $M(A)$.

Proposition 9.5. *If*

$$A \simeq \bigoplus_{j=1}^{r} C_{n_j},$$

then

$$\sum_{j=1}^{r} [\log n_j / \log 2] \leq M(A) \leq \log |A| / \log 2.$$

Proof. The upper bound results from the fact that the number of 0–1 sequences of length M equals 2^M. The lower bound is obtained by considering, for $j = 1, \ldots, r$, the elements $X_j, X_j^2, \ldots, X_j^{2^{k_j}}$ with X_j being the generator of C_{n_j}' and k_j being the largest integer with

$$1 + 2 + \ldots + 2^{k_j} < n_j. \qquad \square$$

Corollary. *If A is either cyclic or a 2-group, then*

$$M(A) = \log|A|/\log 2. \qquad \square$$

§ 2. Quantitative Results

1. In this section we consider the counting functions of irreducible integers and integers with unique factorization. We obtain upper bounds and in certain cases also asymptotic formulas for them.

We start with an auxiliary result concerning the distribution of ideals having a prescribed number of prime ideal divisors in a given class of $H(K)$.

If X is a given set of ideals, we denote by $\omega_X(I)$ the number of distinct prime ideals belonging to X which divide the ideal I, and by $\Omega_X(I)$ the number of those prime ideals counted with multiplicities. In other words, we put $\Omega_X(\mathfrak{P}^m) = m$ for a prime ideal $\mathfrak{P} \in X$, $\Omega_X(\mathfrak{P}^m) = 0$ for a prime ideal not belonging to X and extend Ω_X to all ideals I by additivity; i.e.

$$\Omega_X(I) = \sum \Omega_X(\mathfrak{P}^m),$$

the sum running over all prime ideal powers $\mathfrak{P}^m \| I$.

Theorem 9.4. *Let $X_1, \ldots, X_m$ be given distinct classes in $H(K)$, let $c_1, \ldots, c_m$ be given non-negative integers and, for $i = 1, 2, \ldots, m$, let f_i be equal either to ω_{X_i} or to Ω_{X_i}. Denote by $\Phi(x) = \Phi(x; c_1, \ldots, c_m)$ the number of ideals I of R_K satisfying $N(I) \le x$ and $f_i(I) = c_i$ for $i = 1, 2, \ldots, m$. Further, denote by $\Phi_Y(x) = \Phi_Y(x; c_1, \ldots, c_m)$ the number of such ideals lying in a given class Y of a certain group $H_f^*(K)$. Put also $h = h(K)$. Then we have for x tending to infinity*

$$\Phi(x) = \begin{cases} (C+o(1))x(\log x)^{-m/h}(\log\log x)^T & \text{if } m < h, \\ (C_1+o(1))x(\log x)^{-1}(\log\log x)^{T-1} & \text{if } m = h, \end{cases}$$

(in the case of $m = h$ we have to assume that not all c_i vanish), with $T = c_1 + \ldots + c_m$ and certain positive constants C, C_1, which depend on the classes

$X_1, \ldots, X_m$ and the numbers $c_1, \ldots, c_m$ but not on the choice of the functions $f_1, \ldots, f_m$. More precisely, we have

$$C^{-1} = c_1! \ldots c_m! h^T \Gamma(1 - m/h) \gamma$$

and

$$C_1^{-1} = c_1! \ldots c_m! h^T T^{-1} \gamma,$$

where γ is a non-zero number, depending only on $X_1, \ldots, X_m$. Moreover, if $m < h/2$, then

$$\Phi_Y(x) = (h_f^*(K)^{-1} + o(1)) \Phi(x).$$

For any character χ of $H_f^*(K)$ consider the function

$$H_\chi(s; z_1, \ldots, z_m) = \sum_I \left(\prod_{j=1}^m z_j^{f_j(I)} \right) N(I)^{-s} \qquad (9.7)$$

defined for $\operatorname{Re} s > 1$ and $|z_i| \leq 1$ ($i = 1, 2, \ldots, m$). By Lemma 7.1 we can write

$H_\chi(s; z_1, \ldots, z_m)$

$$= \prod_{\mathfrak{p}} \left(1 + \sum_{j=1}^\infty \left(\prod_{i=1}^m z_i^{f_i(\mathfrak{p}^j)} \right) \chi(\mathfrak{p})^j N(\mathfrak{p})^{-js} \right)$$

$$= \exp \left\{ \sum_{\mathfrak{p}} \log \left(1 + \sum_{j=1}^\infty \left(\prod_{i=1}^m z_i^{f_i(\mathfrak{p}^j)} \right) \chi(\mathfrak{p})^j N(\mathfrak{p})^{-js} \right) \right\} = ABC,$$

where

$$A = \exp \left\{ \sum_{\mathfrak{p}} \left(\prod_{i=1}^m z_i^{f_i(\mathfrak{p})} \right) \chi(\mathfrak{p}) N(\mathfrak{p})^{-s} \right\},$$

$$B = \exp \left\{ \sum_{\mathfrak{p}} \sum_{j \geq 2} \left(\prod_{i=1}^m z_i^{f_i(\mathfrak{p}^j)} \right) \chi(\mathfrak{p})^j N(\mathfrak{p})^{-js} \right\}$$

and

$$C = \exp \left\{ \sum_{\mathfrak{p}} \sum_{k \geq 2} (-1)^{k+1} k^{-1} \left(\sum_{j=1}^\infty \left(\prod_{i=1}^m z_i^{f_i(\mathfrak{p}^j)} \right) \chi(\mathfrak{p})^{-j} N(\mathfrak{p})^{-sj} \right)^k \right\}.$$

Since $f_j(\mathfrak{p}) = 1$ for $\mathfrak{p} \in X_j$ and $f_j(\mathfrak{p}) = 0$ otherwise, we get, using Corollary 6 to Proposition 7.9,

$$A = \exp \left\{ \sum_{A \in H_f^*(K)} \chi(A) \left(\sum_{j=1}^m z_j \sum_{\mathfrak{p} \in X_j \cap A} N(\mathfrak{p})^{-s} + \sum_{\mathfrak{p} \in A \setminus \bigcup_{j=1}^m X_j} N(\mathfrak{p})^{-s} \right) \right\}$$

$$= \exp\left\{\sum_{A \in H_{\bar{1}}^*(K)} \chi(A)\left(\sum_{j=1}^{m} a(A, X_j)z_j + h_{\bar{1}}^*(K)^{-1}\right.\right.$$

$$\left.\left. - \sum_{j=1}^{m} a(A, X_j)\right)\log\frac{1}{s-1} + g_\chi(s; z_1, \ldots, z_m)\right\},$$

where

$$a(A, X_i) = \begin{cases} h_{\bar{1}}^*(K)^{-1} & \text{if } A \subset X_i, \\ 0 & \text{otherwise} \end{cases}$$

and g_χ is a function regular for $\operatorname{Re} s \geq 1$ and $|z_j| \leq 1$. Observe that g_χ does not depend on the choice of the functions $f_1, \ldots, f_m$; in particular, the value $\alpha = \exp g_{\chi_0}(1, 0, \ldots, 0)$ depends only on the classes $X_1, \ldots, X_m$ in the case of the principal character χ_0.

Now, put

$$d_i(\chi) = \sum_{A \in H_{\bar{1}}^*(K)} \chi(A) a(A, X_i);$$

observe that $d_i(\chi_0) = h^{-1}$ and note that the product BC is equal to a function regular in $\operatorname{Re} s \geq 1$ and $|z_i| \leq 1$. Hence

$$H_{\chi_0}(s; z_1, \ldots, z_m)$$

$$= \prod_{j=1}^{m} \exp\left(z_i h^{-1}\log\frac{1}{s-1}\right)(s-1)^{m/h-1} h_{\chi_0}(s; z_1, \ldots, z_m) \quad (9.8)$$

with h_{χ_0} regular for $\operatorname{Re} s \geq 1$ and $|z_i| \leq 1$.

We need to know more about the value of h_{χ_0} at $s = 1$, $z_1 = z_2 = \ldots = z_m = 0$. Using the fact that for $\chi = \chi_0$ the product BC attains at that point the value

$$\beta = \exp\left\{\sum_{\substack{p \notin \bigcup_{j=1}^m X_j}}\left(\sum_{j \geq 2} N(\mathfrak{p})^{-j} + \sum_{k \geq 2}(-1)^{k+1}k^{-1}\left(\sum_{j=1}^{\infty} N(\mathfrak{p})^{-j}\right)^k\right)\right\},$$

we get $h_{\chi_0}(1; 0, \ldots, 0) = \alpha\beta \neq 0$. Observe also that the product $\alpha\beta$ does not depend on the choice of $f_1, \ldots, f_m$.

Further,

$$H_\chi(s; z_1, \ldots, z_m)$$

$$= \prod_{i=1}^{m} \exp\left(z_i d_i(\chi)\log\frac{1}{s-1}\right)(s-1)^{R(\chi)} h_\chi(s; z_1, \ldots, z_m) \quad (9.9)$$

with

$$R(\chi) = \sum_{i=1}^{m} d_i(\chi).$$

Expanding the right-hand sides of (9.8) and (9.9) into power series in $z_1, \ldots, z_m$ and comparing coefficients in (9.7) we arrive at the following equalities, valid for $\operatorname{Re} s > 1$:

$$\sum_{\substack{I \\ (I, \mathfrak{f}) = 1 \\ f_i(I) = c_i \\ (i=1,2,\ldots,m)}} N(I)^{-s}$$

$$= (s-1)^{m/h-1} \sum_{\substack{r_1 + j_1 = c_1 \\ \cdots\cdots\cdots \\ r_m + j_m = c_m}} A^{(\chi_0)}_{r_1, \ldots, r_m}(s) h^{-J} \log^J \frac{1}{s-1} \prod_{i=1}^{m} (j_i!)^{-1} + g_{\chi_0}(s), \quad (9.10)$$

$$\sum_{\substack{I \\ (I, \mathfrak{f}) = 1 \\ f_i(I) = c_i \\ (i=1,2,\ldots,m)}} \chi(I) N(I)^{-s}$$

$$= (s-1)^{R(\chi)} \sum_{\substack{r_1 + j_1 = c_1 \\ \cdots\cdots\cdots \\ r_m + j_m = c_m}} A^{(\chi)}_{r_1, \ldots, r_m} d_1^{j_1}(\chi) \ldots d_m^{j_m}(\chi) \log^J \frac{1}{s-1} \prod_{i=1}^{m} (j_i!)^{-1} + g_\chi(s)$$

where $J = j_1 + \ldots + j_m$, g_χ is regular for $\operatorname{Re} s \geq 1$, and the functions $A^{(\chi)}_{r_1, \ldots, r_m}$ are defined by

$$\sum_{r_1, \ldots, r_m} A^{(\chi)}_{r_1, \ldots, r_m}(s) z_1^{r_1} \ldots z_m^{r_m} = h_\chi(s; z_1, \ldots, z_m).$$

(Note that $A^{(\chi_0)}_{0, \ldots, 0}(1) = \alpha\beta$.)

Applying to equality (9.10) (in the case of $\mathfrak{f} = R_K$) Theorem I of Appendix II, we get the first two assertions of the theorem, with $\gamma = \alpha\beta$.

To obtain the third assertion we also use (9.10). Write for $\operatorname{Re} s > 1$

$$\sum_{\substack{I \in Y \\ f_i(I) = c_i \\ (i=1,2,\ldots,m)}} N(I)^{-s} = h_\mathfrak{f}^*(K)^{-1} \sum_{\chi} \overline{\chi(Y)} \sum_{\substack{I \\ f_i(I) = c_i \\ (i=1,2,\ldots,m)}} \chi(I) N(I)^{-s}$$

$$= (s-1)^{m/h-1} P_0 \left(\log \frac{1}{s-1} \right) + \sum_{\chi \neq \chi_0} (s-1)^{R(\chi)} P_\chi \left(\log \frac{1}{s-1} \right) + g(s),$$

where P_0, P_χ are polynomials over the ring of functions regular in $\operatorname{Re} s \geq 1$ and g is regular there. The degree of P_0 equals $c_1 + \ldots + c_m = T$ and its leading coefficient does not vanish at $s = 1$. Finally, we have

$$\operatorname{Re}(-R(\chi)) = -\operatorname{Re}\left(\sum_{i=1}^{m} \sum_{A \in H_\mathfrak{f}^\bullet(K)} \chi(A) a(A, X_i) \right)$$

$$\leqslant \sum_{i=1}^{m} \sum_{A \in H_{\mathfrak{f}}^{*}(K)} a(A, X_i) = m/h < 1 - m/h,$$

and we again may apply the Tauberian theorem from Appendix II. □

Note that in the third part of the theorem one cannot omit the condition $m < h/2$. For instance, if $h = 2$, $\mathfrak{f} = R_K$, $Y = E$ and $X_1 \neq E$, then for $f_1 = \Omega_{X_1}$ and odd c_1 we get $F_Y(x) = 0$ for all x.

Corollary 1. *If $A(x, n)$, $B(x, n)$ denote the number of ideals I with $N(I) \leqslant x$ which have n prime ideal divisors, resp. n prime ideal factors counted according to their multiplicities, then*

$$A(x, n) = (C + o(1)) x (\log \log x)^{n-1} / \log x$$

with a certain constant $C = C(n, K) > 0$ and

$$B(x, n) = (1 + o(1)) A(x, n).$$

Proof. Observe that for $m = h$ (with $f_i = \omega_i$ resp. $f_i = \Omega_i$ for all i) the sum

$$\sum_{\substack{c_1, \ldots, c_h \\ c_1 + \ldots + c_h = n}} \Phi(x, c_1, \ldots, c_h)$$

equals $A(x, n)$, resp. $B(x, n)$ and apply the theorem. □

Corollary 2. *For the number of principal ideals I with $N(I) \leqslant x$ which have n prime ideal factors counted with their multiplicities we get the asymptotics*

$$(C_1 + o(1)) x (\log \log x)^{n-1} / \log x.$$

Proof. If $X_1, \ldots, X_h$ are all classes of $H(K)$, then the number in question equals

$$\sum_{\substack{c_1, \ldots, c_h \\ c_1 + \ldots + c_h = n \\ X_1^{c_1} \ldots X_h^{c_h} = E}} \Phi(x, c_1, \ldots, c_h)$$

where in the definition of Φ only the functions Ω_{X_i} are involved. Now it suffices to apply the theorem. □

Corollary 3. *Let $X_1, \ldots, X_h$ be all classes of $H(K)$ and let $c_1, \ldots, c_h$ be given, not all vanishing. Then the number of ideals $I \in X_j$ with $N(I) \leqslant x$ and $\Omega_{X_i}(I) = c_i$ for $i = 1, 2, \ldots, h$ equals*

$$(C_2 + o(1)) x (\log \log x)^{T-1} / \log x$$

with positive C_2 and $T = c_1 + \ldots + c_h$ if $\prod_{i=1}^{h} X_i^{c_i} = X_j$ and equals zero otherwise.

Proof. Observe that if $X_1^{c_1} \ldots X_h^{c_h} = X_j$ then every ideal I satisfying $\Omega_{X_i}(I) = c_i$ for $i = 1, 2, \ldots, h$ lies in X_j and apply the theorem. □

Note finally that the same method as used in the proof of Theorem 9.4 leads to the proof of the following result:

Proposition 9.6. *Let $\Pi_1, \ldots, \Pi_m$ be disjoint sets of prime ideals, which are regular and have positive densities $a_1, \ldots, a_m$. Let for each $i = 1, 2, \ldots, m$ a function $f_i(I)$ be given which equals either ω_{Π_i} or Ω_{Π_i}. Further, let $c_1, \ldots, c_m$ be given non-negative integers with non-vanishing sum T and let $F(x, c_1, \ldots, c_m)$ be the number of ideals I with $N(I) \leq x$ and $f_i(I) = c_i$ for $i = 1, 2, \ldots, m$. Then*

$$F(x, c_1, \ldots, c_m) = (C + o(1)) x (\log\log x)^M (\log x)^{-a}$$

where

$$a = a_1 + \ldots + a_m, \qquad M = \begin{cases} T & \text{if } a < 1, \\ T-1 & \text{if } a = 1, \end{cases}$$

and C is a positive number, which does not depend on the choice of the functions f_i.

We leave the needed modifications to the reader. □

2. Our first application of Theorem 9.4 concerns irreducible integers:

Theorem 9.5. *Let $F(x)$ be the number of irreducible integers in K whose norms do not exceed x in absolute value, and where from each set of associated integers only one is counted. Then we have for x tending to infinity*

$$F(x) = (C + o(1)) x (\log\log x)^{D-1} / \log x$$

with a suitable $C > 0$ and $D = D(H(K))$, the Davenport constant of $H(K)$.

Proof. If I is a principal ideal, $I = \mathfrak{P}_1 \ldots \mathfrak{P}_s$ its factorization into not necessarily distinct prime ideals and X_i is the class of $\mathfrak{P}_i$ in $H(K)$, then we denote by $b(I)$ the block $(X_1, X_2, \ldots, X_s)$ in $B(H(K))$. Now let $b_1, \ldots, b_T$ be all irreducible blocks in $B(H(K))$ and note that according to Proposition 9.2 we have $T \leq 1 + D(H(K))^{h-1}$. Now Proposition 9.1 implies

$$F(x) = \sum_{j=1}^{T} \sum_{\substack{I \in E \\ N(I) \leq x \\ b(I) = b_j}} 1.$$

Moreover, if for a class X we denote by $c_j(X)$ the number of appearances of X

in b_j, then $b(I) = b_j$ holds iff $\Omega_X(I) = c_j(X)$ for every class X. This shows that, in the notation of Theorem 9.4,

$$\sum_{\substack{I \in E \\ N(I) \leq X \\ b(I) = b_j}} 1 = \Phi_E(x; \{c_j(X): X \in H(K)\})$$

$$= (C + o(1)) x (\log \log x)^{t_j - 1} (\log x)^{-1},$$

in view of Corollary 3 to Theorem 9.4 with $t_j = \sum_{X \in H(K)} c_j(X)$ being the length of the block b_j. An application of Proposition 9.2 leads to our assertion. □

One can also ask for rational primes which remain irreducible in K. Such primes exist rather seldom, as the following easy result shows:

Proposition 9.6. *Let K/Q be normal and assume that there are rational primes which do not ramify in K/Q and which remain irreducible in K. Then the Galois group of K/Q contains a cyclic subgroup of index not exceeding $D(H(K))$.*

Proof. If p does not ramify in K/Q, then we have $pR_K = \mathfrak{P}_1 \ldots \mathfrak{P}_s$ with distinct prime ideals $\mathfrak{P}_i$. If X_i is the class of $\mathfrak{P}_i$ in $H(K)$, then by Proposition 9.1 the block $(X_1, \ldots, X_s)$ is irreducible and so, by Proposition 9.2, we have $s \leq D(H(K))$. However, by Corollary to Proposition 6.5, s is the index of the decomposition group of $\mathfrak{P}_1$ in $\text{Gal}(K/Q)$ and since the decomposition group is cyclic, the result follows. □

Corollary 2 to Theorem 7.10 shows that if K/Q is cyclic, then infinitely many rational primes generate prime ideals in R_K and Theorem 7.11* implies that the density of the set of all these primes equals $\varphi(n)/n$ with $n = [K:Q]$, because a cyclic group of order n has $\varphi(n)$ generators. There may be also other primes which remain irreducible; we can now obtain a formula expressing the asymptotic behaviour of their counting function in the special case when K/Q is cyclic of a prime degree:

Proposition 9.7. *Let K/Q be cyclic of prime degree q, and denote by $P(x)$ the number of rational primes $p \leq x$ which remain irreducible in K. Then*

$$P(x) = (c + o(1)) x / \log x,$$

where

$$c = 1 - 1/q + \left(\sum_{b \in A} r(b)\right) h(K)^{-1},$$

A denotes the set of all irreducible blocks $b \in B(H(K))$ which are invariant

under the action of $G = \text{Gal}(K/Q)$ and for every such block the number $r(b)$ is defined by

$$r(b) = \begin{cases} 1/q & \text{if all classes in } b \text{ coincide,} \\ 1 & \text{otherwise.} \end{cases}$$

Proof. As in the proof of Theorem 9.5, we have

$$P(x) = \sum_{\substack{b \in A}} \sum_{\substack{p \leq x \\ b(pR_K) = b}} 1 + \sum_{\substack{p \leq x \\ pR_K \text{ is prime}}} 1.$$

In order to evaluate the inner sum in the first term observe that if $b = (X_1, \ldots, X_r)$ and $X_1 = X_2$, then necessarily $X_1 = X_2 = \ldots = X_r$, G being cyclic of a prime degree; thus b consists either of distinct classes, or $b = (X, X, \ldots, X)$. Using Corollary 4 to Proposition 7.10 we get in the first case

$$\sum_{\substack{p \leq x \\ b(pR_K) = b}} 1 = \sum_{\substack{\mathfrak{p} \in X_1 \\ N(\mathfrak{p}) \leq x}} 1 = (1/h(K) + o(1))x/\log x = (r(b) + o(1))x/\log x$$

and in the second case

$$\sum_{\substack{p \leq x \\ b(pR_K) = b}} 1 = q^{-1} \sum_{\substack{\mathfrak{p} \in X_1 \\ N(\mathfrak{p}) \leq x}} 1 = (1/qh(K) + o(1))x/\log x = (r(b) + o(1))x/\log x.$$

Adding these equalities and noting that

$$\sum_{\substack{p \leq x \\ pR_K \text{ is prime}}} 1 = (1 - q^{-1} + o(1))x/\log x,$$

in virtue of Corollary 5 to Proposition 7.9, we obtain our assertion. □

3. Now we turn to numbers with unique factorization. In this subsection $F(x)$ denotes the number of non-associated integers of K with unique factorization, whose norms are absolutely bounded by x. We shall determine the right order of magnitude of $F(x)$; however, we will not prove the corresponding asymptotic equality, whose available proof is rather technical.

Theorem 9.6. *Let K be an algebraic number field with $h = h(K) \geq 2$. Then there exist two positive constants C_1, C_2 such that, for x tending to infinity,*

$$(C_1 + o(1))x(\log\log x)^{a_1}(\log x)^{1 - 1/h} \leq F(x)$$
$$\leq (C_2 + o(1))x(\log\log x)^{a_1}(\log x)^{1 - 1/h}$$

with $a_1 = a_1(H(K))$, as defined in subsection 5 of the preceding section.

Proof. Let $A = \{b_1, \ldots, b_T\}$ be the set of all blocks in $B(H(K))$ which have unique factorization and do not contain the unit class E. Since the length of a block in A does not exceed $a_1(A)$, the set A is finite. If now $X_1, \ldots, X_{h-1}$ denote the non-unit classes of $H(K)$ and $c_j(X_i)$ denotes, as in the proof of Theorem 9.5, the number of occurrences of X_i in the block b_j, then by Proposition 9.3 and Theorem 9.4 we get

$$F(x) \geq \sum_{j=1}^{T} \sum_{\substack{I \in E \\ N(I) \leq x \\ \Omega_{X_i}(I) = c_j(X_i) \\ (i=1,2,\ldots,h-1)}} 1 = \sum_{j=1}^{T} \sum_{\substack{N(I) \leq x \\ \Omega_{X_i}(I) = c_j(X_i) \\ (i=1,2,\ldots,h-1)}} 1$$

$$= (C_1 + o(1)) x (\log\log x)^{a_1(H(K))} (\log x)^{1/h - 1}$$

with a suitable $C_1 = C_1(K) > 0$. This proves the first part of the assertion. To prove the second, we utilize the Corollary to Proposition 9.3, from which we obtain immediately (with $a_1 = a_1(H(K))$)

$$F(x) \leq \sum_{\substack{c_1,\ldots,c_{h-1} \\ c_1+\ldots+c_{h-1} \leq a_1}} \sum_{\substack{I \in E \\ N(I) \leq x \\ \Omega_{X_i}(I) = c_i \\ (i=1,2,\ldots,h-1)}} 1 \leq \sum_{\substack{c_1,\ldots,c_{h-1} \\ c_1+\ldots+c_{h-1} \leq a_1}} \sum_{\substack{N(I) \leq x \\ \Omega_{X_i}(I) = c_i \\ (i=1,2,\ldots,h-1)}} 1$$

$$= (C_2 + o(1)) x (\log\log x)^{a_1} (\log x)^{1/h - 1}$$

with a certain $C_2 = C_2(K)$, the final equality being a consequence of Theorem 9.4. □

A set M of elements of R_K consisting of full classes of associated integers is said to *contain almost all integers of K* if its counting function

$$|\{a \in M: a \text{ pairwise non-associated, } |N(a)| \leq x\}|$$

satisfies

$$\lim_{x \to \infty} T(x)/I(x) = 1,$$

where $I(x)$ denotes the number of pairwise non-associated integers a with $|N(a)| \leq x$. Clearly $I(x)$ equals the number of principal ideals of R_K with norms not exceeding x, and thus by Theorem 7.6 we have $I(x) = \varkappa x + o(x)$ with $\varkappa$ defined by (6.12). Thus M contains almost all integers iff $M(x)/x$ tends to $\varkappa$ when x tends to infinity. Similarly, we say that M contains almost no integers, if $M(x) = o(x)$.

In this terminology, we state:

Corollary. *If $h(K) \geq 2$, then almost no integer of K has a unique factorization.* □

4. It is possible to obtain similar results concerning rational integers with unique factorization in a given field K. Unfortunately, the argument in the general case relies on class-field theory and so we treat here only the case of a cyclic extension of prime degree.

Theorem 9.7. *Let K/Q be cyclic of a prime degree q and $h = h(K) \geqslant 2$. Denote by $F_0(x)$ the number of positive rational integers $n \leqslant x$ which have unique factorization in K. Then*

$$F_0(x) = (C + o(1)) x (\log\log x)^M (\log x)^{(1-h)/hq}$$

holds with a certain positive constant C, where M is a non-negative integer depending on the action of the Galois group G of K/Q on $H(K)$, not exceeding the number of orbits $\neq (E, E, \ldots, E)$. If $q = 2$, then $M = M(H(K))$, as defined in subsection 6 of the preceding section.

Proof. Let $O_1, \ldots, O_r$ be all the orbits $\neq (E, \ldots, E)$ of $H(K)$ under the action of G. With every orbit $O_i = (X_1, \ldots, X_q)$ we associate the set $\mathfrak{P}_i$ of primes consisting of all primes p unramified in K/Q for which

$$p R_K = \mathfrak{P}_1 \ldots \mathfrak{P}_q$$

holds with $\mathfrak{P}_i \in X_i$ ($i = 1, 2, \ldots, q$). Assume that the orbits $O_1, \ldots, O_t$ have the form $(X, X, \ldots, X)$ and the remaining $r - t$ orbits consist of distinct classes. By Corollary 6 to Proposition 7.9 we infer that all sets $\mathfrak{P}_i$ are regular and of density $1/hq$ ($i = 1, 2, \ldots, t$) resp. $1/h$ ($i = t+1, \ldots, r$). Observe that if n has a unique factorization of K, then for $i = 1, 2, \ldots, t$ we have $\omega_{\mathfrak{P}_i}(n) \leqslant 1$. Indeed, supposing that n has two divisors p_1, p_2 from $\mathfrak{P}_i$, we see that

$$p_1 R_K = \mathfrak{P}_1 \ldots \mathfrak{P}_q$$

and

$$p_2 R_K = \mathfrak{P}'_1 \ldots \mathfrak{P}'_q$$

holds with suitable prime ideals $\mathfrak{P}_j, \mathfrak{P}'_j \in X_j$ and thus $p_1 p_2$ would have two factorizations arising from

$$p_1 p_2 R_K = (\mathfrak{P}_1 \ldots \mathfrak{P}_q)(\mathfrak{P}'_1 \ldots \mathfrak{P}'_q) = (\mathfrak{P}'_1 \mathfrak{P}_2 \ldots \mathfrak{P}_q)(\mathfrak{P}_1 \mathfrak{P}'_2 \mathfrak{P}'_3 \ldots \mathfrak{P}'_q).$$

Let A be the set of all sequences $\{a_1, \ldots, a_r\}$ with the property that there exists a square-free number n with unique factorization in K for which $\omega_{\mathfrak{P}_i}(n) = a_i$ holds for $i = 1, 2, \ldots, r$. Put

$$M = \max\left\{\sum_{j=1}^{r} a_j : \{a_1, \ldots, a_r\} \in A\right\}.$$

(The preceding remark shows that $0 \leqslant a_i \leqslant 1$ and thus $M \leqslant r$.)

Finally, note that if n satisfies $\Omega_{\mathfrak{P}_i}(n) = a_i$ for $i = 1, 2, \ldots, r$ and $(a_1, \ldots, a_r) \in A$, then, owing to Proposition 9.3, n has unique factorization in K. This leads to

$$\sum_{(a_1,\ldots,a_r) \in A} \sum_{\substack{n \leqslant x \\ \Omega_{\mathfrak{P}_i}(n) = a_i \\ (i=1,2,\ldots,r)}} 1 \leqslant F_0(x) \leqslant \sum_{(a_1,\ldots,a_r) \in A} \sum_{\substack{n \leqslant x \\ \omega_{\mathfrak{P}_i}(n) = a_i \\ (i=1,2,\ldots,r)}} 1$$

and an application of Proposition 9.6 gives the assertion. If $q = 2$, then the Corollary to Proposition 9.4 gives $M = M(H(K))$. □

In the general case we prove only the following simple result:

Proposition 9.8. *If K/Q is normal of degree N and $h = h(K) \geqslant 2$, then we have*

$$F_0(x) = O(x(\log\log x)^B (\log x)^{(1-h)/hN})$$

with a suitable $B \geqslant 0$.

Proof. Fix a class $X \neq E$ of $H(K)$. The argument used in the proof of the foregoing theorem shows that if n has a unique factorization, then it cannot be divisible by two distinct primes each of which has a prime ideal divisor belonging to X. Let P be the set of all rational primes which do not ramify in K/Q and which have a non-principal prime ideal factor of degree 1 in K. In view of

$$\sum_{p \in P} p^{-s} = N^{-1} \sum_{\substack{\mathfrak{P} \notin E \\ f_{K/Q}(\mathfrak{P}) = 1 \\ e_{K/Q}(\mathfrak{P}) = 1}} N(\mathfrak{P})^{-s}$$

(for $\operatorname{Re} s > 1$) the set P is regular and of density $(h-1)/hN$, by Corollary 6 to Proposition 7.9. If n has unique factorization in K, then the previous remark implies $\omega_P(n) \leqslant B$, with B denoting the number of distinct orbits $\neq (E, \ldots, E)$ of $H(K)$ under $\operatorname{Gal}(K/Q)$. Now, Proposition 9.6 implies our assertion. □

§ 3. Notes to Chapter 9

1. Theorem 9.1 is due to L. Carlitz [60]. This theorem is not true for arbitrary Dedekind domains; suitable counterexamples can be constructed using the results of L. Claborn [68]. See Skula [76], Zaks [76], [80] for domains in which the length of a factorization depends only on the factorized element.

Other characterizations of algebraic number fields with small class-groups have been given in Czogała [81], Feng Keqin [85], Salce, Zanardo [82]. For characterization of fields with particular class-groups see also Kaczorowski [81a], Krause [84].

Theorem 9.2 was proved by J. Kaczorowski [84]. A variant of it was given in Halter-Koch [83], where also another description of $H(K)$ was presented. Yet another description was given in Rush [83]. Cf. Halter-Koch [85].

Absolutely irreducible elements were studied by J. Kaczorowski [81b].

2. Theorem 9.3 was proved by J. E. Olson [69] and S. Schanuel. For further results concerning $D(A)$ see Baayen [65], [68a], [68b], [69], Baayen, Emde Boas, Kruyswijk [69], Emde Boas [68], [69], Emde Boas, Kruyswijk [67], Zame [67].

The constant $M(A)$ appears, in the special case of $A = C_k^n$, already in Shannon [56]. It is easy to see that $M(C_3^n) = n$ and it was shown in Mead, Narkiewicz [82] that $M(C_5^n) = 2n$. In the last mentioned paper the value of $M(C_m^2)$ was found for certain integers m. In the case of $A = C_k^n$ the bound in Proposition 9.5 was improved by S. K. Stein [77].

Other combinatorial constants occurring in § 1 have been considered in Narkiewicz [79] and Narkiewicz, Śliwa [82]. Cf. also Skula [76], Śliwa [76a], [82b].

3. Theorems 9.4 and 9.5 are due to P. Rémond [64], [66]. Cf. Lardon [71]. The proof of Theorem 9.4 given here does not lead to any non-trivial evaluation of the error term, and therefore we were unable to get any information on error terms in its applications in § 2. An evaluation of this term, and even an asymptotic expansion was given by J. Kaczorowski [83]; with use of it he also obtained corresponding expansions in Theorems 9.5 and 9.7 (for all fields).

An extension of Theorem 9.4 was given by J. Śliwa [76a]. For an analogue of Corollary 1 to Theorem 9.4 see Nakamura [59].

4. The condition given in Proposition 9.6 is necessary but in general not sufficient for the existence of unramified primes which remain irreducible in K. In the case of a normal extension K/Q a necessary and sufficient condition was given by J. Śliwa [77].

In Theorem 9.6 one has in fact $C_1 = C_2$; hence, there is an asymptotic equality for $F(x)$. This was proved in Narkiewicz [72]. The Corollary to Theorem 9.6 was proved for the field $Q(i5^{1/2})$ by E. Fogels [43] and in the general case in Narkiewicz [64].

Theorem 9.7 was in case $q = 2$ obtained in Narkiewicz [66]; and R. W. K. Odoni [76] obtained asymptotics for $F_0(x)$ for arbitrary, not necessarily normal fields. For previous work see Allen [75], Narkiewicz [66], [73]. An analogous result for numbers with at most m factorizations was obtained in Śliwa [76b].

It was shown in Narkiewicz [66] that in the case of $h(K) \geqslant 3$ almost all integers of K have factorizations of distinct lengths and the same applies to rational integers. Asymptotics in both cases was obtained in Śliwa [76a]. (Cf. Śliwa [82b].) For previous evaluations see Allen [75], Narkiewicz [73] and for the quadratic case cf. Narkiewicz [66], [68], [81].

J. Rosiński and J. Śliwa [76] settled in the negative a question of P. Turán, who asked whether the function $f(n)$ counting the factorizations of a rational integer n in a given field K can have a non-decreasing normal order when $h(K) \neq 1$. (Recall that F is called a *normal order* for f, if the inequality

$$|f(n) - F(n)| < \varepsilon F(n)$$

holds for all positive ε and almost all n.) For a special case see Narkiewicz [67].

However, it has turned out that the function $\log f(n)$ has a non-decreasing normal order, equal to $C(K) \log \log n \log \log \log n$ with a certain positive $C(K)$ (Narkiewicz [80]). Similarly, the function $C_1(K) \log \log n$ (with a certain positive $C_1(K)$) serves as a normal order for the number of factorizations of distinct length of n in K, provided $h(K) \geqslant 3$. This was proved independently in Allen, Pleasants [80] and Narkiewicz, Śliwa [78].

The mean value for $f(n)$ was found by Rémond [64], [66].

Another approach to counting irreducible integers, resp. integers with at most m factorizations, was presented by H. Weber [84], who used a method of Siegel [36].

For other results concerning factorizations, irreducible integers and similar questions see Bumby [67], Bumby, Dade [67], Butts, Pall [67], Lettl [87], Pall [45].

5. The remaining part of this section will be devoted to a brief survey of three topics which were not touched upon in the main text, viz. arithmetic functions, additive theory and Diophantine equations. We commence with arithmetic functions.

Dirichlet convolution in the ring of functions defined on ideals of R_K has been considered in Hayashi [75], Matsuda [70b], Mauclaire [76], Takahashi [73], K. Yamamoto [58].

The main results of the theory of additive functions in Z can be carried over to algebraic number fields. On this topic see Baibulatov [67], [68], [69], Danilov [65], [67b], Fluch [69], Jushkis [64], Krompiewska [77], de Kroon [65], Prachar [52], Rieger [62c], Wowk [75].

Analogues of the classical evaluations of summatory functions of various divisor functions have been considered in Axer [04], Eda [55], Hasse, Suetuna [31], Iseki [53], Kanemitsu [78], Lai Dyk Thin [62], [65], Landau [03a], [25], Pergel [67], Rieger [57a], Suetuna [25a], [25b], [28a], [28b], [29], [31], Anna Walfisz [64], Arnold Walfisz [25].

In particular, if $d_m(I)$ denotes the number of representations of an ideal I as a product of m ideals, then

$$\sum_{N(I) \leqslant x} d_m(I) = x P_{m-1}(\log x) + O(x^{1-2/(1+mn)+\varepsilon})$$

(with a polynomial P_{m-1} of degree $m-1$, $n = [K:Q]$ and arbitrary $\varepsilon > 0$). This follows from a general result of E. Landau [12a]; the error term cannot be replaced by $O(x^{a-\varepsilon})$ with $a = (mn-1)/2mn$ and $\varepsilon > 0$ (Landau [24b]). This has been slightly improved in Szegö, Walfisz [27]. Cf. also Berndt [69c], [71c], [75], Chandrasekharan, Narasimhan [62a], [62b], [64], Danilov [67a], Hafner [83], Karatsuba [72], Landau [12b].

For the study of other functions see Cohen [52a], Grotz [79], [80], Lenskoi [63], Matsuda [70a].

The behaviour of various sets of ideals defined by multiplicative conditions has been inspected in Archibald [36], Faddeev [55], Friedlander [72], Gillett [70], Hasse, Tornier [28], Hazlewood [75], [77], Jordan [65], Rieger [57b], [59c], Worley [70], Zaikina [57a].

Sieve methods in algebraic number fields have been considered in Hinz [81], Huxley [68], Levin, Tulyaganova [66], Rieger [58e], [60b], [61c], Schaal [68], [70], A. I. Vinogradov [64a], [64b] and Wilson [69]. For a survey see Schaal [84].

Applications of sieve methods to the study of polynomial values with small number of prime ideal factors have been given in Hinz [82a], [82b] and Sarges [76].

Sequences uniformly distributed with respect to an ideal have been considered by S. K. Lo, H. Niederreiter [75] and Niederreiter, Lo [75], and in the special case of $K = Q(i)$ in Burke, Kuipers [76] and Kuipers, Niederreiter, Shiue [75].

6. The research on Waring's problem in algebraic number fields started with the investigations of O. Meissner [04], [05], who proved that in a real quadratic field every totally positive element is a sum of at most 5 squares. D. Hilbert stated that in every algebraic number field all totally positive elements are sums of four squares and this was established by C. L. Siegel [21b] after E. Landau [19c] and L. J. Mordell [21] settled the case of quadratic resp. cubic fields.

The analogous question of representing totally positive integers as sums of squares of numbers with bounded denominators was considered in Siegel [22c], who showed that also in this problem four squares are sufficient. Much later O. Körner [73] showed that if K is not totally real, then all totally positive integers which are sums of squares are sums of 4 squares of integers. If moreover $d(K)$ is odd, then all totally positive integers of K are sums of 4 squares; this was proved by M. Peters [73]. Cf. Fomenko [82], [83], Peters [74], Scharlau [80b], [80c], Siegel [45b].

As regards the case of higher powers, C. L. Siegel [44c], [45b] showed that every totally positive integer of K with sufficiently large norm which is a sum of kth powers of integers of K is a sum of at most $kn(2^{k-1}+n)+1$ such powers, with $n = [K:Q]$. Further improvements have been achieved by Eda [71], [75], Körner [61c], [62] and Tatuzawa [58], [73d], [75]. For an elementary treatment see Rieger [56], [62a], [63a], [64b]. Siegel conjectured that one can obtain a bound for the number of required summands, which is independent of K. To confirm this, B. J. Birch [64a] and C. P. Ramanujam [63a] proved the analogous statement for p-adic fields, and the analytical part of the problem was settled in Birch [61] and Körner [62]. (Birch's proof requires certain amendments.) Cf. also Bateman, Stemmler [62], Bhaskaran [66], J. H. E. Cohn [71], [72], Eda [67], Joly [65], [66], [68], [70b], Kalinka [63], Körner [70], Niven [41a], [41b], Revoy [79a], Tsunekawa [61].

Bounds for Waring's constants in p-adic fields have been considered in Bhaskaran [69], Bovey [76], Dodson [73], [82], C. Moser [73a], [74], Ramanujam [63a], Riehm [64], Siegel [37].

7. Numerous authors have treated the representation of integers in a given field as sums of squares. The case of 2 squares has been studied for various quadratic fields in Eljoseph [54], Hardy [68], Kalinka [61], Leahey [65], Mordell [67], Nagell [53], [61], [62b], [63a], Niven [40], Pall [51], Schaal [62], H. Schmidt [58], K. S. Williams [67], [73]. For relative-quadratic fields see Lenz [53] and for arbitrary totally real fields see Schaal [65].

Sums of 3 squares have been considered in H. Cohn [61b], Donkar [77], Estes, Hsia [83] and Maass [41] and sums of 4 squares in the papers of H. Cohn [60b], [62c], [64], Cohn, Pall [62], Götzky [28], Kirmse [24]. For numerical results see H. Cohn [59]. A formula for the number of representations by sums of q squares in a quadratic field has been found by J. Dzewas [60] in the case when the form $x_1^2 + \ldots + x_q^2$ has a single class per genus. He showed also that this happens only for finitely many fields and $q \leq 4$.

8. The level (Stufe) $s(R)$ of a ring R is defined as the smallest integer $n \geq 1$ (if it exists) such that -1 is a sum of n squares of elements of R. A. Pfister [65] proved the beautiful result, which says that if R is a field, then $s(R)$, if it exists, must be a power of 2 and to each such power he constructed an appropriate field with that level. There are no such restrictions if R is not a field. See Dai, Lam, Peng [80].

It follows from Satz 14 in Hasse [24a] that for an algebraic number field K with $s(K)$ finite one has $s(K) = 1$ iff $i \in K$, $s(K) = 4$ iff all prime ideal factors of $2R_K$ have their ramification indices and degrees odd and $s(K) = 2$ otherwise.

R. Baeza [79] showed that if R is a Dedekind domain with quotient field K, then $s(R) \leq 1 + s(K)$; in the case of $R = R_K$ this had been earlier established in Peters [72] in a slightly stronger form leading to $s(R_K) \leq 4$.

For the determination of the level in various particular cases see Barnes [72], P. Chowla [69a], [69b], Chowla, Chowla [70], Connell [72], Fein, Gordon, Smith [71], Fujisaki [73], Landau [06], C. Moser [70], [71], [73b], Nagell [72a], [72b], [72d], [73], Plotkin [72], Rajwade [75], [76], Singh [76], J. H. Smith [75], Szymiczek [74].

An analogue of the concept of level for 4-th power has been considered in Parnami, Agrawal, Rajwade [81].

9. The "easier Waring's problem" in which one considers representations $a = \pm x_1^k \pm x_2^k \pm \ldots \pm x_s^k$ has been dealt with by R. M. Stemmler [61]. Cf. Bhaskaran [69], Chinburg [79], Joly [66], [70b], Revoy [79b].

An analogue of the Prouhet–Tarry problem has been studied in Eda, Kitayama [69].

10. The analogue of the binary Goldbach problem in algebraic number fields concerns representations of even integers with large norm as a sum of two numbers generating prime ideals. An algebraic integer is called even if the ideal generated by it is divisible by all prime ideals of the first degree lying over $2Z$. Note that for $K = Q$ this leads to the equation $n = p_1 \pm p_2$ and hence is not a true generalization of the classical Goldbach problem. The first result concerning this question was obtained by H. Rademacher [23], [24], who proved (assuming ERH) an asymptotic formula for a function closely connected with the solutions of
$$a = x_1 + \ldots + x_r \qquad (9.11)$$
with $r \geq 3$ fixed, a totally positive and x_i being totally positive generators

of prime ideals. Under the same assumption A. L. Whiteman [40] showed that in a real quadratic field almost all even integers are sums of two primes (i.e. generators of prime ideals) and later T. Mitsui [60] proved this unconditionally.

The analogue of Schnirelman's theorem was proved by T. Tatuzawa [55]. The ternary Goldbach problem was settled for algebraic number fields by O. Körner [60], [61a], [61b], who showed that every totally positive integer having sufficiently large norm and satisfying certain natural congruence conditions is a sum of three primes. See Mitsui [60] for an asymptotic formula for the number of solutions of (9.11).

For related problems see Andrukhaev [64], Ayoub [53], E. Cohen [56a], Hinz [82c], [84], Kløve [75], Körner [62], [63], [64], Rieger [56], [61b], [62d], Tulyaganova [63], A. I. Vinogradov [64a], [64b].

For other additive questions see Cheema [56], Dress, Scharlau [82], Friedrich [31], Meinardus [53], Mitsui [78], Passi [71], Rademacher [50], Rausch [82], Schaal [77].

11. D. Hilbert [97] showed that Fermat's equation $x^p + y^p = z^p$ has no solutions in the pth cyclotomic field for regular primes p. Cf. Inkeri [49], [56], Nagell [58], Plemelj [12], Vandiver [20]. Apart from this case the equation of Fermat was studied mostly in quadratic fields. W. Burnside [14] dealt with the cubic case and proved the existence of solutions in infinitely many quadratic fields. Cf. Aigner [52a], [52b], [56a], [56b], Fueter [13], Mirimanoff [34]. For other exponents see Aigner [34], [57], Bruckner [72], Christy [72], Fogels [37], Hao, Parry [84], Mordell [68]. For fields of larger degree see Dénes [54].

From Faltings [83] it follows that in a given field Fermat's equation can have only finitely many pairwise non-proportional solutions.

12. The equation $X^2 - aY^2 = 1$ in the field $Q(i)$ has been solved already by Dirichlet [42]. I. Niven [42] proved that it has infinitely many solutions in an imaginary quadratic field K provided a is not a square in K, and also in a real quadratic field, provided $-a$ is not totally positive. Cf. Fjellstedt [54], Niven [43], Skolem [45a], [45b], Walfisz [51].

For other quadratic equations see Hemer [52], Lubelski [31], [39b], Nagell [54], Samet [52], Siegel [73], Szymiczek [72]. Quadratic congruences were considered by E. Cohen [52b], [53], who extended his results to higher degrees in [56b], [56c].

For other diophantine equations in algebraic number fields see Ankeny, Chowla [68], Brindza [84], Chabauty [43], P. Chowla, S. Chowla, Ts'ao [77], Fogels [37], Gordon, Mohanty [77], Hausner [61], Jacobsthal [13], Lal,

McFarland, Odoni [77], Laska [82], Leveque [64], Loxton [74b], Mordell [69a], [69b], Nagell [72c], Rieger [62b], Rosenthall [44], Sexauer [66], [68], Skolem [46].

13. We conclude with a survey of results concerning zeros of forms over algebraic number fields or over p-adic fields. E. Artin has put forward the conjecture that any form of degree n in at least $1+n^2$ variables over Q_p has a non-trivial zero there. This was shown to be false by G. Terjanian [66], who exhibited a form of degree 4 in 18 variables over Q_2 without non-trivial zeros. If we denote by $a(k)$ the minimal value of m such that every form of degree d over a field k in more than d^m variables has a non-trivial zero in k, then Terjanian's result gives $a(Q_2) > 2$. This was improved by J. Browkin [66] to $a(Q_v) \geqslant 3$ for all p (cf. Browkin [69]) and finally G. I. Arkhipov and A. A. Karatsuba [81], [82a], [82b] showed that $a(Q_p)$ is infinite for every p. For further work on this topic see Browkin [83], Brownawell [84], Lewis, Montgomery [83], Terjanian [78], [80].

Artin's conjecture is nevertheless true for quadratic forms (Hasse [24a]) and for cubic forms (Dem'yanov [50], Lewis [52]). It holds also if p is sufficiently large in comparison with the degree and an analogue holds also for systems of forms (Ax, Kochen [65], Greenleaf [65]). (Cf. Armbrust [71], Prestel [78], Ribenboim [68], Tai [67].) For other cases in which Artin's conjecture or its generalization to systems of forms is valid see Birch, Lewis [65], Birch, Lewis, Murphy [62], Chowla [60b], Davenport, Lewis [56], Dem'yanov [56], F. Ellison [73], Maxwell [70], Stevenson [82]. For similar questions see Birch, Lewis [59], [62], Chowla [63], Chowla, Shimura [63], Davenport, Lewis [66], Dem'yanov [55], Laxton, Lewis [65], Lewis [57a], Medvedev [64].

B. J. Birch [64b] showed that to every d there corresponds an integer $C(d)$ such that every form of degree d in at least $C(d)$ variables over Q_p has a non-trivial zero. Cf. Davenport, Lewis [63], Dodson [82], Gray [58], Leep [84], Leep, Schmidt [83], Lewis [57a], W. M. Schmidt [80], [82].

The analogue of the result of Birch holds also for systems of forms of an odd degree over an algebraic number field (Birch [57]). (Cf. R. Brauer [45], Lewis [57b], Peck [49].) The minimal number of variables needed to ensure a non-trivial zero is known only for a system consisting of one cubic form, in which case it equals 10 in the case where the form is non-singular and the field is Q, and equals 16 in the general case (Heath-Brown [83], Pleasants [75]). For previous evaluations see Ramanujam [63b], Ryavec [69].

Pairs of cubic forms have been considered in Davenport, Lewis [66].

An exposition of these questions can be found in the book of M. J. Greenberg [69].

EXERCISES TO CHAPTER 9

1. (Czogała [81].) A field K satisfies condition V_n if any equality of type $a_1 a_2 = b_1 \ldots b_k$ with a_i, b_j being irreducible integers of K can hold only for $k \leqslant n$. K satisfies condition W_n if, moreover, any such equality with $a_1 = a_2$ implies $k = 2$. Prove that $h(K) = 3$ iff K satisfies W_3 but not V_2 and $h(K) = 4$ iff K satisfies either W_4 or V_3 but does not satisfy W_3.

2. (Feng Keqin [85].) Prove that a field K satisfies V_n iff $D(H(K)) \leqslant n$.

3. (Krause [84].) (a) If A is a finite Abelian group then the cross-number $k(A)$ of A is defined as the maximal value of the sum

$$\sum_{a \in b} o^{-1}(a)$$

(where $o(a)$ denotes the order of $a \in A$) for all irreducible blocks $b \in B(A)$. Prove that if A is a direct sum of cyclic groups of orders $n_1, n_2, \ldots, n_s \geqslant 2$, then

$$k(A) \geqslant s - \sum_{i=1}^{s} 1/n_i + e^{-1}$$

where $e = e(A)$ is the exponent of A, i.e. $e = \mathrm{LCM}\{n_i\}$.
 (b) Prove that if $k(A) = 1$, then A is cyclic of a prime-power order.
 (c) Prove the converse of (b).
 (d) Show that a field K has its class-group cyclic of a prime-power order iff there exists an integer N such that the Nth power of any irreducible element is a product of at most N absolutely irreducible elements.

4. (Halter-Koch [84b].) Show that if e is the exponent of $H(K)$ then for every $n = 2, 3, \ldots, e$ one can find non-associated irreducible elements $a_0, a_1, \ldots, a_n$ such that $a_0^n = a_1 \ldots a_n$.

5. (Halter-Koch [83].) Prove that an irreducible integer c is absolutely irreducible iff, for every a, b in R_K, condition $c|ab$ forces that c divides either a^2 or b^2.

6. Prove that $M(C_3^n) = n$ and $M(C_5^n) = 2n$.

7. Show that if $h(K) \geqslant 2$, then for any given positive integer N almost all integers of K have at least N distinct factorizations into irreducibles.

8. Show that if $h(K) \geqslant 3$, then almost all integers of K have factorizations into irreducibles of different lengths.

9. Let K be a quadratic field with $h(K) = 3$. For any positive integer n in K let $f(n)$ be the number of factorizations of n of different lengths. Determine a non-decreasing normal order for the function f.

10. Prove that if K/Q is normal and $h(K) \geqslant 3$, then almost all rational integers have in K factorizations of different lengths.

Appendix I. Locally Compact Abelian Groups

1. In this appendix we collect definitions and results from the theory of locally compact Abelian groups which we use in this book. Theorems are quoted in general without proofs which can be found e.g. in Hewitt, Ross [63] and Rudin [62].

A *locally compact Abelian Group* (LCA) is an Abelian group G with a Hausdorff topology in which G is a locally compact topological space. Moreover, it is required that the map $G \times G \to G$ defined by $f(x, y) = x - y$ be continuous in this topology.

Every such group is a homogeneous space since the map $x \to ax$ is a homeomorphism for every fixed $a \in G$. Moreover, to ensure the local compactness of G it suffices to have an open neighbourhood of the zero element which has a compact closure.

A continuous homomorphism of G into T, the group of complex numbers with unit absolute value, with the usual topology of the circle, is called a *character* of G. In the set $\hat{G}$ of characters the structure of a group is induced by $(fg)(x) = f(x)g(x)$ $(f, g \in \hat{G}, x \in G)$. The family $U(\varepsilon, K) = \{f : |f(x) - 1| < \varepsilon$ for x in $K\}$, where ε is positive and K is a compact subset of G, defines a topology in $\hat{G}$ under which it becomes an LCA, the group of characters or the dual group of G.

Every element of G induces a character on this dual group, namely $g(f) = f(g)$ $(f \in \hat{G}, g \in G)$ and this defines a map $t \colon G \to \hat{\hat{G}}$.

Theorem I (Duality theorem). *The map t is a topological isomorphism of G onto $\hat{\hat{G}}$.*

Theorem II. *The dual group of a discrete group is compact and the dual group of a compact group is discrete.*

Theorem III. *The dual group of $G_1 \times G_2$ equals the product $\hat{G}_1 \times \hat{G}_2$.*

Theorem IV. *Let $G_1, G_2, \ldots$ be a sequence of compact Abelian groups and for $i = 2, 3, \ldots$ let $f_i: G_i \to G_{i-1}$ be a continuous surjective homomorphism. Let $G = \liminv G_i$. Then the group G is compact, and the dual groups $\hat{G}_i$ form, in a natural way, a direct system of discrete groups, whose limit is topologically isomorphic to $\hat{G}$.*

Proof. G is compact, since every projective limit of compact groups is so. For $i \geqslant j$ let a map $f_{ij}: G_i \to G_j$ be defined by

$$f_{ij} = f_{j+1} \circ f_{j+2} \circ \ldots \circ f_i.$$

If X_j is a character of G_j, then $X_j \circ f_{ij}$ is a character of G_i and if we now define $g_{ij}: \hat{G}_j \to \hat{G}_i$ (for $j \leqslant i$) by

$$g_{ij}: X_j \mapsto X_j \circ f_{ij},$$

then the system $\langle G_i; g_{ij} \rangle$ will be direct. Moreover, all g_{ij} are injective and hence the direct limit of this system can be identified with the union of all groups $\hat{G}_i$ if we cease to distinguish between X_j and $g_{ij}(X_j)$. No topological questions arise here, since by Theorem II all groups involved are discrete. Now, every element of this direct limit defines a character χ_i of G_i for i large enough, and if $x = [x_1, x_2, \ldots] \in G$ $(x_i \in G_i)$, then

$$X(x) = X_i(x_i) \quad (i \text{ large enough})$$

defines unambiguously a character of G. Conversely, one can easily see that every character of G is obtainable in this way, and finally it has to be observed that this correspondence preserves multiplication. □

Theorem V. *Let G be a compact Abelian group and $G = G_0 \supset G_1 \supset G_2 \ldots$ an infinite sequence of open subgroups of G with only the unit element in common. If we now define for $i \geqslant j$ mappings $f_{ij}: G/G_i \to G/G_j$ by $f_{ij}(xG_i) = xG_j$, then the inverse limit of the resulting system is topologically isomorphic with G.*

Proof. Let $H = \liminv G/G_i$. Since all G_i's are open and G is compact, the quotients G/G_i are finite, and hence H is compact. Let $f: G \to H$ be given by $f(a) = [aG_i]_i$. This mapping is obviously continuous, and it is injective since $f(a) = e$ implies $a \in \bigcap G_i = \{e\}$. Hence, in view of the compactness of G, the mapping f is a topological isomorphism of G onto a closed subgroup of H. It remains to show that $f(G)$ is dense in H. To this aim take $y = [x_i G_i]_i \in H$. For $j \leqslant i$ we have $x_i G_i = x_j G_j G_i = x_j G_i$ and hence for $y_N = f(x_N)$ we obtain

$$y_N = [x_N G_i]_i = [x_1 G_1, \ldots, x_{N-1} G_{N-1}, x_N G_N, x_N G_{N+1}, \ldots],$$

and so y_N tends to y. □

2. We will also need the concept of quasicharacters of a locally compact Abelian group, i.e. its continuous homomorphisms into the multiplicative group of complex numbers. In the same way as for characters, one introduces group structure and a locally compact topology into the set $\tilde{G}$ of quasicharacters of G. We now prove some simple facts about quasicharacters and give some examples.

Theorem VI. (i) *The group $\tilde{G}$ is topologically isomorphic with the direct product of $\hat{G}$ and the group H of positive quasicharacters of G.*

(ii) *Every bounded quasicharacter is a character.*

(iii) *If G is compact or if every element of G is of finite order, then every quasicharacter of G is necessarily a character.*

Proof. (i) Only the trivial character can be positive, and hence $G \cap H = (e)$. Moreover, if $p(x)$ is a quasicharacter, then $p(x)/|p(x)|$ is a character and $|p(x)|$ is a positive quasicharacter; thus $\hat{G}H = \tilde{G}$. The topologies of $\tilde{G}$ and $\hat{G} \times H$ coincide, for each of them is the topology of uniform convergence on compact sets.

(ii) Obvious.

(iii) If G is compact, then every quasicharacter is bounded and we can apply (ii). If every element of G is of finite order, then $x^{n(x)} = 1$ holds for $x \in G$; thus, for every quasicharacter $q(x)$ we have $1 = q(x^{n(x)}) = q(x)^{n(x)}$, whence $|q(x)| = 1$. □

Note also that just as in the case of characters one can prove that $\widetilde{G_1 \times G_2}$ and $\tilde{G}_1 \times \tilde{G}_2$ are topologically isomorphic.

Now we present some examples. First, let G be the infinite cyclic group with discrete topology. The group $\tilde{G}$ is, in this case, topologically isomorphic with the multiplicative group of the field of complex numbers. To see this, consider the mapping $q \mapsto q(a)$, where a is a fixed generator of G.

Now let G be the additive group R^+ of real numbers with the usual topology. As G is self-dual, we only have to find its positive quasicharacters, and hence to solve $q(x+y) = q(x)q(y)$ under the conditions that $q(x)$ be positive and continuous. It is easily seen that $q(x) = \exp ax$ with a suitable real a, and so $\tilde{G}$ is topologically isomorphic with $R^+ \times R^+$. Every quasicharacter $q(x)$ has the form $q(zx)$ with a certain complex z.

Similarly one can see that for the additive group $G = \mathfrak{Z}^+$ of complex numbers we have $\tilde{G} = \mathfrak{Z}^+ \times \mathfrak{Z}^+$ and every quasicharacter has the form $q(x) = \exp(z_1 \operatorname{Re} x + z_2 \operatorname{Im} x)$ with $z_1, z_2 \in \mathfrak{Z}^+$.

This example shows that the analogue of Pontryagin's Theorem I does not hold for the groups of quasicharacters, since $\tilde{\tilde{\mathfrak{Z}}}^+ \simeq \mathfrak{Z}^+ \times \mathfrak{Z}^+ \times \mathfrak{Z}^+ \times \mathfrak{Z}^+ \not= \mathfrak{Z}^+$.

The next two examples concern the multiplicative groups of real and complex numbers. First let $G = R^*$ with the usual topology. Since $R^* \simeq C_2 \times R^+$, we get $\tilde{R}^* \simeq C_2 \times \tilde{3}^+$. Moreover, every quasicharacter $q(x)$ of R^* has the form

$$q(x) = (\operatorname{sgn} x)^\varepsilon \exp(z \log |x|) = (\operatorname{sgn} x)^\varepsilon |x|^z,$$

where $\varepsilon = 0, 1$ and z is a complex number.

Finally, the multiplicative group 3^* of complex numbers can be written as $T \times R^+$ because each of its elements z has the form

$$z = (\exp x)(\exp it),$$

where $x = \log|z|$ and $t = \operatorname{Arg} z$. Thus $\tilde{3}^* \simeq \tilde{T} \times \tilde{R}^+ = C_\infty \times \tilde{3}^+$ and every quasicharacter has the form

$$q(x) = \exp(in \operatorname{Arg} x)|x|^z$$

for an integer n and a complex z.

Note that the last two examples can be written in a uniform way, namely,

$$q(x) = (x/|x|)^n |x|^z.$$

3. We will also use the Haar integral and Haar measure in locally compact Abelian groups. If G is an LCA, then by a *Haar measure* on G we understand any Borel measure $m(E)$ on G which is translation invariant, i.e. such that $m(x+E) = m(E)$ holds for every x in G and every Borel subset E of G, and which is positive on every open and non-void subset of G. In every LCA there is such a measure and it is essentially unique, i.e., if $m(E)$ and $m_1(E)$ are two Haar measures on G, then $m(E) = cm_1(E)$ holds for every Borel E with a constant c independent of E. It is convenient to choose the Haar measure in a compact group G so that $m(G) = 1$ and in a discrete group so that every set consisting of one element has unit measure.

The *Haar integral* is the integral taken with respect to a Haar measure. For a thorough study of its properties see E. Hewitt and K. Ross [63].

The linear space of measurable complex-valued functions on G whose pth power has a finite Haar integral will be denoted, as usual, by $L_p(G)$. For functions in $L_1(G)$ one can define the Fourier transform by

$$\hat{f}(X) = \int_G f(x) X(-x) \, dm(x)$$

($f \in L_1(G)$, $X \in \hat{G}$), which is a continuous function on the dual group of G.

In a similar way one defines the *Mellin transform* on G. If $f(x)$ is a complex-valued measurable function on G and $q(x)$ is a quasicharacter of G, then the integral

$$\tilde{f}(q) = \int_G f(x) q(x) \, dm(x)$$

may happen to be finite. In this case we speak about the value of the Mellin transform of $f(x)$ at q. By $A(f)$ we denote the set of quasicharacters q for which $\tilde{f}(q)$ is finite.

We now present some examples of the Mellin transform.

(a) G = the cyclic infinite group with discrete topology. Here

$$\tilde{f}(q) = \tilde{f}(z) = \sum_{n=-\infty}^{\infty} f(n) z^n,$$

since every quasicharacter is described by an element z of $\mathfrak{Z}^*$.

(b) $G = R^+$ with the usual topology. Here the quasicharacters are described by elements of $\mathfrak{Z}^+$ and we have

$$\tilde{f}(q) = \tilde{f}(z) = \int_{-\infty}^{\infty} f(x) e^{zx} dx = 2\pi i L(f; -z),$$

where $L(f; w)$ denotes the value of the classical Laplace transform at the point w.

(c) $G = \mathfrak{Z}^+$ with the usual topology. Here

$$\tilde{f}(q) = \tilde{f}(z_1, z_2) = \int_{-\infty}^{\infty} \int_{-\infty}^{\infty} f(x+iy) e^{z_1 x + z_2 y} dx\, dy.$$

(d) $G = R^*$ with the usual topology. For the Haar measure $dm(x)$ we may take $dx/|x|$, where dx is the Lebesgue measure on the real line, and we obtain

$$\tilde{f}(q) = \tilde{f}(\varepsilon, z) = \int_{|x|>0} (\operatorname{sgn} x)^\varepsilon |x|^{z-1} f(x) \, dx.$$

(e) $G = \mathfrak{Z}^*$ with the usual topology. For the Haar measure we may take $dx\,dy/(x^2+y^2)$, where $dx\,dy$ is the plane Lebesgue measure and we obtain

$$\tilde{f}(q) = \tilde{f}(n, z) = \iint_{x^2+y^2>0} \exp\{in \operatorname{Arg}(x+iy)\} |x^2+y^2|^{z/2-1} f(x+iy) \, dx\, dy.$$

We shall also use the following result:

Inversion Theorem. *If G is a locally compact Abelian group and dx is a Haar measure on it, then there exists a unique Haar measure $d\hat{x}$ on the dual group $\hat{G}$ of G such that for every continuous and integrable function $f(x)$ on G such that also its Fourier transform $\hat{f}(X)$ is integrable the following formula holds:*

$$f(x) = \int_{\hat{G}} \hat{f}(X) X(x) \, d\hat{x} = \hat{\hat{f}}(-x).$$

In particular, if G is a self-dual group, then one can choose the Haar measure on it so that the inversion formula holds with the same measure in G. In the case of $G = R^+$ the usual Lebesgue measure has this property, and in the case of $G = 3^+$ it is the Lebesgue plane measure multiplied by two, provided we establish the following correspondence between the elements of the group and of its dual group: In the case of $G = R^+$ every $a \in R^+$ determines a character $X_a(t) = \exp(-2\pi i a t)$ and in the case of $G = 3^+$ every $a \in 3^+$ determines a character

$$X_a(t) = \exp(-4\pi i \operatorname{Re}(ax)).$$

This assertion results immediately from the following equalities which are easily verifiable:

$$\int_{-\infty}^{\infty} \exp(-x^2) \exp(2\pi i y x) \, dx = \pi^{1/2} \exp(-\pi^2 y^2),$$

$$\int_{-\infty}^{\infty} \pi^{1/2} \exp(-\pi^2 y^2) \exp(-2\pi i y x) \, dy = \exp(-y^2),$$

$$2 \int_{-\infty}^{\infty} \int_{-\infty}^{\infty} \exp(-x^2 - y^2) \exp(4\pi i (xx_1 - yy_1)) \, dx \, dy = 2\pi \exp(-4\pi^2(x_1^2 + y_1^2)),$$

$$2 \int_{-\infty}^{\infty} \int_{-\infty}^{\infty} 2\pi \exp(-4\pi^2(x_1^2 + y_1^2)) \exp(-4\pi i (xx_1 - yy_1)) \, dx_1 \, dy_1 = \exp(-x^2 - y^2).$$

4. Now we prove the analogue of Theorem 5.17 for the real and complex field:

Theorem VII. *Let K be either the real or the complex number field and let f be a complex-valued function defined on K, which is continuous and integrable with respect to the Haar measure in K^+. Assume, moreover, that its Fourier transform is also integrable and that for all positive t the functions $f(x) v(x)^t$ and $\hat{f}(y) v(y)^t$ are in $L_1(K^*)$. Here $v(x) = |x|$ for real K and $v(x) = |x|^2$ for complex K. For every quasicharacter $q(x) = X(x) v(x)^s$ of K^* (where $X(x) = (\operatorname{sgn} x)^\varepsilon$, $\varepsilon = 0, 1$ for K real, $X(x) = (x/|x|)^n$, $n \in Z$, for K complex) define $\hat{q}(x)$ by $\hat{q}(x) = q^{-1}(x) v(x)$. Then for the zeta-function*

$$Z(f, q) = \int_{K^*} f(x) q(x) \, dx^*$$

the functional equation

$$Z(f, q) = \varrho(q) Z(\hat{f}, \hat{q})$$

holds for $0 < \operatorname{Re} s < 1$, where

$$\varrho(q) = \begin{cases} \dfrac{\pi^{-s/2}\Gamma(s/2)}{\pi^{-(1-s)/2}\Gamma((1-s)/2)} & (\varepsilon = 0), \\[1em] \dfrac{\pi^{-(1+s)/2}\Gamma((1+s)/2)}{i\pi^{-(2-s)/2}\Gamma((2-s)/2)} & (\varepsilon = 1), \end{cases}$$

for real K and

$$\varrho(q) = (-i)^{|n|} \frac{(2\pi)^{1-s}\Gamma(s+|n|/2)}{(2\pi)^s \Gamma(1-s+|n|/2)}$$

for complex K. Moreover, for a fixed equivalence class of quasicharacters the zeta-function is a regular function of s for $\operatorname{Re} s > 0$ and can be continued analytically to a meromorphic function. (Here Xv^s, $X_1 v^s$ are called equivalent if $X = X_1$.)

Proof. We proceed exactly in the same way as in the proof of Theorem 5.17; however, we have to make an appropriate choice of the function $g(x)$, which will be substituted into equality (5.10) to yield the required functional equation. If K is real and $\varepsilon = 0$, then we put $g(x) = \exp(-\pi x^2)$, in which case plainly

$$\hat{g}(y) = g(y) \quad \text{and} \quad Z(g, q) = \pi^{-s/2}\Gamma(s/2);$$

thus

$$Z(\hat{g}, \hat{q}) = \pi^{-(1-s)/2}\Gamma((1-s)/2).$$

If K is real and $\varepsilon = 1$, then we put $g(x) = x\exp(-\pi x^2)$. In this case $\hat{g}(y) = ig(y)$, and to compute the zeta-functions we write

$$Z(g, q) = \int_{-\infty}^{0} x\exp(-\pi x^2)|x|^{s-1}dx - \int_{0}^{\infty} x\exp(-\pi x^2)|x|^{s-1}dx$$

$$= 2\int_{0}^{\infty} \exp(-\pi x^2)x^s dx = \pi^{-(s+1)/2}\Gamma((s+1)/2),$$

and similarly

$$Z(\hat{g}, \hat{q}) = i\pi^{-(2-s)/2}\Gamma((2-s)/2).$$

Finally, if K is complex, then we put $g_n(x) = \bar{x}^n \exp(-2\pi|x|^2)$ in the case of non-negative n, and $g_n(x) = x^{-n}\exp(-2\pi|x|^2)$ for negative n. In this case, for $n = 0$ and $\hat{x} = x+iy$, we obtain

$$\hat{g}_0(\hat{x}) = 2\int_{-\infty}^{\infty}\int_{-\infty}^{\infty} \exp(-2\pi(u^2+v^2))\exp(4\pi i(xu-yv))\,du\,dv$$

$$= 2\int_{-\infty}^{\infty} \exp(-2\pi u^2 + 4\pi i xu)\,du \int_{-\infty}^{\infty} \exp(-2\pi v^2 - 4\pi i yv)\,dv$$

$$= \exp(-2\pi(x^2+y^2)) = g_0(\hat{x}).$$

In the general case we assert that $g_n(x) = i^{|n|}g_{-n}(\hat{x})$. This being true for $n = 0$, let us assume that it holds for a certain $n \geq 0$. Writing this down explicitly and applying the operator $(4\pi i)^{-1}\left(\dfrac{\partial}{\partial x} + i\dfrac{\partial}{\partial y}\right)$, we immediately conclude the truth of our assertion for $n+1$. Having established this for all non-negative n's, we immediately obtain the general case with the use of the inversion theorem.

For the zeta-functions we obtain without difficulty the equalities
$$Z(g_n, q) = (2\pi)^{1-s+|n|/2}\Gamma(s+|n|/2)$$
and
$$Z(\hat{g}_n, \hat{q}) = i^{|n|}(2\pi)^{s+|n|/2}\Gamma(1-s+|n|/2).$$

Once this is done, we can proceed in the same way as in the proof of the said theorem from Chapter 5 and safely arrive at our assertions. □

(*Note.* The first to prove this theorem was J. Tate [50], and the proof given here is due to him.)

5. In this subsection we state and prove the Poisson formula, but only for the special case, which is dealt with in Chapter 6. Let G be a locally compact Abelian group, H its discrete subgroup and assume that the factor group G/H is compact in the quotient topology. Its dual group will be denoted by K. Note that K can be identified with the subgroup of $\hat{G}$ consisting of all those characters which are trivial on H. Assume, moreover, that there exists an open set D in G such that every element $g \in G$ can be uniquely written in the form $g = hd$, where $h \in H$ and $d \in D$. This assumption allows us to consider the mapping $f: G/H \to D$ defined by $f(A) = x$, where x is a representative of the coset A lying in D. This map is continuous and hence D is compact. If we choose in G a Haar measure $m(E)$ so that $m(D) = 1$, then it induces a Haar measure $\mu(E)$ in G/H via $\mu(E) = m(f(E))$, under which G/H acquires the unit measure. Finally, let $\nu(E)$ be the Haar measure in H which gives the unit measure to every point. Note that with this choice of measures we have for $f \in L_1(G)$ the equality

$$\int_G f(x)\,dm(x) = \int_H \int_D f(x+h)\,dm(x)\,d\nu(h) = \sum_{h \in H} \int_D f(x+h)\,dm(x).$$

We now prove

Theorem VIII. *Let f be a continuous function in $L_1(G)$ and assume that*
(i) *the series $\sum_{h \in H} f(x+h)$ is uniformly convergent for $x \in D$ and*

(ii) $\sum_{y \in K} |\hat{f}(y)|$ converges.
Then
$$\sum_{h \in H} f(h) = \sum_{y \in K} \hat{f}(y).$$

Proof. Define $g(x)$ by
$$g(x) = \begin{cases} \sum_{h \in H} f(x+h), & x \in D, \\ 0, & x \notin D, \end{cases}$$
and let $X \in K$. We identify X with a character of G, which is trivial on H. Then
$$\hat{g}(X) = \int_D \sum_{h \in H} f(x+h) X(-x) \, dm(x) = \sum_{h \in H} \int_D f(x+h) X(-x) \, dm(x)$$
$$= \sum_{h \in H} \int_{h+D} f(x) X(h-x) \, dm(x) = \sum_{h \in H} X(h) \int_{h+D} f(x) X(-x) \, dm(x)$$
$$= \sum_{h \in H} \int_{h+D} f(x) X(-x) \, dm(x) = \int_G f(x) X(-x) \, dm(x) = \hat{f}(X).$$

We can regard $g(x)$ as a function on G/H owing to the biunique correspondence between G/H and D. One can easily see that the Fourier transform of this function coincides with $\hat{g}(X)$, and by the inversion formula we get
$$g(x) = \sum_{X \in K} \hat{g}(X) X(x) \quad \text{for } x \in D;$$
hence
$$\sum_{h \in H} f(x+h) = \sum_{X \in K} \hat{f}(X) X(x),$$
and if we put $x = 0$, we obtain the assertion of the theorem. □

6. In this subsection we develop the fundamental properties of restricted direct products of topological Abelian groups, which are needed in Chapter 6. We start with the definition:

Let $\{G_v\}_{v \in V}$ be a family of locally compact Abelian groups and assume that in every G_v with the exception of finitely many v's a compact and open subgroup H_v is selected. The *restricted product* of the G_v's with respect to H_v's is the subgroup G of the direct product of all G_v's consisting of all elements (g_v) such that $g_v \in H_v$ holds for almost all (i.e. all except finitely many) indices v. In G we define a topology by taking for a fundamental system of neighbourhoods of the unit element the system of all neighbourhoods of the unity in the product topology in the group

$$G_S = \mathop{\mathbf{P}}_{v \in S} G_v \times \mathop{\mathbf{P}}_{v \notin S} H_v,$$

where S is an arbitrary fixed finite set of indices containing every index v for which the group H_v is not defined. One can without difficulty see that this topology does not depend on the particular choice of the set S and that all groups G_v are open in G. It is also easy to see that the family $\{\mathbf{P}_v U_v\}$, where, for every v, U_v is a neighbourhood of the unity in G_v and for almost all v's we have $U_v = H_v$, can also serve as a fundamental system of neighbourhoods of the unity in G.

We now have:

Lemma 1. *The restricted product G is a locally compact topological group.*

Proof. The continuity of operations is immediate and the local compactness follows from the local compactness of G_S with arbitrary finite $S \subset V$ and the observation that G_S is open in G. □

Characters and quasicharacters in a restricted product are described by the following

Theorem IX. (i) *The group $\hat{G}$ of characters of G is topologically isomorphic with the restricted product of the character groups $\hat{G}_v$ with respect to the subgroups A_v of $\hat{G}_v$ which annihilate H_v (i.e. consist of characters trivial on H_v). Moreover, for any character X of G we have the equality*

$$X((g_v)) = \prod_v X_v(g_v),$$

where X_v is the restriction of X to G_v, considered as a subgroup of G via the embedding $g_v \mapsto [1, 1, \ldots, 1, g_v, 1, 1, \ldots]$.

(ii) *The same result holds for the group of quasicharacters; however, this time the isomorphism is only algebraical and not necessarily topological.*

Proof. Let $X(t)$ be any quasicharacter of G. We show that for almost all v's the quasicharacter X_v trivializes on H_v. In fact, if U is a neighbourhood of unity in the multiplicative group of complex numbers which does not contain non-trivial subgroups and $O = \mathbf{P}_v O_v$ is a neighbourhood of unity in G such that $X(O) \subset U$, then $X(t)$ is trivial on every subgroup of O. However, for almost all v's we have $O_v = H_v \subset O$; hence, on any such H_v, $X(t)$ is trivial. Now let S be the set of those indices $v \in V$ for which either H_v is undefined or $X(t)$ is not trivial on H_v. Let also $g = (g_v)$ be an arbitrary element of G, $T = \{v \notin S : g_v \notin H_v\}$ and $S_1 = \{v \notin S \cup T : H_v \neq O_v\}$. We can write

$$g = \prod_{v \in S} g_v \prod_{v \in T} g_v \prod_{v \in S_1} g_v \cdot g_0,$$

where $g_0 \in P_v O_v$. This gives

$$X(g) = \prod_{v \in S \cup T \cup S_1} X(g_v) X(g_0) = \prod_{v \in S \cup T \cup S_1} X(g_v) = \prod_{v \in V} X(g_v),$$

because for v outside $S \cup T \cup S_1$ we have $X(g_v) = 1$.

Now note that if for every v we have a quasicharacter X_v of G_v and for almost every v this quasicharacter is trivial on H_v, then by the formula $X((x_v)) = \prod_{v \in V} X_v(x_v)$ one defines a quasicharacter of G. The multiplicative property of $X(t)$ is evident and to obtain continuity denote by S the set of indices v for which $X_v|_{H_v}$ is not trivial and let N be the number of elements of S. For every neighbourhood U of unity in the complex plane choose another neighbourhood U_1 so that $U_1^N \subset U$. If U_v are chosen so that $X_v(U_v) \subset U_1$ holds for $v \in S$ and $U_v = H_v$ for $v \notin S$, then the quasicharacter $X(t)$ maps the product $P_v U_v$ into U and so it is continuous.

Further, consider the map

$$\varphi: (X_v)_v \mapsto X = \prod_v X_v$$

of the restricted product (without topology) of the groups $\tilde{G}_v$ with respect to their subgroups annihilating H_v into $\tilde{G}$. It is obviously a monomorphism and the previous argument shows that it is also surjective. This proves the part of our theorem concerning quasicharacters.

Observe that φ maps the restricted product of the groups $\hat{G}_v$ with respect to their subgroups annihilating H_v onto the group of characters of G. We now prove that it is a homeomorphism.

A subset C of G has a compact closure iff it lies in $P_v C_v$, where C_v are compact in G_v and, for almost every v, C_v equals H_v. Indeed, every such set has a compact closure; on the other hand, if $\bar{C}$ is compact, then for a certain finite $S \subset V$ we have $C \subset G_S$ because $\bar{C} \subset G = \bigcup G_S$, and so a finite system of open subgroups G_S covers $\bar{C}$; it remains to observe that a finite sum of sets G_S also has this form.

Let U be a neighbourhood of unity in the group $\hat{G}$ of the form $U = \{X: |X(t)-1| < \varepsilon \text{ for } t \in P_v C_v\}$, where $\varepsilon < 1/2$ and $S = \{v: C_v \neq H_v\}$ is finite. If $X \in U$, then for $t_v \in C_v$ we have $|X_v(t_v)-1| < \varepsilon$ and for v outside S we even have $X_v(t_v) = 1$, because then $C_v = H_v$ is a group and $X_v(H_v)$ lies in the disc $|z-1| < 1/2$.

If $U_v = \{\psi \in \tilde{G}_v: |\psi(t)-1| < \varepsilon \text{ for } t \text{ in } C_v\}$, then $X \in U$ implies $X_v \in U_v$ for v in S and $X_v(H_v) = 1$ for $v \notin S$. Conversely, if X_v satisfies the last two conditions, then for t in $P_{v \in S} C_v \times P_{v \notin S} H_v$ we have $|X(t)-1| < \varepsilon'$ for suitable ε' depending on ε. □

7. Now we turn to the Haar measure in a group G, which — as in the previous subsection — is the restricted product of G_v's with respect to H_v. We retain all the notation from that subsection. Let us choose in every G a Haar measure which gives the unit measure to the groups H_v in the case where they are defined. It is also possible to allow finitely many exceptions to that condition. Denote by dx_v the chosen measure in G_v and for every finite $S \subset V$ denote by dx^S the product measure in the compact group $P_{v \notin S} H_v$. Then

$$dx_S = \prod_{v \in S} dx_v \times dx^S$$

is a Haar measure in the group G_S.

If $f(x)$ is a continuous function with a compact support A on G, then for suitable S we have $A \subset G_S$, and so the linear functional defined by

$$F(f) = \int_{G_S} f(x) dx_S$$

is well defined and independent of the choice of S. This functional induces, in a well-known way, an invariant integral on G; thus, there exists a Haar measure dx in G, for which

$$F(f) = \int_G f(x) dx.$$

This measure depends on the choice of Haar measures in those groups G_v for which H_v is not defined; however, in all our applications there will be a standard choice for those measures. The next two lemmas are useful for effective integration on G and for finding the Fourier transform.

Lemma 2. *Suppose that, for every $v \in V$, we have a continuous function $f_v(x_v)$ on G_v, which is integrable and whose restriction to H_v, for almost all v's, equals 1. Let $f(x)$ be the function on G defined by*

$$f(x) = \prod_v f_v(x_v) \quad \text{for } x = (x_v).$$

Then $f(x)$ is continuous and, for every finite set $S \subset V$ containing all indices for which either H_v is undefined or $f_v|_{H_v} \neq 1$, we have

$$\int_{G_S} f(x) dx = \prod_{v \in S} \int_{G_v} f_v(x_v) dx_v.$$

Proof. The continuity of $f(x)$ results from the observation that the groups G_S are open and cover G, and on each of them $f(x)$ is continuous; the equality between the integrals is a consequence of

$$\int_{G_S} f(x)\,dx = \int_{G_S} f(x)\,dx_S$$

and the definition of dx_S. □

Corollary. *Under the assumption of the lemma, if*

$$\sup_S \prod_{v\in S} \int_{G_v} |f_v(x_v)|\,dx_v < \infty,$$

then $f(x)$ is integrable on G and

$$\int_G f(x)\,dx = \prod_v \int_{G_v} f_v(x_v)\,dx_v.$$ □

Lemma 3. *For every $v \in V$, let the function $f_v(x_v)$ be continuous and integrable. Suppose that, for almost every v, it is equal to the characteristic function of H_v. If $f((x_v)) = \prod_v f_v(x_v)$ is the induced function on G, then its Fourier transform $\hat{f}(X)$ satisfies*

$$\hat{f}(X) = \prod_v \hat{f}_v(X_v),$$

where $X = \prod_v X_v$ is the factorization of a character X as given by Theorem IX.

Proof. First let us have a look at the Fourier transform of the characteristic function $h_v(x)$ of H_v. We have

$$\hat{h}_v(X_v) = \int_{G_v} h_v(x) X_v(x^{-1})\,dx_v = \int_{H_v} X_v(x^{-1})\,dx_v$$

$$= \begin{cases} \int_{H_v} dx_v & \text{if } X_v \text{ is trivial on } H_v, \\ 0 & \text{otherwise,} \end{cases}$$

and thus $\hat{h}_v(X_v)$ coincides with the characteristic function of the subgroup of G_v which annihilates H_v. We see that, for almost every v, $\hat{h}_v(X_v) = 1$ for any given character $X = \prod_v X_v$ of G. Moreover, for almost all v we have

$$\int_{G_v} |f_v(x_v) X_v(x^{-1})|\,dx_v = 1,$$

and so it is possible to apply the corollary to the previous lemma, which yields

$$\hat{f}(X) = \int_G f(x) X(x^{-1})\,dx = \prod_v \int_{G_v} f_v(x_v) X_v(x_v^{-1})\,dx_v = \prod_v \hat{f}_v(X_v).$$ □

The next lemma describes the Haar measure in the character group $\hat{G}$ which is dual to the Haar measure in G used above.

Lemma 4. *For every v let $d\hat{x}_v$ be the Haar measure in $\hat{G}_v$ which is dual to the measure dx_v; this means that the inversion formula holds for the pair dx_v, $d\hat{x}_v$. Then for every v for which H_v is defined the subgroup of $\hat{G}_v$ annihilating H_v acquires measure 1, and the measure $d\hat{x}$ defined in $\hat{G}$ by the measures $d\hat{x}_v$ in the same way as dx has been constructed from the measures dx_v is dual to dx.*

Proof. If $f_v(x_v)$ is any function integrable and continuous on G_v, whose Fourier transform $\hat{f}_v(X_v)$ is integrable, then

$$f(y^{-1}) = \int_{\hat{G}_v} \hat{f}_v(X_v) y(X_v^{-1}) dX_v.$$

Putting $f_v(x_v) = h_v(x_v)$, where $h_v(x_v)$ is the characteristic function of H_v, we obtain

$$\int_{\hat{G}_v} \hat{h}_v(X_v) X_v^{-1}(y) dX_v = h_v(y^{-1}) = h_v(y)$$

and

$$h_v(y) = \int_{H_v^*} \overline{X_v(y)} dX_v,$$

where H_v^* is the annihilator of H_v. If we put here $y = 1$, then the equality

$$\int_{H_v^*} dx_v = 1$$

results. Therefore we can apply the process of establishing the Haar measure in $\hat{G}$, and to check that it is dual to dx does not present any difficulties. □

Corollary. *If $f(x)$ satisfies the conditions of Lemma 3 and, moreover, for all v the Fourier transform $f_v(X_v)$ is integrable, then also the Fourier transform of $f(x)$ is integrable.*

Proof. Let S_1 be the set of indices at which $f_v \neq h_v$. Then the lemma implies the equality

$$f(X) = \prod_{v \in S_1} f_v(X_v) \prod_{v \notin S_1} h_v(X_v).$$

Now for v outside S_1 we have

$$\int_{\hat{G}_v} |h_v(X_v)| dX_v = 1;$$

hence

$$\sup_S \prod_{v \in S} \int_{\hat{G}_v} |\hat{f}_v(\hat{x}_v)| d\hat{x}_v \leq \prod_{v \in S_1} \int_{\hat{G}_v} |\hat{f}_v(\hat{x}_v)| d\hat{x}_v.$$

This is a finite number, and so we may apply Lemma 3. □

Appendix II. Dirichlet Series

Dirichlet series have been introduced by Dirichlet in his investigations concerning rational primes in arithmetic progressions and are still of utmost importance for the number theory. In this appendix we collect the results concerning those series, which are used in this book. The proofs are omitted.

We start with the definition. Any series of the form

$$\sum_{n=1}^{\infty} a_n n^{-s} \qquad (1)$$

(where a_n are complex coefficients and s is a complex variable) is called a *Dirichlet series*. The domain of its convergence is described by

Proposition I. *If series (1) converges at a point s_0, then it converges also in the open half-plane $\operatorname{Re} s > \operatorname{Re} s_0$, convergence being uniform in every angle $\arg(s-s_0) \leqslant c < \pi/2$. Thus (1) defines a function regular in $\operatorname{Re} s > \operatorname{Re} s_0$. If at some point s_1 the convergence is absolute, then (1) converges absolutely in the open half-plane $\operatorname{Re} s > \operatorname{Re} s_1$, convergence being uniform.*

This proposition allows us to speak about the half-plane of convergence and absolute convergence of (1).

We are mainly interested in the asymptotic behaviour of the coefficient sum $S(x) = \sum_{n \leqslant x} a_n$ of series (1). A very general result regarding this question is the following Tauberian theorem of Delange–Ikehara, proved by S. Ikehara in the special case where (1) defines a function regular for $\operatorname{Re} s > 1$ with a simple pole at $s = 1$ and by H. Delange [54] in the general case.

Theorem I. *Assume that the series (1) has all its coefficients real and non-negative and that it converges in $\operatorname{Re} s > 1$, defining a function $f(s)$ regular there. Assume, moreover, that in the same half-plane we can write*

$$f(s) = g_0(s)\left(\log\frac{1}{s-1}\right)^{b_0} \bigg/ (s-1)^{\alpha_0} + \sum_{j=1}^{q} g_j(s)\left(\log\frac{1}{s-1}\right)^{b_j} \bigg/ (s-1)^{\alpha_j} + g(s),$$

where $g(s)$, $g_0(s)$, ..., $g_q(s)$ are regular in the closed half-plane $\operatorname{Re} s \geq 1$, $b_0, b_1, ..., b_q$ are non-negative rational integers, $\alpha_1, ..., \alpha_q$ are complex numbers whose real parts are smaller than α_0, which is a real number not equal to zero or to a negative integer, and finally $g_0(1) \neq 0$.

Then for the summatory function

$$S(x) = \sum_{n \leq x} a_n$$

of our series we have, for x tending to infinity, the asymptotic expression:

$$S(x) = (g_0(1)\Gamma(\alpha_0)^{-1} + o(1))x(\log x)^{\alpha_0 - 1}(\log\log x)^{b_0}.$$

However, if $f(s)$ satisfies the same assumptions with the following change: $\alpha_0 = 0$, $b_0 \neq 0$, then we get

$$S(x) = (b_0 g_0(1) + o(1))x(\log\log x)^{b_0 - 1}/(\log x).$$

Note that in H. Delange [54] the α_j's are assumed to be real; however, this assumption is not used in the proof and one can easily see that the theorem holds true without it.

We need also a result of Landau on Dirichlet series. For its proof cf. Narkiewicz [84], Th. 3.5 or Prachar [57].

Theorem II. *If $\operatorname{Re} s > C$ is the half-plane of convergence of a Dirichlet series with non-negative coefficients, defining a function $f(s)$, then the point $s = C$ is a singularity for $f(s)$.*

We also include here for reference two theorems from the theory of analytic functions, not connected with Dirichlet series. Their proof can be found in K. Prachar [57].

Theorem III. *Assume that $f(s)$ is regular in $|s - s_0| \leq r$ and satisfies $|f(s)/f(s_0)| < M$ there. Moreover, assume that it does not vanish in $|s - s_0| \leq r/2$, $\operatorname{Re}(s - s_0) > 0$. Then, provided z is a zero of $f(s)$ in $|s - s_0| \leq r/2$, $\operatorname{Re}(s - s_0) < 0$ and h is its order, we have the following evaluations:*

(i) $\operatorname{Re}(f'(s_0)/f(s_0)) \geq -(4/r)\log M$,

(ii) $\operatorname{Re}(f'(s_0)/f(s_0)) \geq -(4/r)\log M + \operatorname{Re}(h/(s_0 - z))$.

Theorem IV. *If $f(s)$ is regular in $|s - s_0| \leq R$ and $0 < r < R$, then for $|s - s_0| \leq r$ we have*

$$|f(s)| \leq 2|f(s_0)|R/(R-r) + (2r/(R-r)) \max_{|s - s_0| \leq R} \{\operatorname{Re} f(s)\}.$$

Appendix III. Geometry of Numbers

1. In this appendix we recall certain results from the geometry of numbers needed in Chapter 2. For their proofs see e.g. Cassels [57], [59b], Narkiewicz [84].

We begin with definitions. A discrete subgroup of the real n-space R^n containing n linearly independent points is called a *lattice*, and if a discrete subgroup of R^n contains $r < n$, but not more, linearly independent points, then we call it an *r-dimensional lattice*, although according to the preceding definition it is not a lattice. It becomes a lattice, however, in a suitable r-dimensional subspace of R^n.

If $x_i = (x_{i1}, \ldots, x_{in})$ $(i = 1, 2, \ldots, n)$ is a set of generators of a lattice L (such set must exist, since L is clearly isomorphic to Z^n), then the absolute value of the determinant $\det[x_{ij}]$ is called the *determinant* of L and is denoted by $d(L)$. It is easy to see that $d(L)$ equals the volume of the set

$$\left\{ \sum_{j=1}^{n} a_j x_j : 0 \leqslant a_j \leqslant 1 \right\}.$$

Theorem I (*Minkowski's convex body theorem*. Minkowski [96a]). *If C is a convex subset of R^n, symmetric about the origin and of volume $V(C)$, which may be infinite, and L is a lattice in R^n satisfying $d(L) < 2^{-n} V(C)$, then there exists in C a point of L distinct from the origin.*

Applying this theorem to sets defined by a system of n linear inequalities we obtain the following important corollary:

Theorem II (*Minkowski's theorem on linear forms*). *Let*

$$L_j(x_1, \ldots, x_n) = \sum_{i=1}^{n} a_{ij} x_i \quad (j = 1, 2, \ldots, n)$$

be a set of linear forms with complex coefficients, containing with a form L

also its complex conjugate $\bar{L}$. Let M be a lattice in R^n and let $D = |\det[a_{ij}]|$. Assume further that $c_1, \ldots, c_n$ are given positive numbers such that $L_i = \bar{L}_j$ implies $c_i = c_j$, and that $c_1 \ldots c_n \geqslant Dd(M)$.

Then there exists a point $(X_1, \ldots, X_n)$ in M, distinct from the origin, for which

$$|L_1(X_1, \ldots, X_n)| \leqslant c_1$$

and

$$|L_j(X_1, \ldots, X_n)| < c_j \quad (j = 2, 3, \ldots, n)$$

holds.

2. Finally we state Kronecker's theorem on diophantine approximations utilized in the proof of Theorem 2.4.

Theorem III. (See Cassels [57], Ch.III.) *Let $L_i(x_1, \ldots, x_m)$ $(i = 1, 2, \ldots, n)$ be linear forms and let $a_1, \ldots, a_n$ be a set of real numbers. Then the following conditions are equivalent:*

(i) *To every positive ε one can find suitable rational integers $A_1, \ldots, A_m$ such that for $i = 1, 2, \ldots, n$*

$$\|L_i(A_1, \ldots, A_m) - a_i\| < \varepsilon,$$

where by $\|t\|$ we denote the distance between t and the nearest rational integer.

(ii) *If $u_1, \ldots, u_n$ are rational integers such that the linear form*

$$\sum_{i=1}^{n} u_i L_i(x_1, \ldots, x_m)$$

has integral coefficients, then the number $u_1 a_1 + \ldots + u_n a_n$ is a rational integer.

Appendix IV. Baker's Method

In the proof of Theorem 8.11 we have made use of a result of A. Baker [77] giving a lower bound for linear forms in logarithms. To state this result we denote by $H(a)$, for an algebraic number a, its height, defined as the maximal absolute value of the coefficients of the minimal polynomial of a with rational integral coefficients.

Theorem. *Let $a_1, \ldots, a_n$ be given positive rational numbers and let*
$$L(X_1, \ldots, X_n) = X_1 \log a_1 + \ldots + X_n \log a_n.$$
Suppose that the numbers $A_1, \ldots, A_n$ satisfy $A_j \geq 4$ and
$$H(a_j) \leq A_j \quad \text{for } j = 1, 2, \ldots, n.$$
If now $M \geq 4$, then for all rational integers $x_1, \ldots, x_n$ satisfying $|x_i| \leq M$ for $i = 1, 2, \ldots, n$ we have either
$$L(x_1, \ldots, x_n) = 0$$
or
$$|L(x_1, \ldots, x_n)| \geq \exp(-CD(\log D')(\log M))$$
where C is a constant depending only on n (one may take $C = (16n)^{200n}$), D denotes the product of the logarithms of $A_1, \ldots, A_n$ and $D' = D/\log A_n$.

If the numbers $a_1, \ldots, a_n$ are algebraic, $H(a_j) \leq A_j$ ($j = 1, 2, \ldots, n$), and the x_j's are also algebraic with $H(x_j) \leq M$, then either
$$L(x_1, \ldots, x_n) = 0$$
or
$$|L(x_1, \ldots, x_n)| \geq \exp(-C_1 D(\log D)(\log D')(\log M))$$
where D, D' are as above and $C_1 = (16dn)^{200n}$, where d is the degree of the field generated by $a_1, \ldots, a_n, x_1, \ldots, x_n$.

Unsolved Problems

In the first edition we gave a list of 35 unsolved problems with short comments on their status. In the meantime certain of them have been solved either completely or partially. Here we reproduce this list giving appropriate comments and also add certain other problems. The newly added problems have numbers ≥ 36.

1. (D. H. Lehmer [33a]). Prove that for every positive ε one can find an integer a with
$$1 < M(a) = \prod \max(1, |a^{(i)}|) < 1+\varepsilon,$$
where the product runs over all conjugates $a^{(i)}$ of a.

Actually, Lehmer asked whether there exists such an a, but the folklore made his question to be stronger. The answer is negative in the case when the minimal polynomial P of a is not reciprocal, i.e. $P(x) \neq x^n P(x^{-1})$ with n being the degree of P (Smyth [71]). Now it is rather believed that the answer is negative also in the general case and the best result in this direction is due to E. Dobrowolski [79], who proved that if $M(a) > 1$, then
$$M(a) \leq 1 + (1-\varepsilon)(\log\log n/\log n)^3$$
for $n =$ degree of a, sufficiently large. The constant here was improved to $2-\varepsilon$ independently in Cantor, Straus [82] and Rausch [85] and to $2.45-\varepsilon$ in Louboutin [83].

2. (Robinson [62]). Show that if $b-a > 4$, then the interval $[a, b]$ contains a full system of conjugates of an integer of degree n, provided n is sufficiently large.

Solved by V. Ennola [75a].

3. (Samet [53]). Let U be the set of all those non-real integers whose all conjugates except two lie in $|z| \leq 1$ and there are some on $|z| = 1$. Determine the closure of U.

4. (Robinson [69]). Determine all circles $|z-x| = R$ with an irrational centre x which contain infinitely many full systems of conjugates of algebraic integers.

Circles of this type with rational centres have been described in Robinson [69]. One now knows all such circles with totally real centres (Ennola [73a]) and with centres of degree 3 or 4 (Ennola, Smyth [74]). Moreover, all algebraic integres whose conjugates lie on a circle were described by V. Ennola, C. J. Smyth [76], however a description of relevant circles is not known.

5. Determine all numbers which are discriminants of finite extensions of Q.

For Abelian extensions of Q with a given Galois group the conductor-discriminant formula leads to an answer. In the general case nothing seems to be known except Stickelbergers theorem.

6. Find a good necessary and sufficient condition for a field to have index 1.

It follows from Uchida [77b] that this happens iff there exists an integral generator of the field whose minimal polynomial does not lie in the square of a maximal ideal in $Z[X]$. On the other hand, K. Győry [73] gave an effective procedure to check whether a given field is of index 1.

7. Let $N_k(x)$ denote the number of fields of degree k whose discriminants do not exceed x in absolute value. Prove the existence of the limit $\lim_{x \to \infty} N_k(x)x^{-1}$. For $k = 2$ this is trivial and for $k = 3$ this has been proved in Davenport, Heilbronn [69].

Lower and upper bounds for $N_4(x)$ have been obtained in Baily [80].

8. Give a necessary and sufficient condition for a normal extension L/Q or, more generally, for L/K to have a normal integral basis.

9. Can one determine reasonable classes of extensions L/Q for which tame ramification is sufficient for the existence of a normal integral basis?

There has been an enormous progress on these two problems. We refer the reader to the book of A. Fröhlich [83a] on that subject and to the literature quoted in Notes to Chapter 4, subsection 6.

10. Characterize fields K (not necessarily algebraic over Q) having the property that every polynomial P over K for which there exists an infinite subset X of K with $P(X) = X$ must be linear. Algebraic number fields have this property (Narkiewicz [62]) and, more generally, all finitely generated extensions of a prime field (Lewis [72], Liardet, Ph. D. thesis, Marseille 1970). See Kubota, Liardet [76], Liardet [71], [72], [75].

11. Assume that K has the property indicated in the preceding problem. Prove that if f is a rational function over K mapping an infinite subset of K onto itself, then $f(t) = (at+b)/(ct+d)$.

This is known for a large class of fields (Liardet, Ph.D. thesis).

12. Let G be a finite Abelian group. Show that there is an algebraic number field with $H(K) = G$. The same problem may be also stated for $H^*(K)$.

13. Give a simple criterion for $N_{K/Q}(\varepsilon) = -1$ in real quadratic fields K.

A rather quick algoritm for testing this was presented by J. C. Lagarias [80a].

14. Which real fields have a system of totally positive fundamental units?

By Theorem 3.8 this is equivalent to $h^*(K) = 2^{r_1-1}h(K)$. See Armitage, Fröhlich [67], where this possibility was ruled out for a large class of fields.

15. Characterize fields with a system of real fundamental units.

16. Give a convenient necessary and sufficient condition for a normal field to have a conjugated system of fundamental units.

By Theorem 3.10 this happens for all cyclic fields of prime degree $\leqslant 19$.

17. In which fields has the group $U^+(K)$ a conjugated system of generators?

For fields with group C_p, $p = 3$ or 5, this was proved in Hasse [48a], resp. Morikawa [68].

18. (Jacobson [64]). In which fields is every integer a sum of distinct units?

There are only two such quadratic fields (with $d(K) = 5, 8$) (Jacobson [64], Śliwa [74]), but infinitely many cubic and quartic fields (Belcher [74], [75]).

19. Prove that only a finite number of units in a field can have the same discriminant.

The assertion was proved in Birch, Merriman [72] and Győry [73], [83]. Győry's proof is effective.

20. (Heilbronn [50]). Prove the existence of infinitely many cubic Euclidean fields.

21. (Gauss). Show that infinitely many real quadratic fields have class-number 1.

For an heuristical approach see Takhtayan, Vinogradov [80], [82].

22. Give an explicit formula for the highest power of a given prime which divides the index of a field K.

For non-ramified primes this problem was solved in Śliwa [82a]. See also Nart [85].

23. (Shafarevich [62]). Let K be a quadratic field with a prime discriminant. Prove that the number of generators of $H(K)$ is bounded by a constant independent of K.

24. (Leopoldt [62]). Find the rank of the p-adic regulator matrix.

In fact Leopoldt conjectured that this rank equals $r_1 + r_2 - 1$. This is true for all Abelian fields and also in certain other cases. See Washington [82], Chapter V.

25. Determine those Dedekind domains in which Theorem 6.8 holds.

Cf. Fröhlich, Serre, Tate [62].

26. (Artin). Show that if $K \subset L$, then the quotient $\zeta_L(s)/\zeta_K(s)$ is entire.

This is true if L/K is normal (Aramata [31], [33], R. Brauer [47a]). Cf. the literature quoted in Chapter 7, § 4, subsection 8.

27. Prove the analogue of Proposition 7.11 for non-normal extensions.

This has been done by R. W. K. Odoni [75a]. (Cf. Odoni [73b], [75b].)

28. Determine the structure of R_L as an $R_K[G]$-module for a given normal extension L/K.

This problem is closely connected with problems 8 and 9 and we refer the reader to the literature mentioned there.

29. Characterize Dedekind domains in which all factorizations into irreducibles of a given element have the same length.

For rings of integers this is Theorem 9.1 of L. Carlitz. In the general case a characterization was given in Skula [76]. Cf. Zaks [76], [80].

30. (Davenport). Let G be a finite Abelian group. Evaluate $D(G)$.
Cf. the literature given in Notes to Chapter 9.

31. Let $f(n)$ denote the number of distinct factorizations into irreducibles of the natural number n in an algebraic number field K with $h(K) \neq 1$. Prove that f cannot have a non-decreasing normal order.

This problem was posed by P. Turán in a conversation and solved by J. Rosiński and J. Śliwa [76].

32. Characterize arithmetically fields with a given class-number $\neq 1, 2$.

This has been done by J. Kaczorowski [84] and D. E. Rush [83]. See Theorem 9.2 and the literature quoted in Notes to Chapter 9.

33. Prove that there are only finitely many totally imaginary fields of a given degree with class-number 1.

34. Prove that for any positive integer N there are infinitely many fields of given degree and signature with class-number divisible by N.

This was shown for not totally real fields by T. Azuhata, H. Ichimura [84] and by S. Nakano [84a] in the general case.

35. Give an elementary proof of Dirichlet's class-number formulas.

For imaginary quadratic fields this was achieved by H. L. S. Orde [78].

Problems added in the second edition:

36. (H. C. Williams [81a]). Prove that the length of the period of the continued fraction for $D^{1/2}$ is $O(D^{1/2} \log \log D)$.

37. (Browkin, Schinzel). Prove that if there exists an infinite sequence of integers of K such that for every ideal I the first $N(I)$ terms of this sequence are distinct $(\mod I)$, then $K = Q$. Evaluate the maximal length of any finite sequence with this property.

See Latham [73], Wantuła [74], Wasén [74], [76], [77].

38. (Boyd [77b]). Is the union of the sets of PV-numbers and Salem numbers closed?

39. (Lenstra [77a]). Evaluate $L(K)$, the Lenstra constant of K, defined as the cardinality of the largest set A of integers of K with the property that for all $a \neq b$ in A, $a-b$ is a unit.

40. Characterize fields K in which every totally positive unit is a square.

Cf. Armitage, Fröhlich [67], Garbanati [76], Hasse [52a], Hughes, Mollin [83].

41. For a finite group G determine those $Z[G]$-modules which are isomorphic to $U(K)/E(K)$ for suitable normal K/Q with Galois group G.

See Notes to Chapter 3, subsection 12.

42. (Newman [74a]). Which rational integers are sums of two units from the pth cyclotomic field?

43. (Newman [74a], [74b]). Find the maximal number $N = N(p)$ such that there is a unit ε in K_p with $\varepsilon, \varepsilon+1, \ldots, \varepsilon+N-1$ all being units.

It is easily seen that $4 \leq N \leq p-1$. Possibly $N = 4$ for all primes p.

44. (Jehne [77b]). Prove that the only Kronecker class of one element consists of the base field.

45. (Perlis [78]). Do arithmetically equivalent fields have the same class-number?

46. (Baker, Schinzel [71]). Prove that if K is an imaginary quadratic field of discriminant d, then in every genus there is an ideal of norm $O(|d|^\varepsilon)$, with any $\varepsilon > 0$.

The best known evaluation is due to D. R. Heath-Brown [79], viz. $O(|d|^{1/4+\varepsilon})$.

47. Obtain analogues of the Frobenius–Rabinowitsch theorem (Proposition 8.9) for other values of h and other classes of fields.

See Hendy [74b], Kutsuna [80], Möller [76b].

48. Evaluate the constants $a_1(A)$ and $M(A)$ associated with a finite Abelian group A.

49. (Chowla, Kessler, Livingston [77]). Prove that if p is a prime congruent to 1 (mod 4) and for all x in $[1, (p-1)/2]$ we have

$$\sum_{n=1}^{x} \left(\frac{n}{p}\right) \geq 0,$$

then $p = 5, 13$ or 37.

This inequality does not hold for other primes less than 43 000.

Bibliography

Apart from the usually applied abbreviations we shall also utilize the following shortenings for the titles of certain journals and series:

AA	—Acta Arithmetica
AJM	—American Journal of Mathematics
AM	—Acta Mathematica
AMM	—American Mathematical Monthly
AnM	—Annals of Mathematics
ASENS	—Annales de l'École Normale Superieure
BAMS	—Bulletin of the American Mathematical Society
CJM	—Canadian Journal of Mathematics
CR	—Comptes Rendus Hebdomadaires, Academie des Sciences, Paris
DAN	—Doklady Akademii Nauk SSSR
GN	—Nachrichten der Akademie der Wissenschaften, Göttingen. (Formerly: ... der Gesellschaft ...)
Hbg	—Abhandlungen Math. Seminar Hamburg
IAN	—Izvestiya Akademii Nauk SSSR, ser. mat.
JLMS	—Journal of the London Mathematical Society
JNT	—Journal of the Number Theory
JRAM	—Journal für die reine und angewandte Mathematik
LN	—Lecture Notes in Mathematics, Springer Verlag
MA	—Mathematische Annalen
MC	—Mathematics of Computation
MN	—Mathematische Nachrichten
MZ	—Mathematische Zeitschrift
PAMS	—Proceedings of the American Mathematical Society
PAUS	—Proceedings of the National Academy of Sciences, USA
PCPS	—Proceedings of the Cambridge Philosophical Society
PJA	—Proceedings of the Japan Academy
PLMS	—Proceedings of the London Mathematical Society
PSPM	—Proceedings of the Symposia in Pure Mathematics
QJM	—Quarterly Journal of Mathematics, Oxford Series
TAMS	—Transactions of the American Mathematical Society

The items are numbered by the two last digits of their year of appearance, items from the same year distinguished by an added letter. There are some exceptions, regarding recent papers (1984, 1985) which at the time of preparing the bibliography were still at print or not yet available to the author in the printed form.

ABRASHKIN, V. A. [74] Determination of imaginary quadratic fields with even discriminant and class-number two by Heegner's method, *Mat. Zametki*, **15**, 1974, 241–246 (Russian).
ADACHI, N. [73] Generalization of Kummer's criterion for divisibility of class numbers, *JNT*, **5**, 1973, 253–265.
ADIBAEV, E. K. [72] A certain invariant of a quadratic field, *Kazakh. Gos. Ped. Inst.*, 1972, 3–7 (Russian).
AGOU, S. [71] Remarques sur la détermination des nombres premiers décomposés dans les corps de nombres du 3ème degré, *Publ. Dept. Math. Lyon*, **8**, 1971, 2, 93–100.
AGRAWAL, M. K. et al. (See J. C. Parnami et al.)
AHERN, P. R. [64] The asymptotic distribution of prime ideals in ideal classes, *Indag. Math.*, **26**, 1964, 10–14.
[65] Elementary methods in the theory of primes, *TAMS*, **118**, 1965, 221–242.
AIGNER, A. [34] Über die Möglichkeit von $x^4+y^4 = z^4$ in quadratischen Körpern, *Jahresber. Deutsch. Math. Verein.*, **43**, 1934, 226–229.
[52a] Weitere Ergebnisse über $x^3+y^3 = z^3$ in quadratischen Körpern, *Monatsh. Math.*, **56**, 1952, 240–252.
[52b] Ein zweiter Fall der Unmöglichkeit von $x^3+y^3 = z^3$ in quadratischen Körpern mit durch 3 teilbarer Klassenzahl, *Monatsh. Math.*, **56**, 1952, 335–338.
[53] Zur einfacher Bestimmung der Klassengruppe eines imaginär quadratischen Körpers, *Arch. Math.*, **4**, 1953, 408–411.
[56a] Die kubische Fermatgleichung in quadratischen Körpern, *JRAM*, **195**, 1956, 3–17.
[56b] Unmöglichkeitskernzahlen der kubischen Fermatgleichung mit Primfaktoren der Art $3n+1$, *JRAM*, **195**, 1956, 175–179.
[57] Die Unmöglichkeit von $x^6+y^6 = z^6$ und $x^9+y^9 = z^9$ in quadratischen Körpern, *Monatsh. Math.*, **61**, 1957, 147–150.
AKIZUKI, Y. [32] Bemerkungen über den Aufbau des Nullideals, *Proc. Phys. Math. Soc. Japan*, **14**, 1932, 253–262.
ALBERT, A. A. [30a] The integers of normal quartic fields, *AnM*, **31**, 1930, 381–418.
[30b] A determination of the integers of all cubic fields, *AnM*, **31**, 1930, 550–566.
[37a] Normalized integral bases of algebraic number fields, I, *AnM*, **38**, 1937, 923–957.
[37b] A note on matrices defining totally real fields, *BAMS*, **43**, 1937, 242–244.
[40] On p-adic fields and rational division algebras, *AnM*, **41**, 1940, 674–693.
ALBIS-GONZÁLEZ, V. S. [73] A remark on primitive roots and ramification, *Rev. Colomb. Mat.*, **7**, 1973, 93–98.
ALBU, T. [70a] Une caractérisation des anneaux de Dedekind, *CR*, **270**, 1970, 699–702.
[70b] On the cohomology of adèles, *Stud. Cerc. Mat.*, **22**, 1970, 383–390 (Rumanian).
[79] On a paper of Uchida concerning simple finite extensions of Dedekind domains, *Osaka J. Math.*, **16**, 1979, 65–69.
ALIBAEV, E. K., URAZBAEV, B. M. (See B. M. Urazbaev, E. K. Alibaev)
ALLEN, S. [75] On the factorization of natural numbers in an algebraic number field, *JLMS*, **11**, 1975, 294–300.
ALLEN, S., PLEASANTS, P. A. B. [80] The number of different lengths of irreducible factorization of a natural number in an algebraic number field, *AA*, **36**, 1980, 59–86.
AMANO, K. [77] On the Galois cohomology groups of algebraic tori and Hasse's norm theorem, *Bull. Fac. Gen. Ed. Gifu Univ.*, **13**, 1977, 185–190.
[79] On the Hasse norm principle for a separable extension, *Bull. Fac. Gen. Ed. Gifu Univ.*, **15**, 1979, 10–13.
AMANO, S. [58] A note on p-adic fields, *Keio Univ. Cent. Mem. Publ.*, Tokyo 1958, 881–892 (Japanese).

[71] Eisenstein equations of degree p in a p-adic field, *J. Fac. Sci. Tokyo*, **18**, 1971, 1–21.

AMARA, H. [81] Groupe des classes et unité fondamentale des extensions quadratiques relatives à un corps quadratique imaginaire principal, *Pacific J. Math.*, **96**, 1981, 1–12.

AMARA, M. [65] Sur un ensemble remarquable de nombres algébriques, *CR*, **260**, 1965, 1052–1054.

[66] Ensembles fermés de nombres algébriques, *ASENS*, (3), **83**, 1966, 215–270.

[79] Sur le produit des conjugués, extérieurs au disque unité, de certains nombres algébriques, *AA*, **34**, 1979, 307–314.

AMBERG, E. J. [97] *Über den Körper, dessen Zahlen sich rational aus zwei Quadratwurzeln zusammensetzen*, Diss., Zürich 1897.

AMERBAEV, V., PAK, I. [69] On the question concerning full residue systems with respect to complex moduli, *IAN Kazakh. SSR*, 1969, 3, 1–5 (Russian).

AMICE, Y. [79] Fonction Γ p-adique associée à un caractère de Dirichlet, *Study Group on Ultrametric Analysis*, **7–8**, 1979/81, Exp. 17.

AMICE, Y., FRESNEL, J. [72] Fonctions zêta p-adiques des corps de nombres abeliens réels, *AA*, **20**, 1972, 353–384.

ANDERSON, D. D. [76] Multiplication ideals, multiplication rings, and the ring $R(X)$, *CJM*, **28**, 1976, 760–768.

[80a] Noetherian rings in which every ideal is a product of primary ideals, *Canad. Math. Bull.*, **23**, 1980, 457–459.

[80b] Some remarks on multiplication ideals, *Math. Japon.*, **25**, 1980, 463–469.

ANDOZHSKII, I. V. [68] The subgroups of Demushkin's group, *Mat. Zametki*, **4**, 1968, 349–354 (Russian).

ANDRIANOV, A. N. [68a] Continuability and functional equation for zeta-functions with nonabelian characters of simple algebras over number fields, *DAN*, **181**, 1968, 775–777 (Russian).

[68b] Zeta-functions of simple algebras with nonabelian characters, *Uspekhi Mat. Nauk*, **23**, 1968, No 4, 3–66 (Russian).

ANDRUKHAEV, H. M. [64] The problem of addition of primes and almost primes in algebraic number fields, *DAN*, **159**, 1964, 1207–1209 (Russian).

[69] Generalization of Kloosterman sums to the Gaussian field and their evaluations, *Nauch. Tr. Krasnodar. Gos. Ped. Inst.*, **118**, 1969, 29–40 (Russian).

ANFERT'EVA, E. A., CHUDAKOV, N. G. [68] On the minimas of the norm-function in imaginary quadratic fields, *DAN*, **183**, 1968, 255–256 (Russian).

[70] Effective estimates from below of the norms of ideals of an imaginary quadratic field, *Mat. Sbornik*, **82**, 1970, 55–66 (Russian).

ANGELL, I. O. [73] A table of complex cubic fields, *Bull. London Math. Soc.*, **5**, 1973, 37–38.

[76] A table of totally real cubic fields, *MC*, **30**, 1976, 184–187.

ANKENY, N. C. [51] An improvement of an inequality of Minkowski, *PAUS*, **37**, 1951, 711–716.

[52a] A generalization of a theorem of Suetuna on Dirichlet series, *PJA*, **28**, 1952, 289–295.

[52b] Representation of primes by quadratic forms, *AJM*, **74**, 1952, 913–919.

ANKENY, N. C., ARTIN, E., CHOWLA, S. [52] The class-number of real quadratic fields, *AnM*, **56**, 1952, 479–493.

ANKENY, N. C., BRAUER, R., CHOWLA, S. [56] A note on the class numbers of algebraic number fields, *AJM*, **78**, 1956, 51–61.

ANKENY, N. C., CHOWLA, S. [49] The class number of the cyclotomic field, *PAUS*, **35**, 1949, 529–532.

[50] The relation between the class number and the distribution of primes, *PAMS*, **1**, 1950, 775–776.
[51] The class number of the cyclotomic field, *CJM*, **3**, 1951, 486–491.
[55] On the divisibility of the class number of quadratic fields, *Pacific J. Math.*, **5**, 1955, 321–324.
[60] A note on the class number of real quadratic fields, *AA*, **6**, 1960, 145–147.
[62] A further note on the class number of real quadratic fields, *AA*, **7**, 1962, 271–272.
[68] Diophantine equations in cyclotomic fields, *JLMS*, **43**, 1968, 67–70.
ANKENY, N. C., CHOWLA, S., HASSE, H. [65] On the class number of the maximal real subfield of a cyclotomic field, *JRAM*, **217**, 1965, 217–220.
ANKENY, N. C., ROGERS, C. A. [51a] A condition for a real lattice to define a zeta function, *PAUS*, **37**, 1951, 159–163.
[51b] A conjecture of Chowla, *AnM*, **53**, 1951, 541–555; *ibid.* **58**, 1953, 591.
ANTONIADIS, J. A. [83] Über die Kennzeichnung zweiklassiger imaginär-quadratischer Zahlkörper durch Lösungen diophantischer Gleichungen, *JRAM*, **339**, 1983, 27–81.
APOSTOL, T. M. [70a] Resultants of cyclotomic polynomials, *PAMS*, **24**, 1970, 457–463.
[70b] Euler's Φ-function and separable Gauss sums, *PAMS*, **24**, 1970, 482–485.
[70c] Dirichlet's *L*-functions and character power sums, *JNT*, **2**, 1970, 223–234.
[72] Dirichlet *L*-functions and primitive characters, *PAMS*, **31**, 1972, 384–386.
[75] The resultant of the cyclotomic polynomials $F_m(ax)$ and $F_n(bx)$, *MC*, **29**, 1975, 1–6.
APOSTOL, T. M., SKLAR, A. [57] The approximate functional equation of Hecke's Dirichlet series, *TAMS*, **86**, 1957, 446–462.
APPELGATE, H., ONISHI, H. [82] Periodic expansions of modules and its relation to units, *JNT*, **15**, 1982, 283–294.
ARAI, M. [81] On Voronoj's theory of cubic fields, I, *PJA*, **57**, 1981, 226–229; II, *PJA*, **57**, 1981, 281–283.
ARAL, H. [39] *Simultane diophantische Approximationen in imaginären quadratischen Zahlkörpern*, Diss., München 1939.
ARAMATA, H. [31] Über die Teilbarkeit der Zetafunktionen gewisser algebraischer Zahlkörper, *Proc. Imp. Acad. Tokyo*, **7**, 1931, 334–336.
[33] Über die Teilbarkeit der Dedekindschen Zetafunktionen, *Proc. Imp. Acad. Tokyo*, **9**, 1933, 31–34.
[39] Über die Eindeutigkeit der Artinschen *L*-Funktionen, *Proc. Imp. Acad. Tokyo*, **15**, 1939, 124–126.
ARCHIBALD, R. G. [36] Highly composite ideals, *Trans. Roy. Soc. Canada*, s. III, **30**, 1936, 41–47.
ARCHINARD, G. [74] Extensions cubiques cycliques de Q dont l'anneau des entiers est monogène, *Enseign. Math.*, **20**, 1974, 179–203.
[84] Submodules of a torsion-free and finitely generated module over a Dedekind ring, *Colloq. Math.*, **48**, 1984, 193–204.
ARF, C. [39] Untersuchungen über reinverzweigte Erweiterungen diskret bewerteter perfekter Körper, *JRAM*, **181**, 1939, 1–44.
ARKHIPOV, G. I., KARATSUBA, A. A. [81] Local representation of zero by a form, *IAN*, **45**, 1981, 948–961 (Russian).
[82a] On the representation of zero by a form in a *p*-adic number field, *DAN*, **262**, 1982, 11–13 (Russian).
[82b] On a problem of the theory of congruences, *Uspekhi Mat. Nauk*, **37**, 1982, No 5, 161–162 (Russian).
ARMBRUST, M. [71] On the theorem of Ax and Kochen, *Arch. Math.*, **22**, 1971, 55–58.

ARMITAGE, J. V. [67] On a theorem of Hecke in number fields and function fields, *Invent. Math.*, **2**, 1967, 238–246.
[72] Zeta functions with a zero at $s = 1/2$, *Invent. Math.*, **15**, 1972, 199–205.

ARMITAGE, J. V., FRÖHLICH, A. [67] Classnumbers and unit signatures, *Mathematika* **14**, 1967, 94–98.

ARNAUDON, M. [76] Étude des normes dans les extensions galoisiennes de corps de nombres, *CR*, **283**, 1976, 269–272.

ARNDT, F. [58a] Über die Anzahl der Genera der quadratischen Formen, *JRAM*, **56**, 1858, 72–78.
[58b] Einfacher Beweis für die Irreduzibilität einer Gleichung in der Kreisteilung, *JRAM*, **56**, 1858, 178–181.

ARNOLD, J. T. [69] On the ideal theory of the Kronecker function ring and the domain $D(X)$, *CJM*, **21**, 1969, 558–563.

ARPAIA, P. J. [68] A note on quadratic Euclidean domains, *AMM*, **75**, 1968, 864–865.

ARTIN, E. [23] Über die Zetafunktionen gewisser algebraischer Zahlkörper, *MA*, **89**, 1923, 147–156.
[24] Über eine neue Art von L-Reihen, *Hbg*, **3**, 1924, 89–108.
[27] Beweis des allgemeinen Reziprozitätsgesetzes, *Hbg*, **5**, 1927, 353–363.
[30a] Idealklassen in Oberkörpern und allgemeines Reziprozitätsgesetz, *Hbg*, **7**, 1930, 46–51.
[30b] Zur Theorie der L-Reihen mit allgemeinen Gruppencharakteren, *Hbg*, **8**, 1930, 292–306.
[31] Die gruppentheoretische Struktur der Diskriminanten algebraischer Zahlkörper, *JRAM*, **164**, 1931, 1–11.
[32a] Über Einheiten relativgaloischer Zahlkörper, *JRAM*, **167**, 1932, 153–156.
[32b] Über die Bewertungen algebraischer Zahlkörper, *JRAM*, **167**, 1932, 157–159.
[50a] Questions de base minimale dans la théorie des nombres algébriques, *Alg. Th. des Nombres, Coll. Internat. CNRS*, **24**, 1950, 19–20.
[50b] Remarques concernant la théorie de Galois, *Alg. Th. des Nombres, Coll. Internat. CNRS*, **24**, 1950, 161–162.
[56] Representatives of the connected component of the idèle class group, *Proc. Int. Symp. Alg. N. Th.*, Tokyo 1956, 51–54.
[59] *Theory of Algebraic Numbers*, Göttingen 1959.
[67] *Algebraic Numbers and Algebraic Functions*, New York 1967.

ARTIN, E., ANKENY, N. C., CHOWLA, S. (See N. C. Ankeny, E. Artin, S. Chowla)

ARTIN, E., TATE, J. [61] *Class Field Theory*, Harvard 1961.

ARTIN, E., WHAPLES, G. [45] Axiomatic characterization of fields by the product formula for valuations, *BAMS*, **51**, 1945, 469–492.
[46] A note on axiomatic characterization of fields, *BAMS*, **52**, 1946, 245–247.

ARUTYUNYAN, L. Z. [77] Generators of the group of principal units of a cyclic p-extension of a regular local field, *Zap. Nauch. Sem. LOMI*, **71**, 1977, 16–23 (Russian).

ARWIN, A. [26] Einige periodische Kettenbruchentwicklungen, *JRAM*, **155**, 1926, 111–128.
[29] On cubic fields, *AnM*, **30**, 1929, 1–11.

ASANO, K. [50] Über Moduln und Elementarteilertheorie im Körper, in dem Arithmetik definiert ist, *Japan. J. Math.*, **20**, 1950, 55–71.
[51] Über kommutative Ringe, in denen jedes Ideal als Produkt von Primidealen darstellbar ist, *J. Math. Soc. Japan*, **3**, 1951, 82–90.

Asenov, E. K. [76] Distribution of absolutely abelian fields of type $(l, l, \ldots, l) - 8$ times, *IAN Kazakh. SSR*, 1976, 3, 65–68 (Russian).

Auslander, L., Tolimieri, R., Winograd, S. [82] Hecke's theorem in quadratic reciprocity, finite nilpotent groups and the Cooley–Tuckey algorithm, *Adv. Math.*, 43, 1982, 122–172.

Auslander, M., Buchsbaum, D. A. [59] On ramification theory in noetherian rings, *AJM*, 81, 1959, 749–765.

Avanesov, É. T. [79] On the fundamental units of algebraic fields of degree n, *AA*, 35, 1979, 175–185 (Russian).

Avanesov, É. T., Billevich, K. K. [81] Fundamental units of cubic fields of positive discriminants, *Mat. Zametki*, 29, 1981, 801–812, 955 (Russian).

Ax, J. [62] The intersection of norm groups, *TAMS*, 105, 1962, 462–474.
[65] On the units of an algebraic number field, *Illinois J. Math.*, 9, 1965, 584–589.

Ax, J., Kochen, S. [65] Diophantine problems over local fields, I, *AJM*, 87, 1965, 605–630; II, *AJM*, 87, 1965, 631–648.

Axer, A. [04] Zahlentheoretische Funktionen und deren asymptotische Werte im Gebiete der aus den dritten Einheitswurzeln gebildeten ganzen komplexen Zahlen, *Monatsh. M.-Phys.*, 15, 1904, 239–291.

Ayoub, C. [68] On the units in certain integral domains, *Arch. Math.*, 19, 1968, 43–46.
[69] On finite primary rings and their groups of units, *Compos. Math.*, 21, 1969, 247–252.
[72] On the group of units of certain rings, *JNT*, 4, 1972, 383–403.
[81] On certain chain rings and their groups of units, *Comm. Algebra*, 9, 1981, 323–338.

Ayoub, R. G. [53] On the Waring–Siegel theorem, *CJM*, 5, 1953, 439–450.
[55] On Selberg's lemma for algebraic fields, *CJM*, 7, 1955, 138–143.
[58] A mean value theorem for quadratic fields, *Pacific J. Math.*, 8, 1958, 23–27.
[64] On L-series with real characters, *Illinois J. Math.*, 8, 1964, 550–555.
[67a] On L-functions, *Monatsh. Math.*, 71, 1967, 193–202.
[67b] A note on the class number of imaginary quadratic fields, *MC*, 21, 1967, 442–445.
[68] On the coefficients of the zeta-function of an imaginary quadratic field, *AA*, 13, 1968, 375–381.

Ayoub, R. G., Chowla, S. [81] On Euler's polynomial, *JNT*, 13, 1981, 443–445.

Azuhata, T., Ichimura, H. [84] On the divisibility problem of the class numbers of algebraic number fields, *J. Fac. Sci. Univ. Tokyo*, 30, 1984, 579–585.

Baayen, P. C. [65] Een combinatorisch vermoeden bevestigt voor $C_2 \oplus C_2 \oplus C_2 \oplus C_6$, *Math. Centzum*, Amsterdam, WN 25, 1965.
[68a] Een combinatorisch problem voor eindige Abelse groepen, MC Syllabus 5, *Coll. Dishrete Wisk.* Cap. 3, 1968.
[68b] Een geral een structurproblem voor Abelse groepen, *Math. Centrum*, Amsterdam, WN 24, 1968.
[69] $(C_2 \oplus C_2 \oplus C_2 \oplus C_{2n})!$ is true for odd n, *Math. Centrum*, Amsterdam, ZW 1969-006.

Baayen, P. C., Emde Boas, P. van, Kruyswijk, D. [69] A combinatorial problem on finite abelian groups, III, *Math. Centrum*, Amsterdam, ZW 1969-008.

Babaev, G. [60] The distribution of integral points on certain surfaces defined by norms, *DAN*, 134, 1960, 13–15 (Russian).
[63] Asymptotical geometrical properties of the set of integral points on certain hyperbolas, *Izv. VUZ*, 1963, No 1, 3–7 (Russian).
[71] An arithmetic proof of the infinity of prime ideals of degree $\geqslant 2$, *DAN Tadzhik. SSR*, 14, 1971, No 9, 3–6 (Russian).

BABAEV, G., KORCHAGINA, V. I. [72] The distribution of prime ideal numbers, *DAN Tadzhik. SSR*, **15**, 1972, 8–9 (Russian).

BABAITSEV, V. A. [76] Certain questions in the theory of Γ-extensions of algebraic number fields, *IAN*, **40**, 1976, 477–487; II, *ibid.*, 715–726 (Russian).

[80] On the boundedness of Iwasawa's μ-invariant, *IAN*, **44**, 1980, 3–23 (Russian).

[81] On the linear character of the behaviour of the Iwasawa μ invariant, *IAN*, **45**, 1981, 691–703 (Russian).

BACHMAN, G. [66] The decomposition of a rational prime ideal in cyclotomic fields, *AMM*, **73**, 1966, 494–497.

BACHMANN, P. [64] *De unitatum complexarum theoria*, Berlin 1864, 23 pp.

BAEZA, R. [79] Über die Stufe von Dedekind Ringen, *Arch. Math.*, **33**, 1979/80, 226–231.

BAIBULATOV, R. S. [67] The Erdös–Wintner theorem for normed semigroups, *DAN Uzbek. SSR*, 1967, No 8, 3–5 (Russian).

[68] Value distribution of certain classes of additive functions in algebraic number fields, *Mat. Zametki*, **4**, 1968, 63–73 (Russian).

[69] A certain generalization of the Levin–Faĭnleĭb theorem, *DAN Uzbek. SSR*, 1969, No 4, 3–5 (Russian).

BAILY, A. M. [80] On the density of discriminants of quartic fields, *JRAM*, **315**, 1980, 190–210.

[81] On octic fields of exponent 2, *JRAM*, **328**, 1981, 33–38.

BAKER, A. [66] Linear forms in the logarithms of algebraic numbers, *Mathematika*, **13**, 1966, 204–216.

[69] A remark on the class number of quadratic fields, *Bull. London Math. Soc.*, **1**, 1969, 98–102.

[71a] On the class-number of imaginary quadratic fields, *BAMS*, **77**, 1971, 678–684.

[71b] Imaginary quadratic fields with class number 2, *AnM*, **94**, 1971, 139–152.

[77] The theory of linear forms in logarithms, in: *Transcendence Theory*, Academic Press, 1977, 1–27.

BAKER, A., SCHINZEL, A. [71] On the least integers represented by the genera of binary quadratic forms, *AA*, **18**, 1971, 137–144.

BALDISSERI, N. [75] Sulla lunghezza dell'algoritmo euclideo nei campi quadratici euclidici complessi, *Boll. Un. Mat. Ital.*, (5), **12**, 1975, 333–347.

BALLIEU, R. [54] Factorisation des polynomes cyclotomiques modulo un nombre premier, *Ann. Soc. Sci. Bruxelles*, Sér. I, **68**, 1954, 140–144.

BAMBAH, R. P., LUTHAR, I. S., MADAN, M. L. [61] On the existence of a certain type of basis in a totally real field, *Res. Bull. Panjab Univ.*, **12**, 1961, 135–137.

BANIUK, L., BARRUCAND, P., WILLIAMS, H. C. (See P. Barrucand, H. C. Williams, L. Baniuk)

BARBAN, M. B. [62] The "large sieve" of Ju. V. Linnik and limit theorem for the class-number of an imaginary quadratic field, *IAN*, **26**, 1962, 563–580 (Russian).

[64] On a theorem of P. Bateman, S. Chowla and P. Erdös, *Magyar Tud. Akad. Mat. Fiz. Oszt. Közl.*, **9**, 1964, 429–435 (Russian).

[67] On the density hypothesis of E. Bombieri, *DAN*, **172**, 1967, 999–1000 (Russian).

BARBAN, M. B., GORDOVER, G. [66] On the moments of the class-numbers of purely radical quadratic forms of negative determinant, *DAN*, **167**, 1966, 267–269 (Russian).

BARBAN, M. B., LEVIN, B. V. [68] Multiplicative functions on "shifted" prime numbers, *DAN*, **181**, 1968, 778–780 (Russian).

BARKAN, P. [75] Sur des sommes de caractères liées aux nombres de classes des corps abéliens imaginaires, *CR*, **281**, 1975, A887–890.

BARNER, K. [67] Zur Reziprozität quadratischer Charaktersummen in algebraischen Zahlkörpern, *Monatsh. Math.*, **71**, 1967, 369–384.
[68] Über die quaternäre Einheitsform in total reellen algebraischen Zahlkörpern, *JRAM*, **229**, 1968, 194–208.
[69] Über die Werte der Ringklassen-L-Funktionen reell-quadratischer Zahlkörper an natürlichen Argumentstellen, *JNT*, **1**, 1969, 28–64.
[81] On A. Weil's explicit formulas, *JRAM*, **323**, 1981, 139–152.
BARNES, E. S. [50] Non-homogeneous binary quadratic forms, *QJM*, **1**, 1950, 199–210.
BARNES, E. S., SWINNERTON-DYER, H. P. F. [52] The inhomogeneous minima of binary quadratic forms, I, *AM*, **87**, 1952, 259–323.
BARNES, F. W. [72] On the Stufe of an algebraic number field, *JNT*, **4**, 1972, 474–476.
BARRUCAND, P. [71] Quelques propriétés des coefficients des séries L associées aux corps cubiques, *CR*, **273**, 1971, 960–963.
BARRUCAND, P., COHN, H. [69] Note on primes of type x^2+32y^2, class number and residuacity, *JRAM*, **238**, 1969, 67–70.
[70] A rational genus, class number divisibility and unit theorem for pure cubic fields, *JNT*, **2**, 1970, 7–21.
[71] Remarks on principal factors in a relative cubic field, *JNT*, **3**, 1971, 226–239.
[73] On some class-fields related to primes of type x^2+32y^2, *JRAM*, **262/3**, 1973, 400–414.
BARRUCAND, P., LAUBIE, F. [82] Ramification modérée dans les corps de nombres de degré premier, *Sém. Th. Nombr.*, Bordeaux, 1981/82, exp. 13, 1–6.
BARRUCAND, P., SATGÉ, P. (See P. Satgé, P. Barrucand)
BARRUCAND, P., WILLIAMS, H. C., BANIUK, L. [76] A computational technique for determining the class number of a pure cubic field, *MC*, **30**, 1976, 312–323.
BARSKY, D. [73] Fonctions k-lipschitziennes sur un anneau local et polynômes à valeurs entières, *Bull. Soc. Math. France*, **101**, 1973, 397–411.
[77a] On Morita's p-adic Γ-function, *Gr. Etud. Anal. Ultrametr.*, **5**, 1977/8, exp. 3, 1–6.
[77b] Fonctions zeta p-adiques d'une classe de rayon de corps totalement réels, *Gr. Etud. Anal. Ultrametr.*, **5**, 1977/78, exp. 16, 1–23.
[78] Transformations de Cauchy p-adique et algèbre d'Iwasawa, *MA*, **232**, 1978, 255–266.
BARTELS, H. J. [80] Über Normen algebraischer Zahlen, *MA*, **251**, 191–212.
[81a] Zur Arithmetik von Diedergruppenerweiterungen, *MA*, **256**, 1981, 465–473.
[81b] Zur Arithmetik von Konjugationsklassen in algebraischen Gruppen, *J. Algebra*, **70**, 1981, 179–199.
BARTZ, K. [76] On some estimates in the theory of the Dedekind zeta-function, *Funct. et Approx.*, **4**, 1976, 27–35.
[78] On a theorem of Sokolovskiĭ, *AA*, **34**, 1978, 113–126.
[80] An estimate for $\zeta(s, \chi)$-functions, *Funct. et Approx.*, **9**, 1980, 99–106.
BASHMAKOV, M. I., KESHTOV, R. A. [76] The localization of the units of a number field, *Zap. Nauch. Sem. LOMI*, **64**, 1976, 5–11 (Russian).
BASS, H. [62] Torsion free and projective modules, *TAMS*, **102**, 1962, 319–327.
[65] A remark on an arithmetic theorem of Chevalley, *PAMS*, **16**, 1965, 875–878.
[66] The Dirichlet unit theorem, induced characters and Whitehead groups of finite group, *Topology*, **4**, 1966, 391–410.
[68] *Algebraic K-theory*, New York 1968.
[69] K_2 and symbols, in: *Algebraic K-theory and its Geometric Applications*, LN 108, 1–11, Springer, 1969.
[71] K_2 des corps globaux [d'après J. Tate, H. Garland, ...], *Sém. Bourbaki*, **23**, 1970/71, exp. 394, 233–255, LN 244, 1971.

BATEMAN, P. T. [49] Note on the coefficients of the cyclotomic polynomial, *BAMS*, **55**, 1949, 1180–1181.
[59] Theorems implying the non-vanishing of $\Sigma\chi(m)m^{-1}$ for real residue-characters, *J. Indian Math. Soc.*, **23**, 1959, 101–115.
BATEMAN, P. T., CHOWLA, S., ERDÖS, P. [50] Remarks on the size of $L(1,\chi)$, *Publ. Math. (Debrecen)*, **1**, 1950, 165–182.
BATEMAN, P. T., GROSSWALD, E. [62] Imaginary quadratic fields with unique factorization, *Illinois J. Math.*, **6**, 1962, 187–192.
BATEMAN, P. T., POMERANCE, C., VAUGHAN, R. C., [85] On the size of the coefficients of the cyclotomic polynomials, in: *Classical Problems in Number Theory*, North-Holland, 1984.
BATEMAN, P. T., STEMMLER, R. M. [62] Waring's problem for algebraic number fields and primes of the form $(p^r-1)(p^d-1)$, *Illinois J. Math.*, **6**, 1962, 142–156.
BAUER, H. [69] Numerische Bestimmung der Klassenzahlen reeller zyklischer Zahlkörper, *JNT*, **1**, 1969, 161–162.
[71] Zur Berechnung der 2-Klassenzahl der quadratischen Zahlkörper mit genau zwei verschiedenen Diskriminantenprimteilern, *JRAM*, **248**, 1971, 42–46.
[72] Die 2-Klassenzahl spezieller quadratischer Zahlkörper, *JRAM*, **252**, 1972, 79–81.
BAUER, M. [04a] Über Kreisteilungsgleichungen, *Arch. Math. Phys.*, (3), **6**, 1904, 220.
[04b] Über zusammengesetzte Körper, *Arch. Math. Phys.*, (3), **6**, 1904, 221–222.
[07] Über die ausserwesentliche Diskriminantenteiler einer Gattung, *MA*, **64**, 1907, 572–576.
[16a] Zur Theorie der algebraischen Zahlkörper, *MA*, **77**, 1916, 353–356.
[16b] Über zusammengesetzte Zahlkörper, *MA*, **77**, 1916, 357–361.
[17] Zur Theorie der arithmetischer Progression, *Arch. Math. Phys.*, (3), **25**, 1917, 131–134.
[19a] Bemerkungen über die Differente des algebraischen Zahlkörpers, *MA*, **79**, 1919, 321–322.
[19b] Zur Theorie der Fundamentalgleichung, *JRAM*, **149**, 1919, 86–96.
[20] Bemerkung über die Zusammensetzung der algebraischen Zahlkörper, *JRAM*, **150**, 1920, 185–188.
[21a] Über relativ-Galoische Zahlkörper, *MA*, **83**, 1921, 70–73.
[21b] Über die Differente eines algebraischen Zahlkörpers, *MA*, **83**, 1921, 74–76.
[21c] Beweis von einigen bekannten Sätzen über zusammengesetzte Körper ohne Anwendung der Idealtheorie, *Jahresber. Deutsch. Math. Verein.*, **30**, 1921, 186–188.
[22] Die Theorie der p-adischen bzw. p-adischen Zahlen und die gewöhnlichen algebraischen Zahlkörpern, *MZ*, **14**, 1922, 244–249; *ibid.* **20**, 1924, 94–97.
[23] Verschiedene Bemerkungen über die Differente und die Diskriminante eines algebraischen Zahlkörpers, *MZ*, **16**, 1923, 1–12.
[24] Über die Erweiterung des Körpers der p-adischen Zahlen zu einem algebraisch abgeschlossenen Körper, *MZ*, **19**, 1924, 308–312.
[27] Über die Norm eines Ideals, *Hbg.*, **5**, 1927, 184.
[36] Bemerkungen zum Hensel–Oreschen Hauptsatze, *Acta Sci. Math. (Szeged)*, **8**, 1936, 64–67.
[37] Bemerkungen über die Galoische Gruppe einer Gleichung, *MA*, **114**, 1937, 352–354.
[39] Zur Theorie der Kreiskörper, *Acta Sci. Math. (Szeged)*, **9**, 1939, 110–112.
[40a] Über zusammengesetzte relativ Galoische Körper, *Acta Sci. Math. (Szeged)*, **9**, 1940, 206–211.
[40b] Über die Zusammensetzung algebraischer Zahlkörper, *Acta Sci. Math. (Szeged)*, **9**, 1940, 212–217.

BAUER, M., CHEBOTAREV, N. G. [28] *p*-adischer Beweis des zweiten Hauptsatzes von Herrn Ore, *Acta Sci. Math. (Szeged)*, **4**, 1928, 56–57.

BAYER, P., NEUKIRCH, J. [79] On values of zeta functions and *l*-adic Euler characteristics, *Invent. Math.*, **50**, 1978/9, 35–64.

BAZYLEWICZ, A. [76] On the product of the conjugates outside the unit circle of an algebraic integer, *AA*, **30**, 1976/7, 43–61.

[82] Traces of monomials in algebraic numbers, *AA*, **41**, 1982, 101–116.

BEACH, B. D., WILLIAMS, H. C., ZARNKE, C. R. [71] Some computer results on units in quadratic and cubic fields, *Proc. 25 Summer Meeting Can. Math. Congr.*, 1971, 609–648.

BEAUMONT, R. A., PIERCE, R. S. [61] Subrings of algebraic number fields, *Acta Sci. Math. (Szeged)*, **22**, 1961, 202–216.

BEBBE, E. [66] Utilisation d'idéaux réduits dans la théorie du corps quadratique, *Ann. Univ. Besançon*, 1966, Nr. 2.

BEDOCCHI, E. [78] Perfect numbers in a real quadratic field, *Boll. Un. Mat. Ital.*, (5), **15**, 1978, 94–103 (Italian).

[80] Sums of divisors in the ring of integers of the field $Q(\sqrt{2})$, *Boll. Un. Mat. Ital.*, Suppl. 1980, No. 2, 15–29 (Italian).

BEEGER, N. G. W. H. [19] Über die Teilkörper des Kreiskörpers $K(\zeta_n)$, *Proc. Akad. Wet. Amsterdam*, **21**, 1919, 454–465, 758–773, 774–779.

[20] Bestimmung der Klassenzahl der Ideale aller Unterkörper des Kreiskörpers der ζ_m, wo *m* durch mehr als eine Primzahl teilbar ist, *Proc. Akad. Wet. Amsterdam*, **22**, 1920, 331–350, 395–414; corr., ibid., **23**, 1922, 1399–1401.

BEHNKE, H. [23] Ueber analytische Funktionen und algebraische Zahlen, *Hbg.*, **2**, 1923, 81–111.

BEHRBOHM, H., RÉDEI, L. [36] Der Euklidische Algorithmus in quadratischen Körpern, *JRAM*, **174**, 1936, 192–205.

BEITER, M. [68] Magnitude of the coefficients of the cyclotomic polynomial $F_{pqr}(x)$, *AMM*, **75**, 1968, 370–372.

[78] Coefficients of the cyclotomic polynomial F_{3qr}, *Fibonacci Quart.*, **16**, 1978, 302–306.

BELCHER, P. [74] Integers expressible as sums of distinct units, *Bull. London Math. Soc.*, **6**, 1974, 66–68.

[75] A test for integers being sums of distinct units applied to cubic fields, *JLMS*, **12**, 1975/6, 141–148.

BENDER, E. A. [68a] Classes of matrices and quadratic fields, *Linear Algebra Appl.*, **1**, 1968, 195–201.

[68b] Characteristic polynomials of symmetric matrices, *Pacific J. Math.*, **25**, 1968, 433–441.

BENNETT, A. A. [23] Some algebraic analogies in matric theory, *AnM*, **23**, 1923, 91–96.

BENSON, C. T., WEBER, B. T. [73] Computing units in certain orders of algebraic integers, *JNT*, **5**, 1973, 99–107.

BENZAGHOU, B. [68] Sur les suites d'unités algébriques vérifiant une relation de récurrence lineaire, *CR*, **267**, 1968, 913–915.

[69] Suites d'unités algébriques satisfaisant a une relation de recurrence lineaire, *Bull. Soc. Math. France*, Mém. **25**, 1969, 29–31.

BERG, E. [35] Über die Existenz eines Euklidischen Algorithmus in quadratischen Zahlkörpern, *Fysiogr. Sällsk. Lund Förh.*, **5**, 1935, 1–6.

BERGÉ, A.-M. [72] Sur l'arithmétique d'une extension diédrale, *Ann. Inst. Fourier*, **22**, 1972, No 2, 31–59.

[75] Sur l'arithmétique d'une extension cyclique totalement ramifiée d'un corps local, *CR*, **281**, 1975, 67–70.

[78] Arithmétique d'une extension galoisienne à groupe d'inertie cyclique, *Ann. Inst. Fourier*, **28**, 1978, No 4, 17–44.

[81] A propos du genre de l'anneau des entiers d'une extension, *Publ. Math. Fac. Sci. Besançon, Th. de Nombres* 1979/80 et 1980/81, 9 pp.

BERGER, A.   [91] Recherches sur les nombres et les fonctions de Bernoulli, *AM*, **14**, 1890/1, 249–304.

BERGER, R.   [60] Über verschiedene Differentenbegriffe, *S.-B. Heidelberg. AW, Math. Nat. Kl.*, 1960/1, 1–44.

[62] Ausdehnung von Derivationen und Schachtelung der Differente, *MZ*, **78**, 1962, 97–115.

[64] Differenten regulärer Ringe, *JRAM*, **214-215**, 1964, 441–442.

BERGER, T. R., REINER, I.   [75] A proof of the normal basis theorem, *AMM*, **82**, 1975, 915–918.

BERGMANN, G.   [65] Untersuchungen zur Einheitengruppe in den totalkomplexen algebraischen Zahlkörpern sechstes Grades (über P) im Rahmen der "Theorie der Netze", *MZ*, **161**, 1965, 349–364.

[66a] Über Eulers Beweis des grossen Fermatschen Satzes für den Exponenten 3, *MA*, **164**, 1966, 159–175.

[66b] Zur numerischen Bestimmung einer Einheitenbasis, *MA*, **166**, 1966, 103–105.

[66c] Beispiel numerischer Einheitenbestimmung, *MA*, **167**, 1966, 143–168.

BERGSTRÖM, H.   [37a] Über die Methode von Woronoj zur Berechnung einer Basis eines kubischen Zahlkörpers, *Arkiv Mat., Astr. Fys.*, **25B**, 1937, Nr. 26, 1–8.

[37b] Zur Theorie der biquadratischen Zahlkörper. Die Arithmetik auf klassenkörpertheoretischer Grundlage, *Nova Acta Soc. Sci. Uppsala*, IV, **10**, 1937, Nr. 8, 1–56.

[44] Die Klassenzahlformel für reelle quadratische Zahlkörper mit zusammengesetzter Diskriminante als Produkt verallgemeinerter Gaussscher Summen, *JRAM*, **186**, 1944–45, 91–115.

BERGUM, G. E.   [71] Distribution of primes in $Z(\omega)$, *Proc. Washington State Univ. Conf. Numb. Th.*, 1971, 207–216.

[72] The distribution of k-th powers residues and nonresidues in the integral domain $Z(\sqrt{-2})$, *Duke Math. J.*, **39**, 1972, 19–24.

[78] Complete residue systems in the quadratic domain $Z(e^{2\pi i/3})$, *Int. J. Math. Sci.*, **1**, 1978, 75–85.

BERLEKAMP, E. R.   [67] Factoring polynomials over finite fields, *Bell Syst. Tech. J.*, **46**, 1967, 1853–1859.

[70] Factoring polynomials over large finite fields, *MC*, **24**, 1970, 713–735.

BERNDT, B. C.   [69a] Identities involving the coefficients of a class of Dirichlet series. I, *TAMS*, **137**, 1969, 349–359; II, *ibid.*, 361–374.

[69b] Arithmetical identities and Hecke's functional equation, *Proc. Edinburgh Math. Soc.*, **16**, 1969, 221–226.

[69c] A note on the number of integral ideals of bounded norm in a quadratic number field, *BAMS*, **75**, 1969, 1283–1285.

[70] On the zeros of a class of Dirichlet series. I, *Illinois J. Math.*, **14**, 1970, 244–258; II, *ibid.*, 678–691.

[71a] The number of zeros of the Dedekind zeta-function on the critical line, *JNT*, **3**, 1971, 1–6.

[71b] The functional equation of some Dirichlet series, *PAMS*, **29**, 1971, 457–460.

[71c] On the average order of a class of arithmetical functions, *JNT*, **3**, 1971, 184–203; II, *ibid.*, 288–305.

[73a] A new proof of the functional equation of Dirichlet L-functions, *PAMS*, **37**, 1973, 355–357.
[73b] Character transformation formulae similar to those of Dedekind eta-function, *PSPM*, **24**, 1973, 9–30.
[75] Identities involving the coefficients of a class of Dirichlet series, VII, *TAMS*, **201**, 1975, 247–261.
BERNDT, B. C., EVANS, R. J. [77] Dedekind sums and class numbers, *Monatsh. Math.*, **84**, 1977, 265–273.
[81] The determination of Gauss sums, *BAMS*, **5**, 1981, 107–129.
BERNSTEIN, L. [67] Ein neuer Algorithmus für absteigende Basispotenzen im kubischen Körper, *MN*, **33**, 1967, 257–272.
[70] Einheitenberechnung in kubischen Körpern mittels des Jacobi–Perronschen Algorithmus aus der Rechenanlage, *JRAM*, **244**, 1970, 201–220.
[71] *The Jacobi–Perron Algorithm, its Theory and Application*, LN 207, Springer, 1971.
[72] On units and fundamental units, *JRAM*, **257**, 1972, 129–145.
[74a] Fundamental units from the preperiod of a generalized Jacobi–Perron algorithm, *JRAM*, **268/9**, 1974, 391–409.
[74b] Units from periodic Jacobi–Perron algorithms in algebraic number fields of degree $n > 2$, *Manuscr. Math.*, **14**, 1974, 249–261.
[75a] Units and periodic Jacobi–Perron algorithms in real algebraic number fields of degree 3, *TAMS*, **212**, 1975, 295–300.
[75b] Truncated units in infinitely many algebraic number fields of degree $n \geqslant 4$, *MA*, **213**, 1975, 275–279.
[75c] Units and their norm equation in real algebraic number fields of any degree, *Symposia Math.*, **15**, 307–340, Academic Press, 1975.
[76a] Fundamental units and cycles in the period of real quadratic number fields. I, *Pacific J. Math.*, **63**, 1976, 37–61; II, *ibid.*, 63–78.
[76b] Zeros of combinatorial functions and combinatorial identities, *Houston J. Math.*, **2**, 1976, 9–16.
[77] Gaining units from units, *CJM*, **29**, 1977, 93–106.
[78a] An algorithm for Halter-Koch units, *Michigan J. Math.*, **25**, 1978, 371–377.
[78b] Applications of units, *JNT*, **10**, 1978, 354–383.
BERNSTEIN, L., HASSE, H. [65] Einheitenberechnung mittels des Jacobi–Perronschen Algorithmus, *JRAM*, **218**, 1965, 51–69.
[69] An explicit formula for the units of an algebraic number field of degree $n \geqslant 2$, *Pacific J. Math.*, **30**, 1969, 293–365.
[75] Ein formales Verfahren zur Herstellung parameterabhängiger Scharen quadratischer Grundeinheiten, *JRAM*, **276**, 1975, 206–212.
BERTIN, M. J. [74] Caractérisation de l'ensemble dérivé de l'ensemble S_q, *CR*, **279**, 1974, 251–254.
[76] Familles fermés de n-uples irréductibles, *Sém. DPP*, **18**, 1976/7, fasc. 1, exp. 2.
[78] Étude des ensembles $\sum_{q,h}$ et de leurs ensembl es dérivés: application à l'étude des ensembles dérivés S des nombres de Pisot, *CR*, **287**, 1978, A381–382.
[80] Ensembles dérivés des ensembles $\sum_{q,h}$ et de l'ensemble S des PV-nombres, *Bull. Sci. Math.*, (2), **104**, 1980, 3–17.
[81] Familles fermées de nombres algébriques, *AA*, **39**, 1981, 207–240.
BERTNESS, C. H., MCCULLOH, L. R. [72] Corresponding residue systems in nonnormal extensions of prime degree, *JNT*, **5**, 1972, 24–42.

BERTRANDIAS, F. [65] Ensembles remarquables d'adèles algébriques, *Bull. Soc. Math. France, Mém.* **4**, 1965.
[78] Entiers d'une p-extension cyclique d'un corps local, *CR*, **286**, 1978, A1083–1086.
[79] Decomposition du Galois-module des entiers d'une extension cyclique de degré premier d'un corps de nombres ou d'un corps local, *Ann. Inst. Fourier*, **29**, 1979, 1, 33–48.
BERTRANDIAS, F., BERTRANDIAS, J.-P., FERTON, M. J. [72] Sur l'anneau des entiers d'une extension cyclique de degré premier d'un corps local, *CR*, **274**, 1972, A1388–1391.
BERTRANDIAS, F., FERTON, M. J. [72] Sur l'anneau des entiers d'une extension cyclique de degré premier d'un corps local, *CR*, **274**, 1972, A1330–1333.
BERTRANDIAS, F., PAYAN, J. J. [72] Γ-extensions et invariants cyclotomiques, *ASENS*, (4), **5**, 1972, 517–543.
BERTRANDIAS, J.-P., BERTRANDIAS, F., FERTON, M. J. (see F. Bertrandias *et al.*)
BERWICK, W. E. H. [13] The classification of ideal numbers that depend on a cubic irrationality, *PLMS*, (2), **12**, 1913, 393–429.
[24] On cubic fields with a given discriminant, *PLMS*, (2), **23**, 1924, 359–378.
[27] *Integral Bases*, Cambridge 1927; repr. Stechert–Hafner, 1964.
[32] Algebraic number fields with two independent units, *PLMS*, (2), **34**, 1932, 360–378.
[34] The classification of ideal numbers in a cubic field, *PLMS*, (2), **38**, 1934, 217–242.
BESICOVITCH, A. S. [40] On the linear independence of fractional powers of integers, *JLMS*, **15**, 1940, 3–6.
BESSIS, D., MOUSSA, P., GERONIMO, J. S. (See P. Moussa *et al.*)
BEYER, G. [54] Über eine Klasseneinteilung aller kubischer Restcharaktere, *Hbg.*, **19**, 1954, 115–116.
BHANDARI, S. K., NANDA, V. C. [79] Ideal matrices for relative extensions, *Hbg.*, **49**, 1979, 3–17.
BHASKARAN, M. [66] Sums of m-th powers in algebraic and Abelian number fields, *Arch. Math.*, **17**, 1966, 497–504.
[69] Sums of p-th powers in a P-adic ring, *AA*, **15**, 1969, 217–219.
[71] A new proof for the law of decomposition in a general cyclotomic field, *Arch. Math.*, **22**, 1971, 62–64.
[74] Some remarks on the decomposition of a rational prime in a Galois extension, *AA*, **26**, 1974, 101–104.
BICKMORE, C. E. [93] Tables connected with the Pellian equation, *British Association Report*, **53**, 1893, 73–120.
BILHAN, M. [81] Théorème de Bauer dans les corps globaux, *Bull. Sci. Math.*, (2), **105**, 1981, 299–303.
BILLEVICH, K. K. [49a] On the equivalence of two ideals in a field of degree n, *Tr. Sev. Kavk. Gorn. Inst.*, **7**, 1949, 43–48 (Russian).
[49b] Determination of the class number of ideals in an algebraic field of degree n, *Tr. Sev. Kavk. Gorn. Inst.*, **7**, 1949, 49–51 (Russian).
[56] On units of algebraic fields of degree 3 and 4, *Mat. Sbornik*, **40**, 1956, 123–136; corr., *ibid.*, **48**, 1959, 256 (Russian).
[62] On the equivalence of two ideals of an algebraic field of degree n, *Mat. Sbornik*, **58**, 1962, 17–28 (Russian).
[64] Theorem on units of algebraic fields of degree n, *Mat. Sbornik*, **64**, 1964, 145–152 (Russian).
BILLEVICH, K. K., AVANESOV, É. T. (See É. T. Avanesov, K. K. Billevich)
BILLEVICH, K. K., DELAUNAY, B. N., SOMINSKII, J. (See B. N. Delaunay, J. Sominskii, K. K. Billevich)

BIRCH, B. J. [57] Homogeneous forms of odd degree in a large number of variables, *Mathematika*, **4**, 1957, 102–105.
[61] Waring's problem in algebraic number fields, *Proc. Cambridge Phil. Soc.*, **57**, 1961, 444–459.
[64a] Waring's problem for p-adic number fields, *AA*, **9**, 1964, 169–176.
[64b] Diagonal equations over p-adic fields, *AA*, **9**, 1964, 291–300.
[69a] K_2 of global fields, *PSPM*, **20**, 1969, 87–95.
[69b] Weber's class invariants, *Mathematika*, **16**, 1969, 283–294.

BIRCH, B. J., LEWIS, D. J. [59] p-adic forms, *J. Indian Math. Soc.*, **23**, 1959, 11–32.
[62] On p-adic forms, *Michigan Math. J.*, **9**, 1962, 53–57.
[65] Systems of three quadratic forms, *AA*, **10**, 1964/5, 423–442.

BIRCH, B. J., LEWIS, D. J., MURPHY, T. G. [62] Simultaneous quadratic forms, *AJM*, **84**, 1962, 110–115.

BIRCH, B. J., MERRIMAN, J. R. [72] Finiteness theorems for binary forms with given discriminant, *PLMS*, (3), **24**, 1972, 385–394.

BIRCH, B. J., SWINNERTON-DYER, H. P. F. [65] Notes on elliptic curves, II, *JRAM*, **218**, 1965, 79–108.
[75] The Hasse problem for rational surfaces, *JRAM*, **274/5**, 1975, 164–174.

BIRD, R. F., PARRY, C. J. [76] Integral bases for bicyclic biquadratic fields over quadratic subfields, *Pacific J. Math.*, **66**, 1976, 29–36.

BITIMBAEV, T. S. [68] Korkin's sum and its connection with the class-number of imaginary quadratic fields, *IAN Kazakh. SSR*, 1968, 1, 1–6 (Russian).
[77] Enumeration of all imaginary quadratic fields of prime discriminant $d = -p$, $p \leqslant 1\,000\,763$ with a prescribed number of ideal classes $h(-p)\ldots$, *Theor. and Appl. Probl. Mat. Mech.*, 70–79, Alma-Ata 1977 (Russian).

BLANKSBY, P. E. [69] A note on algebraic integers, *JNT*, **1**, 1969, 155–160.
[70] A metric inequality associated with valuated fields, *AA*, **17**, 1970, 217–225.
[75] Sums of powers of conjugates of algebraic numbers, *PAMS*, **49**, 1975, 28–32.

BLANKSBY, P. E., LOXTON, J. H. [78] A note on the characterization of *CM*-fields, *J. Austral. Math. Soc., ser. A.*, **26**, 1978, 26–30.

BLANKSBY, P. E., MONTGOMERY, H. L. [71] Algebraic integers near the unit circle, *AA*, **18**, 1971, 355–369.

BLICHFELDT, H. F. [36] A new upper bound for the minimum value of the sum of linear homogeneous forms, *Monatsh. M.-Phys.*, **43**, 1936, 410–414.
[39] Note on the minimum value of the discriminant of an algebraic field, *Monatsh. M.-Phys.*, **48**, 1939, 531–533.

BLOOM, D. M. [68] On the coefficients of the cyclotomic polynomial, *AMM*, **75**, 1968, 372–377.

BLOOM, J. R. [79] On the invariants of some Z_l-extensions, *JNT*, **11**, 1979, 239–256.

BLOOM, J. R., GERTH, F. III [81] The Iwasawa invariant μ in the composite of two Z_l-extensions, *JNT*, **13**, 1981, 262–267.

BOBROVSKII, V. P. [71a] On admissible cells of groups of special elementary-Abelian fields of the type (l, l, l, l, l), *IAN Kazakh. SSR*, 1971, 1, 82–83 (Russian).
[71b] The structure of admissible cells of conductors of special elementary-Abelian fields of type (l, l, l, l, l), *IAN Kazakh. SSR*, 1971, 1, 83–84 (Russian).
[72] The algebra of special elementary-Abelian fields of type (l, l, l, l, l), *Rep. Sem. Alg. Kazakh. Gos. Ped. Inst.*, 1972, 1, 7–21 (Russian).

BOBROVSKII, V. P., RAMAZANOV, R. G. (See R. G. Ramazanov. V. P. Bobrovskii)

BOCHKAREV, D. P. [49] On the properties of the equation of the division of the circle into n equal parts, *Uchen. Zap. Mordovsk. Ped. Inst.*, 1949 (Russian).
[54] On certain properties of the polynomials of the division of the circle, *Uchen. Zap. Mordovsk. Ped. Inst.*, 1954 (Russian).
BOHR, H., LANDAU, E. [10] Über das Verhalten von $\zeta(s)$ und $\zeta_K(s)$ in der Nähe der Geraden $\sigma = 1$, *GN*, 1910, 303–330.
BÖLLING, R. [79] Bemerkungen über Klassenzahlen und Summen von Jacobi-Symbolen, *MN*, **90**, 1979, 159–172.
BOND, R. J. [79] Some results on p-extensions of local and global fields, *AA*, **35**, 1979, 25–32.
[81] Some results on the capitulation problem, *JNT*, **13**, 1981, 246–254.
BOREL, A. [72] Cohomologie réelle stable des groupes arithmétiques classiques, *CR*, **274**, 1972, 1700–1702.
[74] Stable real cohomology of arithmetic groups, *ASENS*, (4), **7**, 1974, 235–272.
[77] Cohomologie de SL_n et valeurs de fonctions zeta aux points entiers, *Ann. Scuola Norm. Sup. Pisa*, (4), **4**, 1977, 613–636.
BOREL, A., CHOWLA, S., HERZ, C. S., IWASAWA, K., SERRE, J. P. [66] *Seminar on Complex Multiplication*, LN 21, Springer, 1966.
BOREVICH, Z. I. [56] On extensions without simple ramification of a regular local field, *Vestnik LGU*, **11**, 1956, nr. 19, 41–47 (Russian).
[57] On the demonstration of the principal ideal theorem, *Vestnik LGU*, **12**, 1957, nr 13, 5–8 (Russian).
[64] Multiplicative group of a regular local field with a cyclic operator group, *IAN*, **28**, 1964, 707–712 (Russian).
[65a] On the multiplicative group of cyclic p-extensions of a local field, *Tr. Mat. Inst. Steklov*, **80**, 1965, 16–29 (Russian).
[65b] On the group of principal units of a normal p-extension of a regular local field, *Tr. Mat. Inst. Steklov*, **80**, 1965, 30–44 (Russian).
[65c] On the multiplicative group of cyclic p-extensions of a local field, II, *Vestnik LGU*, **20**, 1965, nr. 13, 5–12 (Russian).
[66] Cohomology of the unit group of a normal extension of a local field, *Vestnik LGU*, **21**, 1966, nr. 13, 124–125 (Russian).
[67] On groups of principal units of p-extensions of a local field, *DAN*, **173**, 1967, 253–255 (Russian).
BOREVICH, Z. I., EL-MUSA, A. J. [73] Completion of the multiplicative group of a p-extension of an irregular local field, *Zap. Nauch. Sem. LOMI*, **31**, 1973, 6–23 (Russian).
BOREVICH, Z. I., GERLOVIN, É. L. [76] Structure of the group of principal units of a cyclic p-extension of a local field, *Zap. Nauch. Sem. LOMI*, **57**, 1976, 51–63 (Russian).
BOREVICH, Z. I., SHAFAREVICH, I. R. [64] *Number Theory*, Moskva 1964 (Russian); 3rd ed., Moskva 1985; English transl., New York 1966.
BOREVICH, Z. I., SKOPIN, A. I. [65] Extensions of a local field with a normal basis for the principal units, *Tr. Mat. Inst. Steklov*, **80**, 1965, 45–50 (Russian).
BOREVICH, Z. I., VOSTOKOV, S. V. [73] The ring of integral elements of an extension of prime degree of a local field as a Galois module, *Zap. Nauch. Sem. LOMI*, **31**, 1973, 24–37 (Russian).
BOUGAUT, B. [80] Algorithme explicite pour la recherche du P.G.C.D. dans certains anneaux principaux d'entiers de corps de nombres, *Theoret. Comput. Sci.*, **11**, 1980, 207–220.

BOUVIER, L. [71] Sur le 2-groupe des classes au sens restreint de certaines extensions biquadratiques de Q, *CR*, **272**, 1971, A193–196.

[72] Table des 2-rang, 4-rang et 8-rang du 2-groupe des classes d'idéaux au sens restreint de $Q(\sqrt{m})$, m étant un entier relatif sans facteur carré tel que $1 < |m| < 10000$, *Enseign. Math.*, **18**, 1972, 37–45.

BOUVIER, L., PAYAN, J. J. [75] Modules sur certains anneaux de Dedekind, *JRAM*, **274/5**, 1975, 278–286.

[79] Sur la structure galoisienne du groupe des unités d'un corps abélien de type (p, p), *Ann. Inst. Fourier*, **29**, 1979, 1, 171–187.

BOVEY, J. D. [76] A note on Waring's problem in p-adic fields, *AA*, **29**, 1976, 343–351.

BOYD, D. W. [77a] Small Salem numbers, *Duke Math. J.*, **44**, 1977, 315–328.

[77b] Pisot sequences which satisfy no linear recurrence, *AA*, **32**, 1977, 89–98.

[78] Pisot and Salem numbers in intervals of the real line, *MC*, **32**, 1978, 1244–1260.

[79a] On the successive derived sets of the Pisot numbers, *PAMS*, **73**, 1979, 154–156.

[79b] Pisot sequences, Pisot numbers and Salem numbers, *Astérisque*, **61**, 1979, 35–42.

[81a] Speculations concerning the range of Mahler's measure, *Canad. Math. Bull.*, **24**, 1981, 453–469.

[81b] Kronecker's theorem and Lehmer's problem for polynomials in several variables, *JNT*, **13**, 1981, 116–121.

[82] Families of Pisot and Salem numbers, *Sém. DPP*, 1980/81, 19–33, Birkhäuser 1982.

BOYD, D. W., KISILEVSKY, H. [72] On the exponent of the ideal class groups of complex quadratic fields, *PAMS*, **31**, 1972, 433–436.

BRANDAL, W. [79] *Commutative Rings whose Finitely Generated Modeles Decompose*, LN 723, Springer, 1979.

BRANDIS, A. [65] Über die multiplikative Struktur von Körpererweiterungen, *MZ*, **87**, 1965, 71–73.

BRANDLER, J. A. [73] Residuacity properties of real quadratic units, *JNT*, **5**, 1973, 271–286.

BRATTSTRÖM, G. [82] Jacobi-sum Hecke characters of a totally real abelian field, *Sem. Th. Nombr. Bordeaux*, 1981/2, exp. 22.

BRAUER, A. [40] On the non-existence of the Euclidean algorithm in certain quadratic number fields, *AJM*, **62**, 1940, 697–716.

[51] On algebraic equations with all but one root in the interior of the unit circle, *MN*, **4**, 1951, 250–257.

BRAUER, R. [45] A note on systems of homogeneous algebraic equations, *BAMS*, **51**, 1945, 749–755.

[47a] On the Zeta-function of algebraic number fields, *AJM*, **69**, 1947, 243–250; II, *ibid.*, **72**, 1950, 739–746.

[47b] On Artin's L-series with general group characters, *AnM*, **48**, 1947, 502–514.

[51] Beziehungen zwischen Klassenzahlen von Teilkörpern eines galoischen Körpers, *MN*, **4**, 1950/1, 158–174.

[73] A note on zeta-functions of algebraic number fields, *AA*, **24**, 1973, 325–327.

BRAUER, R., ANKENY, N. C., CHOWLA, S. (See N. C. Ankeny, R. Brauer, S. Chowla)

BREDIKHIN, B. M. [58] Free semigroups of numbers with power densities, *Mat. Sbornik*, **46**, 1958, 143–158 (Russian).

BREMNER, A. [78] Some cubic surfaces with no rational points, *Math. Proc. Cambridge Philos. Soc.*, **84**, 1978, 219–223.

BRENTJES, A. J. [81] A two-dimensional continued fraction algorithm for best approximations with an application in cubic number fields, *JRAM*, **326**, 1981, 18–44.

BRIGGS, W. E., CHOWLA, S. [54] On discriminants of binary quadratic forms with a single class in each genus, *CJM*, **6**, 1954, 463–470.

BRILL, A. [77] Ueber die Discriminante, *MA*, **12**, 1877, 87–89.

BRILLHART, J., GERST, I. (See I. Gerst, J. Brillhart)

BRILLHART, J., ZIERLER, N. (See N. Zierler, J. Brillhart)

BRINDZA, S. [84] On S-integral solutions of the equation $y^m = f(x)$, *Acta Math. Acad. Sci. Hungar.*, **44**, 1984, 133–139.

BRINKHUIS, J. [81a] *Embedding problems and Galois modules*, Diss., Leiden 1981.

[81b] Symmetries d'un module Galoisien, *Sem. Th. Nombr. Bordeaux*, 1981/2, exp. 44.

[83] Normal integral bases and embedding problems, *MA*, **264**, 1983, 537–543.

[84a] K_2 and Galois extensions of fields, in: *Algebraic K-theory, Number Theory, Geometry and Analysis*, 13–28, LN 1046, Springer, 1984.

[84b] Galois modules and embedding problems, *JRAM*, **346**, 1984, 141–165.

BRIZOLIS, D. [74] On the ratios of integer-valued polynomials over any algebraic number field, *AMM*, **81**, 1974, 997–999.

[75] Hilbert rings of integral-valued polynomials, *Comm. Algebra*, **3**, 1975, 1051–1081.

[76] Ideals in rings of integer valued polynomials, *JRAM*, **285**, 1976, 28–52.

[79] A theorem on ideals in Prüfer rings of integral-valued polynomials, *Comm. Algebra*, **7**, 1979, 1065–1077.

BROERE, J., WILLIAMS, H. C. (See H. C. Williams, J. Broere)

BROWKIN, J. [63a] On the generalized class field tower, *Bull. Acad. Pol. Sci. sér. sci. math. astr. phys.*, **11**, 1963, 143–145.

[63b] Examples of maximal 3-extensions with two ramified places, *IAN*, **27**, 1963, 613–620 (Russian).

[66] On forms over p-adic field, *Bull. Acad. Pol. Sci. sér. sci. math. astr. phys.*, **14**, 1966, 489–492.

[69] On zeros of forms, *Bull. Acad. Pol. Sci. sér. sci. math. astr. phys.* **17**, 1969, 611–616.

[82a] Elements of small order in $K_2 F$, in: *Algebraic K-Theory*, I, 1–6, LN 966, Springer, 1982.

[82b] The functor K_2 for the ring of integers of a number field, in: *Universal Algebra and Applications*, Banach Center Publications, t. 9, 187–195, Warszawa 1982.

[83] On systems of congruences, *Bull. Acad. Pol. Sci., sér. math.*, **31**, 1983, 219–226.

BROWKIN, J., HOFFMANN, B., HETTLING, K. F. (See B. Hoffmann et al.)

BROWKIN, J., HURRELBRINK, J. [84] On the generation of $K_2(G)$ by symbols, in: *Algebraic K-theory, Number Theory, Geometry and Analysis*, 29–31, LN 1046, Springer, 1984.

BROWKIN, J., SCHINZEL, A. [82] On Sylow 2-subgroups of $K_2 O_F$ for quadratic number fields F, *JRAM*, **331**, 1982, 104–113.

BROWN, E. [72a] Binary quadratic forms of determinant $-pq$, *JNT*, **4**, 1972, 408–410.

[72b] The class number of $Q(\sqrt{-p})$ for $p \equiv 1 \pmod 8$ a prime, *PAMS*, **31**, 1972, 381–383.

[73] The power of 2 dividing the class-number of a binary quadratic discriminant, *JNT*, **5**, 1973, 413–419.

[74a] Class number of complex quadratic fields, *JNT*, **6**, 1974, 185–191.

[74b] Class numbers of real quadratic number fields, *TAMS*, **190**, 1974, 99–107.

[75] Class numbers of quadratic fields, *Symposia Math.*, **15**, 403–411, Academic Press, 1975.

[81] The class-number of $Q(\sqrt{-pq})$, for $p \equiv -q \equiv 1 \pmod 4$ primes, *Houston J. Math.*, **7**, 1981, 497–505.

[83] The class number and fundamental unit of $Q(\sqrt{2p})$ for $p \equiv 1 \pmod{16}$ a prime, *JNT*, **16**, 1983, 95–99.

BROWN, E., PARRY, C. J. [73] Class numbers of imaginary quadratic fields having exactly three discriminantial divisors, *JRAM*, **260**, 1973, 31–34.
[74] The imaginary bicyclic biquadratic fields with class-number 1, *JRAM*, **266**, 1974, 118–120.
[77] The 2-class group of biquadratic fields, *JRAM*, **295**, 1977, 61–71; II, *Pacific J. Math.*, **78**, 1978, 11–26.

BROWN, K. S. [74] Euler characteristic of discrete groups and G-spaces, *Invent. Math.*, **27**, 1974, 229–264.

BROWNAWELL, W. D. [84] On p-adic zeros of forms, *JNT*, **18**, 1984, 342–349.

BRUCKNER, G. [66] Charakterisierung der galoisschen Zahlkörper, deren zerlegte Primzahlen durch binäre quadratische Formen gegeben sind, *MN*, **32**, 1966, 317–326.
[68] Eine Charakterisierung der in algebraischen Zahlkörpern voll zerlegten Primzahlen, *MN*, **36**, 1968, 153–159.
[72] Die Fermatsche Vermutung über $Q(\sqrt{5})$ und die Fibonacci-Folge, *MN*, **52**, 1972, 255–257.

BRUMER, A. [65] Ramification and class towers of number fields, *Michigan Math. J.*, **12**, 1965, 129–131.
[66] Galois groups of extensions of algebraic number fields with given ramification, *Michigan Math. J.*, **13**, 1966, 33–40.
[67] On the units of algebraic number fields, *Mathematika*, **14**, 1967, 121–124.
[69] On the group of units of an absolutely cyclic number field of prime degree, *J. Math. Soc. Japan*, **21**, 1969, 357–358.

BRUMER, A., ROSEN, M. [63] Class number and ramification in number fields, *Nagoya Math. J.*, **23**, 1963, 97–101.

BRUNOTTE, H. [78] Vektoren minimaler Länge in Gittern mit einem Automorphismus, *Monatsh. Math.*, **86**, 1978, 89–100.
[80] Bemerkungen zu einer metrischen Invarianten algebraischer Zahlkörper, *Monatsh. Math.*, **90**, 1980, 171–184.
[82] The computation of a certain metric invariant of an algebraic number field, *MC*, **38**, 1982, 627–632.

BRUNOTTE, H., HALTER-KOCH, F. [79] Zur Einheitenberechnung in totalreellen kubischen Zahlkörpern nach Godwin, *JNT*, **11**, 1979, 552–559.
[81a] Metrische Kennzeichnung von Erzeugenden für Einheitengruppen vom Rang 1 oder 2 in algebraischen Zahlkörpern, *JNT*, **13**, 1981, 320–333.
[81b] Grundeinheitensysteme algebraischer Zahlkörper mit vorgegebener Verteilung der Konjugiertenbeträge, *Arch. Math.*, **37**, 1981, 512–513.

BRUNOTTE, H., KLINGEN, J., STEURICH, M. [77] Einige Bemerkungen zu Einheiten in reinen kubischen Körpern, *Arch. Math.*, **29**, 1977, 154–157.

BUCHNER, P. [26] Annäherung beliebiger komplexer Grössen durch ganze Zahlen des Körpers $\sqrt{-2}$, *JRAM*, **155**, 1926, 37–61.

BUCHSBAUM, D. A., AUSLANDER, M. (See M. Auslander, D. A. Buchsbaum)

BUELL, D. A. [76] Class groups of quadratic fields, *MC*, **30**, 1976, 610–623.
[77] Small class numbers and extreme values of L-functions of quadratic fields, *MC*, **31**, 1977, 786–796.

BUELL, D. A., LEONARD, P. A., WILLIAMS, K. S. [81] Note on the quadratic character of a quadratic unit, *Pacific J. Math.*, **92**, 1981, 35–38.

BUELL, D. A., WILLIAMS, H. C., WILLIAMS, K. S. [77] On the imaginary bicyclic biquadratic fields with class-number 2, *MC*, **31**, 1977, 1034–1042.

BUHLER, J. P. [78] *Icosahedral Galois Representations*, LN 654, Springer, 1978.
BUHR, P. A., WILLIAMS, H. C. (See H. C. Williams, P. A. Buhr)
BULENOV, A. [69] Count of the number of special Abelian fields of type (l, l, l, l), *IAN Kazakh. SSR*, 1969, 1, 87–88 (Russian).
BULLIG, G. [36] Die Berechnung der Grundeinheit in den kubischen Körpern mit negativer Diskriminante, *MA*, **112**, 1936, 325–394.
[38] Ein periodisches Verfahren zur Berechnung eines Systems von Grundeinheiten in den total reellen kubischen Körpern, *Hbg*, **12**, 1938, 369–411.
[39] Zur Zahlengeometrie in den total rellen kubischen Körpern, *MZ*, **45**, 1939, 511–532.
BULOTA, K. [62] Approximate functional equation of Hecke's Z-functions of an imaginary quadratic field, *Litovsk. Mat. Sb.* **2**, 1962, 2, 39–82 (Russian).
[63] Some theorems on the density of zeros of Hecke's Z-functions, *Litovsk. Mat. Sb.*, **3**, 1963, 1, 29–50 (Russian).
[64a] On the approximate functional equation of Hecke's Z-functions, *Litovsk. Mat. Sb.*, **4**, 1964, 183–196 (Russian).
[64b] On Hecke's Z-functions and the distribution of primes in an imaginary quadratic field, *Litovsk. Mat. Sb.*, **4**, 1964, 309–328 (Russian).
BUMBY, R. T. [67] Irreducible integers in Galois extensions, *Pacific J. Math.*, **22**, 1967, 221–229.
BUMBY, R. T., DADE, E. C. [67] Remark on a problem of Niven and Zuckerman, *Pacific J. Math.*, **22**, 1967, 15–18.
BUNDSCHUH, P., HOCK, A. [69] Bestimmung aller imaginär-quadratischen Zahlkörper der Klassenzahl Eins mit Hilfe eines Satzes von Baker, *MZ*, **111**, 1969, 191–204.
BUNGERS, R. [36] Über Zahlkörper mit gemeinsamen ausserwesentlichen Diskriminantenteilern, *Jahresber. Deutsch. Math. Verein.*, **46**, 1936, 93–96; ibid. **47**, 1937, 56.
BURGESS, D. A. [66] Estimating $L_\chi(1)$, *Norske Vid. Selsk. Forh.* **39**, 1966, 101–108.
BURKE, J. R., KUIPERS, L. [76] Asymptotic distribution and independence of sequences of Gaussian integers, *Simon Stevin*, **50**, 1976/7, 3–21.
BURNSIDE, W. [14] On the rational solutions of the equation $X^3+Y^3+Z^3 = 0$ in quadratic fields, *PLMS*, (2), **14**, 1914, 1–4.
BUSHNELL, C. J. [77a] Integrality of Galois Jacobi sums, *JLMS*, (2), **15**, 1977, 35–40.
[77b] Norms of normal integral generators, *JLMS*, (2), **15**, 1977, 199–209.
[79] Norm distribution in Galois orbits, *JRAM*, **310**, 1979, 81–99.
[83] Diophantine approximation and norm distribution in Galois orbits, *Illinois J. Math.*, **27**, 1983, 145–157.
BUSHNELL, C. J., REINER, I. [80] Zeta-functions of arithmetic orders and Solomon's conjectures, *MZ*, **173**, 1980, 135–161.
[81a] L-functions of arithmetic orders and asymptotic distribution of ideals, *JRAM*, **327**, 1981, 156–183.
[81b] Functional equations for L-functions of arithmetical orders, *JRAM*, **329**, 1981, 88–124.
[81c] Zeta-functions of orders, in: *Integral Representations and Applications*, 159–173, LN 882, Springer, 1981.
BÜSSER, A. H. [44] *Über die Primidealzerlegung in Relativkörpern mit der Relativgruppe G_{168}*, Diss., Zürich 1944.
BUTTS, H. S. [64] Unique factorization of ideals into nonfactorable ideals, *PAMS*, **15**, 1964, 21.
[65] Quasi-invertible prime ideals, *PAMS*, **16**, 1965, 291–292.
BUTTS, H., ESTES, D. [68] Modules and binary quadratic forms over integral domains, *Linear Algebra Appl.*, **1**, 1968, 153–180.

BUTTS, H. S., GILBERT, J. R. JR.   (See J. R. Gilbert, Jr., H. S. Butts)
BUTTS, H. S., GILMER, R. W. JR.   [66] Primary ideals and prime power ideals, *CJM*, **18**, 1966, 1183–1195.
BUTTS, H. S., MANN, H. B.   [56] Corresponding residue systems in algebraic number fields, *Pacific J. Math.*, **6**, 1956, 211–224.
BUTTS, H. S., PALL, G.   [67] Factorization in quadratic rings, *Duke Math. J.*, **34**, 1967, 139–146.
[68] Modules and binary quadratic forms, *AA*, **15**, 1968, 23–44.
BUTTS, H. S., PHILLIPS, R. C.   [65] Almost multiplication rings, *CJM*, **17**, 1965, 267–277.
BUTTS, H. S., SMITH, W. W.   [66] On the integral closure of a domain, *J. Sci. Hiroshima Univ.*, **30**, 1966, 117–127.
BUTTS, H. S., WADE, L.   [66] Two criteria for Dedekind domains, *AMM*, **73**, 1966, 14–21.
CAHEN, P. J.   [72] Polynômes à valeurs entiers, *CJM*, **24**, 1972, 747–754.
CAHEN, P. J., CHABERT, J. L.   [71] Coefficients et valeurs d'un polynôme, *Bull. Sci. Math.*, (2), **95**, 1971, 295–304.
CALLAHAN, T.   [74] The 3-class groups of non-Galois fields, I, *Mathematika*, **21**, 1974, 72–89; II, *ibid.*, 168–188.
[76] Dihedral field extensions of order $2p$ whose class numbers are multiples of p, *CJM*, **28**, 1976, 429–439.
CALLAHAN, T., NEWMAN, M., SHEINGORN, M.   [77] Fields with large Kronecker constants, *JNT*, **9**, 1977, 182–186.
CALLAHAN, T., SMITH, R. A.   [76] L-functions of a quadratic form, *TAMS*, **217**, 1976, 297–309.
ÇALLIAL, P. F.   [80] Non-nullité des fonctions zèta des corps quadratiques réels pour $0 < s < 1$, *CR*, **291**, 1980, A623–625.
CALLOWAY, J.   [55] On the discriminant of arbitrary algebraic number fields, *PAMS*, **6**, 1955, 482–489.
CAMION, P., LEVY, L. S., MANN, H. B.   [73] Prüfer rings, *JNT*, **5**, 1973, 132–138.
CANALS, I., ORTIZ, J. J.   [70] The minimal basis of an algebraic number field, *Bol. Soc. Mat. Mexicana*, **15**, 1970, 14–21.
CANDIOTTI, A.   [74] Computations of Iwasawa invariants and K_2, *Compos. Math.*, **29**, 1974, 89–111.
CANDIOTTI, A., KRAMER, K.   (See K. Kramer, A. Candiotti)
CANTOR, D. G.   [62] On sets of algebraic integers whose remaining conjugates lie in the unit circle, *TAMS*, **105**, 1962, 391–406.
[65a] On powers of real numbers (mod 1), *PAMS*, **16**, 1965, 791–793.
[65b] On the elementary theory of diophantine approximation over the ring of adeles, *Illinois J. Math.*, **9**, 1965, 677–700.
[76] On certain algebraic integers and approximation by rational functions with integral coefficients, *Pacific J. Math.*, **67**, 1976, 323–338.
[77] On power series with only finitely many coefficients (mod 1): Solution of a problem of Pisot and Salem, *AA*, **34**, 1977/8, 43–55.
[80] On an extension of the definition of transfinite diameter and some applications, *JRAM*, **316**, 1980, 160–207.
CANTOR, D. G., ROQUETTE, P.   [84] On Diophantine equations over the ring of all algebraic integers, *JNT*, **18**, 1984, 1–26.
CANTOR, D G., STRAUS, E. G.   [82] On a conjecture of D. H. Lehmer, *AA*, **42**, 1982, 97–100; corr., *ibid.*, 325.

CARLITZ, L. [33] On abelian fields, *TAMS*, **35**, 1933, 122–136.
[52] A note on common index divisors, *PAMS*, **3**, 1952, 688–692.
[53a] The class number of an imaginary quadratic field, *Comment. Math. Helv.*, **27**, 1953, 338–345.
[53b] Note on the class number of real quadratic fields, *PAMS*, **4**, 1953, 535–537.
[53c] A theorem of Stickelberger, *Math. Scand.*, **1**, 1953, 82–84.
[54a] The first factor of the class number of a cyclic field, *CJM*, **6**, 1954, 23–26.
[54b] Note on irregular primes, *PAMS*, **5**, 1954, 329–331.
[55] Note on the class number of quadratic fields, *Duke Math. J.*, **22**, 1955, 589–593.
[56] A note on Gauss's sum, *PAMS*, **7**, 1956, 910–911.
[59a] Arithmetic properties of generalized Bernoulli numbers, *JRAM*, **202**, 1959, 174–182.
[59b] Some arithmetical properties of generalized Bernoulli numbers, *BAMS*, **65**, 1959, 68–69.
[60] A characterization of algebraic number fields with class number two, *PAMS*, **11**, 1960, 391–392.
[61] A generalization of Maillet's determinant and a bound for the first factor of the class number, *PAMS*, **12**, 1961, 256–261.
[62] Binomial coefficients in an algebraic number field, *AA*, **7**, 1962, 381–388; corr., *ibid.*, **11**, 1966, 489.
[63] Some arithmetic properties of a special sequence of polynomials in the Gaussian field, *Duke Math. J.*, **30**, 1963, 15–24.
[66a] The number of terms in the cyclotomic polynomial $F_{pq}(x)$, *AMM*, **73**, 1966, 979–981.
[66b] A remark on cubics, *AMM*, **73**, 1966, 1111–1112.
[67] The sum of squares of the coefficients of the cyclotomic polynomial, *Acta Math. Acad. Sci. Hungar.*, **18**, 1967, 297–304.
[68a] A congruence for the second factor of the class number of a cyclotomic field, *AA*, **14**, 1968, 27–34; corr., *ibid.*, **16**, 1970, 437.
[68b] A note on Gauss's sum, *Matematiche*, **23**, 1968, 147–150.
CARLITZ, L., OLSON, F. R. [55] Maillet's determinant, *PAMS*, **6**, 1955, 265–269.
CARROLL, J. E. [73] On the torsion in K_2 of local fields, in: *Algebraic K-Theory*, II, 464–473, LN 342, Springer, 1973.
[75] On determining the quadratic subfields of Z_2-extensions of complex quadratic fields, *Compos. Math.*, **30**, 1975, 259–271.
CARROLL, J. E., KISILEVSKY, H. [76] Initial layers of Z_2-extensions of complex quadratic fields, *Compos. Math.*, **32**, 1976, 157–168.
[81] On Iwasawa's λ-invariant for certain Z_l-extensions, *AA*, **40**, 1981, 1–8.
CARTIER, P., ROY, Y. [73] Certains calculs numériques relatifs à l'interpolation p-adique des séries de Dirichlet, in: *Modular Functions of One Variable*, III, 269–349, LN 350, Springer, 1973.
[74] On the enumeration of quintic fields with small discriminants, *JRAM*, **268/9**, 1974, 213–215.
CASSELS, J. W. S. [48] The lattice properties of asymmetric hyperbolic regions, II. On a theorem of Davenport, *PCPS*, **44**, 1948, 145–154.
[50] The rational solutions of the diophantine equation $y^2 = x^3 - D$, *AM*, **82**, 1950, 243–273.
[52] The inhomogeneous minimum of binary quadratic, ternary cubic and quaternary quartic forms, *PCPS*, **48**, 1952, 72–86; add., *ibid.*, 519–520.

[57] *An Introduction to Diophantine Approximation*, Cambridge 1957.
[59a] Note on quadratic forms over the rational field, *PCPS*, **55**, 1959, 267–270.
[59b] *An Introduction to the Geometry of Numbers*, Springer, 1959.
[66] On a problem of Schinzel and Zassenhaus, *J. Math. Sci.*, **1**, 1966, 1–8.
[69a] On the determination of generalized Gauss sums, *Arch. Math.*, **5**, 1969, 79–84.
[69b] On a conjecture of R. M. Robinson about sums of roots of unity, *JRAM*, **238**, 1969, 112–131.
[70] On Kummer sums, *PLMS*, (3), **3**, 1970, 19–27.
[76] An embedding theorem for fields, *Bull. Austral. Math. Soc.*, **14**, 1976, 193–198; add., *ibid.*, 479–480.

CASSELS, J. W. S., FRÖHLICH, A. (editors) [67] *Algebraic Number Theory*, Academic Press, 1967.

CASSELS, J. W. S., GUY, M. J. T., [66] On the Hasse principle for cubic surfaces, *Mathematika*, **13**, 1966, 111–120.

CASSELS, J. W. S., LEDERMANN, W., MAHLER, K. [51] Farey sections in $k(i)$ and $k(\rho)$, *Philos. Trans. Roy. Soc. London*, *A*, **243**, 1951, 585–626.

CASSELS, J. W. S., WALL, G. E. [50] The normal basis theorem, *JLMS*, **25**, 1950, 259–264.

CASSOU-NOGUÈS, P. [78] Fonctions L p-adiques des corps de nombres totalement réels, *Sém. DPP*, **19**, 1977/8, fasc. 2. exp. 33.
[79] Valeurs aux entiers négatifs des fonctions zêta et fonctions zêta p-adiques, *Invent. Math.*, **51**, 1979, 29–59.

CASSOU-NOGUÈS, P., FRESNEL, J. [79] Le résidu des fonctions zêta de Shintani, *Sem. Th. Nombr. Bordeaux*, 1978/9, exp. 2.

CASSOU-NOGUÈS, PH. [77] Quelques théorèmes de base normale, *Astérisque*, **41/2**, 1977, 183–189.
[78] Quelques théorèmes de base normale d'entiers, *Ann. Inst. Fourier*, **28**, 1978, 3, 1–33.
[79] Structure galoisienne des anneaux d'entiers, *PLMS*, (3), **38**, 1979, 545–576.
[81] Module de Frobenius et structure galoisienne des anneaux d'entiers, *J. Algebra*, **71**, 1981, 268–289.

CASSOU-NOGUÈS, PH., QUEYRUT, J. [82] Structure galoisienne des anneaux d'entiers d'extensions sauvagement ramifiées, II, *Ann. Inst. Fourier*, **32**, 1982, 1, 7–27.

CASSOU-NOGUÈS, PH., TAYLOR, M. J. [83] Constante de l'équation fonctionnelle de la fonction L d'Artin d'une représentation symplectique et modérée, *Ann. Inst. Fourier*, **33**, 1983, 2, 1–17.

CASTELA, C. [78] Nombre de classes d'idéaux d'une extension diédrale d'un corps de nombres, *CR*, **287**, 1978, A 483–486.

CAUCHY, A. [40] Méthode simple et nouvelle pour la détermination complète des sommes alternées formées avec les racines primitives des équations binomes, *CR*, **10**, 1840, 560–572 = *J. de Math.*, **5**, 1840, 154–168 = *Oeuvres*, (1), 5, 152–166, Paris 1885.

CAVIOR, S. R. [64] Exponential sums related to polynomials over $GF(p)$, *PAMS*, **15**, 1964, 175–178.

CAYLEY, A. [62] Tables des formes quadratiques binaires pour les déterminants négatifs depuis $D = -1$ jusqu'a $D = -100$, pour les déterminants positifs non carrés depuis $D = 2$ jusqu'a $D = 99$ et pour les treize déterminants négatifs irréguliers qui se trouvent dans le premier million, *JRAM*, **60**, 1862, 357–372.

CHABAUTY, C. [36] Sur certaines équations diophantiques ternaires, *CR*, **202**, 1936, 2117–2119.

[37] Sur les unités d'un corps de nombres algébriques, qui sont sourmises à des conditions algébriques, *CR*, **205**, 1937, 944–946.

[38] Démonstration d'un théorème de Thue, indépendante de la théorie des approximations diophantiennes, *CR*, **208**, 1938, 1196–1198.

[43] Sur les solutions de certaines équations diophantiennes en nombres algébriques, en particulier en entiers algébriques, de degré borné, *CR*, **217**, 1943, 127–129.

CHABERT, J. L.   [71] Anneaux de "polynômes à valeurs entières" et anneaux de Fatou, *Bull. Soc. Math. France*, **99**, 1971, 273–283.

[77] Les idéaux premiers de l'anneau des polynômes à valeurs entières, *JRAM*, **293/4**, 1977, 275–283.

[78] Polynômes à valeurs entières et propriété de Skolem, *JRAM*, **303/4**, 1978, 366–378.

[79a] Anneaux de Skolem, *Arch. Math.*, **32**, 1979, 555–568.

[79b] Polynômes à valeurs entiers ainsi que leurs dérivées, *Ann. Sci. Univ. Clermont, Math.*, **18**, 1979, 47–64.

CHABERT, J. L., CAHEN, P. J.   (See P. J. Cahen, J. L. Chabert)

CHAMFY, C.   [57] Valeur minima du module pour un ensemble fermé d'indices algébriques, *CR*, **244**, 1957, 1992–1994.

CHANDRASEKHARAN, K., GOOD, A.   [83] On the number of integral ideals in Galois extensions, *Monatsh. Math.*, **95**, 1983, 99–109.

CHANDRASEKHARAN, K., JORIS, H.   [73] Dirichlet series with functional equations and related arithmetical identities, *AA*, **24**, 1973, 165–191.

CHANDRASEKHARAN, K., NARASIMHAN, R.   [60] Sur l'ordre moyen de quelques fonctions arithmétiques, *CR*, **251**, 1960, 1333–1335.

[61a] On Hecke's functional equation, *BAMS*, **67**, 1961, 182–185.

[61b] Hecke's functional equation and arithmetical identities, *AnM*, **74**, 1961, 1–23.

[61c] Hecke's functional equation and the average order of arithmetical functions, *AA*, **6**, 1961, 487–503.

[62a] Functional equations with multiple gamma factors and the average order of arithmetical functions, *AnM*, **76**, 1962, 93–136.

[62b] The average order of arithmetical functions, and the approximate functional equation for a class of zeta-functions, *Rend. Mat. Appl.*, (5), **21**, 1962, 354–363.

[63] The approximate functional equation for a class of zeta-functions, *MA*, **152**, 1963, 30–64.

[64] On the mean value of the error term for a class of arithmetical functions, *AM*, **112**, 1964, 41–67.

[68a] Zeta-functions of ideal classes in quadratic fields and their zeros on the critical line, *Comment. Math. Helv.*, **43**, 1968, 18–30.

[68b] An approximate reciprocity formula for some exponential sums, *Comment. Math. Helv.*, **43**, 1968, 296–300.

CHANG, J. A., GODWIN, H. J.   [69] A table of irreducible polynomials and their exponents, *PCPS*, **65**, 1969, 513–522.

CHANG, S. M.   [77] *Capitulation problems in algebraic number fields*, Thesis, Toronto 1977.

CHANG, S. M., FOOTE, R.   [80] Capitulation in class field extensions of type (p, p), *CJM*, **32**, 1980, 1229–1243.

CHAO, N. L.   [51] Discrete-valued complete fields with residue class fields of characteristic p, *J. Chinese Math. Soc.*, **1**, 1951, 377–394.

CHARIN, V. S.   [54] On groups of automorphisms of nilpotent groups, *Ukrain. Mat. Zh.*, **6**, 1954, 295–304 (Russian).

CHASE, S. U. [60] Direct product of modules, *TAMS*, **97**, 1960, 457–473.
CHATELAIN, D. [70] Bases normales de l'anneau des entiers de certaines extensions abéliennes de Q, *CR*, **270**, 1970, A557–560.
[73] Bases des entiers des corps composés par des extensions quadratiques, *Ann. Univ. Besançon, Math.*, 1973, fasc. 6.
[81] Modules d'entiers d'une extension abélienne sur son ordre maximal, *Publ. Math. Fac. Sci. Besançon, Th. de Nombres*, 1979/80 et 1980/81.
CHÂTELET, A. [11] Sur certaines ensembles de tableaux et leur application à la théorie des nombres, *ASENS*, (2), **28**, 1911, 105–202.
[46] Arithmétique des corps abéliens du troisième degré, *ASENS*, (2), **63**, 1946, 109–160.
[50] Utilisation des matrices dans l'algèbre et l'arithmétique des corps de nombres algébriques, *Alg. Th. des Nombres, Coll. Internat. CNRS*, **24**, 1950, 21–25.
[62] *L'arithmétique des corps quadratiques*, Monographies de "L'Enseignement Mathématique", Génève 1962.
CHATLAND, H. [49] On the euclidean algorithm in quadratic number fields, *BAMS*, **55**, 1949, 948–953.
CHATLAND, H., DAVENPORT, H. [50] Euclid's algorithm in real quadratic fields, *CJM*, **2**, 1950, 289–296.
CHATLAND, H., MANN, H. B. [49] Integral extensions of a ring, *BAMS*, **55**, 1949, 592–594.
CHEBOTAREV, N. G. (Tschebotareff, N. G., Tschebotaröw, N. G.) [23a] On a theorem of Hilbert, *Visti VUAN*, 1923, 3–7 (Russian).
[23b] Determination of the density of the set of primes corresponding to a given class of permutations, *IAN*, **17**, 1923, 205–230, 231–250 (Russian).
[24a] Proof of the Kronecker–Weber theorem on Abelian fields, *Mat. Sbornik*, **31**, 1924, 302–309 (Russian).
[24b] Eine Verallgemeinerung des Minkowski'schen Satzes mit Anwendung auf die Betrachtung der Körperidealklassen, *Zh. N. I. Odessa*, **4**, 1924, 17–20.
[25] On the foundations of ideal theory (according to Zolotarev), *Izv. Fiz.-Mat. Obshch. Kazan'*, **25**, 1925, 1–14 (Russian).
[26] Die Bestimmung der Dichtigkeit einer Menge von Primzahlen, welche zu einer gegebener Substitutionsklasse gehören, *MA*, **95**, 1926, 191–228.
[27] Studien über Primzahldichtigkeiten, I, *Izv. Fiz.-Mat. Obshch. Kazan'*, **2**, 1927, 14–20; II, *ibid.* **3**, 1928, 1–17.
[29] Zur Gruppentheorie des Klassenkörpers, *JRAM*, **161**, 1929, 179–193; corr., *ibid.*, **164**, 1931, 196.
[30] The foundations of the ideal theory of Zolotarev, *AMM*, **37**, 1930, 117–118.
[37a] *Foundations of the Galois Theory*, II, Moskva–Leningrad, 1937 (Russian).
[37b] Kurzer Beweis des Diskriminantensatzes, *AA*, **1**, 1937, 78–82.
[37c] Eine Aufgabe aus der algebraischen Zahlentheorie, *AA*, **2**, 1937, 221–229.
[47] On the foundations of the ideal theory, *Uspekhi Mat. Nauk*, **2**, 1947, 6, 52–67 (Russian).
CHEBOTAREV, N. G., BAUER, M. (See M. Bauer, N. G. Chebotarev)
CHEEMA, M. S. [56] *Tables of partitions of Gaussian integers, giving the number of partitions of $n+im$*, New Delhi 1956.
CHEN-JING-RUN [62] The number of lattice points in a given region, *Acta Math. Sinica*, **12**, 1962, 408–420 (Chinese); English translation: *Chinese Math.*, **3**, 1963, 439–452.
[63] Improvement on the asymptotic formulas for the number of lattice points in a region of three dimensions, *Sci. Sinica*, **12**, 1963, 151–161.

CHEO, L. [51] On the density of sets of Gaussian integers, *AMM*, **58**, 1951, 618–620.
CHEVALLEY, C. [31] Relation entre le nombre de classes d'un sous-corps et celui d'un surcorps, *CR*, **192**, 1931, 257–258.
[33a] Sur la théorie du corps de classes dans les corps finis et les corps locaux, *J. Fac. Sci. Univ. Tokyo*, **2**, 1933, 365–476.
[33b] La théorie du symbole de restes normiques, *JRAM*, **169**, 1933, 140–157.
[36a] *L'arithmétique dans les algèbres de matrices*, Paris 1936.
[36b] Généralisation de la théorie du corps de classes pour les extensions infinies, *J. Math. Pures Appl.*, (9), **15**, 1936, 359–371.
[40] La théorie du corps de classes, *AnM*, **41**, 1940, 394–418.
[51] Deux théorèmes d'arithmétique, *J. Math. Soc. Japan*, **3**, 1951, 36–44.
[54] *Class Field Theory*, Nagoya 1954.
CHEW, K. L., LAWN, S. [70] Residually finite rings, *CJM*, **22**, 1970, 92–101.
CHIDAMBARASWAMY, J. [67] Integral part and factorial in certain Euclidean domains, *AMM*, **74**, 1967, 41–43.
CHILDS, L. N. [77] The group of unramified Kummer extensions of prime degree, *PLMS*, (3), **35**, 1977, 407–422.
[80] Stickelberger relations on tame Kummer extensions of prime degree, *Proc. Queen's N. Th. Conf. 1979*, 249–256, Kingston 1980.
[81] Stickelberger relations and tame extensions of prime degree, *Illinois J. Math.*, **25**, 1981, 258–266.
CHINBURG, T. [79] 'Easier' Waring problems for commutative rings, *AA*, **35**, 1979, 303–331.
[83a] Stark's conjecture for *L*-functions with first-order zeroes at $s = 0$, *Adv. Math.*, **48**, 1983, 82–113.
[83b] On the Galois structure of algebraic integers and *S*-units, *Invent. Math.*, **74**, 1983, 321–349.
CHOWLA, P. [68] On the class-number of real quadratic fields, *JRAM*, **230**, 1968, 51–60.
[69a] On the representation of -1 as sum of squares in a cyclotomic field, *JNT*, **1**, 1969, 208–210.
[69b] On the representation of -1 as sum of two squares of cyclotomic integers, *Norske Vid. Selsk. Forh.*, **42**, 1969, 51–52.
[74] On the nonvanishing of a certain *L*-series at $s = 1/2$, *JNT*, **6**, 1974, 158–159.
CHOWLA, P., CHOWLA, S. (See also S. Chowla, P. Chowla)
[68] Formulae for the units and class-numbers of real quadratic fields, *JRAM*, **230**, 1968, 61–65.
[70] Determination of the Stufe of certain cyclotomic fields, *JNT*, **2**, 1970, 271–272.
[72] Problems on periodic simple continued fractions, *PAUS*, **69**, 1972, 3745.
[73] On Hirzebruch sums and a theorem of Schinzel, *AA*, **24**, 1973, 223–224.
CHOWLA, P., CHOWLA, S., TS'AO, L. C. [77] Some Diophantine equations in cyclotomic fields, *Chinese J. Math.*, **5**, 1977, 11–14.
CHOWLA, S. [34a] The class-number of binary quadratic forms, *QJM*, **5**, 1934, 302–303.
[34b] An extension of Heilbronn's class-number theorem, *QJM*, **5**, 1934, 304–307.
[34c] On the *k*-analogue of a result in the theory of the Riemann zeta function, *MZ*, **38**, 1934, 483–487.
[34d] Heilbronn's class-number theorem, I, *J. Indian Math. Soc.*, **1**, 1934, 66–68; II, *Proc. Nat. Acad. Sci. India, A*, **1**, 1934, 145–146.
[34e] The class-number of binary quadratic forms, *Proc. Nat. Acad. Sci. India, A*, **1**, 1934, 387–389.

[43] On the *K*-analogue of a result in the theory of Riemann zeta function, *Proc. Benares Math. Soc.*, **5**, 1943, 23–27.
[47] On the class-number of the corpus $P(\sqrt{-k})$, *Proc. Nat. Inst. Sci. India*, **13**, 1947, 197–200.
[49] Improvement of a theorem of Linnik and Walfisz, *PLMS*, **50**, 1949, 423–429.
[50] A new proof of a theorem of Siegel, *AnM*, **51**, 1950, 120–122.
[60a] Some results in number theory, *Norske Vid. Selsk. Forh.*, **33**, 1960, 43–44.
[60b] On a conjecture of J. F. Gray, *Norske Vid. Selsk. Forh.*, **33**, 1960, 58–59.
[61a] On the class number of real quadratic fields, *PAUS*, **47**, 1961, 878.
[61b] A remarkable solution of the Pellian equation $X^2 - pY^2 = -4$ in the case when $p \equiv 1 \pmod 4$ and the class-number of $R(\sqrt{p})$ is 1, *J. Indian Math. Soc.*, **25**, 1961, 43–46.
[61c] Proof of a conjecture of Julia Robinson, *Norske Vid. Selsk. Forh.*, **34**, 1961, 100–101.
[62] On Gaussian sums, *Norske Vid. Selsk. Forh.*, **35**, 1962, 66–67.
[63] On a conjecture of Artin, I, II, *Norske Vid. Selsk. Forh.*, **36**, 1963, 125–141.
[64] Bounds for the fundamental unit of a real quadratic field, *Norske Vid. Selsk. Forh.*, **37**, 1964, 88–90.
[65a] Note on the units of a real quadratic field, *PAMS*, **16**, 1965, 551.
[65b] Application of a theorem of A. Weil to improvements on bounds for class-numbers of quadratic fields, *Norske Vid. Selsk. Forh.*, **38**, 1965, 84–85.
[66] On a conjecture of Marshall Hall, *PAUS*, **56**, 1966, 417–418.
[67a] Stark's series expressed by theta-functions, *Norske Vid. Selsk. Forh.*, **40**, 1967, 31–33.
[67b] Observation on a theorem of Stark, *Norske Vid. Selsk. Forh.*, **40**, 1967, 34–36.
[70a] The Heegner–Stark–Baker–Deuring–Siegel theorem, *JRAM*, **241**, 1970, 47–48.
[70b] Leopoldt's criterion for real quadratic fields with class-number 1, *Hbg*, **35**, 1970, 32.

CHOWLA, S., ANKENY, N. C. (See N. C. Ankeny, S. Chowla)
CHOWLA, S., ANKENY, N. C., ARTIN, E. (See N. C. Ankeny, E. Artin, S. Chowla)
CHOWLA, S., ANKENY, N. C., BRAUER, R. (See N. C. Ankeny, R. Brauer, S. Chowla)
CHOWLA, S., ANKENY, N. C., HASSE, H. (See N. C. Ankeny, S. Chowla, H. Hasse)
CHOWLA, S., AYOUB, R. G. (See R. G. Ayoub, S. Chowla)
CHOWLA, S., BATEMAN, P. T., ERDÖS, P. (See P. T. Bateman, S. Chowla, P. Erdös)
CHOWLA, S., BOREL, A., HERZ, C. S., IWASAWA, K., SERRE, J. P. (See A. Borel *et al.*)
CHOWLA, S., BRIGGS, W. E. (See W. E. Briggs, S. Chowla)
CHOWLA, S., CHOWLA, P. (See also P. Chowla, S. Chowla) [73] "Metodo rapido"for finding real quadratic fields of class-number 1, *PAUS*, **70**, 1973, 395.
CHOWLA, S., CHOWLA, P., TS'AO, L. C. (See P. Chowla, S. Chowla, L. C. Ts'ao)
CHOWLA, S., COWLES, J., COWLES, M. [80] On the number of conjugacy classes in SL(2, Z), *JNT*, **12**, 1980, 372–377.
CHOWLA, S., COWLES, M. [77] On the coefficients c_n in the expansion
$$x \prod_1^\infty (1-x^n)^2 (1-x^{11n})^2 = \sum_1^\infty c_n x^n, \; JRAM, 292, 1977, 115-116.$$
CHOWLA, S., ERDÖS, P. [51] A theorem on the distribution of the values of *L*-functions, *J. Indian Math. Soc.*, **15**, 1951, 11–18.
CHOWLA, S., FRIEDLANDER, J. B. [76] Some remarks on *L*-functions and class numbers, *AA*, **28**, 1976, 413–417.
CHOWLA, S., GOLDFELD, D. M. [76] A remark on certain Hecke *L*-series which are non-negative on the real axis, *AA*, **30**, 1976, 1–3.
CHOWLA, S., HARTUNG, P. [74a] Congruence properties of class numbers of quadratic fields, *JNT*, **6**, 1974, 136–137.

[74b] A note on the hypothesis that $L(s, \chi) > 0$ for all real non-principal characters χ and for all $s > 0$, *JNT*, **6**, 1974, 271–275.

CHOWLA, S., KESSLER, I., LIVINGSTON, M. [77] On character sums and the non-vanishing for $s > 0$ of Dirichlet L-series belonging to real odd characters χ, *AA*, **33**, 1977, 81–87.

CHOWLA, S., de LEON, M. J. [74] A note on the Hecke hypothesis and the determination of imaginary quadratic fields with class-number 1, *JNT*, **6**, 1974, 261–263.

CHOWLA, S., de LEON, M. J., HARTUNG, P. [73] On a hypothesis implying the non-vanishing of Dirichlet's L-series $L(s, \chi)$ for $s > 0$ and real odd characters, *JRAM*, **262/3**, 1973, 415–419.

CHOWLA, S., MORDELL, L. J. [61] Note on the nonvanishing of $L(1)$, *PAMS*, **12**, 1961, 283–284.

CHOWLA, S., SELBERG, A. (See A. Selberg, S. Chowla)

CHOWLA, S., SHIMURA, G. [63] On the representation of zero by a linear combination of k-th powers, *Norske Vid. Selsk. Forh.*, **36**, 1963, 169–176.

CHOWLA, S., VIJAYARAGHAVAN, T. [44] The complete factorization $(\bmod p)$ of the cyclotomic polynomial of order p^2-1, *Proc. Nat. Acad. Sci. India*, A, **14**, 1944, 101–105.

CHRISTOFFERSON, S. [57] Über eine Klasse von kubischen Gleichungen mit drei Unbekannten, *Arkiv Mat.*, **3**, 1957, 355–364.

CHRISTY, D. [72] L'équation $x^4+y^4 = z^4$ dans les corps de nombres, *CR*, **274**, 1972, 1193–1196.

CHUDAKOV, N. G. (Tchudakoff, N. G.) [42] On Siegel's theorem, *IAN*, **6**, 1942, 135–142 (Russian).

[47a] *Introduction to the Theory of L-functions*, Moskva–Leningrad 1947 (Russian).

[47b] On Goldbach–Vinogradov's theorem, *AnM*, **48**, 1947, 515–545.

[69] Upper bound for the discriminant of the tenth imaginary quadratic field with class number one, *Issled. T. Chisel*, **3**, 75–77, Saratov 1969 (Russian).

CHUDAKOV, N. G., ANFERT'EVA, E. A. (See E. A. Anfert'eva, N. G. Chudakov)

CHUDAKOV, N. G., FELDMAN, N. I. (See N. I. Feldman, N. G. Chudakov)

CHULANOVSKII, I. V. [56] Elementary proof of the law of distribution of primes in the Gaussian field, *Vestnik LGU*, **11**, 1956, 13, 43–62 (Russian).

CHURCH, R. [35] Tables of irreducible polynomials for the first four prime moduli, *AnM*, **36**, 1935, 198–207.

CIOFFARI, V. G. [79] The euclidean condition in pure cubic and complex quartic fields, *MC*, **33**, 1979, 389–398.

CLABORN, L. [65] Dedekind domains and rings of quotients, *Pacific J. Math.*, **15**, 1965, 59–64.

[66] Every abelian group is a class group, *Pacific J. Math.*, **18**, 1966, 219–222.

[68] Specified relations in the ideal group, *Michigan Math. J.*, **15**, 1968, 249–255.

CLABORN, L., FOSSUM, R. [68] Generalizations of the notion of class-group, *Illinois J. Math.*, **12**, 1968, 228–253.

CLARKE, L. E. [51] On the product of three non-homogeneous linear forms, *PCPS*, **47**, 1951, 260–265.

COATES, J. [72a] On K_2 and some classical conjectures, *AnM*, **95**, 1972, 99–116.

[72b] On Iwasawa's analogue of the Jacobian for totally real number fields, *PSPM*, **24**, 1972, 51–61.

[73] Research problems: arithmetic questions in K-theory, in: *Algebraic K-theory*, II, 521–523, LN 342, Springer, 1973.

[77] p-adic L-functions and Iwasawa's theory, in: *Algebraic Number Fields, Proc. Durham Symp.*, 269–353, London 1977.

[81] The work of Mazur and Wiles on cyclotomic fields, *Sem. Bourbaki*, 1980/81, exp. 575, 220–242, LN 901, Springer, 1981.

COATES, J., LICHTENBAUM, S. [73] On *l*-adic zeta functions, *AnM*, **98**, 1973, 498–550.

COATES, J., SINNOTT, W. [74a] An analogue of Stickelbergers theorem for the higher *K*-groups, *Invent. Math.*, **24**, 1974, 149–161.

[74b] On *p*-adic *L*-functions over real quadratic fields, *Invent. Math.*, **25**, 1974, 253–279.

[77] Integrality properties of the values of partial zeta functions, *PLMS*, (3), **34**, 1977, 365–384.

COATES, J., WILES, A. [77] Kummer's criterion for Hurwitz numbers, in: *Algebraic number theory* (Kyoto Int. Symp.), 9–23, Tokyo 1977.

[78] On *p*-adic *L*-functions and elliptic units, *J. Austral. Math. Soc.*, ser. A, **26**, 1978, 1–25.

COFRÉ-MATTA, A., SHAPIRO, H. N. [77] On the distribution of reduced residues in algebraic number fields, *Comm. Pure Appl. Math.*, **30**, 1977, 13–39.

COHEN, E. [52a] Sur les fonctions arithmétiques relatives aux corps algébriques, *CR*, **234**, 1952, 787–788.

[52b] Sur les congruences du deuxième degré dans les corps algébriques, *CR*, **235**, 1952, 1358–1360.

[53] Congruence representations in algebraic number fields, *TAMS*, **75**, 1953, 444–470; II, *CJM*, **10**, 1958, 561–571.

[56a] The finite Goldbach problem in algebraic number fields, *PAMS*, **7**, 1956, 500–506.

[56b] Congruences in algebraic number fields involving sums of similar powers, *TAMS*, **83**, 1956, 547–556.

[56c] Binary congruences in algebraic number fields, *PAUS*, **42**, 1956, 120–122.

COHEN, G. L. [75a] Boundary conditions for expanding domains, *AA*, **26**, 1975, 213–216.

[75b] Selberg formulae for Gaussian integers, *AA*, **26**, 1975, 385–400.

COHEN, H. [74] Variations sur un thème de Siegel–Hecke, *Publ. Math. Un. Bordeaux*, 1973/4, nr. 5, 1–45.

[76] Variations sur un thème de Siegel et Hecke, *AA*, **30**, 1976, 63–93.

[83] Sur la distribution asymptotique des groupes de classes, *CR*, **296**, 1983, 245–247.

COHEN, H., LENSTRA, H. W. JR. [84] Heuristic on class groups of number fields, in: *Number Theory*, Noordwijkerhout 1983, 33–62, LN 1068, Springer, 1984.

COHEN, I. S. [50] Commutative rings with restricted minimum condition, *Duke Math. J.*, **17**, 1950, 27–42.

COHEN, I. S., KAPLANSKY, I. [51] Rings for which every module is a direct sum of cyclic modules, *MZ*, **54**, 1951, 97–101.

COHN, H. [52a] Note on fields of small discriminants, *PAMS*, **3**, 1952, 713–714.

[52b] A periodic algorithm for cubic forms, *AJM*, **74**, 1952, 821–833; II, ibid. **76**, 1954, 904–914.

[54] The density of abelian cubic fields, *PAMS*, **5**, 1954, 476–477.

[55] A numerical study of quintics of small discriminants, *Comm. Pure Appl. Math.*, **8**, 1955, 377–386.

[56a] A device for generating fields of even class number, *PAMS*, **7**, 1956, 595–598.

[56b] Some algebraic number theory estimates based on Dedekind Eta-functions, *AJM*, **78**, 1956, 791–796.

[57] A numerical study of Dedekind's cubic class number formula, *J. Res. Nat. Bur. Standards*, **59**, 1957, 265–271.

[58] A computation of some bi-quadratic class-numbers, *Math. Tables Aids Comp.*, **12**, 1958, 213–217.

[59] Numerical study of the representation of a totally positive quadratic integer as the sum of quadratic integral squares, *Numer. Math.*, **1**, 1959, 121–134.
[60a] A numerical study of Weber's real class number, *Numer. Math.*, **2**, 1960, 347–362.
[60b] Decomposition into four integral squares in the fields of $2^{1/2}$ and $3^{1/2}$, *AJM*, **82**, 1960, 301–322.
[61a] Proof that Weber's normal units are not perfect powers, *PAMS*, **12**, 1961, 964–966.
[61b] Calculation of class numbers by decomposition into three integral squares in the fields of $2^{1/2}$ and $3^{1/2}$, *AJM*, **83**, 1961, 33–56.
[62a] Some illustrative computations in algebraic number theory, in: *Survey of Numerical Analysis*, 543–549, McGraw Hill, 1962.
[62b] A numerical study of the relative class numbers of real quadratic integral domains, *MC*, **16**, 1962, 127–140.
[62c] Cusp forms arising from Hilbert's modular functions for the field of $3^{1/2}$, *AJM*, **84**, 1962, 283–305.
[64] On theta functions for certain quadratic fields, *AA*, **9**, 1964, 53–66.
[71] A numerical study of units in composite real quartic and octic fields, in: *Computers in Number Theory*, 153–165, Academic Press, 1971.
[76] Dyadotropic polynomials, *MC*, **30**, 1976, 854–862; II, *ibid.*, **33**, 1979, 359–367.
[77] Rational compositum genus for a pure cubic field, *JNT*, **9**, 1977, 244–257.
[79a] Quaternionic compositum genus, *JNT*, **11**, 1979, 399–411.
[79b] Cyclic-sixteen class fields for $Q((-p)^{1/2})$ by modular arithmetic, *MC*, **33**, 1979, 1307–1316.
[81a] The explicit Hilbert 2-cyclic class field for $Q(\sqrt{-p})$, *JRAM*, **321**, 1981, 64–77.
[81b] Iterated ring class fields and the icosahedron, *MA*, **255**, 1981, 107–122.
[83] Some examples of Weber–Hecke ring class field theory, *MA*, **265**, 1983, 83–100.
COHN, H., BARRUCAND, P. (See P. Barrucand, H. Cohn)
COHN, H., COOKE, G. E. [76] Parametric form of an eight class field, *AA*, **30**, 1976, 367–377.

COHN, H., GORN, S. [57] A computation of cyclic cubic units, *J. Res. Nat. Bur. Standards*, **59**, 1957, 155–168.

COHN, H., LAGARIAS, J. C. [83] On the existence of fields governing the 2-invariants of the class-group of $Q(\sqrt{dp})$ as p varies, *MC*, **41**, 1983, 711–730.

COHN, H., PALL, G. [62] Sums of four squares in a quadratic ring, *TAMS*, **105**, 1962, 536–556.

COHN, J. H. E. [71] Sums of cubes of gaussian integers, *PAMS*, **29**, 1971, 426.
[72] Waring's problem in quadratic number fields, *AA*, **20**, 1972, 1–16; add., *ibid.*, **23**, 1973, 417–418.
[77] The length of the period of the simple continued fraction of $\sqrt{d}$, *Pacific J. Math.*, **71**, 1977, 21–32.

COLLIOT-THÉLÈNE, J. L., CORAY, D., SANSUC, J. J. [80] Descente et principe de Hasse pour certaines variétés rationnelles, *JRAM*, **320**, 1980, 150–191.

COLLIOT-THÉLÈNE, J. L., SANSUC, J. J. [82] Sur le principe de Hasse et l'approximation faible, et sur une hypothèse de Schinzel, *AA*, **41**, 1982, 33–53.

CONNELL, I. G. [62] On algebraic number fields with unique factorization, *Canad. Math. Bull.*, **5**, 1962, 151–156.
[65] Elementary generalizations of Hilbert's theorem 90, *Canad. Math. Bull.*, **8**, 1965, 747–757.
[72] The Stufe of number fields, *MZ*, **124**, 1972, 20–22.

CONNELL, I., SUSSMAN, D. [70] The p-dimension of class groups of number fields, *JLMS*, (2), **2**, 1970, 525–529.

CONNER, P. E., PERLIS, R. [84] *Survey of Trace Forms of Algebraic Number Fields*, Singapore 1984.

CONWAY, J. H. [68] A tabulation of some information concerning finite fields, in: *Computers in Mathematical Research*, 37–50, North Holland, 1968.

CONWAY, J. H., JONES, A. J. [76] Trigonometric diophantine equations (On vanishing sums of roots of unity), *AA*, **30**, 1976, 229–240.

COOKE, G. E. [76] A weakening of the Euclidean property for integral domains and applications to algebraic number theory, *JRAM*, **282**, 1976, 133–156; II, *ibid.*, **283/4**, 1976, 71–85.

COOKE, G. E., COHN, H. (See H. Cohn, G. E. Cooke)

COOKE, G., WEINBERGER, P. J. [75] On the construction of division chains in algebraic number rings, with applications to SL_2, *Comm. Alg.*, **3**, 1975, 481–524.

COOPER, A. E. [25] Tables of quadratic forms, *AnM*, **26**, 1925, 309–316.

COQUET, J. [77] Remarques sur les nombres de Pisot-Vijayaraghavan, *AA*, **32**, 1977, 79–87.

CORAY, D., COLLIOT-THÉLÈNE, J. L., SANSUC, J. J. (See J. L. Colliot-Thélène, D. Coray, J. J. Sansuc)

CORMACK, G., WILLIAMS, H. C., SEAH, E. (See H. C. Williams, G. Cormack, E. Seah)

CORNELL, G. [71] Abhyankar's lemma and the class group, in: *Number Theory, Carbondale 1971*, 82–88, LN 751, Springer, 1971.

[82] On the construction of relative genus fields, *TAMS*, **271**, 1982, 501–511.

[83a] Exponential growth of the l-rank of the class group of the maximal real subfield of cyclotomic fields, *BAMS*, **8**, 1983, 55–58.

[83b] Relative genus theory and the class group of l-extensions, *TAMS*, **277**, 1983, 421–429.

CORNELL, G., ROSEN, M. [80] Cohomological analysis of the class group extension problem, *Proc. Queen's N. Th. Conf. 1979*, 287–308, Kingston 1980.

[81] Group-theoretic constraints on the structure of the class group, *JNT*, **13**, 1981, 1–11.

[84] The l-rank of the real class-group of cyclotomic fields, *Compos. Math.*, **53**, 1984, 133–141.

CORNELL, G., WASHINGTON, L. [85] Class numbers of cyclotomic fields, *JNT*, **21**, 1985, 260–274.

CORPUT, J. G., VAN DER, [23] Neue Zahlentheoretische Untersuchungen, *MA*, **89**, 1923, 215–254.

COUGNARD, J. [72] Sur les extensions galoisiennes non abéliennes de degré pq (p et q premiers) des rationnels, *CR*, **274**, 1972, 936–939.

[73] Sur l'anneau des entiers des extensions galoisiennes à groupe de Galois quaternionien d'ordre $4p$, *Publ. Math. Un. Bordeaux*, 1973/4, nr. 1, 1–21.

[74] Sur l'anneau des entiers des extensions galoisiennes non abeliennes de degré pq des rationnels, *Bull. Soc. Math. France, Mém.* **37**, 1974, 33–34.

[75] Sur l'anneau des entiers des p-extensions, *Astérisque*, **24/5**, 1975, 15–20.

[76] Propriétés galoisiennes des anneaux d'entiers des p-extensions, *Compos. Math.*, **33**, 1976, 303–336.

[77] Un contre-exemple à une conjecture de J. Martinet, in: *Algebraic Number Fields, Proc. Durham Symp.*, 539–560, London 1977.

[80] Une propriété de l'anneau des entiers des extensions galoisiennes non abéliennes de degré pq des rationnels, *Compos. Math.*, **40**, 1980, 407–415.

[82] Propriétés locales et globales de certaines extensions métacycliques, *Ann. Inst. Fourier*, **32**, 1982, 2, 1–12.

[83a] Une remarque sur l'anneau des entiers du corps des racines septiemes de l'unité, *Publ. Math. Fac. Sci. Besançon, Th. de Nombres*, 1981/82 et 1982/83.

[83b] Les travaux de A. Fröhlich, Ph. Cassou-Noguès et M. J. Taylor sur les bases normales, *Sém. Bourbaki*, **35**, 1982/3, exp. 598, *Astérisque* **105/6**, 1983, 25–38.

COWLES, J., CHOWLA, S., COWLES, M. (See S. Chowla, J. Cowles, M. Cowles)

COWLES, M. J. [80] On the divisibility of the class number of imaginary quadratic fields, *JNT*, **12**, 1980, 113–115.

COWLES, M., CHOWLA, S. (See S. Chowla, M. Cowles)

COWLES, M., CHOWLA, S., COWLES, J. (See S. Chowla, J. Cowles, M. Cowles)

CRAIG, M. [73] A type of class group for imaginary quadratic fields, *AA*, **22**, 1973, 449–459.

[77] A construction for irregular discriminants, *Osaka Math. J.*, **14**, 1977, 365–402; corr., *ibid.*, **15**, 1978, 461.

CROSS, J. T. [83] The Euler φ-function in the Gaussian integers, *AMM*, **90**, 1983, 518–528.

CUGIANI, M. [48] Observations on the question of existence of an Euclidean Algorithm in quadratic fields, *Bol. Un. Mat. Ital.*, (3), **3**, 1948, 136–141 (Italian).

[50] Quadratic fields with Euclidean Algorithm, *Period. Mat.*, **28**, 1950, 52–62, 114–129 (Italian).

CUOCO, A. A. [80] The growth of Iwasawa invariants in a family, *Compos. Math.*, **41**, 1980, 415–437.

[82] Relations between invariants in Z_p^2-extensions, *MZ*, **181**, 1982, 197–200.

CUOCO, A. A., MONSKY, P. [81] Class numbers in Z_p^d-extensions, *MA*, **255**, 1981, 235–258.

CURRIE, J. D., WILLIAMS, K. S. (See K. S. Williams, J. D. Currie)

CUSICK, T. W. [82] Finding fundamental units in cubic fields, *Math. Proc. Cambridge Phil. Soc.*, **92**, 1982, 385–389.

[84a] Lower bounds for regulators, in: *Number Theory, Noordwijkerhout 1983*, 63–73, LN 1068, Springer, 1984.

[84b] Finding fundamental units in totally real fields, *Math. Proc. Cambridge Phil. Soc.* **96**, 1984, 191–194.

CVETKOV, V. M. [80] Examples of extensions with a Demushkin group, *Zap. Nauch. Sem. LOMI*, **103**, 1980, 146–149 (Russian).

[83] On the Stickelberger-Voronoi theorem, *Zap. Nauch. Sem. LOMI*, **121**, 1983, 171–175 (Russian).

CZARNOWSKI, R. [82] On the zeros of Dedekind zeta-functions on the line $\sigma = 1/2$, *Funct. et Approx.*, **13**, 1982, 149–154.

CZOGAŁA, A. [81] Arithmetic characterization of algebraic number fields with small class numbers, *MZ*, **176**, 1981, 247–253.

DADE, E. C. [63] Algebraic integral representations by arbitrary forms, *Mathematika*, **10**, 1963, 96–100; corr., *ibid.*, **11**, 1964, 89–90; add., *ibid.*, **28**, 1981, 87.

DADE, E. C., BUMBY, R. T. (See R. T. Bumby, E. C. Dade)

DADE, E. C., TAUSSKY, O. [64] On the different in orders in an algebraic number field and special units connected with it, *AA*, **9**, 1964, 47–51.

DADE, E. C., TAUSSKY, O., ZASSENHAUS, H. [61] On the semigroup of ideal classes in an order of an algebraic number field, *BAMS*, **67**, 1961, 305–308.

[62] On the theory of orders, in particular on the semi-group of ideal classes and genera of an order in an algebraic number field, *MA*, **148**, 1962, 31–64.

DAI, Z. D., LAM, T. Y., PENG, C. K. [80] Levels in algebra and topology, *BAMS*, **3**, 1980, 845–848.

DALEN, K. [55] On a theorem of Stickelberger, *Math. Scand.*, **3**, 1955, 124–126.

DAMEY, P., PAYAN, J. J. [70] Existence et construction des extensions galoisiennes et non-abéliennes de degré 8 d'un corps de caractéristique differente de 2, *JRAM*, **244**, 1970, 37–54.

DANILOV, A. N. [65] On the sequence of values of additive arithmetical functions, defined on the set of ideals of a field K of degree n over the rationals, *Uchen. Zap. Leningr. Gos. Ped. Inst.*, **274**, 1965, 59–70 (Russian).

[67a] Asymptotical evaluations of the sums from the divisor problem for algebraic integers, *Uchen. Zap. Leningr. Gos. Ped. Inst.*, **302**, 1967, 3–8 (Russian).

[67b] Integral and local distribution laws for additive functions, defined on the sets of ideals of an algebraic field K, *Uchen. Zap. Leningr. Gos. Ped. Inst.*, **302**, 1967, 9–31 (Russian).

DANILOV, G. V., DANILOV, L. V. [75] On the evaluation of the length of the period of a quadratic irrationality, *Izv. VUZ*, **160**, 1975, 19–24 (Russian).

DAVENPORT, H. [39] On the product of three homogeneous linear forms, III, *PLMS*, (2), **45**, 1939, 98–125.

[47] On the product of three non-homogeneous linear forms, *PCPS*, **43**, 1947, 137–152.

[49] Sur les corps cubiques à discriminants négatifs, *CR*, **228**, 1949, 883–885.

[50a] Euclid's algorithm in cubic fields of negative discriminant, *AM*, **84**, 1950, 159–179.

[50b] Euclid's algorithm in certain quartic fields, *TAMS*, **68**, 1950, 508–532.

[51] Indefinite binary forms and Euclid's algorithm in real quadratic fields, *PLMS*, (2), **53**, 1951, 65–82.

[52] Linear forms associated with an algebraic number field, *QJM*, (2), **3**, 1952, 32–41.

[55] On a theorem of Furtwängler, *JLMS*, **30**, 1955, 186–195.

[66] Eine Bemerkung über Dirichlet's L-Funktionen, *GN*, 1966, 203–212.

DAVENPORT, H., CHATLAND, H. (See H. Chatland, H. Davenport)

DAVENPORT, H., HASSE, H. [34] Die Nullstellen der Kongruenzzetafunktionen in gewissen zyklischen Fällen, *JRAM*, **172**, 1934, 151–182.

DAVENPORT, H., HEILBRONN, H. [36] On the zeros of certain Dirichlet series, *JLMS*, **11**, 1936, 181–185, 307–312.

[69] On the density of discriminants of cubic fields, *Bull. London Math. Soc.*, **1**, 1969, 345–348; II, *Proc. Roy. Soc. London, A*, **322**, 1971, 405–420.

DAVENPORT, H., LEWIS, D. J. [63] Homogeneous additive equations, *Proc. Roy. Soc. London, A*, **274**, 1963, 443–460.

[66] Cubic equations of additive type, *Philos. Trans. Roy. Soc. London, A*, **261**, 1966, 97–163.

DAVENPORT, H., SCHINZEL, A. [67] Diophantine approximations and roots of unity, *MA*, **169**, 1967, 118–135.

DAVID, M. [49] Sur un algorithme voisin de celui de Jacobi, *CR*, **229**, 1949, 965–967.

[50] Sur trois algorithmes associés à l'algorithme du Jacobi, *CR*, **230**, 1950, 1445–1446.

[51] Caractérisation algorithmique des irrationnelles cubiques, *CR*, **232**, 1951, 1795–1798.

[56] Contribution à l'étude algorithmique des approximations rationelles simultanées de deux irrationels, Application au cas cubique, *Publ. Sci. Un. Alger., A*, **3**, 1956, 1–102.

DAVID, P. [78] Détermination de la structure du groupe des unités principales d'une extension abélienne K d'un corps local régulier, considéré comme un $Z_p(G(K/k))$-module, *CR*, **286**, 1978, A985–986.

DAVIES, D. [61] The computation of the zeros of Hecke zeta functions in the Gaussian field, *Proc. Roy. Soc. London, A*, **264**, 1961, 496–502.

[65] An approximate functional equation for Dirichlet L-functions, *Proc. Roy. Soc. London, A*, **284**, 1965, 224–236.

DAVIS, D. [78] Computing the number of totally positive circular units which are squares, *JNT*, **10**, 1978, 1–9.

DAVIS, E. D. [64] Overrings of commutative rings, II, *TAMS*, **110**, 1964, 196–212.
[65] Rings of algebraic numbers and functions, *MN*, **29**, 1965, 1–7.

DAVIS, H. T. [35] *Tables of Higher Mathematical Functions, II*, Bloomington 1935.

DAVIS, R. D., WISHART, E. F. [71] Galois extensions and the ramification sequence of some wildly ramified π-adic fields, *PAMS*, **30**, 1971, 212–216.

DAVIS, R. W. [76] Class number formulae for imaginary quadratic fields, *JRAM*, **286**/7, 1976, 369–379; II, *ibid.*, **299/300**, 1978, 247–255.

DECOMPS-GUILLOUX, A. [65a] Ensembles d'éléments algébriques dans les adèles, *CR*, **261**, 1965, 1929–1931.

[65b] Ensemble d'éléments définis dans les adèles comme limite de certaines suites de rationelles, *CR*, **261**, 1965, 3925–3926.

[70] Généralisation des nombres de Salem aux adèles, *AA*, **16**, 1970, 265–314.

DEDEKIND, R. [57] Beweis für die Irreductibilität der Kreistheilungsgleichungen, *JRAM*, **54**, 1857, 27–30 = *Ges. math. Werke*, *I*, 68–71, Vieweg, 1930.

[71] Über die Theorie der ganzen algebraischen Zahlen, XI Suppl. to Dirichlet's *"Vorlesungen über Zahlentheorie"*, 2nd ed. (1871), 3rd ed. (1879), 4th ed. (1894) = *Ges. math. Werke*, *III*, 1–314, Vieweg, 1932.

[77] Über die Anzahl der Ideal-Classen in den verschiedenen Ordnungen eines endlichen Körpers, *Festschr. zur Saecularfeier C. F. Gauß*, 1–55, Braunschweig 1877 = *Ges. math. Werke*, *I*, 105–108, Vieweg, 1930.

[78] Über den Zusammenhang zwischen der Theorie der Ideale und der Theorie der höheren Kongruenzen, *Abhandl. Kgl. Ges. Wiss. Göttingen*, **23**, 1878, 1–23 = *Ges. math. Werke*, *I*, 202–232, Vieweg, 1930.

[82] Über die Diskriminanten endlicher Körper, *Abhandl. Kgl. Ges. Wiss. Göttingen*, **29**, 1882, 1–56 = *Ges. math. Werke*, *I*, 351–397, Vieweg, 1930.

[92] Über einen arithmetischen Satz von Gauss, *Mitt. Deutsch. Math. Ges. Prag*, 1892, 1–11 = *Ges. math. Werke*, *II*, 28–39, Vieweg, 1931.

[95] Über die Begründung der Idealtheorie, *GN*, 1895, 106–113 = *Ges. math. Werke*, *II*, 50–58, Vieweg, 1931.

[00] Über die Anzahl der Idealklassen in reinen kubischen Zahlkörpern, *JRAM*, **121**, 1900, 40–123 = *Ges. math. Werke*, *II*, 148–233, Vieweg, 1931.

[31] Charakteristische Eigenschaft einklassiger Körper Ω, *Ges. math. Werke*, *II*, 373–375, Vieweg, 1931.

DEDEKIND, R., WEBER, H. [82] Theorie der algebraischer Funktionen einer Veränderlichen, *JRAM*, **92**, 1882, 181–290 = Dedekind, *Ges. math. Werke*, *I*, 238–249, Vieweg, 1930.

DEGEN, C. F. [17] *Canon Pellianus*, Havniae, 1817.

DEGERT, G. [58] Über die Bestimmung der Grundeinheit gewisser reell-quadratischer Zahlkörper, *Hbg*, **22**, 1958, 92–97.

DELANGE, H. [54] Généralisation du théorème de Ikehara, *ASENS*, (3), **71**, 1954, 213–242.

DELAUNAY, B. N. (Delone, B. N.) [23] Zur Bestimmung algebraischer Zahlkörper durch Kongruenzen; eine Anwendung auf die Abelsche Gleichungen, *JRAM*, **152**, 1923, 120–123.

[28] Vollständige Lösung der unbestimmten Gleichung $X^3q+Y^3 = 1$ in ganzen Zahlen, *MZ*, **28**, 1928, 1–9.

[30a] Über die Darstellung der Zahlen durch die binären kubischen Formen von negativer Diskriminante, *MZ*, **31**, 1930, 1–26.

[30b] Bemerkung über die Abhandlung von Herrn Trygve Nagell "Darstellung...", *MZ*, **31**, 1930, 27–28.

[54] On the growth of discriminants of algebraic number fields of given degree, *DAN*, **96**, 1954, 233-236 (Russian).

DELAUNAY, B. N., BILLEVICH, K. K., SOMINSKII, I. S. [35] Table of totally real fields of degree 4, *IAN*, **10**, 1935, 1267-1310 (Russian).

DELAUNAY, B. N., FADDEEV, D. K. [40] The theory of irrationalities of third degree, *Tr. Mat. Inst. Steklov*, **11**, 1940, 1-340 (Russian); English translation: Providence 1964.

[44] Investigations in the geometry of Galois theory, *Mat. Sbornik*, **15**, 1944, 243-284 (Russian).

DELIGNE, P. [73] Les constantes des equations fonctionnelles des fonctions *L*, in: *Modular Functions in One Variable, II*, 501-597, LN 349, Springer, 1973; corr., *ibid.*, *IV*, 149, LN 476, Springer, 1975.

[76] Les constantes locales de l'équation fonctionnelle de la fonction *L* d'Artin d'une représentation orthogonale, *Invent. Math.*, **35**, 1976, 299-316.

DELIGNE, P., RIBET, K. A., [80] Values of abelian *L*-functions at negative integers over totally real fields, *Invent. Math.*, **59**, 1980, 227-286.

DELONE, B. N. (See B. N. Delaunay)

DELSARTE, S. [48] Fonctions de Möbius sur les groupes abéliens, *AnM*, (2), **49**, 1948, 600-609.

DEMUSHKIN, S. P. [59] The group of maximal *p*-extension of a local field, *DAN*, **128**, 1959, 657-660 (Russian).

[61] The group of maximal *p*-extension of a local field, *IAN*, **25**, 1961, 329-346 (Russian).

[63] On 2-extensions of a local field, *Sibir. Mat. Zh.*, **4**, 1963, 951-955 (Russian).

DEM'YANOV, V. B. [50] On cubic forms in discretely normed fields, *DAN*, **74**, 1950, 889-891 (Russian).

[55] On the representation of zero by forms of the type $\sum_{i=1}^{m} \alpha_i x_i^n$, *DAN*, **105**, 1955, 203-205 (Russian).

[56] Pairs of quadratic forms over a field with a discrete norm and finite residue class field, *IAN*, **20**, 1956, 307-324 (Russian).

DENENBERG, C. G. [75] Periodic expansions and units in quadratic and cubic number fields, *JRAM*, **278/9**, 1975, 266-267.

DÉNES, P. [51] Über Einheiten von algebraischen Zahlkörpern, *Monatsh. Math.*, **55**, 1951, 161-163.

[52a] Über relativ-zyklische Körper vom Primzahlgrade, *Publ. Math. (Debrecen)*, **2**, 1952, 64-65.

[52b] Beweis einer Vandiverschen Vermutung bezüglich des zweiten Falles des letzten Fermatschen Satzes, *Acta Sci. Math. (Szeged)*, **14**, 1952, 197-202.

[53a] Über irreguläre Kreiskörper, *Publ. Math. (Debrecen)*, **3**, 1953, 17-23.

[53b] Über Grundeinheitensysteme der irregulären Kreiskörper von besonderen Kongruenzeigenschaften, *Publ. Math. (Debrecen)*, **3**, 1953, 195-204.

[54] Über den letzten Fermatschen Satz in relativzyklischen Zahlkörpern, *Ann. Polon. Math.*, **1**, 1954, 77-80.

[55] Über den zweiten Faktor der Klassenzahl und Irregularitätsgrad der irregulären Kreiskörper, *Publ. Math. (Debrecen)*, **4**, 1955/6, 163-170.

DENNIS, R. K., STEIN, M. R. [72] A new exact sequence for K_2 and some consequences for rings of integers, *BAMS*, **78**, 1972, 600-603.

[75] K_2 of valuation rings, *Adv. Math.*, **18**, 1975, 182-238.

DESCOMBES, R., POITOU, G. [50] Sur l'approximation dans $R(i\sqrt{11})$, *CR*, **231**, 1950, 264-266.

Despujols, P. [45] Norme de l'unité fondamentale du corps quadratique absolu, *CR*, **221**, 1945, 684–685.

Deuring, M. [31] Verzweigungstheorie bewerteter Körper, *MA*, **105**, 1931, 277–307.
[32] Galoische Theorie und Darstellungstheorie, *MA*, **107**, 1932, 140–144.
[33] Imaginäre quadratische Zahlkörper mit der Klassenzahl 1, *MZ*, **37**, 1933, 405–415.
[35a] Über den Tschebotareffschen Dichtigkeitssatz, *MA*, **110**, 1935, 414–415.
[35b] Neuer Beweis des Bauerschen Satzes, *JRAM*, **173**, 1935, 1–4.
[37] On Epstein's zeta function, *AnM*, **38**, 1937, 584–593.
[52] Die Struktur der elliptischen Funktionen-Körper und die Klassenkörper der imaginären quadratischen Zahlkörper, *MA*, **124**, 1952, 393–426.
[68] Imaginäre quadratische Zahlkörper mit der Klassenzahl Eins, *Invent. Math.*, **5**, 1968, 169–179.
[69] Analytische Klassenzahlformeln, in: *Number Theory and Analysis*, 55–75, New York 1969.

Diamond, J. [77] The p-adic log-gamma function and p-adic Euler constants, *TAMS*, **233**, 1977, 321–337.
[79a] The p-adic gamma measures, *PAMS*, **75**, 1979, 211–217.
[79b] On the values of p-adic L-functions at positive integers, *AA*, **35**, 1979, 223–237.

Diaz y Diaz, F. [74] Sur les corps quadratiques imaginaires dont le 3-rang du groupe des classes est supérieur à 1, *Sém. DPP*, **15**, 1973/4, fasc. 2, G 15.
[78a] On some families of imaginary quadratic fields, *MC*, **32**, 1978, 637–650.
[78b] Sur le 3-rang des corps quadratiques, *Publ. Math. d'Orsay*, **78**, 1978, 11.
[82] Sur les discriminants minimaux, *Sem. Th. de Nombr. Bordeaux*, 1981/2, exp. 14.
[83] Valeurs minima du discriminant des corps de degré 7 ayant une seule place réelle, *CR*, **296**, 1983, 137–139.

Diaz y Diaz, F., Shanks, D., Williams, H. C. [79] Quadratic fields with 3-rank equal to 4, *MC*, **33**, 1979, 836–840.

Dickson, L. E. [10] On the negative discriminants for which there is a single class of positive primitive binary quadratic forms, *BAMS*, **17**, 1910/1, 534–537.
[19] *History of the Theory of Numbers*, Washington 1919–1923, reprinted by Chelsea 1952 and 1966.
[23] *Algebras and Their Arithmetic*, Chicago 1923.

Dickson, L. E. et al. [23] *Algebraic numbers*, Report of the Committee on algebraic numbers of the N.R.C., 1923, 1928. Reprinted by Chelsea 1967.

Diederichsen, F. E. [40] Über die Ausreduktion ganzzahliger Gruppendarstellungen bei arithmetischer Äquivalenz, *Hbg*, **13**, 1940, 357–402.

Diekert, V. [84] Über die absolute Galoisgruppe dyadischer Zahlkörper, *JRAM*, **350**, 1984, 152–172.

Dinghas, A. [52] Sur un théorème de Schur concernant les racines d'une classe des équations algébriques, *Norske Vid. Selsk. Forh.*, **25**, 1952, 17–20.

Dirichlet, P. G. Lejeune- [28] Mémoire sur l'impossibilité de quelques équations indéterminées du cinquième degré, *JRAM*, **3**, 1828, 354–355 = *Werke*, *I*, 21–46, Berlin 1889.
[32a] Démonstration d'une propriété analogue à la loi de réciprocité qui existe entre deux nombres premiers quelconques, *JRAM*, **9**, 1832, 379–389 = *Werke*, *I*, 173–188, Berlin 1889.
[32b] Démonstration du théorème de Fermat pour le cas de 14ièmes puissances, *JRAM*, **9**, 1832, 390–393 = *Werke*, *I*, 191–194, Berlin 1889.
[35] Ueber eine neue Anwendung bestimmter Integrale auf die Summation endlicher oder unendlicher Reihen, *Abh. Kgl. Preuß. AW*, 1835, 391–407 = *Werke*, *I*, 237–256, Berlin 1889.

[38] Sur l'usage des séries infinies dans la théorie des nombres, *JRAM*, **18**, 1838, 259–274 = *Werke*, *I*, 357–374, Berlin 1889.
[39] Recherches sur diverses applications de l'analyse infinitésimale à la théorie des nombres, *JRAM*, **19**, 1839, 324–369; ibid. **21**, 1840, 1–12, 134–155 = *Werke*, *I*, 411–496, Berlin 1889.
[40] Sur la théorie des nombres, *CR*, **10**, 1840, 285–288 = *Werke*, *I*, 619–623, Berlin 1889.
[41a] Untersuchungen über die Theorie der complexen Zahlen, *JRAM*, **22**, 1841, 375–378 = *Werke*, *I*, 503–508, Berlin 1889.
[41b] Untersuchungen über die Theorie der complexen Zahlen, *Abh. Kgl. Preuß. AW*, 1841, 141–161 = *Werke*, *I*, 509–532, Berlin 1889.
[41c] Einige Resultate von Untersuchungen über eine Classe homogener Functionen des dritten und der höheren Grade, *Verhandl. Kgl. Preuß. AW*, 1841, 280–285 = *Werke*, *I*, 625–632, Berlin 1889.
[42] Recherches sur les formes quadratiques à coefficients et à indéterminées complexes, *JRAM*, **24**, 1842, 291–371 = *Werke*, *I*, 535–618, Berlin 1889.
[46] Zur Theorie der complexen Einheiten, *Verhandl. Kgl. Preuß. AW*, 1846, 103–107 = *Werke*, *I*, 639–644, Berlin 1889.

DISSE, A. [25] Über die Beziehungen zwischen Logarithmus und Numerus in einem p-adischen algebraischen Körper, *JRAM*, **154**, 1925, 178–198.
[26] Das Fundamentalsystem für die Logarithmen eines p-adischen algebraischen Körpers und sein Regulator, *JRAM*, **155**, 1926, 225–250.

DIVIŠ, B. [77] On the degrees of the sum and product of two algebraic elements, in: *Number Theory and Algebra*, 19–27, New York 1977.

DOBROWOLSKI, E. [78] On the maximal modulus of conjugates of an algebraic integer, *Bull. Acad. Pol. Sci., sér. sci. math. astr. phys.*, **26**, 1978, 291–292.
[79] On a question of Lehmer and the number of irreducible factors of a polynomial, *AA*, **34**, 1979, 391–401.

DOBROWOLSKI, E., LAWTON, W., SCHINZEL, A. [83] On a problem of Lehmer, in: *Studies in Pure Mathematics* (to the memory of Paul Turán), 135–144, Birkhäuser, 1983.

DODSON, M. M. [73] On Waring's problem in p-adic fields, *AA*, **22**, 1972/3, 315–327.
[82] Some estimates for diagonal equations over p-adic fields, *AA*, **40**, 1982, 117–124.

DONKAR, E. N. [77] On sums of three integral squares in algebraic number fields, *AJM*, **99**, 1977, 1297–1328.

DOUBRÉRE, M. [55] Sur les points limites d'un ensemble remarquable d'entiers algébriques imaginaires, *CR*, **240**, 1955, 2111–2113.

DRAKOKHRUST, Ya. A., PLATONOV, V. P. (See V. P. Platonov, Ya. A. Drakokhrust)

DRAXL, P. K. J. [70] Remarques sur le groupe de classes du composé de deux corps de nombres linéairement disjoints, *Sem. DPP*, **12**, 1970/71, exp. 24.
[71] L-Funktionen algebraischer Tori, *JNT*, **3**, 1971, 444–467.

DRESS, A. [64] Zu einem Satz aus der Theorie der algebraischen Zahlen, *JRAM*, **216**, 1964, 218–219.
[65] Eine Bemerkung über Teilringe globaler Körper, *Hbg*, **28**, 1965, 133–138.

DRESS, A., SCHARLAU, R. [82] Indecomposable totally positive numbers in real quadratic orders, *JNT*, **14**, 1982, 292–306.

DRESS, F. [68] Familles de séries formelles et ensembles de nombres algébriques, *ASENS*, (4), **1**, 1968, 1–44.

DRIBIN, D. M. [37] Quartic fields with the symmetric group, *AnM*, **38**, 1937, 739–746.

DRINFEL'D, V. G. [74] Elliptic modules, *Mat. Sbornik*, **94**, 1974, 594–627 (Russian).

DUBOIS, D. W., STEGER, A. [58] A note on division algorithmus in imaginary quadratic number fields, *CJM*, **10**, 1958, 285–286.

DUBOIS, E., PAYSANT-LE-ROUX, R. '[71] Développement périodique par l'algorithme de Jacobi–Perron et nombre de Pisot–Vijayaraghavan, *CR*, **272**, 1971, A649–652.

[75] Algorithme de Jacobi–Perron dans les extensions cubiques, *CR*, **280**, 1975, A183–186.

DUFFIN, R. J. [78] Algorithms for localizing roots of a polynomial and the Pisot–Vijayaraghavan numbers, *Pacific J. Math.*, **74**, 1978, 47–56.

DUFRESNOY, J., PISOT, C. [53a] Sur un ensemble fermé d'entiers algébriques, *ASENS*, (3), **70**, 1953, 105–133.

[53b] Sur les dérivés successifs d'un ensemble fermé d'entiers algébriques, *Bull. Sci. Math.*, **77**, 1953, 129–136.

[55a] Sur les éléments d'accumulation d'un ensemble fermé d'entiers algébriques, *Bull. Sci. Math.*, **79**, 1955, 54–64.

[55b] Étude de certaines fonctions méromorphes bornées sur le cercle unité. Application à un ensemble fermé d'entiers algébriques, *ASENS*, (3), **72**, 1955, 69–92.

DUJČEV, J. [56] On prime ideals of degree 1, *Acta Math. Acad. Sci. Hungar.*, **7**, 1956, 71–73.

DUMMIT, D. S., KISILEVSKY, H. [77] Indices in cubic fields, in: *Number Theory and Algebra*, 29–42, New York 1977.

DUNWOODY, M. J. [76] K_2 of a euclidean ring, *J. Pure Appl. Algebra*, **7**, 1976, 53–58.

DÜRBAUM, H., KOWALSKY, H.-J. [53] Arithmetische Kennzeichnung von Körpertopologien, *JRAM*, **191**, 1953, 135–152.

DUVAL, D. [81] Sur la structure galoisienne du groupe des unités d'un corps abélien réel de type (p, p), *JNT*, **13**, 1981, 228–245.

DWORK, B. [55] The local structure of the Artin root number, *PAUS*, **41**, 1955, 754–756.

[56] On the Artin root number, *AJM*, **78**, 1956, 444–472.

DWYER, W. G., FRIEDLANDER, E. M. [82] Étale K-theory and arithmetic, *BAMS*, **6**, 1982, 453–455.

DZEWAS, J. [60] Quadratsummen in reell-quadratischen Zahlkörpern, *MN*, **21**, 1960, 233–284.

DZHIEMURATOV, U. [68a] The number of ideals with estimates that are uniform in the parameters of a field, *DAN Uzbek. SSR*, 1968, nr. 5, 3 (Russian).

[68b] On ideals of algebraic number fields, *IAN Uzbek. SSR*, **12**, 1968, 9–14 (Russian).

EAKIN, P. M., JR. [68] The converse to a well-known theorem on Noetherian rings, *MA*, **177**, 1968, 278–282.

EAKIN, P., HEINZER, W. [73] More noneuclidean PID's and Dedekind domains with prescribed class group, *PAMS*, **40**, 1973, 66–68.

EARNEST, A. G., ESTES, D. R. [81] An algebraic approach to the growth of class numbers of binary quadratic lattices, *Mathematika*, **28**, 1981, 160–168.

EARNEST, A. G., KÖRNER, O. H. [82] On ideal class groups of 2-power exponents, *PAMS*, **86**, 1982, 196–198.

EDA, Y. [53] On the canonical basis of ideals, *Sci. Rep. Kanazawa Univ.*, **2**, 1953, 1, 15–21.

[55] A note on the general divisor problem, *Sci. Rep. Kanazawa Univ.*, **3**, 1955, 1, 5–9.

[67] On the mean-value theorem in an algebraic number field, *Japan. J. Math.*, **36**, 1967, 5–21.

[71] On the Waring problem in an algebraic number field, *Sem. modern methods in N. Th.*, paper 10, Tokyo 1971.

[75] On Waring's problem in algebraic number fields, *Rev. Colombiana Mat.*, **9**, 1975, 29–73.

EDA, Y., KITAYAMA, T. [69] Note on the generalized Prouhet–Tarry problem in an algebraic number field, *Sci. Rep. Kanazawa Univ.*, **14**, 1969, 21–28.

EDA, Y., NAKAGOSHI, N. [67] An elementary proof of the prime ideal theorem with remainder term, *Sci. Rep. Kanazawa Univ.*, **12**, 1967, 1, 1–12.

EDGAR, H. M. [66] Classes of equations of the type $y^2 = x^3 + k$ having no rational solutions, *Nagoya Math. J.*, **28**, 1966, 49–58.

[79] A number field without an integral basis, *Math. Mag.*, **52**, 1979, 248–251.

EDGAR, H., PETERSON, B. [80] Some contributions to the theory of cyclic quartic extensions of the rationals, *JNT*, **12**, 1980, 77–83.

EDGOROV, ZH., LAVRIK, A. F. (See A. F. Lavrik, Zh. Edgorov)

EDWARDS, H. M. [77] *Fermat's Last Theorem*, Springer, 1977.

[80] The genesis of ideal theory, *Arch. Hist. Ex. Sci.*, **23**, 1980, 321–378.

[83] Dedekind's invention of ideals, *Bull. London Math. Soc.*, **15**, 1983, 8–17.

EDWARDS, H., NEUMANN, O., PURKERT, W. [82] Dedekinds "Bunte Bemerkungen" zu Kronecker's "Grundzüge", *Arch. Hist. Ex. Sci.*, **27**, 1982, 49–85.

EGAMI, S. [79] Euclid's algorithm in pure quartic fields, *Tokyo J. Math.*, **2**, 1979, 379–385.

[80] The distribution of residue classes modulo $\mathfrak{A}$ in an algebraic number field, *Tsukuba J. Math.*, **4**, 1980, 9–13.

[81] Average version of Artin's conjecture in an algebraic number fields, *Tokyo J. Math.*, **4**, 1981, 203–212.

EGGAN, L. C., MAIER, E. A. [63] On complex approximation, *Pacific J. Math.*, **13**, 1963, 497–502.

EICHLER, M. [55] On the class number of imaginary quadratic fields and the sums of divisors of natural numbers, *J. Indian Math. Soc.*, **19**, 1955, 153–180.

[56] Der Hilbertsche Klassenkörper eines imaginärquadratischen Zahlkörpers, *MZ*, **64**, 1956, 229–242; corr., *ibid.*, **65**, 1956, 214.

[63] *Einführung in die Theorie der algebraischen Zahlen und Funktionen*, Birkhäuser, 1963.

[65] Eine Bemerkung zur Fermatschen Vermutung, *AA*, **11**, 1965, 129–131.

EISENBEIS, H., FREY, G., OMMERBORN, B. [78] Computation of the 2-rank of pure cubic fields, *MC*, **32**, 1978, 559–569.

EISENSTEIN, G. [44a] Beweis der Reciprocitätssatzes für die cubischen Reste in der Theorie der aus dritten Wurzeln der Einheit zusammengesetzten complexen Zahlen, *JRAM*, **27**, 1844, 289–310.

[44b] Über die Anzahl der quadratischen Formen in den verschiedenen complexen Theorien, *JRAM*, **27**, 1844, 311–316.

[44c] Allgemeine Untersuchungen über die Formen dritten Grades mit drei Variablen, welche der Kreisteilung ihre Entstehung verdanken, *JRAM*, **28**, 1844, 289–374; ibid. **29**, 1845, 9–53.

ELJOSEPH, N. [54] On the representation of a number as a sum of squares, *Riveon lematematika*, **7**, 1954, 38–43 (Hebrew).

ELLIOTT, P. D. T. A. [69] On the size of $L(1, \chi)$, *JRAM*, **236**, 1969, 26–36.

[70a] The distribution of power residues and certain related results, *AA*, **17**, 1970, 141–159.

[70b] The distribution of the quadratic class number, *Litovsk. Mat. Sb.*, **10**, 1970, 189–197.

[73] On the distribution of the values of quadratic L-series in the half-plane $\sigma > 1/2$, *Invent. Math.*, **21**, 1973, 319–338.

ELLISON, F. [73] Three diagonal quadratic forms, *AA*, **23**, 1973, 137–151.

ELLISON, W. J., PESEK, J., STALL, D. S., LUNDON, W. F. [71] A postscript to a paper of A. Baker, *Bull. London Math. Soc.*, **3**, 1971, 75–78.

EL-MUSA, A. J., BOREVICH, Z. I. (See Z. I. Borevich, A. J. El-Musa)
ELSNER, L., HASSE, H. [67] Numerische Ergebnisse zum Jacobischen Kettenbruchalgorithmus in rein-kubischen Zahlkörpern, *MN*, **34**, 1967, 95–97.
EMDE BOAS, P., VAN [68] Some ALGOL 60 algorithms for the verification of combinatorial conjectures on finite abelian groups, *Math. Centrum*, Amsterdam, ZW 014, 1968.
[69] A combinatorial problem on finite abelian groups, II, *Math. Centrum*, Amsterdam, ZW 007, 1969.
EMDE BOAS, P. VAN, BAAYEN, P. C., KRUYSWIJK, D. (See P. C. Baayen, P. van Emde Boas, D. Kruyswijk)
EMDE BOAS, P., VAN, KRUYSWIJK, D. [67] A combinatorial problem on finite abelian groups, *Math. Centrum*, Amsterdam, ZW 009, 1967.
ENDLER, O. [63a] Bewertungstheorie, *Bonn Math. Schriften*, **15**, 1963.
[63b] Über einen Existenzsatz der Bewertungstheorie, *MA*, **150**, 1963, 54–65.
ENDÔ, A. [73a] On the 2-rank of the ideal class groups of quadratic number fields, *Mem. Fac. Sci. Kyushu Univ.*, **27**, 1973, 7–12.
[73b] On the 2-class number of certain quadratic number fields, *Mem. Fac. Sci. Kyushu Univ.*, **27**, 1973, 111–120.
[76] On divisibility of the class number of $Q(\sqrt[9]{n})$ by 3, *Mem. Fac. Sci. Kyushu Univ.*, **30**, 1976, 299–311.
[78] Fundamental units of certain cubic number fields with negative discriminants, *Kumamoto J. Sci.*, **13**, 1978/9, 1, 24–36.
ENDO, M. [74] On the coefficients of cyclotomic polynomials, *Comment. Math. Univ. St. Pauli*, **23**, 1974/5, 121–126.
ENGSTROM, H. T. [30a] On the common index divisors of an algebraic field, *TAMS*, **32**, 1930, 223–237.
[30b] The theorem of Dedekind in the ideal theory of Zolotarev, *TAMS*, **32**, 1930, 879–887.
[30c] An example of the ideal theory of Zolotarev, *AMM*, **37**, 1930, 128–129.
ENNOLA, V. [58a] On the first inhomogeneous minimum of indefinite binary quadratic forms and Euclid's algorithm in real quadratic fields, *Ann. Univ. Turku*, **28**, 1958, 1–58.
[58b] Two elementary proofs concerning simple quadratic fields, *Nordisk Mat. Tidskr.*, **6**, 1958, 114–117.
[71] Some particular relations between cyclotomic units, *Ann. Univ. Turku*, **147**, 1971, 1–23.
[72] On relations between cyclotomic units, *JNT*, **3**, 1972, 236–247.
[73a] Conjugate algebraic integers on a circle with irrational center, *MZ*, **134**, 1973, 337–350.
[73b] Proof of a conjecture of Morris Newman, *JRAM*, **264**, 1973, 203–206.
[75a] Conjugate algebraic integers in an interval, *PAMS*, **53**, 1975, 259–261.
[75b] A note on a cyclotomic diophantine equation, *AA*, **28**, 1975, 157–159.
[75c] Solution of a cyclotomic diophantine equation, *JRAM*, **272**, 1975, 73–91.
[78] *J*-fields generated by roots of cyclotomic integers, *Mathematika*, **25**, 1978, 242–250.
[80] On the representations of units by cyclotomic polynomials, *AA*, **36**, 1980, 165–170.
ENNOLA, V., SMYTH, C. J. [74] Conjugate algebraic numbers on a circle, *Ann. Acad. Sci. Fenn.*, *AI*, **582**, 1974, 1–31.
[76] Conjugate algebraic numbers on a circles, *AA*, **29**, 1976, 147–157.
EPSTEIN, P. [03] Zur Theorie allgemeiner Zeta-funktionen, *MA*, **56**, 1903, 615–644; *ibid.*, **63**, 1907, 205–216.
[34] Zur Auflösbarkeit der Gleichung $x^2 - Dy^2 = -1$, *JRAM*, **171**, 1934, 243–252.
ERDÖS, P. [46] On the coefficients of the cyclotomic polynomial, *BAMS*, **52**, 1946, 179–184.

[57] On the growth of the cyclotomic polynomial in the interval (0, 1), *Proc. Glasgow Math. Ass.*, **3**, 1957, 102–104.
ERDÖS, P., BATEMAN, P. T., CHOWLA, S.   (See P. T. Bateman, S. Chowla, P. Erdös)
ERDÖS, P., CHOWLA, S.   (See S. Chowla, P. Erdös)
ERDÖS, P., KO, CHAO   [38] Note on the Euclidean algorithm, *JLMS*, **13**, 1938, 3–8.
ERDÖS, P., VAUGHAN, R. C.   [74] Bounds for the r-th coefficients of cyclotomic polynomials, *JLMS*, (2), **8**, 1974, 393–400.
ERNVALL, R.   [75] On the distribution mod 8 of the E-irregular primes, *Ann. Acad. Sci. Fenn.*, *AI*, **1**, 1975, 195–198.
[79] Generalized Bernoulli numbers, generalized irregular primes, and class number, *Ann. Univ. Turku*, **178**, 1979, 1–72.
ERNVALL, R., METSÄNKYLÄ, T.   [78] Cyclotomic invariants and E-irregular primes, *MC*, **32**, 1978, 617–629; corr. *ibid.*, **33**, 1979, 433.
ESTERMANN, T.   [45] On the sign of the Gaussian sums, *JLMS*, **20**, 1945, 66–67.
[48] On Dirichlet's L-functions, *JLMS*, **23**, 1948, 275–279.
ESTES, D., BUTTS, H.   (See H. Butts, D. Estes)
ESTES, D. R., EARNEST, A. G.   (See A. G. Earnest, D. R. Estes)
ESTES, D. R., HSIA, J. S.   [83] Sums of three integer squares in complex quadratic fields, *PAMS*, **89**, 1983, 211–214.
EVANS, R. J.   [77] Generalization of a theorem of Chowla, *Houston J. Math.*, **3**, 1977, 343–349.
[84] The octic and bioctic character of certain quadratic units, *Utilitas Math.*, **25**, 1984, 153–157.
EVANS, R. J., BERNDT, B. C.   (See B. C. Berndt, R. J. Evans)
EVANS, R. J., ISAACS, I. M.   [77] Fields generated by linear combinations of roots of unity, *TAMS*, **229**, 1977, 249–258; II, *Math. Scand.*, **43**, 1978/9, 26–34.
EVERTSE, J. H.   [84] On equations in S-units and the Thue–Mahler equation, *Invent. Math.*, **75**, 1984, 561–584.
EVTEEV, V. P.   [70] Integral ideals of norm z^2+a of a quadratic field, *DAN Tadzhik. SSR*, **13**, 1970, 11, 7–8 (Russian).
[73] Integral ideals of norm az^2+bz+c, *DAN Tadzhik. SSR*, **16**, 1973, 2, 3–5 (Russian).
FADDEEV, D. K.   [34] Tabularisation of fields and rings of degree three, *Tr. Mat. Inst. Steklov*, **5**, 1934, 19–24 (Russian).
[55] On an arithmetical formula, *Uspekhi Mat. Nauk*, **10**, 1955, 1, 169–171 (Russian).
[60] On the structure of reduced multiplicative group of a cyclic extension of a local field, *IAN*, **24**, 1960, 145–152 (Russian).
[74] On representations of algebraic numbers by matrices, *Zap. Nauch. Sem. LOMI*, **96**, 1974, 89–91 (Russian).
FADDEEV, D. K., DELONE, B. N.   (See B. N. Delaunay, D. K. Faddeev)
FADDEEV, D. K., POTAPKIN, V. K.   [59] On a totally real quintic extension of the rational field with the minimal discriminant, *Trudy 3 Mat. S'ezda*, **4**, 1959, 7 (Russian).
FADDEEV, D. K., SKOPIN, A. I.   [59] On the proof of a theorem of Kawada, *DAN*, **127**, 1959, 529–530 (Russian).
FAINLEIB, A. S.   [69] On the limit theorem for the class-number of purely radical quadratic forms with negative determinant, *DAN*, **184**, 1969, 1048–1049 (Russian).
[72] The distribution of classes of positive quadratic forms, *N. Tr. Tashkent Gos. Univ.*, **418**, 1972, 272–279 (Russian).
FAINLEIB, A. S., SAPARNIYAZOV, O.   [75] Dispersion of the sums of real characters and moments of $L(1, \chi)$, *IAN Uzbek. SSR*, 1975, 6, 24–29.

FAITH, C. [66] On Köthe rings, *MA*, **164**, 1966, 207–212.

FALTINGS, G. [83] Endlichkeitssätze für abelsche Varietäten über Zahlkörpern, *Invent. Math.*, **73**, 1983, 349–366; corr., *ibid.*, **75**, 1984, 381.

FANTA, E. [01] Beweis, dass jede lineare Funktion, deren Koeffizienten dem kubischen Kreisteilungskörper entnommene ganze teilerfremde Zahlen sind, unendlich viele Primzahlen dieses Körpers darstellt, *Monatsh. M.-Phys.*, **12**, 1901, 1–44.

FAVARD, J. [29] Sur les formes décomposables et les nombres algébriques, *Bull. Soc. Math. France*, **57**, 1929, 50–71.

[30] Sur les nombres algébriques, *Mathematica*, **4**, 1930, 109–113.

FEDERER, L. J., GROSS, B. H. [81] Regulators and Iwasawa modules, *Invent. Math.*, **62**, 1981, 443–457.

FEIN, B., GORDON, B., SMITH, J. H. [71] On the representation of -1 as a sum of two squares in an algebraic number field, *JNT*, **3**, 1971, 310–315.

FEIT, W. [59] On p-regular extensions of local fields, *PAMS*, **10**, 1959, 592–595.

FEKETE, M. [23] Über die Verteilung der Wurzeln bei gewisser algebraischen Gleichungen mit ganzzahligen Koeffizienten, *MZ*, **17**, 1923, 228–249.

FEKETE, M., SZEGÖ, G. [55] On algebraic equations with integral coefficients whose roots belong to a given point set, *MZ*, **63**, 1955, 158–172.

FELDMAN, N. I., CHUDAKOV, N. G. [72] On a theorem of Stark, *Mat. Zametki*, **11**, 1972, 329–340 (Russian).

FELSCH, V., SCHMIDT, E. [68] Über Perioden in den Koeffizienten der Kreisteilungspolynome $F_{np}(x)$, *MZ*, **106**, 1968, 267–272.

FENG, K. [82a] The rank of group of cyclotomic units in abelian fields, *JNT*, **14**, 1982, 315–326.

[82b] On the first factor of the class number of a cyclotomic field, *PAMS*, **84**, 1982, 479–482.

[82c] An elementary criterion on parity of class number of cyclic number field, *Sci. Sinica, A*, **25**, 1982, 1032–1041.

[85] Arithmetic characterization of algebraic number fields with given class group, *Acta Math. Sinica*, **1**, 1985, 47–54.

FENG, K., ZHANG, X. [83] On relative integral bases of quartic cyclic number fields, *Kexue Tongbao*, **28**, 1983, 456–457.

FERGUSON, LE BARON, O. [70] Algebraic kernels of plane sets, *Duke Math. J.*, **37**, 1970, 225–230.

FERRERO, B. [77] An explicit bound for Iwasawa's λ-invariant, *AA*, **33**, 1977, 405–408.

[78] Iwasawa invariants of abelian number fields, *MA*, **234**, 1978, 9–24.

[80] The cyclotomic Z_2-extension of imaginary quadratic fields, *AJM*, **102**, 1980, 447–459.

FERRERO, B., GREENBERG, R. [78] On the behavior of p-adic L-functions at $s = 0$, *Invent. Math.*, **50**, 1978/9, 91–102.

FERRERO, B., WASHINGTON, L. C. [79] The Iwasawa invariant μ_p vanishes for abelian number fields, *AnM*, **109**, 1979, 377–395.

FERTON, M. J. [72] Sur l'anneau des entiers d'une extension diédrale de degré $2p$ d'un corps local, *CR*, **274**, 1972, A1529–1532.

[73] Sur les idéaux d'une extension cyclique de degré premier d'un corps local, *CR*, **276**, 1973, A1483–1486.

[74] Sur l'anneau des entiers d'extensions cycliques d'un corps local, *Bull. Soc. Math. France*, Mém. **37**, 1974, 69–74.

[75] Sur l'anneau des entiers de certaines extensions cycliques d'un corps local, *Astérisque*, **24/5**, 1975, 21–28.

FERTON, M. J., BERTRANDIAS, F. (See F. Bertrandias, M. J. Ferton)

FERTON, M. J., BERTRANDIAS, F., BERTRANDIAS, J. P.   (See F. Bertrandias, J. P. Bertrandias, M. J. Ferton)
FIELDS, J. C.   [24] A foundation for the theory of ideals, *Proc. ICM*, Toronto 1924, 245-298.
FISCHER, W.   [52] Über die Zetafunktion des reell-quadratischen Zahlkörpers, *MZ*, **57**, 1952, 94-115.
FITTING, H.   [36] Die Determinantenideale eines Moduls, *Jahresber. Deutsch. Math, Verein.*, **46**, 1936, 195-228.
FJELLSTEDT, L.   [54] On a class of Diophantine equations of the second degree in imaginary quadratic fields, *Arkiv Mat.*, **2**, 1952/4, 435-461.
[55] Bemerkungen über gleichzeitige Lösbarkeit von Kongruenzen, *Arkiv Mat.*, **3**, 1955, 193-198.
FLANDERS, H.   [53a] The norm function of an algebraic field extension, *Pacific J. Math.*, **3**, 1953, 103-113; *ibid.*, **5**, 1955, 519-528.
[53b] Generalization of a theorem of Ankeny and Rogers, *AnM*, **57**, 1953, 392-400.
[60] The meaning of the form calculus in classical ideal theory, *TAMS*, **95**, 1960, 92-100.
FLEISCHER, I.   [53] Sur les corps topologiques et les valuations, *CR*, **236**, 1953, 1320-1322.
FLETT, T. M.   [51] On a coefficient problem of Littlewood and some trigonometrical sums, *QJM*, (2), **2**, 1951, 26-52.
FLEXOR, M., GAUTHERON, V.   (See V. Gautheron, M. Flexor)
FLUCH, W.   [64] Zur Abschätzung von $L(1, \chi)$, *GN*, 1964, 101-102.
[69] Über einen Satz von Hardy-Ramanujan, *Monatsh. Math.*, **73**, 1969, 31-35.
FOGELS, E. (Fogel, E. K.)   [37] Über die Möglichkeit einiger Diophantischer Gleichungen 3 und 4 Grades in quadratischen Körpern, *Comment. Math. Helv.*, **10**, 1937/8, 263-269.
[43] Zur Arithmetik quadratischer Zahlenkörper, *Wiss. Abh. Univ. Riga, Kl. Math.*, **1**, 1943, 23-47.
[50] A finite proof of the Gauss-Dirichlet formula, *Latvijas PSR Zin. Akad. Vēstis*, 1950, **9**, 117-125 (Russian).
[52] Finite proofs of some result of the analytic theory of numbers, *Latvijas PSR Zin. Akad. Fiz. Inst. R.* **3**, 1952, 49-63 (Russian).
[61a] On the distribution of primes, *DAN*, **140**, 1961, 1029-1032 (Russian).
[61b] On prime numbers in sequences of ideal norms, *Latvijas PSR Zin. Akad. Vēstis*, 1961, **12**, 29-34 (Russian).
[62a] On the zeros of Hecke's L-functions, *AA*, **7**, 1962, 87-106, 131-147, 225-240.
[62b] On the distribution of prime ideals, *AA*, **7**, 1962, 255-269.
[63] Über die Ausnahmenullstelle der Heckeschen L-Funktionen, *AA*, **8**, 1963, 307-309.
[65] On the zeros of L-functions, *AA*, **11**, 1965, 69-96.
[66a] On the abstract theory of primes, *AA*, **10**, 1964, 137-182, 333-358; *ibid.* **11**, 1966, 293-331.
[66b] On primes represented by a binary quadratic form, *Latvijas PSR Zin. Akad Vēstis*, 1966, **1**, 71-79 (Russian).
[69] Approximate functional equation for Hecke's L-functions of quadratic field, *AA*, **16**, 1969, 161-178.
[71] On the zeros of a class of L-functions, *AA*, **18**, 1971, 153-164.
[72] A mean value theorem of Bombieri's type, *AA*, **21**, 1972, 137-151.
FOMENKO, O. M.   [61] On the problem of Gauss, *AA*, **6**, 1961, 277-284.
[72] Continuability to the whole plane and a functional equation of the scalar product of the Hecke L-series of two quadratic fields, *Tr. Mat. Inst. Steklov*, **128**, 1972, 232-241 (Russian).

[82] On certain diophantine systems, *Zap. Nauch. Sem. LOMI*, **116**, 1982, 155–160 (Russian).
[83] Sums of squares in imaginary quadratic fields, *Zap. Nauch. Sem. LOMI*, **125**, 1983, 184–197 (Russian).

FONTAINE, J. M. [71] Groupes de ramification et représentations d'Artin, *ASENS*, (4), **4**, 1971, 337–392.

FOOTE, R., CHANG, S. M. (See S. M. Chang, R. Foote)

FORD, L. R. [18] Rational approximations to irrational complex numbers, *TAMS*, **19**, 1918, 1–42.
[25] On the closeness of approach of complex rational fractions to a complex irrational number, *TAMS*, **27**, 1925, 146–154.

FORMAN, W., SHAPIRO, H. N. [54] Abstract prime number theorems, *Comm. Pure Appl. Math.*, **7**, 1954, 587–619.

FOSSUM, R. M. [66] The Noetherian different of projective orders, *JRAM*, **224**, 1966, 207–218.
[73] *The Divisor Class Group of a Krull Domain*, Springer, 1973.

FOSSUM, R. M., CLABORN, L. (See L. Claborn, R. M. Fossum)

FOSTER, L. L. T. [70] On the number fields $Q(\vartheta)$, where $\vartheta^{2p}+p^2(\vartheta+1) = 0$, *MN*, **44**, 1970, 145–149.

FRAENKEL, A. [12] Axiomatische Begründung von Hensel's p-adischen Zahlen, *JRAM*, **141**, 1912, 43–76.
[16] *Über gewisse Teilbereiche und Erweiterungen von Ringen*, Habilitationsschrift, Leipzig 1916.

FRANZ, W. [34] Elementarteilertheorie in algebraischen Zahlkörpern, *JRAM*, **171**, 1934, 149–161.

FREI, G. [81a] Fundamental systems of units in number fields $Q(\sqrt{D^2+d}, \sqrt{D^2+4d})$ with $d|D$, *Arch. Math.*, **36**, 1981, 137–144.
[81b] Fundamental systems of units in bicubic parametric number fields, *Arch. Math.*, **36**, 1981, 524–536.
[82] Fundamental systems of units in biquadratic parametric number fields, *JNT*, **15**, 1982, 295–303.

FREI, G., LEVESQUE, C. [79] Independent system of units in certain algebraic number fields, *JRAM*, **311/2**, 1979, 116–144.
[80] On an independent system of units in the field $K = Q(\sqrt[n]{D^n \pm d})$ where $d|D^n$, *Hbg*, **50**, 1980, 162–165.

FREITAG, A. [77] Einige Bemerkungen zur Verteilung der Gaussschen Primzahlen, *Arch. Math.*, **29**, 1977, 93–95.

FRESNEL, J. [66] Nombres de Bernoulli généralisés et fonctions L p-adiques, *CR*, **263**, 1966, A337–340.
[67] Nombres de Bernoulli et fonctions L p-adiques, *Ann. Inst. Fourier*, **17**, 1967, 2, 281–333.
[71] Valeurs des fonctions zeta aux entiers negatifs, *Sem. Th. Nombr. Bordeaux* 1970/1, exp. 27.

FRESNEL, J., AMICE, Y. (See Y. Amice, J. Fresnel)

FRESNEL, J., CASSOU-NOGUÈS, P. (See P. Cassou-Noguès, J. Fresnel)

FREY, G., EISENBEIS, H., OMMERBORN, B. (See H. Eisenbeis, G. Frey, B. Ommerborn)

FREY, G., GEYER, W. D. [72] Über die Fundamentalgruppe von Körpern mit Divisorentheorie, *JRAM*, **254**, 1972, 110–122.

FRIED, E. [54] On linear combinations of roots, *Magyar Tud. Akad. Mat. Fiz. Oszt. Közl.*, **3**, 1954, 155–162.

FRIEDLANDER, E. M., DWYER, W. G. (See W. G. Dwyer, E. M. Friedlander)

FRIEDLANDER, J. B. [72] On the number of ideals free from large prime divisor, *JRAM*, **255**, 1972, 1-7.
[73a] On the least *k*-th power non-residue in an algebraic number field, *PLMS*, (3), **26**, 1973, 19-34.
[73b] On characters and polynomials, *AA*, **25**, 1973, 31-37.
[74] Character sums in quadratic fields, *PLMS*, (3), **28**, 1974, 99-111.
[76] On the class numbers of certain quadratic extensions, *AA*, **28**, 1976, 391-393.
[78] Small values of zeta functions of quadratic fields, *Math. Scand.*, **42**, 1978, 161-168.
[80] Estimates for prime ideals, *JNT*, **12**, 1980, 101-105.
FRIEDLANDER, J. B., CHOWLA, S. (See S. Chowla, J. B. Friedlander)
FRIEDMAN, E. C. [82a] Ideal class groups in basic $Z_{p_1} \times \ldots \times Z_{p_s}$-extensions of abelian number fields, *Invent. Math.*, **65**, 1982, 425-440.
[82b] Iwasawa theory for several primes and connection to Wieferich's criterion, in: *Developments related to Fermat's Last Theorem*, 269-274, Birkhäuser, 1982.
FRIEDRICH, R. [31] Über die Zerfällung einer Zahl in Summanden in beliebigen algebraischen Zahlkörpern, *Mitt. Math. Ges. Hamburg*, **7**, 1931, 31-58.
FROBENIUS, G. [96] Ueber Beziehungen zwischen den Primidealen eines algebraischen Körpers und den Substitutionen seiner Gruppe, *S. B. Kgl. Preuß. AW*, 1896, 689-703.
[12] Über quadratische Formen, die viele Primzahlen darstellen, *S. B. Kgl. Preuß. AW*, 1912, 966-980.
FRÖBERG, C.-E. [74] New results on the Kummer conjecture, *Nordisk Tidskr. Inform.*, **14**, 1974, 117-119.
FRÖHLICH, A. [52] On the class group of relatively abelian fields, *QJM*, (2), **3**, 1952, 98-106.
[54a] On the absolute class-group of abelian fields, *JLMS*, **29**, 1954, 211-217; ibid., **30**, 1955, 72-80.
[54b] The generalization of a theorem of L. Rédei's, *QJM*, (2), **5**, 1954, 130-140.
[54c] On fields of class two, *PLMS*, (3), **4**, 1954, 235-256.
[54d] A note on the class field tower, *QJM*, (2), **5**, 1954, 141-144.
[54e] A remark on the class number of abelian fields, *JLMS*, **29**, 1954, 498.
[57] On a method for the determination of class number factors in number fields, *Mathematika*, **4**, 1957, 113-121.
[59] The genus field and genus group in finite number fields, *Mathematika*, **6**, 1959, 40-46, 142-146.
[60a] A prime decomposition symbol for certain non Abelian number fields, *Acta Sci. Math. (Szeged)*, **21**, 1960, 229-246.
[60b] Discriminants of algebraic number fields, *MZ*, **74**, 1960, 18-28.
[60c] Ideals in an extension field as modules over the algebraic integers in a finite number field, *MZ*, **74**, 1960, 29-38.
[60d] The discriminants of relative extensions and the existence of integral bases, *Mathematika*, **7**, 1960, 15-22.
[61] Discriminants and module invariants over a Dedekind domain, *Mathematika*, **8**, 1961, 170-172.
[62a] On non-ramified extensions with prescribed Galois group, *Mathematika*, **9**, 1962, 133-134.
[62b] The module structure of Kummer extensions over Dedekind domains, *JRAM*, **209**, 1962, 39-53.
[62c] On the *l*-classgroup of the field $P(\sqrt[l]{m})$, *JLMS*, **37**, 1962, 189-192.
[62d] A remark on the classfield tower of the field $P(\sqrt[l]{m})$, *JLMS*, **37**, 1962, 193-194.

[72] Artin root numbers and normal integral bases for quaternion fields, *Invent. Math.*, **17**, 1972, 143–166.

[74a] Module invariants and root numbers for quaternion fields of degree $4l^v$, *PCPS*, **76**, 1974, 393–399.

[74b] Artin root numbers, conductors, and representation for generalized quaternion groups, *PLMS*, (3), **28**, 1974, 402–438.

[74c] Galois module structure and Artin L-functions, *Proc. ICM Vancouver*, 1974, 351–356.

[75] Artin root-number for quaternion characters, *Symposia Math.*, **15**, 1975, 353–363.

[76a] A normal integral basis theorem, *J. Algebra*, **39**, 1976, 131–137.

[76b] Arithmetic and Galois module structure for tame extensions, *JRAM*, **286/7**, 1976, 380–440.

[77a] Galois module structure, in: *Algebraic Number Fields, Proc. Durham Symp.*, 133–191, London 1977.

[77b] Stickelberger without Gauss sums, in: *Algebraic Number Fields, Proc. Durham Symp.*, 589–607, London 1977.

[77c] Rings of integers in tame extensions as Galois modules, *Astérisque*, **41/2**, 1977, 31–33.

[77d] Non-abelian Jacobi sums, in: *Number Theory and Algebra*, 71–75, New York 1977.

[78] Some problems of Galois module structure for wild extensions, *PLMS*, (3), **37**, 1978, 193–212.

[80] Galois module structure and rootnumbers for quaternion extensions of degree 2^n, *JNT*, **12**, 1980, 499–518.

[83a] *Galois module structure of algebraic integers*, Springer, 1983.

[83b] Central extensions, Galois groups and ideal class groups of number fields, *Contemp. Math.*, **24**, 1983.

[83c] Gauss'sche Summen, I, II, *Math. Inst. Univ. Köln.* (not dated).

FRÖHLICH, A., ARMITAGE, J. V. (See J. V. Armitage, A. Fröhlich)

FRÖHLICH, A., CASSELS, J. W. S. (See J. W. S. Cassels, A. Fröhlich)

FRÖHLICH, A., KEATING, M., WILSON, S. [74] The classgroup of dihedral 2-groups, *Mathematika*, **21**, 1974, 64–71.

FRÖHLICH, A., QUEYRUT, J. [73] On the functional equation of the Artin L-function for characters of real representations, *Invent. Math.*, **20**, 1973, 125–138.

FRÖHLICH, A., SERRE, J. P., TATE, J. [62] A different with an odd class, *JRAM*, **209**, 1962, 6–7.

FRÖHLICH, A., TAYLOR, M. J. [80] The arithmetic theory of local Gauss sums for tame characters, *Philos. Trans. Roy. Soc. London, A*, **298**, 1980/1, 141–181.

FRYSKA, T. [79] Über die obere Grenze der reellen Teile der Nullstellen der Dedekindschen Zetafunktionen, *Funct. et Approx.*, **7**, 1979, 91–99.

FUCHS, L. [63] Ueber die Perioden, welche aus den Wurzeln der Gleichung $w^n = 1$ gebildet sind, wenn n eine zusammengesetzte Zahl ist, *JRAM*, **61**, 1863, 374–386.

[66] Ueber die aus Einheitswurzeln gebildeten complexen Zahlen von periodischem Verhalten, insbesondere die Bestimmung der Klassenzahl derselben, *JRAM*, **65**, 1866, 74–111.

FUCHS, L. [48] A theorem on the relative norm of an ideal, *Comment. Math. Helv.*, **21**, 1948, 29–43.

[49] Some theorems on algebraic rings, *AM*, **81**, 1949, 285–289.

[54] On the fundamental theorem of commutative ideal theory, *Acta Math. Acad. Sci. Hungar.*, **5**, 1954, 95–99.

FUCHS, L., LOONSTRA, F. [71] On the cancellation of modules in direct sums over Dedekind domains, *Indag. Math.*, **33**, 1971, 163–169.

FUETER, R. [05] Die Theorie der Zahlstrahlen, *JRAM*, **130**, 1905, 197–237.

[07] Die Klassenzahl der Körper der komplexen Multiplikation, *GN*, 1907, 288–298.
[10] Die verallgemeinerte Kroneckersche Grenzformel und ihre Anwendung auf die Berechnung der Klassenzahl, *Rend. Circ. Mat. Palermo*, **29**, 1910, 380–395.
[11] Die Klassenkörper der komplexen Multiplikation und ihr Einfluss auf die Entwicklung der Zahlentheorie, *Jahresber. Deutsch. Math. Verein.*, **20**, 1911.
[13] Die diophantische Gleichung $\xi^3 + \eta^3 + \zeta^3 = 0$, *S. B. Heidelberg. A. W.* 1913.
[14] Abelsche Gleichungen in quadratisch-imaginären Zahlkörpern, *MA*, **75**, 1914, 177–255.
[17] Die Klassenzahl zyklischer Körper vom Primzahlgrad, deren Diskriminante nur eine Primzahl enthält, *JRAM*, **147**, 1917, 174–183.
[22] Kummers Kriterium zum letzten Theorem von Fermat, *MA*, **85**, 1922, 11–20.
[24] *Vorlesungen über die singulären Moduln und die komplexe Multiplikation der Elliptischen Funktionen*, Berlin 1924–27.
[33] Ein Satz über die Ring- und Strahlklassenzahlen in algebraischen Zahlkörpern, *Comment. Math. Helv.*, **5**, 1933, 319–322.
[45a] Über die Normalbasis in einem absolut Abelschen Zahlkörper, *Festschrift zum 60 Geburtst. von A. Speiser*, 141–152, Zürich 1945.
[45b] Abelsche Gleichungen in algebraischen Zahlkörpern, *Comment. Math. Helv.*, **17**, 1945, 108–127.

FUJII, A. [77] A remark on the zeros of some *L*-functions, *Tôhoku Math. J.*, (2), **29**, 1977, 417–426.

FUJISAKI, G. [57] On an example of an unramified Galois extension, *Sûgaku*, **9**, 1957/8, 97–99 (Japanese).
[62] On *L*-functions of simple algebras over the field of rational numbers, *J. Fac. Sci. Univ. Tokyo*, I, **9**, 1962, 293–311.
[73] Remark on the Stufe of fields, *Sci. Papers College Gen. Ed. Univ. Tokyo*, **23**, 1973, 1–3.
[74] Note on a paper of E. Artin, *Sci. Papers College Gen. Ed. Univ. Tokyo*, **24**, 1974, 93–98.

FUJIWARA, M. [72] Hasse principle in algebraic equations, *AA*, **22**, 1972, 267–276.

FUJIWARA, M., SUDO, M. [76] Some forms of odd degree for which the Hasse principle fails, *Pacific J. Math.*, **67**, 1976, 161–169.

FUNAKURA, T. [79] Characters and Artin *L*-functions, *Comment. Math. Univ. St. Pauli*, **27**, 1978/9, 185–197; II, *ibid.*, **28**, 1980, 81–85.

FURTWÄNGLER, PH. [04] Über die Reziprozitätsgesetze zwischen *l*-ten Potenzresten in algebraischen Zahlkörpern, wenn *l* eine ungerade Primzahl bedeutet, *MA*, **58**, 1904, 1–50.
[07] Allgemeiner Existenzbeweis für den Klassenkörper eines beliebigen algebraischen Zahlkörpers, *MA*, **63**, 1907, 1–37.
[08] Über die Klassenzahl Abelscher Zahlkörper, *JRAM*, **134**, 1908, 91–94.
[09] Die Reziprozitätsgesetze für Potenzreste mit Primzahlexponenten, *MA*, **67**, 1909, 1–31; *ibid.*, **72**, 1912, 346–386; *ibid.*, **74**, 1913, 413–429.
[11] Über die Klassenzahl der Kreisteilungskörper, *JRAM*, **140**, 29–32.
[16] Über das Verhalten der Ideale des Grundkörper im Klassenkörper, *Monatsh. M.-Phys.*, **27**, 1916, 1–15.
[19] Über die Führer der Zahlringen, *Ber. AW Wien*, **128**, 1919, 239–245.
[21] Punktgitter und Idealtheorie, *MA*, **82**, 1921, 256–279.
[27] Über die simultane Approximation von Irrationalzahlen, *MA*, **96**, 1927, 169–175
[30] Beweis des Hauptidealsatzes für die Klassenkörper algebraischer Zahlkörper, *Hbg*, **7**, 1930, 14–36.

[32] Über eine Verschärfung des Hauptidealsatzes für algebraische Zahlkörper, *JRAM*, **167**, 1932, 379–387.
FURUTA, Y. [58] A reciprocity law of the power residue symbol, *J. Math. Soc. Japan*, **10**, 1958, 46–54.
[59a] Norm of units of quadratic fields, *J. Math. Soc. Japan*, **11**, 1959, 139–145.
[59b] On meta-abelian fields of a certain type, *Nagoya Math. J.*, **14**, 1959, 193–199.
[61] A property of meta-abelian extensions, *Nagoya Math. J.*, **19**, 1961, 169–187.
[66] The notion of restricted idèles with applications to some extension fields, *Nagoya Math. J.*, **27**, 1966, 121–132; II, *J. Math. Soc. Japan*, **18**, 1966, 247–252.
[67] The genus field and genus number in algebraic number fields, *Nagoya Math. J.*, **29**, 1967, 281–285.
[70] Über das Geschlecht und die Klassenzahl eines relativ-Galoisches Zahlkörpers vom Primzahlpotenzgrade, *Nagoya Math. J.*, **37**, 1970, 197–200.
[71] Über die zentrale Klassenzahl eines Relativ-galoisschen Zahlkörpers, *JNT*, **3**, 1971, 318–322.
[72] On class field towers and the rank of ideal class groups, *Nagoya Math. J.*, **48**, 1972, 147–157.
[76] On nilpotent factors of congruent ideal class group of Galois extensions, *Nagoya Math. J.*, **62**, 1976, 13–28.
[77] Note on class number factors and prime decomposition, *Nagoya Math. J.*, **66**, 1977, 167–182.
FURUTA, Y., KAPLAN, P. [81] On quadratic and quartic characters of quadratic units, *Sci. Rep. Kanazawa Univ.*, **26**, 1981, 27–30.
FURUTA, Y., SAWADA, Y. [68] On the Galois cohomology group of the ring of integers in a global field and its adele ring, *Nagoya Math. J.*, **32**, 1968, 247–252.
FURUYA, H. [71] On divisibility by 2 of the relative class numbers of imaginary number fields, *Tôhoku Math. J.*, (2), **23**, 1971, 207–218.
[77] Principal ideal theorems in the genus field for absolutely Abelian extensions, *JNT*, **9**, 1977, 4–15.
[82] Ambiguous numbers over $P(\zeta_3)$ of absolutely abelian extensions of degree 6, *Tokyo J. Math.*, **5**, 1982, 457–462.
GADIA, S. K., GANDHI, J. M. (See J. M. Gandhi, S. K. Gadia)
GAIGALAS, E. [75] On the scalar product of Hecke's *L*-series of quadratic fields, *Litovsk. Mat. Sb.*, **15**, 1975, 4, 41–52 (Russian).
[77] The scalar product of Hecke *L*-series of certain algebraic fields, *Litovsk. Mat. Sb.*, **17**, 1977, 1, 65–74 (Russian).
[79] The distribution of primes in two quadratic imaginary fields, *Litovsk. Mat. Sb.*, **19**, 1979, 45–60; II, *ibid.*, 69–76 (Russian).
GALKIN, V. M. [72] The first factor of the class number of ideals of a cyclotomic field, *Uspekhi Mat. Nauk*, **27**, 1972, 6, 233–234 (Russian).
[73] Zeta-functions of certain one-dimensional rings, *IAN*, **37**, 1973, 3–19 (Russian).
GALLAGHER, V. P. [85] The trace form on Galois field extensions, to appear.
GANDHI, J. M., GADIA, S. K. [78] A simple proof of the infinity of irregular primes, *Utilitas Math.*, **20**, 1978, 379–382.
GARAKOV, G. A. [70] Tables of irreducible polynomials over the field $GF(p)$ ($p \leq 11$), *Trudy Vychisl. Tsentra Akad. Nauk Armyan. SSR; Mat. Voprosy Kibernet. Vychisl. Tekhn*, **6**, 1970, 112–142.
GARBANATI, D. A. [75a] Unit signatures, and even class numbers, and relative class numbers, *JRAM*, **274/5**, 1975, 376–384.

[75b] Extensions of the Hasse norm theorem, *BAMS*, **81**, 1975, 583–586.

[75c] Extensions of the Hasse norm theorem, II, *Linear and Multilinear Algebra*, **3**, 1975/6, 143–145.

[76] Units with norm -1 and signatures of units, *JRAM*, **283/4**, 1976, 164–175.

[77] The Hasse norm theorem for *l*-extensions of the rationals, in: *Number Theory and Algebra*, 77–90, New York 1977.

[78a] Invariants of the ideal class group and the Hasse norm theorem, *JRAM*, **297**, 1978, 159–171.

[78b] The Hasse norm theorem for noncyclic extensions of the rationals, *PLMS*, (3), **37**, 1978, 143–164.

[80] An algorithm for finding an algebraic number whose norm is a given rational number, *JRAM*, **316**, 1980, 1–13.

[81] Class field theory summarized, *Rocky Mountain J. Math.*, **11**, 1981, 195–225.

GARLAND, H.   [71] A finiteness theorem for K_2 of a number field, *AnM*, **94**, 1971, 534–548.

GASSMANN, F.   [26] Bemerkungen zu der vorstehender Arbeit von Hurwitz, *MZ*, **25**, 1926, 665–675.

GAUSS, C. F.   [01] *Disquisitiones Arithmeticae*, Göttingen 1801; English translation: Yale Univ. Press, 1966; German translation: New York 1965; Russian translation: Moskva 1959.

[11] Summatio quarumdam serierum singularium, *Comm. Soc. Reg. Sc. Gotting.*, **1**, 1811 = *Werke*, *II*, 9–45, Göttingen 1863.

[28] Letter to Dirichlet of 30 May 1828 = Dirichlet, *Werke*, *II*, 378–380, Berlin 1897.

[32] Theoria residuorum biquadraticorum, *Comm. Soc. Reg. Sc. Gotting.*, **7**, 1832 = *Werke*, *II*, 93–198, Göttingen 1863.

[63] De nexu inter multitudinem classium in quae formae binariae secundi gradus distribuntur earumque determinantem, *Werke*, *II*, 269–291, Göttingen 1863.

GAUTHERON, V., FLEXOR, M.   [69] Un exemple de détermination des entiers d'un corps de nombres, *Bull. Sci. Math.*, (2), **93**, 1969, 3–13; corr., *ibid.*, (2), **96**, 1972, 177–179.

GAUTHIER, F.   [78] Ensembles de Kronecker et représentations des nombres premiers par une forme quadratique binaire, *Bull. Sci. Math.*, (2), **102**, 1978, 129–143.

GAY, D., VÉLEZ, W. Y.   [81] The torsion group of a radical extension, *Pacific J. Math.*, **92**, 1981, 317–327.

GEBHART, H.-M.   [77] Zur Berechnung des Funktors K_2 von einigen euklidischen Ringen, *Schriftenreihe Math. Inst. Univ. Münster*, (2), **13**, 1977, 1–69.

GECHTER, J.   [76] Artin root numbers for real characters, *PAMS*, **57**, 1976, 35–38.

GELBART, S.   [77] Automorphic forms and Artin's conjecture, in: *Modular Forms of One Variable*, VI, 241–276, LN 627, Springer, 1977.

[84] An elementary introduction to the Langlands program, *BAMS*, **10**, 1984, 177–219.

GELFOND, A. O.   [40] Sur la divisibilité de la différence des puissances de deux nombres entiers par une puissance d'un idéal premier, *Mat. Sbornik*, **7**, 1940, 7–25.

[53] On an elementary approach to some problems from the field of distribution of prime numbers, *Vestnik MGU*, **8**, 1953, 2, 21–26 (Russian).

[60] Some functional equations implied by equations of Riemann type, *IAN*, **24**, 1960, 469–474 (Russian).

GELFOND, A. O., LINNIK, YU. V.   [48] On a method of Thue in the problem of effectivization in quadratic fields, *DAN*, **61**, 1948, 773–776.

GÉRARDIN, P., LABESSE, J. P.   [79] The solution of a base change problem for $GL(2)$, *PSPM*, **33**, 1979, II, 115–133.

GERIG, S. [67] Sequences of integers satisfying congruence relations and Pisot–Vijayaraghavan numbers, *J. Austral. Math. Soc.*, **7**, 1967, 508–512.

GERLOVIN, É. L., [69] The completion of the multiplicative group of a cyclic p-extension of a local field, *Vestnik LGU*, **22**, 1969, 7, 14–22.

GERLOVIN, É. L., BOREVICH, Z. I. (See Z. I. Borevich, É. L. Gerlovin)

GERONIMO, J. S., MOUSSA, P., BESSIS, D. (See P. Moussa *et al.*)

GERST, I. [70] On the theory of nth power residues and a conjecture of Kronecker, *AA*, **17**, 1970, 121–139.

GERST, I., BRILLHART, J. [71] On the prime divisors of polynomials, *AMM*, **71**, 1971, 250–266.

GERTH, F., III [73] Ranks of Sylow 3-subgroups of ideal class group of certain cubic fields, *BAMS*, **79**, 1973, 521–525.

[75a] On 3-class groups of pure cubic fields, *JRAM*, **278/9**, 1975, 52–62.

[75b] Number fields with prescribed l-class groups, *PAMS*, **49**, 1975, 284–288.

[75c] A note on the l-class groups of number fields, *MC*, **29**, 1975, 1135–1137.

[76a] On l-class groups of certain number fields, *Mathematika*, **23**, 1976, 116–123.

[76b] Cubic fields whose class numbers are not divisible by 3, *Illinois J. Math.*, **20**, 1976, 486–493.

[76c] Ranks of 3-class groups of non-Galois cubic fields, *AA*, **30**, 1976, 307–322.

[76d] On 3-class groups of cyclic cubic extensions of certain number fields, *JNT*, **8**, 1976, 84–98.

[77a] The Hasse norm principle in metacyclic extensions of number fields, *JLMS*, (2), **16**, 1977, 203–208.

[77b] The Hasse norm principle for abelian extensions of number fields, *BAMS*, **83**, 1977, 264–266.

[77c] Structure of l-class groups of certain number fields and Z_l-extensions, *Mathematika*, **24**, 1977, 16–33.

[78] The Hasse norm principle in cyclotomic fields, *JRAM*, **303/4**, 1978, 249–252.

[79a] Upper bounds for an Iwasawa invariant, *Compos. Math.* **39**, 1979, 3–10.

[79b] The Iwasawa invariant μ for quadratic fields, *Pacific J. Math.*, **80**, 1979, 131–136.

[80] The ideal class group of two cyclotomic fields, *PAMS*, **78**, 1980, 321–323.

[82] Counting certain number fields with prescribed l-class numbers, *JRAM*, **337**, 1982, 195–207.

[83a] Asymptotic results for class number divisibility in cyclotomic fields, *Canad. Math. Bull.*, **26**, 1983, 464–472.

[83b] An application of matrices over finite fields to algebraic number theory, *MC*, **41**, 1983, 229–234.

[83c] Asymptotic behavior of number fields with prescribed l-class numbers, *JNT*, **17**, 1983, 191–203.

[83d] Sufficiency of genus theory for certain number fields, *Expos. Math.*, **1**, 1983, 357–359.

[86] Densities for 3-class rank of pure cubic fields, *AA*, **46**, 1986, 227–242.

GERTH, F., III, BLOOM, J. R. (See J. R. Bloom, F. Gerth, III)

GERTH, F., III, GRAHAM, S. W. [84] Application of a character sum estimate to a 2-class number density, *JNT*, **19**, 1984, 239–247.

[85] Counting certain number fields with l-class numbers 1 or l, to appear.

GEYER, W. D. [78] The automorphism group of the field of all algebraic numbers, *Proc. 5th School of Algebra*, 167–199, Rio de Janeiro 1978.

GEYER, W. D., FREY, G. (See G. Frey, W. D. Geyer)
GILBARG, D. [42] The structure of the group of p-adic 1-units, *Duke Math. J.*, **9**, 1942, 262–271.
GILBERT, J. R. JR., BUTTS, H. S. [68] Rings satisfying the three Noether axioms, *J. Sci. Hiroshima Univ.*, **32**, 1968, 211–224.
GILBERT, W. J. [81] Radix representations of quadratic fields, *J. Math. Anal. Appl.*, **83**, 1981, 264–274.
GILLARD, R. [76] Remarques sur certaines extensions prodiédrales de corps de nombres, *CR*, **282**, 1976, A13–15.
[77] Sur le groupe des classes des extensions abéliennes réelles, *Sém. DPP*, **18**, 1976/7, exp. 10.
[79a] Unités elliptiques et unités cyclotomiques, *MA*, **243**, 1979, 181–189.
[79b] Remarques sur les unités cyclotomiques et les unités elliptiques, *JNT*, **11**, 1979, 21–48.
[79c] Unités cyclotomiques, unités semilocales et Z_l-extensions, *Ann. Inst. Fourier*, **29**, 1979, 1, 49–79; II, *ibid.*, 4, 1–15.
[79d] Formulations de la conjecture de Leopoldt et étude d'une condition suffisante, *Hbg*, **48**, 1979, 125–138.
[80a] Unités elliptiques et unités de Minkowski, *J. Soc. Math. Japan*, **32**, 1980, 697–701.
[80b] Unités elliptiques et fonctions L p-adiques, *Compos. Math.*, **42**, 1980/1, 57–88.
GILLARD, R., ROBERT, G. [79] Groupes d'unités elliptiques, *Bull. Soc. Math. France*, **107**, 1979, 305–317.
GILLETT, J. R. [70] On the largest prime divisors of ideals in fields of degree n, *Duke Math. J.*, **37**, 1970, 589–600.
GILMER, R. W. JR. [63a] On a classical theorem of Noether in ideal theory, *Pacific J. Math.*, **13**, 1963, 579–583.
[63b] Finite rings having a cyclic multiplicative group of units, *AJM*, **85**, 1963, 447–452.
[64] Integral domains which are almost Dedekind, *PAMS*, **15**, 1964, 813–818.
[67a] A note on two criteria for Dedekind domain, *Enseign. Math.*, **13**, 1967, 253–256.
[67b] Some applications of the Hilfssatz von Dedekind-Mertens, *Math. Scand.*, **20**, 1967, 240–244.
[68] *Multiplicative Ideal Theory*, Kingston 1968; 2nd ed., New York 1972.
[69] A note on generating sets for invertible ideals, *PAMS*, **22**, 1969, 426–427.
[72] On commutative rings of finite rank, *Duke Math. J.*, **39**, 1972, 381–383.
[73] The n-generator property for commutative rings, *PAMS*, **38**, 1973, 477–482.
GILMER, R. W. JR., BUTTS, H. S. (See H. S. Butts, R. W. Gilmer, Jr.)
GILMER, R. W. JR., MOTT, J. L. [68] On proper overrings of integral domains, *Monatsh. Math.*, **72**, 1968, 61–71.
GILMER, R. W. JR., OHM, J. [64] Integral domains with quotient overrings, *MA*, **153**, 1964, 97–103.
GIORGIUTTI, J. [60] Modules projectifs sur les algèbres de groupes finis, *CR*, **250**, 1960, 1419–1420.
GIRSTMAIR, K. [79] Elementare Berechnung von kubischen Diskriminanten, *Arch. Math.*, **32**, 1979, 341–343.
GLAISHER, J. W. L. [03] On the expression for the number of classes of a negative discriminant, *QJM*, **34**, 1903, 178–204.
GODEMENT, R., JACQUET, H. [72] *Zeta Functions of Simple Algebras*, LN 260, Springer, 1972.
GODWIN, H. J. [56] Real quartic fields with small discriminants, *JLMS*, **31**, 1956, 478–485.
[57a] On totally complex quartic fields with small discriminant, *PCPS*, **53**, 1957, 1–4.

[57b] On quartic fields with signature one with small discriminants, *QJM*, (2), **8**, 1957, 214–222.
[58] The determination of fields of small discriminant with a given subfield, *Math. Scand.*, **6**, 1958, 40–46.
[60] The determination of units in totally real cubic fields, *PCPS*, **56**, 1960, 318–321.
[61] The determination of the class-numbers of totally real cubic fields, *PCPS*, **57**, 1961, 728–730.
[62] On relations between cubic and quartic fields, *QJM*, (2), **13**, 1962, 206–212; corr., *ibid.*, (2), **26**, 1975, 511–512.
[65a] On the inhomogeneous minima of totally real cubic norm-forms, *JLMS*, **40**, 1965, 623–627.
[65b] On Euclid's algorithm in some quartic and quintic fields, *JLMS*, **40**, 1965, 699–704.
[67] On Euclid's algorithm in some cubic fields with signature one, *QJM*, (2), **18**, 1967, 333–338.
[71] Computations relating to cubic fields, in: *Computers in Number Theory*, 225–229, Academic Press, 1971.
[83] The calculation of large units in cyclic cubic fields, *JRAM*, **338**, 1983, 216–220.
[84] A note on Cusick's theorem on units in totally real cubic fields, *Math. Proc. Cambridge Philos. Soc.*, **95**, 1984, 1–2.

GODWIN, H. J., CHANG, J. A.   (See J. A. Chang, H. J. Godwin)
GODWIN, H. J., SAMET, P. A.   [59] A table of real cubic fields, *JLMS*, **34**, 1959, 108–110; corr., *ibid.*, (2), **9**, 1974/5, 624.

GOGIA, S. K., LUTHAR, I. S.   [78a] Quadratic unramified extensions of $Q(\sqrt{d})$, *JRAM*, **298**, 1978, 108–111.
[78b] The Brauer–Siegel theorem for algebraic function fields, *JRAM*, **299/300**, 1978, 28–37.
[79] Real characters of the ideal class-group and the narrow ideal class-group of $Q(\sqrt{d})$, *Colloq. Math.*, **41**, 1979/80, 153–159.

GOLD, R.   [72] Γ-extensions of imaginary quadratic fields, *Pacific J. Math.*, **40**, 1972, 83–88; II, *JNT*, **8**, 1976, 415–419.
[74a] Examples of Iwasawa invariants, *AA*, **26**, 1974, 21–32; II, *ibid.*, 233–240.
[74b] The nontriviality of certain Z_l-extensions, *JNT*, **6**, 1974, 369–373.
[75] Genera in abelian extensions, *PAMS*, **47**, 1975, 25–28.
[76a] Genera in normal extensions, *Pacific J. Math.*, **63**, 1976, 397–400.
[76b] Z_3-invariants of real and imaginary quadratic fields, *JNT*, **8**, 1976, 420–423.
[77] The principal genus and Hasse's norm theorem, *Indiana Univ. Math. J.*, **26**, 1977, 183–189.
[81] Local class field theory via Lubin–Tate groups, *Indiana Univ. Math. J.*, **30**, 1981, 795–798.

GOLD, R., MADAN, M. L.   [78] Some applications of Abhyankar's lemma, *MN*, **82**, 1978, 115–119.

GOLDFELD, D. M.   (Goldfeld, M., Goldfeld, D.)
[73] A large sieve for a class of nonabelian *L*-functions, *Israel J. Math.*, **14**, 1973, 39–49.
[74] A simple proof of Siegel's theorem, *PAUS*, **71**, 1974, 1055.
[75] An asymptotic formula relating the Siegel zero and the class number of quadratic fields, *Ann. Scuola Norm. Sup. Pisa*, (4), **2**, 1975, 611–615.
[76] The class number of quadratic fields and the conjectures of Birch and Swinnerton-Dyer, *Ann. Scuola Norm. Sup. Pisa* (4), **3**, 1976, 623–663.
[77] The conjectures of Birch and Swinnerton-Dyer and the class numbers of quadratic fields, *Astérisque*, **41/2**, 1977, 219–227.

GOLDFELD, D. M., CHOWLA, S. (See S. Chowla, D. M. Goldfeld)
GOLDFELD, D. M., SCHINZEL, A. [75] On Siegel's zero, *Ann. Scuola Norm. Sup. Pisa*, (4), **2**, 1975, 571–583.
GOLDMAN, O. [64] On a special class of Dedekind domains, *Topology*, 3, 1964, suppl. 1, 113–118.
GOLDSTEIN, L. J. [68] Analogues of Artin's conjecture, *BAMS*, **74**, 1968, 517–519.
[70a] A generalization of the Siegel–Walfisz theorem, *TAMS*, **149**, 1970, 417–429.
[70b] Analogues of Artin's conjecture, *TAMS*, **149**, 1970, 431–442.
[71a] Density questions in algebraic number theory, *AMM*, **78**, 1971, 342–351.
[71b] A generalization of Stark's theorem, *JNT*, **3**, 1971, 323–346.
[71c] *Analytic Number Theory*, Prentice-Hall 1971.
[72a] On prime discriminants, *Nagoya Math. J.*, **45**, 1972, 119–127.
[72b] Relative imaginary quadratic fields of class number 1 or 2, *TAMS*, **165**, 1972, 353–364.
[72c] Relative imaginary quadratic fields of low class number, *BAMS*, **78**, 1972, 80–81.
[72d] Imaginary quadratic fields of class number 2, *JNT*, **4**, 1972, 286–301.
[73a] Some remarks on arithmetic density questions, *PSPM*, **24**, 1973, 103–110.
[73b] On a conjecture of Hecke concerning elementary class number formulas, *Manuscr. Math.*, **9**, 1973, 245–305.
[73c] On the class numbers of cyclotomic fields, *JNT*, **5**, 1973, 58–63.
[74] On a formula of Hecke, *Israel J. Math.*, **17**, 1974, 283–301.
GOLDSTEIN, L. J., RAZAR, M. [76] A generalization of Dirichlet's class number formula, *Duke Math. J.*, **43**, 1976, 349–358.
GOLDSTEIN, L., TORRE, P. DE LA [75] On a function analogous to $\log \eta(\tau)$, *Nagoya Math. J.*, **59**, 1975, 169–198.
GOLDSTINE, H. H., NEUMANN, J. V. (See J. v. Neumann, H. H. Goldstine)
GOLOD, E. S., SHAFAREVICH, I. R. [64] On the class-field tower, *IAN*, **28**, 1964, 261–272 (Russian).
GOLOMB, S. W. [78] Cyclotomic polynomials and factorization theorems, *AMM*, **85**, 1978, 734–737.
GOLUBEVA, E. P. [84] On the length of the period of a quadratic irrationality, *Mat. Sbornik*, **123**, 1984, 120–129 (Russian).
GOOD, A., CHANDRASEKHARAN, K. (See K. Chandrasekharan, A. Good)
GORDEEV, N. L. [77] Infinity of the number of relations in the Galois group of the maximal p-extension with bounded ramification of a local field, *DAN*, **233**, 1977, 1031–1034 (Russian).
GORDON, B., FEIN, B., SMITH, J. H. (See B. Fein, B. Gordon, J. H. Smith)
GORDON, B., MOHANTY, S. P. [77] On a theorem of Delaunay and some related results, *Pacific J. Math.*, **68**, 1977, 399–409.
GORDOVER, G., BARBAN, M. B. (See M. B. Barban, G. Gordover)
GORN, S., COHN, H. (See H. Cohn, S. Gorn)
GORSHKOV, D. S. [41] Cubic fields and symmetrical matrices, *DAN*, **31**, 1941, 842–844.
GOSS, D. [80] The algebraist's upper half-plane, *BAMS*, **2**, 1980, 391–415.
GÖTZKY, F. [28] Über eine zahlentheoretische Anwendung von Modulfunktionen zweier Veränderlichen, *MA*, **100**, 1928, 411–437.
GOUT, G. [70a] Sur les fonctions exponentielles des corps locaux, *CR*, **271**, 1970, A984–986.
[70b] Sur les fonctions exponentielles des corps locaux, *Publ. Dep. Math. Lyon*, **7**, 1970, 3, 105–119.
GRAHAM, S. W., GERTH, F. III (See F. Gerth, III, S. W. Graham)

GRAMM, S. L. [60] Expansion of complex numbers in generalized continued fractions in certain imaginary quadratic fields, *Tr. Inzh. Stroit. Inst. Rostov n/D.*, **10**, 1960, 46–60 (Russian).

GRANDET-HUGOT, M. (Grandet, M., see also M. Hugot) [61] Sur un ensemble d'entiers algébriques, *CR*, **252**, 1961, 1542–1543.

[62] Sur les dérivés d'un ensemble d'entiers algébriques, *CR*, **254**, 1962, 2905–2906.

[65a] Ensembles fermés d'entiers algébriques, *ASENS*, (3), **82**, 1965, 1–35.

[65b] Étude de certaines suites $[\lambda \alpha^n]$ dans les I-adèles, *CR*, **261**, 1965, 4943–4945.

[66] Étude de certaines suites $\{\lambda \alpha^n\}$ dans les adèles, *ASENS*, (3), **83**, 1966, 171–185.

[70] P.-V. éléments dans un corps de nombres algébriques, *Sém. DPP*, **12**, 1970/71, exp. 16.

[72] P. V. éléments dans un corps de nombres algébriques, *AA*, **20**, 1972, 203–214.

[75a] Équirépartition dans les adèles, *CR*, **280**, 1975, 873–876.

[75b] Quelques résultats concernant l'équirépartition dans l'anneau des adèles d'un corps de nombres algébriques, *Bull. Sci. Math.*, (2), **99**, 1975, 91–111; corr., *ibid.*, 243–247.

GRANDJOT, K. [24] Über die Irreduzibilität der Kreisteilungsgleichung, *MZ*, **19**, 1924, 128–129.

GRANT, H. S. [34] Concerning powers of certain classes of ideals in a cyclotomic realm which give the principal class, *AnM*, **35**, 1934, 220–238.

GRAS, G. [72a] Remarques sur la conjecture de Leopoldt, *CR*, **274**, 1972, 377–380.

[72b] Sur le groupe des classes des extensions cycliques de degré premier l, *CR*, **274**, 1972, A1145–1148.

[72c] Extensions abéliennes non ramifiées de degré premier d'un corps quadratique, *Bull. Soc. Math. France*, **100**, 1972, 177–193.

[73a] Sur les l-classes d'idéaux dans les extensions cycliques relatives de degré premier l, *CR*, **276**, 1973, A507–510.

[73b] Sur les l-classes d'idéaux dans les extensions cycliques relatives de degré premier l, *Ann. Inst. Fourier*, **23**, 1973, 3, 1–48.

[74a] Problèmes relatifs aux l-classes d'idéaux dans les extensions cycliques relatives de degré premier l, *Bull. Soc. Math. France, Mém.* **37**, 1974, 91–100.

[74b] Sur les l-classes d'idéaux des extensions non galoisiennes de Q de degré premier impair l à clôture galoisienne diédrale de degré $2l$, *J. Math. Soc. Japan*, **26**, 1974, 677–685.

[75] Critère de parité du nombre de classes des extensions abéliennes réelles de Q de degré impair, *Bull. Soc. Math. France*, **103**, 1975, 177–190.

[77a] Étude d'invariants relatifs aux groupes des classes des corps abéliens, *Astérisque*, **41/2**, 1977, 35–53.

[77b] Classes d'idéaux des corps abéliens et nombres de Bernoulli généralisés, *Ann. Inst. Fourier*, **27**, 1977, 1, 1–66.

[78] Nombre de φ-classes invariantes, application aux classes des corps abéliens, *Bull. Soc. Math. France*, **106**, 1978, 337–364.

[79a] Sur l'annulation en 2 des classes relatives des corps abéliens, *C.R. Math. Rep. Acad. Sci. Canada*, **1**, 1979, 107–110.

[79b] Annulation du groupe des l-classes généralisées d'une extension abélienne réelle de degré premier à l, *Ann. Inst. Fourier*, **29**, 1979, 1, 15–32.

[79c] Sur la construction des fonctions L p-adiques abéliennes, *Sém. DPP*, **20**, 1978/9, exp. 22.

[82a] Groupe de Galois de la p-extension abélienne p-ramifiée maximale d'un corps de nombres, *JRAM*, **333**, 1982, 86–132.

[82b] Canonical divisibilities of values of p-adic L-functions, in: *Journées Arithmétiques 1980*, 291–299, Cambridge Univ. Press, 1982.

[82c] Logarithme p-adique, p-ramification abélienne et K_2, *Sem. Th. de Nombr. Bordeaux*, 1982/3, exp. 12.

[83a] Logarithme p-adique et groupes de Galois, *JRAM*, **343**, 1983, 64–80.

[83b] Sur les Z_2-extensions d'un corps quadratique imaginaire, *Ann. Inst. Fourier*, **33**, 1983, 4, 1–18.

[86] Remarks on K_2 of number fields, *JNT*, **23**, 1986, 322–331.

GRAS, G., GRAS, M.-N. [75] Signature des unités cyclotomiques et parité du nombre de classes des extensions cycliques de Q de degré premier impair, *Ann. Inst. Fourier*, **25**, 1975, 1, 1–22.

[79] Calcul du nombre de classes et des unités des extensions abéliennes réelles de Q, *Bull. Sci. Math.*, (2), **101**, 1979, 97–129.

GRAS, M. N. (See also M. N. Montouchet) [73] Sur les corps cubiques cycliques dont l'anneau des entiers est monogène, *Ann. Univ. Besançon*, (3), 1973, fasc. 6, 26 pp.

[74] Sur les corps cubiques cycliques dont l'anneau des entiers est monogène, *CR*, **278**, 1974, 59–62.

[75] Méthodes et algorithmes pour le calcul numérique du nombre de classes et des unités des extensions cubiques cycliques de Q, *JRAM*, **277**, 1975, 89–116.

[77] Calcul de nombre de classes par dévissage des unités cyclotomiques, *Bull. Soc. Math. France*, Mem. **49/50**, 1977, 109–112.

[79] Classes et unités des extensions cycliques réelles de degré 4 de Q, *Ann. Inst. Fourier*, **29**, 1979, 1, 107–124.

[80] Note à propos d'une conjecture de H. J. Godwin sur les unités des corps cubiques, *Ann. Inst. Fourier*, **30**, 1980, 4, 1–6.

[81] Z-bases d'entiers 1, ϑ, ϑ^2, ϑ^3 dans les extensions cycliques de degré 4 de Q, *Publ. Math. Fac. Sci. Besançon, Th. de Nombres*, 1979/80 et 1980/81.

[83] Bases d'entiers dans les extensions cycliques de degré 4 de Q, *Sem. Th. de Nombr. Bordeaux*, 1982/3, exp. 11.

[86] Non monogénéité de l'anneau des entiers des extensions cycliques de Q de degré premier $l \geq 5$, *JNT*, **23**, 1986, 347–353.

GRAS, M. N., GRAS, G. (See G. Gras, M. N. Gras)

GRAS, M. N., MOSER, N., PAYAN, J. J. [73] Approximation algorithmique du groupe des classes de certains corps cubiques cycliques, *AA*, **23**, 1973, 295–300.

GRAVE, D. A. [24a] On fundamentals of the theory of ideal numbers, *Mat. Sbornik*, **32**, 1924, 134–151 (Russian).

[24b] On decomposition of prime numbers into ideal factors, *Mat. Sbornik*, **32**, 1924, 542–561 (Russian).

GRAY, J. [58] *Diagonal forms of prime degrees*, Ph. D. Thesis, Univ. Notre Dame, 1958.

GREBENYUK, D. G. [58] On the theory of algebraic integers depending on a root of an irreducible equation of fourth degree, *IAN Uzbek. SSR*, 1958, nr. 6, 27–47 (Russian).

GREENBERG, M. J. [69] *Lectures on Forms in Many Variables*, Benjamin, 1969.

[74] An elementary proof of the Kronecker–Weber theorem, *AMM*, **81**, 1974, 601–607; corr., *ibid.*, **82**, 1975, 803.

GREENBERG, R. [73a] A generalization of Kummer's criterion, *Invent. Math.*, **21**, 1973, 247–254.

[73b] The Iwasawa invariants of Γ-extensions of a fixed number field, *AJM*, **95**, 1973, 204–214.

[75] On p-adic L-functions and cyclotomic fields, Nagoya Math. J., **56**, 1975, 61–77; II, ibid., **67**, 1977, 139–158.
[76] On the Iwasawa invariants of totally real number fields, AJM, **98**, 1976, 263–284.
[78a] On 2-adic L-functions and cyclotomic invariants, MZ, **159**, 1978, 37–45.
[78b] A note on K_2 and the theory of Z_p-extensions, AJM, **100**, 1978, 1235–1245.
[78c] On the structure of certain Galois groups, Invent. Math., **47**, 1978, 85–99.
GREENBERG, R., FERRERO, B. (See B. Ferrero, R. Greenberg)
GREENLEAF, N. [65] Irreducible subvarieties and rational points, AJM, **87**, 1965, 25–31.
GREITER, G. [78] A simple proof of a theorem of Kronecker, AMM, **85**, 1978, 756–757.
[80] Explizite Formeln für Einheiten algebraischer Zahlkörper und eine Familie irreduzibler Polynome über Funktionenkörpern, Hbg, **50**, 1980, 157–161.
GRELL, H. [27] Zur Theorie der Ordnungen in algebraischen Zahl- und Funktionenkörpern, MA, **97**, 1927, 524–558.
[35] Verzweigungstheorie in allgemeinen Ordnungen algebraischer Zahlkörper, MZ, **40**, 1935, 629–657.
[36] Über die Gültigkeit der gewöhnlicher Idealtheorie in endlichen algebraischen Erweiterungen erster und zweiter Art, MZ, **40**, 1936, 503–505.
GRIFFIN, M. P. [74] Multiplication rings via their total quotient rings, CJM, **26**, 1974, 430–449.
GRIMM, G. [32] Über die reellen Nullstellen Dirichletscher L-Reihen, Diss., Zürich 1932.
GRISHIN, A. V. [73] The multiplicative group of a field, Uspekhi Mat. Nauk, **28**, 1973, 6, 201–202 (Russian).
GRÖLZ, W. [69] Primteiler von Polynomen, MA, **181**, 1969, 134–136.
GRONWALL, T. H. [13] Sur les séries de Dirichlet correspondant a des caractères complexes, Rend. Circ. Mat. Palermo, **35**, 1913, 145–159.
GROSS, B. H. [80] Arithmetic on Elliptic Curves with Complex Multiplication, LN 776, Springer, 1980.
[81] P-adic L-series at $s = 0$, J. Fac. Sci. Univ. Tokyo, IA, **28**, 1981, 979–994.
GROSS, B. H., FEDERER, L. J. (See L. J. Federer, B. H. Gross)
GROSS, B., ZAGIER, D. [83] Points de Heegner et dérivées de fonctions L, CR, **297**, 1983, 85–87.
[86] Heegner points and derivatives of L-series, Invent. Math. **84**, 1986, 225–320.
GROSSMAN, E. H. [74a] On the prime ideal divisors of $a^n - b^n$, Pacific J. Math., **54**, 1974, 73–83.
[74b] Units and discriminants of algebraic number fields, Comm. Pure Appl. Math., **27**, 1974, 741–747.
[76] On the solutions of diophantine equations in units, AA, **30**, 1976, 137–143.
[77] Sums of roots of unity in cyclotomic fields, JNT, **9**, 1977, 321–329.
GROSSWALD, E. [63] Negative discriminants of binary quadratic forms with one class in each genus, AA, **8**, 1963, 295–306.
[66] L-functions and quadratic fields with unique factorization, Duke Math. J., **33**, 1966, 169–185.
[73] Relations between the values of zeta and L-functions at integral arguments, AA, **24**, 1973, 369–378.
GROSSWALD, E., BATEMAN, P. T. (See P. T. Bateman, E. Grosswald)
GROTZ, W. [79] Mittelwert der Eulerschen φ-Funktion und des Quadrates der Dirichletschen Teilerfunktion in algebraischen Zahlkörpern, Monatsh. Math., **88**, 1979, 219–228.
[80] Einige Anwendungen der Siegelschen Summenformel, AA, **38**, 1980/1, 69–95.

GRUBE, F. [74] Ueber einige Euler'sche Sätze aus der Theorie der quadratischen Formen, *Zeitschr. Math. Phys.* (5), **19**, 1874, 492–519.

GRUNWALD, W. [33] Ein allgemeines Existenztheorem für algebraische Zahlkörper, *JRAM*, **169**, 1933, 103–107.

GUAN, CHI-WEN [63] On the structure of multiplicative group of discretely valued complete fields, *Sci. Sinica*, **12**, 1963, 1079–1103.

GUDIEV, A. H. [57] The determination of the class-number of ideals in algebraic fields of degree 4, *Uchen. Zap. Sev.-Oset. Ped. Inst.*, **21**, 1957, 1, 154–180 (Russian).

GUÉHO, M. F. (See also M. F. Vignéras) [72a] Corps de quaternions et fonctions zêta au point -1, *CR*, **274**, 1972, A296–298.

[72b] *Corps de quaternions sur un corps de nombres algébriques*, Thése, Bordeaux 1972.

[74a] Le théorème d'Eichler sur le nombre de classes d'idéaux d'un corps de quaternions totalement défini et la mesure de Tamagawa, *Bull. Soc. Math. France*, *Mém.* **37**, 1974, 107–114.

[74b] Sur les corps de quaternions, *Publ. M. Univ. Bordeaux*, 1973/4, 1, 35–56.

GUERRIER, W. J. [68] The factorization of the cyclotomic polynomial mod p, *AMM*, **75**, 1968, 46.

GUNDLACH, K.-B. [65] Die Bestimmung der Funktionen zu einigen Hilbertschen Modulgruppen, *JRAM*, **220**, 1965, 109–153.

[72] *Einführung in die Zahlentheorie*, Mannheim–Wien–Zürich 1972.

[73] Die Berechnung von Zetafunktionen mit Vorzeichencharakter an der Stelle 1, *AA*, **24**, 1973, 201–221.

GUNJI, H., MCQUILLAN, D. L. [70] On a class of ideals in an algebraic number field, *JNT*, **2**, 1970, 207–222.

[75] On rings with a certain divisibility property, *Michigan Math. J.*, **22**, 1975, 289–299.

[78] Polynomials with integral values, *Proc. Roy. Irish Acad.*, *A*, **78**, 1978, 1–7.

GUPTA, H. [42] On the class-numbers of binary quadratic forms, *Rev. Univ. Tucumán*, *A*, **3**, 1942, 283–299.

GURAK, S. [77a] Consecutive integers in an algebraic number field, *JRAM*, **290**, 1977, 168–179.

[77b] Ideal-theoretic description of the relative genus field, *JRAM*, **296**, 1977, 119–124.

[78a] On the Hasse norm principle, *JRAM*, **299/300**, 1978, 16–27.

[78b] The Hasse norm principle in non-abelian extensions, *JRAM*, **303/4**, 1978, 314–318.

[80] The Hasse norm principle in a compositum of radical extensions, *JLMS*, (2), **22**, 1980, 385–397.

GUREVICH, M. M. [71] On the determination of L-series by their functional equations, *Mat. Sbornik*, **85**, 1971, 538–552 (Russian).

GUT, M. [29] Die Zeta-Funktion, die Klassenzahl und die Kroneckersche Grenzformel eines beliebigen Kreiskörpers, *Comment. Math. Helv.*, **1**, 1929, 160–226.

[32] Über die Primidealzerlegung in gewisser relativ-ikosaedrischen Zahlkörpern, *Comment. Math. Helv.*, **4**, 1932, 219–229.

[33] Weitere Untersuchungen über die Primidealzerlegung in gewissen relativ-ikosaedrischen Zahlkörpern, *Comment. Math. Helv.*, **6**, 1933, 47–75.

[34a] Über die Primideale in Wurzelkörper einer Gleichung, *Comment. Math. Helv.*, **6**, 1934, 185–191.

[34b] Über die Gradteilerzerlegung in gewissen relativ-ikosaedrischen Zahlkörpern, *Comment. Math. Helv.*, **7**, 1934, 103–130.

[37] Über Erweiterungen von unendlichen algebraischen Zahlkörpern, *Comment. Math. Helv.*, **9**, 1937, 136–155.

[39] Folgen von Dedekindschen Zetafunktionen, *Monatsh. M.-Phys.*, **48**, 1939, 153-160.
[40] Mittel aus Dirichlet-Reihen mit reellen Restcharakteren, *Vierteljschr. Naturforsch. Ges. Zürich*, **85**, Beibl. **32**, 1940, 214-224.
[43a] Zur Theorie der Strahlklassenkörper der quadratisch reellen Zahlkörper, *Comment. Math. Helv.*, **15**, 1943, 37-59.
[43b] Zur Theorie der Klassenkörper der Kreiskörper, insbesondere der Strahlklassenkörper der quadratisch imaginären Zahlkörper, *Comment. Math. Helv.*, **15**, 1943, 81-119.
[51a] Eulersche Zahlen und Klassenanzahl des Körpers der $4l$-ten Einheitswurzeln, *Comment. Math. Helv.*, **25**, 1951, 43-63.
[51b] Kubische Klassenkörper über quadratisch-imaginären Grundkörpern, *Nieuw. Arch. Wisk.*, (2), **23**, 1951, 185-189.
[54] Relativquadratische Zahlkörper, deren Klassenzahl durch eine vorgegebene ungerade Primzahl teilbar ist, *Comment. Math. Helv.*, **28**, 1954, 270-277.
[63] Abschätzungen für die Klassenzahlen der quadratischen Körper, *AA*, **8**, 1963, 113-122.
[73] Erweiterungskörper von Primzahlgrad mit durch diese Primzahl teilbarer Klassenzahl, *Enseign. Math.*, **19**, 1973, 119-123.

GUT, M., STÜNZI, M. [66] Kongruenzen zwischen Koeffizienten trigonometrischen Reihen und Klassenzahlen quadratisch imaginärer Körper, *Comment. Math. Helv.*, **41**, 1966/7, 287-302.

GÜTING, R. [77] Positive inverse Einheiten in komplexen kubischen Zahlkörpern, *AA*, **34**, 1977/8, 1-7.

GUY, M. J. T., CASSELS, J. W. S. (See J. W. S. Cassels, M. J. T. Guy)

GYÖRY, K. [73] Sur les polynômes à coefficients entiérs et de discriminant donné, *AA*, **23**, 1973, 419-426; II, *Publ. Math. (Debrecen)*, **21**, 1974, 125-144; III, *ibid.*, **23**, 1976, 141-165; IV, *ibid.*, **25**, 1978, 155-167; V, *Acta Math. Acad. Sci. Hungar.*, **32**, 1978, 175-190.
[75] Sur une classe des corps de nombres algébriques et ses applications, *Publ. Math. (Debrecen)*, **22**, 1975, 151-175.
[79a] Corps de nombres algébriques d'anneau d'entiers monogène, *Sém. DPP*, **20**, 1978/79, fasc. 2, exp. 26.
[79b] On the number of solutions of linear equations in units of an algebraic number field, *Comment. Math. Helv.*, **54**, 1979, 583-600.
[80a] Explicit upper bounds for the solutions of some diophantine equations, *Ann. Acad. Sci. Fenn.*, *AI*, **5**, 1980, 3-12.
[80b] On certain graphs composed of algebraic integers of a number field and their applications, I, *Publ. Math. (Debrecen)*, **27**, 1980, 229-242.
[81a] On S-integral solutions of norm form, discriminant form and index form equations, *Stud. Sci. Math. Hungar.*, **16**, 1981, 149-161.
[81b] On discriminants and indices of integers of an algebraic number field, *JRAM*, **324**, 1981, 114-126.
[83] Bounds for the solutions of norm form, discriminant form and index form equations in finitely generated integral domains, *Acta Math. Acad. Sci. Hungar.*, **42**, 1983, 45-80.
[84] Effective finiteness theorems for polynomials with given discriminant and integral elements with given discriminants over finitely [generated] domains, *JRAM*, **346**, 1984, 54-100.

GYÖRY, K., LEAHEY, W. [77] A note on Hilbert class fields of algebraic number fields, *Acta Math. Acad. Sci. Hungar.*, **29**, 1977, 241-254.

GYÖRY, K., PAPP, Z. Z. [77] On discriminant form and index form equations, *Stud. Sci. Math. Hungar.*, **12**, 1977, 47-60.

GYÖRY, K., PETHÖ, A. [75] Sur la distribution des solutions des équations du type "normeforme", *Acta Math. Acad. Sci. Hungar.*, **26**, 1975, 135-142.
[77] Über die Verteilung der Lösungen von Normformen Gleichungen, II, *AA*, **32**, 1977, 349-363; III, *ibid.*, **37**, 1980, 143-165.

HABERLAND, K. [74] Über die Anzahl der Erweiterungen eines algebraischen Zahlkörpers mit einer gegebenen abelschen Gruppe als Galoisgruppe, *AA*, **26**, 1974, 153-158.
[78] *Galois Cohomology of Algebraic Number Fields*, Berlin 1978.

HAFNER, J. L. [83] The distribution and average order of the coefficients of Dedekind ζ functions, *JNT*, **17**, 1983, 183-190.

HAFNER, P. [68] Automorphismen von binären quadratischen Formen, *Elem. Math.*, **23**, 1968, 25-30.

HAGGENMÜLLER, R. [82] Signaturen von Einheiten und unverzweigte quadratische Erweiterungen von total-reeller Zahlkörper, *Arch. Math.*, **39**, 1982, 312-321.

HALL, M. [37] Indices in cubic fields, *BAMS*, **43**, 1937, 104-108.

HALL, N. A. [37] Binary quadratic discriminants with a single class of reduced forms in each genus, *PAUS*, **23**, 1937, 414-415.
[39] Binary quadratic discriminants with a single class of forms in each genus, *MZ*, **44**, 1939, 85-90.

HALTER-KOCH, F. [67] Arithmetische Kennzeichnung der Spur des Einsdivisors, *JRAM*, **228**, 1967, 217-219.
[71a] Algebraische Zahlen mit Konjugierten auf dem Einheitskreis, *Arch. Math.*, **22**, 1971, 161-164.
[71b] Arithmetische Theorie der Normalkörper vom 2-Potenzgrad mit Diedergruppe, *JNT*, **3**, 1971, 412-443.
[71c] Geschlechtertheorie der Ringklassenkörper, *JRAM*, **250**, 1971, 107-108.
[72a] Abgeschlossene Mengen algebraischer Zahlen, *Hbg.* **38**, 1972, 65-79.
[72b] Einseinheitengruppen und prime Restklassengruppen in quadratischen Zahlkörpern, *JNT*, **4**, 1972, 70-77.
[72c] Ein Satz über die Geschlechter relativ-zyklischer Zahlkörper von Primzahlgrad und seine Anwendung auf biquadratisch-bizyklische Körper, *JNT*, **4**, 1972, 144-156.
[75] Unabhängige Einheitensysteme für eine allgemeine Klasse algebraischer Zahlkörper, *Hbg*, **43**, 1975, 85-91.
[76] Eine Bemerkung über kubische Einheiten, *Arch. Math.*, **27**, 1976, 593-595.
[77] Einheiten und Divisorenklassen in Galois'schen algebraischen Zahlkörpern mit Diedergruppe der Ordnung 2*l* für eine ungerade Primzahl *l*, *AA*, **33**, 1977, 353-364.
[78a] Die Struktur der Einheitengruppe für eine Klasse metazyklischer Erweiterungen algebraischer Zahlkörper, *JRAM*, **301**, 1978, 147-160.
[78b] Eine allgemeine Geschlechtertheorie und ihre Anwendung auf Teilbarkeitsaussagen für Klassenzahlen algebraischer Zahlkörper, *MA*, **233**, 1978, 55-63.
[78c] Über den *p*-Rang der Klassengruppe des Kompositums algebraischer Zahlkörper, *MA*, **238**, 1978, 119-122.
[79] Zur Geschlechtertheorie algebraischer Zahlkörper, *Arch. Math.*, **31**, 1978/9, 137-142.
[80] Über Radikalerweiterungen, *AA*, **36**, 1980, 43-58.
[81] Grosse Faktoren in der Klassengruppe algebraischer Zahlkörper, *AA*, **39**, 1981, 33-47.
[82] Metrische Theorie der Einheiten algebraischer Zahlkörper, *Mitt. Math. Ges. Hamburg*, **11**, 1982, 131-141.
[83] Factorization of algebraic integers, *Ber. math.-stat. Sekt. Forsch. Graz*, **191**, 1983, 24 pp.
[84a] Über den 4-Rang der Klassengruppe quadratischer Zahlkörper, *JNT*, **19**, 1984, 219-227.

[84b] On the factorization of algebraic integers into irreducibles, in: *Topics in Classical Number Theory*, 699–707, North-Holland, 1984.
[86] Binäre quadratische Formen und rationale Zerlegungsgesetze, I, *JNT*, **22**, 1986, 249–270.
HALTER-KOCH, F., BRUNOTTE, H.   (See H. Brunotte, F. Halter-Koch)
HALTER-KOCH, F., LORENZ, F.   [81] Ein Normalbasissatz für Einheiten algebraischer Zahlkörper, *MA*, **257**, 1981, 335–339.
HALTER-KOCH, F., MOSER, N.   [78] Sur le nombre de classes de certaines extensions métacycliques sur Q ou sur un corps quadratique imaginaire, *J. Math. Soc. Japan*, **30**, 1978, 237–248.
HALTER-KOCH, F., STENDER, H. J.   [74] Unabhängige Einheiten für die Körper $K = Q(\sqrt[n]{D^n+d})$ mit $d|D^n$, *Hbg*, **42**, 1974, 33–40.
HAMAMURA, M.   [81] On absolute class fields of certain algebraic number fields, *Nat. Sci. Rep. Ochanomizu Univ.*, **32**, 1981, 23–34.
HAMBURGER, H.   [21] Über die Riemannsche Funktionalgleichung der ζ-Funktion, *MZ*, **10**, 1921, 240–254; *ibid.*, **11**, 1921, 224–245; *ibid.*, **13**, 1922, 283–311.
HANCOCK, H.   [25] Trigonometric realms of rationality, *Rend. Circ. Mat. Palermo*, **49**, 1925, 263–276.
HANEKE, W.   [69] Darstellung von Primzahlen durch quadratische Formen, *MZ*, **110**, 1969, 10–14.
[73] Über die reellen Nullstellen der Dirichletschen L-Reihen, *AA*, **22**, 1973, 391–421; corr., *ibid.*, **31**, 1976, 99–100.
HANLY, V., MANN, H. B.   [58] A note to the paper "On integral bases" by H. B. Mann, *PAMS*, **9**, 1958, 173–176.
HAO, F. H., PARRY, C. J.   [84] The Fermat equation over quadratic fields, *JNT*, **19**, 1984, 115–130.
HAPPLE, W., REHM, H. P.   (See H. P. Rehm, W. Happle)
HARDMAN, N. R., JORDAN, J. H.   [67] A minimum problem connected with complete residue systems in the Gaussian integers, *AMM*, **74**, 1967, 559–561.
[69] The distribution of quadratic residues in fields of order p^2, *Math. Mag.*, **42**, 1969, 12–17.
HARDY, G. H.   [14] Sur les zéros de la fonction $\zeta(s)$ de Riemann, *CR*, **158**, 1914, 1012–1014.
[19] A problem of diophantine approximation, *J. Indian Math. Soc.*, **11**, 1919, 162–166.
HARDY, G. H., LITTLEWOOD, J. E.   [21] The zeros of Riemann's zeta function on the critical line, *MZ*, **10**, 1921, 283–317.
[23] The approximate functional equation in the theory of zeta-function with applications to the divisor problems of Dirichlet and Piltz, *PLMS*, (2), **21**, 1923, 39–74.
[29] The approximate functional equation for $\zeta(s)$ and $\zeta^2(s)$, *PLMS*, (2), **29**, 1929, 81–97.
HARDY, G. H., WRIGHT, E. M.   [60] *An Introduction to the Theory of Numbers*, Oxford 1938, 4th ed. 1960.
HARDY, J.   [68] Sums of two squares in a quadratic ring, *AA*, **14**, 1968, 357–369.
HARRIS, B., SEGAL, G.   [75] K_i groups of rings of algebraic integers, *AnM*, **101**, 1975, 20–33.
HARTUNG, P.   [74a] Proof of the existence of infinitely many imaginary quadratic fields whose class number is not divisible by 3, *JNT*, **6**, 1974, 276–278.
[74b] Explicit construction of a class of infinitely many imaginary quadratic fields whose class number is divisible by 3, *JNT*, **6**, 1974, 279–281.
HARTUNG, P., CHOWLA, S.   (See S. Chowla, P. Hartung)
HARTUNG, P., CHOWLA, S., DE LEON, M. J.   (See S. Chowla, P. Hartung, M. J. de Leon)

HASELGROVE, C. B. [51] Some theorems in the analytic theory of numbers, *JLMS*, **26**, 1951, 273–277.

HASSE, H. [23a] Über die Darstellbarkeit von Zahlen durch quadratische Formen im Körper der rationalen Zahlen, *JRAM*, **152**, 1923, 129–148.

[23b] Über die Äquivalenz quadratischer Formen im Körper der rationalen Zahlen, *JRAM*, **152**, 1923, 205–224.

[24a] Darstellbarkeit von Zahlen durch quadratische Formen in einem beliebigen algebraischen Zahlkörper, *JRAM*, **153**, 1924, 113–130.

[24b] Äquivalenz quadratischer Formen in einem beliebigen algebraischen Zahlkörper, *JRAM*, **153**, 1924, 158–162.

[26a] Zwei Existenztheoreme über algebraische Zahlkörper, *MA*, **95**, 1926, 229–238.

[26b] Ein weiteres Existenztheorem in der Theorie der algebraischer Zahlkörper, *MZ*, **24**, 1926, 149–160.

[26c] Bericht über neuere Untersuchungen und Probleme aus der Theorie der algebraischen Zahlkörper, *Jahresber. Deutsch. Math. Verein.*, **35**, 1926, 1–55; *ibid.*, **36**, 1927, 233–311; *ibid.*, VI Erg. Bd., 1930; reprinted Wien 1965, 1970.

[27] Neue Begründung der komplexen Multiplikation, *JRAM*, **157**, 1927, 115–139; *ibid.*, **165**, 1931, 64–88.

[28] Über eindeutige Zerlegung in Primelemente oder in Primhauptideale in Integritätsbereichen, *JRAM*, **159**, 1928, 3–12.

[30a] Ein Satz über relativ-Galoische Zahlkörper und seine Anwendung auf relativ-Abelsche Zahlkörper, *MZ*, **31**, 1930, 559–564.

[30b] Arithmetische Theorie der kubischen Zahlkörper auf klassenkörpertheoretischer Grundlage, *MZ*, **31**, 1930, 565–582.

[30c] Die Normenresttheorie relativ-Abelscher Zahlkörper als Klassenkörpertheorie im Kleinen, *JRAM*, **162**, 1930, 145–154.

[30d] Führer, Diskriminante und Verzweigungskörper relativ-Abelscher Zahlkörper, *JRAM*, **162**, 1930, 169–184.

[31a] Zum Hauptidealsatz der komplexen Multiplikation, *Monatsh. M.-Phys.*, **38**, 1931, 315–322.

[31b] Ein Satz über die Ringklassenkörper der komplexen Multiplikation, *Monatsh. M.-Phys.*, **38**, 1931, 323–330.

[31c] Das Zerlegungsgesetz für die Teiler des Moduls in den Ringklassenkörper der komplexen Multiplikation, *Monatsh. M.-Phys.*, **38**, 1931, 331–334.

[31d] Beweis eines Satzes und Widerlegung einer Vermutung über das allgemeine Normenrestsymbol, *GN*, 1931, 64–69.

[32] Zwei Bemerkungen zu der Arbeit "Zur Arithmetik der Polynome" von U. Wegner in den Math. Ann. 105, S. 628–631, *MA*, **106**, 1932, 455–456.

[33] Explizite Konstruktion zyklischer Klassenkörper, *MA*, **109**, 1933, 191–195.

[34] Normenresttheorie Galoisscher Zahlkörper mit Anwendungen auf Führer und Diskriminante algebraischer Zahlkörper, *J. Fac. Sci. Univ. Tokyo*, **2**, 1934, 477–498.

[37] Über die Diskriminante auflösbarer Körper mit Primzahlgrad, *JRAM*, **176**, 1937, 12–17.

[40] Produktformeln für verallgemeinerte Gausssche Summen und ihre Anwendungen auf die Klassenzahlformel für reelle quadratische Zahlkörper, *MZ*, **46**, 1940, 303–314.

[47] Invariante Kennzeichnung relativabelscher Zahlkörper mit vorgegebener Galoisgruppe über einem Teilkörper des Grundkörpers, *Abh. Deutsch. Akad. Wiss.*, 1947, 8, 5–56.

[48a] Arithmetische Bestimmung von Grundeinheit und Klassenzahl in zyklischen kubischen und biquadratischen Zahlkörpern, *Abh. Deutsch. Akad. Wiss.*, 1948, 2, 3–95.

[48b] Die Einheitengruppe in einem total-reellen nichtzyklischen kubischen Zahlkörper und in zugehörigen bikubischen Normalkörper, *Arch. Math.*, **1**, 1948, 42–46; *Miscel. Acad. Berol.*, **1**, 1950, 1–24.

[49] *Zahlentheorie*, Akademie-Verlag, 1949, 3rd ed. 1969.

[50a] *Vorlesungen über Zahlentheorie*, Springer, 1950, 2nd ed. 1964.

[50b] Zum Existenzsatz von Grunwald in der Klassenkörpertheorie, *JRAM*, **188**, 1950, 40–64.

[51a] Allgemeine Theorie der Gaussschen Summen in algebraischen Zahlkörpern, *Abh. Deutsch. Akad. Wiss.*, 1951, 1, 1–23.

[51b] Sopra la formula analitica per il numero delle classi su corpi quadratici immaginari e reali, *Rend. Mat. Appl.*, (5), **10**, 1951, 84–95.

[51c] Zur Geschlechtertheorie in quadratischen Zahlkörpern, *J. Math. Soc. Japan*, 3, 1951, 45–51.

[51d] Über das Problem der Primzerlegung in Galoischen Zahlkörpern, *S.-B. Berlin. Math. Ges.*, 1951/2, 8–27.

[52a] *Über die Klassenzahl Abelscher Zahlkörper*, Akademie-Verlag, 1952.

[52b] Gaussche Summen zu Normalkörpern über endlich-algebraischen Zahlkörpern, *Abh. Deutsch. Akad. Wiss.*, 1952, 1, 1–19.

[54a] Artinsche Führer, Artinsche L-Funktionen und Gausssche Summen über endlich-algebraischen Zahlkörpern, *Acta Salamant., Mat.*, **4**, 1954, 113 pp.

[54b] Zetafunktionen und L-Funktionen zu einem arithmetischen Funktionenkörper vom Fermatschen Typus, *Abh. Deutsch. Akad. Wiss.*, 1954, 4, 5–70.

[55] Die dyadische Einseinheitenoperatorengruppe zum Körper der 2^n-ten Einheitswurzeln nebst Anwendung auf die Klassenzahl seines grössten reellen Teilkörpers, *Rev. Fac. Sci. Univ. Istanbul*, **20**, 1955, 7–126.

[62a] Kurt Hensels entscheidender Anstoss zur Entdeckung des Lokal-Global Prinzips, *JRAM*, **209**, 1962, 3–4.

[62b] Sulla generalizzacione di Leopoldt dei numeri di Bernoulli e sua applicazione alla divisibilitá del numero delle classi nei corpi numerici abeliani, *Rend. Mat. Appl.*, (5), **21**, 1962, 9–27.

[64] Über den Klassenkörper zum quadratischen Zahlkörper mit der Diskriminante -47, *AA*, **9**, 1964, 419–434.

[65] Über mehrklassige, aber eingeschlechtige reell-quadratische Zahlkörper, *Elem. Math.*, **20**, 1965, 49–59.

[66] Vandiver's congruence for the relative class number of the p-th cyclotomic field, *J. Math. Anal. Appl.*, **15**, 1966, 87–90.

[67] *Vorlesungen über Klassenkörpertheorie*, Würzburg 1967.

[69a] A supplement to Leopoldt's theory of genera in abelian number fields, *JNT*, **1**, 1969, 4–7.

[69b] Über die Klassenzahl des Körpers $P(\sqrt{-2p})$ mit einer Primzahl $p \neq 2$, *JNT*, **1**, 1969, 231–234.

[69c] Über die Klassenzahl des Körpers $P(\sqrt{-p})$ mit einer Primzahl $p \equiv 1 \pmod{2^3}$, *Aequat. Math.*, 3, 1969, 165–169.

[69d] Eine Folgerung aus H.-W. Leopoldts Theorie der Geschlechter abelscher Zahlkörper, *MN*, **42**, 1969, 261–262.

[70a] Über die Teilbarkeit durch 2^3 der Klassenzahl imaginärquadratischer Zahlkörper mit genau zwei verschiedenen Diskriminantenprimteilern, *JRAM*, **241**, 1970, 1–6.

[70b] Über Teilbarkeit durch 2^3 der Klassenzahl der quadratischer Zahlkörper mit genau zwei verschiedenen Diskriminantenprimteilern, *MN*, **46**, 1970, 61–70.

[75] An algorithm for determining the structure of the 2-Sylow-subgroups of the divisor group of a quadratic number field, *Symposia Math.*, **15**, 341–352, Academic Press, 1975.
HASSE, H., ANKENY, N. C., CHOWLA, S. (See N. C. Ankeny, S. Chowla, H. Hasse)
HASSE, H., BERNSTEIN, L. (See L. Bernstein, H. Hasse)
HASSE, H., DAVENPORT, H. (See H. Davenport, H. Hasse)
HASSE, H., ELSNER, L. (See L. Elsner, H. Hasse)
HASSE, H., HENSEL, K. [23] Über die Normenreste eines relativ-zyklischen Körpers vom Primzahlgrad l nach einem Primteiler $\mathfrak{L}$ von l, *MA*, **90**, 1923, 262–278.
HASSE, H., LIANG, J. [69] Über den Klassenkörper zum quadratischen Zahlkörper mit der Diskriminante -47 (Fortsetzung), *AA*, **16**, 1969, 89–97.
HASSE, H., SCHMIDT, F. K. [33] Die Struktur diskret bewerteter Körper, *JRAM*, **170**, 1933, 4–63.
HASSE, H., SUETUNA, Z. [31] Ein allgemeines Teilerproblem der Idealtheorie, *J. Fac. Sci. Univ. Tokyo*, **2**, 1931, 133–154.
HASSE, H., TORNIER, E. [28] Über die Dichte der quadratfreien Zahlen und ähnliche Dichten in einem algebraischen Zahlkörper, *Leopoldina*, **3**, 1928, 9–16.
HATADA, K. [79] On the values at rational integers of the p-adic L functions, *J. Math. Soc. Japan*, **31**, 1979, 7–27.
HATTORI, A. [57] On Prüfer rings, *J. Math. Soc. Japan*, **9**, 1957, 381–385.
HAUSMAN, M., SHAPIRO, H. N. [76] Perfect ideals over Gaussian integers, *Comm. Pure Appl. Math.*, **29**, 1976, 323–341.
HAUSNER, A. [61] Algebraic number fields and the diophantine equation $m^n = n^m$, *AMM*, **68**, 1961, 856–861.
HAYASHI, H. [77] Note on the class numbers of the quadratic number fields $Q(\sqrt{-p})$, $Q(\sqrt{-2p})$ and $Q(\sqrt{2p})$ with a prime number $p \equiv 1 \pmod{2^2}$, *Mem. Fac. Gen. Ed. Kumamoto Univ.*, **13**, 1977, 1, 1–8.
[84] On elliptic units and class number of a certain dihedral extension of degree $2l$, *AA*, **44**, 1984, 35–45.
HAYASHI, M. [75] On lifting of the convolution rings associated with the algebraic number fields, *Mem. Osaka Inst. Techn.*, A, **19**, 1974/5, 143–147.
HAYES, D. R. [74] Explicit class field theory for rational function fields, *TAMS*, **189**, 1974, 77–91.
[79] Explicit class field theory in global function fields, in: *Studies in Algebra and Number Theory*, 173–217, New York 1979.
HAYS, J. H. [73] Reductions of ideals in commutative rings, *TAMS*, **177**, 1973, 51–63.
HAZEWINKEL, M. [69] *Abelian Extensions of Local Fields*, Groningen 1969.
[75] Local class field theory is easy, *Adv. Math.*, **18**, 1975, 148–181.
HAZLEWOOD, D. G. [75] On sums over Gaussian integers, *TAMS*, **209**, 1975, 295–309.
[77] On ideals having only small prime factors, *Rocky Mountain J. Math.*, **7**, 1977, 753–768.
HEATH-BROWN, D. R. [77] On the density of the zeros of the Dedekind zeta-function, *AA*, **33**, 1977, 169–181.
[79] On a paper of Baker and Schinzel, *AA*, **35**, 1979, 203–207.
[83] Cubic forms in ten variables, *PLMS*, (3), **47**, 1983, 225–257.
HEATH-BROWN, D. R., PATTERSON, S. J. [79] The distribution of Kummer sums at prime arguments, *JRAM*, **310**, 1979, 111–130.
HECKE, E. [10] Über nicht-reguläre Primzahlen und den Fermatschen Satz, *GN*, 1910, 420–424 = *Math. Werke*, 59–63, Göttingen 1959.

[12] Zur Theorie der Modulfunktionen von zwei Variablen und ihre Anwendung auf die Zahlentheorie, *MA*, **71**, 1912, 1–37 = *Math. Werke*, 21–57, Göttingen 1959.

[13] Über die Konstruktion relativ-abelscher Zahlkörper durch Modulfunktionen von zwei Variablen, *MA*, **74**, 1913, 465–510 = *Math. Werke*, 69–114, Göttingen 1959.

[17a] Über die Zetafunktion beliebiger algebraischer Zahlkörper, *GN*, 1917, 77–89 = *Math. Werke*, 159–171, Göttingen 1959.

[17b] Über eine neue Anwendung der Zetafunktion auf die Arithmetik der Zahlkörper, *GN*, 1917, 90–95 = *Math. Werke*, 172–177, Göttingen 1959.

[17c] Über die *L*-Funktionen und den Dirichletschen Primzahlsatz für einen beliebigen Zahlkörper, *GN*, 1917, 299–318 = *Math. Werke*, 178–197, Göttingen 1959.

[17d] Über die Kroneckersche Grenzformel für reelle quadratische Körper und die Klassenzahl relativ-abelscher Körper, *Verh. Naturforsch. Ges. Basel*, **28**, 1917, 363–372 = *Math. Werke*, 198–207, Göttingen 1959.

[18] Eine neue Art von Zetafunktionen und ihre Beziehungen zur Verteilung der Primzahlen, *MZ*, **1**, 1918, 357–376; II, *ibid.*, **6**, 1920, 11–51 = *Math. Werke*, 215–234, 249–289, Göttingen 1959.

[19] Reziprozitätsgesetz und Gausssche Summen in quadratischen Zahlkörpern, *GN*, 1919, 265–278 = *Math. Werke*, 235–248, Göttingen 1959.

[21a] Analytische Funktionen und algebraische Zahlen, *Hbg*, **1**, 1921, 102–126; II, *ibid.*, **3**, 1924, 213–236 = *Math. Werke*, 336–360, 381–404, Göttingen 1959.

[21b] Bestimmung der Klassenzahl einer neuen Reihe von algebraischen Zahlkörpern, *GN*, 1921, 1–23 = *Math. Werke*, 290–312, Göttingen 1959.

[23] *Vorlesungen über die Theorie der algebraischen Zahlen*, Leipzig 1923, 2nd ed. 1954, reprinted by Chelsea 1970.

[25] Darstellung der Klassenzahlen als Perioden von Integralen 3 Gattung aus dem Gebiet der elliptischen Modulfunktionen, *Hbg*, **4**, 1925, 211–223 = *Math. Werke*, 405–417, Göttingen 1959.

[30] Über das Verhalten der Integrale 1. Gattung bei Abbildungen, insbesondere in der Theorie der Elliptischen Modulfunktionen, *Hbg*, **8**, 1930, 271–281 = *Math. Werke*, 548–558, Göttingen 1959.

[36] Über die Bestimmung Dirichletscher Reihen durch ihre Funktionalgleichung, *MA*, **112**, 1936, 644–699 = *Math. Werke*, 591–626, Göttingen 1959.

[37] Über Dirichlet-Reihen mit Funktionalgleichung und ihre Nullstellen auf der Mittelgeraden, *S.-B. Bayer. Akad. Wiss.*, 1937, 73–95 = *Math. Werke*, 708–730, Göttingen 1959.

[39] Die Klassenzahl imaginär-quadratischer Körper in der Theorie der elliptischen Modulfunktionen, *Monatsh. M.-Phys.*, **48**, 1939, 75–83 = *Math. Werke*, 773–781, Göttingen 1959.

[44] Herleitung des Euler-Produktes der Zeta-funktion und einiger *L*-Reihen aus ihrer Funktionalgleichung, *MA*, **119**, 1944, 266–287.

HEEGNER, K. [52] Diophantische Analysis und Modulfunktionen, *MZ*, **56**, 1952, 227–253.

HEIDER, F. P. [80] Strahlknoten und Geschlechterkörper mod m, *JRAM*, **320**, 1980, 52–67.

[81] Zahlentheoretische Knoten unendlicher Erweiterungen, *Arch. Math.*, **37**, 1981, 341–352.

[84] Kapitulationsprobleme und Knotentheorie, *Manuscr. Math.*, **46**, 1984, 229–272.

HEIDER, F. P., SCHMITHALS, B. [82] Zur Kapitulation der Idealklassen in unverzweigten primzyklischen Erweiterungen, *JRAM*, **336**, 1982, 1–25.

HEILBRONN, H. [34] On the class-number in imaginary quadratic fields, *QJM*, **5**, 1934, 150–160.
[38a] On Euclid's algorithm in real quadratic fields, *PCPS*, **34**, 1938, 521–526.
[38b] On Dirichlet series which satisfy a certain functional equation, *QJM*, **9**, 1938, 194–195.
[50] On Euclid's algorithm in cubic self-conjugated fields, *PCPS*, **46**, 1950, 377–382.
[51] On Euclid's algorithm in cyclic fields, *CJM*, **3**, 1951, 257–268.
[72] On real simple zeros of Dedekind ζ-functions, *Proc. Number Theory Conf.*, Boulder 1972, 108–110.
[73] On real zeros of Dedekind ζ-functions, *CJM*, **25**, 1973, 870–873.
HEILBRONN, H., DAVENPORT, H. (See H. Davenport, H. Heilbronn)
HEILBRONN, H., LINFOOT, E. H. [34] On the imaginary quadratic corpora of class-number one, *QJM*, **5**, 1934, 293–301.
HEINZER, W., EAKIN, P. (See P. Eakin, W. Heinzer)
HELLER, A., REINER, I. [65] Grothendieck group of integral group rings, *Illinois J. Math.*, **9**, 1965, 349–360.
HEMER, O. [52] On the solvability of the Diophantine equation $ax^2+by^2+cz^2 = 0$ in imaginary Euclidean quadratic fields, *Arkiv Mat.*, **2**, 1952-54, 57–82.
HENDY, M. D. [74a] Applications of a continued fraction algorithm to some class number problems, *MC*, **28**, 1974, 267–277.
[74b] Prime quadratics associated with complex quadratic fields of class number two, *PAMS*, **43**, 1974, 253–260.
[75] The distribution of ideal class numbers of real quadratic fields, *MC*, **29**, 1975, 1129–1134; corr., *ibid.*, **30**, 1976, 679.
HENDY, M. D., JEANS, N. S. (See N. S. Jeans, M. D. Hendy)
HENNIART, G. [77] Représentations de degré 2 de Gal($\overline{Q_2}/Q_2$), *CR*, **284**, 1977, 1329–1332.
HENSEL, K. [84] *Arithmetische Untersuchungen über Discriminanten und ihre ausserwesentliche Theiler*, Diss., Berlin 1884.
[87] Untersuchungen der ganzen algebraischen Zahlen eines gegebenes Gattungsbereiches für einen beliebigen algebraischen Primdivisor, *JRAM*, **101**, 1887, 99–141.
[88] Ueber die Darstellung der Zahlen eines Gattungsbereiches für einen beliebigen Primdivisor, *JRAM*, **103**, 1888, 230–237.
[89] Ueber Gattungen, welche durch Composition aus zwei anderen Gattungen entstehen, *JRAM*, **105**, 1889, 329–344.
[93] Über Darstellung der Determinante eines Systems welches aus zwei anderen componiert ist, *AM*, **14**, 1893, 317–319.
[94a] Untersuchung der Fundamentalgleichung einer Gattung für eine reelle Primzahl als Modul und Bestimmung der Theiler ihrer Discriminante, *JRAM*, **113**, 1894, 61–83.
[94b] Arithmetische Untersuchungen über die gemeinsamen Discriminantentheiler einer Gattung, *JRAM*, **113**, 1894, 128–160.
[97a] Ueber die Bestimmung der Discriminante eines algebraischen Körpers, *GN*, 1897, 247–253.
[97b] Ueber die Fundamentalgleichung und die ausserwesentliche Discriminantentheiler eines algebraischen Körpers, *GN*, 1897, 254–260.
[97c] Über eine neue Begründung der Theorie der algebraischen Zahlen, *Jahresber. Deutsch. Math. Verein.*, **6**, 1897, 83–88.
[97d] Ueber die Fundamentaltheiler algebraischer Gattungsbereiche, *JRAM*, **117**, 1897, 333–345.

[97e] Ueber die Elementartheiler zweier Gattungen, von denen die eine unter der anderen enthalten ist, *JRAM*, **117**, 1897, 346–355.
[99] Ueber diejenigen algebraischen Körper, welche aus zwei anderen componiert sind, *JRAM*, **120**, 1899, 99–108.
[02] Ueber die Entwicklung der algebraischen Zahlen in Potenzreihen, *MA*, **55**, 1902, 305–336.
[04] Neue Grundlagen der Arithmetik, *JRAM*, **127**, 1904, 51–84.
[05a] Über eine neue Begründung der Theorie der algebraischen Zahlen, *JRAM*, **128**, 1905, 1–32.
[05b] Über die zu einem algebraischen Körper gehörigen Invarianten, *JRAM*, **129**, 1905, 68–85.
[07] Über die arithmetische Eigenschaften der Zahlen, *Jahresber. Deutsch. Math. Verein.*, **16**, 1907, 299–319; *ibid.*, 386–393; *ibid.*, 473–496.
[08] *Theorie der algebraischen Zahlen*, Leipzig–Berlin 1908.
[09] Über die zu einer algebraischen Gleichung gehörigen Auflösungskörper, *JRAM*, **136**, 1909, 183–209.
[13] *Zahlentheorie*, Berlin–Leipzig 1913.
[14a] Über die Grundlagen einer neuen Theorie der quadratischer Zahlkörper, *JRAM*, **144**, 1914, 57–70.
[14b] Die Exponentialdarstellung der Zahlen eines algebraischen Zahlkörpers für den Bereich eines Primdivisors, *Festschr. H. A. Schwarz*, Berlin 1914.
[15] Untersuchungen der Zahlen eines algebraischen Körpers für den Bereich eines beliebigen Primteilers, *JRAM*, **145**, 1915, 92–113.
[16a] Die multiplikative Darstellung der algebraischen Zahlen für den Bereich eines beliebigen Primteilers, *JRAM*, **146**, 1916, 189–215.
[16b] Untersuchung der Zahlen eines algebraischen Körpers für eine beliebige Primteilerpotenz als Modul, *JRAM*, **146**, 1916, 216–228.
[17] Allgemeine Theorie der Kongruenzklassgruppen und ihrer Invarianten in algebraischen Körpern, *JRAM*, **147**, 1917, 1–15.
[18] Eine neue Theorie der algebraischen Zahlen, *MZ*, **2**, 1918, 433–452.
[21a] Über die Zerlegung der Primteiler in relativ cyklischen Körpern, nebst einer Anwendung auf die Kummerschen Körper, *JRAM*, **151**, 1921, 112–120.
[21b] Die Zerlegung der Primteiler eines beliebigen Zahlkörpers in einem auflösbaren Oberkörper, *JRAM*, **151**, 1921, 200–209.
[21c] Zur multiplikativen Darstellung der algebraischen Zahlen für den Bereich eines Primteilers, *JRAM*, **151**, 1921, 210–212.
[27] Die Exponentialdarstellung der rationalen Zahlen für den Bereich einer Primzahl, *S.-B. Marburg*, **63**, 1927.
[37] Über den Zusammenhang zwischen der Kongruenzgruppen eines algebraischen Körpers für alle Potenzen eines Primteilers als Modul, *JRAM*, **177**, 1937, 82–93.

HENSEL, K., HASSE, H. (See H. Hasse, K. Hensel)

HENSLEY, D. [76] An asymptotic inequality concerning primes in contours for the case of quadratic number fields, *AA*, **28**, 1975/6, 69–79.
[77] Polynomials which take Gaussian integer values at Gaussian integers, *JNT*, **9**, 1977, 510–524.

HERBRAND, J. [30a] Détermination des groupes de ramification d'un corps a partir de ceux d'un sur-corps, *CR*, **191**, 1930, 980–982.
[30b] Nouvelle démonstration et généralisation d'un théorème de Minkowski, *CR*, **191**, 1930, 1282–1285.

[31a] Sur la théorie des groupes de décomposition, d'inertie et de ramification, *J. Math. Pures Appl.*, **10**, 1931, 481–498.
[31b] Sur les unités d'un corps algébrique, *CR*, **192**, 1931, 24–27; corr., *ibid.*, 188.
[31c] Sur la théorie des corps des nombres de degré infini, *CR*, **193**, 1931, 504–506.
[32a] Sur les classes des corps circulaires, *J. Math. Pures Appl.*, **11**, 1932, 417–441.
[32b] Théorie arithmétique des corps des nombres de degré infini, *MA*, **106**, 1932, 473–501; II, *ibid.*, **108**, 1933, 699–717.
[32c] Sur les théorèmes du genre principal et des idéaux principaux, *Hbg*, **9**, 1932, 84–92.
[32d] Une propriété de discriminant des corps algébriques, *ASENS*, (3), **49**, 1932, 105–112.

HERGLOTZ, G.   [22] Über einen Dirichletschen Satz, *MZ*, **12**, 1922, 255–261.
[23] Über die Kroneckersche Grenzformel für reelle, quadratische Körper, I, *Ber. Verh. Sächs. Akad. Wiss. Leipzig*, **75**, 1923, 3–14; II, *ibid.*, 31–37.

HERMITE, C.   [50] Sur different objects de la théorie des nombres, (Lettres de M. Hermite à M. Jacobi), *JRAM*, **40**, 1850, 261–278, 279–315 = *Oeuvres, I*, 100–163, Paris 1905.
[57] Sur le nombre limité d'irrationalités auxelle se réduisent les racines des équations à coefficients entiers complexes d'un degré et d'un discriminant donnés (Extrait d'une lettre à M. Borchardt), *JRAM*, **53**, 1857, 182–192 = *Oeuvres, I*, 414–428, Paris 1905.

HERZ, C. S.   [66] Construction of class fields, in: *Seminar on Complex Multiplication*, LN 21, Springer, 1966.

HERZ, C. S. *et al.*   (See A. Borel *et al.*)

HETTLING, K. F.   [85] *On K_2 of rings of integers of totally real number fields*, Ph. D. thesis, Louisiana State Univ. at Baton Rouge, 1985.

HETTLING, K. F., HOFFMANN, B., BROWKIN, J.   (See B. Hoffmann, K. F. Hettling, J. Browkin)

HEUPEL, W.   [68] Die Verteilung der ganzen Zahlen, die durch quadratische Formen dargestellt werden, *Arch. Math.*, **19**, 1968, 162–166.

HEWITT, E., ROSS, K.   [63] *Abstract Harmonic Analysis*, Springer, 1963.

HEY, F.   [29] *Analytische Zahlentheorie in Systemen hyperkomplexen Zahlen*, Diss., Hamburg 1929.

HICKERSON, D. R.   [73] Length of period of simple continued fraction expansion of $\sqrt{d}$, *Pacific J. Math.*, **46**, 1973, 429–431.

HIDA, H.   [78] On the values of Hecke's *L*-functions at non-positive integers, *J. Math. Soc. Japan*, **30**, 1978, 249–278.

HIGHTOWER, C. J.   [75] Approximation to complex numbers by certain biquadratic numbers, *JNT*, **7**, 1975, 293–309.

HIGMAN, G.   [40] The units of group rings, *PLMS*, (2), **46**, 1940, 231–248.

HIJIKATA, H.   [63] Hasse's principle on quaternionic anti-hermitian forms, *J. Math. Soc. Japan*, **15**, 1963, 165–175.

HILANO, T.   [74a] On the zeros of Hecke's *L*-functions, *PJA*, **50**, 1974, 23–28.
[74b] On the zeros of Hecke's *L*-functions, *Sci. Papers College Gen. Ed. Univ. Tokyo*, **24**, 1974, 9–24.

HILBERT, D.   [90] Ueber die Theorie der algebraischen Formen, *MA*, **36**, 1890, 473–534.
[94a] Grundzüge einer Theorie des Galoisschen Zahlkörpers, *GN*, 1894, 224–236.
[94b] Über die Zerlegung der Ideale eines Zahlkörpers in Primideale, *MA*, **44**, 1894, 1–8.
[94c] Über den Dirichletschen biquadratischen Zahlkörper, *MA*, **45**, 1894, 309–340.
[96] Ein neues Beweis des Kroneckerschen Fundamentalsatzes über Abelsche Zahlkörper, *GN*, 1896, 29–39.
[97] Die Theorie der algebraischer Zahlkörper, *Jahresber. Deutsch. Math. Verein.*, **4**, 1897, 175–546.
[99] Über die Theorie des relativquadratischen Zahlkörpers, *MA*, **51**, 1899, 1–127.

HINZ, J. G. (Hinz, J.) [76a] Régions libres de zéros des fonctions dzéta's de Hecke et la distribution des idéaux premiers, *CR*, **283**, 1976, 919–920.
[76b] Über Nullstellen der Heckeschen Zetafunktionen in algebraischen Zahlkörpern, *AA*, **31**, 1976, 167–193.
[77] Über Nullstellen der m-ten Ableitung der Dedekindschen Zetafunktion, *JNT*, **9**, 1977, 535–560.
[79] A mean value theorem for the Dedekind zeta function of a quadratic number field, *Monatsh. Math.*, **87**, 1979, 229–239.
[80] Eine Erweiterung des nullstellenfreien Bereiches der Heckeschen Zetafunktion und Primideale in Idealklassen, *AA*, **38**, 1980/1, 209–254.
[81] On the theorem of Barban and Davenport–Halberstam in algebraic number fields, *JNT*, **13**, 1981, 463–484.
[82a] Potenzfreie Werte von Polynomen in algebraischen Zahlkörpern, *JRAM*, **332**, 1982, 134–150.
[82b] Eine Anwendung der Selbergschen Siebmethode in algebraischen Zahlkörpern, *AA*, **41**, 1982, 223–254.
[82c] On the representation of even integers as sums of two almost-primes in algebraic number fields, *Mathematika*, **29**, 1982, 93–108.
[83a] Character sums in algebraic number fields, *JNT*, **17**, 1983, 52–70.
[83b] Character sums and primitive roots in algebraic number fields, *Monatsh. Math.*, **95**, 1983, 275–286.
[83c] The average order of magnitude of least primitive roots in algebraic number fields, *Mathematika*, **30**, 1983, 11–25.
[84] Some applications of sieve methods in algebraic number fields, *Manuscr. Math.*, **48**, 1984, 117–137.
[86] A note on Artin's conjecture in algebraic number fields, *JNT*, **22**, 1986, 334–349.
HIRAMATSU, T. [82] Higher reciprocity laws and modular forms of weight one, *Comment. Math. Univ. St. Pauli*, **31**, 1982, 75–85.
HIRONAKA-KOBAYASHI, Y. [76] On the Galois group of the maximal p-extension of algebraic number fields, *Natur. Sci. Rep. Ochanomizu Univ.*, **27**, 1976, 99–105.
HIRST, K. E. [72] The length of periodic continued fractions, *Monatsh. Math.*, **76**, 1972, 428–435.
HIRZEBRUCH, F. [73] The Hilbert modular group and some algebraic surfaces, *Tr. Mat. Inst. Steklov*, **132**, 1973, 55–66.
[76] Hilbert modular surfaces and class numbers, *Astérisque*, **32/3**, 1976, 151–164.
HLAWKA, E. [48] Über Folgen von Quadratwurzeln komplexer Zahlen, *S.-B. Österr. Ak. Wiss.*, **156**, 1948, 255–262.
HOCHSCHILD, G. [50] Local class field theory, *AnM*, **51**, 1950, 331–347.
HOCHSCHILD, G., NAKAYAMA, T. [52] Cohomology in class field theory, *AnM*, **55**, 1952, 348–366.
HOCK, A., BUNDSCHUH, P. (See P. Bundschuh, A. Hock)
HODGES, W. [74] Six impossible rings, *J. Algebra*, **31**, 1974, 218–244.
HOECHSMANN, K. [66] Über die Gruppe der maximalen l-Erweiterung eines globalen Körpers, *JRAM*, **222**, 1966, 142–147.
HOFFMANN, B., HETTLING, K. F., BROWKIN, J. [84] On the group generated by symbols in K_2O_F for real quadratic fields, *Res. Math.*, **7**, 1984, 63–64.
HOFFSTEIN, J. [79a] Some results related to minimal discriminants, in: *Number Theory, Carbondale 1979*, 185–194, LN 751, Springer, 1979.

[79b] Some analytic bounds for zeta functions and class numbers, *Invent. Math.*, **55**, 1979, 37–47.

[80] On the Siegel–Tatuzawa theorem, *AA*, **38**, 1980, 167–174.

HOFREITER, N. [35a] Quadratische Zahlkörper ohne euklidischen Algorithmus, *MA*, **110**, 1935, 194–196.

[35b] Über die Approximation von komplexen Zahlen, *Monatsh. M.-Phys.*, **42**, 1935, 401–416.

[35c] Quadratische Körper mit und ohne Euklidischen Algorithmus, *Monatsh. M.-Phys.*, **42**, 1935, 397–400.

[37] Diophantische Approximationen in imaginär quadratischen Zahlkörpern, *Monatsh. M.-Phys.*, **45**, 1937, 175–190.

[40a] Über das Produkt von Linearformen, *Monatsh. M.-Phys.*, **49**, 1940, 295–298.

[40b] Diophantische Approximationen komplexer Zahlen, *Monatsh. M.-Phys.*, **49**, 1940, 299–302.

[52] Über die Approximation von komplexen Zahlen durch Zahlen des Körpers $K(i)$, *Monatsh. Math.*, **56**, 1952, 61–74.

HÖLDER, O. [35] Zur Theorie der Gaussschen Summen, *Ber. Sächs. Ges. Wiss. Leipzig*, **87**, 1935, 27–36.

[36] Zur Theorie der Kreisteilungsgleichung $K_m(x) = 0$, *Prace Mat.-Fiz.*, **43**, 1936, 13–23.

HOLZAPFEL, R. P. [67] Eine Bemerkung zur Zahlentheorie von H. Hasse, *Wiss. Z. Humboldt Univ. Berlin*, **16**, 1967, 317.

HOLZER, L. [50] Zur Klassenzahl in reinen Zahlkörpern von ungeraden Primzahlgrad, *AM*, **83**, 1950, 327–348.

[58] *Zahlentheorie*, Leipzig 1958–1965.

[66] *Klassenkörpertheorie*, Leipzig 1966.

HONDA, T. [60a] On absolute class fields of certain algebraic number fields, *JRAM*, **203**, 1960, 80–89.

[60b] On the absolute ideal class group of relatively metacyclic number fields of a certain type, *Nagoya Math. J.*, **17**, 1960, 171–179.

[68] On real quadratic fields whose class numbers are multiples of 3, *JRAM*, **233**, 1968, 101–102.

[71] Pure cubic fields whose class numbers are multiples of three, *JNT*, **3**, 1971, 7–12.

[75] A few remarks on class numbers of imaginary quadratic number fields, *Osaka J. Math.*, **12**, 1975, 19–21.

HOOLEY, C. [84] On the Pellian equation and the class number of indefinite binary quadratic forms, *JRAM*, **353**, 1984, 98–131.

HORIE, K., KIMURA, T. (See T. Kimura, K. Horie)

HORIE, M. [83] On the genus field in algebraic number fields, *Tokyo J. Math.*, **6**, 1983, 363–380.

HORNFECK, B. [70] Primteiler von Polynome, *JRAM*, **243**, 1970, 120.

HSIA, J. S. [73] On the representation of cyclotomic polynomials as sums of squares, *AA*, **25**, 1973/4, 115–120.

HSIA, J. S., ESTES, D. R. (See D. R. Estes, J. S. Hsia)

HSÜ, C. S. [62] Theorems on direct sums of modules, *PAMS*, **13**, 1962, 540–542.

HSÜ, T.-N. [63] Über den Hauptidealsatz für imaginär-quadratische Zahlkörper, *JRAM*, **212**, 1963, 49–62.

HUA, LOO-KENG [42] On the least solution of Pell's equation, *BAMS*, **48**, 1942, 731–735.

[44] On the distribution of quadratic non-residues and the Euclidean algorithm in real quadratic fields, I, *TAMS*, **56**, 1944, 537–546.

[51] On exponential sums over an algebraic number field, *CJM*, **3**, 1951, 44–51.

HUA, LOO-KENG, MIN, SZU-HOA [44] On the distribution of quadratic non-residues and the Euclidean algorithm in real quadratic fields, II, *TAMS*, **56**, 1944, 547–569.

HUA, LOO-KENG, SHIH, W. T. [45] On the lack of an Euclidean algorithm in $R(\sqrt{61})$, *AJM*, **67**, 1945, 209–211.

HUARD, J. G. [79] Cyclic cubic fields that contain an integer of given index, in: *Number Theory, Carbondale 1979*, 195–199, LN 751, Springer, 1979.

HUCKABA, J. A., PAPICK, I. J. [81] A localization of $R[X]$, *CJM*, **33**, 1981, 103–115.

HUGHES, I., MOLLIN, R. [83] Totally positive units and squares, *PAMS*, **87**, 1983, 613–616.

HUGOT, M. (See M. Grandet-Hugot)

HUGOT, M., PISOT, CH. [58] Sur certaines entiers algébriques, *CR*, **246**, 1958, 2831–2833.

HULL, R. [35] A determination of all cyclotomic quintic fields, *AnM*, **36**, 1935, 366–372.

HUMBERT, P. [40] Sur les nombres de classes de certains corps quadratiques, *Comment. Math. Helv.*, **12**, 1939/40, 233–245; add., *ibid.*, **13**, 1940/1, 67.

HUNTER, J. [56a] A generalization of the inequality of the arithmetic-geometric mean, *Proc. Glasgow Math. Ass.*, **2**, 1956, 149–158.

[56b] A note on integer solutions of the diophantine equation $x^2 - dy^2 = 1$, *Proc. Glasgow Math. Ass.*, **3**, 1956, 55–56.

[57] The minimum discriminant of quintic fields, *Proc. Glasgow Math. Ass.*, **3**, 1957, 57–67.

HURRELBRINK, J. [82a] On the size of certain K-groups, *Comm. Algebra*, **10**, 1982, 1873–1889.

[82b] $K_2(O)$ for two totally real fields of degree three and four, in: *Algebraic K-Theory, I*, 112–114, LN 966, Springer, 1982.

[83] On the wild kernel, *Arch. Math.*, **40**, 1983, 316–318.

HURRELBRINK, J., BROWKIN, J. (See J. Browkin, J. Hurrelbrink)

HURRELBRINK, J., KOLSTER, M. [86] On the 2-primary part of the Birch–Tate conjecture for cyclotomic fields, *Proc. AMS Conference, Boulder 1983, Contemporary Math.* **55**, 1986, 519–528.

HURWITZ, A. [82] Einige Eigenschaften der Dirichletschen Functionen $F(s) = \sum \left(\frac{D}{n}\right)\frac{1}{n^s}$, die bei der Bestimmung der Classenanzahlen binärer quadratischer Formen auftreten, *Zeitschr. Math. Phys.*, **27**, 1882, 86–101.

[87] Über die Entwicklung complexer Grösse in Kettenbrüche, *AM*, **11**, 1887/8, 187–200.

[94] Über die Theorie der Ideale, *GN*, 1894, 291–298.

[95a] Über einen Fundamentalsatz der arithmetischer Theorie der algebraischen Grössen, *GN*, 1895, 230–240.

[95b] Die unimodularen Substitutionen in einem algebraischen Zahlenkörper, *GN*, 1895, 244–268.

[95c] Zur Theorie der algebraischer Zahlen, *GN*, 1895, 324–331.

[95d] Über die Anzahl der Classen binärer quadratischer Formen von negativer Determinante, *AM*, **19**, 1895, 389–397.

[99] Über die Entwicklungskoeffizienten der lemniskatischen Funktionen, *MA*, **51**, 1899, 196–226.

[19] Der Euklidische Divisionssatz in einem endlichen algebraischen Zahlkörper, *MZ*, **3**, 1919, 123–126.

[26] Über Beziehungen zwischen den Primidealen eines algebraischen Körpers und den Substitutionen seiner Gruppe, *MZ*, **25**, 1926, 661–665.

HURWITZ, J. [02] Über die Reduction der binären quadratischen Formen mit complexen Coeffizienten und Variablen, *AM*, **25**, 1902, 231–290.

HUSHVAKTOV, M. [72] Distribution of absolutely Abelian cyclic fields of degree l^k and of special absolutely Abelian fields of given type, *Nauchn. Tr. Tashkent. Gos. Univ.*, **418**, 1972, 338-351 (Russian).
[75] The distribution of absolutely Abelian cyclic fields of degree q^h, ($h \geqslant 2$), *DAN Uzbek. SSR*, 1975, **5**, 5-6 (Russian).
[76] Distribution of the discriminants of absolutely Abelian cyclic fields of degree q^h, ($h \geqslant 2$), *IAN Uzbek. SSR*, 1976, **1**, 34-39 (Russian).
[77a] On the distribution of certain finite Abelian extensions with small discriminants over the field of rational numbers, *IAN Uzbek. SSR*, 1977, **6**, 47-52 (Russian).
[77b] Distribution of discriminants of regular Abelian extensions of degree n of the field of rational numbers, *Nauchn. Tr. Tashkent. Gos. Univ.*, **548**, 1977, 109-114 (Russian).
[77c] On a theorem on the distribution of discriminants of cyclic fields of prime degree, *IAN Uzbek. SSR*, 1977, **4**, 73-74 (Russian).
HUXLEY, M. N. [68] The large sieve inequality for algebraic number fields, *Mathematika*, **15**, 1968, 178-187; II, *PLMS*, (3), **21**, 1970, 108-128; III, *JLMS*, (2), **3**, 1971, 233-240.
HYYRÖ, S. [67] Über eine Determinantenidentität und den ersten Faktor der Klassenzahl des Kreiskörpers, *Ann. Acad. Sci. Fenn.*, *AI*, **398**, 1967, 1-7.
ICHIMURA, H. [82] On 2-rank of the ideal class groups of totally real number fields, *PJA*, **58**, 1982, 329-332.
ICHIMURA, H., AZUHATA, T. (See T. Azuhata, H. Ichimura)
IDT, J. [74] Calcul der ordres de groupes de ramification et des nombres de ramification des extensions $k^{(i)}/k$ de Chafarévitch, *CR*, **278**, 1974, 669-670.
IIMURA, K. [71] A criterion for the class number of a pure quintic field to be divisible by 5, *JRAM*, **292**, 1971, 201-210.
[79a] Dihedral extensions of Q of degree $2l$ which contain non-Galois extensions with class number not divisible by l, *AA*, **35**, 1979, 385-394.
[79b] On 3-class groups of non-Galois cubic fields, *AA*, **35**, 1979, 395-402.
[80] On the unit group of certain sextic number fields, *Hbg*, **50**, 1980, 32-39.
[81a] On the l-class group of an algebraic number field, *JRAM*, **322**, 1981, 136-144.
[81b] A note on the Stickelberger ideal of conductor level, *Arch. Math.*, **36**, 1981, 45-52.
IKEDA, M. [75a] On the group automorphisms of the absolute Galois group of the rational number field, *Arch. Math.*, **26**, 1975, 250-252.
[75b] Completeness of the absolute Galois group of the rational number field, *Arch. Math.*, **26**, 1975, 602-605.
[77] Completeness of the absolute Galois group of the rational number field, *JRAM*, **291**, 1977, 1-22.
INABA, E. [35] Über die Klassenzahlen abelscher Zahlkörper, *Proc. Imper. Acad. Japan. Tokyo*, **11**, 1935, 81-82.
[37] Über die absoluten Idealklassengruppen algebraischer Zahlkörper, *Japan. J. Math.*, **13**, 1937, 81-84.
[40] Über die Struktur der l-Klassengruppe zyklischer Zahlkörper vom Primzahlgrad l, *J. Fac. Sci. Univ. Tokyo*, **4**, 1940, 61-115.
[41] Klassenkörpertheoretische Deutung der Struktur der Klassengruppe des zyklischen Zahlkörpers, *Proc. Imper. Acad. Japan. Tokyo*, **17**, 1941, 125-128.
[52] Note on relative complete fields, *Natur. Sci. Rep. Ochanomizu Univ.*, **3**, 1952, 5-9.
INCE, E. L. [34] *Cycles of Reduced Ideals in Quadratic Fields*, London 1934.
INGHAM, A. E. [30] Note on Riemann's ζ-function and Dirichlet's L-functions, *JLMS*, **5**, 1930, 107-112.

INKERI, K. [47] Über den Euklidischen Algorithmus in quadratischen Zahlkörpern, *Ann. Acad. Sci. Fenn., AI*, **41**, 1947, 1–35.
 [48] Neue Beweise für einige Sätze zum Euklidischen Algorithmus in quadratischen Zahlkörpern, *Ann. Univ. Turku, A*, **9**, 1948, 1.
 [49] On the second case of Fermat's last theorem, *Ann. Acad. Sci. Fenn., AI*, **60**, 1949, 1–32.
 [56] Über die Klassenanzahl des Kreiskörpers der l-ten Einheitswurzeln, *Ann. Univ. Turku, A*, **23**, 1956, 1–16.
INOUE, H. [42] Eine Eigenschaft der Norm, *Tôhoku Math. J.*, **49**, 1942, 60–68.
IRELAND, K. F., ROSEN, M. I. [82] *A Classical Introduction to Modern Number Theory*, Springer, 1982.
IRFAN, M., QUADRI, M. A. (See M. A. Quadri, M. Irfan)
ISAACS, I. M. [70] Degrees of sums in a separable field extension, *PAMS*, **25**, 1970, 638–641.
ISAACS, I. M., EVANS, R. J. (See R. J. Evans, I. M. Isaacs)
ISEKI, K. [51a] Über die imaginär-quadratischen Zahlkörper der Klassenzahl Eins oder Zwei, *PJA*, **27**, 1951, 621–622.
 [51b] On the imaginary quadratic fields of class-number one or two, *Japan. J. Math.*, **21**, 1951, 145–162.
 [52] Über die negativen Fundamentaldiskriminanten mit der Klassenzahl Zwei, *Natur. Sci. Rep. Ochanomizu Univ.*, **3**, 1952, 23–29.
 [53] On a general divisor problem in algebraic number fields, *Natur. Sci. Rep. Ochanomizu Univ.*, **5**, 1953, 1–21.
ISHIDA, M. [57] On the divisibility of Dedekind's zeta-functions, *PJA*, **33**, 1957, 293–297.
 [69] A note on class numbers of algebraic number fields, *JNT*, **1**, 1969, 65–69.
 [70] Class numbers of algebraic number fields of Eisensteinian type, *JNT*, **2**, 1970, 404–413; II, *ibid.*, **6**, 1974, 99–104.
 [71] On algebraic number fields with even class-numbers, *JRAM*, **247**, 1971, 118–122.
 [73] Fundamental units of certain algebraic number fields, *Hbg*, **39**, 1973, 245–250.
 [74] Some unramified abelian extensions of algebraic number fields, *JRAM*, **268/9**, 1974, 165–173.
 [75a] On the genus field of an algebraic number field of odd prime degree, *J. Math. Soc. Japan*, **27**, 1975, 289–293.
 [75b] On 2-rank of the ideal class groups of algebraic number fields, *JRAM*, **273**, 1975, 165–169.
 [76] *The Genus Field of Algebraic Number Fields*, LN 555, Springer, 1976.
 [77] An algorithm for constructing the genus field of an algebraic number field of odd prime degree, *J. Fac. Sci. Univ. Tokyo, IA*, **24**, 1977, 61–75.
 [80] On the genus field of pure number fields, *Tokyo J. Math.*, **3**, 1980, 163–171; II, *ibid.*, **4**, 1981, 213–220.
 [82] On the index $(R:U)$ and the genus number of an abelian number field, *Arch. Math.*, **39**, 1982, 546–550.
ISHIKAWA, T. [69] On Dedekind rings, *J. Math. Soc. Japan*, **11**, 1969, 83–84.
ISKOVSKIKH, V. A. [71] A counterexample to the Hasse principle for a system of two quadratic forms in five variables, *Mat. Zametki*, **10**, 1971, 253–257 (Russian).
ITO, H. [77] A note on the law of decomposition of primes in certain Galois extension, *PJA*, **53**, 1977, 115–118.
IWASAKI, K. [52] Simple proof of a theorem of Ankeny on Dirichlet series, *PJA*, **28**, 1952, 555–557.

IWASAWA, K. [53a] On the rings of valuation vectors, *AnM*, **57**, 1953, 331–356.
[53b] On solvable extensions of algebraic number fields, *AnM*, **58**, 1953, 548–572.
[53c] A note on Kummer extensions, *J. Math. Soc. Japan*, **5**, 1953, 253–262.
[55a] A note on class numbers of algebraic number fields, *Hbg*, **20**, 1955, 257–258.
[55b] On Galois groups of local fields, *TAMS*, **80**, 1955, 448–469.
[56] A note on the group of units of an algebraic number field, *J. Math. Pures Appl.*, **35**, 1956, 189–192.
[58] On some invariants of cyclotomic fields, *AJM*, **80**, 1958, 773–783; corr., *ibid.*, **81**, 1959, 280.
[59a] On Γ-extensions of algebraic number fields, *BAMS*, **65**, 1959, 183–226.
[59b] Sheaves for algebraic number fields, *AnM*, **69**, 1959, 408–413.
[59c] On some properties of Γ-finite modules, *AnM*, **70**, 1959, 291–312.
[59d] On the theory of cyclotomic fields, *AnM*, **70**, 1959, 530–561.
[60] On local cyclotomic fields, *J. Math. Soc. Japan*, **12**, 1960, 16–21.
[62] A class number formula for cyclotomic fields, *AnM*, **76**, 1962, 171–179.
[65] Some modules in the theory of cyclotomic fields, *PSPM*, **8**, 1965, 66–69.
[66a] A note on ideal class group, *Nagoya Math. J.*, **27**, 1966, 239–247.
[66b] Some modules in local cyclotomic fields, in: *Les tendances géometriques en algèbre et theorie des nombres*, 87–96, Paris 1966.
[69] On p-adic L-functions, *AnM*, **89**, 1969, 198–205.
[72a] *Lectures on p-adic L-functions*, Princeton 1972.
[72b] On the μ-invariants of cyclotomic fields, *AA*, **21**, 1972, 99–101.
[73a] On Z_l-extensions of algebraic number fields, *AnM*, **98**, 1973, 248–326.
[73b] On the μ-invariants of Z_l-extensions, in: *Number Theory, Algebraic Geometry and Commutative Algebra*, 1–11, Tokyo 1973.
[75] A note on Jacobi sums, *Symposia Math.*, **15**, 447–459, Academic Press, 1975.
[76] A note on cyclotomic fields, *Invent. Math.*, **36**, 1976, 115–123.
[77] Some remarks on Hecke characters, in: *Algebraic Number Theory*, (Kyoto), 99–108, Tokyo 1977.
[80] *Local Class Field Theory*, Iwanami Shoten 1980 (Japanese); Russian transl.: Moskva 1983.
[81] Riemann–Hurwitz formula and p-adic Galois representations for number fields, *Tôhoku Math. J.*, (2), **33**, 1981, 263–288.
[83] On cohomology groups of units for Z_p-extensions, *AJM*, **105**, 1983, 189–200.
IWASAWA, K. *et al.* (See A. Borel *et al.*)
IWASAWA, K., SIMS, C. S. [66] Computation of invariants in the theory of cyclotomic fields, *J. Math. Soc. Japan*, **18**, 1966, 86–96.
IWATA, H. [72] Algebraic number fields embedded in Q_p, *Bull. Fac. Sci. Ibaraki Univ.*, *A*, **4**, 1972, 21–28.
IYANAGA, S. [31] Über den allgemeinen Hauptidealsatz, *Japan. J. Math.*, **7**, 1931, 315–333.
[34a] Zum Beweis des Hauptidealsatzes, *Hbg*, **10**, 1934, 349–357.
[34b] Zur Theorie der Geschlechtermoduln, *JRAM*, **171**, 1934, 12–18.
[35] *Sur les classes d'idéaux dans les corps quadratiques*, Paris 1935.
[39] Über die allgemeinen Hauptidealformeln, *Monatsh. M. -Phys.*, **48**, 1939, 400–407.
[75] *The Theory of Numbers*, Amsterdam–Oxford–New York 1975.
IYANAGA, S., TAMAGAWA, T. [51] Sur la théorie du corps de classes sur le corps des nombres rationels, *J. Math. Soc. Japan*, **3**, 1951, 220–227.
JACOB, G. [76] Polynômes représentant la fonction nulle sur un anneaux commutatif unitaire, *CR*, **283**, 1976, A421–424.

JACOBI, C. G. J. [32] Observatio arithmetica de numero classium divisorum quadraticorum formae $aa+Azz$, designante a numerum primum formae $4n+3$, *JRAM*, **9**, 1832, 189–192.

[39] Ueber die complexen Primzahlen, welche in der Theorie der 5-ten, 8-ten und 12-ten Potenzen zu betrachten sind, *JRAM*, **19**, 1839, 314–318.

[46] Über die Kreisteilung und ihre Anwendung auf die Zahlentheorie, *JRAM*, **30**, 1846, 166–182.

[68] Allgemeine Theorie der Kettenbruchähnlichen Algorithmen, in welchen jede Zahl aus drei vorhergehenden gebildet wird, *JRAM*, **69**, 1868, 29–64.

JACOBINSKI, H. [61] Verzweigungsgruppen und Verzweigungskörper, *JRAM*, **208**, 1961, 113–143.

[63] Über die Hauptordnung eines Körpers als Gruppenmodul, *JRAM*, **213**, 1963, 151–164.

JACOBSON, B. [64] Sums of distinct divisors and sums of distinct units, *PAMS*, **15**, 1964, 179–183.

JACOBSTHAL, E. [13] Diophantische Gleichungen im Bereich aller ganzen algebraischen Zahlen, *MA*, **74**, 1913, 31–65.

[58] Zur Theorie der Einheitswurzeln, *Norsk Vid. Selsk. Forh.*, **31**, 1958, 125–129, 130–137.

JACQUET, H., GODEMENT, R.   (See R. Godement, H. Jacquet)

JAKOVLEV, A. V. [68] Galois group of the algebraic closure of a local field, *IAN*, **32**, 1968, 1283–1322; corr., *ibid.*, **42**, 1978, 212–213 (Russian).

[70] Homological determination of p-adic representations of rings with a power basis, *IAN*, **34**, 1970, 1000–1014 (Russian).

[75] On the theory of symplectic spaces with operators, *Sibir. Mat. Zh.*, **16**, 1975, 169–174 (Russian).

[78a] Abstract characterization of the Galois group of the algebraic closure of a local field, *Zap. Nauch. Sem. LOMI*, **75**, 1978, 179–193 (Russian).

[78b] Structure of the multiplicative group of a tamely ramified extension of a local field of odd degree, *Mat. Sbornik*, **107**, 1978, 304–316 (Russian).

JANSSEN, U. [82] Über Galoisgruppen lokaler Körper, *Invent. Math.*, **70**, 1982/3, 53–69.

JANSSEN, U., WINGBERG, K. [79] Die p-Vervollständigung der multiplikativen Gruppe einer p-Erweiterung eines irregulären p-adischen Zahlkörpers, *JRAM*, **307/8**, 1979, 399–410.

[80] Einbettungsprobleme und Galoisstruktur lokaler Körper, *JRAM*, **319**, 1980, 196–212.

[82] Die Struktur der absoluten Galoisgruppe p-adischer Zahlkörper, *Invent. Math.*, **70**, 1982/3, 71–98.

JANUSZ, G. [73] *Algebraic Number Fields*, Academic Press, 1973.

JARDEN, M. [74] On Chebotarev sets, *Arch. Math.*, **25**, 1974, 495–497.

JARDEN, M., RITTER, J. [79] On the characterization of local fields by their absolute Galois groups, *JNT*, **11**, 1979, 1–13.

[80] Normal automorphisms of absolute Galois groups of p-adic fields, *Duke Math. J.*, **47**, 1980, 47–56.

JAULENT, J. F. [79] *Structures galoisiennes dans les extensions métabeliennes*, Thèse, Besançon 1979.

[81a] Remarques sur la structure galoisienne des entiers d'une extension métacyclique de Q, *CR*, **293**, 1981, 231–233.

[81b] Sur la l-structure galoisienne des idéaux ambiges dans une extension metacyclique de degré nl sur le corps des rationnels, *Publ. Math. Fac. Sci. Besançon*, 1979/80 et 1980/81.

[81c] Sur la théorie des genres dans une extension cyclique de degré l^m d'un corps de nombres metabelienne sur un sous-corps, *Publ. Math. Fac. Sci. Besançon*, 1979/80 et 1980/81.

[81d] Théorie d'Iwasawa des tours métabeliennes, *Sem. Th. de Nombr. Bordeaux*, 1980/1, exp. 21.

[81e] Unités et classes dans les extensions métabéliennes de degré nl^s sur un corps de nombres algébriques, *Ann. Inst. Fourier*, 31, 1981, 1, 39–62.

[82] Sur la théorie des genres dans les tours métabéliennes, *Sem. Th. de Nombr. Bordeaux*, 1981/2, exp. 24.

[83] Introduction au K_2 des corps de nombres, *Publ. Math. Fac. Sci. Besançon*, 1981/82 et 1982/83.

JEANS, N. S., HENDY, M. D. [78] Determining the fundamental unit of a pure cubic field given any unit, *MC*, 32, 1978, 925–935.

JEHNE, W. [54] Zur moderner Klassenkörpertheorie, *S.-B. Deutsch. Akad. Wiss.*, 1954, 3, 1–8.

[59] Bemerkung über die p-Klassengruppe des p-ten Kreiskörpers, *Arch. Math.*, 10, 1959, 422–427.

[61] Zur Verschärfung des F. K. Schmidtschen Einheitssatzes, *MZ*, 77, 1961, 439–452.

[77a] Über die Einheiten- und Divisorenklassengruppe von reellen Frobeniuskörpern von Maximaltyp, *MZ*, 152, 1977, 223–252.

[77b] Kronecker classes of algebraic number fields, *JNT*, 9, 1977, 279–320.

[77c] On Kronecker classes of atomic extensions, *PLMS*, (3), 34, 1977, 32–64.

[79] On knots in algebraic number theory, *JRAM*, 311/312, 1979, 215–254.

[82] Der Hassesche Normensatz und seine Entwicklung, *Mitt. Math. Ges. Hamburg*, 11, 1982, 143–153.

JENKINS, E. D., MCDUFFEE, C. C. (See C. C. McDuffee, E. D. Jenkins)

JENNER, W. [69] On Zeta-functions of number fields, *Duke Math. J.*, 36, 1969, 669–671.

JENSEN, C. U. [60] Über die Führer einer Klasse Heckescher Grössencharaktere, *Math. Scand.*, 8, 1960, 81–96.

[62a] Über eine Klasse nicht-Pellscher Gleichungen, *JRAM*, 209, 1962, 36–38.

[62b] On the solvability of a certain class of non-Pellian equations, *Math. Scand.*, 10, 1962, 71–84.

[62c] On the Diophantine equation $\xi^2 - 2m^2\eta^2 = -1$, *Math. Scand.*, 11, 1962, 58–62.

[63] On characterizations of Prüfer rings, *Math. Scand.*, 13, 1963, 90–98.

[64a] A remark on relative integral bases for infinite extensions of finite number fields, *Mathematika*, 11, 1964, 64–66.

[64b] A remark on arithmetical rings, *PAMS*, 15, 1964, 951–954.

[66] A remark on the distributive law for an ideal in a commutative ring, *Proc. Glasgow Math. Ass.*, 7, 1966, 193–198.

JENSEN, K. L. [15] Numbertheoretical properties of Bernoulli numbers, *Nyt Tidsskr. Mat.*, 26, 1915, B, 73–83 (Danish).

JOHNSON, D. [79] Mean values of Hecke L-functions, *JRAM*, 305, 1979, 195–205.

JOHNSON, E. W., LEDIAEV, J. P. [71] A new characterization of Dedekind domains, *PAMS*, 28, 1971, 63–64.

JOHNSON, W. [73] On the vanishing of the Iwasawa invariant μ_p for $p < 8000$, *MC*, 27, 1973, 387–396.

[75] Irregular primes and cyclotomic invariants, *MC*, 29, 1975, 113–120.

JOLY, J. R. [65] Sur les puissances d-ièmes des éléments d'un anneau commutatif, *CR*, 261, 1965, 3259–3262.

[66] Sur le problème de Waring pour un exposant premier dans certains anneaux local, *CR*, **262**, 1966, A1438-1441.
[68] Constantes de Waring des corps commutatifs, *CR*, **266**, 1968, 516-518.
[70a] Note relative aux théorèmes des S-unités et des S-classes, *Enseign. Math.*, **16**, 1970, 247-254.
[70b] Sommes de puissances d-ièmes dans un anneau commutatif, *AA*, **17**, 1970, 37-114.
JOLY, J. R., MOSER, C. [79] Ordre de grandeur de $L(1, \chi)$ et de $L'(1, \chi)$, *Ann. Inst. Fourier*, **29**, 1979, 1, 125-135.
JONES, A. J. [68] Sums of three roots of unity, *PCPS*, **64**, 1968, 673-682; II, *ibid.*, **66**, 1969, 43-59.
JONES, A. J., CONWAY, J. H. (See J. H. Conway, A. J. Jones)
JONES, B. W. [49] The composition of quadratic binary forms, *AMM*, **56**, 1949, 380-391.
JORDAN, J. H. [65] The divisibility of Gaussian integers by large Gaussian primes, *Duke Math. J.*, **32**, 1965, 503-509.
[67a] Covering classes of residues, *CJM*, **19**, 1967, 514-519.
[67b] A covering class of residues with odd moduli, *AA*, **13**, 1967/8, 335-338.
[67c] Character sums in $Z(i)/(p)$, *PLMS*, (3), **17**, 1967, 1-10.
[68] The distribution of k-th power residues and non-residues in the Gaussian integers, *Tôhoku Math. J.*, **20**, 1968, 498-510.
[69] Consecutive residues or non-residues in the Gaussian integers, *JNT*, **1**, 1969, 477-485.
JORDAN, J. H., HARDMAN, N. R. (See N. R. Hardman, J. H. Jordan)
JORDAN, J. H., PORTRATZ, C. J. [65] Complete residue system in the Gaussian integers, *Math. Mag.*, **38**, 1965, 1-12.
JORDAN, J. H., RABUNG, J. R. [70] A conjecture of Paul Erdös concerning Gaussian primes, *MC*, **24**, 1970, 221-223.
[76] Local distribution of Gaussian primes, *JNT*, **8**, 1976, 43-51.
JORDAN, J. H., SCHNEIDER, D. G. [71] Covering classes of residues in $Z(\sqrt{-2})$, *Math. Mag.*, **44**, 1971, 257-261.
JORIS, H. [70] Un Ω-théorème pour la fonction des idéaux d'un corps de nombres algébriques, *CR*, **270**, 1970, A1713.
[72] Ω-Sätze für zwei arithmetische Funktionen, *Comment. Math. Helv.*, **47**, 1972, 220-248.
[77] On the evaluation of Gaussian sums for non-primitive Dirichlet characters, *Enseign. Math.*, **23**, 1977, 13-18.
JORIS, H., CHANDRASEKHARAN, K. (See K. Chandrasekharan, H. Joris)
JOSHI, P. T. [70] The size of $L(1, \chi)$ for real nonprincipal residue characters χ with prime modulus, *JNT*, **2**, 1970, 7-21.
JURICIC, H. [65] *Tables des nombre de classes de formes quadratiques binaires arithmétiques primitives pour les déterminants $\Delta = b^2 - 4ac$ négatifs jusqu'a cent milles*, Marseille 1965.
JUSHKIS, Z. [64] Limit theorems for additive functions, defined on ordered semigroups with regular norms, *Litovsk. Mat. Sb.*, **4**, 1964, 565-603 (Russian).
JUSTIN, J. [69] Bornes des coefficients du polynôme cyclotomique et des certains autres polynômes, *CR*, **268**, 1969, A 995-997.
JUTILA, M. [73] On character sums and class numbers, *JNT*, **5**, 1973, 203-214.
[77] Zero-density estimates for L-functions, *AA*, **32**, 1977, 55-62.
KACZOROWSKI, J. [81a] A pure arithmetical characterization for certain fields with a given class group, *Colloq. Math.*, **45**, 1981, 327-330.
[81b] Completely irreducible numbers in algebraic number fields, *Funct. et Approx.*, **11**, 1981, 95-104.

[83] Some remarks on factorization in algebraic number fields, *AA*, **43**, 1983, 53–68.
[84] A pure arithmetical definition of the classgroup, *Colloq. Math.*, **48**, 1984, 265–267.

KÄHLER, E. [53] Algebra und Differentialrechnung, *Ber. Math. Tagung*, Berlin 1953, 58–163.

KALINKA, V. [61] On the representation of Gaussian integers as a sum of squares, *Litovsk. Mat. Sb.*, **1**, 1961, 1–2, 370–371 (Russian).
[63] A generalization of a lemma of Loo-Keng Hua to algebraic numbers, *Litovsk. Mat. Sb.*, **3**, 1963, 1, 149–155 (Russian).

KALLEN, W. VAN DER [81] Stability for K_2 of Dedekind rings of arithmetic type, in: *Algebraic Number Theory, Evanston 1980*, 217–248, LN 854, Springer, 1981.

KALNIN, I. M. [65a] Distribution of prime numbers representable by a quadratic form, *Rigas Politekhn. Inst., Zin. R. Izl.*, **19**, 1965, 5–45 (Russian).
[65b] On prime numbers of a quadratic imaginary field in sectors, *IAN Latv. SSR*, 1965, 83–92 (Russian).
[68] On Hecke's zeta-function, in: *Differentsial'nye uravneniya i primenen.*, 73–102, Riga 1968 (Russian).

KAMBAYASHI, T. [75] On certain algebraic groups attached to local number fields, *JRAM*, **273**, 1975, 41–48.

KANEIWA, R., SHIOKAWA, I., TAMURA, J. I. (See I. Shiokawa, R. Kaneiwa, J. I. Tamura)

KANEMITSU, S. [77] On some bounds for the value of Dirichlet's L-function $L(s, \chi)$ at the point $s = 1$, *Mem. Fac. Sci. Kyushu Univ., A*, **31**, 1977, 15–23.
[78] A note on the general divisor problem, *Mem. Fac. Sci. Kyushu Univ., A*, **32**, 1978, 211–221.

KANNO, T. [73] Automorphisms of the Galois group of the algebraic closure of the rational number field, *Kōdai Math. Sem. Rep.*, **25**, 1973, 446–448.

KANOLD, H. J. [49] Sätze über Kreisteilungspolynome und ihre Anwendungen auf einige zahlentheoretische Probleme, *JRAM*, **187**, 1949–50, 169–182; II, *ibid.*, **188**, 1950, 129–146.
[52] Abschätzungen bei Kreisteilungspolynomen und daraus hergeleitete Bedingungen für die kleinsten Primzahlen gewisser arithmetischer Folgen, *MZ*, **55**, 1952, 284–287.

KANTZ, G. [55a] Über den Typus eines Zerlegungsringes, *Monatsh. Math.*, **59**, 1955, 104–110.
[55b] Über Integritätsbereiche mit eindeutiger Primelementzerlegung, *Arch. Math.*, **6**, 1955, 397–402.

KAPLAN, P. [72] Divisibilité par 8 du nombre des classes des corps quadratiques reéls dont le 2-sous-groupe des classes est cyclique, *CR*, **275**, 1972, A887–890.
[73a] Divisibilité par 8 du nombre des classes des corps quadratiques dont le 2-groupe des classes est cyclique et réciprocité biquadratique, *J. Math. Soc. Japan*, **25**, 1973, 596–608.
[73b] 2-groupe des classes et facteurs principaux de $Q(\sqrt{pq})$ ou $p \equiv -q \equiv 1 \pmod 4$, *CR*, **276**, 1973, 89–92.
[74] Comparaison des 2-groupes des classes d'idéaux au sense large et au sense étroit d'un corps quadratique réel, *PJA*, **50**, 1974, 688–693.
[76] Sur le 2-groupe de classes d'idéaux des corps quadratiques, *JRAM*, **283/4**, 1976, 313–363.
[77a] Unités de norme -1 de $Q(\sqrt{p})$ et corps de classes de degré 8 de $Q(\sqrt{-p})$ où p est un nombre premier congru à 1 modulo 8, *AA*, **32**, 1977, 239–243.
[77b] Cycles d'ordre au moins 16 dans le 2-groupe des classes d'idéaux de certain corps quadratiques, *Bull. Soc. Math. France, Mém.* **49/50**, 1977, 113–124.

[81] Nouvelle démonstration d'une congruence modulo 16 entre les nombres de classes d'idéaux de $Q(\sqrt{-2p})$ et $Q(\sqrt{2p})$ pour p premier $\equiv 1 \pmod 4$, *PJA*, **57**, 1981, 507–509.

KAPLAN, P., FURUTA, Y. (See Y. Furuta, P. Kaplan)

KAPLAN, P., WILLIAMS, K. S. [82a] On the class numbers of $Q(\sqrt{\pm 2p})$ modulo 16, for $p \equiv 1 \pmod 8$ a prime, *AA*, **40**, 1982, 289–296.

[82b] Congruences modulo 16 for the class numbers of the quadratic fields $Q(\sqrt{\pm p})$ and $Q(\sqrt{\pm 2p})$ for p a prime congruent to 5 modulo 8, *AA*, **40**, 1982, 375–397.

KAPLANSKY, I. [47] Topological methods in valuation theory, *Duke Math. J.*, **14**, 1947, 527–541.

[52] Modules over Dedekind rings and valuation rings, *TAMS*, **72**, 1952, 327–340.

[54] *Infinite Abelian Groups*, Ann Arbor 1954.

[60] A characterization of Prüfer rings, *J. Indian Math. Soc.*, **24**, 1960, 279–281.

[68] Composition of binary quadratic forms, *Studia Math.*, **31**, 1968, 523–530.

KAPLANSKY, I., COHEN, I. S. (See I. S. Cohen, I. Kaplansky)

KARATSUBA, A. A. [72] Dirichlet's divisor problem in number fields, *DAN*, **204**, 1972, 540–541 (Russian).

KARATSUBA, A. A., ARKHIPOV, G. I. (See G. I. Arkhipov, A. A. Karatsuba)

KARIBAEV, S. K. [72a] Asymptotic formulae for the distribution of the elementary abelian fields of degree l^6 and type (l, l, l, l, l, l), *Rep. Alg. Sem. Kazakh. Gos.-Ped. Inst. Alma-Ata*, 1972, 1, 21–32 (Russian).

[72b] Asymptotic formulae for the distribution of the elementary Abelian fields of type (l, l, l, l, l, l) in the classes Q_v, *Rep. Alg. Sem. Kazakh. Gos.-Ped. Inst. Alma-Ata*, 1972, 2, 12–20 (Russian).

KÁTAI, I., KOVÁCS, B. [80] Kanonische Zahlensysteme in der Theorie der quadratischen algebraischen Zahlen, *Acta Sci. Math. (Szeged)*, **42**, 1980, 99–107.

[81] Canonical number systems in imaginary quadratic fields, *Acta Math. Acad. Sci. Hungar.*, **37**, 1981, 159–164.

KÁTAI, I., SZABÓ, J. [75] Canonical number-systems for complex integers, *Acta Sci. Math. (Szeged)*, **37**, 1975, 255–260.

KATAOKA, T. [79] Some types of ideals of group rings and its applications to algebraic number fields, *J. Fac. Sci. Tokyo*, **26**, 1979, 443–452.

[80] On the integer ring of the compositum of algebraic number fields, *Nagoya Math. J.*, **77**, 1980, 25–31.

KATAYAMA, K. [66] Kronecker's limit formulas and their applications, *J. Fac. Sci. Tokyo*, **13**, 1966, 1–44.

[76] On the values of ray-class L-functions for real quadratic fields, *J. Math. Soc. Japan*, **28**, 1976, 455–482.

[81] On the Galois cohomology groups of C_K/D_K, *PJA*, **57**, 1981, 378–380.

KATZ, N. M. [76] P-adic interpolation of real analytic Eisenstein series, *AnM*, **104**, 1976, 459–571.

[77] The Eisenstein measure and p-adic interpolation, *AJM*, **99**, 1977, 238–311.

[78] P-adic L-functions for CM fields, *Invent. Math.*, **49**, 1978, 199–297.

[81] Another look at p-adic L-functions for totally real fields, *MA*, **255**, 1981, 33–43.

KAUFMAN, R. M. [77] The geometric aspect of Linnik's theorem on the least prime, *Litovsk. Mat. Sb.*, **17**, 1977, 1, 111–114 (Russian).

[78a] An estimate of Hecke's L-functions of the Gaussian field on the line $\operatorname{Re} s = 1/2$, *DAN Belorus. SSR*, **22**, 1978, 25–82 (Russian).

[78b] A. F. Lavrik truncated equations, *Zap. Nauch. Sem. LOMI*, **76**, 1978, 124–158 (Russian).
[79] An evaluation of Hecke's *L*-functions on the critical line, *Zap. Nauch. Sem. LOMI*, **91**, 1979, 40–51 (Russian).

KAWADA, Y. [51] On the derivations in number fields, *AnM*, **54**, 1951, 302–314.
[53] On the ramification theory of infinite algebraic extensions, *AnM*, **58**, 1953, 24–47.
[54] On the structure of the Galois group of some infinite extensions, *J. Fac. Sci. Tokyo*, **7**, 1954, 1–18; II, ibid., 87–106.
[68] A remark on the principal ideal theorem, *J. Math. Soc. Japan*, **20**, 1968, 166–169.

KAZANDZIDIS, G. S. [63a] On the cyclotomic polynomial: coefficients, *Bull. Soc. Math. Grèce*, **4**, 1963, 1–11.
[63b] On the cyclotomic polynomial: morphology, estimates, *Bull. Soc. Math. Grèce*, **4**, 1963, 50–73.
[64a] On the cyclotomic polynomial $\Phi_m(z)$ (Estimates of $|\Phi_m(z)|/\Phi_m(\pm r)$, $r = |z|$), *Bull. Soc. Math. Grèce*, **5**, 1964, 11–17.
[64b] On the cyclotomic polynomial Φ_N (Upper estimates of $|\Phi_N(z)|$ on the *z*-plane), *Bull. Soc. Math. Grèce*, **5**, 1964, 18–36.

KEATING, M., FRÖHLICH, A., WILSON, S. (See A. Fröhlich, M. Keating, S. Wilson)

KELLER, O. H. [39] Eine Bemerkung zur Berechnung der Diskriminante imprimitiver Gleichungen, insbesondere der Ikosaedergleichung, *MA*, **116**, 1939, 456–462.

KELLY, J. B. [50] A closed set of algebraic integers, *AJM*, **72**, 1950, 565–572.

KEMPFERT, H. [62] Zum allgemeinen Hauptidealsatz, *JRAM*, **210**, 1962, 38–64; II, ibid., **223**, 1966, 28–55.
[69] On the factorization of polynomials, *JNT*, **1**, 1969, 116–120.

KENKU, M. A. [70] Determination of the even discriminants of complex quadratic fields with class-number 2, *PLMS*, (3), **22**, 1970, 734–746.
[75] On the *L*-function of quadratic forms, *JRAM*, **276**, 1975, 36–43.

KENNEDY, R. E. [80] Krull rings, *Pacific J. Math.*, **89**, 1980, 131–136.

KENZHEBAEV, S. [70] Distribution of special absolutely Abelian fields of the type (l, q), *IAN Kazakh. SSR*, 1970, 5, 85–87 (Russian).
[71] Distribution of special cyclic fields of degree lq, *IAN Kazakh. SSR*, 1971, 1, 84–86 (Russian).
[72] The algebra of special absolutely Abelian fields of type (l, l, q), *Rep. Alg. Sem. Kazakh. Gos. Ped. Inst. Alma-Ata*, 1972, 1, 32–53 (Russian).
[79] Calculation of the number of certain absolutely Abelian fields of the type (l, l, q), in: *The theory of nonregular curves* ..., 60–62, Alma-Ata 1979 (Russian).

KERSEY, D. [80] Modular units inside cyclotomic units, *AnM*, **112**, 1980, 361–380.

KESHTOV, R. A., BASHMAKOV, M. I. (See M. I. Bashmakov, R. A. Keshtov)

KESSLER, I., CHOWLA, S., LIVINGSTON, M. (See S. Chowla, I. Kessler, M. Livingston)

KIDA, Y. [80] *l*-extensions of *CM*-fields and cyclotomic invariants, *JNT*, **12**, 1980, 519–528.
[82] Cyclotomic Z_2-extensions of *J*-fields, *JNT*, **14**, 1982, 340–352.

KIMURA, N. [79a] Kummersche Kongruenzen für die verallgemeinerte Bernoullischen Zahlen, *JNT*, **11**, 1979, 171–187.
[79b] On the class number of real quadratic fields $Q(\sqrt{p})$ with $p \equiv 1 \pmod 4$, *Tokyo J. Math.*, **2**, 1979, 387–396.

KIMURA, T., HORIE, K. [82] On the Stickelberger ideal and the relative class number, *PJA*, **58**, 1982, 170–171.

KING, H. [68] Analytic representation for the L-series for certain ray class characters, *Comm. Pure Appl. Math.*, **21**, 1968, 523–533.

KINOHARA, A. [52] On the derivations and the relative differents in algebraic number fields, *J. Sci. Hiroshima Univ.*, **16**, 1952, 261–266.

[53] On the derivations and the relative different in commutative fields, *J. Sci. Hiroshima Univ.*, **16**, 1953, 441–456.

[54] On the different theorem in complete fields with respect to a discrete valuation, *J. Sci. Hiroshima Univ.*, **18**, 1954, 9–12.

KIRMSE, J. [24] Zur Darstellung total positiver Zahlen als Summen von vier Quadraten, *MZ*, **21**, 1924, 195–202.

KISELEV, A. A. [48] The expression of the class-number of ideals of real quadratic fields by Bernoulli numbers, *DAN*, **61**, 1948, 777–779 (Russian).

[55a] On the class-number of ideals in cubic fields, *Uchen. Zap. Leningr. Gos. Ped. Inst.*, **14**, 1955, 46–51 (Russian).

[55b] On a congruence relating the class-number of ideals in two quadratic fields, whose discriminants differ by a factor, *Uchen. Zap. Leningr. Gos. Ped. Inst.*, **14**, 1955, 52–56 (Russian).

[59] On certain congruences for the class-number of ideals of real quadratic fields, *Uchen. Zap. LGU*, **16**, 1959, 20–31 (Russian).

KISELEV, A. A., SLAVUTSKII, I. SH. [59] On the class-number of ideals in a quadratic fields and its rings, *DAN*, **126**, 1959, 1191–1194 (Russian).

[62] On certain congruences for the number of representations by sums of an odd number of squares, *DAN*, **143**, 1962, 272–274 (Russian).

[64] Transformation of Dirichlet's formulae and arithmetical computation of the class-number of ideals in quadratic fields, *Trudy IV Mat. S'ezda*, **2**, 1964, 105–112 (Russian).

KISILEVSKY, H. [70] Some results related to Hilbert's theorem 94, *JNT*, **2**, 1970, 199–206.

[76] Number fields with class number congruent to 4(mod 8) and Hilbert's theorem 94, *JNT*, **8**, 1976, 271–279.

[82] The Rédei-Reichardt theorem: another proof, in: *Ternary Quadratic Forms and Norms*, 1–4, M. Dekker, 1982.

KISILEVSKY, H., BOYD, D. W. (See D. W. Boyd, H. Kisilevsky)

KISILEVSKY, H. H., CARROLL, J. E. (See J. E. Carroll, H. H. Kisilevsky)

KISILEVSKY, H., DUMMIT, D. S. (See D. S. Dummit, H. Kisilevsky)

KITAOKA, Y. [71a] A simple proof of the functional equation of a certain L-function, *JNT*, **3**, 1971, 155–158.

[71b] A note on Hecke operators and theta series, *Nagoya Math. J.*, **42**, 1971, 189–195.

[77] Scalar extensions of quadratic lattices, *Nagoya Math. J.*, **66**, 1977, 139–149.

KITAYAMA, T., EDA, Y. (See Y. Eda, T. Kitayama)

KLEBOTH, H. [55] *Untersuchung über Klassenzahl und Reziprozitätsgesetz im Körper der l-ten Einheitswurzeln und die Diophantische Gleichung $X^{2l}+3^l Y^{2l} = Z^{2l}$ für eine Primzahl l grösser als* 3, Diss., Zürich 1955.

KLEIMAN, H. [71] Totally real subfields of p-adic fields having the symmetric group as Galois group, *Canad. Math. Bull.*, **14**, 1971, 441–442.

[72] Normal integral bases and the Galois groups of Galois extensions of number fields, *JRAM*, **255**, 1972, 83–84.

[73] A class of irreducible soluble polynomials in $Q[X]$, *JLMS*, (2), **7**, 1973, 571–576.

KLINGEN, H. [62] Über die Werte der Dedekindschen Zetafunktion, *MA*, **145**, 1961/62, 265–277.

KLINGEN, N. [78] Zahlkörper mit gleicher Primzerlegung, *JRAM*, **299/300**, 1978, 342–384.
[79] Atomare Kronecker-Klassen mit speziellen Galoisgruppen, *Hbg*, **48**, 1979, 42–53.
[80] Über schwache quadratische Zerlegungsgesetze, *Comment. Math. Helv.*, **55**, 1980, 645–651.
[83] Allgemeine Primstellen und Kroneckerklassen unendlicher Körpererweiterungen, *Res. Math.*, **6**, 1983, 183–193.
KLINGEN, J., BRUNOTTE, H., STEURICH, M. (See H. Brunotte, J. Klingen, M. Steurich)
KLOBE, W. [49] Über eine untere Abschätzung der n-ten Kreisteilungspolynome $g_n(z) = \prod_{d\mid n} (z^d-1)^{\mu(n/d)}$, *JRAM*, **187**, 1949, 68–69.
KLOOSTERMAN, H. D. [30] Thetareihen in total-reellen algebraischen Zahlkörpern, *MA*, **103**, 1930, 279–299.
KLOSS, K. E. [65] Some number-theoretic calculations, *J. Res. Nat. Bur. Standards*, **69**, 1965, 335–336.
KLØVE, T. [75] Sums of distinct Gaussian primes, *Nordisk Mat. Tidskr.*, **23**, 1975, 20–22.
KNAPOWSKI, S. [68] On Siegel's theorem, *AA*, **14**, 1968, 417–424.
[69] On a theorem of Hecke, *JNT*, **1**, 1969, 235–251.
KNEBUSCH, M., SCHARLAU, W. [71] Quadratische Formen und quadratische Reziprozitätsgesetze über algebraischen Zahlkörpern, *MZ*, **121**, 1971, 346–368.
KNESER, H. [42] Zur Stetigkeit der Wurzeln einer algebraischen Gleichung, *MZ*, **48**, 1942, 101–104.
KNESER, M. [75] Lineare Abhängigkeit von Wurzeln, *AA*, **26**, 1974/5, 307–308.
[82] Composition of binary quadratic forms, *JNT*, **15**, 1982, 406–413.
KNUTH, D. E. [69] *The Art of Computer Programming*, vol. II, Reading 1969.
KO CHAO, ERDÖS, P. (See P. Erdös, Chao Ko)
KOBAYASHI, M. [83] The connected component of the idèle class group of an algebraic number field, *Pacific J. Math.*, **106**, 1983, 129–134.
KOBAYASHI, S. [71] On the l-dimension of the ideal class group of Kummer extensions of a certain type, *PJA*, **18**, 1971, 399–404.
[73] On the 3-rank of the ideal class groups of certain pure cubic fields, *J. Fac. Sci. Univ. Tokyo*, **20**, 1973, 209–216; II, *ibid.*, **21**, 1974, 263–270.
[74] On the l-class rank in some algebraic number fields, *J. Math. Soc. Japan*, **26**, 1974, 668–676.
[77] Complete determination of the 3-class rank in pure cubic fields, *J. Math. Soc. Japan*, **29**, 1977, 373–384.
[79] Divisibility of class numbers of real cyclic extensions of degree 4 over Q, *JRAM*, **307/8**, 1979, 365–372.
[80] Divisibilité du nombre de classes des corps abéliens réels, *JRAM*, **320**, 1980, 142–149.
[82] L'indice de l'idéal de Stickelberger l-adique, *MZ*, **179**, 1982, 453–464.
KOBAYASHI, Y., MORIYA, M. [41a] Eine notwendige Bedingung für die eindeutige Primfaktorzerlegung der Ideale in einem kommutativen Ring, *Proc. Imp. Acad. Tokyo*, **17**, 1941, 129–133.
[41b] Eine hinreichende Bedingung für die eindeutige Primfaktorzerlegung der Ideale in einem kommutativen Ring, *Proc. Imp. Acad. Tokyo*, **17**, 1941, 134–138.
KOBLITZ, N. [77] *P-adic Numbers, p-adic Analysis and Zeta-functions*, Springer, 1977.
[78] Interpretation of the p-adic log gamma function and Euler constants using the Bernoulli measure, *TAMS*, **242**, 1978, 261–269.

[79] A new proof of certain formulas for p-adic L-functions, *Duke Math. J.*, **46**, 1979, 455–468.

KOCH, H. [61] Galois group of a local field, *DAN*, **137**, 1961, 1291–1294 (Russian).
[62] Über galoische Erweiterungen p-adischer Zahlkörper, *JRAM*, **209**, 1962, 8–11.
[63a] Über maximale l-Erweiterungen mit vorgeschriebenen Verzweigungsstellen, *Mon. Ber. DAW*, **5**, 1963, 341–344.
[63b] Über Darstellungsräume und die Struktur der multiplikativen Gruppe eines p-adischen Zahlkörpers, *MN*, **26**, 1963, 67–100.
[64a] Über Halbkörpern, die in algebraischen Zahlkörpern enthalten sind, *Acta Math. Acad. Sci. Hungar.*, **15**, 1964, 439–444.
[64b] Über den 2-Klassenkörperturm eines quadratischen Zahlkörpers, I, *JRAM*, **214/5**, 1964, 201–206.
[65a] l-Erweiterungen mit vorgeschriebenen Verzweigungsstellen, *JRAM*, **219**, 1965, 30–61.
[65b] l-Erweiterungen mit zwei Verzweigungsstellen, *Mon. Ber. DAW*, **7**, 1965, 616–623.
[65c] Über Galoische Gruppen von p-adischen Zahlkörpern, *MN*, **29**, 1965, 77–111.
[66] Zur Begründung der Arithmetik in algebraischen Zahl- und Funktionenkörpern, *Wiss. Z. Humboldt Univ. Berlin*, **15**, 1966, 203–206.
[67] Beweis einer Vermutung von Höchsmann aus der Theorie der l-Erweiterungen, *JRAM*, **225**, 1967, 203–206.
[68] Über die Dimension der Galoisschen Gruppen maximaler p-Erweiterungen, *Mon. Ber. DAW*, **10**, 1968, 5–8.
[69] Zum Satz von Golod–Schafarewitsch, *MN*, **42**, 1969, 321–333.
[70] *Galoissche Theorie der p-Erweiterungen*, Akademie-Verlag, 1970.
[72] Über das Normenrestsymbol einer lokalen unverzweigten Erweiterung von 2-Potenzgrad, *MN*, **52**, 1972, 355–369.
[74] Zur Galoischen Theorie der maximalen p-Erweiterungen mit vorgegebenen Verzweigungstellen, *MN*, **61**, 1974, 47–50.
[75] Zum Satz von Golod–Schafarewitsch, *JRAM*, **274/5**, 1975, 240–243.
[77] Fields of class two and Galois cohomology, in: *Algebraic Number Fields*, Proc. Durham Symp., 609–624, London 1977.
[78] Galois group of a p-closed extension of a local field, *DAN*, **238**, 1978, 19–22 (Russian).
KOCH, H., VENKOV, B. B. (See B. B. Venkov, H. Koch)
KOCH, H., ZINK, W. [72] Über die 2-Komponente der Klassengruppe quadratischer Zahlkörper mit zwei Diskriminantenteilern, *MN*, **54**, 1972, 309–333.
KOCHEN, S., AX, J. (See J. Ax, S. Kochen)
KOKSMA, J. F., MEULENBELD, B. [41a] Simultane Approximationen in imaginären quadratischen Zahlkörpern, *Proc. Akad. Wet. Amsterdam*, **44**, 1941, 310–323.
[41b] Diophantische Approximationen homogener Linearformen in imaginären quadratischen Zahlkörpern, *Proc. Akad. Wet. Amsterdam*, **44**, 1941, 426–434.
KOLSTER, M. [84] On the Birch–Tate conjecture for maximal real subfields of cyclotomic fields, in: *Algebraic K-theory, Number Theory, Geometry and Analysis*, 229–234, LN 1046, Springer, 1984.
KOLSTER, M., HURRELBRINK, J. (See J. Hurrelbrink, M. Kolster)
KOMATSU, K. [74] The Galois group of the algebraic closure of an algebraic number field, *Kōdai Math. Sem. Rep.*, **26**, 1974/5, 44–52.
[75] Integral bases in algebraic number fields, *JRAM*, **278/9**, 1975, 137–144.
[76a] Discriminants of certain algebraic number fields, *JRAM*, **285**, 1976, 114–125.
[76b] An integral basis of the algebraic number field $Q(\sqrt[l]{a}, \sqrt[l]{1})$, *JRAM*, **288**, 1976, 152–153.

[76c] On the adele rings of algebraic number fields, *Kōdai Math. Sem. Rep.*, **28**, 1976, 78–84.
[78] On the adele rings and zeta-functions of algebraic number fields, *Kōdai Math. J.*, **1**, 1978, 394–400.
[82] On a certain property of profinite groups, *PJA*, **58**, 1982, 319–322.
[84] On adele rings of arithmetically equivalent fields, *AA*, **43**, 1984, 93–95.

KONDAKOVA, L. F. [71] Application of the large sieve to the solution of additive problems of mixed type, *Mat. Zametki*, **10**, 1971, 73–81 (Russian).

KÖNIG, R. [13] Über quadratische Formen und Zahlkörper, sowie zwei Gruppensätze, *Jahresber. Deutsch. Math. Verein.*, **22**, 1913, 239–254.

KONNO, S. [65] On Kronecker's limit formula in a totally imaginary quadratic field over a totally real algebraic number field, *J. Math. Soc. Japan*, **17**, 1965, 412–424.

KORCHAGINA, V. I. [79] The distribution of prime ideal numbers in a quadratic field, *DAN Tadzhik. SSR*, **22**, 1979, 152–154 (Russian).

KORCHAGINA, V. I., BABAEV, G. (See G. Babaev, V. I. Korchagina)

KÖRNER, O. [60] Übertragung des Goldbach–Vinogradovschen Satzes auf reell-quadratische Zahlkörper, *MA*, **141**, 1960, 343–360.
[61a] Erweiterter Goldbach–Vinogradovscher Satz in beliebigen algebraischen Zahlkörpern, *MA*, **143**, 1961, 344–378.
[61b] Zur additiver Primzahltheorie algebraischer Zahlkörper, *MA*, **144**, 1961, 97–104.
[61c] Über das Waringsche Problem in algebraischen Zahlkörpern, *MA*, **144**, 1961, 224–238.
[62] Über Mittelwerte trigonometrischer Summen und ihre Anwendung in algebraischen Zahlkörpern, *MA*, **147**, 1962, 205–239; corr., *ibid.*, **149**, 1963, 462.
[63] Ganze algebraische Zahlen als Summen von Polynomwerten, *MA*, **149**, 1963, 97–104.
[64] Darstellung ganzer Grössen durch Primzahlpotenzen in algebraischen Zahlkörpern, *MA*, **155**, 1964, 204–245.
[70] Über durch Potenzen erzeugte Ringe und Gruppen in algebraischen Zahlkörpern, *Manuscr. Math.*, 3, 1970, 157–174.
[73] Quadratsummen und Kongruenzlösungsdichten in p-adischen Zahlkörpern, *MN*, **57**, 1973, 15–38; II, *ibid.*, **59**, 1974, 63–73.

KÖRNER, O. H., EARNEST, A. G. (See A. G. Earnest, O. H. Körner)

KÖRNYEI, I. [62] Eine Bemerkung zur Theorie der durch quadratischen Formen darstellbaren Primzahlen, *Ann. Univ. Budapest*, **5**, 1962, 95–108.

KOSHLYAKOV, N. S. [37] On Kronecker's fundamental limit formula in the theory of quadratic fields, *Mitt. Forsch. Inst. Math. Mech. Univ. Tomsk*, **1**, 1937, 237–241.

KOSTANDI, G. [43] Une propriété des équations irréductibles de la division de cercle, *Bull. Éc. Polytech. Bucarest*, **14**, 1943, 10–18.

KOSTRIKIN, A. J. [65] Defining groups by generators and relations, *IAN*, **29**, 1965, 1119–1122 (Russian).

KÖTHE, G. [35] Verallgemeinerte Abelsche Gruppen mit hyperkomplexen Operatorenring, *MZ*, **39**, 1935, 31–44.

KOVÁCS, B. [81] Canonical number systems in algebraic number fields, *Acta Math. Acad. Sci. Hungar.*, **37**, 1981, 405–407.

KOVÁCS, B., KÁTAI, I. (See I. Kátai, B. Kovács)

KOVAL'CHIK, F. B. [74] Density theorems and the distribution of primes in sectors and progressions, *DAN*, **219**, 1974, 31–34 (Russian).
[75a] On a generalization of the Halász–Montgomery method, *Litovsk. Mat. Sb.*, **15**, 1975, 3, 139–149 (Russian).

[75b] Density theorems for sectors and progressions, *Litovsk. Mat. Sb.*, **15**, 1975, 4, 133–151 (Russian).

KOWALSKY, H.-J., DÜRBAUM, H. (See H. Dürbaum, H.-J. Kowalsky)

KOYAMA, T., NISHI, M., YANAGIHARA, H. [74] On characterizations of Dedekind domains, *Hiroshima Math. J.*, **4**, 1974, 71–74.

KRAKOWSKI, F. [65a] A remark on the lemma of Gauss, *Pacific J. Math.*, **15**, 1965, 917–920.

[65b] On the content of polynomials, *PAMS*, **16**, 1965, 810–812.

KRAMER, K., CANDIOTTI, A. [78] On K_2 and Z_l-extensions of number fields, *AJM*, **100**, 1978, 177–196.

KRASNER, M. [34] Sur le premier cas du théorème de Fermat, *CR*, **199**, 1934, 256–258.

[35] Sur la théorie de la ramification des idéaux, *CR*, **200**, 1935, 1813–1815.

[36] Sur la représentation multiplicative dans les corps de nombres P-adiques relativement galoisiens, *CR*, **203**, 1936, 907–908.

[37a] Définition de certains anneaux non commutatifs. Classification des extensions primitives des corps à valuation discrète, *CR*, **205**, 1937, 772–774; corr., *ibid.*, 1111, 1347–1348; *ibid.*, **206**, 1938, 288.

[37b] Le nombre des surcorps d'un degré donné d'un corps de nombres p-adiques, *CR*, **205**, 1937, 1026–1028; corr., *ibid.*, 1267.

[37c] Sur la primitivité des corps P-adiques, *Mathematica*, **13**, 1937, 72–191.

[37d] Sur la théorie de ramification des idéaux de corps de nombres algébriques, *Mém. Acad. Belg.*, **20**, 1937, 4, 1–110.

[38] Le nombre de surcorps primitifs d'un degré donné et le nombre des surcorps métagaloisiens d'une degré donné d'un corps de nombres p-adiques, *CR*, **206**, 1938, 876–878; corr., *ibid.*, 1152.

[39] Sur la représentation exponentielle dans les corps relativement galoisiens de nombres p-adiques, *AA*, **3**, 1939, 133–173.

[40] La loi de Jordan–Hölder dans les hypergroupes et les suites génératrices des corps de nombres p-adiques, *Duke Math. J.*, **6**, 1940, 120–140; II, *ibid.*, **7**, 1940, 121–135.

[44] Théorie de la ramification dans les extensions finies des corps valués, *CR*, **219**, 1944, 539–541; *ibid.*, **220**, 1945, 28–30, 761–763; *ibid.*, **221**, 1945, 737–739.

[46] Théorie non abélienne des corps de classes pour les extensions finies et séparables des corps valués complets, *CR*, **222**, 1946, 626–628, 984–986, 1370–1372; *ibid.*, **224**, 1947, 173–175, 434–436.

[47] Théorie non-abélienne des corps de classes pour les extensions galoisiennes des corps de nombres algébriques, *CR*, **225**, 1947, 785–787, 973–975, 1113–1115; *ibid.*, **226**, 1948, 535–537, 1231–1233, 1656–1658.

[49] Le produit complet et la théorie de la ramification, *CR*, **229**, 1949, 1103–1105, 1287–1289; *ibid.*, **230**, 1950, 162–164.

[50] Quelques méthodes nouvelles dans la théorie des corps valués complets, in: *Alg. Th. des Nombres, Coll. Internat. CNRS*, **24**, 1950, 29–39.

[52] Généralisation nonabélienne de la théorie locale des corps de classes, *Proc. ICM Cambridge*, Mass., 1950, II, 71–76, Providence 1952.

[62] Nombre des extensions d'un degré donné d'un corps p-adique, *CR*, **254**, 1962, 3470–3472; *ibid.*, **255**, 1962, 224–226, 1682–1684, 2342–2344, 3095–3097.

[66] Nombre des extensions d'un degré donné d'un corps p-adique, in: *Les tendances géométriques en algèbre et théorie des nombres*, 143–169, Paris 1966.

[79] Remarques au sujet d'une note de J. P. Serre, *CR*, **288**, 1979, A863–865.

KRASNER, M., RANULAC, B. [37] Sur une propriété des polynômes de la division du cercle, *CR*, **204**, 1937, 397-399.

KRÄTZEL, E. [67] Kubische und biquadratische Gaussche Summen, *JRAM*, **228**, 1967, 159-165.

KRAUSE, U. [84] A characterization of algebraic number fields with cyclic class groups of prime power order, *MZ*, **186**, 1984, 143-148.

KROMPIEWSKA, E. [77] On some sequences of integral ideals, *Funct. et Approx.*, **5**, 1977, 65-68.

KRONECKER, L. [45a] *De unitatibus complexis*, Diss., Berlin 1845 = *Werke*, *I*, 5-73, Leipzig 1895.

[45b] Beweis, dass für jede Primzahl p die Gleichung $1+x+x^2+ \ldots +x^{p-1} = 0$ irreducibel ist, *JRAM*, **29**, 1845, 280 = *Werke*, *I*, 3-4, Leipzig 1895.

[53] Über die algebraisch auflösbaren Gleichungen, *Mon. Ber. Kgl. Preuß. AW*, 1853, 365-374; ibid., 1856, 203-215 = *Werke*, *IV*, 1-11, 25-37, Leipzig-Berlin 1929.

[54] Mémoire sur les facteurs irréductibles de l'expression x^n-1, *J. Math. Pures Appl.*, **19**, 1854, 177-192 = *Werke*, *I*, 75-92, Leipzig 1895.

[56a] Sur une formule de Gauss, *J. Math. Pures Appl.*, (2), **1**, 1856, 392-395 = *Werke*, *IV*, 171-175, Leipzig-Berlin 1929.

[56b] Démonstration d'un théorème de M. Kummer, *J. Math. Pures Appl.*, (2), **1**, 1856, 396-398 = *Werke*, *I*, 93-97, Leipzig 1895.

[56c] Démonstration de l'irréductibilité de l'équation $x^{n-1}+x^{n-2}+ \ldots +1 = 0$ ou n désigne un nombre premier, *J. Math. Pures Appl.*, (2), **1**, 1856, 399-400 = *Werke*, *I*, 99-102, Leipzig 1895.

[57a] Zwei Sätze über Gleichungen mit ganzzahligen Coefficienten, *JRAM*, **53**, 1857, 173-175 = *Werke*, *I*, 103-108, Leipzig 1895.

[57b] Über complexe Einheiten, *JRAM*, **53**, 1857, 176-181 = *Werke*, *I*, 109-118, Leipzig 1895.

[63] Über die Klassenzahl der aus Wurzeln der Einheit gebildeten complexen Zahlen, *Mon. Ber. Kgl. Preuß. AW*, 1863, 340-341 = *Werke*, *I*, 123-131, Leipzig 1895.

[64] Über den Gebrauch der Dirichletschen Methoden in der Theorie der quadratischen Formen, *Mon. Ber. Kgl. Preuß. AW*, 1864, 285-303 = *Werke*, *IV*, 227-244, Leipzig-Berlin 1929.

[77] Über Abelsche Gleichungen, *Mon. Ber. Kgl. Preuß. AW*, 1877, 845-851 = *Werke*, *IV*, 63-71, Leipzig-Berlin 1929.

[80] Über die Irreduktibilität der Gleichungen, *Mon. Ber. Kgl. Preuß. AW*, 1880, 155-163 = *Werke*, *II*, 85-93, Leipzig 1897.

[82] Grundzüge einer arithmetischer Theorie der algebraischen Grössen, *JRAM*, **92**, 1882, 1-122 = *Werke*, *II*, 237-387, Leipzig 1897.

[83] Sur les unités complexes, *CR*, **96**, 1883, 93-98, 148-152, 216-222 = *Werke*, III_1, 1-20, Leipzig 1899.

[84] Additions au mémoire sur les unités complexes, *CR*, **99**, 1884, 765-771 = *Werke*, III_1, 21-30, Leipzig 1899.

[85] Zur Theorie der elliptischen Funktionen, *SB Kgl. Preuß. AW*, 1885, 761-784; ibid., 1889, 123-135 = *Werke*, *IV*, 363-389, 482-495, Leipzig-Berlin 1929.

[89] Summirung der Gaussischen Reihen $\sum_{h=0}^{h=n-1} e^{\frac{2h^2\pi i}{n}}$, *JRAM*, **105**, 1889, 267-268 = *Werke*, *IV*, 295-300, Leipzig-Berlin 1929.

[90] Über die Dirichletsche Methode der Wertbestimmung der Gauss'schen Reihen, *Festschr. Math. Ges. Hamburg*, 1890, 32-36 = *Werke*, *IV*, 301-307, Leipzig-Berlin 1929.

KROON, J. P. M. DE [65] The asymptotic behaviour of additive functions in algebraic number theory, *Compos. Math.*, **17**, 1965, 207–261.

KRULL, W. [25] Über Multiplikationsringe, *SB Heidelberg. AW, Mat. Nat. Kl.*, 1925, **5**, 13–18.

[28a] Zur Theorie der allgemeinen Zahlringen, *MA*, **98**, 1928, 51–70.

[28b] Idealtheorie in unendlichen Zahlkörpern, *MZ*, **29**, 1928, 42–54; II, *ibid.*, **31**, 1930, 527–557.

[28c] Galoissche Theorie der unendlichen algebraischen Erweiterungen, *MA*, **100**, 1928, 687–698.

[30a] Ein Hauptsatz über umkehrbare Ideale, *MZ*, **31**, 1930, 558.

[30b] Galoissche Theorie bewerteter Körper, *SB Bayer. AW*, 1930, 225–238.

[31a] Über die Zerlegung der Hauptideale in allgemeinen Ringen, *MA*, **105**, 1931, 779–785.

[31b] Allgemeine Bewertungstheorie, *JRAM*, **167**, 1931, 160–196.

[32] Matrizen, Moduln und verallgemeinerte abelsche Gruppen im Bereich der ganzen algebraischen Zahlen, *SB Heidelberg. AW, Math. Nat. Kl.*, 1932, **2**, 1–38.

[35] *Idealtheorie*, Springer, 1935.

[51] Zur Arithmetik der endlichen diskreten Hauptordnungen, *JRAM*, **189**, 1951, 118–128.

[52] Über unendliche algebraische Erweiterungen bewerteter Körper, *Rend. Circ. Mat. Palermo*, (2), **1**, 1952, 164–169.

[59] Über ein Existenzsatz der Bewertungstheorie, *Hbg*, **23**, 1959, 29–35.

KRUYSWIJK, D., BAAYEN, P. C., EMDE BOAS, P. van (See P. C. Baayen, P. van Emde Boas, D. Kruyswijk)

KRUYSWIJK, D., EMDE BOAS, P. VAN (See P. van Emde Boas, D. Kruyswijk)

KUBERT, D. [85] Jacobi sums and Hecke characters, *AJM*, **107**, 1985, 253–280.

KUBERT, D. S., LANG, S. [79] Modular units inside cyclotomic units, *Bull. Soc. Math. France*, **107**, 1979, 161–178.

[81] *Modular Units*, Springer, 1981.

KUBERT, D., LICHTENBAUM, S. [83] Jacobi-sum Hecke characters and Gauss sum identities, *Compos. Math.*, **48**, 1983, 55–87.

KUBILIUS, I. P. [50] Distribution of prime numbers of the Gaussian fields in sectors and contours, *Uchen. Zap. LGU*, **37**, 1950, 19, 40–52 (Russian).

[51] The decomposition of prime numbers into two squares, *DAN*, **77**, 1951, 791–794 (Russian).

[52] On certain problems of the geometry of prime numbers, *Mat. Sbornik*, **31**, 1952, 507–542.

[55] On a problem in the n-dimensional analytic theory of numbers, *Vilniaus Valst. Univ. Mokslu Darbai*, **4**, 1955, 5-43 (Lithuanian).

KUBO, K. [40] Über die Noethersche fünf Axiome in kommutativen Ringen, *J. Fac. Sci. Hiroshima Univ.*, **10**, 1940, 77–84.

KUBOKAWA, Y. [77] The genus field for composite of quadratic fields, *J. Saitama Univ. Fac. Ed. Math. Natur. Sci.*, **26**, 1977, 1–3.

KUBOTA, K. K. [72a] Note on a conjecture of W. Narkiewicz, *JNT*, **4**, 1972, 181–190.

[72b] Factors of polynomials under composition, *JNT*, **4**, 1972, 587–595.

[73] Image sets of polynomials, *AA*, **23**, 1973, 183–194.

KUBOTA, K. K., LIARDET, P. [76] Réfutation d'une conjecture de W. Narkiewicz, *CR*, **282**, 1976, A1261–1264.

KUBOTA, T. [53] Über die Beziehungen der Klassenzahlen der Unterkörper des bizyklischen biquadratischen Zahlkörpers, *Nagoya Math. J.*, **6**, 1953, 119–127.

[56a] Density in a family of abelian extensions, *Proc. Int. Symp. Alg. N. Th.*, 77–91, Tokyo 1956.
[56b] Über den bizyklischen biquadratischen Zahlkörper, *Nagoya Math. J.*, **10**, 1956, 65–85.
[57] Galois group of the maximal abelian extension over an algebraic number field, *Nagoya Math. J.*, **12**, 1957, 177–189.
[60] Local relation of Gauss sums, *AA*, **6**, 1960/1, 285–294.
[61] Über quadratische Charaktersummen, *Nagoya Math. J.*, **19**, 1961, 15–25.
[63] Über eine Verallgemeinerung der Reziprozität der Gausssschen Summen, *MZ*, **82**, 1963, 91–100.
[64] Some arithmetical applications of an elliptic function, *JRAM*, **214/5**, 1964, 141–145.
[68] On a special kind of Dirichlet series, *J. Math. Soc. Japan*, **20**, 1968, 193–207.
KUBOTA, T., LEOPOLDT, H. W. [64] Eine p-adische Theorie der Zeta-Werte, I, *JRAM*, **214/5**, 1964, 328–339.
KUDO, A. [72] On the reflection theorem in prime cyclotomic fields, *Mem. Fac. Sci. Kyushu Univ.*, *A*, **26**, 1972, 333–337.
[75a] On a generalization of a theorem of Kummer, *Mem. Fac. Sci. Kyushu Univ.*, *A*, **29**, 1975, 255–261.
[75b] On a class number relation of imaginary abelian fields, *J. Math. Soc. Japan*, **27**, 1975, 150–159.
[78] Generalized Bernoulli numbers and the basic Z_p-extension of imaginary quadratic number fields, *Mem. Fac. Sci. Kyushu Univ.*, *A*, **32**, 1978, 191–198.
KÜHNOVA, J. [79] Maillet's determinant $D_{p^{n+1}}$, *Arch. Math. (Brno)*, **15**, 1979, 209–212.
KUIPERS, L., BURKE, J. R. (See J. R. Burke, L. Kuipers)
KUIPERS, L., NIEDERREITER, H., SHIUE, J. S. [75] Uniform distribution of sequences in the ring of Gaussian integers, *Bull. Inst. Math. Acad. Sinica*, **3**, 1975, 311–325.
KULKARNI, R. S. [67] On a theorem of Jensen, *AMM*, **74**, 1967, 960–961.
KUMMER, E. E. [42] Eine Aufgabe, betreffend die Theorie der cubischen Reste, *JRAM*, **23**, 1842, 285–286 = *Coll. Papers*, *I*, 143–144, Springer, 1975.
[46] De residuis cubicis disquisitiones nonnullae analyticae, *JRAM*, **32**, 1846, 341–359 = *Coll. Papers*, *I*, 145–163, Springer, 1975.
[47a] Zur Theorie der complexen Zahlen, *JRAM*, **35**, 1847, 319–325 = *Coll. Papers*, *I*, 203–210, Springer, 1975.
[47b] Über die Zerlegung der aus Wurzeln der Einheit gebildeten complexen Zahlen in ihre Primfaktoren, *JRAM*, **35**, 1847, 327–367 = *Coll. Papers*, *I*, 211–251, Springer, 1975.
[50a] Bestimmung der Anzahl nicht Äquivalenter Classen für die aus λ-ten Wurzeln der Einheit gebildeten complexen Zahlen und die ideale Faktoren derselben, *JRAM*, **40**, 1850, 93–116 = *Coll. Papers*, *I*, 299–322, Springer, 1975.
[50b] Zwei besondere Untersuchungen über die Classen Anzahl und über die Einheiten der aus λ-ten Wurzeln der Einheit gebildeten complexen Zahlen, *JRAM*, **40**, 1850, 117–129 = *Coll. Papers*, *I*, 323–335, Springer, 1975.
[50c] Allgemeiner Beweis des Fermatschen Satzes, dass die Gleichung $x^\lambda + y^\lambda = z^\lambda$ durch ganze Zahlen unlösbar ist, für alle diejenige Potenz-Exponenten λ, welche ungerade Primzahlen sind und in den Zählern der ersten $\frac{1}{2}(\lambda-3)$ Bernoulli'schen Zahlen als Factoren nicht vorkommen, *JRAM*, **40**, 1850, 130–138 = *Coll. Papers*, *I*, 336–344, Springer, 1975.
[51] Mémoire sur la théorie des nombres complexes composés de racines de l'unité et des nombres entiers, *J. Math. Pures Appl.*, **16**, 1851, 377–498 = *Coll. Papers*, *I*, 363–484, Springer, 1975.

[52] Über die Ergänzungssätze zu den allgemeinen Reziprozitätsgesetzen, *JRAM*, **44**, 1852, 93–146 = *Coll. Papers, I*, 485–538, Springer, 1975.

[53] Über die Irregularität der Determinanten, *Mon. Ber. Kgl. Preuß. AW*, 1853, 194–200 = *Coll. Papers, I*, 539–545, Springer, 1975.

[55] Über eine besondere Art aus complexen Einheiten gebildeten Ausdrücke, *JRAM*, **50**, 1855, 212–232 = *Coll. Papers, I*, 552–572, Springer, 1975.

[56] Theorie der idealen Primfaktoren der complexen Zahlen, welche aus der Wurzeln der Gleichung $\omega^n = 1$ gebildet sind, wenn n eine zusammengesetzte Zahl ist, *Abh. Kgl. Preuß. AW*, 1856, 1–47 = *Coll. Papers, I*, 583–629, Springer, 1975.

[57] Einige Sätze über die aus den Wurzeln der Gleichung $\alpha^\lambda = 1$ gebildeten complexen Zahlen für den Fall, dass die Classenzahl durch λ teilbar ist, nebst Anwendung derselben auf einen weiteren Beweis des letzten Fermatschen Lehrsatzes, *Abh. Kgl. Preuß. AW*, 1857, 41–74 = *Coll. Papers, I*, 639–672, Springer, 1975.

[59] Über die allgemeinen Reziprozitätsgesetze unter den Resten und Nichtresten der Potenzen, deren Grad eine Primzahl ist, *Abh. Kgl. Preuß. AW*, 1859, 19–159 = *Coll. Papers, I*, 699–839, Springer, 1975.

[61] Über die Classenanzahl der aus n-ten Einheitswurzeln gebildeten complexen Zahlen, *Mon. Ber. Kgl. Preuß. AW*, 1861, 1051–1053 = *Coll. Papers, I*, 883–885, Springer, 1975.

[63] Über die Classenanzahl der aus zusammengesetzten Einheitswurzeln gebildeten idealen complexen Zahlen, *Mon. Ber. Kgl. Preuß. AW*, 1863, 21–28 = *Coll. Papers, I*, 887–894, Springer, 1975.

[70] Über eine Eigenschaft der Einheiten der aus den Wurzeln der Gleichung $\alpha^\lambda = 1$ gebildeten complexen Zahlen und über den zweiten Faktor der Classenzahl, *Mon. Ber. Kgl. Preuß. AW*, 1870, 855–880 = *Coll. Papers, I*, 919–944, Springer, 1975.

KUNERT, D. [35] Ein neuer Beweis für die Reziprozitätsformel der Gaussschen Summen in beliebigen algebraischen Zahlkörpern, *MZ*, **40**, 1935, 326–347.

KUNIYOSHI, H. [58] Cohomology theory and different, *Tôhoku Math. J.*, **10**, 1958, 313–337.

KUNIYOSHI, H., TAKAHASHI, S. [53] On the principal genus theorem, *Tôhoku Math. J.*, **5**, 1953, 128–131.

KUNZ, E. [60] Die Primidealteiler der Differenten in allgemeinen Ringen, *JRAM*, **204**, 1960, 165–182.

[68] Vollständige Durchschnitte und Differenten, *Arch. Math.*, **19**, 1968, 47–58.

KURMANALIN, H. [66] Completely critical cyclic fields of degree l^h, *Uchen. Zap. Kazakh. Gos. Ped. Inst.*, **23**, 1966, 19–23 (Russian).

KURODA, S. [43a] Über den Dirichletschen Körper, *J. Fac. Sci. Univ. Tokyo*, **4**, 1943, 383–406.

[43b] Über die Pellsche Gleichung, *Proc. Imp. Acad. Tokyo*, **19**, 1943, 611–612.

[50] Über die Klassenzahlen algebraischer Zahlkörper, *Nagoya Math. J.*, **1**, 1950, 1–10.

[51] Über die Zerlegung rationaler Primzahlen in gewissen nicht-abelschen galoisschen Körpern, *J. Math. Soc. Japan*, **3**, 1951, 148–156.

KURODA, S. N. [62] On a theorem of Minkowski, *Sûgaku*, **14**, 1962/3, 171–172 (Japanese).

[64a] On the class number of imaginary quadratic number fields, *PJA*, **40**, 1964, 365–367.

[64b] Über die Klassenzahl eines relativzyklischen Zahlkörpers vom Primzahlgrade, *PJA*, **40**, 1964, 623–626.

[68] Some results on Γ-extensions of algebraic number fields, *J. Math. Soc. Japan*, **20**, 1968, 208–222.

[70a] Idealgruppen und Dirichletsche Reihen in algebraischen Zahlkörpern, *J. Math. Soc. Japan*, **22**, 1970, 353–387.
[70b] Über den allgemeinen Spiegelungssatz für Galoische Zahlkörper, *JNT*, **2**, 1970, 282–297.
[71] Kapitulation von Idealklassen in einer Γ-Erweiterung, in: *Sem. Modern Methods in N.Th.*, 4 pp., Tokyo 1971.
[74] Über eine Klasse von L-Funktionen algebraischer Zahlkörper, *Nagoya Math. J.*, **55**, 1974, 151–159.

KUROKAWA, N. [78a] On the meromorphy of Euler products, *PJA*, **54**, 1978, 163–166.
[78b] On Linnik's problem, *PJA*, **54**, 1978, 167–169.

KÜRSCHAK, J. [13] Über Limesbildung und allgemeine Körpertheorie, *JRAM*, **142**, 1913, 211–253.

KURSHAN, R. P., ODLYZKO, A. M. [81] Values of cyclotomic polynomials at roots of unity, *Math. Scand.*, **49**, 1981, 15–35.

KUTSUNA, M. [74] On the fundamental units of real quadratic fields, *PJA*, **50**, 1974, 580–583.
[80] On a criterion for the class number of a quadratic number field to be one, *Nagoya Math. J.*, **79**, 1980, 123–129.

KUZMIN, L. V. [69] Homologies of profinite groups, the Schur multiplicator, and class field theory, *IAN*, **33**, 1969, 1220–1254 (Russian).
[72] The Tate module of algebraic number fields, *IAN*, **36**, 1972, 267–327 (Russian).
[75a] Cohomological dimension of certain Galois groups, *IAN*, **39**, 1975, 487–495 (Russian).
[75b] Local extensions associated with l-extensions with given ramification, *IAN*, **39**, 1975, 739–772 (Russian).
[81] Some remarks on an l-adic Dirichlet theorem and the l-adic regulator, *IAN*, **45**, 1981, 1203–1240 (Russian).

KWON, S. H. [84] Corps de nombres de degré 4 de type alterné, *CR*, **299**, 1984, 41–43.

LABESSE, J. P., GÉRARDIN, P. (See P. Gérardin, J. P. Labesse)

LABUTE, J. [65] Classification des groupes de Demuškin, *CR*, **260**, 1965, 1043–1046.
[66a] Le groupe de Demuškin de rang dénombrable, *CR*, **262**, 1966, A4–7.
[66b] Demuškin groups of rank $\aleph_0$, *Bull. Soc. Math. France*, **94**, 1966, 211–244.
[67] Classification of Demushkin groups, *CJM*, **19**, 1967, 106–132.

LAGARIAS, J. C. [78] Signatures of units and congruences (mod 4) in certain real quadratic fields, *JRAM*, **301**, 1978, 142–146; II, *ibid.*, **320**, 1980, 115–126.
[80a] On the computational complexity of determining the solvability or unsolvability of the equation $X^2 - DY^2 = -1$, *TAMS*, **260**, 1980, 485–508.
[80b] Signatures of units and congruences (mod 4) in certain totally real fields, *JRAM*, **320**, 1980, 1–5.
[80c] On determining the 4-rank of the ideal class group of a quadratic field, *JNT*, **12**, 1980, 191–196.

LAGARIAS, J. C., COHN, H. (See H. Cohn, J. C. Lagarias)

LAGARIAS, J. C., LENSTRA, H. W. JR. [81] Problem A 6341, *AMM*, **88**, 1981, 294.

LAGARIAS, J. C., MONTGOMERY, H. L., ODLYZKO, A. M. [79] A bound for the least prime ideal in the Chebotarev density theorem, *Invent. Math.*, **54**, 1979, 271–296.

LAGARIAS, J. C., ODLYZKO, A. M. [77] Effective versions of the Chebotarev density theorem, in: *Algebraic Number Fields*, Proc. Durham Symp., 409–464, London 1977.
[79] On computing Artin L-functions in the critical strip, *MC*, **33**, 1979, 1081–1085.

LAGRANGE, J. L. [66] Solution d'un problème d'arithmétique, *Misc. Taurinensia*, **4**, 1766/69 = *Oeuvres, I*, 671–731, Paris 1867.

[73] Recherches d'arithmétique, *N. Mém. Acad. Roy. Sci. Bel. Lettr. de Berlin*, 1773, 1775 = *Oeuvres, III*, 695–795, Paris 1869.

LAI DYK THIN' [62] On the number of divisors in an angle, *DAN*, **143**, 1962, 28–30; *ibid.*, **149**, 1963, 513–515 (Russian).

[65] On the number of divisors in angles, *Mat. Sbornik*, **67**, 1965, 345–365 (Russian).

LAKEIN, R. B. [69] A Gauss bound for a class of biquadratic fields, *JNT*, **1**, 1969, 108–112.

[71] Class numbers and units of complex quartic fields, in: *Computers in Number Theory*, 167–172, Academic Press, 1971.

[72] Euclid's algorithm in complex quartic fields, *AA*, **20**, 1972, 393–400.

[74a] Bounds for consecutive k-th power residues in the Eisensteinian integers, *JNT*, **6**, 1974, 318–323.

[74b] Computation of the ideal class group of certain complex quartic fields, *MC*, **28**, 1974, 839–846; *ibid.*, **29**, 1975, 137–144.

LAKKIS, K. [66a] Die galoisschen Gauss'schen Summen von Hasse, *Bull. Soc. Math. Grèce*, **7**, 1966, 183–371.

[66b] Die verallgemeinerten Gaussschen Summen, *Arch. Math.*, **17**, 1966, 505–509.

[67] Die lokalen verallgemeinerten Gauss'schen Summen, *Bull. Soc. Math. Grèce*, **8**, 1967, 143–150.

[69] Über die Weilsche Wurzelzahl, *JRAM*, **234**, 1969, 197–206.

LAL, S., MCFARLAND, R. L., ODONI, R. W. K. [77] On $|\alpha|^2 + |\beta|^2 = p^t$ in certain fields, in: *Number Theory and Algebra*, 195–197, New York 1977.

LAM, T. Y., DAI, Z. D., PENG, C. K. (See Z. D. Dai, T. Y. Lam, C. K. Peng)

LAM, T. Y., SIU, M. K. [75] K_0 and K_1—an introduction to algebraic K-theory, *AMM*, **82**, 1975, 329–363.

LAMPRECHT, E. [53] Allgemeine Theorie der Gaussschen Summen in endlichen kommutativen Ringen, *MN*, **9**, 1953, 149–196.

[57] Struktur und Relationen allgemeiner Gaussscher Summen in endlichen Ringen, I, *JRAM*, **197**, 1957, 1–26; II, *ibid.*, 27–48.

[67] Existenz von Zahlkörpern mit nicht abbrechenden Klassenkörperturm, *Arch. Math.*, **18**, 1967, 140–152.

LÁNCZI, E. [65] Unique prime factorization in imaginary quadratic number fields, *Acta Math. Acad. Sci. Hungar.*, **16**, 1965, 453–466; add., *Ann. Univ. Budapest*, **26**, 1983, 195–196.

LANDAU, E. [03a] Ueber die zu einem algebraischen Zahlkörper gehörige Zetafunction und die Ausdehnung der Tschebyscheffschen Primzahlentheorie auf das Problem der Verteilung der Primideale, *JRAM*, **125**, 1903, 64–188.

[03b] Neuer Beweis des Primzahlsatzes und Beweis des Primidealsatzes, *MA*, **56**, 1903, 645–670.

[03c] Über die Klassenzahl der binären quadratischen Formen von negativer Discriminante, *MA*, **56**, 1903, 671–676.

[04] Über die Darstellung der Anzahl der Idealklassen eines algebraischen Körpers durch eine unendliche Reihe, *JRAM*, **127**, 1904, 167–174.

[06] Über die Darstellung definiter Funktionen durch Quadrate, *MA*, **62**, 1906, 272–285.

[07] Über die Verteilung der Primideale in den Idealklassen eines algebraischen Zahlkörpers, *MA*, **63**, 1907, 145–204.

[08] Über die Einteilung der positiven ganzen Zahlen in vier Klassen nach der Mindestanzahl der zu ihrer additiver Zusammensetzung erforderlichen Quadrate, *Arch. Math. Phys.*, (3), **13**, 1908, 305–312.

[12a] Über die Anzahl der Gitterpunkte in gewisser Bereichen, *GN*, 1912, 687–771.

[12b] Über eine idealtheoretische Funktion, *TAMS*, **13**, 1912, 1–21.
[15] Über die Hardysche Entdeckung unendlich vieler Nullstellen der Zetafunktion mit reellem Teil 1/2, *MA*, **76**, 1915, 212–243.
[18a] Abschätzungen von Charaktersummen, Einheiten und Klassenzahlen, *GN*, 1918, 79–97.
[18b] Über imaginärquadratische Zahlkörper mit gleicher Klassenzahl, *GN*, 1918, 277–284.
[18c] Über die Klassenzahl imaginärquadratischer Zahlkörper, *GN*, 1918, 285–296.
[18d] Verallgemeinerungen eines Pólyaschen Satzes auf algebraische Zahlkörper, *GN*, 1918, 478–488.
[18e] *Einführung in die elementare und analytische Theorie der algebraischen Zahlen und der Ideale*, Leipzig 1918, 2nd ed. 1927, reprinted by Chelsea 1949.
[18f] Über Ideale und Primideale in Idealklassen, *MZ*, **2**, 1918, 52–154.
[19a] Zur Theorie der Heckeschen Zetafunktionen, welche komplexen Charakteren entsprechen, *MZ*, **4**, 1919, 152–162.
[19b] Über die Wurzeln der Zetafunktion eines algebraischen Zahlkörpers, *MA*, **79**, 1919, 388–401.
[19c] Über die Zerlegung total positiver Zahlen in Quadrate, *GN*, 1919, 392–396.
[22] Der Minkowskische Satz über die Körperdiskriminante, *GN*, 1922, 80–82.
[24a] Über die Wurzeln der Zetafunktion, *MZ*, **20**, 1924, 98–104.
[24b] Über die Anzahl der Gitterpunkte in gewisser Bereichen, IV, *GN*, 1924, 137–150.
[25] Bemerkungen zu der Arbeit des Herrn Walfisz: "Über das Piltzsche Problem in algebraischen Zahlkörpern", *MZ*, **22**, 1925, 189–205.
[27a] *Vorlesungen über Zahlentheorie*, Leipzig 1927; reprinted by Chelsea 1969.
[27b] Über Dirichletsche Reihen mit komplexen Charakteren, *JRAM*, **157**, 1927, 26–32.
[28] Über das Vorzeichen der Gaussschen Summe, *GN*, 1928, 19–20.
[29] Über die Irreduzibilität der Kreisteilungsgleichung, *MZ*, **29**, 1929, 462.
[36] Bemerkungen zum Heilbronnschen Satz, *AA*, **1**, 1936, 1–18.

LANDAU, E., BOHR, H.   (See H. Bohr, E. Landau)

LANDHERR, W.   [36] Äquivalenz Hermitescher Formen über einem beliebigen algebraischen Zahlkörper, *Hbg*, **11**, 1936, 245–248.

LANDSBERG, G.   [93] Zur Theorie der Gaussschen Summen und der linearen Transformation der Thetafunktionen, *JRAM*, **111**, 1893, 234–253.
[97] Ueber das Fundamentalsystem und die Diskriminante der Gattungen algebraischer Zahlen welche aus Wurzelgrössen gebildet sind, *JRAM*, **117**, 1897, 140–147.

LANG, H.   [68] Über eine Gattung elementararithmetischer Klasseninvarianten reell-quadratischer Zahlkörper, *JRAM*, **233**, 1968, 123–175.
[72] Über Anwendung höheren Dedekindscher Summen auf die Struktur elementar-arithmetischer Klasseninvarianten reell-quadratischer Zahlkörper, *JRAM*, **254**, 1972, 17–32.
[73a] Über Bernoullische Zahlen in reell-quadratischen Zahlkörpern, *AA*, **22**, 1973, 423–437.
[73b] Über verallgemeinerte Bernoullische Zahlen und die Klassenzahl reell-quadratischer Zahlkörper, *AA*, **23**, 1973, 13–18.
[73c] Über die Klassenzahlen eines imaginären bizyklischen biquadratischen Zahlkörpers und seines reell-quadratischen Teilkörpers, *JRAM*, **262/3**, 1973, 18–40; II, *ibid.*, **267**, 1974, 175–178.
[75] Eine Invariante modulo 8 von Geschlechtern in reell-quadratischen Zahlkörpern, *MA*, **217**, 1975, 263–265.
[76] Über einfache periodische Kettenbrüche und Vermutungen von P. Chowla und S. Chowla, *AA*, **28**, 1975/6, 419–428.

[77] Über die Klassenzahl der Ringklassenkörper mit einem reellquadratischen Grundkörper, *MA*, **227**, 1977, 127–133.

LANG, H., SCHERTZ, R. [74] Bemerkungen über die Klassenzahl eines imaginären bizyklischen Zahlkörpers, *Hbg*, **42**, 1974, 212–216.

[76] Kongruenzen zwischen Klassenzahlen quadratischer Zahlkörper, *JNT*, **8**, 1976, 352–365.

LANG, S. [64] *Algebraic Numbers*, Reading 1964.

[70] *Algebraic Number Theory*, Reading–London 1970.

[71] On the zeta function of number fields, *Invent. Math.*, **12**, 1971, 337–345.

[78] *Cyclotomic Fields, I, II*, Springer, 1978, 1980.

[82] Units and class groups in number theory and algebraic geometry, *BAMS*, **6**, 1982, 253–316.

LANG, S., KUBERT, D. S. (See D. S. Kubert, S. Lang)

LANG, S. D. [77] Note on the class number of the maximal real subfield of a cyclotomic field, *JRAM*, **290**, 1977, 70–72.

LANGLANDS, R. P. [70] On Artin's *L*-functions, in: *Complex Analysis, Rice Univ. St.*, **56**, 1970, 23–28.

[80] Base change for GL(2), *Ann. of Math. Stud.*, **96**, 1980.

LARDON, R. [71] Évaluations asymptotiques concernant la fonction Ω dans certains semi-groupes normés à factorisation unique, *CR*, **273**, 1971, 76–79.

LASKA, M. [82] Solving the equation $x^3 - y^2 = v$ in number fields, *JRAM*, **333**, 1982, 73–85.

LASKER, E. [05] Zur Theorie der Moduln und Ideale, *MA*, **60**, 1905, 20–116.

[16] Über eine Eigenschaft der Diskriminante, *SB Math. Ges. Berlin*, **15**, 1916, 176–178.

LATHAM, J. [73] On sequences of algebraic integers, *JLMS*, (2), **6**, 1973, 555–560.

LATIMER, C. G. [29] On the prime ideals of the general cubic Galois field, *AJM*, **51**, 1929, 295–304.

[33a] Note on the invariants of the class group of a cyclic field, *AnM*, **34**, 1933, 872–874.

[33b] On the class-number of a cyclic field and a subfield, *BAMS*, **39**, 1933, 115–118.

[34] On the units in a cyclic field, *AJM*, **56**, 1934, 69–74.

LATIMER, C. G., MCDUFFEE, C. C. [33] A correspondence between classes of ideals and classes of matrices, *AnM*, **34**, 1933, 313–316.

LAUBIE, F. [80] Sur les nombres de ramification des extensions abéliennes des corps locaux, *CR*, **291**, 1980, A483–484.

LAUBIE, F., BARRUCAND, P. (See P. Barrucand, F. Laubie)

LAVRIK, A. F. [59] On the problem of distribution of the values of the class-numbers of purely radical quadratic forms with negative determinant, *IAN Uzbek. SSR*, 1959, 81–90 (Russian).

[65a] The abbreviated functional equation for the *L*-function of Dirichlet, *IAN Uzbek. SSR*, **9**, 1965, 4, 17–22.

[65b] On the divisor problem in segments of arithmetical progressions, *DAN*, **164**, 1965, 1232–1234 (Russian).

[66a] On the functional equation of Dirichlet's functions, *DAN*, **171**, 1966, 278–280 (Russian).

[66b] Functional equation for Dirichlet's *L*-functions and the divisor problem in arithmetical progressions, *IAN*, **30**, 1966, 433–448 (Russian).

[67a] On functional equations of Dirichlet's functions, *IAN*, **31**, 1967, 431–442 (Russian).

[67b] Approximate functional equation of Hecke's zeta-function of an imaginary quadratic field, *Mat. Zametki*, **2**, 1967, 475–482 (Russian).

[68a] On the approximate equation of Dirichlet's L-functions, *Tr. Moskov. Mat. Obshch.*, **18**, 1968, 91–104 (Russian).

[68b] Approximate functional equations of Dirichlet functions, *IAN*, **32**, 1968, 134–185 (Russian).

[70a] A note on the Siegel–Brauer theorem concerning parameters of algebraic number fields, *Mat. Zametki*, **8**, 1970, 259–263 (Russian).

[70b] On $L(1, \chi)$ with real Dirichlet character on sparse sets of values of the character modulus, *DAN*, **190**, 1970, 1286–1288 (Russian).

[71a] On moments of the class-number of purely radical quadratic forms with negative determinant, *DAN*, **197**, 1971, 32–35 (Russian).

[71b] A method of evaluation of double sums with a real quadratic character and its applications, *IAN*, **35**, 1971, 1189–1207 (Russian).

LAVRIK, A. F., EDGOROV, ZH. [75] The product of the number of divisor classes by the regulator of algebraic number fields, *IAN Uzbek. SSR*, 1975, 2, 77–79 (Russian).

LAWN, S., CHEW, K. L. (See K. L. Chew, S. Lawn)

LAWTON, W. [75] Heights of algebraic numbers and Szegö's theorem, *PAMS*, **49**, 1975, 47–50.

[77] A generalization of a theorem of Kronecker, *J. Sci. Fac. Chiangmai Univ.*, **4**, 1977, 15–23.

[83] A problem of Boyd concerning geometric means of polynomials, *JNT*, **16**, 1983, 356–362.

LAWTON, W., DOBROWOLSKI, E., SCHINZEL, A. (See E. Dobrowolski, W. Lawton, A. Schinzel)

LAXTON, R. R., LEWIS, D. J. [65] Forms of degree 7 and 11 over p-adic fields, *PSPM*, **8**, 1965, 16–21.

LAZAMI, F., [78] Sur les élements de $S \cap [1, 2[$, *Sém. DPP*, **20**, 1978/9, 1, exp. 3.

LEAHEY, W. J. [65] A note on a theorem of I. Niven, *PAMS*, **16**, 1965, 1130–1131.

[66] A new proof of a theorem of Kummer, *PAMS*, **17**, 1966, 497–498.

LEAHEY, W., GYÖRY, K. (See K. Györy, W. Leahey)

LEBESGUE, V. A. [59] Démonstration de l'irréductibilité de l'équation aux racines primitives de l'unité, *J. Math. Pures Appl.*, (2), **4**, 1859, 105–110.

LEDERMANN, W., CASSELS, J. W. S., MAHLER, K. (See J. W. S. Cassels, W. Ledermann, K. Mahler)

LEDIAEV, J. P., JOHNSON, E. W. (See E. W. Johnson, J. P. Lediaev)

LEDNEV, N. A. [39] On units of relatively cyclic algebraic fields, *Mat. Sbornik*, **6**, 1939, 227–261 (Russian).

LEE, K. C. [79] On the average order of characters in totally real algebraic number fields, *Chinese J. Math.*, **7**, 1979, 77–90.

LEE, M. P., MADAN, M. L. [69] On the Galois cohomology of the ring of integers in an algebraic number field, *PAMS*, **20**, 1969, 405–408.

LEEDHAM-GREEN, C. R. [72] The classgroup of Dedekind domains, *TAMS*, **163**, 1972, 493–500.

LEEP, D. [84] Systems of quadratic forms, *JRAM*, **350**, 1984, 109–116.

LEEP, D., SCHMIDT, W. M. [83] Systems of homogeneous equations, *Invent. Math.*, **71**, 1983, 539–549.

LEGENDRE, A. M., [98] *Théorie des nombres*, Paris 1798.

LEHMER, D. H. [26a] On the indeterminate equation $t^2 - p^2 Du^2 = 1$, *AnM*, **27**, 1926, 471–476.

[26b] A list of errors in tables of the Pell equation, *BAMS*, **32**, 1926, 545–550.

[32] Quasi-cyclotomic polynomials, *AMM*, **39**, 1932, 383-389.
[33a] Factorization of certain cyclotomic functions, *AnM*, **34**, 1933, 461-479.
[33b] On imaginary quadratic fields whose class-number is unity, *BAMS*, **39**, 1933, 360.
[36] An extension of the table of Bernoulli numbers, *Duke Math. J.*, **2**, 1936, 460-464.
[66] Some properties of cyclotomic polynomial, *J. Math. Anal. Appl.*, **15**, 1966, 105-117.
[77] Prime factors of cyclotomic class numbers, *MC*, **31**, 1977, 599-607.

LEHMER, D. H., MASLEY, J. M. [78] Table of the cyclotomic class number $h^*(p)$ and their factors for $200 < p < 521$, *MC*, **32**, 1978, 577-582.

LEHMER, E. [30] A numerical function applied to cyclotomy, *BAMS*, **36**, 1930, 291-298.
[36] On the magnitude of the coefficients of the cyclotomic polynomial, *BAMS*, **42**, 1936, 389-392.
[56] On the location of Gauss sums, *Math. Tables Aids Comp.*, **10**, 1956, 194-202.
[71] On the quadratic character of some quadratic surds, *JRAM*, **250**, 1971, 42-48.
[72] On some special quartic reciprocity law, *AA*, **21**, 1972, 367-372.
[73] On the cubic character of quadratic units, *JNT*, **5**, 1973, 385-389.
[74] On the quartic character of quadratic units, *JRAM*, **268/9**, 1974, 294-301.

LEMMLEIN, V. G. [54] On Euclidean rings and principal ideal rings, *DAN*, **97**, 1954, 585-587 (Russian).

[68] On the distribution of class-numbers h of real quadratic fields $K(\sqrt{p})$ with a prime discriminant $p \equiv 1 \pmod 4$ in the residue-classes $\{4k+1\}$ and $\{4k+3\}$, *DAN*, **179**, 1968, 1050-1053 (Russian).

LENSKOI, D. N. [63] On upper bounds for certain arithmetical functions in algebraic number fields, *DAN*, **150**, 1963, 251-253 (Russian).

LENSTRA, H. W. JR. [75] Euclid's algorithm in cyclotomic fields, *JLMS*, **10**, 1975, 457-465.
[76] K_2 of a global field consists of symbols, in: *Algebraic K-theory*, 69-73, LN 551, Springer, 1976.
[77a] Euclidean number fields of large degree, *Invent. Math.*, **38**, 1977, 237-254.
[77b] On Artin's conjecture and Euclid's algorithm in global fields, *Invent. Math.*, **42**, 1977, 201-224.
[78] Quelques examples d'anneaux euclidiens, *CR*, **286**, 1978, 683-685.
[79] Euclidean ideal classes, *Astérisque*, **61**, 1979, 121-131.
[82] On the calculation of regulators and class numbers of quadratic fields, in: *Journées Arithmétiques 1980*, 123-150, Cambridge Univ. Press, 1982.

LENSTRA, H. W. JR., COHEN, H. (See H. Cohen, H. W. Lenstra, Jr.)
LENSTRA, H. W. JR., LAGARIAS, J. C. (See J. C. Lagarias, H. W. Lenstra, Jr.)
LENZ, H. [53] Zur Quadratsummendarstellung in relativ-quadratischen Zahlkörpern, *S. B. Bayer. AW*, 1953, 283-288.

LEON, M. J. DE, CHOWLA, S. (See S. Chowla, M. J. de Leon)
LEON, M. J. DE, CHOWLA, S., HARTUNG, P. (See S. Chowla, M. J. de Leon, P. Hartung)
LEONARD, P. A. [69] On factoring quartics $(\bmod p)$, *JNT*, **1**, 1969, 113-115.
LEONARD, P. A., BUELL, D. A., WILLIAMS, K. S. (See D. A. Buell, P. A. Leonard, K. S. Williams)

LEONARD, P. A., WILLIAMS, K. S. [73] Representability of binary quadratic forms over a Bézout domain, *Duke Math. J.*, **40**, 1973, 533-539.
[74] Forms representable by integral binary quadratic forms, *AA*, **26**, 1974, 1-9.
[77] The quadratic and quartic character of certain quadratic units, I, *Pacific J. Math.*, **71**, 1977, 101-106; II, *Rocky Mountain J. Math.*, **9**, 1979, 683-692.
[80] The quartic characters of certain quadratic units, *JNT*, **12**, 1980, 106-109.

[82] On the divisibility of the class numbers of $Q(\sqrt{-p})$ and $Q(\sqrt{-2p})$ by 16, *Canad. Math. Bull.*, **25**, 1982, 200–206.

LEOPOLDT, H. W. [53a] Zur Geschlechtertheorie in abelschen Zahlkörpern, *MN*, **9**, 1953, 350–362.

[53b] Über Einheitengruppe und Klassenzahl reeller abelscher Zahlkörper, *Abh. Deutsch. Akad. Wiss.*, 1953, 2, 1–48.

[56] Über ein Fundamentalproblem der Theorie der Einheiten algebraischer Zahlkörper, *S. B. Bayer. AW*, 1956, 41–48.

[58a] Eine Verallgemeinerung der Bernoullischen Zahlen, *Hbg*, **22**, 1958, 131–140.

[58b] Zur Struktur der l-Klassengruppe galoisscher Zahlkörper, *JRAM*, **199**, 1958, 165–174.

[59a] Über die Hauptordnung der ganzen Elemente eines abelschen Zahlkörpers, *JRAM*, **201**, 1959, 119–149.

[59b] Über Klassenzahlprimteiler reeller abelscher Zahlkörper als Primteiler verallgemeinerter Bernoullischer Zahlen, *Hbg*, **23**, 1959, 36–47.

[60] Über Fermatquotienten von Kreiseinheiten und Klassenzahlformeln modulo p, *Rend. Circ. Mat. Palermo*, (2), **9**, 1960, 39–50.

[61] Zur Approximation des p-adischen Logarithmus, *Hbg*, **25**, 1961/2, 77–81.

[62] Zur Arithmetik in abelschen Zahlkörpern, *JRAM*, **209**, 1962, 54–71.

[75] Eine p-adische Theorie der Zetawerte, II, *JRAM*, **274/5**, 1975, 224–239.

LEOPOLDT, H. W., KUBOTA, T. (See T. Kubota, H. W. Leopoldt)

LEPISTÖ, T. [63] The first factor of the class number of the cyclotomic field $k(e^{2\pi i/p^n})$, *Ann. Univ. Turku, AI*, **70**, 1963, 1–7.

[66] On the first factor of the class-number of the cyclotomic field and Dirichlet's L-functions, *Ann. Acad. Sci. Fenn.*, **387**, 1966, 1–53.

[67] On the first factor of the class number of the cyclotomic field $k(\exp 2\pi i/p^u)$, *Ann. Univ. Turku, AI*, **108**, 1967, 1–8.

[68a] An upper bound for the first factor of the class number of the cyclotomic field $k\left(\exp\left(\dfrac{2\pi i}{p^u}\right)\right)$, *Ann. Univ. Turku, AI*, **116**, 1968, 1–8.

[68b] Two remarks concerning Dirichlet's L-functions and the class-number of the cyclotomic field, *Ann. Acad. Sci. Fenn.*, **433**, 1968, 1–8.

[68c] On the product of the regulator and the class-number of the cyclotomic field, *Ann. Univ. Turku, AI*, **118**, 1968, 1–7.

[69] On the class number of the cyclotomic field $k(\exp(2\pi i/p^n))$, *Ann. Univ. Turku, AI*, **125**, 1969, 1–13.

[70] An estimate for the class number of the Abelian field, *Ann. Acad. Sci. Fenn., AI*, **473**, 1970, 1–15.

[74] On the growth of the first factor of the class number of the prime cyclotomic field, *Ann. Acad. Sci. Fenn.*, **577**, 1974, 1–21.

LEPTIN, H. [55] Die Funktionalgleichung der Zeta-funktion einer einfachen Algebra, *Hbg*, **19**, 1955, 198–220.

LERCH, M. [03] Über die arithmetische Gleichung $Cl(-\varDelta) = 1$, *MA*, **57**, 1903, 568–571.

[05] Essai sur le calcul du nombre des classes de formes quadratiques binaires aux coefficients entiers, *AM*, **29**, 1905, 333–424; II, *ibid.*, **30**, 1906, 203–294.

LESEV, V. D. [64] On the multiplicative group of a regular local field with cyclic higher ramification, *Uchen. Zap. Kabard.-Balk. Univ.*, **22**, 1964, 117–121 (Russian).

[65] On the group of principal units of certain classes of normal extensions of a regular local field, *Uchen. Zap. Kabard.-Balk. Univ.*, **24**, 1965, 267–270 (Russian).

[66a] On the group of principal units of fully ramified extensions of a regular local field, *Uchen. Zap. Kabard.-Balk. Univ.*, 30, 1966, 85–91 (Russian).

[66b] On a theorem of E. Noether, *Uchen. Zap. Kabard.-Balk. Univ.*, 30, 1966, 92–95 (Russian).

[73] The p-adic completion of the multiplicative group of a normal extension of a regular local field with cyclic ramification subfield, *Algebra and Number Theory*, 1, 1973, 27–31. *Kabard.-Balk. Univ.* (Russian).

LETTL, G. [87] Characterization of irreducible integers by their norms, *Colloq. Math.*, 54, 1987, 325–332.

LEUTBECHER, A. [77] Euklidisches Algorithmus und die Gruppe GL_2, *MA*, 231, 1977/8, 269–285.

[85] Euclidean fields having a large Lenstra constant, *Ann. Inst. Fourier*, 35, 1985, 2, 83–106.

LEUTBECHER, A., MARTINET, J. [82a] Lenstra's constant and Euclidean number fields, *Astérisque*, 94, 1982, 87–131.

[82b] Constante de Lenstra et corps de nombres euclidiens, *Sem. Th. Nombr. Bordeaux*, 1981/2, exp. 4.

LEVEQUE, W. J. [52a] Geometric properties of Farey sections in $k(i)$, *Indag. Math.*, 14, 1952, 415–426.

[52b] Continued fractions and approximations in $k(i)$, *Indag. Math.*, 14, 1952, 526–535.

[64] On the equation $y^m = f(x)$, *AA*, 9, 1964, 209–219.

LEVESQUE, C. [79] A class of fundamental units and some classes of Jacobi–Perron algorithms in pure cubic fields, *Pacific J. Math.*, 81, 1979, 447–466.

[81] Systèmes fondamentaux d'unités de certains composés de deux corps quadratiques, I, *CJM*, 33, 1981, 937–945.

[82] Systèmes fondamentaux d'unités de certains corps de degré 4 et de degré 8 sur Q, *CJM*, 34, 1982, 1059–1090.

LEVESQUE, C., FREI, G. (See G. Frei, C. Levesque)

LEVI, F. [31] Zur Irreduzibilität der Kreisteilungspolynome, *Compos. Math.*, 1, 1931, 303–304.

LEVIN, B. V., BARBAN, M. B. (See M. B. Barban, B. V. Levin)

LEVIN, B. V., TULYAGANOVA, M. I. [66] A. Selberg's sieve in algebraic number fields, *Litovsk. Mat. Sb.*, 6, 1966, 59–73 (Russian).

LEVITZ, K. B. [72] A characterization of general Z.P.I.-rings, *PAMS*, 32, 1972, 376–380; II, *Pacific J. Math.*, 42, 1972, 147–151.

LEVITZ, K. B., MOTT, J. L. [72] Rings with finite norm property, *CJM*, 24, 1972, 557–565.

LEVY, L. S., CAMION, P., MANN, H. B. (See P. Camion, L. S. Levy, H. B. Mann)

LÉVY, P. [37] L'arithmétique des lois de probabilité, *CR*, 204, 1937, 80–82.

LEWIN, J. [67] Subrings of finite index in finitely generated rings, *J. Algebra*, 5, 1967, 84–88.

LEWIS, D. J. [52] Cubic homogeneous polynomials over p-adic number fields, *AnM*, 56, 1952, 473–478.

[57a] Cubic congruences, *Michigan Math. J.*, 4, 1957, 85–95.

[57b] Cubic forms over algebraic number fields, *Mathematika*, 4, 1957, 97–101.

[72] Invariant sets of morphisms in projective and affine number spaces, *J. Algebra*, 20, 1972, 419–434.

LEWIS, D. J., BIRCH, B. J. (See B. J. Birch, D. J. Lewis)

LEWIS, D. J., BIRCH, B. J., MURPHY, T. G. (See B. J. Birch, D. J. Lewis, T. G. Murphy)

LEWIS, D. J., DAVENPORT, H. (See H. Davenport, D. J. Lewis)

LEWIS, D. J., LAXTON, R. R. (See R. R. Laxton, D. J. Lewis)

LEWIS, D. J., MONTGOMERY, H. L.   [83] On zeros of p-adic forms, *Michigan Math. J.*, **30**, 1983, 83–87.

LEWIS, D. J., SCHINZEL, A., ZASSENHAUS, H.   [66] An extension of the theorem of Bauer and polynomials of certain special type, *AA*, **11**, 1966, 345–352.

LEWITTES, J.   [83] Characters and decomposition of a representation in a number field, *JNT*, **16**, 1983, 31–48.

LIANG, J. J.   [72] On relations between units of normal algebraic number fields and their subfields, *AA*, **20**, 1972, 331–344.

[73] On discriminants and maximal orders, *Acta Math. Acad. Sci. Hungar.*, **24**, 1973, 41–57.

[76] On the integral basis of the maximal real subfield of a cyclotomic field, *JRAM*, **286/7**, 1976, 223–226.

LIANG, J. J., HASSE, H.   (See H. Hasse, J. J. Liang)

LIANG, J. J., TORO, E.   [80] On the periods of the cyclotomic fields, *Hbg*, **50**, 1980, 127–134.

LIANG, J. J., ZASSENHAUS, H.   [69] On a problem of Hasse, *MC*, **23**, 1969, 515–519.

[77] The minimum discriminant of sixth degree totally complex algebraic number fields, *JNT*, **9**, 1977, 16–35.

LIARDET, P.   [71] Sur les transformations polynomiales et rationnelles, *Sem. Th. Nombr. Bordeaux*, 1971/2, exp. 29.

[72] Sur une conjecture de W. Narkiewicz, *CR*, **274**, 1972, 1836–1838.

[75] Stabilité algébrique et topologies hilbertiennes, *Sém. DPP*, **17**, 1975/6, exp. 8.

LIARDET, P., KUBOTA, K. K.   (See K. K. Kubota, P. Liardet)

LIARDET, P., VENTADOUX, M.   (See M. Ventadoux, P. Liardet)

LICHTENBAUM, S.   [72] On the values of zeta and L-functions, I, *AnM*, **96**, 1972, 338–360.

[73] Values of zeta-functions, étale cohomology and algebraic K-theory, in: *Algebraic K-theory*, II, 489–501, LN 342, Springer, 1973.

[75] Values of zeta and L-functions at zero, *Astérisque*, **24/5**, 1975, 133–138.

[80] On p-adic L-functions associated to elliptic curves, *Invent. Math.*, **56**, 1980, 19–55.

LICHTENBAUM, S., COATES, J.   (See J. Coates, S. Lichtenbaum)

LICHTENBAUM, S., KUBERT, D.   (See D. Kubert, S. Lichtenbaum)

LIENEN, H. V.   [78] The quadratic form x^2-2py^2, *JNT*, **10**, 1978, 10–15.

LINDEN, F. J. VAN DER   [82a] Class numbers of real abelian number fields of small conductor, in: *Journées Arithmétiques 1980*, 350–359, Cambridge Univ. Press, 1982.

[82b] *Euclidean rings in imaginary quadratic fields with two infinite primes*, Report 82-03, Univ. Amsterdam.

[84] Euclidean rings of integers of fourth degree fields, in: *Number Theory, Noordwijkerhout 1983*, 139–148, LN 1068, Springer, 1984.

LINFOOT, E., HEILBRONN, H.   (See H. Heilbronn, E. Linfoot)

LINNIK, YU. V.   [42] On a conditional theorem of J. E. Littlewood, *DAN*, **37**, 1942, 142–144 (Russian).

[43] The "analogy property" of L-series and Siegel's theorem, *DAN*, **38**, 1943, 115–117 (Russian).

[46] On the density of zeros of L-series, *IAN*, **10**, 1946, 35–46 (Russian).

[50] An elementary proof of Siegel's theorem based on the method of I. M. Vinogradov, *IAN*, **14**, 1950, 327–342 (Russian).

[54] Asymptotical distribution of integral points on a sphere, *DAN*, **96**, 1954, 909–912 (Russian).

LINNIK, YU. V., GELFOND, A. O.   (See A. O. Gelfond, Yu. V. Linnik)

LINNIK, YU. V., VINOGRADOV, A. I. [66] Hyperelliptic curves and the smallest prime quadratic residue, *DAN*, **168**, 1966, 259–261 (Russian).
LIPPMANN, R. A. [63] Note on irregular discriminats, *JLMS*, **38**, 1963, 385–386.
LITTLEWOOD, J. E. [28] On the class-number of the corpus $P(\sqrt{-k})$, *PLMS*, (2), **27**, 1928, 358–372.
LITTLEWOOD, J. E., HARDY, G. H. (See G. H. Hardy, J. E. Littlewood)
LITVER, E. L. [49] On the number of ideal classes in certain special fields, *DAN*, **66**, 1949, 335–338 (Russian).
[55] Fundamental basis of a composite of quadratic fields, *Uchen. Zap. Univ. Rostov*, **32**, 1955, 29–36 (Russian).
[59] On the group of ideal classes of certain algebraic fields of degree p^n, *Uchen. Zap. Univ. Rostov*, **43**, 1959, 137–145 (Russian).
LIVINGSTON, M., CHOWLA, S., KESSLER, I. (See S. Chowla, I. Kessler, M. Livingston)
LLORENTE, P., NART, E. [83] Effective determination of the decomposition of the rational primes in a cubic field, *PAMS*, **87**, 1983, 579–585.
LLORENTE, P., NART, E., VILA, N. [84] Discriminants of number fields defined by trinomials, *AA*, **43**, 1984, 368–373.
LLORENTE, P., ONETO, A. V. [82] On the real cubic field, *MC*, **39**, 1982, 689–692.
LLOYD-SMITH, C. W. [84] On a problem of Favard concerning algebraic integers, *Bull. Austral. Math. Soc.*, **29**, 1984, 111–121.
LO, S. K., NIEDERREITER, H. (See also H. Niederreiter, S. K. Lo) [75] Banach-Buck measure, density and uniform distribution in rings of algebraic integers, *Pacific J. Math.*, **61**, 1975, 191–208.
LOH, H. K. [77] On a group structure of the set of solutions to $f(x_1, ..., x_s) \equiv 1 (\mod n)$, *Chinese J. Math.*, **5**, 1977, 51–69.
LOHOUÉ, N. [70] Nombres de Pisot et synthèse harmonique dans les algèbres $A_p(R)$, *CR*, **270**, 1970, A1676–1678.
LONG, R. L. [71] Steinitz classes of cyclic extensions of prime degree, *JRAM*, **250**, 1971, 87–98.
[72] The module structure of some tamely ramified extensions of algebraic number fields, in: *Proc. Numb. Theory Conf., Boulder*, 139–141, Boulder, 1972.
[75] Steinitz classes of cyclic extensions of degree l^r, *PAMS*, **49**, 1975, 297–304.
[77] *Algebraic Number Theory*, Basel 1977.
LOONSTRA, F., FUCHS, L. (See L. Fuchs, F. Loonstra)
LORENZ, F. [80] Über eine Verallgemeinerung des Hasseschen Normensatzes, *MZ*, **173**, 1980, 203–210.
[82] Zur Theorie der Normenreste, *JRAM*, **334**, 1982, 157–170.
LORENZ, F., HALTER-KOCH, F. (See F. Halter-Koch, F. Lorenz)
LOUBOUTIN, R. [83] Sur la mesure de Mahler d'un nombre algébrique, *CR*, **296**, 707–708.
LOW, M. E. [68] Real zeros of the Dedekind zeta function of an imaginary quadratic field, *AA*, **14**, 1968, 117–140.
LOXTON, J. H. [72] On the maximum modulus of cyclotomic integers, *AA*, **22**, 1972, 69–85.
[74a] Products related to Gauss sums, *JRAM*, **268/9**, 1974, 53–67.
[74b] On a cyclotomic diophantine equation, *JRAM*, **270**, 1974, 164–168.
[74c] On two problems of R. M. Robinson about sums of roots of unity, *AA*, **26**, 1974/5, 159–174.
[78] Some conjectures concerning Gauss sums, *JRAM*, **297**, 1978, 153–158.

LOXTON, J. H., BLANKSBY, P. E. (See P. E. Blanksby, J. H. Loxton)
LU, H. W. [79] On the class number of real quadratic fields, *Sci. Sinica*, 1979, Special Issue, II, 118–130.
LUBELSKI, S. [31] Beweis und Verallgemeinerung eines Waring–Legendreschen Satzes, *MZ*, **33**, 1931, 321–349.
[36a] Über Klassenzahlrelationen quadratischer Formen in quadratischen Körpern, *JRAM*, **174**, 1936, 160–184.
[36b] Zur Gaussschen Kompositionstheorie der binären quadratischen Formen, *JRAM*, **176**, 1936, 56–60.
[39a] Zur Arithmetizierung des Beweises des Minkowskischen Diskriminanten und Kronecker–Weberschen Einbettungssatzes, *AA*, **3**, 1939, 235–254.
[39b] Zur Erweiterung des Legendreschen Satzes auf algebraische Zahlkörper, *Mathematica*, **15**, 1939, 125–134.
LUBIN, J. [81] The local Kronecker–Weber theorem, *TAMS*, **267**, 1981, 133–138.
LUBIN, J., TATE, J. [65] Formal complex multiplication in local fields, *AnM*, **81**, 1965, 380–387.
LUNDON, W. F. *et al.* (See W. J. Ellison *et al.*)
LUTHAR, I. S. [66] A note on a result of Mahler's, *J. Austral. Math. Soc.*, **6**, 1966, 399–401.
[67] A generalization of a theorem of Landau, *AA*, **12**, 1967, 223–228.
LUTHAR, I. S., BAMBAH, R. P., MADAN, M. L. (See R. P. Bambah, I. S. Luthar, M. L. Madan)
LUTHAR, I. S., GOGIA, S. K. (See S. K. Gogia, I. S. Luthar)
MAASS, H. [41] Über die Darstellung total positiver Zahlen des Körpers $R(\sqrt{5})$ als Summe von drei Quadraten, *Hbg*, **14**, 1941, 185–191.
MACAULAY, F. S. [13] On the resolution of a given modular system into primary systems including some properties of Hilbert numbers, *MA*, **74**, 1913, 66–121.
MADAN, M. L. [70] On class numbers of algebraic number fields, *JNT*, **2**, 1970, 116–119.
[72] Class groups of global fields, *JRAM*, **252**, 1972, 171–177.
MADAN, M. L., BAMBAH, R. P., LUTHAR, I. S. (See R. P. Bambah, I. S. Luthar, M. L. Madan)
MADAN, M. L., GOLD, R. (See R. Gold, M. L. Madan)
MADAN, M. L., LEE, M. P. (See M. P. Lee, M. L. Madan)
MADAN, M. L., PAL, S. [77] Abelian varieties and a conjecture of Robinson, *JRAM*, **291**, 1977, 78–91.
MADAN, M. L., QUEEN, C. S. [73] Euclidean function fields, *JRAM*, **262/3**, 1973, 271–277.
MADDEN, D. J., VÉLEZ, W. Y. [80] A note on the normality of unramified abelian extensions of quadratic extensions, *Manuscr. Math.*, **30**, 1980, 343–349.
MAGNUS, W. [34] Über den Beweis des Hauptidealsatzes, *JRAM*, **170**, 1934, 235–240.
MAHLER, K. [34] On Hecke's theorem on the real zeros of the L-functions and the class number of quadratic fields, *JLMS*, **9**, 1934, 298–302.
[36] Über Pseudobewertungen, II, *AM*, **67**, 1936, 51–80.
[50] On algebraic relations between two units of an algebraic field, in: *Alg. Th. des Nombres, Coll. Internat. CNRS*, **24**, 1950, 47–50.
[64a] Inequalities for ideal bases in algebraic number fields, *J. Austral. Math. Soc.*, **4**, 1964, 425–448.
[64b] An inequality for the discriminant of a polynomial, *Michigan Math. J.*, **11**, 1964, 257–262.
[73] *Introduction to p-adic Numbers and their Functions*, Cambridge 1973.

MAHLER, K., CASSELS, J. W. S., LEDERMANN, W.   (See J. W. S. Cassels, W. Ledermann, K. Mahler)
MAIER, E. A., EGGAN, L. C.   (See L. C. Eggan, E. A. Maier)
MAIKOTOV, N. R.   [69] On the classification of special Abelian fields of type (3, 3, 3), *IAN Kazakh. SSR*, 1969, 1, 88 (Russian).
MÄKI, S.   [80] *The Determination of Units in Real Cyclic Sextic Fields*, LN 797, Springer, 1980.
[85] On the density of Abelian number fields, *Ann. Acad. Sci. Fenn.*, *AI*, **54**, 1985, 104 pp.
MAKNYS, M. (Maknis, M.)   [75a] On Hecke Z-functions of an imaginary quadratic field, *Litovsk. Mat. Sb.*, **15**, 1975, 1, 157–172 (Russian).
[75b] Zeros of Hecke Z-functions and the distribution of prime numbers of an imaginary quadratic field, *Litovsk. Mat. Sb.*, **15**, 1975, 1, 173–184 (Russian).
[76] Density theorems of Hecke Z-functions and the distribution of prime numbers of an imaginary quadratic field, *Litovsk. Mat. Sb.*, **16**, 1976, 1, 173–180 (Russian).
[80] The "large sieve" in quadratic fields, *Litovsk. Mat. Sb.*, **20**, 1980, 2, 79–86 (Russian).
MALLIK, A.   [79] A note on Friedlander's paper: "On the class numbers of certain quadratic extensions", *AA*, **35**, 1979, 53–54.
[81a] New formulations of the class number one problem, *AA*, **39**, 1981, 361–364.
[81b] Bounding L-functions by class numbers, *AA*, **39**, 1981, 365–368.
MANIN, J. I.   [71] Le groupe de Brauer-Grothendieck et géométrie diophantienne, *Proc. ICM Nice*, *I*, 401–411, Paris 1971.
[76] Non-Archimedean integration and p-adic Jacquet–Langlands functions, *Uspekhi Mat. Nauk*, **31**, 1976, 1, 5–54 (Russian).
MANIN, YU. I., VISHIK, M. M.   (See M. M. Vishik, Yu. I. Manin)
MANN, H. B.   [50] On the field of origin of an ideal, *CJM*, **2**, 1950, 16–21.
[54] A generalization of a theorem of Ankeny and Rogers, *Rend. Circ. Mat. Palermo*, (2), **3**, 1954, 476–477.
[55] *Introduction to Algebraic Number Theory*, Columbus 1955.
[58] On integral bases, *PAMS*, **9**, 1958, 167–172.
MANN, H. B., BUTTS, H. S.   (See H. S. Butts, H. B. Mann)
MANN, H. B., CAMION, P., LEVY, L. S.   (See P. Camion, L. S. Levy, H. B. Mann)
MANN, H. B., CHATLAND, H.   (See H. Chatland, H. B. Mann)
MANN, H. B., HANLY, V.   (See V. Hanly, H. B. Mann)
MANN, H. B., VÉLEZ, W. Y.   [76] Prime ideal decomposition in $F(\sqrt[n]{\mu})$, *Monatsh. Math.*, **81**, 1976, 131–139.
MANN, H. B., YAMAMOTO, K.   [67] On canonical bases of ideals, *J. Combin. Theory*, **2**, 1967, 71–76.
MARCUS, D. A.   [77] *Number Fields*, Springer, 1977.
MARININA, S. F.   [56] An evaluation of the number of irregular functions of an imaginary quadratic field, *Ukr. Mat. Zh.*, **8**, 1956, 319–324 (Russian).
MARKANDA, R.   [75] Euclidean rings of algebraic numbers and functions, *J. Algebra*, **37**, 1975, 425–440.
MARKE, P. W.   [37] Über die Bestimmung Dirichletscher Reihen durch ihre Funktionalgleichung, *MA*, **114**, 1937, 29–56.
MARKOFF, A.   [82] Sur les nombres entiers dépendent d'une racine cubique d'un nombre entier, *Mém. Acad. Imp. Sci. St. Petersburg*, (7), **38**, 1882, nr. 9.
[91] Sur une classe de nombres complexes, *CR*, **92**, 1891, 780–782, 1049–1050, 1123–1124.
MARKSHAITIS, G. N.   [63] On p-extensions with one critical number, *IAN*, **27**, 1963, 463–466 (Russian).

MARSH, R. W. [57] *Tables of irreducible polynomials over GF*(2) *through degree 19*, U.S. Dept. of Commerce, Washington 1957.

MARSHALL, M. A. [71] Ramification groups of abelian local field extensions, *CJM*, **23**, 1971, 271–281.

MARTEL, B. [74] Sur l'anneau des entiers d'une extension biquadratique d'un corps 2-adique, *CR*, **278**, 1974, 117–120.

MARTIN, J. N. [83] Calcul du résidu en $s = 1$ de la fonction dzéta p-adique de corps cubiques cycliques, *Publ. Math. Fac. Sci. Besançon, Th. de Nombres*, 1981/2 et 1982/3.

MARTINET, J. [68] Sur les extensions galoisiennes non abéliennes de degré $2p$ des rationnels, *CR*, **266**, 1968, A959–962.

[69] Sur l'arithmétique des extensions galoisiennes à groupe de Galois diédral d'ordre $2p$, *Ann. Inst. Fourier*, **19**, 1969, 1, 1–80.

[71a] Modules sur l'algèbre du groupe quaternionien, *ASENS*, (4), **4**, 1971, 399–408.

[71b] Anneau des entiers d'une extension galoisienne considéré comme module sur l'algèbre du groupe de Galois, *Bull. Soc. Math. France, Mém.* **25**, 1971, 123–126.

[72] Sur les extensions à groupe de Galois quaternionien, *CR*, **274**, 1972, 933–935.

[73] Bases normales et constante de l'équation fonctionnelle des fonctions L d'Artin, *Sém. Bourbaki*, **26**, 1973/4, exp. 450.

[77a] Character theory and Artin L-functions, in: *Algebraic Number Fields*, Proc. Durham Symp., 1–87, London 1977.

[77b] H_8, in: *Algebraic Number Fields*, Proc. Durham Symp., 525–538, London 1977.

[78] Tours de corps de classes et estimations de discriminants, *Invent. Math.*, **44**, 1978, 65–73.

[79a] Sur la constante de Lenstra des corps de nombres, *Sem. Th. Nombr. Bordeaux*, 1979/80, exp. 17.

[79b] Petits discriminants, *Ann. Inst. Fourier*, **29**, 1979, 1, 159–170.

[82] Petits discriminants des corps de nombres, in: *Journées Arithmétiques 1980*, 151–193, Cambridge Univ. Press, 1982.

MARTINET, J., LEUTBECHER, A. (See A. Leutbecher, J. Martinet)

MARTINET, J., PAYAN, J. J. [67] Sur les extensions cubiques non-Galoisiennes des rationnels et leur clôture Galoisienne, *JRAM*, **228**, 1967, 15–37.

[68] Sur les bases d'entiers des extensions galoisiennes et non abéliennes de degré 6 des rationnels, *JRAM*, **229**, 1968, 29–33.

MARUNO, T. [76] On the ramification numbers of a tower of the maximal Abelian extension of p-adic number fields with exponent p^m, *Rep. Fac. Sci. Kagoshima Univ.*, **9**, 1976, 43–49.

MASLEY, J. M. [75a] On Euclidean rings of integers in cyclotomic fields, *JRAM*, **272**, 1975, 45–48.

[75b] Solution of the class number two problem for cyclotomic fields, *Invent. Math.*, **28**, 1975, 243–244.

[76] Solution of small class number problems for cyclotomic fields, *Compos. Math.*, **33**, 1976, 179–186.

[77] Odlyzko bounds and class number problems, in: *Algebraic Number Fields*, Proc. Durham Symp., 465–474, London 1977.

[78a] Class number of real cyclic number fields with small conductor, *Compos. Math.*, **37**, 1978, 297–319.

[78b] On the first factor of the class number of prime cyclotomic fields, *JNT*, **10**, 1978, 273–290.

[79] Where are number fields with small class numbers? in: *Number Theory, Carbondale 1979*, 221–242, LN 751, Springer, 1979.

MASLEY, J. M., LEHMER, D. H.   (See D. H. Lehmer, J. M. Masley)

MASLEY, J. M., MONTGOMERY, H. L.   [76] Cyclotomic fields with unique factorization, *JRAM*, **286**/7, 1976, 248–256.

MASSY, R., NGUYEN-QUANG-DO, T.   [75] Extensions galoisiennes non abéliennes de degré p^3 d'un corps $\mathfrak{P}$-adique, *CR*, **280**, 1975, A1345–1347.

MASUDA, K.   [57] Certain subgroups of the idèle group, *PJA*, **33**, 1957, 70–72.

[59] An application of the generalized norm residue symbol, *PAMS*, **10**, 1959, 245–252.

[61] Note on characters of the group of units of algebraic number fields, *Tôhoku Math. J.*, (2), **13**, 1961, 248–252.

MATHEWS, G. B.   [93] On the algebraic integers derived from an irreducible cubic equation, *PLMS*, **24**, 1893, 327–336.

MATLIS, E.   [59] Injective moduls over Prüfer rings, *Nagoya Math. J.*, **15**, 1959, 57–69.

[68] Reflexive domains, *J. Algebra*, **8**, 1968, 1–33.

[70] The two-generator problem for ideals, *Michigan Math. J.*, **17**, 1970, 157–265.

MATSUDA, R.   [70a] On a generalization of Euler's φ-function, *Bull. Fac. Sci. Ibaraki Univ.*, **2-2**, 1970, 19–21.

[70b] The ring of arithmetic functions of many variables, *Bull. Fac. Sci. Ibaraki Univ.*, **2-2**, 1970, 41–43.

MATSUMOTO, H.   [69] Sur les sous-groupes arithmétiques des groupes semi-simples déployés, *ASENS*, (4), **2**, 1969, 1–62.

MATSUMURA, N.   [72] On the cohomology group of the unit group of an algebraic number field, *Mem. Fac. Sci. Kyushu Univ.*, A, **26**, 1972, 279–283.

[77] On the class field tower of an imaginary quadratic number field, *Mem. Fac. Sci. Kyushu Univ.*, A, **31**, 1977, 165–171.

MATTHEWS, C. R.   [79] Gauss sums and elliptic functions, I, *Invent. Math.* **52**, 1979, 163–185; II, *ibid.*, **53**, 1979, 23–52.

MATULJAUSKAS, A.   [69] Approximate functional equation for Hecke's ζ-function of a real quadratic field, *Litovsk. Mat. Sb.*, **9**, 1969, 291–321 (Russian).

[71] The Hecke's ζ-function of a real quadratic field, *Litovsk. Mat. Sb.*, **11**, 1971, 597–605 (Russian).

MATUSITA, K.   [44] Über ein bewertungstheoretisches Axiomensystem für die Dedekind-Noethersche Idealtheorie, *Japan. J. Math.*, **19**, 1944, 97–110.

MAUCLAIRE, J. L.   [76] On the extension of a multiplicative arithmetical function in an algebraic number field, *Math. Japan.*, **21**, 1976, 337–342.

[83] Sommes de Gauss modulo p^α, *PJA*, **59**, 1983, 109–112, 161–163.

MAURER, D.   [73] The trace-form of an algebraic number field, *JNT*, **5**, 1973, 379–384.

[78a] Invariants of the trace-form of a number field, *Linear and Multilinear Algebra*, **6**, 1978, 33–36.

[78b] A matrix criterion for normal integral bases, *Illinois J. Math.*, **22**, 1978, 672–681.

[79] Arithmetic properties of the idèle discriminant, *Pacific J. Math.*, **85**, 1979, 393–401.

MAUS, E.   [67] Arithmetisch disjunkte Körper, *JRAM*, **226**, 1967, 184–203.

[68] Die gruppentheoretische Struktur der Verzweigungsgruppenreihen, *JRAM*, **230**, 1968, 1–28.

[72] Über die Verteilung der Grundverzweigungszahlen von wild verzweigten Erweiterungen p-adischer Zahlkörper, *JRAM*, **257**, 1972, 47–79.

[73] Relationen in Verzweigungsgruppen, *JRAM*, **258**, 1973, 23–50.

MAUTNER, F. I.   [53] On congruence characters, *Monatsh. Math.*, **57**, 1953, 307–316.

MAXWELL, G. [70] A note on Artin's diophantine conjecture, *Canad. Math. Bull.*, **13**, 1970, 119–120.
MAY, W. [70] Unit groups of infinite abelian extensions, *PAMS*, **25**, 1970, 680–683.
[72] Multiplicative groups of fields, *PLMS*, (3), **24**, 1972, 295–306.
[79] Multiplicative groups under field extension, *CJM*, **31**, 1979, 436–440.
MAYER, J. [29] Die absolut-kleinsten Diskriminanten der biquadratischer Zahlkörper, *S.B. AW Wien*, IIa, **138**, 1929, 733–742.
MAZUR, B., WILES, A. [84] Class fields of abelian extensions of Q, *Invent. Math.*, **76**, 1984, 179–330.
MCCARTHY, P. J. [66] *Algebraic Extensions of Fields*, Waltham–Toronto–London 1966.
MCCLUER, C. R. [68] A reduction of the Čebotarev density theorem to the cyclic case, *AA*, **15**, 1968, 45–47.
[71a] Common divisors of values of polynomials, *JNT*, **3**, 1971, 33–34.
[71b] Non-principal divisors among the values of polynomials, *AA*, **19**, 1971, 319–320.
MCCLUER, C. R., PARRY, C. J. [75] Units of modulus 1, *JNT*, **7**, 1975, 371–375.
MCCULLOH, L. R. [63] Integral bases in Kummer extensions of Dedekind fields, *CJM*, **15**, 1963, 755–765.
[66] Cyclic extensions without relative integral bases, *PAMS*, **17**, 1966, 1191–1194.
[69] Cyclic extensions of prime power degree and corresponding residue systems, *JNT*, **1**, 1969, 459–466.
[71] Frobenius groups and integral bases, *JRAM*, **248**, 1971, 123–126.
[77] A Stickelberger condition on Galois module structure for Kummer extensions of prime degree, in: *Algebraic Number Fields, Proc. Durham Symp.*, 561–588, London 1977.
[82] Stickelberger relations in class groups and Galois module structure, in: *Journées Arithmétiques 1980*, 194–201, Cambridge Univ. Press, 1982.
MCCULLOH, L. R., BERTNESS, C. H. (See C. H. Bertness, L. R. McCulloh)
MCCULLOH, L. R., STOUT, W. T. JR. [69] Corresponding residue systems in cyclic extensions of prime degree over algebraic number fields, *JNT*, **1**, 1969, 312–325.
MCDANIEL, W. L. [74a] Perfect Gaussian integers, *AA*, **25**, 1974, 137–144.
[74b] On multiple prime divisors of cyclotomic polynomials, *MC*, **28**, 1974, 847–850.
MCDUFFEE, C. C. [31a] A method for determining the canonical basis of an ideal, *MA*, **105**, 1931, 663–665.
[31b] Ideals in linear algebras, *BAMS*, **37**, 1931, 845–853.
MCDUFFEE, C. C., JENKINS, E. D. [35] A substitute for the Euclid algorithm in algebraic fields, *AnM*, **36**, 1935, 40–45.
MCDUFFEE, C. C., LATIMER, C. G. (See C. G. Latimer, C. C. McDuffee)
MCELIECE, R. J. [69] Factorization of polynomials over finite fields, *MC*, **23**, 1969, 861–867.
MCFARLAND, R. L., LAL, S., ODONI, R. W. K. (See S. Lal, R. L. McFarland, R. W. K. Odoni)
MCFEAT, R. B. [71] Geometry of numbers in adele spaces, *Dissert. Math.*, **88**, 1971, 49 pp.
MCGETTRICK, A. D. [72a] On the biquadratic Gauss sum, *PCPS*, **71**, 1972, 79–83.
[72b] A result in the theory of Weierstrass elliptic functions, *PLMS*, (3), **25**, 1972, 41–54.
MCKENZIE, R. E. [52] Class group relations in cyclotomic fields, *AJM*, **74**, 1952, 759–763.
MCKENZIE, R., SCHEUNEMANN, J. [71] A number field without a relative integral basis, *AMM*, **78**, 1971, 882–883.

McKenzie, R. E., Whaples, G. [56] Artin-Schreier equations in characteristic zero, *AJM*, **78**, 1956, 473–485.
McLane, S. [35] The ideal-decomposition of rational primes in terms of absolute values, *PAUS*, **21**, 1935, 663–667.
[36] A construction for prime ideals as absolute values of an algebraic field, *Duke Math. J.*, **2**, 1936, 492–510.
[39a] Steinitz towers for modular fields, *TAMS*, **46**, 1939, 23–45.
[39b] Subfields and automorphism groups of p-adic fields, *AnM*, **40**, 1939, 423–442.
[40] Note on the relative structure of p-adic fields, *AnM*, **41**, 1940, 751–753.
McQuillan, D. L. [62] A generalization of a theorem of Hecke, *AJM*, **84**, 1962, 306–316.
[72a] Modules over algebraic integers, *Sem. Th. Nombr. Bordeaux*, 1972/3, exp. 4.
[72b] Réseaux sur les anneaux d'entiers algébriques, *Sém. DPP*, **14**, 1972/3, exp. 25.
[73] A remark on Hilbert's Theorem 92, *AA*, **22**, 1973, 125–128.
[76] On the Galois cohomology of Dedekind rings, *JNT*, **8**, 1976, 438–445.
McQuillan, D. L., Gunji, H. (See H. Gunji, D. L. McQuillan)
Mead, D. G., Narkiewicz, W. [82] The capacity of C_5 and free sets in C_m^2, *PAMS*, **84**, 1982, 308–310.
Medvedev, P. A. [64] On the representation of zero by a cubic form in the field of p-adic numbers, *Uspekhi Mat. Nauk*, **19**, 1964, 6, 187–190 (Russian).
Megibben, C. [70] Absolutely pure modules, *PAMS*, **26**, 1970, 561–566.
Meinardus, G. [53] Über das Partitionsproblem eines reell-quadratischen Zahlkörpers, *MA*, **126**, 1953, 343–361.
Meissner, G. [11] Bemerkungen zur Bestimmung der nächsten ganzen Zahl im Gebiete der komplexen Zahlen $a+b\varrho$, *Mitt. Math. Ges. Hamburg*, **4**, 1911, 441–444.
Meissner, O. [04] Über die Darstellung der Zahlen einiger algebraischen Zahlkörper als Summen von Quadratzahlen des Körpers, *Arch. Math. Phys.*, (3), **7**, 1904, 266–268.
[05] Über die Darstellbarkeit der Zahlen quadratischer und kubischer Zahlkörper als Quadratsummen, *Arch. Math. Phys.*, (3), **9**, 1905, 202–203.
Mendès-France, M. [67] Deux remarques concernant l'équirépartition des suites, *AA*, **14**, 1967/8, 163–167.
[76] A characterization of Pisot numbers, *Mathematika*, **23**, 1976, 32–34.
Merriman, J. R., Birch, B. J. (See B. J. Birch, J. R. Merriman)
Mertens, F. [74] Ueber einige asymptotische Gesetze der Zahlentheorie, *JRAM*, **77**, 1874, 289–338.
[94] Über die Fundamentalgleichung eines Gattungsbereiches algebraischer Zahlen, *S.B. Kais. AW Wien*, **103**, 1894, 5–40.
[96] Über die Gaussischen Summen, *S.B. Kgl. Preuß. AW*, 1896, 217–219.
[05] Ein Beweis des Satzes, dass jede Klasse von ganzzahligen primitiven binären quadratischen Formen des Hauptgeschlechts durch Duplikation entsteht, *JRAM*, **129**, 1905, 181–186.
[06] Über zyklische Gleichungen, *JRAM*, **131**, 1906, 87–112.
Mestre, J. F. [81] Corps euclidiens, unités exceptionnelles et courbes elliptiques, *JNT*, **13**, 1981, 123–137.
[83] Courbes elliptiques et groupe de classes d'idéaux de certains corps quadratiques, *JRAM*, **343**, 1983, 23–35.
Metsänkylä, T. [67a] Bemerkungen über den ersten Faktor der Klassenzahl des Kreiskörpers, *Ann. Univ. Turku*, AI, **105**, 1967, 1–15.
[67b] Über den ersten Faktor der Klassenzahl des Kreiskörpers, *Ann. Acad. Sci. Fenn.*, **416**, 1967, 1–48.

[68a] Über die Teilbarkeit der Relativklassenzahl des Kreiskörpers durch zwei, *Ann. Univ. Turku, AI*, **118**, 1968, 1–8.
[68b] Über die Teilbarkeit des ersten Faktors der Klassenzahl des Kreiskörpers, *Ann. Univ. Turku, AI*, **124**, 1968, 1–6.
[69a] Congruences modulo 2 for class number factors in cyclotomic fields, *Ann. Acad. Sci. Fenn.* **453**, 1969, 1–12.
[69b] Calculation of the first factor of the class number of the cyclotomic field, *MC*, **23**, 1969, 533–537.
[70a] A congruence for the class number of a cyclic field, *Ann. Acad. Sci. Fenn.*, **472**, 1970, 1–11.
[70b] Estimations for L-functions and the class numbers of certain imaginary cyclic fields, *Ann. Univ. Turku, A I*, **140**, 1970, 1–11.
[71a] Note on the distribution of irregular primes, *Ann. Acad. Sci. Fenn.*, **492**, 1971, 1–6.
[71b] On prime factors of the relative class number of cyclotomic fields, *Ann. Univ. Turku, A I*, **149**, 1971, 1–8.
[72] On the growth of the first factor of the cyclotomic class number, *Ann. Univ. Turku, A I*, **155**, 1972, 1–12.
[73] A class number congruence for cyclotomic fields and their subfields, *AA*, **23**, 1973, 107–116.
[74] Class numbers and μ-invariants of cyclotomic field, *PAMS*, **43**, 1974, 299–300.
[75a] On the cyclotomic invariants of Iwasawa, *Math. Scand.*, **37**, 1975, 61–75.
[75b] On the Iwasawa invariants of imaginary abelian fields, *Ann. Acad. Sci. Fenn., ser. math.*, **1**, 1975, 2, 343–353.
[76] Distribution of irregular prime numbers, *JRAM*, **282**, 1976, 126–130.
[77] A short proof for the nonvanishing of a character sum, *JNT*, **9**, 1977, 507–509.
[78a] Note on certain congruences for generalized Bernoulli numbers, *Arch. Math.*, **30**, 1978, 595–598.
[78b] Iwasawa invariants and Kummer congruences, *JNT*, **10**, 1978, 510–522.
[83] An upper bound for the λ-invariant of imaginary abelian fields, *MA*, **264**, 1983, 5–8.
[84] Maillet's matrix and irregular primes, *Ann. Univ. Turku, AI*, **186**, 1984, 72–79.
METSÄNKYLÄ, T., ERNVALL, R. (See R. Ernvall, T. Metsänkylä)
MEULENBELD, B., KOKSMA, J. F. (See J. F. Koksma, B. Meulenbeld)
MEYER, C. [57] *Die Berechnung der Klassenzahl Abelscher Körper über quadratischen Zahlkörpern*, Akademie-Verlag, 1957.
[67] Über die Bildung von elementar-arithmetischen Klasseninvarianten in reell-quadratischen Zahlkörpern, in: *Alg. Zahlentheorie, Oberwolfach 1964*, 165–215, Mannheim 1967.
[70] Bemerkungen zum Satz von Heegner-Stark über die imaginär-quadratischen Zahlkörper mit der Klassenzahl Eins, *JRAM*, **242**, 1970, 179–214.
[75] Imaginäre bizyklische biquadratische Zahlkörper als Klassenkörper, *Symposia Math.*, **15**, 1975, 365–387, Academic Press.
MEYER, W. [78] Konvexe Untergruppen in algebraischen Zahlkörpern, *MA*, **233**, 1978, 275–283.
MEYER, W., PERLIS, R. [79] On the genus of norm forms, *MA*, **246**, 1979/80, 117–119.
MEYER, Y. [68] Une caractérisation des nombres de Pisot, *CR*, **266**, 1968, A63–64.
[70a] Nombres algébriques et analyse harmonique, *ASENS*, (4), **3**, 1970, 75–110.
[70b] Les nombres de Pisot et la synthèse harmonique, *ASENS*, (4), **3**, 1970, 235–346.
[70c] *Nombres de Pisot, nombres de Salem et analyse harmonique*, LN 117, Springer, 1970.
MIDGARDEN, B., WIEGAND, S. [81] Commutative rings of bounded module type, *Comm. Algebra*, **9**, 1981, 1001–1025.

MIGNOTTE, M. [77] Entiers algébriques dont les conjugués sont proches du cercle unité, *Sém. DPP*, **19**, 1977/8, exp. 39.
[84] Sur les conjugés des nombres de Pisot, *CR*, **298**, 1984, 21.
MIKI, H. [76] On some Galois cohomology groups of a local field and its application to the maximal p-extension, *J. Math. Soc. Japan*, **28**, 1976, 114-122.
[77] A note on Maus' ramification theorem, *Tôhoku Math. J.*, (2), **29**, 1977, 61-68.
[78a] On the maximal l-extension of a finite algebraic number field with given ramification, *Nagoya Math. J.*, **70**, 1978, 183-202.
[78b] On Grunwald-Hasse-Wang theorem, *J. Math. Soc. Japan*, **30**, 1978, 313-325.
[81] On the ramification numbers of cyclic p-extensions over local fields, *JRAM*, **328**, 1981, 99-115.
MILGRAM, R. J. [81] Odd index subgroups of units in cyclotomic fields and applications, in: *Algebraic K-theory, Evanston 1980*, 269-298, LN 854, Springer, 1981.
MILLS, W. H. [63] Characters with preassigned values, *CJM*, **15**, 1963, 169-171.
MILNOR, J. [71] *Introduction to Algebraic K-theory*, Princeton 1971.
MIN, S. H. [47] On the euclidean algorithm in real quadratic fields, *JLMS*, **22**, 1947, 88-90.
MIN, S. H., HUA, L. K. (See L. K. Hua, S. H. Min)
MINES, R., RICHMAN, F. [81] Dedekind domains, in: *Constructive mathematics*, 16-30, LN 873, Springer, 1981.
[84] Valuation theory: a constructive view, *JNT*, **19**, 1984, 40-62.
MINKOWSKI, H. [91a] Über die positiven quadratischen Formen und über kettenbruchähnlichen Algorithmen, *JRAM*, **107**, 1891, 278-297 = *Ges. Abh. I*, 244-260, Leipzig-Berlin 1911.
[91b] Théorèmes arithmétiques, *CR*, **112**, 1891, 209-212 = *Ges. Abh., I*, 261-263, Leipzig-Berlin 1911.
[96a] *Geometrie der Zahlen*, Leipzig 1896; reprinted by Johnson 1968.
[96b] Zur Theorie der Kettenbrüche, *ASENS*, (2), **13**, 1896, 41-60 = *Ges. Abh. I*, 278-292, Leipzig-Berlin 1911.
[00] Zur Theorie der Einheiten in den algebraischen Zahlkörpern, *GN*, 1900, 90-93 = *Ges. Abh. I*, 316-319, Leipzig-Berlin 1911.
[07] *Diophantische Approximationen*, Leipzig 1907; reprints: Chelsea 1957, Physica-Verlag, 1961.
MIRIMANOFF, D. [91] Sur une question de la théorie des nombres, *JRAM*, **99**, 1891, 82-88.
[34] L'équation $\xi^3+\eta^3+\zeta^3 = 0$ et la courbe $x^3+y^3 = 1$, *Comment. Math. Helv.*, **6**, 1934, 192-198.
MITCHELL, H. H. [26] On classes of ideals in a quadratic field, *AnM*, **27**, 1926, 297-314.
MITSUI, T. [56] Generalized prime number theorem, *Japan. J. Math.*, **26**, 1956, 1-42.
[60] On the Goldbach problem in an algebraic number field, *J. Math. Soc. Japan*, **12**, 1960, 290-324, 325-372.
[68] On the prime ideal theorem, *J. Math. Soc. Japan*, **20**, 1968, 233-247.
[78] On the partition problem in an algebraic number field, *Tokyo J. Math.*, **1**, 1978, 189-236.
MIYAKE, K. [80a] On the general principal ideal theorem, *PJA*, **56**, 1980, 171-174.
[80b] On the structure of the idèle group of an algebraic number field, *Nagoya Math. J.*, **80**, 1980, 117-127; II, *Tôhoku Math. J.*, (2), **34**, 1982, 101-112.
[82] On the units of an algebraic number field, *J. Math. Soc. Japan*, **34**, 1982, 515-525.
MIYATA, T. [80] A normal integral basis theorem for dihedral groups, *Tôhoku Math. J.*, (2), **32**, 1980, 49-62,

MIYATA, Y. [68] Remark on Yokoi's theorem concerning the basis of algebraic integers and tame ramification, *PJA*, **44**, 1968, 987–989.
[71] On a characterization of the first ramification group as the vertex of the ring of integers, *Nagoya Math. J.*, **43**, 1971, 151–156.
[74] On the module structure of the ring of all integers of a p-adic number field, *Nagoya Math. J.*, **54**, 1974, 53–59.
[79] On the module structure in a cyclic extension over a p-adic number field, *Nagoya Math. J.*, **73**, 1979, 61–68.
[80] On the module structure of a *p*-extension over a p-adic number field, *Nagoya Math. J.*, **77**, 1980, 13–23.
MOHANTY, S. P., GORDON, B. (See B. Gordon, S. P. Mohanty)
MOINE, J. M. [72] Quelques problèmes concernant les classes ambiges des corps quadratiques, *Ann. Univ. Besançon*, 1972, 4, 1–63.
MOLK, J. [83] Sur les unités complexes, *Bull. Sci. Math.*, (2), **7**, 1883, 133-136 = L. KRONECKER, *Werke*, *V*, 504–506, Leipzig–Berlin 1930.
MÖLLER, H. [70] Über die *i*-ten Koeffizienten der Kreisteilungspolynome, *MA*, **188**, 1970, 26–38.
[71] Über die Koeffizienten des *n*-ten Kreisteilungspolynom, *MZ*, **119**, 1971, 33–40.
[76a] Imaginär-quadratische Zahlkörper mit einklassigen Geschlechtern, *AA*, **30**, 1976, 179–186.
[76b] Verallgemeinerung eines Satzes von Rabinowitsch über imaginär-quadratische Zahlkörper, *JRAM*, **285**, 1976, 100–113.
MOLLIN, R. A. [83a] Class numbers and a generalized Fermat theorem, *JNT*, **16**, 1983, 420–429.
[83b] On the cyclotomic polynomial, *JNT*, **17**, 1983, 165–175.
MOLLIN, R., HUGHES, I. (See I. Hughes, R. Mollin)
MONSKY, P. [83] *p*-ranks of class groups in Z_p^d-extensions, *MA*, **263**, 1983, 509–514.
MONSKY, P., CUOCO, A. A. (See A. A. Cuoco, P. Monsky)
MONTGOMERY, H. L. [65] Distribution of irregular primes, *Illinois J. Math.*, **9**, 1965, 553–558.
MONTGOMERY, H. L., BLANKSBY, P. E. (See P. E. Blanksby, H. L. Montgomery)
MONTGOMERY, H. L., LAGARIAS, J. C., ODLYZKO, A. M. (See J. C. Lagarias, H. L. Montgomery, A. M. Odlyzko)
MONTGOMERY, H. L., LEWIS, D. J. (See D. J. Lewis, H. L. Montgomery)
MONTGOMERY, H. L., MASLEY, J. M. (See J. M. Masley, H. L. Montgomery)
MONTGOMERY, H. L., ROHRLICH, D. E. [82] On the *L*-functions of canonical Hecke characters of imaginary quadratic fields, II, *Duke Math. J.*, **49**, 1982, 937–942.
MONTGOMERY, H. L., SCHINZEL, A. [77] Some arithmetic properties of polynomials in several variables, in: *Transcendence Theory; Advances and Applications*, 195–203, Academic Press, 1977.
MONTGOMERY, H. L., WEINBERGER, P. J. [74] Notes on small class numbers, *AA*, **24**, 1974, 529–542.
[77] Real quadratic fields with large class number, *MA*, **225**, 1977, 173–176.
MONTOUCHET, M. N. (See also M. N. Gras) [71] Sur le nombre de classes de sous-corps cyclique de $Q^{(p)}$, $p \equiv 1(\mathrm{mod}\,3)$, *PJA*, **47**, 1971, 585–586.
MOORE, C. [69] Group extensions of *p*-adic and adelic linear groups, *Publ. IHES*, **35**, 1969, 5–74.
MOORE, M. E. [75] A strong complement property of Dedekind domains, *Czechoslov. Math. J.*, **25**, 1975, 282–283.

MORDELL, L. J.   [18a] The class number for definite binary quadratics, *Messeng. Math.*, **47**, 1918, 138–142.

[18b] On a simple summation of the series $\sum_{s=0}^{n-1} e^{2s^2\pi i/n}$, *Messeng. Math.*, **48**, 1918, 54–56.

[21] On the representation of algebraic numbers as a sum of four squares, *PCPS*, **20**, 1921, 250-256.

[22a] On the reciprocity formula for the Gauss's sums in the quadratic field, *PLMS*, (2), **20**, 1922, 289–296.

[22b] On trigonometric series involving algebraic numbers, *PLMS*, (2), **21**, 1922, 493–496.

[29] Kronecker's fundamental limit formula in the theory of numbers and elliptic functions and similar theorems, *Proc. Roy. Soc. London*, **125**, 1929, 262–276.

[31] On Hecke's modular functions, zeta functions, and some other analytic functions in the theory of numbers, *PLMS*, (2), **32**, 1931, 501–556.

[34] On the Riemann hypothesis and imaginary quadratic fields with a given class number, *JLMS*, **9**, 1934, 289–298.

[53] On the linear independence of algebraic numbers, *Pacific J. Math.*, **3**, 1953, 625–630.

[60a] On a Pellian equation conjecture, *AA*, **6**, 1960, 137–144; II, *JLMS*, **36**, 1961, 282–288.

[60b] On recurrence formulae for the number of classes of definite binary quadratic forms, *J. Indian Math. Soc.*, **24**, 1960, 367–378.

[61] The congruence $[\frac{1}{2}(p-1)]! \equiv \pm 1 (\mathrm{mod} p)$, *AMM*, **68**, 1961, 145–146.

[62a] The sign of the Gaussian sum, *Illinois J. Math.*, **6**, 1962, 177–180.

[62b] On a cyclotomic resolvent, *Arch. Math.*, **13**, 1962, 486–487.

[63] On a cyclotomic diophantine equation, *J. Math. Pures Appl.*, (9), **42**, 1963, 205–208.

[64] On Lerch's class number formulae for binary quadratic forms, *Arkiv. Mat.* **5**, 1964, 97–100.

[65] On the conjecture for the rational points on a cubic surface, *JLMS*, **40**, 1965, 149–158.

[67] The representation of a Gaussian integer as a sum of two squares, *Math. Mag.*, **40**, 1967, 209.

[68] The diophantine equation $x^4 + y^4 = 1$ in algebraic number fields, *AA*, **14**, 1968, 347–355.

[69a] The integer solutions of the equation $ax^2 + by^2 + c = 0$ in quadratic fields, *Bull. London Math. Soc.*, **1**, 1969, 43–44.

[69b] The diophantine equation $y^2 = x^4 \pm 1$ in quadratic fields, *JLMS*, **44**, 1969, 112–114.

[69c] A norm ideal bound for a class of biquadratic fields, *Norske Vid. Selsk. Forh.*, **42**, 1969, 53–55.

MORDELL, L. J., CHOWLA, S.   (See S. Chowla, L. J. Mordell)

MORENO, C. J.   [74] Sur le problème de Kummer, *Enseign. Math.*, **20**, 1974, 45–51.

[80] The higher reciprocity laws; an example, *JNT*, **12**, 1980, 57–70.

MORI, S.   [32] Axiomatische Begründung des Multiplikationsringes, *J. Sci. Hiroshima Univ.*, A, **3**, 1932, 45–49.

[33] Über allgemeine Multiplikationsringe, I, *J. Sci. Hiroshima Univ.*, A, **4**, 1933, 1–26; II, *ibid.*, 33–109.

[40] Allgemeine Z. P. I.-Ringe, *J. Sci. Hiroshima Univ.*, A, **10**, 1940, 117–136.

MÔRI, Y.   [33] Zum Fundamentalsatze der Idealtheorie, *Proc. Phys. Math. Soc. Japan*, (3), **15**, 1933, 225–226.

MORIKAWA, R.   [68] On the unit group of an absolutely cyclic number field of degree five, *J. Math. Soc. Japan*, **20**, 1968, 263–265.

[72] On units of real quadratic fields, *JNT*, **4**, 1972, 503–507.

[74] On units of certain cubic number fields, *Hbg*, **42**, 1974, 72–77.

[79] On units of real quadratic number fields, *J. Math. Soc. Japan*, **31**, 1979, 245–250.

MORISHIMA, T. [33] Über die Einheiten und Idealklassen des Galoischen Zahlkörpers und die Theorie der Kreiskörper der l^ν-ten Einheitswurzeln, *Japan. J. Math.*, **10**, 1933, 83–126.

[34] Über die Theorie der Kreiskörper der l^ν-ten Einheitswurzeln, *Japan. J. Math.*, **11**, 1934, 225–240.

[66] On the second factor of the class-number of the cyclotomic field, *J. Math. Anal. Appl.*, **15**, 1966, 141–153.

MORITA, Y. [75] A p-adic analogue of the Γ-function, *J. Fac. Sci. Univ. Tokyo*, I A, **22**, 1975, 255–266.

[77] On the Hurwitz–Lerch L-functions, *J. Fac. Sci. Univ. Tokyo*, I A, **24**, 1977, 29–43.

[78] A p-adic integral representation of the p-adic L-function, *JRAM*, **302**, 1978, 71–95.

[79] On the radius of convergence of the p-adic L-function, *Nagoya Math. J.*, **75**, 1979, 177–193.

MORIYA, M. [30] Ueber die Klassenzahl eines relativ-zyklischen Zahlkörpers vom Primzahlgrad, *Proc. Imp. Acad. Tokyo*, **6**, 1930, 245–247; *Japan. J. Math.* **10**, 1933, 1–18.

[33] Eine Bemerkung über Einheiten im Kreiskörper, *Proc. Imp. Acad. Tokyo*, **9**, 1933, 199–200.

[34a] Über die Konstruktion algebraischer Zahlkörper unendlichen Grades, *J. Fac. Sci. Hokkaido Univ.*, **2**, 1934, 119–128.

[34b] Eine Bemerkung über die Klassenzahl des absoluten Klassenkörper, *Proc. Imp. Acad. Tokyo*, **10**, 1934, 623–625.

[36] Über einem Satz von Herbrand, *J. Fac. Sci. Hokkaido Univ.*, **4**, 1936, 181–194.

[40] Bewertungstheoretischer Aufbau der multiplikativen Idealtheorie, *J. Fac. Sci. Hokkaido Univ.*, **8**, 1940, 109–144.

[50] Rein arithmetischer Beweis über die Unendlichkeit der Primideale 1 Grades aus einem endlichen algebraischen Zahlkörper, *J. Fac. Sci. Hokkaido Univ.*, **11**, 1950, 164–166.

[53] Theorie der Derivationen und Körperdifferenten, *Math. J. Okayama Univ.*, **2**, 1953, 111–148.

MORIYA, M., KOBAYASHI, Y. (See Y. Kobayashi, M. Moriya)

MOROZ, B. Z. [63] Analytic continuation of the scalar product of Hecke series of two quadratic fields and its application, *DAN*, **150**, 1963, 752–754 (Russian).

[64] On the continuability of the scalar product of Hecke series of two quadratic fields, *DAN*, **155**, 1964, 1265–1267 (Russian).

[65a] Composition of binary quadratic forms and scalar product of Hecke series, *Tr. Mat. Inst. Steklov*, **80**, 1965, 102–109 (Russian).

[65b] The distribution of pairs of prime divisors of two quadratic fields, I, *Vestnik LGU*, **20**, 1965, 19, 47–57; II, *ibid.*, **21**, 1966, 1, 64–79 (Russian).

[68] On zeta-functions of algebraic number fields, *Mat. Zametki*, **4**, 1968, 333–339 (Russian).

[80] On the convolution of Hecke L-functions, *Mathematika*, **27**, 1980, 312–320.

[82] Scalar products of L-functions with grössencharacters: its meromorphic continuation and natural boundary, *JRAM*, **332**, 1982, 99–117.

[84] On the distribution of integral and prime divisors with equal norm, *Ann. Inst. Fourier*, **34**, 1984, 4, 1–17.

MORTON, P. [79] On Rédei's theory of the Pell equation, *JRAM*, **307/8**, 1979, 373–398.

[82a] Density results for the 2-classgroups and fundamental units of real quadratic fields, *Studia Sci. Math. Hungar.*, **17**, 1982, 21–43.

[82b] Density results for the 2-classgroups of imaginary quadratic fields, *JRAM*, **323**, 1982, 156–187.

[83] The quadratic number fields with cyclic 2-classgroups, *Pacific J. Math.*, **108**, 1983, 165–175.

MOSER, C. [70] Représentation de −1 par une somme de carrés dans certains corps locaux et globaux, et dans certains anneaux d'entiers algébriques, *CR*, **271**, 1970, A1200–1203.

[71] Représentation de −1 comme somme de carrés d'entiers dans un corps quadratique imaginaire, *Enseign. Math.*, **17**, 1971, 279–287.

[73a] Sommes de carrés d'entiers dans un corps dyadique, *CR*, **277**, 1973, A571–574.

[73b] Représentation de −1 comme somme de carrés dans un corps cyclotomique quelconque, *JNT*, **5**, 1973, 139–141.

[74] Somme de carrés d'entiers d'un corps p-adique, *Enseign. Math.*, **20**, 1974, 299–322.

[81] Nombre de classes d'une extension cyclique réelle de Q, de degré 4 ou 6 et de conducteur premier, *MN*, **102**, 1981, 45–52.

MOSER, C., JOLY, J. R. (See J. R. Joly, C. Moser)

MOSER, C., PAYAN, J. J. [81] Majoration du nombre de classes d'un corps cubique cyclique de conducteur premier, *J. Math. Soc. Japan*, **33**, 1981, 701–706.

MOSER, N. [75] Unités et nombre de classes d'une extension diédrale de Q, *Astérisque*, **24/5**, 1975, 29–35.

[78] *Contraintes galoisiennes sur le groupe des unités de certaines extensions de Q—applications arithmétiques*, Thése, Grenoble 1978.

[79a] Unités et nombre de classes d'une extension galoisienne diédrale de Q, *Hbg*, **48**, 1979, 54–75.

[79b] Sur les unités d'une extension galoisienne non abélienne de degré pq du corps des rationnels, p et q nombres premiers impairs, *Ann. Inst. Fourier*, **29**, 1979, 1, 137–158.

[83] Théorème de densité de Tchebotareff et monogénéité de modules sur l'algèbre d'un groupe métacyclique, *AA*, **42**, 1983, 311–323.

MOSER, N., GRAS, M. N., PAYAN, J. J. (See M. N. Gras, N. Moser, J. J. Payan)

MOSER, N., HALTER-KOCH, F. (See F. Halter-Koch, N. Moser)

MOSSIGE, S. [72] Table of irreducible polynomials over GF[2] of degrees 10 through 20, *MC*, **26**, 1972, 1007–1009.

MOSTOWSKI, A. [55] Determination of the degree of certain algebraic numbers, *Prace Mat.*, **1**, 1955, 239–252 (Polish).

MOTODA, Y. [75] On biquadratic fields, *Mem. Fac. Sci. Kyushu Univ.*, **29**, 1975, 263–268.

[77] On units of a real quadratic field, *Mem. Fac. Gen. Ed. Kumamoto Univ.*, **13**, 1977, 1, 9–13.

MOTOHASHI, Y. [70] A note on the mean value of the Dedekind Zeta-function of the quadratic field, *MA*, **188**, 1970, 123–127.

MOTT, J. L. [64] Equivalent conditions for a ring to be a multiplication ring, *CJM*, **16**, 1964, 429–434.

[66] Integral domains with quotient overrings, *MA*, **166**, 1966, 229–232.

MOTT, J. L., GILMER, R. W. (See R. W. Gilmer, J. L. Mott)

MOTT, J. L., LEVITZ, K. B. (See K. B. Levitz, J. L. Mott)

MOTZKIN, T. [45] Sur l'équation irréductible $z^n + a_1 z^{n-1} + \ldots + a_n = 0$, $n > 1$, à coefficients complexes entiers, dont toutes les racines sont sur une droite. Les 11 classes de droites admissibles, *CR*, **221**, 1945, 220–222.

[47] From among n conjugate algebraic integers $n-1$ can be approximately given, *BAMS*, **53**, 1947, 156–162.

[49] The euclidean algorithm, *BAMS*, **55**, 1949, 1142–1146.

MOUSSA, P., GERONIMO, J. S., BESSIS, D. [84] Ensembles de Julia et propriétés de localisation des familles itérées d'entiers algébriques, *CR*, **299**, 1984, 281–284.

MULHOLLAND, H. P.   [60] On the product of n complex homogeneous linear forms, *JLMS*, **35**, 1960, 241–250.
MÜLLER, H.   [78] A calculation of class-numbers of imaginary quadratic number-fields, *Tamkang J. Math.*, **9**, 1978, 121–128.
MÜNTZ, C.   [23] Der Summensatz von Cauchy in beliebigen algebraischen Zahlkörpern und die Diskriminante derselben, *MA*, **90**, 1923, 279–291.
[24] Allgemeine Begründung der Theorie der höheren ζ-Funktionen, *Hbg*, **3**, 1924, 1–11.
MURPHY, T. G., BIRCH, B. J., LEWIS, D. J.   (See B. J. Birch, D. J. Lewis, T. G. Murphy)
NAGATA, M.   [53] On the theory of Henselian rings, *Nagoya Math. J.*, **5**, 1953, 45–57; II, *ibid.*, **7**, 1954, 1–19.
[54] Note on integral closures of Noetherian domains, *Mem. Coll. Sci. Univ. Kyoto*, **28**, 1954, 121–124.
[68] A type of subrings of a noetherian ring, *J. Math. Kyoto Univ.*, **8**, 1968, 465–467.
[78] On Euclid algorithm, in: *C. P. Ramanujam—a tribute*, 175–186, Springer, 1978.
NAGATA, M., NAKAYAMA, T., TUZUKU, T.   [53] On an existence lemma in valuation theory, *Nagoya Math. J.*, **6**, 1953, 59–61.
NAGELL, T.   [19] Le discriminant de l'équation de la division du cercle, *Norsk Mat. Tidsskr.*, **1**, 1919, 99–101.
[22] Über die Klassenzahl imaginär-quadratischer Zahlkörper, *Hbg*, **1**, 1922, 140–150.
[23] Über die Einheiten in reinen kubischen Zahlkörpern, *Vidensk. Skr. Christiania*, 1922.
[26] Über einige kubische Gleichungen mit zwei Unbestimmten, *MZ*, **24**, 1926, 422–447.
[28] Darstellung ganzer Zahlen durch binäre kubische Formen mit negativer Diskriminante, *MZ*, **28**, 1928, 10–29.
[30] Zur Theorie der kubischen Irrationalitäten, *AM*, **55**, 1930, 33–65.
[31] Zur Theorie der algebraischen Ringe, *JRAM*, **164**, 1931, 80–84.
[32a] Bemerkungen über numerisches Rechnen mit algebraischen Zahlen, *JRAM*, **167**, 1932, 70–72.
[32b] Sätze über algebraische Ringe, *MZ*, **34**, 1932, 179–182.
[32c] Zur algebraischen Zahlentheorie, *MZ*, **34**, 1932, 183–193.
[33] Die Bestimmung der Ringe mit gegebener Diskriminante in einem algebraischen Zahlkörper, *Norsk. Mat. Foren. Skr.* **1/12**, 1933, 69–72.
[37] Bemerkungen über zusammengesetzte Zahlkörper, *Avh. Norske Vid. Akad. Oslo*, 1937, **4**, 1–26.
[38] Bemerkung über die Klassenzahl reell-quadratischer Zahlkörper, *Norske Vid. Selsk. Forh.*, **11**, 1938, 7–10.
[39] Bestimmung des Grades gewisser relativ-algebraischer Zahlen, *Monatsh. M.-Phys.*, **48**, 1939, 61–74.
[53] On the representations of integers as the sum of two integral squares in algebraic, mainly quadratic fields, *Nova Acta Soc. Sci. Uppsala*, **15**, 1953, 11, 1–73.
[54] On a special class of Diophantine equations of the second degree, *Arkiv Mat.*, **3**, 1954, 51–65.
[55] Contributions to the theory of a category of diophantine equations of the second degree with two unknowns, *Nova Acta Soc. Sci. Uppsala*, **16**, 1955, 2, 1–38.
[58] Sur l'équation $x^5+y^5 = z^5$, *Arkiv Mat.*, **3**, 1958, 511–514.
[59] Les points exceptionnels rationnels sur certaines cubiques du premier genre, *AA*, **5**, 1959, 333–357.
[60] Les points exceptionnels sur les cubiques $ax^3+by^3+cz^3 = 0$, *Acta Sci. Math. (Szeged)*, **21**, 1960, 173–180.

[61] On the sum of two integral squares in certain quadratic fields, *Arkiv Mat.*, **4**, 1961, 267–286.

[62a] Sur quelques questions dans le théorie des corps biquadratiques, *Arkiv Mat.*, **4**, 1962, 347–376.

[62b] On the number of representations of an A-number in an algebraic field, *Arkiv Mat.*, **4**, 1962, 467–478.

[63a] On the A-numbers in the quadratic fields $K(\sqrt{\pm 37})$, *Arkiv Mat.*, **4**, 1963, 511–521.

[63b] Sur les sous-corps des corps métacycliques du sixième degré, *Arkiv Mat.*, **5**, 1963, 43–54.

[64a] Contributions à la théorie des corps et des polynômes cyclotomiques, *Arkiv Mat.*, **5**, 1964, 153–192.

[64b] Sur une propriété des unités d'un corps algébrique, *Arkiv Mat.*, **5**, 1964, 343–356.

[65] Contributions à la théorie des modules et des anneaux algébriques, *Arkiv Mat.*, **6**, 1965, 161–178.

[66] Quelques résultats sur les diviseurs fixes de l'index des nombres entiers d'un corps algébrique, *Arkiv Mat.*, **6**, 1966, 269–289.

[67] Sur les discriminants des nombres algébriques, *Arkiv Mat.*, **7**, 1967, 265–282.

[68a] Sur les diviseurs premiers des polynômes, *AA*, **15**, 1968/9, 235–244.

[68b] Sur les unités dans les corps biquadratiques primitifs du premier rang, *Arkiv Mat.*, **7**, 1968, 359–394.

[68c] Quelques propriétés des nombres algébriques du quatrième degré, *Arkiv Mat.*, **7**, 1968, 517–525.

[69a] Quelques problèmes relatifs aux unités algébriques, *Arkiv Mat.*, **8**, 1969, 115–127.

[69b] Sur un type particulier d'unités algébriques, *Arkiv Mat.*, **8**, 1969, 163–184.

[72a] Sur la résolubilité de l'équation $x^2+y^2+z^2 = 0$ dans un corps quadratique, *AA*, **21**, 1972, 35–43.

[72b] Über die Darstellung der Zahlen ± 1 als die Summe von zwei Quadraten in algebraischen Zahlkörpern, *Arch. Math.*, **23**, 1972, 25–29.

[72c] Sur quelques équations diophantiennes de degré supérieur à plusieurs variables, *Norske Vid. Selsk. Skr.*, 1972, 5, 1–6.

[72d] Sur la représentabilité de zéro par certaines formes quadratiques, *Norske Vid. Selsk. Skr.*, 1972, 6, 1–7.

[73] Sur la représentation de zéro par une somme de carrés dans un corps algébrique, *AA*, **24**, 1973, 379–383.

NAITO, H. [82] The p-adic Hurwitz L-functions, *Tôhoku Math. J.*, (2), **34**, 1982, 553–558.

NAKAGOSHI, N. [75] On indices of unit groups related to the genus number of Galois extensions, *Sci. Rep. Kanazawa Univ.*, **20**, 1975, 7–13.

[79] The structure of the multiplicative group of residue classes modulo $\mathfrak{p}^{N+1}$, *Nagoya Math. J.*, **73**, 1979, 41–60.

[81] On the class number relations of abelian extensions whose Galois groups are of the type (p, p), *Math. Rep. Toyama Univ.*, **4**, 1981, 91–106.

[84a] A note on l-class groups of certain algebraic number fields, *JNT*, **19**, 1984, 140–147.

[84b] A construction of unramified Abelian l-extensions of regular Kummer extensions, *AA*, **44**, 1984, 47–58.

NAKAGOSHI, N., EDA, Y. (See Y. Eda, N. Nakagoshi)

NAKAHARA, T. [70] On the determination of the fundamental units of certain real quadratic fields, *Mem. Fac. Sci. Kyushu Univ.*, **24**, 1970, 300–304.

[73] Examples of algebraic number fields which have not unramified extensions, *Rep. Fac. Sci. Saga Univ.*, **1**, 1973, 1–8.

[74] On the fundamental units and an estimate of the class numbers of real quadratic fields, *Rep. Fac. Sci. Saga Univ.*, **2**, 1974, 1–13.
[78] On real quadratic fields whose ideal class groups have a cyclic *p*-subgroup, *Rep. Fac. Sci. Saga Univ.*, **6**, 1978, 15–26.
[82] On cyclic biquadratic fields related to a problem of Hasse, *Monatsh. Math.*, **94**, 1982, 125–132.
[83] On the indices and integral bases of noncyclic but abelian biquadratic fields, *Arch. Math.*, **41**, 1983, 504–508.

NAKAMULA, K. [77] On a fundamental domain of R_+^3 for the action of the group of totally positive units of a cyclic cubic field, *J. Fac. Sci. Univ. Tokyo*, *I A*, **24**, 1977, 3, 701–713.
[79] An explicit formula for the fundamental units of a real pure sextic number field and its Galois closure, *Pacific J. Math.*, **83**, 1979, 463–471.
[80] On the group of units of a non-Galois quartic or sextic number field, *PJA*, **56**, 1980, 77–81.
[81] Class number calculation and elliptic unit, I. Cubic case, *PJA*, **57**, 1981, 56–59; II, Quartic case, *ibid.*, 117–120; III, Sextic case, *ibid.*, 363–366.
[82a] A construction of the groups of units of some number fields from certain subgroups, *Tokyo J. Math.*, **5**, 1982, 85–106.
[82b] Class number calculation of a cubic field from the elliptic unit, *JRAM*, **331**, 1982, 114–123.

NAKAMURA, Y. [59] On the distribution of ideals with exactly *r* different prime divisors in an ideal class of an algebraic number field, *Sci. Rep. Tokyo Kyoiku Daigaku*, *A*, **6**, 1959, 241–257.
[74] Degrees of Galois extensions and norms of prime ideals of algebraic number fields, *Math. Japon.*, **19**, 1974, 135–138.

NAKANO, N. [43] Über die Umkehrbarkeit der Ideale in Integritätsbereichen, *Proc. Imp. Acad. Tokyo*, **19**, 1943, 230–234.
[53] Idealtheorie in einem speziellen unendlichen algebraischen Zahlkörper, *J. Sci. Hiroshima Univ.*, **16**, 1953, 425–439.

NAKANO, S. [83a] Class numbers of pure cubic fields, *PJA*, **59**, 1983, 263–265.
[83b] On the construction of certain number fields, *Tokyo J. Math.*, **6**, 1983, 389–395.
[84a] On ideal class groups of algebraic number fields, *PJA*, **60**, 1984, 74–77.
[84b] On the 2-rank of the ideal class groups of pure number fields, *Arch. Math.*, **42**, 1984, 53–57.
[85] On ideal class groups of algebraic number fields, *JRAM*, **358**, 1985, 61–75.
[86] Ideal class groups of cubic cyclic fields, *AA*, **46**, 1986, 297–300.

NAKATSUCHI, S. [68] A note on certain properties of algebraic number fields, *Mem. Osaka Univ.*, **17**, 1968, 1–10.
[70] On a relation between Kronecker's assertion and Gassmann's theorem, *Mem. Osaka Univ.*, **19**, 1970, 97–105.
[72] A note on regular domains of algebraic number fields, *Mem. Osaka Univ.*, **21**, 1972, 205–211.
[73] A note on Kronecker's "Randwertsatz", *J. Math. Kyoto Univ.*, **13**, 1973, 129–137.
[75] On Čebotarev-sets of normal number fields, *Math. Japon.*, **20**, 1975, 183–206; suppl., *ibid.*, **21**, 1976, 105–109.

NAKAYAMA, T. [40] Note on uni-serial and generalized uni-serial rings, *Proc. Imp. Acad. Tokyo*, **16**, 1940, 285–289.
[51] Factor system approach to the isomorphism and reciprocity theorem, *J. Math. Soc. Japan*, **3**, 1951, 52–58.

[52] Idèle-class factor sets and class field theory, *AnM*, **55**, 1952, 73–84.
NAKAYAMA, T., HOCHSCHILD, G.   (See G. Hochschild, T. Nakayama)
NAKAYAMA, T., NAGATA, M., TUZUKU, T.   (See M. Nagata, T. Nakayama, T. Tuzuku)
NANDA, V. C., BHANDARI, S. K.   (See S. K. Bhandari, V. C. Nanda)
NARASIMHAN, R.   [68] Une remarque sur $\zeta(1+it)$, *Enseign. Math.*, **14**, 1968, 189–191.
NARASIMHAN, R., CHANDRASEKHARAN, K.   (See K. Chandrasekharan, R. Narasimhan)
NARKIEWICZ, W.   [62] On polynomial transformations, *AA*, **7**, 1962, 241–249; II, *ibid.*, **8**, 1962/3, 11–19.
  [63] Remark on rational transformations, *Colloq. Math.*, **10**, 1963, 139–142.
  [64] On algebraic number fields with non-unique factorization, *Colloq. Math.*, **12**, 1964, 59–67; II, *ibid.*, **15**, 1966, 49–58.
  [65] On polynomial transformations in several variables, *AA*, **11**, 1965, 163–168.
  [66] On natural numbers having unique factorization in a quadratic number field, *AA*, **12**, 1966, 1–22; II, *ibid.*, **13**, 1967, 123–129.
  [67] Factorization of natural numbers in some quadratic number fields, *Colloq. Math.*, **16**, 1967, 257–268.
  [68] A note on factorizations in quadratic fields, *AA*, **15**, 1968, 19–22.
  [69] On a theorem of A. Weil on derivations in number fields, *Colloq. Math.*, **20**, 1969, 57–58.
  [72] Numbers with unique factorization in an algebraic number field, *AA*, **21**, 1972, 313–322.
  [73] A note on numbers with good factorization properties, *Colloq. Math.*, **27**, 1973, 275–276.
  [79] Finite abelian groups and factorization problems, *Colloq. Math.*, **42**, 1979, 319–330.
  [80] Normal order for a function associated with factorization into irreducibles, *AA*, **37**, 1980, 77–84.
  [81] Numbers with all factorizations of the same length in a quadratic number field, *Colloq. Math.*, **45**, 1981, 71–74.
  [84] *Number Theory*, Singapore 1984.
  [86] *Classical Problems in Number Theory*, Warszawa 1986.
NARKIEWICZ, W., MEAD, D. G.   (See D. G. Mead, W. Narkiewicz)
NARKIEWICZ, W., SCHINZEL, A.   [69] Ein einfacher Beweis des Dedekindschen Satzes über die Differente, *Colloq. Math.*, **20**, 1969, 65–66.
NARKIEWICZ, W., ŚLIWA, J.   [78] Normal order for certain functions associated with factorizations in number fields, *Colloq. Math.*, **38**, 1978, 323–328.
  [82] Finite abelian groups and factorization problems, II, *Colloq. Math.*, **46**, 1982, 115–122.
NART, E.   [85] On the index of a number field, *TAMS*, **289**, 1985, 171–183.
NART, E., LLORENTE, P.   (See P. Llorente, E. Nart)
NART, E., LLORENTE, P., VILA, N.   (See P. Llorente, E. Nart, P. Vila)
NECHAEV, V. I., STEPANOVA, L. L.   [65] Distribution of non-residues and primitive roots in recurrent sequences over an algebraic number field, *Uspekhi Mat. Nauk*, **20**, 1965, 3, 197–203 (Russian).
NEHRKORN, H.   [33] Über die absolute Idealklassengruppe und Einheiten in algebraischen Zahlkörpern, *Hbg*, **9**, 1933, 318–334.
NEILD, C., SHANKS, D.   [74] On the 3-rank of quadratic fields and the Euler product, *MC* **28**, 1974, 279–291.
NEISS, F.   [31] Darstellung relativ-Abelscher Zahlkörper durch Primkörper und Einheitskörper, *JRAM*, **166**, 1931, 30–53.

NETTO, E. [83] Notiz über Gleichungen, deren Discriminante ein Quadrat ist, *JRAM*, **95**, 1883, 237–239.
[84] Über die Factorenzerlegung der Discriminanten algebraischer Gleichungen, *MA*, **24**, 1884, 579–587.
NEUBRAND, M. [78] Einheiten in algebraischen Funktionen- und Zahlkörpern, *JRAM*, **303/4**, 1978, 170–204.
[81] Scharen quadratischer Zahlkörper mit gleichgebauten Einheiten, *AA*, **39**, 1981, 125–132.
NEUKIRCH, J. [67a] Zur Differententheorie, *Arch. Math.*, **18**, 1967, 241–249.
[67b] Klassenkörpertheorie, *Bonn Math. Schriften*, **26**, 1967.
[68] Über eine algebraische Kennzeichnung der Henselkörper, *JRAM*, **231**, 1968, 75–81.
[69a] Kennzeichnung der p-adischen und der endlichen algebraischen Zahlkörper, *Invent. Math.*, **6**, 1969, 296–314.
[69b] *Klassenkörpertheorie*, Mannheim 1969.
[69c] Kennzeichnung der endlich-algebraischen Zahlkörper durch die Galoisgruppe der maximal auflösbaren Erweiterungen, *JRAM*, **238**, 1969, 135–147.
[73] Über das Einbettungsproblem der algebraischen Zahlentheorie, *Invent. Math.*, **21**, 1973, 59–116.
[74a] Über die absolute Galoisgruppe algebraischer Zahlkörper, *Jahresber. Deutsch. Math. Verein.*, **76**, 1974, 18–37.
[74b] Eine Bemerkung zum Existenzsatz von Grunwald–Hasse–Wang, *JRAM*, **268/9**, 1974, 315–317.
[74c] On an existence theorem of Grunwald's type, *Bol. Soc. Brasil. Mat.*, **5**, 1974, 79–83.
[77] Über die absoluten Galoisgruppen algebraischer Zahlkörper, *Astérisque*, **41/2**, 1977, 67–79.
[84] Neubegründung der Klassenkörpertheorie, *MZ*, **186**, 1984, 557–574.
NEUKIRCH, J., BAYER, P. (See P. Bayer, J. Neukirch)
NEUMANN, J. V. [26] Zur Prüferschen Theorie der idealen Zahlen, *Acta Sci. Math. (Szeged)*, **2**, 1926, 193–227.
NEUMANN, J. V., GOLDSTINE, H. H. [53] A numerical study of a conjecture of Kummer, *Math. Tables Aids Comput.*, **7**, 1953, 133–134.
NEUMANN, O. [73] Relativ-quadratische Zahlkörper, deren Klassenzahlen durch 3 teilbar sind, *MN*, **56**, 1973, 281–306.
[75] On p-closed algebraic number fields with bounded ramification, *IAN*, **39**, 1975, 259–271 (Russian).
[77a] On p-closed number fields and an analogue of Riemann's existence theorem, in: *Algebraic Number Fields*, Proc. Durham Symp., 625–647, London 1977.
[77b] On maximal p-extensions, class numbers and unit signatures, *Astérisque*, **41/2**, 1977, 239–246.
[81a] Über die Anstösse zu Kummers Schöpfung der "idealen complexen Zahlen", in: *Mathematical Perspectives*, 179–199, Academic Press, 1981.
[81b] Two proofs of the Kronecker–Weber theorem "according to Kronecker and Weber", *JRAM*, **323**, 1981, 105–126.
NEUMANN, O., EDWARDS, H., PURKERT, W. (See H. Edwards, O. Neumann, W. Purkert)
NEWMAN, M. [65] Bounds for class numbers, *PSPM*, **8**, 1965, 70–77.
[70] A table for the first factor for prime cyclotomic fields, *MC*, **24**, 1970, 215–219.
[71] Units in cyclotomic fields, *JRAM*, **250**, 1971, 3–11.
[74a] Diophantine equations in cyclotomic fields, *JRAM*, **265**, 1974, 84–89; corr., *ibid.*, **280**, 1976, 211–212.

[74b] Units in arithmetic progression in an algebraic number field, *PAMS*, **43**, 1974, 266–268.
[82] Cyclotomic units and Hilbert's Satz 90, *AA*, **41**, 1982, 353–357.

NEWMAN, M., CALLAHAN, T., SHEINGORN, M. (See T. Callahan, M. Newman, M. Sheingorn)

NEWMAN, M., TAUSSKY, O. [58] On a generalization of the normal basis in abelian algebraic number fields, *Comm. Pure Appl. Math.*, **9**, 1958, 85–91.

NGUYEN-QUANG-DO, T. [75] Ramification et dénombrement dans les p-extensions des corps locaux, *CR*, **280**, 1975, 401–402.
[76] Filtration de K^*/K^{*p} et ramification sauvage, *AA*, **30**, 1976, 323–340.
[82] Sur la structure galoisienne des corps locaux et la théorie d'Iwasawa, I, *Compos. Math.*, **46**, 1982, 85–119; II, *JRAM*, **333**, 1982, 133–143.

NGUYEN-QUANG-DO, T., MASSY, R. (See R. Massy, T. Nguyen-Quang-Do)

NIEDERREITER, H., KUIPERS, L., SHIUE, J. S. (See L. Kuipers, H. Niederreiter, J. S. Shiue)

NIEDERREITER, H., LO, S. K. (See also S. K. Lo, H. Niederreiter)
[75] Uniform distribution of sequences of algebraic integers, *Math. J. Okayama Univ.*, **18**, 1975, 13–29.

NISHI, M., KOYAMA, T., YANAGIHARA, H. (See T. Koyama, M. Nishi, H. Yanagihara)

NIVEN, I. [40] Integers of quadratic fields as sums of squares, *TAMS*, **48**, 1940, 405–417.
[41a] Sums of n-th powers of quadratic integers, *Duke Math. J.*, **8**, 1941, 441–451.
[41b] Sums of fourth powers of gaussian integers, *BAMS*, **47**, 1941, 923–926.
[42] Quadratic Diophantine equations in the rational and quadratic fields, *TAMS*, **52**, 1942, 1–11.
[43] The Pell equation in quadratic fields, *BAMS*, **49**, 1943, 413–416.
[51] A class of algebraic integers, *AMM*, **58**, 1951, 27–29.

NOETHER, E. [19] Die arithmetische Theorie der algebraischen Funktionen einer Veränderlichen in ihrer Beziehung zu den übrigen Theorie und zu der Zahlkörpertheorie, *Jahresber. Deutsch. Math. Verein.*, **28**, 1919, 182–203.
[21] Idealtheorie in Ringbereichen, *MA*, **83**, 1921, 24–66.
[27a] Abstrakter Aufbau der Idealtheorie in algebraischen Zahl- und Funktionenkörpern, *MA*, **96**, 1927, 26–61.
[27b] Der Diskriminantensatz für die Ordnungen eines algebraischen Zahl- oder Funktionenkörpers, *JRAM*, **157**, 1927, 82–104.
[30] Idealdifferentiation und Differente, *Jahresber. Deutsch. Math. Verein.*, **39**, 1930, 17 kursiv.
[32] Normalbasis bei Körpern ohne höhere Verzweigung, *JRAM*, **167**, 1932, 147–152.
[33] Der Hauptgeschlechtsatz für relativ-galoische Zahlkörper, *MA*, **108**, 1933, 411–419.
[50] Idealdifferentiation und Differente, *JRAM*, **188**, 1950, 1–21.

NORDHOFF, H. U. [74] Explizite Darstellungen von Einheiten und ihre Anwendung auf Mehrklassigkeitsfragen bei reell-quadratischen Zahlkörpern, I, *JRAM*, **268/9**, 1974, 131–149; II, *ibid.*, **280**, 1976, 37–60.

NORRIS, M. J., VÉLEZ, W. Y. [80] Structure theorems for radical extensions of fields, *AA*, **38**, 1980/1, 111–115; corr., *ibid.*, **42**, 1983, 427–428.

NORTHCOTT, D. G. [55] A note on classical ideal theory, *PCPS*, **51**, 1955, 766–767.

NOTARI, C. [78] Sur le produit des conjugués à l'extérieur du cercle unité d'un nombre algébrique, *CR*, **286**, 1978, A313–315.

NOVIKOV, A. P. [62] On class-number of fields of complex multiplication, *IAN*, **26**, 1962, 677–686 (Russian).

[67] On class number of fields, which are abelian over an imaginary quadratic field, *IAN*, **31**, 1967, 717–726 (Russian).

[69] On the regularity of prime divisors of first degree in an imaginary quadratic field, *IAN*, **33**, 1969, 1059–1079 (Russian).

[80] The Kronecker's limit formula in a real quadratic field, *IAN*, **44**, 1980, 886–917 (Russian).

Nowlan, F. S. [26] Representations of integers by certain ternary cubic forms, *BAMS*, **32**, 1926, 374–380.

Nyberg, N. [33] Bemerkungen über die Berechnung einer Körperbasis in einem gegebenen algebraischen Zahlkörper, *Norsk Mat. Tidsskr.*, **15**, 1933, 50–54.

Nymann, J. [67] A Minkowskian type bound for a class of relative quadratic fields, *Duke Math. J.*, **34**, 1967, 53–55.

Odlyzko, A. [75] Some analytic estimates of class numbers and discriminants, *Invent. Math.*, **29**, 1975, 275–286.

[76] Lower bounds for discriminants of number fields, *AA*, **29**, 1976, 275–297; II, *Tôhoku Math. J.*, **29**, 1977, 209–216.

[77] On conductors and discriminants, in: *Algebraic Number Fields, Proc. Durham Symp.*, 377–407, London 1977.

Odlyzko, A. M., Kurshan, R. P. (See R. P. Kurshan, A. M. Odlyzko)

Odlyzko, A. M., Lagarias, J. C. (See J. C. Lagarias, A. M. Odlyzko)

Odlyzko, A. M., Lagarias, J. C., Montgomery, H. L. (See J. C. Lagarias, H. L. Montgomery, A. M. Odlyzko)

Odoni, R. W. K. [73a] On Gauss sums $(\bmod p^n)$, *Bull. London Math. Soc.*, **5**, 1973, 325–327.

[73b] The Farey density of norm subgroups of global fields, I, *Mathematika*, **20**, 1973, 155–169.

[75a] On the norms of algebraic integers, *Mathematika*, **22**, 1975, 71–80.

[75b] On norms of integers in a full module of an algebraic number field and the distribution of values of binary integral quadratic forms, *Mathematika*, **22**, 1975, 108–111.

[76] On a problem of Narkiewicz, *JRAM*, **288**, 1976, 160–167.

[77a] Some global norm density results from an extended Čebotarev density theorem, in: *Algebraic Number Fields, Prop. Durham Symp.*, 485–495, London 1977.

[77b] A new equidistribution property of norms of ideals in given classes, *AA*, **33**, 1977, 53–63.

[78] Representation of algebraic integers by binary quadratic forms and norm forms from full modules of extension fields, *JNT*, **10**, 1978, 324–333.

Odoni, R. W. K., Lal, S., McFarland, R. L. (See S. Lal, R. L. McFarland, R. W. K. Odoni)

Ohm, J., Gilmer, R. W. Jr. (See R. W. Gilmer, Jr., J. Ohm)

Ohta, K. [72] On the relative class number of a relative Galois number field, *J. Math. Soc. Japan*, **24**, 1972, 552–557.

[78] On the p-class groups of a Galois number field and its subfields, *J. Math. Soc. Japan*, **30**, 1978, 763–770.

[81] On algebraic number fields whose class numbers are multiples of 3, *Bull. Fac. Gen. Ed. Gifu Univ.*, 1981, 51–54.

Ojala, T. [75] Sums of roots of unity, *Math. Scand.*, **37**, 1975, 83–101.

[77] Euclid's algorithm in the cyclotomic field $Q(\zeta_{16})$, *MC*, **31**, 1977, 268–273.

Okada, T. [80] Normal bases of class fields over Gauss's number field, *JLMS*, (2), **22**, 1980, 221–225.

OKAMOTO, T. [76] A remark on the relative class number of certain algebraic number fields, *TRU Math.*, **12**, 1976, 2, 1–3.
OKUTSU, K. [82a] Construction of integral basis, I, *PJA*, **58**, 1982, 47–49; II, *ibid.*, 87–89; III, *ibid.*, 117–119; IV, *ibid.*, 167–169.
[82b] Integral basis of the field $Q(\sqrt[n]{a})$, *PJA*, **58**, 1982, 219–222.
OLIWA, G. [53] Über die Approximation von komplexen Zahlen durch Zahlen des Körpers $K(i\sqrt{2})$, *Anz. Österr. AW*, **90**, 1953, 171–173.
OLSON, F. R., CARLITZ, L. (See L. Carlitz, F. R. Olson)
OLSON, J. E. [69] A combinatorial problem on finite Abelian groups, *JNT*, **1**, 1969, 8–10; II, *ibid.*, 195–199.
OMAROV, R. [68] On the distribution of special elementary Abelian fields of type (l, l), *IAN Kazakh. SSR*, 1968, 5, 73–76 (Russian).
O'MEARA, O. T. [56] Basis structure of modules, *PAMS*, **7**, 1956, 965–974.
[59] Infinite dimensional quadratic forms over algebraic number fields, *PAMS*, **10**, 1959, 55–58.
[63] *Introduction to Quadratic Forms*, Springer, 1963.
[65] On the finite generation of linear groups over Hasse domains, *JRAM*, **217**, 1965, 79–108.
OMMERBORN, B., EISENBEIS, H., FREY, G. (See H. Eisenbeis, G. Frey, B. Ommerborn)
ONABE, M. [76] On the isomorphisms of the Galois groups of the maximal Abelian extensions of imaginary quadratic fields, *Natur. Sci. Rep. Ochanomizu Univ.*, **27**, 1976, 155–161.
[78] On idèle class groups of imaginary quadratic fields, *Natur. Sci. Rep. Ochanomizu Univ.*, **29**, 1978, 37–42.
ONETO, A. V., LLORENTE, P. (See P. Llorente, A. V. Oneto)
ONISHI, H., APPELGATE, H. (See H. Appelgate, H. Onishi)
ONO, T. [63] On the Tamagawa number of algebraic tori, *AnM*, **78**, 1963, 47–73.
[70] Gauss transforms and zeta-functions, *AnM*, **91**, 1970, 332–361.
ONUKI, M., YAMAMOTO, K. (See K. Yamamoto, M. Onuki)
OOZEKI, K. [78] On truncated units, *TRU Math.*, **14**, 1978, 1–3.
[79] On some truncated units in algebraic number fields of degree $n \geq 3$, *Monatsh. Math.*, **87**, 1979, 310–312.
OOZEKI, K., YAMAGUCHI, I. (See I. Yamaguchi, K. Oozeki)
OPOLKA, H. [80a] Zur Auflösung zahlentheoretischer Knoten, *MZ*, **173**, 1980, 95–103.
[80b] Auflösung zahlentheoretischer Knoten in Galoiserweiterungen von Q, *Arch. Math.*, **34**, 1980, 416–420.
[81] Geschlechter von zentralen Erweiterungen, *Arch. Math.*, **37**, 1981, 418–424.
[82] Some remarks on the Hasse norm theorem, *PAMS*, **84**, 1982, 464–466.
OPPENHEIM, A. [34] Quadratic fields with and without Euclid's algorithm, *MA*, **109**, 1934, 349–352.
[37] Diophantische Approximationen in imaginär-quadratischen Zahlkörpern, *Monatsh. M.-Phys.*, **46**, 1937, 196.
ORDE, H. L. S. [78] On Dirichlet's class number formula, *JLMS*, (2), **18**, 1978, 409–420.
ORE, O. [23] Zur Theorie der algebraischen Körper, *AM*, **44**, 1923, 219–314.
[24] Zur Theorie der Eisensteinschen Gleichungen, *MZ*, **20**, 1924, 267–279.
[25a] Weitere Untersuchungen zur Theorie der algebraischen Körper, *AM*, **45**, 1925, 145–160.
[25b] Bestimmung der Diskriminanten algebraischer Körper, *AM*, **45**, 1925, 303–344.
[25c] Bestimmung der Differente eines algebraischen Zahlkörpers, *AM*, **46**, 1925, 363–392.

[26a] Bemerkungen zur Theorie der Differente, *MZ*, **25**, 1926, 1-8.
[26b] Über zusammengesetzte algebraische Körper, *AM*, **49**, 1926, 379-396.
[26c] Existenzbeweise für algebraische Körper mit vorgeschriebenen Eigenschaften, *MZ*, **25**, 1926, 474-489.
[26d] Über die Bedeutung der Fundamentalgleichung in der Theorie der algebraischen Körper, *MA*, **95**, 1926, 239-246.
[27] Über den Zusammenhang zwischen den definierenden Gleichungen und der Idealtheorie in algebraischen Körpern, *MA*, **96**, 1927, 313-352; II, *ibid.*, **97**, 1927, 569-598.
[28a] Newtonsche Polygone in der Theorie der algebraischer Körper, *MA*, **99**, 1928, 84-117.
[28b] Abriss einer arithmetischer Theorie der Galoisschen Körper, *MA*, **100**, 1928, 650-673; II, *ibid.*, **102**, 1930, 283-304.
[34] Les corps algébriques et la théorie des idéaux, *Mém. Sci. Math.*, **64**, 1934, 1-72.

ORIAT, B. [72] Étude arithmétique des corps cycliques de degré p^r sur le corps des nombres rationnels, *Enseign. Math.*, (2), **18**, 1972, 57-104.
[76] Relation entre les 2-groupe des classes d'idéaux au sens ordinaire et restraint de certains corps de nombres, *Bull. Soc. Math. France*, **104**, 1976, 301-307.
[77] Relation entre le 2-groupe des classes d'idéaux des extensions quadratiques $k(\sqrt{d})$ et $k(\sqrt{-d})$, *Ann. Inst. Fourier*, **27**, 1977, 2, 37-59.
[78] Sur la divisibilité par 8 et 16 des nombres de classes d'idéaux des corps quadratiques $Q(\sqrt{2p})$ et $Q(\sqrt{-2p})$, *J. Math. Soc. Japan*, **30**, 1978, 279-285.
[81] Annulation de groupe de classes réelles, *Nagoya Math. J.*, **81**, 1981, 45-56.

ORIAT, B., SATGÉ, P. [79] Un essai de généralisation du "Spiegelungssatz", *JRAM*, **307/8**, 1979, 134-159.

ORTIZ, J. J., CANALS, I. (See I. Canals, J. J. Ortiz)

OSGOOD, C. F. [66] Some theorems on diophantine approximation, *TAMS*, **123**, 1966, 64-87.

OSIPOV, J. V. [79] P-adic zeta-functions, *Uspekhi Mat. Nauk*, **34**, 1979, 3, 209-210 (Russian).
[80] P-adic zeta-functions and Bernoulli numbers, *Zap. Nauch. Sem. LOMI*, **93**, 1980, 192-203 (Russian).

OSTMANN, H. H. [68] *Additive Zahlentheorie*, Springer, 1968.

OSTROWSKI, A. [13] Über einige Fragen der allgemeinen Körpertheorie, *JRAM*, **143**, 1913, 255-284.
[17] Über sogenannte perfekte Körper, *JRAM*, **147**, 1917, 191-204.
[18] Über einige Lösungen der Funktionalgleichung $\varphi(x)\varphi(y) = \varphi(xy)$, *AM*, **41**, 1918, 271-284.
[19] Über ganzwertige Polynome in algebraischen Zahlkörpern, *JRAM*, **149**, 1919, 117-124.
[35] Untersuchungen zur arithmetischer Theorie der Körper, *MZ*, **39**, 1935, 269-320; II-III, *ibid.*, 321-404.

PAGE, A. [35] On the number of primes in an arithmetical progression, *PLMS*, (2), **39**, 1935, 116-141.

PAJUNEN, S. [76] Computations on the growth of the first factor for prime cyclotomic fields, *Nordisk Tidskr. Inform.*, **16**, 1976, 85-87; II, *ibid.*, **17**, 1977, 113-114.

PAK, I., AMERBAEV, V. (See V. Amerbaev, I. Pak)

PAL, S., MADAN, M. L. (See M. L. Madan, S. Pal)

PALL, G. [45] Note on factorization in a quadratic field, *BAMS*, **51**, 1945, 771-775.
[51] Sums of two squares in a quadratic field, *Duke Math. J.*, **18**, 1951, 399-409.
[69] Discriminantal divisors of binary quadratic forms, *JNT*, **1**, 1969, 525-533.

PALL, G., BUTTS, H. S.   (See H. S. Butts, G. Pall)
PALL, G., COHN, H.   (See H. Cohn, G. Pall)
PAN, CHENG-DUNG   [63] On k-th moments of the number of classes of imaginary quadratic fields, *Sci. Sinica*, **12**, 1963, 737–738 (Chinese).
PANELLA, G.   [66] Un teorema di Golod–Šafarevič e alcune sue conseguenze, *Confer. Sem. Math. Univ. Bari*, **104**, 1966, 1–17.
PAPICK, I. J., HUCKABA, J. A.   (See J. A. Huckaba, I. J. Papick)
PAPKOV, P. S.   [38] On imaginary quadratic realms with a given group of ideal classes, *Uchen. Zap. Univ. Rostov*, **2**, 1938, 8–14 (Russian).
[44] On imaginary quadratic realms admitting only ambiguous classes, *Soobshch. AN Gruz. SSR*, **5**, 1944, 588–592 (Russian).
PAPP, Z. Z., GYÖRY, K.   (See K. Györy, Z. Z. Papp)
PARIS, C.   [72a] Approximations p-adiques de certains entiers de degré 3, *CR*, **274**, 1972, 289–291.
[72b] Bases sur Z des idéaux canoniques d'un corps cubique cyclique, *CR*, **274**, 1972, 610–611.
[83] Calculs numériques dans les extensions de Q, cycliques de degré 5, *Publ. Math. Fac. Sci. Besançon, Th. de Nombres*, 1981/82 et 1982/83.
PARNAMI, J. C., AGRAWAL, M. K., RAJWADE, A. R.   [81] On the 4-power Stufe of a field, *Rend. Circ. Mat. Palermo*, (2), **30**, 1981, 245–254.
PARRY, C. J.   [71a] On a problem of Schinzel concerning principal divisors in arithmetic progressions, *AA*, **19**, 1971, 215–222.
[71b] Algebraic number fields with the principal ideal condition, *AA*, **19**, 1971, 409–413.
[71c] A further note on principal divisors in arithmetic progressions, *JNT*, **3**, 1971, 182–183.
[75a] Units of algebraic number fields, *JNT*, **7**, 1975, 385–388; corr., *ibid.*, **9**, 1977, 278.
[75b] Class number relations in pure quintic fields, *Symposia Math.*, **15**, 1975, 475–485, Academic Press.
[75c] Pure quartic fields whose class numbers are even, *JRAM*, **272**, 1975, 102–112.
[75d] Class number relations in pure sextic fields, *JRAM*, **274/5**, 1975, 360–375.
[76] On a conjecture of Brandler, *JNT*, **8**, 1976, 492–495.
[77a] Class number formulae for bicubic fields, *Illinois J. Math.*, **21**, 1977, 148–163.
[77b] Real quadratic fields with class numbers divisible by five, *MC*, **31**, 1977, 1019–1029.
[78] On the class number of relative quadratic fields, *MC*, **32**, 1978, 1261–1270.
[80] A genus theory for quartic fields, *JRAM*, **314**, 1980, 40–71.
[81] A unit relationship for pure sextic fields, *Arch. Math.*, **37**, 1981, 210–221.
PARRY, C. J., BIRD, R. F.   (See R. F. Bird, C. J. Parry)
PARRY, C. J., BROWN, E.   (See E. Brown, C. J. Parry)
PARRY, C. J., HAO, F. H.   (See F. H. Hao, C. J. Parry)
PARRY, C. J., MCCLUER, C. R.   (See C. R. McCluer, C. J. Parry)
PARRY, C. J., WALTER, C. D.   [76] The class number of pure fields of prime degree, *Mathematika*, **23**, 1976, 220–226; corr., *ibid.*, **24**, 1977, 133.
PASSI, H. A.   [71] An asymptotic formula in partition theory, *Duke Math., J.*, **38**, 1971, 327–337.
PATHIAUX, M.   [69] Répartition modulo 1 de la suite $\lambda \alpha^n$, *Sém. DPP*, **11**, 1969/70, exp. 13.
[73] Sur le produit des conjugués d'un nombre algébrique situés à l'extérieur du disque unité, *Sém. DPP*, **15**, 1973/74, exp. 66.
[77] Familles fermées de nombres algébriques dont la réunion est l'ensemble des nombres algébriques de module supérieur à 1, tous leurs conjugués étant de module inférieur à 1, *CR*, **284**, 1977, 1319–1320.

PATTERSON, S. J. [78] On the distribution of Kummer sums, *JRAM*, **303/4**, 1978, 126-143.
[81] The distribution of general Gauss sums at prime arguments, in: *Progress in Analytic Number Theory*, II, 171-182, New York 1981.
PATTERSON, S. J., HEATH-BROWN, D. R. (See D. R. Heath-Brown, S. J. Patterson)
PATZ, W. [41] *Tafel der regelmässiger Kettenbrüche für Quadratwurzeln aus den natürlichen Zahlen von 1–10 000*, Leipzig 1941.
PAVLOVA, I. V. [68] Local fields of degree 5, *Uchen. Zap. Dalnevost. Gos. Univ.*, **16**, 1968, 148-151 (Russian).
PAYAN, J. J. [62a] Construction des corps abéliens de degré 5, *CR*, **254**, 1962, 3618-3620.
[62b] Entiers des corps abéliens de degré 5, *CR*, **255**, 1962, 2345-2347.
[65] Contribution a l'étude des corps abéliens absolus de degré premier impair, *Ann. Inst. Fourier*, **15**, 1965, 2, 133-199.
[73] Sur les classes ambiges et les ordres monogénes d'une extension cyclique de degré premier impair sur Q ou sur un corps quadratique imaginaire, *Arkiv Mat.*, **11**, 1973, 239-244.
[81] Remarques sur la structure galoisienne des unités des corps de nombres, *AA*, **39**, 1981, 77-82.
PAYAN, J. J., BERTRANDIAS, F. (See F. Bertrandias, J. J. Payan)
PAYAN, J. J., BOUVIER, L. (See L. Bouvier, J. J. Payan)
PAYAN, J. J., DAMEY, P. (See P. Damey, J. J. Payan)
PAYAN, J. J., GRAS, M. N., MOSER, N. (See M. N. Gras, N. Moser, J. J. Payan)
PAYAN, J. J., MARTINET, J. (See J. Martinet, J. J. Payan)
PAYAN, J. J., MOSER, C. (See C. Moser, J. J. Payan)
PAYSANT-LE-ROUX, R., DUBOIS, E. (See E. Dubois, R. Paysant-le-Roux)
PEARSON, K. R., SCHNEIDER, J. E. [70] Rings with a cyclic group of units, *J. Algebra*, **16**, 1970, 243-251.
PECK, L. G. [49] Diophantine equations in algebraic number fields, *AJM*, **71**, 1949, 387-402.
PELLET, A. [78] Sur la décomposition d'une fonction entiére en facteurs irréductibles suivant un module premier, *CR*, **86**, 1878, 1071-1072.
PEN, A. S., SKUBENKO, B. F. [69] Upper bound for the period of a quadratic irrationality, *Mat. Zametki*, **5**, 1969, 413-418 (Russian).
PENG, C. K., DAI, Z. D., LAM, T. Y. (See Z. D. Dai, T. Y. Lam, C. K. Peng)
PERGEL, J. [67] Generalizations of Linnik's asymptotic formula for the additive problem of divisors to Gaussian numbers, *Studia Sci. Math. Hungar.*, **2**, 1967, 133-151.
PERLIS, R. [77a] On the equation $\zeta_K(s) = \zeta_{K'}(s)$, *JNT*, **9**, 1977, 342-360.
[77b] A remark about zeta functions of number fields of prime degree, *JRAM*, **293/4**, 1977, 435-436.
[78] On the class numbers of arithmetically equivalent fields, *JNT*, **10**, 1978, 489-509.
PERLIS, R., CONNER, P. E. (See P. E. Conner, R. Perlis)
PERLIS, R., MEYER, W. (See W. Meyer, R. Perlis)
PERLIS, R., SCHINZEL, A. [79] Zeta functions and the equivalence of integral norms, *JRAM*, **309**, 1979, 176-182.
PERLIS, S. [42] Normal bases of cyclic fields of prime-power degree, *Duke Math. J.*, **9**, 1942, 507-517.
PEROTT, J. [88] Sur l'équation $t^2 - Du^2 = -1$, *JRAM*, **102**, 1888, 185-223.
PERRON, O. [07] Grundlagen fuer eine Theorie des Jacobischen Kettenbruchalgorithmus, *MA*, **64**, 1907, 1-76.
[14] Abschätzung der Lösung der Pellschen Gleichung, *JRAM*, **144**, 1914, 71-73.

[30] Über die Approximation einer komplexer Zahl durch Zahlen des Körpers $K(i)$, *MA*, **103**, 1930, 534–544; II, *ibid.*, **105**, 1931, 160–164.
[31] Über einen Approximationssatz von Hurwitz und über die Approximation einer komplexen Zahl des Körpers der dritten Einheitswurzeln, *S.B. Bayer. AW*, 1931, 129–154.
[32] Quadratische Körper mit Euklidischem Algorithmus, *MA*, **107**, 1932, 489–495.
[33] Diophantische Approximationen in imaginären quadratischen Zahlkörpern, insbesondere im Körper $K(i\sqrt{2})$, *MZ*, **37**, 1933, 749–767.
[49] Diophantische Ungleichungen in imaginären quadratischen Körpern, *Mat. Tidsskr.*, B, 1949, 1–7.
PESEK, J. et al. (See W. J. Ellison et al.)
PETERS, M. [72] Die Stufe der Ordnungen ganzer Zahlen in algebraischen Zahlkörpern, *MA*, **195**, 1972, 309–314.
[73] Quadratische Formen über Zahlringen, *AA*, **24**, 1973, 157–164.
[74] Summen von Quadraten in Zahlringen, *JRAM*, **268/9**, 1974, 318–323.
PETERSON, B., EDGAR, H. (See H. Edgar, B. Peterson)
PETERSSON, H. [55] Über die Zerlegung des Kreisteilungspolynom von Primzahlordnung, *MN*, **14**, 1955, 361–375.
[59] Über Darstellungsanzahlen von Primzahlen durch Quadratsummen, *MZ*, **71**, 1959, 289–307.
PETHÖ, A. [74] Über die Darstellung der rationalen Zahlen durch Normformen, *Publ. Math. (Debrecen)*, **21**, 1974, 31–38.
PETHÖ, A., GYÖRY, K. (See K. Györy, A. Pethö)
PETR, K. [35] Basis der ganzen Zahlen in algebraischen Zahlkörpern, *Časopis mat.-fys.*, **64**, 1935, 62–72.
PFISTER, A. [65] Zur Darstellung von -1 als Summe von Quadraten in einem Körper, *JLMS*, **40**, 1965, 159–165.
PHILLIPS, R. C., BUTTS, H. S. (See H. S. Butts, R. C. Phillips)
PHRAGMÉN, E. [92] Sur la distribution des nombres premiers, *CR*, **114**, 1892, 337–340.
PIEPER, H. [72] Die Einheitengruppe einer zahm-verzweigten galoisschen lokalen Körpers als Galois-Modul, *MN*, **54**, 1972, 173–210.
[73] Die Einseinheitengruppen höheren Stufen einer zerfallenden zahmverzweigten Erweiterung als Galoismoduln, *MN*, **58**, 1973, 193–200.
PIERCE, R. S., BEAUMONT, R. A. (See R. A. Beaumont, R. S. Pierce)
PIERCE, S. [74] Steinitz classes in quartic fields, *PAMS*, **43**, 1974, 39–41.
PIERCE, T. A. [26] An approximation to the least root of a cubic equation with application to the determination of units in pure cubic fields, *BAMS*, **32**, 1926, 263–268.
PINTZ, J. [71] On a certain point in the theory of Dirichlet's L-functions, I, *Mat. Lapok*, **22**, 1971, 143–148; II, *ibid.*, 331–335 (Hungarian).
[74a] On Siegel's theorem, *AA*, **24**, 1973/4, 543–551.
[74b] On the Brauer–Siegel theorem, *Proc. Coll. Number Theory*, Debrecen 1974, 259–265.
[76a] Elementary methods in the theory of L-functions, I, Hecke's theorem, *AA*, **31**, 1976, 53–60.
[76b] —, II, On the greatest real zero of a real L-function, *AA*, **31**, 1976, 273–289.
[76c] —, IV, The Heilbronn phenomenon, *AA*, **31**, 1976, 419–429.
[77a] —, V, The theorems of Landau and Page, *AA*, **32**, 1977, 163–171.
[77b] —, VII, Upper bound for $L(1,\chi)$, *AA*, **32**, 1977, 397–406; corr., *ibid.*, **33**, 1977, 293–295.
[77c] —, VIII, Real zeros of real L-functions, *AA*, **33**, 1977, 89–98.
PIRTLE, E. B., JR. [70] A note on almost-Dedekind domains, *Publ. Math. (Debrecen)*, **17**, 1970, 243–247.

Pisot, C. [36] Sur une propriété de certains entiers algébriques, *CR*, **202**, 1936, 892–894.

[38] La répartition modulo 1 et les nombres algébriques, *Ann. Scuola Norm. Sup. Pisa* (2), **7**, 1938, 205–248.

[63a] *Quelques aspects de la théorie des entiers algébriques*, Montreal 1963; 2nd ed. 1966.

[63b] Ensembles fermés de nombres algébriques et familles normales des fractions rationnelles, *CR*, **256**, 1963, 1418–1419.

[64] Familles compactes de fractions rationnelles et ensembles fermés de nombres algébriques, *ASENS*, **81**, 1964, 165–188.

[66] Ensembles remarquables de nombres algébriques, in: *Les tendances géometriques en algèbre et théorie des nombres*, 191–199, Paris 1966.

Pisot, C., Dufresnoy, J. (See J. Dufresnoy, C. Pisot)

Pisot, C., Hugot, M. (See M. Hugot, C. Pisot)

Pisot, C., Salem, R. [64] Distribution modulo 1 of the powers of real numbers larger than 1, *Compos. Math.*, **16**, 1964, 164–168.

Pitti, C. [70] Élements de norme 1 des corps cubiques non galoisiens. Généralisations, *Sém. Algèbre et Th. de Nombres*, Paris 1969/70, exp. 3.

[71] Sur un problème diophantien relatif au corps diédraux, *CR*, **272**, 1971, A92–94.

[72] Étude des normes dans les extensions à groupe de Klein, *CR*, **274**, 1972, A1433–1435.

[73a] Étude des normes dans les extensions à groupe de Klein, *Sém. DPP*, **14**, 1972/3, exp. 7.

[73b] Éléments de norme 1 dans les extensions a groupe de Klein, *CR*, **277**, 1973, A273–275.

Pizer, A. [76] On the 2-part of the class-number of imaginary quadratic number fields, *JNT*, **8**, 1976, 184–192.

Platonov, V. P., Drakokhrust, Ya. A. [85] On Hasse's principle for algebraic number fields, *DAN*, **281**, 1985, 793–797 (Russian).

Pleasants, P. A. B. [74] The number of generators of the integers of a number field, *Mathematika*, **21**, 1974, 160–167.

[75] Cubic polynomials over algebraic number fields, *JNT*, **7**, 1975, 310–344.

Pleasants, P. A. B., Allen, S. (See S. Allen, P. A. B. Pleasants)

Plemelj, J. [12] Die Unlösbarkeit von $x^5+y^5+z^5 = 0$ im Körper $k(\sqrt{5})$, *Monatsh. M.-Phys.*, **23**, 1912, 305–308.

[33] Die Irreduzibilität der Kreisteilungsgleichung, *Publ. Math. Univ. Belgrade*, **2**, 1933, 164–165.

Plotkin, N. [72] The solvability of the equation $ax^2+by^2 = c$ in quadratic fields, *PAMS*, **34**, 1972, 337–339.

Podsypanin, E. V. [79] On the length of the period of a quadratic irrationality, *Zap. Nauch. Sem. LOMI*, **82**, 1979, 95–99 (Russian).

Podsypanin, V. D. [49] On algebraic units in normal rings of fourth degree, *Uchen. Zap. Leningr. Gos. Ped. Inst.*, **86**, 1949, 183–194 (Russian).

Pohst, M. [75a] Über biquadratische Zahlkörper gleicher Diskriminante, *Hbg.*, **43**, 1975, 192–197.

[75b] Berechnung kleiner Diskriminanten total reeller algebraischer Zahlkörper, *JRAM*, **278/9**, 1975, 278–300.

[75c] Berechnung unabhängiger Einheiten und Klassenzahlen in total reellen biquadratischen Zahlkörpern, *Computing*, **14**, 1975, 67–78.

[76] Invarianten des total reellen Körpers siebten Grades mit Minimaldiskriminante, *AA*, **30**, 1976, 199–207.

[77a] The minimum discriminant of seventh degree totally real algebraic number fields, in: *Number Theory and Algebra*, 235–240, New York 1977.

[77b] Regulatorabschätzungen für total reelle algebraische Zahlkörper, *JNT*, **9**, 1977, 459–492.
[78] Eine Regulatorabschätzung, *Hbg*, **47**, 1978, 95–106.
[82] On the computation of number fields of small discriminants including the minimum discriminants of sixth degree fields, *JNT*, **14**, 1982, 99–117.

POHST, M., WEILER, P., ZASSENHAUS, H. [82] On effective computation of fundamental units, II, *MC*, **38**, 1982, 293–329.

POHST, M., ZASSENHAUS, H. [77] An effective number geometric method of computing the fundamental units of an algebraic number field, *MC*, **31**, 1977, 754–770.
[79] On unit computation in real quadratic fields, in: *Symbolic and Algebraic Computation*, 140–152, Lecture Notes in Comp. Sci., 72, Springer, 1979.
[82] On effective computation of fundamental units, I, *MC*, **38**, 1982, 275–291.

POINCARÉ, H. [92] Extension aux nombres premiers complexes des théorèmes de M. Tchebicheff, *J. Math. Pures Appl.*, (4), **8**, 1892, 25–68.

POITOU, G. [53] Sur l'approximation des nombres complexes par les nombres des corps imaginaires quadratiques dénués d'idéaux non principaux particuliérement lorsque vaut l'algorithme d'Euclide, *ASENS*, (3), **70**, 1953, 199–265.
[76] Sur les petits discriminants, *Sém. DPP*, **18**, 1976/7, exp. 6.
[77] Minorations de discriminants (d'aprés Odlyzko), *Sém. Bourbaki*, 136–153, exp. 479, LN 567, Springer, 1977.

POITOU, G., DESCOMBES, R. (See R. Descombes, G. Poitou)

POL, B. VAN DER, SPEZIALI, P. [51] The primes in $k(\varrho)$, *Indag. Math.*, **13**, 1951, 9–15.

POLLACZEK, F. [24] Über die irregulären Kreiskörper der l-ten und l^2-ten Einheitswurzeln, *MZ*, **21**, 1924, 1–38.
[29] Über die Einheiten relativabelscher Zahlkörper, *MZ*, **30**, 1929, 520–551.
[46] Relation entre les dérivées logarithmiques de Kummer et des logarithmes π-adiques, *Bull. Sci. Math.*, (2), **70**, 1946, 199–218.

POLLÁK, G. [59] On types of euclidean norms, *Acta Sci. Math. (Szeged)*, **20**, 1959, 252–268 (Russian).
[62] Mengentheoretische Betrachtung der euklidischen und Hauptidealringe, *Magyar Tud. Akad. Mat. Fiz. Oszt. Közl.*, **7**, 1962, 323–333.

PÓLYA, G. [19] Über ganzwertige Polynome in algebraischen Zahlkörpern, *JRAM*, **149**, 1919, 97–116.

POMERANCE, C., BATEMAN, P. T., VAUGHAN, R. C. (See P. T. Bateman, C. Pomerance, R. C. Vaughan)

POORTEN, A. J. VAN DER, [68] Transcendental entire functions mapping every algebraic number field into itself, *J. Austral. Math. Soc.*, **8**, 1968, 192–193.

POPKEN, J. [66] Algebraic independence of certain zêta functions, *Indag. Math.*, **28**, 1966, 1–5.

POPOVICI, C. P. [55] On the uniqueness of factorization into prime factors in the ring of Gaussian integers, *Bul. St. Acad. Rep. Pop. Rom.*, **7**, 1955, 518–528 (Rumanian).
[57] On the uniqueness of decomposition into prime factors in rings of integral quadratic numbers, *Bull. Soc. Math. Rep. Pop. Rom.*, **1**, 1957, 99–120 (Russian).
[59] Integral polynomials irreducible (mod p), *Rev. Mat. P. Appl.* **4**, 1959, 369–379 (Russian).

PORTRATZ, C. J., JORDAN, J. H. (See J. H. Jordan, C. J. Portratz)

PORUSCH, J. [33] Die Arithmetik in Zahlkörpern, deren zugehörige Galoische Körper spezielle metabelsche Gruppen besitzen, auf klassenkörpertheoretischer Grundlage, *MZ*, **37**, 1933, 134–160.

POSTNIKOVA, L. P., SCHINZEL, A. [68] On the primitive divisors of $a^n - b^n$ in algebraic number fields, *Mat. Sbornik*, **75**, 1968, 171–177 (Russian).

POTAPKIN, V. K., FADDEEV, D. K. (See D. K. Faddeev, V. K. Potapkin)

POTTER, H. S. A. [36] Approximate equation for the Epstein zeta-function, *PLMS*, (2), **36**, 1936, 501–515.

POTTER, H. S. A., TITCHMARSH, E. C. [35] The zeros of Epstein Zetafunctions, *PLMS*, (2), **35**, 1935, 372–384.

PRACHAR, K. [52] Verallgemeinerung eines Satzes von Hardy und Ramanujan auf algebraische Zahlkörper, *Monatsh. Math.*, **56**, 1952, 229–232.

[57] *Primzahlverteilung*, Springer, 1957.

PRENAT, M. [75] Sur des ensembles fermés de nombres algébriques contenant les ensembles S_q, *CR*, **280**, 1975, A487–488.

PRESTEL, A. [78] Artin's conjecture on p-adic number fields, in: *Proc. 5th School of Algebra*, 79–109, Rio de Janeiro 1978.

PROSKURIN, N. V., VENKOV, A. B. (See A. B. Venkov, N. V. Proskurin)

PRÜFER, H. [25] Neue Begründung der algebraischen Zahlentheorie, *MA*, **94**, 1925, 198–243.

[32] Untersuchungen über Teilbarkeitseigenschaften in Körpern, *JRAM*, **168**, 1932, 1–36.

PUMPLÜN, D. [63] Über Zerlegungen des Kreisteilungspolynoms, *JRAM*, **213**, 1963, 200–220.

[65] Über die Klassenzahl imaginär-quadratischer Zahlkörper, *JRAM*, **218**, 1965, 23–30.

[66] Eine Bemerkung über das Kompositum von Dedekindringen über Hauptidealringen, *JRAM*, **222**, 1966, 214–220.

[68] Über die Klassenzahl und die Grundeinheit des reell-quadratischen Zahlkörpers, *JRAM*, **230**, 1968, 167–210.

PURKERT, W., EDWARDS, H., NEUMANN, O. (See H. Edwards, O. Neumann, W. Purkert)

PYATETSKII-SHAPIRO, I. I. [52] On the question of unicity of the development of a function into trigonometric series, *DAN*, **85**, 1952, 497–500 (Russian).

QUADRI, M. A., IRFAN, M. [79] A characterization of Dedekind domains, *Tamkang J. Math.*, **10**, 1979, 165–167.

QUEEN, C. S. [73a] Euclidean subrings of global fields, *BAMS*, **79**, 1973, 437–439.

[73b] Euclid's algorithm in global fields, *BAMS*, **79**, 1973, 1229–1232.

[74] Arithmetic euclidean rings, *AA*, **26**, 1974, 105–113.

[76] A note on class numbers of imaginary quadratic number fields, *Arch. Math.*, **27**, 1976, 295–298.

[77] The existence of p-adic Abelian L-functions, in: *Number Theory and Algebra*, 263–288, New York 1977.

QUEEN, C. S., MADAN, M. L. (See M. L. Madan, C. S. Queen)

QUEYRUT, J. [72] Extensions quaternioniennes généralisées et constante de l'équation fonctionnelle des séries L d'Artin, *Publ. Math. Un. Bordeaux*, 1972/3, 4, 91–119; add., ibid., 1973/4, 1, 71–72.

[81a] Structure galoisienne des anneaux d'entiers d'extensions sauvagement ramifiées, I, *Ann. Inst. Fourier*, **31**, 1981, 3, 1–35.

[81b] S-groupes de Grothendieck et structure galoisienne des anneaux d'entiers, in: *Integral Representations and Applications*, 219–239, LN 882, Springer, 1981.

[82] S-groupes des classes d'un ordre arithmétique, *J. Algebra*, **76**, 1982, 234–260.

QUEYRUT, J., CASSOU-NOGUÈS, PH. (See Ph. Cassou-Noguès, J. Queyrut)

QUEYRUT, J., FRÖHLICH, A. (See A. Fröhlich, J. Queyrut)

QUILLEN, D. [73] Finite generation of the groups K_i of rings of algebraic integers, in: *Algebraic K-theory, I*, 179–210, LN 341, Springer, 1973.

RABINOWITSCH, G. [13] Eindeutigkeit der Zerlegung in Primzahlfaktoren in quadratischen Zahlkörpern, *JRAM*, **142**, 1913, 153–164.

RABUNG, J. R. [70] Preassigned character values in the gaussian integers, *JNT*, **2**, 1970, 329–332.

RABUNG, J. R., JORDAN, J. H. (See J. H. Jordan, J. R. Rabung)

RADEMACHER, H. [23] Über die Anwendung der Viggo Brunscher Methode auf die Theorie der algebraischen Zahlkörpern, *S.B. Preuß. AW*, 1923, 211–218.

[24] Zur additiver Primzahltheorie algebraischer Zahlkörper, *Hbg*, **3**, 1924, 109–163; II, ibid., 331–378; III, *MZ*, **27**, 1928, 319–426.

[35] Primzahlen reell-quadratischer Zahlkörper in Winkelräumen, *MA*, **111**, 1935, 209–228.

[36a] Über die Primzahlen eines reell-quadratischen Zahlkörpers, *AA*, **1**, 1936, 67–77.

[36b] On prime numbers of real quadratic fields in rectangles, *TAMS*, **39**, 1936, 380–386.

[50] Additive algebraic number theory, *Proc. ICM Cambridge*, 1950, I, 356–362.

RADFORD, D. E. [71] On the convergence of the zeta-function of a commutative ring, *Duke Math. J.*, **38**, 1971, 521–526.

RADOS, G. [90] Zur Theorie der Determinanten, *Math. Nat. Ber. Ungarn.* **8**, 1890, 60.

[06] Die Diskriminante der allgemeinen Kreisteilungsgleichung, *JRAM*, **131**, 1906, 49–55.

[30] Über die Verallgemeinerung eines Kroneckerschen Determinantensatzes, *JRAM*, **162**, 1930, 198–202.

RAGHAVENDRAN, R. [70] A class of finite rings, *Compos. Math.*, **22**, 1970, 49–57.

RAIKOV, D. A. [37] On a property of the cyclotomic polynomials, *Mat. Sbornik*, **2**, 1937, 379–382 (Russian).

RAJU, N. S. [76] Periodic Jacobi–Perron algorithm and fundamental units, *Pacific J. Math.*, **64**, 1976, 241–251.

RAJWADE, A. R. [75] A note on the Stufe of quadratic fields, *Indian J. Pure Appl. Math.*, **6**, 1975, 725–726.

[76] Note sur le théorème des trois carrés, *Enseign. Math.*, (2), **22**, 1976, 171–173.

RAJWADE, A. R., M. K. AGRAWAL, J. C. PARNAMI (See J. C. Parnami, M. K. Agrawal, A. R. Rajwade)

RAM MURTY, M. [84] An analogue of Artin's conjecture for abelian extensions, *JNT*, **18**, 1984, 241–248.

RAMACHANDRA, K. [64] Some applications of Kronecker's limit formula, *AnM*, **80**, 1964, 104–148.

[66] On the units of cyclotomic fields, *AA*, **12**, 1966, 165–173.

[69] On the class number of relative abelian fields, *JRAM*, **236**, 1969, 1–10.

[75] On a theorem of Siegel, *GN*, 1975, 5, 43–47.

[80] One more proof of Siegel's theorem, *Hardy–Ramanujan J.*, **3**, 1980, 25–40.

RAMANATHAN, K. G. [59] The zeta function and discriminant of a division algebra, *AA*, **5**, 1959, 277–288.

RAMANUJAM, C. P. [63a] Sums of m-th powers in p-adic rings, *Mathematika*, **10**, 1963, 137–146.

[63b] Cubic forms over algebraic number fields, *PCPS*, **59**, 1963, 683–705.

RAMAZANOV, R. G., BOBROVSKII, V. P. [72] Classes of special elementary Abelian fields of type (2, 2, 2, 2, 2), *Rep. Alg. Sem. Kazakh. Gos. Ped. Inst. Alma-Ata*, 1972, 1, 54–72 (Russian).

RANULAC, B., KRASNER, M. (See M. Krasner, B. Ranulac)

RANUM, A. [10] The group of classes of congruent quadratic integers with respect to a composite ideal modulus, *TAMS*, **11**, 1910, 172–198.

Rausch, U. [82] *Geometrische Reihen in algebraischen Zahlkörpern*, Diss., Marburg/Lahn 1982.

[85] On a theorem of Dobrowolski about the product of conjugate numbers, *Colloq. Math.*, 50, 1985, 137–142.

Rauzy, G. [64] Répartition modulo 1 pour des suites partielles d'entiers. Développements en série de Taylor donnés sur des suites partielles, *CR*, **258**, 1964, 4881–4884.

[69] Transformations rationelles pour lesquelles l'ensemble des nombres de Pisot–Vijayaraghavan est stable, *CR*, **268**, 1969, 305–307.

Rayner, F. J. [57] Hensel's lemma, *QJM*, (2), **8**, 1957, 307–311.

[58] Relatively complete fields, *Proc. Edinburgh Math. Soc.*, **11**, 1958/9, 131–133.

Razar, M. J. [77] Central and genus class fields and Hasse norm theorem, *Compos. Math.*, **35**, 1977, 281–298.

Razar, M., Goldstein, L. J. (See L. J. Goldstein, M. Razar)

Rédei, L. [28] Über die Klassenzahl des imaginären quadratischen Zahlkörpers, *JRAM*, **159**, 1928, 210–219.

[32a] On the class-number and the fundamental unit of a real quadratic number field, *Mat. term. Értes.*, **48**, 1932, 648–682 (Hungarian).

[32b] On the class-number of a quadratic field, *Mat. term. Értes.*, **48**, 1932, 683–707 (Hungarian).

[34a] Arithmetischer Beweis des Satzes über die Anzahl der durch 4 teilbaren Invarianten der absoluten Klassengruppe im quadratischen Zahlkörper, *JRAM*, **171**, 1934, 55–60.

[34b] Eine obere Schranke der Anzahl der durch 4 teilbaren Invarianten der absoluten Klassengruppe im quadratischen Zahlkörper, *JRAM*, **171**, 1934, 61–64.

[34c] Über die Grundeinheit und die durch 8 teilbaren Invarianten der absoluten Klassengruppe in quadratischen Zahlkörpern, *JRAM*, **171**, 1934, 131–148.

[35] Über die Pellsche Gleichung $t^2 - du^2 = -1$, *JRAM*, **173**, 1935, 193–211.

[36] Über einige Mittelwertfragen in quadratischen Zahlkörpern, *JRAM*, **174**, 1936, 15–55.

[38] Ein neues zahlentheoretisches Symbol mit Anwendungen auf die Theorie der quadratischen Zahlkörper, *JRAM*, **180**, 1938, 1–43.

[41a] Zur Frage des Euklidischen Algorithmus in quadratischen Zahlkörpern, *MA*, **118**, 1941/3, 588–608.

[41b] Über den Euklidischen Algorithmus in reell quadratischen Zahlkörpern, *JRAM*, **183**, 1941, 183–192.

[44] Über die Klassengruppen und Klassenkörper algebraischer Zahlkörper, *JRAM*, **186**, 1944/5, 80–90.

[47] Zwei Lückensätze über Polynome in endlichen Primkörpern mit Anwendung auf die endlichen Abelschen Gruppen und die Gaussischen Summen, *AM*, **79**, 1947, 273–290.

[53a] Bedingtes Artinsches Symbol mit Anwendung in der Klassenkörpertheorie, *Acta Math. Acad. Sci. Hungar.*, **4**, 1953, 1–29.

[53b] Die 2-Ringklassengruppe des quadratischen Zahlkörpers und die Theorie der Pellschen Gleichung, *Acta Math. Acad. Sci. Hungar.*, **4**, 1953, 31–87.

[54] Über das Kreisteilungspolynom, *Acta Math. Acad. Sci. Hungar.*, **5**, 1954, 27–28.

[58] Über die algebraisch zahlentheoretische Verallgemeinerung eines elementarzahlentheoretisches Satzes von Zsigmondy, *Acta Sci. Math. (Szeged)*, **19**, 1958, 98–126.

[59] Natürliche Basen des Kreisteilungskörpers, *Hbg*, **23**, 1959, 180–200; II, *ibid.*, **24**, 1960, 12–40.

[60] Über die quadratischen Zahlkörper mit Primzerlegung, *Acta Sci. Math. (Szeged)*, **21**, 1960, 1–3.

Rédei, L., Behrbohm, H. (See H. Behrbohm, L. Rédei)

RÉDEI, L., REICHARDT, H. [34] Die Anzahl der durch 4 teilbaren Invarianten der Klassengruppe eines beliebigen quadratischen Zahlkörpers, *JRAM*, **170**, 1934, 69–74.

REHM, H. P. [75] Über die gruppentheoretische Struktur der Relationen zwischen Relativnormabbildungen in endlichen Galoischen Körpererweiterungen, *JNT*, **7**, 1975, 49–70.

REHM, H. P., HAPPLE, W. [74] Zur gruppentheoretischer Abschätzung von Idealklassenexponenten galoischer Zahlkörper, durch Exponenten geeigneter Teilkörper, *JRAM*, **268/9**, 1974, 439–440.

REICHARDT, H. [33] Arithmetische Theorie der kubischen Körper als Radikalkörper, *Monatsh. M.-Phys.*, **40**, 1933, 323–350.

[34] Zur Struktur der absoluten Idealklassengruppe im quadratischen Zahlkörper, *JRAM*, **170**, 1934, 75–82.

[58] Ein Beweis des Hauptidealsatzes für imaginär-quadratische Zahlkörper, *MN*, **17**, 1958/9, 318–329.

[70] Über die 2-Klassengruppe gewisser quadratischer Zahlkörper, *MN*, **46**, 1970, 71–80.

[72] Über die Idealklassengruppe des Dirichletschen biquadratischen Zahlenkörpers, *AA*, **21**, 1972, 323–327.

REICHARDT, H., RÉDEI, L. (See L. Rédei, H. Reichardt)

REICHARDT, H., WEGNER, U. [37] Arithmetische Charakterisierung von algebraisch auflösbaren Körpern und Gleichungen von Primzahlgrad, *JRAM*, **178**, 1937, 1–10.

REID, L. W. [01] A table of class numbers for cubic number fields, *AJM*, **23**, 1901, 68–84.

REIDEMEISTER, K. [22] Über die Relativklassenzahl gewisser relativquadratischer Zahlkörper, *Hbg*, **1**, 1922, 27–48.

REINER, I. [45] On genera of binary quadratic forms, *BAMS*, **51**, 1945, 909–912.

[56] Unimodular complements, *AMM*, **63**, 1956, 246–247.

[76] *Class Groups and Picard Groups of Group Rings and Orders*, Providence 1976.

REINER, I., BERGER, T. R. (See T. R. Berger, I. Reiner)

REINER, I., BUSHNELL, C. J. (See C. J. Bushnell, I. Reiner)

REINER, I., HELLER, A. (See A. Heller, I. Reiner)

REJEB, S. [77] Sur les points d'accumulations d'une classe remarquable d'entiers algébriques, *CR*, **284**, 1977, 1321–1323.

RELLA, T. [19] Die Zerlegungsgesetze für die Primideale eines beliebigen algebraischen Zahlkörpers im Körper der l-ten Einheitswurzeln, *MZ*, **5**, 1919, 11–16.

[20] Über die multiplikative Darstellung von algebraischen Zahlen eines Galoischen Zahlkörpers für den Bereich eines beliebigen Primteilers, *JRAM*, **150**, 1920, 157–174.

[24a] Bemerkungen zu Herrn Hensels Arbeit "Die Zerlegung der Primteiler eines beliebigen Zahlkörpers in einem auflösbaren Oberkörper", *JRAM*, **153**, 1924, 108–110.

[24b] Zur Newtonschen Approximationsmethode in der Theorie der p-adischen Gleichungswurzeln, *JRAM*, **153**, 1924, 111–112.

REMAK, R. [13] Abschätzung der Lösung der Pellschen Gleichung im Anschluss an den Dirichletschen Existenzsatz, *JRAM*, **143**, 1913, 250–254.

[31] Elementare Abschätzungen von Fundamentaleinheiten und des Regulators eines algebraischen Zahlkörpers, *JRAM*, **165**, 1931, 159–171.

[32] Über die Abschätzung des absoluten Betrages des Regulators eines algebraischen Zahlkörpers nach unten, *JRAM*, **167**, 1932, 360–378.

[34] Über den Euklidischen Algorithmus in reell-quadratischen Zahlkörpern, *Jahresber. Deutsch. Math. Verein.*, **44**, 1934, 238–250.

[52] Über Grössenbeziehungen zwischen Diskriminante und Regulator eines algebraischen Zahlkörpers, *Compos. Math.*, **10**, 1952, 245–285.

[54] Über algebraische Zahlkörper mit schwachem Einheitsdefekt, *Compos. Math.*, **12**, 1954, 35–80.

RÉMOND, P. [64] Évaluations asymptotiques dans certains semigroupes, *CR*, **258**, 1964, 4179–4181; *ibid.*, **260**, 1965, 6250–6251; *ibid.*, **262**, 1966, A271–273.

[66] Étude asymptotique de certaines partitions dans certaines semi-groupes, *ASENS*, (3), **83**, 1966, 343–410.

RESHETUKHA, I. V. [70] A question of the theory of cubic residues, *Mat. Zametki*, **7**, 1970, 464–476 (Russian).

[75] The analytic definition of a product with a cubic character, *Ukr. Mat. Zh.*, **27**, 1975, 193–201 (Russian).

RÉVESZ, S. G. [83] Irregularities in the distribution of prime ideals, I, *Studia Sci. Math. Hungar.*, **18**, 1983, 57–67.

REVOY, P. [79a] Sommes de bicarrés dans $Z[\sqrt{-1}]$ et $Z[\sqrt[3]{-1}]$, *Enseign. Math.*, (2), **25**, 1979, 257–260.

[79b] Sur les sommes de carrés dans un anneau, *Ann. Univ. Besançon*, III, **11**, 1979, 3–8.

REVUZ, G. [74] Ordre et indice modulo les puissances d'idéal premier, *Sém. DPP*, **15**, 1973/74, exp. 68.

RIBENBOIM, P. [59] Remarques sur le prolongement des valuations de Krull, *Rend. Circ. Mat. Palermo*, (2), **8**, 1959, 152–159.

[62a] An existence theorem for fields with Krull valuations, *TAMS*, **105**, 1962, 278–294.

[62b] Résultats et problèmes de la théorie des corps de classes, *Rev. Un. Mat. Argentina*, **20**, 1962, 125–145.

[66] *Tópicos de teoria dos números*, Rio de Janeiro 1966.

[68] La conjecture d'Artin sur les équations diophantiennes, *Queen's Papers in Pure Appl. Math.*, **14**, 1968.

[72a] *L'Arithmétique des corps*, Paris 1972.

[72b] *Algebraic Numbers*, Wiley-Interscience, 1972.

[79] *13 Lectures on Fermat's Last Theorem*, Springer, 1979.

RIBET, K. A. [75] P-adic interpolation via Hilbert modular forms, *PSPM*, **29**, 1975, 581–592.

[76] A modular construction of unramified p-extensions of $Q(\mu_p)$, *Invent. Math.*, **34**, 1976, 151–162.

[78] P-adic L-functions attached to characters of p-power order, *Sém. DPP*, **19**, 1977/8, exp. 9.

[79a] Fonctions L p-adiques et théorie d'Iwasawa, *Publ. Math. d'Orsay*, 1979, 1–149.

[79b] Report on p-adic L-functions over totally real fields, *Astérisque*, **61**, 1979, 177–192.

RIBET, K. A., DELIGNE, P. (See P. Deligne, K. A. Ribet)

RICHARDS, I. [74] An application of Galois theory to elementary arithmetic, *Adv. Math.*, **13**, 1974, 268–273.

RICHAUD, C. [66] Sur la résolution des équations $x^2-Ay^2 = \pm 1$, *Atti Acad. Pontif. Nuovi Lincei*, 1866, 177–182.

RICHERT, H. E. [57] Über Dirichletreihen mit Funktionalgleichung, *Publ. Inst. Math. Acad. Sci. Serbe*, **11**, 1957, 73–124.

RICHERT, N. [81] A canonical form for planar Farey sets, *PAMS*, **83**, 1981, 259–262.

RICHMAN, F., MINES, R. (See R. Mines, F. Richman)

RICHTER, B. [72] Die Primfaktorzerlegung der Werte der Kreisteilungspolynome, I, *JRAM*, **254**, 1972, 123–132; II, *ibid.*, **267**, 1974, 77–89.

[74] Eine Abschätzung der Werte der Kreisteilungspolynome für reelles Argument, I, *JRAM*, **267**, 1974, 74–76.

RIDEOUT, D. E. [73] A simplification of the formula for $L(1, \chi)$ where χ is a totally imaginary Dirichlet character of a real quadratic field, *AA*, **23**, 1973, 329–337.

RIEGER, G. J. [56] Zum Waringschen Problem für algebraische Zahlen und Polynome, *JRAM*, **195**, 1956, 108–121.

[57a] Über die Anzahl der Teiler der Ideale in einem algebraischen Zahlkörper, *Arch. Math.*, **8**, 1957, 162–165.

[57b] Über die Verteilung gewisser Idealmengen in einem algebraischen Zahlkörper, *Arch. Math.*, **8**, 1957, 401–404.

[58a] Verallgemeinerung der Selbergschen Formel auf Idealklassen mod f in algebraischen Zahlkörpern, *MZ*, **69**, 1958, 183–194.

[58b] Ein weiterer Beweis der Selbergschen Formel für Idealklassen mod f in algebraischen Zahlkörpern, *MA*, **134**, 1958, 403–407.

[58c] Über die Anzahl der Ideale in einer Idealklasse mod f eines algebraischen Zahlkörpers, *MA*, **135**, 1958, 444–466.

[58d] Einige Sätze über Ideale in algebraischen Zahlkörpern, *MA*, **136**, 1958, 339–341.

[58e] Verallgemeinerung der Siebmethode von A. Selberg auf algebraische Zahlkörper, *JRAM*, **199**, 1958, 208–214; II, *ibid.*, **201**, 1959, 157–171; III, *ibid.*, **208**, 1961, 79–90.

[59a] A numbertheoretic application of a theorem of M. Riesz, *Rev. Cienc. (Lima)*, **61**, 1959, 30–33.

[59b] Zur Wienerschen Methode in der Zahlentheorie, *Arch. Math.*, **10**, 1959, 258–260.

[59c] Verallgemeinerung eines Satzes von Romanov und anderes, *MN*, **20**, 1959, 107–122.

[60a] Ramanujansche Summen in algebraischen Zahlkörpern, *MN*, **22**, 1960, 371–377.

[60b] Zum Sieb von Linnik, *Arch. Math.*, **11**, 1960, 14–22.

[61a] On the prime ideals of smallest norm in an ideal class mod f of an algebraic number field, *BAMS*, **67**, 1961, 314–315.

[61b] Verallgemeinerung zweier Sätze von Romanov aus der additiven Zahlentheorie, *MA*, **144**, 1961, 49–55.

[61c] Das grosse Sieb von Linnik für algebraische Zahlen, *Arch. Math.*, **12**, 1961, 184–187.

[61d] Eine Selbergsche Identität für algebraische Zahlen, *MA*, **145**, 1961/2, 77–80.

[62a] Elementare Lösung des Waringschen Problems für algebraische Zahlkörper mit der verallgemeinerter Linnikschen Methode, *MA*, **148**, 1962, 83–88.

[62b] Über die Anzahl der Lösungen der Diophantischen Gleichung $\xi_1\zeta_1+\xi_2\zeta_2 = \nu$ unterhalb gewisser Schranken in algebraischen Zahlkörpern, *JRAM*, **211**, 1962, 54–64.

[62c] Über die Anzahl der Primfaktoren algebraischen Zahlen und das Gausssche Fehlergesetz, *MN*, **24**, 1962, 77–89.

[62d] Solution of the Waring-Goldbach problem for algebraic number fields, *BAMS*, **68**, 1962, 234–236.

[63a] Über die Darstellung ganzer algebraischer Zahlen durch Quadrate, *Arch. Math.*, **14**, 1963, 22–28.

[63b] Some theorems on prime ideals in algebraic number fields, *Pacific J. Math.*, **13**, 1963, 687–692.

[64a] Über die multiplikative Halbgruppe der Restklassen nach einem ganzen Ideal in einem algebraischen Zahlkörper und ihre Halbcharaktere, *MA*, **156**, 1964, 192–197.

[64b] Ein Dreiquadratesatz für algebraische Zahlkörper, *Arch. Math.*, **15**, 1964, 310–315.

RIEHM, C. [64] On the integral representations of quadratic forms over local fields, *AJM*, **86**, 1964, 25–62.

RIM, D. S. [57] Relatively complete fields, *Duke Math. J.*, **24**, 1957, 197–200.

[59] Modules over finite groups, *AnM*, **69**, 1959, 700–712.

RISMAN, L. J.   [76a] On the order and degree of solutions to pure equations, *PAMS*, **55**, 1976, 261–266.
[76b] A counterexample to a conjecture on multinomial degree, *AA*, **31**, 1976, 271–272.
RITTER, J.   [78] $\mathfrak{P}$-adic fields having the same type of algebraic extensions, *MA*, **238**, 1978, 281–288.
RITTER, J., JARDEN, M.   (See M. Jarden, J. Ritter)
ROBERT, A.   [74] Des adèles; pourquoi? *Enseign. Math.*, (2), **20**, 1974, 133–145.
ROBERT, G.   [73a] Unités elliptiques et formules pour le nombre de classes, *CR*, **277**, 1973, A1143–1146.
[73b] Unités elliptiques et formules pour le nombre de classes des extensions abéliennes d'un corps quadratique, *Bull. Soc. Math. France, Mém.* **36**, 1973, 5–77.
[74] Nombres de Hurwitz et régularité des idéaux premiers, *Sém. DPP*, **16**, 1974/5, exp. 21.
[78] Nombres de Hurwitz et unités elliptiques, *ASENS*, (4), **11**, 1978, 297–389.
[79] Caractéres exceptionnels, *JNT*, **11**, 1979, 161–170.
ROBERT, G., GILLARD, R.   (See R. Gillard, G. Robert)
ROBINSON, R. M.   [62] Intervals containing infinitely many sets of conjugate algebraic integers, in: *Studies in Mathematical Analysis*, 305–315, Stanford 1962.
[64a] Conjugate algebraic integers in real point sets, *MZ*, **84**, 1964, 415–427.
[64b] Algebraic equations with span less than 4, *MC*, **18**, 1964, 547–559.
[64c] Intervals containing infinitely many sets of conjugate algebraic units, *AnM*, **80**, 1964, 411–428.
[65] Some conjectures about cyclotomic integers, *MC*, **19**, 1965, 210–217.
[67] On the distribution of certain algebraic integers, *MZ*, **99**, 1967, 28–41.
[69] Conjugate algebraic integers on a circle, *MZ*, **110**, 1969, 41–51.
[77] Conjugate algebraic units in a special interval, *MZ*, **154**, 1977, 31–40.
RODEMICH, E. R., RUMSEY, H. JR.   [68] Primitive trinomials of high degree, *MC*, **22**, 1968, 863–865.
RODOSSKII, K. A.   [56] On the exceptional zero, *IAN*, **20**, 1956, 667–672 (Russian).
[80] Euclidean rings, *DAN*, **273**, 1980, 819–822 (Russian).
ROGERS, C. A.   [50] The product of n real homogeneous linear forms, *AM*, **83**, 1950, 185–208.
ROGERS, C. A., ANKENY, N. C.   (See N. C. Ankeny, C. A. Rogers)
ROGERS, K.   [65] Cyclotomic polynomials and division rings, *Monatsh. Math.*, **69**, 1965, 239–242.
ROGGENKAMP, K., SCOTT, L.   [82] Hecke actions on Picard groups, *J. Pure Appl. Algebra*, **26**, 1982, 85–100.
ROHRLICH, D. E.   [80a] The nonvanishing of certain Hecke L-functions at the center of the critical strip, *Duke Math. J.*, **47**, 1980, 223–232.
[80b] On the L-functions of canonical Hecke characters of imaginary quadratic fields, *Duke Math. J.*, **47**, 1980, 547–557.
[80c] Galois conjugacy of unramified twists of Hecke characters, *Duke Math. J.*, **47**, 1980, 695–703.
[82] Root numbers of Hecke L-functions of CM fields, *AJM*, **104**, 1982, 517–543.
[84a] On L-functions of elliptic curves and anticyclotomic towers, *Invent. Math.*, **75**, 1984, 383–408.
[84b] On L-functions of elliptic curves and cyclotomic towers, *Invent. Math.*, **75**, 1984, 409–423.
ROHRLICH, D. E., MONTGOMERY, H. L.   (See H. L. Montgomery, D. E. Rohrlich)

ROQUETTE, P. [57] Einheiten und Divisorklassen in endlich erzeugbaren Körpern, *Jahresber. Deutsch. Math. Verein.*, **60**, 1957, 1–21.
[67] On class field towers, in: *Algebraic Number Theory, Brighton 1965*, 231–249, Academic Press, 1967.
ROQUETTE, P., CANTOR, D. G. (See D. G. Cantor, P. Roquette)
ROQUETTE, P., ZASSENHAUS, H. [69] A class rank estimate for algebraic number fields, *JLMS*, **44**, 1969, 31–38.
ROSEN, K. H. [80] L-series for genera at $s = 1$, *Mathematika*, **27**, 1980, 10–16.
ROSEN, M. [66] Two theorems on Galois cohomology, *PAMS*, **17**, 1966, 1183–1185.
[81] An elementary proof of the local Kronecker–Weber theorem, *TAMS*, **265**, 1981, 599–605.
ROSEN, M., BRUMER, A. (See A. Brumer, M. Rosen)
ROSEN, M., CORNELL, G. (See G. Cornell, M. Rosen)
ROSEN, M., IRELAND, K. F. (See K. F. Ireland, M. I. Rosen)
ROSENBAUM, K. [66] On the multiplicative group of cyclic extensions of a local field, *Vestnik LGU*, **21**, 1966, 1, 80–92 (Russian).
[70] Über irreguläre zyklische Erweiterungen lokaler Körper, *MN*, **43**, 1970, 143–159.
[78] On the structure of p-adic closure of cyclic extensions of local fields, *Indian J. Math.*, **20**, 1978, 255–264.
ROSENBLÜTH, E. [34] Die arithmetische Theorie und die Konstruktion der Quaternionenkörper auf klassenkörpertheoretischer Grundlage, *Monatsh. M.-Phys.*, **41**, 1934, 85–125.
ROSENTHALL, E. [44] On some special Diophantine equations, *BAMS*, **50**, 1944, 753–758.
ROSIŃSKI, J., ŚLIWA, J. [76] The number of factorizations in an algebraic number field, *Bull. Acad. Polon. Sci. sér. sci. math. astr. phys.*, **24**, 1976, 821–826.
ROSS, K., HEWITT, E. (See E. Hewitt, K. Ross)
ROSSER, B. [49] Real roots of Dirichlet series, *BAMS*, **55**, 1949, 906–913; *J. Res. Nat. Bur. Standards*, **45**, 1950, 505–514.
ROTH, R. L. [71] On extensions of Q by square roots, *AMM*, **78**, 1971, 392–393.
ROTMAN, J. [60] A characterization of fields among integral domains, *An. Acad. Brasil. Ciênc.*, **32**, 1960, 193–194.
ROY, Y., CARTIER, P. (See P. Cartier, Y. Roy)
RUDIN, W. [61] Unique factorization of gaussian integers, *AMM*, **68**, 1961, 907–908.
[62] *Fourier Analysis on Groups*, New York–London 1962.
RUD'KO, V. P. [68] On integral representations of the Galois group of a cyclotomic field, *Dopovidi AN URSR*, 1968, A 6, 529–531 (Ukrainian).
RUDMAN, R. J. [73] On the fundamental unit of purely cubic field, *Pacific J. Math.*, **46**, 1973, 253–256.
RUDMAN, R. J., STEINER, R. (See also R. Steiner, R. Rudman) [78] A generalization of Berwick's unit algorithm, *JNT*, **10**, 1978, 16–34.
RUMP, S. M. [79] Polynomial minimum root separation, *MC*, **33**, 1979, 327–336.
RUMSEY, H. JR., RODEMICH, E. R. (See E. R. Rodemich, H. Rumsey, Jr.)
RUSH, D. E. [83] An arithmetic characterization of algebraic number fields with a given class group, *Math. Proc. Cambridge Philos. Soc.*, **94**, 1983, 23–28.
RUTHINGER, M. [07] *Die Irreduzibilitätsbeweise der Kreisteilungsgleichung*, Diss., Strassburg 1907.
RYAVEC, C. [69] Cubic forms over algebraic number fields, *PCPS*, **66**, 1969, 323–333.

Rychlik, K. [24] Zur Bewertungstheorie der algebraischer Körper, *JRAM*, **153**, 1924, 94–107.
Sachs, H. [56] Untersuchungen über das Problem der eigentlichen Teiler, *Wiss. Z. Univ. Halle*, **6**, 1956/7, 223–259.
Salce, L., Zanardo, P. [82] Arithmetical characterization of rings of algebraic integers with cyclic ideal class group, *Boll. Un. Mat. Ital.*, (6), **1**, 1982, 117–122.
Salem, R. [43] Sets of uniqueness and sets of multiplicities, *TAMS*, **54**, 1943, 218–228; II, *ibid.*, **56**, 1944, 32–49; corr., *ibid.*, **63**, 1948, 595–598.
[44] A remarkable class of algebraic integers. Proof of a conjecture of Vijayaraghavan, *Duke Math. J.*, **11**, 1944, 103–108.
[45] Power series with integral coefficients, *Duke Math. J.*, **12**, 1945, 153–172.
[63] *Algebraic Numbers and Fourier Analysis*, Boston 1963.
Salem, R., Pisot, C. (See C. Pisot, R. Salem)
Salem, R., Zygmund, A. [55] Sur un théorème de Piatetcki-Shapiro, *CR*, **240**, 1955, 2040–2042.
Saltman, D. J. [82] Generic Galois extensions and problems in field theory, *Adv. Math.*, **43**, 1982, 250–283.
[84] Retract rational fields and cyclic extensions, *Israel J. Math.*, **47**, 1984, 165–215.
Samet, P. A. [52] An equation in Gaussian integers, *AMM*, **59**, 1952, 448–452.
[53] Algebraic integers with two conjugates outside the unit circle, *PCPS*, **49**, 1953, 421–436; II, *ibid.*, **50**, 1954, 346.
Samet, P. A., Godwin, H. J. (See H. J. Godwin, P. A. Samet)
Samko, G. P. [49] A study of an algebraic field obtained by composing two pure cubic fields, *Uchen. Zap. Ped. Inst. Rostov*, **1**, 1949, 15–42 (Russian).
[55] A basis for the field defined by a root of an odd prime degree of a rational integer, *Uchen. Zap. Ped. Inst. Rostov*, **3**, 1955, 29–37 (Russian).
[57] Construction of a basis of the field defined by two cubic radicals of Gaussian integers, *Uchen. Zap. Ped. Inst. Rostov*, **4**, 1957, 3–24 (Russian).
Samuel, P. [66] À propos du théorème des unités, *Bull. Sci. Math.*, (2), **90**, 1966, 84–96.
[67] *Théorie algébrique des nombres*, Paris 1967; English transl., Boston 1970.
[71] About euclidean rings, *J. Algebra*, **19**, 1971, 282–301.
Samuel, P., Zariski, O. (See O. Zariski, P. Samuel)
Sansuc, J. J., Colliot-Thélène, J. L. (See J. L. Colliot-Thélène, J. J. Sansuc)
Sansuc, J. J., Colliot-Thélène, J. L., Coray, D. (See J. L. Colliot-Thélène, D. Coray, J. J. Sansuc)
Saparniyazov, O. [65] Asymptotic equalities for the class number of ideals in an imaginary quadratic field, *Litovsk. Mat. Sb.*, **5**, 1965, 303–305 (Russian).
Saparniyazov, O., Fainleib, A. S. (See A. S. Fainleib, O. Saparniyazov)
Sarbasov, G. [66] On the remainder term in the asymptotical formula for the distribution of elementary Abelian fields of degree l^2 and type (l, l), *Uchen. Zap. Kazakh. Gos. Ped. Inst.*, **23**, 1966, 15–18 (Russian).
[67] Improvement of the remainder term in the asymptotical formula for the distribution of cyclic fields of prime degree, *IAN Kazakh. SSR*, **1**, 1967, 61–62 (Russian).
Sarbasov, G., Urazbaev, G. M. [66] On the remainder term of the asymptotical formula for the distribution of cyclic fields of prime degree, *IAN Kazakh. SSR*, **1**, 1966, 14–19, (Russian).
Sarges, H. [76] Eine Anwendung des Selbergschen Siebes auf algebraische Zahlkörper, *AA*. **28**, 1976. 433–455,

SARGES, H., SCHAAL, W. [82] Least quadratic non-residues in algebraic number fields, *JNT*, **15**, 1982, 275–281.

SATAKE, I. [52] On a generalization of Hilbert's theory of ramification, *Sci. Papers College Gen. Ed. Univ. Tokoy*, **2**, 1952, 25–39.

SATGÉ, P. [74] Lois de réciprocité et lois de décomposition, *Sém. Th. de Nombr. Bordeaux*, 1974/75, exp. 9.

[77] Décomposition des nombres premiers dans les extensions non abéliennes, *Ann. Inst. Fourier*, **27**, 1977, 4, 1–8.

[79a] Corps résolubles et divisibilité de nombres de classes d'idéaux, *Enseign. Math.*, (2), **25**, 1979, 165–188.

[79b] Divisibilité du nombre de classes de certains corps cycliques, *Astérisque*, **61**, 1979, 193–203.

[81] Corps cubiques de discriminant donné, *AA*, **39**, 1981, 295–301.

SATGÉ, P., BARRUCAND, P. [76] Une classe de corps résolubles, les corps tchebycheviens, *CR*, **282**, 1976, A947–949.

SATGÉ, P., ORIAT, B. (See B. Oriat, P. Satgé)

SATO, K. [77] On Artin's *L*-functions, *J. College Eng. Nihon Univ.*, B, **18**, 1977, 21–22.

[81] A remark concerning the prime decomposition in $Q(\zeta, \sqrt[5]{n})$, *J. College Eng. Nihon Univ.*, B, **22**, 1981, 9–15.

[82] On the divisibility of $\zeta_{K_1 K_2}(s) \zeta_{K_1 \cap K_2}(s) | \zeta_{K_1}(s) \zeta_{K_2}(s)$, *J. College Eng. Nihon Univ.*, B, **23**, 1982, 41–46.

SAWADA, Y., FURUTA, Y. (See Y. Furuta, Y. Sawada)

SCHAAL, W. [62] Übertragung des Kreisproblems auf reell-quadratische Zahlkörper, *MA*, **145**, 1962, 273–284.

[65] On the expression of a number as a sum of two squares in totally real algebraic number fields, *PAMS*, **16**, 1965, 529–537.

[68] Obere und untere Abschätzungen in algebraischen Zahlkörpern mit Hilfe des linearen Selbergschen Siebes, *AA*, **13**, 1968, 267–313.

[70] On the large sieve method in algebraic number fields, *JNT*, **2**, 1970, 249–270.

[77] Der Satz von Erdös und Fuchs in reell-quadratischen Zahlkörpern, *AA*, **32**, 1977, 147–156.

[84] Siebmethoden in algebraischen Zahlkörpern, *Überblicke Math.*, 1984, 37–53.

SCHAAL, W., SARGES, H. (See H. Sarges, W. Schaal)

SCHACHER, M., STRAUS, E. G. [74] Some applications of a non-Archimedean analogue of Descartes' rule of signs, *AA*, **25**, 1974, 353–357.

SCHÄFER, W. [29] *Beweis des Hauptidealsatzes der Klassenkörpertheorie für den Fall der komplexen Multiplikation*, Diss., Halle 1929.

SCHAFFSTEIN, K. [28] Tafel der Klassenzahlen der reellen quadratischen Zahlkörper mit Primzahldiskriminante unter 12 000 und zwischen 100 000–101 000 und 1 000 000–1 001 000, *MA*, **98**, 1928, 745–748.

SCHARLAU, R. [80a] The fundamental unit in quadratic extensions of imaginary quadratic fields, *Arch. Math.*, **34**, 1980, 534–537.

[80b] Darstellbarkeit von ganzen Zahlen durch Quadratsummen in einigen totalreellen Zahlkörpern, *MA*, **249**, 1980, 49–54.

[80c] On the Pythagoras number of orders in totally real number fields, *JRAM*, **316**, 1980, 208–210.

SCHARLAU, R., DRESS, A. (See A. Dress, R. Scharlau)

SCHARLAU, W. [73] Eine Invariante endlicher Gruppen, *MZ*, **130**, 1973, 291–296.

SCHARLAU, W., KNEBUSCH, M. (See M. Knebusch, W. Scharlau)

SCHENKMAN, E. [64] On the multiplicative group of a field, *Arch. Math.*, **15**, 1964, 282–285.
SCHERTZ, R. [73] *L*-Reihen in imaginär-quadratischen Zahlkörpern und ihre Anwendung auf Klassenzahlprobleme bei quadratischen und biquadratischen Zahlkörpern, I, *JRAM*, 262/3, 1973, 120–133; II, *ibid.*, 270, 1974, 195–212.
[74a] Arithmetische Ausdeutung der Klassenzahlformel für einfach reelle kubische Zahlkörper, *Hbg*, **41**, 1974, 211–223.
[74b] Über die Klassenzahl gewisser nicht galoischer Körper 6-ten Grades, *Hbg*, **42**, 1974, 217–227.
[76] Die singulären Werte der Weberschen Funktionen $\mathfrak{f}$, $\mathfrak{f}_1$, $\mathfrak{f}_2$, γ_2, γ_3, *JRAM*, **286**/7, 1976, 46–74.
[77] Die Klassenzahl der Teilkörper abelscher Erweiterungen imaginär-quadratischer Zahlkörper, *JRAM*, **295**, 1977, 151–168; II, *ibid.*, **296**, 1977, 58–79.
[78a] Zur Theorie der Ringklassenkörper über imaginär-quadratischen Zahlkörpern, *JNT*, **10**, 1978, 70–82.
[78b] Teilkörper relativ abelscher Erweiterungen imaginärquadratischer Zahlkörper, deren Klassenzahl durch Primteiler des Körpergrades teilbar ist, *JRAM*, **302**, 1978, 59–69.
[79] Über die analytische Klassenzahlformel für reelle abelsche Zahlkörper, *JRAM*, **307**/8, 1979, 424–430.
[81] Über die Klassenzahl einfach reeller kubischer Zahlkörper, *AA*, **39**, 1981, 369–379.
SCHERTZ, R., LANG, H. (See H. Lang, R. Schertz)
SCHERTZ, R., STENDER, H. J. [79] Eine Abschätzung der Klassenzahl gewisser reiner Zahlkörper sechsten Grades, *JRAM*, **311**/2, 1979, 347–355.
SCHEUNEMANN, J., MCKENZIE, R. (See R. McKenzie, J. Scheunemann)
SCHILLING, O. F. G. [40] Regular normal extensions over complete fields, *TAMS*, **47**, 1940, 440–454.
[43] Normal extensions of relatively complete fields, *AJM*, **65**, 1943, 309–334.
[50] *The Theory of Valuations*, New York 1950.
[61] On local class field theory, *J. Math. Soc. Japan*, **13**, 1961, 234–245.
SCHINZEL, A. [66a] On a theorem of Bauer and some of its applications, *AA*, **11**, 1966, 333–344; II, *ibid.*, **22**, 1973, 221–231.
[66b] On sums of roots of unity, *AA*, **11**, 1966, 419–432.
[68] Remarque sur le travail précédent de T. Nagell, *AA*, **15**, 1968/9, 245–246.
[73] On the product of conjugates outside the unit circle of an algebraic number, *AA*, **24**, 1973, 385–399; add., *ibid.*, **26**, 1974/5, 329–331.
[74a] Primitive divisors of the expression $A^n - B^n$ in algebraic number fields, *JRAM*, **268**/9, 1974, 27–33.
[74b] On two conjectures of P. Chowla and S. Chowla concerning continued fractions, *Ann. Mat. Pura Appl.*, (4), **98**, 1974, 111–117.
[75a] Traces of polynomials in algebraic numbers, *Norske Vid. Selsk. Skr.*, **6**, 1975, 1–3.
[75b] On linear dependence of roots, *AA*, **28**, 1975, 161–175.
SCHINZEL, A., BAKER, A. (See A. Baker, A. Schinzel)
SCHINZEL, A., BROWKIN, J. (See J. Browkin, A. Schinzel)
SCHINZEL, A., DAVENPORT, H. (See H. Davenport, A. Schinzel)
SCHINZEL, A., DOBROWOLSKI, E., LAWTON, W. (See E. Dobrowolski, W. Lawton, A. Schinzel)
SCHINZEL, A., GOLDFELD, D. M. (See D. M. Goldfeld, A. Schinzel)
SCHINZEL, A., LEWIS, D. J., ZASSENHAUS, H. (See D. J. Lewis, A. Schinzel, H. Zassenhaus)
SCHINZEL, A., MONTGOMERY, H. L. (See H. L. Montgomery, A. Schinzel)

SCHINZEL, A., NARKIEWICZ, W. (See W. Narkiewicz, A. Schinzel)
SCHINZEL, A., PERLIS, R. (See R. Perlis, A. Schinzel)
SCHINZEL, A., POSTNIKOVA, L. P. (See L. P. Postnikova, A. Schinzel)
SCHINZEL, A., ZASSENHAUS, H. [65] A refinement of two theorems of Kronecker, *Michigan Math. J.*, **12**, 1965, 81–85.
SCHIPPER, R. [77] On the behavior of ideal classes in cyclic unramified extensions of prime degree, in: *Number Theory and Algebra*, 303–309, New York 1977.
SCHMID, L. H. [36] Relationen zwischen verallgemeinerten Gaussschen Summen, *JRAM*, **176**, 1936, 189–191.
SCHMID, L. P., SHANKS, D. [66] Variations on a theorem of Landau, I, *MC*, **20**, 1966, 551–564.
SCHMIDT, A. L. [67] Farey triangles and Farey quadrangles in the complex plane, *Math. Scand.*, **21**, 1967, 241–295.
[75] Diophantine approximation of complex numbers, *AM*, **134**, 1975, 1–85.
[78] Diophantine approximation in the field $Q(i\,11^{1/2})$, *JNT*, **10**, 1978, 151–176.
[83] Diophantine approximation in the Eisensteinian field, *JNT*, **16**, 1983, 169–204.
SCHMIDT, C. G. [78] Die Relationen von Gaussschen Summen und Kreiseinheiten, *Arch. Math.*, **31**, 1978/9, 457–463.
[79] Grössencharaktere und relativ-Klassenzahl abelscher Zahlkörper, *JNT*, **11**, 1979, 128–159.
[80a] Die Relationenfaktorengruppen von Stickelberger-Elementen und Kreiszahlen, *JRAM*, **315**, 1980, 60–72.
[80b] Über die Führer von Gaussschen Summen als Grössencharaktere, *JNT*, **12**, 1980, 283–310.
[82] On ray class annihilators of cyclotomic fields, *Invent. Math.*, **66**, 1982, 215–230.
SCHMIDT, E., FELSCH, V. (See V. Felsch, E. Schmidt)
SCHMIDT, F. K. [30] Zur Klassenkörpertheorie im Kleinen, *JRAM*, **162**, 1930, 155–168.
[33] Mehrfach perfekte Körper, *MA*, **108**, 1933, 1–25.
[36] Über die Erhaltung der Kettensätze der Idealtheorie bei beliebigen endlichen Körpererweiterungen, *MZ*, **41**, 1936, 443–450.
SCHMIDT, F. K., HASSE, H. (See H. Hasse, F. K. Schmidt)
SCHMIDT, H. [58] Über spezielle mehrgradige Gleichungen und Quadratsummendarstellungen in imaginär quadratischen Körpern, *MN*, **19**, 1958, 323–330.
SCHMIDT, K. [80] On periodic expansions of Pisot numbers and Salem numbers, *Bull. London Math. Soc.*, **12**, 1980, 269–278.
SCHMIDT, W. M. [80] Simultaneous zeros of quadratic forms, *Monatsh. Math.*, **90**, 1980, 45–65.
[82] On cubic polynomials, III. Systems of p-adic equations, *Monatsh. Math.*, **93**, 1982, 211–223.
SCHMIDT, W. M., LEEP, D. (See D. Leep, W. M. Schmidt)
SCHMITHALS, B. [80a] Konstruktion imaginärquadratischer Körper mit unendlichen Klassenkörperturm, *Arch. Math.*, **34**, 1980, 307–312.
[80b] Eine Verallgemeinerung der Klassenrangabschätzung für Zahlkörper von Roquette und Zassenhaus, *Arch. Math.*, **34**, 1980, 412–415.
SCHMITHALS, B., HEIDER, F. P. (See F. P. Heider, B. Schmithals)
SCHMITZ, T. [16] Abschätzung der Lösung der Pellschen Gleichung, *Arch. M.-Phys.*, **24**, 1916, 87–88.
SCHNEIDER, D. G., JORDAN, J. H. (See J. H. Jordan, D. G. Schneider)
SCHNEIDER, J. E., PEARSON, K. R. (See K. R. Pearson, J. E. Schneider)

SCHNEIDER, P. [79] Über gewisse Galoiscohomologiegruppen, *MZ*, **168**, 1979, 181–205.
SCHOENBERG, I. J. [64] A note on the cyclotomic polynomial, *Mathematika*, **11**, 1964, 131–136.
SCHOLZ, A. [29] Zwei Bemerkungen zum Klassenkörperturm, *JRAM*, **161**, 1929, 201–207.
[30] Über das Verhältnis von Idealklassen und Einheitengruppe in Abelschen Körpern vom Primzahlgrad, *S.B. Heidelberg AW, Math. Nat. Kl.*, **3**, 1930, 31–55.
[31] Die Abgrenzungssätze für Kreiskörper und Klassenkörper, *S.B. Preuß. AW*, **20/1**, 1931, 417–426.
[32] Über die Beziehungen der Klassenzahlen quadratischer Körper zueinander, *JRAM*, **166**, 1932, 201–203.
[33] Idealklassen und Einheiten in kubischen Körpern, *Monatsh. M.-Phys.*, **40**, 1933, 211–222.
[35] Über die Lösbarkeit der Gleichung $t^2 - Du^2 = -4$, *MZ*, **39**, 1935, 95–111.
[36] Totale Normenreste, die keine Normen sind, als Erzeuger nichtabelscher Körpererweiterungen, *JRAM*, **175**, 1936, 100–107; II, *ibid.*, **182**, 1940, 217–234.
[38] Minimaldiskriminanten algebraischer Zahlkörper, *JRAM*, **179**, 1938, 16–21.
[43] Zur Idealtheorie in unendlichen algebraischen Zahlkörpern, *JRAM*, **185**, 1943, 113–126.
SCHOLZ, A., TAUSSKY, O. [34] Die Hauptideale der kubischen Klassenkörper imaginärquadratischer Zahlkörper: ihre rechnerische Bestimmung und ihr Einfluss auf das Klassenkörperturm, *JRAM*, **171**, 1934, 19–41.
SCHÖNEMANN, T. [46] Grundzüge einer allgemeiner Theorie der höheren Congruenzen, deren Modul eine reelle Primzahl ist, *JRAM*, **31**, 1846, 269–325.
SCHOOF, R. J. [83] Class groups of complex quadratic fields, *MC*, **41**, 1983, 295–302.
SCHREIER, O. [27] Über eine Arbeit von Herrn Tschebotareff, *Hbg*, **5**, 1927, 1–6.
SCHRUTKA V. RECHTENSTAMM, G. [64] Tabelle der (relativ-)Klassenzahlen der Kreiskörper, *Abh. Deutsch. Akad. Wiss.*, **2**, 1964, 1–64.
SCHULZ-ARENSTORFF, R. [57] Über Verteilung der Primzahlen reell-quadratischer Zahlkörper in Restklassen, *JRAM*, **198**, 1957, 204–220.
SCHULZE, V. [72] Die Primteilerdichte von ganzzahligen Polynomen, *JRAM*, **253**, 1972, 175–185; II, *ibid.*, **256**, 1972, 153–162; III, *ibid.*, **273**, 1975, 144–145.
[73] Über die Zerlegung von Primzahlen bestimmter arithmetischer Progressionen in algebraischen Zahlkörpern, *JRAM*, **264**, 1973, 147–148.
[76a] Polynome mit nicht durch Restklassen beschreibbaren Primteilermengen, *AA*, **31**, 1976, 195–197.
[76b] Die Verteilung der Primteiler von Polynomen auf Restklassen, I, *JRAM*, **280**, 1976, 122–133; II, *ibid.*, **281**, 1976, 126–148.
[81] Kronecker-äquivalente Körpererweiterungen und p-Ränge, *JRAM*, **328**, 1981, 9–21.
SCHUMANN, H. G. [37] Zum Beweis des Hauptidealsatzes, *Hbg*, **12**, 1937, 42–47.
SCHUR, I. [18a] Über die Verteilung der Wurzeln bei gewisser algebraischen Gleichungen mit ganzzahligen Koeffizienten, *MZ*, **1**, 1918, 377–402.
[18b] Einige Bemerkungen zu der vorstehender Arbeit des Herrn Pólya: Über die Verteilung der quadratischen Reste und Nichtreste, *GN*, 1918, 30–36.
[21] Über die Gaussschen Summen, *GN*, 1921, 147–153.
[29a] Zur Irreduzibilität der Kreisteilungsgleichung, *MZ*, **29**, 1929, 463.
[29b] Elementarer Beweis eines Satzes von L. Stickelberger, *MZ*, **29**, 1929, 464.
[32] Einige Bemerkungen über die Diskriminante eines algebraischen Zahlkörpers, *JRAM*, **167**, 1932, 264–269.

SCHUSTER, I.   [38] Reelquadratische Zahlkörper ohne Euklidischen Algorithmus, *Monatsh. M.-Phys.*, **47**, 1938, 117–127.

SCHWARZ, Š.   [61] On the number of irreducible factors of a given polynomial over a finite field, *Czechoslov. Math. J.*, **11**, 1961, 213–225 (Russian).

SCOTT, L., ROGGENKAMP, K.   (See K. Roggenkamp, L. Scott)

SEAH, E., WASHINGTON, L. C., WILLIAMS, H. C.   [83] The calculation of a large cubic class number with an application to real cyclotomic fields, *MC*, **41**, 1983, 303–305.

SEAH, E., WILLIAMS, H. C., CORMACK, G.   (See H. C. Williams, G. Cormack, E. Seah)

SEGAL, G., HARRIS, B.   (See B. Harris, G. Segal)

SEGAL, R.   [68] Generalized Bernoulli numbers and the theory of cyclotomic fields, *AM*, **121**, 1968, 49–75.

SEIDENBERG, A.   [84] On the Lasker–Noether decomposition theorem, *AJM*, **106**, 1984, 611–638.

SELBERG, A., CHOWLA, S.   [49] On Epstein's zeta function, I, *PAUS*, **35**, 1949, 371–374.

[67] On Epstein's zeta function, *JRAM*, **227**, 1967, 86–110.

SELMER, E. S.   [51] The diophantine equation $ax^3+by^3+cz^3 = 0$, *AM*, **85**, 1951, 203–362.

[53] Sufficient congruence conditions for the existence of rational points on certain cubic surfaces, *Math. Scand.*, **1**, 1953, 113–119.

[55] Tables for the purely cubic field $K(\sqrt[3]{m})$, *Avh. Norske Vid.Akad. Oslo*, I, 1955, 5, 1–38.

[56] The rational solutions of the Diophantine equation $\eta^2 = \xi^3 - D$ for $|D| \leq 100$, *Math. Scand.*, **4**, 1956, 281–286.

SELUCKÝ, K., SKULA, L.   [81] Irregular imaginary fields, *Arch. Math. (Brno)*, **17**, 1981, 95–112.

SEN, S.   [69] On automorphisms of local fields, *AnM*, **90**, 1969, 33–46.

SEN, S., TATE, J.   [63] Ramification groups of local fields, *J. Indian Math. Soc.*, **27**, 1963, 147–202.

SENGE, H. G.   [67] Closed sets of algebraic numbers, *Duke Math. J.*, **34**, 1967, 307–323.

SENGE, H. G., STRAUS, E. G.   [71] PV-numbers and sets of multiplicity, in: *Proc. Washington State Univ. Conf. Numb. Th.*, 55–67, 1971.

[73] PV-numbers and sets of multiplicity, *Period. Mat.*, **3**, 1973, 93–100.

SERAFIN, R., SHANKS, D.   (See D. Shanks, R. Serafin)

SERGEEV, É. A.   [73] An integral basis of algebraic fields, *Mat. Zametki*, **13**, 1973, 229–234 (Russian).

[78] An integral basis for pure cubic extensions of one-class quadratic fields, *Mat. Zametki*, **24**, 1978, 175–181 (Russian).

SERRE, J. P.   [62] *Corps locaux*, Paris 1962; 2nd ed. 1968; English transl. Springer, 1979.

[66] Existence de tour infinies de corps de classes d'aprés Golod et Safarevic, in: *Les tendances géometriques en algèbre et théorie des nombres*, 231–238, Paris 1966.

[68] *Abelian l-adic representations and elliptic curves*, New York, Amsterdam 1968.

[71a] Conducteurs d'Artin des caractères réels, *Invent. Math.*, **14**, 1971, 173–183.

[71b] Cohomologie des groupes discrets, *Annals of Math. Studies*, **70**, 1971, 77–169.

[73] Formes modulaires et fonctions zêta p-adiques, in: *Modular functions of one variable*, III, 191–268, LN 350, Springer, 1973; corr., ibid., IV, 149–150, LN 476, Springer, 1975.

[78a] Une "formule de masse" pour les extensions totalement ramifiées de degré donné d'un corps local, *CR*, **286**, 1978, A1031–1036.

[78b] Sur le résidu de la fonction zêta p-adique d'un corps de nombres, *CR*, **287**, 1978, A183–188.

[81] Quelques applications du théorème de densité de Chebotarev, *Publ. IHES*, **54**, 1981, 323–401.

Serre, J. P., Borel, A. et al. (See A. Borel et al.)
Serre, J. P., Fröhlich, A., Tate, J. (See A. Fröhlich, J. P. Serre, J. Tate)
Setzer, B. [78] Units in totally complex S_3 fields, JNT, **10**, 1978, 244–249.
[80a] The determination of all imaginary, quartic, abelian fields with class number 1, MC, **35**, 1980, 1383–1386.
[80b] Units over totally real $C_2 \times C_2$ fields, JNT, **12**, 1980, 160–175.
Sexauer, N. E. [66] Pythagorean triples over Gaussian domains, AMM, **73**, 1966, 829–834.
[68] Pythagorean triples over Gaussian domains with fundamental units, AMM, **75**, 1968, 278–279.
Shafarevich, I. R. [43] On introduction of a norm in topological fields, DAN, **40**, 1943, 133–135 (Russian).
[46] On Galois groups of p-adic fields, DAN, **53**, 1946, 15–16 (Russian).
[47] On p-extensions, Mat. Sbornik, **20**, 1947, 351–363 (Russian).
[51] A new proof of Kronecker–Weber theorem, Tr. Mat. Inst. Steklov, **38**, 1951, 382–387 (Russian).
[54] On an existence theorem in the theory of algebraic numbers, IAN, **18**, 1954, 327–334 (Russian).
[62] Fields of algebraic numbers, Proc. ICM, Stockholm 1962, 163–176 (Russian).
[63] Extensions with given ramifications, Publ. IHES, **18**, 1963, 71–95.
Shafarevich, I. R., Borevich, Z. I. (See Z. I. Borevich, I. R. Shafarevich)
Shafarevich, I. R., Golod, E. S. (See E. S. Golod, I. R. Shafarevich)
Shanks, D. [64] The second-order term in the asymptotic expansion of $B(x)$, MC, **18**, 1964, 75–86.
[69a] On Gauss's class-number problem, MC, **23**, 1969, 151–163.
[69b] Class number, a theory of factorization, and genera, PSPM, **20**, 1969, 415–440.
[71] Gauss's ternary form reduction and the 2-Sylow subgroup, MC, **25**, 1971, 837–853.
[72a] The infrastructure of a real quadratic field and its applications, Proc. Number Theory Conf. Boulder, 1972, 217–224.
[72b] Five number-theoretic algorithms, Proc. II Manitoba Conf. Numer. Math., 1972, 51–70.
[72c] New types of quadratic fields having three invariants divisible by 3, JNT, **4**, 1972, 537–556.
[73] Systematic examination of Littlewood's bounds on $L(1, \chi)$, PSPM, **24**, 1973, 267–283.
[74] The simplest cubic fields, MC, **28**, 1974, 1137–1152.
[76a] A survey of quadratic, cubic and quartic algebraic number fields (from a computational point of view), Proc. VII Southeastern Conf., 15–40, Congr. Numer. **17**, 1976.
[76b] Class group of the quadratic fields found by F. Diaz y Diaz, MC, **30**, 1976, 173–178.
Shanks, D., Diaz y Diaz, F., Williams, H. C. (See F. Diaz y Diaz, D. Shanks, H. C. Williams)
Shanks, D., Neild, C. (See C. Neild, D. Shanks)
Shanks, D., Schmid, L. P. (See L. P. Schmid, D. Shanks)
Shanks, D., Serafin, R. [73] Quadratic fields with four invariants divisible by 3, MC, **27**, 1973, 183–187; corr., ibid., 1012.
Shanks, D., Weinberger, P. [72] A quadratic field of prime discriminant requiring three generators for its class group, and related theory, AA, **21**, 1972, 71–87.
Shanks, D., Williams, H. C. (See H. C. Williams, D. Shanks)
Shanks, D., Wrench, J. W. Jr. [63] The calculation of certain Dirichlet series, MC, **17**, 1963, 136–154.

SHANNON, C. E. [56] The zero error capacity of a noisy channel, *IRF Trans. Inform. Theory*, *IT*-2, 1956, 8–19.
SHAPIRO, H. N. [49] An elementary proof of the prime ideal theorem, *Comm. Pure Appl. Math.*, **2**, 1949, 309–323.
SHAPIRO, H. N., COFRÉ-MATTA, A. (See A. Cofré-Matta, H. N. Shapiro)
SHAPIRO, H. N., FORMAN, W. (See W. Forman, H. N. Shapiro)
SHAPIRO, H. N., HAUSMAN, M. (See M. Hausman, H. N. Shapiro)
SHAPIRO, H. N., SPARER, G. H. [73] On the units of cubic fields, *Comm. Pure Appl. Math.*, **26**, 1973, 819–835.
SHEINGORN, M., CALLAHAN, T., NEWMAN, M. (See T. Callahan, M. Newman, M. Sheingorn)
SHIH, W. T., HUA, L. K. (See L. K. Hua, W. T. Shih)
SHIMURA, G. [62] On the class-fields obtained by complex multiplication of abelian varieties, *Osaka Math. J.*, **14**, 1962, 33–44.
[66] A reciprocity law in non-solvable extensions, *JRAM*, **221**, 1966, 209–220.
[68] *Automorphic Functions and Number Theory*, LN 54, Springer, 1968.
[71a] *Introduction to the Arithmetic Theory of Automorphic Functions*, Iwanami Shoten—Princeton Univ. Press 1971.
[71b] Class fields over real quadratic fields in the theory of modular functions, in: *Several Complex Variables*, 169–188, LN 185, Springer, 1971.
[72] Class fields over real quadratic fields and Hecke operators, *AnM*, **95**, 1972, 130–190.
SHIMURA, G., CHOWLA, S. (See S. Chowla, G. Shimura)
SHIMURA, G., TANIYAMA, Y. [61] *Complex Multiplication of Abelian Varieties and its Application to Number Theory*, Tokyo 1961.
SHINTANI, T. [76a] On Kronecker limit formula for real quadratic fields, *PJA*, **52**, 1976, 355–358.
[76b] On evaluation of zeta functions of totally real algebraic number fields at nonpositive integers, *J. Fac. Sci. Univ. Tokyo, IA*, **23**, 1976, 393–417.
[77a] On a Kronecker limit formula for real quadratic fields, *J. Fac. Sci. Univ. Tokyo, IA*, **24**, 1977, 167–199.
[77b] On values at $s = 1$ of certain L functions of totally real algebraic number fields, in: *Algebraic Number Theory* (*Kyoto*), 201–212, Tokyo 1977.
[77c] On certain ray class invariants of real quadratic fields, *PJA*, **53**, 1977, 128–131.
[78] On certain ray class invariants of real quadratic fields, *J. Math. Soc. Japan*, **30**, 1978, 139–167.
[80] A proof of the classical Kronecker limit formula, *Tokyo J. Math.*, **3**, 1980, 191–199.
[81] A remark on zeta functions of algebraic number fields, in: *Automorphic Forms, Representation Theory and Arithmetic*, 255–260, Tata Inst. 1981.
SHIOKAWA, I., KANEIWA, R., TAMURA, J. I. [75] A proof of Perron's theorem on Diophantine approximation of complex numbers, *Keio Engrg. Rep.*, **28**, 1975, 131–147.
SHIRAI, S. [75] Central class numbers in central class field towers, *PJA*, **51**, 1975, 389–393.
[78] On the central class field mod $\mathfrak{m}$ of Galois extension of an algebraic number field, *Nagoya Math. J.*, **71**, 1978, 61–85.
[79] On the central ideal class group of cyclotomic fields, *Nagoya Math. J.*, **75**, 1979, 133–143.
SHIRATANI, K. [64a] Bemerkungen zur Theorie der Kreiskörper, *Mem. Fac. Sci. Kyushu Univ., A*, **18**, 1964, 121–126.
[64b] On the Gauss–Hecke sums, *J. Math. Soc. Japan*, **16**, 1964, 32–38.

[64c] On some relations between Bernoulli numbers and class numbers of cyclotomic fields, *Mem. Fac. Sci. Kyushu Univ.*, A, **18**, 1964, 127–135.

[67] Ein Satz zu den Relativklassenzahlen der Kreiskörpern, *Mem. Fac. Sci. Kyushu Univ.*, A, **21**, 1967, 132–137.

[71] A generalization of Vandiver's congruences, *Mem. Fac. Sci. Kyushu Univ.*, A, **25**, 1971, 144–151.

[72] Kummer's congruence for generalized Bernoulli numbers and its application, *Mem. Fac. Sci. Kyushu Univ.*, A, **26**, 1972, 119–138.

[74] On certain values of p-adic L-functions, *Mem. Fac. Sci. Kyushu Univ.*, A, **28**, 1974, 59–82.

[77] On a formula for p-adic L-functions, *J. Fac. Sci. Univ. Tokyo*, I A, **24**, 1977, 45–53.

SHIUE, J. S., KUIPERS, L., NIEDERREITER, H. (See L. Kuipers, H. Niederreiter, J. S. Shiue)

SHYR, J. M. [75] On relative class numbers of certain quadratic extensions, *BAMS*, **81**, 1975, 500–502.

[79] Class numbers of binary quadratic forms over algebraic number fields, *JRAM*, 307/8, 1979, 353–364.

SIEGEL, C. L. [21a] Approximation algebraischer Zahlen, *MZ*, **10**, 1921, 173–213 = *Ges. Abh.*, I, 6–46, Springer, 1966.

[21b] Darstellung total positiver Zahlen durch Quadrate, *MZ*, **11**, 1921, 246–275 = *Ges. Abh.*, I, 47–76, Springer, 1966.

[22a] Über die Diskriminanten total reeller Körper, *GN*, 1922, 17–24 = *Ges. Abh.*, I, 157–164, Springer, 1966.

[22b] Neuer Beweis für die Funktionalgleichung der Dedekindschen Zetafunktion, *MA*, **85**, 1922, 123–128; II, *GN*, 1922, 25–31 = *Ges. Abh.*, I, 113–118, 173–179, Springer, 1966.

[22c] Additive Theorie der Zahlkörper, *MA*, **87**, 1922, 1–35; II, *ibid.*, **88**, 1923, 184–210 = *Ges. Abh.*, I, 118–153, 180–206, Springer, 1966.

[22d] Bemerkungen zu einem Satz von Hamburger über die Funktionalgleichung der Riemannscher Zetafunktion, *MA*, **86**, 1922, 276–279 = *Ges. Abh.*, I, 154–156, Springer, 1966.

[32] Über Riemanns Nachlaβ zur analytischen Zahlentheorie, *Quellen u. Studien zur Geschichte der Math., Astr. Phys.*, **2**, 1932, 45–80 = *Ges. Abh.*, I, 275–310, Springer, 1966.

[35] Über die Classenzahl quadratischer Zahlkörper, *AA*, **1**, 1935, 83–86 = *Ges. Abh.*, I, 406–409, Springer, 1966.

[36] Mittelwerte arithmetischer Funktionen in Zahlkörpern, *TAMS*, **39**, 1936, 219–224 = *Ges. Abh.*, I, 453–458, Springer, 1966.

[37] Über die analytische Theorie der quadratischen Formen, III, *AnM*, **38**, 1937, 212–291 = *Ges. Abh.* I, 469–548, Springer, 1966.

[41] Equivalence of quadratic forms, *AJM*, **63**, 1941, 658–680 = *Ges. Abh.*, II, 217–239, Springer, 1966.

[44a] Algebraic integers whose conjugates lie in the unit circle, *Duke Math. J.*, **11**, 1944, 597–602 = *Ges. Abh.*, II, 467–472, Springer, 1966.

[44b] The average measure of quadratic forms with given determinant and signature, *AnM*, **45**, 1944, 667–685 = *Ges. Abh. II*, 473–491, Springer, 1966.

[44c] Generalization of Waring's problem to algebraic number fields, *AJM*, **66**, 1944, 122–136 = *Ges. Abh. II*, 406–420, Springer, 1966.

[45a] The trace of totally positive and real algebraic integers, *AnM*, **46**, 1945, 302–312 = *Ges. Abh.*, III, 1–11, Springer, 1966.

[45b] Sums of mth powers of algebraic integers, *AnM*, **46**, 1945, 313-339 = *Ges. Abh.*, *III*, 12-38, Springer, 1966.
[61] *Lectures on Advanced Analytic Number Theory*, Tata Inst. 1961; 3rd ed., 1980.
[64] Zu zwei Bemerkungen Kummers, *GN*, 1964, 51-57 = *Ges. Abh.*, *III*, 436-442, Springer, 1966.
[68a] Bernoullische Polynome und quadratische Zahlkörper, *GN*, 1968, 7-38 = *Ges. Abh.*, *IV*, 9-40, Springer, 1979.
[68b] Zum Beweise des Starkschen Satzes, *Invent. Math.*, **5**, 1968, 180-191 = *Ges. Abh.*, *IV*, 41-52, Springer, 1979.
[69a] Abschätzung von Einheiten, *GN*, 1969, 71-86 = *Ges. Abh.*, *IV*, 66-81, Springer, 1979.
[69b] Berechnung von Zetafunktionen an ganzzahligen Stellen, *GN*, 1969, 87-102 = *Ges. Abh.*, *IV*, 82-97, Springer, 1979.
[70] Über die Fourierschen Koeffizienten der Modulformen, *GN*, 1970, 15-56 = *Ges. Abh.*, *IV*, 98-139, Springer, 1979.
[72a] Wurzeln Heckescher Zetafunktionen, *GN*, 1972, 11-20 = *Ges. Abh.*, *IV*, 214-223, Springer, 1979.
[72b] Algebraische Abhängigkeit von Wurzeln, *AA*, **21**, 1972, 59-64 = *Ges. Abh.*, *IV*, 167-172, Springer, 1979.
[73] Normen algebraischer Zahlen, *GN*, 1973, 197-215 = *Ges. Abh.*, *IV*, 250-268, Springer, 1979.
[75] Zur Summation von *L*-Reihen, *GN*, 1975, 269-292 = *Ges. Abh.*, *IV*, 305-328, Springer, 1979.

SIERPIŃSKI, W. [64] *Elementary Theory of Numbers*, Warszawa 1964, 2nd ed. 1987.

SIMS, C. S., IWASAWA, K. (See K. Iwasawa, C. S. Sims)

SINGH, S. [76] The Stufe of some integral domains, *Delta (Waukesha)*, **6**, 1976, 34-42.

SINNOTT, W. [78] On the Stickelberger ideal and the circular units of a cyclotomic field, *AnM*, **108**, 1978, 107-134.
[80] On the Stickelberger ideal and the circular units of an abelian field, *Invent. Math.*, **62**, 1980/1, 181-234.
[84] On the μ-invariant of the Γ-transform of a rational function, *Invent. Math.*, **75**, 1984, 273-282.

SINNOTT, W., COATES, J. (See J. Coates, W. Sinnott)

SIROVICH, C. [69] On the distribution of elements belonging to certain subgroups of algebraic numbers, *TAMS*, **141**, 1969, 93-98.

SIU, M. K., LAM, T. Y. (See T. Y. Lam, M. K. Siu)

SKLAR, A., APOSTOL, T. M. (See T. M. Apostol, A. Sklar)

SKOLEM, T. [23] Integritätsbereiche in algebraischen Zahlkörpern, *Skr. Oslo*, 1923, nr. 21.
[35] Lösung gewisser Gleichungen in ganzen algebraischen Zahlen, insbesondere in Einheiten, *Skr. Norske Vid. Akad. Oslo*, 1935, 1-19.
[45a] A theorem on the equation $\zeta^2 - \delta\eta^2 = 1$ where δ, ζ, η are integers in an imaginary quadratic field, *Avh. Norske Vid. Akad. Oslo*, *I*, 1945, 1, 1-13.
[45b] A remark on the equation $\zeta^2 - \delta\eta^2 = 1$, $\delta > 0$, δ', δ'', ... < 0 where δ, ζ, η belong to a totally real number field, *Avh. Norske Vid. Akad. Oslo*, *I*, 1945, 12, 1-15.
[46] Solutions of the equation $axy+bx+cy+d = 0$ in algebraic integers, *Avh. Norske Vid. Akad. Oslo*, *I*, 1946, 3, 1-8.
[48] On the existence of a multiplicative basis for an algebraic number field, *Norske Vid. Selsk. Forh.*, **20**, 1948, 4-7.

[49] A proof of the irreducibility of the cyclotomic equation, *Norsk Mat. Tidsskr.*, **31**, 1949, 116–120.
[52] On a certain connection between the discriminant of a polynomial and the number of its irreducible factors (mod p), *Norsk Mat. Tidsskr.*, **34**, 1952, 81–85.
SKOPIN, A. I. [55] p-extensions of a local field, containing the p^M-th roots of unity, *IAN*, **19**, 1955, 445–470 (Russian).
SKOPIN, A. I., BOREVICH, Z. I. (See Z. I. Borevich, A. I. Skopin)
SKOPIN, A. I., FADDEEV, D. K. (See D. K. Faddeev, A. I. Skopin)
SKUBENKO, B. F. [62] Asymptotical distribution of integral points on one sheet hyperboloid and ergodic theorems, *IAN*, **26**, 1962, 721–752 (Russian).
SKUBENKO, B. F., PEN, A. S. (See A. S. Pen, B. F. Skubenko)
SKULA, L. [70] Divisorentheorie einer Halbgruppe, *MZ*, **114**, 1970, 113–120.
[72] Eine Bemerkung zu dem ersten Fall der Fermatschen Vermutung, *JRAM*, **253**, 1972, 1–14.
[75] Über pseudoreguläre Primzahlen, *JRAM*, **277**, 1975, 37–39.
[76] On c-semigroups, *AA*, **31**, 1976, 247–257.
[77] Non-possibility to prove infinity of regular primes from certain theorems, *JRAM*, **291**, 1977, 162–181.
[80] Index of irregularity of a prime, *JRAM*, **315**, 1980, 92–106.
[81] Another proof of Iwasawa's class number formula, *AA*, **39**, 1981, 1–6.
SKULA, L., SELUCKÝ, K. (See K. Selucký, L. Skula)
SLAVUTSKII, I. SH. (Slavutsky, I. Sh.) [60] On the class-number of a real quadratic field, *Izv. VUZ*, 1960, 4, 173–177 (Russian).
[61] On the class-number of ideals of a real quadratic field with a prime discriminant, *Uchen. Zap. Leningr. Gos. Ped. Inst.*, **218**, 1961, 179–189 (Russian).
[63] On the problem of irregular prime numbers, *AA*, **8**, 1963, 123–125 (Russian).
[65a] On Mordell's theorem, *AA*, **11**, 1965, 57–66.
[65b] Upper bound and arithmetical determination of the class number of ideals in real quadratic fields, *Izv. VUZ*, 1965, 2, 161–165 (Russian).
[66] Generalized Voronoi's congruence and the class number of ideals in an imaginary quadratic field, II, *Izv. VUZ*, 1966, 4, 118–126 (Russian).
[69a] L-functions of a local field and the real quadratic field, *Izv. VUZ*, 1969, 2, 99–105 (Russian).
[69b] The simplest proof of Vandiver's theorem, *AA*, **15**, 1969, 117–118.
[72a] Generalized Bernoulli numbers that belong to unequal characters and an extension of Vandiver's theorem, *Uchen. Zap. Leningr. Gos. Ped. Inst.*, **496**, 1972, 1, 61–68 (Russian).
[72b] Local properties of Bernoulli numbers and a generalization of the Kummer–Vandiver theorem, *Izv. VUZ*, 1972, 3, 61–69 (Russian).
[75] Square-free numbers and the quadratic field, *Colloq. Math.*, **32**, 1975, 291–300 (Russian).
SLAVUTSKII, I. SH., KISELEV, A. A. (See A. A. Kiselev, I. Sh. Slavutskii)
ŚLIWA, J. [74] Sums of distinct units, *Bull. Acad. Pol. Sci., sér. sci. math. astr. phys.*, **22**, 1974, 11–13.
[76a] Factorizations of distinct lengths in algebraic number fields, *AA*, **31**, 1976, 399–417.
[76b] A note on factorizations in algebraic number fields, *Bull. Acad. Pol. Sci., sér. sci. math. astr. phys.*, **24**, 1976, 313–314.
[77] Primes which remain irreducible in a normal field, *Colloq. Math.*, **37**, 1977, 159–165.
[82a] On the nonessential discriminant divisor of an algebraic number field, *AA*, **42**, 1982, 57–72.

[82b] Remarks on factorizations in algebraic number fields, *Colloq. Math.*, **46**, 1982, 123–130.

ŚLIWA, J., NARKIEWICZ, W.   (See W. Narkiewicz, J. Śliwa)

ŚLIWA, J., ROSIŃSKI, J.   (See J. Rosiński, J. Śliwa)

SMADJA, R.   [73] Sur le groupe des classes des corps de nombres, *CR*, **276**, 1973, A1639–1641.

[77] Utilisation des ordinateurs dans les calculs sur les idéaux des corps de nombres algébriques, *Astérisque*, **41/2**, 1977, 277–282.

SMITH, H. I. S.   [94] Report on the theory of numbers, *Collected Math. Papers*, *I*, 38–364, Oxford 1894.

SMITH, J.   [65] On class 2 extensions of algebraic number fields, *AJM*, **87**, 1965, 537–550.

[69] A remark on class numbers of number field extensions, *PAMS*, **20**, 1969, 388–390.

SMITH, J. H.   [69] A remark on fields with unramified composition, *JLMS*, (2), **1**, 1969, 1–2.

[70a] On S-units almost generated by S-units of subfields, *Pacific J. Math.*, **34**, 1970, 803–805.

[70b] A result of Bass on cyclotomic extension fields, *PAMS*, **24**, 1970, 394–395.

[75] Representability by certain norm forms over algebraic number fields, *AA*, **28**, 1975, 223–227.

SMITH, J. H., FEIN, B., GORDON, B.   (See B. Fein, B. Gordon, J. H. Smith)

SMITH, J. R.   [69] On Euclid's algorithm in some cyclic cubic fields, *JLMS*, **44**, 1969, 577–582.

[71] The inhomogeneous minima of some totally real cubic fields, in: *Computers in Number Theory*, 223–224, Academic Press, 1971.

SMITH, R. A., CALLAHAN, T.   (See T. Callahan, R. A. Smith)

SMITH, W. W., BUTTS, H. S.   (See H. S. Butts, W. W. Smith)

SMYTH, C. J.   [70] Closed sets of algebraic numbers in complete fields, *Mathematika*, **17**, 1970, 199–205.

[71] On the product of the conjugates outside the unit circle of an algebraic integer, *Bull. London Math. Soc.*, **3**, 1971, 169–175.

[73] Problem A 5931, *AMM*, **80**, 1973, 949; solution: *ibid.*, **82**, 1975, 86.

[80] On the measure of totally real algebraic integers, *J. Austral. Math. Soc.*, **30**, 1980/1, 137–149; II, *MC*, **37**, 1981, 205–208.

[81] A Kronecker-type theorem for complex polynomials in several variables, *Canad. Math. Bull.*, **24**, 1981, 447–452; corr., *ibid.*, **25**, 1982, 504.

[82] Conjugate algebraic numbers on conics, *AA*, **40**, 1982, 333–346.

SMYTH, C. J., ENNOLA, V.   (See V. Ennola, C. J. Smyth)

SNAITH, V.   [82] A topological "proof" of a theorem of Ribet, in: *Current Trends in Algebraic Topology*, *I*, 43–47, Providence 1982.

SOKOLOVSKII, A. V.   [66a] On zeros of Dedekind zeta-function, *DAN Uzbek. SSR*, **10**, 1966, 1, 40–50 (Russian).

[66b] Density theorems for a class of zeta functions, *IAN Uzbek. SSR*, **10**, 1966, 3, 33–40 (Russian).

[67] Distance between "neighbouring" prime ideals, *DAN*, **172**, 1967, 1273–1275 (Russian).

[68] Theorems on zeros of Dedekind zeta-function and the distance between "neighbouring" prime ideals, *AA*, **13**, 1968, 321–334 (Russian).

[71] On small differences between "neighbouring" prime ideals, *DAN*, **196**, 1971, 53–56 (Russian).

SOLDERITSCH, J. J.   [85] Quadratic fields with special class groups, to appear.

SOLOMON, L. [77] Zeta functions and integral representation theory, *Adv. Math.*, **26**, 1977, 306–326.

SOMINSKII, I. S., BILLEVICH, K. K. DELAUNAY, B. (See B. Delaunay, K. K. Billevich, I. S. Sominskii)

SONN, J. [74] Epimorphisms of Demuškin groups, *Israel J. Math.*, **17**, 1974, 176–190.
[79] Classgroups and Brauer groups, *Israel J. Math.*, **34**, 1979, 97–105.
[83] Direct summands of class groups, *JNT*, **17**, 1983, 343–349.

SOULÉ, C. [79] K-théorie des anneaux d'entiers de corps de nombres et cohomologie étale, *Invent. Math.*, **55**, 1979, 251–295.
[82] On K-theory and values of zeta functions, in: *Current Trends in Algebraic Topology, I*, 49–58, Providence 1982.

SPARER, G. H., SHAPIRO, H. N. (See H. N. Shapiro, G. H. Sparer)

SPÄTH, H. [27] Über die Irreduzibilität der Kreisteilungsgleichung, *MZ*, **26**, 1927, 442–444.

SPEISER, A. [16] Gruppendeterminante und Körperdiskriminante, *MA*, **77**, 1916, 546–562.
[19] Die Zerlegungsgruppe, *JRAM*, **149**, 1919, 174–188.
[22] Die Zerlegung von Primzahlen in algebraischen Zahlkörpern, *TAMS*, **23**, 1922, 173–178.
[32] Über die Minima Hermitescher Formen, *JRAM*, **167**, 1932, 88–97.
[39] Die Funktionalgleichung der Dirichletschen L-Funktionen, *Monatsh. M.-Phys.*, **48**, 1939, 240–244.

SPENCER, J. [77] An elementary proof of Kronecker's theorem, *Fibonacci Quart.*, **15**, 1977, 9–10.

SPEZIALI, P., POL, van der B. (See B. van der Pol, P. Speziali)

SPIRA, R. [61] The complex sums of divisors, *AMM*, **68**, 1961, 120–124.

SPRINDZHUK, V. G. (Sprindžuk, V. G.) [74a] The distribution of the fundamental units of real quadratic fields, *AA*, **25**, 1974, 405–409.
[74b] "Almost every" algebraic number-field has a large class-number, *AA*, **25**, 1974, 411–413.
[74c] Algebraic number fields with a large number of classes, *IAN*, **38**, 1974, 971–982 (Russian).
[76] The hyperelliptical diophantine equation and the class-number of ideals, *AA*, **30**, 1976, 95–108 (Russian).
[77] The arithmetic structure of polynomials and the number of classes of ideals, *Tr. Mat. Inst. Steklov*, **143**, 1977, 152–174 (Russian).

SPRINGER, T. A. [57] Note on quadratic forms over algebraic number fields, *Indag. Math.*, **19**, 1957, 39–43.

STAHNKE, W. [73] Primitive binary polynomials, *MC*, **27**, 1973, 977–980.

STALL, D. S. *et al.* (See W. J. Ellison *et al.*)

STANKUS, E. [75] Distribution of Dirichlet L-functions with real characters in the half-plane Re$s > 1/2$, *Litovsk. Mat. Sb.*, **15**, 1975, 4, 199–214 (Russian).
[76] On the moments of $L(1, \chi_m)$ on arithmetical progressions, *Litovsk. Mat. Sb.*, **16**, 1976, 1, 207–226 (Russian).

STANTON, R. G., SUDLER, C. JR., WILLIAMS, H. C. JR. [76] An upper bound for the period of the simple continued fraction for $\sqrt{D}$, *Pacific J. Math.*, **67**, 1976, 525–536.

STANTON, R. G., WILLIAMS, H. C. [76] An application of combinatorics in number theory, *Ars Combin.*, **1**, 1976, 321–330.

STARK, H. M. [66] On complex quadratic fields with class number equal to one, *TAMS*, **122**, 1966, 112–119.
[67a] There is no tenth complex quadratic field with class number one, *PAUS*, **57**, 1967, 216–221.
[67b] A complete determination of the complex quadratic fields of class-number one, *Michigan Math. J.*, **14**, 1967, 1–27.
[68] *L*-functions and character sums for quadratic forms, I, *AA*, **14**, 1968, 35–50; II, *ibid.*, **15**, 1969, 307–317.
[69a] On the "gap" in a theorem of Heegner, *JNT*, **1**, 1969, 16–27.
[69b] A historical note on complex quadratic fields with class-number one, *PAMS*, **21**, 1969, 254–255.
[69c] The role of modular functions in a class-number problem, *JNT*, **1**, 1969, 252–260.
[69d] On the problem of unique factorization in complex quadratic fields, *PSPM*, **12**, 1969, 41–56.
[71a] Class number problems in quadratic fields, in: *Actes Congr. Internat. Math. Nice 1970*, *I*, 511–518, Paris 1971.
[71b] Values of *L*-functions at $s = 1$; I. *L*-functions for quadratic forms, *Adv. Math.*, **7**, 1971, 301–343.
[71c] A transcendence theorem for class-number problems, *AnM*, **94**, 1971, 153–173; II, *ibid.*, **96**, 1972, 174–209.
[71d] Recent advances in determining all complex quadratic fields of a given classnumber, *PSPM*, **20**, 1971, 415–440.
[73] Class-numbers of complex quadratic fields, in: *Modular Functions of One Variable*, *I*, 153–174, LN 320, Springer, 1973.
[74] Some effective cases of the Brauer–Siegel theorem, *Invent. Math.*, **23**, 1974, 135–152.
[75a] *L*-functions at $s = 1$; II. Artin *L*-functions with rational characters, *Adv. Math.*, **17**, 1975, 60–92.
[75b] On complex quadratic fields with class-number two, *MC*, **29**, 1975, 289–302.
[75c] The analytic theory of algebraic numbers, *BAMS*, **81**, 1975, 961–972.
[76a] *L*-functions at $s = 1$; III. Totally real fields and Hilbert's twelfth problem, *Adv. Math.*, **22**, 1976, 64–84.
[76b] The genus theory of number fields, *Comm. Pure Appl. Math.*, **29**, 1976, 805–811.
[77a] Class fields for real quadratic fields and *L*-series at 1, in: *Algebraic Number Fields*, *Proc. Durham Symp.*, 355–375, London 1977.
[77b] Class fields and modular forms of weight one, in: *Modular Functions of One Variable*, *V*, 277–287, LN 601, Springer, 1977.
[80] *L*-functions at $s = 1$; V. First derivative at $s = 0$, *Adv. Math.*, **35**, 1980, 197–235.
STAŚ, W. [59] Über eine Anwendung der Methode von Turán, auf die Theorie des Restgliedes in Primidealsatz, *AA*, **5**, 1959, 179–195.
[60] Über einige Abschätzungen in Idealklassen, *AA*, **6**, 1960, 1–10.
[61a] Über das Verhalten der Riemannschen ζ-Funktion und einiger verwandter Funktionen, in der Nähe der Geraden $\sigma = 1$, *AA*, **7**, 1961/2, 217–224.
[61b] On a certain evaluation of the remainder in the theorem on the distribution of prime ideals, *Prace Mat.*, **5**, 1961, 53–60 (Polish).
[76] On the order of Dedekind zeta-functions in the critical strip, *Funct. et Approx.*, **4**, 1976, 19–26.
[79] On the order of Dedekind zeta-functions near the line $\sigma = 1$, *AA*, **35**, 1979, 195–202.
STAŚ, W., WIERTELAK, K. [73] Some estimates in the theory of Dedekind zeta-functions, *AA*, **23**, 1973, 127–135.

[75a] Certain evaluations in the theory of Landau's L-functions, *Discuss. Math.*, **21**, 1975, 17–25 (Polish).
[75b] On some estimates in the theory of $\zeta(s, \chi)$ functions, *AA*, **26**, 1974/5, 293–301.
[76a] An equivalence in ideal classes of algebraic number fields, *Funct. et Approx.*, **2**, 1976, 219–232.
[76b] Further applications of Turán's methods to the distribution of prime ideals in ideal classes mod f, *AA*, **31**, 1976, 153–165.
STAUFFER, R. [36] The construction of a normal basis in a separable extension field, *AJM*, **58**, 1936, 585–597.
STECKEL, H. D. [82a] Abelsche Erweiterungen mit vorgegebenen Zahlknoten, *JRAM*, **330**, 1982, 93–99.
[82b] Arithmetik in Frobeniuserweiterungen, *Manuscr. Math.*, **39**, 1982, 359–386.
[83] Dichte von Frobeniuskörpern bei fixiertem Kernkörper, *JRAM*, **343**, 1983, 36–63.
STEGER, A., DUBOIS, D. W. (See D. W. Dubois, A. Steger)
STEIN, A. [27] Die Gewinnung der Einheiten in gewissen relativquadratischen Zahlkörpern durch das J. Hurwitzsche Kettenbruchverfahren, *JRAM*, **156**, 1927, 69–92.
STEIN, M. R., DENNIS, R. K. (see R. K. Dennis, M. R. Stein)
STEIN, S. K. [77] Modified linear dependence and the capacity of a cyclic graph, *Linear Algebra Appl.*, **17**, 1977, 191–195.
STEINBACHER, F. [11] Abelsche Körper als Kreisteilungskörper, *JRAM*, **139**, 1911, 85–100.
STEINER, R. [77] On the units in algebraic number fields, *Proc. VI Manitoba Conf. Numer. Math.*, 413–435, 1977.
STEINER, R., RUDMAN, R. (See also R. Rudman, R. P. Steiner) [76] On an algorithm of Billevich for finding units in algebraic fields, *MC*, **30**, 1976, 598–609.
STEINIG, J. [66] On Euler's idoneal numbers, *Elem. Math.*, **21**, 1966, 73–88.
STEINITZ, E. [10] Algebraische Theorie der Körper, *JRAM*, **137**, 1910, 167–309.
[12] Rechteckige Systeme und Moduln in algebraischen Zahlkörpern, *MA*, **71**, 1912, 328–354; II, *ibid.*, **72**, 1912, 297–345.
STEMMLER, R. M. [61] The easier Waring problem in algebraic number fields, *AA*, **6**, 1961, 447–468.
STEMMLER, R. M., BATEMAN, P. T. (See P. T. Bateman, R. M. Stemmler)
STENDER, H. J. [69] Über die Grundeinheit für spezielle unendliche Klassen rein kubischer Zahlkörper, *Hbg*, **33**, 1969, 203–215.
[72] Einheiten für eine allgemeine Klasse total reeller algebraischer Zahlkörper, *JRAM*, **257**, 1972, 151–178.
[73] Grundeinheiten für einige unendliche Klassen reiner biquadratischer Zahlkörper mit einer Anwendung auf die diophantische Gleichung $x^4 - ay^4 = \pm c$ ($c = 1, 2, 4$ oder 8), *JRAM*, **264**, 1973, 207–220.
[74] Über die Einheitengruppe der reinen algebraischen Zahlkörper sechsten Grades, *JRAM*, **268/9**, 1974, 78–93.
[75] Eine Formel für Grundeinheiten in reinen algebraischen Zahlkörpern dritten, vierten und sechsten Grades, *JNT*, **7**, 1975, 235–250.
[77] Lösbare Gleichungen $ax^n - by^n = c$ und Grundeinheiten für einige algebraische Zahlkörper vom Grade $n = 3, 4, 6$, *JRAM*, **290**, 1977, 24–62.
[78] "Verstümmelte" Grundeinheiten für biquadratische und bikubische Zahlkörper, *MA*, **232**, 1978, 55–64.
[79a] Zur Parametrisierung reell-quadratischer Zahlkörper, *JRAM*, **311/2**, 1979, 291–301.

[79b] Über die Grundeinheit der reell-quadratischen Zahlkörper $Q(\sqrt{AN^2+BN+C})$, JRAM, **311**/2, 1979, 302–306.

[83] Einheitenbasen für parametrisierte Zahlkörper vierten und achten Grades mit beliebigen reell-quadratischen Teilkörpern, JNT, **17**, 1983, 246–269.

STENDER, H. J., HALTER-KOCH, F. (See F. Halter-Koch, H. J. Stender)
STENDER, H. J., SCHERTZ, R. (See R. Schertz, H. J. Stender)
STEPANOVA, L. L., NECHAEV, V. I. (See V. I. Nechaev, L. L. Stepanova)
STEPHENS, P. J. [72] Optimizing the size of $L(1, \chi)$, PLMS, (3), **24**, 1972, 1–14.
STEURICH, M., BRUNOTTE, H., KLINGEN, J. (See H. Brunotte, J. Klingen, M. Steurich)
STEVENSON, E. [82] The Artin conjecture for three diagonal cubic forms, JNT, **14**, 1982, 374–390.
STEWART, C. L. [78] Algebraic integers whose conjugates lie near the unit circle, Bull. Soc. Math. France, **106**, 1978, 169–176.
STEWART, I., TALL, D. [79] *Algebraic Number Theory*, London–New York 1979.
STICKELBERGER, L. [90] Über eine Verallgemeinerung der Kreisteilung, MA, **37**, 1890, 321–367.

[97] Über eine neue Eigenschaft der Diskriminanten algebraischer Zahlkörper, *I. Math. Kongress*, Zürich, 1897, 182–193.

STIEMKE, E. [26] Über unendliche algebraische Zahlkörper, MZ, **25**, 1926, 9–39.
STOLL, W. [50] Eine Bemerkung über die Dirichletschen L-Reihen, MZ, **52**, 1950, 307–309.
STOLT, B. [52] On the diophantine equation $u^2 - Dv^2 = \pm 4N$, Arkiv Mat., **2**, 1952, 1–23; II, ibid., 251–268; III, ibid., **3**, 1955, 117–132.
STORRER, H. H. [69] A characterization of Prüfer domains, Canad. Math. Bull., **12**, 1969, 809–812.
STOUT, W. T. JR. [73] Corresponding residue systems in normal extensions, JNT, **5**, 1973, 116–122.
STOUT, W. T. JR., MCCULLOH, L. R. (See L. R. McCulloh, W. T. Stout, Jr.)
STRASSMANN, R. [26] *Zur Theorie der π-adischen Zahlen*, Diss., Göttingen 1926.
STRAUS, E. G., CANTOR, D. G. (See D. G. Cantor, E. G. Straus)
STRAUS, E. G., SCHACHER, M. (See M. Schacher, E. G. Straus)
STRAUS, E. G., SENGE, H. G. (See H. G. Senge, E. G. Straus)
STROOKER, J. R. [66] A remark on Artin–van der Waerden equivalence of ideals, MZ, **93**, 1966, 241–242.
STÜNZI, M., GUT, M. (See M. Gut, M. Stünzi)
SUDAN, G. [67] Sur les suites interdites pour les fractions continues dont les éléments appartiennent aux corps $k(\sqrt{-7})$ et $k(\sqrt{-11})$, Rev. Roumaine Math. Pures Appl., **12**, 1967, 1391–1398.
SUDLER, C. JR., STANTON, R. G., WILLIAMS, H. C. JR. (See R. G. Stanton, H. C. Williams, Jr., C. Sudler, Jr.)
SUDO, M., FUJIWARA, M. (See M. Fujiwara, M. Sudo)
SUETUNA, Z. [24] On the mean value of L-functions, Japan. J. Math., **1**, 1924, 69–82.

[25a] Über die Maximalordnung einiger Funktionen in der Idealtheorie, J. Fac. Sci. Univ. Tokyo, **3**, 1925, 105–153.

[25b] On the product of L-functions, Japan. J. Math., **2**, 1925, 19–32.

[25c] The zeros of the L-functions on the critical line, Tôhoku Math. J., **24**, 1925, 313–331.

[28a] Über die Idealnormen eines algebraischen Körpers, J. Fac. Sci. Univ. Tokyo, **1**, 1928, 417–434.

[28b] Bemerkung zu meiner Arbeit "Über die Maximalordnung einiger Funktionen in der Idealtheorie", *J. Fac. Sci. Univ. Tokyo* **1**, 1928, 435–437.
[29] Über die Anzahl der Idealfaktoren von n in einem algebraischen Zahlkörper, *J. Fac. Sci. Univ. Tokyo*, **2**, 1929, 1–24.
[30] Über die Nullstellen der Dedekindschen Zetafunktionen, *MZ*, **32**, 1930, 190–191.
[31] Über die Anzahl der Idealteilern, *J. Fac. Sci. Univ. Tokyo*, **2**, 1931, 155–177.
[32] Über die approximative Funktionalgleichung für Dirichletsche L-Funktionen, *Japan. J. Math.*, **9**, 1932, 111–116.
[35] Über die L-Funktionen in einem kubischen Körper, *PJA*, **11**, 1935, 132–134.
[36] Über die L-Funktionen in gewissen algebraischen Zahlkörpern, *Japan. J. Math.*, **13**, 1936, 27–28.
[37] Abhängigkeit der L-Funktionen in gewissen algebraischen Zahlkörpern, *JRAM*, **177**, 1937, 6–12; II, *J. Fac. Sci. Univ. Tokyo*, **3**, 1937, 223–252.
SUETUNA, Z., HASSE, H. (See H. Hasse, Z. Suetuna)
SUEYOSHI, Y. [78] On ramification of p-extensions of p-adic number fields, *Mem. Fac. Sci. Kyushu Univ.*, **32**, 1978, 199–204.
SUKALLO, A. A. [55] On the question of determining the index of an algebraic number field, *Uchen. Zap. Un. Rostov*, **32**, 1955, 4, 37–42 (Russian).
SUNLEY, J. S. [72a] On the class numbers of totally imaginary quadratic extensions of totally real fields, *BAMS*, **78**, 1972, 74–76.
[72b] Remarks concerning generalized prime discriminants, *Proc. Conf. Number Theory Boulder*, 233–237, 1972.
[73] Class numbers of totally imaginary quadratic extensions of totally real fields, *TAMS*, **175**, 1973, 209–232.
[79] Prime discriminants in real quadratic fields of narrow class number one, in: *Number Theory, Carbondale 1979*, 294–301, LN 751, Springer, 1979.
SUSSMAN, D., CONNELL, I. G. (See I. G. Connell, D. Sussman)
SUTHANKAR, N. S. [73] On Grimm's conjecture in algebraic number fields, *Indag. Math.*, **35**, 1973, 475–484; II, *ibid.*, **37**, 1975, 13–25.
SVED, M. [70] Units in pure cubic number fields, *Ann. Univ. Budapest*, **13**, 1970, 141–149.
SWAN, R. G. [60] Induced representations and projective modules, *AnM*, **71**, 1960, 552–578.
[62] Factorization of polynomials over finite fields, *Pacific J. Math.*, **12**, 1962, 1099–1106.
[63] The Grothendieck ring of a finite group, *Topology*, **2**, 1963, 85–110.
SWIFT, J. D. [48] Note on discriminants of binary quadratic forms with a single class in each genus, *BAMS*, **54**, 1948, 560–561.
SWINNERTON-DYER, H. P. F. [62] Two special cubic surfaces, *Mathematika*, **9**, 1962, 54–56.
[71] On the product of three homogeneous linear forms, *AA*, **18**, 1971, 371–385.
SWINNERTON-DYER, H. P. F., BARNES, E. S. (See E. S. Barnes, H. P. F. Swinnerton-Dyer)
SWINNERTON-DYER, H. P. F., BIRCH, B. J. (See B. J. Birch, H. P. F. Swinnerton-Dyer)
SZABÓ, J., KÁTAI, I. (See I. Kátai, J. Szabó)
SZEGÖ, G., FEKETE, M. (See M. Fekete, G. Szegö)
SZEGÖ, G., WALFISZ, A. [27] Über das Piltzsche Teilerproblem in algebraischen Zahlkörpern, *MZ*, **26**, 1927, 138–156; II, *ibid.*, 467–486.
SZEKERES, G. [74] On the number of divisors of x^2+x+A, *JNT*, **6**, 1974, 434–442.
SZYMICZEK, K. [72] The integer solutions of the equation $ax^2+by^2+cz^2 = 0$ in quadratic fields, *Pr. Nauk. Uniw. Śl.*, 1972, 12, 91–94.
[74] Note on a paper of T. Nagell, *AA*, **25**, 1974, 313–314.

TAI, Y. S. [67] On a conjecture of Lang, in: *Hung-ching Chow 65th Anniversary Volume*, 121–123, Taipei 1967.

TAKAGI, T. [20] Über eine Theorie des relativ-Abelschen Zahlkörpers, *J. Coll. Sci. Tokyo*, **41**, 1920.

[27] Zur Theorie des Kreiskörpers, *JRAM*, **157**, 1927, 230–238.

[48] *Algebraic Number Theory. Generalities and Class Field Theory*, Tokyo 1948 (Japanese).

TAKAHASHI, SHŌICHI [64] An explicit representation of the generalized principal ideal theorem for the rational ground field, *Tôhoku Math. J.*, (2), **16**, 1964, 176–182.

[65] On Tannaka-Terada's principal ideal theorem for rational ground field, *Tôhoku Math. J.*, (2), **17**, 1965, 87–104.

TAKAHASHI, SHÔICHI [73] The number theoretic functions, *Tôhoku Math. J.*, (2), **25**, 1973, 375–382.

TAKAHASHI, SHUICHI [53] Homology groups in class field theory, *Tôhoku Math. J.*, (2), **5**, 1953, 8–11.

TAKAHASHI, SHUICHI, KUNIYOSHI, H. (See H. Kuniyoshi, S. Takahashi)

TAKAHASHI, T. [68] On extensions with given ramification, *PJA*, **33**, 1968, 771–775.

TAKAKU, A. [71] Units of real quadratic fields, *Nagoya Math. J.*, **44**, 1971, 51–55.

[75] Elementary proof of "The class number of $Q(\sqrt{l})$ is odd when l is a prime", *Bull. Sci. Univ. Ryukyus*, **19**, 1975, 27–31.

TAKASE, K. [82] Some remarks on the relative genus fields, *Kōdai Math. J.*, **5**, 1982, 482–494.

TAKENOUCHI, T. [13] On the classes of congruent integers in an algebraic körper, *J. Coll. Sci. Tokyo*, **36**, 1913, 1–28.

TAKETA, K. [32] Neuer Beweis eines Satzes von Herrn Furtwängler über die metabelschen Gruppen, *Japan. J. Math.* **9**, 1932, 199–218.

TAKEUCHI, H. [81] On the class number of the maximal real subfield of a cyclotomic field, *CJM*, **33**, 1981, 55–58.

TAKEUCHI, T. [79] Note on the class field towers of cyclic fields of degree l, *Tôhoku Math. J.*, **31**, 1979, 301–307.

[80] On the l-class field towers of cyclic fields of degree l, *Sci. Rep. Niigata Univ.*, A, **17**, 1980, 23–25.

[82] Genus number and l-rank of genus group of cyclic extensions of degree l, *Manuscr. Math.*, **39**, 1982, 99–109.

TAKHTAYAN, L. A., VINOGRADOV, A. I. (Tahtadžjan, L. A.) [80] On the Gauss-Hasse hypothesis for real quadratic fields, *DAN*, **255**, 1980, 1306–1309 (Russian).

[82] The Gauss-Hasse hypothesis on real quadratic fields with class number one, *JRAM*, **335**, 1982, 40–86.

TALL, D., STEWART, I. (See I. Stewart, D. Tall)

TALMOUDI, F. [77] Sur les nombres de $S \cap [1, 2]$, *CR*, **285**, 1977, A969–971; ibid., **287**, 1978, A739–741.

TAMAGAWA, T. [51] On the theory of ramification groups and conductors, *Japan. J. Math.*, **21**, 1951, 197–215.

[53] On the functional equation of the generalized L-function, *J. Fac. Sci. Univ. Tokyo*, **6**, 1953, 421–428.

[63] On the ζ-function of a division algebra, *AnM*, **77**, 1963, 387–405.

TAMAGAWA, T., IYANAGA, S. (See S. Iyanaga, T. Tamagawa)

TAMURA, J. I. et al. (See I. Shiokawa et al.)

Taniyama, Y. [57] *L*-functions of number fields and zeta functions of abelian varieties, *J. Math. Soc. Japan*, **9**, 1957, 330–336.

Taniyama, Y., Shimura, G.   (See G. Shimura, Y. Taniyama)

Tannaka, T.   [33a] Über einem Satz von Herrn Artin, *Proc. Imp. Acad. Tokyo*, **9**, 1933, 197–198.

[33b] Ein Hauptidealsatz relativ-Galoischer Zahlkörper und ein Satz über den Normenrest, *Proc. Imp. Acad. Tokyo*, **9**, 1933, 355–356; *Japan. J. Math.* **10**, 1934, 183–189.

[34] Einige Bemerkungen zu den Arbeiten über den allgemeinen Hauptidealsatz, *Japan. J. Math.*, **10**, 1934, 163–167.

[49] An alternative proof of a generalized principal ideal theorem, *PJA*, **25**, 1949, 26–31.

[50] Some remarks concerning principal ideal theorem, *Tôhoku Math., J.*, (2), **1**, 1950, 270–278.

[56] On the generalized principal ideal theorem, *Proc. Int. Symp. Alg. N. Th.*, Tokyo 1956, 65–77.

[58] A generalized principal ideal theorem and a proof of a conjecture of Deuring, *AnM*, **67**, 1958, 574–589.

Tannaka, T., Terada, F.   [49] A generalization of the principal ideal theorem, *PJA*, **25**, 1949, 7–8.

Tano, F.   [89] Sur quelques théorèmes de Dirichlet, *JRAM*, **105**, 1889, 160–169.

Tasaka, T.   [70] Remarks on the validity of Hasse's norm theorem, *J. Math. Soc. Japan*, **22**, 1970, 330–341.

Tate, J.   [50] *Fourier analysis in number fields and Hecke's zeta function*, Thesis, Princeton 1950; reproduced in Cassels, Fröhlich [67].

[52] The higher dimensional cohomology groups of class field theory, *AnM*, **56**, 1952, 294–297.

[62] Duality theorems in Galois cohomology, *Proc. ICM Stockholm 1962*, 288–295.

[70] Symbols in arithmetic, *Proc. ICM Nice 1970*, I, 201–211.

[76] Relations between K_2 and Galois cohomology, *Invent. Math.* **36**, 1976, 257–274.

[77a] On the torsion in K_2 of fields, in: *Algebraic Number Theory (Kyoto)*, 243–261, Tokyo 1977.

[77b] Local constants, in: *Algebraic Number Fields, Proc. Durham. Symp.*, 89–131, London 1977.

[81a] On Stark's conjectures on the behavior of $L(s, \chi)$ at $s = 0$, *J. Fac. Sci. Univ. Tokyo*, IA, **28**, 1981, 963–978.

[81b] Brume–Stark–Stickelberger, *Sém. Th. Nombr. Bordeaux*, 1980/1, exp. 24.

Tate, J., Artin, E.   (See E. Artin, J. Tate)

Tate, J., Fröhlich, A., Serre, J. P.   (See A. Fröhlich, J. P. Serre, J. Tate)

Tate, J., Lubin, J.   (See J. Lubin, J. Tate)

Tate, J., Sen, S.   (See S. Sen, J. Tate)

Tateyama, K.   [82a] On the ideal class groups of some cyclotomic fields, *PJA*, **58**, 1982, 333–335.

[82b] Maillet's determinant, *Sci. Papers College Gen. Ed. Univ. Tokyo*, **32**, 1982, 97–100.

Tatuzawa, T.   [51] On a theorem of Siegel, *Japan. J. Math.*, **21**, 1951, 163–178.

[52] The approximate functional equation for Dirichlet's *L*-series, *Japan. J. Math.*, **22**, 1952, 19–25.

[53] On the product of $L(1, \chi)$, *Nagoya Math. J.*, **5**, 1953, 105–111.

[55] Additive prime number theory in an algebraic number field, *J. Math. Soc. Japan*, **7**, 1955, 409–423.

[58] On the Waring problem in an algebraic number field, *J. Math. Soc. Japan*, **10**, 1958, 322–341.
[60] On the Hecke–Landau *L*-series, *Nagoya Math. J.*, **16**, 1960, 11–20.
[73a] On the number of integral ideals in algebraic number fields, whose norms not exceed x, *Sci. Papers College Gen. Ed. Univ. Tokyo*, **23**, 1973, 73–86.
[73b] On the extended Hecke theta-formula, *Tr. Mat. Inst. Steklov*, **132**, 1973, 206–210.
[73c] On the conductor-discriminant formula, *JRAM*, **262/3**, 1973, 436–440.
[73d] On Waring's problem in algebraic number fields, *AA*, **24**, 1973, 37–60.
[75] Fourier analysis used in analytic number theory, *AA*, **28**, 1975, 263–272.
[77] On the number of integral ideals whose norms belong to some norm residue classes mod q, *Sci. Papers College Gen. Ed. Univ. Tokyo*, **27**, 1977, 1–8.

TAUSSKY, O. [32] Über eine Verschärfung des Hauptidealsatzes für algebraische Zahlkörper, *JRAM*, **168**, 1932, 193–210.
[37a] A remark on the class field tower, *JLMS*, **12**, 1937, 82–85.
[37b] A remark on unramified class fields, *JLMS*, **12**, 1937, 86–88.
[49] On a theorem of Latimer and MacDuffee, *CJM*, **1**, 1949, 300–302.
[51] Classes of matrices and quadratic fields, *Pacific J. Math.*, **1**, 1951, 127–132; II, *JLMS*, **27**, 1952, 237–239.
[57] On matrix classes corresponding to an ideal and its inverse, *Illinois J. Math.*, **1**, 1957, 108–113.
[60] Matrices of rational integers, *BAMS*, **66**, 1960, 327–345.
[62] Ideal matrices, *Arch. Math.*, **13**, 1962, 275–282; II, *MA*, **160**, 1963, 218–225.
[68] The discriminant matrices of an algebraic number field, *JLMS*, **43**, 1968, 152–154.
[69] A remark concerning Hilbert's theorem 94, *JRAM*, **239/240**, 1969, 435–438.
[71] Hilbert's theorem 94, in: *Computers in Number Theory*, 65–71, Academic Press, 1971.
[77a] Norms from quadratic fields and their relation to non-commuting 2×2 matrices, II, The principal genus, *Houston J. Math.*, **3**, 1977, 543–547.
[77b] —, III, A link between the 4-rank of the ideal class groups on $Q(\sqrt{m})$ and $Q(\sqrt{-m})$, *MZ*, **154**, 1977, 91–95.
[80] Some facts concerning integral representations of the ideals in an algebraic number field, *Linear Algebra Appl.*, **31**, 1980, 245–248.

TAUSSKY, O., DADE, E. C. (See E. C. Dade, O. Taussky)
TAUSSKY, O., DADE, E. C., ZASSENHAUS, H. (See E. C. Dade, O. Taussky, H. Zassenhaus)
TAUSSKY, O., NEWMAN, M. (See M. Newman, O. Taussky)
TAUSSKY, O., SCHOLZ, A. (See A. Scholz, O. Taussky)
TAUSSKY, O., TODD, J. [40] A characterization of algebraic numbers, *Proc. Roy. Irish Acad.*, *A*, **46**, 1940, 1–8.
[60] Some discrete variable computations, *Proc. Sympos. Appl. Math.*, **10**, 1960, 201–209.

TAYLOR, E. M. [76] Euclid's algorithm in cubic fields with complex conjugates, *JLMS*, (2), **14**, 1976, 49–54.

TAYLOR, M. J. [75] Galois module structure of classgroups and units, *Mathematika*, **22**, 1975, 156–160.
[78a] On the self-duality of a ring of integers as a Galois module, *Invent. Math.*, **46**, 1978, 173–177.
[78b] Galoismodule structure of integers of relative abelian extensions, *JRAM*, **303/4**, 1978, 97–101.
[79] Adams operations, local root numbers and the Galois module structure of rings of integers, *PLMS*, (3), **39**, 1979, 147–175.
[80a] Galois module structure of rings of integers, *Ann. Inst. Fourier*, **30**, 1980, 3. 11–48.

[80b] Galois module structure of rings of integers in Kummer extensions, *Bull. London Math. Soc.*, **12**, 1980, 96–98.
[81a] On Fröhlichs conjecture for rings of integers of tame extensions, *Invent. Math.*, **63**, 1981, 41–79.
[81b] Monomial representations and rings of integers, *JRAM*, **324**, 1981, 127–135.
[81c] Galois module type congruences for values of L-functions, *JLMS*, (2), **24**, 1981, 441–448.
[82a] Galois module structure of rings of integers, in: *Journées arithmétiques 1980*, 218–225, Cambridge Univ. Press, 1982.
[82b] Group laws and rings with normal bases, *JRAM*, **337**, 1982, 121–141.
[83] Relative Galois module structure of rings of integers and elliptic functions, *Math. Proc. Cambridge Philos. Soc.*, **94**, 1983, 389–397.

TAYLOR, M. J., CASSOU-NOGUÈS, PH. (See Ph. Cassou-Noguès, M. J. Taylor)
TAYLOR, M. J., FRÖHLICH, A. (See A. Fröhlich, M. J. Taylor)
TCHUDAKOFF, N. G. (See N. G. Chudakov)
TEEGE, H. [11] Beweis, dass die unendliche Reihe $\sum_{n=1}^{\infty} \left(\frac{n}{p}\right)\frac{1}{n}$ einen positiven von Null verschiedenen Wert hat, *Mitt. Math. Ges. Hamburg*, **4**, 1911, 1–11.
[21] Über den Zusammenhang von $f \equiv 1 \cdot 2 \cdot 3 \cdot \ldots \cdot \frac{1}{2}(p-1) \pmod{p}$ mit der Klassenzahl der binären quadratischen Formen von positiver Determinante $+p$, *Mitt. Math. Ges. Hamburg*, **6**, 1921–30, 87–100.
TEICHMÜLLER, O. [36] Über die Struktur diskret bewerteter Körper, *GN*, 1936, 151–161.
[37] Diskret bewertete perfekte Körper mit unvollkommenen Restklassenkörper, *JRAM*, **176**, 1937, 141–152.
TENA AYUSO, J. [77] Euclidean subrings of real quadratic fields, *Rev. Mat. Hisp.-Amer.*, (4), **37**, 1977, 43–50 (Spanish).
TERADA, F. [50] On a generalization of the principal ideal theorem, *Tôhoku Math. J.*, (2), **1**, 1950, 229–269.
[52] On the principal genus theorem concerning the abelian extensions, *Tôhoku Math. J.*, (2), **4**, 1952, 141–152.
[53] A note on the principal genus theorem, *Tôhoku Math. J.*, (2), **5**, 1953, 211–213.
[54a] Complex multiplication and principal ideal theorem, *Tôhoku Math. J.*, (2), **6**, 1954, 21–25.
[54b] On a generalized principal ideal theorem, *Tôhoku Math. J.*, (2), **6**, 1954, 95–100.
[55] A generalization of the principal ideal theorem, *J. Math. Soc. Japan*, **7**, 1955, 530–536.
[71] A principal ideal theorem in the genus field, *Tôhoku Math. J.*, (2), **23**, 1971, 697–718.
TERADA, F., TANNAKA, T. (See T. Tannaka, F. Terada)
TERJANIAN, G. [66] Un contre-exemple à une conjecture d'Artin, *CR*, **262**, 1966, A 612.
[78] Sur la dimension diophantienne des corps p-adiques, *AA*, **34**, 1977/8, 127–130.
[80] Formes p-adiques anisotropes, *JRAM*, **313**, 1980, 217–220.
TERRAS, A. [76] The Fourier expansion of Epstein's zeta function for totally real algebraic number fields and some consequences for Dedekind's zeta function, *AA*, **30**, 1976, 187–197.
[77a] The Fourier expansion of Epstein's zeta function over an algebraic number field and its consequences for algebraic number theory, *AA*, **32**, 1977, 37–53.
[77b] A relation between $\zeta_K(s)$ and $\zeta_K(s-1)$ for any algebraic number field K, in: *Algebraic Number Fields, Proc. Durham Symp.*, 475–483, London 1977.

THÉROND, J. D. [77] Inégalités sur les coordonnées des unités de $Z[\sqrt{d}]$, *Bull. Sci. Math.*, (2), **101**, 1977, 249–253.

THIELMANN, M. v. [26] Zur Pellscher Gleichung, *MA*, **95**, 1926, 635–640.

THOMAS, E. [79] Fundamental units for orders in certain cubic number fields, *JRAM*, **310**, 1979, 33–55.

THOMPSON, R. C. [62] Normal matrices and the normal basis in abelian number fields, *Pacific J. Math.*, **12**, 1962, 1115–1124.

THOMPSON, W. R. [31] On the possible forms of discriminants of algebraic fields, *AJM*, **53**, 1931, 81–90; II, *ibid.*, **55**, 1933, 111–118.

THUE, A. [12] Über eine Eigenschaft, die keine Transzendente Grösse haben kann, *Kra. Vidensk. Selsk. Skr. I*, 1912, nr. 20 = *Selected Math. Papers*, 479–491, Oslo 1977.

THURNHEER, P. [83] Kleine Pisot–Vijayaraghavan Zahlen und die Fibonacci-Folge, *Monatsh. Math.*, **95**, 1983, 321–331.

THURSTON, H. S. [43] The solution of p-adic equations, *AMM*, **50**, 1943, 142–148.

TIETZE, H. [44] Über die Herstellung einer Basis für die ganzen Zahlen eines algebraischen Zahlkörpers, *SB Bayer. AW*, 1944, 147–162.

TITCHMARSH, E. C. [38] The approximate functional equation for $\zeta^2(s)$, *QJM*, **9**, 1938, 109–114.

TITCHMARSH, E. C., POTTER, H. S. A. (See H. S. A. Potter, E. C. Titchmarsh)

TODD, J., TAUSSKY, O. (See O. Taussky, J. Todd)

TOEPKEN, H. [37] Zur Irreduzibilität der Kreisteilungsgleichung, *Deutsche Math.*, **2**, 1937, 631–633.

TOLIMIERI, R., AUSLANDER, L., WINOGRAD, S. (See L. Auslander *et al.*)

TOMÁS, F. [73] On the normalized groups among the decomposition groups, *An. Inst. Mat. Univ. Nac. Autónoma México*, **13**, 1973, 187–208 (Spanish).

TORELLI, G. [01] Sulla totalitá dei numeri primi fino ad un limite assegnato, *Atti. Accad. Napoli*, (2), **11**, 1901.

[24] Determinazione dei coefficientti dell'equaz. $X_m = 0$ arente per radici tutti e sole le radici primitive dell'equazione $x^m - 1 = 0$, *Gaz. Mat.*, **62**, 1924, 215–226.

TORNHEIM, L. [55] Minimal basis and inessential discriminant divisors for a cubic field, *Pacific J. Math.*, **5**, 1955, 623–631.

TORNIER, E., HASSE, H. (See H. Hasse, E. Tornier)

TORO, E. [80] Integral bases in p-adic cyclotomic fields, *Bull. Calcutta Math. Soc.*, **72**, 1980, 315–317.

TORO, E., LIANG, J. J. (See J. J. Liang, E. Toro)

TORRE, P. DE LA, GOLDSTEIN, L. (See L. Goldstein, P. de la Torre)

TOWBER, J. [80] Composition of oriented binary quadratic form-classes over commutative rings, *Adv. Math.*, **36**, 1980, 1–107.

TÔYAMA, H. [55] A note on the different of the composed field, *Kōdai Math. Sem. Rep.*, **7**, 1955, 43–44.

TOYOIZUMI, M. [81] Formulae for the values of zeta and L-functions at half integers, *Tokyo J. Math.*, **4**, 1981, 193–201.

[82] On the values of the Dedekind zeta function of an imaginary quadratic field at $s = 1/3$, *Comment. Math. Univ. St. Pauli*, **31**, 1982, 159–161.

TRAN QUANG PHUNG [77] Construction d'une extension L/K de ramification triviale, *An. Univ. Bucureşti Mat.*, **26**, 1977, 75–78.

TRELINA, L. A. [77a] On algebraic integers with discriminants containing fixed prime divisors, *Mat. Zametki*, **21**, 1977, 289–296 (Russian).

[77b] The least prime divisor of an index form, *DAN Beloruss. SSR*, **21**, 1977, 975–976 (Russian).

TROTTER, H. F. [69] On the norm of units in quadratic fields, *PAMS*, **22**, 1969, 198–201.

TS'AO, L. C. et al. (See P. Chowla, S. Chowla, L. C. Ts'ao)

TSCHEBOTAREW, N. G. (See N. G. Chebotarev)

TSUNEKAWA, M. [61] Sum of two fourth powers of integers, *Nagoya Math. J.*, **18**, 1961, 53–61.

TULYAGANOVA, M. I. [63] On the Goldbach–Euler problem for imaginary quadratic field, *IAN Uzbek. SSR*, 1963, 1, 11–17 (Russian).

TULYAGANOVA, M. I., LEVIN, B. V. (See B. V. Levin, M. I. Tulyaganova)

TUNNELL, J. [81] Artin's conjecture for representations of octahedral type, *BAMS*, **5**, 1981, 173–175.

TURÁN, P. [50] On the remainder-term of the prime-number formula, *Acta Math. Acad. Sci. Hungar.*, **1**, 1950, 155–166.

[53] *Eine neue Methode in der Analysis und deren Anwendungen*, Budapest 1953; English edition, J. Wiley, 1984.

[59] A note on the real zeros of Dirichlet's L-functions, *AA*, **5**, 1959, 309–314.

TURGANALIEV, R. [62] On the fundamental basis of algebraic integers, depending on a root of an irreducible equation of fifth degree, *Tr. II Resp. Konf. Mat. Mekh.*, Alma-Ata, 1962, 93–98 (Russian).

TURNBULL, H. W. [41] On certain modular determinants, *Edinburgh Math. Notes*, **32**, 1941, 23–30.

TUZUKU, T., NAGATA, M., NAKAYAMA, M. (See M. Nagata, M. Nakayama, T. Tuzuku)

UCHIDA, K. [69] On Tate's duality theorems in Galois cohomology, *Tôhoku Math. J.*, (2), **21**, 1969, 92–101.

[70] Unramified extensions of quadratic number fields, *Tôhoku Math. J.*, (2), **22**, 1970, 138–141; II, *ibid.*, 220–224.

[71] Class numbers of imaginary abelian number fields, *Tôhoku Math. J.*, (2) **23**, 1971, 97–104; II, *ibid.*, 335–348; III, *ibid.*, 573–580.

[72] Imaginary abelian number fields with class number one, *Tôhoku Math. J.*, (2), **24**, 1972, 487–499.

[73] Relative class numbers of normal CM-fields, *Tôhoku Math. J.*, (2), **25**, 1973, 347–353.

[74] Class number of cubic cyclic fields, *J. Math. Soc. Japan*, **26**, 1974, 447–453.

[75] On Artin L-functions, *Tôhoku Math. J.*, (2), **27**, 1975, 75–81.

[76a] Isomorphisms of Galois groups, *J. Math. Soc. Japan*, **28**, 1976, 617–620.

[76b] On a cubic cyclic field with discriminant 163^2, *JNT*, **8**, 1976, 346–349.

[77a] Isomorphisms of Galois groups of algebraic number fields, in: *Algebraic Number Theory (Kyoto)*, 263–266, Tokyo 1977.

[77b] When is $Z[\alpha]$ the ring of integers? *Osaka J. Math.*, **14**, 1977, 155–157.

[82] Galois groups of unramified solvable extensions, *Tôhoku Math. J.*, (2), **34**, 1982, 311–317.

UDRESCU, V. S. [69] On the zetafunction of a totally positive definite quadratic form, *Rev. Roumaine Math. Pures Appl.*, **14**, 1969, 1629–1632.

UEHARA, T. [75] Vandiver's congruence for the relative class number of an imaginary abelian field, *Mem. Fac. Sci. Kyushu Univ.*, **29**, 1975, 249–254.

[76] Bernoulli numbers in real quadratic fields (a remark on a work of H. Lang (Acta Arith. 22 (1973), 423–437)), *Rep. Fac. Sci. Saga Univ.*, **4**, 1976, 1–5.

[78] Fermat's conjecture and Bernoulli numbers, *Rep. Fac. Sci. Saga Univ.*, **6**, 1978, 9–14.

[82] On some congruences for generalized Bernoulli numbers, *Rep. Fac. Sci. Saga Univ.*, **10**, 1982, 1–8.

ULLOM, S. [69a] Normal bases in Galois extensions of number fields, *Nagoya Math. J.*, **34**, 1969, 153–167.

[69b] Galois cohomology of ambiguous ideals, *JNT*, **1**, 1969, 11–15.

[70] Integral normal bases in Galois extensions of local fields, *Nagoya Math. J.*, **39**, 1970, 141–148.

[74a] Integral representations afforded by ambiguous ideals in some abelian extensions, *JNT*, **6**, 1974, 32–49.

[74b] The nonvanishing of certain character sums, *PAMS*, **45**, 1974, 164–166.

[80] Galois module structure for intermediate extensions, *JLMS*, (2), **22**, 1980, 204–214.

[81] Ratios of rings of integers as Galois modules, in: *Integral Representations and Applications*, 240–246, LN 882, Springer, 1981.

URAZBAEV, B. M. [50] On indexes of algebraic equations, *IAN Kazakh. SSR*, 1950, 4, 33–41 (Russian).

[54a] On an asymptotical formula in algebra, *DAN*, **95**, 1954, 935–938 (Russian).

[54b] Asymptotical formula for the growth of the number of abelian fields of degree l^2, *DAN*, **95**, 1954, 1145–1147 (Russian).

[55] Asymptotical formula for the growth of the number of abelian fields of type $(l, l, ..., l)$, *DAN*, **105**, 1955, 659–661 (Russian).

[62] On the growth of the number of special abelian fields, *DAN*, **142**, 1962, 42–45 (Russian).

[64] On certain asymptotical evaluations and their applications to the problem of distribution of absolutely abelian fields, *IAN Kazakh. SSR*, 1964, 2, 16–21 (Russian).

[67] Asymptotical evaluations of arithmetical sums, *IAN Kazakh. SSR*, 1967, 1, 7–16 (Russian).

[69] The distribution of the discriminants of cyclic fields of prime degree, *IAN Kazakh. SSR*, 1969, 1, 8–14 (Russian).

[72] On the growth of the number of discriminants of cyclic fields of degree l^h, *Rep. Alg. Sem. Kazakh. Gos. Ped. Inst. Alma-Ata*, 1972, 92–109 (Russian).

[77] The distribution of cyclic fields, *IAN Kazakh. SSR*, 1977, 5, 66–70 (Russian).

[81] On a class of special abelian fields, *IAN Kazakh. SSR*, 1981, 3, 33–38 (Russian).

URAZBAEV, B. M., ALIBAEV, E. K. [72] Table of the class number of imaginary quadratic fields for primes of form $4n+3$, *Rep. Alg. Sem. Kazakh. Gos. Ped. Inst., Alma-Ata*, 1972, 26–47 (Russian).

URAZBAEV, B. M., SARBASOV, G. (See G. Sarbasov, B. M. Urazbaev)

URBANOWICZ, J. [84] On the 2-primary part of a conjecture of Birch and Tate, *AA*, **43**, 1984, 69–81.

URBELIS, I. [64] Distribution of primes in the real quadratic field $K(\sqrt{2})$, *Litovsk. Mat. Sb.*, **4**, 1964, 409–427 (Russian).

[65a] Distribution of primes in a totally real algebraic number field, *Litovsk. Mat. Sb.*, **5**, 1965, 307–324 (Russian).

[65b] Distribution of algebraic prime numbers, *Litovsk. Mat. Sb.*, **5**, 1965, 504–516 (Russian).

URSELL, H. D. [74] The degrees of radical extensions, *Canad. Math. Bull.*, **17**, 1974, 615–617.

USPENSKY, J. V. [06] A note on integers depending on the fifth root of unity, *Mat. Sbornik*, **26**, 1906, 1–17 (Russian).

[09] Note sur les nombres dépendant d'une racine cinquième de l'unité, *MA*, **66**, 1909, 109–112.

[31] A method for finding units in cubic orders of a negative discriminant, *TAMS*, **33**, 1931, 1–31.

UZKOV, A. I.　[63] On the decomposition of modules over a commutative ring in direct sums of cyclic submodules, *Mat. Sbornik*, **62**, 1963, 469–475 (Russian).

VAL'FISH, A. Z.　(See Arnold Walfisz)

VANDIVER, H. S.　[18] On the first factor of the class number of a cyclotomic field, *BAMS*, **25**, 1918/9, 458–461.

[19] A property of cyclotomic integers and its relation to Fermat's last theorem, *AnM*, **21**, 1919/20, 73–80; II, ibid., **26**, 1925, 217–232.

[20] On Kummer's memoir of 1857 concerning Fermat last theorem, *PAUS*, **6**, 1920, 266–269; II, *BAMS*, **28**, 1922, 400–407.

[25] On the power characters of units in a cyclotomic field, *AJM*, **47**, 1925, 140–147.

[26] Applications of the theory of relative cyclic fields to both cases of Fermat's last theorem, *TAMS*, **28**, 1926, 554–560; II, ibid., **29**, 1927, 154–162.

[29a] On a theorem of Kummer's concerning power characters of units in a cyclotomic field, *AnM*, **30**, 1929, 487–491.

[29b] A theorem of Kummer's concerning the second factor of the class number of a cyclotomic field, *BAMS*, **35**, 1929, 333–335.

[29c] On Fermat's Last Theorem, *TAMS*, **31**, 1929, 613–642.

[30a] On power characters of singular integers in a properly irregular cyclotomic field, *TAMS*, **32**, 1930, 391–408.

[30b] Some properties of a certain system of independent units in a cyclotomic field, *AnM*, **31**, 1930, 123–125.

[34a] On power characters in cyclotomic fields, *BAMS*, **40**, 1934, 111–117.

[34b] Fermat's last theorem and the second factor in the cyclotomic class number, *BAMS*, **40**, 1934, 118–126.

[34c] A note on units in super-cyclic fields, *BAMS*, **40**, 1934, 855–858.

[39a] On basis systems for group of ideal classes in a properly irregular cyclotomic field, *PAUS*, **25**, 1939, 586–591.

[39b] On the composition of the group of ideal classes in a properly irregular cyclotomic field, *Monatsh. M.-Phys.*, **48**, 1939, 369–380.

[41] On improperly irregular cyclotomic fields, *PAUS*, **27**, 1941, 77–83.

[55] On divisors of the second factor of the class number of a cyclotomic field, *PAUS*, **41**, 1955, 780–783.

VÄRMON, J.　[30] Über die Klassenzahl Abelscher Körper, *Arkiv Mat. Astr. Fys.*, **22**, 1930, 13, 1–47.

VARNAVIDES, P.　[52] The euclidean real quadratic fields, *Indag. Math.*, **14**, 1952, 111–122.

VASCONCELOS, W. V.　[67] Ideals and cancellation, *MZ*, **102**, 1967, 353–355.

VASSILIOU, P.　[32] Über den Grad eines Primideals in einem komponierten Körper, *Rend. Circ. Mat. Palermo*, **56**, 1932, 446–448.

[33] Bestimmung der Führer der Verzweigungskörper relativabelscher Zahlkörper. Beweis der Produktformel für den Führer-Diskriminanten-Satz, *JRAM*, **169**, 1933, 131–139.

VAUGHAN, R. C.　[74] Bounds for the coefficients of cyclotomic polynomials, *Michigan Math. J.*, **21**, 1974, 289–295.

VAUGHAN, R. C., BATEMAN, P. T., POMERANCE, C.　(See P. T. Bateman, C. Pomerance, R. C. Vaughan)

VAUGHAN, R. C., ERDÖS, P.　(See P. Erdös, R. C. Vaughan)

VELDKAMP, G. R.　[60] Remark on Euclidean rings, *Nieuw Tid. Wisk.*, **48**, 1960/1, 268–270 (Dutch).

VÉLEZ, W. Y. [77] Prime ideal decomposition in $F(\mu^{1/m})$, II, in: *Number Theory and Algebra*, 331–338, New York 1977.
[78] Prime ideal decomposition in $F(\mu^{1/p})$, *Pacific J. Math.*, **75**, 1978, 589–600.
VÉLEZ, W. Y., GAY, D.   (See D. Gay, W. Y. Vélez)
VÉLEZ, W. Y., MADDEN, D. J.   (See D. J. Madden, W. Y. Vélez)
VÉLEZ, W. Y., MANN, H. B.   (See H. B. Mann, W. Y. Vélez)
VÉLEZ, W. Y., NORRIS, M. J.   (See M. J. Norris, W. Y. Vélez)
VEL'MIN, V. P. [51] Determination of fundamental units and the class-group of cubic fields, *Mat. Sbornik*, **5**, 1951, 53–58 (Russian).
VENKOV, A. B., PROSKURIN, N. V.,   [82] Automorphic functions and Kummer's problem, *Uspekhi Mat. Nauk*, **37**, 1982, 3, 143–165 (Russian).
VENKOV, B. A. [28] On the class-number of binary quadratic forms with negative determinants, *IAN*, 1928, 375–392; II, *ibid.*, 455–480 = *Issled. po t. chisel*, 91–125, Leningrad 1981 (Russian).
[31] Über die Klassenzahl positiver binären quadratischen Formen, *MZ*, **33**, 1931, 350–374.
[34] Construction of cubic fields with given discriminant, *Tr. Inst. Inzh.-zh.-dor. Transp.*, Leningrad, **9**, 1934, 1–107; Summary in: *Tr. II Vsesoj. S'ezda, Leningrad 1934*, II, 28–31, Leningrad–Moskva 1936 = *Issled. po t. chisel*, 154–157, Leningrad 1981 (Russian).
VENKOV, B. B., KOCH, H. [74] The p-tower of class-fields for an imaginary quadratic field, *Zap. Nauch. Sem. LOMI*, **46**, 1974, 5–13 (Russian).
VENTADOUX, M., LIARDET, P. [69] Transformations rationelles laissant stables certains ensembles de nombres algébriques, *CR*, **269**, 1969, 181–183.
VIGNÉRAS, M. F.   (See also M. F. Guého)   [74] Partie fractionnaire de zêta au point -1, *CR*, **279**, 1974, 359–361.
[75a] Quaternions et applications, *Astérisque*, **24/5**, 1975, 47–56.
[75b] Nombre de classes d'un ordre d'Eichler et valeur au point -1 de la fonction zêta d'un corps quadratique réel, *Enseign. Math.*, (2), **21**, 1975, 69–105.
VIJAYARAGHAVAN, T. [27] Periodic simple continued fractions, *PLMS*, (2), **26**, 1927, 403–414.
[40] On the fractional parts of the powers of a number, *JLMS*, **15**, 1940, 159–160; II, *PCPS*, **37**, 1941, 349–357; III, *JLMS*, **17**, 1942, 137–138; IV, *J. Indian Math. Soc.*, **12**, 1948, 33–39.
VIJAYARAGHAVAN, T., CHOWLA, S.   (See S. Chowla, T. Vijayaraghavan)
VILA, N., LLORENTE, P., NART, E.   (See P. Llorente, E. Nart, N. Vila)
VINBERG, E. B. [65] On the theorem on infinite dimensionality of an associative algebra, *IAN*, **29**, 1965, 209–214 (Russian).
VINOGRADOV, A. I. [62] On the class-number, *DAN*, **146**, 1962, 274–276 (Russian).
[63a] On the number of classes of ideals and the group of divisor classes, *IAN*, **27**, 1963, 561–576 (Russian).
[63b] On Siegel's zeros, *DAN*, **151**, 1963, 479–481 (Russian).
[64a] Lower bounds with sieve methods in algebraic number fields, *DAN*, **154**, 1964, 13–15 (Russian).
[64b] Sieve methods in algebraic fields. Lower bounds, *Mat. Sbornik*, **64**, 1964, 52–78 (Russian).
[65] On the continuability into the left half-plane of the scalar product of Hecke L-series with Grössencharacters, *IAN*, **29**, 1965, 485–492 (Russian).
[71] Artin's L-series and the adèle groups, *Tr. Mat. Inst. Steklov*, **112**, 1971, 105–122 (Russian).

[73] Artin's conjectures and reciprocity law, *Tr. Mat. Inst. Steklov*, **132**, 1973, 35–43 (Russian).

VINOGRADOV, A. I., LINNIK, YU. V. (See Yu. V. Linnik, A. I. Vinogradov)

VINOGRADOV, A. I., TAKHTAYAN, L. A. (See L. A. Takhtayan, A. I. Vinogradov)

VINOGRADOV, I. M. [18] On the mean value of the number of classes of purely radical forms of negative determinant, *Zap. Mat. Obshch. Kharkov*, (2), **16**, 1918, 10–38 (Russian).

[49] Improvement of the remainder term in an asymptotical formula, *IAN*, **13**, 1949, 97–110 (Russian).

[55] Improvement of asymptotical formulas for the number of integral points in three-dimensional regions, *IAN*, **19**, 1955, 3–10 (Russian).

[60] On the question of the number of integral points in a given region, *IAN*, **24**, 1960, 777–786 (Russian).

[63] On the number of integral points in a three-dimensional region, *IAN*, **27**, 1963, 3–8 (Russian).

VISHIK, M. M. [77] The p-adic zeta function of an imaginary quadratic field, and the Leopoldt regulator, *Mat. Sbornik*, **102**, 1977, 173–181 (Russian).

VISHIK, M. M., MANIN, YU. I. [74] p-adic Hecke series of imaginary quadratic fields, *Mat. Sbornik*, **95**, 1974, 357–383 (Russian).

VORONIN, S. M. [75] On functional independence] of Dirichlet's L-functions, *AA*, **27**, 1975, 493–503 (Russian).

VORONOI, G. F. [96] *On a Generalization of the Continued Fractions Algorithm*, Warszawa 1896 (Russian).

[04] Sur une propriété du discriminant des fonctions entières, *Verhandl. III Internat. Kongr., Heidelberg*, 1904, 186–189.

VOSTOKOV, S. V. [74] Ideals of an Abelian p-extension of an irregular local field as Galois modules, *Zap. Nauch. Sem. LOMI*, **46**, 1974, 14–35 (Russian).

[76a] Ideals of an abelian p-extension of a local field as Galois modules, *Zap. Nauch. Sem. LOMI*, **57**, 1976, 64–84 (Russian).

[76b] A normal basis of an ideal of a local field, *Zap. Nauch. Sem. LOMI*, **64**, 1976, 64–68 (Russian).

[77] The ring of integral elements of an algebraic number field as a Galois module, *Zap. Nauch. Sem. LOMI*, **71**, 1977, 80–84 (Russian).

VOSTOKOV, S. V., BOREVICH, Z. I. (See Z. I. Borevich, S. V. Vostokov)

VULAH, L. J. [77] Hurwitz constants, *Izv. VUZ*, 1977, 2, 21–23 (Russian).

WAALL, R. W. VAN DER, [73] Remarks on the Artin L-functions of the groups $GL_2(F_3)$ and $SL_2(F_3)$, *Indag. Math.*, **35**, 1973, 41–46.

[74a] On splitting properties of primes by means of Artin conductors, *Indag. Math.*, **36**, 1974, 82–88; corr., *ibid.*, 411.

[74b] A remark on the zeta-function of an algebraic number field, *JRAM*, **266**, 1974, 159–162.

[75] On a conjecture of Dedekind on zeta-functions, *Indag. Math.*, **37**, 1975, 83–86.

[82] Some results connected to Dedekind's zeta-function, *Abh. Braunschw. Wiss. Ges.*, **33**, 1982, 243–251.

WADA, H. [66] On the class number and the unit group of certain algebraic number fields, *J. Fac. Sci. Univ. Tokyo*, **13**, 1966, 201–209.

[70a] On cubic galois extensions of $Q(\sqrt{-3})$, *PJA*, **46**, 1970, 397–400.

[70b] A table of ideal class groups of imaginary quadratic fields, *PJA*, **46**, 1970, 401–403.

[70c] A table of fundamental units of purely cubic fields, *PJA*, **46**, 1970, 1135–1140.

WADE, L., BUTTS, H. S.   (See H. S. Butts, L. Wade)
WAERDEN, B. L. VAN DER,   [28] Ein logarithmenfreier Beweis des Dirichletschen Einheitssatzes, *Hbg*, **6**, 1928, 259–262.
[30] *Moderne Algebra*, Springer, 1930–1931; 3rd ed., 1950.
[34] Die Zerlegungs- und Trägheitsgruppe als Permutationsgruppen, *MA*, **111**, 1934, 731–733.
WAGNER, G. B.   [69] Ideal matrices and ideal vectors, *MA*, **183**, 1969, 241–249.
WAGSTAFF, S.   [78] The irregular primes to 125 000, *MC*, **32**, 1978, 583–591.
WAHLIN, G. E.   [15a] A new development of the theory of algebraic numbers, *TAMS*, **16**, 1915, 502–508.
[15b] The equation $x^l - A \equiv 0\,(p)$, *JRAM*, **145**, 1915, 114–138.
[16] On the principal units of an algebraic domain $k(\mathfrak{p}, \alpha)$, *BAMS*, **23**, 1916/7, 450–455.
[22] The factorization of the rational primes in a cubic domain, *AJM*, **44**, 1922, 191–203.
[32] The multiplicative representation of the principal units of a relative cyclic field, *JRAM*, **167**, 1932, 122–128.
WALDSCHMIDT, M.   [81] Transcendance et exponentielles en plusieurs variables, *Invent. Math.*, **63**, 1981, 97–127.
WALFISZ, ANNA   [64] Über die summatorischen Funktionen einiger Dirichletschen Reihen, II, *AA*, **10**, 1964, 71–118.
WALFISZ, ARNOLD (VAL'FISH, A. Z.)   [25] Über das Piltzsche Teilerproblem in algebraischen Zahlkörpern, *MZ*, **22**, 1925, 153–188; II, *ibid.*, **26**, 1927, 487–494.
[26] Über die Idealfunktion quadratischer Zahlkörper, *Prace Mat.-Fiz.*, **34**, 1925/6, 35–47.
[27] Beiträge zur Theorie der Dedekindschen Zetafunktion, I. Abschätzung von $\zeta_K(1+it)$, *MA*, **97**, 1927, 629–634.
[32] Wertevorrat der Dedekindschen Zetafunktion um $\sigma = 1$, *Jahresber. Deutsch. Math. Verein.*, **42**, 1932, 62–68.
[36] Zur additiven Zahlentheorie, II, *MZ*, **40**, 1936, 592–607.
[42] On the class-number of binary quadratic forms, *Tr. Mat. Inst. Gruz. SSR*, **11**, 1942, 57–72; II, *ibid.*, 173–186.
[51] Pell's equation in imaginary quadratic fields, *Tr. Mat. Inst. Gruz. SSR*, **18**, 1951, 133–151 (Russian).
WALFISZ, A., SZEGÖ, G.   (See G. Szegö, A. Walfisz)
WALL, G. E., CASSELS, J. W. S.   (See J. W. S. Cassels, G. E. Wall)
WALTER, C. D.   [77] A class number relation in Frobenius extensions of number fields, *Mathematika*, **24**, 1977, 216–225.
[79a] The ambiguous class group and the genus group of certain non-normal extensions, *Mathematika*, **26**, 1979, 113–124.
[79b] Brauer's class number relations, *AA*, **35**, 1979, 33–40.
[79c] Kuroda's class number relation, *AA*, **35**, 1979, 41–51.
[80] Pure fields of degree 9 with class number prime to 3, *Ann. Inst. Fourier*, **30**, 1980, 2, 1–15.
WALTER, C. D., PARRY, C. J.   (See C. J. Parry, C. D. Walter)
WALTON, L. F.   [50] Ideal numbers over integral domains having non-maximal prime ideals, *Duke Math. J.*, **17**, 1950, 285–289.
WALTON, R. A.   [71] A criterion for Dedekind domains, *JLMS*, (2), **3**, 1971, 539–543.
WANG, K.   [84] On Maillet determinant, *JNT*, **18**, 1984, 306–312.
WANG, S.   [48] A counterexample to Grunwald's theorem, *AnM*, **49**, 1948, 1008–1009.
[50a] On Grunwald's theorem, *AnM*, **51**, 1950, 471–484.

[50b] An existence theorem for abelian extension over algebraic number fields, *Sci. Rec.*, **3**, 1950, 25–27.
WANG, Y. [64] Estimation and application of character sums, *Shuxue Jinhan*, **7**, 1964, 78–83 (Chinese).
WANTUŁA, B. [74] Problem of Browkin on quadratic fields, *Zesz. Nauk. Polit. Śl.*, 1974, no 386, 173–178.
WARLIMONT, R. [67] Eine Bemerkung über Dirichletreihen mit Funktionalgleichung, *JRAM*, **228**, 1967, 144–158.
[71] Über die k-ten Mittelwerte der Klassenzahlen primitiven binären quadratischer Formen negativer Diskriminante, *Monatsh. Math.*, **75**, 1971, 173–179.
WASÉN, R. [74] On sequences of algebraic numbers in pure extensions of prime degree, *Colloq. Math.*, **30**, 1974, 89–104.
[76] Remark on a problem of Schinzel, *AA*, **29**, 1976, 425–426.
[77] On additive relations between algebraic integers of bounded norm, in: *Journées de théorie additive des nombres*, 153–160, Bordeaux 1977.
WASHINGTON, L. C. [74] On the self-duality of Q_p, *AMM*, **81**, 1974, 369–371.
[75] Class numbers and Z_p-extensions, *MA*, **214**, 1975, 177–193.
[76a] Relative integral bases, *PAMS*, **56**, 1976, 93–94.
[76b] A note on p-adic L-functions, *JNT*, **8**, 1976, 245–250.
[76c] The class number of the field of 5^nth roots of unity, *PAMS*, **61**, 1976, 205–208.
[77] The calculation of $L_p(1, \chi)$, *JNT*, **9**, 1977, 175–178.
[78a] Euler factors for p-adic L-functions, *Mathematika*, **25**, 1978, 68–75.
[78b] The non-p-part of the class number in a cyclotomic Z_p-extension, *Invent. Math.*, **49**, 1978, 87–97.
[79a] Kummer's calculation of $L_p(1, \chi)$, *JRAM*, **305**, 1979, 1–8.
[79b] Units of irregular cyclotomic fields, *Illinois J. Math.*, **23**, 1979, 635–647.
[81a] The derivative of p-adic L-functions, *AA*, **40**, 1981, 109–115.
[81b] p-adic L-functions at $s = 0$ and $s = 1$, in: *Analytic Number Theory*, 166–170, LN 899, Springer, 1981.
[82] *Introduction to Cyclotomic Fields*, Springer, 1982.
WASHINGTON, L., CORNELL, G. (See G. Cornell, L. Washington)
WASHINGTON, L. C., FERRERO, B. (See B. Ferrero, L. C. Washington)
WASHINGTON, L., SEAH, E., WILLIAMS, H. C. (See E. Seah, L. Washington, H. C. Williams)
WATABE, M. [78] On class numbers of some cyclotomic fields, *JRAM*, **301**, 1978, 212–215; corr., *ibid.*, **329**, 1981, 176.
[82] On certain Diophantine equations in algebraic number fields, *PJA*, **58**, 1982, 410–412.
[83] On certain cubic fields, I, *PJA*, **59**, 1983, 66–69; II, *ibid.*, 107–108; III, *ibid.*, 260–262, IV, *ibid.*, 387–389.
WATANABE, S. I. [77] The Galois group of local fields, *Tôhoku Math. J.*, (2), **29**, 1977, 385–416.
WATERHOUSE, W. C. [70] The sign of the Gaussian sum, *JNT*, **2**, 1970, 363.
[73] Pieces of eight in class groups of quadratic fields, *JNT*, **5**, 1973, 95–97.
[76] Pairs of quadratic forms, *Invent. Math.*, **37**, 1976, 157–164.
[77] A nonsymmetric Hasse–Minkowski theorem, *AJM*, **99**, 1977, 755–759.
[78] A probable Hasse principle for pencils of quadrics, *TAMS*, **242**, 1978, 297–306.
[79] The normal basis theorem, *AMM*, **86**, 1979, 212.
WATSON, E. J. [62] Primitive polynomials (mod 2), *MC*, **16**, 1962, 368–369.
WATSON, G. L. [63] A problem of Dade on quadratic forms, *Mathematika*, **10**, 1963, 101–106.

WEBER, B. T., BENSON, C. T.   (See C. T. Benson, B. T. Weber)
WEBER, H.   [86] Theorie der Abelscher Zahlkörper, *AM*, **8**, 1886, 193–263; II, *ibid.*, **9**, 1886/7, 105–130.
[96a] Über einen in der Zahlentheorie angewandten Satz der Integralrechnung, *GN*, 1896, 275–281.
[96b] *Lehrbuch der Algebra*, vol. II, III, 2nd ed., Braunschweig 1896, 1908.
[97] Über Zahlengruppen in algebraischen Körpern, *MA*, **48**, 1897, 433–473; II, *ibid.*, **49**, 1897, 83–100; III, *ibid.*, **50**, 1898, 1–26.
[05] Über komplexe Primzahlen in Linearformen, *JRAM*, **129**, 1905, 35–62.
[07] Über zyklische Zahlkörper, *JRAM*, **132**, 1907, 167–188.
[09] Zur Theorie der zyklischen Zahlkörper, *MA*, **67**, 1909, 32–60; II, *ibid.*, **70**, 1911, 459–470.
WEBER, H., DEDEKIND, R.   (See R. Dedekind, H. Weber)
WEBER, H., WELLSTEIN, J.   [13] Der Minkowskische Satz über die Körperdiskriminante, *MA*, **73**, 1913, 275–285.
WEBER, H.   [84] Über die Verteilung ganzer Zahlen mit ausgezeichneten Eigenschaften der Faktorzerlegung in algebraischen Zahlkörpern, *AA*, **44**, 1984, 215–239.
WEBER, W.   [31] Umkehrbare Ideale, *MZ*, **34**, 1931, 131–157.
WEGNER, U.   [31] Über ein algebraisches Problem, *MA*, **105**, 1931, 779–785.
[32a] Zur Theorie der auflösbaren Gleichung von Primzahlgrad, I, *JRAM*, **168**, 1932, 176–192.
[32b] Ein Satz über die Zerlegung von Primzahlen bestimmter arithmetischer Progressionen in algebraischen Zahlkörpern, *JRAM*, **168**, 1932, 231–232.
[35] Zur Theorie der affektlosen Gleichungen, *MA*, **111**, 1935, 738–742.
[37] Bestimmung eines auflösbaren Körpers von Primzahlgrad aus der Form seiner Diskriminante, *JRAM*, **176**, 1937, 1–11.
WEGNER, U., REICHARDT, H.   (See H. Reichardt, U. Wegner)
WEIL, A.   [36] Remarques sur des résultats recents de C. Chevalley, *CR*, **203**, 1936, 1208–1210 = *Coll. Papers, I*, 145–146, Springer, 1979.
[39] Sur l'analogie entre les corps de nombres algébriques et les corps de fonctions algébriques, *Revue Scient.*, **77**, 1939, 104–106 = *Coll. Papers, I*, 236–240, Springer, 1979.
[43] Differentiation in algebraic number fields, *BAMS*, **49**, 1943, 41 = *Coll. Papers, I*, 329, Springer, 1979.
[51] Sur la théorie de corps de classes, *J. Math. Soc. Japan*, **3**, 1951, 1–35 = *Coll. Papers, I*, 483–517, Springer, 1979.
[52a] Jacobi sums as "Grössencharaktere", *TAMS*, **73**, 1952, 487–495 = *Coll. Papers, II*, 63–71, Springer, 1979.
[52b] Sur les "formules explicites" de la théorie des nombres premiers, *Medd. Lunds Univ. Math. Sem.*, 1952, 252–265 = *Coll. Papers, II*, 48–61, Springer, 1979.
[56] On a certain type of characters of the idèle-class group of an algebraic number field, in: *Proc. Int. Symp. Alg. N. Th.*, 1–7, Tokyo 1956.
[66] Fontions zeta et distributions, *Sém. Bourbaki*, 1966, exp. 312 = *Coll. Papers, III*, 158–163, Springer, 1979.
[67a] *Basic Number Theory*, Springer, 1967.
[67b] Über die Bestimmung Dirichletscher Reihen durch Funktionalgleichungen, *MA*, **168**, 1967, 149–156 = *Coll. Papers, III*, 165–172, Springer, 1979.
[71] *Dirichlet Series and Automorphic Forms*, LN 189, Springer, 1971.
[72] Sur les formules explicites de la théorie des nombres, *IAN*, **36**, 1972, 3–18 = *Coll. Papers, III*, 249–264, Springer, 1979.

[74] Sommes de Jacobi et caractéres de Hecke, *GN*, 1974, 1–14 = *Coll. Papers, III*, 329–342, Springer, 1979.
WEILER, P., POHST, M., ZASSENHAUS, H.   (See M. Pohst, P. Weiler, H. Zassenhaus)
WEINBERGER, P. J.   [69] *A proof of a conjecture of Gauss on class-number two*, Thesis, Berkeley 1969.
[72a] On euclidean rings of algebraic integers, *PSPM*, **24**, 1972, 321–332.
[72b] The cubic character of quadratic units, *Proc. Number Theory Conference, Boulder 1972*, 241–242.
[72c] A counterexample to an analogue of Artin's conjecture, *PAMS*, **35**, 1972, 49–52.
[73a] Exponents of the class groups of complex quadratic fields, *AA*, **22**, 1973, 117–124.
[73b] Real quadratic fields with class numbers divisible by n, *JNT*, **5**, 1973, 237–241.
[75] On small zeros of Dirichlet L-functions, *MC*, **29**, 1975, 319–328.
WEINBERGER, P. J., COOKE, G.   (See G. Cooke, P. J. Weinberger)
WEINBERGER, P. J., MONTGOMERY, H. L.   (See H. L. Montgomery, P. J. Weinberger)
WEINBERGER, P. J., SHANKS, D.   (See D. Shanks, P. J. Weinberger)
WEINERT, H. J.   [63] Unterhalbkörper quadratischer Zahlkörper, *Wiss. Z. Pädagog. Hochsch. Potsdam*, **8**, 1963/4, 83–86.
WEINSTEIN, L.   [77] The mean value of the derivative of the Dedekind zeta-function of a real quadratic field, *Mathematika*, **24**, 1977, 226–236.
[79] The zeros of the Artin L-series of a cubic field on the critical line, *JNT*, **11**, 1979, 279–284.
[80] The mean value of the Artin L-series and its derivative of a cubic field, *Glasgow Math. J.*, **21**, 1980, 9–18.
WEISNER, L.   [28] Quadratic fields over which cyclotomic polynomials are reducible, *AnM*, **29**, 1928, 377–381.
WEISS, A.   [83] The least prime ideal, *JRAM*, **338**, 1983, 56–94.
WEISS, E.   [63] *Algebraic Number Theory*, New York 1963.
WEISS, M. J.   [36] Fundamental systems of units in normal fields, *AJM*, **58**, 1936, 249–254.
WELLSTEIN, J., WEBER, H.   (See H. Weber, J. Wellstein)
WESTLUND, J.   [03] On the class number of the cyclotomic number field, *TAMS*, **4**, 1903, 201–212.
[10] On the fundamental number of the algebraic number field $k(\sqrt[p]{m})$, *TAMS*, **11**, 1910, 388–392.
[12] Primitive roots of ideals in algebraic number-fields, *MA*, **71**, 1912, 246–250.
[13] On the factorization of rational primes in cubic cyclotomic number fields, *Jahresber. Deutsch. Math. Verein.*, **22**, 1913, 135–140.
WEYL, H.   [40] *Algebraic Theory of Numbers*, Princeton 1940.
WHAPLES, G.   [42] Non-analytic class-field theory and Grunwald's theorem, *Duke Math. J.*, **9**, 1942, 455–473.
[47] On a conjecture about infinite class fields, *BAMS*, **53**, 1947, 377–380.
WHAPLES, G., ARTIN, E.   (See E. Artin, G. Whaples)
WHAPLES, G., MCKENZIE, R. E.   (See R. E. McKenzie, G. Whaples)
WHITEMAN, A. L.   [40] Additive prime number theory in real quadratic fields, *Duke Math. J.*, **7**, 1940, 208–232.
WHITFORD, E. E.   [12] *The Pell Equation*, New York 1912.
WIEBELITZ, R.   [52] Über approximative Funktionalgleichungen der Potenzen der Riemannschen Zetafunktion, *GN*, 1952, 263–270.
WIECZORKIEWICZ, J. K.   [79] Some explicit estimates for the Dedekind zeta function, *Funct. et Approx.*, **7**, 1979, 9–12.

[80] Some remarks on a result of A. V. Sokolovskij, *Funct. et Approx.*, **8**, 1980, 49–58.

WIEGAND, R., WIEGAND, S. [77] Commutative rings whose finitely generated modules are direct sums of cyclics, in: *Abelian Group Theory*, 406–423, LN 616, Springer, 1977.

WIEGAND, S., MIDGARDEN, B. (See B. Midgarden, S. Wiegand)

WIEGAND, S., WIEGAND, R. (See R. Wiegand, S. Wiegand)

WIEGANDT, R. [59] On the general theory of Möbius inversion formula and Möbius product, *Acta Acad. Sci. (Szeged)*, **20**, 1959, 164–180.

WIERTELAK, K. [78] On the density of some sets of primes, II, *AA*, **34**, 1977/8, 197–210.

WIERTELAK, K., STAŚ, W. (See W. Staś, K. Wiertelak)

WIĘSŁAW, W. [82] *Topological Fields*, Wrocław 1982.

WILES, A. [80] Modular curves and the class group of $Q(\zeta_p)$, *Invent. Math.*, **58**, 1980, 1–35.

WILES, A., COATES, J. (See J. Coates, A. Wiles)

WILES, A., MAZUR, B. (See B. Mazur, A. Wiles)

WILLIAMS, H. C. [76] Some results on fundamental units in cubic fields, *JRAM*, **286/7**, 1976, 75–85.

[77] Certain pure cubic fields with class-number one, *MC*, **31**, 1977, 578–580; corr., *ibid.*, **33**, 1979, 847–848.

[80] Improving the speed of calculating the regulator of certain pure cubic fields, *MC*, **35**, 1980, 1423–1434.

[81a] A numerical investigation into the length of the period of the continued fraction expansion of $\sqrt{D}$, *MC*, **36**, 1981, 593–601.

[81b] Some results concerning Voronoï's continued fraction over $Q(\sqrt[3]{D})$, *MC*, **36**, 1981, 631–652.

WILLIAMS, H. C., BARRUCAND, P., BANIUK, L. (See P. Barrucand, H. C. Williams, L. Baniuk)

WILLIAMS, H. C., BEACH, B. D., ZARNKE, C. R. (See B. D. Beach, H. C. Williams, C. R. Zarnke)

WILLIAMS, H. C., BROERE, J. [76] A computational technique for evaluating $L(1, \chi)$ and the class number of a real quadratic field, *MC*, **30**, 1976, 887–893.

WILLIAMS, H. C., BUELL, D. A., WILLIAMS, K. S. (See D. A. Buell, H. C. Williams, K. S. Williams)

WILLIAMS, H. C., BUHR, P. A. [79] Calculation of the regulator of $Q(\sqrt{D})$ by use of the nearest integer continued fraction algorithm, *MC*, **33**, 1979, 369–381.

WILLIAMS, H. C., CORMACK, G., SEAH, E. [80] Calculation of the regulator of a pure cubic field, *MC*, **34**, 1980, 567–611.

WILLIAMS, H. C., DIAZ Y DIAZ, F., SHANKS, D. (See F. Diaz y Diaz, D. Shanks, H. C. Williams)

WILLIAMS, H. C., SEAH, E., WASHINGTON, L. (See E. Seah, L. Washington, H. C. Williams)

WILLIAMS, H. C., SHANKS, D. [79] A note on class-number one in pure cubic fields, *MC*, **33**, 1979, 1317–1320.

WILLIAMS, H. C., STANTON, R. G. (See R. G. Stanton, H. C. Williams)

WILLIAMS, H. C. JR., STANTON, R. G., SUDLER, C. JR. (See R. G. Stanton, C. Sudler, Jr., H. C. Williams, Jr.).

WILLIAMS, K. S. [67] On a theorem of Niven, *Canad. Math. Bull.*, **10**, 1967, 573–587.

[70] Integers of biquadratic fields, *Canad. Math. Bull.*, **13**, 1970, 519–526.

[73] Another proof of a theorem of Niven, *Math. Mag.*, **46**, 1973, 39.

[75] Note on non-Euclidean principal ideal domains, *Math. Mag.*, **48**, 1975, 176–177.

[76] Note on a result of Barrucand and Cohn, *JRAM*, **285**, 1976, 218–220.

[79] The class number of $Q(\sqrt{-p})$ modulo 4, for $p \equiv 3 \pmod{4}$ a prime, *Pacific J. Math.*, 83, 1979, 565-570.
[80] On the evaluation of $(\varepsilon_{q_1 q_2}/p)$, *Rocky Mountain J. Math.*, 10, 1980, 559-573.
[81a] On the class number of $Q(\sqrt{-p})$ modulo 16, for $p \equiv 1 \pmod{8}$ a prime, *AA*, 39, 1981, 381-398.
[81b] The class number of $Q(\sqrt{-2p})$ modulo 8, for $p \equiv 5 \pmod{8}$ a prime, *Rocky Mountain J. Math.*, 11, 1981, 19-26.
[81c] The class number of $Q(\sqrt{p})$ modulo 4, for $p \equiv 5 \pmod{8}$ a prime, *Pacific J. Math.*, 92, 1981, 241-248.
[82] Congruences modulo 8 for the class numbers of $Q(\sqrt{\pm p})$, $p \equiv 3 \pmod{4}$ a prime, *JNT*, 15, 1982, 182-198.
WILLIAMS, K. S., BUELL, D. A., LEONARD, P. A. (See D. A. Buell, P. A. Leonard, K. S. Williams)
WILLIAMS, K. S., BUELL, D. A., WILLIAMS, H. C. (See D. A. Buell, H. C. Williams, K. S. Williams)
WILLIAMS, K. S., CURRIE, J. D. [82] Class numbers and biquadratic reciprocity, *CJM*, 34, 1982, 969-988.
WILLIAMS, K. S., KAPLAN, P. (See P. Kaplan, K. S. Williams)
WILLIAMS, K. S., LEONARD, P. A. (See P. A. Leonard, K. S. Williams)
WILSON, N. R. [27] Integers and basis of a number field, *TAMS*, 29, 1927, 111-126.
[29] On finding ideals, *AnM*, 30, 1929, 411-428.
[31] Constructing bases for an algebraic number field, *Trans. Roy. Soc. Canada, III*, (3), 25, 1931, 171-184.
WILSON, R. J. [69] The large sieve in algebraic number fields, *Mathematika*, 16, 1969, 189-204.
WILSON, S. M. J. [80] Extensions with identical wild ramifications, *Sem. Th. Nombr. Bordeaux*, 1980/1, exp. 20.
WILSON, S., FRÖHLICH, A., KEATING, M. (See A. Fröhlich, M. Keating, S. Wilson)
WILTON, J. R. [30] An approximate functional equation for the product of two ζ-functions, *PLMS*, (2), 31, 1930, 11-17.
WIMAN, A. [99] Über die Ideale in einem algebraischen Zahlkörper, nach denen Primitivzahlen existieren, *Ofversikt Svenska Vet. Akad. Förhandl.*, 56, 1899.
WINGBERG, K. [79] Die Einseinheitengruppe von p-Erweiterungen regulärer p-adischer Zahlkörper als Galoismodul, *JRAM*, 305, 1979, 206-214.
[82] Der Eindeutigkeitssatz für Demuškinformationen, *Invent. Math.*, 70, 1982/3, 99-113.
[83] Freie Produktzerlegungen von Galoisgruppen und Iwasawa-Invarianten für p-Erweiterungen von Q, *JRAM*, 341, 1983, 111-129.
WINGBERG, K., JANSSEN, U. (See U. Janssen, K. Wingberg)
WINOGRAD, S. et al. (See L. Auslander et al.)
WINTENBERGER, J. P. [80] Structure galoisienne de limites projectives d'unités locales, *Compos. Math.*, 42, 1980/1, 89-103.
WINTER, D. J. [72] A generalization of the normal basis theorem, *MN*, 54, 1972, 75-77.
WINTNER, A. [45] The densities of ideal classes and the existence of unities in algebraic number fields, *AJM*, 67, 1945, 235-238.
[46a] The values of the norms in algebraic number fields, *AJM*, 68, 1946, 223-229.
[46b] A factorization of the densities of ideals in algebraic number fields, *AJM*, 68, 1946, 273-284.
WISHART, E. F., DAVIS, R. D. (See R. D. Davis, E. F. Wishart)

WITT, E. [35a] Über ein Gegenbeispiel zum Normensatz, *MZ*, **39**, 1935, 462–467.
[35b] Riemann-Rochscher Satz und *Z*-Funktion im Hyperkomplexen, *MA*, **110**, 1935, 12–28.
[36] Zyklische Körper und Algebren der Charakteristik p von Grade p^n, *JRAM*, **176**, 1936, 126–140.
WÓJCIK, J. [69] On prime ideals with prescribed values of characters of prime degree, *Colloq. Math.*, **20**, 1969, 261–263.
[75] A purely algebraic proof of special cases of Tchebotarev's theorem, *AA*, **28**, 1975, 137–145.
WOLFF, G. [05] *Über die Gruppen der Reste eines beliebigen Moduls in algebraischen Zahlkörpern*, Diss., Giessen 1905.
WOLKE, D. [69] Moments of the number of classes of primitive quadratic forms with negative discriminant, *JNT*, **1**, 1969, 502–511.
[71] Momente der Klassenzahlen, II, *Arch. Math.*, **22**, 1971, 65–69.
[72] Moments of class numbers, III, *JNT*, **4**, 1972, 523–531.
WORLEY, T. R. [70] On a result of Cassels, *J. Austral. Math. Soc.*, **11**, 1970, 191–194.
WOWK, C. [75] On the asymptotical densities of certain ideals in algebraic number fields, *Discuss. Math.*, **1**, 1975, 27–29.
WRENCH, J. W. JR., SHANKS, D. (See D. Shanks, J. W. Wrench, Jr.)
WRIGHT, E. M., HARDY, G. H. (See G. H. Hardy, E. M. Wright)
WUNDERLICH, M. C. [71] Some computer results related to the Gaussian primes on the line Im $s = 1$, *Proc. Washington State Univ. Conf. Numb. Th.*, 92–95, 1971.
[73] On the Gaussian primes on the line Im $X = 1$, *MC*, **27**, 1973, 399–400.
WYMAN, B. F. [69] Wildly ramified gamma extensions, *AJM*, **91**, 1969, 135–152.
YAGER, R. I. [82] A Kummer criterion for imaginary quadratic fields, *Compos. Math.*, **47**, 1982, 31–42.
YAGI, A. [72] Explicit formulas for some functions associated with the *L*-series with some Hecke characters, *Bull. Yamagata Univ.*, **8**, 1972, 1, 17–27.
YAHAGI, O. [78] Construction of number fields with prescribed *l*-class groups, *Tokyo J. Math.*, **1**, 1978, 275–283.
YAMAGATA, S. [76] A counterexample for the local analogy of a theorem of Iwasawa and Uchida, *PJA*, **52**, 1976, 276–278.
YAMAGUCHI, I. [71] On a property of the irregular class group in a properly irregular *l*-th cyclotomic field, *TRU Math.*, **7**, 1971, 21–24.
[75] On the class-number of the maximal real subfield of a cyclotomic field, *JRAM*, **272**, 1975, 217–220.
[77] On the units in a l^ν-th cyclotomic field, *TRU Math.*, **13**, 1977, 2, 1–12.
YAMAGUCHI, I., OOZEKI, K. [72] On the class number of the real quadratic field, *TRU Math.*, **8**, 1972, 13–14.
YAMAMOTO, K. [58] Arithmetic linear transformations in an algebraic number field, *Mem. Fac. Sci. Kyushu, Univ.*, A, **12**, 1958, 41–66.
[65] On Gaussian sums with biquadratic residue characters, *JRAM*, **219**, 1965, 200–213.
[66] On a conjecture of Hasse concerning multiplicative relations of Gaussian sums, *J. Combin. Theory*, **1**, 1966, 476–489.
YAMAMOTO, K., MANN, H. B. (See H. B. Mann, K. Yamamoto)
YAMAMOTO, K., ONUKI, M. [75] On Kronecker's theorem about Abelian extensions, *Sci. Rep. Tokyo Woman's Christian Coll.*, **35–38**, 1975, 415–418.
YAMAMOTO, S. [68] On a property of Hasse's function in the ramification theory, *Mem. Fac. Sci. Kyushu, Univ.*, A, **22**, 1968, 96–109.

[72] On the rank of the p-divisor class group of Galois extensions of algebraic number fields, *Kumamoto J. Sci.*, **9**, 1972, 33-40.

YAMAMOTO, Y. [70] On unramified Galois extensions of quadratic number fields, *Osaka J. Math.*, **7**, 1970, 57-76.

[71] Real quadratic number fields with large fundamental units, *Osaka J. Math.*, **8**, 1971, 261-270.

YANAGIHARA, H. *et al.* (See T. Koyama *et al.*)

YOKOI, H. [60a] On unit groups of absolute abelian number fields of degree pq, *Nagoya Math. J.*, **16**, 1960, 73-81.

[60b] On the ring of integers in an algebraic number field as a representation module of Galois group, *Nagoya Math. J.*, **16**, 1960, 83-90.

[62a] On the Galois cohomology group of the ring of integers in an algebraic number field, *AA*, **8**, 1962/3, 243-250.

[62b] On an isomorphism of Galois cohomology groups $H^m(G, O_k)$ of integers in an algebraic number field, *PJA*, **38**, 1962, 499-501.

[64] A note on the Galois cohomology group of the ring of integers in an algebraic number field, *PJA*, **40**, 1964, 245-246.

[66] A cohomological investigation of the discriminant of a normal algebraic number field, *Nagoya Math. J.*, **27**, 1966, 207-211.

[67] On the class number of a relatively cyclic field, *Nagoya Math. J.*, **29**, 1967, 31-44.

[68a] On real quadratic fields containing units with norm -1, *Nagoya Math. J.*, **33**, 1968, 139-152.

[68b] On the divisibility of the class number in an algebraic number field, *J. Math. Soc. Japan*, **20**, 1968, 411-418.

[70a] On the fundamental unit of real quadratic fields with norm 1, *JNT*, **2**, 1970, 106-115.

[70b] Units and class-numbers of real quadratic fields, *Nagoya Math. J.*, **37**, 1970, 61-65.

[74] The diophantine equation $x^3+dy^3 = 1$ and the fundamental unit of a pure cubic field $Q(\sqrt[3]{d})$, *JRAM*, **268/9**, 174-179.

[75] On the distribution of irregular primes, *JNT*, **7**, 1975, 71-76.

YOKOYAMA, A. [64] On the Gaussian sum and the Jacobi sum with its application, *Tôhoku Math. J.*, (2), **16**, 1964, 142-153.

[65] On class numbers of finite algebraic number fields, *Tôhoku Math. J.*, (2), **17**, 1965, 349-357.

[66] Über die Relativklassenzahl eines relativ-galoisschen Zahlkörpers von Primzahlpotenzgrad, *Tôhoku Math. J.*, (2), **18**, 1966, 318-324.

[67] On the relative class number of finite algebraic number fields, *J. Math. Soc. Japan*, **19**, 1967, 179-184.

YOSHIDA, H. [77] On Artin L-functions, *Japan. J. Math.*, **3**, 1977, 369-380.

ZAGIER, D. [75a] A Kronecker limit formula for real quadratic fields, *MA*, **213**, 1975, 153-184.

[75b] Nombre de classes et fractions continues, *Astérisque*, **24/5**, 1975, 81-97.

[76] On the values at negative integers of the zeta-function of a real quadratic field, *Enseign. Math.*, (2), **22**, 1976, 55-95.

[77] Valeurs des fonctions zêta des corps quadratiques réels aux entiers négatifs, *Astérisque*, **41/2**, 1977, 135-151.

[81] *Zetafunktionen und quadratische Körper*, Springer, 1981.

ZAGIER, D., GROSS, B. (See B. Gross, D. Zagier)

ZAIKINA, N. G. [57a] The distribution of numbers of an imaginary quadratic field, composed of small prime ideals, *Uchen. Zap. Mosk. Ped. Inst.*, **108**, 1957, 261-272 (Russian).

[57b] The distribution of n-th non-residues with respect to a prime ideal in an imaginary quadratic field, *Uchen. Zap. Mosk. Ped. Inst.*, **108**, 1957, 273–282 (Russian).

ZAKS, A. [76] Half-factorial domains, *BAMS*, **82**, 1976, 721–723.

[80] Half-factorial domains, *Israel J. Math.*, **37**, 1980, 281–302.

ZAME, A. [67] Note on a paper by M. F. Tinsley, *Duke Math. J.*, **34**, 1967, 231.

ZANARDO, P., SALCE, L. (See L. Salce, P. Zanardo)

ZANTEMA, H. [82] Integer valued polynomials over a number field, *Manuscr. Math.*, **40**, 1982, 155–203.

ZARISKI, O., SAMUEL, P. [58] *Commutative Algebra*, New York 1958–1960.

ZARNKE, C. R. et al. (See B. D. Beach et al.)

ZASSENHAUS, H. [54] Über eine Verallgemeinerung des Henselschen Lemmas, *Arch. Math.*, **5**, 1954, 317–325.

[65] Ein Algorithmus zur Berechnung einer Minimalbasis über gegebener Ordnung, in: *Funktionalanalysis, Approx. Th., Numer. Math., Oberwolfach 1965*, 90–103, Stuttgart 1965.

[68] On a theorem of Kronecker, *Delta (Waukesha)*, **1**, 1968/9, 1–14.

[72] On the units of orders, *J. Algebra*, **20**, 1972, 368–395.

ZASSENHAUS, H., DADE, E. C., TAUSSKY, O. (See E. C. Dade, O. Taussky, H. Zassenhaus)

ZASSENHAUS, H., LEWIS, D. J., SCHINZEL, A. (See D. J. Lewis, A. Schinzel, H. Zassenhaus)

ZASSENHAUS, H., LIANG, J. (See J. Liang, H. Zassenhaus)

ZASSENHAUS, H., POHST, M. (See M. Pohst, H. Zassenhaus)

ZASSENHAUS, H., POHST, M., WEILER, P. (See M. Pohst, P. Weiler, H. Zassenhaus)

ZASSENHAUS, H., ROQUETTE, P. (See P. Roquette, H. Zassenhaus)

ZASSENHAUS, H., SCHINZEL, A. (See A. Schinzel, H. Zassenhaus)

ZAUPPER, T. [83] A note on unique factorization in imaginary quadratic fields, *Ann. Univ. Budapest*, **26**, 1983, 197–203.

ZEINALOV, B. A. [65] On units of a cyclic real field, *Sb. Nauch. Soobshch. Dagestan. Univ.*, 1965, 21–23 (Russian).

ZEITLIN, D. [68] On coefficients identities for cyclotomic polynomials $F_{pq}(x)$, *AMM*, **75**, 1968, 976–980.

ZELINSKY, D. [48] Topological characterization of fields with valuations, *BAMS*, **54**, 1948, 1145–1150.

ZELVENSKII, I. G. [72] The algebraic closure of a local field when $p = 2$, *IAN*, **36**, 1972, 933–946 (Russian).

[73] Extensions without tame ramification of a local field, *Zap. Nauch. Sem. LOMI*, **31**, 1973, 102–105 (Russian).

[78] The maximal extension without simple ramification of a local field, *IAN*, **42**, 1978, 1385–1400 (Russian).

ZHANG XIANKE [82] On number fields of type $(2, 2, \ldots, 2)$, *J. Univ. Sci. Techn. China*, **12**, 1982, 4, 29–41 (Chinese).

[84a] Note on a paper by A. A. Albert, *J. Univ. Sci. Techn. China*, **14**, 1984, 171–177.

[84b] Density of number fields of type $(2, 2, \ldots, 2)$, *Sci. Sinica*, **27**, 1984, 345–347.

[84c] On number fields of type $(l, l, \ldots, l)$, *Sci. Sinica*, A **27**, 1984, 1018–1026.

[84d] Cyclic quartic fields and genus theory of their subfields, *JNT*, **18**, 1984, 350–355.

[84e] Relative integral bases and units of C_l^n fields, *J. Univ. Sci. Techn. China*, **14**, 1984, 427–428.

ZHANG XIANKE, FENG K. (See K. Feng, Zhang Xianke)

ZIERLER, N. [69] Primitive trinomials whose degree is a Mersenne prime, *Inform. and Control*, **15**, 1969, 67–69.

[70] On x^n+x+1 over GF(2), *Inform. and Control*, **16**, 1970, 502–505.
ZIERLER, N., BRILLHART, J. [68] On primitive trinomials (mod 2), *Inform. and Control*, **13**, 1968, 541–554; II, *ibid.*, **14**, 1969, 566–569.
ZIMMER, H. G. [72] *Computational Problems, and Results in Algebraic Number Theory*, LN 262, Springer, 1972.
ZIMMERT, R. [80] Ideale kleiner Norm in Idealklassen und eine Regulatorabschätzung, *Invent. Math.*, **62**, 1980, 367–380.
ZINK, E. W. [75] Zum Hauptidealsatz von Tannaka-Terada, *MN*, **67**, 1975, 317–325.
ZINK, W., KOCH, H. (See H. Koch, W. Zink)
ZLEBOV, E. D. [66] The Pisot-Vijayaraghavan numbers and fundamental units of algebraic fields, *IAN Belorus. SSR*, 1966, 110–112 (Russian).
ZOLOTAREV, E. [69] *On an indeterminate cubic equation*, Diss., St. Petersburg 1869 (Russian).
[74] Théorie des nombres entiers complexes avec une application au calcul intégral, *Bull. Acad. Sci. St. Petersburg*, 1874.
[80] Sur la théorie des nombres complexes, *J. Math. Pures Appl.*, (3), **6**, 1880, 51–84.
ZYGMUND, A., SALEM, R. (See R. Salem, A. Zygmund)
ŻYLIŃSKI, E. [13] Zur Theorie der ausserwesentlichen Diskriminantenteiler algebraischer Körper, *MA*, **73**, 1913, 273–274.
[32] Zur Begründung der Idealtheorie, *C.R. Soc. Sci. Varsovie*, **24**, 1932, 87–92.

List of important symbols

(In order of appearance)

$(A, B), [A, B]$ 8
$a \equiv b \pmod{I}$ 9
$N(I)$ 12
$\varphi(I)$ 13
R_n, P_n 18
$\text{Ann}(m)$ 31
$M(P)$ 31
$\text{Ann}(M)$ 32
$H(R)$ 35
$\deg a$ 42
$\deg_K a$ 42
R_K 42
$r_1(K), r_2(K)$ 43
Sgn 43
$\text{Sgn}(K)$ 43
$a \geqslant 0$ 43
$N_{L/K}(x)$ 45
$T_{L/K}(x)$ 45
$|a|$ 45
$T_n(x)$ 48
$d_{L/K}(v_1, \ldots, v_n)$ 52
$d_{L/K}(a)$ 53
$d_{K/Q}(M)$ 57
$d(K)$ 57
K_n 63
$M(r_1, r_2)$ 69
$M(a)$ 74
$\varepsilon(K)$ 74
$d(a, b)$ 75
$C(d)$ 83
$G(K), P(K)$ 94
$H(K)$ 94
$D(K)$ 94
$G_+(K)$ 94
$H^*(K)$ 94
$G_{\mathfrak{f}}(K)$ 95
$G_{\mathfrak{f}}^+(K)$ 95

$H_{\mathfrak{f}}(K)$ 95
$H_{\mathfrak{f}}^*(K)$ 95
$h(K), h^*(K)$ 97
$h_{\mathfrak{f}}(K), h_{\mathfrak{f}}^*(K)$ 97
$U(K)$ 99
$E(K)$ 100
S_∞ 101
$U_S(K)$ 101
$R(u_1, \ldots, u_r)$ 106
$R(K)$ 106
$U(K, \mathfrak{f})$ 110
$U^+(K, \mathfrak{f})$ 110
$U^+(K)$ 110
$Z[G]$ 114
$U_0(K)$ 114
$\Gamma(R), K_i(R)$ 128
$C(r_1, r_2)$ 129
$U'(K), U'_S(K)$ 136
$C(K_p)$ 136
K^+ 136
$C_0(K)$ 137
$I(K)$ 143
$i_{L/K}$ 143
$e_{L/K}(\mathfrak{P}_i)$ 144
$f_{L/K}(\mathfrak{P})$ 145
$N_{L/K}(I)$ 149
A^* 156
$D_{L/K}(A)$ 157
$D_{L/K}$ 157
$d(L/K)$ 159
$\delta_{L/K}(a)$ 160
$\mathfrak{f}_A$ 160
$F_m(x)$ 169
$i(K)$ 178
$i_{L/K}^*$ 188
$H_p(K)$ 190
$\text{Am}(L/K)$ 191

List of Important Symbols

$\mathrm{Am}_p(L/K)$ 191
$K_\mathfrak{p}, K_v$ 207
Q_p 207
Z_p 208
$e(L_\mathfrak{P}/K_\mathfrak{p})$ 218
$f(L_\mathfrak{P}/K_\mathfrak{p})$ 218
$\partial(L_\mathfrak{P}/K_\mathfrak{p})$ 219
$U_1(K)$ 223
$E_1(K)$ 223
$U_m(K)$ 223
G_i 240
$\mathfrak{f}_\chi$ 247
$e(q)$ 251
$Z(f, q)$ 251
$\tau_0(X)$ 254
$\varrho(q)$ 254
$G_i(\mathfrak{P})$ 271
$G(I)$ 274
$\tau_a(\chi)$ 280
$\tau(\chi)$ 280
A_K 295
A_S 295
I_K 296
I_S 296
U_K 296
A_0, I_0 296
$V(x)$ 296
J_K 296
$C(K)$ 297
$I_\mathfrak{f}, I_\mathfrak{f}^*$ 298
$D(K)$ 300
$\partial(L/K)$ 303
Q_0 307
$\mathrm{An} J_K$ 307
J_K' 308
$\tilde{f}$ 309

$\varkappa$ 310
$\zeta_K(s)$ 324
$\mu(I)$ 326
$G(K; S)$ 335
$\zeta(s, \chi)$ 343
$L(s, \chi)$ 348
$\left[\dfrac{L/K}{\mathfrak{P}}\right], s_\mathfrak{P}$ 379
$F_{L/K}$ 380
$P(L/K)$ 391
$\left(\dfrac{a}{\mathfrak{P}}\right)_p$ 394
$C_K(L)$ 397
$\mathrm{Cl}(M)$ 399
$f(K)$ 421
$X(K)$ 425
$h(d)$ 436
$L_d(s)$ 436
h_p^+, h_p^- 437
$d(f)$ 448
$\mathfrak{G}(K)$ 455
$g(K)$ 455
h_n^+, h_n^- 473
$\mathrm{ord}\,a$ 491
$B(A)$ 493
$D(A)$ 494
$a_1(A)$ 498
$M(A)$ 499
$\omega_x(I)$ 501
$\Omega_x(I)$ 501
$\hat{G}$ 520
$\tilde{G}$ 522
$\hat{f}$ 523
$\tilde{f}$ 524
$H(a)$ 538

Subject Index

Abelian fields 80, 134, 137, 289, 421-488
— — with given Galois group 431, 471
— — with $h = 1$ 446, 480
Abhyankar's lemma 236
absolute discriminant 143
— extension 143
— norm 12
absolutely irreducible element 491
additive function 513
— problems 514-517
adele 295, 321
— class group 297
— group 295
—, principal 296
— ring 295, 321
admissible homomorphism 335
algebraic integer 42, 71
— number 42, 72
— —, degree 42, 77
— —, minimal polynomial 42
algebraic number field 42
almost all integers 509
almost Dedekind domain 38
almost no integers 509
ambiguous ideal class 191
annihilator 31
approximate functional equation 409
approximation by quadratic integers 70, 82, 83
— theorem, strong 302
— —, weak 20
Archimedean valuation 16
arithmetic functions 513, 514
arithmetically equivalent fields 405
Artin-Hecke functions 408
Artinian ring 36
Artin's conjecture 407

— L-functions 196, 197, 407
— reciprocity theorem 389, 425
— root number 196, 408
— symbol 380, 415
associated elements 99

Baker's method 538
Bauerian extension 391, 416
Bernoulli numbers 133, 473, 475
— —, generalized 412, 476
binary quadratic forms 448, 481
— — —, classes 448, 481
block 493
—, equivalent factorizations 497
—, irreducible 493
—, length 493
—, with unique factorization 497

cancellation law for ideals 38, 40
— — for modules 40
capitulation 203
central ideal class group 202
character, Hecke's 334, 404
— —, conductor 340, 341
— —, exceptional set 334
— —, normalized 334
— —, primitive 347
— —, proper 334
character of $G(I)$ 274
— — —, conductor 274
— — —, even 275
— — —, odd 275
— — —, primitive 274
— — —, real primitive 277, 283
— of I , normalized 308
— of locally compact Abelian group 520
— group 520

716 Subject Index

— — associated with an Abelian field 426
Chinese remainder-theorem 10
circular units 136, 137
class-field theory over Q 425
— tower 203
class-group mod $\mathfrak{f}$ 95
— of a field 94, 127, 128, 491
— — —, arithmetic description 491, 512
— — —, behaviour under extensions 188-192
— — —, determination 185-188, 200, 201
— — —, finiteness 96, 127
— — —, exponent 485
— — —, invariants 455, 462, 463, 483-485
— of Dedekind domains 35, 40, 41
— of group rings 197
— relations 487
class-number 97
—, bounds 153, 441, 451, 481-483
—, congruences 473, 474
—, divisibility 459, 485, 486, 488
— formula for Abelian fields 436, 471, 472
— — for quadratic fields 440, 472
— mod $\mathfrak{f}$ 97, 111
—, narrow 97, 98, 111
— of imaginary quadratic fields, asymptotics 446, 454, 478, 480
— — — — —, divisibility 459, 485
— — — — —, mean value 482
— one, fields with 124, 446, 463, 465, 479, 480, 488
CM-fields 73, 134, 480
codifferent 156
cohomology groups 135
combinatorial constants 494, 498, 499, 512
common non-essential discriminatial divisor 65, 179, 180, 198
completion 87
complex embedding 43
— multiplication 204
conductor of an Abelian field 421, 428
— of a character of $G(I)$ 274
— of a character of U_1 247
— of a Hecke character 340, 341
— of a quasicharacter 248
— of a ring 160, 193
conductor-discriminant formula 428, 471
congruences for $h(K)$ 473, 474

conjugated fields 43
— elements 43
— ideals 148
continued fraction of $D^{1/2}$ 119, 132
convex body theorem 536
critical interval 138
cubic fields, discriminant 62, 78, 79
— —, Euclidean 140, 141
— —, pure, integral basis 62
— —, units 133
cyclic fields 107, 115
cyclotomic fields K_n 63, 72
— —, class number 188, 436-438, 441, 473-477, 480
— —, discriminant 63, 169
— —, factorization of primes 182
— —, integral basis 63, 169, 175
— —, units 134, 136, 170
— — with $h = 1$ 188, 480
— polynomials 171, 194, 195
— units 136, 137
— Z_p-extensions 476

Davenport's constant 494, 495, 512
decomposition field 272
— group 272
— — for cyclotomic fields 423
Dedekind domain 3, 35, 36
— —, characterizations 5, 36, 37
— —, class group 34, 40, 41
— —, congruences 9
— —, extensions 13, 39
— zeta-function 324, 404, 405
— —, analogues 405
— —, analytic continuation 326, 404
— —, approximate functional equation 409
— —, functional equation 326, 404
— — of an Abelian field 427
— —, values at integers 412
— —, zeros 325, 363, 369, 409-411
defect of units 107, 134
degree of a number 42, 77
— of a prime ideal 145
density, Dirichlet's 355
— hypothesis 410
— theorem 382, 383, 415
derivation 163
—, essential 163

different of an element 160
— of a fractional ideal 157
— of an extension 157, 243, 267, 273, 322
— theorem 166, 194, 233, 270
diophantine equations 138, 517
Dirichlet's class-number formula 440, 472
— convolution 513
— density 355
— L-functions, approximate functional equation 348, 404
— —, functional equation 348, 404
— —, value at $s = 1$ 439, 451, 481, 482
— —, zeros 409, 478
— series 534
— unit theorem 101, 129, 130
discrete valuation 18
discriminant $\partial(L/K)$ 219, 303, 322
— of a field, absolute 57-64, 66-69, 78-80
— — —, evaluations 66-69, 81, 82
— — —, of Abelian fields 428
— — —, of composite fields 167, 168, 194, 269
— — —, of cyclotomic fields 63, 169
— — —, of pure cubic fields 62
— — —, of quadratic fields 61
— — —, sign of 59
— of an integer 53
— of a system of numbers 52, 53
— of a Z-module 56, 57
—, regular 485
—, relative 159, 304
— theorem 167, 194
divisor functions 514
— group 94
divisors, finite 94
—, positive principal 94
—, principal 94
—, real infinite 94
dual group of K^+ 245
— — of U_1 247

easier Waring's problem 516
Eisensteinian polynomial 181, 232
elliptic units 137
embedding, complex 42
—, imaginary 42
—, real 42
equivalent quasi-characters 251
— valuations 16

ERH (see Extended Riemann Hypothesis)
essential derivation 163
Euclidean algorithm 120
— domain 120
— field 120, 140, 141
— —, quadratic 121, 122, 140
— ideal 141
Euler product 325
even character 275
Existence Theorem 425
exponent of a field 18
— of a quasi-character 251
— of the class-group 485
exponent-ring 19
Extended Riemann Hypothesis 82, 123, 136, 141, 410, 478, 479, 481, 485, 516
extension, absolute 143
—, fully ramified 229
—, Galois structure 171-175
—, infinite 205, 260, 261
—, of p-adic field 229-244, 260-262
—, ramified 145, 229
—, relative 143
—, tame (tamely ramified) 145, 229
— —, characterization 268
—, totally ramified 229
—, unramified 145, 229
—, — at infinity 145
—, wildly ramified 229

factorization of prime ideals 177, 181-185, 198-200
— — — in Abelian fields 423
— — — in cyclotomic fields 182
— — — in Kummerian extensions 184
— — — in quadratic fields 181
FCS (see full conjugated sets)
Fermat's Last Theorem 71, 125, 517
FGC-rings 39
finite divisors 94
finiteness of $H(K)$ 96, 127
first factor of the cyclotomic class-number 437, 438, 441, 473-475
FLT (see Fermat's Last Theorem)
FN property 12, 37, 39
forms, quadratic 448
—, zeros of 518
Fourier transform 248
fractional ideal 1

—, norm of 12
Frobenius automorphism 379
— symbol 379
full conjugated sets (FCS), of integers 47, 75
— — —, of units 138
fully ramified extension 229
— — prime ideal 148
functional equation of Dedekind zeta-function 326, 404
— — of Dirichlet's L-functions 348, 404
— — of Hecke's zeta-functions 343, 404
— — of zeta-functions of quasicharacters 254, 309
fundamental system of S-units 105
— units 105, 130-134
— — in cubic fields 133
— — in quartic fields 134
— — in real quadratic fields 118, 119, 131-133
— — in sextic fields 134

Galois structure of an extension 171-175
— — of p-adic units 262, 263
— — of rings of integers 173-175, 195-198
— — — — of p-adic integers 262
— — of the class-group 191, 192, 201, 202
— — of units 113-116, 135, 376
Gaussian sums 279, 319, 320, 408
— —, of quadratic characters 283, 319
generalized valuation 93
genus 455, 483, 485
— field 483
— group 455
Goldbach problem 516
Goldstein's conjecture 417
group of ambiguous ideal classes 191
— of ideal classes 34, 40, 41, 94-98, 127, 128, 188-193, 200, 201, 455, 458, 461, 483-487
— — — —, in the narrow sense 94
— — — — mod $\mathfrak{f}$ 95
— of ray classes (mod $\mathfrak{f}$) 95
— — — —, narrow 95
— of residue classes, multiplicative 274, 275, 318
— of roots of unity 100

Haar measure in adeles, ideles 300
— — in p-adic fields 248

Hasse Norm Theorem (HNT) 317, 318
— principle 317
Hecke character 334, 401
— —, conductor 340, 341
— —, normalized 334
— —, of type A, A_0 406
— —, proper 334
— zeta-function 343, 404
— —, functional equation 343, 404
— —, zeros 353, 363, 369
Hensel's Lemma 211, 259
Hilbert class-field 202, 203
HNT (see Hasse Norm Theorem)
Hurwitz constant 82

ideal classes 94-98
— class-group 35, 40, 41
— numbers 125
Ideal Theorem 361, 414, 415
ideals, conjugated 148
—, divisibility 8
—, fractional 1
—, greatest common divisor 8
—, invertible 3
— —, characterization 26
—, lattice of 37
—, least common multiple 8
—, norm 12, 149
—, number of generators 10, 38, 39
— of small norm in a class 96, 129
—, primary 38
—, principal 3
—, product 1
—, unique factorization 7, 36, 37
idele 295, 321
— class-group 297, 321
— group 295
—, principal 296
—, volume 296
index of a field 178, 198
— of an integer 60
— of irregularity of a field 226
— — — of a prime 475
inertia field 272
— group 240, 271
— — of a cyclotomic field 423
infinite prime divisor 94
integers of a field 42
—, p-adic 208

integral basis 55, 78-81
— — for cyclotomic fields 63, 169
— — for pure cubic fields 62
— — for quadratic fields 61
— —, normal (NIB) 174, 195-198
— —, relative (RIB) 55, 401, 417
— closure 5
— element 4
integrally closed domain 5
inversion theorem 524
invertible ideals 3
— —, characterization 26
irreducible element 35, 506
irregular prime 475
irregularity index of a field 226
— — of a prime 475
Iwasawa coefficients 477
— theory 476, 477

Jacobi sum 406
J-field (see CM-field)

Kaplansky's test-problems 40
Köthe ring 40
Krasner's Lemma 214, 260
Kronecker class 416
— equivalence 416
Kronecker's character 437
— constant 74
— limit formula 411
— theorem on diophantine approximations 537
K-theory 128, 129
Kummerian extension 184
Kummer's conjecture 319

Langland's program 204
lattice 536
— determinant 536
LCA (= locally compact abelian group) 520
length of a block 493
— of a factorization 490
Lenstra constant 140
Leopoldt's conjecture 322
level 516
L-function, Artin's 196, 197, 407
—, Dirichlet's 347, 348, 409, 439, 451, 481, 482

—, p-adic 262, 418
lying above 144
— below 144

Main Conjecture 128, 413
matrices 127
Mellin transform 250, 309, 524
minimal polynomial 42
Minkowski constant 98, 129, 140
— unit 115, 135, 375
— —, strong 115
Minkowski's convex body theorem 436
— theorem on linear forms 436
module, annihilator 32
—, Noetherian 2
—, projective 23
—, —, characterizations 25
modules over Dedekind domains, structure 23, 29, 32, 39
monic polynomial 42
monodromy theorem 290
monogenic field 80, 81
multinomial degree 77
multiplication ideal 38
— ring 38

narrow class group 94
NIB (see normal integral basis)
Noetherian module 2
— ring 1, 36-39
non-Archimedean valuation 16, 90
norm, absolute 12
norm of an element 45, 72, 73, 266
— of an ideal 12, 149
— of an ideal class 189
normal integral basis (NIB) 174, 195-198, 422
— order 513
normalized Hecke character 334
— idele character 308
— valuation 193
norm-form 73
number-knot 318

odd character 275
overring 37

p-adic fields 207
— —, characterization 216
— —, extensions 229
— —, —, biquadratic 239
— —, —, fully ramified 229, 232, 234
— —, —, normal 240-244
— —, —, quadratic 237
— —, —, tamely ramified (tame) 229, 234, 235
— —, —, totally ramified 229, 232, 234
— —, —, unramified 229-232
— —, —, wildly ramified 229, 233
— L-functions 262, 418
— logarithm 228, 261
— regulator 322
P-adic topology 18
Pellian equation 118, 131, 517
PID (see principal ideal domain)
p-independent elements 394
Poisson formula 527
Pólya field 83
polynomial, integral-valued 83
— map 83
—, minimal 42
—, monic 42
—, reciprocal 74
power integral basis 64, 80, 81
— residue symbol 394
primary ideal 38
prime ideal, at most tamely ramified 145
— —, degree 145
— —, fully ramified 148
— —, ramified 144
— —, splitting 148
— —, tamely ramified 145
— —, totally ramified 148
— —, unramified 144
— —, wildly ramified 145
Prime Ideal Theorem 360, 369, 414
— — — for ideal classes 360, 369, 414
primitive character of $G(I)$ 274
— — — —, real 277, 283
— —, Hecke 347
— quadratic form 448
— root 275, 318
principal adele 296
— ideal 3
— — domain (PID) 31, 35, 40, 124
Principal Ideal Theorem 203

principal idele 296
— units 223
— —, Galois structure 262, 263
— —, module structure 226
product formula 93, 126
projective module 23
— —, characterization 25
proper Hecke character 334
Prouhet–Tarry problem 516
Prüfer domain 37
pseudo-regular primes 476
pure fields, integral bases 61, 62, 79
PV-numbers 76, 77, 136
—, analogues 77, 322

quadratic fields, class-group 455, 483-485
— —, class-number, asymptotical behaviour 446, 454, 478, 480
— — —, bounds 451, 481
— — —, congruences 473, 474
— — —, divisibility 459, 485, 486
— — —, formula 436, 440, 492
— — —, mean value 482
— —, discriminant 61
— —, Euclidean 121-123, 140
— —, integral basis 61
— —, units 117-120, 131-133
— — with $h = 1$ 185, 200, 446, 463, 465, 479, 488
— — with $h = 2$ 479, 480
— forms, binary 448, 481
— — —, classes 448, 481
quartic fields, discriminant 79
— —, integral basis 79
— —, units 134
quasi-character 521
— of idele group 306
— — — —, exponent 307
— of p-adic fields 247
— — — —, equivalence 251
— — — —, exponent 251
— — — —, ramification degree 248
— — — —, unramified 248
— of R and $\mathfrak{J}$ 522

Ramanujan sum 320
ramification degree of a quasicharacter 248
— field 272
— groups 240, 271, 318

— — of cyclotomic fields 423
— index 144
ramified extension 145, 229
— prime ideal 144
rank of units 106
ray classes 95
— class-field 204
real embedding 43
— primitive character 277, 283
reciprocal polynomial 74
regular discriminant 485
— field 226
— prime 138, 475
— set of prime ideals 355
regulator 106
— of a field 106, 139
— — —, bounds 139, 333, 404
—, p-adic 322
relative discriminant 159
— extension 143
— integral basis (RIB) 55, 401, 417
— units 135
relatively complete field 259
restricted product 528
RIB (see relative integral basis)
ring of integers 42
— — —, Galois structure 173-175, 195-198
— — —, module structure 55, 398, 417
— — —, p-adic 208
— — — —, Galois structure 262
— — — —, module structure 219
root number 196, 408
roots of unity 100, 138

Salem numbers 77
scalar product of zeta-functions 406
second factor of the cyclotomic class-number 437, 438, 473, 474, 476
Serre's conjecture 412
sieves 514
signature group 43
— map 43
— of a field 43, 51
— of an element 43
splitting field 272
— prime ideals 148, 180, 199, 357
— primes in Abelian fields 424
Steinitz class 397, 418

strong approximation theorem 302
sums of squares 44, 514, 515, 516
— of units 138
S-unit 101

tamely ramified (tame) extension 145, 229
— — prime ideal 145
Tauberian theorem 534
test-problems of Kaplansky 40
Theorem of Ankeny–Brauer–Chowla 454, 482
— of Artin 389
— of Chebotarev 382, 415, 416
— of Dirichlet 101, 129, 130
— of Dirichlet–Chevalley–Hasse 101, 130
— of Dobrowolski 46, 74
— of Euler 12
— of Fermat (small) 12
— of Frobenius–Rabinowitsch 463, 488
— of Golod–Shafarevich 203
— of Grunwald–Hasse–Wang 321
— of Hermite 69
— of Ikehara–Delange 534
— of Kronecker 45, 46, 73
— of Kronecker–Weber 289, 320
— of Motzkin 49
— of Noether 171
— of Ostrowski 90
— of Pellet–Stickelberger 153, 193
— of Perron 70
— of Robinson 47
— of Schur 47, 75
— of Siegel–Brauer 442, 477, 478
totally complex field 43
— — number 43
— imaginary field 43
— — number 43
— positive number 43
— ramified extension 229
— — prime ideal 148
— real field 43
— — number 43
trace 45, 72, 73, 173, 266
—, surjectivity 173, 194, 234, 268
trivial valuation 16

UFD (see unique factorization domain)
uniform distribution 514
unique factorization 35, 497, 498, 500, 508, 510, 513

— — domain (UFD)   35
— — of rational integers   498, 510, 513
unit   99
— ideles   296
— rank   106
— theorem   101, 129, 130
units, defect of   107, 134
—, FCS   138
—, fundamental   105, 130-134
—, Galois structure   113-116, 135
—, independent conjugated   113, 115, 135
—, p-adic   208, 233, 261-263
— —, principal   223, 262, 263
—, quadratic fields   117-120, 131-133
—, real   107, 109, 134
—, signatures   111, 134
unramified extension   145, 229
— prime ideal   144

valuation   15, 39
—, Archimedean   16
—, discrete   18
—, generalized   93
— ideal   19
—, induced by an exponent   18
—, non-Archimedean   16
—, normalized   18, 193
— ring   19

valuations, equivalent   16
— of algebraic number fields   90
— of the rational field   22
—, product formula   93, 126
Vandiver's conjecture   476
volume of an idele   296

Waring's problem   514, 515
— —, "Easier"   516
weak approximation theorem   20
wildly ramified extension   229
— — prime ideal   145

zeta-function of an ideal class   411, 412
— — — —, values at integers   412, 413
— of Artin–Hecke   408
— of Dedekind   324, 404
— — —, analytic continuation   326, 404
— — —, functional equation   326, 404
— — — —, approximate   409
— — —, values at integers   412
— — —, zeros   325, 363, 369, 409-411
— of Hecke   343, 404
— — —, functional equation   343, 404
— — —, zeros   353, 363, 369
— of quasicharacters   251
Z_p-extension   476
Z.P.I. rings   37

Author Index

Abrashkin, V. A. 480
Adachi, N. 475, 488
Adibaev, E. K. 472
Agou, S. 199
Agrawal, M. K. 516
Ahern, P. R. 414
Aigner, A. 200, 517
Akizuki, Y. 38
Albert, A. A. 78, 79, 127, 260, 262
Albis-González, V. S. 318
Albu, T. 37, 81, 321
Alibaev, E. K. 201
Allen, S. 513
Amano, K. 318
Amano, S. 262
Amara, H. 134, 200
Amara, M. 74, 77
Amberg, E. J. 79, 487
Amerbaev, V. 77
Amice, Y. 262, 418
Anderson, D. D. 38
Andozhskii, I. V. 261
Andrianov, A. N. 405
Andrukhaev, H. M. 77, 517
Anfert'eva, E. A. 129
Angell, I. O. 78, 133, 201
Ankeny, N. C. 72, 82, 131, 132, 136, 415, 416, 471, 474, 476, 481, 482, 485, 486, 488, 517
Antoniadis, J. A. 480
Apostol, T. M. 195, 320, 404, 409
Appelgate, H. 133
Arai, M. 78, 199
Aral, H. 83
Aramata, H. 407, 408, 541
Archibald, R. G. 514
Archinard, G. 39, 80
Arf, C. 262
Arkhipov, G. I. 518
Armbrust, M. 518

Armitage, J. V. 135, 136, 137, 196, 322, 408, 411, 485, 540, 542
Arnaudon, M. 318
Arndt, F. 194, 483
Arnold, J. T. 37
Arpaia, P. J. 141
Artin, E. 20, 39, 72, 126, 127, 130, 132, 135, 194, 203, 205, 321, 407, 408, 415, 416, 417, 471, 474, 518, 541
Arutyunyan, L. Z. 263
Arwin, A. 83, 127, 133
Asano, K. 37, 39
Asenov, E. K. 471
Auslander, L. 320
Auslander, M. 193
Avanesov, É. T. 130, 133
Ax, J. 318, 322, 518
Axer, A. 514
Ayoub, C. 130, 261, 319
Ayoub, R. G. 404, 414, 415, 479, 482, 488, 517
Azuhata, T. 486, 487, 542

Baayen, P. C. 512
Babaev, G. 413, 415
Babaitsev, V. A. 477
Bachman, G. 199
Bachmann, P. 130, 487
Baeza, R. 516
Baibulatov, R. S. 514
Baily, A. M. 79, 471, 540
Baker, A. 479, 485, 538, 543
Baldisseri, N. 141
Ballieu, R. 195
Bambah, R. P. 80
Baniuk, L. 201
Barban, M. B. 413, 482
Barkan, P. 472
Barner, K. 320, 406, 412
Barnes, E. S. 140

Barnes, F. W. 516
Barrucand, P. 80, 136, 200, 201, 204, 408, 483, 486
Barsky, D. 83, 262, 418
Bartels, H. J. 72, 318
Bartz, K. 409, 410, 413
Bashmakov, M. I. 263
Bass, H. 39, 128, 129, 130, 138
Bateman, P. T. 74, 195, 409, 479, 482, 515
Bauer, H. 201, 484, 486
Bauer, M. 80, 193, 194, 195, 198, 200, 260, 262, 317, 416
Bayer, P. 413
Bazylewicz, A. 73, 74
Beach, B. D. 133, 201
Beaumont, R. A. 81
Bebbe, E. 483
Bedocchi, E. 77
Beeger, N. G. W. H. 471
Behnke, H. 73
Behrbohm, H. 140, 488
Beiter, M. 195
Belcher, P. 138, 541
Bender, E. A. 127
Bennett, A. A. 127
Benson, C. T. 130
Benzaghou, B. 139
Berg, E. 140
Bergè, A.-M. 197, 262
Berger, A. 412
Berger, R. 193
Berger, T. R. 195
Bergmann, G. 71, 133, 134
Bergström, H. 78, 84, 320, 472
Bergum, G. E. 77
Berlekamp, E. R. 198
Berndt, B. C. 319, 320, 404, 405, 411, 415, 472, 514
Bernstein, L. 131, 133, 134
Bertin, M. J. 77
Bertness, C. H. 77
Bertrandias, F. 197, 262, 322
Bertrandias, J.-P. 262
Berwick, W. E. H. 78, 79, 133, 134, 136
Besicovitch, A. S. 417
Bessis, D. 73, 84
Beyer, G. 319
Bhandari, S. K. 127
Bhaskaran, M. 199, 200, 515, 516

Bickmore, C. E. 131
Bilhan, M. 416, 420
Billevich, K. K. 79, 127, 130, 133, 134, 200
Birch, B. J. 81, 129, 317, 406, 479, 515, 518, 541
Bird, R. F. 417
Bitimbaev, T. S. 201, 472
Blanksby, P. E. 73, 74, 75, 84, 134
Blichfeldt, H. F. 82
Bloom, D. M. 195
Bloom, J. R. 477
Bobrovskii, V. P. 471
Bochkarev, D. P. 195
Bohr, H. 409
Bölling, R. 472
Bond, R. J. 203, 205
Borel, A. 128, 204, 412
Borevich, Z. I. 72, 73, 126, 203, 261, 262, 263, 481
Bougaut, B. 141
Bouvier, L. 135, 201, 487
Bovey, J. D. 515
Boyd, D. W. 74, 76, 77, 485, 542
Brandal, W. 39
Brandis, A. 131
Brandler, J. A. 132
Brattström, G. 406
Brauer, A. 77, 140
Brauer, R. 193, 407, 408, 477, 482, 487, 518, 541
Bredikhin, B. M. 414
Bremner, A. 317
Brentjes, A. J. 133
Briggs, W. E. 485
Brill, A. 59
Brillhart, J. 198, 416
Brindza, S. 517
Brinkhuis, J. 129, 195, 196, 197
Brizolis, D. 83
Broere, J. 131, 201
Browkin, J. 129, 203, 205, 518, 542
Brown, E. 132, 200, 480, 486
Brown, K. S. 413, 475
Brownawell, W. D. 518
Bruckner, G. 199, 200, 517
Brumer, A. 82, 135, 203, 205, 322, 486, 487
Brunotte, H. 74, 75, 133, 135, 136
Buchner, P. 83

Buchsbaum, D. A.   193
Buell, D. A.   132, 201, 480
Buhler, J. P.   407
Buhr, P. A.   131
Bulenov, A.   471
Bullig, G.   75, 133
Bulota, K.   409, 410, 415
Bumby, R. T.   513
Bundschuh, P.   465, 479
Bungers, R.   198
Burgess, D. A.   481
Burke, J. R.   514
Burnside, W.   517
Bushnell, C. J.   195, 405, 408
Büsser, A. H.   200
Butts, H. S.   37, 38, 77, 481, 513

Cahen, P. J.   83
Callahan, T.   74, 84, 479, 486, 487
Çallial, P. F.   411, 488
Calloway, J.   81
Camion, P.   37, 38, 41
Canals, I.   78
Candiotti, A.   322, 477
Cantor, D. G.   74, 75, 77, 138, 139, 317, 322, 539
Carlitz, L.   77, 79, 81, 133, 193, 195, 198, 319, 412, 470, 473, 474, 475, 476, 490, 511, 541
Carroll, J. E.   263, 477
Cartier, P.   79, 200, 413, 418
Cassels, J. W. S.   72, 74, 76, 82, 83, 127, 133, 138, 140, 195, 199, 201, 205, 260, 263, 299, 317, 318, 319, 320, 321, 404, 407, 416, 471, 536, 537
Cassou-Noguès, P.   404, 413, 418
Cassou-Noguès, Ph.   196, 197
Castela, C.   487
Cauchy, A.   319
Cavior, S. R.   319
Cayley, A.   201
Chabauty, C.   138, 517
Chabert, J. L.   83
Chamfy, C.   77
Chandrasekharan, K.   405, 409, 411, 415, 514
Chang, J. A.   198
Chang, S. M.   203
Chao, N. L.   259

Charin, V. S.   131
Chase, S. U.   36, 39
Chatelain, D.   197
Châtelet, A.   72, 79, 84, 127
Chatland H.   122, 140, 201
Chebotarev, N. G.   126, 194, 198, 317, 320, 415, 416, 417, 488
Chebyshev, P. L.   414
Cheema, M. S.   517
Chen-Jing-Run   482
Cheo, L.   77
Chevalley, C.   39, 72, 101, 130, 138, 195, 201, 202, 263, 318, 321
Chew, K. L.   39
Chidambaraswamy, J.   77
Childs, L. N.   197, 202
Chinburg, T.   135, 197, 408, 516
Chowla, P.   411, 474, 488, 516, 517
Chowla, S.   127, 131, 132, 133, 136, 138, 139, 195, 200, 319, 409, 411, 474, 476, 478, 479, 480, 481, 482, 483, 485, 486, 488, 516, 517, 518, 543
Christofferson, S.   130
Christy, D.   517
Chudakov, N. G.   129, 404, 409, 478, 479
Chulanovskii, I. V.   415
Church, R.   198
Cioffari, V. G.   140
Claborn, L.   40, 41, 511
Clarke, L. E.   140
Coates, J.   128, 129, 137, 413, 418, 475, 477
Cofrè-Matta, A.   77, 84
Cohen, E.   514, 517
Cohen, G. L.   414, 415
Cohen, H.   201, 412
Cohen, I. S.   36, 37, 38, 40
Cohn, H.   79, 82, 131, 133, 134, 136, 200, 201, 204, 404, 474, 476, 479, 481, 483, 484, 486, 515
Cohn, J. H. E.   132, 139, 515
Colliot-Thélène, J. L.   317
Connell, I. G.   139, 486, 488, 516
Conner, P. E.   73
Conway, J. H.   195, 198
Cooke, G. E.   141, 204, 474
Cooper, A. E.   201
Coquet, J.   76
Coray, D.   317
Cormack, G.   133

Cornell, G. 128, 201, 202, 203, 476, 483, 485, 486, 487
Corput, J. G. van der 415
Cougnard, J. 196, 197
Cowles, J. 127, 483
Cowles, M. J. 127, 200, 483, 485
Craig, M. 484
Cresse, G. H. 481
Cross, J. T. 318
Cugiani, M. 140
Cuoco, A. A. 477
Currie, J. D. 474
Cusick, T. W. 133, 139
Cvetkov, V. M. 193, 205
Czarnowski, R. 411
Czogała, A. 512, 519

Dade, E. C. 81, 127, 138, 513
Dai, Z. D. 516
Dalen, K. 193
Damey, P. 487
Danilov, A. N. 514
Danilov, G. V. 132
Danilov, L. V. 132
Davenport, H. 79, 82, 122, 129, 138, 140, 141, 320, 471, 478, 479, 494, 518, 540, 542
David, M. 134
David, P. 263
Davies, D. 409, 411
Davis, D. 137
Davis, E. D. 37, 81
Davis, H. T. 473
Davis, R. D. 262
Davis, R. W. 472
Decomps-Guilloux, A. 322
Dedekind, R. 35, 38, 64, 72, 78, 79, 80, 123, 125, 126, 127, 141, 176, 193, 194, 198, 199, 404, 407, 414, 472, 481
Degen, C. F. 131
Degert, G. 131
Delange, H. 324, 534, 535
Delaunay, B. N. 73, 77, 78, 79, 82, 84, 126, 129, 133, 135, 138, 320
Deligne, P. 408, 413, 418
Delsarte, S. 471
Demushkin, S. P. 261
Dem'yanov, V. B. 518
Denenberg, C. G. 131
Dénes, P. 134, 139, 201, 202, 476, 517

Dennis, R. K. 129, 263
Descombes, R. 83
Despujols, P. 132
Deuring, M. 195, 204, 262, 416, 472, 478, 479
Diamond, J. 262, 418
Diaz y Diaz, F. 82, 484
Dickson, L. E. 72, 131, 140, 472, 479, 481
Diederichsen, F. E. 195
Diekert, V. 261
Dinghas, A. 73
Dirichlet, P. G. Lejeune- 71, 72, 99, 101, 129, 130, 318, 319, 404, 409, 413, 472, 486, 487, 489, 517
Disse, A. 261
Diviš, B. 77
Dobrowolski, E. 46, 74, 539
Dodson, M. 515, 518
Donkar, E. N. 515
Doubrére, M. 77
Drakokhrust, Ya. A. 318
Draxl, P. K. J. 201, 407
Dress, A. 199, 317, 517
Dress, F. 77
Dribin, D. M. 84, 200
Drinfel'd, V. G. 320
Dubois, D. W. 141
Dubois, E. 134
Duffin, R. J. 76
Dufresnoy, J. 76, 77
Dujčev, J. 199
Dummit, D. S. 80
Dunwoody, M. J. 129
Dürbaum, H. 259
Duval, D. 135
Dwork, B. 408
Dwyer, W. G. 129
Dzewas, J. 515
Dzhiemuratov, U. 415

Eakin, P. 39, 40
Earnest, A. G. 485
Eda, Y. 78, 414, 514, 515, 516
Edgar, H. M. 404, 417
Edgorov, Zh. 404
Edwards, H. M. 71, 125, 126
Egami, S. 129, 140, 319
Eggan, L. C. 83
Eichler, M. 72, 204, 472, 475

Eisenbeis, H. 201
Eisenstein, G. 71, 127, 404
Eljoseph, N. 515
Elliott, P. T. D. A. 417, 482
Ellison, F. 518
Ellison, W. J. 479
El-Musa, A. J. 263
Elsner, L. 134
Emde Boas, P. van 512
Endler, O. 39, 199
Endô, A. 133, 484, 486
Endo, M. 195
Engstrom, H. T. 198, 199
Ennola, V. 75, 77, 134, 137, 138, 140, 195, 488, 539
Epstein, P. 132, 479
Erdös, P. 140, 195, 411, 479, 482
Ernvall, R. 476
Estermann, T. 319, 478
Estes, D. R. 481, 485, 515
Euler, L. 71, 194, 412, 485
Evans, R. J. 77, 132, 319, 320, 472
Evertse, J. H. 81, 138
Evteev, V. P. 415

Faddeev, D. K. 73, 77, 78, 82, 84, 126, 127, 129, 133, 135, 138, 261, 263, 514
Fainleib, A. S. 482
Faith, C. 40
Faltings, G. 517
Fanta, E. 413, 414
Favard, J. 75
Federer, L. J. 477
Fein, B. 516
Feit, W. 260
Fekete, M. 75
Feldman, N. I. 479
Felsch, V. 195
Feng Keqin 137, 417, 473, 486, 512, 519
Ferguson, L. B. O. 75
Ferrero, B. 418, 477
Ferton, M. J. 262
Fields, J. C. 126
Fischer, W. 409
Fitting, H. 193
Fjellstedt, L. 416, 517
Flanders, H. 72, 126, 193, 416
Fleischer, I. 259
Flett, T. M. 413

Flexor, M. 79
Fluch, W. 482, 514
Fogels, E. 409, 410, 411, 413, 415, 472, 513, 517
Fomenko, O. M. 407, 482, 515
Fontaine, J. M. 262
Foote, R. 203
Ford, L. R. 83
Forman, W. 414
Fossum, R. M. 40, 193
Foster, L. L. T. 80
Fraenkel, A. 36, 259
Franz, W. 39
Frei, G. 134
Freitag, A. 77
Fresnel, J. 404, 412, 413, 418
Frey, G. 126, 128, 201, 260, 486
Fried, E. 77
Friedlander, E. M. 129
Friedlander, J. B. 77, 413, 414, 480, 514
Friedman, E. C. 477
Friedrich, R. 517
Frobenius, G. 390, 415, 463, 488
Fröberg, C.-E. 319
Fröhlich, A. 72, 127, 128, 135, 136, 137, 196, 197, 199, 200, 201, 202, 203, 205, 260, 263, 299, 318, 321, 322, 407, 408, 416, 417, 471, 483, 485, 486, 487, 488, 540, 541, 542
Fryska, T. 410
Fuchs, L. (XIX C.) 125, 471
Fuchs, L. (XX C.) 37, 40, 78
Fueter, R. 127, 137, 196, 204, 320, 471, 472, 475, 517
Fujii, A. 411
Fujisaki, G. 194, 405, 417, 516
Fujiwara, M. 317
Funakura, T. 408
Furtwängler, Ph. 82, 126, 191, 193, 201, 203, 317, 413, 414, 488
Furuta, Y. 132, 200, 202, 203, 205, 321, 483, 486
Furuya, H. 202, 203, 486

Gadia, S. K. 476
Gaigalas, E. 407
Galkin, V. M. 405, 473
Gallagher, V. P. 73, 322
Gandhi, J. M. 475
Garakov, G. A. 198

Garbanati, D. A. 72, 134, 137, 142, 200, 202, 318, 542
Garland, H. 129
Gassmann, F. 405, 416, 417
Gauss, C. F. 40, 71, 127, 129, 194, 198, 199, 200, 319, 448, 455, 478, 482, 483, 485, 486, 488, 541
Gautheron, V. 79
Gauthier, F. 199
Gay, D. 131
Gebhart, H.-M. 128
Gechter, J. 408
Gelbart, S. 204, 407
Gelfond, A. O. 77, 409, 479, 482
Gérardin, P. 407
Gerig, S. 77
Gerlovin, É. L. 263
Geronimo, J. S. 73, 84
Gerst, I. 416, 417
Gerth, F. III 128, 202, 318, 473, 476, 477, 484, 485, 486
Geyer, W. D. 126, 128, 205, 260, 486
Gilbarg, D. 263
Gilbert, J. R. 38
Gilbert, W. J. 81
Gillard, R. 137, 322, 477
Gillett, J. R. 514
Gilmer, R. W. 37, 38, 39, 319
Giorgiutti, J. 40
Girstmair, K. 78
Glaisher, J. W. L. 486
Godement, R. 405
Godwin, H. J. 75, 78, 79, 82, 133, 135, 141, 198, 201
Gogia, S. K. 204, 478, 483
Gold, R. 317, 318, 320, 477, 483, 486
Goldfeld, D. M. 408, 411, 478, 480
Goldman, O. 37
Goldstein, L. J. 72, 80, 412, 413, 414, 417, 472, 474, 480, 483
Goldstine, H. H. 319
Golod, E. S. 82, 203
Golomb, S. W. 195
Golubeva, E. P. 132
Good, A. 409, 415
Gordeev, N. L. 262
Gordon, B. 516, 517
Gordover, G. 482
Gorn, S. 133

Gorshkov, D. S. 127
Goss, D. 320
Götzky, F. 515
Gout, G. 261
Graham, S. W. 486
Gramm, S. L. 83
Grandet-Hugot, M. (see also M. Hugot) 76, 77, 322
Grandjot, K. 171, 194
Grant, H. S. 202
Gras, G. 129, 134, 137, 200, 202, 204, 205, 322, 418, 477, 484, 485, 486, 487
Gras, M. N. (see also M. N. Montouchet) 75, 80, 133, 134, 135, 137, 197, 200, 201, 203, 485, 486
Grave, D. A. 126
Gray, J. 518
Grebenyuk, D. G. 79
Greenberg, M. J. 320, 518
Greenberg, R. 137, 413, 418, 475, 477
Greenleaf, N. 518
Greiter, G. 73, 134
Grell, H. 39, 193
Griffin, M. P. 38
Grimm, G. 479
Grishin, A. V. 131
Grölz, W. 416
Gronwall, T. H. 478
Gross, B. H. 406, 418, 477, 480
Grossman, E. H. 77, 138, 139
Grosswald, E. 412, 478, 479, 485
Grotz, W. 514
Grube, F. 485
Grunwald, W. 320, 321
Guan, C. W. 261
Gudiev, A. H. 201
Guého, M. F. (see also M. Vigneras) 413
Guerrier, W. J. 195
Gundlach, K.-B. 72, 412
Gunji, H. 83
Gupta, H. 201
Gurak, S. 318, 415, 483
Gurevich, M. M. 321, 405
Gut, M. 126, 200, 204, 261, 405, 471, 474, 476, 481, 486
Güting, R. 133
Guy, M. J. T. 317
Györy, K. 72, 73, 80, 81, 134, 138, 204, 540, 541

Haberland, K. 205, 471
Hafner, J. L. 514
Hafner, P. 133
Haggenmüller, R. 132
Hall, M. 80
Hall, N. A. 485
Halter-Koch, F. 75, 77, 133, 134, 135, 136, 194, 195, 200, 261, 318, 417, 483, 484, 485, 487, 512, 519
Hamamura, M. 204, 483
Hamburger, H. 404, 405
Hancock, H. 80
Haneke, W. 478, 488
Hanly, V. 417
Hao, F. H. 517
Happle, W. 487
Hardman, N. R. 77
Hardy, G. H. 76, 199, 409, 411
Hardy, J. 515
Harris, B. 129
Hartung, P. 411, 480, 486
Haselgrove, C. B. 411
Hasse, H. 72, 78, 79, 80, 84, 101, 123, 128, 131, 133, 134, 135, 136, 137, 141, 199, 200, 201, 203, 204, 259, 261, 262, 263, 317, 319, 320, 321, 404, 407, 408, 412, 415, 416, 471, 472, 473, 476, 483, 484, 486, 488, 514, 516, 518, 541, 542
Hatada, K. 418
Hattori, A. 37
Hausman, M. 77
Hausner, A. 517
Hayashi, H. 200, 474, 486
Hayashi, M. 513
Hayes, D. R. 320
Hays, J. H. 37
Hazewinkel, M. 263, 320
Hazlewood, D. G. 514
Heath-Brown, D. R. 319, 410, 485, 518, 543
Hecke, E. 72, 125, 194, 201, 204, 302, 316, 319, 320, 322, 329, 341, 404, 405, 410, 411, 412, 413, 414, 415, 417, 420, 472, 473, 475, 478, 481, 487
Heegner, K. 479
Heider, F. P. 203, 318
Heilbronn, H. 79, 122, 140, 407, 411, 471, 478, 479, 540, 541
Heinzer, W. 40

Heller, A. 128
Hemer, O. 517
Hendy, M. D. 132, 133, 201, 485, 488, 543
Henniart, G. 260
Hensel, K. 65, 80, 126, 193, 194, 195, 198, 199, 200, 259, 260, 261, 262, 264, 317, 318
Hensley, D. 83, 415
Herbrand, J. 126, 135, 193, 200, 203, 262, 318, 322, 475, 488
Herglotz, G. 412, 487
Hermite, C. 52, 69, 81, 130, 133, 487
Herz, C. S. 204, 483
Hettling, K. F. 128, 129
Heupel, W. 415
Hewitt, E. 520
Hey, F. 405
Hickerson, D. R. 132
Hida, H. 412
Hightower, C. J. 83
Higman, G. 196
Hijikata, H. 317
Hilano, T. 410
Hilbert, D. 36, 72, 80, 84, 125, 126, 136, 138, 139, 193, 194, 196, 199, 203, 204, 205, 262, 317, 318, 320, 417, 470, 475, 487, 514, 517
Hinz, J. G. 319, 409, 410, 413, 414, 514, 517
Hiramatsu, T. 200
Hironaka-Kobayashi, Y. 205
Hirst, K. E. 132
Hirzebruch, F. 413, 488
Hlawka, E. 83
Hochschild, G. 205, 263
Hock, A. 465, 479
Hodges, W. 36
Hoechsmann, K. 205
Hoffmann, B. 129
Hoffstein, J. 404, 411, 474, 478, 480, 481, 482
Hofreiter, N. 82, 83, 140
Hölder, O. 195, 320
Holzapfel, R. P. 261
Holzer, L. 72, 486
Honda, T. 204, 486, 488, 489
Hooley, C. 139, 482
Horie, K. 202
Horie, M. 483
Hornfeck, B. 416

Hsia, J. S. 195, 515
Hsü, C. S. 40
Hsü, T. N. 203
Hua, Loo-Keng 77, 132, 139, 140, 451, 481
Huard, J. G. 80
Huckaba, J. A. 37
Hughes, I. 135, 542
Hugot, M. (See also M. Grandet-Hugot) 77
Hull, R. 84
Humbert, P. 485
Hunter, J. 73, 75, 82, 131
Hurrelbrink, J. 128, 129
Hurwitz, A. 82, 96, 126, 127, 129, 404, 412, 415, 474
Hurwitz, J. 83
Hushvaktov, M. 471
Huxley, M. N. 410, 514
Hyyrö, S. 474

Ichimura, H. 486, 487, 542
Idt, J. 262
Iimura, K. 128, 134, 202, 486, 487
Ikeda, M. 205
Ikehara, S. 534
Inaba, E. 200, 260, 485, 487, 488
Ince, E. L. 201
Ingham, A. E. 410
Inkeri, K. 140, 473, 488, 517
Inoue, H. 72
Ireland, K. F. 72
Irfan, M. 37
Isaacs, I. M. 77
Iseki, K. 480, 514
Ishida, M. 133, 135, 136, 137, 408, 483, 486, 487
Ishikawa, T. 37
Iskovskikh, V. A. 317
Ito, H. 200
Iwasaki, K. 471
Iwasawa, K. 130, 131, 135, 201, 202, 205, 261, 263, 321, 322, 406, 418, 472, 473, 476, 477, 487, 488
Iwata, H. 260
Iyanaga, S. 72, 133, 203, 483

Jacob, G. 83
Jacobi, C. G. J. 71, 133, 134, 406, 489
Jacobinski, H. 197, 262

Jacobson, B. 138, 541
Jacobsthal, E. 75, 138, 195, 517
Jacquet, H. 405
Jakovlev, A. V. 261, 263
Janssen, U. 261, 263
Janusz, G. 72
Jarden, M. 261, 416
Jaulent, J. F. 129, 135, 196, 197, 198, 202, 477, 487
Jeans, N. S. 133
Jehne, W. 138, 203, 317, 318, 409, 416, 417, 477, 487, 543
Jenkins, E. D. 141
Jenner, W. 405
Jensen, C. U. 37, 38, 78, 132, 133, 406
Jensen, K. L. 475
Johnson, D. 410
Johnson, E. W. 37
Johnson, W. 477
Joly, J. R. 130, 482, 515, 516
Jones, A. J. 138, 195
Jones, B. W. 481
Jordan, J. H. 77, 413, 514
Joris, H. 320, 405, 415
Joshi, P. T. 482
Juricic, H. 201
Jushkis, Z. 514
Justin, J. 195
Jutila, M. 410, 482

Kaczorowski, J. 491, 512, 542
Kähler, E. 193
Kalinka, V. 515
Kallen, W. van der 129
Kalnin, I. M. 409, 413, 414, 415
Kambayashi, T. 261
Kaneiwa, R. 83
Kanemitsu, S. 481, 514
Kanno, T. 205
Kanold, H. J. 195
Kantz, G. 141
Kaplan, P. 132, 134, 204, 474, 484, 486
Kaplansky, I. 36, 37, 39, 40, 259, 481
Karatsuba, A. A. 415, 514, 518
Karibaev, S. K. 471
Kátai, I. 81
Kataoka, T. 131, 194
Katayama, K. 321, 412
Katz, N. M. 418

Kaufman, R. M. 409, 415
Kawada, Y. 194, 203, 205, 261, 262
Kazandzidis, G. S. 195
Keating, M. 196
Keller, O. H. 194
Kelly, J. B. 76
Kempfert, H. 199, 203
Kenku, M. A. 479
Kennedy, R. E. 40
Kenzhebaev, S. 471
Kersey, D. 137
Keshtov, R. A. 263
Kessler, I. 543
Kida, Y. 477
Kimura, N. 412, 474
Kimura, T. 202
King, H. 473
Kinohara, A. 194
Kirmse, J. 515
Kiselev, A. A. 472, 474
Kisilevsky, H. 80, 203, 204, 477, 484, 485
Kitaoka, Y. 201, 479, 485
Kitayama, T. 516
Kleboth, H. 476
Kleiman, H. 81, 197, 198, 260
Klingen, H. 412
Klingen, J. 136
Klingen, N. 405, 417
Klobe, W. 195
Kloosterman, H. D. 320
Kloss, K. E. 488
Kløve, T. 517
Knapowski, S. 415, 478
Knebusch, M. 322
Kneser, H. 77
Kneser, M. 417, 481
Knuth, D. E. 199
Ko, Chao 140
Kobayashi, M. 321
Kobayashi, S. 202, 484, 486, 487
Kobayashi, Y. 37
Koblitz, N. 261, 262, 418
Koch, H. 81, 126, 203, 204, 205, 261, 263, 486
Kochen, S. 518
Koksma, J. F. 83
Kolster, M. 128
Komatsu, K. 79, 80, 205, 321, 405, 417
Kondakova, L. F. 410

König, R. 481
Konno, S. 412
Korchagina, V. I. 415
Körner, O. 485, 515, 517
Környei, I. 415
Koshlyakov, N. S. 412
Kostandi, G. 195
Kostrikin, A. J. 203
Köthe, G. 40
Kovács, B. 81
Koval'chik, F. B. 410, 415
Kowalsky, H.-J. 259
Koyama, T. 37
Krakowski, F. 126
Kramer, K. 322, 477
Krasner, M. 195, 204, 260, 262, 263, 318, 475
Krätzel, E. 319
Krause, U. 512, 519
Krompiewska, E. 514
Kronecker, L. 45, 59, 73, 81, 125, 126, 127, 130, 142, 194, 195, 198, 319, 320, 390, 411, 416, 438, 472, 475, 483
Kroon, J. P. M. de 514
Krull, W. 36, 38, 39, 41, 81, 126, 141, 199, 262
Kruyswijk, D. 512
Kubert, D. 137, 406, 472
Kubilius, I. P. 411, 415
Kubo, K. 37
Kubokawa, Y. 483
Kubota, K. K. 83, 540
Kubota, T. 134, 136, 205, 262, 319, 320, 412, 418, 471, 487
Kudo, A. 474, 475, 477, 487
Kühnova, J. 473
Kuipers, L. 514
Kulkarni, R. S. 78
Kummer, E. E. 40, 72, 125, 127, 130, 134, 136, 137, 138, 176, 198, 199, 202, 319, 438, 471, 472, 473, 474, 475, 488
Kunert, D. 320
Kuniyoshi, H. 193, 203, 205
Kunz, E. 193
Kurmanalin, H. 471
Kuroda, S. 129, 131, 134, 136, 200
Kuroda, S. N. 194, 200, 202, 408, 477, 483, 485, 487
Kurokawa, N. 406

Kürschak, J. 39, 126, 259, 260
Kurshan, R. P. 77
Kutsuna, M. 131, 141, 488, 543
Kuzmin, L. V. 203, 205, 261, 322, 477
Kwon, S. H. 79

Labesse, J. P. 407
Labute, J. 261
Lagarias, J. C. 132, 138, 200, 408, 413, 416, 484, 540
Lagrange, J. L. 127, 131, 319
Lai Dyk Thin' 514
Lakein, R. B. 77, 129, 134, 141, 200, 201
Lakkis, K. 408, 409
Lal, S. 517
Lam, T. Y. 128, 516
Lamprecht, E. 203, 319
Lánczi, E. 200
Landau, E. 72, 81, 127, 139, 152, 193, 194, 319, 404, 409, 410, 411, 413, 414, 415, 472, 478, 479, 481, 514, 515, 516
Landherr, W. 317
Landsberg, G. 79, 320
Lang, H. 133, 412, 474, 487, 488
Lang, S. 72, 136, 137, 202, 404, 415, 471, 472
Lang, S. D. 476
Langlands, R. P. 407, 408
Lardon, R. 512
Laska, M. 518
Lasker, E. 36, 38, 193
Latham, J. 542
Latimer, C. G. 127, 134, 135, 199, 200, 488
Laubie, F. 200, 262
Lavrik, A. F. 139, 404, 409, 482
Lawn, S. 39
Lawton, W. 74
Laxton, R. R. 518
Lazami, F. 76
Leahey, W. J. 138, 204, 515
Lebesgue, V. A. 194
Ledermann, W. 83
Lediaev, J. P. 37
Lednev, N. A. 134, 135
Lee, K. C. 413
Lee, M. P. 205
Leedham-Green, C. R. 40, 41
Leep, D. 518

Legendre, A. M. 131
Lehmer, D. H. 73, 74, 131, 195, 473, 479, 539
Lehmer, E. 132, 195, 319, 486
Lemmlein, V. G. 141, 488
Lenskoi, D. N. 514
Lenstra, H. W. Jr. 129, 138, 140, 141, 200, 201, 416, 542
Lenz, H. 515
Leon, M. J. de 411, 480
Leonard, P. A. 132, 199, 481, 486
Leopoldt, H. W. 80, 134, 135, 196, 197, 261, 262, 320, 322, 412, 418, 472, 483, 487, 541
Lepistö, T. 473, 474, 478, 480
Leptin, H. 405
Lerch, M. 473, 474, 479
Lesev, V. D. 262, 263
Lettl, G. 513
Leutbecher, A. 140, 141
Leveque, W. J. 83, 518
Levesque, C. 134
Levi, F. 194
Levin, B. V. 413, 478, 514
Levitz, K. B. 38, 39
Levy, L. S. 37, 38, 41
Lévy, P. 195
Lewin, J. 81
Lewis, D. J. 83, 416, 518, 540
Lewittes, J. 319
Liang, J. J. 80, 81, 82, 136, 194, 204
Liardet, P. 76, 83, 540
Lichtenbaum, S. 129, 406, 413, 418
Lienen, H. v. 132
Linden, F. J. van der 136, 141, 201
Linfoot, E. 479
Linnik, Yu. V. 409, 415, 478, 479, 482, 488
Lippmann, R. A. 485
Littlewood, J. E. 409, 482
Litver, E. L. 79, 487
Livingston, M. 543
Llorente, P. 78, 79, 80, 199
Lloyd-Smith, C. W. 75
Lo, S. K. 514
Loh, H. K. 318
Lohoué, N. 76
Long, R. L. 72, 418
Loonstra, F. 40
Lorenz, F. 195, 318

Louboutin, R. 74, 539
Low, M. E. 411, 479
Loxton, J. H. 73, 84, 134, 138, 319, 518
Lu, H. W. 488
Lubelski, S. 81, 320, 481, 487, 517
Lubin, J. 263, 320
Luthar, I. S. 78, 80, 204, 415, 478, 483

Maass, H. 515
Macaulay, F. S. 36
Madan, M. L. 80, 139, 141, 205, 486, 487
Madden, D. J. 204
Magnus, W. 203
Mahler, K. 74, 78, 83, 127, 130, 138, 261, 478, 481
Maier, E. A. 83
Maikotov, N. R. 471
Mäki, S. 134, 471
Maknys, M. 411, 415
Mallik, A. 480, 483, 488
Manin, J. I. (Manin, Yu. I.) 262, 317, 418
Mann, H. B. 37, 38, 41, 72, 77, 78, 193, 200, 201, 416, 417
Marcus, D. A. 72
Marinina, S. F. 411
Markanda, R. 141
Marke, P. W. 405
Markoff, A. 79, 84, 199
Markshaitis, G. N. 205
Marsh, R. W. 198
Marshall, M. A. 262
Martel, B. 262
Martin, J. N. 418
Martinet, J. 78, 79, 80, 82, 140, 196, 197, 198, 199, 203, 407, 408, 417, 481, 487
Maruno, T. 262
Masley, J. M. 141, 473, 474, 480
Massy, R. 260
Masuda, K. 318, 321, 406
Mathews, G. B. 78
Matlis, E. 37, 38
Matsuda, R. 513, 514
Matsumoto, H. 129
Matsumura, N. 135, 203
Matthews, C. R. 319
Matuljauskas, A. 409
Matusita, K. 36, 37
Mauclaire, J. L. 320, 513
Maurer, D. 73, 79, 198, 200, 322

Maus, E. 200, 262
Mautner, F. I. 415
Maxwell, G. 518
May, W. 130, 131
Mayer, J. 82
Mazur, B. 128, 137, 413, 475
McAuley, M. J. 75
McCarthy, P. J. 72
McCluer, C. R. 83, 134, 200, 416
McCulloh, L. R. 77, 197, 206, 318, 403, 417
McDaniel, W. L. 77, 195
McDuffee, C. C. 78, 127, 141
McEliece, R. J. 198, 199
McFarland, R. L. 518
McFeat, R. B. 322
McGettrick, A. D. 319
McKenzie, R. E. 202, 262, 417
McLane, S. 126, 200, 259, 260
McQuillan, D. L. 83, 135, 205, 473
Mead, D. G. 512
Medvedev, P. A. 518
Megibben, C. 37
Meinardus, G. 517
Meissner, G. 78
Meissner, O. 514
Mendès-France, M. 76
Merriman, J. R. 81, 541
Mertens, F. 126, 319, 320, 482, 483
Mestre, J. F. 140, 484
Metsänkylä, T. 418, 473, 474, 475, 476, 477, 480, 482, 486
Meulenbeld, B. 83
Meyer, C. 412, 472, 479, 480
Meyer, W. 73, 131, 321
Meyer, Y. 76
Midgarden, B. 39
Mignotte, M. 74, 77
Miki, H. 205, 261, 262, 321, 477, 488
Milgram, R. J. 137
Mills, W. H. 417
Milnor, J. 128, 129
Min, S. H. 140
Mines, R. 126
Minkowski, H. 67, 81, 82, 96, 113, 127, 129, 130, 133, 135, 536
Mirimanoff, D. 138, 476, 517
Mitchell, H. H. 481, 488
Mitsui, T. 410, 414, 415, 517

Miyake, K. 203, 321, 322
Miyata, T. 196
Miyata, Y. 80, 262
Mohanty, S. P. 517
Moine, J. M. 484
Molk, J. 130
Möller, H. 195, 485, 488, 543
Mollin, R. A. 135, 195, 486, 542
Monsky, P. 477
Montgomery, H. L. 74, 411, 416, 474, 476, 479, 480, 483, 518
Montouchet, M. N. (see also M. N. Gras) 137, 201, 486
Moore, C. 129
Moore, M. E. 41
Mordell, L. J. 81, 129, 132, 133, 138, 317, 319, 320, 409, 417, 472, 473, 478, 479, 489, 515, 517, 518
Moreno, C. J. 200, 319
Mori, S. 36, 37, 38
Môri, Y. 129
Morikawa, R. 133, 135, 541
Morishima, T. 475, 476
Morita, Y. 262, 418
Moriya, M. 37, 126, 139, 194, 199, 202, 204, 416, 483, 485, 486
Moroz, B. Z. 407
Morton, P. 132, 484, 486
Moser, C. 476, 482, 485, 515, 516
Moser, N. 135, 136, 201, 376, 487
Mossige, S. 198
Mostowski, A. 77, 417
Motoda, Y. 133, 194
Motohashi, Y. 409
Mott, J. L. 37, 38, 39
Motzkin, T. 49, 75, 141
Moussa, P. 73, 84
Mulholland, H. P. 82, 129
Müller, H. 201
Müntz, C. 81, 404
Murphy, T. G. 518

Nagata; M. 39, 141, 199, 260, 262
Nagell, T. 77, 78, 79, 80, 81, 84, 130, 132, 133, 134, 138, 195, 198, 199, 403, 404, 416, 417, 459, 483, 485, 488, 515, 516, 517, 518
Naito, H. 262
Nakagoshi, N. 318, 414, 483, 487
Nakahara, T. 81, 131, 194, 486

Nakamula, K. 130, 133, 134, 135, 136, 137, 200
Nakamura, Y. 198, 512
Nakano, N. 36, 39
Nakano, S. 484, 486, 487, 542
Nakatsuchi, S. 416
Nakayama, T. 40, 199, 205
Nanda, V. C. 127
Narasimhan, R. 405, 409, 410, 411, 415, 514
Narkiewicz, W. 83, 194, 451, 472, 512, 513, 535, 536, 540
Nart, E. 79, 80, 198, 199, 541
Nechaev, V. I. 319
Nehrkorn, H. 203, 487
Neild, C. 484
Neiss, F. 320
Netto, E. 318
Neubrand, M. 131, 134
Neukirch, J. 72, 194, 205, 206, 260, 263, 321, 413
Neumann, J. v. 126, 319
Neumann, O. 125, 126, 134, 206, 320, 486
Newman, M. 74, 84, 138, 139, 193, 196, 206, 473, 481, 542, 543
Nguyen-Quang-Do, T. 260, 262, 263
Niederreiter, H. 514
Nishi, M. 37
Niven, I. 77, 515, 517
Noether, E. 36, 38, 39, 171, 193, 194, 195, 196, 262, 317
Nordhoff, H. U. 131
Norris, M. J. 130
Northcott, D. G. 39
Notari, C. 74
Novikov, A. P. 137, 412, 472, 475
Nowlan, F. S. 199
Nyberg, N. 78
Nymann, J. 129

Odlyzko, A. 77, 81, 82, 408, 413, 416, 480
Odoni, R. W. K. 320, 415, 416, 513, 518, 541
Ohm, J. 37
Ohta, K. 201, 486
Ojala, T. 138, 141
Okada, T. 195
Okamoto, T. 201
Okutsu, K. 78, 79, 260

Oliwa, G.   83
Olson, F. R.   473
Olson, J. E.   494, 512
Omarov, R.   471
O'Meara, O. T.   40, 141, 317
Ommerborn, B.   201
Onabe, M.   205, 321
Oneto, A. V.   78
Onishi, H.   133
Ono, T.   318, 322
Onuki, M.   320
Oozeki, K.   134, 488
Opolka, H.   317, 318
Oppenheim, A.   83, 140
Orde, H. L. S.   472, 542
Ore, O.   72, 78, 80, 126, 130, 193, 194, 198, 199, 200, 262, 317
Oriat, B.   80, 84, 202, 484, 485, 486, 487
Ortiz, J. J.   78
Osgood, C. F.   83
Osipov, J. V.   418
Ostmann, H. H.   415
Ostrowski, A.   39, 83, 90, 126, 215, 259, 260

Page, A.   478
Pajunen, S.   473, 474
Pak, I.   77
Pal, S.   139
Pall, G.   132, 481, 486, 513, 515
Pan, C. D.   482
Panella, G.   203
Papick, I. J.   37
Papkov, P. S.   485, 488
Papp, Z. Z.   73, 80, 81
Paris, C.   78, 79
Parnami, J. C.   516
Parry, C. J.   132, 134, 136, 200, 417, 480, 486, 487, 517
Passi, H. A.   517
Pathiaux, M.   74, 77
Patterson, S. J.   319
Patz, W.   131
Pavlova, I. V.   260
Payan, J. J.   78, 79, 80, 84, 135, 198, 199, 201, 202, 260, 322, 417, 476, 485, 487
Paysant-Le-Roux, R.   134
Pearson, K. R.   319
Peck, L. G.   518
Pellet, A.   193

Pen, A. S.   132
Peng, C. K.   516
Pergel, J.   514
Perlis, R.   73, 321, 405, 543
Perlis, S.   195
Perott, J.   132
Perron, O.   70, 82, 83, 134, 139, 140
Peters, M.   515, 516
Peterson, B.   417
Petersson, H.   195
Pethö, A.   72, 73
Petr, K.   78
Pfister, A.   516
Phillips, R. C.   38
Phragmén, E.   414
Pieper, H.   263
Pierce, R. S.   81
Pierce, S.   417
Pierce, T. A.   133
Pintz, J.   478, 479, 481, 482
Pirtle, E. B. Jr.   38
Pisot, C.   76, 77, 322
Pitti, C.   139
Pizer, A.   474
Platonov, V. P.   318
Pleasants, P. A. B.   81, 513, 518
Plemelj, J.   194, 517
Plotkin, N.   516
Podsypanin, E. V.   132
Podsypanin, V. D.   134
Pohst, M.   79, 82, 130, 131, 134, 139, 200, 201
Poincaré, H.   414
Poitou, G.   82, 83
Pol, B. van der   199
Pollaczek, F.   135, 203, 261, 475, 487
Pollák, G.   141
Pólya, G.   83
Pomerance, C.   195
Poorten, A. J. van der   84
Popken, J.   405
Popovici, C. P.   198, 200, 488
Portratz, C. J.   77
Porusch, J.   84
Postnikova, L. P.   77
Potapkin, V. K.   82
Potter, H. S. A.   409, 411
Prachar, K.   404, 514, 535
Prenat, M.   77

Prestel, A. 518
Proskurin, N. V. 319
Prüfer, H. 126
Pumplün, D. 132, 194, 195, 474
Purkert, W. 126
Pyatetskii-Shapiro, I. I. 76

Quadri, M. A. 37
Queen, C. S. 141, 418, 486
Queyrut, J. 196, 197, 408
Quillen, D. 128

Rabinowitsch, G. 141, 463, 488
Rabung, J. R. 77, 417
Rademacher, H. 415, 516, 517
Radford, D. E. 405
Rados, G. 80, 138, 194, 195
Raghavendran, R. 319
Raikov, D. A. 195
Raju, N. S. 134
Rajwade, A. R. 516
Ram Murty, M. 417
Ramachandra, K. 137, 204, 320, 412, 472, 478, 479
Ramanathan, K. G. 81, 405, 472
Ramanujam, C. P. 515, 518
Ramazanov, R. G. 471
Ranulac, B. 195
Ranum, A. 318
Rausch, U. 74, 517, 539
Rauzy, G. 76, 77
Rayner, F. J. 260
Razar, M. J. 318, 472
Rédei, L. 77, 80, 132, 133, 140, 195, 200, 204, 319, 474, 484, 485, 488
Rehm, H. P. 73, 487
Reichardt, H. 79, 84, 133, 199, 200, 203, 484, 486, 487
Reid, L. W. 78, 133, 201
Reidemeister, K. 472
Reiner, I. 41, 128, 195, 197, 405, 483
Rejeb, S. 77
Rella, T. 199, 200, 259, 261, 317
Remak, R. 134, 139, 140
Rémond, P. 512, 513
Reshetukha, I. V. 319
Révesz, S. G. 414
Revoy, P. 515, 516
Revuz, G. 78
Ribenboim, P. 72, 199, 476, 518

Ribet, K. A. 413, 418, 475
Richards, I. 417
Richaud, C. 131
Richert, H. E. 415
Richert, N. 83
Richman, F. 126
Richter, B. 195
Rideout, D. E. 473
Rieger, G. J. 129, 318, 320, 413, 414, 415, 514, 515, 517, 518
Riehm, C. 515
Riemann, B. 316, 404, 409
Rim, D. S. 128, 260
Risman, L. J. 77
Ritter, J. 261
Robert, A. 321
Robert, G. 137, 472, 475
Robinson, R. M. 47, 74, 75, 138, 139, 539
Rodemich, E. R. 198
Rodosskii, K. A. 141, 478
Rogers, C. A. 72, 82, 129, 416
Rogers, K. 195
Roggenkamp, K. 405
Rohrlich, D. E. 411
Roquette, P. 130, 138, 203, 317, 486, 487
Rosen, K. H. 479
Rosen, M. 72, 135, 201, 202, 203, 205, 320, 486, 487
Rosenbaum, K. 263
Rosenblüth, E. 84
Rosenthall, E. 518
Rosiński, J. 513, 542
Ross, K. 520
Rosser, B. 411, 479
Roth, R. L. 417
Rotman, J. 36
Roy, Y. 79, 200, 413, 418
Rudin, W. 200, 520
Rud'ko, V. P. 197
Rudman, R. J. 130, 133
Rump, S. M. 75
Rumsey, H. Jr. 198
Rush, D. E. 512, 542
Ruthinger, M. 194
Ryavec, C. 518
Rychlik, K. 261

Sachs, H. 77
Salce, L. 512

Salem, R. 76, 77
Saltman, D. J. 321
Samet, P. A. 77, 78, 201, 517, 539
Samko, G. P. 79, 80
Samuel, P. 40, 72, 130, 141
Sansuc, J. J. 317
Saparniyazov, O. 482
Sarbasov, G. 471
Sarges, H. 78, 514
Satake, I. 262
Satgé, P. 79, 80, 200, 486, 487
Sato, K. 200, 407, 408
Sawada, Y. 205
Schaal, W. 78, 514, 515, 517
Schacher, M. 77
Schäfer, W. 203
Schaffstein, K. 201
Schanuel, S. 512
Scharlau, R. 134, 515, 517
Scharlau, W. 201, 322
Schenkman, E. 131
Schertz, R. 204, 472, 474, 479, 486, 487
Scheunemann, J. 417
Schilling, O. F. G. 39, 260, 262, 263
Schinzel, A. 46, 73, 74, 77, 129, 138, 194, 198, 405, 416, 417, 478, 485, 488, 542, 543
Schipper, R. 201
Schmid, L. H. 320
Schmid, L. P. 415
Schmidt, A. L. 83
Schmidt, C. G. 137, 202, 406
Schmidt, E. 195
Schmidt, F. K. 39, 138, 259, 260, 263
Schmidt, H. 515
Schmidt, K. 77
Schmidt, W. M. 518
Schmithals, B. 203, 487
Schmitz, T. 139
Schneider, D. G. 77
Schneider, J. E. 319
Schneider, P. 322, 413
Schoenberg, I. J. 75, 195
Scholz, A. 82, 126, 132, 136, 203, 318, 416, 484, 487
Schönemann, T. 198
Schoof, R. J. 484
Schreier, O. 415, 416
Schrutka v. Rechtenstamm, G. 473
Schulz-Arenstorff, R. 415

Schulze, V. 199, 416, 417
Schumann, H. G. 203
Schur, I. 47, 73, 80, 81, 82, 139, 194, 319
Schuster, I. 140
Schwarz, Š. 198
Scott, L. 405
Seah, E. 133, 476
Segal, G. 129
Segal, R. 472
Seidenberg, A. 38
Selberg, A. 479
Selmer, E. S. 133, 199, 317, 404
Selucký, K. 475
Sen, S. 262
Senge, H. G. 76, 322
Serafin, R. 484
Serebrennikov 473
Sergeev, É. A. 78, 80
Serre, J. P. 196, 200, 203, 205, 207, 260, 263, 322, 408, 412, 416, 418, 541
Setzer, B. 134, 135, 136, 480
Sexauer, N. E. 518
Shafarevich, I. R. 72, 73, 82, 126, 203, 206, 259, 260, 261, 289, 320, 417, 481, 541
Shanks, D. 127, 131, 200, 201, 205, 412, 415, 482, 484, 485, 487
Shannon, C. E. 512
Shapiro, H. N. 77, 84, 139, 414
Sheingorn, M. 74, 84
Shih, W. T. 140
Shimura, G. 200, 204, 518
Shintani, T. 130, 408, 412, 472, 473
Shiokawa, I. 83
Shirai, S. 202
Shiratani, K. 320, 412, 418, 474, 475, 477, 487
Shiue, J. S. 514
Shyr, J. M. 481, 483, 487
Siegel, C. L. 44, 72, 73, 76, 81, 82, 137, **138**, 139, 193, 317, 404, 405, 408, 409, 411, **412**, 417, 472, 473, 474, 476, 477, 479, 482, 513, 515, 517
Sierpiński, W. 119
Sims, C. S. 477
Singh, S. 516
Sinnott, W. 129, 137, 202, 413, 418, 477
Sirovich, C. 138
Siu, M. K. 128
Sklar, A. 409

Skolem, T. 81, 131, 138, 142, 193, 194, 517, 518
Skopin, A. I. 261, 263
Skubenko, B. F. 132
Skula, L. 126, 202, 473, 475, 476, 511, 512, 541
Slavutskii, I. Sh. 133, 418, 473, 474, 476, 481, 486
Śliwa, J. 138, 198, 512, 513, 541, 542
Smadja, R. 78, 79, 200
Smith, H. I. S. 72
Smith, J. 206, 487
Smith, J. H. 136, 138, 194, 318, 516
Smith, J. R. 141
Smith, R. A. 479
Smith, W. W. 37
Smyth, C. J. 74, 75, 76, 84, 539
Snaith, V. 475
Sokolovskii, A. V. 409, 410, 414
Solderitsch, J. J. 484
Solomon, L. 405
Sominskii, I. S. 79
Sonn, J. 128, 261
Soulé, C. 129, 413
Sparer, G. H. 139
Späth, H. 194
Speiser, A. 83, 196, 200, 318, 320, 404
Spencer, J. 73
Speziali, P. 199
Spira, R. 78
Sprindzhuk, V. G. 133, 139, 483
Springer, T. A. 317
Stahnke, W. 198
Stankus, E. 482
Stanton, R. G. 132
Stark, H. M. 200, 408, 478, 479, 483
Staś, W. 409, 414
Stauffer, R. 195
Steckel, H. D. 84, 318, 471
Steger, A. 141
Stein, A. 134
Stein, M. R. 129, 263
Stein, S. K. 512
Steinbacher, F. 81, 320
Steiner, R. 130, 133
Steinig, J. 485
Steinitz, E. 36, 39, 138
Stemmler, R. M. 515, 516
Stender, H. J. 133, 134, 487

Stepanova, L. L. 319
Stephens, P. J. 139, 481
Steurich, M. 136
Stevenson, E. 518
Stewart, C. L. 74
Stewart, I. 72
Stickelberger, L. 58, 80, 193, 202
Stiemke, E. 78, 126
Stoll, W. 410
Stolt, B. 133
Storrer, H. H. 37
Stout, W. T. Jr. 77, 78
Strassmann, R. 260
Straus, E. G. 74, 76, 77, 539
Strooker, J. R. 40
Stünzi, M. 474
Sudan, G. 83
Sudler, C. Jr. 132
Sudo, M. 317
Suetuna, Z. 407, 408, 409, 414, 514
Sueyoshi, Y. 263
Sukallo, A. A. 198
Sunley, J. S. 80, 132, 480
Sussman, D. 486
Suthankar, N. S. 78
Sved, M. 133
Swan, R. G. 40, 128, 193
Swift, J. D. 485
Swinnerton-Dyer, H. P. F. 140, 141, 317, 406
Szabó, J. 81
Szegö, G. 75, 415, 514
Szekeres, G. 488
Szymiczek, K. 516, 517

Tai, Y. S. 518
Takagi, T. 72, 320, 487
Takahashi, Shōichi 203
Takahashi, Shôichi 513
Takahashi, Shuichi 205
Takahashi, T. 205
Takaku, A. 139, 483
Takase, K. 483
Takenouchi, T. 318
Taketa, K. 203
Takeuchi, H. 488
Takeuchi, T. 203, 483
Takhtayan, L. A. 488, 541
Tall, D. 72

Talmoudi, F. 76
Tamagawa, T. 262, 405, 408, 409, 483
Tamura, J. I. 83
Taniyama, Y. 204, 406
Tannaka, T. 203
Tano, F. 132
Tartakovskii, V. A. 138
Tasaka, T. 318
Tate, J. 72, 129, 205, 245, 262, 263, 318, 320, 321, 322, 404, 405, 406, 408, 527, 541
Tateyama, K. 473
Tatuzawa, T. 129, 404, 406, 409, 415, 471, 474, 478, 479, 482, 515, 517
Taussky, O. 73, 81, 127, 133, 194, 196, 202, 203, 204, 206, 487
Taylor, E. M. 141
Taylor, M. J. 136, 196, 197, 198, 408, 413, 485
Teege, H. 133, 410
Teichmüller, O. 259
Tena Ayuso, J. 141
Terada, F. 203
Terjanian, G. 518
Terras, A. 413
Thérond, J. D. 133
Thielmann, M. v. 131
Thomas, E. 133
Thompson, R. C. 196
Thompson, W. R. 80, 194
Thue, A. 76, 138
Thurnheer, P. 76
Thurston, H. S. 260
Tietze, H. 79
Titchmarsh, E. C. 409, 411
Todd, J. 127, 133
Toepken, H. 194
Tolimieri, R. 320
Tomás, F. 318
Torelli, G. 195, 414
Tornheim, L. 78, 198
Tornier, E. 514
Toro, E. 80, 260
Torre, P. de la 472
Towber, J. 481
Tôyama, H. 194
Toyoizumi, M. 413
Tran, Q. P. 262
Trelina, L. A. 80, 81
Trotter, H. F. 132

Ts'ao, L. C. 517
Tsunekawa, M. 515
Tulyaganova, M. I. 478, 514, 517
Tunnell, J. 407
Turán, P. 414, 479, 513, 542
Turganaliev, R. 79
Turnbull, H. W. 473
Tuzuku, T. 199

Uchida, K. 81, 194, 205, 206, 407, 474, 480, 486, 540
Udrescu, V. S. 479
Uehara, T. 412, 418, 474, 475, 486
Ullom, S. 194, 198, 205, 262, 408, 473
Urazbaev, B. M. 80, 201, 471, 481
Urbanowicz, J. 128
Urbelis, I. 415
Ursell, H. D. 417
Uspensky, J. V. 133, 141
Uzkov, A. I. 39

Vandiver, H. S. 137, 138, 473, 475, 476, 517
Värmon, J. 487
Varnavides, P. 140
Vasconcelos, W. V. 40
Vassiliou, P. 200, 471
Vaughan, R. C. 195
Veldkamp, G. R. 120
Vélez, W. Y. 130, 131, 200, 204
Vel'min, V. P. 133, 200
Venkov, A. B. 319
Venkov, B. A. 79, 472
Venkov, B. B. 204
Ventadoux, M. 76
Vignéras, M. F. (see also M. F. Guého) 413
Vijayaraghavan, T. 76, 132, 195
Vila, N. 80
Vinberg, E. B. 203
Vinogradov, A. I. 406, 407, 415, 478, 488, 514, 517, 541
Vinogradov, I. M. 482
Vishik, M. M. 418
Voronin, S. M. 405
Voronoi, G. F. 78, 133, 193
Vostokov, S. V. 198, 262
Vulah, L. J. 83

Waall, L. J. 200, 407, 408
Wada, H. 133, 134, 201, 417, 486

Wade, L. 37
Waerden, B. L. van der 36, 130, 318, 416
Wagner, G. B. 127
Wagstaff, S. 475, 476
Wahlin, G. E. 199, 261, 263, 317
Waldschmidt, M. 322, 406
Walfisz, Anna 514
Walfisz, Arnold 409, 410, 413, 414, 415, 478, 482, 514, 517
Wall, G. E. 195
Walter, C. D. 202, 486, 487
Walton, L. F. 126
Walton, R. A. 37
Wang, K. 473
Wang, S. 321
Wang, Y. 139
Wantuła, B. 542
Warlimont, R. 415, 482
Wasén, R. 542
Washington, L. C. 72, 136, 139, 202, 261, 263, 412, 417, 418, 472, 473, 476, 477, 480, 541
Watabe, M. 133, 138, 483, 486
Watanabe, S. I. 261
Waterhouse, W. C. 171, 195, 284, 317, 319, 484
Watson, E. J. 198
Watson, G. L. 138
Weber, B. T. 130
Weber, H. (XIX C.) 35, 72, 81, 84, 126, 127, 134, 204, 320, 414, 415, 425
Weber, H. (XX C.) 513
Weber, W. 39
Wegner, U. 79, 195, 199, 200, 318, 416
Weil, A. 72, 163, 194, 321, 322, 323, 404, 405, 406, 408, 409, 410, 420
Weiler, P. 130
Weinberger, P. J. 123, 132, 141, 417, 479, 483, 484, 485, 486
Weinert, H. J. 81
Weinstein, L. 408, 409
Weisner, L. 195
Weiss, A. 413
Weiss, E. 72
Weiss, M. J. 134, 135
Wellstein, J. 81
Westlund, J. 79, 199, 318, 474
Weyl, H. 72, 126

Whaples, G. 20, 126, 127, 130, 262, 321, 417
Whiteman, A. L. 517
Whitford, E. E. 131
Wiebelitz, R. 409
Wieczorkiewicz, J. K. 409, 410
Wiegand, R. 39
Wiegand, S. 39
Wiegandt, R. 471
Wiertelak, K. 414
Więsław, W. 259
Wiles, A. 128, 137, 413, 418, 475
Williams, H. C. 131, 132, 133, 201, 480, 484, 542
Williams, K. S. 79, 132, 141, 474, 476, 480, 481, 486, 515
Wilson, N. R. 78
Wilson, R. J. 514
Wilson, S. M. J. 196, 197
Wilton, J. R. 409
Wiman, A. 318
Wingberg, K. 261, 263, 322
Winograd, S. 320
Wintenberger, J. P. 263
Winter, D. J. 195
Wintner, A. 404, 415
Wirsing, E. 415
Wishart, E. F. 262
Witt, E. 259, 317, 405
Wójcik, J. 416, 417
Wolff, G. 318
Wolke, D. 482
Worley, T. R. 514
Wowk, C. 514
Wrench, J. W. Jr. 412
Wright, E. M. 199
Wunderlich, M. C. 78
Wyman, B. F. 262

Yager, R. I. 475
Yagi, A. 406
Yahagi, O. 128
Yamagata, S. 261
Yamaguchi, I. 134, 476, 488
Yamamoto, K. 78, 319, 320, 414, 513
Yamamoto, S. 262, 475
Yamamoto, Y. 132, 139, 194, 483, 484, 486
Yanagihara, H. 37

Yokoi, H.   80, 131, 133, 139, 142, 194, 197, 201, 205, 476, 488
Yokoyama, A.   201, 319, 486, 488
Yoshida, H.   407

Zagier, D.   412, 480, 488
Zaikina, N. G.   78, 514
Zaks, A.   511, 541
Zame, A.   512
Zanardo, P.   512
Zantema, H.   83
Zariski, O.   40
Zarnke, C. R.   133, 201
Zassenhaus, H.   46, 73, 74, 78, 82, 127, 130, 131, 204, 260, 320, 416, 486, 487
Zaupper, T.   200
Zeinalov, B. A.   135
Zeitlin, D.   195
Zelinsky, D.   259
Zelvenskii, I. G.   261
Zhang Xianke   79, 417, 471
Zierler, N.   198
Zimmer, H. G.   72, 201
Zimmert, R.   129
Zink, E. W.   203
Zink, W.   486
Zlebov, E. D.   136
Zolotarev, E.   126, 133, 198
Zygmund, A.   76
Żyliński, E.   126, 198

Addendum

(*Added in January 1989*)

We give here a few addenda to the comments sections.

CHAPTER 2

p. 75. The problem considered by Favard has been solved by M. Langevin, R. Reyssat, G. Rhin (*Ann. Inst. Fourier*, **38**, 1988) who have shown that if a is an algebraic integer, $a = a_1, \ldots, a_n$ are its conjugates over Q and $e(a) = \max|a_i - a_j|$, then $e(a) \geq 3^{1/2}$. (This estimate is best possible in view of the polynomial $X^2 + X + 1$.) They also showed that if the degree of a is sufficiently large, then $e(a) \geq 1.8819$, and later G. Langevin showed (*Ann. Inst. Fourier*, **38**, 1988) that the last number can be replaced by any number smaller than 2. The last evaluation is also true when one replaces Q by any imaginary quadratic field (G. Langevin, *C.R. Acad. Sci. Paris*, **307**, 1988, 427–429).

p. 76. A method for an effective determination of all PV-numbers in the neighbourhood of a limit point gave D. W. Boyd (*JNT*, **21**, 1985, 17–43).

p. 77. Corrections to Boyd [77b] appeared in *AA*, **48**, 1987, 191–195.

p. 82. $M(3,1) = 612\,233$ (F. Diaz y Diaz, *AIF*, **34**, 1984, 3, 29–38), $M(0,4) = 1\,257\,728$ (F. Diaz y Diaz, *JNT*, **25**, 1987, 34–52).

p. 83. K. Rogers and E. G. Straus (*Pacific J. Math.*, **118**, 1985, 507–522) characterized polynomials which together with all their derivatives up to a given order map Z_K in Z_K.

An analogue of the result of Brizolis [74] for several variables has been obtained by D. J. Lewis and P. Morton (*J. Fac. Sci. Tokyo*, IA, **28**, 1981, 813–822).

CHAPTER 3

p. 133. A bibliography on Bernoulli numbers containing over 1300 entries has been prepared by L. Skula and I. Sh. Slavutskii: *Bernoulli numbers, Bibliography 1713–1983*, J. E. Purkyne Univ., Brno).

p. 139. It has been shown by J. H. Silverman (*JNT*, **19**, 1984, 437–442) that for fields of fixed degree N one has $R(K) > c(N)(\log d(K))^{r(K)-r'(K)}$, where

$r(K)$ is the unit rank of K, $r'(K)$ is the maximal unit rank of proper subfields of K and $c(N) > 0$.

p. 140. S. Egami (*Tokyo J. Math.*, **7**, 1984, 183–196) showed the finiteness of the set of Euclidean fields lying in the following classes of fields:

(i) Cyclic fields of a fixed degree N with totally ramified prime factors of the discriminant. (In case of a prime N every cyclic field of degree N has this property.)

(ii) Pure quintic fields.

p. 141. It has been proved in Gupta, Ram Murty, Kumar Murty (*Proc. CMS Conf. Number Theory*, 189–201, AMS 1987) that if K is real, $|S| \geq \max\{5, 2[K:Q]-1\}$ and A_S is PID then it must be Euclidean. An analogous result holds also for non-real K, but one has to assume that K contains sufficiently many (depending on S) roots of unity.

CHAPTER 4

p. 198. Parts II and III of Taylor [85] appeared: *AnM* (2), **121**, 1985, 519–535; *PLMS* (3), **51**, 1985, 415–431.

p. 200. An efficient algorithm for determining $H(K)$ has been given in Pohst, Zassenhaus (*JRAM*, **361**, 1985, 50–72).

p. 202. For a new kind of annihilator of the p-part of the class-group in case of abelian fields see Rubin (*Invent. Math.*, **89**, 1987, 511–526) and F. Thaine (*Ann. Math.*, **128**, 1988, 1–18).

p. 203. R. Schoof (*JRAM*, **372**, 1986, 209–220) produced infinitely many real and imaginary quadratic fields K with $d(K)$ being divisible by two primes and infinite class-field tower.

CHAPTER 5

p. 259. See the new book of J. W. S. Cassels (*Local fields*, Cambridge Univ. Press, 1986) for a good introduction into that theory.

A new edition of Więsław [82] was published by Marcel Dekker in 1988.

CHAPTER 6

p. 317. A. Schinzel (*Studia Math.* **77**, 1984, 103–109) proved that the Hasse principle holds for zeros of systems of ternary quadratic forms over Q and showed it fails for zeros of $x^4 - 17y^4 - 2(z_1^2 + \ldots + z_n^2)$.

p. 318. Proofs for Platonov, Drakokhrust [85] appeared in *IAN*, **50**, 1986, 946–968.

p. 322. Leopoldt's conjecture has been established for imaginary fields with Galois group A_4 in Emsalem, Kisilevsky, Wales (*JNT*, **19**, 1984, 384–391).

CHAPTER 7

p. 404. Cohn's result for pure cubic fields has been improved to $h(K)R(K) < (d^{1/2}\log d)/6.3^{1/2}$ (with $d = |d(K)|$) in Barrucand, Loxton, H. C. Williams (*Pacific J. Math.*, **128**, 1987, 209–222).

p. 406. R. Perlis (*Canad. Math. Bull.*, **28**, 1985, 422–430) proved that arithmetically equivalent fields have rationally equivalent trace forms $T_{L/K}(x^2)$.

p. 407. subsect. 7. Proofs of Kurokawa [78a, b] appeared in *PLMS* (3), **53**, 1986, 1–47; 209–236. The book of Moroz (*Analytic arithmetic in algebraic number fields*, LN 1205, Springer, 1986) treats the question of analytic continuation of scalar products.

p. 408. A discussion of Stark's conjectures is contained in J. Tate's lectures (*Progress in Math.*, **47**, 1984).

CHAPTER 8

subsect. 5. T. Kimura, K. Horie (*TAMS*, **302**, 1987, 727–739) showed that for any fixed N and almost all n, N divides h_n^-.

subsect. 12. Another way of determining imaginary quadratic fields with $h = 1$ was shown by J. M. Cherubini and R. V. Walliser (*MC*, **49**, 1987, 295–299).

subsect. 14. An exposition of the theorem of Gross–Zagier gave J. Oesterlé (*Astérisque*, **121/2**, 1985, 309–312). There one can find the following explicit lower bound for $h(K)$:

$$h(K) > C\log D \prod_{\substack{p|D \\ p<D}} (1 - [2p^{1/2}]/(1+p)),$$

where $D = |d(K)|$ and $C = 1/7000$. Later work of J. F. Mestre, J. Oesterlé and J. P. Serre showed that one can put $C = 1/55$. This implies that all imaginary quadratic fields K with $h(K) = 3$ satisfy $|d(K)| \leq 907$.

A survey of work on imaginary quadratic fields with a given class number gave D. Goldfeld (*BAMS*, **13**, 1985, 23–37).

subsect. 20. J. Quer (*CR*, **305**, 1987, 215–218) found three imaginary quadratic fields with $e_3(H) = 6$.

CHAPTER 9

subsect. 11. The analogue of Chen's theorem was for totally real fields obtained by J. Hinz (to appear).

subsect. 12. The Catalan equation $x^p - y^q = 1$ (with x, y in R_K, not roots of unity, $p, q > 1$) has been treated by B. Brindza, K. Györy, R. Tijdeman

(*JRAM*, **367,** 1986, 90–102) who showed that in case $pq > 5$ it has only finitely many solutions which can all be effectively determined.

subsect. 13. The analogue of the result of Arkhipov and Karatsuba was obtained for finite extensions of Q_p by Y. Alemu (*AA*, **45,** 1985, 163–171) and V. T. Vilchinskii (*DAN Belorus. SSR*, **29,** 1985, 972–975).